Progress in Fluid Flow Research: Turbulence and Applied MHD

Herman Branover
Yeshajahu Unger
Ben Gurion University of the Negev
Beer–Sheva, Israel

Volume 182
PROGRESS IN ASTRONAUTICS AND AERONAUTICS

Paul Zarchan, Editor-in-Chief
The Charles Stark Draper Laboratory, Inc.
Cambridge, Massachusetts

Published by the
American Institute of Aeronautics and Astronautics, Inc.
1801 Alexander Bell Drive, Reston, Virginia 20191-4344

ISBN 1-56347-284-8

Table of Contents

VII. SOLIDIFICATION 577

VIII. INCOMPRESSIBLE FLOW 619

Preface

This volume contains a collection of papers devoted to progress in the research of fluid flows, presented at the Eighth Beer-Sheva International Seminar on Magnetohydrodynamic (MHD) Flows and Turbulence.

The Beer-Sheva Seminars have been held every three years since 1975, when a group of researchers from different countries, interested in both MHD flows and different aspects of turbulence in electro-conductive fluids as well as nonconductive fluids, came together for a week-long seminar. All participants felt that the meeting was very fruitful in detecting promising areas of research and in developing new ideas. Therefore, the seminar was repeated in 1978 and ultimately became an ongoing event, taking place once every three years. After a thorough review and editing, the papers presented at each Beer-Sheva Seminar are published. The first two volumes were published by John Wiley and Sons and the subsequent volumes by AIAA (Vols. 84, 100, 111, 112, 148, 149, and 162 in the Progress in Astronautics and Aeronautics Series).

The Beer-Sheva Seminars became well known in the international scientific community and the number of researchers willing to participate kept growing steadily. This forced the organizers, who desired to preserve the special intimate atmosphere of the seminars and to keep very substantial time for informal discussions, to introduce more and more severe criteria in the paper selection process. The intention was to keep the number of participants in each event at approximately 100. The eighth seminar, held February 25–29, 1996, under the auspices of the HYDROMAG association, hosted 139 participants from 19 countries. The number of studies presented at the seminar was 100. After a thorough review 66 papers related to transition, turbulent flows, energy conversion and hydrodynamic (HD) and MHD flows have been selected for this volume. They cover many important contemporary trends in both experimental and theoretical research and give a concise and comprehensive picture of the present status in most of the areas in which extensive studies are occurring. The authors of many of the articles in this volume are recognized leaders in their research area.

To better acquaint the reader with the content of the volume, we will briefly describe here some experimental and theoretical papers.

This volume opens with two chapters on transition to turbulence. Seifert and Tumin studied, both experimentally and numerically, the linear as well as nonlinear evolution of artificially introduced localized disturbances in an adverse pressure gradient boundary layer. No assumption was made as to the self-similarity of the boundary layer, even though it can also be treated as such. Very good agreement was found between measured amplitudes and integrated eigenvalues. The deviation of the harmonic point source (HPS) disturbances from linearity in the close proximity of the source is attributed to receptivity and away from the source to nonlinearity. An amplitude threshold of 0.3–0.4% was identified as the trigger of rapid second harmonic amplification. Only at the lowest HPS amplitude simultaneous amplification of the subharmonic accompanied. The wave packet nonlinearity was characterized primarily by subharmonic amplification. Nonparallel effects were not found at the moderate frequencies and Reynolds numbers tested.

Hot-wire measurements were carried out by Konig et al. in a two-dimensional two-stream mixing layer with an initial velocity ratio $U_1/U_2 = 0.5$, subjected to an adverse pressure gradient in streamwide direction. The extension of the Abramovich–Sabin relation on slightly nonparallel flows was found to be valid. The bulk quantities show similar distributions when normalized with the local width. The maximum energy values indicate enhanced mixing. Effects of various forcing amplitudes at single frequency forcing as well as of forcing with fundamental and subharmonic frequency have been investigated.

Various chapters are related to turbulence. Sukoriansky et al. are concerned with a problem in computational fluid mechanics. This is how to model correctly the transfer of energy through the scales in two-dimensional turbulence. Much of the energy resides in the unresolved scales. A difficulty arises with the representation of the viscosity in the numerical simulation; if it is kept constant, it appears that the numerical scheme becomes unstable. This was overcome by allowing the viscosity to depend on the flowfield.

Godeferd et al. presented a stochastic model of turbulent flows in a rotating frame and in a stable stratified field. The model is based on a generalized EDQNM and kinematic simulation models and is developed to account for flow anisotropy and waves. Theoretical results were compared to the results of a high resolution direct numerical simulation.

Juttner and Thess discussed a theory of equilibrium statistical mechanics for inertial organization in a forced two-dimensional turbulence. New predictions were presented for the case of a forced turbulence with a stationary forcing balanced by a linear friction.

Ecket et al. summarized results of a long term experimental program concerning MHD turbulence measurements in a sodium channel flow. Mean velocities, turbulence intensities, and turbulent kinetic energy spectra were presented for different flow situations and for a wide range of flow parameters.

Without going into detail we should mention here that a number of chapters deal with the results of direct numerical simulation of both hydrodynamic and magnetohydrodynamic turbulence.

Turbulence in the atmosphere is studied by Moiseev et al. Extensive work on the role of fluid helicity in the generation of turbulence is shown. A key result was the discovery that the existence of mean helicity was sufficient to destabilize helical disturbances above a certain amplitude, providing the flow was either compressible or subject to shear or similar variations. Understanding the connection between turbulence and helicity is an important area of research in atmospheric physics. A practical application of the work is to identify the precursors of typhoons.

Golbraikh et al. studied the instability of turbulent MHD flow in a uniform magnetic field with respect to weak large-scale disturbances. It is shown that helical turbulence is unstable. The limiting wave number of the instability range linear depends on the interaction parameter N for the weak magnetic field, which agrees with experimental data.

Sideridis et al. report on a new experimental flowfield. A flat plate at zero incidence was placed in the test-section of a low-speed, suction-type wind tunnel. A circular cylinder with a deliberately rough surface was placed parallel to the plate, just outside the boundary layer and at right angles to the flow. The results are of importance to those who develop models for the Reynolds stresses as well as to turbulent heat flux.

Schwarz is dealing with the problem of pollutant propagation in the atmospheric boundary layer over the scales of tens of kilometers. The atmospheric boundary layer is described within "shallow water approximation," which defines the distribution of parameters averaged over the altitude in a certain way. To take into account the vertical structure of the velocity field required for computing pollutant transfer, the author applies well-known velocity distribution in Eckman's layer, which is connected with mean flow parameters.

The chapter by Dolginov, in the section on astrophysics, is mentioned here due to new data from Jupiter presented in it.

In the field of measuring techniques, two developments are presented. Reshotko et al. report on microfabricated shear stress sensors based on polycrystalline silicon lateral resonant structures with capacitive pickup and integrated electronics, which hold a promise to become cost effective devices to be calibrated in situ under certain conditions and can be used for laminar, transitional, or turbulent flow.

Fautrelle et al. report on the mass transfer through interfaces, which plays a role in many chemical purification processes. This mass transfer has been investigated prototypically through measurement of the Faradic electric current, which is caused by the oxidation of amalgamated indium by an aqueous solution of sodium chloride. The technique allows for determining mass transfer coefficients in different stirring conditions of the interface by an analysis of the mean current and its superimposed fluctuations.

Two-phase flow measurement of liquid considered as conducting continuous phase and gas as insulating dispersed phase with an induced voltage electromagnetic flowmeter is introduced by Opara and Bajsic. The flowmeter output and pressure drop are related to the volumetric quality and total gas–liquid volumetric flow. The validity of these relations and the influence of flow regime are experimentally confirmed by the master–slave method in an air–water system.

Lykoudis, in his chapter "Tau Theorem" (an alternative to Buckingham's Pi-Theorem), presents a very interesting scheme to deal with physical processes of various kinds, where everything is expressed in terms of time, hence the name Tau-Theorem. It is a novel approach in methodology of evident merits and overcomes the weakness of Buckingham's Pi-Theorem, when the choice of properties and parameters of a problem is not correct or complete.

Magnetohydrodynamic energy conversion is presented by different investigators.

In the field of seawater MHD, three chapters are presented. Meng, using experimental data, establishes a conceptual phenomenology of microturbulence, including spatial and temporal patterns. The author aims primarily at developing an effective methodology for the control of drag and noise in a turbulent flow. For a natural turbulent boundary layer with zero pressure gradient, the simplified Chapman–Kolmogorov equations for the transition probability are solved for the limiting probabilities.

When a Lorentz pressure gradient is imposed, the transition probability is modified. With the presence of externally applied electromagnetic fields, the limiting probabilities no longer apply. Under these new conditions the author successfully analyzes the solution of the Chapman–Kolmogorov equation.

Lastly, a specific method for electromagnetic control of turbulent boundary layers, based upon both local shear stress and wall pressure sensors, is proposed.

Lee et al., in their chapter, provide contributions of real data of seawater MHD flows at low Reynolds number at 0.6 m/s flow speed and low MHD interaction parameter in the order of 0.0003. Their chapter provides the first concurrent measurements of bubble and flow turbulence data in a seawater MHD duct. This chapter used fiberoptic laser Doppler velocimetry (LDV) for turbulence measurement. This contribution is novel and discovered the existence of inflectional point in the flow velocity and high turbulence intensity near the wall.

Miroshnikov and Shishko consider a steady creeping electrolyte flow past an electromagnetic (EM) cylinder moving with a uniform speed, when the EM force helps to resolve the well-known Stokes paradox of the ordinary hydrodynamics forming the nonslip flow pattern. An exact solution of the three-dimensional Navier–Stokes equations forced electromagnetically is derived via the modified Bessel's functions, which describe exponentially vanishing solutions in the cylindrical case. For the upstream velocity, the nonslip velocity distribution without any separation points is possible for the critical EM Hartmann number because of an effective EM shaping of the cylindrical body, Depending on β, which characterizes the ratio of the cylinder perimeter to the period of the EM system, the EM cylinder effectively takes a form of a streamlined body with drop-shaped cutters upstream and downstream of the cylinder, and an optimal β value is determined.

Sheindlin and Matveyeva discuss MHD power generation in their chapter "On Perspectives of Plasma MHD Installations." They say that in spite of previous dead-end perspectives in the MHD power generation, there is now some hope that under some special conditions it should be possible to get interesting results in this field. Most of the problems that remain to be solved are of a technological nature.

A direct numerical modeling of kinematic dynamo effect based on the finite element method is presented by Plunian et al. The resolution is done in potentials A and φ, for a given velocity field. The model is then used to predict any eventual dynamo effect in a geometry corresponding to the core of the French fast breeder reactor Phenix. The direct simulation leads to a core's critical minimum size necessary to self-excite and shows the importance of electromagnetic properties heterogeneousness.

Phenomena in metal solidification are addressed by Lehman et al. In their chapter they present the effect of a magnetic field applied during the dendritic solidification of a Cu 60% mass Ag alloy. It is shown that an electrical current density exists in the dendritic network due to thermoelectric effects. The magnetic field creates a Lorentz force. The alloy is solidified horizontally so that this force directly opposes or reinforces the buoyancy forces. Asymptotic regimes are studied for both high magnetic field values (MHD braking) and low magnetic field (thermoelectric motion).

Mikelsons et al. present overcooling as a possible cause of thermoelectric currents and magnetic fields in the case of metal solidification.

The work presented by Vives concerns a new method of transmitting vibration to solidifying aluminum alloy melts. This vibration is generated by the use of an alternative electromagnetic field of force obtained by the cross of a static magnetic field with an alternative electric current. When the intensity of the field of force is strong enough, it is able to produce cavitation inside the liquid metal, which, after collapse of the vapor hole when the force cycle is in its upper part, produces brake and dispersion of the solid grains of metal. Then the solidification obtained in such a situation exhibits a very fine and homogeneous microstructure with smaller and less rounded grains than those produced above cavitation.

Different aspects of incompressible fluid flow subjected to a magnetic field are discussed by various authors. Davidson et al. argue that Ekman pumping is an essential component of many confined, MHD flows. This pumping occurs whenever the axis of a forced, columnar vortex intersects a solid boundary. The weak recirculation associated with Ekman pumping is important for two reasons. First, the recirculation provides an efficient mechanism of removing the energy supplied by the Lorentz force, by flushing all streamlines through a boundary layer. Second, the Coriolis acceleration associated with the secondary flow can be used to balance the azimuthal component of the Lorentz force. Ekman pumping is an efficient mechanism of destroying mechanical energy, and so such flows have a lower kinetic energy than the equivalent poloidal flow that might exist in the absence of a small azimuthal force. This behavior has been observed in weld pools.

Numerical simulation of flow driven by a rotating magnetic field (RMF) in a rotating cylindrical vessel is presented by Grants and Gelfgat. The obtained results conclude that the imposed rotation of the vessel is a powerful means to control flow patterns induced by the RMF, therefore providing opportunities to choose optimal ones for every particular technological process.

Liquid metal jets through electromagnetic nozzle is characterized due to theoretical model predicting electromagnetic striction. A new model of the flow is proposed by Moreau et al. aiming at clarifying the understanding of this phenomenon and at predicting both the position of the separation and the free surface shape downstream of the separation. The new ingredients of this model are 1) the inertial overpressure due to the curvature of the streamlines and 2) the contact angle of the separation due to the surface tension; hence, a good agreement with experimental data is achieved.

Lindsay and Davidson in their chapter discuss the stability of interfacial gravity waves in aluminium reduction cells. Specifically, existing models of cell stability are reviewed, showing that the two primary candidates are quite distinct in their physical origins. The importance of the cell boundaries in selecting the instability mode is discussed.

Gelfgat et al. present several unusual variations of RMF action on a liquid; at asymmetric position of an RMF inductor and cylindrical vessel along the axis of finite length and at the action of two counter-rotating magnetic fields on a liquid cylinder. An original numerical technique for electromagnetic fields calculation accounting for all changes of geometrical parameters as well as magnetic fields of frontal parts of coils of an RMF inductor, thickness and electrical conductivity of side and end walls confiding a liquid volume, value of linear current load and number of pole pairs, etc., has been specially developed. On the basis of the above technique, the distribution of electromagnetic forces in a fluid and their action on velocity patterns, temperature distribution, and shape of the solidification front at different values of the Taylor, Grashof, and Prandtl numbers are considered.

Shatrov et al. investigated stability of the time-dependent, three-dimensional, incompressible flow around a circular cylinder exposed to an external magnetic field. The analysis is restricted to a magnetic field being aligned with the oncoming flow, confirming that in the (Re,N)-plane, parameter regions above the two-dimensional neutral stability curve exist where the flow is three-dimensional unstable but two-dimensional stable (steady) because the magnetic field acts in a different way on two-dimensional and three-dimensional instabilities.

Yang et al. present a two-dimensional model to compute the steady-state flow of an aluminum (Al) melt in the case of centrifugal casting with electromagnetic stirring. The distributions of flow velocity caused by electromagnetic stirring, body forces, and pressure in the melt are studied by means of numerical simulations.

The paper "Bifurcations in a Magnetofluid with Helical Forcing" by Seehafer et al. studies bifurcations in incompressible MHD flows on the basis of ABC flow. As a result of numerical modeling, at a maximum helicity the dynamo effect arises due to Hopf's bifurcation.

A recent experimental investigation of the MHD free convection inside a differentially heated horizontal mercury-filled cylinder whose end walls are regulated to the temperatures T_1 and T_2 ($T_1 < T_2$) is presented by Davoust et al. By measuring temperatures both at the wall and within mercury, the influence of a uniform vertical magnetic field Bo either on the steady or the unsteady regimes has been studied. Buoyancy effects as well as the influence of nonuniformity of the magnetic field are discussed and compared to predictions.

Patel and El-Kaddah examine the secondary MHD flow around a nonconducting spherical droplet in uniform crossed electric and magnetic fields at low Reynolds numbers. The flow inside the droplet is also considered. Analytical expressions are presented for the two characteristic velocity fields.

Kapusta and Shamota consider steady quasilaminar and turbulent flow of a viscous incompressible conductive liquid in cylindrical vessel, using a new simple semiempirical model of external friction. The authors managed to derive a sufficiently simple formula for the average angular velocity. Theoretical results obtained in this chapter are compared with experimental data revealing good agreement.

Application of plasma MHD is presented by Abramova et al. They describe a new experimental ability concerning the confinement and heating of plasma. This ability is obtained for the first time, using a rotating plasma operation in spiral coil "Tornado" magnetic traps. In this apparatus, 20% of the plasma rotation energy was converted to thermal energy. Provided that sufficiently high ion and electron temperatures can be reached, this system may be operated in the future at higher plasma densities and smaller dimensions than those of other existing systems.

The chapter by Girka and Lapshin is devoted to the theory of resonant absorption of satellite Alfven waves in a radially inhomogeneous plasma column. Such satellite waves can be produced due to the modulation of the main Alfven wave by an external magnetic field that varies periodically in the axial direction. The authors consider the ion–cyclotron turbulence excited by Alfven waves as a dominant mechanism of the wave energy absorption in the resonant layer.

In their totality, the chapters referred to as well as those not mentioned here present a quite complete picture of the contemporary trends in those fields. They present a substantial amount of the most recent results. The extensive references given in most chapters of this volume should be an additional aid to the reader who wishes to make himself familiar with the present status of the field.

The authors and editors are very pleased that AIAA is continuing to publish the proceedings of the Beer-Sheva International Seminars on MHD Flows and Turbulence in their Progress in Astronautics and Aeronautics Series.

The editors would like to express their most sincere gratitude to Paul Zarchan, Editor-in-Chief of the AIAA Progress in Astronautics and Aeronautics Series as well as to the entire team of the AIAA Progress in Astronautics and Aeronautics Series. We are also grateful to Victor Halebski of the Joint Israeli–Russian Laboratory and Shulamith Friedmann of the Center for MHD studies for their selfless assistance enabling the completion of this volume.

We are now looking forward to the next seminar scheduled for 2000 to be held in Paris under the auspices of the HYDROMAG Association. It will be joined by the PAMIR (Grenoble, France), the Riga Conference (Institute of

Physics, Latvian Academy of Sciences), and the International Beer-Sheva Seminar (Center for MHD Studies, Israel).

Herman Branover
Yeshajahu Unger
October 1998

I. Transition

Chapter 1

Nonlinear Localized Disturbances in an Adverse Pressure Gradient Boundary-Layer Transition: Experiment and Linear Stability Analysis

A. Seifert and A. Tumin

Department of Fluid Mechanics and Heat Transfer, Faculty of Engineering, Tel-Aviv University, Ramat-Aviv, 69978, Israel

Introduction

LOCALIZED disturbances evolving in a transitional boundary layer (BL) represent much more realistically the natural transition process rather than the periodic, in space and time, disturbances commonly studied. They also present a much more difficult task for prediction and simulation. The harmonic point source (HPS) is simple because only one continuous spectrum is present (i.e., in the spanwise wave number). A pulsed localized disturbance complicates the analysis by generating also a continuous frequency (or streamwise wave number) spectrum. Even the linear stage of HPS evolution in Blasius flow is not successfully predicted when one tries to simulate the measured amplitudes to summation of eigenvalues,[1–4] whereas discrepancies are partly attributed to experimental uncertainties, unknown input spectra, and nonparallel effects. Certainly, the importance of controlling and documenting the base flow conditions for stability experiments cannot be underestimated.

Qualitative agreement was found in controlled experiments[5–7] when a decomposition of the localized disturbance into normal modes was performed and compared to the eigenmodes of linear stability theory (LST). The effect of boundary-layer growth on the wave pattern prediction is in some dispute.[7] It was recently shown[8] that in Blasius flow the nonparallel effects are very weak and in adverse pressure gradient BLs they are stronger but still can be neglected if one considers the inner maximum of the amplitude distribution as the criteria for instability. However, this is not a general statement. The critical *Re* number and the evolution of high frequencies at low *Re* numbers are certainly more sensitive to nonparallel effects than the amplification of low frequencies at moderate *Re*. Because most of the experimental results were presented elsewhere,[9] we would concentrate here on the comparison between the parallel LST and the experiment. The effect of an adverse pressure gradient on the stability of localized disturbances is expected to be significant, as was recently found for a turbulent spot.[10] The presence of an inflection point in the velocity profile, near $U/Ue = 0.4$, should enable easier identification of linear as well as nonlinear and nonparallel effects due to the higher amplification rates expected from the LST.

A detailed description of the experiment could be found in previous publications.[9,10] It should be emphasized that all the HPS characteristics were measured across the entire BL because nonparallel effects alter both the amplitude distribution across the BL and its growth rate. To detect these effects, one should measure the complete HPS disturbance. The localized disturbances were introduced through a single 0.5-mm hole located in $X =$ 50 cm, where the reference freestream velocity was $Ur = 7.2$ m/s ($R_{\delta^* p} = 830$, *Re* based on δ^* at the source).

Base Flow

It is critically important to control and document the base flow in BL transition experiments. The acquisition of repeatable results and the development of reliable prediction tools are impossible without this step. Therefore, a major effort was made to produce a self-similar, adverse pressure gradient. This was achieved by measuring the evolution of the freestream in the X direction and many velocity profiles along X and Z. The local pressure gradient parameter, β (Hartree parameter), was determined by fitting the closest theoretical velocity profile to the measured ones. The resolution of the fitting procedure was $\Delta\beta = \pm 0.01$. The geometry of the flexible insert imposing the pressure gradient was altered until the desired flow was obtained. The evolution of the freestream and the BL δ^* (displacement thickness) and θ (momentum thickness) along the plate are presented in Fig. 1 along with the calculations according to Ref. 11.

The measurements also agree with the theoretical predictions for a laminar adverse pressure gradient BL with $\beta = -0.1$ (± 0.02) over the range $50 < X < 110$ cm, even though no assumption about the self-similarity of the flow was made in solving the boundary-layer equations for the stability analysis. They were solved with the code given in Ref. 11 using only the external velocity distribution $U_c(X)$ measured in the experiment. The velocity distribution near the leading edge could not be measured with sufficient resolution and accuracy, and, therefore, we could not start our calculations from $X = 0$. To match the calculations and the experimental data for δ^* and θ (Fig. 1), we assumed that the boundary layer originates from $X_0 = 12$ cm, as it was established,[9] assuming self-similar flow. At $X > 120$ cm (and $Ur =$ 7.2 m/s), transition takes place. The short stretch of laminar flow and the explosive transition indicate the highly unstable nature of this BL even in a low-disturbance environment.

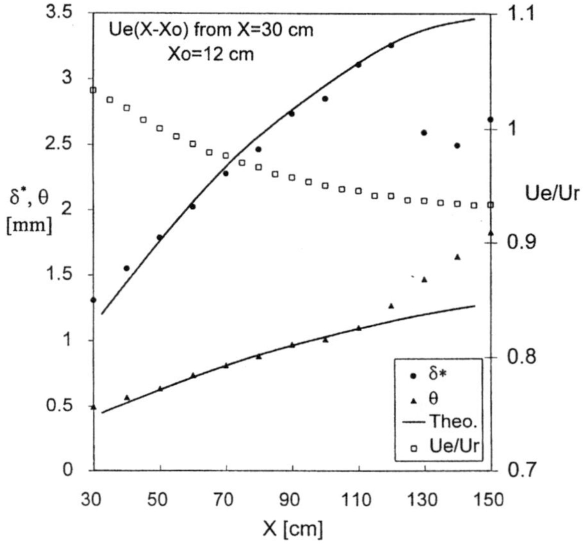

Fig. 1 Measured and calculated boundary-layer parameters.

In Fig. 2a we present the comparison between the measured and predicted mean velocity profiles at $X = 50$ cm. The profiles calculated assuming Hartree parameter, $\beta = -0.1$, and according to the external flow distribution (i.e., Cebeci) are indistinguishable. The experimental data are in very good agreement with these calculations. Figure 2b presents the dimensional mean velocity profiles as measured and calculated in $X = 70$ cm.

During the transition (e.g., $X > 120$ cm at $Ur = 7.2$ m/s), the freestream velocity ceased to decelerate, the displacement thickness decreased, and the mean velocity profile deviated from the laminar velocity prediction.

Streamwise Evolution of the Harmonic Point Source

The analysis was performed by direct summation of inclined Tollmien–Schlichting–Schubauer (TS) waves of equal amplitudes at the source. The transversal wave number was changed in the band $-0.45 < \beta$ [1/mm] < 0.45 (the maximum inclination angle was about 80 deg.). Streamwise integration was performed every 2.5 cm. Extensive comparison between the evolution of the artificially introduced disturbances and parallel BL, LST calculations was performed in the intermediate region, i.e., after the transient decayed and before nonlinearity takes place, as a check for the linearity of the initial stages of disturbances evolution and also as a first step in a future nonlinear analysis.

Experimentally, the authors found that the early stages of localized disturbance evolution are dominated by transient effects.[9] Even at the weakest HPS input used, about four wavelengths were required for the disturbance to obtain the predicted TS form and start following the LST prediction. This distance is about 30% shorter than in Blasius flow.

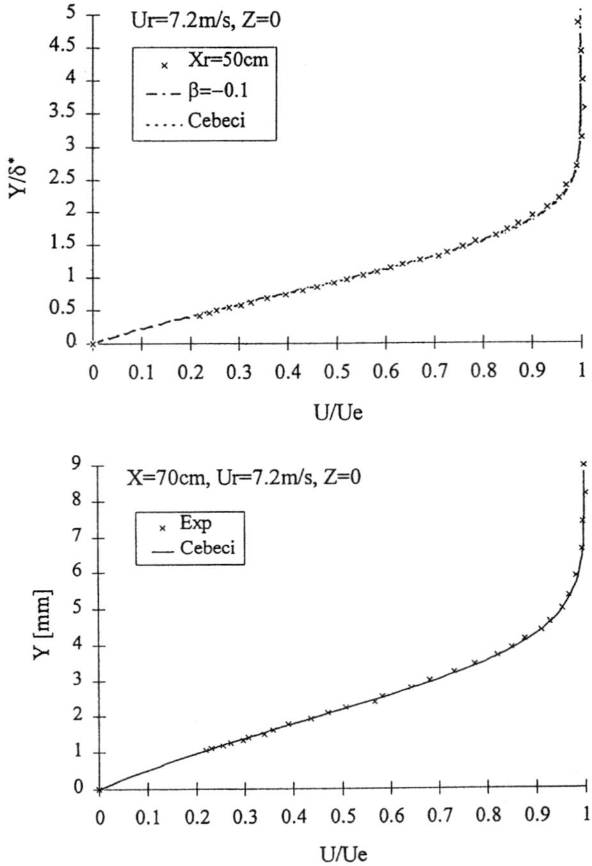

Fig. 2 Measured and calculated boundary-layer velocity profiles.

Figure 3 presents the comparison between the centerline normalized HPS amplitudes and the LST calculations, for f = 60 Hz (F = 113 at X = 50 cm, $F = 2\pi f\nu/U_\infty^2$) and A_0 = 0.03%. The reference amplitude is the maximum HPS amplitude measured at Xs = 15 cm ($R_{\delta *}$ = 1000). The two amplitudes are matched at X = 70 cm and agree very well until X = 110 cm. At this location the fundamental (F) saturates, indicating nonlinearity.

Figure 4 presents the comparison between the measured and calculated HPS evolution for three different frequencies when all initial amplitudes are about 0.1%. A quantitative agreement can be observed in the linear range. The highest frequency departs from linearity when the linear calculation exits branch II of the neutral stability curve. The lower frequencies saturate despite the LST prediction for further amplification.

Figure 5 presents the nature of the fundamental (F) amplitude saturation for f = 60 Hz at four excitation amplitudes. All of the amplitudes are

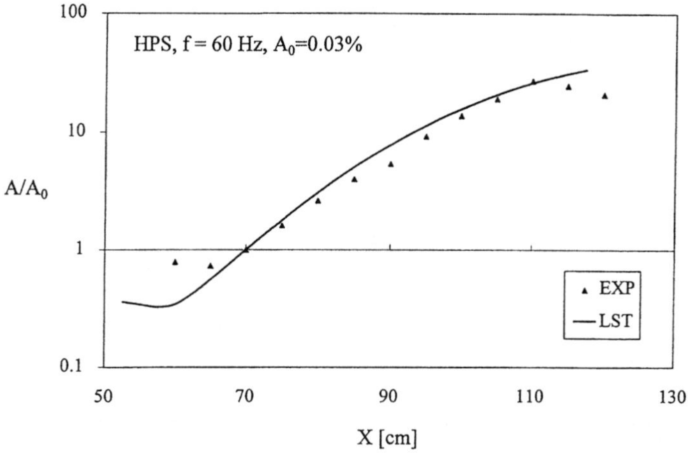

Fig. 3 Measured and calculated HPS maximum amplitudes, *Fr* = 113 and A_0 = 0.03%.

normalized by the amplitude at X = 70 cm (A_0) and matched there to the amplitude of the LST. As the forcing amplitude increases, the amplitude saturates at lower $R_{\delta*}$.

Figure 6a presents the same data, but the ordinate is dimensional. The saturation of F does not happen in a constant amplitude. The fundamental

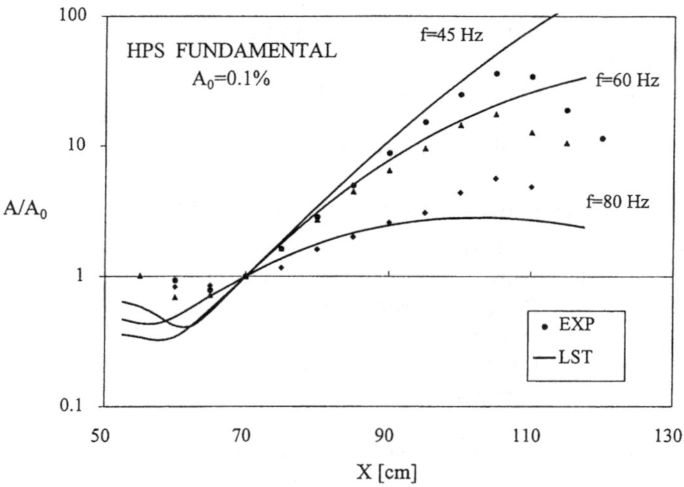

Fig. 4 Measured and calculated HPS maximum amplitudes, *Fr* = 85, 113, 150, and A_0 = 0.1%.

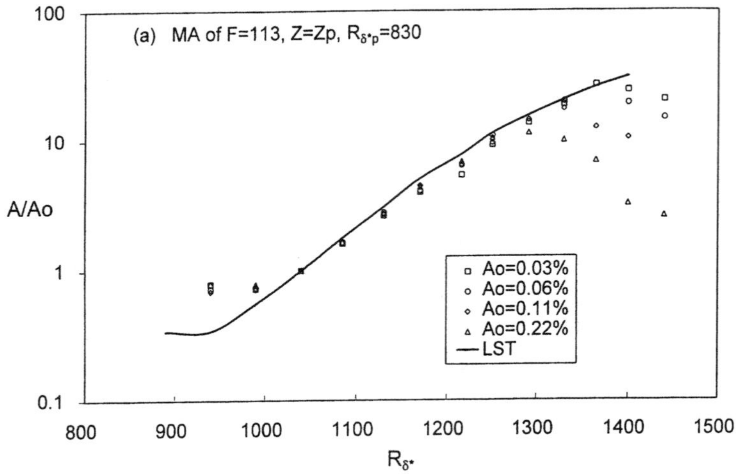

Fig. 5 Initial amplitude effect on the HPS saturation amplitudes, *Fr* = 113.

saturates further downstream and at a lower amplitude as A_0 decreases. From these findings it is clear that the saturation of the fundamental does not depend on a critical amplitude nor on a certain amplification factor or even on the Reynolds number. What is evident, when one considers the amplification of the second harmonic (2F) above the noise level (see Fig. 6b), is that 2F begins to be amplified as the fundamental increases beyond 0.3–0.4%. The amplification of the subharmonic (F/2 presented in Fig. 6c) occurs only further downstream. Only at the lowest HPS amplitude used (i.e., $A_0 = 0.03\%$) is the subharmonic amplified simultaneously with the second harmonic.

This threshold level could be connected to the known results of shifting between K-type and HN-type[12] (for Herbert *and* Novosibyrsk) of transition in Blasius BL at an amplitude of about 0.4%.

Streamwise Evolution of the Wave Pocket

The pulsed localized disturbance, generated by impulsive jet ejected from a hole in the wall, permits many combinations of nonlinear interactions if the disturbance is sufficiently strong. Transient and probably nonlinear receptivity calculations will be required to predict this behavior. Even at the lowest wave pocket (WP) input used, the BL transition occurs within less than 300 $\delta^* r$ (displacement thickness at the source) downstream of the disturbance origin.

In the LST of wave packets, we have considered a direct summation of inclined waves of equal amplitudes at the source with frequencies in the band 20 Hz $< f <$ 140 Hz.

Figures 7a–7c present the comparison between the measured and predicted WP disturbance at various streamwise locations downstream of the

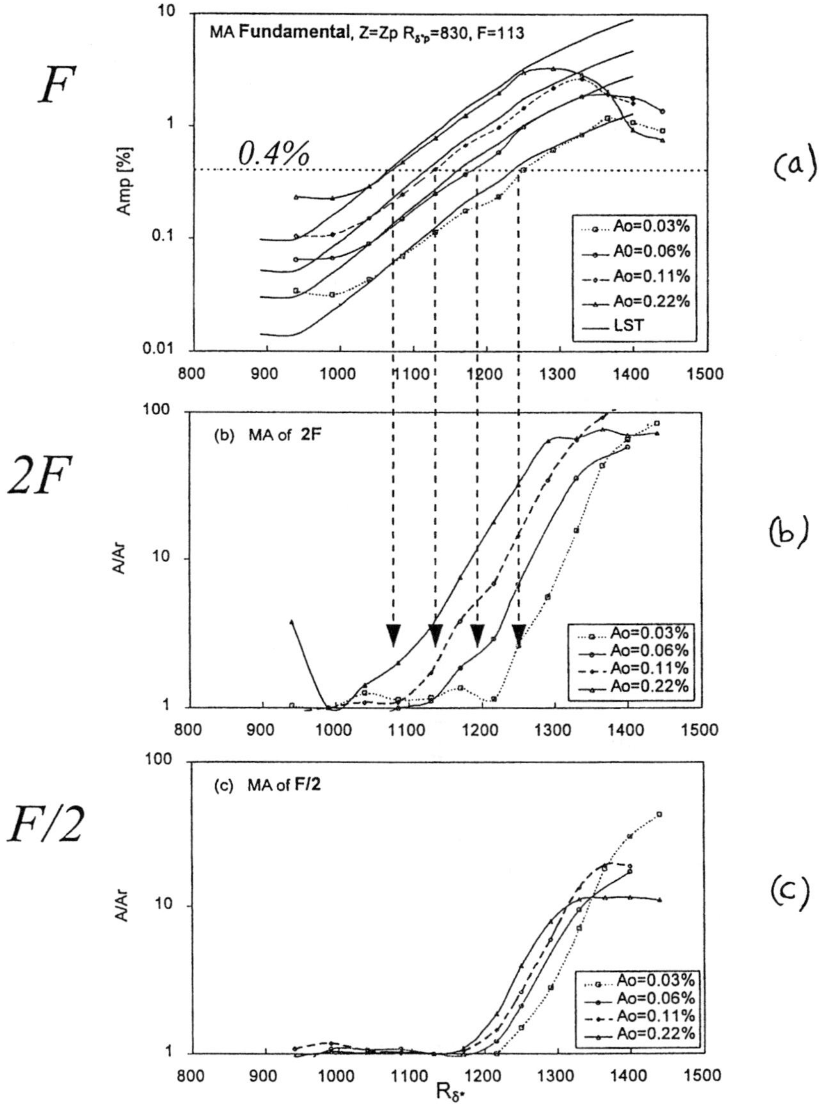

Fig. 6 Harmonic point source nonlinearity.
a) Same data as in Fig. 5 but in dimensional amplitudes
b) Amplitude of the second harmonic (2F)
c) Amplitude of the subharmonic (F/2)

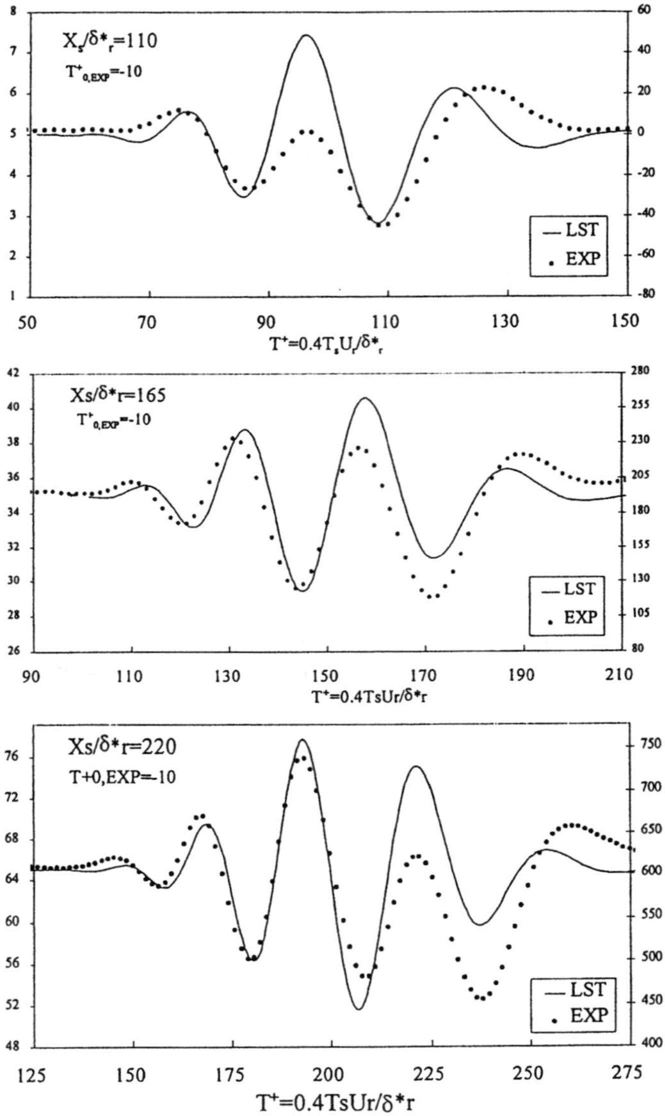

Fig. 7 Streamwise evolution of the wave packet.

source. The ordinate in these figures is arbitrary, but the abscissa is the WP travel distance (assuming convection velocity of $0.4Ur$) normalized by δ^*r. These data were measured at Y corresponding to $0.35Ur$. A time shift of -10 ms was assumed as a "virtual production time" of the WP. At $X = 70$ cm ($Xs/\delta^*r = 110$), the transient part of the disturbance has almost decayed, and

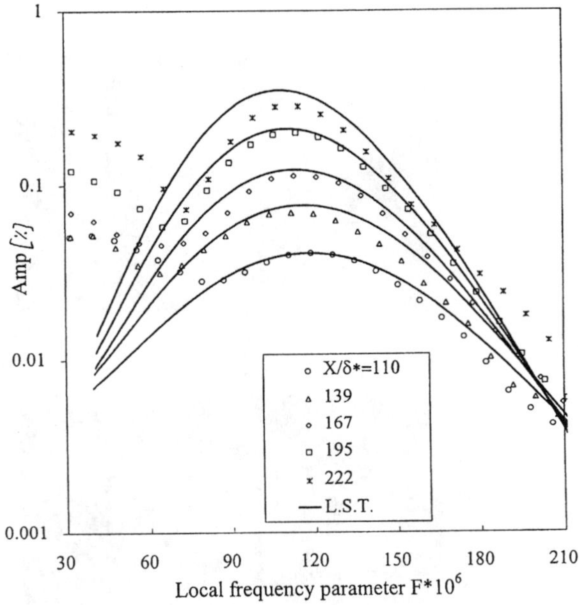

Fig. 8 Wave packet amplitude spectra.

the resemblance to LST is fair. At $X = 80$ cm (Fig. 7b $Xs/\delta^*r = 165$), the agreement is even better, indicating that the WP is mostly linear at this X station. A weak low frequency modulation can be seen at the rear part of the WP. Another 10 cm downstream (i.e., Fig. 7c), the forward part of the WP is in good agreement with the prediction of the LST, but the rear part presents a stronger low frequency modulation.

Figure 8 presents the measured and predicted WP spectra. The only matching performed was in the first X station where the maximum amplitude of the theory and experiment were equaled. The agreement in the three first X stations is excellent over most of the linearly amplified frequencies. The deviation in the lower end of the frequency range is due to the initial transient disturbance. The low frequencies are the first to be amplified because of nonlinearity. In the last X station presented here the deviation from the LST is evident in the low and high frequency ranges and also in the fundamental (linearly amplified) band of frequencies.

Figure 9 presents the measured and predicted spanwise wave pattern at Y corresponding to $0.35Ur$ and $Xs/\delta^*r = 220$. Scales are dimensional to demonstrate the general agreement. The differences are seen most clearly in the rear part of the experimental WP, where a low frequency modulation can be seen mostly away from the centerline (only one side of it is presented).

Further downstream a turbulent spot is formed when the most probable location for initial breakdown to turbulence is on the WP centerline.

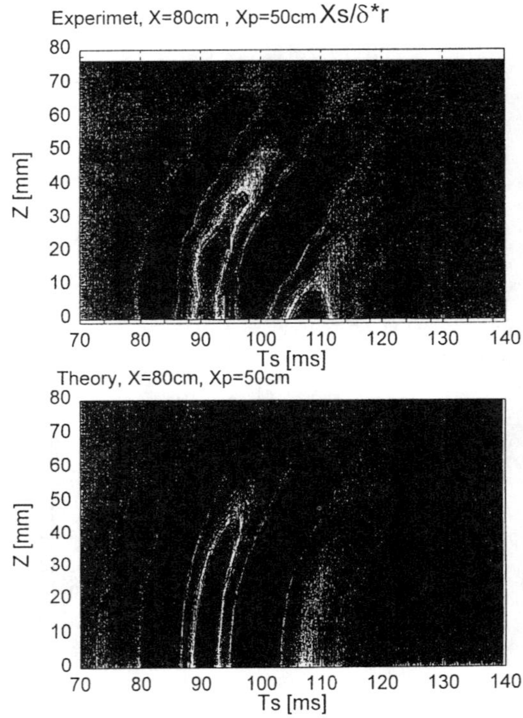

Fig. 9 Spanwise distribution of the wave packet velocity perturbation, experiment and theory.

Conclusions

Very good agreement was found between the LST and the linear stages of the harmonic point source as well as wave packet disturbances evolution on the centerline. This confirms the usefulness of the approach taken presently as far as controlling and documenting the base flow, performing local boundary-layer calculations, and using real localized disturbances both in space and in time. The results also indicate that nonparallel effects are very weak, even in this adverse pressure gradient flow, at least for the low F and moderate *Re* tested.

The nonlinearity of the HPS is characterized by simultaneous amplification of F/2, 3F/2, and 2F, with 2F being the most sensitive to the amplitude of F, as long as it exceeds 0.3–0.4%. The saturation amplitude is lower as the initial amplitude is smaller or as the HPS frequency increases. The nonlinearity of the WP is led by amplification of F/2 and 3F/2, whereas 2F is amplified only when the F band saturates. The initial breakdown to turbulence occurs first—for the HPS as well as for the WP—on the disturbance centerline.

The next step of the investigation will include weakly nonlinear expansion of the linear solution to predict some of the measured effects prior to breakdown to turbulence. The experimental findings also demonstrate transient nonlinear characteristics, waiting for proper theoretical modeling.

References

[1]Mack, L. M., and Kendall, J. M., "Wave Pattern Produced by a Localized Harmonic Source in a Blasius Boundary Layer," AIAA Paper 83-0046, 1983.

[2]Seifert, A., "On the Interaction of Low Amplitude Disturbances Emanating from Discrete Points in a Blasius Boundary Layer," Ph.D. Thesis, Tel-Aviv Univ., Ramat-Aviv, Israel, 1990.

[3]Seifert, A., and Wygnanski, I., "On the Interaction of Wave Trains Emanating from Two Point Sources in a Blasius Boundary Layer," *Proceedings of the Boundary Layer Transition and Control Conference*, Royal Astronomical Society, Cambridge, England, UK, 1991.

[4]Mack, L. M., and Herbert, Th., "Linear Wave Motion from Concentrated Harmonic Sources in Blasius Flow," AIAA Paper 95-0774, 1995.

[5]Gaster, M., and Grant, I., "An Experimental Investigation of the Formation and Development of a Wave Packet in a Laminar Boundary Layer," *Proceedings of the Royal Society of London, A.* 347, London, 1975, p. 253.

[6]Gaster, M., "A Theoretical Model for the Development of a Wave Packet in a Laminar Boundary," *Proceedings of the Royal Society of London, A.* 347, London, 1975, p. 271.

[7]Kachanov, Y. S., and Michalke, A., "Three Dimensional Instability of Flat Plate Boundary Layers: Theory and Experiment," *Eur. J. Mech., B/Fluids*, Vol. 13, No. 4, 1994, pp. 401–422.

[8]Govindarajan, R., and Narasimha, R., "Stability of Spatially Developing Boundary Layers in Pressure Gradients," *Journal of Fluid Mechanics*, Vol. 300, 1995, pp. 117–147.

[9]Seifert, A., "Non-Linear Evolution of Point-Source Disturbances in an Adverse Pressure Gradient Laminar Boundary Layer," *Proceedings of the IUTAM Symposium on Nonlinear Stability of 3D Boundary Layers 1995*, Edited by P. Duck, Kluwer Academic Press (to be published).

[10]Seifert, A., and Wygnanski, I., "On Turbulent Spots in a Laminar Boundary Layer Subjected to a Self-Similar Adverse Pressure Gradient," *Journal of Fluid Mechanics*, Vol. 296, Aug. 10, 1995, pp. 185–209.

[11]Cebeci, T., and Bradshow, P., *Physical and Computational Aspects of Convective Heat Transfer*, Springer–Verlag, 1984.

[12]Kachanov, Yu. S., "Physical Mechanism of Laminar Boundary Layer Transition," *Annual Review of Fluid Mechanics,* Vol. 26, 1994, p. 411–482.

Chapter 2

Turbulent Mixing Layer in Adverse Pressure Gradient

O. König,* J. Schulülter,†, and H. E. Fiedler‡
Technische Universitä Berlin, Berlin, Germany

Introduction

TURBULENT mixing layers are found wherever two flows of different properties mix. They are therefore considered a basic flow module for the general investigation into the nature of turbulence. For many decades plane and axisymmetric mixing layers in constant pressure flows have been investigated, mostly between fluids of equal properties but different velocities; whereas, on the other hand, only little is known about mixing layers undergoing a pressure gradient in streamwise direction. Sabin[1] introduced the local width of the flow as a similarity variable, and Rebollo[2] pointed out that decelerated flow is not self-similar for $\rho_1 = \rho_2$. More recent investigations by Fiedler et al.[3] and by Böttcher[4] indicate enhanced mixing and intensification of structural dynamics in a positive pressure gradient.

The one-stream mixing layer has the strongest spread rate and, therefore, the best mixing conditions. (The width is often taken as an indicator for mixing since there is no generally agreed upon criterion to quantify mixing.) In practical applications, however, it is often necessary to mix two individual streams of different species at a defined mass ratio. This mixing can be done in a two-stream layer. Enhancement of the mixing process in this configuration, and thereby reduction of the necessary length of the mixing regime, is then achieved by introducing a longitudinal positive pressure gradient. This process causes a deceleration mainly of the low-speed stream and hence an increase of the flow parameter λ in x. The opposite, namely reduced mixing, is achieved by a negative pressure gradient. Table 1 summarizes some important relationships.

*Scientist, Hermann-Föttinger-Institut für Strömungsmechanik.
†Student, Hermann-Föttinger-Institut für Strömungsmechanik.
‡Professor, Hermann-Föttinger-Institut für Strömungsmechanik.

Table 1 Shear layer types

Shear layer type	Velocity ratio, U_1/U_2	Spread rate, db/dx	Mixing
One stream	0	$\cong 0.17$	Maximum, but uncontrolled
Two stream			
$d_p/d_x = 0$	$1 \div 0$	$0 \div 0.17$	Below maximum
$d_p/d_x < 0$	$\to 1$	$\to 0$	Reduced
$d_p/d_x > 0$	$\to 0$	$\to 0.17$	Enhanced

Forcing a shear layer in zero pressure gradient (referred to as CSL in the following) increases the local spread rate and, therefore, the mixing region is partly widened. Although the exact effect of this widening on mixing intensification is not known, a certain effect may be assumed.

Theory

The growth of an unforced mixing layer without pressure gradient is proportional to the ratio of a global diffusion velocity ($\propto \delta U_{1,2}$) to mean convection velocity ($\propto \Sigma U_{1,2}$). Assuming the validity of the so-called Abramovich–Sabin relation for slightly nonparallel flows,[3] we write

$$\frac{\mathrm{d}b(x)}{\mathrm{d}x} \propto \lambda(x) = \frac{\Delta U(x)}{\Sigma U(x)} = \frac{U_2(x) - U_1(x)}{U_2(x) + U_1(x)} \tag{1}$$

where $b(x)$ is the local width, $U_1(x)$ and $U_2(x)$ are the local freestream velocities, and $\lambda(x)$ is the shear layer parameter. The Bernoulli equation is satisfied with an exponential law for diffusion and convection velocity, viz.,

$$\Delta U(x) \propto (\hat{x} - x_0)^k \tag{2}$$

$$\Sigma U(x) \propto (\hat{x} - x_0)^n \tag{3}$$

where $x = \hat{x} - x_0$ and x_0 denoting a virtual origin. Neglecting streamline curvature in potential flow and assuming constant pressure across the mixing layer, the governing velocities are

$$\Delta U \propto (\hat{x} - x_0)^{-n} \propto \Sigma U^{-1} \tag{4}$$

With these results Eq. (1) yields an expression for the width of the shear layer in streamwise direction as

$$b = \frac{C}{1 - 2n} \lambda x + b_0 \tag{5}$$

where C is a constant and b_0 denotes the virtual width at the trailing edge of the splitter plate. The value of n characterizes the different shear layer types: $n = 0$ stands for the CSL, $n > 0$ stands for the mixing layer in favorable pressure gradient, and $n < 0$ stands for the mixing layer in adverse pressure gradient. Classical analysis yields relations for various bulk and other quantities, e.g., the characteristic frequency, the Taylor microscale, etc.

Experimental Setup

In adverse pressure gradient the low-speed side of a mixing layer decelerates stronger than the high-speed side. Our test section (Fig. 1) has, therefore, only one variable perforated wall on the low-speed side to control the pressure gradient. Suction on the low-speed side is used to finely adjust the freestream velocities. The shape of the test section with suction produces an approximately hyperbolic decrease of the freestream velocities. The initial velocities are $U_{1_0} = 14$m/s and $U_{2_0} = 28$m/s; hence, $\lambda_0 \approx 0.33$. Maximum deceleration is 54% at low and 13% at high-speed side, bringing the velocity parameter to a maximum value of $\lambda_{max} = 0.58$ at the end of the test section. This value was found to be the limit, beyond which separation occurs on the

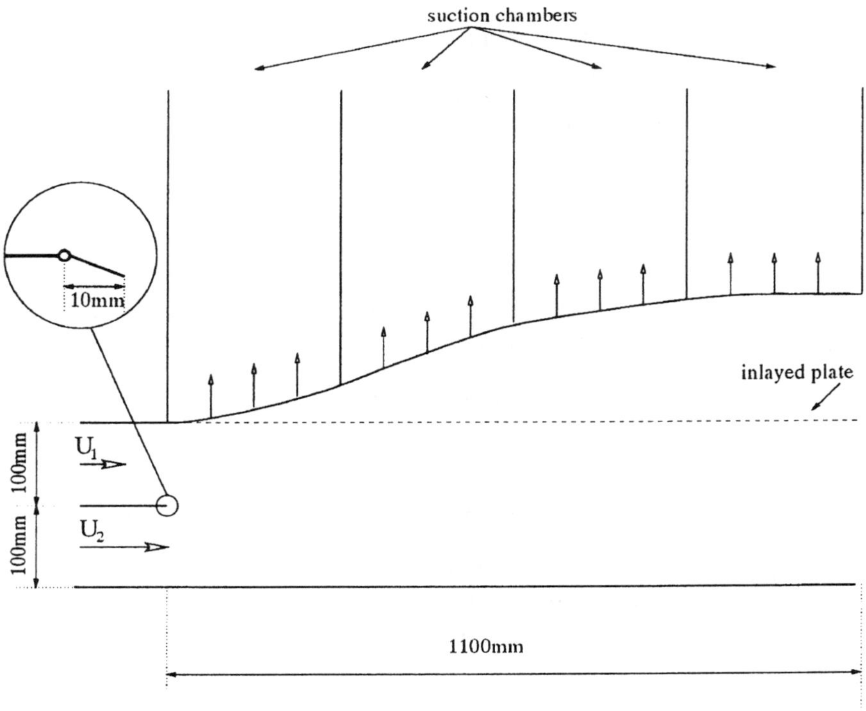

Fig. 1 Test section.

side walls. The turbulence level in the potential streams is less than 0.4% of the initial velocity difference; both boundary layers on the splitter plate are tripped.

A spanwise flap driven by two loudspeakers outside the test section is mounted at the trailing edge of the splitter plate. A computer generated amplified sinusoidal signal forces the shear layer with amplitudes $u'_{periodic}/\Delta U_0 < 1\%$, measured at the end of the wake region, i.e, $x \approx 80$ mm. The forcing frequencies were chosen as $f = 100$ Hz and 200 Hz for the single frequency and double frequency modes.

To compare the results with a CSL, measurements were done with a flat plate placed into the test section where then the velocity difference was constant within 2% along x.

Measurements were carried out with conventional hot-wire technique. X-wire probes were used for the mean velocity quantities and a six-wire probe[5] was used to measure the vorticity. To analyze the energy budget, the signals were triple-decomposed into the mean value, a phase-dependent coherent constituent, and a time-dependent random part, viz.,

$$f(t) = \bar{f} + \tilde{f}(\varphi) + f_R(t) \tag{6}$$

For phase averaging the forcing signal was used as a trigger. Averages were taken at 60 phases per period for up to six periods of the fundamental forcing frequency. The phase-averaged energy equations[6] could be evaluated and the quantities related to the fundamental or the subharmonic frequency separated.

Results

Natural Flow

All averaged quantities (Figs. 2–4) show excellent similarity when the y coordinate is nondimensionalized with the local shear layer width. The width $b = y_{0.05} - y_{0.95}$ is evaluated from the mean streamwise velocity profile (Fig. 2). All velocities were found to scale with the local velocity difference.

The normalized shear stress (Fig. 3) reaches a maximum value of 0.0095. The maxima of the normal stresses (Fig. 4) are 15, 13, and 15% (rms-value), which are slightly lower than the maxima in a CSL with identical initial conditions (17, 15, and 17%, respectively). Although the nondimensionalized maximum values are lower, their absolute values increase in streamwise direction due to the increasing velocity difference.

Figure 5 shows the streamwise development of width, vorticity thickness, and momentum thickness. By curve fitting we found

$$b = 4.66 + \frac{0.188}{1 - 2n} \lambda x \tag{7}$$

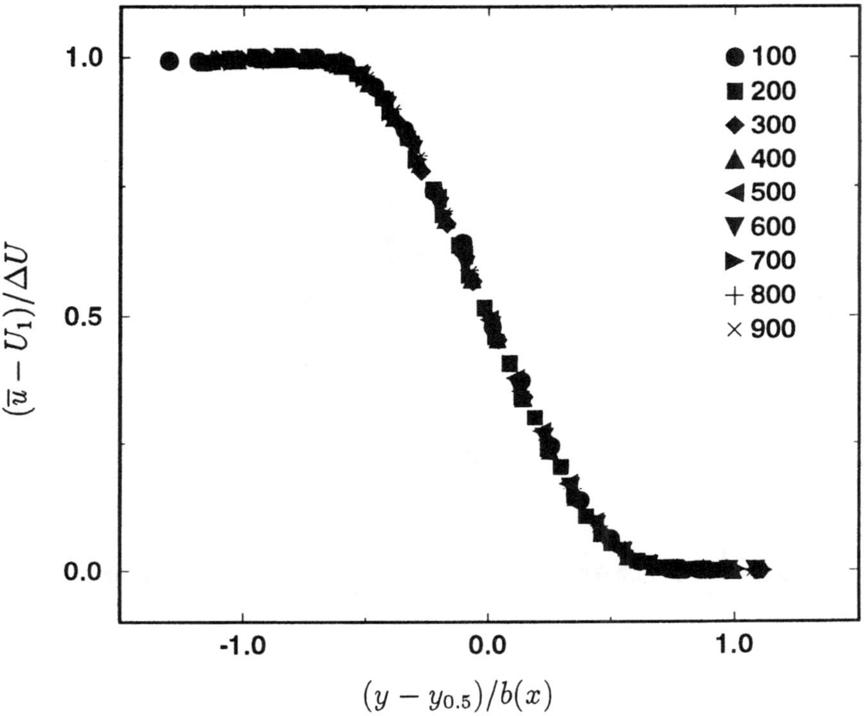

Fig. 2 Mean velocity profiles.

$$\delta = 1.27 + \frac{0.178}{1 - 2n}\lambda x \tag{8}$$

$$\theta = 0.81 + \frac{0.034}{1 - 2n}\lambda x \tag{9}$$

where $n = -0.26$ was evaluated from Eq. (4). Compared to the width of a CSL (dotted line in Fig. 5), the mixing region is considerably expanded, which was expected according to the preceding. The spread rate (solid line in Fig. 7) related to the local shear layer parameter is indeed constant, proving the Abramovich–Sabin relation also for slightly nonparallel flows. This rate is slightly higher than that one of the CSL ($\approx$0.176), but in the range of values found in the literature.[7] The streamwise development of the Taylor microscale, the turbulent Reynolds number, and the characteristic frequency is well described by the appropriate relations.

Figure 6 shows enstrophy components nondimensionalized with the Taylor microscale λ_f and the velocity difference. This normalization was chosen because the vorticity fluctuations are components of fluctuating derivatives

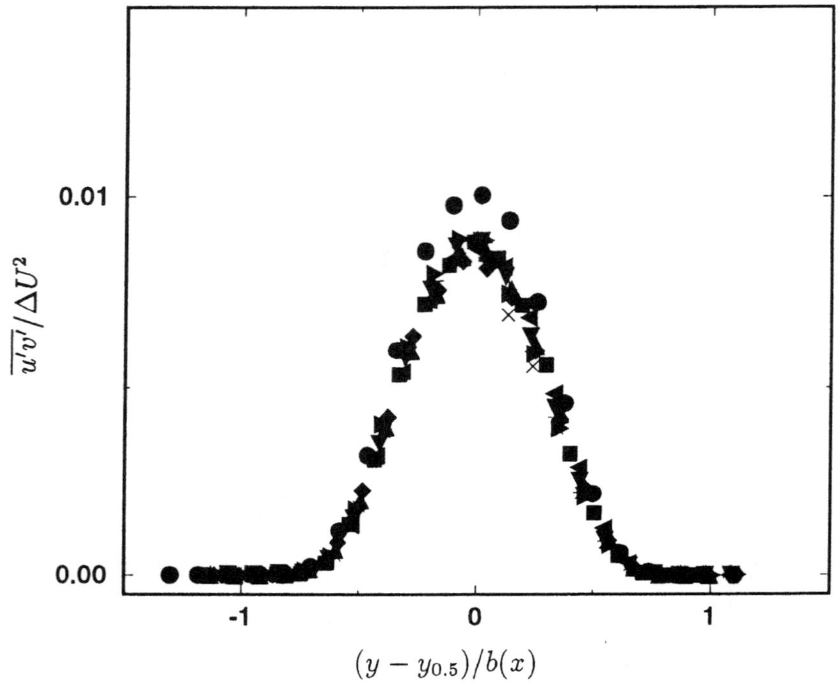

Fig. 3 Shear stress, symbols see Fig. 2.

and are therefore correlated to the small scale turbulence. Measurements in a CSL[8] show the vorticity scaled with the global width increases downstream. Comparing the streamwise development of width Eq. (5) and the Taylor microscale,

$$\lambda_f \propto (\hat{x} - x_0)^{(1-n)/2} \tag{10}$$

The reason for this is obvious: the ratio of Taylor microscale and width increases with downstream distance.

Whereas the properties of the streamwise component are similar, the spanwise and crosswise components increase along x. This statement indicates that the structures are compressed in streamwise direction and therefore grow in spanwise and mostly lateral direction. Also, they reach slightly higher values than in the CSL. In agreement with Ballint and Wallace[9] and Kim,[6] the maximum value of the streamwise component was found to be higher than those of the normal components.

Forced Flow

Forcing the shear layer with a single frequency or in combination with its subharmonic leads to organized formation, amalgamation, and collapse of

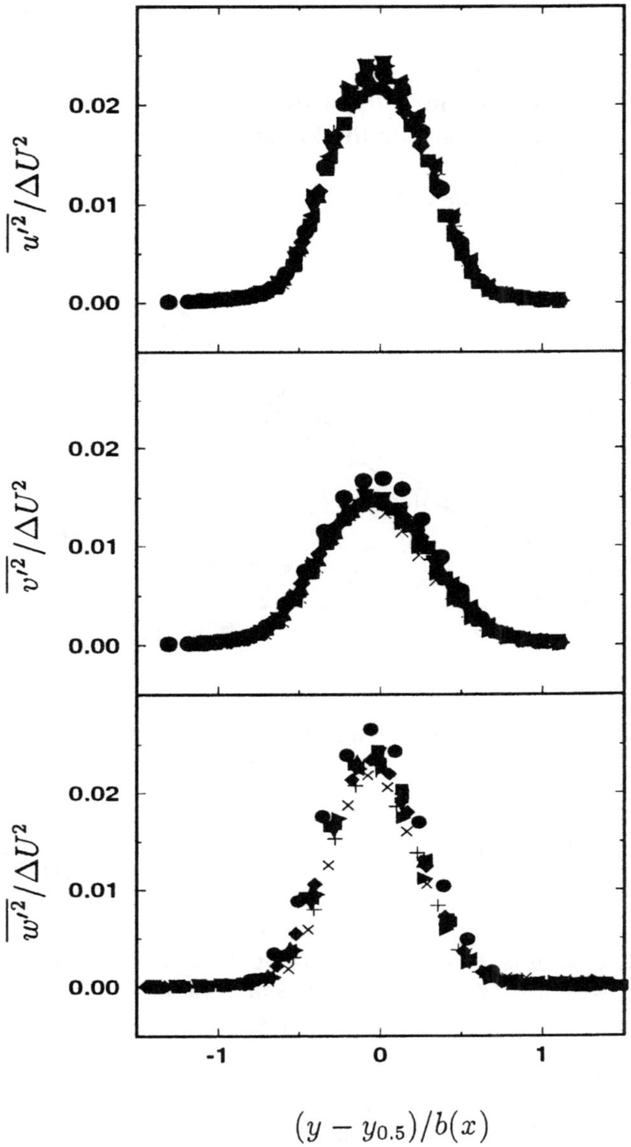

Fig. 4 Normal stresses, symbols see Fig. 2.

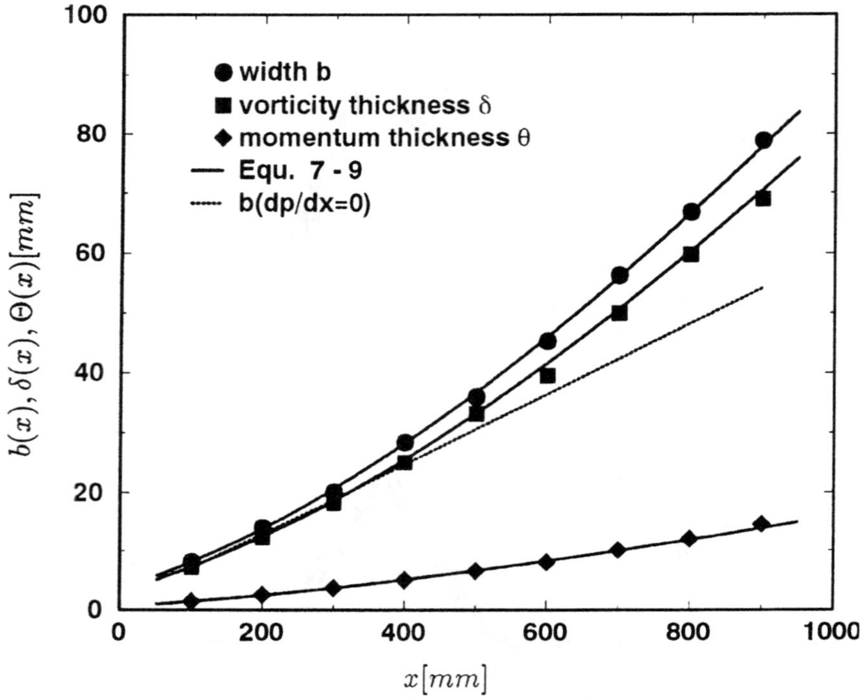

Fig. 5 Mixing layer development in streamwise direction.

the large structures in the flow. The size of the structures formed and, therefore, the width of the shear layer depends on the forcing amplitude. In Fig. 7 the widths for different forcing amplitudes at $f = 200$ Hz are shown. Once the shear layer is enlarged, its width will not return to the unforced situation further downstream as in a CSL. The spread rate is the same as in the unforced case, as is seen in Fig. 8, showing the width along x of the mixing layer forced with a fundamental and its subharmonic. For comparison the width of the unforced and the single frequency forced shear layer is included.

Figures 5 and 7 are interpreted as follows: whereas the turbulent structure of a CSL is fed with a constant amount of energy per unit length due to the constant velocity difference, the structure develops until it reaches a state where energy production and dissipation are in equilibrium. After this the shear layer is self-similar in the classical sense. In an adverse pressure gradient, where the velocity difference increases in streamwise direction, the amount of energy fed into the mixing region increases along x, and the turbulent structure will not reach a state of equilibrium corresponding to the local velocity difference. Forcing this shear layer causes an increased energy flux into the flow, which pushes the turbulent structure into a state corresponding to the local velocity difference, but because this difference is

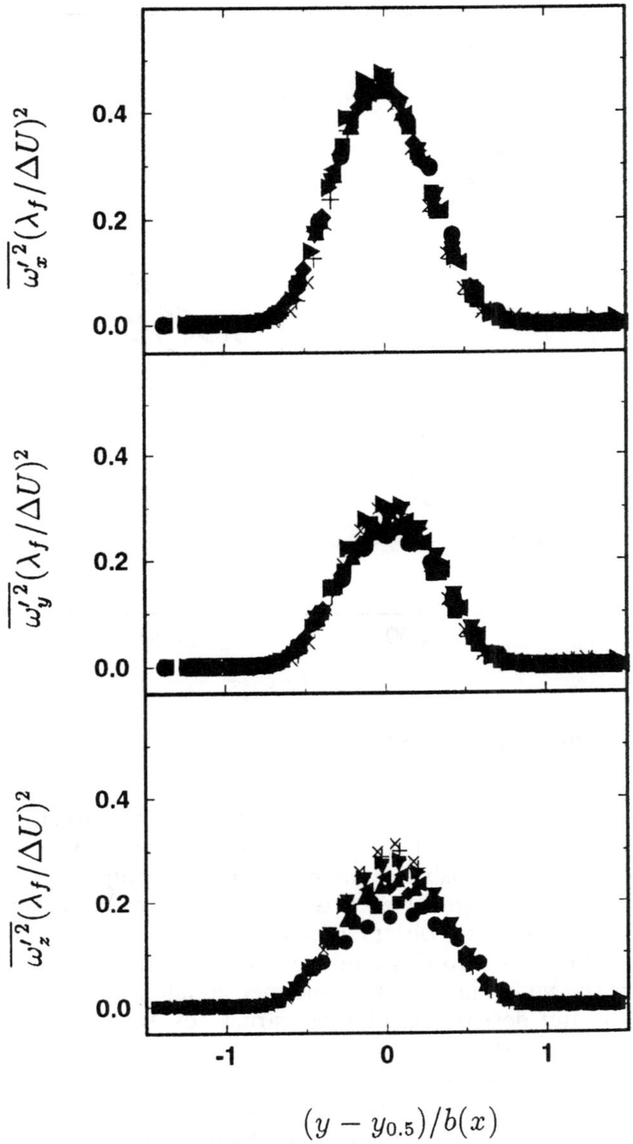

Fig. 6 Normal enstrophy components, symbols see Fig. 2.

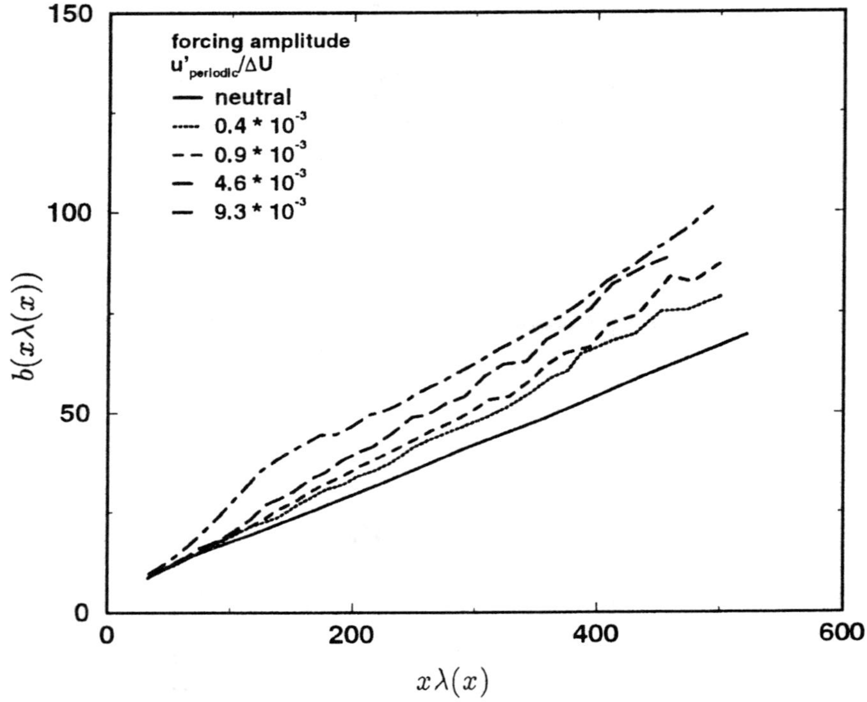

Fig. 7 Streamwise development of forced mixing layer f = 200 Hz.

increasing with downstream distance, the mixing region starts to grow naturally again thereafter. Probably only strong periodic perturbations, i.e., large forcing amplitudes, can push the mixing layer into a saturated state where the width remains constant or even decreases for a short distance until it starts increasing again, as is observed in a CSL.

The different effects of forcing a mixing layer in zero and adverse pressure gradients are shown in Fig. 9 by the nondimensionalized energy production: the formation and destruction of large structures are compressed into a shorter streamwise distance.

In both configurations the energy from the mean motion to the turbulent is transferred to 80% via its coherent part. Also, the energy transferred from the organized turbulent structures, once they are formed, back to the mean flow is only done by the coherent part, while the production of uncorrelated energy is even decreasing. The production of uncorrelated energy by the coherent motion is one order of magnitude smaller.

A quantitative look into the mixing layer forced with fundamental (f_f = 200 Hz) and subharmonic (f_s = 100 Hz) frequencies is given in Fig. 10a by the phase-locked vorticity $\langle\omega_z\rangle = \overline{\omega}_z + \tilde{\omega}_z$. The first roll up occurs at x = 300 mm, while the amalgamation of two structures takes place between x = 500 and 700 mm. The measurement technique enables separation of the con-

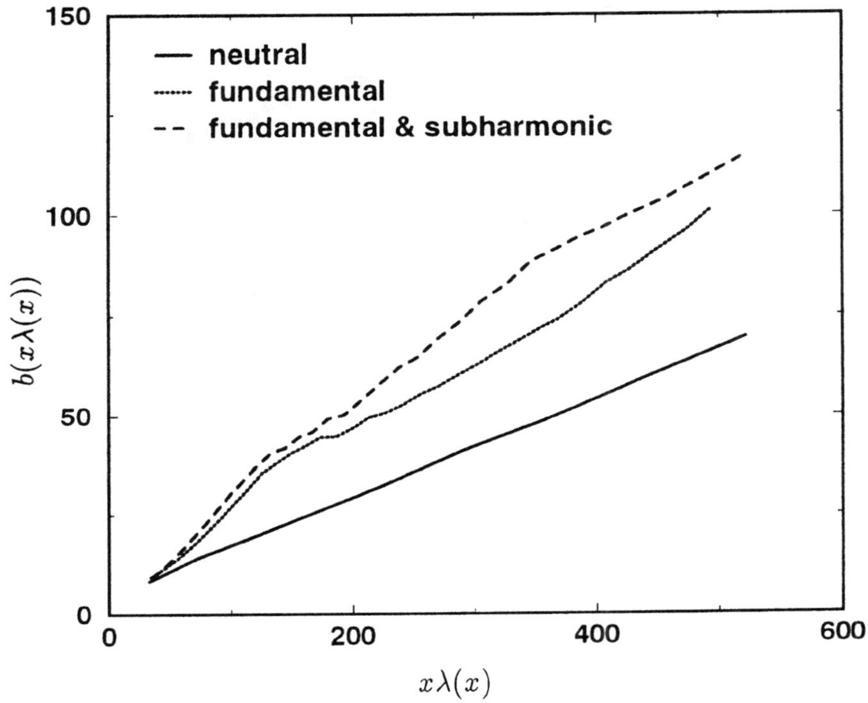

Fig. 8 Streamwise mixing layer development, fundamental (f_f = 200 Hz) and subharmonic forcing (f_s = 100 Hz).

stituents of the vorticity due to the fundamental and the subharmonic frequency (Figs. 10b and 10c). Whereas the fundamental part is mainly found in the upstream region of the test section where the first large structures form, the subharmonic part is found further downstream where two fundamental structures amalgamate into subharmonic. For comparison the phase-locked vorticity of the mixing layer forced with single frequency is shown in Figs. 10d and 10e. Figures 10b and 10c and Figs. 10d and 10e appear similar, so that the bifrequential forced mixing layer appears to be merely a superposition of the two effects. This is also shown in Fig. 11, where the coherent energy is plotted in streamwise direction. The energy amount of the bifrequential forced shear layer is roughly the sum of the structures in single-frequency-forced layers. Different behaviors at various phase shifts have not been found so far.

Conclusions

The investigation shows the validity of the Abramovich–Sabin relation for slightly nonparallel flows. The increasing spreading rate indicates enhanced mixing, whereas the increasing shear stress accounts for the increasing tur-

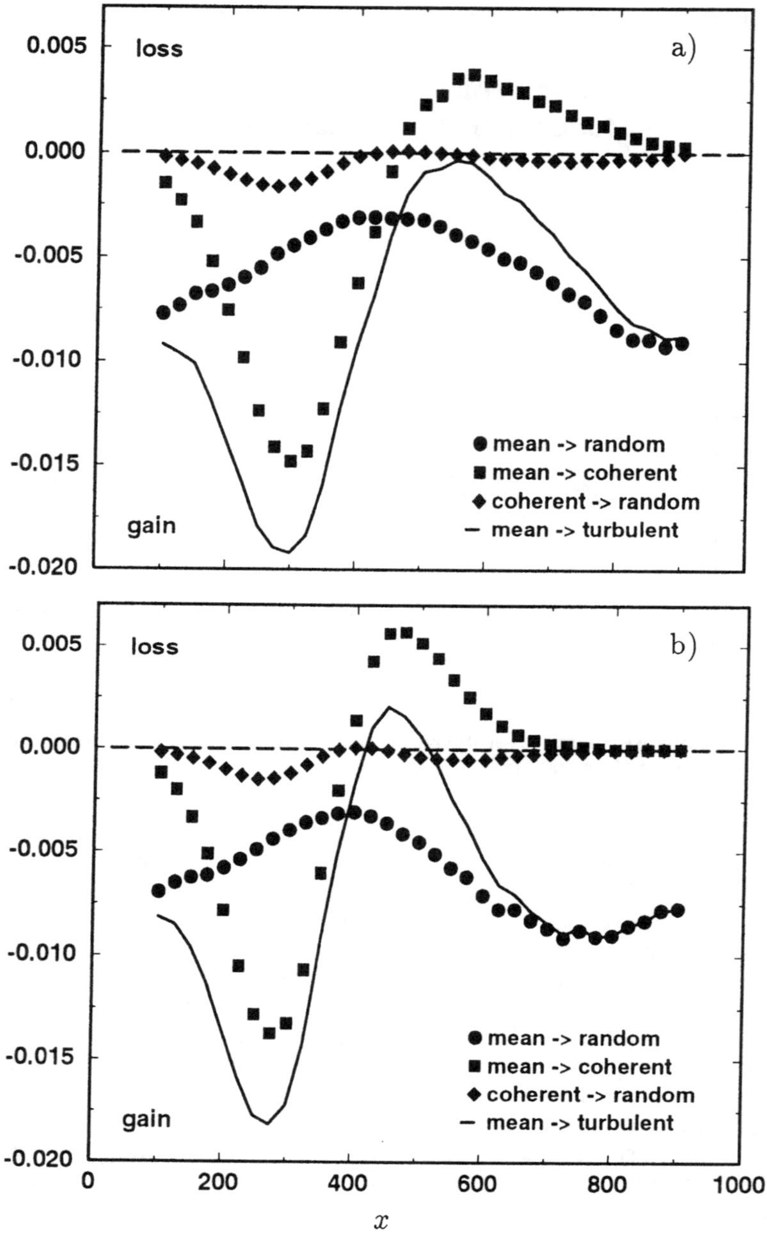

Fig. 9 Comparison of streamwise development of triple-decomposed, integrated energy production of a forced mixing layer (f = 200 Hz): a) d_p/d_x = 0 and b) d_p/d_x > 0.

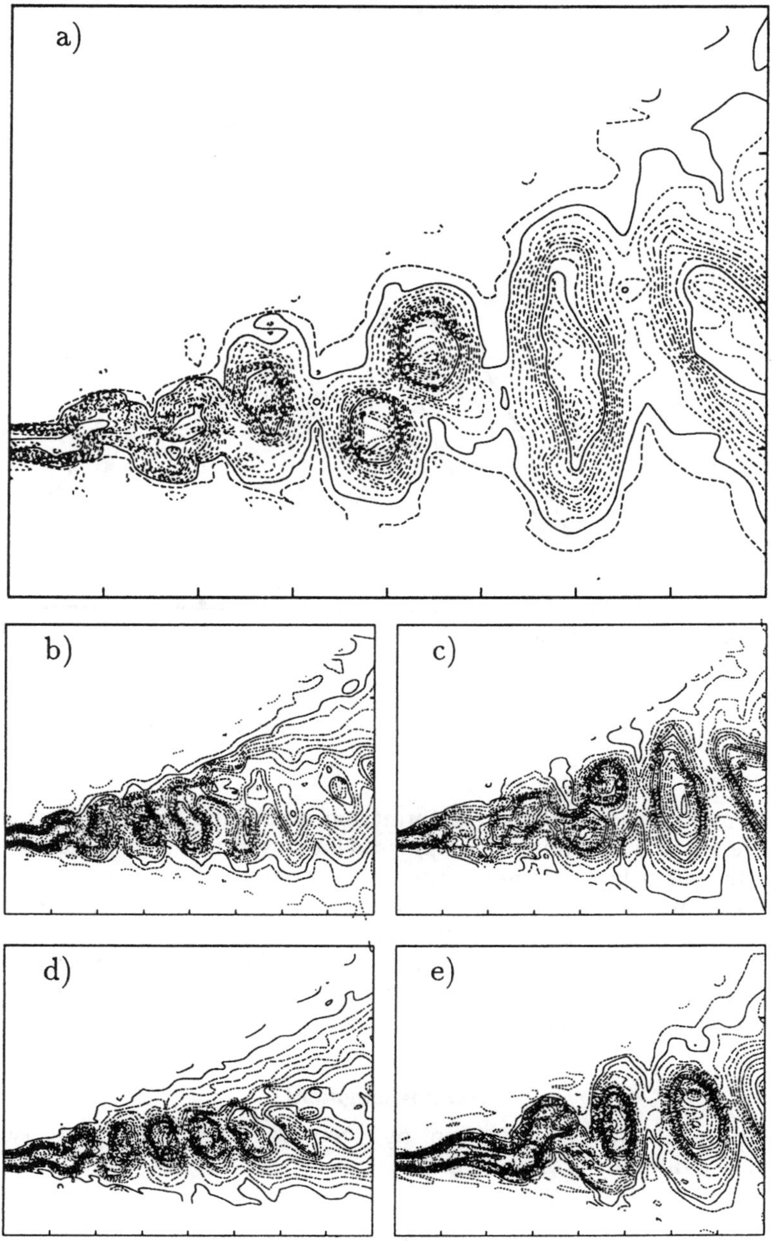

Fig. 10 Phase-locked vorticity of forced mixing layer: bifrequential forcing (f_f = 200 Hz, f_s = 100 *Hz*); a) fundamental and subharmonic part $\langle\omega_z\rangle$, b) fundamental part $\langle\omega_z\rangle_{200\ \mathrm{Hz}}$ of a, c) subharmonic part $\langle\omega\rangle_{z100\ \mathrm{Hz}}$ of a, and single frequency forcing: d) f = 200 Hz, e) f = 100 Hz.

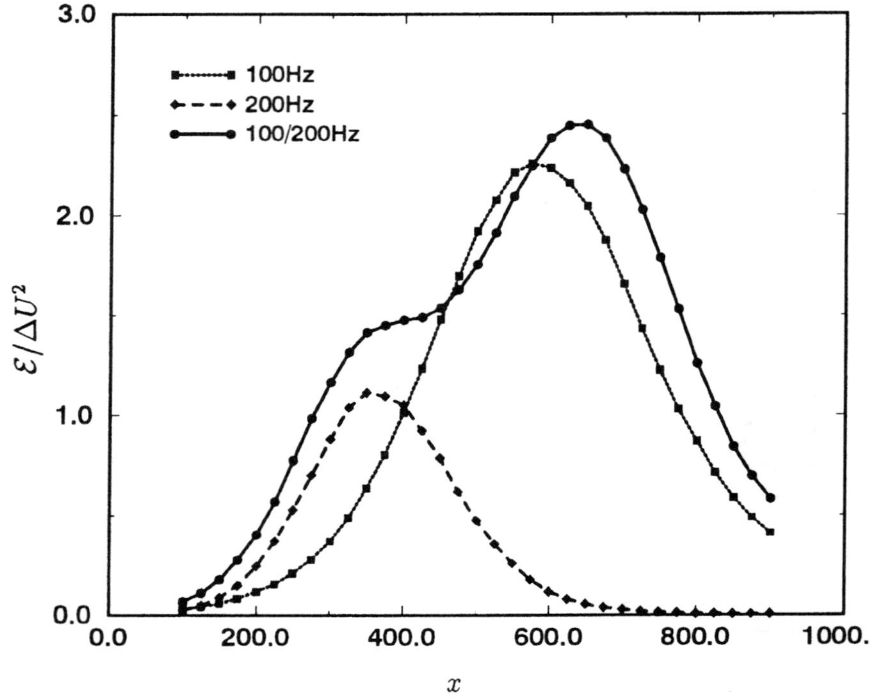

Fig. 11 Streamwise development of integrated coherent energy.

bulent energy and vorticity. The formation, amalgamation, and destruction of large vortical structures is compressed into a shorter streamwise distance when the shear layer is forced. The energy content of a shear layer forced with fundamental and subharmonic frequency appears as a simple superposition of the energy contents of the single frequency forced layers.

Although a quantitative criterion of scalar mixing is still lacking, a positive pressure gradient in combination with forcing apparently is an effective tool to control the mixing efficiency in a shear layer.

Acknowledgment

Support of this investigation by Deutsche Forschungsgemeinschaft (DFG) is gratefully acknowledged.

References

[1]Sabin, C. M., "An Analytical and Experimental Study of the Plane, Incompressible, Turbulent Free Shear Layer with Arbitrary Velocity Ratio and Pressure Gradient," Mech. Eng., No. MD9, Stanford Univ., Palo Alto, CA, 1963.

[2]Rebollo, R. M., "Analytical and Experimental Investigation of a Turbulent Mix-

ing Layer of Different Gases in a Pressure Gradient," Ph.D. Thesis, Cal. Inst. of Tech., CA, 1972.

[3]Fiedler, H. E., Köpp, N., Kim, J.-H., "The Spatially Accelerated Mixing Layer," *Eur. J. Mech., B/Fluids*, Vol. 10, No. 4, 1991.

[4]Böttcher, M., "Numerische Simulation von Scherschichten Mit der Discrete-Vortex Methode," Dissertation, Techn. Univ. Berlin, Berlin, 1991.

[5]Kim, J.-H., "Wirbelstärkemessungen in Einer Turbulenten Scherschicht," Dissertation, Techn. Univ. Berlin, Berlin, 1989.

[6]Hussain, A. K. M. F., "Coherent Structures—Reality and Myth," *Physics of Fluids,* Vol. 26, No. 10, 1983.

[7]Oster, D., and Wygnanski, I., "The Forced Mixing Layer Between Parallel Streams," *Journal of Fluid Mechanics,* Vol. 123, 1982.

[8]Lang, D. B., "Laser Doppler Velocity and Vorticity Measurementis in Turbulent Shear Layers," Ph.D. Thesis, Cal. Inst. of Tech., CA, 1985.

[9]Ballint, J.-L., and Wallace, J. M., "The Statistical Properties of the Vorticity Field of a Two-stream Turbulent Mixing Layer," *Proceedings of Advances in Turbulence 2*, edited by H.-H. Fernholz and H. E. Fiedler, Springer–Verlag, 1989.

II. Turbulence

Chapter 3

Stabilized Negative Viscosity as Subgrid-Scale Representation of Two-Dimensional Turbulence

Semion Sukoriansky*
Ben Gurion University of the Negev, Beer Sheva, Israel
Alexey Chekhlov†
Cambridge Hydrodynamics, Inc., Princeton, New Jersey
and
Boris Galperin‡
University of South Florida, St. Petersburg, Florida

Introduction

MANY natural flows that span a large number of scales possess features of both three- and two-dimensional turbulence and can be classified somewhere between the purely three-dimensional and two-dimensional extremes. Thus, even though the focus of this paper is isotropic two-dimensional turbulence, one should keep in mind that applications of the results to quasi-two-dimensional flows are thought. Quasi-two-dimensional turbulent flows are widely found in geophysics and engineering. Although under normal circumstances all flows are unstable to three-dimensional instabilities,[2] there exist natural situations when a flow may attain a quasi-two-dimensional configuration or even become quasi-two-dimensionalized on certain scales. There are two major factors that may cause a flow to become quasi-two-dimensional: geometry of the flow boundaries and/or certain body forces (or "extra strains") whose action leads to smoothing of the velocity fluctuations in a preferred direction. While in the geophysical context, both factors are equally important (i.e., the small aspect ratio, density

*Senior Lecturer, Department of Mechanical Engineering and the Perlstone Center for the Aeronautical Engineering Studies.
†Research Scientist.
‡Associate Professor, Department of Marine Science.

stratification, rotation), in the engineering context the second factor usually is the dominant one (for instance, the so-called mechanism of "magnetic friction" in magnetohydrodynamic flows with low magnetic Reynolds number[3]).

Despite the specific peculiarities of quasi-two-dimensional flows related to the energy and vorticity dynamics, their subgrid scale representation has not received sufficient attention so far. Moreover, large-eddy simulation (LES) of quasi-two-dimensional flows in the energy transfer subrange has never been attempted, despite the fact that such flows would bear strong analogy to large-scale oceanic and atmospheric circulation and that direct numerical simulation (DNS) of such flows cannot be expected in the foreseeable future. Thus, there exists a need to improve our understanding of the subgrid scale (SGS) processes in the energy transfer subrange of quasi-two-dimensional flows and to successfully simulate such flows when their energy sources reside in the SGS region. Both of these issues were addressed recently by Sukoriansky et al.[4] and are reviewed in the present paper.

Specific difficulties of LES of two-dimensional turbulence in the energy transfer subrange are that the SGS representation must simultaneously be consistent with the conservation laws for both energy and enstrophy. The enstrophy conservation prevents a cascade of energy from large to small scales because such a cascade would increase enstrophy.[5–7] Thus, the energy is transferred up scales. Mathematically, this important feature is illustrated by the Fjørtoft theorem.[8] However, the direct cascade of enstrophy from large to small scales is possible. It results in molecular dissipation of large-scale vorticity at small scales. Drawing analogy to eddy viscosity in three-dimensional flows, one can infer that small scale processes in two-dimensional flows generate an effective or eddy viscosity for the vorticity of large scales. However, introducing an eddy viscosity concept in two-dimensional flows seems to be intrinsically inconsistent and self-defeating because the dissipation of enstrophy will be accompanied by the dissipation of energy, which is physically incorrect.

In addition, in LES of the inverse energy cascade, the wave number modes of the energy source lie in the subgrid scales, so that the SGS representation must assume the function of the forcing. A possible resolution to this problem can involve the replacement of the SGS forcing by force located in the explicitly resolved region near k_c. However, this solution is not only quite cumbersome but also significantly distorts the explicit scales near k_c. In addition, this approach is difficult for implementation in physical space, particularly for bounded systems and/or systems with spatially nonuniform energy sources.

Another solution would be to introduce a negative eddy viscosity, first extensively discussed by Starr,[9] as an SGS parameterization of the unresolved energy source. Although such an SGS representation could address the issue of the inverse energy cascade, it would not satisfy the constraint of the zero enstrophy flux. In addition, equations of motion with negative viscosity produce ill-posed problems. It appears therefore that addressing the issue of SGS representation for quasi-two-dimensional flows in self-consistent and comprehensive way would require full consideration of energy and enstrophy dynamics and should be based upon the corresponding transport equations. Such an approach was first outlined by Kraichnan[1] who introduced the notion of two-parametric viscosity. This approach and its implications will be elaborated in the next section.

Two-Parametric Viscosity and Its Implementation for Large Eddy Simulation of Two-Dimensional Turbulence

Two-dimensional incompressible turbulent flows are described by the vorticity equation

$$\frac{\partial \zeta}{\partial t} + \frac{\partial(\nabla^{-2}\zeta,\zeta)}{\partial(x,y)} = \nu_0 \nabla^2 \zeta + f \tag{1}$$

where ζ is fluid vorticity, ν_0 is molecular viscosity, and f is external force.

The introduction of the classical eddy viscosity concept for LES with Eq. (1) implies that there is a distinct scale separation between the resolvable and SGS modes. Indeed, only if such a separation exists, the eddy viscosity would be k-independent and a function of the cutoff wave number k_c only. However, the assumption of scale separation fails in all turbulent flows, particularly in two-dimensional flows, such that, strictly speaking, an SGS representation should depend on two parameters, k and k_c. Such two-parametric viscosity, denoted by $\nu(k|k_c)$, was first introduced by Kraichnan.[1] It describes the energy exchange between given resolved vorticity mode with the wave number k and all SGS modes with $k > k_c$. The two-parametric viscosity is derived from the evolution equation for the spectral enstrophy density $\Omega(k,t) \equiv (4\pi)^{-1}k\langle\zeta(\boldsymbol{k},t)\zeta(-\boldsymbol{k},t)\rangle$, where $\langle\ldots\rangle$ denotes ensemble averaging:

$$(\partial_t + 2\nu k^2)\Omega(k,t) = \mathcal{T}_\Omega(k,t) \tag{2}$$

The enstrophy transfer function in Eq. (2), $\mathcal{T}_\Omega(k,t)$, is given by

$$\mathcal{T}_\Omega(k,t) = \frac{k}{2\pi}\Re\left\{\int_{\boldsymbol{p}+\boldsymbol{q}=\boldsymbol{k}} \frac{\boldsymbol{p}\times\boldsymbol{q}}{p^2}\langle(\zeta\boldsymbol{p},t)\zeta(\boldsymbol{q},t)\zeta(-\boldsymbol{k},t)\rangle \frac{\mathrm{d}\boldsymbol{p}\,\mathrm{d}\boldsymbol{q}}{(2\pi)^2}\right\} \tag{3}$$

For a system in statistical steady state, the two-parametric transfer $\mathcal{T}_\Omega(k|k_c)$ and viscosity $\nu(k|k_c)$ are calculated from Eq. (3) by extending integration only over all such triangles $(\boldsymbol{k},\boldsymbol{p},\boldsymbol{q})$ that $|k+p| < q < k-p$ and p and/or q are greater than k_c:

$$\nu(k|k_c) = -\frac{\mathcal{T}_\Omega(k|k_c)}{2k^2\Omega(k)} \tag{4}$$

In a wide class of quasi-normal approximations, $\mathcal{T}_\Omega(k|k_c)$ in two dimensions is given by

$$\mathcal{T}_\Omega(k|k_c) = \frac{2k}{\pi}\iint_\Delta \Theta_{-k,p,q}(p^2 - q^2)\sin\alpha\left[\frac{p^2-q^2}{p^3q^3}\Omega(p)\Omega(q) - \frac{k^2-q^2}{k^3q^3}\Omega(q)\Omega(k) + \frac{k^2-p^2}{k^3p^3}\Omega(p)\Omega(k)\right]\mathrm{d}p\,\mathrm{d}q \tag{5}$$

where $\Theta_{-k,p,q}$ is the triad relaxation time. Also, the angle α is formed by the vectors $\boldsymbol{p}$ and $\boldsymbol{q}$, and $\iint_\Delta$ denotes integration over the area defined above.

Different spectral closure models provided different specification of $\Theta_{-k,p,q}$. Chekhlov et al.[10] compared $\nu(k|k_c)$ calculated from their DNS data with those evaluated by Kraichnan[1] using his test field model (TFM) and obtained from the renormalization group (RG) theory.[11] Very good agreement between all three results was observed. The RG-based normalized two-parametric viscosity,

$$N(k/k_c) \equiv \nu(k|k_c)/|\nu(0|k_c)| \tag{6}$$

is plotted in Fig. 1. One can see that for the large scale modes, for which $k \ll k_c$ and scale separation exists, the effect of the SGS modes is represented by a negative and constant viscosity, such that these modes gain energy from

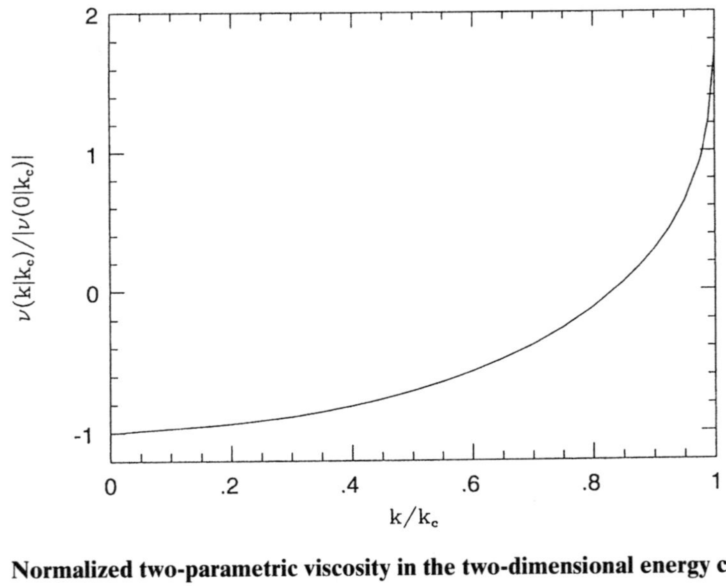

Fig. 1 Normalized two-parametric viscosity in the two-dimensional energy cascade range.

their SGS counterparts by means of the inverse transfer. On the other hand, $\nu(k|k_c) > 0$ for $k \to k_c$ such that the modes close to k_c lose their energy to the SGS modes. The difference between the large scale gain and the small scale loss is equal to ε, the rate of the energy input due to the forcing f in Eq. (1). If enstrophy balance is considered, recall that the enstrophy dissipation is most efficient at small scales, such that the resulting balance for the resolvable scales turns out to be zero,[6] i.e., the enstrophy is conserved. This explains how the two-parametric viscosity resolves the controversy of the inverse cascade of energy and conservation of enstrophy in the energy subrange of two-dimensional flows. It appears therefore that the only physically correct way to represent SGS processes in two-dimensional turbulence would be through the two-parametric viscosity.

To test the two-parametric viscosity-based SGS parameterization in the energy transfer subrange, a series of LES of two-dimensional turbulence in Fourier space was designed. These LES were based upon Eq. (1) in which all subgrid scale processes including the forcing were represented by the two-parametric viscosity $\nu(k|k_c)$,

$$\frac{\partial(\boldsymbol{k})}{\partial \boldsymbol{t}} + \int_{|p|,|k-p|<k_c} \frac{\boldsymbol{p}\times\boldsymbol{k}}{p^2}\,\zeta(p)\,\zeta(k-p)\frac{\mathrm{d}\boldsymbol{p}}{(2\pi)^2} = -\nu(k|k_c)k^2\zeta(k)\zeta\, 0 < k < k_c \quad (7)$$

It is important to reiterate that in LES of two-dimensional turbulence in the energy subrange, the source of energy resides on the unresolved scales, such that Eq. (7) appears unforced. However, as was explained earlier, the negative part of $\nu(k|k_c)$ serves as the only energy source for the resolved modes. In the course of the present LES, numerical results critically depend on the way $\nu(k|k_c)$ is introduced in the solver. Thus, a series of simulations was designed with the purpose of understanding the nature of the problems associated with the implementation of the two-parametric viscosity and of identifying the most viable and robust ways to use this viscosity in LES of two-dimensional flows.

A detailed description of numerical method is given in Ref. 4. We solved Eq. (7) in a periodic box $2\pi \times 2\pi$ using 162^2 Fourier modes with complete dealiasing based on the two-thirds rule. The cutoff wave number was set at $k_c = 50$, which is about half of the resolution used in DNS of Chekhlov et al.[10] The initial flowfield was set to zero everywhere except for a narrow band of wave numbers in the middle part of the spectrum where it was assigned random Gaussian distribution. The DNS inferred value of $\bar{\varepsilon}$ was about 5.19×10^{-10}; the same $\bar{\varepsilon}$ was used in LES. The time step in LES was set to $\delta t = 0.5$ satisfying both convective and viscous necessary conditions for linear stability. Based upon the size of the largest energy containing eddy with the wave number $k_{\min}$, $2\pi/k_{\min}$, and the total energy of the steady-state $\overline{E}$, the maximum large-scale eddy turnover time defined as $\tau_{tu} = 2\pi/(k_{\min}\sqrt{2\overline{E}})$ was about $\tau_{tu} \simeq 900$ for LES where $k_{\min}$ was set to 4.

With $\nu(k|k_c)$ known, Eq. (7) can be solved directly. As shown in Ref. 4,

$$\nu(k|k_c) = 0.327\bar{\varepsilon}^{-1/3}k_c^{-4/3}N(k/k_c) \tag{8}$$

where $N(k|k_c)$ is given by Eq. (6), and since $\bar{\varepsilon}$ is a constant, $\nu(k|k_c)$ is a function of k and k_c only. It was assumed that thus defined $\nu(k|k_c)$ would be capable of supporting inverse energy cascade with constant SGS energy input ε. To verify this assumption, one needs to examine the evolution of total energy and enstrophy of the resolved modes, $\overline{E}(t)$ and $\overline{\Omega}(t)$, respectively.

The SGS formulation (8) with constant $\bar{\varepsilon}$ was used in the first LES. Both $\overline{E}(t)$ and $\overline{\Omega}(t)$ exhibit nonlinear growth indicating that not only the rate of the energy transfer to the resolvable scales $\bar{\varepsilon}_{\text{LES}} \neq$ const, but also the rate of the enstrophy transfer $\bar{\eta}_{\text{LES}} \neq 0$. The roots of the problem are revealed when one calculates the rate of the energy input into all resolved models $\bar{\varepsilon}_{\text{LES}}$. The energy equation derived from the definition of spectral energy density and Eq. (7) yields

$$\bar{\varepsilon}_{\text{LES}}(t) \equiv \frac{\partial \overline{E}(t)}{\partial t} = -2\int_0^{k_c} \nu(k|k_c)E(k,t)k^2 \, \mathrm{d}k \tag{9}$$

Because in the first series of LES $\nu(k|k_c)$ depends on k and k_c only and $E(k,t)$ is a dynamic variable that depends on the evolution of the flowfield, $\bar{\varepsilon}_{\text{LES}}(t)$ also turns out to be time dependent. This is not only in direct conflict with the requirement that $\bar{\varepsilon}_{\text{LES}} = \bar{\varepsilon} =$ const, but also leads to a positive feedback between the energy input and total energy of the system, which results in numerical instability. To correct this problem, one must ensure that $\bar{\varepsilon}_{\text{LES}} =$ const. This result can be achieved by allowing $\nu(k|k_c)$ to become time dependent and related to resolved variables. The philosophy of using actual flowfield characteristics to determine energy input and dissipation would be analogous to a standard practice of three-dimensional LES.

Inspection of Eq. (9) reveals that $\bar{\varepsilon}_{\text{LES}}(t)$ is proportional to the total enstrophy of the resolvable field, such that stipulating $\bar{\varepsilon}_{\text{LES}} =$ const would require $\nu(k|k_c)$ to become time dependent and inversely proportional to $\overline{\Omega}(t)$. Then, following Sukoriansky et al.,[4] Eq. (8) may be rewritten in the form

$$\nu(k|k_c) = \frac{\bar{\varepsilon}}{0.8\overline{\Omega}(t)}N(k/k_c) \tag{10}$$

(the time dependency of $\nu(k|k_c)$ will be implied in the following discussion but suppressed in notations). The negative feedback between energy input and enstrophy of the resolved modes is the mechanism that stabilizes numerical process. Note that the SGS formulation based upon Eq. (10) complicates Eq. (7) because its right-hand side now depends on the functional of the solution $\overline{\Omega}(t)$. However, the SGS representation (10), though complicated, is in line with the eddy viscosity approach, in which the eddy viscosity coefficient is usually solution dependent.[12–15]

In a LES that employs the formulation (10) together with a specially designed large-scale drag that effectively removes energy that reaches the largest computational scales and prevents generation of coherent structures,[16,17] it was found that simulations can be carried out nearly indefinitely (being terminated after about 200 turnover times), while a very robust Kolmogorov $k^{-5/3}$ energy spectrum was established. Figures 2a and 2b show that $\overline{E}(t)$ and $\overline{\Omega}(t)$ slightly oscillate around their steady-state values (the source of these oscillations is probably the self-adjustment of the numerical scheme to the mismatch between the small scale energy forcing and large scale withdrawal). The instantaneous energy spectrum, Fig. 3, exhibits steady and nearly perfect Kolmogorov scaling.

An important characteristic of two-dimensional turbulence in the energy transfer subrange is the energy flux $\Pi_E(k) = \int_0^k \mathcal{T}_\Omega(n) n^{-2}\, dn$, which theoretically should be equal to $\bar{\varepsilon}$ for any k. Note that the enstrophy transfer function $\mathcal{T}_\Omega(n)$ is defined by Eq. (3) where integration is extended over all triangles $\boldsymbol{k} + \boldsymbol{p} + \boldsymbol{q} = 0$ including the SGS ones for which k and/or p and/or $q > k_c$.

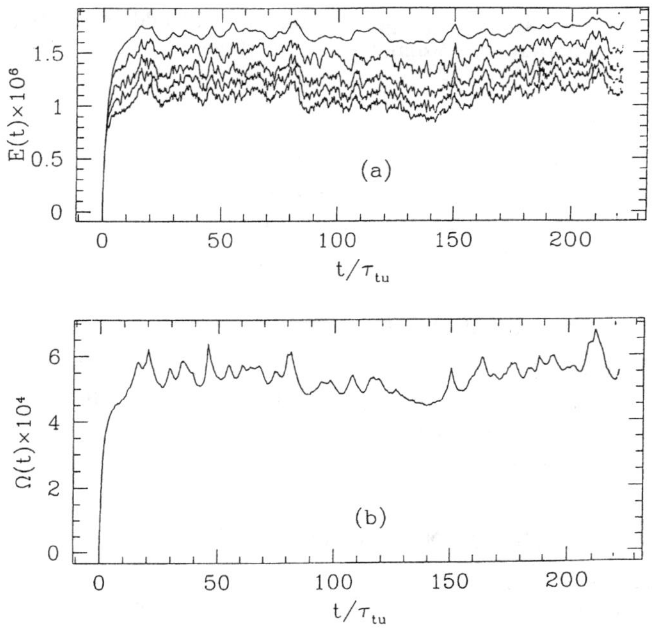

Fig. 2 Evolution: a) the total energy $\overline{E}(t)$ and b) total enstrophy $\overline{\Omega}(t)$ in LES with eddy viscosity given by Eq. (10). Figure 2a also shows the evolution of $\overline{E}(t)$ with the energy of the fourth, fifth, sixth, and seventh modes removed.

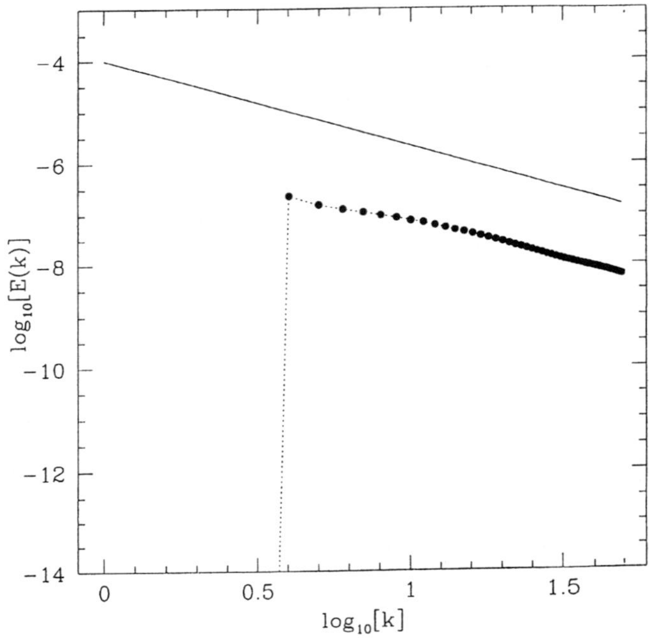

Fig. 3 The time-averaged energy spectrum for LES with eddy viscosity given by Eq. (10). The solid line shows the Kolmogorov −5/3 slope.

However, if $\Pi_E(k)$ is calculated for LES results, then the SGS triangles should be excluded such that the integration area in Eq. (3) is reduced. On the other hand, the energy flux into the interval $0 < n < k$ in LES consists of two contributions, $\int_0^k \mathcal{T}'_\Omega(n)n^{-2}\, dn$ [$\mathcal{T}'_\Omega$ implies the reduced integration area in Eq. (3)], and the direct flux from the SGS modes, $-2\int_0^k \nu(n|k_c)n^2E(n)\,dn$. Figure 4 shows that thus calculated $\Pi_E(k)$ remains nearly constant for $k > 16$ and is equal to about 5×10^{-10}, which is very close to the specified value of $\bar{\varepsilon}$ obtained from DNS by Chekhlov et al.[10] The decrease of $\Pi_E(k)$ at small k is due to the large-scale energy removal.

Stabilized Negative Viscosity Formulation

For practical implementation of SGS formulation [Eq. (10)], it is convenient to approximate $N(k/k_c)$ in Eq. (6) by a series in powers of k^2. Even the first two terms of this series,

$$N(k/k_c) \simeq -1 = \alpha(k/k_c)^2 \tag{11}$$

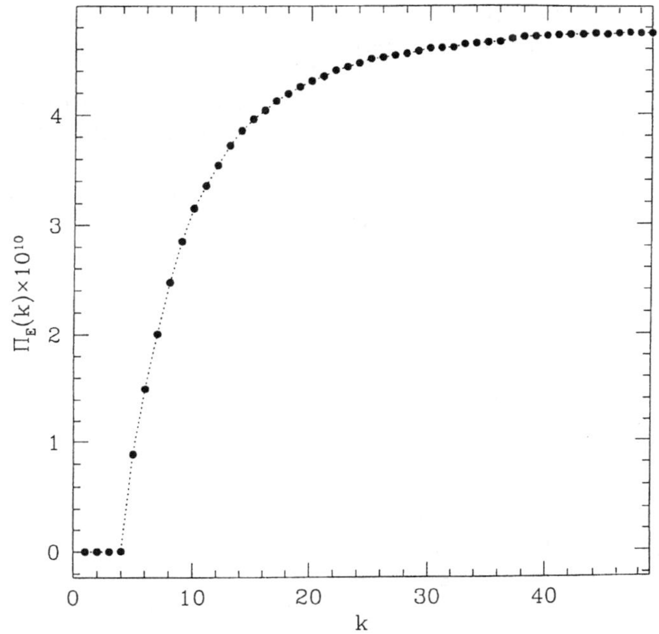

Fig. 4 The energy flux $\Pi_E(k)$.

where α is a constant, are sufficient to perform successful LES of two-dimensional turbulence. To find α, recall that representation (11) must ensure zero enstrophy input in the energy subrange,

$$\bar{\eta}_{\rm LES} = 2\int_0^{k_c} \nu(k|k_c)E(k,t)k^4\, \mathrm{d}k = 0 \tag{12}$$

Substituting Eq. (11) into Eq. (12) and assuming that $E(k,t)$ is Kolmogorovian, one finds that α = 8/5. This leads to the two-term SGS parameterization for LES of two-dimensional turbulence in which the eddy viscosity takes the form

$$\nu(k|k_c) = \frac{25}{18}\frac{\bar{\varepsilon}}{\bar{\Omega}(t)}\left[-1+\frac{8}{5}\left(\frac{k}{k_c}\right)^2\right] \tag{13}$$

The first term in the right-hand side of Eq. (13) accounts for the energy flux from the unresolved modes and the inverse cascade of this energy, whereas the second term represents the energy dissipation near the cutoff.[4]

Simulations performed with the formulation (13) exhibit very little difference compared to the LES that employed the full curve $N(k/k_c)$ given by Eq.

(6). One infers therefore that Eq. (13) is a viable two-term SGS representation for LES of two-dimensional turbulence; obviously, formulation (13) is significantly simpler than Eq. (10).

For practical purposes, it would be most appealing to use formulation (13) with the dissipation term constant. A numerical experiment was conducted with the energy source in Eq. (13) not changed, but in the dissipation term. $\overline{\Omega}(t)$ was replaced by its value calculated for the Kolmogorov spectrum. Such an approach yields the eddy viscosity in the form

$$\nu(k|k_c) = -\frac{25}{18}\frac{\bar{\varepsilon}}{\overline{\Omega}(t)} + Ak^2 \tag{14}$$

where A is a constant given by

$$A = 0.511\varepsilon^{-1/3}k_c^{-10/3} \tag{15}$$

The results of this LES are presented in Figs. 5 and 6; there is a very good agreement with the corresponding results obtained with the full curve $N(k/k_c)$ up to $t \simeq 40\ \tau_{tu}$ after which the solution begins to deteriorate. The Kolmogorov constant C_K is close to 5.8, which is close to the DNS value.[10]

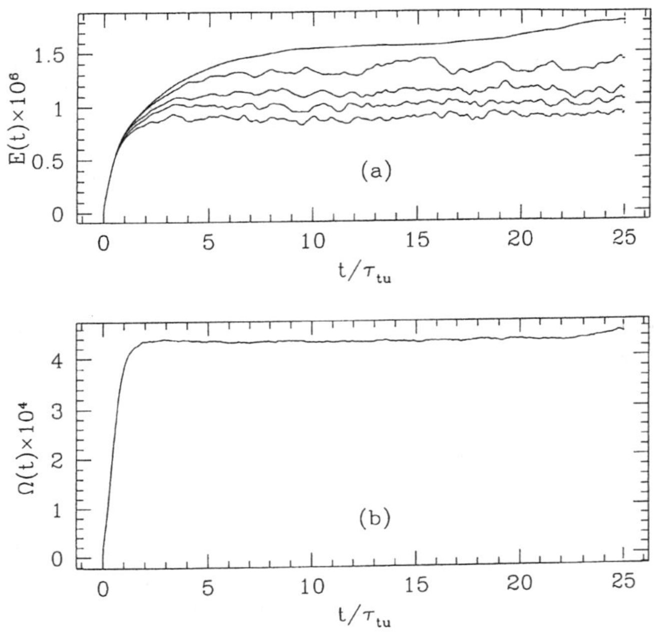

Fig. 5 Same as Fig. 2 but for LES with eddy viscosity given by Eq. (14).

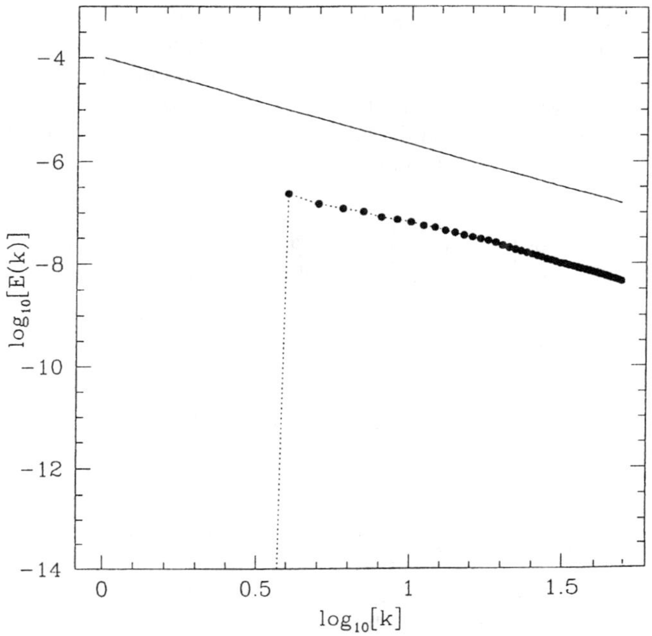

Fig. 6 Same as Fig. 3 but for LES with eddy viscosity given by Eq. (14).

One can therefore infer that Eqs. (14) and (15) are probably the simplest SGS representation for LES of two-dimensional turbulence possible.

Clearly, the two term SGS representation (14) is also simpler to implement in physical space. In simulations of homogeneous flows with space independent resolution, a straightforward SGS model based upon Eq. (14) yields

$$-\frac{\partial}{\partial x_i}\left(A_2\frac{\partial}{\partial x_i}\zeta(x)\right)-A_4\frac{\partial^4}{\partial x_i^2\partial x_j^2}\zeta(x) \tag{16}$$

where

$$A_2=\frac{25}{18}\frac{\bar{\varepsilon}}{\overline{\Omega}(x)} \tag{17}$$

$$A_4=A=0.511\varepsilon^{-1/3}(\Delta(x)/2\pi)^{10/3}=\text{const} \tag{18}$$

and where Δ is the grid resolution and $\overline{\Omega}(x)$ denotes the enstrophy averaged over the computational domain [note that the Laplacian term in Eq. (16) is written in the conservative form]. In inhomogeneous flows, $\overline{\Omega}(x)$ should represent the enstrophy averaged over an area adjacent to the grid cell x. Finally, when a grid of variable resolution is used, a conservative representation of the biharmonic term with space dependent coefficient A_4 is required. From tensorial considerations, such a representation can be constructed by analogy to the Laplacian term in Eq. (16), where $\zeta(x)$ is replaced by $\nabla^2\zeta(x)$, such that Eqs. (16) and (18) become

$$-\frac{\partial}{\partial x_i}\left(A_2(x)\frac{\partial}{\partial x_i}\zeta(x)\right)-\frac{\partial}{\partial x_i}\left(A_4(x)\frac{\partial}{\partial x_i}\nabla^2\zeta(x)\right) \tag{16a}$$

and

$$A_4(x)=0.511\varepsilon^{-1/3}[\Delta(x)/2\pi]^{10/3} \tag{18a}$$

respectively.

Equations (16) and (16a) thus include two terms, the negative Laplacian and positive (in the sense of dissipation) biharmonic, and structurally resemble the Kuramoto–Sivashinsky equation widely known from combustion theory[18] and flows with chemical reactions.[19,20] However, the SGS representation (16–18) combined with the explicit equation for the resolved scales produces far more complicated equations than the Kuramoto–Sivashinsky equation because generally its coefficients are not constant but, as in the eddy viscosity approach, are functions of the flow. There have been previous attempts to use formulations similar to (16–18) but with constant coefficients (dubbed the Kuramoto–Sivashinsky–Navier–Stokes equation)[21] to perform LES of two-dimensional turbulence. However, they were not overly successful even in reproducing the Kolmogorov spectrum, mostly because they used constant "eddy viscosity" coefficients. Because, on the one hand, SGS representation (16–18) includes a negative Laplacian viscosity term and a positive, stabilizing, dissipation term, but, on the other hand, it is quite different from the Kuramoto–Sivashinsky equation in which "viscosity" coefficients are constant, which will be referred to as the stabilized negative viscosity (SNV) formulation. The SNV formulation is expected to be particularly useful in simulations of atmospheric and oceanic flows where large scale motions are quasi-two-dimensional by their nature while small-scale forcing is significant.[22]

Discussion and Conclusions

The central subject of this paper has been the derivation of an appropriate SGS representation for LES of homogeneous and isotropic two-dimensional turbulence in the energy transfer subrange. Conservation of energy and enstrophy in two-dimensional turbulent flows leads to coexistence of two spectral transfer processes, upscale for energy and downscale for enstrophy. Conventional eddy viscosity formulations are purely dissipative and thus fail to accommodate both transfers simultaneously. The proper SGS parameterization for LES of two-dimensional flows is given by the two-parametric viscosity $\nu(k|k_c)$ that accounts for the energy (or enstrophy) exchange between given resolved and all SGS modes and which includes negative and positive branches. The negative branch of $\nu(k|k_c)$ represents the unresolved, small-scale forcing and inverse cascade of energy, whereas the positive one represents the dissipation. Negative and positive parts of the two-parametric viscosity play a vital role in ensuring that all conservation and spectral transfer laws of two-dimensional turbulence are satisfied. Then, $\nu(k|k_c)$ was used in LES of two-dimensional turbulence in the energy transfer subrange; the negative part of $\nu(k|k_c)$ was the only energy source. If $\nu(k|k_c)$ is flow independent, then a positive feedback is established between the forcing and the total energy of the system leading to numerical instability. Thus, another scheme was designed in which $\nu(k|k_c)$ was flow dependent, but the rate of the energy input was kept constant. This LES exhibited a very robust behavior and produced results consistent with analytical theories and DNS. Then, a simplified SGS representation was advanced in which only two terms were retained: one negative and proportional to k^2, and one positive and proportional to k^4. The viscosity coefficient in the negative term served as the energy source and had to be flow dependent, to ensure that the energy input remained constant. However, because the k^4 term is mostly active at high wave number region near the cutoff wave number k_c and its main role is energy and enstrophy dissipation at $k \to k_c$, this term did not need to be flow dependent and was kept constant evaluated from analytical theories of two-dimensional turbulence. Indeed, LES with constant and flow dependent k^4 terms gave similar results.

The implementation of the two term SGS parameterization in physical space results in the so-called stabilized negative viscosity (SNV) representation that includes negative Laplacian and positive biharmonic viscosities. A conservative version of SGS SNV representation was developed for spatially varying viscosities.

By its very design, the negative viscosity term, which is a nonlinear functional of the solution, causes the SNV scheme to differ substantially from the Kuramoto–Sivashinsky equation where viscosity coefficients are constant. In fact, the SNV representation is a peculiar case of the eddy viscosity approach. Following the common sense drawn from simulations of three-dimensional turbulence, one would expect that if all scales of a turbulent flow are resolved, then eddy viscosity would simply become the molecular viscosity. In two-dimensional turbulence, however, resolution of all scales must be accompanied by restoration of the explicit small-scale forcing, disappearance of both negative Laplacian and positive biharmonic eddy viscosities, and appearance of a single, positive, molecular, Laplacian viscosity.

The SNV representation should be especially useful in simulations of the large-scale behavior of quasi-two-dimensional flows in which a considerable amount of the forcing energy resides on the unresolved scales. Such flows are typical in geophysical fluid dynamics where, due to the topographic and other constraints, large-scale flows are quasi-two-dimensional and where the small-scale forcing is a major source of energy. In fact, in such situations, the use of the SNV scheme should be preferable because then, all the unresolved forcing factors, which are very difficult to account for otherwise, are included in a single, simple, negative Laplacian eddy viscosity term.

Acknowledgments

This research has been partially supported by ONR Grants N00014-92-J-1363, N00014-92-C-0118, N00014-92-C-0089, and NSF Grant OCE 9010851. Stimulating discussions with Ilya Staroselsky are appreciated.

References

[1]Kraichnan, R. H., "Eddy Viscosity in Two and Three Dimensions," *Journal of the Atmospheric Sciences,* Vol. 33, 1976, pp. 1521–1536.

[2]Batchelor, G. K., "Computation of the Energy Spectrum in Homogeneous Two-Dimensional Turbulence," *Physics of Fluids Supplement II*, Vol. 12, 1969, pp. 233–238.

[3]Sommeria, J., and Moreau, R., "Why, How and When, MHD Turbulence Becomes Two-Dimensional," *Journal of Fluid Mechanics*, Vol. 118, 1982, pp. 507–518.

[4]Sukoriansky, S., Chekhlov, A., Orszag, S. A., Galperin, B., and Staroselsky, I., "Large Eddy Simulation of Isotropic Two-Dimensional Turbulence in the Energy Subrange," *Journal of Scientific Computing*, Vol. 11, No. 1, 1996, pp. 13–45.

[5]Kraichnan, R. H., "Inertial Ranges in Two-Dimensional Turbulence," *Physics of Fluids*, Vol. 10, 1967, pp. 1417–1423.

[6]Kraichnan, R. H., "Inertial-Range Transfer in Two- and Three-Dimensional Turbulence," *Journal of Fluid Mechanics*, Vol. 47, 1971, pp. 525–535.

[7]Kraichnan, R. H., and Montgomery, D., "Two-Dimensional Turbulence," *Reports on Progress in Physics*, Vol. 43, 1980, pp. 547–619.

[8]Lesieur, M., *Turbulence in Fluids*, 2nd ed., Kluwer, Dordrecht, The Netherlands, 1990.

[9]Starr, V. P., *Physics of Negative Viscosity Phenomena*, McGraw–Hill, New York, 1968.

[10]Chekhlov, A., Orszag, S. A., Sukoriansky, S., Galperin, B., and Staroselsky, I., "Direct Numerical Simulation Tests of Eddy Viscosity in Two Dimensions," *Physics of Fluids*, Vol. 6, 1994, pp. 2548–2550.

[11]Staroselsky, I., and Sukoriansky, S., "Renormalization Group Approach to Two-Dimensional Turbulence and the ε-Expansion for the Vorticity Equation," *Advances in Turbulence Studies*, edited by H. Branover and Y. Unger, Vol. 149, Progress in Astronautics and Aeronautics, AIAA, Washington, DC, 1993, pp. 150–164.

[12]Smagorinsky, J., "General Circulation Experiments with the Primitive Equations," *Monthly Weather Review*, Vol. 91, 1963, pp. 99–164.

[13]Smagorinsky, J., "Some Historical Remarks on the Use of Nonlinear Viscosities," *Large Eddy Simulation of Complex Engineering and Geophysical Flows*, edited by B. Galperin and S. A. Orszag, Cambridge Univ. Press, New York, 1993, pp. 3–36.

[14]Yakhot, V., and Orszag, S. A., "Renormalization Group Analysis of Turbulence. I. Basic Theory," *Journal of Scientific Computing*, Vol. 1, No. 1, 1986, pp. 3–51.

[15]Yakhot, A., Orszag, S. A., Yakhot, V., and Israeli, M., "Renormalization Group

Formulation of Large-Eddy Simulations," *Journal of Scientific Computing*, Vol. 4, No. 2, 1989, pp. 139–158.

[16]Smith, L., and Yakhot, V., "Bose Condensation and Small-Scale Structure Generation in a Random Force Driven 2D Turbulence," *Physical Review Letters*, Vol. 71, 1993, pp. 352–355.

[17]Smith, L., and Yakhot, V., "Finite-Size Effects in Forced Two-Dimensional Turbulence," *Journal of Fluid Mechanics*, 274, 1994, pp. 115–138.

[18]Sivashinsky, G. I., "Nonlinear Analysis of Hydrodynamic Instability in Laminar Flames—I. Derivation of Basic Equations," *Acta Astronautica*, Vol. 4, 1977, pp. 1177–1206.

[19]Kuramoto, Y., and Tsuzuki, T., "Persistent Propagation of Concentration Waves in Dissipative Media Far from Thermal Equilibrium," *Progress of Theoretical Physics*, Vol. 55, 1976, pp. 365–369.

[20]Kuramoto, Y., *Chemical Oscillations, Waves, and Turbulence*, Springer–Verlag, 1984.

[21]Gama, S., Frisch, U., and Scholl, H., "The Two-Dimensional Navier–Stokes Equations with a Large-Scale Instability of the Kuramoto–Sivashinsky Type: Numerical Exploration on the Connection Machine," *Journal of Scientific Computing*, Vol. 6, 1991, pp. 425–452.

[22]Monin, A. S., "Stratification and Circulation of the Ocean," *Synoptic Eddies in the Ocean*, edited by V. M. Kamenkovich, M. N. Koshlyakov, and A. S. Monin, D. Riedel Publishing Co., 1986, pp. 1–33.

Chapter 4

Stochastic Modeling of Turbulence in the Presence of Gravity or Inertial Waves: from Kinematic to Dynamic Simulation

F. S. Godeferd, N. A. Malik, and C. Cambon
Centre National de la Recherche Scientifique, 5509, École Centrale de Lyon, BP 163 - 69131 Ecully Cedex, France

I. Introduction

TURBULENT flows in a rotating frame and/or in a stably stratified field can develop inertiogravity waves, which affect the statistical anisotropy and the energy cascade. On the one hand, two-point Eulerian closure models have been developed to take into account the effects of wave dynamics on both linear and nonlinear interactions (see, e.g., Refs. 1 and 2) in a statistical way. This approach is based on a generalized eddy damped quasi-normal Markovian (EDQNM2) model.[3] On the other hand, given the two-point second-order correlation tensor, kinematic simulation (KS)[4] is a Lagrangian-based model using unsteady random Fourier modes and capable of simulating realizations of a full spatio-temporal homogeneous turbulent velocity (and coupled temperature) field. The statistical properties of the ensemble of realizations are frozen in time, but a particular realization is not. The KS spatio-temporal field can be treated like a turbulent field in order to compute trajectories directly and hence to obtain Lagrangian statistics, like the dispersion of particles. In comparison to a local random flight model, KS ensures the solenoidal property (divergence-free) for the incompressible velocity field. Because it generates structured flowfields explicitly, both the Eulerian and Lagrangian statistics obtained from KS are sensitive to the scale and structure of these eddies and thus can improve Lagrangian prediction.[4,5] But KS can also be used to study some fundamental concepts such as scaling laws of particle diffusion and the role that streamline topology plays in the process of structural diffusion.[6] It is simpler, however, than the stochastic Langevin models,[7,8] which are consistent with two-point statistical theories (EDQNM, DIA, TFM) that underlie them.

Because KS was initially developed with the additional assumption of statistical isotropy and did not include organized wave-like time dependency, in this paper we generalize the procedure and attempt to reconcile both approaches that are previously mentioned (EDQNM2 and KS). For this purpose the KS model is made consistent with a time-dependent anisotropic second-order spectral tensor $\hat{U}_{ij}$ $(\boldsymbol{k}, t)$ resulting from the time-dependent EDQNM2 equations. Linear dynamics is introduced in any realization by adding to the random frequency the deterministic frequency due to gravity or inertial waves according to their dispersion law. The latter procedure amounts to making the KS consistent with the so-called rapid distortion theory (RDT) at any point and at any time. In addition to this explicit deterministic time-dependency, which reflects linear dynamics and can be considered as rapid, nonlinear dynamics is also implicitly reflected in the anisotropic structure of the second-order spectral tensor. In the above mentioned papers about EDQNM2, it is shown that the rise of directional and polarization anisotropy in $\hat{U}_{ij}$ $(\boldsymbol{k}, t)$ reflects an irreversible transition (which cannot be captured by RDT). The transition evolves toward a two-dimensional velocity field in the case of rotating turbulence and toward a three-dimensional two-component (collapse of vertical motion, sometime wrongly described as two-dimensional) state in the case of stably stratified turbulence (see also Ref. 9, previous Beer-Sheva seminar).

II. Equations of the Coupled Model

The Navier–Stokes equations for the fluctuating velocity field $\boldsymbol{u}$ and the fluctuating temperature field τ are used within the Boussinesq hypothesis, with a vertical gradient γ of mean temperature. Depending on the sign of γ, one can distinguish the unstable stratified case ($\gamma > 0$) and the stably stratified one ($\gamma < 0$), the latter of which we consider here. The system of equations under the Boussinesq approximation then reads

$$\left(\frac{\partial}{\partial t} + u_j \frac{\partial}{\partial x_j} - \nu\nabla^2\right) u_i + \frac{\partial p}{\partial x_i} = \beta g \tau \delta_{i3}$$
$$\left(\frac{\partial}{\partial t} + u_j \frac{\partial}{\partial x_j} - \nu\nabla^2\right) \tau = -\gamma u_3 \tag{1}$$

The temperature diffusivity is assumed to be equal to the kinematic viscosity ν in the temperature equation, therefore yielding a unit Prandtl number. The Brünt–Väisälä frequency $N = (\beta\gamma g)^{1/2}$ that appears in the nondimensionalized version of these equations represents the inverse timescale of the internal waves regime arising from the stratification effect in the linear limit. β is the thermometric expansivity, and g is the acceleration of gravity. The system (1) is treated in spectral space by Fourier transforming all quantities (denoted by a ^), and the velocity equation is projected onto the direction of the wave vector $\boldsymbol{\kappa}$ in order to get rid of the term involving the pressure p, in virtue of the divergenceless property of the fluctuating velocity field.

A useful and illuminating treatment of the fields in this situation (and also in the case of rotating turbulence) is the use of a special frame of reference, namely the Craya–Herring frame, in which an orthonormal eigenframe associated with the linear regime is readily derived.[10,11] Any solenoidal (divergence-free) velocity field in spectral space can be decomposed into two modes that are orthogonal unit vectors in the plane normal to the wave vector $\boldsymbol{\kappa}$, or:

$$
\begin{aligned}
\hat{u}_i &= \rho^1 e_i^1 + \rho^2 e_i^2 \qquad (2)\\
\boldsymbol{e}^1(\boldsymbol{\kappa}) &= (\boldsymbol{\kappa} \times \boldsymbol{n}) / |\boldsymbol{\kappa} \times \boldsymbol{n}|\\
\boldsymbol{e}^2(\boldsymbol{\kappa}) &= (\boldsymbol{\kappa} \times \boldsymbol{e}^1) / |\boldsymbol{\kappa} \times \boldsymbol{e}^1|\\
u_i &= \int (\varphi^1 e_i^1 + \varphi^2 e_i^2) \exp(I\boldsymbol{\kappa} \cdot \boldsymbol{x}) d^3\boldsymbol{\kappa} \qquad (3)
\end{aligned}
$$

The auxiliary (fixed) unit vector $\boldsymbol{n}$ is chosen as the upward vector $n_i = -g_i/g = \delta_{i3}$ due to the symmetry properties of the physical geometry of the flow-field. The components $\varphi^1 = \varphi^1(\kappa, \theta)$ and $\varphi^2 = \varphi^2(\kappa, \theta)$ are the energy spectral densities associated with the respective directions, where θ is the angle between the wave vector $\boldsymbol{\kappa}$ and the upward vertical. To avoid distinguishing the kinematic (velocity) components from the temperature field, and taking advantage of the property $\hat{\boldsymbol{u}} \perp \boldsymbol{\kappa}$, the temperature is scaled as a velocity by the factor $\beta g/N$ and considered as a third component $\varphi^3 \equiv \tau^* \propto \tau$ of a global temperature–velocity field.

Therefore, the projection of the two-point correlations Φ_{ij} of its components (in a fixed frame) onto the different axes of the local Craya–Herring frame gives the different correlation spectra for the components of the temperature–velocity in the local frame, such that

$$
\frac{1}{2} e_i^u \Phi_{ij} e_j^v = \begin{pmatrix} \Phi_1 & 0 & 0 \\ 0 & \Phi_2 & \Psi^* \\ 0 & \Psi & \Phi_3 \end{pmatrix} \qquad (4)
$$

which depends on the wave mode $k = |\boldsymbol{\kappa}|$, and the angle θ_k which $\hat{\boldsymbol{\kappa}}$ makes with the vertical $\boldsymbol{n}$. Whence we obtain the equations for the energy spectra $\Phi_i(\boldsymbol{\kappa}, t) = \langle \hat{\varphi}^{i*} \hat{\varphi}^i \rangle / 2$ and for the correlation spectrum $\Psi(\boldsymbol{\kappa}, t) = \langle \hat{\varphi}^{2*} \hat{\varphi}^3 \rangle / 2$, whose integration in the vertical direction yields the vertical heat flux

$$
\frac{1}{2} \langle u_3 \tau \rangle = -\frac{N}{\beta g} \int \Psi_R(\boldsymbol{\kappa}) \sin \theta_\kappa \, d^3\boldsymbol{\kappa} \qquad (5)
$$

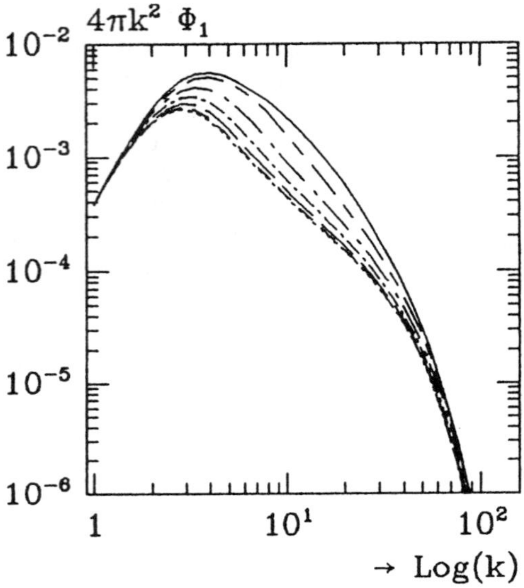

Fig. 1 Anisotropic Φ^1 spectrum as given by the EDQNM2 model. The energy spectrum is distributed over several angles of the wave vectors; from top to bottom, *k* parallel to the gravity (polar direction) and *k* perpendicular to it (equatorial direction).

where the subscript R stands for "real part." The physical meaning of the three other spectra stems directly from the component φ^i they are associated with. Namely, Φ_1 is the spectrum of kinetic energy contained in the horizontal vortical motion in the flow; Φ_2 contains the kinetic energy for the wave motion; and Φ_3 represents the potential energy associated with the waves.

A. Dynamic Eddy Damped Quasinormal Markovian Model

For the sake of brevity, the detailed equations for the fully anisotropic EDQNM model will not be given here, nor will all the features of the closure be discussed. The reader is referred to Refs. 1 and 12, for an extensive presentation of the model. We shall discuss hereafter the part of the statistical closure that is relevant to the coupling to the kinematic model.

The corresponding exact equations for the energy spectra in Eq. (4) and the different energy transfer terms associated to them are closed by the EDQNM2 model, allowing the computation of time-varying spectra of a freely decaying stably stratified homogeneous turbulence. Because the EDQNM2 model takes into account the complete (modeled) dynamics of the flow, the statistical Eulerian spectra reproduce very well the behavior of the real flowfield. The accuracy of the obtained results have been checked against high resolution direct numerical simulations.[13] Figure 1 shows the developed spectra that are picked up at a given time in the decay of turbulence and will be assumed to be almost frozen from the point of view of the nonlinear energy transfers (with respect to the rapid timescale of the kinematic simulation) during the KS computations.

B. Stochastic Kinematic Simulation Coupled to the Eddy Damped Quasinormal Markovian

The eddy damped quasinormal Markovian model gives three time-varying spectra associated with the horizontal (two-component) modes φ^1 ($\boldsymbol{\kappa}$), the kinematic energy modes φ^2 ($\boldsymbol{\kappa}$), and the potential energy modes $\varphi^3(\boldsymbol{\kappa})$. Particular care is taken of the fact that the spectra are distributed anisotropically in wave number space, and so they have a dependence on θ, the angle between the wave number $\boldsymbol{\kappa}$ and the vertical direction. In the KS stratified models this aspect is made easier to incorporate by reference to the Craya–Herring frame. We can define a complex velocity field $\boldsymbol{\upsilon}$ whose Fourier component $\hat{\boldsymbol{\upsilon}}(\boldsymbol{\kappa}, t)$ is given by

$$\hat{\upsilon}(\kappa,t) = \varphi^1(\kappa,t)e_i^1(\boldsymbol{\kappa}) + \varphi^2(\kappa,t)e_i^2(\boldsymbol{\kappa}) + I\varphi^3(\kappa,t)e_i^3(\boldsymbol{\kappa}) \tag{6}$$

where e_i, $i = 1, 2, 3$ are the unit vectors in the Craya–Herring frame, and $I^2 = -1$. The solenoidal part of $\boldsymbol{V}$ gives the actual velocity field $\boldsymbol{u}$. The spectral density of the total energy (kinetic + potential) is equal to

$$\frac{1}{2}\hat{v}_i^*\hat{v}_i = \left(\frac{1}{2}\hat{u}_i^*\hat{u}_i + \frac{1}{2}\left(\frac{\beta g}{N}\right)^2 \hat{\tau}*\hat{\tau}\right)$$

Internal Gravity Waves

From studies of homogeneous stratified turbulence in spectral space, it has been observed that from a physical point of view the linear regime (ignoring energy transfer terms) consists of vortical contributions that are unaffected by the stratification and internal gravity waves, generated by some perturbations. The EDQNM2 model includes these exact effects and also takes advantage of such a physical decomposition in different contributions to the velocity field. The nonlinear regime, expressed in terms of the eigenmodes of the linear regime, can thus be viewed as the superposition of infinitely many internal waves propagating in all directions (see Ref. 1). It is therefore possible to introduce the exact linear solution in the KS model by means of the eigenmodes of the linear stratification operator, namely the internal gravity waves. The latter arise directly by solving the Green's function problem associated to this operator: in the linearized limit, the dynamical equations of motion yield an explicit expression for the Fourier components $\hat{\boldsymbol{\upsilon}}(\boldsymbol{\kappa}, t)$ from an initial $\hat{\boldsymbol{\upsilon}}(\boldsymbol{\kappa}, t_0)$ at time t_0 via a Green's operator $G(\boldsymbol{\kappa}, t, t_0)$, (Ref. 1), viz.,

$$\hat{v}_i(\boldsymbol{\kappa},t) = G_{ij}(\boldsymbol{\kappa},t,t_0)\hat{v}_j \tag{7}$$

$$G_{ij}^{\varepsilon}(\boldsymbol{\kappa},t) = \sum\nolimits_{\varepsilon\in\{0,\pm 1\}} N_i^{\varepsilon} N_j^{\varepsilon} e^{I\varepsilon \sin\theta_k N t} \tag{8}$$

where

$$N_i^{\varepsilon}(\boldsymbol{\kappa}) = \left[e_i^{|\varepsilon|}(\boldsymbol{\kappa}) + \varepsilon e_i^{3}(\boldsymbol{\kappa})\right] / \sqrt{|\varepsilon| + 1}$$

is the eigenmode associated with the eigenvalue $-\varepsilon INe_3^2(\boldsymbol{\kappa}) = \varepsilon IN\sin(\theta_k)$. In Eq. (8) there appears the phase (when multiplied by the Fourier mode $e^{I\boldsymbol{\kappa}\cdot\boldsymbol{x}}$) $\varphi_w = \boldsymbol{\kappa}\cdot\boldsymbol{x} \pm N\sin\theta_k t$ of the gravity wave of frequency $\omega_k = N\sin\theta_k$. These internal wave modes are then easily introduced in the KS model, so that the linear solution for stratified homogeneous turbulence is reproduced, as shown in Section III.

The mathematical formulation of the KS flowfield for isotropic turbulence is given in detail in Ref. 4. Most of this formulation is valid also for generating the stratified kinematic simulation (KSTRAT) flowfield. We begin here with the basic expression for the complex velocity field in the discretized KS approximation [c.f., Eq. (2.2) in Ref. 4], at initial time $t = 0$, viz.,

$$\boldsymbol{v}(\boldsymbol{x},0) = \sum_{n=1}^{N_k}\left(\sum_{m=1}^{N_\theta}(A_{nm}^1 e^{I\chi_{nm}^1}\boldsymbol{e}_1 + A_{nm}^2 e^{I\chi_{nm}^2}\boldsymbol{e}_2 + A_{nm}^3 e^{I\chi_{nm}^3}\boldsymbol{e}_3) \times \exp(I\boldsymbol{\kappa}_{nm}\cdot\boldsymbol{x} + I\omega_{nm}t)\right) \tag{9}$$

where ω_{nm} are the frequencies due to the intrinsic unsteadiness of eddying motions. The $\chi_{nm}^i, i = 1,2,3$ are random phases selected between $[0, 2\pi]$. At later time t, $\boldsymbol{v}(\boldsymbol{s},t)$ is obtained from $\boldsymbol{v}(x, 0)$ from Eqs. (7) and (8) using the Green's operator G_{ij}. Thus, the phase in the Fourier modes contains an additional time-varying term exp $(\pm IN\sin\theta_k t)$ due to the internal waves in the linearized regime. The actual velocity field $\boldsymbol{u}(\boldsymbol{x}, t)$ is then obtained by a summation of each of the Fourier components $(\hat{\boldsymbol{v}}_{mn})$ projected onto the plane orthogonal to the vector $\boldsymbol{e}_3^{mn}$, and the normalized temperature is the sum of components projected parallel to $\boldsymbol{e}_3^{mn}$.

This form of the KS velocity field ensures incompressibility $(\nabla\cdot u = 0)$. The distribution of wave numbers K_{mn} and frequencies ω_{mn} is given by an energy spectrum, which is known. Note that in contrast to stochastic models such as random flight models, the KS models turbulence with many hundreds, even thousands, of length and (possibly) timescales. This is much more complex than random flight models, but still less complex and less expensive than DNS.

In the KS stratified models, the time-varying spectra are readily discretized and distributed anisotropically in wave number space. The spectra are obtained from the EDQNM2 data of Godeferd and Cambon[1] and the use of Eq. (8) to account for the internal waves. The velocity and temperature fields are correlated through vertical flux equation (5). The random phases χ_{mn}^2 and χ_{mn}^3 are adjusted in the KS model, so that it yields the correct value for the vertical heat flux that is given by the EDQNM2 data for the initial flow $u(t, 0)$ at time $t = 0$.

Lagrangian Motion

Fluid particle trajectories are computed by numerical integration of the particle's Lagrangian velocity $\boldsymbol{v}^L(t) = \boldsymbol{v}^L(t;\boldsymbol{x}_0,t_0)$ given that the particle was at $\boldsymbol{x}_0(t_0)$ at some earlier time t_0, viz.,

$$\frac{\mathrm{d}\boldsymbol{x}(t)}{\mathrm{d}t} = \boldsymbol{v}^L(t) \tag{10}$$

where $\boldsymbol{v}^L(t)$ is related to the instantaneous KS Eulerian velocity at $\boldsymbol{x}(t)$ by $\boldsymbol{v}^L(t) = \boldsymbol{u}(\boldsymbol{x}, t)$. We solve for particle trajectories using a fourth-order Adams–Bashforth predictor–corrector method. (The Lagrangian motion of small inertial particles or bubbles can also be computed in principle, but this is left for subsequent study.) The Lagrangian statistics of particle motion is then obtained by averaging over ensembles of particle trajectories from different flowfield realizations.

III. Results

As far as the stratified case is concerned, we present results from a set of data obtained from the EDQNM2 closure model for a stratification of $N = 16\pi\ s^{-1}$. The spectral data are in the form of $N_k = 37$ wave number shells, distributed on $N_\theta = 19$ discrete angles between $[0, \pi/2]$, Fig. 1.

A consistency check of the effects of the stratification on the model is a plot of the energy in the three energy modes as a function of time. Figure 2

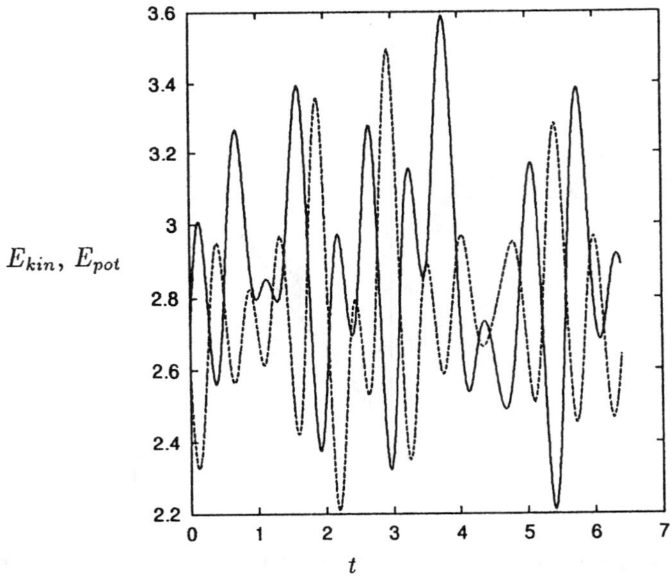

Fig. 2 Oscillating exchange between the kinetic and potential energies.

shows that the potential and kinetic energies oscillate and exchange energy between each other in time and are thus in antiphase. This is the correct behavior in the linearized limit and ensures that the heat flux $\langle u_3 \mathrm{r}\rangle/2$ is correctly modeled.

Lagrangian models are well suited for dispersion study. Figure 3 shows the log-log plot of one-particle diffusion, $< (x - x_0)^2 >$, against time t. The solid line is the horizontal dispersion $< (x - x_0)^2 + (y - y_0)^2 > /2$, and the dotted line is the vertical dispersion $< (z - z_0)^2 >$. The dispersion follows the classic[14] diffusion laws; the $< x^2 > \sim < u^2 > t^2$ law for small times $t \ll T^L$; and $< u^2 > u^2 > \sim Dt$ for long times $t \gg T^L$, where T^L is the Lagrangian (decorrelation) timescale and the diffusivity $D = < u^2 > T^L$. The difference in the horizontal and vertical diffusion reflects the anisotrophy in the flow: $< (u_z')^2 > / < (u_x')^2 > \approx 0.86$; and $D_z/D_x \approx 0.46$. This gives $T_z^L/T_x^L \approx 0.5$. Pearson et al.[15] have conjectured on the basis of a stochastic model that if the stratification is sufficiently high, i.e., $NT \gg 1$, where T is the Eulerian timescale of the flow, then the vertical dispersion will be confined by buoyancy forces to a finite spread, i.e., $< z^2 > \to$ constant as $t \to \infty$. Our KS simulation suggests that we need $NT > NT^L > 16$ for this to happen.

The future work on this model will introduce drag forces on the particles so that we can simulate the motion of inertial solid particles as well. Kinematic simulation stochastic models for rotation, and for stratification and rotation could also be constructed.

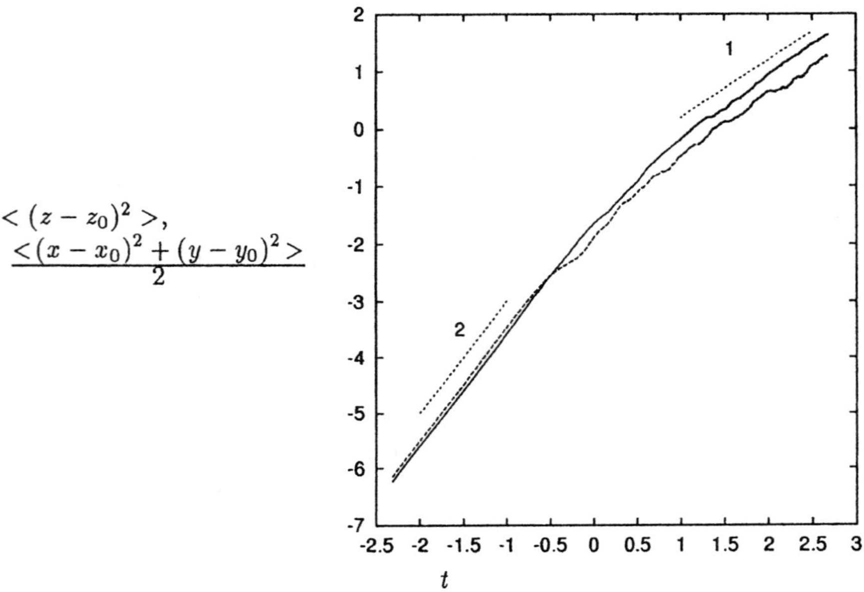

Fig. 3 The one particle diffusion in the horizontal and vertical components. The Taylor[14] diffusion laws for short times ~t^2 and for long times ~t are obtained. This case is for $N = 16\pi$. See text for details.

Acknowledgments

The authors gratefully acknowledge financial support from the European community under contract number ERBCGBI-CT94-1230. The computational resources were made available on the Cray 98 by IDRIS (Institut du Developpement et des Ressources en Informatique Scientifique), which receives major funding from the CNRS.

References

[1]Godeferd, F. S., and Cambon, C., "Detailed Investigation of Energy Transfers in Homogeneous Stratified Turbulence," *Physics of Fluids,* Vol. 6, No. 6, 1994, pp. 2084–2100.

[2]Cambon, C., Mansour, N. N., and Godeferd, F. S., "Energy Transfer in Rotating Turbulence," *Journal of Fluid Mechanics* (to be published).

[3]Orszag, S. A., "Analytical Theories of Turbulence," *Journal of Fluid Mechanics,* Vol. 41, 1970, p. 363.

[4]Fung, F. C. H., Hunt, J. C. R., Malik, N. A., and Perkins, R. J., "Kinematic Simulations in Turbulent Flow Generated by Unsteady Random Fourier Modes," *Journal of Fluid Mechanics,* Vol. 236, 1992, pp. 281–317.

[5]Perkins, R. J., Malik, N. A., and Fung, F. C. H., "Cloud Dispersion Models," *Journal of Applied Sci. Res.,* Vol. 51, 1993, pp. 539–545.

[6]Malik, N. A., "Structural Diffusion of Particles in Two- and Three-Dimensional Random Flows: The Role of Streamline Intermittency," *Journal of Fluid Mechanics* (submitted for publication).

[7]Herring, J. R., and Kraichnan, R. H., "Comparison of Some Approximations for Isotropic Turbulence," *Statistical Models and Turbulence, No. 12 in Lecture Notes in Physics,* Springer-Verlag, 1971.

[8]Leith, C. E., "Atmospheric Predictability and Two-Dimensional Turbulence," *Journal of Atmospheric Sciences,* Vol. 28, 1971, pp. 145–161.

[9]Godeferd, F. S., and Cambon, C., "Inertial Transfers in Rotating, Stably Stratified and MHD Turbulence," *Progress in Turbulence Research,* edited by H. Branover and Y. Unger, Vol. 162, Progress in Astronautics and Aeronautics, AIAA, Washington, DC, 1993, pp. 150–168.

[10]Craya, A., "Contribution à l'Analyse de la Turbulence Associée à des Vitesses Moyennes," Ministère de l'Air, Tech. Rept. 345, 1958.

[11]Herring, J. R., "Approach of Axisymmetric Turbulence to Isotropy," *Physics of Fluids,* Vol. 17, No. 5, 1974, pp. 859–872.

[12]van Haren, L., "Étude Théorique et Modélisation de la Turbulence en Présence D'ondes Internes," Ph.D. Thesis, École Centrale de Lyon, 1993.

[13]Godeferd, F. S., Staquet, C., and Cambon, C., "Stably Stratified Turbulence: A Statistical Two-Point Closure vs. Direct Numerical Simulations," *Tenth Symposium on Turbulent Shear Flows,* Penn State Univ., State College, PA, 1995.

[14]Taylor, G. I., "Diffusion by Continuous Movements," *Proceedings of the London Mathematical Society A,* Vol. 20, 1921, pp. 196–212.

[15]Pearson, H. J., Puttock, J. S., and Hunt, J. C. R., "A Statistical Model of Fluid Element Motions and Vertical Diffusion in a Homogeneous Stratified Turbulent Flow," *Journal of Fluid Mechanics,* Vol. 129, 1983, pp. 219–249.

Chapter 5

Turbulent Liquid-Metal Flow in Rectangular Ducts with Thin Conducting Walls and Strong Magnetic Fields

Sergio Cuevas* and Eduardo Ramos*
Laboratorio de Energía Solar, IIM-UNAM, Mor., 62580, Mexico
and
Basil F. Picologlou†
Argonne National Laboratory, Argonne, IL, 60439, USA

I. Introduction

AT PRESENT, magnetohydrodynamic (MHD) turbulence in ducts is a challenging problem, and although there is a satisfactory understanding of some aspects, there are many others that are still unclear. Specifically, there is a widespread belief that MHD turbulence at low magnetic Reynolds numbers presents a tendency toward two-dimensionality. A uniform magnetic field transverse to the flow favors two-dimensional motion, first by dissipating the components of vorticity perpendicular to the field and second by keeping unaffected pure two-dimensional motion in the plane normal to the field. The picture of fully developed MHD turbulence as a collection of two-dimensional eddies with axes parallel to the magnetic field has been advocated by many authors.[3–5] Sommeria and Moreau[4] stressed that the presence of insulating walls perpendicular to the magnetic field is a necessary condition for the persistence of turbulence and eventually to the development of two-dimensional behavior. These results are related with the fact that the rate of decay of a vortex elongated in the direction of the magnetic field is much lower in insulated channels than in conducting ones. However, some experimental evidence obtained in the late 1980s in Beer-Sheva and Argonne National Laboratory (ANL) points to the existence of persistent nonlaminar flows in conducting wall ducts. The

*Researcher, A.P. 34, Temixco.
†Technology Development Division.

Beer-Sheva experiments, carried out far away from fusion reactor relevant conditions, showed the persistence of turbulence in conducting channels in conditions where the flow should be "laminarized,"[1,5,6] On the other hand, ANL experiments demonstrated the existence of persisting velocity fluctuations in the region near the side walls (parallel to the magnetic field) of a rectangular thin conducting wall duct near fusion blanket relevant operation conditions.[2] It is expected that these kinds of flows may have beneficial effects on the thermohydraulic performance of self-cooled liquid metal blankets through an enhancement of heat transfer rates from the first wall.

The present study is aimed at the theoretical investigation of fully developed turbulent liquid-metal flows in rectangular ducts with thin conducting walls and strong, transverse magnetic fields. In rectangular ducts, the heat transfer characteristics and the stability are determined mainly by the flow in the side layers, whereas the Hartmann layers play a rather passive role. Therefore, the problem is addressed here using the model presented by Cuevas et al.[7] consisting in a composite core-side-layer spectral collocation solution that includes turbulent effects through an eddy viscosity within an iterative scheme. Eddy viscosity models from the renormalization group (RNG)[8,9] and mixing length[10,15] theories are used to analyze the behavior of the flow in situations in which the parameter values coincide with cases studied in the ANL side-layer instability experiments. Hence, in the next section a brief description of experimental conditions and observations is made. In Section III the governing equations are formulated, whereas in Section IV the RNG and mixing length models are presented. Finally, the numerical results are shown in Section V, and discussion and concluding remarks are addressed in Section VI.

II. Survey of Experimental Results

Experiments performed at the Ben-Gurion University investigated turbulent mercury flows, far from fusion reactor relevant conditions, in a rectangular cross-section duct with conducting walls subject to a transverse magnetic field. They show the persistence of turbulence in conditions where the flow should be laminarized.[1,5] The magnetic field appears to suppress the turbulence only in the core of the flow whereas significantly enhances the turbulence intensity in the vicinity of the channel walls without the need of external flow perturbing means. Experiments indicate that the turbulence in conducting channels is continuously created along the channel, probably because of steep velocity gradients associated with M-shaped velocity profiles. The continuous excitation of eddies might lead to the establishment of a significant level of quasi-two-dimensional turbulence and to the enhancement of heat transfer.[6]

Likewise, experiments carried out at ANL demonstrated the existence of an instability in a NaK-flow inside a square duct with thin metal walls under a transverse magnetic field, near fusion reactor relevant operation conditions.[2] The instability developed in a side-layer jet (characteristic of MHD flows in rectangular ducts with thin conducting walls) attached to a side wall parallel to the magnetic field and presented a very strong periodicity and lack of small-scale structure. Although the appearance of the instability has not been satisfactorily explained, the speculation is that transport mechanisms similar to those present in MHD turbulence might be involved. Measurements were performed in the midplane of the duct $y = 0$ (see Fig. 1) and embraced almost the whole extent of the magnet in the axial direction. In the transverse-to-the-field z direction, measurements were obtained in the region $0.95 > z > 0.60$. The instability always appeared well within the region of the homogeneous magnetic field. It begins as a harmonic oscillation in the mean velocity and, after a few characteristic lengths, reaches a somewhat stable size (occupying approximately the region $0.70 < z < 1.0$) that maintains with slow changes over many characteristic lengths in the axial direction. Actually, one of the main effects of the presence of the instability on the mean velocity distributions is the apparent thickening of the wall jet layer. The large amplitude of the instability oscillations with respect to the mean velocity and the strongly periodic character of the velocity signals as the flow proceeds downstream indicate the well-organized structure of the instability in the near-the-wall region. However, oscillations present a fast decrease with distance from the wall, disappearing almost completely at $z = 0.75$. Increasing the interaction parameter N from 2900 to 11,000 (for a Hartmann number M of 5300) leads to a very rapid suppression of the instability. For high N, its effects are more closely confined to the near-the-wall region than in the low interaction parameter case, and for the highest experimental value of $N = 48{,}000$, they can be considered completely nonexistent. Over the ranges of M and N examined, the onset of the instability depends primarily on the Reynolds number. It was concluded that the critical Reynolds number lies in the range $2650 < Re_{cr} < 5100$ and is independent of M in the range $2700 < M < 5400$.

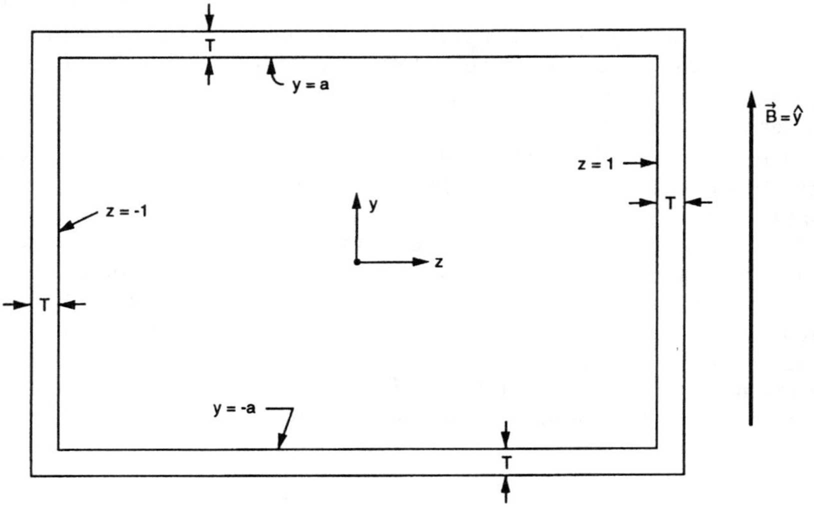

Fig. 1 Dimensionless duct cross section.

III. Governing Equations

To get the governing Reynolds equations for the turbulent regime, we consider the flow variables to consist of a mean part, which will be denoted by an overbar, and a fluctuating part with a zero average, denoted by a prime. Hence, the velocity, pressure, and electric potential fields are expressed by $\boldsymbol{u} = \overline{\boldsymbol{u}} + \overline{\boldsymbol{u}}'$, $p = \overline{p} + p'$, $\varphi = \overline{\varphi} + \varphi'$, respectively, where, by definition, $\overline{\boldsymbol{u}}' = \overline{p}' = \overline{\varphi}' = 0$. Because the induced magnetic field can be neglected in comparison to the applied field, $\boldsymbol{B} = \overline{\boldsymbol{B}}$ (low magnetic Reynolds number approximation). With the previous assumptions and an averaging procedure, the mean flow governing equations can be obtained and expressed in the following dimensionless form[7,11]:

$$N^{-1}(\overline{\boldsymbol{u}}\cdot\nabla)\overline{\boldsymbol{u}} = -\nabla\overline{p} + M^{-2}\nabla\cdot(\nabla\overline{\boldsymbol{u}} - Re\overline{\boldsymbol{u}'\boldsymbol{u}'}) + \boldsymbol{j}\times\overline{\boldsymbol{B}} \tag{1}$$

$$\nabla\cdot\overline{\boldsymbol{u}} = 0, \quad \nabla\cdot\bar{\boldsymbol{j}} = 0 \tag{2a,2b}$$

$$\bar{\boldsymbol{j}} = -\nabla\overline{\varphi} + \overline{\boldsymbol{u}} \times \overline{\boldsymbol{B}} \tag{3}$$

where $\bar{\boldsymbol{j}}$ is the mean electric current density. Mean variables $\overline{\boldsymbol{u}}, \overline{p}, \bar{\boldsymbol{j}}, \overline{\varphi}$, and $\overline{\boldsymbol{B}}$ are normalized by $\overline{U}, \sigma\overline{U}B_o^2 L, \sigma\overline{U}B_o, \overline{U}B_oL$, and B_o, respectively, where $\sigma, \overline{U}$, B_o, and L are the electrical conductivity of the fluid, the average velocity, the applied magnetic field strength, and a characteristic length. In addition, the dimensionless parameters $N = \sigma B_o^2 L/\rho\overline{U}$, $M = B_oL, \sqrt{\sigma/\rho\nu_O}$ and $Re = \overline{U}L/\nu_o$ are the interaction parameter, the Hartmann number, and the Reynolds number, respectively, where ρ and ν_o are the mass density and kinematic viscosity of the liquid, respectively. Notice that only two of these parameters are independent because $N = M^2/Re$. The Navier–Stokes equation (1) includes the electromagnetic body force $\bar{\boldsymbol{j}} \times \overline{\boldsymbol{B}}$, and the turbulent Reynolds stresses term $-Re\boldsymbol{u}'\boldsymbol{u}'$. Conservation of mass and electric charge are expressed through Eqs. (2a) and (2b), respectively, while Eq. (3) represents Ohm's law in a moving media.

We consider a turbulent liquid-metal flow in a constant area rectangular duct whose axis coincides with the x-coordinate axis, in the presence of a uniform, transverse magnetic field, $\overline{\boldsymbol{B}} = \hat{\boldsymbol{y}}$ (see Fig. 1). All length dimensions are normalized by L, half the distance between the side walls, parallel to the magnetic field. Under these conditions the flow is fully developed and the problem becomes two-dimensional. Hence, $\overline{\boldsymbol{u}} = (\overline{u}(y, z), 0, 0)$, $\overline{\varphi} = \overline{\varphi}(y, z)$, and $\nabla\overline{p} = (-K, 0, 0)$, where K is a constant. Therefore, Eq. (1) reduces to

$$M^{-2}\left[\frac{\partial}{\partial y}\left(\frac{\partial\overline{u}}{\partial y} - Re\overline{u'v'}\right) + \frac{\partial}{\partial z}\left(\frac{\partial\overline{u}}{\partial z} - Re\overline{u'w'}\right)\right] - \bar{j}_z - \frac{\partial\overline{p}}{\partial x} = 0 \tag{4}$$

Using Boussinesq's approach, the turbulent Reynolds stresses can be expressed in the form

$$-Re\overline{u'v'} = \nu_{ey}\frac{\partial \bar{u}}{\partial y}, \quad -Re\overline{u'w'} = \nu_{ez}\frac{\partial \bar{u}}{\partial z} \tag{5a,5b}$$

where ν_{ej} ($j = y, z$) is known as the eddy viscosity (normalized by the molecular kinematic viscosity ν_o). Because of the anisotropic effect of the magnetic field, we allow for different eddy viscosities in each direction. Introducing Eqs. (5a) and (5b) into Eq. (4) and definining the total viscous diffusivities as $\nu_{ty} = 1 + \nu_{ey}$ and $\nu_{t2} = 1 + \nu_{ez}$, which account for the molecular and turbulent contributions, the set of Eqs. (1–3) are reduced to the form

$$M^{-2}\left[\frac{\partial}{\partial y}\left(\nu_{ty}\frac{\partial \bar{u}}{\partial y}\right)+\frac{\partial}{\partial z}\left(\nu_{tz}\frac{\partial \bar{u}}{\partial z}\right)\right] - \bar{j}_z - \frac{\partial \bar{p}}{\partial x} = -K \tag{6}$$

$$\bar{j}_y = -\frac{\partial \bar{\varphi}}{\partial y}, \quad \bar{j}_z = -\frac{\partial \bar{\varphi}}{\partial z} + \bar{u} \tag{7a,7b}$$

$$\frac{\partial \bar{j}_y}{\partial y} + \frac{\partial \bar{j}_z}{\partial z} = 0 \tag{8}$$

In addition, the mean variables have also to satisfy Ohm's law and current conservation at the walls:

$$\bar{j}_{yw} = -\left(\frac{\sigma_w}{\sigma}\right)\frac{\partial \bar{\varphi}_w}{\partial y}, \quad \bar{j}_{zw} = -\left(\frac{\sigma_w}{\sigma}\right)\frac{\partial \bar{\varphi}_w}{\partial z} \tag{9a,9b}$$

$$\frac{\partial \bar{j}_{yw}}{\partial y} + \frac{\partial \bar{j}_{zw}}{\partial z} = 0 \tag{10}$$

where the subindex w indicates a variable or property of the wall. At a wall–liquid interface, both φ and $\boldsymbol{j} \cdot \hat{\boldsymbol{n}}$ are continuous, where $\hat{\boldsymbol{n}}$ stands for the normal to the surface. By setting $\nu_t = 1$ and dropping the overbars, Eqs. (6–10) reduce to the laminar set of equations.

Because of symmetry conditions in both y and z directions, we need to consider only one quarter of the duct's cross section, for instance, $y \geq 0$ and $z \geq 0$. Equations (8–10) allow the elimination of the electric current density by introducing electric current stream functions h the fluid, defined as

$$\bar{j}_y = -\frac{\partial h}{\partial z}, \quad \bar{j}_z = \frac{\partial h}{\partial y}$$

From Eq. (7a), $\partial h/\partial z = \partial\varphi/\partial y$; therefore, a suitable potential function $F(y, z)$ can be introduced for h and φ, defined in the following way:

$$h = \frac{\partial F}{\partial y}, \quad \bar{\varphi} = \frac{\partial F}{\partial z}$$

Similarly, functions h_w and F_w are defined in the walls. In general, the electric current density and electric potential function of the fluid, walls, and surrounding medium are coupled. However, by assuming that the walls are thin and the surrounding medium is an electrical insulator, a boundary condition can be derived on the fluid variables only at the inside surface of each wall.[12] This boundary condition was extended in[7,11] where the boundary value problem for the turbulent flow in a rectangular duct with a uniform, transverse magnetic field was formulated in terms of the potential function F, in a duct with walls ranging from thin conducting to insulating. To reduce the computation task, the problem is simplified by considering a joint core-side-layer flow that occupies the region $0 \le y \le \alpha$, $0 \le z \le 1$. Assuming that M is sufficiently large, the Hartmann layers and corner regions are ignored in the sense that they are not resolved numerically. This implies that the non-slip condition at the top (bottom) wall is not satisfied. Nevertheless, the Hartmann layers are taken into account as a return path for the electric currents. The composite core-side-layer flow does not distinguish between the core and the side layer and is governed by a boundary value problem with the Hartmann number and the wall conductance ratio as $O(1)$ parameters. By retaining terms up to $O(M^{-1})$ and neglecting all of the other smaller terms in both the core and the side layer, and assuming that $v_{ty} \approx v_{tz} = O(1)$, the boundary value problem is determined by the following system of equations[7,11]:

$$\bar{u} = K + \frac{\partial^2 F}{\partial z^2} \tag{11}$$

$$\frac{\partial^2 F}{\partial y^2} - M^{-2}\left[v_{tz}\frac{\partial^4 F}{\partial z^4} + \frac{\partial v_{tz}}{\partial z}\frac{\partial^3 F}{\partial z^3}\right] = K \tag{12}$$

for $0 \le y \le a$ and $0 \le z \le 1$. At $y = 0$,

$$\frac{\partial^3 F}{\partial y \partial z^2} = 0, \quad \frac{\partial F}{\partial y} = 0 \tag{13a,13b}$$

At $z = 0$,

$$\frac{\partial^3 F}{\partial z^3} = 0, \quad \frac{\partial F}{\partial z} = 0 \qquad (14a,14b)$$

At $z = 1$,

$$\frac{\partial^2 F}{\partial z^2} = -K, \qquad F + c\frac{\partial F}{\partial z} = 0 \qquad (15a,15b)$$

and at $y = a$,

$$\frac{\partial F}{\partial y} - (c + M^{-1})\frac{\partial^2 F}{\partial z^2} = M^{-1}K \qquad (16)$$

In addition, the volume flux conservation condition has to be satisfied; namely,

$$\int_0^a \int_0^1 \overline{u}\, \mathrm{d}z\, \mathrm{d}y = a \qquad (17)$$

where a is the duct aspect ratio and $c = \sigma_w t_w / \sigma L$ is the wall conductance ratio, t_w being the wall thickness. To close the system of governing equations (12–16), the explicit form of the total viscosity $\nu_{tz} = \nu_t$ must be given. The boundary value problem given by the set of equations (12–16) and the corresponding expression for ν_t was solved with the spectral collocation method,[13,11] expressing the unknown function F as a finite series in terms of even Chebyshev polynomials in y and z and using appropriate Gauss–Lobatto collocation points. The system of simultaneous, linear algebraic equations for the coefficients of the series was solved with Gauss–Jordan elimination method. All of the solutions were normalized by applying condition (17) and corresponded to a square cross-section duct ($a = 1$). Turbulent flow solutions involve an iterative procedure in which *initial* Reynolds number and pressure gradient values are input, being properly rescaled once convergence is reached.[11]

IV. Turbulence Models

Renormalization Group Model

In dimensionless terms, the total viscous diffusivity (molecular + turbulent) given by the RNG theory[8] is

$$\nu_t = [1 + H(C_1 Re^2 \nu_t (\nabla \overline{u})^2 l_o^4 - C_2)]^{1/3} \qquad (18)$$

where the Heaviside function $H(x)$ is defined as $H(x) = x$ when $x > 0$ and $H(x) = 0$ when $x < 0$; $C_1 = 0.0256$ and $C_2 = 110$ are constants of the model used by Talmage[14] for the MHD flow in homopolar devices and l_o is the integral length scale, normalized by L. Because the steepest velocity gradients are in the z direction (perpendicular to the side wall), we can assume $|\nabla \bar{u}| \approx \mathrm{d}\bar{u}/\mathrm{d}z$. The integral length scale l_o is a free parameter and must be specified outside the theory. Here, it was determined through the model proposed by Yakhot et al.,[9] namely

$$l_o = \min\,(z, \gamma\delta^*) \tag{19}$$

where z is the distance to the wall and $\gamma = \gamma_o(1 - \mathcal{H}^{-1})^{-1}$. Here, γ_o is a constant which value was estimated as 0.3 in two limiting non-MHD cases of fully turbulent and transitional boundary layer over a flat plate.[9] For the present MHD flow, γ_o was determined to be 0.4 [Ref. 11]. In addition, $\mathcal{H}^{-1} = \theta/\delta_*$ is a shape factor, where δ_* and θ are the displacement and momentum-loss thickensses, respectively, defined as

$$\delta_* = \int_0^\infty \left(1 - \frac{u}{U_o}\right)\mathrm{d}z, \qquad \theta = \int_0^\infty \frac{u}{U_o}\left(1 - \frac{u}{U_o}\right)\mathrm{d}z \tag{20a,20b}$$

For a duct flow, using definitions (19–20b) the outer integral length scale can be written as

$$l_o^{(\mathrm{outer})} = \gamma_o(1 - \mathcal{H}^{-1})^{-1}\delta_* = \gamma_o \frac{\left[\int_0^\infty (U_o - u)\mathrm{d}z\right]^2}{\int_0^\infty (U_o - u)^2 \mathrm{d}z} \tag{21}$$

The introduction of boundary-layer characteristics in the definition of the integral length provides a better assessment of the factors influencing the turbulence. Under this approach, the magnetic effects are considered by the introduction of the particular structure of the MHD boundary layer in the integral length model. The total viscosity model defined by Eqs. (18–21), closes the system of governing Eqs. (11–16).

Mixing Length Model

One of the most common options for modeling the eddy viscous diffusivity is by means of the Prandtl mixing length theory.[10,15] The mixing length l is defined as the distance traveled by the fluid element in the z direction, which is normal to the direction of bulk flow. The assumption is made that the parcel of fluid moving through the distance l retains the mean momentum from its point of origin upon reaching its destination, so that the instantaneous velocity field can be expressed as

$$u = \overline{u} + u' \approx \overline{u} + l\frac{\partial \overline{u}}{\partial z}$$

that is, the mixing length is small enough to permit the velocity fluctuation to be written as $u' = l\partial\overline{u}/\partial z$. Because $u' \approx w'$, we can write the Reynolds stress term as

$$\overline{u'w'} \approx -l^2\left(\frac{\partial \overline{u}}{\partial z}\right)^2 = -\nu_e\frac{\partial \overline{u}}{\partial z}$$

So, the eddy viscous diffusivity is given by

$$\nu_e = l^2\frac{\partial \overline{u}}{\partial z} \tag{22}$$

and the total viscous diffusivity is

$$\nu_t = 1 + \nu_e \tag{23}$$

Different expressions have been proposed to take account for l. In this study, satisfactory results were found by taking $l = l_o$, where l_o is given by Eqs. (19–21).

V. Numerical Results

The validation of the numerical method was made by obtaining the laminar solution and comparing it with the analytical result of Walker.[12] This is accomplished by setting $\nu_t = 1$. The results obtained with the two methods coincide within 2%, except in a small region near the maximum velocity where the differences are of approximately 10%. Figures 2a and 2b show the axial velocity distribution as a function of the transversal coordinate z and the electric current density as a function of position in the transversal plane. Recall that due to the symmetry of the flow only one-quarter of the transversal plane is solved. Results were calculated for $M = 5300$ and $c = 0.07$. These values, along with $N = 2900$, correspond to the parameter values used in one of the experimental cases observed by Reed and Picologlou.[2] This numerical example is useful for it can be employed to gage the modifications introduced by the turbulence models, although it must be emphasized that the type of flow observed by Reed and Picologlou is not strictly turbulent because large scale structures with a strong periodic behavior were identified. According to Walker's asymptotic analysis,[12] the former values of M and c correspond to the case $c = \alpha M^{-1/2}$, where $\propto$ is a $O(1)$ constant. For this case, in the side layer u is $O(M^{1/2})$, φ and j_y are $O(1)$, and j_z is $O(M^{-1/2})$. In Fig. 2a the velocity profile at $y = 0$ displays a maximum at $z \sim 0.98$, and the

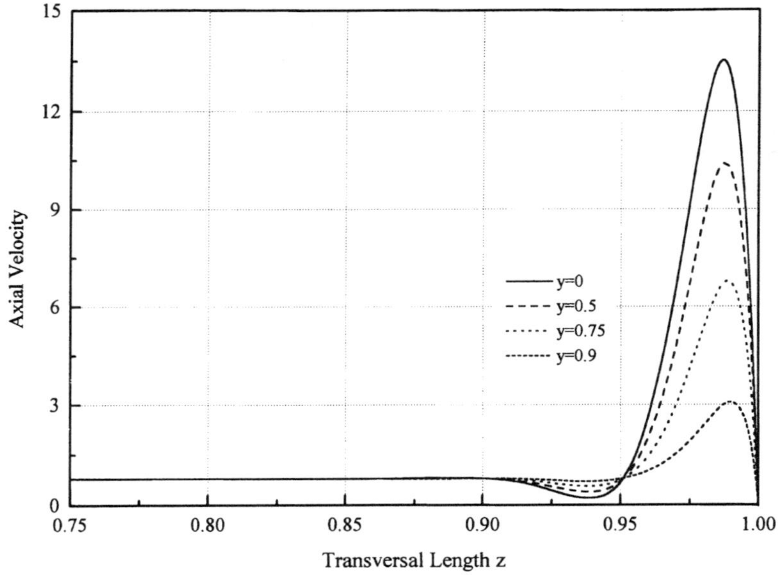

Fig. 2a Axial velocity as a function of transversal length z. Laminar flow. $M = 5300$ and $c = 0.07$.

maximum velocity is approximately an order of magnitude larger than that of the core value. This velocity distribution presents large shear stresses on each side of the velocity maximum that may generate instabilities or turbulence. The side boundary layer presents an $O(M^{-1/2})$ thickness and extends to approximately 6% of the channel width. The qualitative behavior for y = 0.5, 0.75, and 0.9 is similar to that described above for $y = 0$, but the maximum velocity is smaller as the top wall is approached. The electric potential shown in Fig. 3 indicates that the electric boundary layer is confined to the region $z > 0.96$, being wider near the mid-horizontal plane ($y = 0$) and getting narrower in the region close to the top wall ($y = 1$). In the core and up to approximately $z = 0.9$, the electric current, which runs perpendicular to the electric potential lines, moves horizontally in the direction of positive z. In the region next to the side wall ($z = 1$), the electric potential lines are perpendicular to the wall, indicating that the electric current runs parallel to it and in the positive direction of y. The return path for the electric current cannot be described from the plot because Hartmann layers are not resolved.

Renormalization Group Model

The mean velocity profiles obtained with the RNG total viscosity model given in Eqs. (18–21) are shown in Fig. 3a. Most of the qualitative features of the velocity profiles are analogous to those displayed by the laminar

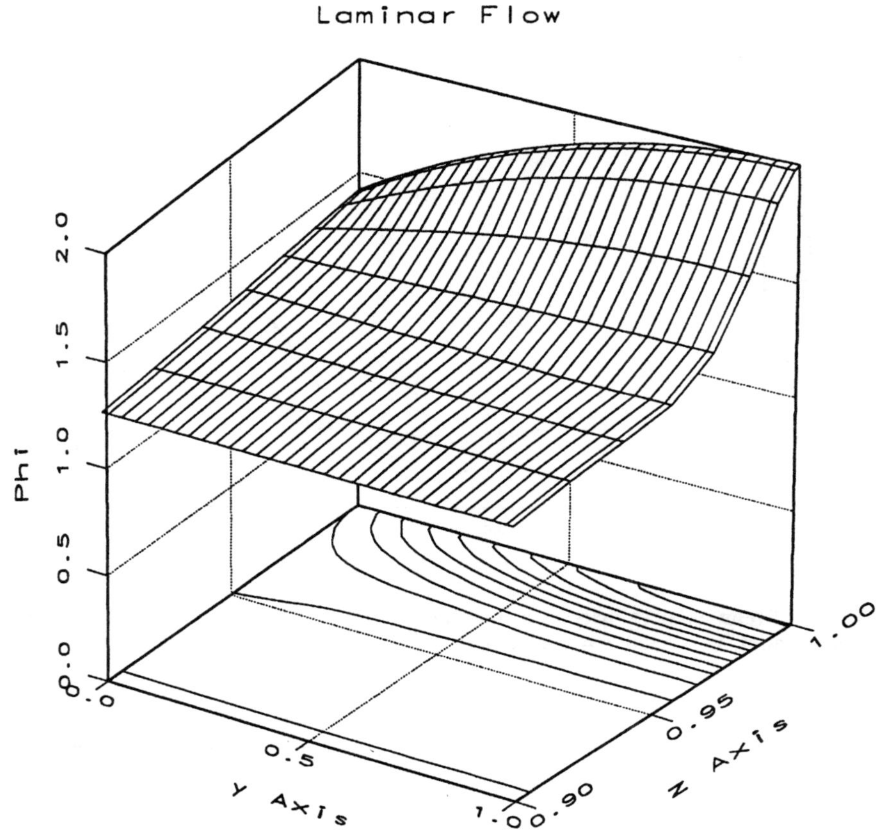

Fig. 2b Electric potential φ as a function of position yz. Laminar flow. $M = 5300$ and $c = 0.07$.

profiles, but the maximum velocity is larger than the core velocity by only a factor of five. Also, the velocity distribution displays a wider side boundary layer as compared to the laminar flow, extending to up to 12% of the channel width. The mean velocity distributions located near the top wall present a change in the profile's slope that makes the maximum of the distribution more clearly defined. The calculated total viscosity is plotted in Fig. 3b. The wavy form of the traces is a consequence of incomplete spectral resolution. The total viscosity is up to a factor of 50 larger than the molecular viscosity, but its effect on reducing the velocity is modulated due to the large Hartmann number used for the calculation [see Eq. (12)]. The electric potential plot of Fig. 3c shows that the electric side boundary layer extends to 5% of the channel width, which is only slightly wider than that of the laminar case. The qualitative features of the potential distribution are similar to those of the former case.

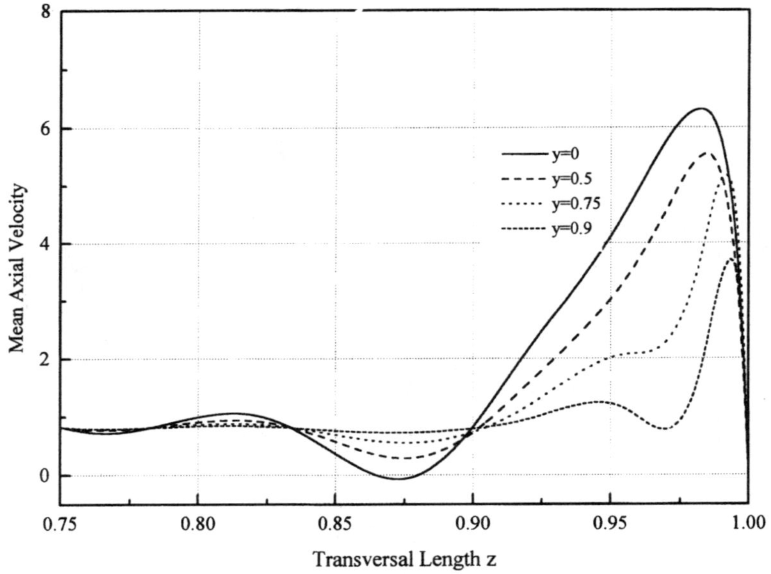

Fig. 3a Mean axial velocity as a function of transversal length *z*. Turbulent flow, RNG model. *M* = 5300, *c* = 0.07, and *N* = 2900.

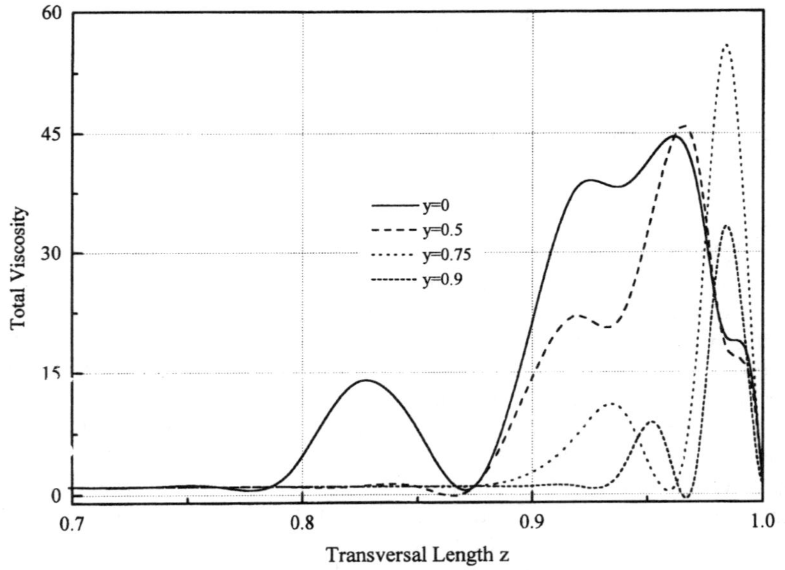

Fig. 3b Total viscosity as a function of transversal length *z*. Turbulent flow, RNG model. *M* = 5300, *c* = 0.07, and *N* = 2900.

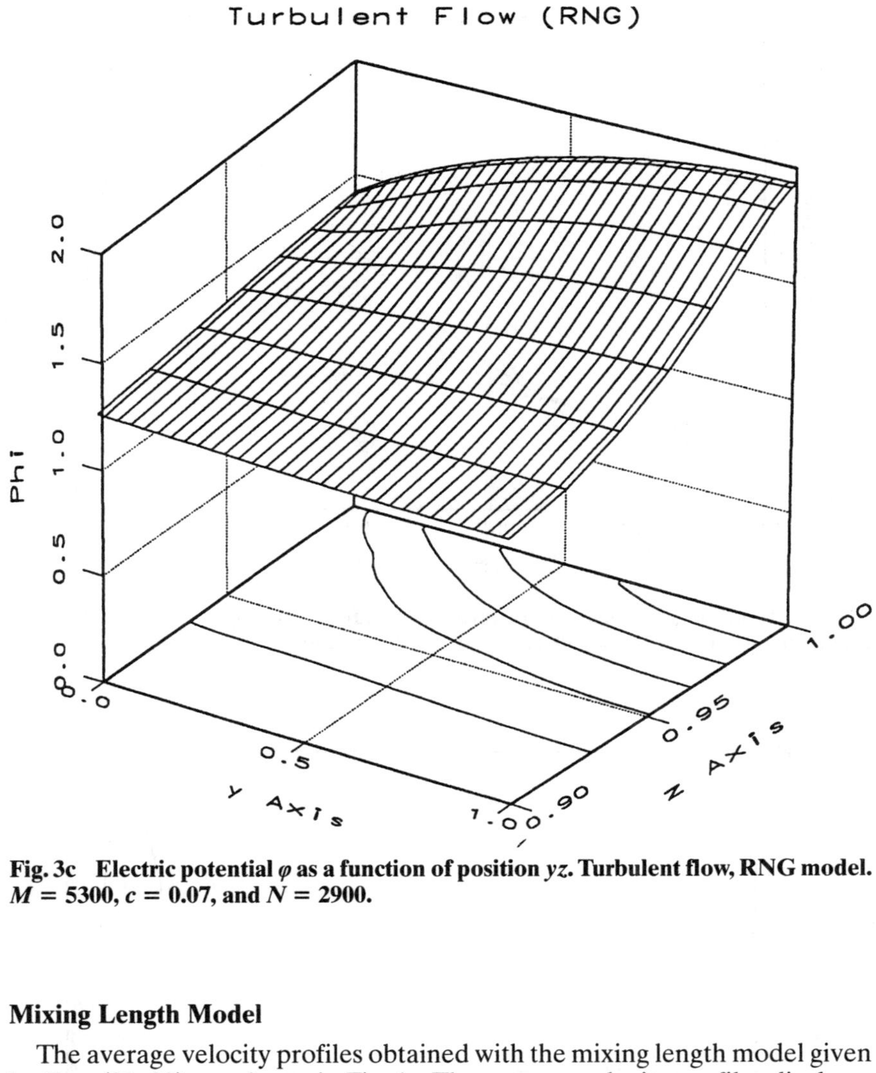

Fig. 3c Electric potential φ as a function of position yz. Turbulent flow, RNG model. $M = 5300$, $c = 0.07$, and $N = 2900$.

Mixing Length Model

The average velocity profiles obtained with the mixing length model given by Eqs. (22–23) are shown in Fig. 4a. The average velocity profiles display a wider side-boundary layer as compared to the laminar flow, extending up to $z = 0.75$, and although the velocity presents a maximum near the side wall at approximately $z = 0.98$, the maximum value is only a factor of four larger than the core value. Turbulence has the effect of smearing out the steep velocity gradients present near the wall. This observation is consistent with the fact that the total viscosity is two orders of magnitude larger than the molecular viscosity and is concentrated in the region $z > 0.8$ (see Fig. 4b). The electric current potential displays a similar distribution as compared to the laminar flow out the side boundary layer extends up to $z = 0.9$, which is approximately twice as wide as that of the laminar flow.

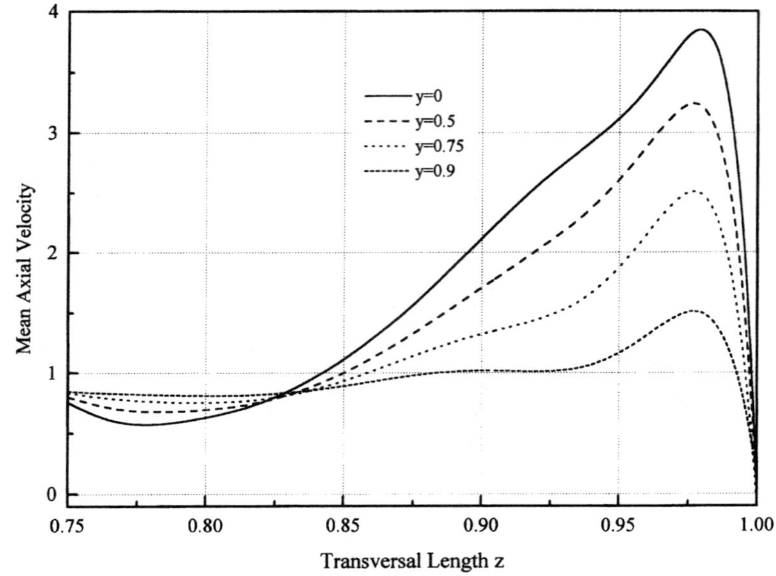

Fig. 4a Mean axial velocity as a function of transversal length z. Turbulent flow, mixing length model. M = 5300, c = 0.07, and N = 2900.

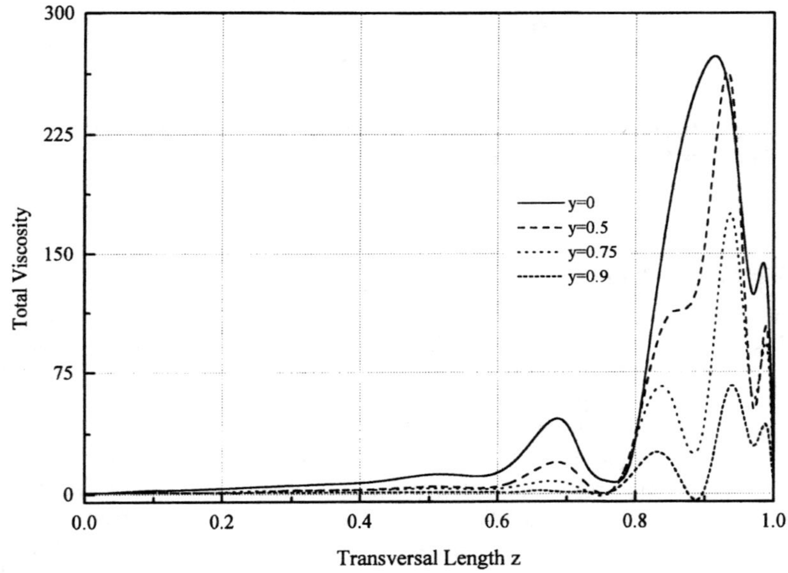

Fig. 4b Total viscosity as a function of transversal length z. Turbulent flow, mixing length model. M = 5300, c = 0.07, and N = 2900.

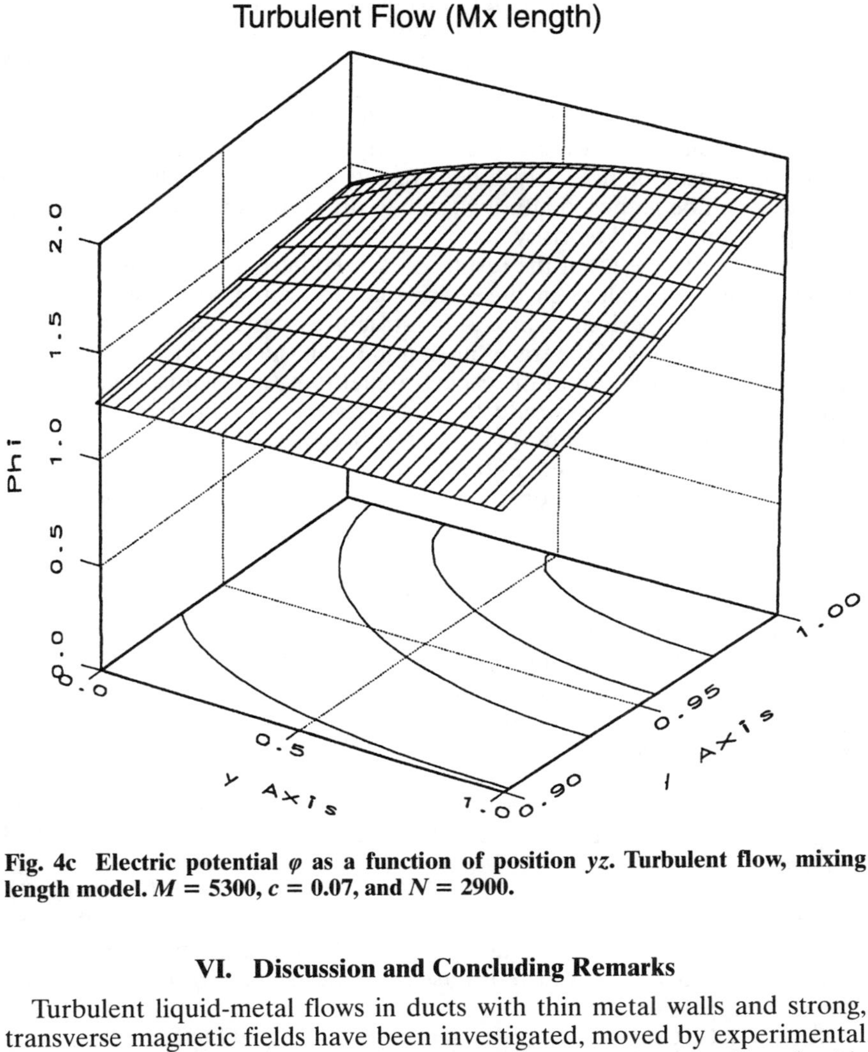

Fig. 4c Electric potential φ as a function of position yz. Turbulent flow, mixing length model. $M = 5300$, $c = 0.07$, and $N = 2900$.

VI. Discussion and Concluding Remarks

Turbulent liquid-metal flows in ducts with thin metal walls and strong, transverse magnetic fields have been investigated, moved by experimental evidences that point to the existence of persistent nonlaminar flows in this kind of duct. In qualitative terms, RNG as well as mixing length turbulence models display properties that are expected from the physical point of view. The steep velocity gradients are smoothed out and the side boundary layer is widened. The considerable increase of the total viscosity in the side layer compared to the laminar case indicates the enhancement of mixing in that region, though the total viscosity correctly approaches the laminar value in the core. On the other hand, as the side layer gets thicker, the electric current paths bend in the field direction farther from the side wall. The former features induced by turbulence are more pronounced when the mixing length mode is used. Numerical results are consistent with a linear stability analysis of the side layer instability flow which indicates that the steady flow in the core is unconditionally stable and is never perturbed by any fluctuations in the side layers.[16] A direct comparison of numerical results against experimental data is not possible at this time because the fully developed turbulent behavior assumed by the models does not correspond to the physical conditions prevailing in the side layer instability experiments. However, some resemblance of turbulent results with the overall experimental behavior can be glimpsed although due to differences in physical condition extrapolations would be risky.

It must be stressed that the results presented here must be interpreted in terms of average values in the sense that according to the experiments, time-dependent, periodic, vortex-like structures have been observed. The detailed oscillations of the traces most notable in the total viscosity plots are due to lack of resolution of the espectral method. An obvious suggestion to overcome this difficulty would be to increase the number of collocation points. This however is not an easy step because the convergence of the algorithm is not guaranteed. Future work will decide whether convergence can be improved or an altogether different numerical method of solution should be used.

References

[1]Branover, H., Sukoriansky, S., Talmage, G., and Greenspan, E. "Turbulence and the Feasibility of Self-Cooled Liquid-Metal Blankets for Fusion Reactors," *Proceedings of the American Nuclear Society Meeting*, Reno, NV, June 1986, pp. 15–19.

[2]Reed, C. B., and Picologlou, B. F., "Sidewall Flow Instabilities in Liquid-Metal MHD Flow under Blankets Relevant Conditions," *Fusion Technology*, Vol. 15, 1989, pp. 705–715.

[3]Alemany, A., Moreau, R., Sulem, P. L., and Frisch, U., "Influence of an External Magnetic Field on Homogeneous MHD Turbulence," *Journal de Mécanique*, Vol. 18, 1979, pp. 277–313.

[4]Sommeria, J., and Moreau, R., "Why, How, and When, MHD Turbulence Becomes Two-Dimensional," *Journal of Fluid Mechanics*, Vol. 118, 1982, pp. 507–518.

[5]Sukoriansky, S., Zilberman, I., and Branover, H., "Experimental Studies of Turbulence in Mercury Flows with Transverse Magnetic Fields," *Experiments in Fluids*, Vol. 4, 1986, pp. 11–16.

[6]Branover, H., Greenspan, E., and Sukoriansky, S., "Liquid-Metal Turbulent Flow Phenomena and Their Implications on Fusion Reactor Blankets Design," *Liquid Metal Magnetohydrodynamics*, edited by J. Lielpeteris and R. Moreau, Kluwer, Dordrecht, The Netherlands, 1989, pp. 63–77.

[7]Cuevas, S., Picologlou, B. F., Walker, J. S., and Talmage, G., "Liquid-Metal Flow in Rectangular Ducts with Thin Conducting or Insulating Walls: Laminar and Turbulent Solutions" (submitted for publication).

[8]Yakhot, V., and Orszag, S. A., "Renormalization Group Analysis of Turbulence. I. Basic Theory," *J. Sci. Comput.*, Vol. 1, 1986, pp. 3–51.

[9]Yakhot, V., Kedar, O., and Orszag, S. A., "An Algebraic-Q4 Turbulent Eddy Viscosity Model: Boundary Layer Flow over a Flat Plate and Flow in a Pipe," *J. Sci. Comput.*, Vol. 7, 1992, pp. 229–239.

[10]Hinze, J. O., *Turbulence*, 2nd ed., McGraw–Hill, New York, 1975.

[11]Cuevas, S., "Liquid-Metal Flow and Heat Transfer under Strong Magnetic Fields," Ph.D. Thesis, Facultad de Ciencias, Univ. Nacional Autónoma de México, 1994.

[12]Walker, J. S., "Magnetohydrodynamic Flows in Rectangular Ducts with Thin Conducting Walls. Part I. Constant Area and Variable Area Ducts with Strong Magnetic Fields," *Journal of Mecanique*, Vol. 20, 1981, pp. 79–112.

[13]Canuto, C., Hussani, M. Y., Quarteroni, A., and Zang, T. A., *Spectral Methods in Fluid Dynamics*, Springer–Verlag, New York, 1987.

[14]Talmage, G., Walker, J. S., Brown, S. H., Sondergaard, N. A., Branover, H., and Sukoriansky, S., "Liquid-Metal Flows in Sliding Electrical Contacts: Solution for Turbulent Primary Azimuthal Velocity," Vol. 148, Progress in Astronautics and Aeronautics, AIAA, Washington, DC, 1993, p. 165.

[15]Tennekes, H., and Lumley, J. L., *A First Course in Turbulence*, MIT Press, Cambridge, MA, 1972.

[16]Ting, A., Walker, J. S., Moon, T. J., Reed, C. B, and Picologlou, B. F., "Linear Stability Analysis for High-Velocity Boundary Layers in Liquid-Metal Magnetohydrodynamic Flows," *Int. J. Engng. Sci.*, Vol. 29, 1991, pp. 939–948.

Chapter 6

Inertial Organization in Forced Two-Dimensional Turbulence

Bernhard Jüttner* and André Thess†
Institut für Strömungsmechanik, Technische Universität Dresden, Germany

Introduction

BOTH atmospheric and oceanic as well as some magnetohydrodynamic flows are able to exhibit long-living large-scale turbulent structures, often called "coherent structure." A well-known example is the Great Red Spot of Jupiter, which has stably existed for more than 300 years in a highly turbulent atmosphere. In laboratory experiments on rotating fluids[1] and on electromagnetically driven flows,[2] coherent structures also can be observed. The experimental observations[2] were confirmed by previously performed numerical simulations.[3] However, due to the high costs of the computations, the authors could not study the long-term evolution of the flow, which is necessary for a quantitative comparison with results from statistical computations. Indeed, the strong anisotropy of the flow causes the existence of the large-scale structures. Although these structures cannot be understood in the frame of fully three-dimensional turbulence, they can be described by a statistical theory of inertial organization of two-dimensional turbulence.[4–6] Originally, this maximum entropy theory (MET) was proposed to describe equilibrium states of the two-dimensional Euler equation. The existence of conserved integral quantities for fluid motion governed by the two-dimensional Euler equation is responsible for the organization of two-dimensional turbulence into large-scale structures. Although results of this theory can be successfully compared with those of direct numerical simulation (DNS) on freely decaying two-dimensional turbulence at high Reynolds numbers,[7–9] there is a large gap. Whereas theoretical results are obtained as solutions of the Euler equation, the natural systems of our interest are subject to forcing and friction. In the case of the Euler equation, the conserved quantities are given a priori; in the latter case, there is a dynamical equilibrium of the

*Researcher
†Researcher

integral quantities. The aim of the present communication is to present some preliminary results supporting the conjecture that the ideas of MET can be successfully carried over to the important case of forced turbulence.

In the next section we present the results of a DNS of forced two-dimensional turbulence. The kind of forcing and the additional linear friction models the experimental situation.[2] Then we will use the measured values of time-averaged integral quantities from the DNS as input in a numerical scheme,[10] which is based on the MET. It calculates coarse-grained vorticity distributions that are related to the most probable macroscopic vorticity distributions under the constraints of fixed integral quantities.

Direct Numerical Simulation

To study the properties of large-scale structures evolving in forced two-dimensional turbulence, we have performed a DNS of the generalized two-dimensional Navier–Stokes equation

$$\frac{\partial \omega}{\partial t} = \frac{\partial \psi}{\partial x}\frac{\partial \omega}{\partial y} - \frac{\partial \psi}{\partial y}\frac{\partial \omega}{\partial x} + \nu \Delta \omega - \mu \omega + g(x, y) \tag{1}$$

in a square domain $(x,y) \in [0,1] \times [0,1]$. In Eq. (1) $\omega = -\Delta\psi$ is the vorticity, and ψ denotes the stream function. The term $-\mu\omega$ describes the friction of the fluid in the boundary layers parallel to the flow, e.g., Hartmann layer(s) in magnetohydrodynamic (MHD) flows or Ekman layer(s) in rotating flow systems. We apply free-slip boundary conditions, that is, $\psi|_{\partial D} = 0$ and $\omega|_{\partial D} = 0$. The function $g(x, y)$ models the stationary forcing of the vorticity field.

We use a numerical scheme that bases on the pseudospectral collocation method[11] including the 2/3 rule for removal of the aliasing errors. Using sine-series for ψ and ω in both spatial directions, the imposed boundary conditions are satisfied automatically. The time-stepping scheme for the nonlinear terms uses the Adams–Bashford–3 method. The linear terms are calculated exactly using the integrating-factor technique. For the results reported in this communication, we have used a spatial resolution of 128 $\times$ 128 collocation points corresponding to 84 $\times$ 84 Fourier modes. The time step was $\Delta t = 0.0025$.

The DNS is performed with reference to an experiment of Sommeria[2] on an electromagnetically forced layer of liquid metal. In these MHD experiments the author has studied forced quasi-two-dimensional turbulence at high Reynolds numbers. He has observed that for high enough values of the bottom friction number $Rh = 1/\mu$ a stable large-scale annular motion in the box has been established. The aim of our numerical study is to investigate the regime where such structures appear. Therefore, we have applied conditions for our DNS that are close to the experimental situation. The forcing of the flow is then given by

$$g = \sin(6\pi x)\sin(6\pi y) \tag{2}$$

which for low Reynolds numbers $Re = 1/\nu$ produces an array of 36 counter-rotating vortices as the basic flow. The parameters of the dissipative terms are chosen as $\nu = 10^{-5}$ and $\mu = 10^{-3}$.

We start our simulation from an initial condition of the vorticity $\omega(x,y,t = 0) = \delta(x,y)$, where $\delta(x,y)$ is a random perturbation with a magnitude $||\delta|| = 0.001$. The small perturbations are used to break the symmetry of the problem. After a comparatively short period of time of the order of 10, the flow becomes turbulent and enters into a statistically stationary regime, henceforth referred to as "dynamical equilibrium." We follow this regime up to time $t_{\text{end}} = 850$.

An essential signature of the dynamical equilibrium is the weak time dependence of the integral quantities in time explained by the plots in Fig. 1. In Fig. 1a the temporal behavior of the energy E is shown. We see, that in the dynamical equilibrium state the mean value of the energy is $\langle E\rangle \approx 0.026 \pm 0.02$ and relative fluctuations are small. Other integral quantities, which we have measured for later use in the statistical theory, are the areas A_i of different vorticity levels q_i in the domain. They are calculated by discretisizing the vorticity in intervals that cover all values from the minimum up to the maximum vorticity. Then we measure the area A_i of vorticity

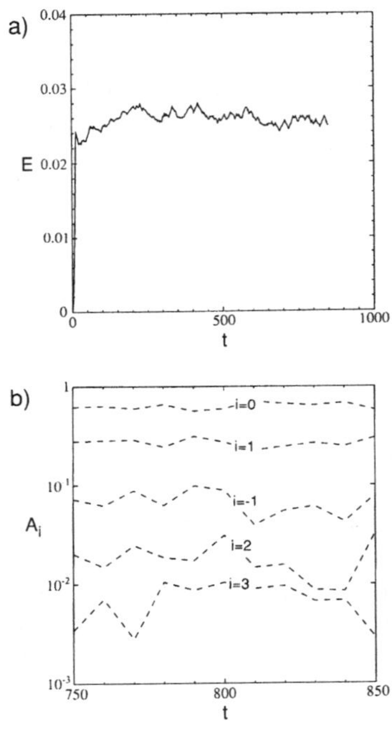

Fig. 1 Temporal behavior: a) of the kinetic energy E and b) of the areas A_i (of the dominating vorticity levels $q_i = iq_0/n$ from the DNS, with $i = -n, \ldots, n$, $q_0 = 22.8$, and $n = 10$).

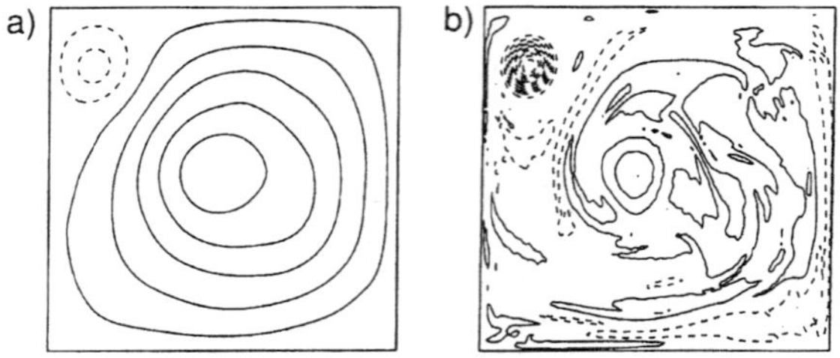

Fig. 2 Isoline plots of a snapshot from the direct numerical simulation at t = 800: a) of the stream function and b) of the vorticity. The isolines are equidistantly distributed. Dashed lines represent negative values.

lying in a special interval. We have done this procedure at 11 equidistantly distributed time steps in the range of $t = 750$ and $t = 850$. In Fig. 1b the area fractions A_i for the dominating vorticity levels q_i are plotted versus time. The areas fluctuate less than approximately 30%. What about the evolving flow structures? Figure 2 shows snapshots of the stream function (Fig. 2a) and of the vorticity (Fig. 2b) at time $t = 800$. In Fig. 2a we see that, like in the experiment, the flow displays large-scale annular motion around the center of the box. In detail we find that positive vorticity spreads around the center of the box whereas negative vorticity is mainly concentrated in the vortex in the upper left corner. This vortex occupies just a small area of the domain. However, the vortex it has a much larger magnitude than the positive vorticity in the center. To model the balanced forced system by a closed set of statistical equations, it is necessary that the main cause for energy dissipation is due to linear friction. This condition is fulfilled if $\mu \approx \tilde{\mu}$ where

$$\tilde{\mu} = \mu + v\int_D \omega^2 \, \mathrm{d}D \Big/ \int_D \psi\omega \, \mathrm{d}D \tag{3}$$

is a total dissipation rate defined with respect to energy dissipation. For our DNS, however, we realize that this total dissipation rate is twice the dissipation rate with respect to the linear friction term μ, as shown in Fig. 3. Therefore we conclude that the condition $\mu \approx \tilde{\mu}$ is not very well fulfilled in our DNS. Furthermore we will make sure that the vorticity produced at a special localization can be transported through the whole domain before it is dissipated by the linear friction. This is the case if the characteristic time of the motion is small in comparison with the dissipative time scale due to the linear friction, that is,

$$\mu \ll \frac{\sqrt{2E}}{L} \tag{4}$$

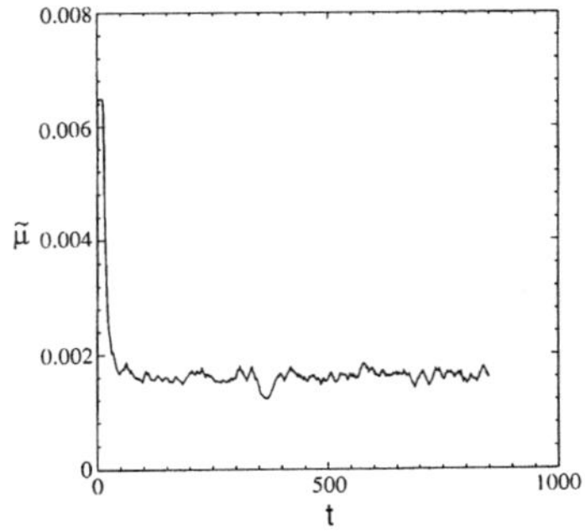

Fig. 3 Temporal behavior of the effective dissipation rate $\tilde{\mu}$ from the DNS for the whole time of observed evolution.

with $L = 1$ is the size of the box. This condition is fulfilled by the parameters chosen in the simulation.

We conclude that our numerical observations reported so far are in qualitative agreement with the experimental findings of Sommeria.[2] In particular the appearing of a nonsymmetric dipolar structure as result of the dynamical equilibrium of forcing and friction in the turbulent regime seems to be a strong sign that some properties of the MHD experiment[2] can be modeled by the two-dimensional Navier–Stokes equation (1) with the inhomogeneity (2). In the following section we will show that these structures can be understood in terms of the statistical theory of equilibrium states of two-dimensional turbulence.

Direct Numerical Simulation—Assisted Maximum Entropy Theory

Self-organization of freely evolving two-dimensional turbulence [i.e., $\nu = \mu = g = 0$ in Eq. (1)] into large-scale coherent structures can be understood in the framework of the MET. In the present section we are going to investigate as to wether the basic ideas of MET can be carried over to the forced case. This theory considers the flow as a set of vorticity patches of different vorticity levels q_i. The theory assumes a strong mixing of these vorticity patches. In a first step, probabilities $p_i(x,y)$ for finding the vorticity level q_i at a fixed point are defined. They have to fulfill the normalization condition

$$\sum_i p_i = 1 \tag{5}$$

at each point. Then the most probable macroscopic distribution of the vorticity levels in the domain maximizes a Boltzmann mixing entropy

$$S = -\frac{1}{A_D}\int_D \sum_i p_i \ell n(p_i)\,\mathrm{d}D \Rightarrow \max \tag{6}$$

From this distribution a coarse-grained vorticity distribution

$$\Omega = \sum_i q_i p_i \tag{7}$$

is calculated. The related coarse-grained stream function is given by the solution of the Poisson equation

$$\Delta\psi = -\Omega \tag{8}$$

and has to fulfill the boundary conditions $\psi|_{\partial D} = 0$. The mixing process takes place under the constraints that the kinetic energy

$$E = \frac{1}{2}\int_D \Psi\Omega \,\mathrm{d}D \tag{9}$$

of the system and the areas

$$A_i = \int_D p_i \,\mathrm{d}D \tag{10}$$

of vorticity patches have to be conserved due to the properties of the two-dimensional Euler equation. Equations (5–10) constitute a closed system of equations for the probabilities p_i and the Lagrange multipliers α_i and α_E that are related to the constraints of A_i and E, respectively. In principle there are multiple solutions that satisfy Eqs. (5–10). To find the global maximum of entropy, one can use either the relaxation equation method of Robert and Sommeria[12] or the iterative algorithm of Whitaker and Turkington.[10] It is now accepted that MET for free turbulence can produce quantitative predictions with reasonable accuracy.[7–9]

Next we turn to the case of forced two-dimensional turbulence. In this case the integral quantities A_i and E are not given a priori but result from an equilibrium of forcing and friction that can be expressed by the requirements

$$\frac{\mathrm{d}A_i}{\mathrm{d}t} = \frac{\mathrm{d}E}{\mathrm{d}t} = 0 \tag{11}$$

Assuming that the dissipation is mainly affected by the linear friction due to μ, we can neglect the effect of inner friction due to ν. Then, according to Kanzantsev et al.,[13] the energy balance is given by

$$E_{\mathrm{eq}} = \frac{1}{2\mu}\int_D g\Psi \,\mathrm{d}D \tag{12}$$

and the areas of the vorticity levels are determined from an equilibrium condition

$$A_{\mathrm{eq},i} = \frac{1}{\mu q_i} \int_D g p_i \, \mathrm{d}D \tag{13}$$

If the forcing function $g(x,y)$ and the linear friction parameter μ are known, MET equations (5–10), (12), and (13) determine the coarse-grained vorticity $\Omega(x,y)$ as well as the energy E and the area fractions A_i. A straightforward way to check the validity of this theory against the DNS of the previous section would be to solve the forced MET with $g = \sin(6\pi x)\sin(6\pi y)$ and $\mu = \tilde{\mu} \approx 0.002$ and to compare E, A_i, Ψ, and Ω with the results of Figs. 1 and 2. Unfortunately, there is as yet no algorithm available that is comparable with the iterative procedure[10] and solves the system (5–10), (12), and (13). Therefore we must have recourse to a restricted approach by taking the time averages of the areas A_i from the DNS as input in the iterative algorithm. The distribution of the averaged values $\langle A_i \rangle$ are shown in Fig. 4. In a next step we have calculated the maximum entropy solutions for all feasible values of the energy. As a result we obtain the bifurcation diagram shown in Fig. 5. We see that, depending on applied energy, two distinct structures evolve: monopolar structures for low energies and nonsymmetric dipolar structures for the high energy range. The diagram in Fig. 5 shows the dependence of the dissipation parameter $\tilde{\mu}$ from the kinetic energy E, necessary to fulfill balance equation (12) if kinetic energy, coarse-grained stream function, and forcing function are known. We see that for a forcing function (2) monopolar structures cannot fulfill the balance condition (12). On the other hand, in the case of a lattice-like forcing (2), dipolar structures can result from an equilibrium of forcing and friction. The most interesting detail of Fig. 5 is the quantitative comparison of the graph $\tilde{\mu}(E)$ with the values from DNS. These values are plotted as clouds of points nearby the maximum of $\tilde{\mu}$. From this excellent agreement we conclude that condition (12) seems to be valid also in cases where the dissipation is caused by both linear friction and kinematic viscosity. Comparing coarse-grained stream function (Fig. 6a)

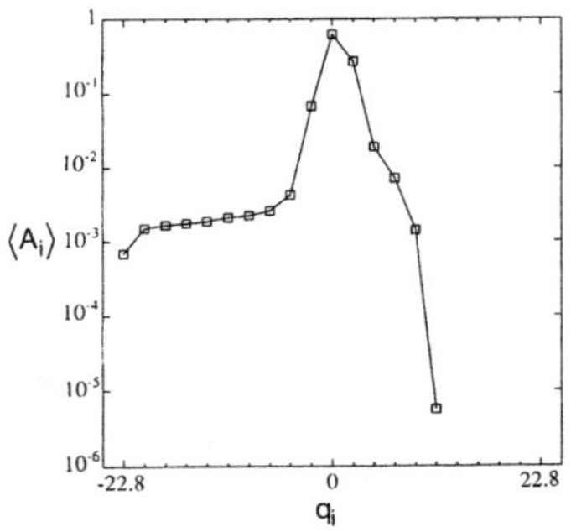

Fig. 4 Time-averaged areas $\langle A_i \rangle$ of the vorticity levels $q_i = iq_0/n$ from the DNS, with $i = -n, \ldots, n$, $q_0 = 22.8$, and $n = 10$ used as input to the iterative solver of the multiconstrained variational problem.

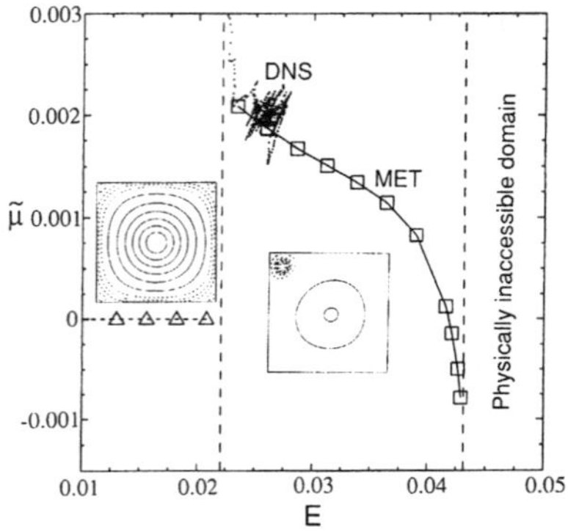

Fig. 5 Bifurcation diagram of the maximum entropy solutions for the distribution $\langle A_i \rangle(q_i)$ showing the necessary effective dissipation rate $\tilde{\mu}$ for the equilibrium of forcing and friction in dependence of the energy E. Solid line: $\tilde{\mu}(E)$ from the MET; cloud of dots: $\tilde{\mu}(E)$ from the DNS.

and coarse-grained vorticity (Fig. 6b) obtained by statistical theory with time-averaged stream function (Fig. 6c) and time-averaged vorticity (Fig. 6d) obtained by DNS, we find the same types of spatial distributions. Now we turn to the most impressive comparison of statistical theory with DNS. Remember that important features of the solutions of the problem (5–10) are coarse-grained functions Ψ and Ω fulfilling a special monotonous relation $\Omega = f(\Psi)$. Therefore look at Fig. 7 where Ω is plotted versus Ψ. The solid line shows the relation as a result of MET. The cloud of points close to the

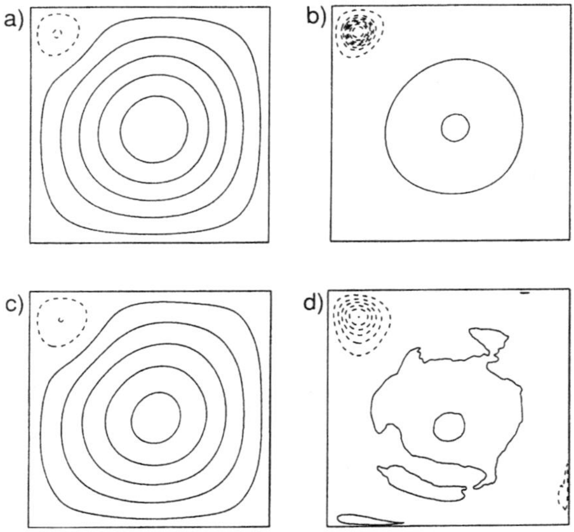

Fig. 6 Isoline plots: a) the coarse-grained stream function from the MET, b) the coarse-grained vorticity from the MET, and c) the time-averaged stream function from the DNS for the time interval $t \in [750,850]$, and d) the time-averaged vorticity from the DNS for the time interval $t \in [750,850]$. The isolines are equidistantly distributed. Dashed lines represent negative values.

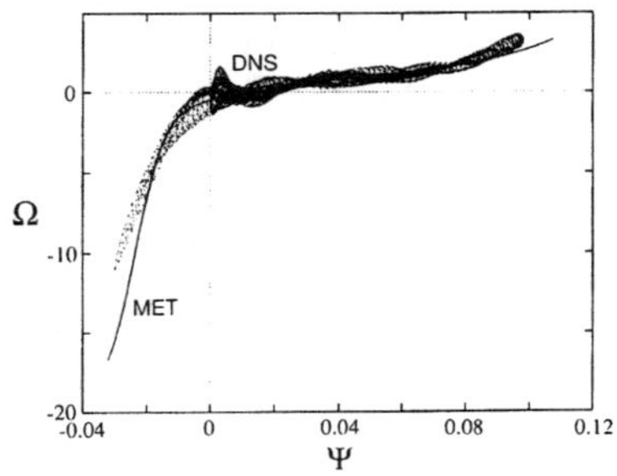

Fig. 7 Scatter plot of the ψ–Ω relation for the time-averaged values from the DNS (cloud of dots) and for the values from the MET (solid line).

solid line comes from the time-averaged values of the stream function and vorticity obtained by DNS. It is shown that, except the values in the core of the small, negative vortex in the upper left corner, the Ψ–Ω relation from the DNS is the same as from the statistical theory.

Conclusion and Outlook

We have studied features of large-scale structures evolving in two-dimensional forced turbulence as result of a mixing process satisfying equilibrium conditions for the kinetic energy of the system and for the areas of given vorticity levels. We find results of the DNS to agree with predictions of the MET. However, we have used an indirect way for the calculation of the maximum entropy structure by using the areas of the vorticity levels from the DNS as input for the statistical theory. Further work should concentrate on investigating the question if the very fast working algorithm of Whitaker and Turkington can also be used for the system (5–10) with equilibrium conditions (12) and (13).

Acknowledgment

This work is supported by the Deutsche Forschungsgemeinschaft under Grant Th 497/4-2.

References

[1]Sommeria, J., Meyers, S. D., and Swinney, H. L., "Laboratory Simulation of Jupiter's Great Red Spot," *Nature*, Vol. 331, 1988, pp. 689–693.

[2]Sommeria, J., "Experimental Study of the Two-dimensional Inverse Energy Cascade in a Square Box," *Journal of Fluid Mechanics*, Vol. 170, 1986, pp. 139–168.

[3]Verron, J., and Sommeria, J., "Numerical Simulations of a Two-Dimensional Turbulence Experiment in Magnetohydrodynamics," *Physics of Fluids*, Vol. 30, 1987, pp. 732–739.

[4]Kuzmin, G. A., "Statistical Mechanics of the Organization into Two-Dimensional Coherent Structures," *Structural Turbulence*, edited by M. A. Goldshtik, Academy of Naouk CCCP Novosibirsk, Inst. of Thermophysics, Novosibirsk, 1982, p. 103.

[5]Miller, J., "Statistical Mechanics of Euler Equations in Two Dimensions," *Physics Rev. Lett.*, Vol. 65, 1990, pp. 2137–2140.

[6]Robert, R., and Sommeria, J., "Statistical Equilibrium States for Two-Dimensional Flows," *Journal of Fluid Mechanics*, Vol. 229, 1991, pp. 291–310.

[7]Sommeria, J., Staquet, C., and Robert, R., "Final Equilibrium States of a Two-Dimensional Shear Layer," *Journal* of *Fluid Mechanics*, Vol. 233, 1991, pp. 661–689.

[8]Thess, A., Sommeria, J., and Jüttner, B., "Intertial Organization of a Two-Dimensional Turbulent Vortex Street," *Physics of Fluids*, Vol. 6, 1994, pp. 2417–2429.

[9]Jüttner, B., Thess, A., and Sommeria, J., "On the Symmetry of Self-Organized Structures in Two-Dimensional Turbulence," *Physics of Fluids*, Vol. 7, 1995, pp. 2108–2110.

[10]Whitaker, N., and Turkington, B., "Maximum Entropy States for Rotating Vortex Patches," *Physics of Fluids*, Vol. 6, 1994, pp. 3963–3973.

[11]Canuto, C., Hussaini, M. Y., Quarteroni, A., and Zang, T. A., *Spectral Methods in Fluids Dynamics*, Springer–Verlag, New York, 1988.

[12]Robert, R., and Sommeria, J., "Relaxation Towards a Statistical Equilibrium State in Two-Dimensional Perfect Fluid Dynamics," *Phys. Rev. Lett.*, Vol. 69, 1992, pp. 2776–2779.

[13]Kazantsev, E., Sommeria, J., and Verron, J., "Subgrid Eddy Parametrisation by Statistical Mechanics in a Barotropic Ocean Model," *J. Phys. Ocean* (submitted for publication).

Chapter 7

Magnetohydrodynamic Turbulence Measurements in a Sodium Channel Flow

Sven Eckert, Gunter Gerbeth, Horst Langenbrunner, and Willy Witke
Research Centre Rossendorf, Dresden, 01314, Germany

Introduction

THE knowledge about properties of the anisotropic turbulence in liquid-metal flows exposed to an external magnetic field is essential for several technological applications such as the concept of a self-cooled liquid metal blanket for thermonuclear fusion reactors, but also of basic interest for turbulence research.

Flow parameters like pressure drop and heat transfer rate are essentially determined by the interaction between the external magnetic field and the liquid-metal flow. Several experimental studies[1–4] revealed that the application of a magnetic field leads not exclusively to a suppression of the turbulent perturbations. Velocity fluctuations remain and demonstrate a distinct anisotropy of the magnetohydrodynamic (MHD) turbulence showing a tendency to become two-dimensional. The properties of the local transport of heat or mass are strongly governed by the anisotropic character of the flow.[5,6] In this context a number of questions arises regarding the origin, the decay time, or the size of the turbulent elements. A typical feature of the two-dimensional turbulence is that the energy becomes concentrated in organized large-scale fluctuations.[7] This fact ensures an intensive heat transfer on longer distances.

Because of the anisotropy of the electromagnetic dissipation term, vortices will be scarcely damped over long distances if their axes are aligned with the magnetic field lines.[8] For practical applications it is important to identify possibilities to promote the formation of such quasi-two-dimensional vortices in order to control the heat or mass transfer rate of the flow. It is well-known that the mentioned perturbations can be generated by rather specific means, for example, mechanical inserts or electrical currents between definitely arranged electrodes.

In this paper we follow the procedure proposed by Branover and Sukoriansky[9,10] to force the turbulence intensity by mechanical means employing a grid of cylindrical bars or flat stripes. Because of favorable material properties of the used liquid sodium, we are able to extend the measurements into the region of high interaction parameters. In this way the obtained results complement the already existing knowledge about the two-dimensional MHD turbulence.

After a brief description of the experimental conditions, the results concerning the measured turbulence intensity as a function of the interaction parameter N and the corresponding power spectral densities will be presented. The measurements took place in two different test sections allowing either the variation of the axial distance between the turbulence promoter and the probe or the installation of special channel inserts in order to suppress the generation of the M-profile of the local velocity and to decrease the wall conductance ratio. The results obtained in both test sections will be compared, and the influence of the altered boundary conditions onto the turbulent structure of the flow will be discussed.

Experimental Configuration and Data Processing

For technical reasons the experiments were performed in two different test sections in the following denoted as version A and B, respectively.

With version A we have the most simple channel configuration shown in Fig. 1. The walls of the rectangular channel have a thickness of 5 mm. This leads to a wall conductance ratio c of 0.03. The corresponding ratio of the electrical resistivities of the Hartmann layers and the channel wall has reached a value greater than unity already for the smallest Ha number in the experiment. That means, in this case we have to deal with electrically conducting walls. The conductivity of the channel walls affects clearly the damping process of the MHD turbulence.[11] If the walls are perfectly conducting, a short-circuit appears directly through the wall and causes strong currents resulting in a rapid damping of the turbulence. On the contrary, in the case of nonconducting walls in the region behind a grid of bars, turbulence intensities were reported[9] exceeding remarkably the values typical for ordinary turbulence without magnetic field at the same Re number. So, if one wants to guarantee optimal conditions for the MHD turbulence, nonconducting walls are required. However, this causes serious practical difficulties to find suitable nonconducting materials for aggressive environments as liquid metals like sodium or lithium.

A special method was employed in the experiments with the test section B with the aim to decrease the wall conductance ratio. To cover the inner channel wall, a U-shaped insert was produced consisting of two thin stainless steel sheets (δ = 0.2 mm). An electrically insulating layer is enclosed by these metal sheets welded by a special technology along the edges. The

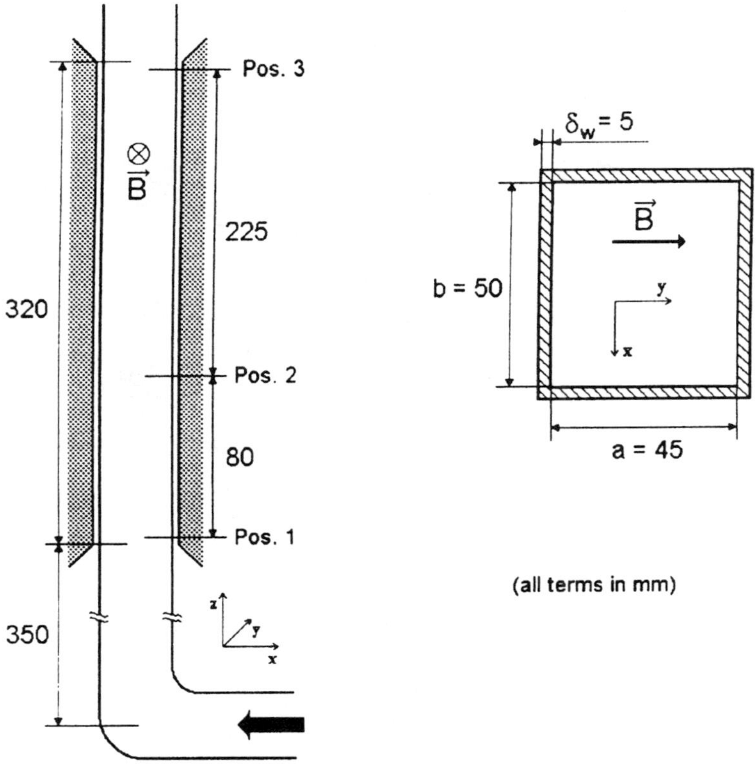

Fig. 1 Sketch of the test section version A.

result is a considerable reduction of the thickness of the effective conducting wall leading to a wall conductance ratio of 0.0012. As a consequence the resistivity ratio of the Hartmann layers and the wall is clearly below unity for $Ha = 250$ and becomes larger than one for the maximum of the magnetic field strength. The reason for the U-shape is caused by the condition to have a free side in order to install local probes into the flow or to perform heat transfer experiments with a heat flux generated at the channel wall.

This U-shape installation belongs to an ensemble of channel inserts proposed and manufactured at the Institute of Physics in Riga/Salaspils. Furthermore the entire set of inserts for the channel version B is completed by the following components (see Fig. 2): 1) two honeycombs for the fringe field regions at the ends of the magnet consisting of flat sheets created according to the same "sandwich" technology as the U-shape insert described before in order to avoid the generation of the typical M-shape velocity profile; 2) a 3-mm copper plate located inside the U for a better homogenization of the flow; 3) a grid of copper bars directed perpendicular

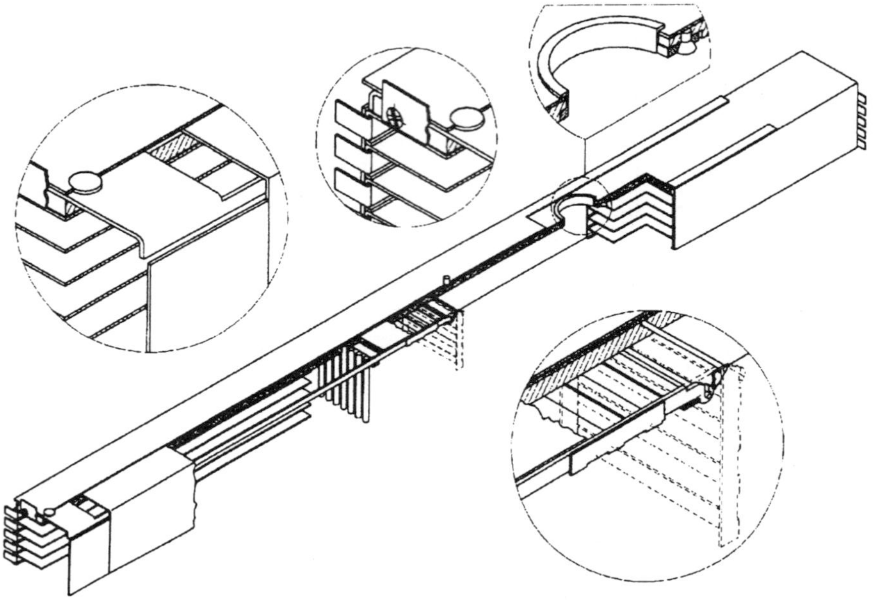

Fig. 2 Sketch of the inserts applied to the test section version B.

to the direction of the external magnetic field in order to suppress the remaining turbulence coming from the outlet of the first honeycomb; and 4) a grid of steel sheets parallel to the field lines acting as turbulence promoter.

In the frame of the measurements performed with the test section A, a second grid was used in order to increase the level of turbulence in the flow (Fig. 3). The material of the installation is stainless steel. The cylinders are

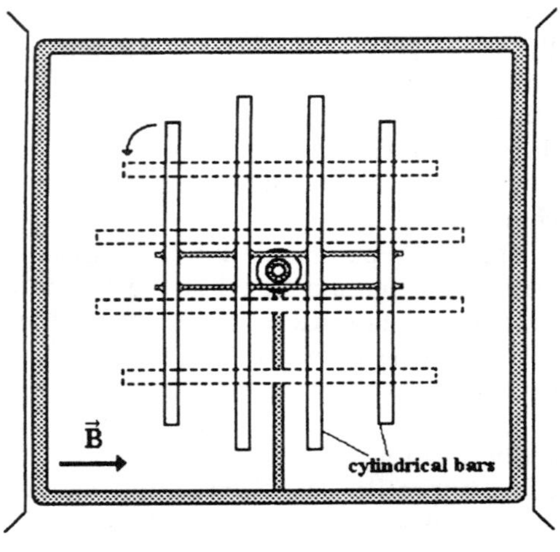

Fig. 3 Grid consisting of four cylindrical bars used for the turbulence promotion in the test section version A.

rotatable mounted, so that any angle from 0 to 90 deg between the direction of the cylinder axes and the magnetic field lines can be selected. This grid was positioned in the core of the flow inside the region of the homogeneous magnetic field.

The axial component of the velocity and their fluctuations were measured by means of two electrode potential probes. The probe was connected with a traversing mechanism that allowed the sensor to move over the channel cross section. Local measurements were possible in a distance from the wall of 3 mm (direction perpendicular to the magnetic field) and 9 mm (direction parallel to the magnetic field). To measure the profiles of the velocity and the turbulence intensities, it was necessary to vary the probe position during the experiment. In other cases, if the turbulent properties of the flow should be investigated as function of the dimensionless parameter N, the location of the probe was fixed in the center of the channel cross section. The chosen distance between the sensitive wires at the probe tip was 2 mm.

In the test section A it is possible to choose three different distances between the turbulence creating grid and the potential probe. Thus, the velocity signal can be measured at distances of 80, 225, and 305 mm behind the grid.

For the signal processing high performance low noise amplifiers were used. The acquisition of the data records is based on an analog–digital converter board installed inside a PC-system. Sample rates of 500 Hz or 2 kHz were chosen. As the relative turbulence intensity tu we define the ratio of the rms value of the fluctuating signal divided by the mean part of the signal. The presented power spectral densities were obtained from the ratio of the fluctuating and the mean signal by direct fast Fourier transform (FFT). For this reason the data records were subdivided in time windows of 8192 points corresponding to measuring times of about 4 and 11 s, respectively, leading to resolutions with respect to the frequency of 0.25 Hz or 0.085 Hz. The power spectral density was finally obtained by an ensemble averaging based on 87 pieces with an overlapping of 75%. Using the well-known Taylor hypothesis, the corresponding wave numbers $k = 2\pi f/\overline{u}$ have been received.

The turbulence is characterized by the slope of the power spectral density in its inertial range. These exponents were also calculated for our experiments. However, here exists a problem arising from the finite spatial resolution of the potential probes. The choice of the distance between the probe wires at the sensitive tip is a compromise between the magnitude of the signals and the ability to resolve the small scales of the spectrum. Nevertheless, in general the probe acts as an additional low-pass filter. Thus, the measured power spectra have to be corrected in order to exclude distortions of the obtained slopes caused by the spatial averaging over the volume between the wires of the sensor. A method to correct the results in this sense has been already proposed by Debray[12] corresponding to the correction procedure applied to our presented power spectra.

Moreover, in the discussion about the spectral slopes presented below one has to take into consideration that the domain of the inertial range used for the fitting procedure is limited by the described correction procedure and covers in most of the cases less than one decade in the wave number space.

Experimental Results

Channel Version A

As a first step the profiles of the axial velocity component and the corresponding turbulence intensity were measured. Typical results are shown in Fig. 4a–4d. The influence of the magnetic field on the velocity distribution is evident. Perpendicular to the field direction, the velocity profiles show the expected M-shaping. Along the same line a distinct maximum of the turbulence intensity was detected in the case of the installed grid. This maximum is especially pronounced if the cylindrical bars are aligned with the field direction. A quite different picture presents itself in the orthogonal direction with rather flat profiles of the mean velocity as well as for the turbulence intensity.

In the case of installed bars perpendicular to B, the velocity in the core seems to be lower compared with the two other cases, where the curves nearly show an overlapping. This fact indicates the larger braking effect of the fluid by flow obstacles aligned perpendicular to the direction of the applied field, which is explainable by the formation of a magnetic wake. In the perpendicular case such a wake is much more pronounced as in the parallel case.

Because the presence of the M-shape velocity profile, the experimental situation has to be considered as a superposition of the turbulent production arising from the grid installed in the core of the flow and from the steep velocity gradients near the channel wall, too. Nevertheless, the profiles of the fluctuations indicate that the turbulence promotion by means of the grid is clearly the dominating process.

The dependence of the turbulence intensity from the interaction parameter is displayed in Fig. 5a for the case without grid and in Fig. 5b–5d for the three different orientations of the turbulence promoting cylinders. The distance between grid and probe was chosen as 80 mm. According to our expectations, the direction of the cylindrical bars has an essential influence on the measured turbulence level. The differences in the displayed curves are especially pronounced in the region $10 < N < 100$. While for the constellation of a parallel alignment a maximum of about 14% ($N \approx 70$) can be noted, the turbulence intensity reaches only values of around 1% for the case without grid and shows a monotonous decrease with increasing N. On the contrary, in the case of perpendicular arrangement of the cylinders toward the field direction, we notice a nearly constant level between 4 and 5%. The "perpendicular" case cannot be considered as an ideal one. Mounting supports of the cylinders are directed here along field lines being also able to promote the origin of vortices with axes parallel to the magnetic field. Thus, the diagonal position of the bars should be further away from the "two-dimensional" configuration. This is confirmed by Fig. 5d showing a slope of the turbulent intensity already for $N > 10$, whereas in both of the other situations with installed grid the intensity decreases for values of N greater than 50.

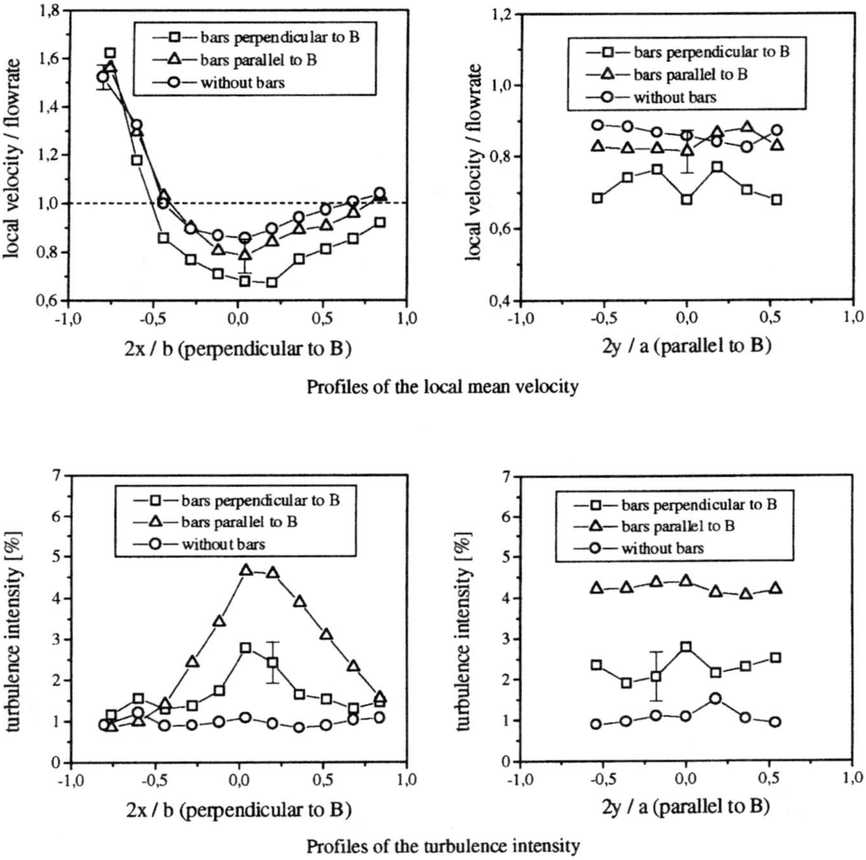

Fig. 4 Measured profiles of the local velocity and the turbulence intensity, respectively, obtained in the test section version A with and without turbulence promoter. (The typical inaccuracy of the measurements is illustrated by error bars.)
a) Perpendicular to the B-field
b) Parallel to the B-field
c) Perpendicular to the B-field
d) Parallel to the B-field

Starting with the same experimental data, the power spectral densities (PSD) at *Re* numbers of 18,600 and 65,100 are obtained. The distinct influence of the cylinder arrangement can also be detected in the resulting spectra. According to the results depicted in Figs. 7–9, the PSD will be reduced drastically in the region of low wave numbers for increasing interaction parameters. This fact indicates clearly the efficient damping mechanism of the third turbulent perturbations by the electromagnetic forces. Such a strong decrease of the PSD in this energy region was not observed in the case with turbulizing cylinders aligned with the magnetic field lines (Fig. 6). More-

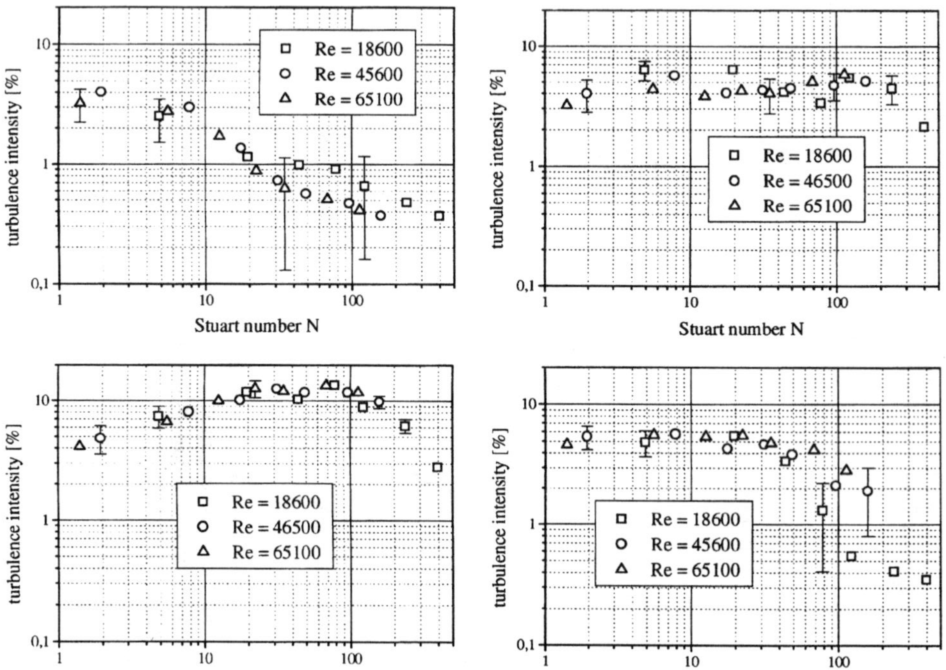

Fig. 5 Turbulence intensity as a function of the Stuart number *N* obtained in the test section version A (distance promoter-probe: 80 mm). (The typical inaccuracy of the measurements is illustrated by error bars.)

a) Smooth channel without any inserts

b) Cylindrical bars directed parallel to the magnetic field lines

c) Cylindrical bars directed perpendicular to the magnetic field lines

d) Cylindrical bars directed diagonal to the magnetic field lines

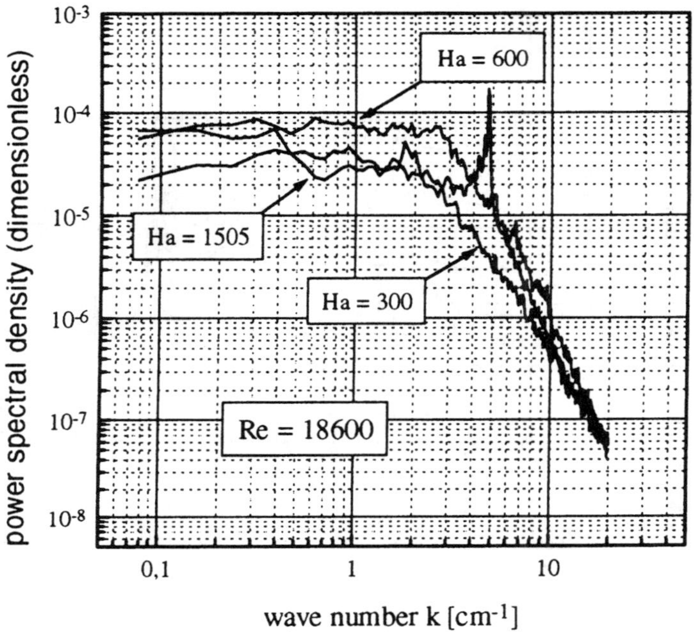

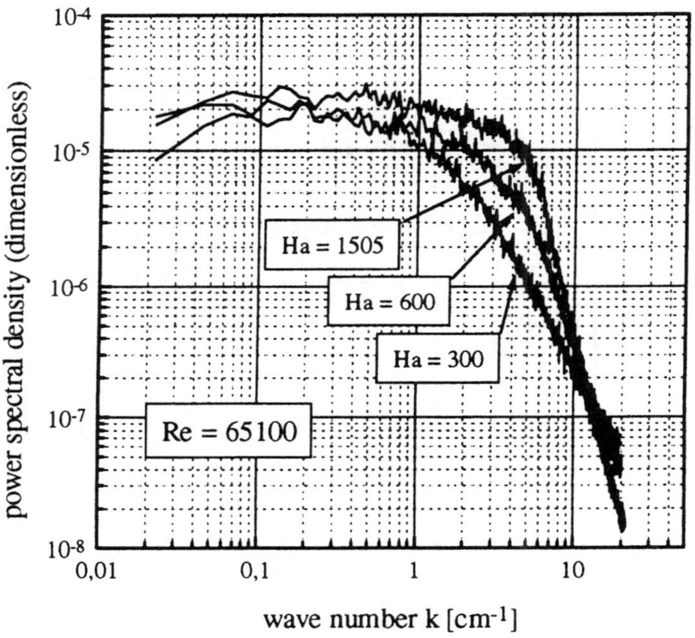

Fig. 6 Power spectral density (corrected) versus wave number *k* (version A, bars directed parallel to the B-field).

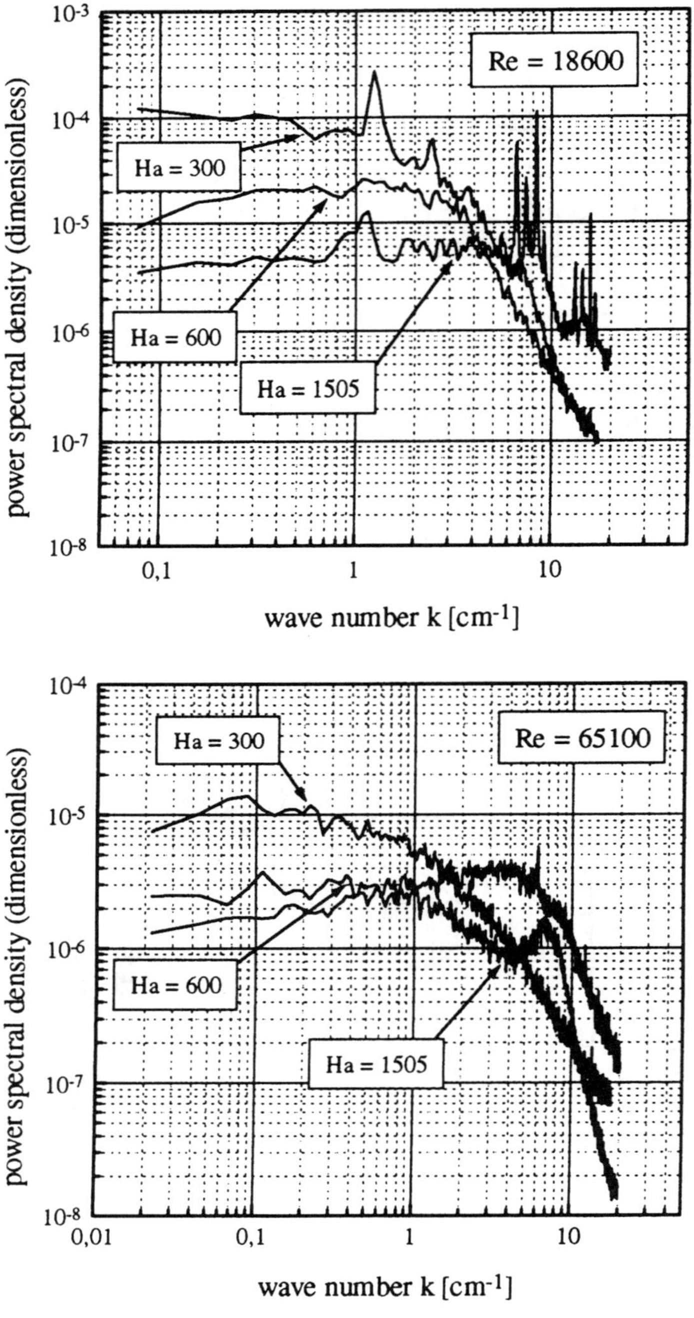

Fig. 7 Power spectral density (corrected) versus wave number *k* (version A, bars directed perpendicular to the B-field).

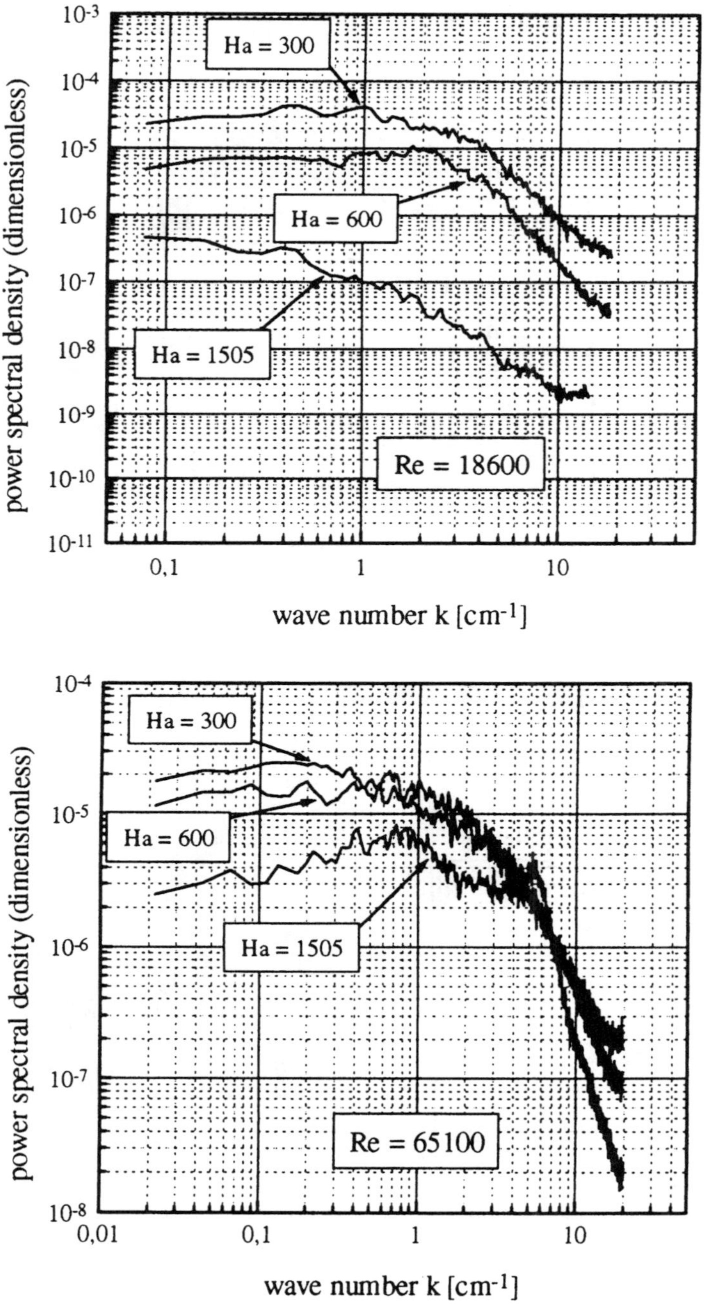

Fig. 8 Power spectral density (corrected) versus wave number *k* (version A, bars directed diagonal to the B-field).

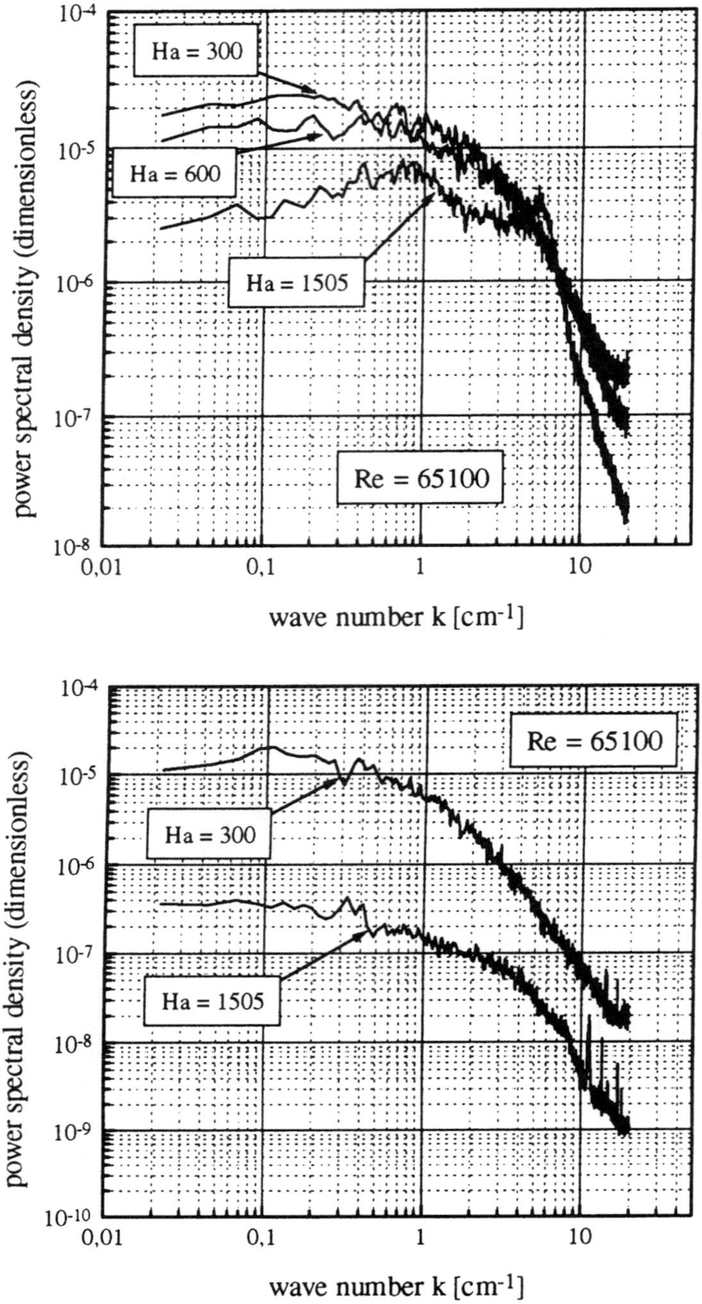

Fig. 9 Power spectral density (corrected) versus wave number *k* (version A, smooth channel without turbulence promoter).

over, the enhancement of the magnetic field strength is accompanied by an increase of the PSD in the k-region between 0.1 . . . 10 cm^{-1}. As a consequence the domain of the energy containing modes will be prolongated toward higher wave numbers. The shift of the transition between energy and inertial range is combined with a change of the slope in the inertial range (see Fig. 14).

Channel Version B

As a next step a similar experimental program was performed in the test section B equipped with the set of flow channel inserts. The function of the ensemble of inserts was tested by an experiment carried out without turbulence promoter.

In Fig. 10 profiles of the velocity and the turbulence intensity are shown along the direction perpendicular to the magnetic field lines. The velocity distribution is not symmetrical because of the velocity jet in the neighborhood of the left channel wall. As reason for this behavior; a defect honeycomb sheet was found. Fortunately, the turbulence intensity shows a flat profile with some small peaks near the walls. Nevertheless, all of the values are clearly below 5%.

The corresponding profiles for the experimental configuration with the turbulence promoter can be seen in Fig. 11. After the substitution of the faulty part of the honeycomb, the maxima of the velocity near the left wall disappear. In the turbulence intensity profiles, no distinct irregularities can be found. Consequently, we conclude that now the vortices created by the turbulizing grid determine the character of the observed turbulence almost solely.

The turbulence intensities measured in the center of the channel cross section are displayed versus the interaction parameter N in Figs. 12a and 12b. Here we have a comparison between the similar experimental constellations of both test sections: 1) smooth channel (version A), without turbulence

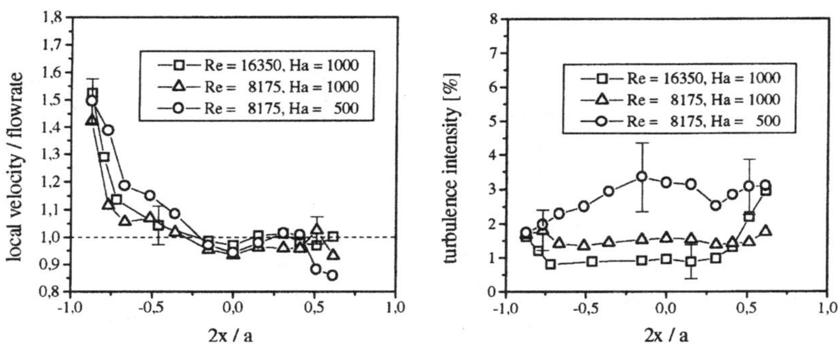

Fig. 10 Measured profiles of the local velocity and the turbulence intensity, respectively, obtained in the test section version B without turbulence promoter. (The typical inaccuracy of the measurements is illustrated by error bars.)

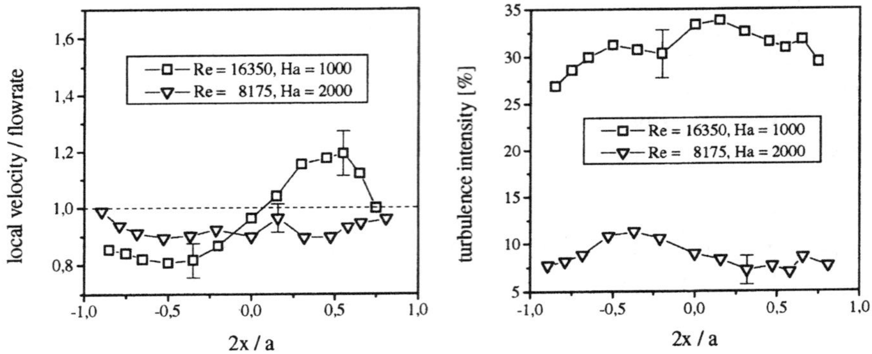

Fig. 11 Measured profiles of the local velocity and the turbulence intensity, respectively, obtained in the test section version B with turbulence promoter. (The typical inaccuracy of the measurements is illustrated by error bars.)

promoter (version B); and 2) installed grid with the cylinders oriented parallel to the magnetic field (version A), with turbulence promoter (version B).

The qualitative courses of the confronted curves are quite similar. In case 1) there are no considerable differences. The turbulence level is below 5% and decreases with increasing *N*. The influence of the electrical wall conductivity becomes clear in the case of the well-defined turbulence promotion. The reduction of the wall conduction ratio in version B by a factor 10 results in an enhancement of the turbulence level by a factor 2 . . . 5. For both curves we find a maximum at values of the interaction parameter around 100. That means the position of this maximum is not a function of the wall conductance ratio *c*.

A selection of PSD functions determined for *Re* numbers of 16,350 and 24,525 is depicted in Fig. 13. Corresponding to the enhancement of the turbulence intensity with increasing *N* as shown in Fig. 12, the amplitude of the power spectral density for low wave numbers (energy range) is growing. The inertial range of the spectrum is characterized by a slope of -3 nearly independent of the magnitude of the interaction parameter. On the other hand, as can be seen in Fig. 14 displaying the comparison of version A and B with respect to the obtained exponents of the slope a rather different influence of the magnetic field on the inertial range of the spectra has been revealed for version A. The reason for this behavior was not clear until now. However, a determining role of the electrical boundary conditions is obvious.

A dependence of the inertial slope from the interaction parameter was also reported by Branover et al.[13] for an experiment performed in a channel with conducting walls made from stainless steel, too. The authors describe the behavior of the spectral exponents versus magnetic field for a range of interaction parameters between 0.15 and 30. Generally, a decrease of the spectral exponents was found from $-5/3$ at $10^3\,Ha/Re \approx 2.5$ up to -4 at $10^3\,Ha/Re \approx 18$. The authors suggest the existence of two quasi-two-dimensional

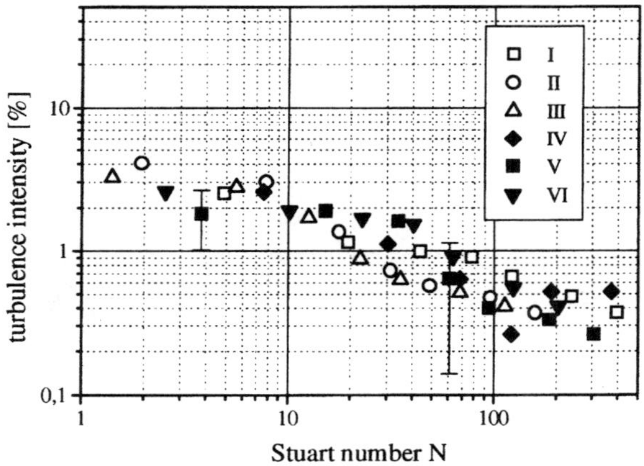

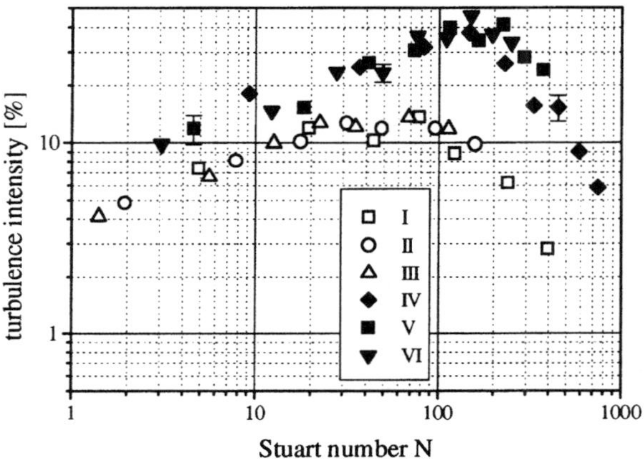

Fig. 12 Turbulence intensity as a function of the Stuart number *N*: I: version A, *Re* = 18,600; II: version A, *Re* = 46,500; III: version A, *Re* = 65,100; IV: version B, *Re* = 8175; V: version B, *Re* = 16,350; and VI: version B, *Re* = 24,525.

(The typical inaccuracy of the measurements is illustrated by error bars.)

a) Test section version A: smooth channel and test section version B: without turbulence promoter

b) Test section version A: cylindrical bars directed parallel to the magnetic field lines and test section version B: with turbulence promoter

modes combined with spectral laws of −7/3 and −11/3, respectively, which is based on the observation that both transitions from Kolmogorov's turbulence to the first quasi-two-dimensional mode (−7/3) and from the latter to the second (−11/3) proceed abruptly at $10^3\, Ha/Re \approx 2.5$ and $10^3\, Ha/Re \approx 4.5$.

In comparison to Branover et al., the advantage of our experiment consists in the possibility to extend the parameter region to distinct larger values for N and Ha/Re, respectively. The Re numbers for both experiments are in

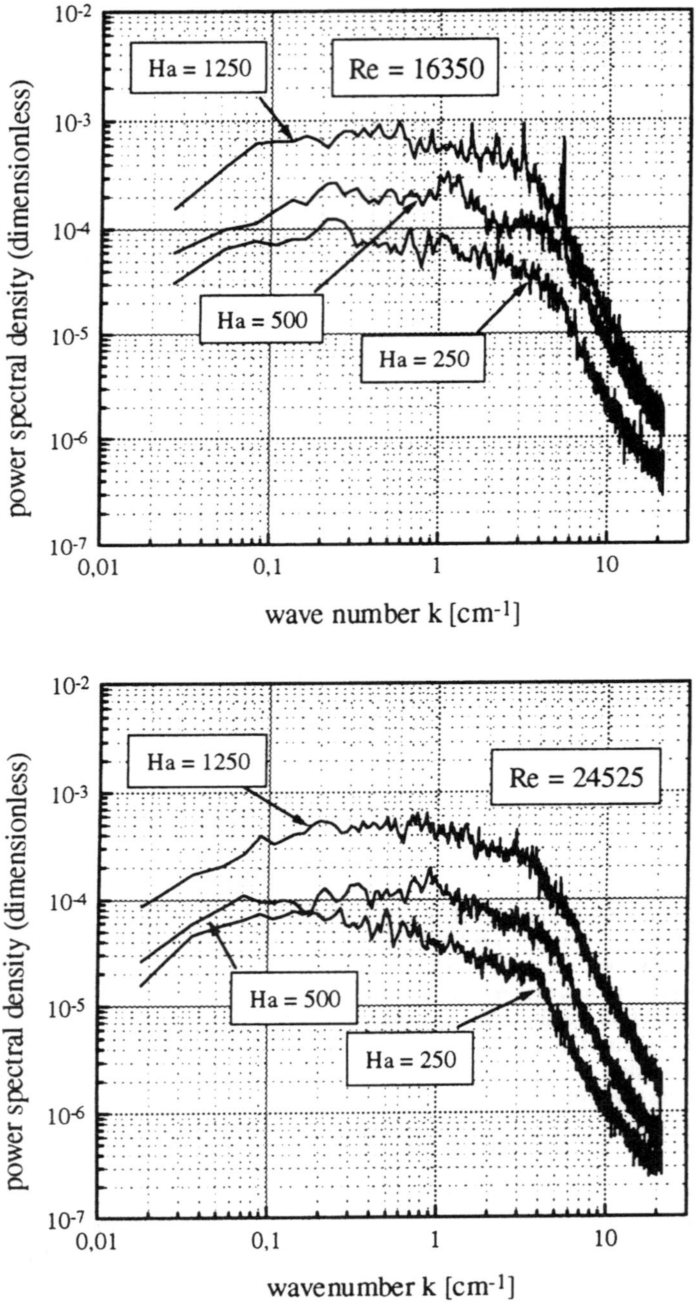

Fig. 13 Power spectral density (corrected) versus wave number *k* (version B, with installed turbulence promoter).

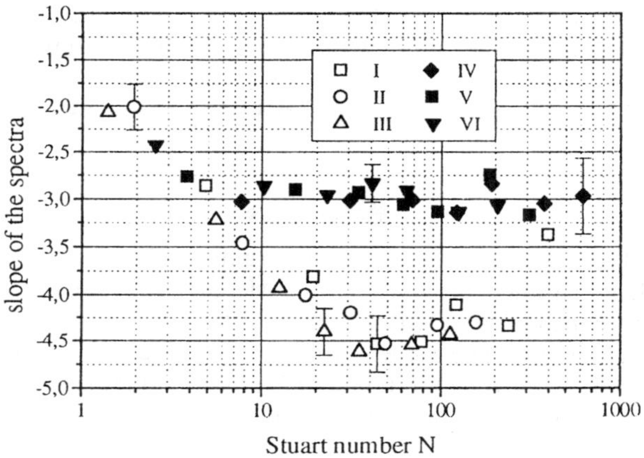

Fig. 14 Exponents obtained from the inertial range of the power spectrum as a function of the Stuart number *N*: I: version A, *Re* = 18,600; II: version A, *Re* = 46,500; III: version A, *Re* = 65,100; IV: version B, *Re* = 8175; V: version B, *Re* = 16,350; and VI: version B, *Re* = 24,525.
(The typical inaccuracy of the measurements is illustrated by error bars.)

the same order of magnitude. Unfortunately, there is no information in the paper of Branover et al. about the exact value of the wall conductance ratio *c*. The spectral exponents extracted from our measurements are displayed in Fig. 14. The same sharp transitions as described by Branover et al. cannot be observed. However, in our case the smallest value for 10^3 *Ha*/*Re* was 4.6. The spectral exponents in the region of small magnetic field values (10^3 *Ha*/*Re* ≈ 5) are about -2.2 ($\approx -7/3$). A minimum of about -4.5 was detected in the region $25<10^3$ *Ha*/*Re*<40. For a further enhancement of the magnetic field intensity, the spectral exponents show a tendency to increase.

Summary

A set of experiments was performed with respect to the two-dimensional behavior of liquid metal MHD channel turbulence for the first time in a range of higher *Ha*/*Re* and *N* values. A suitable installation of a mechanical turbulence promoter drastically changes the turbulence characteristics. If the cylindrical bars of the promoter are elongated parallel to the magnetic field lines, a pronounced turbulence intensification is found.

Two types of test sections were used. Version A corresponds to a simple configuration with essentially electrical conducting walls, whereas Version B represents a much more sophisticated installation as described before. The turbulence intensities due to turbulizing bars parallel to the magnetic field clearly increase for increasing field strength and reach values of 14% in case A, and even 20 . . . 40% for $30<N<100$ in case B (see Fig. 12b). On the contrary, without the turbulence promoter a continous decrease of turbulence intensity below 2% was observed for increasing magnetic field.

It is worthwhile to emphasize these very large intensification rates, particularly for the test section A where it was obtained without sophisticated means just by installing a simple promoter into a given channel flow.

Analysis of the power spectra reveals a stable inertial range exponent of -3 for case B, whereas much steeper slopes of up to -4.5 were found in case A.

References

[1]Branover, H. H., Slyusarev, N. M., and Scherbinin, E. V., "Some Results of Measuring Turbulent Pulsation of Velocity in a Mercury Flow in the Presence of a Transverse Magnetic Field." *Magn. Gidrodin.,* Vol. 1, 1965, p. 33.

[2]Branover, H. H., Gelfgat, Yu. M., Kit, L. G., and Platnick, E. A., "Effect of a Transverse Magnetic Field on the Intensity Profiles of Turbulent Velocity Fluctuations in a Channel of Rectangular Cross Section," *Magn. Gidrodin.*, Vol. 3, 1974, p. 41.

[3]Hua, H. M., and Lykoudis, P. S., "Turbulence Measurements in a Magneto-Fluid-Mechanic Channel," *Nuclear Science and Engineering*, Vol. 54, 1974, p. 445.

[4]Votsish, A. D., and Kolesnikov, Yu. B., "Spatial Correlation and Vorticity in Two-Dimensional Turbulence," *Magn. Gidrodin.*, Vol. 3, 1976, pp. 25–28.

[5]Kolesnikov, Yu. B., and Tsinober, A. B., "An Experimental Study of Two-Dimensional Turbulence Behind an Array," *Mekhanika Zhidkosti i Gasa*, 1974, N., 4, pp. 146–151.

[6]Eckert, S., Gerbeth, G., Langenbrunner, H., and Witke, W., "MHD Turbulence: Use of Gas Bubbles as Local Tracers in Liquid Metal MHD Flows," Rept. FZR-80, Rossendorf, Germany, 1995.

[7]Sommeria, J., "Experimental Study of the Two-Dimensional Inverse Energy Cascade in a Square Box," *Journal of Fluid Mechanics*, Vol. 170, 1986, pp. 139–168.

[8]Sommeria, J., and Moreau, R., "Why, How and When MHD Turbulence Becomes Two-Dimensional," *Journal of Fluid Mechanics,* Vol. 118, 1982, pp. 507–518.

[9]Sukoriansky, S., and Branover, H., "Turbulence Pecularities Caused by Interference of Magnetic Fields with the Energy Transfer Phenomena," *Proceedings of the 5th Beer-Sheva Seminar*, Vol. 112, Progress in Astronautics and Aeronautics, AIAA, Washington, DC, 1988, pp. 87–99.

[10]Branover, H., and Sukoriansky, S., "MHD Turbulence with Inverse Energy Cascades and Enhanced Heat Transfer," *Proceedings of the 25th Symposium Engineering Aspects of Magnetohydrodynamics*, Bethesda, 1987.

[11]Platnieks, I., and Freibergs, J., *Magn. Gidrodin.*, Vol. 2, 1972, p. 29.

[12]Debray, F., *Research Centre Karlsruhe*, Rept.

[13]Branover, H., Eidelmann, A., and Nagorny, M., "Quasi-Two-Dimensional Helical Turbulence in MHD and Geophysical Flows," *Proceedings of Energy Transfer in MHD Flows*, Aussois, 1994, pp. 777–785.

Chapter 8

Origin of Wave-Particle Turbulence in Beam-Plasma Systems

V. A. Buts,* A. N. Kupriyanov,† V. V. Ognivenko,‡
I. N. Onishchenko,§ and A. P. Tolstoluzhskii@
Kharkov Institute of Physics and Technology, Kharkov, 310108, USSR

I. Instability and Dynamical Chaos in a Weak Nonlinear Interaction of Waves

IT IS well established now that nearly all dynamical systems in their extended phase space (a space of all canonical variables and parameters) have regions of chaotic behavior. The stringent analysis of dynamics in such regions is possible only if employing numerical methods. However, the random character of integral curves in these regions enables one to use methods of statistical physics for their description. Thus analytical criteria for defining of such regions are of paramount importance because they not only reveal the regions where the qualitative transition of the dynamics from regular to chaotic occurs and also give an analytical description of a dynamical system in this region by means of methods of statistical physics. There exist a few criteria of transition to global stochasticity (see, e.g., Ref. 1). The most simple and effective is a Chirikov's criterion.[2] Whereas this criterion is based on the definition of a nonlinear resonance, it is well adapted and widely used in the investigation of processes comprising wave-to-particle interaction.

In the present report formulate conditions if being fulfilled the dynamics of the wave-to-wave type of coupling turns out to be chaotic (see also Ref. 3). These conditions are verified via the system that describes coupling of high frequency (HF) and low frequency (LF) waves in nonlinear media.

---

*Head of Laboratory.
†Researcher.
‡Senior Researcher.
§Head of Dept.
@Senior Researcher.

As known, the energy transfer from one wave to others bears the character of an instability. Therefore one may suggest that value of the growth rate Γ of this instability plays the role of width of a nonlinear resonance under wave-to-wave type of coupling. Further it is quite natural to assume that dynamics of this coupling will be chaotic when parameter K becomes greater than unity $K \equiv \Gamma/\delta\omega$, where $\delta\omega$ is the distance between resonances of different waves. Observe that the value of $\delta\omega$ usually corresponds to a minimum frequency.

The above formulated conditions of transition to stochasticity are verified via the nonlinear coupling of HF and LF waves in nonlinear media.

$$
\begin{aligned}
&i\frac{\mathrm{d}}{\mathrm{d}\tau}a_n = ba_{n-1}e^{i\Delta\tau} + b^* a_{n+1}e^{-i\Delta\tau} \\
&\frac{\mathrm{d}^2}{\mathrm{d}\tau^2}b + \Omega^2 b = -\sum_n a^*_{n-1}a_n e^{-i\Delta\tau}
\end{aligned}
\tag{1}
$$

where b is the complex amplitude of an LF wave; a_n is the complex amplitude of an HF wave with number n, $\Delta \equiv \Omega_n - \Omega_{n-1}$; Ω_n is the frequency of an HF wave with number n; and Ω is the free frequency of LF oscillations.

To obtain the set of Eq. (1), we assume that there is a space synchronism between coupling waves, i.e., the following conditions have to be satisfied: $\boldsymbol{k}_n - \boldsymbol{k}_{n-1} = æ$, $\boldsymbol{k}_n$ is the wave number of an HF wave with number n; æ is the wave number of a LF wave. In general we lay emphasis on processes having a growth time less then a period of HF waves. Thereby Eq. (2) of the set (1) for the LF is full, of the second order, i.e., not shortened.

The set (1) describes a variety of nonlinear processes of wave coupling. Let us dwell upon the most simple and important case of three waves coupling, propagating in nonlinear medium. To be certain, we treat the decay of transverse electromagnetic HF wave with the frequency ω_i, and wave number $\boldsymbol{k}_i$, amplitude A_i, which is propagated in a magnetized plasma $\left(\omega_H^2 \gg \omega_p^2\right)$, on the electromagnetic (ω_s, $\boldsymbol{k}_s$, $\boldsymbol{A}_s$) wave and langmuir (ω_p, $\boldsymbol{k}_p$, $\boldsymbol{A}_p$) one. In other words $t \to t + l$. In Eq. (1) the index runs two values: $n = \{0, 1\}$, and the dimensionless variables are expressed in terms of the physical parameters of the problem as follows:

$$
a_o \equiv A_i(\tau)/A_o, \quad a_{-1} \equiv A_s(\tau)/A_o, \quad b \equiv \frac{\delta n}{n_o}\frac{\mu}{\Omega_o}, \quad \tau \equiv \Omega_o e
$$

$$
\Delta \equiv (\omega_i - \omega_s)/\Omega_*, \quad \Omega \equiv \frac{\omega_p}{\Omega_*}\frac{k_{pz}}{k_p}, \quad \Omega_* \equiv \frac{1}{2}\omega_p\left[\left(\frac{V_e}{V_f}\right)^2\frac{\omega_p}{\omega_i}\right]^{1/3}
$$

$$
\mu \equiv \frac{1}{4}\frac{\omega_p^2}{\omega_i}, \quad V_e = \frac{eA_o}{m_e\omega_i}, \quad V_f = \frac{\omega_i}{k_i}
$$

where

$a_o \equiv A_i\,(\tau)/A_o$ is the initial value of the incident of a plasma HF wave
$\omega_p^2 \equiv \dfrac{4\pi n_o e^2}{m}$ is the frequency of the langmuir oscillations of plasma
n_o is the unperturbed plasma density
δ_n are the slow perturbations of this density
m,e are the mass and charge of electron, respectively

On the linear stage of decay, one may get from the set (1) the following expressions for maximum values of the growth rates at small amplitudes of the incident waves ($\Omega^2 \gg 1$)

$$\Gamma \equiv Im\omega = 1/\sqrt{2\Omega}$$

and at large ones ($\Omega^2 \ll 1$)

$$\Gamma \equiv Im\omega = \sqrt{3}/2$$

In the first case parameter $K < 1$ and dynamics of the decay must be regular. At large amplitudes parameter K exceeds unity ($K > 1$); thus, decay must be chaotic. Note that the region of parameters under consideration is related to the so-called modified decay. Therefore it is possible to claim that the modified decay has a chaotic dynamics.

From the expression for the maximum growth rates, one may determine the value of amplitude of electromagnetic wave necessary for the chaos outcome. At $\omega_p = 6 \cdot 10^9$ Hz, $V_f = 3 \cdot 10^9$ cm/s, and $\omega_i/\omega_p = 3$, this value is $1.5 \cdot 10^5$ V/cm. At smaller field values the process of decay will be regular. However even at small amplitude values the chaotic dynamics of decay is possible. It arises when there is a mismatch between the frequencies of coupling waves $\omega_o - \omega_{-1} - \omega_p = \delta$.

The set of Eq. (1) has been solved numerically for the different values of the amplitude of the incident wave. Time-dependent dynamics of amplitudes of coupling waves, spectra of obtained realizations, and autocorrelation functions have been studied. Typical dependencies of these values for the case $K > 1$ are demonstrated in Fig. 1. As it is seen, the ordinary process of three-wave coupling takes place. Amplitudes of the coupling waves periodically alter their value; the spectrum of realization is discrete, and correlation function oscillates.

For the spectral and correlation analysis these formulas were used:

$$g(t) = \frac{1}{\pi}\frac{1}{E}\left|a_T(\omega)\right|^2, \qquad a_T(\omega) = \int_o^E u(t)\exp(-i\omega t)\,\mathrm{d}t$$

$$\rho(\tau) = \int_o^\infty g(\omega)\exp(i\omega\tau)\,\mathrm{d}\omega$$

where $a_T\,(\omega)$ is the Fourier transform of process studied $u(t)$.

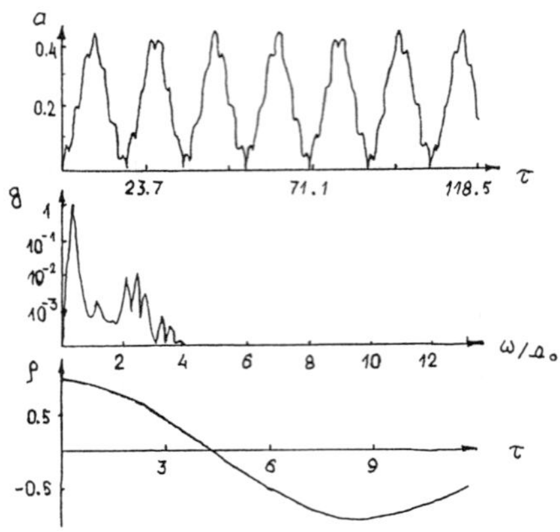

Fig. 1 Dependence of amplitude on time (a_{-1}), power spectrum ($S\omega$), and correlation function (C_f).

Dynamics of the field change quantitatively when the amplitude of the incident wave attains such a value that parameter K becomes greater than unity (see Fig. 2). In this case fields turn out to be chaotic with broad, continuous spectra, and correlation functions sharply drop to zero.

In conclusion it is necessary to observe that the following criterion really defines the range of parameters where the chaotic dynamics of nonintegrable systems reign. But for the integrable ones it does not work at all. So one should be cautious before using the criterion $K > 1$. Indeed, the set (1) at small amplitudes of incident wave is reduced to the set of the equation of the first order,[4,5] which in the case of an infinite number of waves is completely integrable. It is also integrable at the finite number of waves when the quantity of red satellites equals the blue ones.

In addition recall the dynamics of a system consisting of a great number of bounded nonlinear oscillators. As it was found in Ref. 6, the onset of chaotic motion is possible only in the presence of nonlinear self-influence, when the Hamiltonian contains nonlinearity of the fourth order.

If one takes the Hamiltonian restricted to the cubic term and assumes that there are many oscillators with natural frequencies ω_n being bounded one with another via Ω low natural frequency ($\omega_n \gg \Omega$), then the system treated in Ref. 6 can be easily reduced to the set (1). Therefore the preceding-formulated criterion may be employed not only for description of the wave-to-wave type of coupling but enables one to treat various physical processes as well.

II. Stochastization of Langmuir Wave Packets in a Beam–Plasma System

The beam–plasma interaction[7,8] leads to exponential growth of the plasma wave resonant with a beam, which afterwards will be stabilized because of the trapping of the beam particles by the generated wave. The consequent dynamics of this interaction may be represented as a regular energy exchange between wave and beam with a period of trapping oscilla-

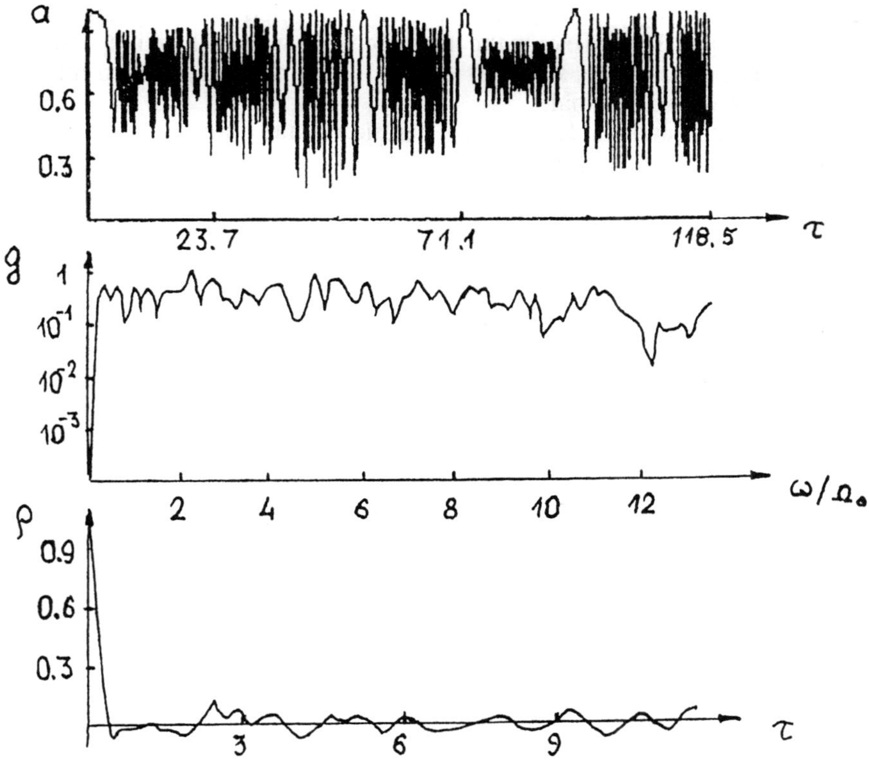

Fig. 2 Dependence of amplitude on time (a_{-1}), power spectrum (S_ω), and correlation function (C_f).

tions. The situation changes drastically if the beam-generated wave is unstable in respect to the decay on langmuir and ion-acoustic waves. In this case the character of interaction of the two high-frequency (langmuir) and low-frequency (ion-acoustic) packets depends on the relationship between the beam increment of the resonant langmuir wave's growth and the decrement of its decay on langmuir and ion-acoustic waves. This process may be varied from regular energy exchange between the packets to their stochastical behavior.

Let us consider the evolution of decay instability under stationary injection of an electron beam modulated with frequency $\omega \simeq \omega_p = \sqrt{4\pi ne^2/m}$, corresponding to the maximum gain, into a plasma. The beam-generated wave having wave number k_1 and frequency $\omega_1(k_1 \sim \omega_p/V_o)$ is unstable[9] with respect to the decay on ion-acoustic wave (ω_s, k_s) and langmuir satellite (ω_2, k_2), which has wave numbers and frequencies determined by the decay conditions:

$$\omega_s = \omega_1 - \omega_2, \qquad \boldsymbol{k}_s = \boldsymbol{k}_1 - \boldsymbol{k}_2 \tag{2}$$

If the condition $k_{2z} \gg \omega_p / V_o(n_1 V_o / n_b V_g)^{1/3}$ is satisfied, one may discount a modified decay; therefore only the main langmuir wave will be in resonance with a beam. Moreover, as it was shown in Ref. 10, the problem may be treated as one-dimensional.

The scaled equations describing generation of the main wave $E_1(t, z) = E_1(z)\cos(\omega_1 z/k_1 - \omega_1 t + a_1)$ and its decay on ion-acoustic wave $E_s(t, z) = E_s(z)\cos(\omega_s z/k_s - \omega_s t + a_s)$ and langmuir satellite $E_2(t, z) = E_2(z)\cos(\omega_2 z/k_2 - \omega_2 t + a_2)$ have the following form[10]:

$$\frac{dE_1}{d\zeta} = -\Lambda E_1 E_s \cos\varphi + \int_{-1/2}^{1/2} d\tau(0)\cos(2\pi\tau + a_1)$$

$$E_1\frac{da_1}{d\zeta} = -\Lambda E_1 E_s \sin\varphi - \int_{-1/2}^{1/2} d\tau(0)\sin(2\pi\tau + a_1)$$

$$\frac{dE_2}{d\zeta} = -\Lambda E_2 E_s \cos\varphi, \qquad E_2\frac{da_2}{d\zeta} = -\Lambda E_2 E_s \sin\varphi \tag{3}$$

$$\frac{dE_s}{d\zeta} = -\Lambda E_1 E_2 \cos\varphi - \tau E_s, \qquad E_s\frac{da_s}{d\zeta} = -\Lambda E_1 E_2 \sin\varphi$$

$$\varphi = a_1 - a_2 - a_3$$

The set (3) becomes complete if one adds an equation of motion of the beam particles:

$$\frac{d\tau}{d\zeta} = \frac{v}{2\pi}, \qquad \frac{dv}{d\zeta} = -E_1\cos(2\pi\tau + a_1) \tag{4}$$

The introduced scaled variables look as follows:

$$E_{1.2} = E_{1.2} \Big/ \sqrt{4\pi n_1 m V_o^2 \left(\frac{n_1 V_o^4}{n_o V_g^4}\right)^{1/3}}$$

$$E_s = E_s \Big/ \sqrt{\frac{T_e}{mV_o^2} 2\pi n_1 m V_o^2 \left(\frac{n_1 V_o^4}{n_o V_g^4}\right)^{1/3}} \cos\theta \tag{5}$$

$$\tau = -\frac{\omega}{2\pi}(t - \frac{z}{V_o}, \qquad \zeta = \frac{\omega_p \gamma}{V_o}, \left(\frac{n_1 V_o}{n_o V_g}\right)^{1/3}$$

where θ is the angle between $\boldsymbol{k}_2$ and the direction of beam motion.

The ratio of decay increment δ_d to beam increment δ_b

$$\Lambda \equiv \frac{\delta_d}{\delta_b} = \frac{m_e^2 V_o^2}{8T_e} \frac{1}{\cos\theta} \frac{V_o}{V_g} \left(\frac{n_1 V_o}{n_o V_g} \right)^{2/3} \tag{6}$$

$$\Gamma \equiv \frac{\sqrt{\pi}}{2^{1/3}} \left(\frac{m_e}{m_i} \right)^{1/2} \frac{k_s}{k_1} \left(\frac{n_1 V_o}{n_o V_g} \right)^{1/3} \tag{7}$$

and Landau damping of the acoustic wave on electrons are parameters of the problem under consideration. In set (3) the first terms determine amplitude and phase variation of the main wave in the process of decay whereas the last ones describe interaction of the main (resonance) wave with the beam particles.

Equations (3) and (4) were integrated using the Runge–Kutta method with the step $h = 0.001$ for $N = 100$ macroparticles.

In Figs. 3 and 4 the results for two values of parameters $\Lambda^2 = 1; 5$ and $\Gamma = 0$ are demonstrated (a is the normalized dependence of amplitudes a_1, a_2, a_3 on coordinate ζ; b is the normalized correlation function; and c is the spectral density).

It is clear that if $\Lambda^2 = 1$, the process of decay becomes significant after a single trapping oscillation. Then the energy exchange between the waves commences on the average stationary level but with a presence of the irregular component this time. The same is true for the correlation function (modulations with a weak damping) and for the spectrum as well (broadening of the base of the main resonant wave and emergence of additional satellites). With parameter Λ (Fig. 4, $\Lambda^2 = 2$) increasing, the frequency of the energy exchange grows, and the modulation depth of the correlation function enhances while in the spectrum a more intensive satellite appears that is separated from the main wave at a greater distance. The spectrum becomes wider and its content more complex. However, realization $E(\zeta)$, correlation, and spectral analyses (two quite distinct lines) show a weak irregularity. The picture of regular trapping oscillations in the absence of decay[12] is replaced by another though nearly periodical process having its own, intrinsic periods. In the phase portrait of the beam (Fig. 5) for $\Lambda^2 = 5$, the mixing of trajectories was revealed,[10] which does not correspond to the onset of stochastization of the wave phase that was proven by means of the correlation analysis.

If one introduces a dissipation ($\Gamma \neq 0$) into the system under investigation, then an irregularity relevant to the strange attractor arises.[11] Actually, for the decay triplet under high-frequency pumping (increment γ_o) and decrement ν for two low-frequency waves, on which the growing high-frequency wave decays, at certain values of parameter ν/γ_o the chaotic state characterized by the strange attractor is realized. The transition to such a state is accomplished via the sequence of doubling bifurcations. In the case

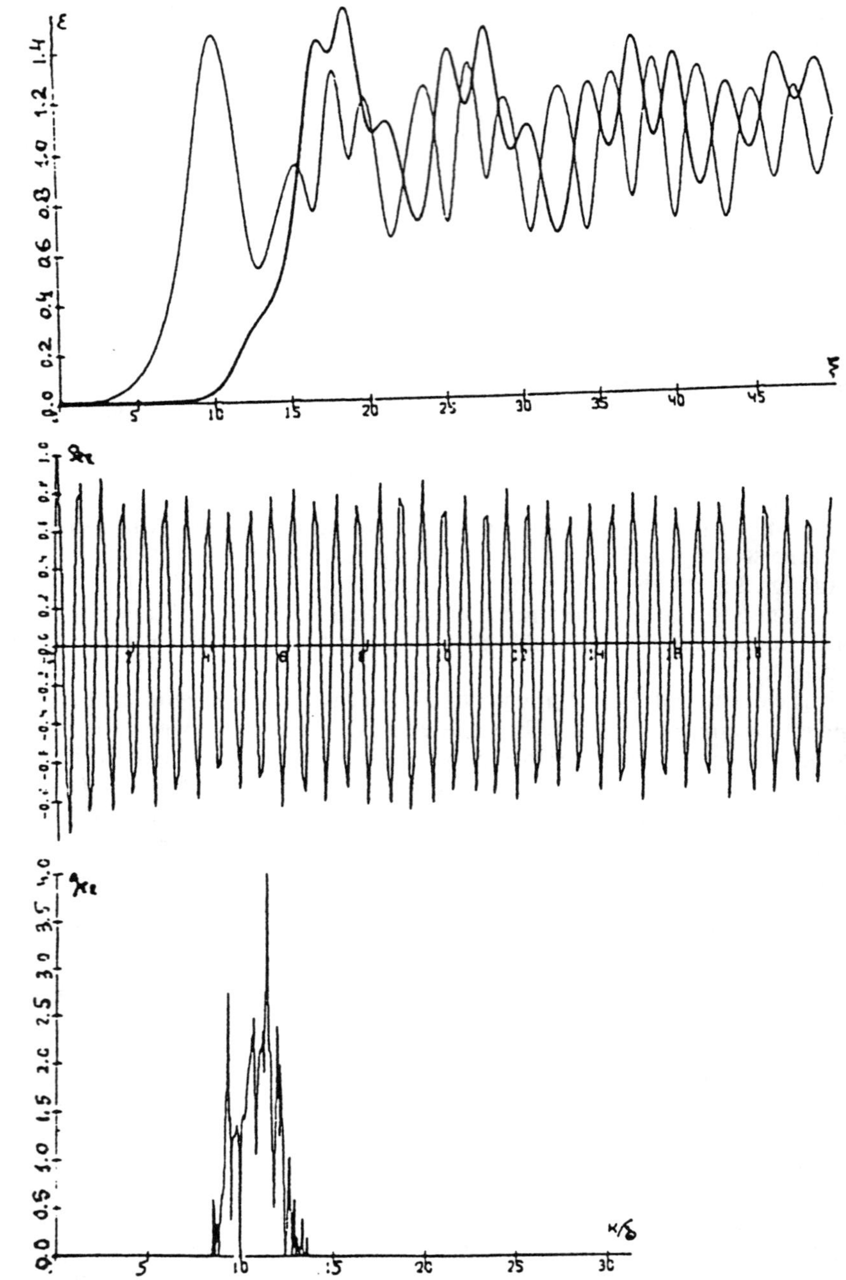

Fig. 3 The dependence of normalized amplitude ε_1, ε_2, and ε_3 on coordinate ζ (*a*), the normalized auto correlation function $\rho_{\xi\xi}$ of ε_1 (*b*) and spectral density $g_{\xi\xi}$ in K-space for the parameter $\Lambda = 1$.

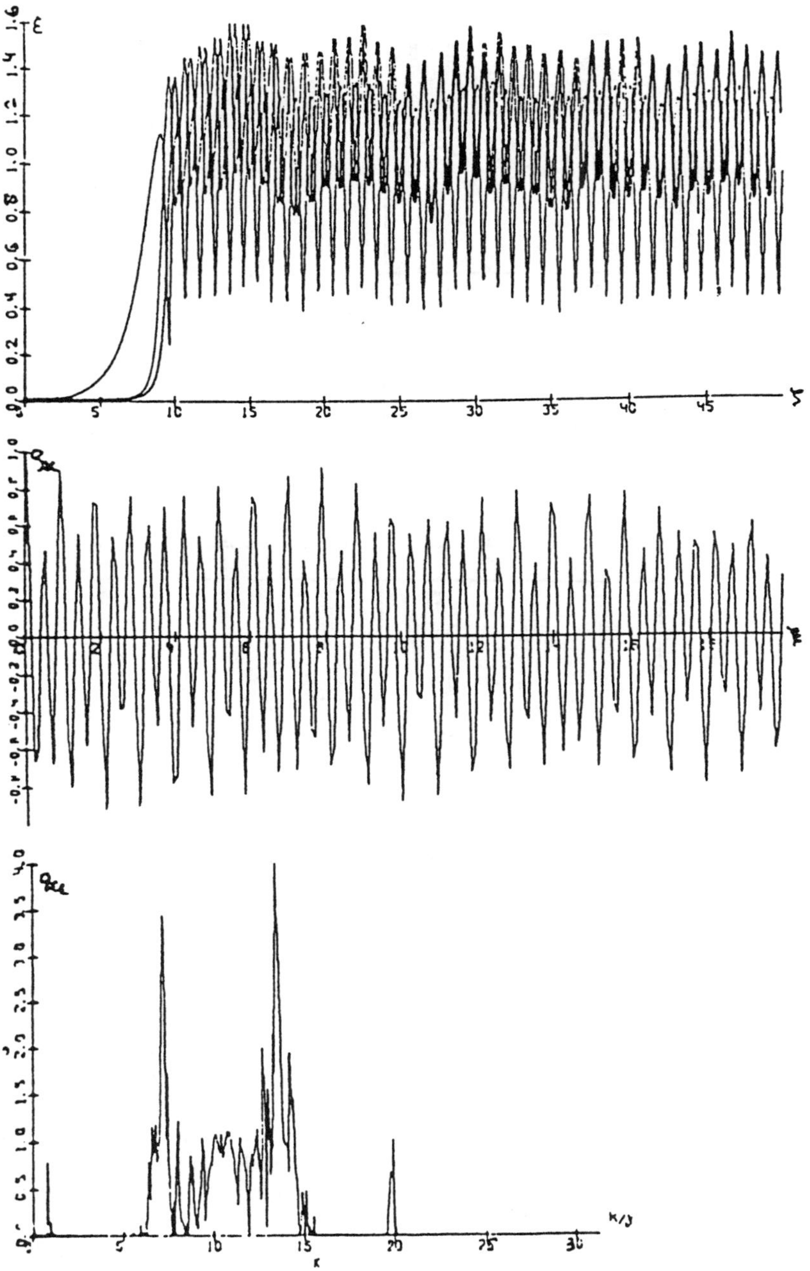

Fig. 4 The dependence of normalized amplitude ε_1, ε_2, and ε_3 on coordinate ζ (a), the normalized autocorrelation function $\rho_{\xi\xi}$ of ε_1 (*b*) and spectral density $g_{\xi\xi}$ in K-space for the parameter $\Lambda = 5$.

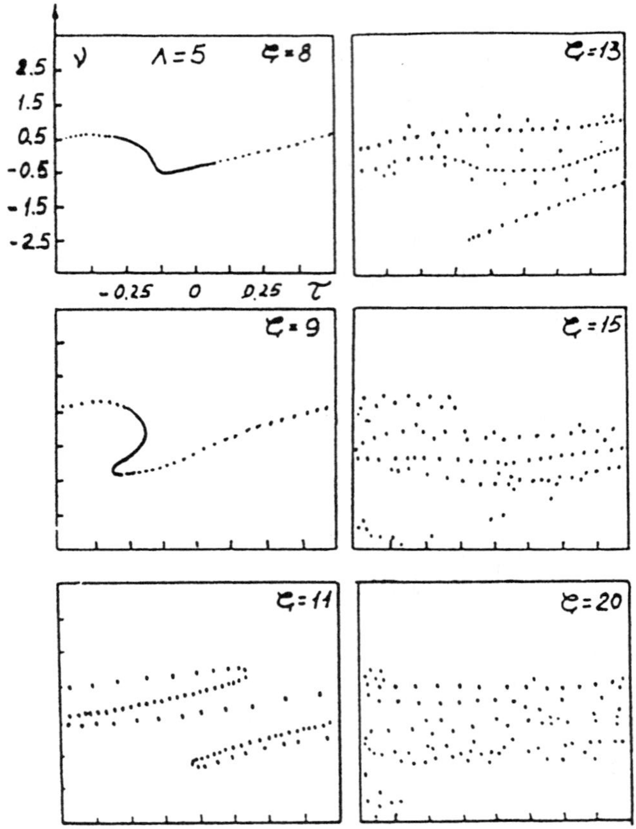

Fig. 5 Phase plane (v, ζ) of beam at various coordinates for Λ^2 = 5.

of the nonlinear increment $\gamma_b = \gamma_o\ (1 + \beta\alpha^2)$ where α is the amplitude of wave and β is the small parameter characterizing a cubic nonlinearity, the strange attractor is destroyed at the same range of parameters.[13] Numerical experiments show that in such a case the dynamical system behaves quite differently, not forming a strange attractor. As distinct from Ref. 11, the system is described not by three equations (for the amplitudes and phases of three waves) but by means of six ones, including those for fields and equations of motion for the beam particles. In addition to parameter $\Gamma = v/\gamma_b$, there is a parameter $\Lambda = \delta_d/\delta_b$. The development of instability goes on as follows. At the first stage, as in the case $\Gamma = 0$ the wave resonant with the beam grows due to the beam–plasma instability. Starting from the threshold value of amplitude of this wave, the decay instability takes place that causes periodical energy exchange between waves.

At $\Gamma \neq 0$, the ion-acoustic wave is damping along with the resonant langmuir wave. Meanwhile the second langmuir attains the stationary level

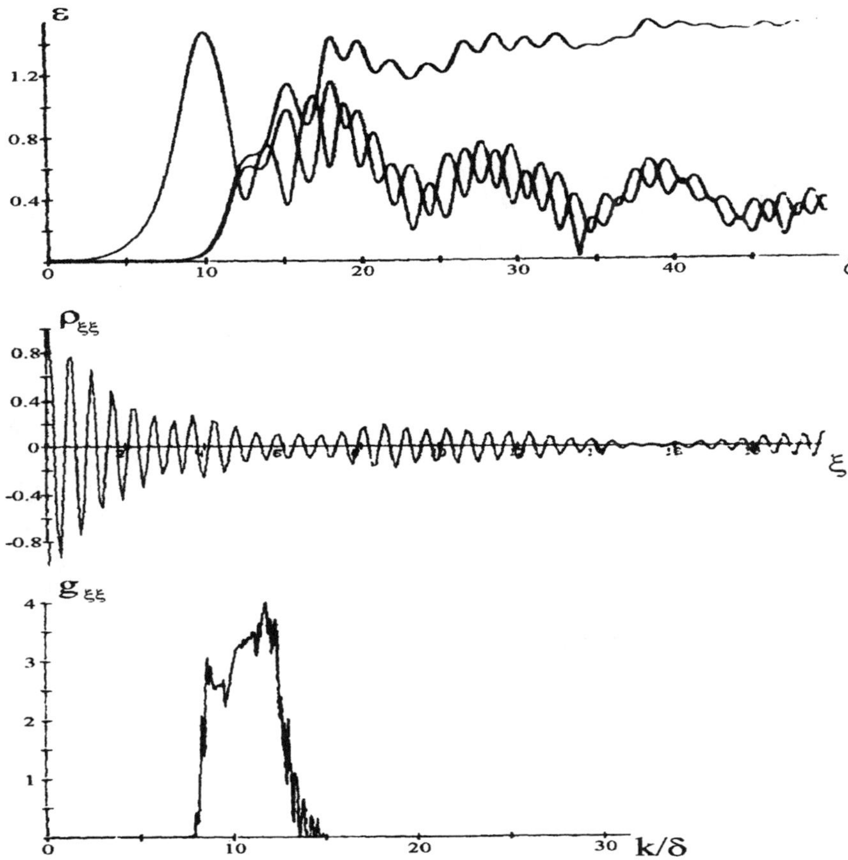

Fig. 6 The dependence of normalized amplitude of langmuir waves ε_1, ε_2, and ion-acoustic wave ε_3 on coordinate ζ (*a*), the normalized autocorrelation function $\rho_{\xi\xi}$ of ε_1 (*b*) and spectral density $g_{\xi\xi}$ in K-space for the parameter $\Lambda = 1$.

(see Fig. 6, for $\Lambda^2 = 1$). The resonant wave is gradually going out the resonance with the beam, the latter being appreciably dispersed at this time. The amplitude and phase of the resonant wave undergo nonlinear oscillations; therefore, the correlation function turns out strongly damping whereas the spectrum of wave numbers are broad and uniform in density. For the large values $\Lambda^2 = 5$ (Fig. 7), further broadening of the spectrum occurs accompanied by the arising of distinct spectral lines. The latter is coherent with emerging modulation of the correlation function.

III. Explosive Beam-Wave Instability

Nonlinear interaction of wave in plasma can lead not only to stochastical behavior of wave but to coherent oscillation. In this section the processes of

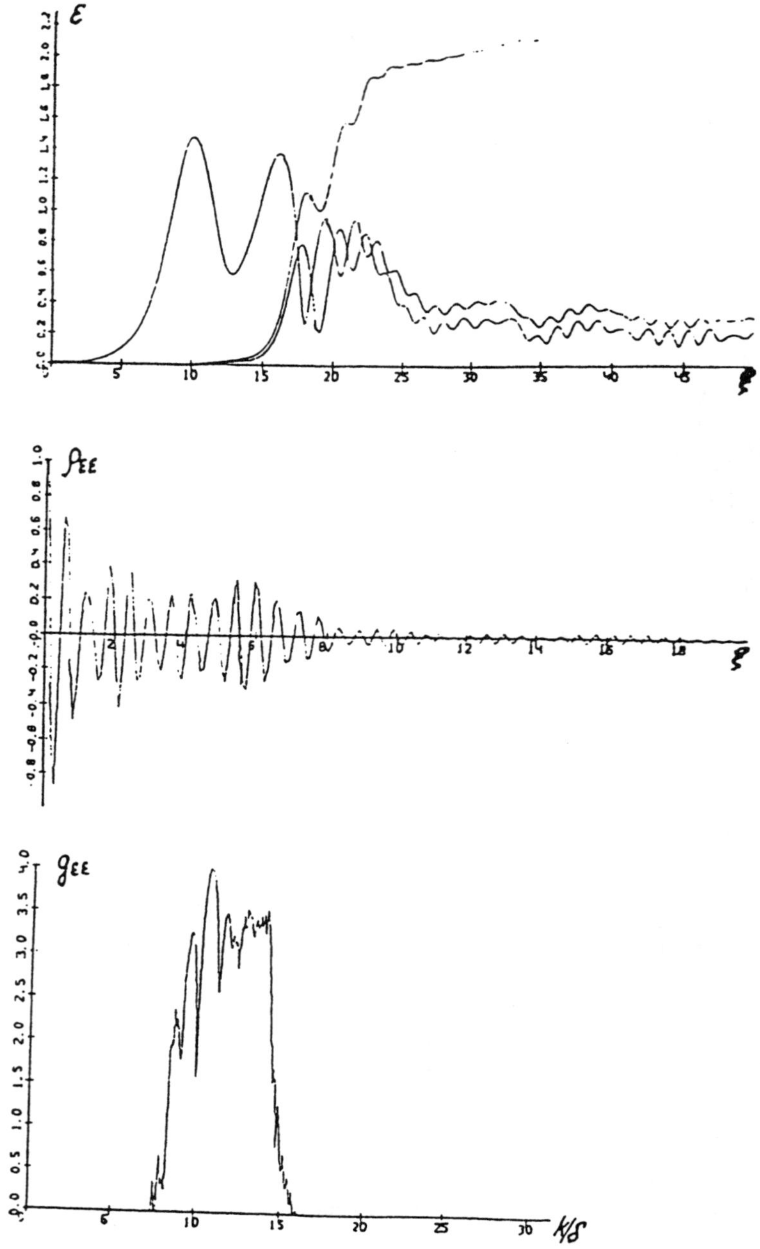

Fig. 7 The dependence of normalized amplitude of langmuir waves ε_1, ε_2, and ion-acoustic wave ε_3 on coordinate ζ (*a*), the normalized autocorrelation function $\rho_{\xi\xi}$ of ε_1 (*b*) and spectral density $g_{\xi\xi}$ in K-space for the parameter $\Lambda = 5$.

explosive interaction of waves in which amplitudes of parametric coupling waves grow simultaneously is present. An example of this parametric instability can be an interaction of a slow electromagnetic wave (pump wave) and a high-frequency (HF) wave with a charged particle beam in a slowing-down medium (plasma).

Let us consider the interaction a slow pump electromagnetic wave (PW) $(\omega_1,\boldsymbol{k}_1)$ and a HF wave $(\omega_2,\boldsymbol{k}_2)$ with the electron beam that is moving along the positive Z axis. The frequencies and wave numbers are satisfying the linear dispersion relation $\omega_2\varepsilon(\omega_2,\boldsymbol{k}_2) = \boldsymbol{k}_2C$ and the phase matching $\omega_3 = \omega_1 + \omega_2$, $\boldsymbol{k}_3 = \boldsymbol{k}_1 + \boldsymbol{k}_2$.

In the case, where two coupling waves are the transverse waves, the set of equations, describing the amplification of waves and the equation of beam particles motion, written in terms of the Lagrange variables is[14,15]:

$$\frac{\mathrm{d}}{\mathrm{d}\xi}A_1 = i\mu A_2^* I, \qquad \frac{\mathrm{d}}{\mathrm{d}\xi}A_2 = iA_1^* I \tag{8a}$$

$$\frac{\mathrm{d}^2}{\mathrm{d}\xi^2}\tau = -V\,\mathrm{Im}[e^{-\vartheta(\xi,q_0)}\,(2\Delta^2 I_b - A_1A_2)] \tag{8b}$$

where

$$I = \langle V(\xi,q_0)e^{i\vartheta(\xi,q_0)}\rangle, \qquad I_b = \langle e^{i\vartheta(\xi,q_0)}\rangle$$

$$V = (1+\lambda\rho)^{-1}, \qquad \xi = æz, \qquad \vartheta(\xi,q_0) = \tau(\xi,q_0) - \Delta\xi$$

$$\tau = \omega_3(t - z/v_0), \qquad \rho = (v/v_0 - 1)/\lambda$$

$$A_n = b_n e^{ia_n}, \qquad b_1 = \frac{E_1}{E_{10}}, \qquad b_2 = \frac{E_2}{\mu E_{10}}\left(\frac{\omega_1 v_2}{\omega_2 v_1}\right)^{1/2}$$

$$\Delta = \frac{\omega_b}{æv_0}, \qquad v_n = (v_g/2c\omega)\frac{\partial}{\partial\omega}\omega^2\,\varepsilon(\omega,k)|\omega = \omega_n$$

$$\lambda = æv_0/\omega_3, \qquad v_g = \frac{\partial\omega}{\partial k}$$

$$æ = k_3(\beta_1\omega_b/2k_3c)^{2/3}\,(\omega_3/\omega_2 v_2)^{1/3}$$

$$\beta_1 = eE_{10}/m\omega_l c, \qquad \mu = \Delta^2\omega_b^2/|4\omega_1\omega_2 v_1 v_2|$$

$$\boldsymbol{E}(z,t) = \vec{e}_x\sum_{n=1,2}E_n(z)\cos[k_n z - \omega_n t + a_n(z)]$$

$\langle\,\rangle$ represents the ensemble average over the initial phases of the electron; ω_b is the beam–plasma frequency; e_x is the unit vector along the X axis; $\varepsilon(\omega,\boldsymbol{k})$ is the dielectric medium susceptibility; and E_{n0}, v_0 are the initial significances of the amplitudes and electron-velocity at $z = 0$.

In that case, the pump wave is a longitudinal slow plasma wave (see, e.g., Ref. 16); the set of self-consistent scaled equations describing explosive parametric instability is the same form as Eq. (8). The peculiarity of the scattering longitudinal plasma wave into the electromagnetic wave propagating parallel to the electron beam consists in that growth rate; æ will be proportionate to the angle between the wave number of the plasma wave and a beam electron-velocity.

The low of conservation of energy electromagnetic waves flux follows from Eqs. (8a):

$$b_1^2 - \mu b_2^2 = \text{const} \tag{9}$$

We can see from this equation that the amplitudes of pump wave and HF wave grow simultaneously.

Figures 8 and 9 show the results of the numerical simulation of Eqs. (8): graphs of the HF wave field envelope versus axial position (Figs. 8a and 9a) and spectral density Y versus scaled wave number X ($X = \boldsymbol{k}_2/æ$) (Figs. 8b and 9b). These graphs have the following parameter values: $\lambda = 0.2$, $b_2(0) = 0.02$, $Re(\boldsymbol{k}_2)/æ = 10$; $\Delta = 0.2, \mu = 10^{-3}$ (Fig. 8); and $\Delta = 1, \mu = 0.1$ (Fig. 9).

From Fig. 8a we can see that in the initial stage the HF wave amplifiers with the spatial grow rate $\approx$ æ. According to conservation law (9), the pump wave amplitude grows also. However, relative increase in its amplitude is small. After the saturation of the instability caused by beam particle trap-

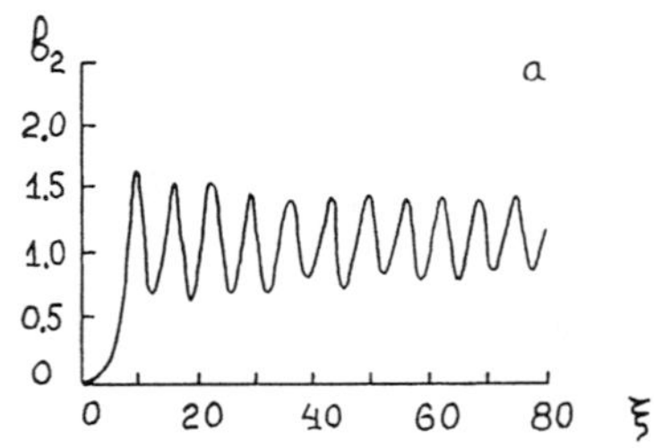

Fig. 8a The dependence of the HF wave field envelope versus axial position.

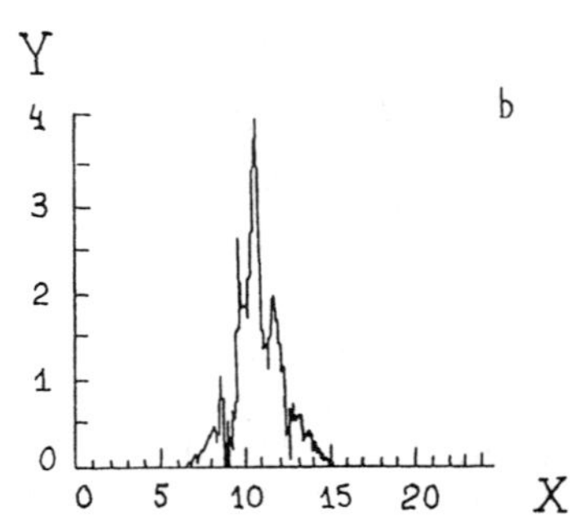

Fig. 8b The spectral density (Y) versus scaled of the wave number (X) for $\Delta = 0.2$ and $\mu = 10^{-3}$.

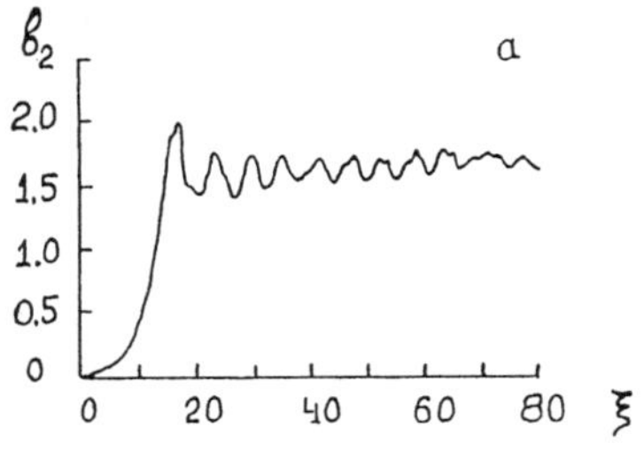

Fig. 9a The dependence of the HF wave field envelope versus axial position.

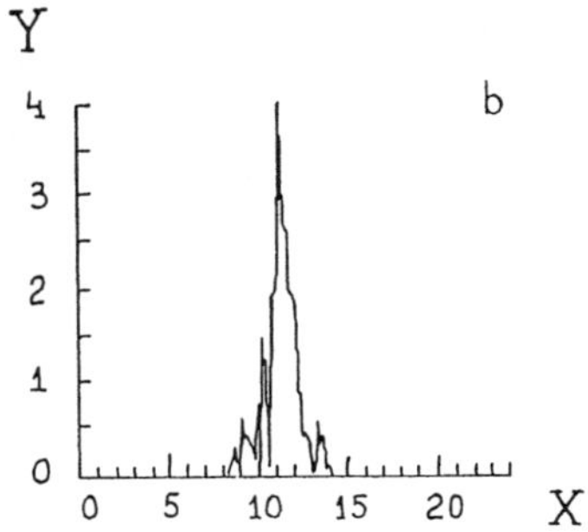

Fig. 9b The spectral density (*Y*) versus scaled of the wave number (*X*) for $\Delta = 1$ and $\mu = 10^{-1}$.

ping in a potential well formed by the combination wave, field envelope of HF wave oscillates with the period of particle trapping oscillation. Figure 8b shows the main distinct line in oscillation spectrum and two satellites due to the HF wave field envelope oscillation.

The results cited at Fig. 9 are received at smaller initial pump wave amplitudes than at Fig. 8. Figure 9a shows that after stabilization of instability, the wave amplitude oscillations are here less deeper than at Fig. 8a, which are dumped along the z axis. That is conditioned by the chaotic particles oscillation, which strapped in a potential well formed by the combination wave and slow space-charge wave of beam. Figure 9b shows that excited oscillation is a narrow spectrum (one distinct line). In this case the spectrum pedestal is less narrow than in the realization cited at Fig. 8b.

References

[1]Lihtenberg, A., and Liberman, M., *Regular and Stochastic Dynamics M.*, Mir., 1984, p. 528.

[2]Zaslavskii, G. M., and Chirikov, B. V., *Soviet Physics–Uspekhi*, Vol. 105, No. 5, 1971, p. 3.

[3]Buts, V. A., Kupriyanov, A. N., and Tolstoluzhskii, A. P., "Current Problems of Statistical Physics," *Proceedings Abstracts of All Union Conference*, Kharkov, 1991, p. 30.

[4]Bakaj, A. S., *Nuclear Fusion*, Vol. 10, 1970, pp. 53–67.

[5]Vodjanitskii, A. A., and Repalov, N. S., *Soviet Physics—Technical Physics,* Vol. 40, No. 1, 1970, p. 32.

[6]Zaslavskii, G. M., and Sagdeev, R. Z., *Introduction in Nonlinear Physics*, M., Nauka, 1988.

[7]Ahiezer, A. I., and Fainberg, Ya. B., *Soviet Physics—Doklady*, Vol. 69, 1949, p. 555.

[8]Bomm, D., and Gross, E. P., *Phys. Rev.*, Vol. 75, 1949, p. 1851.

[9]Oraevskii, V. N., and Sagdeev, R. Z., *Soviet Physics Technical Physics,* Vol. 32, 1962, p. 1291.

[10]Matsiborko, N. G., Onishchenko, I. N., et al., *Soviet Journal of Plasma Physics*, Vol. 1, 1975, p. 765.

[11]Fin, J. M., Wersinger, J. M., and Ott, E., *Physics of Fluids*, Vol. 26, 1980, p. 1142.

[12]Onischchenko, I. N., Linetskii, A. R., et al., *SETP Letters*, Vol. 12, 1970, p. 407.

[13]Balakirev, V. A., Sotnikov, G. V., et al., *Soviet Technical Physics Letters*, Vol. 52, 1982, p. 531.

[14]Ognivenko, V. V., *Radiotehnica i Electronica*, Vol. 27, No. 9, 1982, pp. 1818–1824.

[15]Ognivenko, V. V., Preprint XFTI 83–42, Kharkov, 1983, p. 22.

[16]Kuzelev, M., and Rukhadze, A., *Electrodynamics of Dense Electron Beams in Plasma*, M., Nauka, 1990, p. 325.

Chapter 9

Local Similarity of Reynolds-Stress Transport Processes in Turbulent Rotating Flows

E. Sukhovich
Institute of Physics, Latvian Academy of Sciences, Salaspils 1, LV-2169, Latvia

THE effects of the Earth's gravitational field, rotation, and magnetic field on turbulence have received much attention, concerned with probability that some of these effects are analogous. In the previous work it was found that in the presence of external forces turbulence becomes anisotropic and in specific cases its features tend to the features of the two-dimensional turbulence. The stabilizing and destabilizing influence of gravity in stratified flows and centrifugal forces in rotating flows on the pulsating motion have a similar mechanism. On the other hand qualitative similarity is not yet sufficient to allow definite conclusion about the behavior of turbulence properties in these cases.

In the work reported here the second-order form of the locally anisotropic model of turbulence has been developed. In the first part of this paper, we used the data of direct numerical simulation (DNS) to evaluate some existing models for the pressure-strain correlation. Some of the previously familiar and a number of the new approximations for this term were investigated to be certain that these approximations describe the DNS data.

In the second part of this chapter, the analysis of equations for Reynolds stresses and dissipation-rate functions was carried out. The analysis showed that a limited number of dimensionless parameters can be determined, and they may be considered as local criteria of similarity for processes of turbulent heat and momentum transfer. If the values of the criteria are equal for the same flows in the same point, then the Reynolds stresses and heat flux for these flows would be equal also.

A parametric investigation of the influence of these criteria on Reynolds stresses has been performed. The calculations were carried out for the rotating flows in a wide range of criterion values. The results of the analyses may be used for any one of turbulent rotating flows. The similar method of attack may be used for analysis of alternate types of turbulent flows.

Exact Equations for Second-Order Moments

By applying the standard procedure the transport equations for the second-order moments may be derived from the Navier–Stokes equations. For incompressible turbulent flow the transport equations are given by

$$\frac{\mathrm{d}R_{ij}}{\mathrm{d}t} = F_{ij} + P_{ij} + \Phi_{ij} - Z\varepsilon_{ij} + D_{ij} \tag{1}$$

$$\frac{\mathrm{d}K}{\mathrm{d}t} = F_k + P_k - \varepsilon + D_k \tag{2}$$

where $\frac{\mathrm{d}}{\mathrm{d}t} = \frac{\partial}{\partial_t} + U_k \frac{\partial}{\partial x_k}$, is the mean velocity, $R_{ij} = \langle u_i u_j \rangle$ is the Reynolds stress tensor, u_i is the fluctuating velocity component, t is the time, and $K = \langle u_i u_i \rangle /2$ is the turbulent kinetic energy. The angular brackets denote averaging. In the preceding equations and in the ones following, repeated indices imply summation:

Production of R_{ij} by internal force:

$$F_{ij} = (\langle f_i u_j \rangle + \langle f_j u_i \rangle)/\rho$$

Stress production of R_{ij}:

$$P_{ij} = -(R_{ik} U_{j,k} + R_{jk} U_{i,k})$$

Pressure-strain correlations:

$$\Phi_{ij} = -\frac{1}{\rho}(\langle u_i p_{,j} \rangle + \langle u_j p_{,i} \rangle)$$

Turbulent transport:

$$D_{ij} = -(\langle u_i u_j u_k \rangle - \nu R_{ij,k})_{,k}$$

Dissipation rate of Reynolds stresses, ν denotes kinematic viscosity:

$$\varepsilon_{ij} = \nu \langle u_{i,k} u_{j,k} \rangle$$

Reynolds-Stress Equation

In the context of a full second-order model of turbulence, R_{ij} and ε_{ij} are dependent variables that must be determined from a solution of appropriate differential equations. The terms F_{ij} and P_{ij} are defined by Reynolds stresses, heat flux, and gradient of mean velocity. Therefore these terms are exact. The terms Φ_{ij} and D_{ij} must be represented as empirical functions of the mean velocities, Reynolds stresses, and their derivatives.

Velocity Pressure-Gradient Term Φ_{ij}

This term plays the dominant role in energy redistribution among the components of Reynolds stresses. The expression for the Φ_{ij} has been extensively analyzed and modeled by various groups. In the early works Launder et al.[1] and Hanjalic and Launder[2] split the pressure-strain term into three parts: a return term $[\varphi_{(1)ij}]$, a rapid term $[\varphi_{(2)ij}]$ and a wall term $[\varphi_{(w)ij}]$:

$$\Phi_{ij} = \varphi_{(1)ij} + \varphi_{(2)ij} + \varphi_{(w)ij} \tag{3}$$

Mansour et al.[3] have used a direct numerical simulation to compute a return term, a rapid term, and a Stokes term. These data may be used for model development because they provide a direct means to evaluate closure models.

Term $\Phi_{(1)ij}$

The return-to-isotropy part of the pressure-strain term is usually modeled as the product of the dissipation-rate ε and a function of the Reynolds-stress anisotropy tensor $b_{ij} = R_{ij}/K - 2\delta_{ij}/3$. The simplest, linear form

$$\varphi_{(1)ij} = -C_1 \varepsilon b_{ij} \tag{4}$$

was suggested by Rotta[4] over 45 years ago. More recently Weinstock and Burk[5] showed that coefficient C_1 is not only not constant, but differs for different components ij. If this is the case, the components of C_1 can be written as

$$C_{(1)xx} = -\frac{\varphi_{(1)xx}}{\varepsilon b_{xx}}, \qquad C_{(1)yy} = \frac{\Phi_{(1)yy}}{\varepsilon b_{yy}}, \qquad C_{(1)xy} = \frac{\Phi_{(1)xy}}{\varepsilon b_{xy}}$$

Figure 1 shows the distributions of these components across the channel estimated from DNS data of Mansour et al.[3] $Y^+ = YU_\tau/\nu$ is normal to the wall nondimensional coordinate $[U_\tau = (\nu U_{,y|\text{wall}})^{1/2}$ is the wall-shear velocity]. Coefficient C_1 is not constant, and Eq. (4) should be modified to include the correct behavior return-to-isotropy term in a channel flow.

Lumley[6] emphasized the need to incorporate nonlinear effects and developed a general nonlinear representation for the slow pressure-strain correlation based on invariant tensor representation theory that can be written as follows:

$$C_1 = f(II_b, III_b, Re_t) \tag{5}$$

where $II_b = b_{ik}b_{ki}$ and $III_b = b_{ik}b_{kj}b_{ij}$ are scalar invariants and $Re_t = K^2/\varepsilon\nu$ is the turbulent Reynolds number.

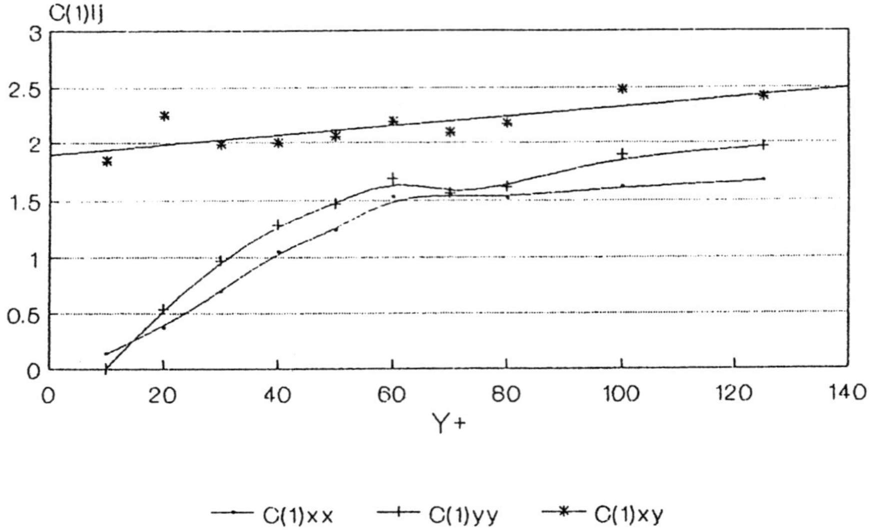

Fig. 1 Distributions of $C(1)ij$ across the channel. DNS data.

The values of $C_{(1)ij}$ are graphed in Fig. 2 as functions of invariants II_b. This figure shows that $C_{(1)xx}$ and $C_{(1)yy}$ vary greatly with anisotropy and that diagonal components of $C_{(1)xy}$ can be written as

$$C_{(1)ii} = -2.25\,(1 - 0.5\,II_b)\; i = 1,2,3 \tag{6}$$

Equation (6) cannot be used for determination of the off-diagonal term $C_{(1)xy}$. This is an indication that Eq. (5) and (6) should be modified to govern the DNS data in channel flow.

The present models for the return-to-isotropy term continue to be based on the idea that the pressure-strain must be a function of the Reynolds-stress anisotropy tensor. These models add the higher-order terms in the anisotropy tensor B_{ij} to the standard Rotta's model. Examples of these type of models are:

Launder et al.[1] model:

$$\frac{\Phi_{(1)ij}}{\varepsilon} = -C_1 b_{ij} \tag{7}$$

where $C_1 = 1.5$.

Shih and Lumley[7] model:

$$\frac{\Phi_{(1)ij}}{\varepsilon} = -\frac{1}{2}\beta b_{ij} + \frac{3}{5}\frac{P_k}{\varepsilon} b_{ij} \tag{8}$$

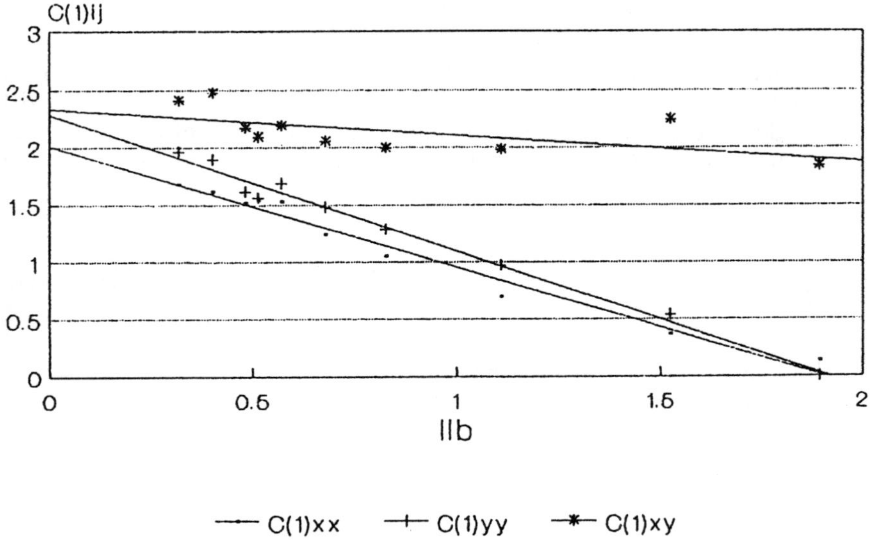

Fig. 2 Dependence of $C(1)ij$ from invariant IIb. DNS data.

where

$$\beta = 2 + \frac{F}{9} \exp\left(-\frac{7.77}{\sqrt{Re_t}}\right)$$

$$\left\{\frac{72}{\sqrt{Re_t}} + 80.1\ \ell n[1 + 62.4(-II_b + 2.3III_b)]\right\}$$

$$Re_t = 4\,K^2/9\varepsilon\nu$$

Speziale et al.[8] model:

$$\frac{\Phi_{(1)ij}}{\varepsilon} = -\left\{C_1 + \frac{C_1^*}{2}\frac{P_k}{\varepsilon}\right\} b_{ij} \; + \; \frac{C_2}{4}(b_{ik}b_{kj} - b_{kl}b_{lk}\delta_{ij}/3) \tag{9}$$

where $C_1 = 1.7$, $C_1^* = 1.8$, and $C_2 = 4.2$.

Launder and Li[9] model:

$$\frac{\Phi_{(1)ij}}{\varepsilon} = -C_1(b_{ij} + C_1 b_{ik}b_{kj} - \delta_{ij}A_2/3) - b_{ij} \tag{10}$$

where $C_1 = 3.1\sqrt{AA_2}$, $A = 1 - 9(A_2 - A_3)/8$, $A_2 = II_b$, $A_3 = III_b$, and $C_1 = 1.2$.

Comparison of the model predictions with the DNS data are shown in Figs. 3 and 4. This comparison shows that the predictions of models (7–10) are inadequate. For a better understanding of disagreement reasons, let us consider in some detail equations for the anisotropy tensor b_{ij}. With Eqs. (1) and (2) we can obtain the next equation:

$$\frac{K}{\varepsilon}\frac{db_{ij}}{dt} = \frac{4}{3}\delta_{ij} - b_{ij}\left(\frac{F_k}{\varepsilon} + \frac{P_k}{\varepsilon} - 1 + \frac{D_k}{\varepsilon}\right) + \frac{\varphi_{ij}}{\varepsilon} - 2d_{ij} + \frac{P_k}{\varepsilon}c_{ij} + \frac{D_k}{\varepsilon}e_{ij} + \frac{F_k}{\varepsilon}m_{ij} \quad (11)$$

The following designations are used:
Anisotropy tensor of stress production:

$$c_{ij} = \left(\frac{P_{ij}}{P_k} - \frac{2}{3}\delta_{ij}\right)$$

Anisotropy tensor of turbulent transport:

$$e_{ij} = \left(\frac{D_{ij}}{D_k} - \frac{2}{3}\delta_{ij}\right)$$

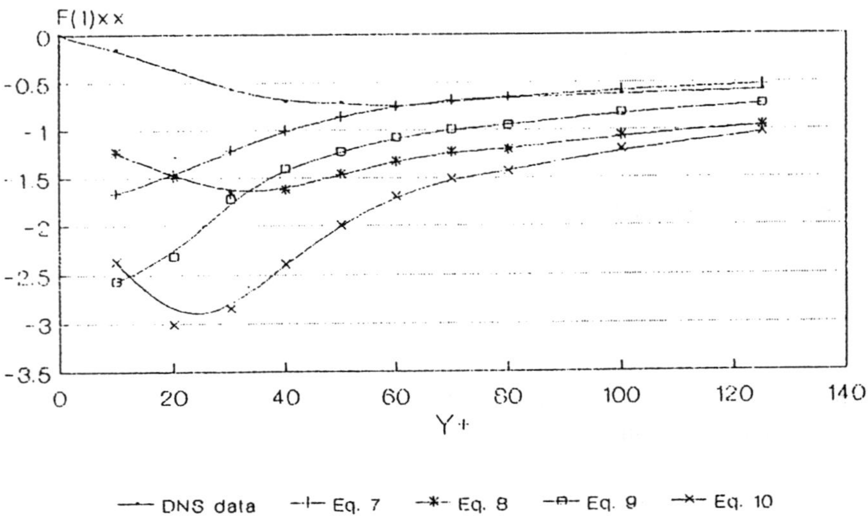

Fig. 3 Comparison of DNS data with calculations of $F(1)xx$.

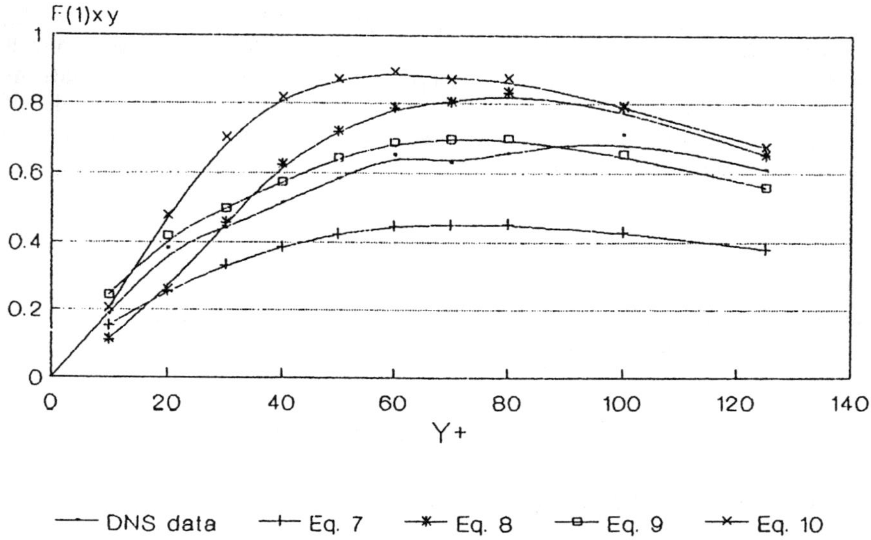

Fig. 4 Comparison of DNS data with calculations of $F(1)xy$.

Anisotropy tensor of internal force production:

$$m_{ij} = \left(\frac{F_{ij}}{F_k} - \frac{2}{3}\delta_{ij} \right)$$

The normalized pressure-strain term in this equation depends on anisotropy tensors b_{ij}, c_{ij}, e_{ij}, m_{ij}, and d_{ij} and the rate of change b_{ij}. The models for the return-to-isotropy term include the anisotropy tensor b_{ij}. The models for rapid term $[\varphi_{(2)ij}]$ include the anisotropy tensors c_{ij}. We believe that the tensor d_{ij} must be included in the approximation for the return-to-isotropy term. This proposition has a physical meaning. The rate of energy transfer between components of Reynolds stresses is bound to increase if the value of anisotropy tensor b_{ij} increases and d_{ij} decreases, all things being equal. It would appear natural that part of the energy stored in the component R_{ij} would be transferred to other components, but the remainder of the energy would be dissipated. This second part would be larger if the dissipation rate ε_{ij} and anisotropy of the dissipation rate d_{ij} are large.

On this basis the general form of model for the return-to-isotropy term can be written as

$$\frac{\Phi_{(1)ij}}{\varepsilon} = -C_1 b_{ij} + C_{12}(b_{ik}b_{kj} - \delta_{ij}II_b/3) + C_{13}(b_{ik}b_{mj}b_{km} - \delta_{ij}III_b/3) + C_{14}d_{ij} \quad (12)$$

where the coefficients C_1, C_{12}, C_{13}, and C_{14} are, in principle, functions of II_b and III_b. We have compared relation (12) with the DNS data of Mansour et al.[3] for the components $\Phi_{(1)xx}$, $\Phi_{(1)yy}$, $\Phi_{(1)zz}$, and $\Phi_{(1)xy}$. Various combinations of the terms on the right of Eq. (12) are used in these tests. The results of these comparisons show that the closest approach to DNS data can be achieved using the next expressions:

$$\frac{\Phi_{(1)ij}}{\varepsilon} = -C_1 b_{ij} + C_{11} II_b b_{ij} \tag{13}$$

where C_1=2.2 and C_{11}= $C_1/2$, $\langle\delta\rangle$ = 19%, and $\langle\delta\rangle$ is the average deviation of computations from DNS data.

$$\frac{\Phi_{(1)ij}}{\varepsilon} = -C_1 b_{ij} + C_{11} II_b b_{ij} + C_{14} d_{ij} \tag{14}$$

where C_1= 3, C_{11}=1, C_{14}= 2.4, and $\langle\delta\rangle$ = 12.8%;

$$\frac{\Phi_{(1)ij}}{\varepsilon} = -C_1 b_{ij} + C_{11} II_b b_{ij} + C_{13}(b_{ik} b_{kj} - \delta_{ij} II_b/3) + C_{14} d_{ij} \tag{15}$$

where C_1= 3, C_{13}= 0.7, C_{13}= 0.65, C_{14}= 2.8, and $\langle\delta\rangle$ 11.2%.

Figures 5 and 6 show the components of the return-to-isotropy term. In a qualitative sense all three models (13), (14), and (15) realistically govern the distributions of $F_{(1)ij}$ across the whole width of the channel. Figure 6 shows

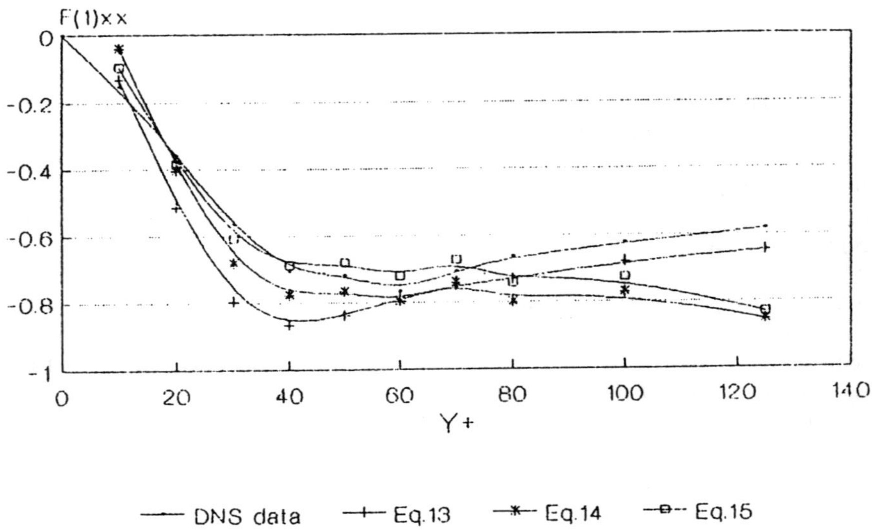

Fig. 5 Comparison of DNS data with calculations.

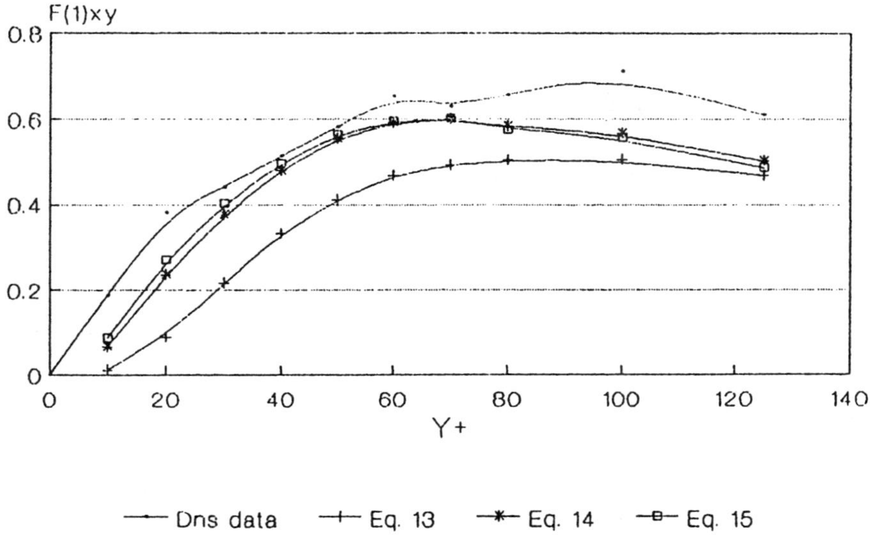

Fig. 6 Comparison of DNS data with calculations.

that the off-diagonal term deviates more than the other terms from the model (13). The remaining two models are very similar. Note that the model (15) is more complicated to calculate. Therefore, at this point preference can be given to the model (14). The eventual conclusion will be made more recently.

Term $\Phi_{(2)ij}$

The expression for the rapid part of the pressure-strain terms has been extensively analyzed and modeled by various groups. The previously derived detailed forms of this term are as follows:

Launder et al.[1] model:

$$\frac{\Phi_{(1)ij}}{\varepsilon} = -\alpha_1 \frac{P_k}{\varepsilon} c_{ij} - \alpha_2 \frac{P_k}{\varepsilon} l_{ij} - 2\alpha_3 \frac{K}{\varepsilon} S_{ij} \tag{16}$$

where $l_{ij} = L_{ij}/P_k - \frac{2}{3}\delta_{ij}$ is the anisotropy tensor of stress production, $L_{ij} = -R_{ik}U_{kj} - R_{jk}U_{k,i}$, $S_{ij} = \frac{1}{2}\{U_{i,j} + U_{j,i}\}, \alpha_1 = C_2 + 8/11 = 0.76, \alpha_2 = 8C_2 - 2/11 = 0.11, \alpha_3 = 30C_2 - 28/55 = 0.18$, and $C_2 = 0.4$.

Shih and Lumley[3] model:

$$\frac{\Phi_{(2)ij}}{\varepsilon} = \frac{4}{5}\frac{K}{\varepsilon}S_{ij} + 6\alpha_5\frac{K}{\varepsilon}(b_{ik}S_{jk} + b_{jk}S_{ik} - \frac{2}{3}b_{kl}S_{kl}\delta_{ij})$$
$$+\frac{2}{3}(2-7\alpha_5)\frac{K}{\varepsilon}(b_{ik}W_{kj} + b_{jk}W_{ik})$$
$$+\frac{1}{5}\frac{K}{\varepsilon}(b_{il}b_{lm}S_{jm} + b_{jl}b_{lm}S_{im} - 2b_{ik}S_{kl}b_{lj} - 3b_{kl}S_{kl}b_{ij})$$
$$+\frac{1}{5}\frac{K}{\varepsilon}(b_{il}b_{lm}W_{jm} + b_{jl}b_{lm}W_{im}) \tag{17}$$

where $W_{ij} = \frac{1}{2}\{U_{i,j} - U_{j,i}\}$, $\alpha_5 = 1/10(1 + 4/5\, F^{1/2})$, $F = 1 + 9II_b + 27III_b$, $II_b = -b_{ij}b_{ij}/8$, and $IIIb = b_{ij}b_{jk}b_{ki}/24$.

Speziale et al.[8] model:

$$\frac{\Phi_{(2)ij}}{\varepsilon} = [C_3 - C_3^*(-2II_b)^{1/2}]\frac{K}{\varepsilon}S_{ij} + C_4\frac{K}{\varepsilon}(b_{ik}S_{jk} + b_{jk}S_{ik}$$
$$-\frac{2}{3}b_{kl}S_{kl}\delta_{ij}) + C_5\frac{K}{\varepsilon}(b_{ik}W_{kj} + b_{jk}W_{ik}) \tag{18}$$

where $C_3 = 4/5$, $C_3^* = 1.30$, $C_4 = 0.625$, and $C_5 = 0.2$.

The model of Launder et al.[1] is the simplest one of the preceding equations. The approximation (16) led to the correct development of the turbulent and mean flowfields over a fairly wide range of conditions. The next two models (17) and (18) for the rapid part of the pressure-strain term were based on a dynamical system analysis, constrained by requiring consistency with rapid-distortion theory and experiments on homogeneous shear flow. The Shih and Lumley model strictly adheres to realizability constraints.

Comparison of the model predictions with the DNS data are shown in Figs. 7 and 8. This comparison shows that the predictions of models (7–10) are inadequate. For good agreement with DNS data, it is necessary to find a workable form of approximation for $F_{(2)ij}$. The general linear form of such an approximation can be written as

$$\frac{\Phi_{(2)ij}}{\varepsilon} = -\alpha_1\frac{P_k}{\varepsilon}c_{ij} - \alpha_2\frac{P_k}{\varepsilon}l_{ij} - 2\alpha_3\frac{K}{\varepsilon}S_{ij} + \alpha_4\frac{\pi_{11}}{\varepsilon}q_{ij} + \frac{\mu_{11}}{\varepsilon}r_{ij} \tag{19}$$

where α_n are the scalar coefficients depending on invariant II_b

$$\alpha_n = \alpha_{n0} + \alpha_{n1}II_b$$

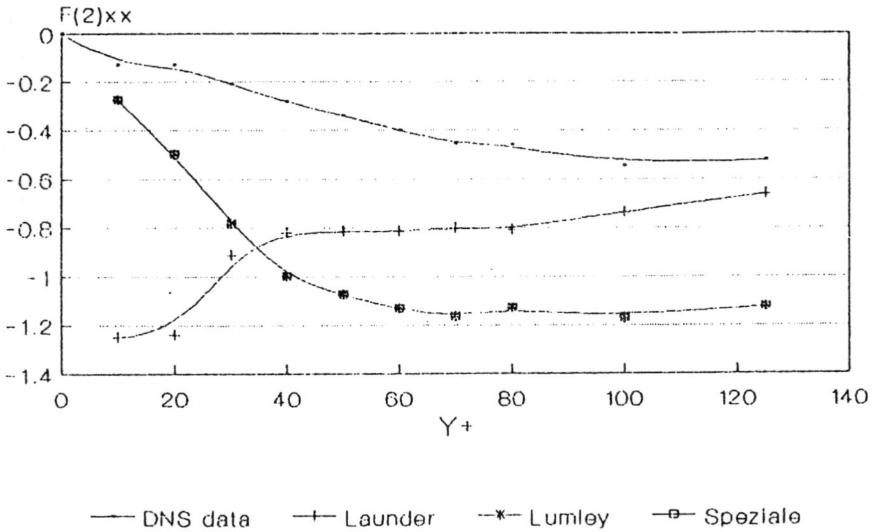

Fig. 7 Comparison of DNS data with calculations of $F(2)xx$.

and

$$q_{ij} = \left(\frac{\pi_{ij}}{\pi_{ll}} - \frac{1}{3}\delta_{ij} \right)$$

$$\pi_{ij} = -\frac{1}{K}(R_{is}R_{sk}U_{j,k} + R_{js}R_{sk}U_{i,k})$$

$$r_{ij} = \left(\frac{\mu_{ij}}{\mu_{ll}} - \frac{1}{3}\delta_{ij} \right)$$

$$\mu_{ij} = -\frac{1}{K}(R_{is}R_{sk}U_{k,j} + R_{js}R_{sk}U_{k,i})$$

We have compared relation (19) with the DNS data of Mansour et al.[3] for the components $\Phi_{(2)xx}$, $\Phi_{(2)yy}$, $\Phi_{(2)zz}$, and $\Phi_{(2)xy}$. Various combinations of the terms on the right side of Eq. (19) are used in these tests. The results of these comparisons show that the closest approach to DNS data can be achieved using the next expressions:

$$\frac{\Phi_{(2)ij}}{\varepsilon} = -\alpha_1 \frac{P_k}{\varepsilon} c_{ij} - \alpha_2 \frac{P_k}{\varepsilon} l_{ij} - 2\alpha_3 \frac{K}{\varepsilon} S_{ij} \tag{20}$$

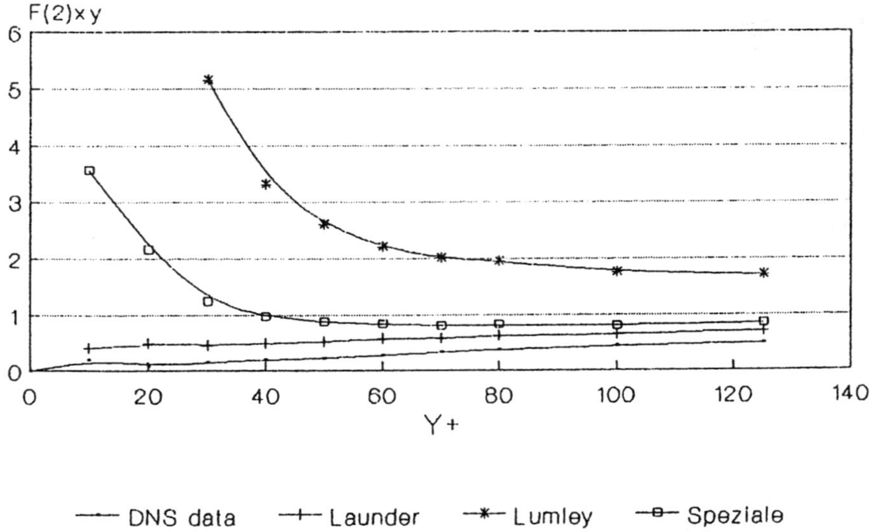

Fig. 8 Comparison of DNS data with calculations of *F*(2)*xy*.

where $\alpha_n = \alpha_{n0} + \alpha_{n1} II_b, \alpha_{10} = 0.71, \alpha_{20} = 0.36, \alpha_{30} = 0.51$, and $\alpha_{n1} = -\alpha_{n0}/2$;

$$\frac{\Phi_{(2)ij}}{\varepsilon} = -\alpha_1 \frac{P_k}{\varepsilon} c_{ij} - \alpha_2 \frac{P_k}{\varepsilon} l_{ij} - 2\alpha_3 \frac{K}{\varepsilon} S_{ij} + \alpha_4 \frac{\pi_{ll}}{\varepsilon} q_{ij} \tag{21}$$

where $\alpha_1 = 1.13, \alpha_2 = 0.25, \alpha_3 = 0.46$, and $\alpha_4 = 0.47$;

$$\frac{\Phi_{(2)ij}}{\varepsilon} = -\alpha_1 \frac{P_k}{\varepsilon} c_{ij} - \alpha_2 \frac{P_k}{\varepsilon} l_{ij} - 2\alpha_3 \frac{K}{\varepsilon} S_{ij} \tag{22}$$

where $\alpha_1 = 0.6 - 0.28\, II_b, \alpha_2 = 0.14$, and $\alpha_3 = 0.23$.

Figures 9 and 10 show the components of the rapid part of the pressure-strain term. Qualitative sense models (21) and (22) realistically describe the distributions of $F_{(1)xx}$, $F_{(1)yy}$, and $F_{(1)xy}$ across the whole width of the channel. The distribution of $F_{(1)yy}$ better describes model (22). Note that model (22) is more simple. Therefore, at this point preference can be given to the model (23).

Term $\Phi_{(w)ij}$

There have been various proposals on how to include near-wall effects into the pressure-strain term. Launder et al.[1] used the wall-normal vector and damping functions based on the wall-normal coordinate direction. In a like manner Shih and Lumley [10] and Launder and Shima[11] used a coordinate

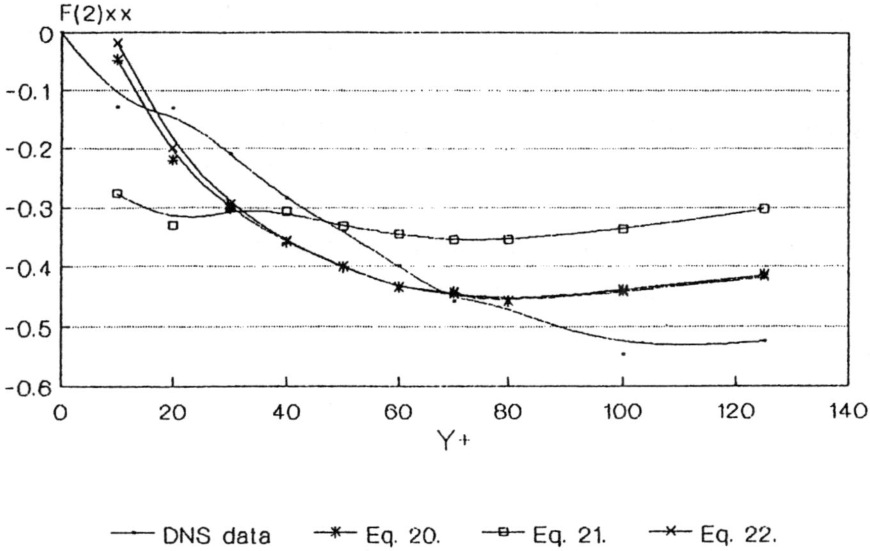

Fig. 9 Comparison of DNS data with calculations of $F(2)xx$.

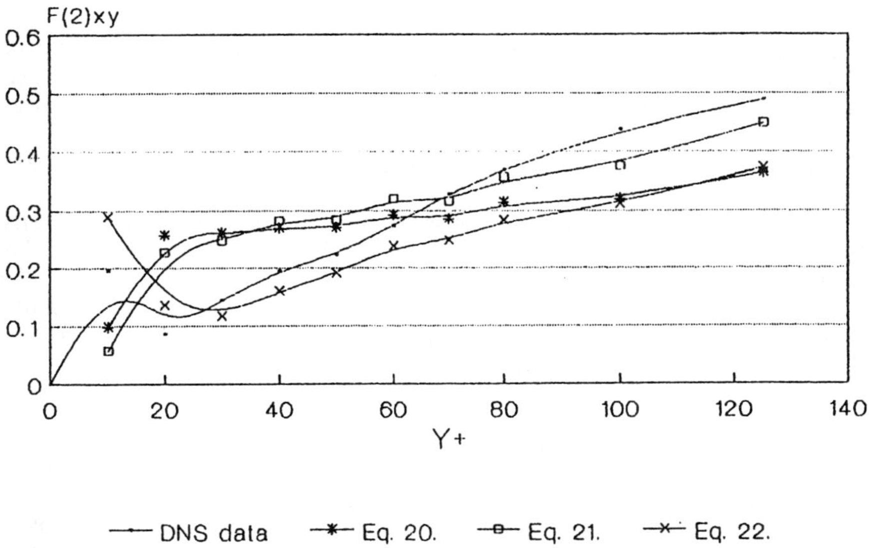

Fig. 10 Comparison of DNS data with calculations of $F(2)xy$.

dependence of $\Phi_{(w)ij}$. However, Mansour et al.[3] with DNS showed that this term is negligible throughout the channel. Because of this, Launder and Li[9] had indicated that in many flows near walls, including flows over surfaces of complex topography, the "wall-reflection" contribution to the pressure-strain process can be eliminated. Thus, we can write

$$\Phi_{(w)ij} = 0$$

Results of Pressure-Strain Modeling

The following model for the pressure-strain is therefore proposed:

$$\Phi_{ij} = \varphi_{(1)ij} + \varphi_{(2)ij} + \varphi_{(w)ij} \tag{23}$$

$$\frac{\Phi_{(1)ij}}{\varepsilon} = -C_1 b_{ij} + C_{11} II_b b_{ij} + C_{14} d_{ij}$$

$$C_1 = 3, \qquad C_{11} = 1, \qquad C_{14} = 2.4, \qquad \langle\delta\rangle = 12.8\%$$

$$\frac{\Phi_{(2)ij}}{\varepsilon} = -\alpha_1 \frac{P_k}{\varepsilon} c_{ij} - \alpha_2 \frac{P_k}{\varepsilon} l_{ij} - 2\alpha_3 \frac{K}{\varepsilon} S_{ij}$$

where $\alpha_1 = 0.6 - 0.28\, II_b$, $\alpha_2 = 0.14$, and $\alpha_3 = 0.23$.

$$\Phi_{(w)ij} = 0$$

Comparison of the model predictions with the DNS data are shown in Figs. 11–13. This comparison shows that the predictions of model (23) are good. For reference, the pressure-strain terms Φ_{ij} calculated by using equations of Launder et al. are also plotted in the same figures.

Based on term-by-term modeling of pressure-strain correlation, the next conclusions may be done:

1) There is no need to use the wall-normal vector and dumping functions for the description of the behavior of pressure-strain correlation near the wall.

2) Nonlinear effects in the energy redistribution among the components of Reynolds stresses can be taken into account by using invariant II_b in approximation for the pressure-strain term.

Local Similarity in Turbulent Flows

In the work reported here the second-order form of the locally anisotropic model of turbulence has been used to give criteria for the description of Reynolds-stress and dissipation-rate transport processes in rotating flows. The full second-order model contains differential transport equations (1) for the Reynolds-stress and dissipation-rate tensor. Sukhovich[12] on this seminar

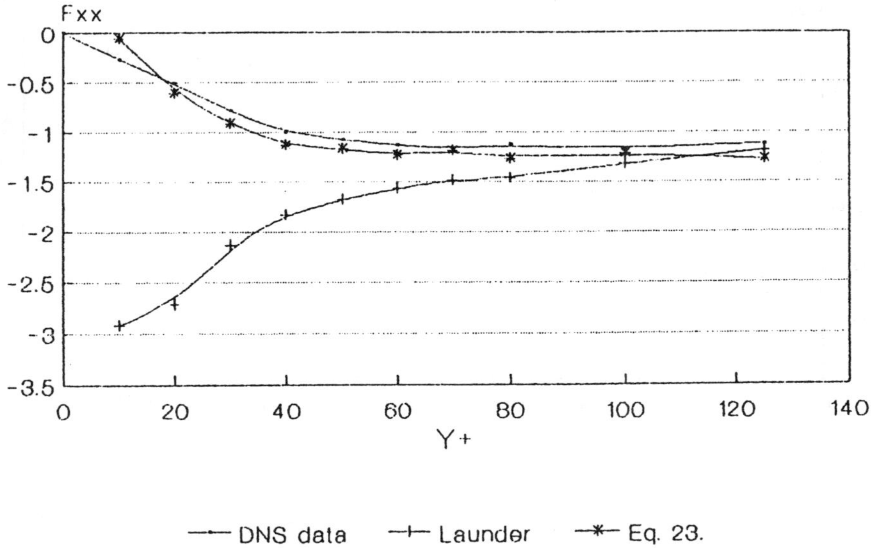

Fig. 11 Comparison of DNS data with calculations of *Fxx*.

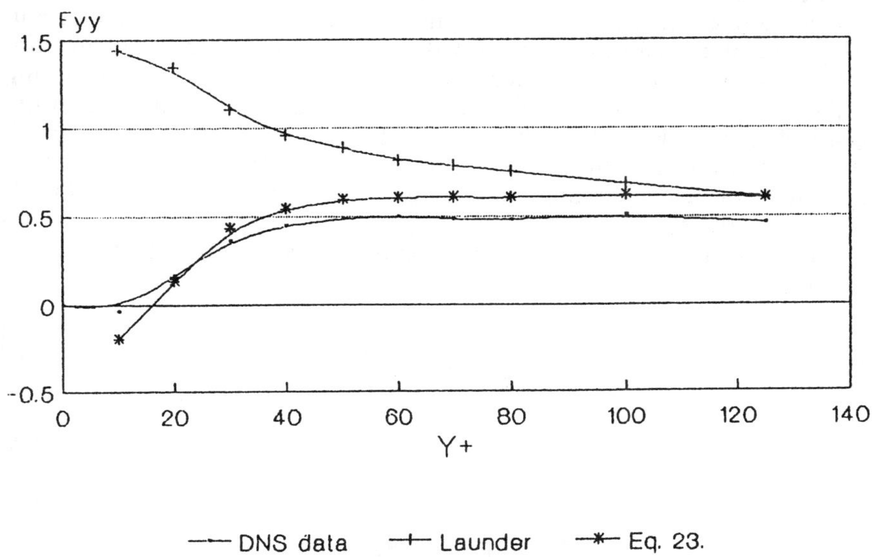

Fig. 12 Comparison of DNS data with calculations of *Fyy*.

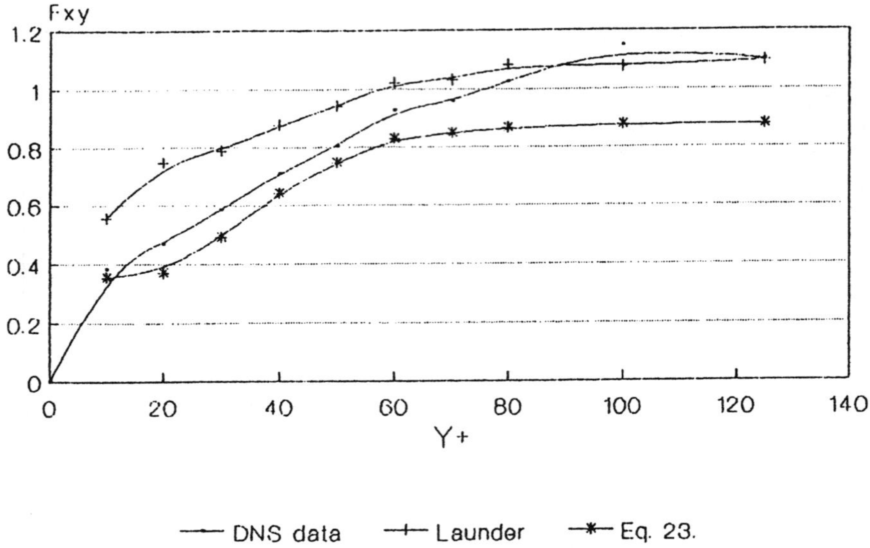

Fig. 13 Comparison of DNS data with calculations of *Fxy*.

proposed the full equation for the tensor dissipation rate. These equations are not convenient for analysis. Therefore the so-called algebraic model has been developed by simplifying the differential transport equations.

Algebraic models are suitable whenever the transport of Reynolds stresses and dissipation rates of Reynolds stresses are not very important. Therefore all effects that are governed by processes of production, pressure strain, and dissipation can be accounted for by the algebraic model. During the last decade several hundred sets of predictions have been made of the test turbulent flows in which different algebraic models have been explored. These models have been successfully used to predict the force effects (buoyancy, rotation, and streamline curvature), nonisotropic strain field, and wall-dumping influence.

The algebraic form of the locally anisotropic second-order model may be written as

$$\frac{Dk}{dt} = P_k - \varepsilon + D_k \tag{24}$$

$$\frac{d\varepsilon}{dt} = F_\varepsilon + P_{\varepsilon 1} + P_{\varepsilon 2} + P_{\varepsilon 3} + P_{\varepsilon 4} + T_\varepsilon + \pi_\varepsilon + D_\varepsilon - Y_\varepsilon \tag{25}$$

$$\frac{R_{ij}}{K} = \frac{F_{ij} + P_{ij} + \Phi_{ij} - 2\varepsilon_{ij}}{F_k + P_k - \varepsilon} \tag{26}$$

$$\frac{\varepsilon_{ij}}{\varepsilon} = \frac{F_{(\varepsilon)ij} + P_{(\varepsilon 1)ij} + P_{(\varepsilon 2)ij} + P_{(\varepsilon 4)ij} - Y_{(\varepsilon)ij}}{F_\varepsilon + P_{\varepsilon 1} + P_{\varepsilon 2} + P_{\varepsilon 4} - Y_\varepsilon} \tag{27}$$

With these equations and with approximations for Φ_{ij}, we can obtain the nondimensional equations for the Reynolds stress and dissipation rate anisotropy tensors:

$$\begin{aligned} b_{ij} &= \varphi_1(b_{ik}X_{j,k} + b_{jk}X_{i,k}) + \varphi_2(b_{ik}X_{k,j} + b_{jk}X_{k,i}) + \varphi_3 S_{(1)ij} \\ &\quad + \varphi_4 d_{ij} + \varphi_5 \frac{F_k}{\varepsilon}\left(\frac{F_{ij}}{F_k} - \frac{2}{3}\delta_{ij}\right) \end{aligned} \tag{28}$$

$$d_{ij} = f_1 S_{(1)ij} + f_2 S_{(2)ij} + f_3 \frac{KF_k}{\varepsilon^2}\left(\frac{F_{(\varepsilon)ij}}{F_k} - \delta_{ij}/3\right) \tag{29}$$

where $X_{ij} = K/\varepsilon\ U_{i,j}$, $S_{(1)ij} = K/\varepsilon\ S_{ij}$, $S_{ij} = (U_{i,j} + U_{k,i})/2$, and $S_{(2)ij} = K^2/\varepsilon^2\ (U_{k,i}S_{jk} + U_{k,j}S_{ik})$.

The analysis of algebraic expressions show that Reynolds stresses and dissipation-rate tensor are functions of the limited number of dimensionless parameters: X_{ij}, $S_{(1)ij}$, and $S_{(2)ij}$, which may be considered as local criteria of similarity for processes of turbulent momentum transfer. To illustrate, for a two-dimensional flow,

$$U_1 > 0, \qquad U_2 \approx U_3 \approx 0, \qquad U_{1,1} \approx U_{1,3} \approx 0$$

there is one nondimensional parameter $\dfrac{K}{\varepsilon}\dfrac{\partial U_x}{\partial y}$, which controls the values b_{ij} and d_{ij}.

For swirling flows:

$$U_z \neq 0, \qquad U_\varphi \neq 0, \qquad U_r \approx 0$$

$$\frac{\partial U_z}{\partial_r} \neq 0, \qquad \frac{\partial U_\varphi}{\partial_r} \neq 0, \qquad \frac{\partial}{\partial_z} \approx \frac{\partial}{\partial_\varphi} \approx 0$$

there are three nondimensional parameters $X_1 = \dfrac{K}{\varepsilon}\dfrac{\partial U_z}{\partial_r}$, $X_2 = \dfrac{K}{\varepsilon}\dfrac{\partial U_\varphi}{\partial_r}$, and $X_3 = \dfrac{K}{\varepsilon}\dfrac{U_\varphi}{r}$, which control the values b_{ij} and d_{ij}.

A parametric investigation of the influence of these criteria on Reynolds stresses has been performed. The calculations were carried out in a wide range of criterion values. As an illustration the areas in which the components of R_{ij} production $P_z = -R_{rz}\partial U_z/\partial_r$ and $P\varphi = -R_r\varphi\ (\partial U_\varphi/\partial r - U_\varphi/r)$ less than zero are shown in Fig. 14. The calculations showed that in rotating flows there may be areas in which occurs the transfer of kinetic energy from pulsating flow to mean flow ($P_z < 0$ and $P_\varphi < 0$) or transfer of kinetic energy

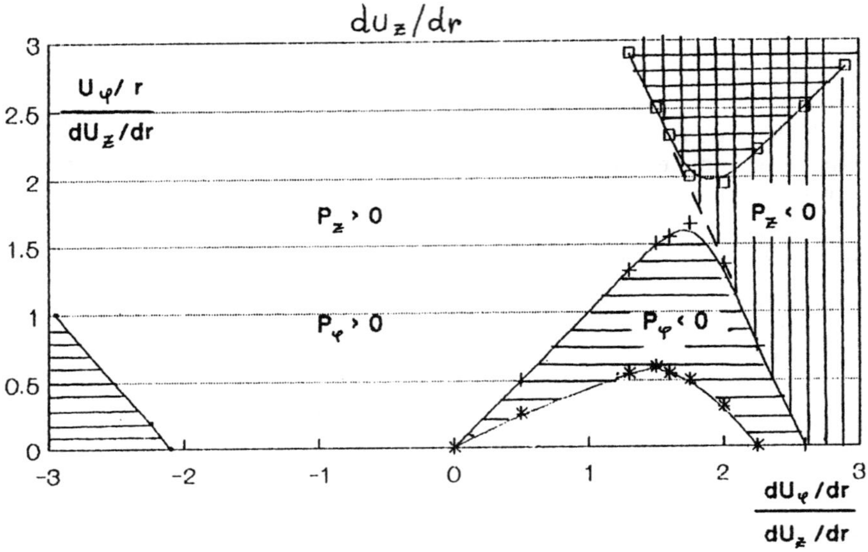

Fig. 14

from mean longitudinal flow into mean rotational flow through pulsating motion ($P_z > 0$ and $P_\varphi < 0$). This process had been found experimentally by Sukhovich.[13]

The results of the analyses may be used for any one of the turbulent rotating flows. If the values of criteria are equal for the same flows in the same point, the Reynolds-stresses for these flows would be equal also. The similar method of attack may be used for analysis of alternate types of turbulent flows.

References

[1]Launder, B. E., Reece, G. J., and Rodi, W., "Progress in the Development of a Reynolds-Stress Turbulence Closure," *Journal of Fluid Mechanics*, Vol. 68, 1975, pp. 537–566.

[2]Hanjalic, K., and Launder, B. E., "A Reynolds Stress Model of Turbulence and Its Application to Thin Shear Flows," *Journal of Fluid Mechanics*, Vol. 52, 1972, pp. 609–638.

[3]Mansour, N. N., Kim, J., and Moin, P., "Reynolds-Stress and Dissipation-Rate Budgets in a Turbulent Channel Flow," *Journal of Fluid Mechanics*, Vol. 194, 1988, pp. 15–44.

[4]Rotta, J. C., "Statistical Theory of Inhomogeneous Turbulence, Part 1," *Zeitschrift fur Physik*, Vol. 129, 1951, pp. 257–572.

[5]Weinstock, J., and Burk, S., "Theoretical Pressure-Strain Term, Experimental Comparison, and Resistance to Large Anisotropy," *Journal of Fluid Mechanics*, Vol. 154, 1985, pp. 429–443.

[6]Lumley, J. L., "Computational Modeling of Turbulent Flows," *Advances in Applied Mechanics*, Vol. 18, 1978, pp. 123–176.

[7]Shih, T. H., and Lumley, J. L., "Modelling of Pressure Correlation Terms in Reynolds Stress and Scalar Flux Equations," *Sibley School Mech. and Aerospace Engng,* Rept. FDA-85-3, Cornell Univ., Ithaca, NY, 1985.

[8]Speziale, C. G., Sarkar, S., and Gatski, T. B., "Modelling the Pressure-Strain Correlation of Turbulence—an Invariant Dynamical Systems Approach," *Journal of Fluid Mechanics*, Vol. 227, 1991, p. 245.

[9]Launder, B. E., and Li, S.-P., "On the Elimination of Wall-Topography Parameters from Second-Moment Closure," *Physics of Fluids*, Vol. 6, 1994, pp. 999–1006.

[10]Shih, T. H., and Lumley, J. L., "Second-Order Modelling of Near-Wall Turbulence," *Physics of Fluids,* Vol. 29, 1986, pp. 971–975.

[11]Launder, B., and Shima, N., "Second-Moment Closure for the Near-Wall Sublayer: Development and Application," *AIAA Journal*, Vol. 27, 1989, pp. 1319–1325.

[12]Sukhovich, E. P., "Dissipation-Rate Modelling with the Aid of Direct Simulation Data," *The Eight Beer-Sheva International Seminar on MHD Flows and Turbulence*, 1996.

[13]Sukhovich, E. P., "Turbulent Kinetic Energy Transfer to Mean Motion in Swirling Flows," *Abstracts of International Symposium "Generation of Large-scale Structures in Continuous Media" (The Nonlinear Dynamics of Structures)*, Perm-Moscow, 1990, pp. 236–237.

Chapter 10

Dissipation-Rate Modeling with the Aid of Direct Simulation Data

E. Sukhovich*
Institute of Physics, Latvian Academy of Sciences Salaspils-1, LV-2169, Latvia

THE instantaneous velocity field in turbulent flow is described by the Navier–Stokes equations. Exact numerical simulation of turbulence at relatively low Reynolds numbers is possible. As the turbulent Reynolds number increases, the ratio of the large to small scale increases, rapidly exceeding the storage capacity of the computers. Therefore turbulent flows cannot be calculated with an exact method, even though the exact equations governing the turbulent motion are known; numerical schemes are available and these equations can be solved in principle. The desire to make practical calculations useful in design and planning led to the development of various models in the study of turbulence.

Among the most successful and popular turbulence models are the various second-order closure schemes. These models attempt to express the physical properties of turbulence in a universal mathematical form. The full second-order model carries a prognostic differential equation for all second moments. For incompressible turbulent flow, the transport equations for the Reynolds stresses are given by

$$\frac{\mathrm{d}R_{ij}}{\mathrm{d}t} = F_{ij} + P_{ij} + \Phi_{ij} - 2\varepsilon_{ij} + D_{ij} \tag{1}$$

where $\frac{\mathrm{d}}{\mathrm{d}t} = \frac{\partial}{\partial t} + U_k \frac{\partial}{\partial x_k}$, U_k is the mean velocity, $R_{ij} = \langle u_i u_j \rangle$ is the Reynolds stress tensor, u_i is the fluctuating velocity component, and t is the time. The

*Salaspils-l.

angular brackets denote averaging. In the preceding and following equations, repeated indices imply summation.

Production of R_{ij} by internal force:

$$F_{ij} = (\langle f_i u_j \rangle + \langle f_j u_i \rangle)/\rho$$

Stress production of R_{ij}:

$$P_{ij} = -(R_{ik}U_{j,k} + R_{jk}U_{i,k})$$

Pressure-strain correlations:

$$\Phi_{ij} = -\frac{1}{\rho}(\langle u_i p_{,j} \rangle + \langle u_j p_{,i} \rangle)$$

Turbulent transport:

$$D_{ij} = -(\langle u_i u_j u_k \rangle - \nu R_{ij,k})_{,k}$$

Dissipation rate of Reynolds stresses, ν denotes kinematic viscosity:

$$\varepsilon_{ij} = \nu \langle u_{i,k} u_{j,k} \rangle$$

In these equations various terms appear other than the dependent second moments. To close the equations and make them determinate, these terms must be modeled. All models contain a number of adjustable numerical coefficients. The form of the modeled terms is subject to many constraints of a mathematical and physical nature. The most obvious is the requirement that the terms have the same tensor structure as the terms they replace, so that they will transform properly in different coordinate systems. This type of modeling was referred to as invariant modeling by Donaldson.[1]

In the context of a full second-order model of turbulence, R_{ij} and ε_{ij} are dependent variables that must be determined from a solution of the relevant equations. The terms F_{ij} and P_{ij} are defined by Reynolds stresses, heat flux, and gradient of mean velocity. Therefore these terms are exact. The terms Φ_{ij} and D_{ij} must be represented as empirical functions of the mean velocities, Reynolds stresses, and their derivatives. The expressions for the Φ_{ij} and D_{ij} have been extensively analyzed and modeled by various groups. A comparison of four models for the turbulent transport diffusion terms and four models for the pressure-strain redistribution terms have been represented by Schwarz and Bradshaw.[2] Term-by-term tests show that the main source of error is the model equations for dissipation. They surmise that because of the way dissipation rate is used in the different terms of models for tensor quantities then different errors will occur for different Reynolds stresses. The alternative is to use the exact equations for the tensor and scalar dissipation rate and to model them as the equation for the Reynolds stresses.

The objective of this work is to present the full locally anisotropic second-order model of turbulence. Some of the previously familiar and a number of the new approximations for all of the terms of the exact equations for dissipation-rate functions were investigated to be certain that these approximations described the DNS data. The system of empirical coefficients has been determined. In the general case the differential equation for tensor dissipation rate is very cumbersome for use in numerical calculations. Therefore this equation may be used for a deduction of a more simple equation.

Modeling Terms in the ε_{ij} Equation with Aid of Direct Numerical Simulation Data

The Kolmogorov's local-isotropy hypothesis, which states that at sufficiently high Reynolds numbers the small-scale structures of turbulent motions are independent of large-scale structures and the mean deformation rate, has been used in most approaches to understanding turbulence. The recent experimental studies and simulations have shown that turbulence does not remain locally isotropic either in the presence of gravity, magnetic, centrifugal forces or in the presence of strong strain fields. However, there are no equations for determination of tensor dissipation rate for locally anisotropic turbulent flows. Usually ε_{ij} are modeled by assuming the dissipative motions to be isotropic in the limit $Re \to \omega$.

$$\varepsilon_{ij} = -\frac{1}{3} - \varepsilon\delta_{ij} \tag{2}$$

Figure 1 shows the profiles of dissipation rate of Reynolds stresses in channel flow. $Y^+ = YU_\tau/\nu$ is normal to the wall nondimensional coordinate

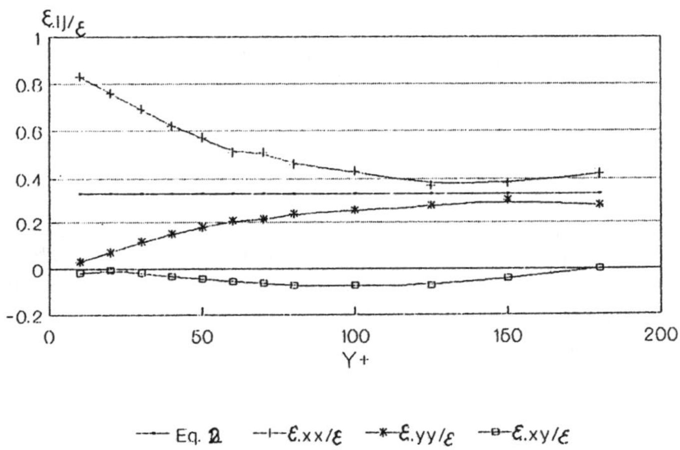

Fig. 1 Dependence of a dissipation rate on distance from the wall.

[$U_\tau = (\nu U_{,y}|_{\text{wall}})^{1/2}$ is the wall-shear velocity]. The ε_{ij} profiles show a consistent disagreement between isotropic value and data of direct numerical simulation of Mansour et al.[3] Rotta[4] argued that in the limit as $Re \to 0$, the dissipation-rate tensor will be aligned with the Reynolds-stress tensor and can be modeled as

$$\varepsilon_{ij}/\varepsilon = f_s R_{ij}/K \tag{3}$$

where $K = R_{ii}/2$ is the turbulent kinetic energy. In fact, if we assume that $f_s = 1$, we find better agreement between the model (3) and DNS data. The off-diagonal component shows the largest difference between the model and the data. The diagonal components show better agreement but would require a different function f_s for the different components to obtain an improvement in the agreement.

There is the distinct possibility that the approximations (2) and (3) do not contain some important effect. Therefore it is necessary to develop the closure models for all of the terms of the equations for tensor and scalar dissipation rate.

In the work reported here the closure models for all of the terms of the exact equation for dissipation rate were investigated, including mixed production, production by mean velocity gradient, gradient production, turbulent production, turbulent transport, pressure transport, viscous diffusion, and dissipation.

Basically three methods may be employed to determine the form of the modeled terms and empirical coefficients in closure approximations: indirect method, direct comparison of the closure models to experimental data for terms of the exact equation, and direct comparison to numerical simulation data.

In the indirect method the adequacy of the models is judged by computing a flow with the model and by comparing the predicted mean velocity and Reynolds stresses with experimental data. Unfortunately, the results of the comparison cannot be interpreted unambiguously. The terms in the exact equation for tensor dissipation rate cannot be measured, and, therefore, information on these terms can be obtained from direct numerical simulation data (DNS) only.

In recent years numerical simulations of turbulent flows have become an important research tool in studying the basic physics of turbulence. Direct simulations of simple inhomogeneous flows have been carried out by Mansour et al., Kim et al., and Spalart.[3,5,6] In the computations the grid resolution was sufficiently fine to resolve the essential turbulent scales. The database generated by such a simulation is of considerable value for the design and testing of turbulence closure models.

A set of the exact equations describing the dynamics of ε_{ij} can be derived from the Navier–Stokes equation:

$$\frac{\mathrm{d}\varepsilon_{ij}}{\mathrm{d}t} = F_{(\varepsilon)ij} + P_{(\varepsilon 1)ij} + P_{(\varepsilon 2)ij} + P_{(\varepsilon 3)ij} + P_{(\varepsilon 4)ij} + \Pi_{(\varepsilon)ij} + D_{(\varepsilon)ij} + T_{(\varepsilon)ij} - Y_{(\varepsilon)ij} \tag{4}$$

where

Production of dissipation rate ε_{ij}:

$$F_{(\varepsilon)ij} = \nu(\langle u_{i,k}f_{j,k}\rangle + \langle u_{j,k}f_{i,k}\rangle)/\rho$$

The production rates attributable to mean shear are

Mixed production of ε_{ij}:

$$P_{(\varepsilon 1)ij} = -(\varepsilon_{ik}U_{j,k} + \varepsilon_{jk}U_{i,k})$$

Production of ε_{ij} by mean velocity gradient:

$$P_{(\varepsilon 2)ij} = -\nu U_{m,k}(\langle u_{i,k}u_{j,m}\rangle + \langle u_{j,k}u_{i,m}\rangle)$$

Gradient production of ε_{ij}:

$$P_{(\varepsilon 3)ij} = -\nu(\langle u_{ui,k}\rangle U_{j,k}\nu_m + \langle u_m u_{j,k}\rangle U_{i,km})$$

Turbulent production:

$$P_{(\varepsilon 4)ij} = -2\nu\langle u_{i,k}u_{j,m}u_{k,m}\rangle$$

The pressure-strain correlations may be written as

$$\Pi_{(\varepsilon)ij} = -\frac{\nu}{\rho}(\langle u_{j,k}p_{,ik}\rangle) + (\langle u_{i,k}p_{,jk}\rangle)$$

The turbulent transport terms take the form:

$$T_{(\varepsilon)ij} = -\nu(\langle u_k u_{i,m}u_{j,m}\rangle - \varepsilon_{ij,k})_{,k}$$

The viscous diffusion term takes the form:

$$D_{(\varepsilon)ij} = \nu\varepsilon_{ij,kk}$$

The destruction terms by viscous action are

$$Y_{(\varepsilon)ij} = -\,2\varepsilon\langle u_{i,km}u_{j,km}\rangle$$

The exact equation for the scalar dissipation rate ε is given by

$$\frac{d\varepsilon}{dt} = F_\varepsilon + P_{\varepsilon 1} + P_{\varepsilon 2} + P_{\varepsilon 3} + P_{\varepsilon 4} + T_\varepsilon + \Pi_\varepsilon + D_\varepsilon - Y_\varepsilon \tag{5}$$

where the terms F_ε, P_ε, T_ε, D_ε, and Y_ε may be derived by convolution of the corresponding terms in Eq. (4).

Mansour et al.[3] and Redi and Mansour[7] have calculated the ε budget from the DNS data for the low-*Re* channel flow. The ε budget has been evaluated from the data for the channel with $Re_t = 180$ and $Re_t = 395$. All terms of Eq. (5) have been made dimensionless with U_τ^6/ν^2. The imbalance in the budget is also given. As can be seen, the imbalance is fairly small relative to the terms in budget, except very close to the wall. This imbalance is small for $y^+ > 8$, so that the ε budget determined from DNS data can be considered accurate for $y^+ > 8$.

As was to be expected from the order-of-magnitude analysis of Tennekes and Lumley,[8] the turbulent production rate $P_{\varepsilon 4}$ due to vortex stretching and the viscous destruction Y_ε dominate the balance equation in the high-Reynolds-number region away from the wall. However, near the wall, the production terms $P_{\varepsilon 1}$ and $P_{\varepsilon 2}$ become equally important; $P_{\varepsilon 3}$, T_ε, and Π_ε are smaller terms. The pressure diffusion term Π_ε can be seen to be negligible everywhere. The term $P_{\varepsilon 3}$ is of the same order of magnitude as the turbulent diffusion T_ε.

To turn the exact Eqs. (4) and (5) into equations that can be used in a turbulence model, all terms at the right side of these equations need to be modeled. Fineburg[9] used a tensor approximation technique and proposed the general form of approximations for all terms of the ε_{ij} equation. We have taken these expressions as a starting point for modeling. Consider term by term the results of modeling.

Results of Modeling

Term $F_{(\varepsilon)ij}$

$$F_{(\varepsilon)ij} = -\frac{\beta \nu}{\nu + a}(g_i \varepsilon_{(\tau)j} + g_j \varepsilon_{(\tau)i})$$

where β is the volumetric expansion coefficient, g_i is the gravitational acceleration vector and $\varepsilon_{(\tau)i}$ is the destruction of the temperature field by molecular action. The term contains the known values, and it has no need of modeling.

Term $P_{(\varepsilon 1)ij}$

$$P_{(\varepsilon 1)ij} = -(\varepsilon_{ik} U_{j,k} + \varepsilon_{jk} U_{i,k}) \tag{6}$$

$$P_{\varepsilon 1} = P_{(\varepsilon 1)ii} = -2\varepsilon_{ik} U_{k,j} \tag{7}$$

In the context of a full second-order model of turbulence, $P_{(\varepsilon 1)ij}$ is the dependent variable that must be determined from a solution of the relevant equations. Figure 2 shows that Eq. (7) describes the DNS data adequately.

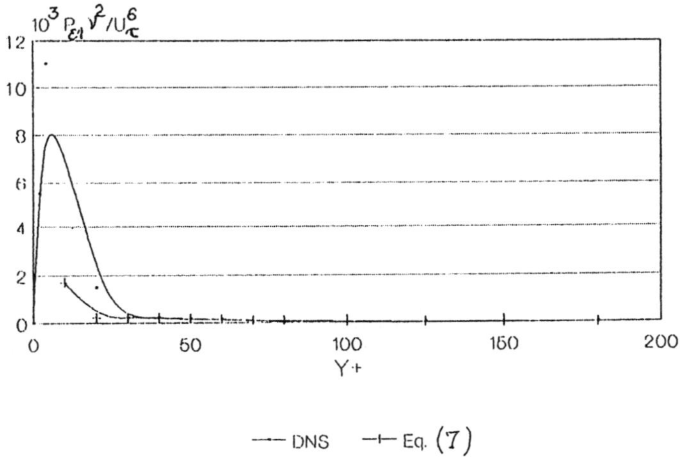

Fig. 2 Distribution of term $P_{\varepsilon 1}$ across a channel.

Term $P_{(\varepsilon 2)ij}$

Fineburg[9] supposed the next approximation for $P_{(\varepsilon 2)ij}$:

$$P_{(\varepsilon 2)ij} = -\alpha_1 \varepsilon (U_{i,j} + U_{j,i}) + \alpha_2 (\varepsilon_{ik} U_{j,k} + \varepsilon_{jk} U_{i,k} + \varepsilon_{ik} U_{k,j} + \varepsilon_{jk} U_{k,i})$$

where a_1 and a_2 are some constants. Convolution of this expression gives

$$P_{\varepsilon 2} = P_{(\varepsilon 2)ii} = 2a_2 P_{\varepsilon 1}$$

Launder et al.[10] proposed that the source term in equation for the dissipation rate of kinetic energy has the form:

$$P_\varepsilon = C_{\varepsilon 1} P_k \varepsilon / K \tag{8}$$

where $C_{\varepsilon 1} = 1.45$. The distribution of P_ε resulting from Eq. (8) is compared in Fig. 3 with DNS data for $P_{\varepsilon 2}$.

The agreement between the expression (8) and distribution of $P_{\varepsilon 2}$ is very good. Thus we can write

$$P_{\varepsilon 2} = C_{\varepsilon 1} P_k \varepsilon / K \tag{9}$$

Because $P_{\varepsilon 2}$ is a trace of tensor $P_{(\varepsilon 2)ij}$, a possible form of the relation for $P_{(\varepsilon 2)ij}$ may be written as

$$P_{(\varepsilon 2)ij} = -\rho_1 \frac{\varepsilon}{K}(R_{ik} U_{j,k} + R_{jk} U_{i,k}) - \rho_2 \frac{\varepsilon}{K}(R_{ik} U_{k,j} + R_{jk} U_{k,i}) - \rho_3 (U_{i,j} + U_{j,i})$$

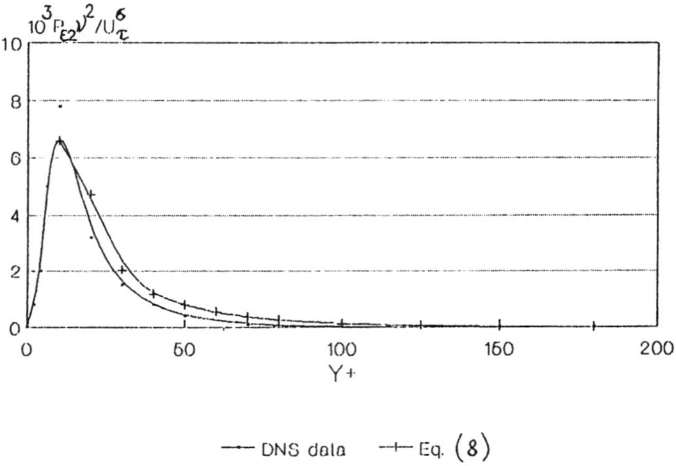

Fig. 3 Dependence of term $P_{\varepsilon 2}$ on distance from the wall.

In the opening stage of modeling, the equation for tensor dissipation rate may be recommended to be a more simple form of this relation

$$P_{(\varepsilon 2)ij} = -\frac{1}{2} C_{\varepsilon 1} \frac{\varepsilon}{K} (R_{ik} U_{j,k} + R_{jk} U_{i,k}) \tag{10}$$

Term $P_{(\varepsilon 3)ij}$

Fineburg[9] proposed the next approximation for $P_{(\varepsilon 3)ij}$:

$$\begin{aligned} P_{(\varepsilon 3)ij} = -\nu[&(\xi K_{,k} + \xi_2 R_{mk,m})(\overline{U}_{i,kj} + \overline{U}_{j,kl}) + (\xi_1 R_{ki,m} \\ &- R_{km,i})\overline{U}_{j,km} + (\xi_1 R_{kj,m} - \xi R_{km,j})\overline{U}_{i,km} + (\xi K_{,i} \\ &+ \xi_2 R_{mi,m})\overline{U}_{j,kk} + (\xi K_{,j} + \xi_2 R_{mj,m})\overline{U}_{i,kk}] \end{aligned} \tag{11}$$

where ξ_i are some constants. Convolution of this expression gives

$$P_{\varepsilon 3} = -2\nu(\xi_1 R_{ki,m} - \xi R_{km,i})\overline{U}_{i,km} - 2\nu(\xi K_{,i} + \xi_2 R_{mi,m})\overline{U}_{i,kk}$$

For developed channel flow this relation has a form

$$P_{\varepsilon 3} = -2\nu(\xi_1 + \xi_2) R_{xy,y} U_{x,yy} \tag{12}$$

Assuming $R_{xy} = \nu_t U_{x,y}$ and $\nu_t = C_\mu K^2/\varepsilon$ in the context of K–ε eddy-viscosity model, the following equation may be obtained from Eq. (12):

$$P_{\varepsilon 3} = -2(\xi_1 + \xi_2)\nu\nu_t U^2_{x,yy} - 2(\xi_1 + \xi_2)\nu U_{x,y} U_{x,yy} \frac{K^2}{\varepsilon}\left(\frac{2}{K} K_{,y} - \frac{1}{\varepsilon}\varepsilon_{,y}\right) \tag{13}$$

By other means Rodi and Mansour[7] obtained the following relation for shear-layer flows:

$$P_{\varepsilon 3} = 2C_{13}\nu\nu_t U^2_{x,yy} + C_{23}\nu \frac{K}{\varepsilon} K_{,y} U_{x,y} U_{x,yy} \tag{14}$$

where $C_{13} = 0.5$ and $C_{23} = 0.006$.

Equation (14) is in good agreement with DNS data. From comparison of Eq. (12) and (13), we obtain

$$\xi_1 + \xi_2 = -0.5$$

It is important to keep in mind that values of $P_{\varepsilon 3}$ are less than 0.002 at $y^+ < 30$ and less than 0.0001 at other points across the boundary layer.

Term $P_{(\varepsilon 4)ij}$

This term is modeled as (see Ref. 9)

$$P_{(\varepsilon 3)ij} = \frac{d_6}{C_d} \cdot \frac{\varepsilon \varepsilon_{ij}}{K} + \frac{d_7}{C_d} \frac{\varepsilon^2}{K} \delta_{ij}$$

$$P_{\varepsilon 4} = \frac{d_6 + 3d_7}{C_d} \frac{\varepsilon^2}{K} \tag{15}$$

where $d_6 + 3d_7/C_d = 3$. The distribution of this term in channel flow is shown in Fig. 4.

Term $T(_\varepsilon)_{ij}$

Following Ref. 9 the expression for this term may be written as

$$T_{(\varepsilon)ij} = C_{s4}\left[\frac{K}{\varepsilon}\left(R_{km}\varepsilon_{ij,m} + R_{im}\varepsilon_{jk,m} + R_{jm}\varepsilon_{ik,m}\right)\right]_{,k}$$

Then the turbulent transport of dissipation rate ε takes the form:

$$T_\varepsilon = C_{s1}\left[\frac{K}{\varepsilon}\left(R_{km}\varepsilon_{,m} + 2R_{im}\varepsilon_{jk,m}\right)\right]_{,k}$$

This equation, with $C_{s4} = 0.15$, yields the disagreement with DNS data and the equation of Rodi and Mansour[7] for this term less than 20%. The values of term T_ε are less than 0.001 in all points across the boundary layer.

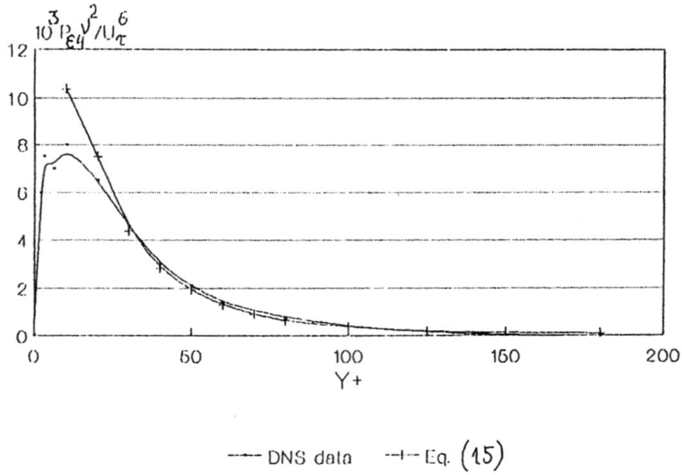

Fig. 4 Dependence of term $P_{\varepsilon 4}$ on distance from the wall.

Term $Y_{(\varepsilon)ij}$

This term is modeled as (see Ref. 9)

$$Y_{(\varepsilon)ij} = 6\mathrm{d}_1\mathrm{d}_5 \frac{\varepsilon_{ij}\varepsilon}{K} + 6\mathrm{d}_1(1-\mathrm{d}_5)\frac{\varepsilon^2}{K}\delta_{ij}$$
$$Y_\varepsilon = 6\mathrm{d}_1 \frac{\varepsilon^2}{K} \tag{16}$$

where $d_1 = 0.8$. The distribution of this term in channel flow is shown in Fig. 5.

Term $\Pi_{(\varepsilon)ij}$

The general form of the approximation for $\Pi(_\varepsilon)_{ij}$ may be written as

$$\Pi_{(\varepsilon)ij} = \alpha_{1\varepsilon}\left(\varepsilon_{ik}\overline{U}_{j,k} + \varepsilon_{jk}\overline{U}_{i,k} - \frac{2}{3}\varepsilon_{mn}\overline{U}_{m,n}\delta_{ij}\right) + \alpha_{2\varepsilon}\left(\varepsilon_{ik}\overline{U}_{k,j} + \varepsilon_{jk}\overline{U}_{k,i} - \frac{2}{3}\varepsilon_{mn}\overline{U}_{n,m}\delta_{ij}\right) + \alpha_{3\varepsilon}\varepsilon\left(\overline{U}_{i,j} + \overline{U}_{j,i}\right)$$

then $\Pi_\varepsilon = 0$. The coefficients of this equation remain unknown. The values of term Π_ε are less then 0.001 in all points across the boundary layer.

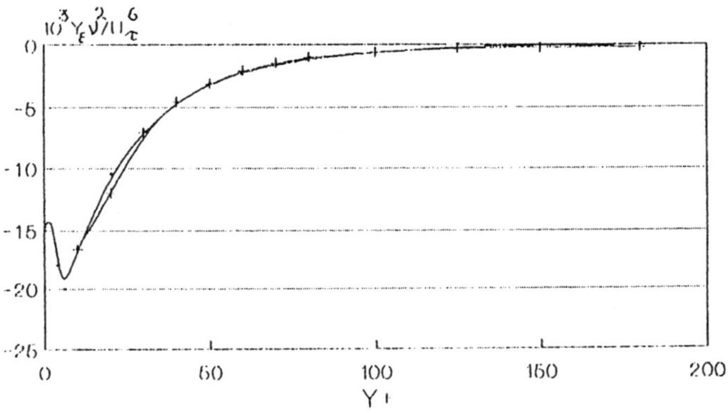

Fig. 5 Dependence of term Y_ε on distance from the wall.

Algebraic Form of Second Order Model

In summary we obtain the closed form of equations for tensor and scalar dissipation functions. In the general case differential equation (4) for ε_{ij} is very cumbersome for using in numerical calculations. It is the author's opinion that Eq. (4) may be used for a justified deduction of more simple equations. For instance, in many of the numerical simulations of turbulent shear flows Rodi's hypothesis has been used:

$$\frac{\mathrm{D}R_{ij}}{\mathrm{d}t} - \mathrm{D}_{ij} = \frac{R_{ij}}{K}\left(\frac{\mathrm{D}K}{\mathrm{d}t} - D_k\right)$$

Incorporating this expression into the R_{ij} equation yields the algebraic expression for R_{ij}:

$$\frac{R_{ij}}{K} = \frac{F_{ij} + P_{ij} + \Phi_{ij} - 2\varepsilon_{ij}}{F_k + P_k - \varepsilon} \tag{17}$$

Analogously, the suggestion has been made that

$$\frac{\mathrm{d}\varepsilon_{ij}}{\mathrm{d}t} - P_{(\varepsilon 3)ij} - \mathrm{D}_{(\varepsilon)ij} - T_{(ij\varepsilon)ij} - \Pi_{(\varepsilon)ij} = \frac{\varepsilon_{ij}}{\varepsilon}\left(\frac{\mathrm{D}\varepsilon}{\mathrm{d}t} - P_{\varepsilon 3} - \mathrm{D}_\varepsilon - T_\varepsilon - \Pi_\varepsilon\right) \tag{18}$$

Substituting this expression into Eq. (4) and (6) gives

$$\frac{\varepsilon_{ij}}{\varepsilon} = \frac{F_{(\varepsilon)ij} + P_{(\varepsilon 1)ij} + P_{(\varepsilon 2)ij} + P_{(\varepsilon 4)ij} - Y_{(\varepsilon)ij}}{F_\varepsilon + P_{\varepsilon 1} + P_{\varepsilon 2} + P_{\varepsilon 4} - Y_\varepsilon} \tag{19}$$

The final form of the locally anisotropic second-order model may thus be written as

$$\frac{\mathrm{D}K}{\mathrm{d}t} = P_k - \varepsilon + \mathrm{D}_k$$

Differential equation (5) for dissipation-rate ε
Algebraic equation (17) for Reynolds stresses R_{ij}
Algebraic equation (19) for tensor dissipation/rate ε_{ij}

Concluding Remarks

Numerical simulations of turbulent flows are not in perfect agreement with experimental data. A certain number of the discrepancies can be attributed with inadequacy in the pressure-strain and dissipation-rate approximations. The report develops proposals for a locally anisotropic model in which the dissipation rates of Reynolds stresses are determined from the corresponding transport equations.

We have used the data of direct numerical simulation of Mansour et al.[3] to derive the approximate expressions for all terms of the equations for tensor and scalar dissipation rate. There is, of course, the distinct possibility that a certain number of the expressions discussed do not contain some important effect. We believe, however, that the suggested form of the equation for the dissipation-rate tensor may be used for an understanding of the behavior of dissipative processes in turbulent flows.

The full locally anisotropic model of turbulence was presented. The second-order model contains transfer equations for the dissipation rate and Reynolds-stress tensors. The empirical coefficients of the turbulence model were determined.

References

[1]Donaldson, D. du R., "Calculation of Turbulent Shear Flows for Atmospheric and Vortex Motions," *AIAA Journal*, Vol. 10, 1972, pp. 4–12.

[2]Schwarz, W. R., and Bradshaw P., "Term-by-Term Tests of Stress-Transport Turbulence Models in a Three-Dimensional Boundary-Layer," *Physics of Fluids*, Vol. 6, 1994, pp. 986–998.

[3]Mansour, N. N., Kim, J., and Moin, P., "Reynolds-Stress and Dissipation-Rate Budgets in a Turbulent Channel Flow," *Journal of Fluid Mechanics*, Vol. 194, 1988, pp. 5–44.

[4]Rotta, J., "Statistical Theory of Inhomogeneous Turbulence, Part 1," *Zeitschrift fur Physik*, Vol. 129, 1951, pp. 257–572.

[5]Kim, J., Moin, P., and Moser, R, "Turbulence Statistics in Fully Developed Channel Flow at Low Reynolds Number," *Journal of Fluid Mechanics* Vol. 177, 1987, pp. 133–166.

[6]Spalart, P. R., "Direct Simulation of Turbulent Boundary Layers up to R_Θ = 1400," *Journal of Fluid Mechanics*, Vol. 187, pp. 61–98.

[7]Rodi, W., and Mansour, N. N., "Low Reynolds Number $K-\varepsilon$ Modeling with the Aid of Direct Simulation Data," *Journal of Fluid Mechanics*, Vol. 250, 1993, pp. 509–529.

[8]Tennekes, H., and Lumley, J. B., *A First Course in Turbulence*, MIT Press, Cambridge, MA, 1972.

[9]Fineburg, G. Z., "Equations for Free-Convection Turbulence in Incompressible Fluids, USSR," *Yyenue zanucku Nepmckozo neauhcmumyma*, Vol. 152, 1976, pp. 100–112 (in Russian).

Chapter 11

Lagrangian Stochastic Models for Two-Particle Relative Turbulent Dispersion

Burkhard M. O. Heppe*
Cambridge, CB3 9EW, England, United Kingdom

I. Introduction

A STATISTICAL theory of turbulent dispersion and mixing that characterizes the fluctuations of the concentration field of a passive diffusive species requires a realistic description and resolution of turbulent convection and molecular diffusion. Turbulent convection is effective at redistributing the scalar field at all length scales above the Kolmogorov scale, whereas the molecular diffusion acts most effectively at the smallest scalar length scales of the flow. An accurate description of mixing then requires a realistic treatment of the flow at the smallest hydrodynamic and scalar length scales.

To formulate a model of turbulent mixing, one must either resolve all relevant length scales or adapt a reduced description of the velocity and/or scalar field, which at present involves ad hoc assumptions to some extent. For references on the large literature on scalar turbulent transport including a rigorous treatment of simple model problems by renormalization methods, see Avellaneda and Majda.[1]

The first and the only fully systematic approach at the present state of knowledge is direct numerical simulation (DNS) of the Navier–Stokes equations and the molecular transport equations. However, this approach is limited with respect to the range of Reynolds and Peclet numbers that can be examined affordably. Nevertheless, within the accessible range of parameters it provides a reliable data supply against which model assumptions and results can be compared. In this paper we concentrate on turbulent dispersion of fluid particles although our approach seems to be promising for the treatment of molecular mixing as well.

---

*Department of Applied Mathematics and Theoretical Physics, Silver Street.

As is well known from Taylor's work, turbulent dispersion properties are conveniently described in terms of Lagrangian statistics. Hence, the basic problem in turbulent dispersion and mixing is to calculate Lagrangian statistics from Eulerian ones that are far better known from experiments.

Our approach to solving this problem is to apply a projector method in the formulation of Mori[2] to the evolution of Lagrangian velocity differences given by the Navier–Stokes equations. The method provides a projection from the dynamical Lagrangian quantities onto its initial values that are Eulerian in nature. The intrinsic nonlinearity of the Navier–Stokes equations is dealt with in a phenomenological way. This is achieved by, on the one hand, employing a short time or Markov approximation for the memory as is usually done in applications of the projector formalism. On the other hand, rather than depending on the initial values, the model coefficients are updated during the evolution in time, i.e., they are assumed to depend on the time dependent separation of the two particles. The approach we will develop in section II makes explicit use of the skewness of the velocity field and the non-Hermitian property of the time evolution operator representing the lack of time reversal symmetry in relative turbulent dispersion.

The Mori–Zwanzig projector formalism was introduced in the context of turbulence by Grossmann and Thomae.[3] Instead of considering stochastic models, their results were used by Grossmann and Procaccia[4] for the development of a mean-field theory of turbulent diffusion, i.e., they derived ordinary differential equations for the mean square separation. The nonlinearity was also treated phenomenologically, but memory effects were neglected completely.

Intermittency effects of the velocity field will be ignored in the following because we are mainly interested in low Reynolds number effects due to inclusion of the viscous subrange. Including finite Reynolds number effects is crucial for comparison with Lagrangian DNS data. However, the intermittency of a passively advected scalar will be addressed in the Appendix where our Lagrangian approach applied to the scalar field recovers the recent closure of Kraichnan.[5] Neglecting memory effects yields anomalous scaling of higher order scalar structure functions.

Because one-particle models are quite well understood even at finite Reynolds numbers,[6] we restrict ourselves here entirely to the *relative* motion of two particles. A more sophisticated approach based on the nonlinear generalized Langevin equation has been developed by the author, which overcomes the phenomenological treatment of the nonlinearity. Furthermore, it has been shown that by taking non-Markovian effects into account one can recover Sawford's model[6] in the far separation or one-particle limit. A non-Markovian one-particle model without the inconsistencies of the Sawford model has been proposed in Heppe,[8] showing an improved agreement with DNS.

This chapter is organized as follows. In the second section we first review the modeling approaches by Thomson and Novikov. Then we use the Mori projector formalism in a phenomenological way to derive stochastic model equations for the relative velocities from the Navier–Stokes equations. The

Lagrangian stochastic model is solved numerically and the results are compared to the DNS data of Yeung[9] in section 3. In the Appendix, we apply the same projection procedure to the concentration differences of a passive scalar.

II. Stochastic Dispersion Models

A. Models of Thomson and Novikov

Two approaches have been suggested to devise stochastic Markov models for the particle separations lying in the inertial range. The well-mixed approach of Thomson[10] and a hierarchy of constraints by Pedrizzetti and Novikov.[11]

The well-mixed approach of Thomson[10] specifies the model in terms of given Eulerian velocity statistics, via the probability distribution of the random velocity at two points. Typically this probability distribution is taken to be Gaussian. Thus the Eulerian statistics are specified in practice by two-point velocity structure functions that are common turbulent field diagnostics. The importance of Thomson's approach (see also Ref. 2) is that it ensures that when the material distribution is uniform the model equations are satisfied by Eulerian flow statistics.

In contrast to Thomson's approach, Pedrizzetti and Novikov[11] do not assume a certain shape of the Eulerian velocity distribution, but rather construct a stochastic process for the Lagrangian velocity differences and derive the corresponding Fokker–Planck equation for three-dimensional velocity increments (velocity differences between two fixed points in space). The Lagrangian model is then devised in consistency with a hierarchy of exact moment constraints. Our Markov model to be devised in section II.F will qualitatively recover the Pedrizzetti and Novikov model in the inertial subrange, however, with different model parameters. Furthermore, our model is applicable in the viscous range, where we use multiplicative forcing, and it satisfies the first- and second-order constraint on all scales.

Roughly speaking, the Novikov approach ensures exact compatibility with the dynamical equations up to second or third order, whereas Thomson constructs his Lagrangian model in consistency with an assumed Eulerian velocity field to all orders (the Gaussian assumption reduces the effort to the first two moments).

In the following we introduce the Mori–Langevin equation in the context of stochastic models for turbulent dispersion reformulating an equivalent approach of Grossmann and Thomae[3] to the evaluation of the dynamic Lagrangian correlation function. The projector formalism of Mori[2] is a standard method in nonequilibrium statistical mechanics and is presented in many textbooks such as Lindenberg and West.[12]

B. Lagrangian Velocities

For the description of turbulent transport, one often chooses the Lagrangian representation. Here the incompressible velocity field $\boldsymbol{u}(\boldsymbol{x}, t)$ with

$\nabla \cdot \boldsymbol{u} = 0$ is represented by a continuum of fluid particles, i.e., points lying on a trajectory $\boldsymbol{X}(\boldsymbol{x}_0, t)$ where

$$\frac{\mathrm{d}}{\mathrm{d}t}\boldsymbol{X}(\boldsymbol{x}_0,t) = \boldsymbol{u}(\boldsymbol{x}_0,t) = \boldsymbol{u}[\boldsymbol{X}(\boldsymbol{x}_0,t), t + t_0] \tag{1}$$

Each particle is identified by its initial position $\boldsymbol{x}_0$ at time t_0. We consider the relative motion of two fluid particles initially (at time t_0) separated by a distance $\boldsymbol{r}_0$ with instantaneous separation $\boldsymbol{r}(t) = \boldsymbol{X}(\boldsymbol{x}_0 + \boldsymbol{r}_0,t) - \boldsymbol{X}(\boldsymbol{x}_0,t)$. The relative velocity or velocity difference between the two particles is initially $\boldsymbol{v}(\boldsymbol{r}_0) \equiv \boldsymbol{v}(\boldsymbol{r}_0,0) = \boldsymbol{u}(\boldsymbol{x}_0 + \boldsymbol{r}_0, t_0) - \boldsymbol{u}(\boldsymbol{x}_0, t_0)$ and at time t $\boldsymbol{v}(\boldsymbol{r}_0, t) = \boldsymbol{u}[\boldsymbol{X}(\boldsymbol{x}_0 + \boldsymbol{r}_0, t), t + t_0] - \boldsymbol{u}[\boldsymbol{X}(\boldsymbol{x}_0, t), t + t_0]$.

The Navier–Stokes equation in terms of Lagrangian velocity differences can be written as

$$\frac{\mathrm{d}}{\mathrm{d}t}v_i(\boldsymbol{r}_0,t) = \nu\partial_{x_m}\partial_{x_m}v_i(\boldsymbol{r}_0,t) - \partial_{x_i}\pi(\boldsymbol{r}_0,t) \tag{2}$$

with the pressure difference $\pi(\boldsymbol{r}_0,t) = p[\boldsymbol{X}(\boldsymbol{x}_0 + \boldsymbol{r}_0,t),t + t_0] - p[\boldsymbol{X}(\boldsymbol{x}_0,t),t + t_0]$ and kinematic viscosity ν, where the notation is meant to first write $\boldsymbol{v}$ and π as differences and then differentiate with respect to t or $\boldsymbol{x}$. Note that the Navier–Stokes equations formulated in Lagrangian quantities are linear in $\boldsymbol{v}$ and π with the nonlinearity entering implicitly via the Lagrangian paths $\boldsymbol{X}(t)$.

C. Eulerian Velocity Field

Describing the statistical state of the flow in terms of $\boldsymbol{v}$ and π, $\langle..\rangle$ denoting an ensemble average, is equivalent to specifying the static structure functions as for instance $D_{ij}(\boldsymbol{r}) = \langle v_i(\boldsymbol{r})v_j(\boldsymbol{r})\rangle$, or $D_{ijk}(\boldsymbol{r}) = \langle v_i(\boldsymbol{r})v_j(\boldsymbol{r})v_k(\boldsymbol{r})\rangle$. The notation already represents the assumption that the ensemble is locally isotropic, homogeneous, and stationary. Therefore the structure functions do not depend on $\boldsymbol{x}$,t_0 and can be decomposed into longitudinal and transversal parts

$$D_{ij}(\boldsymbol{r}) = D_{NN}(r)(\delta_{ij} - n_i n_j) + D_{LL}(r)n_i n_j \tag{3}$$

with $r = |\boldsymbol{r}|$ and $\boldsymbol{n} = \boldsymbol{r}/r$.

With the rate of energy dissipation per unit volume ε, one defines the Kolmogorov length, time, and velocity scales of the turbulence $\eta = (\nu^3/\varepsilon)^{1/4}$, $\tau_\eta = (\nu/\varepsilon)^{1/2}$,$v_\eta = (\nu\varepsilon)^{1/4}$, respectively. The Taylor–Reynolds number Re_λ as a measure of the turbulence intensity is defined with the root-mean-square turbulence velocity u' and the Taylor microscale $\lambda = u'/(\langle\partial_{x1}u_1\rangle)^{1/2}$.

As is well known from the kinematics of homogeneous and isotropic turbulence, transversal structure functions can be expressed in terms of longitudinal ones because of incompressibility. Hence, the most important statistical quantity is the second-order longitudinal structure function

$D_{LL}(r)$, which is very well approximated by a Batchelor type of interpolation function

$$D_{LL}(r) = 2u'^2\left(\frac{r^2}{r_d^2 + r^2}\right)^{2/3}\left(\frac{r^2}{r^2 + L^2}\right)^{1/3} \tag{4}$$

Here, $r_d = (15b_{LL})^{3/4}\eta$ determines the crossover from the viscous subrange (VSR) to the inertial subrange (ISR) and b_{LL} is the Kolmogorov constant in the ISR. The length scale L in Eq. (4) can be related to the integral scale $L_1 = \int_0^\infty [1 - D_{LL}(r)/2u'^2)dr \approx BL$. For large r ($r_d \approx 0$) the integral can be evaluated exactly $B = \Gamma(\frac{5}{6})\Gamma(\frac{1}{2})/\Gamma(\frac{1}{3}) \approx 0.75$ and B increases slightly for finite r_d.

It is well known that the coefficient A (the normalized energy dissipation) defined as $\varepsilon = A(Re_\lambda)u'^3/L_1$ is Reynolds number dependent for $Re_\lambda \leq 100$ (Ref 13). The experimental data suggest a constant asymptotic value $A(\infty) \approx 0.8 - 1.0$ where saturation takes place around $Re_\lambda \approx 50$. The DNS of Yeung and Pope[14] suggests $A(Re_\lambda) \propto Re_\lambda^{-1/2}$ without a clear saturation in their range of Reynolds numbers ($38 \leq Re_\lambda \leq 93$). Because no consistent picture has yet emerged, we specify our numerical model by assuming $A = B = 0.8$, which fixes the Kolmogorov constant $b_{LL} = 2\left(\frac{B}{A}\right)^{2/3} = 2$, a value that is usually taken in high Reynolds number dispersion studies. Similar to Eq. (4), we employ an interpolation for $D_{NN}(r)$ as well.

The pressure structure function $D_{\pi\pi}(r) = \langle\pi(\mathbf{r})\pi(\mathbf{r})\rangle$ depends only on $r = |r|$ due to local isotropy. Taylor expansion for $r \ll \eta$ yields $D_{\pi\pi}(r) = 1/3 \langle(\nabla p)^2\rangle(r/\eta)^2 v_\eta^4 + O(r^4)$. The inertial range behavior follows from dimensional analysis $D_{\pi\pi}(r) = 2C_p\varepsilon^{4/3}r^{4/3}$. The variance of the pressure gradient $a_p = \frac{1}{3}\langle(\nabla p)^2\rangle$ has been found to be strongly Reynolds-number dependent. Pope[15] suggests $a_p = 3(1 - 22/Re_\lambda)$ as a reasonable fit for the DNS results of Yeung and Pope.[14]

D. Mori–Langevin Equation for Lagrangian Velocities

Following Grossmann and Thomae,[3] we choose the Lagrangian velocity differences $\boldsymbol{v}(\boldsymbol{r}_0,t)$ and pressure differences $\pi(\boldsymbol{r}_0,t)$ as the dynamical quantities of interest. Hence, the space of observables is given as

$$\mathcal{L} = \{\boldsymbol{A}(t) = \boldsymbol{a}[\boldsymbol{X}(\boldsymbol{x}_0 + \boldsymbol{r}_0,t),t + t_0] - \boldsymbol{a}[\boldsymbol{X}(\boldsymbol{x}_0,t),t + t_0]: \langle|\boldsymbol{A}(t)|^2\rangle < \infty\} \tag{5}$$

where $\mathbf{a}(\boldsymbol{x},t)$ is a sum of products of $u_i(\boldsymbol{x}, t)$ and $p(\boldsymbol{x},t)$. The differences $\boldsymbol{v}(\boldsymbol{r}_o, t)$ and $\pi(\boldsymbol{r}_0, t)$ are obviously members of $\mathcal{L}$. The inner product in $\mathcal{L}$ is defined as

$$\langle A_i|A_j\rangle = \langle A_iA_j\rangle \tag{6}$$

where the right-hand side has to be interpreted as an average over the underlying ensemble. The subspace of the static variables or initial values is then defined as

$$\mathcal{L}_A = \{\boldsymbol{A}(0) : \boldsymbol{A}(t) \in \mathcal{L}\} \tag{7}$$

The static or Eulerian structure functions D_{ij} $(\boldsymbol{r}_0)$ and D_{ijk} $(\boldsymbol{r}_0)$ are inner products of elements of $\mathcal{L}_A$. The ensemble average is meant to be taken over different $\boldsymbol{x}_0$ but for a fixed separation $\boldsymbol{r}_0$. The projector P is chosen such as to project on the initial velocity differences $v(\boldsymbol{r}_0)$

$$\mathrm{P}(\boldsymbol{r}_0) = \left| v_i(\boldsymbol{r}_0)\rangle D_{ij}^{-1}(\boldsymbol{r}_0)\langle v_j(\boldsymbol{r}_0) \right| \tag{8}$$

It is plausible that $\mathcal{L}_A \neq \mathcal{L}$, i.e., $\mathcal{L}_A^{\perp} \neq \varnothing$ and $\mathrm{P} \neq \mathbf{1}$ because particle pairs in a turbulent flow on average separate in time. Hence, the development of $\boldsymbol{A}(t)$ is unsteady, although the turbulent velocity field $\boldsymbol{u}$ $(\boldsymbol{x}, t)$ is assumed to be stationary.

The exact Mori–Langevin equation (Ref. 12, Chap. 4.2.2) for the Langragian velocity differences can now be formulated as

$$\frac{\mathrm{d}}{\mathrm{d}t}\boldsymbol{v}(\boldsymbol{r}_0,t) = -\gamma(\boldsymbol{r}_0)\boldsymbol{v}(\boldsymbol{r}_0,t) - \int_0^t m(\boldsymbol{r}_0,t-t')\boldsymbol{v}(\boldsymbol{r}_0,t')\,\mathrm{d}t' + \boldsymbol{f}(\boldsymbol{r}_0,t) \tag{9}$$

with the (nonnormalized) relaxation matrix $\Gamma_{ij}(\boldsymbol{r}_0) = \gamma_{ik}(\boldsymbol{r}_0)\, D_{kj}(\boldsymbol{r}_0)$

$$\Gamma_{ij}(\boldsymbol{r}_0) = -\langle \mathrm{L}v_i(\boldsymbol{r}_0) | v_j(\boldsymbol{r}_0)\rangle \tag{10}$$

the memory matrix

$$M_{ij}(\boldsymbol{r}_0, t) = -\langle \mathrm{L}e^{\mathrm{QL}t}\mathrm{QL}v_i(\boldsymbol{r}_0) | v_j(\boldsymbol{r}_0)\rangle \tag{11}$$

and the residual force

$$\boldsymbol{f}(\boldsymbol{r}_0,t) = e^{\mathrm{QL}t}\mathrm{QL}v(\boldsymbol{r}_0) \tag{12}$$

The residual force $\boldsymbol{f}(t)$, the average of which vanishes by construction, is often taken to be Gaussian, zero centered, and hence completely specified by the correlation matrix given (rigorously in case of a Hermitian L) by the generalized fluctuation-dissipation theorem (FDT)

$$\langle f_i(\boldsymbol{r}_0,t) | f_j(\boldsymbol{r}_0,t')\rangle = M_{ij}(\boldsymbol{r}_0, t - t') \tag{13}$$

The equation presumes that the ensemble distribution is stationary; hence, the dependence of M only on the time difference $t - t'$. We will assume the FDT to be approximately valid for the construction of stochastic dispersion models in Section II.F. Only if we assume the FDT to hold, then exact low-order constraints can be satisfied.

E. Model Coefficients

For practical convenience, the ensemble is assumed to be stationary, homogeneous, locally isotropic, and reflection invariant. The relaxation matrix $\Gamma_{ij}(\boldsymbol{r})$ is hence symmetric and can be evaluated exactly (Ref. 3)

$$\Gamma_{ij}(\boldsymbol{r}) = -\frac{1}{2}\frac{\mathrm{d}}{\mathrm{d}r_m} D_{mij}(\boldsymbol{r}) \tag{14}$$

By means of the Kolmogorov structure equation (Ref. 16), the third-order structure function in Eq. (14) can be expressed in terms of $D_{LL}(\boldsymbol{r})$. Using Eq. (4), the normalized relaxation coefficient in the viscous range reads $\gamma_{LL} = \gamma_{NN} = \gamma_{VR}/\tau_\eta \equiv \frac{56}{3}\,(15b_{LL})^{-3/2}/\tau_\eta$. The leading contribution to Γ_* ($* = NN, LL$) in the inertial range is essentially the energy dissipation rate.

To simplify the numerical calculations and to extract the main contributions, we again approximate Γ_* by interpolation functions

$$\Gamma_*(r) = \frac{2}{3}\frac{(r/\eta)^2}{a_{\Gamma *} + (r/\eta)^2} v_\eta^2/\tau_\eta \tag{15}$$

where $a_{\Gamma LL} = 10/\gamma_{VR}$ and $\alpha_{\Gamma NN} = 5/\gamma_{VR}$. Note that this interpolation for γ_* does not include integral scales, where γ_* vanishes.

The memory matrix in the second-order continued fraction expansion[3] decays exponentially

$$m_{ij}(\boldsymbol{r},t) = k_{LL}(r)e^{-\tilde{\gamma}LL(r)t} n_i n_j + k_{NN}(r)e^{-\tilde{\gamma}NN(r)t}(\delta_{ij} - n_i n_j) \tag{16}$$

Dimensional analysis of the K41 type suggests that the leading contribution to the memory amplitude $K_*(r) = k_*(r)D_*(r)$ in the inertial range is given by the variance of the relative acceleration

$$\left\langle \mathrm{L}v_i \middle| \mathrm{L}v_j \right\rangle = v^2 \left\langle \nabla_x^2 v_i \middle| \nabla_x^2 v_j \right\rangle + \left\langle \partial_{x_i}\pi \middle| \partial_{x_j}\pi \right\rangle \tag{17}$$

$$= \left[v^2 \nabla_r^4 D_{ij} - d_{r_i} d_{r_j} D_{\pi\pi} \right]_0^r \tag{18}$$

where the relative acceleration $\mathrm{L}\boldsymbol{V}(\boldsymbol{r})$ has been replaced by the right-hand side of the Navier–Stokes equation (2) and $\nabla_r^2 = \partial_{r_i}\partial_{r_i}$. This yields in the ISR

$$K_*(r) = \left(\frac{2}{3}\gamma_{VR} + 2a_p\right)\frac{v_\eta^2}{\tau_\eta^2} \tag{19}$$

Also one finds that $\tilde{\gamma}_{LL}$ and $\tilde{\gamma}_{NN}$ are constant (r independent) in the VSR as well as in the ISR, but unfortunately not much more is known about their numerical value and their possible Reynolds number dependence.

A non-Markovian model based on the second-order continued fraction approximation recovers Sawford's second-order auto-regressive model in the limit of large separations where the two particles become statistically independent.[7]

F. Markovian Model

The two main assumptions of our modeling approach are the stochastic properties of the forcing in Eq. (9) and the replacement of the dependence on the initial r_0 by the instantaneous separation $\boldsymbol{r}(t)$. The latter is physically sensible because initial conditions should be forgotten after some time and the nonlinearity of the dynamical problem enters via the paths (see also Ref. 7 where a similar model is derived from the nonlinear Mori theory).

The Markov approximation is obtained by formally setting

$$m_{ij}(\boldsymbol{r},t - s) = 2\delta(t - s)\tilde{m}_{ij}(\boldsymbol{r}) \tag{20}$$

Hence, we get the following dispersion model

$$\dot{v}_i(\boldsymbol{r}_0,t) = -\{\gamma_{ij}[\boldsymbol{r}(t)) + \tilde{m}_{ij}(\boldsymbol{r}(t)]\}v_j(\boldsymbol{r}_0,t) + f_i(\boldsymbol{r}_0,t) \tag{21}$$

with a delta-correlated, zero-centered Gaussian forcing

$$\langle f_i(s)f_j(t)\rangle = 2\delta(t - s)\tilde{m}_{ik}D_{kj} \tag{22}$$

Although the Markov approximation results from the limit $\gamma \to \infty$, an estimate of $\tilde{m}$ could be obtained from

$$m(t) = ke^{-\tilde{\gamma}t} = k\tilde{\gamma}^{-1}(\tilde{\gamma}e^{-\tilde{\gamma}t}) \to 2k\tilde{\gamma}^{-1}\delta(t) \quad \text{as} \quad \tilde{\gamma} \to \infty \tag{23}$$

where γ is given by Eq. (15). Because of the lack of knowledge about $\tilde{\gamma}$ but in accordance with K41 dimensional analysis, we assume $\tilde{\gamma}_{NN} = \tilde{\gamma}_{LL} = \text{const.}$ Also by dimensional analysis $k(r)_{ij}$ seems to have the same r dependence as $\gamma_{ij}(r)$. We therefore simply set

$$\tilde{m}_{ij}(\boldsymbol{r}) = a(Re_\lambda)\gamma_{ij}(\boldsymbol{r}) \tag{24}$$

where $\alpha(Re_\lambda)$ is considered a free-model parameter fixed according to the DNS data rather than estimating $\alpha(Re_\lambda)$ from Eq. (23).

Further support for our Markov model can be gained from the hierarchy of exact constraints for not necessarily Gaussian but Markovian models given in Pedrizzetti and Novikov.[11] The first constraint is essentially the incompressibility condition that is for our model trivially satisfied as it is for their model. The second constraint reads in our notation with $\boldsymbol{v}(\boldsymbol{r}_0, t) = \boldsymbol{v}(\boldsymbol{r})$

$$-\frac{\mathrm{d}}{\mathrm{d}r_k}\langle v_k(\boldsymbol{r})v_i(\boldsymbol{r})v_j(\boldsymbol{r})\rangle + \langle v_i(\boldsymbol{r})a_j(\boldsymbol{r},\boldsymbol{v})\rangle + \langle a_i(\boldsymbol{r},\boldsymbol{v})v_j(\boldsymbol{r})\rangle = -\langle m_{ij}(\boldsymbol{r},\boldsymbol{v})\rangle \quad (25)$$

where in our case

$$\langle m_{ij}(\boldsymbol{r},\boldsymbol{v})\rangle = 2\tilde{M}_{ij}(\boldsymbol{r}) \quad (26)$$

and

$$a_i(\boldsymbol{r},\boldsymbol{v}) = -(\gamma_{ij}(\boldsymbol{r}) + \tilde{m}_{ij}(\boldsymbol{r}))\, v_j(\boldsymbol{r}) \quad (27)$$

Because the first term in Eq. (25) is exactly $2\Gamma_{ij}(\boldsymbol{r})$ and all matrices in our model are symmetric, we see immediately that our model satisfies the second constraint.

It is important to note that the second-order constraint does not give any information about the memory. Although we know the memory only approximately, the second-order constraints can be fulfilled exactly because the memory does not contribute. On the other hand, if we had not used the FDT (13), the second-order constraint would fix the force correlation accordingly. In Heppe[7] we prove directly that for an isotropic ensemble the model should satisfy the FDT.

III. Numerical Simulations

The solution of the model in the Markov approximation (21) can be computed affordably provided we know how to handle the multiplicative noise, i.e., the force depending on the stochastic process itself. To find a stochastic force with correlation (22), we employ the decomposition property of isotropic matrices $A = A_{NN}P + A_{LL}N$, where $P_{ij} = \delta_{ij} - n_i n_j$ and $N_{ij} = n_i n_j$. Because of $PN = NP = 0$, we can easily take the square root of A. A Gaussian white noise with correlation $\langle \mathrm{d}f_i \mathrm{d}f_j\rangle = A_{ij}\mathrm{d}t$ can hence be generated as

$$\mathrm{d}f_i = \left(\sqrt{A_{NN}}P_{ij} + \sqrt{A_{LL}}N_{ij}\right)\sqrt{\mathrm{d}t}\,\xi_j \quad (28)$$

where the ξ_i are independent Gaussian random variables with variance one and mean zero.

The well-known forward Euler method is used for numerical time stepping of the stochastic differential equations (SDE) with a variable step-size comparable to the one of Thomson.[10] Note that for SDE with multiplicative noise one builds in the interpretation of the stochastic integrals (Ito or Stratonovich) within the numerical scheme. The Ito–Stratonovich dilemma does not enter the stage here because the so-called spurious drift vanishes.

Mean-Square Dispersion

To compare the statistics of our model with the two-particle DNS data of Yeung[9] at the moderate Reynolds number $Re_\lambda = 91$, we match the long time behavior of the Markov model with the well-known Taylor result for $t \gg T_L$

$$\langle(\boldsymbol{r}(t) - \boldsymbol{r}_0)^2\rangle = 12u'^2 T_L t \tag{29}$$

which yields $12u'^2T_L/(v_\eta^2\tau_\eta) = 36u'^4/(av_\eta'^4)$ and hence $a = \dfrac{3u'^2/v_\eta^2}{T_L/\tau_\eta} \approx 8.0$ where we have used the DNS data $u'^2/v_\eta^2 = 23.2$ and $T_L/\tau_\eta = 8.7$. The integral length scale of the DNS is $L_1/\eta = 56.0$.

There is great uncertainty about the numerical value of the Lagrangian Kolmogorov constant C_0, which is the inertial range coefficient of the Lagrangian structure function $D(\tau) = C_0\varepsilon\tau$. Estimates from DNS, experiments, and random walk dispersion studies vary in the range $C_0 = 2-7$ (Ref. 15) possibly due to a strong Reynolds number dependence. In the inertial range the amplitude of the stochastic force should become $\sqrt{2C_0\varepsilon}$, which gives a reasonable value $2a/3 = C_0 = 5.3$.

Figure 1 shows the growth of the root-mean-square distribution $\langle(\boldsymbol{r}(t) - \boldsymbol{r}_0)^2\rangle^{1/2}$ in Kolmogorov scaling at $Re_\lambda = 91$ compared with DNS data for different initial separations. Over sufficiently small times ($t \ll 10\tau_\eta$) the fluid particles move in straight lines, with no appreciable change in their relative velocity. Here, we simply give all particles the same initial separation $r_0/\sqrt{3}\,(1,1,1)^T$ and velocity $\boldsymbol{v}(\boldsymbol{r}_0, 0) = \sqrt{D_{ii}(\boldsymbol{r}_0)/3}\,(1,1,1)^T$, which approximates the short time behavior of the DNS sufficiently well. For small times $t \lesssim T_L$, the model gives a surprisingly accurate description of the dispersion, also as expected for very long times. In the transition region between viscous and integral scales, however, the Markov model over-predicts the dispersion. The discrepancy is hardly surprising because only a little is known about the (possibly nonuniversal) transition to integral scales.

For intermediate times ($\tau_\eta \ll t \ll T_L$) at high Reynolds numbers, one expects a Richardson law

$$\langle \boldsymbol{r}^2(t)\rangle = G_\Delta \varepsilon t^3 \tag{30}$$

Note that there is no indication of a Richardson law at the moderate Re_λ considered in the DNS. The curves for small initial separations in Fig. 1 show a growth even stronger than $t^{3/2}$. This growth is due to an exponential

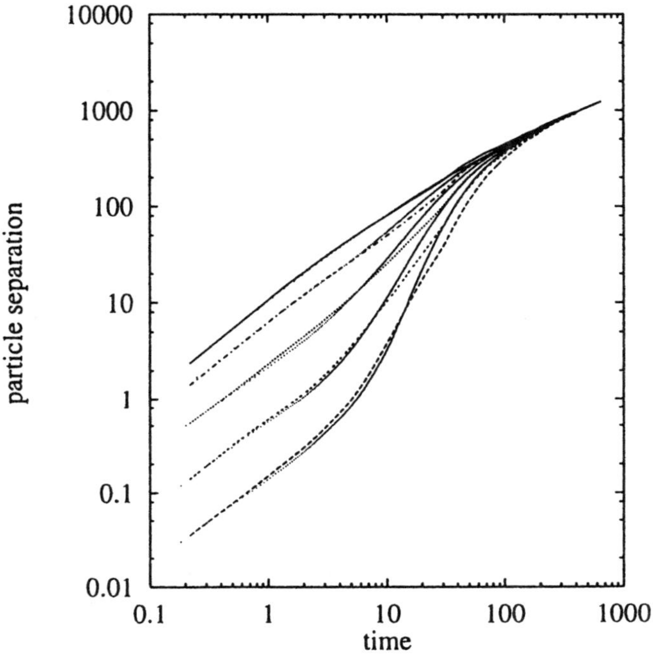

Fig. 1 Relative separation $\langle [r(t) - r_0]^2 \rangle^{1/2}/\eta$ versus time t/τ_η in Kolmogorov scaling for two particles initially separated by 1/4, 1, 4, 16, 32, 64 η (from bottom to top), respectively, $Re_\lambda = 91$. Lines show DNS data, model results are dotted.

transition region between viscous and inertial subrange previously identified by Grossmann and Procaccia.[17]

At a much higher Reynolds number $Re_\lambda = 10^4$ (far beyond the reach of DNS), Fig. 2 shows the $t^{3/2}$ law that extends over roughly two decades with a Richardson constant G_Δ $(a = 8) = 0.45$. For comparison we find G_Δ $(a = 4) = 0.56$ and G_Δ $(a = 12) = 0.31$. This is larger than the experimental value $0.06 - 0.4$, which is highly uncertain.[16] Kinematic simulations that so far do not include a viscous range and are computationally much more expensive than our random flight simulations give $G_\Delta = O(0.1)$ (Ref. 18). However, for a fixed value of C_0, our G_Δ is smaller than those given by well-mixed stochastic models [$O(1)$, see, e.g., Ref. 19] and those of other two-point closure theories such as eddy-damped quasinormal Markovian (EDQNM) ($\approx 1.5 - 3.5$). Note that Grossmann and Procaccia,[17] who also apply the Mori–Zwanzig projector formalism but completely neglect memory effects, obtain a very large $G_\Delta = 13$.

As in the DNS, we find that the particle separation vectors at large times are Gaussian to a very good approximation, which is reflected in a chi-square distribution of the probability density function (PDF) of the magnitude of the separation.

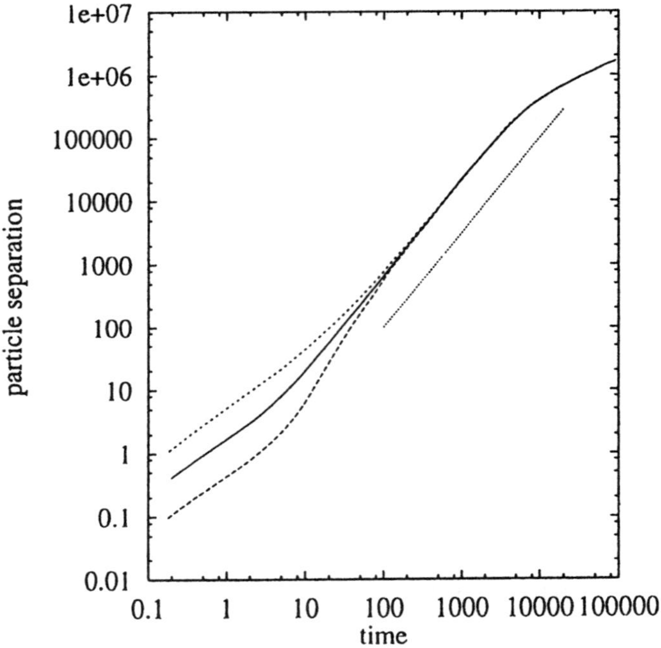

Fig. 2 Relative separation $\langle [r(t) - r]^2 \rangle^{1/2}/\eta$ versus time t/τ_η for two particles initially separated by 1, 4, 16_η (from bottom to top), respectively. For comparison a line with slope 1.5, $Re_\lambda = 10^4$.

Correlation and Alignment Statistics

Further separation and velocity statistics of Eq. (21) are discussed in greater detail in Heppe.[7] Here we concentrate on the failure of the Markov assumption (22) for short times. Figure 3 displays the variance $\langle \Delta v_i \Delta v_i \rangle$ of relative velocity increments $\Delta \boldsymbol{v} = \boldsymbol{v}(t) - \boldsymbol{v}(0)$, which for $r_0 \gg L_1$ should equal the one-particle Lagrangian structure function $2\langle \Delta u_i \Delta u_i \rangle = 6D(t)$. Whereas the long-time behavior of the Markov model coincides with that of the DNS, the short-time behavior proportional to t is qualitatively in error with the physical $D(t) \propto t^2$ of the DNS. In principle, this flaw could be remedied by taking a non-Markovian model based on Eq. (16) (cf., Refs. 7 and 8). The non-Markovian model displayed in Fig. 3 is the one of Sawford.[6]

Finally, we investigate the probability of the alignment between the separation $\boldsymbol{r}(t)$ and velocity vector $\boldsymbol{v}(t)$, which determines whether the particle pairs are instantaneously moving apart or coming closer to each other. An acute alignment angle θ means the particles are moving apart, i.e.,

$$\cos\theta = \frac{1}{2}\frac{\mathrm{d}}{\mathrm{d}t}|\boldsymbol{r}|^2/|\boldsymbol{v}||\boldsymbol{r}| = \boldsymbol{r}\cdot\boldsymbol{u}/|\boldsymbol{v}||\boldsymbol{r}| > 0. \tag{31}$$

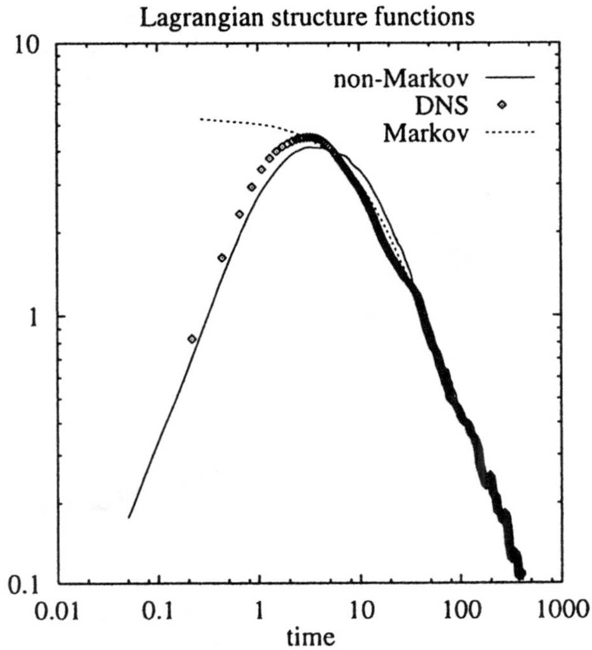

Fig. 3 Variances of relative velocity increments $\langle \Delta v_i \Delta v_i \rangle /(6\varepsilon t)$ versus t/τ_η for particles initially separated by $r_0/\eta = 64$ ($Re_\lambda = 91$) for the Markovian, the non-Markovian stochastic model, and the DNS, respectively.

The PDF of acute alignment $P(0 < \theta < 90 \text{ deg})$ compared to DNS data for $r_0 = 1/4\eta$ is shown in Fig. 4. A preferred alignment begins to develop as the particle pairs separate from each other, particularly strong at small times ($t \leq 10T_L$). The very slow decrease of alignment at large times is well captured.

In conclusion, the Markov model seems to give a reasonable picture of small-scale moderate Reynolds number dispersion. The very short-time behavior could be improved by taking non-Markovian effects into account, which, however, would increase the computational cost considerably. Further improvement might come from the inclusion of intermittency effects, which, however, would introduce further arbitrariness to the model.

IV. Conclusions

We have proposed a novel theory of stochastic Lagrangian models for two-particle turbulent dispersion and mixing based on the Navier–Stokes and the passive scalar field equations. Markovian models of the relative particle motion and the evolution of concentration differences have been suggested.

The advantages of our new models are that they account for finite

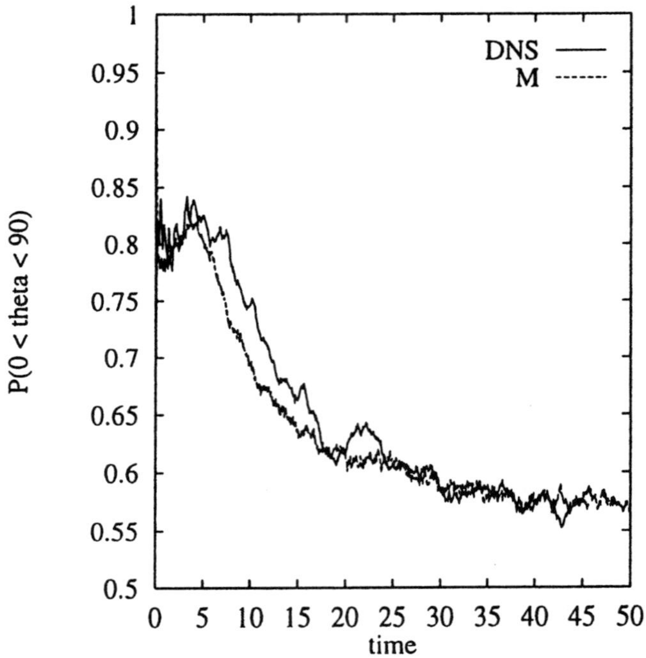

Fig. 4 Probability of acute alignment between separation and relative velocity vectors $P\,(0 < \theta < 90$ deg) for particles initially separated by $r_0 = 1/4\eta$ compared to DNS data ($Re_\lambda = 91$).

Reynolds and Peclet number effects, incompressibility, and other exact constraints. Their validity is not restricted to the case where the particle separation lies in the inertial range. The Markovian dispersion model qualitatively explains observation of two-particle dispersion from direct numerical simulations at moderate Reynolds number, whereas the mixing model sheds new light on a recent closure for higher order scalar structure functions.

Acknowledgments

I am grateful to my Ph.D. supervisor, J. C. Vassilicos, for his comments and support. Several fruitful discussions with J. C. R. Hunt and D. J. Thomson are gratefully acknowledged. The author thanks P. K. Yeung for making available his DNS data. The work has been supported by the Gottlieb Daimler- und Karl Benz-Stiftung, Germany, and the British National Environment Research Council.

Appendix: The Passive Scalar

In this section we sketch how to apply our projector approach to stochastic modeling of Lagrangian concentration differences. Neglecting memory effects completely, we find exactly the scalar probability density closure of Kraichnan,[5] which successfully compares to DNS (Ref. 9) and seems to give a reasonable account even of higher order scalar structure functions showing anomalous (non-Kolmogorov) scaling.

We consider the concentration difference between two fluid particles initially at time t_0 separated by a distance $\boldsymbol{r}_0$ denoted as $\Delta\varphi(\boldsymbol{r}_0, t) = \varphi[\boldsymbol{X}(\boldsymbol{x}_0 + \boldsymbol{r}_0, t)] - \varphi[\boldsymbol{X}(\boldsymbol{x}_0, t)]$, which evolves by the advection-diffusion equation

$$\frac{\mathrm{d}}{\mathrm{d}t}\Delta\varphi(\boldsymbol{r}_0,t) = \kappa\nabla_x^2\Delta\varphi(\boldsymbol{r}_0,t) \tag{A.1}$$

where κ is the molecular diffusivity.

To make analytical progress, we have to choose the simplest model for the convection of fluid particles, namely the Markov model of Section II.F in the diffusion limit.[19] In a given flow, in which u' and ε are specified we identify this diffusion limit with the limit $a \to \infty$. Note that this is only for mathematical convenience; for the implementation of a Monte–Carlo solution algorithm, the full Markov model might give more reliable results without much additional numerical effort.

Neglecting memory effects with $\boldsymbol{r}(t) = \boldsymbol{r}(\boldsymbol{r}_0, t), \Delta\varphi(t) = \Delta\varphi(\boldsymbol{r}_0, t)$, we obtain the mixing model

$$\mathrm{d}r_i(t) = \frac{1}{2}\sqrt{\eta[\boldsymbol{r}(t)]_{kj}}\,\partial_{r_k}\sqrt{\eta[\boldsymbol{r}(t)]_{ij}}\quad \mathrm{d}t + \sqrt{\eta[\boldsymbol{r}(t)]_{ij}}\;\mathrm{d}W_j(t) \tag{A.2}$$

$$\mathrm{d}\Delta\varphi(t) = -\gamma_\varphi[r,t]\Delta\varphi(t)\,\mathrm{d}t \tag{A.3}$$

where in our case the eddy diffusivity $\eta_*(\boldsymbol{r}) \propto \gamma_*{}^{-1}(\boldsymbol{r})D_*(\boldsymbol{r})$. The memory of the scalar differences has been neglected.

Note that once more we have used the closure assumption $\boldsymbol{r}_0 \to \boldsymbol{r}(t)$ (see Section II.F, Ref. 7). The relaxation coefficient $\gamma_\varphi(r)$ is given as

$$\gamma_\varphi(r) = -\frac{\langle\Delta\varphi(\boldsymbol{r})\mathrm{L}_\varphi\Delta\varphi(\boldsymbol{r})\rangle}{\langle[\Delta\varphi(\boldsymbol{r})]^2\rangle} = -\kappa\frac{\nabla_r^2 S_2(r) - \nabla_r^2 S_2(0)}{S_2(r)} \tag{A.4}$$

where $\nabla_r^2 F(r) = \partial_{r_i}\partial_{r_i}F = 1/r^2\partial_r(r^2\partial_r F)$ due to isotropy. This can be seen either from inserting the right-hand side of Eq. (32) in Eq. (35), or by using an explicit representation of the time evolution operator analogously to the procedure for the evaluation of γ. Then, by making use of the Yaglom equation one also arrives at Eq. (35).

The Fokker–Planck equation according to the model (33) and (34) can now be derived for the Lagrangian PDF $P_L(\Delta\varphi,\boldsymbol{r},t|\boldsymbol{r}_0,t_0)$. By means of the relation $P_E(\Delta\varphi|\boldsymbol{r},t) = \int P_L\,(\Delta\varphi,\boldsymbol{r},t|\boldsymbol{r}_0,t_0)d\boldsymbol{r}_0$ (Ref. 11), this Fokker–Planck equation is also obeyed by the Eulerian PDF $P_E(\Delta\varphi|\boldsymbol{r},t)$ of the concentration difference $\Delta\varphi$ at time t between two fixed points at distance $\boldsymbol{r}$

$$\partial_t P_E - /r^2 \partial_r (r^2 \eta_{LL}(r) \partial_r P_E) = \partial_{\Delta\varphi}(\gamma_\varphi \Delta\varphi P_E) \tag{A.5}$$

Isotropy implies that P_E depends only on $r = |\boldsymbol{r}|$. The fact that P_E satisfies Eq. (36) is a direct consequence of the closure assumption $\boldsymbol{r}_0 \rightarrow \boldsymbol{r}$ since then the model coefficients are independent of $\boldsymbol{r}_0$. Multiplying by $(\Delta\varphi)^{2n}$ and integrating over $\Delta\varphi$ gives a differential equation for the even order scalar structure functions $S_{2n}(r) = \langle(\Delta\varphi)^{2n}\rangle$ yielding anomalous inertia range scaling $S_{2n}(r) \propto r^{\zeta_{2n}}$ with $\zeta_{2n} \propto \sqrt{n}$. This is exactly the Kraichnan PDF closure[59] here derived from a Lagrangian approach. At second order the well-approved Batchelor phenomenology for scalar structure functions is recovered.

Note that this deterministic model cannot be valid in the far viscous range where regular scaling $\zeta_{2n} = 2n$ is expected. The correct behavior in the viscous range can be obtained by including memory effects in the mixing model analogous to the dispersion model in Section II.F, which leaves the (exact) second-order equation unaltered. At finite Peclet number the inclusion of memory effects restores Kolmogorov scaling for $S_{2n}(r)$ in the inertial range. However, a closer investigation of the memory suggests a decrease with Peclet number such that anomalous scaling actually reflects the asymptotic inertial scaling. The implication for scalar intermittency requires further investigation.

References

[1]Avellaneda, M., and Majda, A. J., "Simple Examples with Features of Renormalisation of Turbulent Transport," *Philosophical Transactions of the Royal Society of London, Series A*, Vol. 346, 1994, p. 205.

[2]Mori, H., "Transport, Collective Motion, and Brownian Motion," *Progress of Theoretical Physics*, Vol. 33, 1965, p. 423.

[3]Grossmann, S., and Thomae, S., "Correlation Decay of Lagrangian Velocity Differences in Locally Isotropic Turbulence," *Z. Phys. B—Condensed Matter,* Vol. 49, 1982, pp. 253–261.

[4]Kraichnan, R. H., Yakhot, V., and Chen, S., "Scaling Relations for a Randomly Advected Passive Scalar Field," *Physical Review Letters*, Vol. 75, 1995, p. 240.

[5]Kraichnan, R. H., "Anomalous Scaling of a Randomly Advected Passive Scalar," *Physical Review Letters*, Vol. 72, 1994, p. 1016.

[6]Sawford, B. L., "Reynolds Number Effects in Lagrangian Stochastic Models of Turbulent Dispersion," *Physics of Fluids*, Vol. 143, 1991, pp. 1577–1586.

[7]Heppe, B. M. O., "Generalized Langevin Equation for Relative Turbulent Dispersion," *Journal of Fluid Mechanics* (submitted for publication).

[8]Heppe, B. M. O., "Lagrangian Correlations and Stochastic Models in Homogeneous Turbulence," *Phys. Rev. E* (submitted for publication).

[9]Yeung, P. K., "Direct Numerical Simulation of Two-Particle Relative Diffusion in Isotropic Turbulence," *Physics of Fluids*, Vol. 6, 1994, p. 3416.

[10]Thomson, D. J., "A Stochastic Model for the Motion of Particle Pair in Isotropic

High-Reynolds-Number Turbulence, and Its Application to the Problem of Concentration Variance," *Journal of Fluid Mechanics*, Vol. 210, 1990 p. 113.

[11]Pedrizzetti, G., and Novikov, E. A., "On Markov Modelling of Turbulence," *Journal of Fluid Mechanics,* Vol. 280, 1994, pp. 69–93.

[12]Lindenberg, K., and West, B. J., *The Non-Equilibrium Statistical Mechanics of Open and Closed Systems*, VCH Publishers, New York, 1990.

[13]Lohse, D., "Crossover from High to Low Reynolds Number Turbulence," *Physical Review Letters*, Vol. 73, 1995, p. 3223.

[14]Yeung, P. K., and Pope, S. B., "Lagrangian Statistics from Direct Numerical Simulation of Isotropic Turbulence," *Journal of Fluid Mechanics,* Vol. 207, 1989, p. 531.

[15]Pope, S. B., "Lagrangian PDF Methods for Turbulent Flows," *Annual Review of Fluid Mechanics,* Vol. 26, 1994, pp. 23–63.

[16]Monin, A. S., and Yaglom, *Statistical Fluid Mechanics, Vol. II*, MIT Press, Cambridge, MA, 1975.

[17]Grossmann, S., and Procaccia, I., "Unified Theory of Relative Turbulent Diffusion," *Physical Review A; General Physics*, Vol. 29, 1984, p. 1358.

[18]Elliott, F. W., and Majda, A. J. "Pair Dispersion Over an Inertial Range Spanning Many Decades," *Physics of Fluids*, Vol. 8, 1996, p. 1052.

[19]Borgas, M. S., and Sawford, B. L., "A Family of Stochastic Models for Two-Particle Dispersion in Isotropic Homogeneous Stationary Turbulence," *Journal of Fluid Mechanics*, Vol. 279, 1994, p. 69.

Chapter 12

Contribution to the Analysis of Energy Spectrum and Transport Phenomena in a Turbulent Two-Phase Flow

J. Dueck* and Th. Neesse†
University Erlangen-Nuremberg, Erlangen, D-91058, Germany

I. Introduction

According to current knowledge,[1–3] fully developed turbulence can be understood by the overlapping motion of turbulent elements (eddies). The eddies can be arranged in accordance with their size λ and their velocity $u(\lambda)$ relative to the mean flow velocity. Energy is transferred without loss from larger to smaller eddies in the subrange of inertia, the latter being characterized by eddy sizes $\lambda_1 \leq \lambda \leq \lambda_1$. The eddies spectral distribution of velocities can be determined by the help of dimensional analysis[1]:

$$u(l) = u_1(l/l_1)^{1/3} \tag{1}$$

The translatory movement of particles in a turbulent flow has often been investigated.[4,5] Contrarily, little attention has been paid to the influence of particles on the inner structure of turbulence. Bujewich and Gupalo[6] came to the result that damping of turbulence occurs over the whole range of the spectrum and as a consequence of this, the general structure of turbulence does not change. This result could be confirmed by measurements of fine suspensions.[7] However, these results have to be looked at critically. By measuring in the range of high frequencies, certain difficulties occur with the type of measurement technique used in the experiments (piezo-probe). Hot-wire anemometry can be criticized for the same reasons.[8]

Information about a change of turbulent structure in the presence of particles can also be found in the literature.[9,10] Furthermore, it was discov-

*Lehrstuhl für Umweltverfahrenstechnik und Recycling, Schottkystr. 10.
†Professor, Lehrstuhl für Umweltverfahrenstechnik und Recycling, Schottkystr. 10.

ered and explained theoretically that particles present in a turbulent flow not only damp[9,11,12] but also can enhance turbulent energy under certain conditions.[13–18] Jarin and Hetsroni state that "this effect manifests itself . . . also in changing the energy spectra—mostly a decrease in the spectral components at high-frequency and a corresponding change in the distribution of the fluctuations energy."[18] Here particles of different size have different influences on the energy spectrum.[10,17] In a very interesting paper of Al-Taweel and Landau,[16] the energy transfer in the spectrum was described. The energy consumption caused by inertia of the particles was taken into account, but no physical model was created for the spectrum. In addition to this, a work of Schwab[19] can be mentioned, where the influence of particles present in the turbulent flow was considered for the energy analysis of turbulence. The analysis is carried out without using the connection of particle motion and fluid motion. Consequently, no quantitative conclusions can be drawn from the theoretical models.

The following remarks are trying to approach these problems and to overcome some of the difficulties mentioned above. Additional dimensional parameters have to be taken into account for calculations of turbulent flow because of the presence of particles: particle size d_p, particle density ρ_p, and mass concentration m of the particles. Thus the assumption of constant dissipation is no longer sufficient for statements about the structure of turbulence. It is necessary to develop a model that is able to describe the interaction of the two phases. Knowledge about the spectral structure of turbulence is also important for calculation of diffusion processes. With the following remarks, importance is attached to the clearness of the developed model.

II. Cascade Model

A certain cascade is formed by eddies with index n. This means that kinetic energy of eddies with index n is transferred to eddies with index $n+1$. Only the fluid takes part in the energy transfer by pressure fluctuations, whereas the total energy comprises a fluid part $V_f\rho_f u_n^2$ and a particle part $V_p\rho_p w_n^2$. The turnover time $t_n = l_n/u_n$ of a certain eddy n decreases with increasing n. The dissipation for an eddy n that is used to create the energy of an eddy $n+1$ within a certain time $t_{n+1} = l_{n+1}/u_{n+1}$ is $\rho_f V_f u_n^3/l_n$. An energy balance leads to

$$\frac{\rho_f V_f u_n^3}{l_n} t_{n+1} = V_f\rho_f u_{n+1}^2 + V_p\rho_p w_{n+1}^2$$

The relation of two neighboring levels of the cascade is considered in this equation. More complicated formulation with three participating eddies can be found, e.g., in Refs. 20 and 21.

Finally, we get the fundamental equation

$$\frac{u_{n+1}}{u_n} = \left(\frac{l_{n+1}}{l_n}\right)^{1/3} y_{n+1} - \frac{1}{3} \tag{2}$$

where

$$y_n = 1 + \frac{V_p \rho_p}{V_f \rho_f}\left(\frac{w_n}{u_n}\right)^2 \tag{3}$$

can be calculated by solving the particle equation of motion. The particle component w_n is given by

$$\frac{w_n}{u_n} = \frac{1}{\sqrt{1 + (f_n t_p)^2}} \tag{4}$$

where the frequencies can directly be substituted by the eddy sizes:

$$\frac{f_{n+1}}{f_1} = \frac{\lambda_1}{\lambda_{n+1}} \tag{5}$$

The relaxation time t_p for the range of Stokes' law is $t_p = d_p^2 \rho_p / 18 n \rho_f$, where $m = V_p \rho_p / V_f \rho_f$ and $\delta = (f_1 t_p)^2$.

Taking into account Eqs. (2–4), the spectral energy distribution can be calculated with Eq. (2). Presuming logarithmic spacing of eddy sizes leads to

$$\frac{\lambda_{n+1}}{\lambda_n} = q = \text{const} \tag{6}$$

or

$$\frac{\lambda_{n+1}}{\lambda_1} = q^n \tag{7}$$

If we assume splitting of an eddy n into two eddies with index $n + 1$, we obtain $q = \sqrt[3]{0.5} \approx 0.8$. V. Weizsäcker[22] presumed another distribution of eddy sizes, which corresponds to $q = 0.5$. Frisch et al.[23] proposed to introduce a function $q(n)$ with $q(n) \propto a\, q(n-1)$. According to Ref. 23, one obtains $a \approx 0.75$, Eq. (6) results for $a = 1$. With Eq. (7) n is given by

$$n = \ln(\lambda_{n+1}/\lambda_1)/\ln q = -\ln(k_{n+1}/k_1)/\ln q$$

Now Eq. (2) can thus be written as

$$\frac{u_{n+1}}{u_n} = q^{1/3}\left(1 + \frac{m}{1+\delta q^{-2n}}\right)^{-1/3} \tag{8}$$

It is obvious that in a two-phase flow the relation of the velocities of two neighboring eddies of a cascade [right-hand side of Eq. (8)] cannot exceed the coefficient of an unloaded flow.

III. Energy Spectrum

The solution of Eq. (8) for u_{n+1}^2 expressed by u_1 is

$$ln\left(\frac{u_{n+1}}{u_1}\right)^2 = nlnq^{2/3} - \frac{2}{3}\sum_{j=1}^{n} ln\left(1 + \frac{m}{1+\delta q^{-2j}}\right)$$

where the sum can be estimated as

$$nln\left[1 + m\left(1 + \frac{\delta}{q^2(q^2 + n(1-q^2))}\right)^{-1}\right]$$

The energy distribution can now be expressed as follows:

$$\left(\frac{u_{n+1}}{u_1}\right)^2 = q^{2n/3}\left[1 + m\left(1 + \frac{\delta}{q^2[q^2 + n(1-q^2)]}\right)^{-1}\right]^{-2n/3} \tag{9}$$

For a unloaded flow, Eq. (1) results from Eq. (9).

In Fig. 1 the intensity distribution of velocities is depicted as a function of wave numbers. For all curves the mass fraction m is constant and δ, a measure for particle size, changes. The structure of the spectrum is affected merely by small particles. Large particles are not activated by turbulence because of their inertia and consequently no interaction of the two phases occurs.

Taking into account the fact that δ is the relation of eddy turnover time to its relaxation time, gets plausible that damping of turbulence mainly occurs in the high frequency part of the spectrum. Authors[11,17,24] pointed out the great importance of the relation of these two characteristic times for the structure of the spectrum. With measurements of the energy spectrum of particulate two phase flows, Tsuj et al.[10] came to the result that the spectrum is not affected by large particles. Similar results were obtained by authors of theoretical and numerical publications.[25,26]

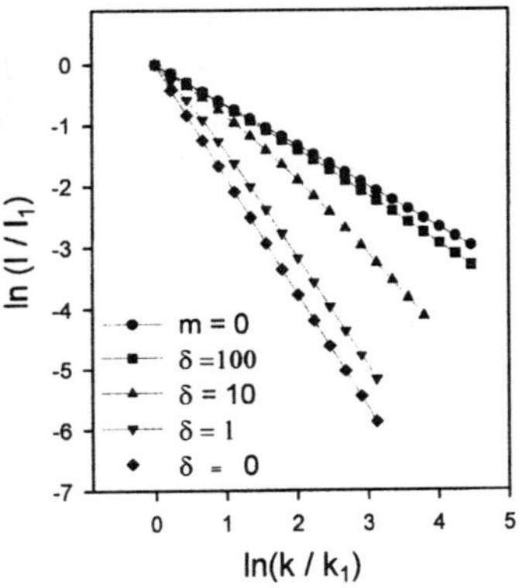

Fig. 1 Spectral intensity distribution, m = 0.5.

The velocity distribution of the carrier fluid (liquid) for δ = const (constant particle size) is depicted in Fig. 2. The higher the particle mass fraction, the higher the decrease of velocities will be. This tendency was expected and is confirmed by some experimental[9,17] and numerical works.[26] For the description of the spectrum with the termini energy density and wave number, the formula for $I = u_{n+1}^2$ can be expressed as follows:

$$I = I_1 k_1^{2/3} k^{-2/3}\left[1 + m\left(1 + \{\delta/q^2\,[q^2 + n(1-q^2)]\}\right)^{-1}\right]^{-2n/3}$$

where $n = -ln(k/k_1)/lnq$. For $m \ll 1$ and $n \ll q^2(1 - q^2)^{-1}$, the following expression is obtained for the energy density $E(k) = |dI/dk|$:

$$lnE(k) \propto -\frac{5}{3}\left(1 + \frac{2m}{5(1+\delta q^{-4})\,ln(1/q)}\right) lnk$$

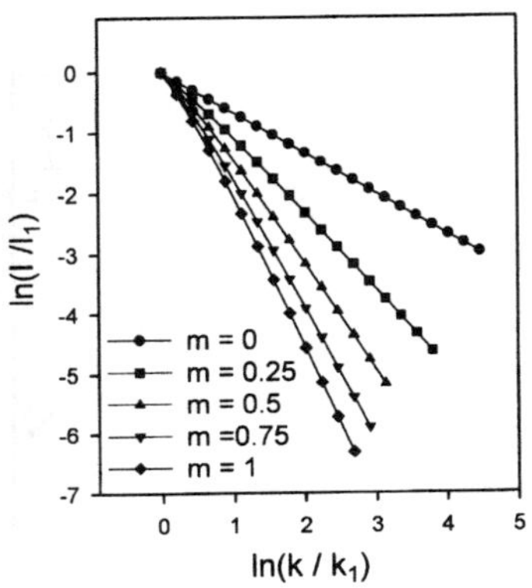

Fig. 2 Spectral intensity distribution, δ = 1.

This means that in a plot *ln(E)* versus *ln(k)* the slopes of the straight lines increase with increasing particle mass fraction. For larger particles on the other hand, the dependency of energy density and wave number is less distinctive. These results are confirmed by experimental investigations.[9]

IV. Diffusion Processes

For the following remarks it is assumed that a dispersing particle is sequentially affected by eddies of different sizes. Consequently, the averaged, squared particle-distance $\langle x^2 \rangle$ from an arbitrary starting point can be calculated by the square of each eddy size $\langle x_n^2 \rangle$. The dispersion coefficient can be calculated according to

$$D_{ep} = \frac{\sum \langle x_n^2 \rangle}{\sum t_n} = \frac{\sum w_n^2 t_n^2}{\sum t_n} = \lambda_1 u_1 \left[\sum_{n=1}^{l} \left(\frac{w_n}{u_n} \right)^2 \left(\frac{\lambda_n}{\lambda_1} \right)^2 \right] \left[\sum_{n=1}^{l} \left(\frac{\lambda_n}{\lambda_1} \right) \left(\frac{u_1}{u_n} \right) \right]^{-1} \quad (10)$$

Here λ_1 can be calculated assuming $Re(\lambda_1) = u_l \lambda_1 / \nu = 1$. Values of $D_{ep}/u_1\lambda_1$ that were calculated by Eq. (4) for $q = 0.8$, $Re_1 = 500$, and different m and δ are depicted in Fig. 3.

A simple formula is obtained for migration of one particle. Assuming $m = 0$ results in

$$\frac{D_{ep}}{u_1 \lambda_1} = \left(\sum_{n=1}^{l} \frac{q^{3(n-1)}}{\delta + q^{2(n-1)}} \right) \cdot \left(\sum_{n=1}^{l} q^{2n/3} \right)^{-1}$$

Substituting the quickly converging series by integration according to $\Sigma Z(n) \approx \int Z(n)\, dn$ leads to the equation

$$\frac{D_{ep}}{u_1 \lambda_1} = \frac{0.4}{1 + 1.6\delta} \quad (11)$$

which is relatively exact and simplifies the following remarks.

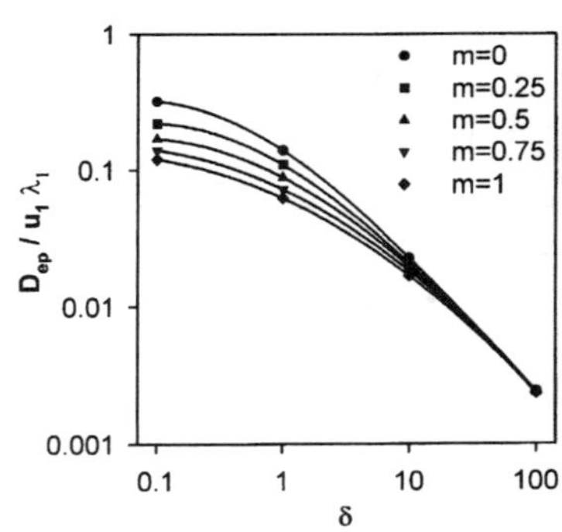

Fig. 3 Values of $D_{ep}/u_1\lambda_1$ for different particle mass fractions *m* and particle sizes *δ*.

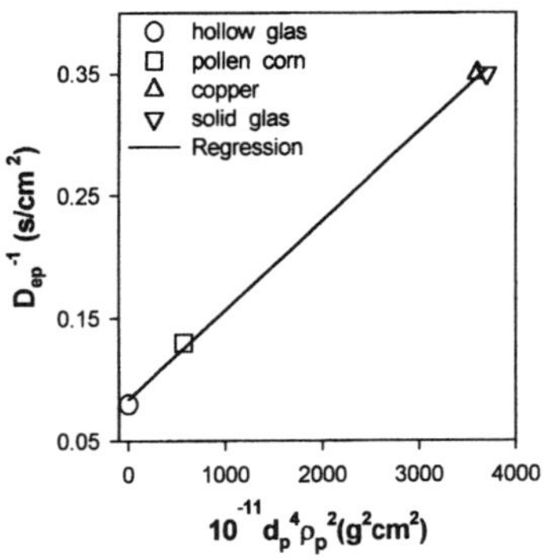

Fig. 4 Data according to Ref. 27.

Snyder and Lumley[27] describe measurements of turbulent dispersion coefficients of single spherical particles with several diameters and densities. In accordance with Eq. (11), D_{ep}^{-1} should be proportional to $d_p^4 \rho_p^2$.

A straight line can be fitted satisfactorily through data presented in Ref. 27 as it is shown in Fig. 4. However, there are results contrary to formula (11). According to experiments of Arnason,[28] the turbulent dispersion coefficient increases with increasing particle size (cited from Ref. 29).

V. Summary and Conclusions

The model of cascade structure of turbulence, well known for turbulence without particle load, has been modified for turbulent two-phase flow. From the two-phase cascade model the presence of suspended particles results in a restriction of the inertia subrange of turbulence. This phenomenon is more significant for high mass fractions of particles and for small particle sizes. Calculations have been carried out concerning the dispersion coefficient of the solid phase in a turbulent flow. Qualitative and quantitative analyses provide the following conclusions:

1) The ratio of the velocities of two eddies of different sizes in the two-phase flow is always more or less as in the unloaded one.

2) The dependence of energy density in a turbulent flow is stronger for heavier particles. The influence of the particle diameter on the turbulent energy density for constant mass fraction grows with decreasing particle size.

3) The dispersion coefficient of particles in a turbulent flow drops if the particle size or mass fraction of the solid phase grows.

Acknowledgment

The authors thank Th. Thaufelder for his help in preparing the paper.

Nomenclature

a = particle area
D_{ep} = effective particle dispersion coefficient
E = energy density

f = frequency
I = eddy intensity
k = wave number
m = particle mass fraction
n = number of cascade in spectrum
Re = Reynolds number
t_p = relaxation time
u = eddy velocity
V_f = volume fraction of fluid
V_p = volume fraction of particle
W = help function
w = particle velocity
y = ratio of total eddy energy to fluid energy
δ = dimensionless eddy size
λ = eddy size
λ_1 = size of smallest eddy
ν = viscosity
ρ_f = fluid density
ρ_p = particle density

References

[1]Landau, L., and Liefschitz, E., *Lehrbuch der Theoretischen Physik*, Berlin, 1996.

[2]Hinze, J. O., *Turbulence*, McGraw–Hill, New York, 1975.

[3]Monin, A. S., and Yaglom, A. M., *Statistical Hydromechanics. The Mechanics of Turbulence. Part 1*, Nauka Press, Moscow, 1965.

[4]Soo, S. L., *Fluid Dynamics of Multiphase Systems*, Toronto/London, 1967.

[5]Nigmatulin, R. I., *Dynamics of Multiphase Media*, Hemisphere, New York/Washington, 1990.

[6]Bujewich, Yu. A., and Gupalo Yu. P., "Influence of the Particles Suspended in a Fluid on the Degeneration of Turbulence," *Prikl. Mech. i Techn. Phis. (PMTF)*, No. 4, 1965, pp. 90–96 (in Russian).

[7]Neeße, Th., "Modellierung der Turbulenten Querstromhydroklassierung," Ph.D. Dissertation, der Bergakademie Freiberg, 1978.

[8]Faeth, G. M., "Mixing, Transport and Combustion in Sprays," *Progress in Energy and Combustion Science*, Vol. 13, pp. 293–345.

[9]Hetsroni, G., and Sokolov, M., "Distribution of Mass, Velocity and Intensity of Turbulence in a Two-Phase Turbulent Jet," *Journal of Applied Mechanics*, Vol. 38, 1971, pp. 315–327.

[10]Tsuji, Y., Morikawa, Y., and Shiomi, H., "LDV Measurements of an Air-Solid Two-Phase Flow in a Vertical Pipe," *Journal of Fluid Mechanics*, Vol. 139, 1982, pp. 417–434.

[11]Owen, P. R., "Pneumatic Transport," *Journal of Fluid Mechanics*, Vol. 39, 1969, pp. 407–432.

[12]Abramovich, G. N., "The Effect of Admixture of Solid Particles or Droplets on the Structure of a Turbulent Gas Jet," *Soviet Physics—Dokla*, Vol. 190, 1970, pp. 1052–1055.

[13]Zisselman, R., and Molerus, O., "Investigation of Solid-Liquid Pipe Flow with

Regard to Turbulence Modification," *Chemical Engineering Journal*, Vol. 18, 1979, p. 233.

[14]Lee, S. L., and Durst, F., "On the Motion of Particles in Turbulent Duct Flows," *International Journal of Multiphase Flow*, Vol. 8, 1982, pp. 125–146.

[15]Tsuji, Y., and Morikawa, Y., "LDV Measurements of an Air-Solid Two-Phase Flow in a Horizontal Pipe," *Journal of Fluid Mechanics*, Vol. 120, 1982, pp. 358–409.

[16]Al-Taweel, A. M., and Landau, J., "Turbulence Modulation in Two-Phase Jets, *International Journal of Multiphase Flow*, Vol. 3, 1977, pp. 341–351.

[17]Hetsroni, G., "Particles-Turbulence Interaction," *International Journal of Multiphase Flow*, Vol. 15, 1989, pp. 735–747.

[18]Jarin, L. P., and Hetsroni, G., "Turbulence Intensity in Dilute Two-Phase Flow-3. The Particles-Turbulence Interaction in Dilute Two-Phase Flow," *International Journal of Multiphase Flow*, Vol. 20, 1994, pp. 27–44.

[19]Schvab, V. A., "Estimation of the Diffusion Processes in a Turbulent Flow of Heterogeneous Media by the Size Method," *Fizika Gorenia i Vsriva*, 1973, pp. 673–682 (in Russian).

[20]Jairazbhoy, V., Tavlarides, L. L., and Lewalle, J., "A Cascade Model for Neutrally Buoyant Dispersed Two-Phase Homogeneous Turbulence—I. Model Formulation," *International Journal of Multiphase Flow*, Vol. 21, 1995, pp. 467–483.

[21]Desnianskii, V. N., and Novikov, E. A., "Simulation of Cascade Processes in Turbulent Flows," *Journal of Applied Mathematics and Mechanics,* Vol. 38, 1974, pp. 468–475.

[22]von Weizsäcker, C. F., "Das Spektrum der Turbulenz Bei Großen Reynoldsschen Zahlen," *Zhurnal Physik*, Vol. 124, 1948, pp. 614–627.

[23]Frisch, U., Sulem, P. L., and Nelkin, M., "A Simple Dynamical Model of Intermittency," *Journal of Fluid Mechanics*, Vol. 87, 1978, pp. 719–736.

[24]Wells, M. W., and Stock, D. E., "The Effects of Crossing Trajectories on the Dispersion of Particles in Turbulent Fluid," *Journal of Fluid Mechanics,* Vol. 136, 1983, pp. 31–62.

[25]Jairazbhoy, V., and Tavlarides, L. L., "A Cascade Model for Neutrally Buoyant Dispersed Two-Phase Homogeneous Turbulence—II. Numerical Solution and Results," *International Journal of Multiphase Flow*, Vol. 21, 1995, pp. 483–499.

[26]Squires, K. D., and Eaton, J. K., "Particle Response and Turbulence Modification in Isotropic Turbulence," *Physics of Fluids*, Vol. A2, No. 1, 1990, pp. 1191–1203.

[27]Snyder, W. H., and Lumley, J. L., "Some Measurements of Particle Velocity Autocorrelation Functions in a Turbulent Flow," *Journal of Fluid Mechanics*, Vol. 48, 1971, pp. 47–71.

[28]Arnason, G., "Measurement of Particle Dispersion in Turbulent Pipe Flow," Ph.D. Thesis, Washington State Univ., Pullman, WA, 1982.

[29]Milojevic, D., "The Effects of Inertia and Crossing Trajectories on the Dispersion of Particles in Turbulence," *Proceedings of the Fourth Workshop on the Two-Phase Flow Predictions*, Erlangen, 1987.

Chapter 13

One Self-Similar Solution of Two-Dimensional Kármán–Howarth Equation: Phase Transition in Two-Dimensional Turbulence

A. L. Tseskis
Scientific Research Institute of the Canning and Vegetable-Drying Industry, Russian Academy of Agricultural Sciences, Vidnoe, Moscow Region, 142700, Russia

I. Introduction

IT IS well known that two-dimensional turbulent motion differs essentially from three-dimensional motion. On the other hand the description of two-dimensional hydrodynamics is based on the same Navier–Stokes (or Euler) equation, so that all of the anomalies in the behavior of two-dimensional turbulent motion are obviously connected with the change in the dimensionality of the space. For instance, in the two-dimensional case at zero viscosity the number of the integrals of motion is infinite so one cannot use an approximation with a finite number of the degrees of freedom; at the same time the conservation of the integral of $(\mathrm{rot}\mathbf{v})^2$ makes it possible to establish some distinguishing properties of the two-dimensional turbulent motion. The most interesting of these properties is the possibility of transfer of energy from the fine-grained to coarse-grained components and the coherent structures generation in the two-dimensional turbulence. The latter corresponds to some exact solution of hydrodynamic equations, as shown in the following section.

II. Self-Similar Solution of Kármán–Howarth Equation Leading to δ-Spectrum

Apparently, the description of isotropic two-dimensional turbulence must be based on the two-dimensional Kármán–Howarth equation [1] analog. This two-dimensional equation was written in Ref. 2,

$$\frac{\partial B_{LL}}{\partial t} = \left(\frac{\partial}{\partial r} + \frac{3}{r}\right) \cdot \left(B_{LL,L} + 2\nu \frac{\partial B_{LL}}{\partial r}\right) \tag{1}$$

and so were the corresponding relations between the correlation scalars $B_{LL}, B_{NN}, B_{LL,L}$ (concerning their meaning see Ref. 1). Equation (1) cannot be solved in any common form; therefore, one should use some additional assumptions for the problem consideration. One of these such assumptions (the so-called Kármán hypothesis) is applicable to the Kármán–Howarth equation of self-similar solution searching. Taking the hypothesis in the form

$$\begin{aligned} B_{LL}(r,t) &= v^2(t) \cdot f\left(\frac{r}{l(t)}\right) \\ B_{LL,L}(r,t) &= v^3(t) \cdot h\left(\frac{r}{l(t)}\right) \end{aligned} \tag{2}$$

where $l(t)$ and $v(t)$ are length and velocity scales, respectively, one obtains

$$f\frac{l}{v^2}\frac{\mathrm{d}v}{\mathrm{d}t} - xf'\frac{1}{2v}\frac{\mathrm{d}l}{\mathrm{d}t} - \frac{1}{2}\left(h' + \frac{3}{x}h\right) - \left(f'' + \frac{3}{x}f\right)\frac{\nu}{vl} = 0 \tag{3}$$

In this equation $x = r/l$ and the stroke corresponds to the x derivative.

All of the self-similar solutions classification for the $d = 3$ case was given by Sedov (for detailed acquaintance, see Ref. 1). Accordingly results for the $d = 2$ situation were described in Refs. 2 and 3. One of the special cases leads from Eq. (3) to the following equations:

$$f'' + \frac{3}{x}f' + f = 0 \tag{4}$$

$$h' + \frac{3}{x}h - \frac{p}{2}xf' = 0 \tag{5}$$

where p is some constant of integration. The first of these equations can be reduced to the Bessel equation:

$$f'' + \frac{1-2\alpha}{x}f' + \left[\left(\beta\gamma x^{\gamma-1}\right)^2 + \frac{\alpha^2 - n^2\gamma^2}{x^2}\right]f = 0 \tag{6}$$

where $\alpha = -1, \beta = \gamma = 1$, and $n = \pm 1$. Limited at $x = 0$ solution of Eq. (6) corresponds to $n = 1$; therefore, one obtains from the known common form.[4]

$$f = x^{\alpha} J_n\left(\beta x^{\gamma}\right)$$

the following result

$$f = \frac{J_1(\kappa r)}{\kappa r} \tag{7}$$

Here $\kappa = 1/l$. Then for the longitudinal correlation scalar, it reads

$$B_{LL} = v^2 \cdot \frac{J_1(\kappa r)}{\kappa r}$$

For the corresponding spectrum obtained one must use the relation[2,3]

$$B_{NN} = B_{LL} + r\frac{\partial B_{LL}}{\partial r}$$

and known recurrence relations for Bessel functions.[4] For B_{NN} it gives

$$B_{NN} = v^2 J_1'(\kappa r)$$

so that

$$B_{ij} = v^2 \left\{ \left[\frac{J_1(\kappa r)}{\kappa r} - J_1' \right] \frac{r_i r_j}{r^2} + J_1' \delta_{ij} \right\}$$

Taking now the trace of this tensor, one obtains

$$B_{ii} = v^2 J_0(\kappa r) \tag{8}$$

So for the spectral energy density $E(k) = 2\pi F(k)$ (F is the spectrum), it reads

$$E = \frac{v^2}{2\pi} \int_0^{\infty} r\mathrm{d}r \int_0^{2\pi} e^{-ikr\cos\varphi} \mathrm{d}\varphi \cdot J_0(\kappa r)$$

Then taking account of Bessel function J_0 integral representation,

$$J_0(z) = \frac{1}{2\pi}\int_0^{2\pi} e^{-ikz\cos\varphi} d\varphi$$

we have the following relations

$$\frac{E}{v^2} = k\int_0^{\infty} r dr \frac{1}{2\pi}\int_0^{2\pi} e^{-ikr\cos\varphi} d\varphi \cdot J_0(\kappa r) = k\int_0^{\infty} J_0(kr)J_0(\kappa r) r dr = \delta(k-\kappa)$$

so that finally

$$E(k,t) = v^2\delta(k-\kappa) \tag{9}$$

and we come to the δ-like spectrum, which was first obtained in Ref. 5 from different considerations.

III. Phase Transition in Two-Dimensional Turbulence

Such a singular form of the spectrum corresponds to some unusual behavior of the correlation functions. Let us consider the so-called Loitsyansky invariant, which, as it was shown in Ref. 6, is connected to the statistical averaged value M^2, where M is the angular momentum of the turbulent fluid. In the case under consideration, this invariant takes the form

$$\Lambda = \int_0^{\infty} B_{LL}(r) \cdot r^3 dr$$

For finite X to within a factor v^2, one obtains

$$\Lambda = \int_0^{X} x^3 B_{LL}(x) dx = \int_0^{X} x^2 J_1(x) dx = -X^2 J_0(X) + 2XJ_1(X) \tag{10}$$

Taking into account the known asymptotic properties of Bessel functions, at large values of X the Loitsynsky invariant absolute value grows as $X^{3/2}$ and Λ changes its sign periodically. But by virtue of $\Lambda \sim \langle M^2 \rangle$, the latter is impossible. Therefore, if the spectrum $E(k)$ approaches δ like one the turbulent motion must change its character. For this character establishing let us consider a certain periodic structure corresponding to a system of plane vortices.[7] The system is given by the relations

$$v_1 = -\frac{\partial}{\partial x_2}(\cos k_1^* x_1 \cos k_2^* x_2)$$
$$v_2 = -\frac{\partial}{\partial x_1}(\cos k_1^* x_1 \cos k_2^* x_2) \tag{11}$$

This system can be tentatively assigned a correlation function with corresponding averaging over $\boldsymbol{x}$, as it was proposed in Ref. 5,

$$B_{ij}(\boldsymbol{r}) = \langle v_i(\boldsymbol{x})\, v_j(\boldsymbol{x}+\boldsymbol{r})\rangle$$

so that, for instance,

$$\langle \varphi \rangle = \lim_{a,b\to\infty} (ab)^{-1} \int_{-a}^{a}\int_{-b}^{b} \varphi(\boldsymbol{x})\,\mathrm{d}x\,\mathrm{d}x_2 \tag{12}$$

After taking derivatives in Eq. (11), the product $v_1 v_2$ appears to be represented in the form

$$v_1(\boldsymbol{x})v_2(\boldsymbol{x}+\boldsymbol{r}) = \frac{k_1^* k_2^*}{2}[\cos(\boldsymbol{k}^*\boldsymbol{r}) - \cos(2\boldsymbol{k}^*\boldsymbol{x} + \boldsymbol{k}^*\boldsymbol{r})]$$

The above-mentioned averaging gives, correspondingly,

$$B_{12}(\boldsymbol{r}) = \frac{k_1^* k_2^*}{4}\sin k_1^* r_1 \sin k_2^* r_2$$

Realizing now Fourier transformation and taking account of known relations,

$$\int_{-\infty}^{\infty} \cos\alpha x \cdot e^{-i\beta x}\mathrm{d}k = \pi[\delta(\alpha-\beta) + \delta(\alpha+\beta)]$$

$$\int_{-\infty}^{\infty} \sin\alpha x \cdot e^{-i\beta x}\mathrm{d}x = -i\pi[\delta(\alpha+\beta) + \delta(\alpha-\beta)]$$

we obtain finally

$$F_{12} = \frac{k_1^* k_2^*}{2} [\delta(\boldsymbol{k} - \boldsymbol{k}^*) + \delta(\boldsymbol{k} + \boldsymbol{k}^*) \tag{13}$$

i.e., the anisotropic δ spectrum, which results in the vector character of the arguments of δ functions in Eq. (13). Such a change in the turbulent motion structure looks like phase transition with its typical violation of the symmetry of the correlation functions. If at some initial instant of time the correlation functions (and the corresponding spectrum) are isotropic, then the appearance of a spectrum of the form of Eq. (13) indicates a change in their symmetry, and the corresponding picture is similar to a phase transition «liquid–crystal» in the sense mentioned.

The solution (11) corresponds to dividing the plane into rectangular cells; in each of the cells a vortex is placed so that the neighboring vortices have opposite signs. Of course, one could find the solutions, corresponding to anisotropic δ-like spectrum as well, which lead to different ones from Eq. (11) but also periodic structures. At all events, such behavior of two-dimensional turbulence demonstrates coherent vortices appearance, which has been established by numerous computer simulations (see Ref. 8) and physical experiments [9] as well.

IV. Conclusions

In this context, one can cast doubts on the reality of two-dimensional turbulence and on the validity of the ordinary description of two-dimensional fluid stochastic motion, based on the statistical consideration. Because different relations among the correlation function are the consequences of the consideration, it makes sense to examine them to establish the validity (or invalidity) of the usual statistical description of two-dimensional turbulence. The answer, as in the case of $d = 3$, may be obtained in the simplest way as in the experimental examination of the kinematic relation mentioned in one of the preceding sections:

$$B_{NN} = B_{LL} + r\frac{\partial B_{LL}}{\partial r}$$

which follows from the continuity equation at $d = 2$. Note in conclusion that the analogous relation for $d = 3$,

$$B_{NN} = B_{LL} + \frac{1}{2}r \cdot \frac{\partial B_{LL}}{\partial r}$$

was examined as far back as 1937 [Ref. 10]; until now such an examination for the cause $d = 2$ has not been realized.

References

[1]Monin, A. S., and Jaglom, A. M., *Statistical Fluid Mechanics*, Vol. II, MIT Press, Cambridge, MA, 1975.

[2]Tseskis, A. L., *Fluid Dynamics*, Vol. 22, No. 5, 1987, p. 670.

[3]Tseskis, A. L., Master's Thesis, Moscow State Univ., Moscow, 1986.

[4]Madelung, E., *Die Mathematischen Hilfsmittel des Physikers*, Springer–Verlag, Berlin/Heidelberg, 1957.

[5]Tseskis, A. L., *JETP Letters*, Vol. 56, July 1982, p. 95.

[6]Landau, L. D., and Lifshitz, E. M., *Fluid Mechanics*, Pergamon, Oxford, England, UK, 1959.

[7]Townsend, A. A., *Structure of Turbulent Shear Flow*, Cambridge Univ. Press, Cambridge, England, UK, 1957.

[8]Matthaues, W. Y., Stribbling, W. T., Martinez, D., Oughton S., and Montgomery, D., *Physica*, Vol. D51, 1991, p. 531.

[9]Couder, Y., *J. Phys. Lett. France*, Vol. 45, 1984, p. 353.

[10]Taylor, G. I., *Journal of Astronautical Science*, Vol. 4, No. 8, 1937, p. 311.

Chapter 14

Turbulent Boundary Layer–Wake Interaction Study

Antonis E. Kourounis,* Demosthenes D. Papailiou,† and Panagiotis Koutmos‡
University of Patras, Rion-Patras, Greece

I. Introduction

KNOWLEDGE of the effects of the interaction of a wake with a boundary layer on its subsequent development is of practical importance to the successful design of high-lift configurations. This interaction phenomenon has also been identified as a problem of fundamental interest in the study of turbulent flows contributing to the better understanding of the transport processes necessary for the development of suitable calculation methods for related flow configurations.

Several experimental and computational studies have been carried out during the past years, focusing on various aspects of the above-flow configuration. Lift and drag computations of an airfoil with flap have been performed, taking into account the mixing of the wake and the boundary layer over the upper surface of the flap[1] as well as the mixing of a slat's wake with a boundary layer developing on the main airfoil element.[2] Methods to attain maximum lift by optimizing parameters like slat deflection and the gap between the slat and the main wing have been also investigated.[2] Detailed mean and turbulent velocity measurements have been conducted for configurations such as those of a flat-plate boundary layer interacting with the wake produced by another flat plate,[3] and of the wake–boundary layer problem for slat-main element arrangements. The objective in these experiments was to provide a database for the development of suitable prediction

*Research Associate, Laboratory of Applied Thermodynamics and Statistical Mechanics, Department of Mechanical and Aeronautical Engineering.

†Professor, Director of the Laboratory of Applied Thermodynamics and Statistical Mechanics, Department of Mechanical and Aeronautical Engineering.

‡Assistant Professor, Laboratory of Applied Thermodynamics and Statistical Mechanics, Department of Mechanical and Aeronautical Engineering.

methods for the disturbed boundary-layer flow.[4,5] Comparative three-dimensional turbulent velocity measurements have been conducted for wakes from a streamlined body (airfoil) and a bluff body (square cylinder) of the same drag interacting with a flat-plate boundary layer. The bluff body produces faster mixing there.[6] Comparison of the interaction of wakes generated by thick airfoils and flat plates with similar drag indicated that the flat-plate wake mixes less rapidly than the airfoil produced wake.[7] Flow visualization study of the embedded shear layers within the wake–boundary layer region has been performed[8] with emphasis placed on the role played by active large eddies in the interaction process. The effects of positive streamwise pressure have been investigated in two-dimensional computations of Ref. 9. Improved algebraic stress models outperformed standard k-ε procedures. However, these extensive comparisons revealed significant inadequacies of currently used turbulence models; the influence of intermittency was considered responsible for these failures. Similarly, the effects of active (time-dependent) large-scale structures have been clearly identified in related experimental results.[9]

In conclusion the works described in the preceding references, although of significant importance in the understanding of the basic problem, cannot offer necessary detailed information regarding the influence of the interaction processes on the turbulence structure and the involved transport mechanisms in the interaction and post-interaction regions that affect the characteristics of the developing boundary layer. Furthermore, information on the alterations occurring in the shear layer structure and the extent of the contribution of active eddies on the flow dynamics is at present insufficient.

Based on the above discussion, the aim of the present tudy is to provide further understanding of the phenomena involved in the turbulent wake–boundary layer interaction process appraising also their effect on the turbulence structure and related transfer mechanisms. In particular the present work focuses attention on the estimation of the distributions of the terms appearing in the turbulent kinetic energy equation obtained from detailed measurements within the interaction region. These results are presented and discussed together with corresponding measurements of the spanwise component of the mean vorticity in an attempt to provide a clearer insight into the physical mechanisms governing the interaction of the two flows.

II. Experimental Arrangement and Procedure

The experiments were conducted in a low-speed, low-turbulence, indraft open wind tunnel, shown in Fig. 1, operating at the Laboratory of Applied Thermodynamics. The dimensions of the tunnel's test section are $0.5 \times 0.3 \times 3$ m. The inflow freestream turbulence level is 0.27% whereas uniformity of the mean inlet axial velocity profile was achieved to within ±0.8% for a mean velocity of 14 m/s.

Detailed measurements of the three components of the mean and turbulent velocity fields were obtained with the use of hot-wire anemometry by suitably placing an X-wire sensor in the x-y and x-z planes. The anemometer output was low-pass filtered to prevent aliasing and improve the signal-to-

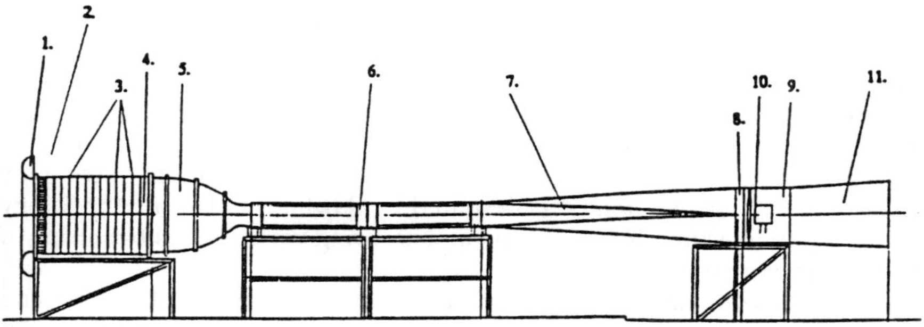

Fg. 1 Open circuit, low-speed wind tunnel.

noise ratio. Subsequently, it was fed to a computer (PC-486) for data processing and analysis through a 12-bit A/D interface card with adjustable gain. The sample size was $1.6*10^4$ events/measurement for each channel. Computer codes have been developed for statistical analysis of the digitized anemometer output and for processing the experimental data to obtain mean velocities and statistical characteristics.

Sources of errors considered that contribute to the uncertainty in the measured and derived quantities are the electronic noise of the anemometer, the probe support displacement, the probe placing in the x-y and x-z planes, the errors due to wire spacing, the derivation of the experimental quantities, the spatial nonuniformity of the flow along the length of the sensing element, the probe's aerodynamic disturbance, the signal interpretation errors in high turbulence intensity regions, and the curve-fit uncertainty of the selected response equation of calibration.

An analysis of the estimated errors is presented in the following:

1) The general error due to electronic noise was found to be approximately 1%.

2) The precision of the probe support displacement is ± 0.1 mm and ± 2 mm for transverse and longitudinal measurements, respectively.

3) The difference in values of $\overline{u^2}$ obtained by placing the probe in the x-y and x-z planes was within 6%.

To estimate the errors due to wire spacing (of the X-wire probe used) in the mean velocities and velocity fluctuations measured by this sensor, we need relations between the measured and the true quantities. Such analytic expressions capable of estimating the error in the mean and rms values of the velocity measured with X-wire probes were derived by Hirota et al.[10] By using these expressions the maximum error in the present experiments was found to be equal to 2% for mean velocity and 2, -6, and -6.5% for the fluctuating components $\overline{u^2}$, $\overline{v^2}$, and $\overline{w^2}$, respectively. These maximum values were found in the vicinity of the wall. Positive errors denote overestimation whereas negative ones denote underestimation of the measured quantities.

There are also uncertainties in the evaluation of the derivatives representing the mean spanwise vorticity and the terms in the turbulent kinetic energy balance equation. A detailed analysis of error estimation for complex measurements of two points is presented by Sawhney.[11] This methodology is employed in estimating the errors in measuring vorticity and the terms of the energy equation.

The spatial nonuniformity of the flow along the length l of the sensing element should also be taken into account. There are two main sources of spatial resolution errors. The first is because of changes in the mean velocity along the length of the wire. The second occurs when the length of the wire is not small compared with the fine scales of the turbulence. Spatial resolution errors are most severe in the near wall region of the turbulent boundary layer. These uncertainties are difficult to estimate; however, errors due to imperfect spatial resolution of the fine scale turbulence are expected to be present only very close to the wall. Other important sources of uncertainties are the probe's aerodynamic disturbance[12] (±2%), the curve-fit uncertainty of the selected response equation of calibration (±0.3%), and the signal interpretation errors in high turbulence intensity regions[13] like near the airfoil's trailing edge and near the wall (4%).

The turbulent boundary layer developed along the bottom wall of the test section while the interacting wake was produced by an airfoil (NACA 0015, Fig. 2) with a trailing-edge flap of varying angle of deflection. The airfoil was positioned at the test section entrance and could be placed at different heights from the bottom wall. Two configurations representing "strong" and "weak" interaction conditions were investigated. In the former the airfoil was placed at a distance of 80 mm from the wall with a flap deflection angle of 25 deg, producing an immediate impingement of the wake on the boundary layer. In the latter the airfoil was positioned at a height of 110 mm with a flap deflection angle of 5 deg resulting in a delayed interaction arrangement.

The Reynolds number based on the airfoil's chord was $1.4*10^5$, and the measurement stations covered a distance from 1.33 to 16 chords. The undisturbed boundary layer was also documented to allow comparisons with the interacting flowfield.

The detailed results allowed the evaluation of the terms appearing in the turbulence kinetic energy equation budgets across the boundary layer. These budgets were performed for the interaction and parts of the pre- and

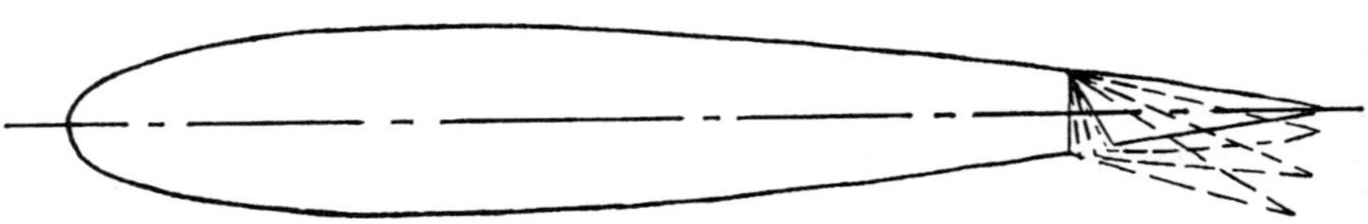

Fig. 2 Wake-producing airfoil.

post-interaction regions and are discussed together with measurements of the corresponding mean spanwise vorticity.

III. Experimental Results

The presented distributions of the various turbulent energy transfer terms appearing in the kinetic energy equation of the undisturbed and disturbed boundary layer constitute part of a broader investigation that includes detailed measurements of the mean and turbulent statistical characteristics of the wake-boundary layer interaction flowfield. An outline of the general character of the structure of this flowfield is presented in the following.

A. Mean and Fluctuating Characteristics of the Turbulent Flowfield

In this part a brief account of the mean and fluctuating structure of the interacting flow will provide the necessary background in discussing observed changes in the turbulent transfer processes as they are represented by the various terms in the kinetic energy equation.

Nondimensionalized mean axial velocity distributions for the wake and the disturbed boundary layer are shown in Fig. 3 for the "strong" interaction case. The undisturbed boundary layer corresponding data are also included for comparison. The distributions show that the potential core between the two interacting flows is maintained up to a distance of $x/c = 6.67$ while further on the inviscid region is affected by the mean shear. Some general characteristics of the flowfield resulting from the interaction of the two flows are discussed below.

In Fig. 4 typical logarithmic velocity profiles are presented at the post-interaction region for both the undisturbed and disturbed boundary layer. Clauser's method has been used for the evaluation of the wall friction velocity. All of the displayed velocity distributions, for the disturbed boundary layer, appear to remain unaffected by the impinging wake logarithmic region near the wall. The logarithmic profile seems to be affected only at the outer part of the disturbed layer, that is, approximately above the buffer layer.

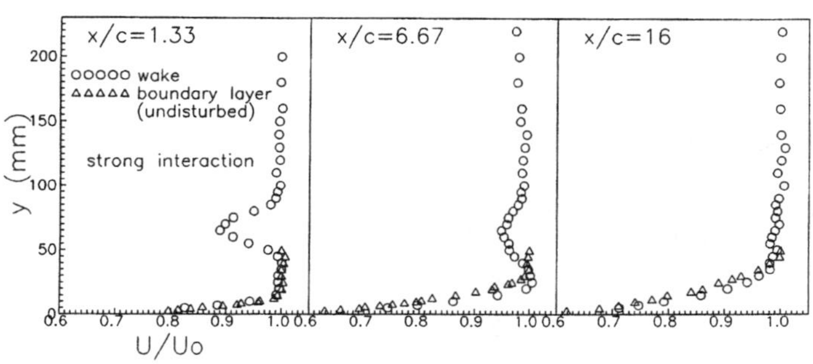

Fig. 3 Mean velocity distributions of the strong interaction along the test section.

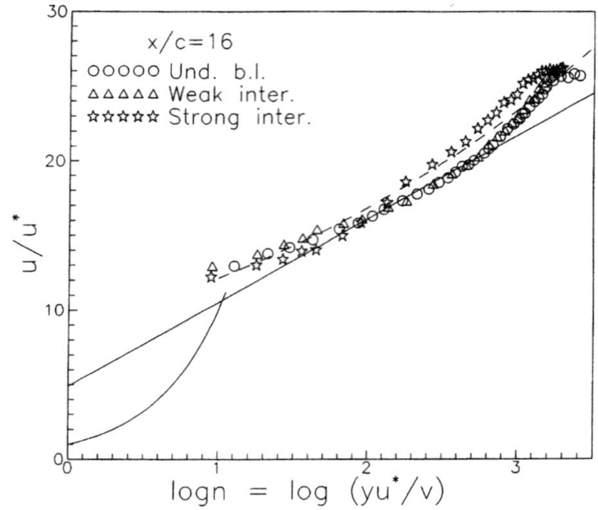

Fig. 4 Logarithmic law profiles of the velocity in the disturbed and the undisturbed boundary layer at station *x/c* = 16.

The axial development of the characteristic thicknesses of the measured boundary layers are shown in Fig. 5a and 5b. A decrease of the displacement and momentum thicknesses is evidenced as the impinging wake prevents the transverse development of the boundary layer. The observed reduction increases with increasing interaction strength. Figure 6 displays comparisons between the three rms components. In the boundary layer the inequality $\overline{u^2} > \overline{w^2} > \overline{v^2}$ is clearly maintained. The distributions of these turbulent quantities appear to follow a wake-like form at the outer edge of the layer while they assume a boundary layer form near the wall.

a.

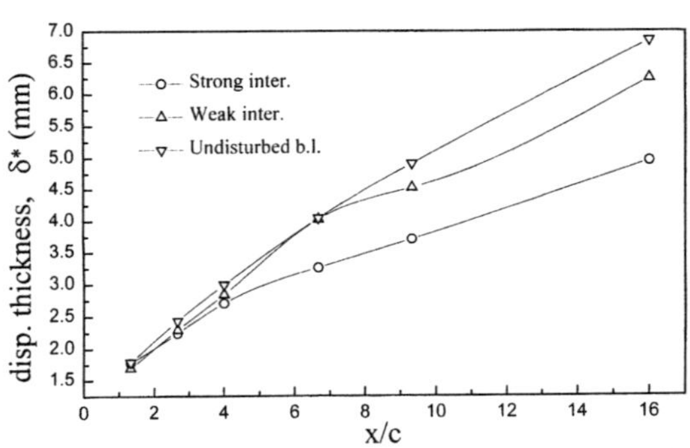

Fig. 5a Displacement thickness distributions of the undisturbed and disturbed boundary layer along the test section.

b.

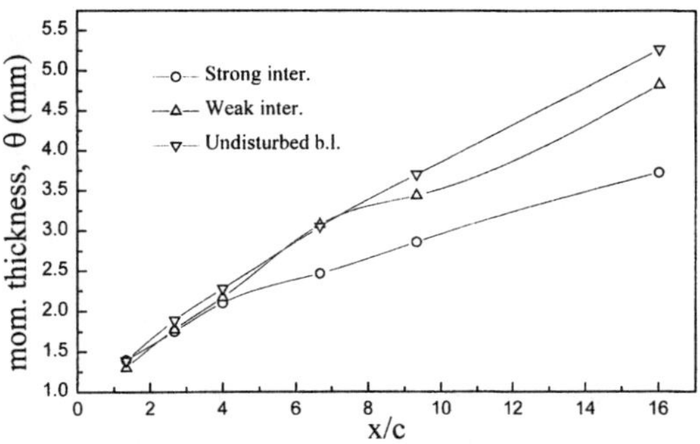

Fig. 5b Momentum thickness distributions of the undisturbed and disturbed boundary layer along the test section.

The presented evidence suggests that, in general, the boundary layer retains its turbulent structure intact up to a certain height approximately extending to the end of the buffer layer, above which the influence of the impinging wake is more pronounced. The dynamic development of the interaction region is dominated by the turbulent structure of the wake. Observations similar to those discussed above have also been reported by Zhou and Squire.[7]

Finally, from the skewness and flatness distributions (presented in the next section), one can see that the turbulent mixing, which increases with increasing interaction strength, is responsible for smoothing out the inhomogeneities that are created in the interaction zone by the impinging wake. Additionally, the interaction strength depends strongly on the airfoil proximity to the wall and the strength of the vorticity shed by the wake producing body (i.e., flap deflection angle).

B. Mean Spanwise Vorticity Distributions

Flow visualization experiments attempting to identify the presence of large-scale vorticity have been conducted for different cases of wake-bound-

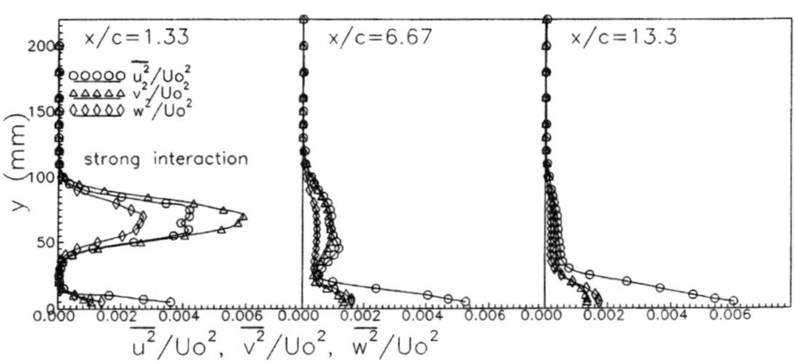

Fig. 6 Distributions of the three-dimensional turbulence velocity field along the test section at strong interaction.

ary layer interaction.[8] In these, "large anisotropic swirling motions" initiated in the wake have been observed to penetrate into the boundary layer dominating its convective characteristics and resulting in increased mixing. It was then suggested that the presence of organized eddy motion constitutes an important feature in the dynamic development of the interaction process. Furthermore Savill and Zhou discussed[8] that the wall region is seriously perturbed by these structures resulting in extra shear stress generation, which crucially determines enhancement of turbulent mixing. In this context the existence of a net motion away from the wall was reported in the outer layer region. This tends to suppress the return of mass and momentum to the wall and reduces skin friction even though bursting may be enhanced.

The present results support and complement the information reported in the investigation[8] mentioned earlier regarding the presence of large vortical structures probably originated by the wake in the interaction region and their effect on this process as inferred from the power spectra and skewness and flatness distributions, presented in Figs. 7 and 8.

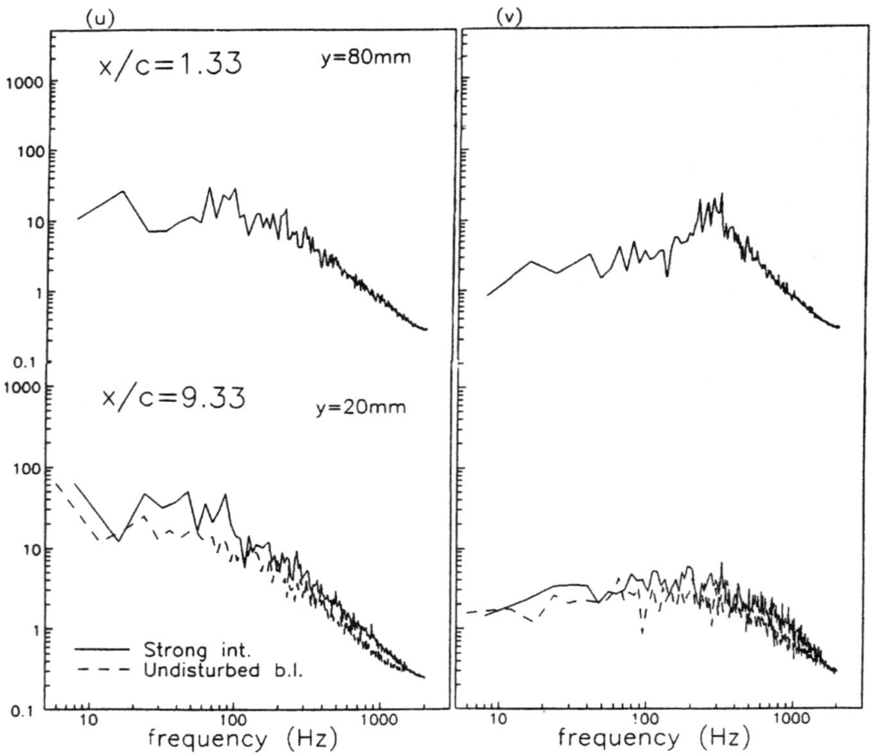

Fig. 7 Power spectra density distributions of the fluctuating turbulent velocity field (u and v components), at the wake-producing body centerline (y = 80 mm, x/c = 1.33) and at the undisturbed and disturbed boundary layer (y = 20 mm, x/c = 9.33).

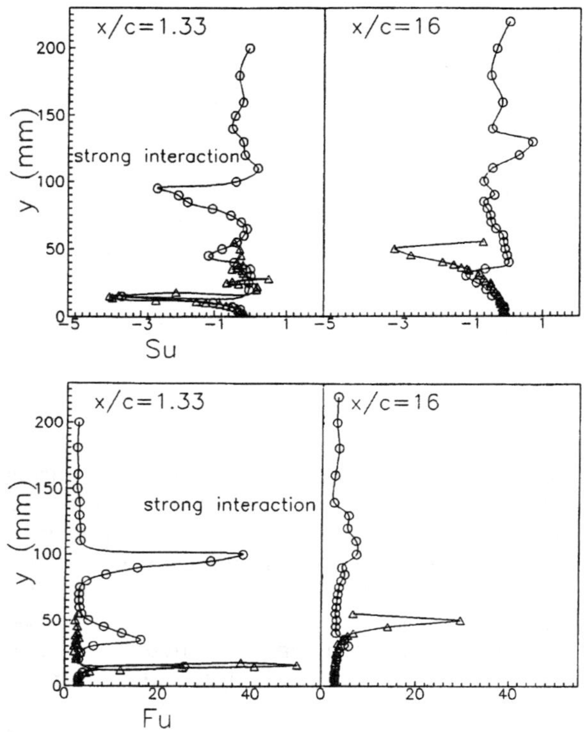

Fig. 8 Typical skewness and flatness factor distributions along the test section at strong interaction.

Direct evidence supporting the presence of a quasiperiodic large-scale structure pattern, probably originating in the near wake field, can be detected in the power spectra displayed in Fig. 7 (x/c = 1.33). The v-turbulent velocity component spectra, in particular, exhibits clear frequency bands representing eddies of different size. These coherent vortical structures certainly undergo a dynamic evolution that might involve phenomena such as vortex breakdown and generation as well as pairing,[14] as they travel and finally merge with the boundary layer interacting with its structure. Their spectral distribution close to the boundary-layer edge is shown in the spectra obtained at x/c = 9.33 where corresponding spectra of the undisturbed layer are also included. By comparing these spectra, it is evident that the interaction resulted in an increase of the turbulent energy budget, enriching also its eddy structure. The provided skewness and flatness distributions corroborate the preceding observations exhibiting strong inhomogeneities in the turbulent field of the disturbed boundary layer. These inhomogeneities are observed to smooth out at downstream stations because of increased turbulent mixing.

The distributions of the mean spanwise vorticity, $\Omega_z = (\partial V/\partial x - \partial U/\partial y)$ were evaluated from velocity measurements and are shown in Figs. 9a and 9b for the undisturbed and disturbed boundary layer. By using the method

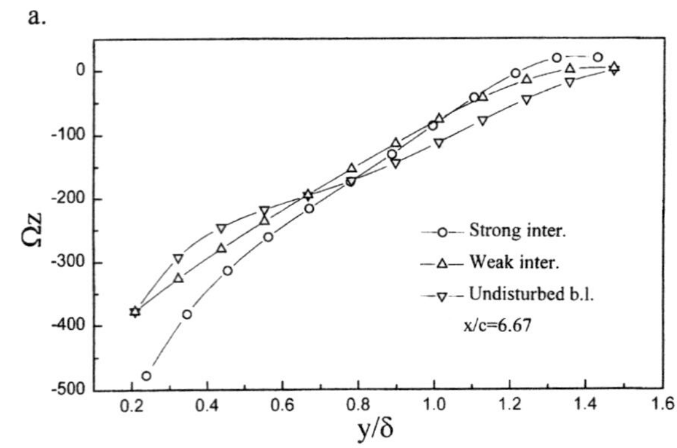

Fig. 9a Distributions of the mean spanwise vorticity in the disturbed and the undisturbed boundary layer at station *x*/*c* = 6.67.

suggested by Sawhney,[11] uncertainty in evaluating the derivatives appearing in the expression of the streamwise vorticity was estimated to be $\pm 4\%$. The negative sign in mean vorticity represents clockwise rotation and appears to dominate in the boundary layer whereas the positive sign, indicating counterclockwise rotation, characterizes the wake. The "border" between these two layers coincides with the region where vorticity changes sign. Because only one vorticity component was evaluated, a complete understanding of the effect of the interaction process on the vorticity budget and its distribution in the boundary layer cannot be achieved particularly in relation to a likely inflow of organized vortical structures. Nonetheless, a general trend can be clearly detected in the presented graphs suggesting a decrease of negative vorticity in the upper part of the boundary layer with a corresponding increase in its lower part closer to the wall. Although limited, the presented information points to the fact that the interaction alters the content and distribution of boundary-layer vorticity with implications on the turbulent kinetic energy budget as discussed below.

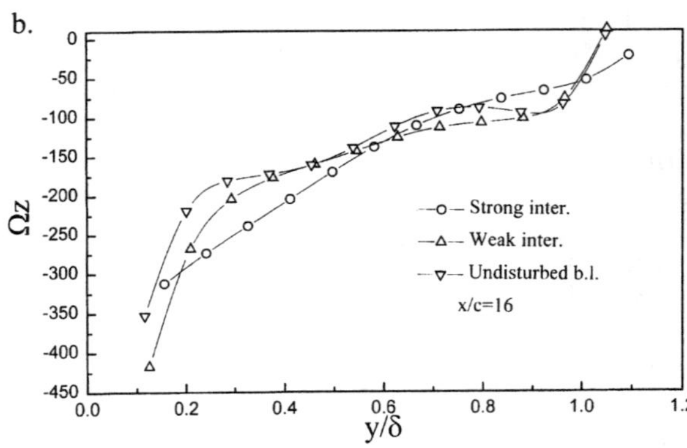

Fig. 9b Distributions of the mean spanwise vorticity in the disturbed and the undisturbed boundary layer at station *x*/*c* = 16.

C. Turbulent Kinetic Energy Budget

The transport equation for the turbulent kinetic energy for two-dimensional boundary layers and wakes (where streamwise gradients can be neglected) may be written as follows:

$$\frac{\delta}{2u_\tau^3}\left(U\frac{\partial\overline{q^2}}{\partial x}+V\frac{\partial\overline{q^2}}{\partial y}\right)=\frac{\delta}{u_\tau^3}\left(-\overline{uv}\frac{\partial U}{\partial y}-(\overline{u^2}-\overline{v^2})\right)\frac{\partial U}{\partial X}$$
$$-\frac{\delta}{u_\tau^3}\frac{\partial}{\partial y}\left(\frac{\overline{vq^2}}{2}+\frac{\overline{vp}}{\rho}\right)-\frac{\delta_\varepsilon}{u_\tau^3} \tag{1}$$

where $\overline{q^2} = \overline{u^2} + \overline{v^2} + \overline{w^2}$. The viscous diffusion terms are neglected because their importance is restricted only to the near-the-wall region. Also, in a zero pressure-gradient boundary layer, as is the present case, production and dissipation are normally the main contributors to the total budget except close the edge of the layer. This has been previously shown[15–19] and is also verified by the present results for both the undisturbed and disturbed boundary layers.

In the present experiments the terms appearing in the above equation have been evaluated directly from the measured data with the exception of the correlations $(\overline{vw^2})$ and $(\overline{pv})$, which contribute to the diffusion processes. In the following each of the evaluated terms is discussed in detail.

The turbulent energy production terms are presented in Fig. 10. Turbulent production is dominated by shear $-\overline{uv}(\partial U/\partial y)$ whereas production due to normal stresses, $-\overline{u^2}-\overline{v^2}\,(\partial U/\partial x)$, which is usually neglected, becomes relatively important in the post-interaction region where it reaches approximately a level of 8% of the production term. Transport of kinetic energy by diffusion receives its main contribution from the term $1/2\ \partial\overline{q^2v}/\partial y$.[15,18,19] To compute this term, an estimate of the correlation $(\overline{vw^2})$ is required. Skåre and Krogstad[18] used the approximation $\overline{vw^2}=k(\overline{u^2v}+\overline{v^3})$ where $k = \overline{w^2}/(\overline{u^2v}+\overline{v^3})$ and reported satisfactory results. As already mentioned, the contribution from the pressure–velocity interaction $(\overline{pv})$ to the diffusion process

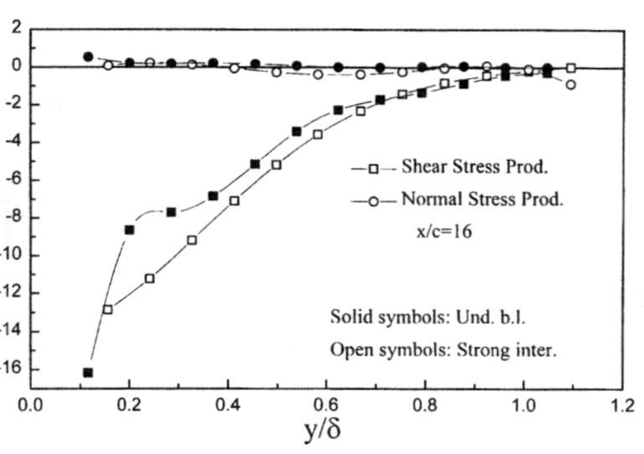

Fig. 10 Distribution of the production terms in the undisturbed and disturbed boundary layer at post interaction region (*x*/*c* = 16).

a.

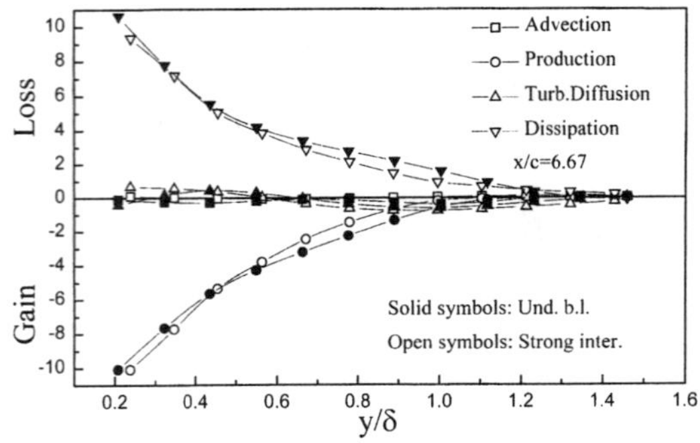

Fig. 11a Comparison between the energy budgets of the undisturbed and disturbed boundary layer at station *x/c* = 6.67.

could not be estimated. However, experience from direct numerical simulations (DNS) indicates that, although the effect of pressure gradients on this is not precisely known, it is of rather minor importance.[17] The advection of kinetic energy by the mean flow is controlled by the two terms at the left-hand side of the turbulent energy equation. These terms have been reported to be of the same order of magnitude and of opposite sign.[18] At zero pressure gradient advection remains significant only in the outer part of the boundary layer.[15,18,19] Finally, most of the methods used to determine the dissipation rate are based on the assumption that the fine scale motion at high wave numbers approaches isotropy. This assumption is not valid in the present case, and the dissipation was estimated as the remainder from all other terms in the kinetic energy equation. The pressure–velocity contribution to the diffusion and any errors that may accrue from the estimation of the various terms are therefore embodied into the dissipation values. There are uncertainties in the turbulent velocity fluctuation results as well as in the locations of the measuring positions. The estimated uncertainty[11] in the present experimental data and their derivatives representing the advection, shear stress production, normal stress production, and turbulent diffusion

b.

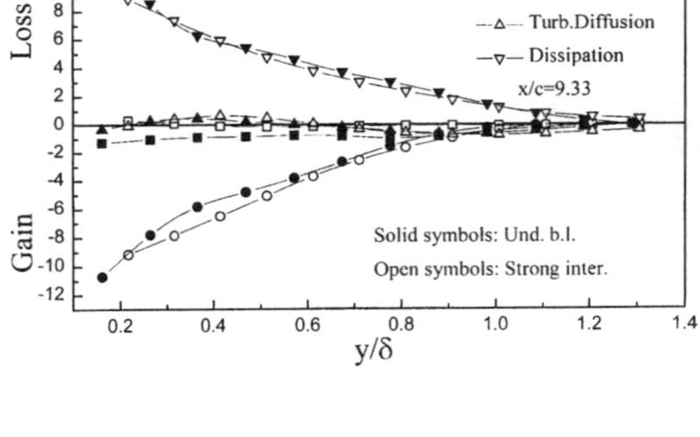

Fig. 11b Comparison between the energy budgets of the undisturbed and disturbed boundary layer at station *x/c* = 9.33.

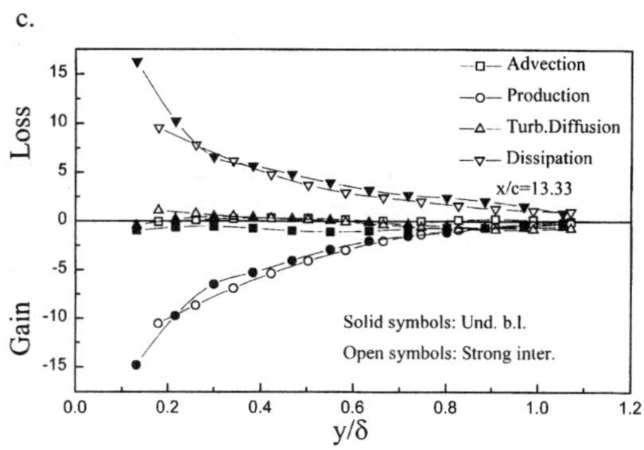

Fig. 11c Comparison between the energy budgets of the undisturbed and disturbed boundary layer at station *x/c* = 13.33.

terms are less than ± 8, ± 8, ± 7, and ± 11, respectively. Nonetheless, it is believed that such an estimate for the dissipation is sufficiently accurate because $\overline{pv}$ contribution is expected to be negligible.

In Figs. 11a–11d the distributions of the above terms are presented for the boundary layer. The terms have been nondimensionalized by δ/u_τ^3, a quantity varying with position in the layer. A comparison is also provided between the disturbed (strong interaction) and the undisturbed boundary layer. Several downstream stations are displayed to demonstrate the variations in the turbulent kinetic energy budget terms in the interaction and post-interaction regions. Negative terms represent a gain of energy whereas positive ones indicate a loss of energy. In line with previous studies,[15,18,19] it was found that close to the wall the main contributors to the energy balance are the production and dissipation terms, both for the undisturbed and disturbed boundary layers. This is in agreement with the suggestion made by Hinze[19]; namely, that part of the energy produced by the shear stress is transported by turbulent diffusion toward the outer region of the boundary layer where it is subsequently dissipated.

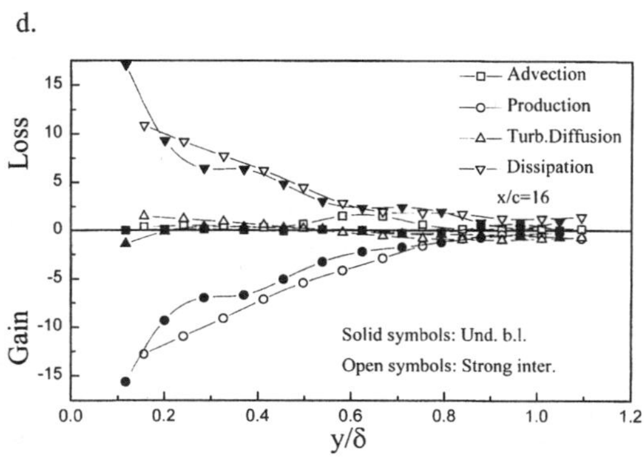

Fig. 11d Comparison between the energy budgets of the undisturbed and disturbed boundary layer at station *x/c* = 16.

Inspection of the estimated budget distributions reveals that shear stress production is somewhat higher for the disturbed layer at stations beyond $x/c = 9.33$. This result may be attributed to the penetration of the wake structures as they are entrained into the layer leading to stronger mean shear at the initial stages of the interaction ($x/c = 6.67$). Increased levels of shear stresses appear near the wall downstream of these stations, as evidenced in some typical distributions shown in Figs. 3 and 12. These variations in the production terms between the disturbed and undisturbed layers increase near the wall at the post-interaction region ($x/c = 16$, Fig. 11d). On the contrary, there are no significant differences in the dissipation terms between those layers with the exception of the post-interaction region (Fig. 11d).

The increased contribution of the dissipation term at the outer edge of both layers is rather surprising. A possible explanation might be that the pressure–velocity correlation acquires significant values in this region causing the inflation of the dissipation estimate. As remarked by Bradshaw,[15] estimates of dissipation derived by neglecting the pressure–velocity correlation term and measuring all others should not be in serious error except near the boundary-layer edge. He also expects that in an irrotational field the dissipation should be negligible and the advection supplied entirely by pressure diffusion, with the result that the latter cannot be neglected in the presence of small intermittency. However in the present measurements that advection term remains indeed negligible confirming the accuracy of the measured dissipation distribution as derived by this difference.

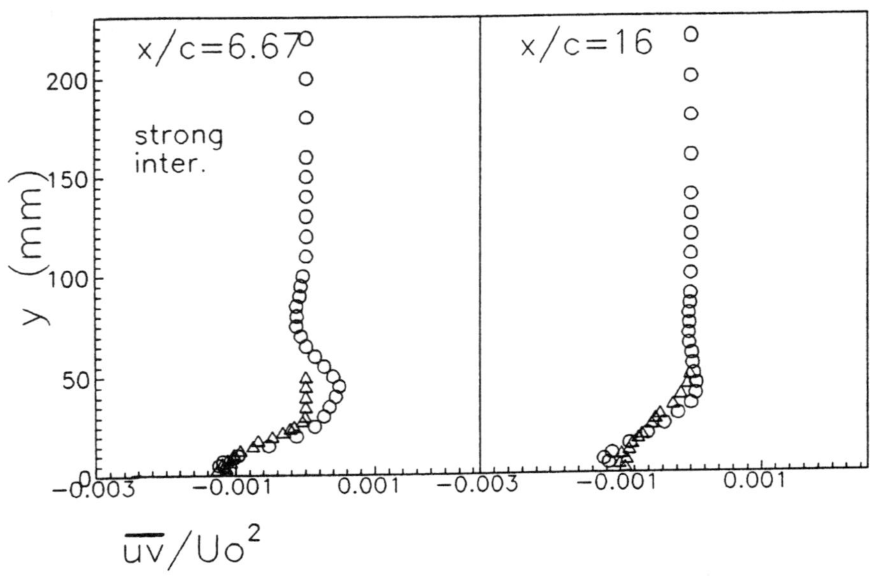

Fig. 12 Typical shear stress distributions along the test section at strong interaction.

In the undisturbed boundary layer, production and dissipation appear to be nearly in balance attaining closer values as the wall is approached (Fig. 11d). The interacted layer exhibits a different behavior. At the outer edge ($y/\delta = 1$), the dissipation term is balanced mainly by turbulent diffusion and partly by production. The diffusion term is slightly larger in this layer probably because of merging with the wake in this region. This mechanism appears to be important at the inflowing outer region of the half-wake as well. The interaction between the two flows results to a supply of energy mainly through turbulent diffusion. Furthermore most of the diffused kinetic energy is locally dissipated, and only a small part is transferred by the mean flow toward the wall (Figs. 11a–11d). It is significant to underline a difference between the preceding conclusion and the views regarding the behavior of equilibrium boundary layers, as expressed by Hinze[19]; namely; diffusion is normally in balance with advection and that dissipation and production are negligible.

In the uninteracted boundary layer the advection term is negligible near the edge while it becomes slightly negative (energy gain) as the inner region ($y/\delta \leq 0.2$) is approached. On the contrary this term becomes positive (energy loss) in the disturbed layer implying an increase of transfer of kinetic energy from upstream regions of the layer and the wake (Figs. 11c and 11d). The turbulent kinetic energy convected by the mean flow is "retarded" by the action of the turbulent shear stresses, decreasing in the inner part of the layer. Therefore the behavior of the advection term is different in the undisturbed and disturbed layers.

In summary the discussed experimental evidence leads to the important conclusion that the inner and outer layer regions of the boundary layer are dominated by different turbulent transfer mechanisms. As a consequence there is a strong energy exchange between the two regions in both the undisturbed and disturbed layers. This consequence suggests a strong interaction between the wall and the outer flow region of the boundary layer. The interaction is additionally strengthened in the case of the disturbed layer due to the inflow of large-scale structures as they are transported from the wake into the outer region where interaction is stronger. Although the identified quantitative differences between the undisturbed and disturbed boundary layer budgets, the trends remain similar in both cases.

There are many similarities between the structure of the outer region of the boundary layer with other turbulent shear flows and especially with wake flows. The turbulence field is characterized in this region of the boundary layer by large eddies elongated at the downstream direction. Their length scale is increasing along the layer. In the present study a more rigorous growth of these scales was identified caused by the effect of the inflow of the wake structures in the interaction region, as shown in Fig. 13.

As suggested by Clauser,[20] there is a clear dependence of the turbulence structure of the outer region by the upstream conditions. The flow at the outer region possesses a strong "memory" of any disturbance or property while the recovery from their influence is very slow. Thus, in the present case,

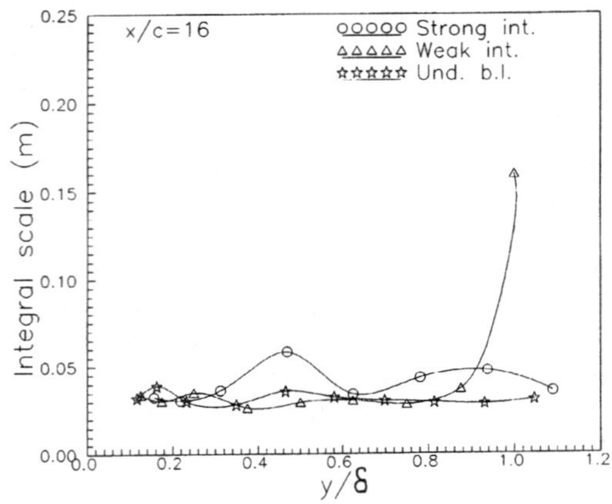

Fig. 13 Distributions of the integral scales in the disturbed and the undisturbed boundary layer at station *x/c* = 16.

significant changes in the terms can only be detected at the post-interaction region (Fig. 11d) as they are subjected to a slow recovery.

IV. Implications for Modeling

Previous attempts to model this important and practical flow have largely failed because of the inadequacy of conventional eddy viscosity procedures within standard steady-state schemes to address properly the effects of active (time-depended) large-scale eddy structures. Recently a computational model that can handle separated flows exhibiting a combination of periodic, quasi-periodic, or flapping motions has been put forward by Koutmos et al.[21] The model has proved successful in the calculation of a range of square cylinder and back-step flows containing the influences of active vortex structure that were present as low or high level contributors to the total fluctuation energy. Such a modeling approach may well be suitable for the representation of the wake–boundary layer interaction flows that have been discussed in detail in the present chapter.

V. Conclusions

In conclusion, the presented experimental results confirm the presence of large-scale vortical structures in the interaction region. These are originated in the wake, and they strongly affect the turbulence field of the boundary layer. More specifically, their presence alters both the vorticity content and its distribution in the boundary layers. As a result an influence of the energy transfer processes occurs manifested by an observed alteration in the distribution of the estimated terms contributing in the turbulent kinetic energy equation.

Production and dissipation are the most affected terms by the interaction with implications on turbulence modeling. Also, the observed alterations persist in the post-interaction region affecting its structure.

Finally an alternative computational procedure has been proposed that is considered suitable for the effective representation of the preceding flows.

Acknowledgment

The presented research work was supported by the BRITE-EURAM/ AERONAUTICS-2076/2032 project of the European Communities.

References

[1]Goradia, S. H., and Colwell, G. T., "Parametric Study of a Two-dimensional Turbulent Wall-Jet in a Moving Stream with Arbitrary Pressure Gradient," *AIAA Journal*, Vol. 9, 1971, pp. 2156–2165.

[2]Omar, E., Zierten, T., Hahn, M., Szpiro, E., and Mahal, A., "Two-Dimensional Wind Tunnel Tests of a NASA Supercritical Airfoil with various High-Lift Systems," NASA CR-2215, 1973.

[3]Pot, P. J., "Measurements in a Two-Dimensional Wake and in a Two-Dimensional Wake Merging into a Boundary Layer," Rept. NLR TR-79063U, 1979.

[4]Bario, F., Charnay, G., and Papailiou, K. D., "An Experiment Concerning the Confluence of a Wake and a Boundary Layer," *Journal of Fluids Engineering*, Vol. 104, 1982, pp. 18–24.

[5]Brune, G. W., and Sikavi, D. A., "Experimental Investigation of the Confluent Boundary Layer of a Multielement Low Speed Airfoil," AIAA Paper 83-0566, 1983.

[6]Tulapurkara, E. G., Ramjee, V., and Rajasekar, R., "Interaction Between Boundary Layer and Wakes of Different Bodies," *AIAA Journal*, Vol. 28, No. 6, 1990, pp. 1032–1037.

[7]Zhou, M. D., and Squire, L. C., "The Interaction of a Wake with a Turbulent Boundary Layer," *Aeronautical Journal*, Vol. 89, 1985, pp. 72–81.

[8]Savill, A. M., and Zhou, M. D., "Wake/Boundary Layer and Wake-Wake Interactions—Smoke Flow Visualization and Modelling," *Proceedings of 2nd Asian Congress of Fluid Mechanics*, 1983, pp. 743–752.

[9]Agoropoulos, D., and Squire, L. C., "Interactions Between Turbulent Wakes and Boundary Layers," *AIAA Journal*, Vol. 26, No. 11, 1987, pp. 1194–1200.

[10]Hirota, M., Fujita, H., and Yokosawa, H., "Influences of Velocity Gradient on Hot-Wire Anemometry with an X-Wire Probe," *Journal of Physics E. Scientific Instruments*, Vol. 21, 1988, pp. 1077–1084.

[11]Sawhney, A. K., *A Course in Electrical and Electronic Measurements and Instrumentation*, Dhanpat Rai & Sons, 1996, Chap. 3.

[12]Strohl, D. E., and Comte-Bellot, G., "Aerodynamic Effects due to Configuration of X-Wire Anemometers," *Journal of Applied Mechanics, ASME*, Vol. 40, 1973, pp. 661–666.

[13]Swaminathan, M. K., Rankin, G. W., and Sridhar, K., "Evaluation of the Basic Systems of Equation for Turbulence Measurements Using the Monte Carlo Technique," *Journal of Fluid Mechanics*, Vol. 170, 1986b, pp. 1–19.

[14]Christakis, D. G., and Papailiou, D. D., "Dynamics of Large Scale Structures in Turbulent Two Dimensional Wakes," *The Global Geometry of Turbulence*, edited by J. Jimenez, Plenum, New York, 1991, pp. 33–49.

[15]Bradshaw, P., “The Turbulence Structure of Equilibrium Boundary Layers,” *Journal of Fluid Mechanics*, Vol. 29, 1967, pp. 625–645.

[16]Townsend, A. A., *The Structure of Turbulent Shear Flow*, Cambridge Univ. Press, Cambridge, MA, 1976.

[17]Spalart, P. R., “Direct Simulation of the Turbulent Boundary Layer up to Re_θ = 1410,” *Journal of Fluid Mechanics*, Vol. 187, 1988, pp. 61–98.

[18]Skåre, P. E., and Krogstad, P. A., “A Turbulent Equilibrium Boundary Layer near Separation,” *Journal of Fluid Mechanics*, Vol. 272, 1994, pp. 319–348.

[19]Hinze, J. O., *Turbulence*, McGraw–Hill, New York, 1975.

[20]Clauser, F. H., “The Turbulent Boundary Layer,” *Advances in Mechanics*, Vol. 4, 1956, pp. 1–51.

[21]Koutmos, P., Mavridis, K., and Papailiou, D. D., “A Study of Unsteady Wake Flows past a 2-D Square Cylinder with and without Jet Injection into the Vortex Formation Region,” *Applied Scientific Research*, Vol. 55, 1996, pp. 187–210.

Chapter 15

Structure of the Velocity Field of a Turbulent Line-Fire Plume

Kalliopi N. Saini* and Demosthenes D. Papailiou†
University of Patras, Rion-Patras, Greece

I. Introduction

THE results of an experimental study on the structure of the velocity field of a turbulent plume formed above a line fire are reported in the present communication. The obtained measurements complement the findings of a similar experiment reported by Verriopoulos and Papailiou,[1] regarding the structure of the corresponding temperature field of the plume. Both experimental efforts have been part of an attempt to improve understanding of the structure and transfer processes in strongly buoyant turbulent plumes developing above forest fires and to provide needed information for their modeling. Particular emphasis was placed on the dynamics of large vortical structures known to exist in fire plumes that appear to participate in the physical mechanisms dominating the development and spreading of these catastrophic events.

Interest on line fires in relation to forest fire research stems from their close resemblance. As experience has shown, under certain circumstances situations encountered in forest fires after time elapses resemble forest fires in the sense of exhibiting approximately constant ratios of burning fuel per unit length.[2]

The main volume of published work on strongly buoyant fire plumes refers to round configurations and mainly consists of visualization studies and velocity–temperature measurements[3–6] establishing the general characteristics of the plume's structure and, also, identifying the presence of large-

*Research Associate, Laboratory of Applied Thermodynamics and Statistical Mechanics, Department of Mechanical and Aeronautical Engineering.

†Professor, Director of the Laboratory of Applied Thermodynamics and Statistical Mechanics, Department of Mechanical and Aeronautical Engineering.

scale buoyant vortices in the ascending column. However, the dynamic behavior, the influence of these structures on the plume's transition to its asymptotic regime, and, most important, their participation to turbulent transport were rather disregarded.

Experiments on the upper asymptotic region of both fire and conventional plumes also have been conducted mainly in round configurations[7–12] whereas only a limited number address line fires[13,14] and generally plane plumes.[15,16] These studies allowed the formulation of expressions regarding nondimensional distributions of the temperature and velocity in these regions.

Existing experimental evidence on fire plumes will also be discussed selectively in the following as compared to related findings of the present work. A more detailed review of previous work on the subject is presented in the mentioned communication on the line-fire plume's temperature field,[1] the main findings and conclusions of which are outlined next in order to provide the necessary background for the interpretation of the presented experimental results.

The obtained measurements in the mentioned investigation on the temperature field[1] covered the intermediate and upper regions of the plume allowing an accurate description of the form of the distribution of both the mean and statistical characteristics of turbulence in those regions confirming also the existence of self-similarity in the latter region. Furthermore, the conducted measurements and visual observation established the presence of large vortical structures in all three regions of the plume, namely, the lower flame region, the strongly buoyant intermediate region, and the self-similar upper region. The dynamic behavior of these structures, including interactions among themselves and the mean flowfield resulting to phenomena such as pairing and merging of vortices of various scales, was found to vary in the different regions of the plume. However, their presence dominated the development of the plume and the turbulent transfer processes such as mixing and especially entrainment.

II. Experiment

The investigated line-fire plumes were produced in an experimental setup that consisted of a line burner 2000 mm long, 20 mm wide, and 15 mm deep made from copper constructed in the Laboratory of Applied Thermodynamics.[1] The burner was cooled by water flowing into jackets attached to its sides while inlet and outlet water temperatures were measured by appropriately located thermocouples. The burner was placed along the centerline of the floor plane of a steel framed parallelepiped enclosure of dimensions 200 × 200 × 350 cm^3. The sides of the enclosure at the two ends of the burner were covered by two vertical asbestos plates. Two more asbestos plates were placed along the long sides of the burner covering the floorplane to prevent air entrainment from below. To protect the flow from air disturbances, two series of side screens were placed parallel to the length of the burner while partitions located appropriately at

150 cm from each screen isolated the experimental facility from the rest of the laboratory space. The fuel supplying the burner was methanol and was deposited in a constant head tank and fed to the burner through four evenly distributed orifices, after passing through a calibrated microflowmeter. A detailed description of the experimental setup is given in Ref. 1.

For measuring the mean and statistical characteristics of the plume's turbulent velocity field, laser-Doppler velocimetry (LDV) was employed, the apparatus consisting of a Spectra Physics He-Ne 15 mW laser source and TSI optics. These instruments were mounted on a specially constructed support with a transversing mechanism allowing movement in the vertical and horizontal directions. Magnesium oxide powder of particles of an average diameter equal to 10 μm was used for the seeding of the plume. The particles were released with a minimal velocity outside the plume's region where measurements were to be conducted and at a distance allowing them to reach an almost motionless state before entrained into the plume by the air's natural circulation.

For the selection of the appropriate sampling rate for LDV measurements, the results of flow visualization obtained from experiments by Verriopoulos and Papailiou[1] conducted under conditions similar to those presently described were used. The obtained information indicated that the prevailing eddies in the plume large-scale structure frequencies were between 10 and 100 Hz while smaller eddies were also detected with frequencies ranging between 100 and 300 Hz. Although the obtained information was adequate for defining an appropriate sampling rate, the measured mean velocity distribution along the vertical axis of the plume also was used to estimate the frequency range of the upward moving eddies by applying the nondimensional analysis suggested by Tennekes and Lumley.[17] The latter estimates indicated frequencies not exceeding the 290 Hz; therefore, a sampling rate of 500 Hz was finally adopted with each sample consisting of 30,000 digitized values.

The velocity measurements were conducted in a constant temperature environment of 20±1°C and covered distances from 4 to 32 burner widths, D, upward in steps of 1 D. Also in order to detect the possible presence of unwanted stratification developing in the vicinity around the plume, the environmental temperature was checked periodically by four thermocouples placed at a distance of 1 m from the apparatus and at heights 0.5, 1.0, 1.5, and 2 m.

The experiments were conducted at two source Grashof numbers, namely, 1.8×10^8 and 2.8×10^8. The source Grashof number is defined as $Gr_{D,Q} = WD^3/p\nu^3$ based on burner width D used as a characteristic length and heat output Q of the source, according to the widely accepted suggestion of Chen and Rodi.[12] In this definition W is the weight deficit, $W = Qg/C_pT_a$ being a quantity characterizing the buoyant strength of the fire. The values of the parameters of the investigated plumes are presented in Table 1. In this table the quantities M_0 and F_0 have been estimated by using the effective area of the source that is the area of the parallelepiped $L \times D$ while the Morton's

Table 1 Flow parameters of the plumes

$Gr_{D,Q}$	F_0, m^4/s	M_0, m^4/s	L_m, m
1.8^8 10^8	0.04	0.025	0.31
2.8^8 10^8	0.05	0.027	0.30

Note. M_0, the specific momentum, $M_0 = 2\pi\int_0^R U^2 r\,dr, R = D/2$; F_0, the buoyant force, $F_0 = 2\pi\int_0^R U_g \frac{\Delta\rho}{\rho} r\,dr$ and L_m, the Morton's length, $L_m = L_m = M_0^{3/4}/F_0^{1/2}$

length $L_m = M_0^{3/4}/F_0^{1/2}$, valid for both plane and line fires, is estimated accordingly. The aspect ratio D/L is not of importance for line fires because of its negligible magnitude. The conditions at the source are dictated here from the characteristic dimensions of the vertical structures.

The obtained measurements in the present experiment consisted of mean and statistical characteristic distributions of the turbulence field. The measured statistical quantities are the turbulent intensity and skewness and flatness factors along the vertical axis of the plume.

III. Experimental Results and Discussion

Considerable insight regarding the upward development of the plume's mean turbulent field can be gained by considering the change of its mean axial velocity U_{CL} in the vertical direction as depicted in nondimensional form in Fig. 1. Based on accumulated experimental evidence on plumes and buoyant jets, it is presently agreed by the majority of researchers that above the nondimensional height $\xi = Z/L_m = 5$, the flowfield attains the structure of a conventional plume exhibiting an asymptotic, self-similar behavior. According to Morton,[2] Kotsovinos and List,[16] and Baker et al.,[18] this region is characterized by a $\xi^{2/3}$ dependence. However, in deference to the preceding behavior the present measurements clearly indicate that this state was established in the investigated plume at considerably shorter distance from the source, namely, at approximately 12 diam corresponding to $\xi = 0.8$. Notwithstanding that for final confirmation of this finding further investigation covering a wider range of Grashof numbers is needed, it appears that self-similar behavior in two-dimensional line-fire plumes is attained earlier than the mentioned accepted distance from the source. The identified difference with previously investigated buoyant flows probably could be attributed to different initial conditions prevailing at the source of the line-fire plume, namely, its negligible specific momentum and probably its geometry. The measurements presented in Fig. 1 allowed the authors also to suggest

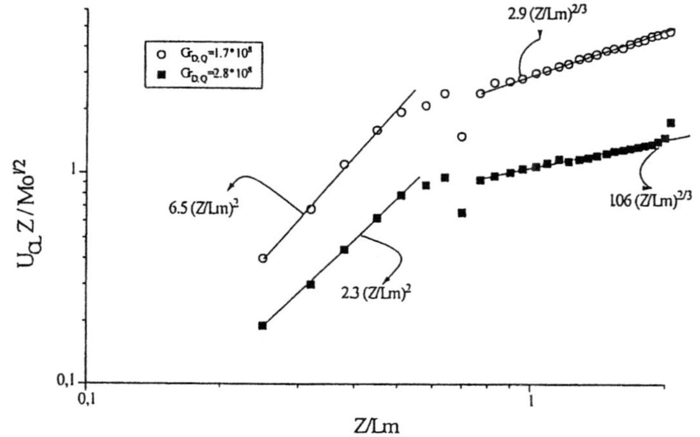

Fig. 1 Nondimensionalized axial mean velocity distribution of the line-fire plume.

the following expression relating the axial velocity with height in the upper region of the plume

$$U_{CL}/M_0^{1/2} = K_p(Z/L_m)^{-2/3} \quad \text{for} \quad \xi > 0.8 \tag{1}$$

where the value of K_p according to the obtained results appears to depend on the Grashof number $Gr_{D,Q}$, its estimated values being $K_p = 2.9$ and $K_p = 1.06$ for the lower and higher Grashof numbers, respectively. Further inspections of Fig. 1 reveal the existence of a region closer to the source, corresponding to the strongly buoyant intermediate region exhibiting a ξ^2 dependence.

The two regions, namely, the intermediate and self-similar upper one, appear separated by a transitional region in which the axial velocity undergoes a sudden dip. For a more convenient inspection of the successive changes of the axial velocity of the plume with upward distance from the source and for comparison with available visualization observation,[1,19] a direct plot of $U_{C,L}$ versus the distance Z is presented in Fig. 2. Aided by the visualization studies regarding the structure of the plume's flowfield,[1,19] an overall physical interpretation of its dynamic development with height covering all of its regions will be attempted in the following.

The obtained distribution of the axial velocity with distance from the source confirms the existence of three regions in the plume's development, as already stated.

In the lower, strongly buoyant region a linear increase of the axial velocity is observed of the form,

$$U_{CL} \approx K_k\,(Z/D) \tag{2}$$

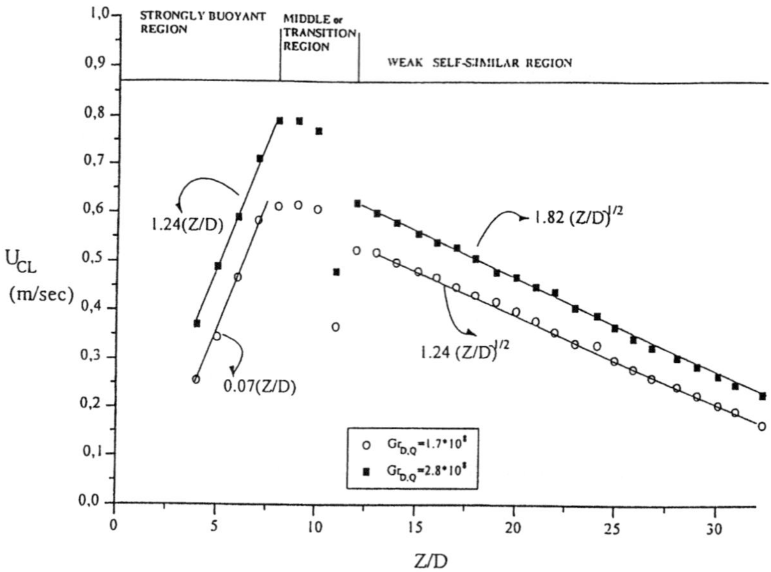

Fig. 2 Change of the mean axial velocity with vertical distance from the source.

where K_k was found to be Grashof number depended. As discussed by Verriopoulos and Papailiou,[1] the measured increase is apparently due to that identified in these experiments as "outer buoyant vortical structures responsible for stretching the flow field of the plume in the transverse direction as they roll upward."

In the same visualization studies the merging and subsequent detachment of these large-scale vortices was observed resulting to the subsequent shedding of large, hot structures into the plume. This observation appears to explain the observed dip of the axial velocity in the transitional region of the plume suggesting its relation to the shedding into this region of the mentioned hot structures.

Finally, further upward in the upper region the velocity follows the behavior of a conventional plume as expressed by the relation

$$U_{CL} \approx K_d \, (Z/D)^{-1/2} \tag{3}$$

where again K_d depends on the Grashof number.

The reported measurements regarding the plume's axial velocity variation with height present new conclusive evidence on the initial stages of its development not offered for either two-dimensional or axisymmetric fire plumes for which, as in conventional plumes, one is mainly referred to the upper self-similar region. Although further investigation covering a wide range of initial conditions at the source is needed to understand the processes involved, the observed axial velocity distribution in the lower strongly buoyant region appears to reflect to the nature of the entrainment mechanism processes as dictated by the large-scale vortical structure.

It should be noted that, as reported,[1] the upward axial mean temperature variation in the self-similar upper region of the line-fire plume was found to be inversely proportional to the distance from the source that is $\Theta_{CL} \sim (Z/D)^{-1}$. This form of distribution is similar to that observed in the corresponding region of two-dimensional conventional plumes. However, the corresponding axial mean velocity measurements obtained in the present experiments shown in Fig. 2 indicate a $(Z/D)^{-1/2}$ dependence with height to characterize the same region that is similar to that observed in two-dimensional turbulent free flows such as wakes and jets as shown in Table 2. In this table all two-dimensional cases follow the same law of axial velocity dependence on distance, that is, $U \sim (Z)^{-1/2}$. This observation probably indicates that turbulent structure similarities exist in the self-preserving region of these flows. According to the discussed experimental evidence, the structure of the temperature and velocity turbulent fields[5] in the upper region of the line-fire plume appear to be developing differently and, also, the temperature field is decaying faster than the corresponding velocity field. It is also plausible to suggest in this case that the observed differences might be associated with corresponding to differences in the transfer processes such as turbulent mixing in which the dynamics of the identified large-scale buoyant structure should be significant.

In Fig. 3 the development of the half-widths of mean velocity profiles of the plume in the vertical direction is presented for the lower Grashof number. They have been estimated for Gaussian distributions of the form,

$$U = U_{CL} \exp\,[-ln2(x/b_{0.5u})^2] \qquad (4)$$

selected to fit the measured velocity distributions.

The development of the velocity distribution half-widths with height is in agreement with the discussed findings of the vertical distribution of the axial velocity according to which a self-similar upper region develops above the distance of 12 diam from the source. The obtained linear dependence of the half-widths on the distance from the source in this region can be expressed as follows:

$$b_{0.5u} = K_{1u}(Z/D + K_{2u}) = 0.12(Z/D - 4) \qquad (5)$$

for $Z/D > 12$, where the experimental constants K_{1u}, K_{2u} were found to depend on Grashof number.

The virtual origin Z_0 of the plume's velocity field was defined from the relation $-Z_0 = K_{2u}/D$. $Z_0 = 2$ cm $= 1$ D. Given that the mean height of the detected flames for the investigated Grashof number is approximately the same, the virtual origin of the velocity field appears to be located above the source at a height comparable to that of the flames. The virtual origin of the corresponding temperature field at the same Grashof number was found to be located below the source.[1] Therefore, in agreement with the

Table 2 Free shear flows: axial velocity distribution[11]

Plane wake	Axisymmetric wake	Plane jet	Axisymmetric jet	Plane plume	Axisymmetric plume	Fire plume
$U \sim Z^{-1/2}$	$U \sim Z^{-2/3}$	$U \sim Z^{-1/2}$	$U \sim Z^{-1}$	$T \sim Z^{-1}$	$U \sim Z^{-1/3}, T \sim Z^{2/3}$	$U \sim Z^{-1/2}, T \sim Z^{-1}$

Note. $Z = Z' + Z_0$ indicates axial distance whereas Z_0 is equal to the virtual origin of the plume.

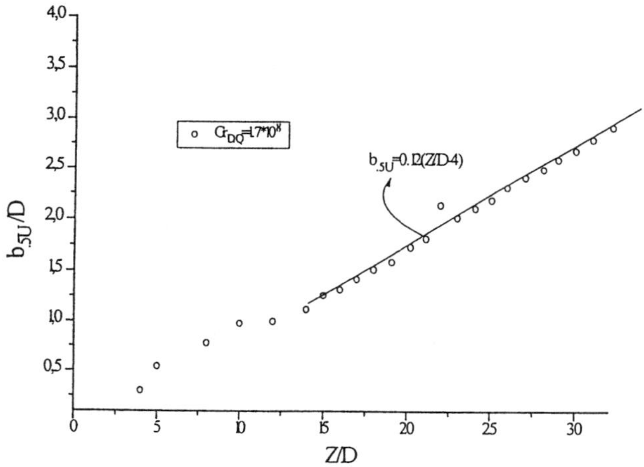

Fig. 3 Change of the half-width of the measured mean velocity distribution with height.

nature of the physical phenomena involved, the velocity field follows the temperature field with its initiation being the result of the buoyant forces generated by it.

The results of an attempt to gain a better insight into the dynamic development of the structure of the turbulent velocity field on the basis of measured axial distributions of a number of turbulent statistical quantities follow: turbulent intensity, as well as the skewness and flatness factors shown in Figs. 4, 5, and 6, respectively.

The axial distribution of the turbulent intensity shown in Fig. 4 follows the discussed pattern of the three successive regions of the plume's development. Contrary to the distribution of the mean axial velocity shown in Fig. 2, the turbulent intensity appears to decrease in the strongly buoyant region followed by a significant increase in the transitional region. Subsequently, it attains the asymptotic value 0.41 at a distance of the source equal to 12 diam corresponding to the initiation of the upper self-preserving region. Similar behavior was observed in the corresponding temperature turbulent intensity, the asymptotic value of which was reported equal to 0.51 (Ref. 1). As indicated in Table 3 where intensity levels in the self-preserving region are presented for different plume configurations, the attained intensities in the line plume are definitely higher than the other cases. This occurrence might be associated with the observed early appearance of the self-preserving region in this two-dimensional configuration.

Further information regarding the nature of the physical processes in the upward evolution of turbulence in the plume can be obtained from the examination of the skewness and flatness factor distributions shown in Figs. 5 and 6. A description of the physical meaning of these factors was given by Townsend.[9] Accordingly, the probability density function of a homogeneous turbulent field is joint-normal corresponding to values of zero for skewness

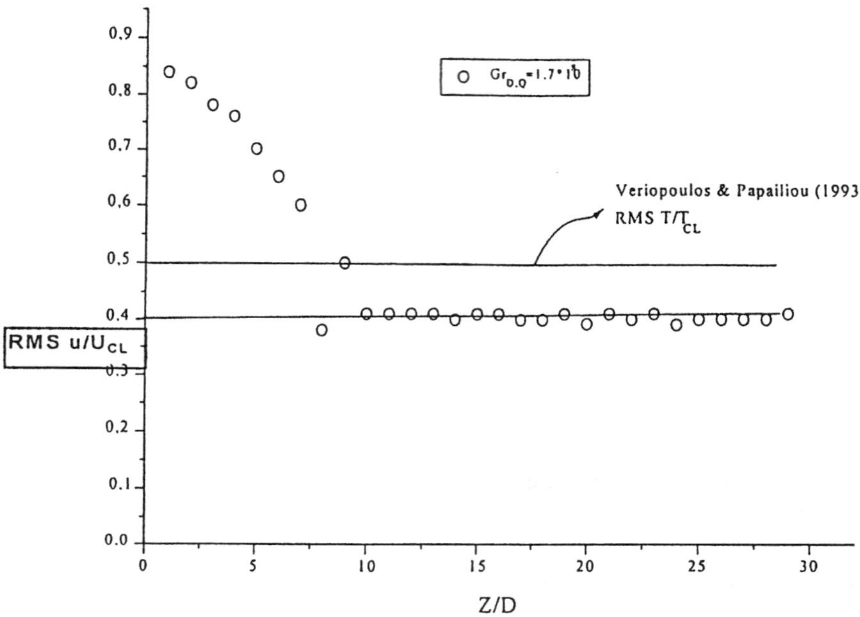

Fig. 4 Axial distribution of the velocity turbulent intensity.

and three for flatness. Deviation from these values indicates the presence of inhomogeneities in the turbulence flowfield with spotty character in the turbulence intensities and, also, transfer of turbulent kinetic energy from high intensity regions to low intensity regions. Furthermore, abnormally high values of these factors are interpreted as indicating the presence of large-scale structures in the flow. According to the preceding discussion, the obtained measurements are in complete agreement with the distributions presented earlier. The observed increased values above their normal level in the strongly buoyant lower region are indicative of severe nonhomogeneity caused by the large vertical structure action accompanied by intense transfer of turbulence energy. This event is followed by the abrupt change in both factors in the transitional region due to large-scale structure shedding with the final asymptotic approach to a state close to homogeneity characterizing the upper region of self-preservation of the plumes.

IV. Conclusions

In conclusion, the presented results complement those previously published by Veriopoulos and Papaliou on the structure of the corresponding temperature field of the line fire plume, offering a better understanding of turbulent strongly buoyant plumes. Besides confirming the existence of three successive regions in the plume's development, these measurements indicated a significantly earlier than previously reported appearance of the

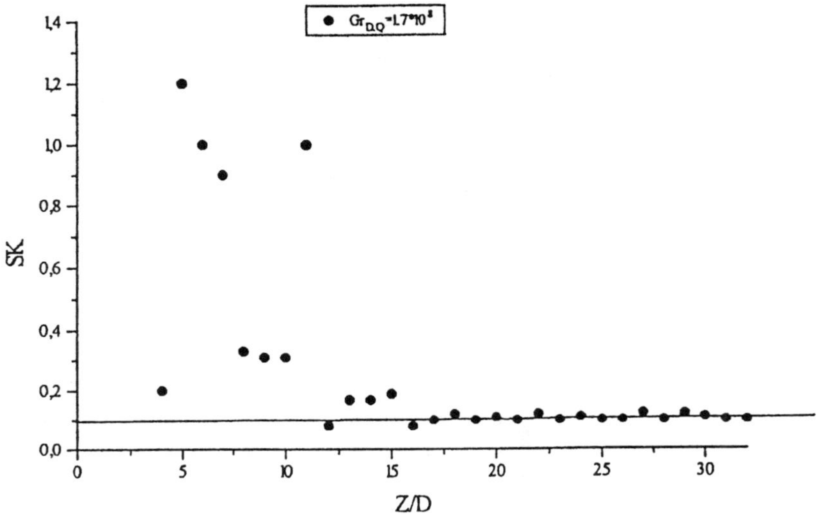

Fig. 5 Distribution of the skewness factor along the axis of the plume.

self-preserving region. The axial velocity decays faster with distance from the source than its temperature counterpart. Furthermore, the obtained distributions of the axial velocity as well as those of the measured statistical characteristics of the turbulent field offer not previously existing information on the structure of the strongly buoyant region and the process leading to the plume's transition towards its self-preserving state.

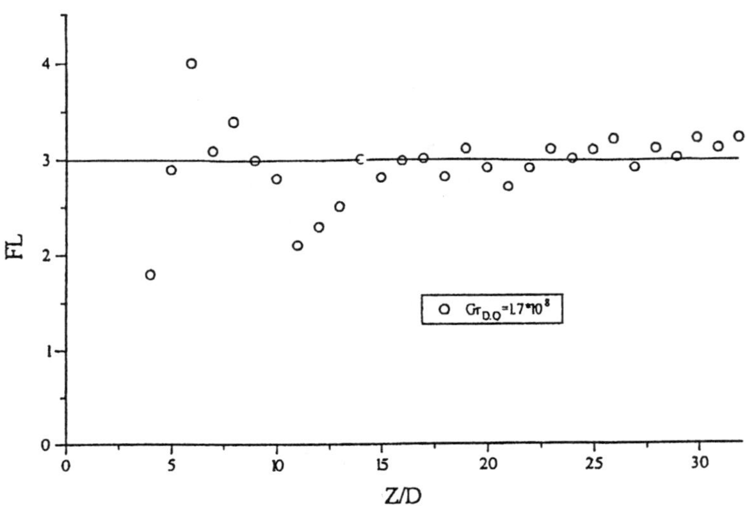

Fig. 6 Distribution of the flatness factor along the axis of the plume.

Table 3 Level of turbulent intensity of axial velocity at self-similar region[11]

Rouse et al.	George et al.	Nakagoma and Hirata	Kotsovinos	Papanicolaou and List	Chen and Rodi	Shabbir and George	Papailiou Group
RMS T/T_{CL} = –	0.38	0.36	0.42	0.42	—	0.40	0.51
RMS u/U_{CL} = –	0.28	0.25	0.38	0.25	—	0.33	0.41

Although further investigation is required, the strong participation of large-scale buoyant structures, present in all identified regions of the plume, is clearly evident in its development. On the basis of the presented turbulent velocity measurements further supported by existing visualization study,[1] the dynamics of the plume's structure as well as that occurring in its turbulent transfer processes are decisively influenced by these structures.

References

[1]Verriopoulos, P. A., and Papailiou, D. D., "Structure of the Turbulent Temperature Field of a Two-Dimensional Fire Plume," *Progress in Turbulent Research*, Vol. 162, Progress in Astronautics and Aeronautics, AIAA, Washington, DC, 1994, pp. 218.

[2]Lee, S. L., and Hellman, J. M., "Heat and Mass Transfer in Fire Research," *Advances in Heat Transfer*, Vol. 10, 1974, pp. 210–284.

[3]Zukoski, E. E., Cetegen, B. M., and Kubota, T., "Visible of Buoyant Diffusion Flames," *Twentieth International Symposium on Combustion*, Combustion Inst., 1984, pp. 361–366.

[4]Fischer, S. J., Harduin-Duparc, B., and Grosshandler, W. L., "The Structure and Radiation of an Ethanol Pool Fire," *Combustion and Flame*, Vol. 70, 1987, pp. 291–306.

[5]Weckman, E. J., and Sobiesac, A., "The Oscillatory Behavior of Medium-Scale Pool Fires," *Twenty-Second International Symposium on Combustion*, Combustion Inst., 1988, pp. 1299–1310.

[6]Davis, R. W., Moore, E. F., Roquemore, W. M., Chen, W. M., Villimpoc, V., and Gross, L. P., "Preliminary Results of a Numerical-Experimental Study of the Dynamic Structure of a Buoyant Jet Diffusion Flame," *Combustion and Flame*, Vol. 83, 1991, pp. 263–270.

[7]George, W. K., Alpert, R. L., and Tamanini, F., "Turbulence Measurements in an Axisymmetric Buoyant Plume," *International Journal of Heat and Mass Transfer*, Vol. 20, 1977, pp. 1145–1154.

[8]Papanicolaou, P. N., and List, E. J., "Statistical Properties of Tracer Concentration in Round Buoyant Jets," *International Journal of Heat and Mass Transfer*, Vol. 30, 1987, pp. 2059–1071.

[9]Papanicolaou, P. N., and List, E. J., "Investigation of Round Vertical Turbulent Buoyant Jets," *Journal of Fluid Mechanics*, Vol. 195, 1989, pp. 341–391.

[10]Papandoniou, D., and List, E. J., "Large Scale Structure in the Far Field of Buoyant Jets," *Journal of Fluid Mechanics*, Vol. 209, 1989, pp. 151–190.

[11]Shabbir, A., and George, W. K., "Experiments on a Round Turbulent Buoyant Plume," *Journal of Fluid Mechanics*, Vol. 275, 1994, pp. 1–32.

[12]Chen, C. J., and Rodi, W., *Vertical Turbulent Buoyant Jets*, Pergamon, New York, 1980.

[13]Rouse, H., Yih, C. S., and Humphreys, H. W., "Gravitational Convection from a Buoyant Source," *Tellus*, Vol. 4, 1952, pp. 200–210.

[14]Lee, S. L., and Emmons, H. W., "A Study of Natural Convection Above Line Fires," *Journal of Fluid Mechanics*, Vol. 11, 1961, pp. 353–368.

[15]Kotsovinos, N. E., "A Study of the Entrainment and Turbulence in a Plane Buoyant Jet," W. M. Keck Lab. of Hydraulics and Water Resources, Rept. No. KH-R-32, Caltech, Pasadena, CA, 1975.

[16]Kotsivinos, N. E., and List, E. J., "Plane Turbulent Buoyant Jets. Part I. Integral Properties," *Journal of Fluid Mechanics*, Vol. 81, 1977, pp. 25–44.

[17]Tennekes, H., and Lumley, J. I., *A First Course in Turbulence*, MIT Press, Cambridge, MA, 1972.

[18]Baker, C. B., Taulbee, D. B., and George, W. K., "An Analysis of Buoyant Jet Heat Transfer," *Proceedings of the 7th Heat Transfer Conferences*, Munich, Germany, 1980.

[19]Verripoulos, P. A., Ph.D. Thesis, Mechanical Engineering Dept., Univ. of Patras, Greece.

[20]Morton, B. R., "Forced Plumes," *Journal of Fluid Mechanics*, Vol. 2, 1959, pp. 151–163.

[21]Townsend, A. A., *The Structure of Turbulent Shear Flow*, Cambridge Univ. Press, London, England, UK, 1965.

III. MHD Turbulence and Atmospheric Turbulence

Chapter 16

Helical Turbulence and Study of Atmospheric and Magnetohydrodynamic Laboratory Flows

S. Moiseev*
Space Research Institute, Moscow, Russia
H. Branover† and A. Eidelman‡
Ben-Gurion University of the Negev, Beer-Sheva, Israel

IN THE present report we give a short review of the impact of small-scale helical turbulence on the processes in a continuous medium. The principal parameters defining these processes are the mean kinetic energy of turbulent fluctuations and the helicity $\langle V^t \text{ rot } V^t \rangle$, V^t being the velocity of turbulent fluctuations and $\langle \ldots \rangle$ the average over the ensemble. Using the transfer rates of the two mentioned parameters, we can define the most characteristic turbulence spectra of incompressible fluid. We also discuss some other spectra connected with helicity fluctuations, taking into account the compressibility of the medium as well. We draw attention to the similarity of theoretical and observation results for atmospheric turbulence and magnetohydrodynamic (MHD) flows.

Further we briefly dwell upon the results relevant to the generation of large-scale structures, mainly for convective media. The generation of a tropical cyclone may be consistent with the theory of helical structures generation. The decrease in turbulent viscosity in flows with helical turbulence is discussed. Experimental results confirming this effect are presented. These peculiarities, as well as the absence of energy transfer in the small-scale helical range ($k = -7/3$), may be, in the long run, the cause of drag reduction in turbulent flows.

---

*Professor, Dept. of Cosmogeophysics, Profsoyuznaya Str. 84/32, Moscow 117810
†Professor, Head of the Center for MHD Studies, POB 653, Beer-Sheva 84105
‡Chief Scientist, Center for MHD Studies, POB 653, Beer-Sheva 84105.

I. Certain Properties of Systems Involving Helical Motion

The Euler equations of inviscid flow admit an integral invariant for every volume bounded by a surface moving with the velocity field, on which $\boldsymbol{\omega} \cdot \boldsymbol{n} = 0$, namely, the helicity

$$H_v = \int \boldsymbol{u} \text{ rot } \boldsymbol{u} \, d\boldsymbol{x} = \text{const} \tag{1}$$

where $\boldsymbol{u}$ is velocity, $\omega = \text{rot } \boldsymbol{u}$. Vortex lines behave in this case like material lines. Therefore, this invariant, the helicity of the flow, is associated with the degree of linkage of vortex lines and has topological significance.[1] Helicity, just as kinetic energy, is a most important characteristic of turbulent media. For certain modes of turbulent motion, both the energy transfer rate $\varepsilon = du^2/dt$ and the helicity transfer rate $\eta = dH/dt$ are crucial parameters defining, for example, spectra formation.

Helical motions can be generated because of various conditions: rotation of a stratified medium (or its parts); charged particles motion in a nonuniform magnetic field; media under the condition of thermal convection; quasistationary vortical motions of an ideal liquid, etc. (see, e.g., Ref. 2). Recent research[3] has shown the possibility of such generation in an electrically conductive fluid under the action of the magnetic field. Thus, helical turbulence exists and determines the properties of a medium in various situations important for practical applications.

The appearance of mean helicity $\langle(\boldsymbol{V}^t \text{ rot } \boldsymbol{V}^t)\rangle$ in a stratified atmosphere rotating with an angular velocity ω is clear from Fig. 1. In fact, let us consider an ensemble of cells taking into account the action of Coriolis force. Cells coming upward broaden and, therefore, rotate. Cells coming downward compresses and rotates, too, but in the opposite direction. Thus, $\boldsymbol{V}^t$ sign reversal leads to rot $\boldsymbol{V}^t$ sign reversal, forming the helicity of the same sign for upward and downward cells. The reason of mean helicity of small-scale turbulence $\langle(\boldsymbol{V}^t \text{ rot } \boldsymbol{V}^t)\rangle \neq 0$ is quite clear now. Helicity estimation for atmospheric motion is

$$H = \langle(\boldsymbol{V}^t \text{ rot } \boldsymbol{V}^t)\rangle = \boldsymbol{V}^t \, \Omega_0 \sin \varphi \tag{2}$$

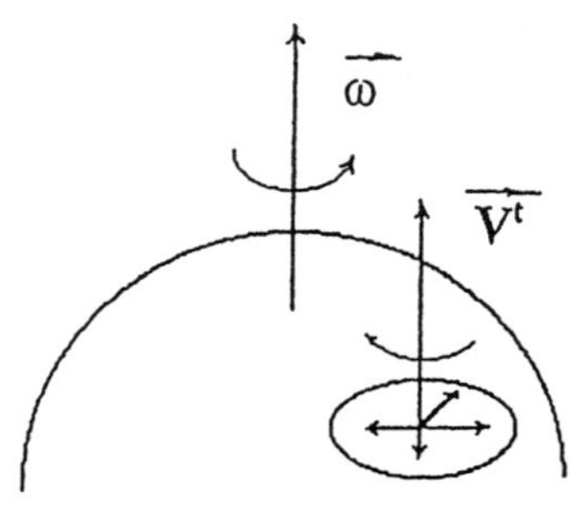

Fig. 1 On the mean helicity generation in the stratified atmosphere under rotation.

where Ω_o is the Earth's rotation angular rate and φ is the geographical latitude. Next is an example of helical motion described by the Euler equations and the equation for rot $\boldsymbol{u}$:

$$\begin{aligned} \partial \boldsymbol{u}/\partial t - [\boldsymbol{u} \cdot \mathrm{rot}\, \boldsymbol{u}] &= -\nabla(p/\rho + u^2/2) \\ \partial\, \mathrm{rot}\, \boldsymbol{u}/\partial t &= \mathrm{rot}\, [\boldsymbol{u} \cdot \mathrm{rot}\, \boldsymbol{u}] \end{aligned} \tag{3}$$

where ρ is the fluid density and p is the pressure. We consider the ensemble of small-scale helical motions with $\langle u \rangle = 0$. First, we restrict ourselves to the case of Beltramy flow rot $\boldsymbol{u} = \lambda \boldsymbol{u}$, where λ is a parameter. From the Euler equation we get the Bernoulli equation:

$$\langle p/\rho \rangle + \langle u^2 \rangle/2 = \mathrm{const} \tag{4}$$

with the constant that is the same over the whole region of the helical motion and not only for streamlines, as is usually. The characteristic property of Eq. (4) is that in some spatial region with increasing helical motion the average pressure decreases.

The following genesis of the large-scale coherent structure can be suggested. Let a quasi-Beltramy's flow described by rot $\boldsymbol{u} = \lambda \boldsymbol{u} + \boldsymbol{b}$ exist in a certain region. The sign-changed term $\langle \int [\boldsymbol{u}\, \boldsymbol{b}] \mathrm{d}\boldsymbol{l} \rangle$ appears at the averaging over the small-scale ensemble in an Eq. (4) type of equation. The Bernoulli equation for helical motion is true within reasonable limits of the smallness of this term. Furthermore, the tendency to a quasistationary state (local Beltramization) will arise in small regions.

Coriolis force twists convective motions in the atmosphere, transforming them into helical ones (see, e.g., Ref. 2). In a medium with helical motion, vortex groups having a tendency to merge are of great importance. Vortex merging exists in helical systems as an inverse transfer effect confined by the external pressure. In the case of helical vortices, the Bernoulli effect makes the tendency to merging stronger. This process will capture more and more new regions with the center near the helical maximum. As a result, the region of the large-scale perturbations with a lower pressure will turn into intense coherent structures. Numerical modeling, for example, that of turbulent dynamics in the solar convection zone by Brummell et al.,[4] shows how large-scale coherent structures and strong mean flows can coexist with intense small-scale turbulence.

Experimental investigations of the effect of vortex merging in a rotating convective medium was conducted by Zimin et al.[5] with $Re = 2\Omega H^2/\nu$, where Ω is the angular velocity of the facility rotation and H is the vertical dimension of the liquid layer. The liquid used was an electrolyte. Horizontal dimension of the convective cell was $L \sim H$ at Rayleigh numbers $Ra < 10^6$. With improving thermal insulation of the electrolyte at $Re \sim 200$–300, comparatively long-living large-scale vortices arise with $L \sim 6H$, looking in a plan like untwisting helices (see Fig. 2). Under horizontally uniform heat

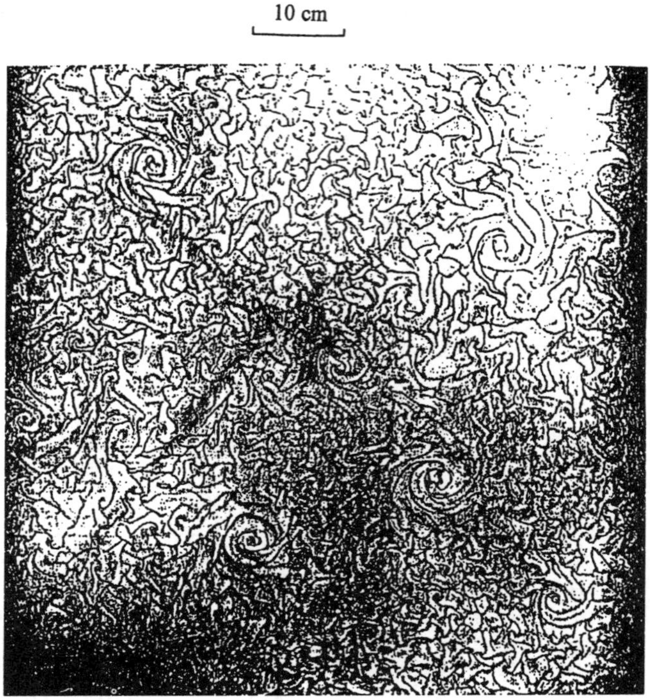

Fig. 2 Effect of vortex merging in a rotating convective medium.[5]

exchange conditions, they cover the entire surface of the layer. Isolated helical vortices are observed in the presence of large-scale horizontal inhomogeneity.

Despite the differences in atmospheric and MHD turbulent flows conditions, the observations show that a number of specific turbulence properties are similar. Our analysis has revealed the following prerequisites for this. Mean helicity of small-scale turbulent field may be considered nonzero in the atmosphere because of Coriolis force and stratification. Under MHD laboratory conditions, besides the flow inhomogeneity, a magnetic field is involved. Therefore, despite different modifications of electromagnetic forces in the equations of motion, Lorenz's force ability "to swirl the motion" manifests itself.

Nonlinear physics results are, as usual, of highly generalized nature. Thus, irrespective of the conditions, the existence of solitary structures is primarily defined by two factors: by dispersion providing for localization effect (solitons in dispersive media) or by the properties of integrals of motion (topological solitons).

Turbulence spectra are formed primarily because of the transfer of one or another physical characteristic (helicity, kinetic energy) from its source to the sink. This basic process inherent both to atmospheric and MHD turbulence is revealed in spectral observations.

Following recent studies conducted in the Space Research Institute, Moscow, the process of generation of large-scale structures (typhoons and other natural disasters) is shown to have several specific properties. The most important among them are the processes of inverse transfer, partial regularization of small-scale random field, and wave enhancement.[6] We may hypothesize that, in general, the process of large-scale structure generation in MHD case and in the atmosphere is similar because of a helical mode of motion.[7]

Here we would like to emphasize certain principal features of helical motions revealed in the examples considered, as well as of those anticipated. In case of helical motion and a respective pressure drop, there exist groups of vortices with a tendency to merge together generating larger vortices. The role of inverse transfer increases leading to the transfer of helicity and kinetic energy toward larger scales. Average turbulent viscosity in an ensemble of vortices drops, and the generated larger vortices are more long-living because their viscous damping is lower. Viscosity decrease, partial (in the ideal case, complete) interruption of energy transfer to smaller vortices, and its dissipation into heat should, particularly, lead to drag reduction in channel or pipe flows, as well as in case of overflow of bodies.

The question of two-dimensional or helical nature of turbulence arises in the attempt to apply one model or another to the description of turbulence having a pronounced quasi-two-dimensional character under constraint. Thus, the question becomes, what is an adequate model for such quasi-two-dimensional turbulence?

We can hardly give an absolutely definite answer to this question now. Indeed, both models lead to an intriguing effect that has been confirmed in numerous natural phenomena—so-called inverse transfer, i.e., formation of larger vortices from smaller ones (see the preceding). In both models turbulent viscosity can become negative. But it is clear that the two-dimensional model's assumption of the absence of motion along one of the coordinates is rather strong. It introduces, on the one hand, enstrophy invariant and an inertial interval of the -3 spectrum, and, on the other hand, interferes, for instance, with the coherent nature of the occurring processes. A qualitative idea that such turbulence is two-dimensional with a bit of three-dimensionality, suggesting the adequacy of the two-dimensional model, has proved to be erroneous.

The preference has been given to the helical model. Obviously, quasi-two-dimensional structures, such as typhoons or vortices in rotating tanks are topologically three-dimensional with knotted streamlines. A large-scale vortex possesses a marked helicity, i.e., its structure corresponds to disturbances resulting from helical instability. Thus, inverse transfer in helical turbulence is a coherent process, whereas in the case of a two-dimensional model it is incoherent.

In the atmosphere, spectra adjacent to Kolmogorov's interval is closer to $k^{-7/3}$ (helical scaling) than to k^{-3} (enstrophy scaling). Arguments testifying to this are also provided by laboratory results on MHD turbulence spectra.[8] Golbraikh et al.[9] have considered, particularly, the behavior of undoubtedly three-dimensional helical turbulence in the magnetic field. They have shown the mechanism of quasi-two-dimensionality formation in such conditions.

II. Spectral Properties of Helical Turbulence

Large-scale atmospheric helical structures are generated in relatively outstanding regions on the Earth: strongly inhomogeneous flows, overflowing of uneven terrain, thermal convection. As for helical scaling of wind spectra, it is almost universal.[10] Spectral slope of helical turbulence in MHD flow was obtained by Moiseev et al.[11] on the basis of group analysis of the equation of characteristic functional. Here we examine scaling properties of helical turbulence in non-stratified and stratified media following Moiseev and Chkhetiani.[12] The problem of helical cascades was qualitatively discussed from a phenomenological viewpoint by Brissaud et al.,[13] but that work did not lead to conclusive evidence of the existence of helicity cascades. Let us emphasize that a group approach allows us to obtain more strict results than usual dimensional consideration.[13] For instance, it becomes possible to substantiate Kolmogorov's spectrum (see, e.g., Ref. 14).

Helical turbulence in the atmosphere is, in particular, connected with a cumulative action of stratification and rotation. The equations of motion, continuity and heat conduction were considered under external random forces following Moiseev et al. (1976) in the Boussinesq's approximation and using assumptions suggested by Sazontov (1979):

$$\partial u_i/\partial t + u_k \partial u_i/x_k = -\partial p'/\partial x_i + g\beta\vartheta e_i + \nu\Delta u_i + f_i\,(\boldsymbol{x},t)$$

$$\partial\vartheta/\partial t + u_k\partial\vartheta/\partial x_k = (\partial T/\partial z)\, u_j e_j + \chi\Delta\vartheta + q(\boldsymbol{x},t) \tag{5}$$

$$\partial u_k/\partial x_k = 0$$

where u_i are the components of the velocity field, p' and ϑ are the deviations of the pressure and the temperature from their mean values, ν is the viscosity, χ is the thermal diffusivity, $g\beta$ is the buoyancy parameter, $\partial T/\partial z$ is the gradient of mean temperature, and $\boldsymbol{e}$ is the unit vector in the vertical direction. The turbulent kinetic energy and the temperature fluctuations are maintained by the external random forces $f_i(\boldsymbol{x},t)$ and $q(\boldsymbol{x},t)$, which we assume to be Gaussian, uniform, and delta-correlated in time with zero mean values:

$$\langle f_i\,(\boldsymbol{x},t)\rangle = \langle q(\boldsymbol{x},t)\rangle = 0$$

$$\langle f_i\,(\boldsymbol{x}_1,t_1)\, f_j\,(\boldsymbol{x}_2,t_2)\rangle = B_{ij}(\boldsymbol{x}_2-\boldsymbol{x}_1, t_2-t_1)\,\delta(t_2-t_1) \tag{6}$$

$$\langle q(\boldsymbol{x}_1,t_1)\, q(\boldsymbol{x}_2,t_2)\rangle = G\,(\boldsymbol{x}_2-\boldsymbol{x}_1, t_2-t_1)\,\delta(t_2-t_1) \tag{7}$$

The statistical properties of the velocity and temperature fields are described by the characteristic functional

$$\Phi = \langle \exp i \int \{y_i(\boldsymbol{x}) u_i(\boldsymbol{x},t) + z(\boldsymbol{x}) \vartheta(\boldsymbol{x},t)\} \, \mathrm{d}x \rangle \tag{8}$$

where $y_i(\boldsymbol{x}), z(\boldsymbol{x})$ are arbitrary smooth functions. It follows from the incompressibility condition that

$$\Phi \{y(\boldsymbol{x}) + \Delta\psi; z(\boldsymbol{x})\} = \Phi \{y(\boldsymbol{x}); z(\boldsymbol{x})\}$$

where ψ is an arbitrary function whose gradient tends rapidly to zero at infinity. Differentiating Eq. (8) with respect to time and taking into account Eq. (5), we obtain

$$\begin{aligned} \partial\Phi/\partial t = {} & i \int y_i(\boldsymbol{x}) \{D_k(\partial D_i/\partial x_k) + ig\beta e_i D_\vartheta\} \Phi \, \mathrm{d}\boldsymbol{x} + i \int z(\boldsymbol{x}) \{D_k(\partial D_\vartheta/\partial_k) \\ & + i(\mathrm{d}T/\mathrm{d}z) e_j D_j - i\chi\Delta D_\vartheta\} \Phi \, \mathrm{d}\boldsymbol{x} + I \end{aligned} \tag{9}$$

where $D_i = \delta/\delta y_i(x)$, $D_\vartheta = \delta/\delta z(x)$ are operators, and I is the source describing the influence of external forces

$$\begin{aligned} I = {} & -0.5 \int \{y_i(\boldsymbol{x}) \langle f_i(\boldsymbol{x}, t) \exp i \int (y_k u_k + z\vartheta) \, \mathrm{d}\boldsymbol{x}' \rangle + z(\boldsymbol{x}) \langle q(\boldsymbol{x}, t) \\ & \exp i \int (y_k u_k + z\vartheta) \, \mathrm{d}\boldsymbol{x}' \rangle\} \, \mathrm{d}\boldsymbol{x} \end{aligned} \tag{10}$$

We assume that the forces start to act at time $t = 0$ and that at $t = 0$ the fluid is assumed to be at rest

$$\boldsymbol{u}|_{t=0} = \vartheta|_{t=0} = 0, \qquad \Phi_{t=0} = 1 \tag{11}$$

Equation (9) for Gaussian forces that are delta-correlated in time can be closed. The source (10) can then be expressed in terms of Φ

$$\begin{aligned} I = {} & -0.5 \iint y_i(\boldsymbol{x}) y_j(\boldsymbol{x}_1) B_{ij}(\boldsymbol{x} - \boldsymbol{x}_1) \Phi \, \mathrm{d}\boldsymbol{x} \, \mathrm{d}\boldsymbol{x}_1 - 0.5 \iint z(\boldsymbol{x}) z(\boldsymbol{x}_1) \\ & G(\boldsymbol{x} - \boldsymbol{x}_1) \Phi \, \mathrm{d}\boldsymbol{x} \, \mathrm{d}\boldsymbol{x}_1 \end{aligned} \tag{12}$$

After averaging the energy balance equation of the fluid and the square of the temperature inhomogeneity, we obtain

$$\langle f_i u_i \rangle = 2\nu \langle (\partial u_i/\partial x_k)(\partial u_k/\partial x_i) \rangle - g\beta \langle \vartheta u_3 \rangle \tag{13}$$

$$\langle q \vartheta \rangle = 2\chi \langle (\partial\vartheta/\partial x_j)^2 \rangle - (\partial T/\partial z) \langle \vartheta u_3 \rangle \tag{14}$$

External random input of the helicity leads to an additional balance condition

$$\langle f_i \omega_i \rangle = \nu \langle (\partial u_i/x_j)(\partial \omega_j/\partial x_i) \rangle - 0.5 \, g\beta \langle \vartheta \, \omega_3 \rangle \tag{15}$$

The left-hand sides of Eqs. (13–15) can be expressed in terms of the corresponding correlators of the external forces according to the Furutsu–Novikov equation[15] yielding

$$0.5\, B_{ij}(0) = \varepsilon - g\beta\, \langle \vartheta u_3 \rangle \tag{16}$$

$$0.5\, G(0) = N - (\partial T/\partial z)\, \langle \vartheta u_3 \rangle \tag{17}$$

$$0.5\, C(0) = \eta - 0.5\, g\beta\, \langle \vartheta \omega_3 \rangle \tag{18}$$

where $\varepsilon = \nu \langle (\partial u_i/\partial x_j)^2 \rangle$ is the mean energy transfer rate; $C(0) = \langle \boldsymbol{f} \operatorname{rot} \boldsymbol{f} \rangle$ specifies the helicity input on the source scale; $\eta = \nu \langle (\partial u_i/\partial x_j)(\partial \omega_j/\partial x_i) \rangle$ is the mean helicity transfer rate; $N = \chi \langle (\partial \vartheta/\partial x_j)^2 \rangle$ is the transfer rate of the temperature inhomogeneities; and $g\beta\langle \vartheta u_3 \rangle$ is the mean work of the Archimedes force in the turbulent motion of the fluid.

We note one important consequence. We turn our attention to the equilibrium condition (15) for the helicity flow. In a stratified fluid the Archimedes force can serve as a helicity source. Attention has been already focused on this circumstance in a nonturbulent case.[16] To generate helicity, an additional source is not needed. A simultaneous coexistence of an intense temperature fluctuations source and a simple nonhelical velocity fluctuation source suffices.

Equation (9) with the source (12) and the initial conditions (11) is invariant under the following group of transformations:

$$\begin{aligned}
&\alpha \boldsymbol{x} = \boldsymbol{x}', \alpha L = L', \alpha L_T = L'_T, \alpha^{1-\gamma} t = t', \alpha^{-(\gamma+3)}\, y(\boldsymbol{x}) = y'(\boldsymbol{x}') \\
&\quad \alpha^{-(\delta+3)}\, z(\boldsymbol{x}) = z'(\boldsymbol{x}')\ \alpha^{1+\gamma}\, \nu = \nu', \alpha^{1+\gamma}\, \chi = \chi', \alpha^{3\gamma-1}\, \varepsilon = \varepsilon', \alpha^{3\gamma-2}\, \eta = \eta' \\
&\quad \alpha^{2\delta+\gamma-1}\, N = N'\ \alpha^{2\gamma-\delta-1}\, g\beta = (g\beta)', \alpha^{\delta-1}\, \partial T/\partial z = (\partial T/\partial z)', \alpha^{3\gamma-1}\, B_{ij} = B_{ij}' \\
&\quad \alpha^{3\gamma-2}\, C = C', \alpha^{2\delta+\gamma-1}\, G = G'
\end{aligned} \tag{19}$$

where L is the external turbulence scale, L_T is the temperature scale, and $\alpha > \gamma > \delta$ are arbitrary parameters ($0 < \alpha < \infty, \gamma, \delta < \infty$). The transformation (19) performs the transition to new measurement scales of the physical quantities. The uniqueness of the solution of the Cauchy problem for Eq. (9) leads to the similarity theorem

$$\begin{aligned}
&\Phi\, \{y(\boldsymbol{x}), z(\boldsymbol{x}), t, \varepsilon, \eta, N, \partial T/\partial z, g\beta, \nu, \chi, L, L_T\} \\
&= \Phi\{y'(\boldsymbol{x}'), z'(\boldsymbol{x}'), t', \varepsilon', \eta', N', (\partial T/\partial z)', (g\beta)', \nu', \chi', L', L'_T\}
\end{aligned} \tag{20}$$

Unlike the similarity hypothesis, Eq. (20) is exact.

We restrict ourselves below to consideration of stationary turbulence spectra for the scales $l^{-1} \gg k \gg L^{-1}$, where l is a certain internal scale that specifies the upper limit of the dissipation range. Let us consider the special case of a vanishing temperature gradient in the medium, i.e., $\partial T/\partial z = 0$. We consider the kinetic energy spectrum

$$E = -(1/8\pi^3) \iint \{D_i D_i \Phi \exp(-i\boldsymbol{kx})\}_{y=z=0}\, \mathrm{d}\boldsymbol{x}\, \mathrm{d}S(\boldsymbol{k}) \tag{21}$$

where $\mathrm{d}S(\boldsymbol{k})$ is an area element of the surface of sphere $|\boldsymbol{k}| = k$. Going over Eq. (21) to the functional (20) that depends on the primed variables and setting $\alpha = k$, we obtain

$$E(k) = k^{-2\gamma-1} f(k^{1+\gamma} \nu, k^{1+\gamma} \chi, k^{3\gamma-1} \varepsilon, k^{3\gamma-2} \eta, k^{2\delta+\gamma-1} N, k^{2\gamma-\delta-1} g\beta) \tag{22}$$

It follows from the arbitrariness of the parameters of the transformation group γ and δ that $\partial E/\partial\gamma = 0$ and $\partial E/\partial\delta = 0$. From the condition $\partial E/\partial\delta = 0$, we obtain the equation

$$2\xi_1 \partial f/\partial\xi_1 - \xi_2 \partial f/\partial\xi_2 = 0$$

where

$$\xi_1 = k^{2\delta+\gamma-1} N, \qquad \xi_2 = k^{2\gamma-\delta-1} g\beta$$

whose solution is

$$\begin{aligned} f &= f(k^{1+\gamma} \nu, k^{1+\gamma} \chi, k^{3\gamma-1} \varepsilon, k^{3\gamma-2} \eta, \xi_1^{1/2} \xi_2) \\ &\equiv f(k^{1+\gamma} \nu, k^{1+\gamma} \chi, k^{3\gamma-1} \varepsilon, k^{3\gamma-2} \eta, g\beta N^{1/2} k^{(5\gamma-3)/2}) \end{aligned}$$

From the condition $\partial E/\partial\gamma = 0$, we obtain the equation

$$6\zeta_1\, \partial f/\partial\zeta_1 + 6\zeta_2\, \partial f/\partial\zeta_2 f_2 + 5\, \zeta_3\, \partial f/\partial\zeta_3 + 2\zeta_4\, \partial f/\partial\zeta_4 + 2\zeta_5\, \partial f/\partial\zeta_5 = 4\mathrm{f}$$

where

$$\zeta_1 = k^{3\gamma-1} \varepsilon, \qquad \zeta_2 = k^{3\gamma-2} \eta, \qquad \zeta_3 = k^{(5\gamma-3)/2} g\beta N^{1/2}$$

$$\zeta_4 = k^{1+\gamma} \nu, \qquad \zeta_5 = k^{1+\gamma} \chi$$

The solution of this and the analogous equations provides spectra of the energy and the square of the temperature fluctuations, as well as a mixed spectrum of the temperature and the vertical component of the velocity.[12] We present classical (nonhelical) Kolmogorov's[18] and helical turbulence energy spectra without any assumptions regarding the characteristics of energy transfer

$$E(k) = \varepsilon^{2/3} k^{-5/3} \Psi_1(kL_{BO}) \tag{23}$$

$$E(k) = \eta^{2/3} k^{-7/3} \Psi_2[(kL_0)^{1/2}, kL_h] \tag{24}$$

In the case of helical turbulence, there is a helicity scale

$$L_h = \varepsilon / \eta \tag{25}$$

that separates the wave number range into two regions. We also have a characteristic scale $L_0 = \eta^{-5}N^3(g\beta)^6$ arising due to the action of stratification. In absence of a direct stratification influence on the spectra formation ($g\beta \to 0$ and $L_0 \to 0$), the spectrum (24) converts into

$$E(k) = \eta^{2/3}k^{-7/3}\Psi(kL_h) \tag{26}$$

Helical spectrum $E(k) \sim k^{7/3}$ is obtained under an additional assumption $\Psi(kL_h) = 1$, corresponding to experimental and observation data.

Let us examine in more detail the mentioned atmospheric observations. Wind spectra (see Fig. 3) computed by Boer and Shepherd[19] from the FGGE-IIIa global data set can be well approximated by a dependence with the exponent $-7/3$ in the global wave number range of 8–25. They have noted a strong up-scale energy transfer over the same scale range. The authors note that the spectral slope is shallower than -3, although they try to interpret the results within the two-dimensional model. Helical turbulence model adequately describes these spectra.

The results of wind velocity spectra measurements fulfilled in the framework of Global Atmospheric Sampling Program (GASP) (Ref. 20) are shown in Fig. 4.

We approximated wind spectra for wavelengths of 150–2400 km and for latitudes of 30 deg S– 60 deg N over two ranges: $-7/3$ in larger scales and $-5/3$ in smaller scales. The spectra correspond to the helical turbulence model. It is of interest that the scale L_η (25) increases at lower latitudes at the expense of decreasing helicity transfer rate η. Such a behavior of η whose magnitude is proportional to the helicity is determined by its dependence on the geographical latitude (2). These observations confirm the validity of the accepted approach and the adequacy of helical turbulence model to atmospheric motions.

Note that helical scaling of $-7/3$ is observed not only in atmospheric or MHD turbulence, but also in numerous laboratory studies: in presence of rotation,[21] stratification,[22] as well as in numerical studies with the appropriate external conditions.

Two more helical spectra were also obtained for helical turbulence under the following specific conditions. The spectral exponent equals -1 in the case of fast changes of structural functions in time and strong helicity fluctuations[23] (Altaisky et al., 1992), and -3 in a helical medium at Mach numbers of the order of unit.[12] Such a shape of the spectrum points to the weakening of the shock sink in the helical case. In the absence of helicity, $E(k) \sim k^{-2}$ (Ref. 24).

III. Generation of Large-Scale Helical Mean Motions

In the previous section we examined incoherent mechanisms of large vortices generation and the respective helical scaling. However, if the helicity $H = \langle (V' \text{rot } V') \rangle \neq 0$, large- scale instability can be developed in the medium and, respectively, a helical vortex is generated with scales depending on the velocity field helicity, turbulence energy, and correlation times of energy and

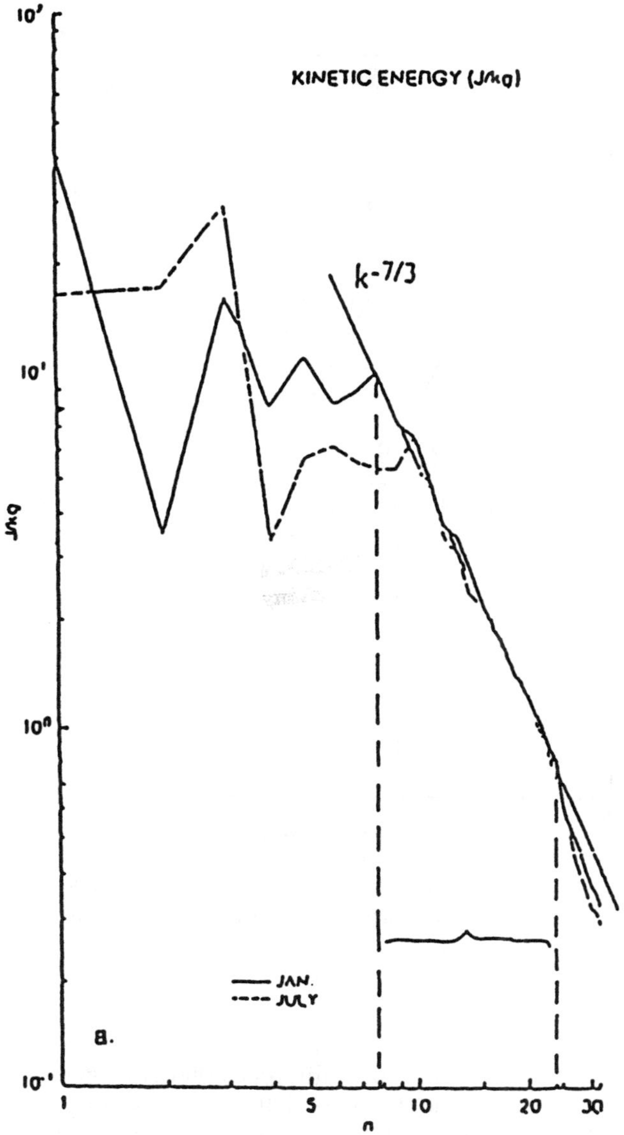

Fig. 3 Integrated kinetic energy spectra for January and July.[19] The straight line with −7/3 slope in the scaling range $n = 8$–24 is drawn by us.

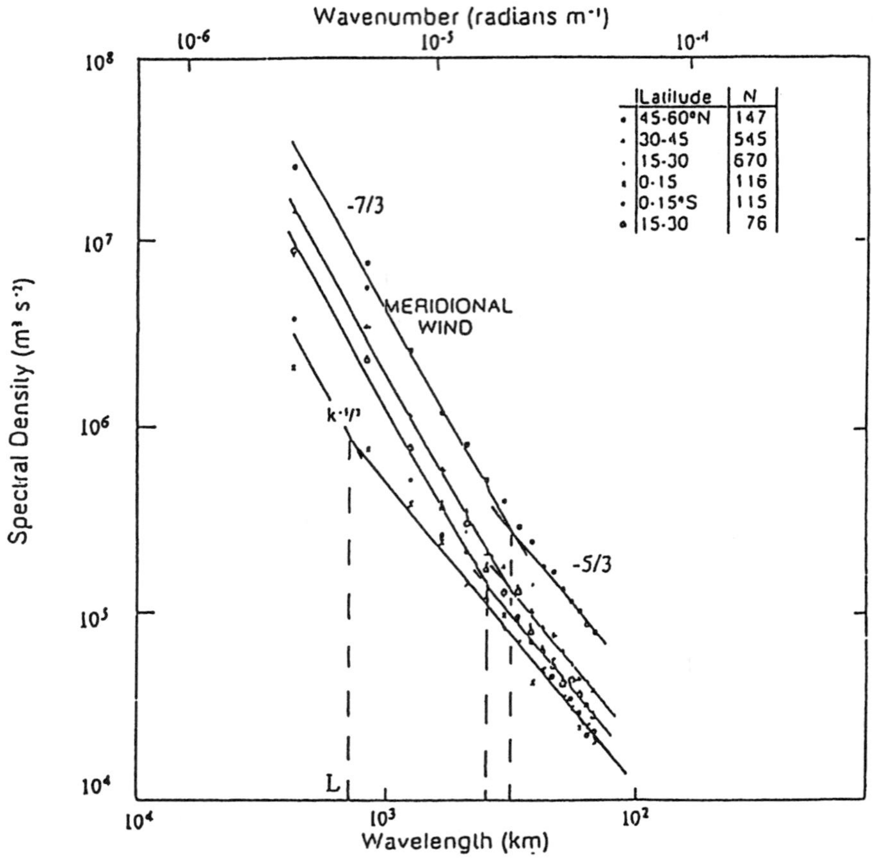

Fig. 4 Meridional wind spectra of GASP data[20] variance by latitude in the trophosphere. The straight lines approximating spectra with helical turbulence slope are drawn by us.

helical characteristics. Such a vortical dynamo is largely similar to a hydromagnetic dynamo with an essential difference that only mean helicity of the field $\boldsymbol{V}^t$ is insufficient for the rise of a coherent process. It is necessary for coherent generation that the structure of nonlinear terms at the interaction of $\langle \boldsymbol{V} \rangle$ and $\boldsymbol{V}^t$ fields were not reduced to a pure symmetrical one; here $\langle \boldsymbol{V} \rangle$ is the generated mean vortex field. Note that for the generation of correlation characteristics, $H \neq 0$ is sufficient if finite correlation time is taken into account.[25]

In the present section we examine only the problem of mean velocity generation. We are summing up a number of known results and discussing some new consequences. The equation for mean velocity $\langle \boldsymbol{V} \rangle$ is obtained as a result of averaging over the small-scale turbulent ensemble. It appears[2] that characteristic features of the process can be described by the model equation:

$$\partial\langle V\rangle/\partial t = \alpha \operatorname{rot}\langle V\rangle + \nu\Delta\langle V\rangle \qquad (27)$$

where $\alpha = \langle V^t \operatorname{rot} V^t\rangle\,\tau = H\tau$ and τ is the correlation time. A new remarkable quality of Eq. (27) is the dissipative "antiviscous" force resulting from the term α rot $V \sim L^{-1}$ providing for the feedback between motions along different axes. This force decreases more slowly that the viscous force ($\sim L^{-2}$) with increasing characteristic scale L, and, hence, Eq. (27) describes instability. Particularly, the perturbations of the form $\langle V\rangle = V_0(\cos qz, \sin qz, 0)\exp(\gamma t)$ increase with an increment $\gamma = \alpha q - \nu q^2$.

Let us discuss qualitatively the causes of the helical structure localization, the medium inhomogeneity being one of them. If the helicity parameter in Eq. (27) is inhomogeneous and, for simplicity, axisymmetric, i.e., $\alpha = \alpha(r)$, then the solution in the vicinity of $\alpha(r)$ maximum α_0 is of the form[26]:

$$\langle V\rangle = \langle V_0\rangle\{0; \sin(R_\alpha r/2L_\alpha); \cos(R_\alpha r/2L_\alpha)\}\exp(\gamma_0 t - R_\alpha r^2/4L_\alpha{}^2) \qquad (28)$$

Here the parameter $R_\alpha = \alpha_0 L_\alpha/\nu \gg 1$, L_α is the characteristic scale of $\alpha(r)$ inhomogeneity, r is the cylindrical radius, and $\gamma_0 = (\alpha_0^2/4\nu)(1 - 4/R_\alpha)$ is the increment. Using the parameters of the tropical atmosphere and Eq. (28), one can obtain the characteristic scale Δ and the evolution time τ of the tropical cyclone (TC). The characteristic scale of mean helicity change is of the order of $L_\alpha \sim 10^3$ km (Ref. 26). From Eq. (2) one obtains $\alpha \sim 2\,\Omega_0 r_0 \sin\varphi$, where $r_0 \sim 10$ km is the scale of convective motion. Assuming a widespread estimate of turbulent viscosity equal to $\nu \sim 10^3$ m^2/s, the following values are obtained: $R_\alpha \approx 1.5\cdot 10^3 \sin\varphi$, $L \approx 50$ km/$(\sin\varphi)^{1/2}$, $\gamma^{-1} = \mathrm{T} \approx 30$ min/$\sin^2\varphi$. For the latitude $\varphi = 10$ deg, they are $L \approx 120$ km and $\gamma^{-1} \approx 1$ day showing good agreement with respective values for natural cyclones.

The similarity of MHD and atmospheric turbulence spectra exists due to the helical character of motions. This similarity has been confirmed in a recent study by Chkhetiani et al.[3] The inhomogeneity of the flow or of the magnetic field leading to nonzero helicity is, in the present case, one of possible conditions of the generation of coherent structures. Because of a relatively high helicity value under MHD laboratory conditions, the interval of disturbances where the helicity is essential is shifted toward small scales. Therefore, the question arises: over what scales is the coherent structure generated? The criterion of the coherent structure generation can be written as $N\,Re_{cr}(l/L) > 1$, where N is the Stewart number, l is the scale of the generated disturbances, and L is the characteristic dimension of the system.

IV. Viscosity and Drag Reduction in Media with Helical Turbulence

Direct computation of turbulent viscosity shows that mean helicity and its fluctuations decrease its value.[2] Indeed, using the representation of the helicity parameter $\alpha = \langle\alpha\rangle + \alpha'$, where $\langle\alpha'\rangle = 0$ and letting $\langle\alpha'(t)\,\alpha'(t')\rangle = 2D\delta(t - t')$, we average Eq. (27) over the helicity fluctuations and obtain the same equation with the substitutions $\alpha \to \langle\alpha\rangle$ and $\nu \to \nu - D$ showing the turbulent viscosity decrease.

We note also that the turbulent viscosity can essentially decrease[27] in helical turbulence with finite correlation time and nonzero mean helicity. It is important that under real conditions when there are neither homogeneity nor isotropy nonnegativity conditions for even moments do not limit the helicity value. Moreover, even under isotropic and homogeneous conditions, but with the account of sufficiently high correlation times of turbulence and intense helicity fluctuations, turbulent viscosity may reverse its sign. The condition of sign reversal in this case is $\tau_1 > \tau$, where τ_1 is the characteristic time of helicity fluctuations and τ is the characteristic correlation time of small-scale turbulence, supposing that the amplitude of helicity fluctuations is close to the average level. Therefore, negative viscosity observed in the experiments on the analysis of MHD flows[28] is well explained from the standpoint of helical turbulence model.

Thus, the principal properties of energy transfer are essentially changed in the helical turbulence mode. In contrast to Kolmogorov's turbulence, where energy losses occur at the expense of energy transfer from the source scale to the sink small-scale region with high dissipation at sufficiently high mean helicity, the usual transfer of turbulent energy into the viscous sink is interrupted.

In fact, the more oscillations that are excited in the system, the better are the chaotization conditions. Roughly, the time of correlation unlinking is $\tau \sim 1/N_0$ where N_0 is the number of excited oscillations.[29] Therefore, under the conditions of a direct transfer (splitting—the number of vortical harmonics grows), the chaotization process is accelerated. However, under the conditions of inverse transfer along the spectrum (merging—the number of vortical harmonics decreases), the chaotization process is slowed down or stops (quasilaminarization).

Drag coefficient for helical turbulent motions in channels and at the overflow of bodies under the mentioned conditions should decrease. Thus, in case of frictional drag (for instance, in channel flows), drag coefficient should approach its laminar value. Indeed, drag reduction observed also in MHD turbulent flows (see Ref. 30) and interpreted as laminarization was obtained in flows with significant turbulent fluctuations. This was revealed in our experiment, and the drag reduction range begins at the parameter Ha/Re value where the helical turbulence mode with the spectrum of $E(k) \sim k^{-7/3}$ is generated. Figure 5a shows friction factor dependence on the Ha/Re parameter obtained by Brouillette and Lykoudis[31] for a broad range of Re numbers in a smooth channel under a transverse magnetic field. Figure 5b presents a dependence of spectral exponent n on Ha/Re parameter obtained by the authors.[32] Apparently, at n values close to $-7/3$, corresponding to the appearance of helical turbulent mode, friction factor becomes close to its laminar value shown in Fig. 5a as Hartmann's solution.

V. Conclusions

1) The principal physical feature of helical media is the appearance of effective attraction for helical vortices and their merging.

2) Large-scale helical vortex structure is generated in a medium with

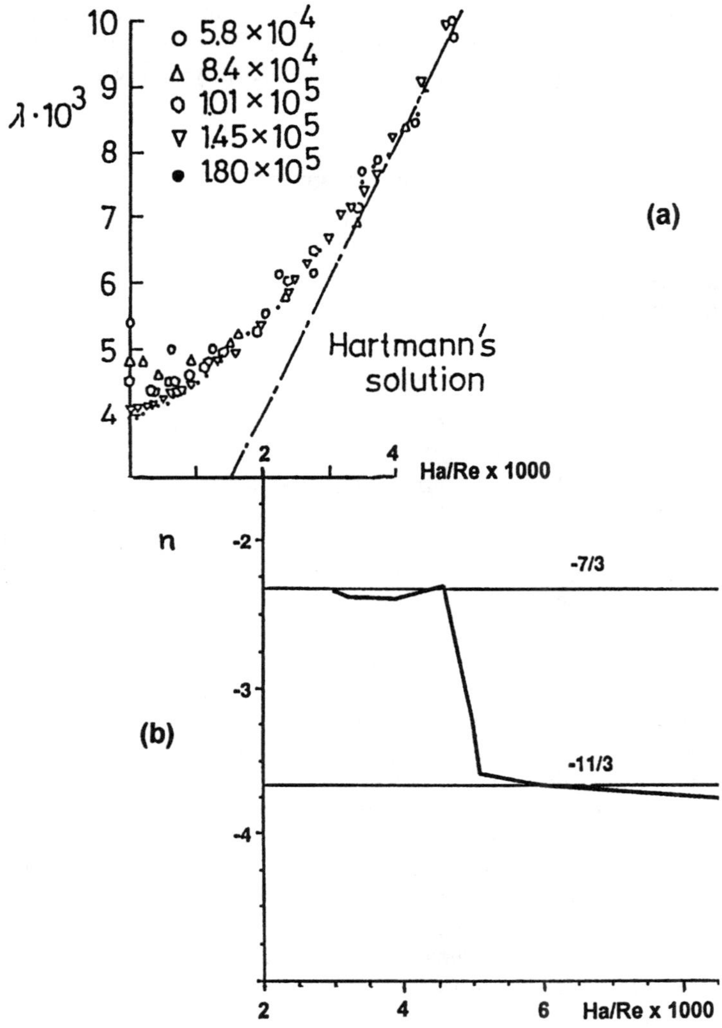

Fig. 5a Dependence of the friction factor λ (Ref. 31) as compared with Hartmann's solution for laminar flow on the parameter *Ha/Re*.
Fig. 5b Dependence of the friction factor λ (Ref. 32) as compared with turbulent spectra exponent n (Ref. 31) on the parameter *Ha/Re*.

small-scale helical turbulence, if this medium is inhomogeneous, or two-component, or else compressible.

3) The spectral scaling of helical turbulence is qualitatively specific: self-similarity regions appear with new exponents, in particular, $-7/3$.

4) Turbulent viscosity can be reduced and even become negative in helical turbulence. At a high enough mean helicity, turbulent energy transfer to a

viscous sink is interrupted. As a result, there must be drag reduction in channel flows and at the overflow of bodies under mentioned conditions.

5) The preceding phenomena exist both in geophysical (e.g., atmospheric) and in MHD flows.

Acknowledgment

This research was partly sponsored by the Ministry of Infrastructure and Energy, Israel (No. 95-11-026), and ECC Commission (AVI-CT92-0017). The authors are grateful to N. Goldbaum for her inestimable assistance in the article preparation.

References

[1]Moffatt, H. K., "The Degree of Knottedness of Tangled Vortex Lines," *Journal of Fluid Mechanics*, Vol. 35, 1969, pp. 1, 117–129.

[2]Moiseev, S. S., "Helical Mechanism of Large-Scale Structure Generation in Continuous Media," *Soviet Journal of Plasma Physics*, Vol. 16, No. 8, 1990, pp. 951–958.

[3]Chkhetiani, O., Moiseev, S., Golbraikh, E., and Eidelman, A., "On the Helicity Generation in Shear Flows in the External Magnetic Field," *Annales de Geophysique Suppl. II*, 1997, p. C607.

[4]Brummell, N., Cattaneo, F., and Toomre, J., "Turbulent Dynamics in the Solar Convection Zone," *Science*, Vol. 269, 1995, pp. 1370–1379.

[5]Zimin, V. D., Startsev, S. E., and Shaidurov, V. G., "Laboratory Modeling of the Early Stage of a Tropical Cyclone," *Chaos*, Vol. 1, No. 2, 1991, pp. 232–236.

[6] Moiseev, S. S., "On Physical Principles of Large-Scale Instabilities Prediction in Geophysics," *Proceedings of International Symposium on Nonlinear Variability in Geophysics,* Inst. d'Etudes Scientifiques de Cargese, 1993, p. 37.

[7]Eidelman, A., Moiseev, S., and Branover, H., "Quasi-Two-Dimensional Turbulence and Large-Scale Structure in MHD Flows," *Proceedings of International Workshop on Coherence and Turbulence,* Kiev, 1994.

[8]Branover, H., Bershadsky, A., Eidelman, A., and Nagorny, M., "Possibility of Simulating Geophysical Flow Phenomena by Laboratory Experiments," *Boundary Layer Meteorology*, Vol. 62, 1993, pp. 117–128.

[9]Golbraikh, E., Eidelman, A., Branover, H., Chkhetiani, O. G., and Moiseev, S. S., "On Helical Instability of MHD Turbulence, (present Proceedings) 1998.

[10]Eidelman, A., Branover, H., and Moiseev, S., "On the Experimental Verification of Helical Scaling," *Annales Geophysicae, Suppl. II*, Pt. 2, Vol. 13, 1995, p. 604.

[11]Moiseev, S., Branover, H., Eidelman, A. and Nagorny, M. 1994. Quasi-two-dimensional helical turbulence in MHD and geophysical flows. Proc. 2nd Intern. Conf. on Energy Transfer in MHD Flows, Aussois, France, v. 2, 777–785.

[12]Moiseev, S. S., and Chkhetiani, O. G., "Helical Scaling in Turbulence," *JETP Letters*, Vol. 83, No. 1, 1996, pp. 192–198.

[13]Brissaud, A., Frisch, U., Leorat, J., Lesieur, M., and Mazure, A., "Helicity Cascades in Fully Developed Isotropic Turbulence," *Physics of Fluids*, Vol. 16, No. 8, 1973, pp. 1366–1367.

[14]Altaisky, M. V., Moiseev, S. S., and Pavlik, S. I., "Scaling and Supersymmetry in

Spectral Problems of Strong Turbulence, *Physics Letters A*, Vol. 147, 1990, pp. 142–146.

[15]Novikov, E. A., "Functionals and Random Forces Method in the Turbulence Theory," *Soviet Physics—JETP*, Vol. 20, 1965.

[16]Lilly, D. K., "The Structure, Energetics and Propagation of Rotating Convective Storms," *Journal of Atmospheric Sciences*, Vol. 43, 1986, pp. 113–140.

[17]Hide, R., "Superhelicity, Helicity and Potential Vorticity," *Geophys. Astrophys. Fluid Dynamics*, Vol. 48, 1989, pp. 69–79.

[18]Sazontov, A. G., Izv. AN SSSR, *Fiz. Atmos. Okeana*, Vol. 15, 1979, pp. 820–828.

[19]Boer, G. J., and Shepherd, T. G., "Large-Scale Two-Dimensional turbulence in the Atmosphere," *Journal of the Atmospheric Sciences*, Vol. 40, 1983, pp. 164–184.

[20]Nastrom, G. D., and Gage, K. S., "A Climatology of Atmospheric Wavenumber Spectra of Wind and Temperature Observed by Commercial Aircraft," *Journal of Atmospheric Sciences*, Vol. 42, No. 9, 1985, pp. 950–960.

[21]Mory, M., and Hopfinger, E. J., "Structure Functions in a Rotationally Dominated Turbulent Flow," *Physics of Fluids*, Vol. 29, No. 7, 1986, pp. 2140–2146.

[22]Istweire, E. C., and Helland, K. N., "Turbulent Mixing and Energy Transfer in Stably Stratified Turbulence," 7th Symposium on Turbulence and Diffusion, Colorado, 1985, pp. 172–175.

[23]Altaisky, M. V., Klepikov, I. N., and Moiseev, S. S., "Experimental and Theoretical Investigations of Spiral Turbulence," *Annales Geophysicae, Suppl.*I, Vol. 10, 1992, p. 138.

[24]Kadomtsev, B. B., and Petviashvili, V. I., *Soviet Physics Doklady*, Vol. 18 1973, p. 115.

[25]Belyan, A. V., Golbraikh, E.I., Moiseev, S. S., and Chkhetiani, O. G., "Instability of Second Moments in Helical Turbulence," *JETP Letters*, Vol. 22, No. 5, 1996, pp. 20–24.

[26]Moiseev, S. S., Sagdeev, R. Z., Tur, A. V., Khomenko, G. A., and Shukurov, A. M., "Physical Mechanism of Amplification of Vortex Disturbances in the Atmosphere," *Soviet Physics Doklady*, Vol. 28, 1983, pp. 925–928.

[27]Belyan, A. V., Moiseev, S. S., and Chkhetiani, O. G., "On Turbulent Viscosity in Helical Turbulence," *Doklady Akademii Nauk* SSR, Vol. 334, 1994, No. 1, pp. 34–36.

[28]Henoch, C., Hoffert, M., Branover, H., and Sukoriansky, S., "Anisotropic Turbulence: Analogies Between Geophysical and Hydromagnetic Flows," *Current Trends in Turbulent Research*, Vol. 149, Progress in Astronautics and Aeronautics, AIAA, Washington, DC, 1993, 190–209.

[29]Zaslavsky, G. M., *Statistical Irreversibility in Nonlinear Systems*, Nauka, Moscow, 1970.

[30]Branover, H., *Magnetohydrodynamic Flow in Ducts*, Wiley, New York, 1978.

[31]Brouillette, E. C., and Lykoudis, P. S., "Magneto-fluid-mechanic Channel Flow. I. Experiment," *Physics of Fluids*, Vol. 10, No. 5, 1967, pp. 995–1001.

[32]Branover, H., Eidelman, A., Nagorny, M., and Kireev, M., "Magnetohydrodynamic Simulation of Quasi-two-dimensional Geophysical Turbulence," *Progress in Turbulence Research*, Edited by H. Branover and Y. Unger, Vol. 162, AIAA, Washington, DC, 1994, 64–79.

Chapter 17

On Helical Instability of Magnetohydrodynamic Turbulence

E. Golbraikh,* A. Eidelman,† and H. Branover‡
Center for MHD Studies, Ben-Gurion University of the Negev, Beer-Sheva, 84105, Israel
O. G. Chkhetiani§ and S. S. Moiseev¶
Space Research Institute, Royal Astronomical Society, Moscow, 117810, Russia

Introduction

THE effect of a steady uniform magnetic field on the turbulence becomes the principal problem in a conductive fluid at low magnetic Reynolds numbers Re_m ($Re_m = \mu_o \sigma UL$, where μ_o is magnetic permeability, σ is electrical conductivity and U and L are characteristic velocity and scale, respectively). Earlier experiments[1] have shown that turbulent velocity fluctuations are not completely suppressed in magnetohydrodynamic (MHD) flow. References 2 and 3 established that a uniform external magnetic field leads to the turbulent energy redistribution between the components of the velocity directed along and across the former. According to experimental data,[4–6] turbulence remains three-dimensional under the applied magnetic field B. Spectral dependence of turbulent energy E on the wave number k is of the form $E \sim k^{-m}$, where the exponent m changes with increasing magnetic field from $-5/3$ to $+7/3$ (at low B values). Then, with growing magnetic field, $m \rightarrow -11/3 \div +4$, and the turbulence becomes highly intermittent.[5]

In the most general case, three-dimensional turbulence is characterized not only by energy, but by the helicity as well. For instance, the correlational velocity tensor of the second order for isotropic turbulence involves a helical

*Chief Scientist, P. O. Box 653.
†Chief Scientist P. O. Box 653.
‡Professor, Head of the Center for MHD Studies, P. O. Box 653.
§Senior Researcher, Profsoyuznaya St., 84/32.
¶Professor, Cosmogeophysics Dept., Profsoyuznaya St., 84/32.

term. Therefore, we are studying helical turbulence in the presence of the magnetic field, i.e., turbulence with nonzero mean helicity.

Helical turbulence occupies a special position among the variety of turbulent motions. Helicity

$$H = \int_V (\boldsymbol{u} \cdot \boldsymbol{w})\, \mathrm{d}V$$

(here $\boldsymbol{u}$ is the velocity and $\boldsymbol{w} = rot\, \boldsymbol{u}$ is the vorticity), being a second invariant of Euler's equation just as energy,[7] has a great influence on the evolution and stability of turbulent and laminar flows. Apparently, helicity plays an important role in the magnetic fields generation and maintenance in astrophysical objects.[8]

In this paper we examine the behavior of helical turbulence under an external uniform magnetic field, particularly, the energy and the helicity of turbulence as a function of the magnetic field value. We assume that helical turbulence already exists in the MHD flow. Helicity generation in a laboratory MHD flow is a separate problem discussed in Ref. 9. As Shown in Ref. 10, in the hydrodynamic case (in the absence of external magnetic field), helical turbulence is unstable with respect to weak large-scale disturbances. Here the helicity plays the crucial role because the instability disappears in its absence. An applied magnetic field essentially changes the instability character.

The present paper involves several sections. In the first one we derive the principal MHD equation for the second moments in the presence of an external magnetic field. Based on this equation, we study in the second section the stability of second moments with respect to weak large-scale disturbances. In the third section we compare the obtained results with the available experimental data and in the conclusion we discuss the obtained results.

Principal Equations

Let us write the system of MHD equations in a dimensionless form for incompressible fluid:

$$\frac{\partial \boldsymbol{u}}{\partial t} + (\boldsymbol{u} \cdot \nabla)\boldsymbol{u} = -\nabla P + \frac{1}{Re}\Delta \boldsymbol{u} + \frac{N}{Re_m}[\mathrm{rot}\, \boldsymbol{B} \times \boldsymbol{B}] + \boldsymbol{F} \tag{1}$$

$$\frac{\partial \boldsymbol{B}}{\partial t} = \mathrm{rot}[\boldsymbol{u} \times \boldsymbol{B}] + \frac{1}{Re_m}\Delta \boldsymbol{B} \tag{2}$$

$$\mathrm{div}\, \boldsymbol{u} = \mathrm{div}\, \boldsymbol{B} = 0$$

where $\boldsymbol{F}$ is an external nonelectromagnetic force. The problem is characterized by three dimensionless numbers: Reynolds number $Re = UL/\nu$, magnetic Reynolds number Re_m, and magnetic interaction parameter $N = \sigma B^2L/\rho U$ (here ρ and ν are fluid density and viscosity). We study the situation with $Re_m \ll 1$ that is usually valid in laboratory MHD flows. In this case, Eq. (2) may be written in the form:

$$\mathrm{rot}[\boldsymbol{u}\times\boldsymbol{B}] = -\frac{1}{Re_m}\Delta\boldsymbol{B} \tag{3}$$

We represent all fields as a sum of averaged and fluctuational values:

$$\boldsymbol{u} = \langle\boldsymbol{u}\rangle + \boldsymbol{u}'; \qquad \boldsymbol{B} = \boldsymbol{B}_o + \boldsymbol{h}; \qquad P = \langle P\rangle + P'; \qquad \boldsymbol{F} = \langle\boldsymbol{F}\rangle + \boldsymbol{F}'$$

$$\boldsymbol{B}_o \gg \boldsymbol{h}; \qquad \langle\boldsymbol{u}'\rangle = \langle\boldsymbol{h}\rangle = \langle P'\rangle = \langle\boldsymbol{F}'\rangle = 0$$

where $\langle\ \rangle$ denotes averaging over an ensemble.

In this case, equations for $\boldsymbol{u}'$ and $\boldsymbol{h}$ take the form:

$$\frac{\partial\boldsymbol{u}'}{\partial t} + (\boldsymbol{u}'\cdot\nabla)\boldsymbol{u}' - \langle(\boldsymbol{u}'\cdot\nabla)\rangle + \boldsymbol{u}' = -\nabla P' + \frac{1}{Re}\Delta\boldsymbol{u}' + \frac{N}{Re_m}[\mathrm{rot}\,\boldsymbol{h}\times\boldsymbol{B}_o] + \boldsymbol{F}' \tag{4}$$

$$\mathrm{rot}[\boldsymbol{u}'\times\boldsymbol{B}_o] = -\frac{1}{Re_m}\Delta\boldsymbol{h} \tag{5}$$

For the sake of simplicity, we have assumed that $\langle\boldsymbol{u}\rangle = 0$ and $\boldsymbol{B}_o = B_o\cdot\boldsymbol{z}$ is constant ($B_o = 1$) and directed along the z axis. In Eq. (5) we keep only terms linear with respect to $\boldsymbol{h}$. In contrast to Ref. 3, we do not neglect the term involving the second moments in Eq. (4).

Let us examine the stability of a turbulent flow under the magnetic field with respect to weak large-scale nonuniform disturbances. In this case we represent turbulent field as a sum of the initial turbulent field $\boldsymbol{u}^{(0)}$ and its disturbance $\boldsymbol{u}^{(1)}$:

$$\boldsymbol{u}' = \boldsymbol{u}^{(0)} + \boldsymbol{u}^{(1)}$$

where $\boldsymbol{u}^{(0)} \gg \boldsymbol{u}^{(1)}$.

Before proceeding further, we derive an expression of the correlation function $< u_i'(\boldsymbol{x},t)u_j'\ (\boldsymbol{x} + \xi, t + \tau) >$. In Eqs. (4) and (5), we pass into the point $\boldsymbol{x} + \xi$, $t + \tau$, multiply them by $u_j'(\boldsymbol{x},t)$, and average them. Then we obtain

$$\langle h_i(\boldsymbol{x}+\xi,t+\tau)u_j'(\boldsymbol{x},t)\rangle = -\frac{Re_m}{\Delta}\langle\mathrm{rot}_i[\boldsymbol{u}'(\boldsymbol{x}+\xi,t+\tau)\times\boldsymbol{B}_o]u_j'(\boldsymbol{x},t)\rangle \tag{6}$$

$$\frac{\partial\langle u_i'(\boldsymbol{x}+\xi,t+\tau)u_j'(\boldsymbol{x},t)\rangle}{\partial\tau}+\nabla_k\langle u_k'(\boldsymbol{x}+\xi,t+\tau)u_i'(\boldsymbol{x}+\xi,t+\tau)u_j'(\boldsymbol{x},t)\rangle$$
$$=-\nabla_i\langle P'(\boldsymbol{x}+\xi,t+\tau)u_j'(\boldsymbol{x},t)\rangle$$
$$+\frac{1}{Re}\Delta u_i'(\boldsymbol{x}+\xi,t+\tau)u_j'(\boldsymbol{x},t)+\frac{N}{Re_m}\langle[\mathrm{rot}\,\boldsymbol{h}(\boldsymbol{x}+\xi,t+\tau)$$
$$\times\boldsymbol{B}_o]_i\,u_j'(\boldsymbol{x},t)\rangle+\langle F_i'(\boldsymbol{x}+\xi,t+\tau)u_j'(\boldsymbol{x},t)\rangle \tag{7}$$

where $rot_i[\ldots]$ is the ith component of the curl, and the operators ∇ and Δ are taken over ξ.

After certain transformations, the correlator in the right-hand side of Eq. (7) involving the magnetic field takes the form:

$$\frac{N}{Re_m}\langle[\mathrm{rot}\,\boldsymbol{h}\times\boldsymbol{B}_o]_i\,u_j'\rangle=\frac{N}{\Delta}\left(\nabla_i\nabla_m\langle u_l'u_j'\rangle-\nabla_m\nabla_l\langle u_i'u_j'\rangle\right)B_{om}B_{ol}$$
$$=\frac{N}{\Delta}\left(\nabla_i\nabla_3\langle u_3'u_j'\rangle-\nabla_3^2\langle u_i'u_j'\rangle\right) \tag{8}$$

where we have taken into consideration that $B_{om}=B_{ok}=B_{o3}$.

Let us introduce notations for correlation functions:

$$Q_{ij}^{00}(\xi,\tau)=\left\langle u_i^{(0)}(\boldsymbol{x},t)u_j^{(0)}(\boldsymbol{x}+\xi,t+\tau)\right\rangle$$
$$Q_{ij}^{10}(\boldsymbol{x},\xi,t,\tau)=\left\langle u_i^{(1)}(\boldsymbol{x},t)u_j^{(0)}(\boldsymbol{x}+\xi,t+\tau)\right\rangle \tag{9}$$

and, in the same manner, for correlators of higher orders.

Assuming that the external force $\boldsymbol{F}'$ maintains the initial helical turbulence, we can derive an equation for $Q_{ij}^{00}=Q_{ij}^{00}\,(\xi,\tau)$:

$$\frac{\partial Q_{ij}^{00}}{\partial t}+\nabla_k Q_{kij}^{000}=-\nabla_i\left\langle P^{(0)}u_j^{(0)}\right\rangle+\frac{1}{Re}\Delta Q_{ij}^{00}-\frac{N}{\Delta}(\nabla_i\nabla_3 Q_{3j}^{10}-\nabla_3^2 Q_{ij}^{10})+\left\langle F_i'u_j^{(0)}\right\rangle$$

In a similar way, we can write an equation for $Q_{ij}^{10}=Q_{ij}^{10}\,(\boldsymbol{x},\xi,t,\tau)$:

$$\frac{\partial Q_{ij}^{10}}{\partial t}-\frac{1}{Re}\Delta Q_{ij}^{10}+\nabla_k(Q_{ikj}^{100}+Q_{kij}^{000})=-\nabla_i\left\langle P^{(1)}u_j^{(0)}\right\rangle-\frac{N}{\Delta}(\nabla_i\nabla_3 Q_{3j}^{10}-\nabla_3^2 Q_{ij}^{10})$$

Excluding pressure from this equation, we obtain the final equation for Q_{ij}^{10}:

$$\frac{\partial Q_{ij}^{10}}{\partial \tau} - \frac{1}{Re} \Delta Q_{ij}^{10} + \Pi_{im} \nabla_k (Q_{mkj}^{100} + Q_{kmj}^{100}) = -N \cos^2 \theta Q_{ij}^{10} \qquad (10)$$

where $\Pi_{im} = (\delta_{im} - \nabla_i \nabla_m / \Delta)$ is the projection operator, $\cos^2 \theta = \nabla_3^2 / \Delta$.

Similarly to Eq. (10), we obtain an equation for third moments $Q_{ijk}^{100} = Q_{ij}^{10}\ (\boldsymbol{x}, \xi, \xi', t, \tau, \tau')$:

$$\left(\frac{\partial}{\partial \tau} - \frac{1}{Re} \Delta + N \cos \theta \right) Q_{ijn}^{100} + \Pi_{im} \nabla_k (Q_{mkjn}^{1000} + Q_{kmjn}^{1000}$$
$$- Q_{mk}^{10} Q_{jn}^{00} - Q_{km}^{10} Q_{jn}^{00}) = 0 \qquad (11)$$

To close the system of equations (10) and (11), we use the results of Ref. 11. In this approximation we obtain from Eq. (11):

$$Q_{ipj}^{100} = -\tilde{\tau} \Pi_{im} Z_{mkpj} \qquad (12)$$

where the three-point correlator

$$Z_{mkpj}(x,t,x,t,\xi,\tau,\xi',\tau') = Q_{mk}^{10}(x,t,x,t) Q_{pj}^{00}(\xi - \xi', \tau - \tau')$$
$$+ Q_{mp}^{10}(x,t,\xi,\tau) Q_{kj}^{00}(\xi',\tau') + Q_{mj}^{10}(x,t,\xi',\tau') Q_{kp}^{00}(\xi,\tau) \qquad (13)$$

$$\tilde{\tau} = \left(\frac{1}{\tau^*} + N \cos \theta \right)^{-1}$$

and the relaxation time $\tau^* \sim L_{\text{tur}} / E_{tur}^{1/2}$ (L_{tur} and E_{tur} being the characteristic scale and average energy of turbulent flow, respectively).

To substitute the expression (13) for the third moment into Eq. (10), one should pass to the limit $\xi' \to \xi, \tau' \to \tau$. Here (see Appendix) one should take into account that $A\delta_{ij} = Q_{ij}^{00}\ (0, 0)$ and $C\varepsilon_{ijp} = \partial Q_{ij}^{00}\ (\xi, \tau)/\partial \xi_p|\xi,\tau \to 0$ depend on B (or N), i.e., on the magnetic field. Passing to the limit $\xi' \to \xi$, $\tau' \to \tau$ in Eq. (13), we obtain the following equation:

$$\left(\frac{\partial}{\partial \tau} - \left(\frac{1}{Re} + A\tilde{\tau} + N \cos\theta \right) \Delta \right) Q_{ij}^{10}(\boldsymbol{x},t,\xi,\tau) - \tilde{\tau} C H_{ij}^{10}$$
$$= \tilde{\tau} \nabla_p \left[\frac{\partial}{\partial x_k} Q_{ip}^{10}(\boldsymbol{x},t,0,0) Q_{kj}^{00}(\xi,\tau) + \frac{\partial}{\partial x_s} Q_{kp}^{10}(\boldsymbol{x},t,0,0) \right.$$
$$\left. \times\ \nabla_k \xi_s Q_{ij}^{00}(\xi,\tau) + Q_{kp}^{10}(\boldsymbol{x},t,0,0) \nabla_k \xi_s Q_{ij}^{00}(\xi,\tau) \right] \qquad (14)$$

where $A = A(0, 0, N)$ and $C = C(0, 0, N)$ are scalar functions of N.

Instability of the Second Moments

To study the stability of the system (14), we apply the operator $\varepsilon_{lmi} \nabla_m$ to it and write the obtained system in a uniform form:

$$\left(\nabla_\tau - \frac{1}{Re}\Delta - \tilde{\tau}A\Delta + N\cos^2\theta\right)Q_{ij}^{10} - \tilde{\tau}CH_{ij}^{10} = 0$$
$$\left(\nabla_\tau - \frac{1}{Re}\Delta - \tilde{\tau}A\Delta + N\cos^2\theta\right)H_{ij}^{10} - \tilde{\tau}C\Delta_{ij}^{10} = 0 \tag{15}$$

An equation for Q_{ij}^{10} follows from Eq. (15):

$$[(\nabla_\tau - \nu_H\Delta + N\cos^2\theta)^2 + \tilde{\tau}^2C^2\Delta]Q_{ij}^{10} = 0 \tag{16}$$

where the effective viscosity $\nu_H = 1/Re + \tilde{\tau}A$. Passing into k space in Eq. (16), we derive an expression of the increment $\gamma = -i\omega$:

$$\gamma = -\nu_H k^2 - N\cos\theta + \tilde{\tau}|Ck|$$

From this expression we obtain, under the condition $\gamma > 0$, the instability condition:

$$\frac{1}{2}k_o\left[1 - \left(1 - \frac{4N\cos^2\theta}{k_o^2\nu_H}\right)^{1/2}\right] < k_b < \frac{1}{2}k_o\left[1 + \left(1 - \frac{4N\cos^2\theta}{k_o^2\nu_H}\right)^{1/2}\right] \tag{17}$$

where $k_o = \tilde{\tau}C/\nu_H$. As follows from Eq. (17), at $4N\cos^2\theta/k_o^2\nu_H \ll 1$ the instability condition acquires the form

$$0 < k < k_o \tag{18}$$

which coincides formally with purely hydrodynamic instability condition (9).

In fact, under the condition $A\tilde{\tau} \gg 1/Re$, $k_o \approx |C|/A$. In case of $Re_m \ll 1$, both energy A and helicity C decrease with growing magnetic field, as noted in Ref. 3. In case of weak magnetic fields, k_o may be represented, up to the term linear in N, as follows (see Appendix):

$$k_o = \frac{C_o}{A_o}\left(1 + \frac{\aleph N}{A_o}\right) \tag{19}$$

where C_o and A_o are turbulence helicity and energy in absence of the magnetic field. It follows from Eq. (19) that the upper limit wave number of the interval in Eq. (18) linearly grows with N.

Comparison with Experimental Results

The behavior of a turbulent flow of a conductive fluid (mercury) in the external uniform transverse magnetic field has been studied in a series of experiments carried out in the Center for MHD Studies of the Ben-Gurion University.[5] During these experiments, turbulence characteristics were studied at different values of the magnetic field and flow velocities. The stainless steel channel had a rectangular cross section of 2.8 × 5.6 cm. Experiments were performed mainly at mean velocity $U = 0.16$ m/s corresponding to $Re = 52 \cdot 10^3$; however, some of the experiments were performed at $U = 0.08$ m/s. Therefore, Hartmann number $Ha = BL(\sigma/\rho\nu)^{0.5}$ varied in the range of 60–1200, and intereaction parameter N varied in the range of 0.15–35, the magnetic Reynolds number being $Re_m \sim 5 \cdot 10^{-3} \ll 1$.

To eliminate penetration of any turbulence from upstream into the test section, a fine honeycomb was placed into the flow upstream of the test section. The flow entrance into a magnetic field is accompanied by a rearrangement of velocity profiles and by the generation of turbulent disturbances. To be efficient, the honeycomb had to be extended into the section with magnetic field. The honeycomb was made of electrically insulated small diameter tubes. The external diameter of the honeycomb tube was 2.4 mm, its wall thickness was 0.3 mm, and the honeycomb solidity was approximately 0.25. The length of the honeycomb in the flow direction was 25 cm.

The dependence of turbulence intensity u'/U on the interaction parameter is rather complicated. A number of turbulence modes can be revealed in the flow. Helical turbulence with a scaling $m = -7/3$ arises in the range of $Ha/Re \cdot 10^3 = 2.5$–4.5 and respective interaction parameter range $N = 0.3$–1.1. Longitudinal velocity fluctuations were measured and their spectra in this range (Fig. 1) were established in the experiment. Therefore, we can compare theoretical [Eq. (19)] and experimental values of the limiting wave number k for streamwise velocity fluctuations.

We assume that the experimental frequency value f_o separating small-scale spectral range formed by helical turbulence corresponds to the inflection point of the energy density dependence on f (see Fig. 1). Then, with growing magnetic field, this value of the frequency for the longitudinal velocity component normal to the magnetic field is shown in Fig. 2a as a dependence on the interaction parameter N. Note that a similar tendency to shift the spectrum of the longitudinal velocity component toward higher frequencies in a low magnetic field was also observed in Ref. 12. This change occurs in the range of helical mode turbulence.

Figure 2b shows spectral scaling m, where $m = -7/3$ corresponds to helical turbulence. The dependence of experimental f_o values on N qualitatively agrees with the expression (19) and allows to estimate the coefficient of N.

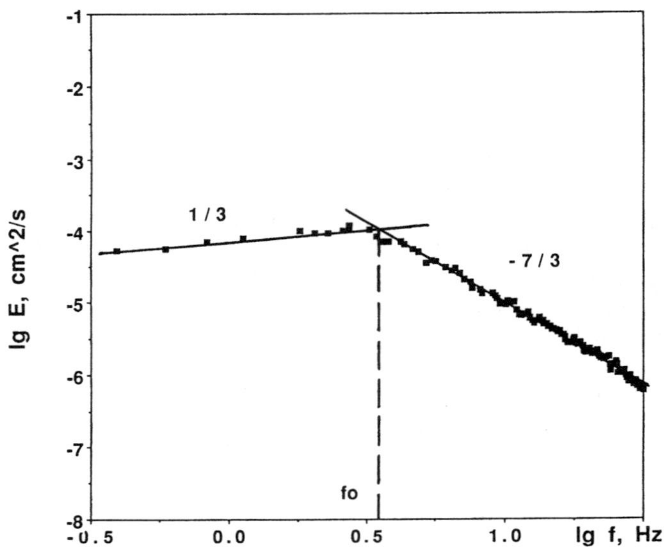

Fig. 1 Helical turbulence velocity spectrum with −7/3 scaling at the frequencies above the limiting value f_o.

Thus, helical turbulence in a uniform magnetic field is unstable with respect to weak large-scale disturbances. This leads to energy redistribution between small and large scales.

Appendix

The influence of an external uniform magnetic field on the behavior of correlations of a turbulent medium has been studied in Ref. 3. In presence of the magnetic field, the second moment of the velocity field acquires the form:

$$\widehat{Q_{ij}}(\boldsymbol{k},\omega) = \widehat{Q'_{ij}}(\boldsymbol{k},\omega) \Bigg/ \left[1 + \frac{(\boldsymbol{k}\cdot\boldsymbol{B}_o)^2}{\mu\rho} \frac{2\eta\nu k^4 - 2\omega^2 + (\boldsymbol{k}\cdot\boldsymbol{B}_o)^2/\mu\rho}{(\eta^2k^4+\omega^2)(\nu^2k^4+\omega^2)} \right]$$

where $\widehat{Q'_{ij}}(\boldsymbol{k},\omega)$ is a correlator in absence of the magnetic field, η, ν, ρ are magnetic and hydrodynamic viscosities and fluid density, respectively. Here we retain, for convenience sake, the notations applied in Ref. 3. Assuming that in the absence of the magnetic field the turbulence is helical and isotropic, we can write $\widehat{Q'_{ij}}(\boldsymbol{k},\omega)$ as follows:

$$\widehat{Q'_{ij}}(\boldsymbol{k},\omega) = \widehat{A_o}(\boldsymbol{k},\omega)\left(\delta_{ij} - \frac{k_ik_j}{k^2}\right) + i\widehat{C_o}(\boldsymbol{k},\omega)\varepsilon_{ijs}k_s$$

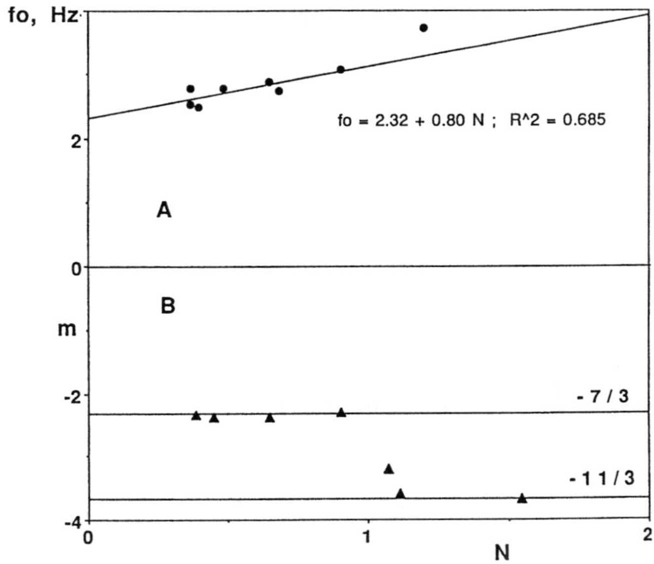

Fig. 2 Limiting frequency f_o (a) and spectral exponent m (b) dependence on the interaction parameter N.

When passing into x space, $Q_{ij}^{00} = \int \widehat{Q_{ij}}(\boldsymbol{k},\omega) d^3\boldsymbol{k}\, d\omega$ in the limit of $\xi = x - x' \to 0, \tau = t - t' \to 0$.

In case of weak magnetic field, the expressions for the energy A and helicity C are of the following form:

$$\frac{1}{2} A(0,0,\boldsymbol{B}) = Q_{ii}^{00} - 2\frac{B_o^2}{\mu\rho} \iint \frac{k_3^2(\eta\nu k^4 - \omega^2)}{(\eta^2 k^4 + \omega^2)(\nu^2 k^4 + \omega^4)} \widehat{Q'_{kk}}(\boldsymbol{k},\omega)\, \mathrm{d}\boldsymbol{k}\, \mathrm{d}\omega$$

$$C(0,0,\boldsymbol{B}) = C_o - \frac{2i}{3}\frac{B_o^2}{\mu\rho} \iint \frac{k_3^2(\eta\nu k^4 - \omega^2)}{(\eta^2 k^4 + \omega^2)(\nu^2 k^4 + \omega^4)} \varepsilon_{ijk} k \widehat{Q'_{jk}}(\boldsymbol{k},\omega)\, \mathrm{d}\boldsymbol{k}\, \mathrm{d}\omega$$

at $\eta \gg L \cdot U($ or $Re_m \ll 1)$.

In this case we obtain in the dimensionless form

$$C = C_o - \frac{2i}{3} N \cdot Re_m \iint \frac{k_3^2 Re^{-1}}{(Re^{-2}k^4 + \omega^4)k^2} \varepsilon_{ijk} k_i \widehat{Q'_{jk}}(\boldsymbol{k},\omega)\, \mathrm{d}\boldsymbol{k}\, \mathrm{d}\omega$$

$$A = A_o - 2N \iint \frac{k_3^2 Re^{-1}}{(Re^{-2}k^4 + \omega^4)} \widehat{Q'_{kk}}(\boldsymbol{k},\omega)\, \mathrm{d}\boldsymbol{k}\, \mathrm{d}\omega$$

We can obtain an expression for k_o in the form

$$k_o = \frac{C}{A} = \frac{C_o - N\aleph_1}{A_o - N\aleph}$$

We derive that

$$k_o \approx \frac{C_o}{A_o}\left(1 + \frac{\aleph N}{A_o} + \ldots\right)$$

is a linear function of N at small N values.

In the limit of $B \to 0$, k_o tends to its hydrodynamic limit.[9]

Acknowledgment

This research was partly sponsored by the Ministry of Infrastructure and Energy, Israel (No. 95-11-026), and ECC Commission (AVI-CT92-0017).

The authors are grateful to N. Goldbaum for her inestimable assistance in the paper preparation.

References

[1]Branover, H. H., Slyusarev, N. M., and Shcherbinin, E. V., "Some Results of Measuring Turbulent Velocity Pulsations in Mercury Flow under Transverse Magnetic Field," *Magn. Gidrodin.*, No. 1, 1965, pp. 33–36.

[2]Moreau, R., "On Magnetohydrodynamic Turbulence," *Proceedings of Symposium on Turbulence of Fluid and Plasma*, Brooklyn Polytech. Inst. Press, 1968, p. 359.

[3]Krause, F., and Radler, K.-H., *Mean-Field Magnetohydrodynamics and Dynamo Theory*, Springer–Verlag, Berlin, 1980.

[4]Alemany, A., Moreau, R., Sulem, P. L., and Frisch, U., "Influence of an External Magnetic Field on Homogeneous MHD Turbulence," *Journal de Mecanique*, Vol. 18, No. 2, 1979, pp. 277–313.

[5]Branover, H., Eidelman, A., Nagorny, M., and Kireev, M., "Magnetohydrodynamic Simulation of Quasi-Two-Dimensional Geophysical Turbulence," *Progress in Turbulence Research*, edited by H. Branover and Y. Unger., Vol. 162, Progress in Astronautics and Aeronautics AIAA, Washington, DC, 1994, pp. 64–79.

[6]Platnieks, I., and Seluto, S. F., "The Effect of Initial and Boundary Conditions upon the Formation and Development of MHD Turbulence Structure," *Liquid Metal Magnetohydrodynamics*, Kluwer Academic, Norwell, MA, 1989, pp. 433–439.

[7]Moffatt, H. K., "The Degree of Knottedness of Tangled Vortex Lines," *Journal of Fluid Mechanics*, Vol. 35, 1969, pp. 1, 117–129.

[8]Vainshtein, S. I., Zeldovich, Ya. B., and Ruzmaikin, A. A., *Turbulent Dynamo in Astrophysics*, Nauka, Moscow, 1980.

[9]Chkhetiani, O., Golbraikh, E., and Moiseev, S., "Helicity Generation in Turbulent and Laminar Magnetohydrodynamic Flow" (to be published).

[10]Belian, A. V., Golbraikh, E. I., Moiseev, S. S., and Chkhetiani, O. G., "Instability

of Second Moments in Helical Turbulence," *SETP Letters*, Vol. 22, No. 5, 1996, pp. 20–24.

[11]Belyan, A. V., Moiseev, S. S., and Chkhetiani, O. G., "Turbulent Viscosity in Helical Turbulence," Preprint Pr-1845, Space Research Inst., RAS, Moscow, 1992.

[12]Gardner, R. A., and Lykoudis, P. S., "Magneto-Fluid-Mechanic Pipe Flow in a Transverse Magnetic Field. Part 1. Isothermal Flow," *Journal of Fluid Mechanics*, Vol. 47, No. 4, 1971, p. 737–764.

Chapter 18

Turbulent Heat Flux Measurements in a Quasi-Two-Dimensional Flow and in Presence of Large-Scale Structures

G. A. Sideridis,* E. G. Kastrinakis,† and S. G. Nychas‡
Aristotle University of Thessaloniki, Thessaloniki, 54006, Greece

I. Introduction

THE term quasi-two-dimensional implies that the flow exhibits to some extent the following features that characterize two-dimensional turbulence: suppression of one fluctuating velocity component and inverse energy transfer, i.e., from smaller to larger scales. This is associated with a scaling region in the kinetic energy spectrum deviating from the $-5/3$ power law, characteristic of isotropic turbulence. Extensive discussions on the differences between two- and three-dimensional turbulent flows can be found in the literature.[1,2] Quasi-two-dimensional turbulent flow occurs in nature at high atmospheric altitudes above the Earth's boundary layer (i.e., at heights greater than 1.5 km), where large-scale eddy structures are also present. As reported in the literature, this type of flow results from body forces due to temperature and density stratification and the Coriolis' effect due to the Earth's rotation.[3–5]

Flows with features similar to those mentioned above have been created in the past by magnetohydrodynamic means. Branover et al.[6] describe an apparatus, with mercury as the working fluid, constrained by an electromagnetic field. The interaction between the currents generated by the turbulent motions and the magnetic fields produces an electromagnetic force that suppresses velocity fluctuations in one direction and generates turbulence with spectral slope equal to $-7/3$. This value has been reported by several investigators and has been generally accepted as characteristic of two-dimensional turbulence.[3,5,7] During the present work, atmospheric flow was simulated by more conventional means: A circular cylinder in a crossflow

---

*Post-doctoral Researcher, Department of Chemical Engineering, Univ. Box 453.
†Assistant Professor, Department of Chemical Engineering, Univ. Box 453.
‡Professor, Department of Chemical Engineering, Univ. Box 453.

configuration was placed over a flat plate, just outside the boundary layer. The von Kármán vortices behind the cylinder interacted with the boundary layer, creating a region with quasi-two-dimensional behavior. Heat was supplied to the present flow in order to simulate mass transport because the Prandtl and Schmidt numbers for gases are both close to one.

The structure of the turbulent wake behind a circular cylinder has attracted the interest of researchers for many years. The strong coherent vortices last for a distance of about 40 diameters downstream of the cylinder. At greater distances the vortices start to break up, producing a flow with different type and degree of organization in the wake.[8,9] The topology and behavior of the velocity field in a cylinder wake, both at close and far away ranges, have been studied in detail. In nonisothermal conditions, however, most publications present temperature measurements in the far wake of a slightly heated cylinder.[10,11] Matsumura and Antonia[12] have looked at the momentum and heat transport in the von Kármán vortex street close to a slightly heated cylinder. They examined the contribution from the coherent motion to the time-averaged momentum and heat fluxes, and they provided a detailed description of the topology of the velocity and temperature fields.

The reverse situation has been examined in the present work. The cylinder used was at ambient temperature. The heat supplied to the flow originated from a line heat source downstream of the cylinder and close to the surface of the plate. Simultaneous measurement of velocity and temperature has been carried out in the area of interaction of the boundary layer with the wake. Analysis and evaluation of results were focused on the heat transport characteristics related to the coherent motion.

II. Experimental Procedure

Measurements for the present work were carried out in an open-return, suction type wind tunnel, driven by a 2 hp electric motor. The test section of the tunnel had dimensions $0.3 \times 0.3 \times 2$ m. Freestream turbulence was less than 0.4%. The configuration used for the present experiments appears in Fig. 1. The surface of the cylinder with diameter $D = 7.6$ mm producing the vortex street was deliberately rough for reasons that will be explained later. The boundary layer was tripped near the leading edge of the plate to ensure that it is fully turbulent in the vicinity of the cylinder. Heat was supplied to the boundary-layer flow by means of the electrically heated wire with diameter $D = 0.1$ mm and an estimated surface temperature of about 160°C.

All measurements reported here were taken using hot-wire anemometry. The triple-wire probe shown in Fig. 1 was used. The X-wire probe (operated as a double hot wire), monitored two velocity components and the single wire probe (operated as a cold wire) recorded temperature variations. All wires were 1.25 mm long and 5 μm in diameter. The effective measuring volume of the triple-wire probe was $2 \text{ mm} \times 1.5 \text{ mm} \times 1 \text{ mm}$. The probe was operated by an AN-1003 anemometer multichannel system of A. A. Lab Systems Ltd. The overheat ratio of the hot wires was 30%, and the current fed to the cold wire was 0.265 mA. The probe output signals were digitized by an 8-channel, 16-bit analog-to-digital converter (type Data Translation

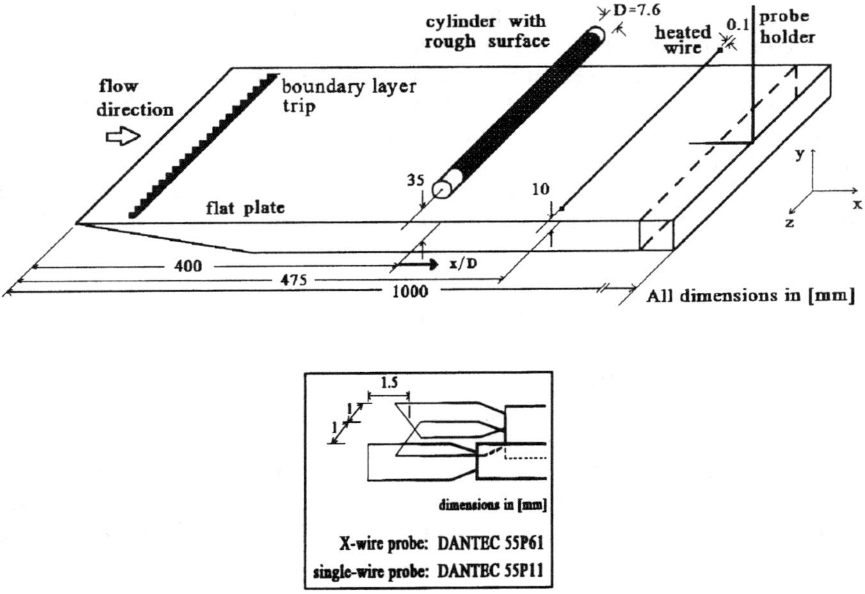

Fig. 1 Schematic presentation of the experimental setup and the triple-wire probe used.

DT-2809) and then fed to an IBM compatible personal computer for processing. Sampling frequency was 4 kHz per channel, and 163,800 points were taken per sample. The probe was calibrated by the technique summarized as follows: A jet of air at constant flow rate was heated in steps. After each increment, the cold wire output voltage was recorded and plotted against the jet temperature provided by a flow thermometer. The jet was allowed to cool off, and then it was heated slowly at constant mass flow rate. The cold and hot wire outputs were recorded simultaneously. The jet temperature was evaluated at several instances using the cold wire calibration curve and then the corresponding value of air density. The latter was used for the estimation of the current value of jet velocity. The procedure was carried out using specially developed computer software. The procedure was repeated at several mass flow rates, providing thus a hot wire calibration data bank.

A series of experiments were performed prior to the present measurements, aimed at the investigation of the two-dimensional character of the wake behind an isolated circular cylinder in crossflow.[13] In the particular case the surface of the cylinder was rough, the velocity power spectrum exhibited a scaling slope in the vicinity of $-7/3$ when $1500 < R_D < 3000$, $|Y_0/D| < 2$, and $x/D < 40$ (y_0 is measured from the cylinder's plane of symmetry, and R_D is the Reynolds number based on cylinder diameter D). During the present measurements, a cylinder with sandpaper glued around its surface was used. A freestream velocity of $U_o = 3.5$ m/s was selected ($R_D = 1820$). The friction velocity U_τ was estimated equal to 0.175 m/s.

The triple wire probe recorded simultaneously the streamwise and normal velocity components $U(t) = \overline{U} + u(t)$, $V(t) = \overline{V} + v(t)$ and temperature $T(t) = \overline{T} + T'(t)$ (an overbar denotes time-averaged values and lower case symbols or a prime denote fluctuations). Time variations of the two heat flux components $uT'(t)$, $vT'(t)$ and the momentum flux $uv(t)$ were evaluated by digital post-processing. The probe was traversed along the y direction from a point well within the freestream, to a distance of 4 mm from the surface of the plate. Four x stations have been considered, at $x/D = 20, 25, 30$, and 35.

III. Results and Discussion

A. Mean Flow Characteristics

Mean and rms profiles of the streamwise and normal velocity components and temperature are shown in Fig. 2. The mean velocity profile is plotted in Fig. 2a against y^+ on a logarithmic scale and compared to a logarithmic law, typical of a fully turbulent boundary-layer flow ($y^+ = U_\tau y/\nu$, where ν is the kinematic viscosity of air). The mean velocity defect behind the cylinder U_d

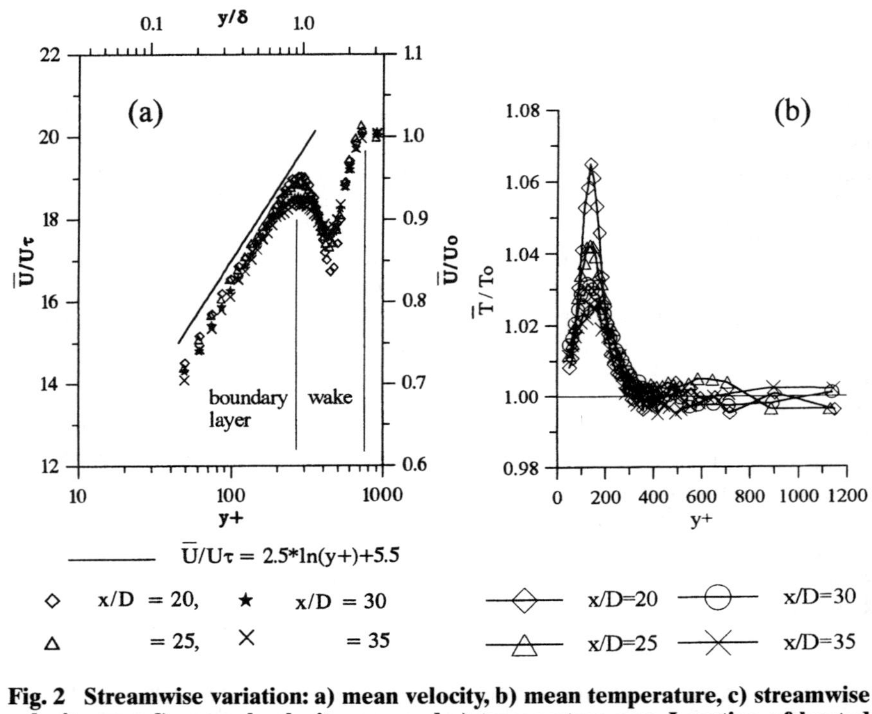

Fig. 2 Streamwise variation: a) mean velocity, b) mean temperature, c) streamwise velocity rms, d) normal velocity rms, and e) temperature rms. Location of heated wire, $y^+ = 120$; of cylinder, $y^+ = 450$.

and the mean temperature excess behind the heated wire T_e remain appreciable at all measuring stations. These parameters, used later for normalization, are defined at each measuring station as follows: $U_d = (U_o - U_{\min})$, where U_o is the freestream velocity and $U_{\min}$ is the minimum velocity in the wake; $T_e = (T_{\max} - T_o)$, where $T_{\max}$ is the maximum temperature behind the heated wire and T_o is the ambient temperature.

The rms velocity profiles in Figs. 2c and 2d indicate that the normal fluctuations dominate in the wake. The rms values of both velocity fluctuation components as well as their difference decrease with distance downstream, having nearly the same value at $x/D = 35$. This behavior indicates the continuous weakening of the vortices along the x direction. Within the boundary layer, however, the flow has a self-preserving character. The rms temperature profiles in Fig. 2e show the progressive spreading of heat along the y direction.

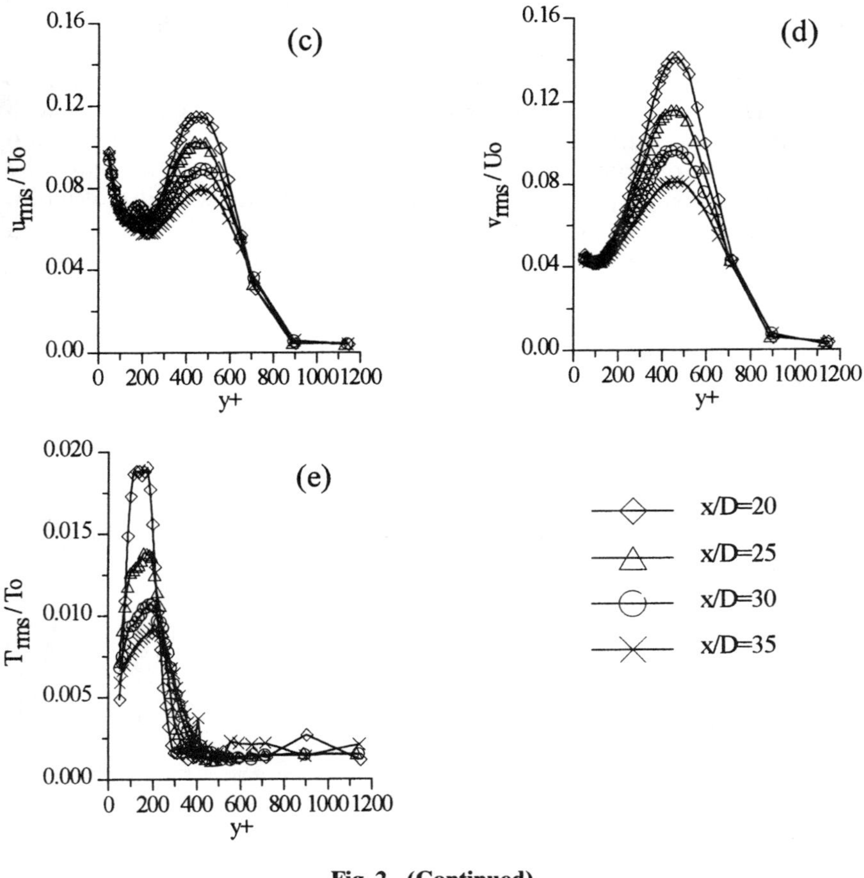

Fig. 2 (Continued)

B. Quasi-Two-Dimensional Characteristics

Velocity Power Spectra

Systematic investigation of the velocity power spectra inside the wake of the cylinder indicated that the spectral slope is in the vicinity of $-7/3$ in the region $180 < y^+ < 600$. Typical frequency power spectra of the streamwise and normal velocity components appear in Figs. 3a and 3b, respectively. A straight line with slope $-7/3$ is also plotted for comparison with the scaling region. Three regions can be identified after the peak at the Strouhal frequency: 97–200 Hz with slope $-5/3$, 200–600 Hz with slope $-7/3$ and over 600 Hz with slope -4.

Velocity Fluctuation Components

To examine whether the present flow fulfills the criterion for two-dimensionality stating that one velocity component ought to be suppressed relative to the other two, the spanwise velocity component profile $W(t)$ was recorded (by rotating the X-wire probe through 90 deg). As shown in Fig. 3, both $u(t)$ and $v(t)$ velocity fluctuation components show clearly contributions from the coherent motion. These contributions were removed by passing the $u(t)$ and $v(t)$ signals through a narrow band-pass digital filter with center frequency at the Strouhal frequency f_{St} ($= 0.21\ U_o/D = 97$ Hz) and subtracting the outcome from the original signals. The rms values of the result, together with w_{rms} (which was found that it did not contain a coherent contribution), have been plotted in Fig. 4. As expected, the u_{rms} and v_{rms} profiles almost coincide. There is a distinct difference between these and the

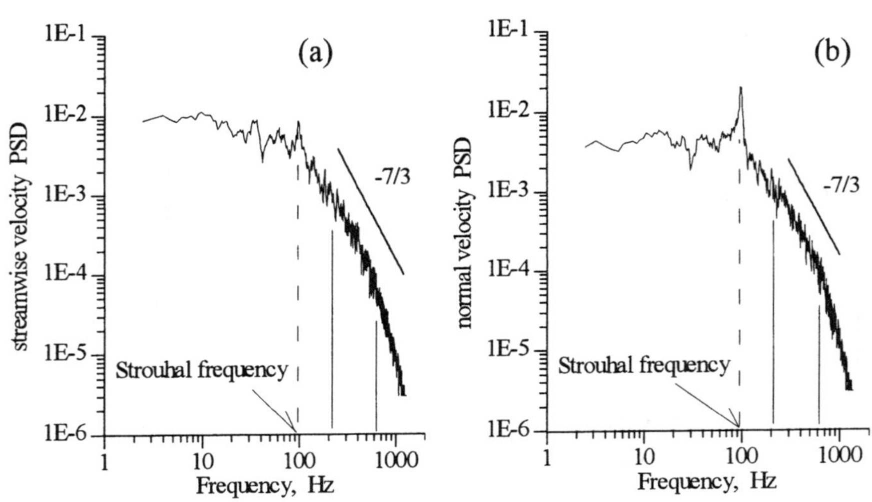

Fig. 3 Typical velocity power spectra for the present flow.

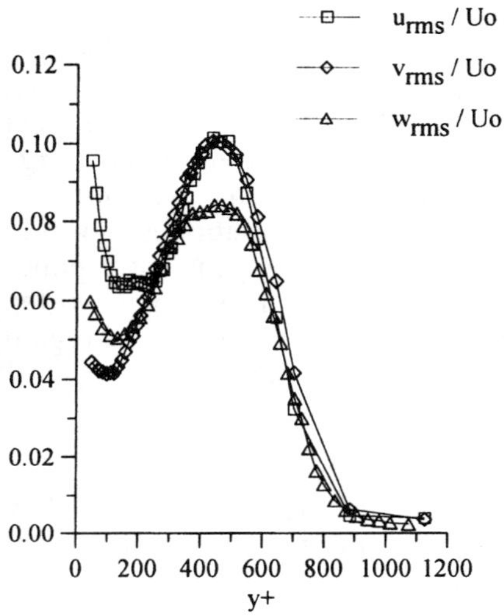

Fig. 4 Comparison of the rms values of all velocity components at x/D = 25.

w_{rms} profile in the region $250 < y^+ < 600$, which is the region where the slope of the velocity spectrum is $-7/3$.

A sketch of the present flowfield based on the results discussed so far appears in Fig. 5.

Turbulent Kinetic Energy Production

The traditional turbulent energy production term $-(\overline{uv})(\partial\overline{U}/\partial y)$ has been evaluated using the present data. The first parameter was determined from the directly measured momentum flux $uv(t)$ and the second from the mean velocity profile. Normally, in turbulent flows with symmetric mean velocity distribution, these parameters have opposite signs and assume zero value at

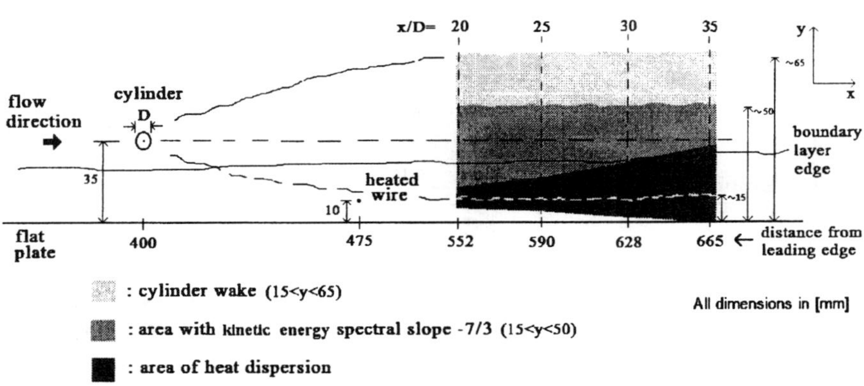

Fig. 5 Schematic presentation of the current flowfield.

the same point. Reported in the literature, however, is that in certain flows with asymmetric mean velocity profile, localized areas may exist at which the energy production term is negative[14] because the points where $(uv) = 0$ and $(\partial \overline{U}/\partial y) = 0$ do not coincide. These areas have been interpreted as zones at which Reynolds stress opposes viscous shear stress, resulting in an energy transfer from small-scale turbulence motion to the large-scale mean motion.

The present turbulent kinetic energy production distribution is shown in Fig. 6. Three distinct areas of intense energy production appear, corresponding to the boundary layer, the lower vortex street, and the upper vortex street. Turbulent energy production acquires negative values in the region of interaction of the boundary layer with the lower vortex street.

C. Time Signals and Probability Density Functions

Sample records of the directly measured parameters of the present flow are shown in Fig. 7. Strong deviation from the normal (Gaussian) distribution is evident in all cases, especially in the temperature and heat fluxes. Consideration of the signs of the tails of the temperature and heat flux probability density functions (pdf) indicate that heat is transported by masses of fluid traveling away from the plate and, relative to a reference frame moving with the mean streamwise velocity, in the upstream direction.

D. Time-Averaged Heat and Momentum Fluxes

Distributions of the time-averaged heat and momentum fluxes $\overline{uT'}$, $\overline{vT'}$, and $\overline{uv}$, normalized by the velocity defect behind the cylinder U_d and the temperature excess relative to the ambient T_e, are shown in Fig. 8.

Streamwise heat flux (Fig. 8a) appears to remain confined to the boundary-layer thickness ($y^+ = 300$) after an initial expansion. This is attributed to

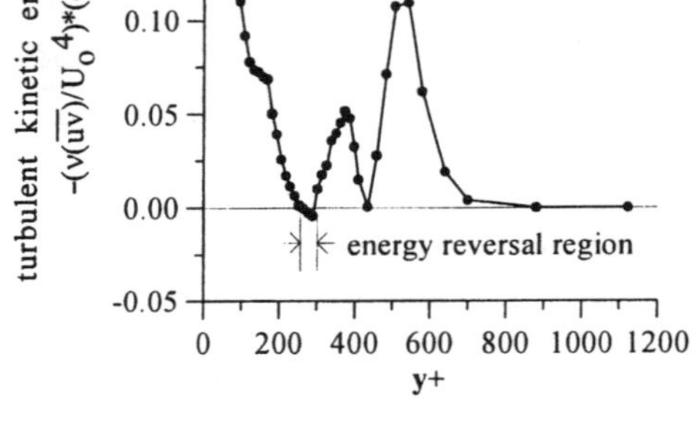

Fig. 6 Turbulent kinetic energy production distribution at *x/D* = 25.

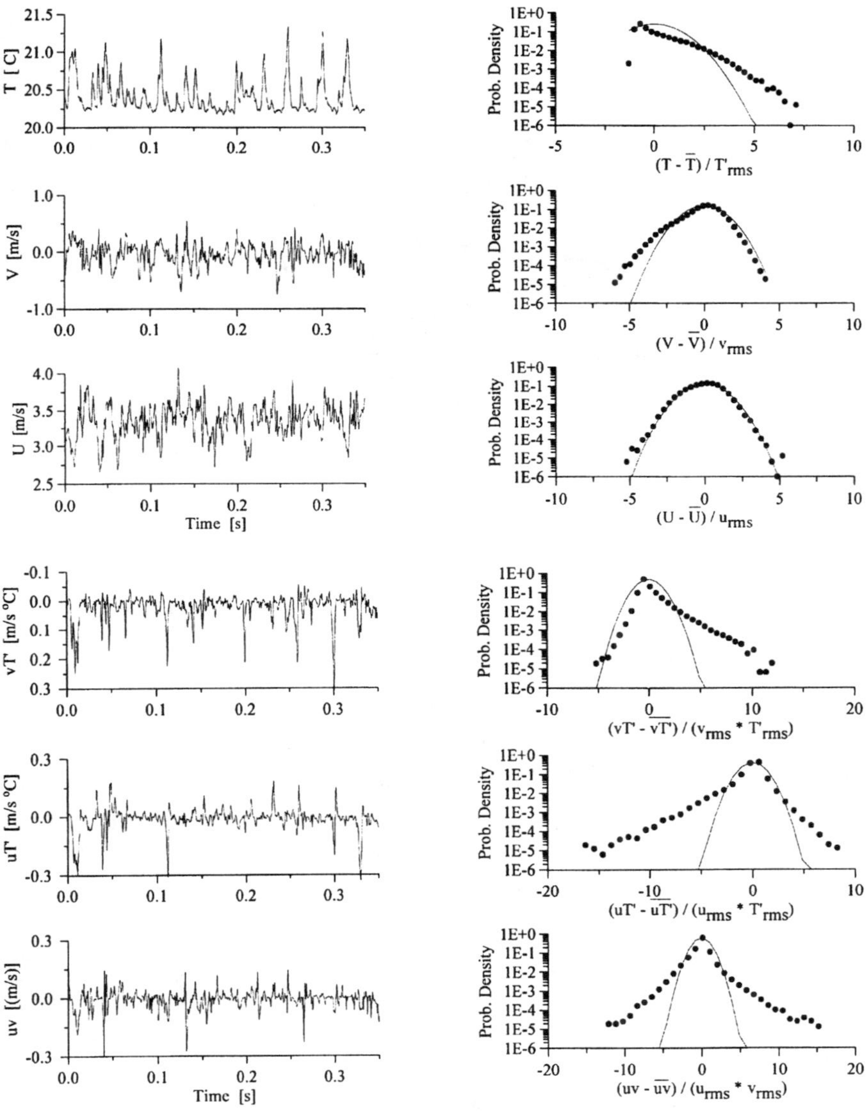

Fig. 7 Sample records of the directly measured statistical parameters at x/D = 25 and y^+ = 220.

the coherent motion that presumably moves hot as well as cold masses of fluid in the streamwise direction in such a manner that there is no net heat flux. On the contrary, coherent motion does not oppose normal heat flux (Fig. 8b) and, consequently, it expands continuously along the y direction.

The eddy diffusivity of heat E_h has been evaluated in the region $180 < y^+$

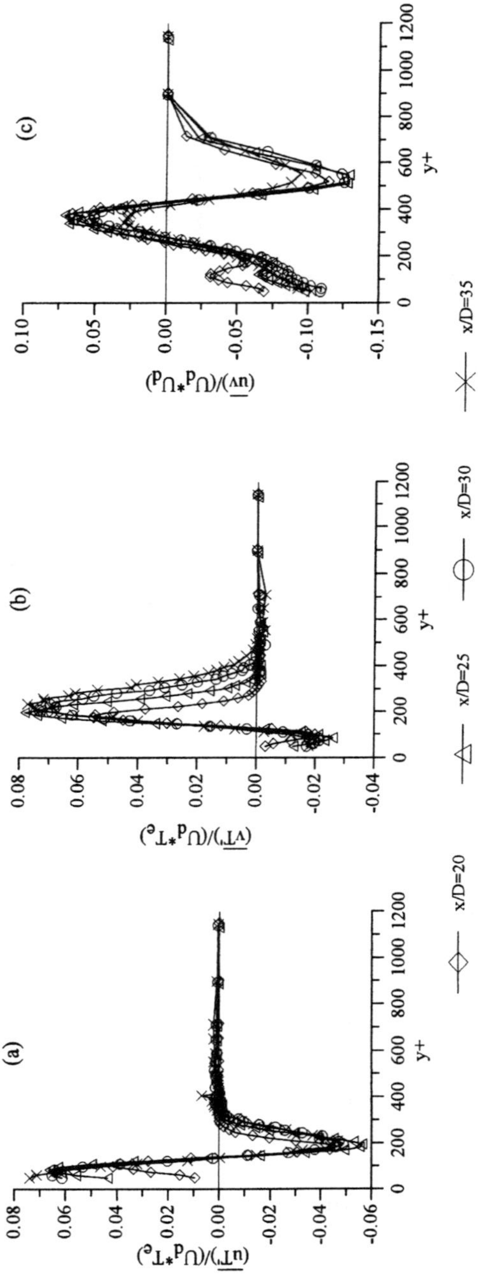

Fig. 8 Conventionally time-averaged heat and momentum flux distributions.

< 300 directly from the present experimental data, using the following relation:

$$E_h = -(\overline{vT'})/(\mathrm{d}\overline{T}/\mathrm{d}y) - \alpha$$

where α = 0.21cm²/s, the thermal diffusivity of air at 20°C. A representative value is E_h = 4 cm²/s. In nondimensional form $E_h/(U_oS)$ = 0.0045 (U_o = freestream velocity and S = scale of von Kármán vortices ≈ boundary layer thickness).

Momentum flux has typical distribution within the wake of the cylinder (Fig. 8c). The flux decreases continuously toward negative values closer to the plate. The following relation has been used for the estimation from the present data of typical values for the eddy diffusivity of momentum:

$$E_m = -(\overline{uv})/(\mathrm{d}\overline{U}/\mathrm{d}y) - \nu$$

where ν = 0.146 cm²/s, the kinematic viscosity of air. Representative values in the wake and the boundary layer are 7.4 cm²/s and 3.5 cm²/s, respectively. In nondimensional form $E_m/(U_oS)$ = 0.009 (wake) and $E_m/(U_oS)$ = 0.004 (boundary layer). Lee and Rodi[15] reported the numerical value of 0.018 for the maximum momentum diffusivity in a line puff (normalized by the puff's maximum velocity and a characteristic length).

E. Conditionally Averaged Heat and Momentum Fluxes Based on Quadrant Splitting Analysis

To obtain a better physical insight into the mechanisms of scalar transport, the quadrant-splitting analysis technique was applied. Developed by Wallace et al.,[16] the technique is based on splitting the instantaneous $uv(t)$ signal into quadrants in accordance with the sign of $u(t)$ and $v(t)$:

$u > 0, v > 0$: 1st quadrant (outward interactions)
$u < 0, v > 0$: 2nd quadrant (ejections)
$u < 0, v < 0$: 3rd quadrant (wallward interactions)
$u > 0, v < 0$: 4th quadrant (sweeps)

The names correspond to visually observed wall structures in fully developed pipe and channel flows.

For the present data, threshold values for each $u(t)$ and $v(t)$ equal to 1.2 times the corresponding rms value have been assumed. Because of that thresholding, it may be argued that for $y^+ > 150$ the four elementary flows of quadrant-splitting analysis are related primarily to the coherent motion. Angular brackets < > have been used to denote conditional averaging.

The $<\overline{uv}>$ is indicative of the relative strength and distribution of the four basic fluid motions. Close to the plate, the dominant motions are sweeps and ejections. Within the wake region, a part of a moderate distortion due to the interaction with the boundary layer, the curves display symmetrical distribution about a line drawn at $y^+ = 450$, the location of the cylinder. The wake data shown in the plot may be interpreted in physical terms as follows: In any single vortex, irrespectively of the direction of rotation, the dominant motion is in an outward (relative to the wake) and upstream (relative to the mean flow) direction. The weakest is in an inward and upstream direction.

Conditional averaging of $uT'(t)$ and $vT'(t)$ evaluated the contribution from each of the four elementary fluid motions. Consideration of the signs of correlated $u(t), v(t), uT'(t)$, and $vT'(t)$ determined the status of the moving masses of fluid regarding velocity and temperature. Representative plots of $<\overline{uT'}>$ and $<\overline{vT'}>$ are shown in Figs. 9b and 9c. The events with major contributions to both heat flux components come from ejections and sweeps. Ejections are associated with hot fluid and sweeps with cold (at locations above the heated wire, i.e., $y^+ > 120$). There is additional contribution to heat flux from the cold fluid moving toward the plate at a streamwise speed smaller than the local average (interactions wallward). This motion is associated to the coherent because it is present only for $y^+ > 150$. Its contribution to the streamwise heat flux is of opposite sign to that of sweeps and ejections, making the overall time-averaged value to be negligible in the wake (refer to Section D). On the contrary, its contribution to normal heat flux is added to that of sweeps and ejections. The contribution of outward interactions associated with hot fluid, although very minor at $x/D = 25$, becomes more significant further downstream. A sketch showing the motions of hot and cold masses of fluid, based on the preceded discussion, appears in Fig. 10.

IV. Conclusions

The heat transport characteristics from a turbulent boundary layer to a turbulent wake of a cylinder interacting with it and displaying a quasi-two-dimensional nature have been investigated in the present work.

Conventionally time-averaged streamwise heat flux was found to be confined to the boundary layer, whereas normal heat flux expanded continuously with distance downstream.

There are discrete hot and cold masses of fluid traveling at different speeds and directions: Hot masses move in a direction away from the plate, with lower streamwise speeds than the local average. Cold fluid moves toward the plate primarily at faster streamwise speeds than the local average. Assuming a reference frame traveling with the large-scale vortices, one may argue that vortices with negative vorticity transport heat away from the boundary layer toward the adjacent vortices with positive vorticity. Cold fluid originating presumably from the freestream is transferred toward the plate mainly under the action of positive vorticity.

A more detailed investigation of these motions is currently under way. A conditional averaging method with the following features has been selected: triple decomposition of the recorded signals into a time-mean value, a coherent part and a random part, and phase averaging of the coherent and random parts relative to the vortex shedding period. These results will be reported elsewhere.

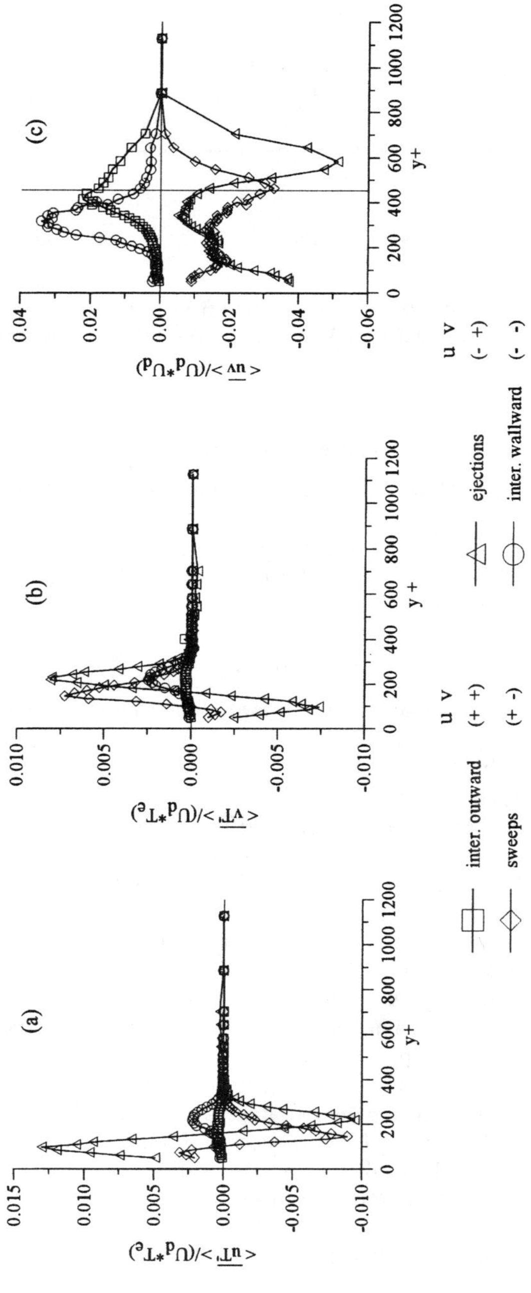

Fig. 9 Conditionally time-averaged heat and momentum fluxes at x/D = 25, according to the quadrant-splitting analysis.

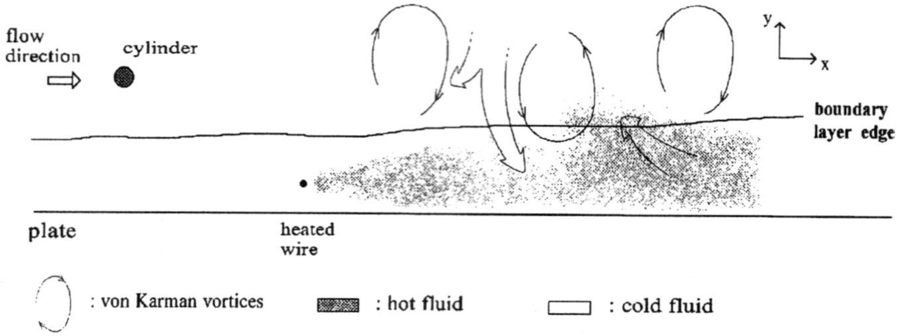

Fig. 10 Schematic presentation of the movement of hot and cold masses of fluid relative to the large-scale structures. View from a reference frame moving with the vortex street.

Acknowledgments

This work was supported by European Communities Commission, under contract AVI-CT92-0017.

Data acquisition and post-processing was carried out using computer software developed in the Chemical Engineering Department by J. N. E. Papaspyros (post-graduate researcher).

References

[1]Kraichnan, R. H., "Inertial Subranges in Two-Dimensional Turbulence," *Physics of Fluids*, Vol. 10, 1967, pp. 1417–1423.

[2]Tennekes, H., "Turbulent Flow in Two and Three Dimensions," *Bulletin of the American Meteorological Society*, Vol. 59, 1978, pp. 22–28.

[3]Boer, G. J., and Shepherd, T. G., "Large-scale Two-Dimensional Turbulence in the Atmosphere," *Journal of Atmospheric Sciences*, Vol. 40, 1983, pp. 165–184.

[4]Wyngaard, J. C., "Atmospheric Turbulence," *Annual Review of Fluid Mechanics*, Vol. 24, 1992, pp. 205–233.

[5]Lilly, D. K., and Petersen, E. L., "Aircraft Measurements of Atmospheric Kinetic Energy Spectra," *Tellus*, Vol. 35A, 1983, pp. 379–382.

[6]Branover, H., Bershadsky, A., Eidelman, A., and Nagorny, M., "Possibility of Simulating Geophysical Flow Phenomena by Laboratory Experiments," *Bound. Layer Meteorol.*, Vol. 62, 1993, pp. 117–128.

[7]Gage, K. S., and Nastrom, G. D., "Theoretical Interpretation of Atmospheric Wave Number Spectra of Wind and Temperature Observed by Commercial Aircraft During GASP," *Journal of Atmospheric Sciences*, Vol. 43, No. 7, 1986, pp. 729–740.

[8]Hussain, A. K. M. F., and Hayakawa, M., "Education of Large-Scale Organized Structures in a Turbulent Plane Wave," *Journal of Fluid Mechanics*, Vol. 180, 1987, pp. 193–229.

[9]Ferre, J. A., Mumford, J. C., Savill, A. M., and Giralt, F., "Three-dimensional Large-Eddy Motions and Fine-Scale Activity in a Plane Turbulent Wake," *Journal of Fluid Mechanics*, Vol. 210, 1990, pp. 371–414.

[10]Fabris, G. "Conditional Sampling Study of the Turbulent Wake of a Cylinder," Part 1, *Journal of Fluid Mechanics*, Vol. 94, Part 4, 1979, pp. 673–709.

[11]Ferre, J. A., and Giralt, F., "Some Topological Features of the Entrainment Process in a Heated Turbulent Wake," *Journal of Fluid Mechanics*, Vol. 198, 1989, pp. 65–78.

[12]Matsumura, M., and Antonia, R. A., "Momentum and Heat Transport in the Turbulent Intermediate Wake of a Cylinder," *Journal of Fluid Mechanics*, Vol. 250, 1993, pp. 651–668.

[13]Sideridis, G. A., Kastrinakis, E. G., and Nychas, S. G., "Experimental Simulation of Air Pollutant Dispersion by Atmospheric Motions," *Air Pollution III*, edited by H. Power et al., Vol. 2, Computational Mechanics Publications, Boston, MA, 1995, pp. 177–184.

[14]Eskinazi, S., and Erian, F. F., "Energy Reversal in Turbulent Flows," *Physics of Fluids*, Vol. 12, Part 10, 1969, pp. 1988–1998.

[15]Lee, J. H. W., and Rodi, W., "Numerical Simulation of Line Puffs," *Recent Research Advances in the Fluid Mechanics of Turbulent Jets and Plumes*, edited by P. A. Davies and M. J. Vlente Neves, Kluwer Academic, Norwell, MA, 1994, pp. 73–88.

[16]Wallace, J. N., Eckelmann, H., and Brodkey, R. S., "The Wall Region in Turbulent Shear Flow," *Journal of Fluid Mechanics*, Vol. 54, 1972, pp. 39–48.

Chapter 19

Mesa-Scale Flows over Large City

K. G. Schwarz*
Perm State University, Perm, 614600, Russia

Introduction

THE modeling of mesa-scale atmospheric processes is developed in present time everywhere. It is associated with the solving of the actual applied problems of meteorology and problems of environment monitoring. For this purpose many different models were created.[1–7] For mesa-scale processes the typical horizontal scale motion L is between 1 and 100 km, the depth of layer D is about 1000 m, and the typical time is between 10^3 s and 10^5 s. The local motion of air in lower atmospheric layer arises background-corrected of large-scale turbulence. But on the another hand the local motion of air is defined by nonuniform surface heating, orography, degree of roughness, and the other small-scale factors. Usually, small-scale processes are parameterized in equations, and large-scale factors are used as initial and boundary conditions. By analogy the interaction of impurity with the underlying surface and the presence of overground pollution sources are accounted by parameterization.

Two-dimensional models, which save the main physical properties of investigating atmospheric processes, are used very often because of the complication of modeling and large computational labor content. In this paper such quasi-two-dimensional, which describes three-dimensional convective processes in thin rotating layer of air, is offered. Its derivation is based on the sufficiently known approach,[8] which was used successfully in geophysical applications.[9,10]

General Theory

In the paper[11] equations of motion and heat transfer equations of this model were derived by methodology.[12] The initial equations are three-dimensional ones of "dry" atmospheric hydrothermodynamics in a Cartesian coordinate system where x axis and y axis direct to the east and to the north,

*Bukireva Str., 15.

and z axis directs upward. In accordance with Refs. 3, 5, and 13, Boussinesq and static approximation for incompressible gas was used for the boundary layer of atmosphere and condition

$$\delta = \frac{D}{L} \ll 1 \tag{1}$$

On the upper boundary in free atmosphere for $z = D$, the "rigid lid" condition is assigned and potential temperature is fixed and equals θ_h-const. Tangential stress of turbulent friction $\boldsymbol{\tau}\ (x, y)$, flow condition, and turbulent heat flow, which propagates by Newton's law, is defined on the lower boundary for $z = H\ (x,y)$. Function H describes the orography of the underlying surface. Moreover, the value of potential temperature of soil surface $\theta_S(x, y)$, the convective heat exchange coefficient $\gamma_T\ (x, y)$, and the components of vector $\boldsymbol{\tau}$ depend on surface type, and they were calculated by formulas that are presented in Refs. 3, 14, and 15.

To consider the mesa-scale processes (1) only, let us suppose that the typical time of equilibrium state evolution is much greater than the air flow relaxation time to the state of equilibrium after the changing of external conditions. It is supposed also that the orography described by function H (x, y) is changed sufficiently smoothly and its height differential of underlying surface is small relatively to the depth D. The "meteorological" part of the model equations, which are represented in Refs. 11 and 14, is written in terms of vortex $\omega(t, x, y)$, stream function $\psi(t, x, y)$, and averaging across the layer of potential temperature $\theta(t, x, y)$ as follows:

$$\begin{aligned} &\frac{\partial \omega}{\partial t} + \frac{1}{D-H}\{\psi, \omega\} + (l+\omega)\left\{\psi, \frac{1}{D-H}\right\} \\ &\quad + 4k_3\left(\frac{D-H}{D}\right)^2 \{\theta, \Delta\theta\} - 4k_{33}\left(\frac{D-H}{D}\right)^2 \operatorname{div}(\Delta\theta\nabla\theta) \\ &\quad = l\left\{\theta, \frac{D-H}{D}\right\} + A_M\left[\Delta\omega - k_4\nabla(\Delta\theta)\frac{\nabla H}{D-H}\right] - k_M \mathrm{rot}_z\left(\frac{\tau_1}{D-H}\right) \end{aligned} \tag{2}$$

$$\frac{\partial}{\partial x}\left(\frac{1}{D-H}\frac{\partial \psi}{\partial x}\right) + \frac{\partial}{\partial y}\left(\frac{1}{D-H}\frac{\partial \psi}{\partial y}\right) = \omega \tag{3}$$

$$\begin{aligned} &\frac{\partial \theta}{\partial t} + \frac{1}{D-H}\{\psi, \theta + 0.5\theta_1\} + \theta\left\{\psi, \frac{1}{D-H}\right\} - 4k_5\theta_1\left\{\theta, \frac{D-H}{D}\right\} \\ &\quad = A_T\left[\Delta\theta - \nabla\theta_1\frac{\nabla H}{D-H}\right] - \frac{k_T}{(D-H)^2}[\gamma_H(\theta - \theta_S) - (1+0.5\gamma_H)\theta_1] \end{aligned} \tag{4}$$

Here $\{A, B\}$ is Jakobian; A_M, k_M are coefficients of horizontal and vertical turbulent impulse exchange; A_T, k_T are analogous coefficients of heat exchange; $\tau_1 = (k_M/\rho_0 k_0)\ \tau$, where ρ_0 is the average density, $k_0\ (x, y)$ is the coefficient of turbulent air viscosity on the underlying surface; l const is the Coriolis parameter, coefficient $k_3 = Re\left\langle f_3^2\right\rangle$, $k_{33} = Im\left\langle f_3^2\right\rangle$, $k_4 = Im[f_3(H)]$, complex function

$$f_3(z) = \left(\frac{z-H}{D-H}\right)^2 - 2\frac{z-H}{D-H} + \frac{2}{3} + 2 * f_2(z) \tag{5}$$

Here

$$f_2(z) = \frac{1}{\lambda^2} - \frac{ch\left[\lambda\left(\frac{z-H}{D-H} - 1\right)\right]}{\lambda sh(\lambda)}$$

where $\lambda = (2E)^{-0.5}(1 + i)$, $E = k_M/lD^2$ is the Ekman number, $\theta_1 = \frac{\gamma_H}{1+\gamma_H}(\theta_h - \theta_S)$, and $\gamma_H = \gamma_T\,(D - H)$.

The derivation of the quasi-two-dimensional equation of impurity transfer and diffusion is presented in this paper. Initially it is used as a three-dimensional equation of passive impurity transfer in atmosphere[16]:

$$\frac{\partial c}{\partial t} + v_x\frac{\partial c}{\partial x} + v_y\frac{\partial c}{\partial y} + (v_z - w_g)\frac{\partial c}{\partial z} = A_p\Delta c + k_p\frac{\partial^2 c}{\partial z^2} + f \tag{6}$$

where c is the impurity concentration, $f\ (t, x, y, z)$ describes the impurity high-altitude sources that are higher than roughness level

$$f = \sum_{i=1}^{m} Q_i(t)\delta(x - x_i)\delta(y - y_i)\delta(z - z_i) \tag{7}$$

where Q_i is the power of the ith source; m is the number of the high-altitude source; $\delta(x)$ is the delta function; x_i, y_i, z_i are the source coordinates; w_g const is the velocity of the heavy particles gravity settling (for light particles $w_g = 0$); A_p, k_p are the coefficients of horizontal and vertical turbulent pollution diffusion; and v_x, v_y, v_z are the components of the velocity vector. For $x = D$,

$$\frac{\partial c}{\partial z} = 0 \tag{8}$$

On the lower boundary for $z = H(x, y)$,

$$\frac{\partial c}{\partial z} = \alpha_\theta(c - c_S) \tag{9}$$

where function $c_S\,(x, y)$ is the impurity distribution on the surface and the coefficient $\alpha_\theta(x, y)$ is calculated by formulas[16] from the impurity balance equation on the surface roughness level. Any average across layer function is denoted as

$$< g > = \frac{1}{D - H} \int_H^D g(t, x, y, z)\, \mathrm{d}z \tag{10}$$

By integration of Eq. (6) across the layer and by using boundary conditions (8) and (9), we have

$$\frac{\partial}{\partial t}[(D - H) < c >] + \frac{\partial}{\partial x}[(D - H) < v_x c >] + \frac{\partial}{\partial t}[(D - H) < v_y c >] - w_g c\Big|_H^D$$
$$= A_p \int_H^D \Delta c \mathrm{d}z - -k_p \alpha_\theta (c\big|_{z=H} - c_S) + (D - H) < f > \tag{11}$$

To complete Eq. (11), we must express all components in terms of averaging velocity and concentration $\langle v_x \rangle$, $\langle v_y \rangle$, and $\langle c \rangle$. Let us use the exact solution of the initial three-dimensional problem, which describes uniform by x, y flux. When θ_S, c_S, and surface pressure are linear functions, H, γ_T, α_θ, Q_i are constant values, and $\boldsymbol{\tau}_1$ is a constant vector. For velocity[11,14]

$$\Phi\,(z) = \langle \Phi \rangle + f_2(z)(D - H)T - f_3(z)U \tag{12}$$

and for impurity concentration

$$c(z) = < c > + [c_0(z) - < c_0 >] + \sum_{i=1}^{m} Q_i [c_i(z) - < c_i >] \delta(x - x_i) \delta(y - y_i) \tag{13}$$

Here complex function $\Phi = v_x + iv_y$, complex values $T = \tau_{1x} + i\tau 1_y$, $U = u_x + iu_y$, and β is buoyancy parameter

$$u_x = -2 \frac{\beta D}{2l} \frac{\partial < \theta >}{\partial y}, \qquad u_y = 2 \frac{\beta D}{2l} \frac{\partial < \theta >}{\partial x}$$
$$c_0(z) = \frac{1}{k_p} \int_H^z e^{-\alpha(z-H)} \int_z^D e^{\alpha(\zeta - H)} \left(\frac{\partial c_S}{\partial x} v_x(\zeta) + \frac{\partial c_S}{\partial y} v_y(\zeta) \right) \mathrm{d}\zeta\, \mathrm{d}z$$

For light impurity

$$c_i(z) = \frac{1}{kp}(z_i + \Delta z_i - z)\,U - (z - z_i)\,U - (z_i + \Delta z_i - z)$$

and for heavy impurity

$$c_i(z) = \frac{1}{w_g} e^{-a\Delta z_i}\left[e^{-a(z - z_i - \Delta z_i)} - 1\right] U - (z - z_i)\,U - (z_i + \Delta z_i - z)$$

where $a = w_g/k_p$, Δz_i is the pollution hoisting height of the ith high-altitude source,[17,18] and step function $U - (\xi) = 0$ for $\xi < 0$ and equals 1 for $\xi > 0$.

By allowing for

$$< v_x > = -\frac{1}{D-H}\frac{\partial \psi}{\partial y}, \quad < v_y > = \frac{1}{D-H}\frac{\partial \psi}{\partial x} \tag{14}$$

the transfer and diffusion equation of averaging impurity concentration follows:

$$\frac{\partial < c >}{\partial t} + \frac{1}{D-H}\{\psi, < c >\} + < c > \left\{\psi, \frac{1}{D-H}\right\} = A_p \Delta < c >$$
$$- \frac{\alpha_\theta k_p}{D-H}(< c > - c_S) + Q_p(x, y) \tag{15}$$

where pollution source

$$Q_p = \sum_{i=1}^{m} \frac{Q_i}{D-H}(1 + \alpha_\theta k_p < c_i >)\delta(x - x_i)\delta(y - y_i)$$
$$+ \frac{\alpha_\theta k_p}{D-H} < c_0 > + \frac{w_g}{D-H} c_0(D) \tag{16}$$

is a function of coordinates, velocity, and potential temperature.

In the limits of the quasi-two-dimensional model, the conception of locally equilibrium solutions is used. Approximate three-dimensional pattern is restored by asymptotical formulas (12–13), where averaging values $\langle \Phi \rangle$, T, U, and $\langle c \rangle$ are already considered as the functions of x, y. They are calculated numerically by two-dimensional equations (2–4), (14–16) with corresponding boundary conditions for averaging fields of velocity, potential temperature, and impurity concentration.

Application

The system (2–4), (15–16) was used to study the dynamical and thermal nonuniformity influence of underlying surface to dynamics of air flux under Perm for faint ageostrophic wind with velocity 0.5 m/s.[11,14] It was solved, by the net-point method of E. L. Tarunin[19] with implicit scheme. The calculations confirmed the fact of dominance of thermal nonuniformity influence to modification of the boundary-layer dynamical characters.

For Berezniky city in Russia, the impurity propagation character from high-altitude industrial sources was investigated. The stationary regimes were studied. It is considered a ground 17.5 km × 23 km. On the side boundaries the Γ value of the velocity is constant and its direction and impurity absence condition are designated with the windward side.

$$<v_x>\big|_\Gamma = <u_0>, \quad <v_y|_\Gamma = <v_0>, \quad <c>|_\Gamma = 0 \tag{17}$$

Velocity components u_0 and v_0 are defined by the Ekman formula[20]:

$$\begin{aligned} u_0 &= [1 - e^{-kz/D}\cos(kz/D)]u_g - e^{-kz/D}\sin(kz/D)v_{g,} \\ v_0 &= e^{-kz/D}\sin(kz/D)u_g + [1 - e^{-kz/D}\cos(kz/D)]v_{g,} \end{aligned}$$

where $k = (2E)^{-0.5}$, $u_g = -c_g\sin(dd)$, $v_g = -c_g\cos(dd)$, c_g is the velocity of ageostrophic wind, and dd is its azimuth. With leeward side

$$\frac{\partial <v_x>}{\partial n}|_\Gamma = \frac{\partial <v_y>}{\partial n}|_\Gamma = 0, \quad A_p\frac{\partial <c>}{\partial n}|_\Gamma = <v_n c>|_\Gamma \tag{18}$$

where n is normal, v_n is the projection of velocity on the normal. The side boundaries are assumed as thermal insulative:

$$\frac{\partial\theta}{\partial n}|_\Gamma = 0 \tag{19}$$

The system (2–4), (15–19) is numerically solved by the alternating directions method. It chose the scheme with difference against the flow. The number of points are 35 × 47 (step is 500 m). The terrain is sufficiently flat (Fig. 1), the largest height difference between the nearest point of net is only

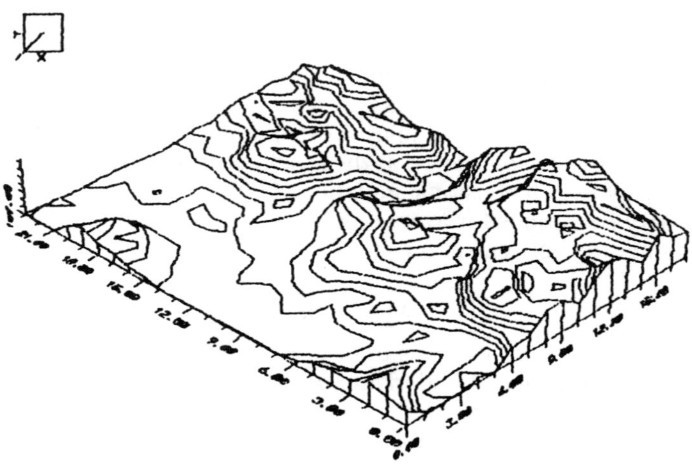

Fig. 1 Surface relief of Berezniky city and its environs.

60 m. On Fig. 2 the results of the numerical calculations for the case when c_g = 3 m/s, the wind direction near Earth equals 300 deg, the depth $D = 300$ m, $A_M = A_T = A_P = 10^3$ m/s, $l = 1.24 \cdot 10^{-4}$1/s, $\beta = 1/30$ m/(s · deg) and $E = 1$ are presumed; consequently, $k_M = k_T = k_P = lD^2$. It separated 17 types of surface for which are defined roughness degree, potential temperature θ_s, and values of coefficients γ_T and α_v. The height of impurity source equals 120 m.

This sufficiently ordinary example illustrates the possibility of investigation by this model of the pollution degree of a large city territory with large

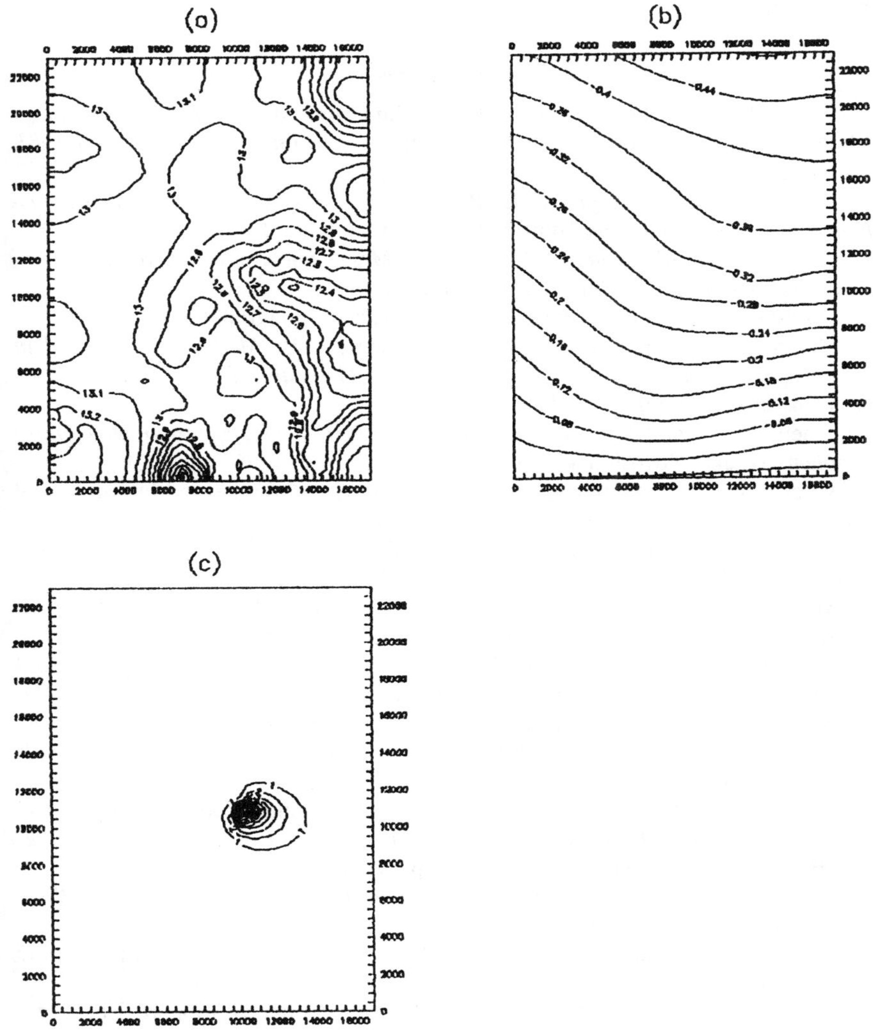

Fig. 2 Isolines: a) temperature, b) stream function, and c) impurity.

space resolution of net point and with a good degree of accuracy for reasonable expenditure of computer resources. The calculations show that by actions of temperature nonuniformity (Fig. 2a), the change of wind direction from northwest to west takes place and in the southeast part of the site (at the foot of one), the wind increases (Fig. 2b). It turns the cloud of impurity from high-altitude source in this case (Fig. 2c).

Acknowledgment

This work was supported by a grant of the Russian Funds of Fundamental Researches N94-01-00487-A.

References

[1]Belov, P. N., and Scherbakov, A. Yu., "Numerical Modelling of Meteorology Elements Daily Variation in Large City," *Meteorologia and Hydrologiya*, No. 7, 1983, pp. 45–53.

[2]Vager, B. G., and Nadegina, E. D., *Boundary Layer of Atmosphere in Horizontal Non-Uniformity Conditions,* Hidrometeoizdat, Leningrad, 1979.

[3]Gavrilov, A. S., *Mathematical Modelling of Mesa-Meteorological Processes,* Leningrad Polytechnical Ins. Publ., Leningrad, 1988.

[4]Landsberg, G. E., *City Climate,* Hidrometeoizdat, Leningrad, 1983.

[5]Marchuk, G. I., Dimnikov, V. P., and Zalesniy, V. B., *Mathematical Model in Geophysical Hydrodynamics and Numerical Methods of Their Realization,* Hidrometeoizdat, Leningrad, 1987.

[6]Penenko, V. V., and Aloyan, A. E., *Models and Methods for Environment Control Problems,* Nauka, Novosibirsk, 1985.

[7]Kibell, I. L., "Hydrodynamic Short-Term Forecast of Weather in Meteorological Problems," *Proceedings of HMC of USSR*, Vol. 48, 1970, pp. 3–18.

[8]Aristov, S. N., and Frik, P. G., "Large-Scale Turbulence in Thin Layer of Non-Isothermal Rotating Fluid," *Fluid Dynamics*, Vol. 23, No. 4, 1988, pp. 522–528.

[9]Aristov, S. N., and Schwarz, K. G., "Evolution of Wind-Circulation in Non-Isothermal Ocean," *Oceanology*, Vol. 30, No. 4, 1990, pp. 562–566.

[10]Aristov, S. N., and Schwarz, K. G., "About Influence of Salt Exchange to Fluid Circulation in Closed Basin," *Marine Hydrophysical Journal*, No. 4, 1990, pp. 38–42.

[11]Schwarz, K. G., "Two-Dimensional Model of Mesa-Scale Atmospheric Processes," *Proceedings of the VIII International Conference on Finite Elements in Fluids* (Barcelona, Spain), 1993.

[12]Aristov, S. N., Schwarz, K. G., "New Two-Dimensional Model of Large-Scale Oceanic Circulation," *Proceedings of the 2nd International Conference on Computer Modelling in Ocean Engineering '91*, 1991, pp. 49–54.

[13]Brown, R. A., *Analytical Methods in Planetary Boundary-Layer Modelling*, Adam Hilger, London, 1974.

[14]Schwarz, K. G., and Shklyaev, V. A., "Modelling of Mesa-Scale Atmospheric Processes over a Large City," *Meteoriligia and Hydrologia*, No. 9, 1994, pp. 29–37.

[15]Tarnapolskiy, L. G., and Schnaidman, V. A., "Parametrization of Baroclinic Boundary Atmosphere Layer," *Proceedings of Hydrometeocentre of the USSR*, Vol. 180, 1976, pp. 33–40.

[16]Marchuk, G. I., *Mathematical Modelling in Environment Problems*, Nauka, Moscow, 1982.

[17]Berlyand, M. E., *Prediction and Regulation of Air Pollution*, Hidrometeoizdat, Leningrad, 1985.

[18]Berlyand, M. E., *Modern Problems of Atmospheric Diffusion and Impurity of Atmosphere*, Hidrometeoizdat, Leningrad, 1975.

[19]Tarunin, E. L., *Numerical Experiment in Problems of Free Convection*, Irkutsk Univ. Publ., Irkutsk, 1990.

[20]Gill, A. E., *Atmosphere-Ocean Dynamics*, Academic, London, 1982.

Chapter 20

Coupled Viscous Solitary Waves in Quasi-Two-Dimensional Magnetohydrodynamical and Geophysical Flows

Victor Miroshnikov*

Laboratoire d'Energétique et de Mécanique Théorique et Appliquée, Centre National de la Recherche Scientifique, Vandoeuvre Cedex, 54504, France

I. Introduction

THE simplest unique description of quasi-two-dimensional laboratory and natural flows is possible up to a sufficient accuracy by the two-dimensional Navier–Stokes equations added by the linear-braking term with a coefficient λ, which was deduced for magnetohydrodynamic (MHD) flows[1] in the case of small magnetic Reynolds numbers, high Hartmann numbers, high interaction parameters, and large-scale geophysical and atmospheric flows[2] in the case of negligible β effect and horizontal compressibility. The emergence of long-lived vortices in two-dimensional viscous incompressible flows, well established through numerical[3] and laboratory[4] experiments, is intrinsically connected with a dependence between the vorticity ω and the stream function Ψ, which might take several forms.[5] The existence of such a dependence means that the considered flow exhibits the properties of steady inviscid flows[6] in the reference frame of the moving vortex. Large enough Reynolds numbers Re_v usually met in laboratory and natural conditions cannot entirely explain the emergence of two-dimensional long-lived vortices as the Reynolds number with respect to the linear friction Re_λ has the order of 10^0–10^1 for laboratory and natural flows[7] as well.

Among various coherent structures studied recently, the two-dimensional symmetric inertial vortex couples primarily studied by Lamb and Chaplygin has a rather long history.[8] The linear dependence $\omega = k^2\Psi$ assumed in their solutions has been fairly confirmed by experiments on liquid metal flows.[3] The current paper is devoted to conditions of a potential friction, which provides a quasi-inviscid steady-state behavior of two-dimensional viscous

*Researcher, Institute of Physics, Academy of Sciences, Salaspils LV-2169, Latvia

vortex flows. As the conditions are fulfilled, the vorticity is related with the stream function for Re_v and $Re\lambda$ of finite values, and the flow possesses an extended form of the Bernoulli integral, which was recently written in the explicit form for two-dimensional inviscid ordinary[9] and MHD[10] flows. One-dimensional channel flows identically obey these conditions resulting in a Poiseuille–Couette flow transformed by the λ braking, and two-dimensional flows with the linear dependence are possible only for imaginary values of $k = iHa_\lambda$, whereas the viscous continuation of the Lamb inertial vortex couple results in an inertial coupled solitary wave. Internal coupled solitary waves also admit a linear with respect to a velocity version that differs in the vorticity distribution. For enough large amplitude the flow inside the coupled solitary wave separates, and the coupled solitary wave transforms into a solitary vortex couple. Permanence of the flow pattern and the propagation velocity of both the solitary vortex couple and the coupled solitary were maintained by the constant pressure drop of the basic flow that compensates the potential friction force.

II. Basic Equations

Under named assumptions the flow splits into the Hartmann (or Ekmann) boundary layers, and the core flow and the equations of two-dimensional motion $\boldsymbol{v}(x_1, x_2, t) = v_1\boldsymbol{i}_1 + v_2\boldsymbol{i}_2$ of the viscous incompressible core become[1,2,7]:

$$\partial_t \boldsymbol{v} + (\boldsymbol{v}\,\text{grad})\boldsymbol{v} = -\text{grad}(p/\rho) + \nu\Delta\boldsymbol{v} - \lambda\boldsymbol{v} + \boldsymbol{f}_b \tag{1}$$

$$\text{div}\ \boldsymbol{v} = 0 \tag{2}$$

where ρ is the density, $\boldsymbol{f}_b = -\text{grad}\ \varphi_b$ is a potential body force, ν is the kinematic viscosity, λ is an inversed Hartmann timescale t_H^{-1} characterizing the electromagnetic braking due to interaction of the induced current and the external magnetic field or an inversed Ekmann timescale t_E^{-1} characterizing the dynamic braking produced by the Ekmann layers. Here, $t_H = (d/Bn)\sqrt{\rho/(\sigma\nu)}$, where n is a number of the Hartmann layers ($n = 1$ for a horizontal layer of depth d with a free upper surface and $n = 2$ for a layer with a rigid lid), B is a uniform magnetic field directed in vertical x_3 direction, σ is an electric conductivity, $t_E = d_a/\sqrt{f_C \nu_t}$, d_a is a free atmosphere's depth, f_C is a Coriolis parameter, and ν_t is a turbulent viscosity. In the reference frame ($X_1 = x_1 - ct, x_2$) moving along $\boldsymbol{i}_1$ at a permanent velocity c, the stream function–vorticity formulation of the problem (1–2) may be written as

$$\text{grad}\ E_B + \omega_3\ \text{grad}\ \Psi_3 + \text{rot}\ (\nu\omega_3 + \lambda\Psi_3|\ \boldsymbol{i}_3) = 0 \tag{3}$$

$$J(\Psi_3, \omega_3) + \Delta\ (\nu\omega_3 + \lambda\Psi_3) = 0 \tag{4}$$

Here conventional definitions for the vorticity $\omega_3 \boldsymbol{i}_3$ = rot $\boldsymbol{V} = -\Delta\Psi_3\, \boldsymbol{i}_3$ and the stream function $\boldsymbol{V} = (V_1, v_2) = \text{rot}\, \Psi_3\, \boldsymbol{i}_3 = (\partial_{x_2} \Psi_3, -\partial_{X_1} \Psi_3)$ are used; J is the Jacobian with respect to X_1 and x_2, and $E_B = |\text{grad}\, \Psi_3|^2/2 + (_p/\rho) + \varphi_b + \lambda\, cX_1$ is the Bernoulli energy (the total head), where the last term means that the fluid might set in the uniform motion $(v_1,v_2) = (c,0)$ by the constant pressure gradient $d_{x_1}\, p = -\lambda c\rho$.

If the considered flows similar to the inviscid Beltrami flows possess a functional dependence (in explicit, implicit, or parametric forms) between the vorticity and the stream function, $J(\Psi_3,\omega_3)$ identically vanishes, and $v\omega + \lambda\Psi_3$ gives a vector potential of the potential friction $f_r = -$ grad φ_r with a scalar potential φ_r, whereas the potentials are connected by the Cauchy-Riemann conditions

$$\partial_{x_2}(v\omega_3 + \lambda\psi_3) = \partial_{X_1}\varphi_r, \quad \partial_{X_1}(v\omega_3 + \lambda\psi_3) = -\partial_{x_2}\varphi_r \tag{5}$$

Then the momentum equation (3) reduces to an ordinary differential equation

$$d_{\psi_3}(E_B + \varphi_r) = -\omega_3 \tag{6}$$

which coincides for $\varphi_r = 0$ with that of Batchelor.[11] Equation (6) means that the vortex lines, the isolines of $E_B + \varphi_r$, and the streamlines form a family of parallel curves because $J(E_B + \varphi_r, \Psi_3) = 0$. Integrating Eq. (6) and expressing φ_r through the vorticity and the stream function, we obtain the Bernoulli integral in the case of potential friction

$$\frac{1}{2}|\boldsymbol{V}|^2 + \int_{\psi_0}^{\psi_3} \omega_3(\chi)\,\mathrm{d}\chi + \lambda c X_1 + \int_{\boldsymbol{R}_0}^{\boldsymbol{R}} (\lambda + v\,\mathrm{d}_{\psi_3}\omega_3)\boldsymbol{V}\cdot \mathrm{d}\boldsymbol{R} + \frac{p}{\rho} + \varphi_b = E_0 \tag{7}$$

which is valid along any curve between points $\boldsymbol{R} = (X_1, x_2)$ and $\boldsymbol{R}_0 = (X_{10}, x_{20})$. Here $E_0 = E_B\,(\Psi_0)$ and $\Psi_0 = \Psi_3(X_{10}, x_{20})$. Under the conditions of potential friction Eq. (5), the two-dimensional flow possesses potentials for the convective terms, the braking, and the viscous friction, respectively. When one moves along a streamline $\Psi_3(X_1, x_2) = \Psi_1$ in the absence of friction, Eq. (7) reduces to the conventional form $\frac{1}{2}|\boldsymbol{V}|^2 + \frac{p}{\rho} + \varphi_b = E(\Psi_1)$, where the constant $E(\Psi_1)$ depends on the chosen streamline $E(\Psi_1) = E_0 - \int_{\Psi_0}^{\Psi_1} \omega_3\, d\chi$.

In the first partial case of $v = \lambda = 0$, it reduces to the explicit form[9,10] of the Bernoulli integral for steady-state inviscid flows without the linear braking. Secondly, integrating Eq. (1) in the case of potential flows ($\omega_3 = 0$), we have the Cauchy integral with the linear braking

$$\partial_t\varphi_v \frac{1}{2}|\boldsymbol{v}|^2 + \frac{p}{\rho} + \varphi_b + \lambda\varphi_v = f(t) \tag{8}$$

where φ_v is the velocity potential $\boldsymbol{v}$ = grad φ_v and $f(t)$ is an arbitrary time-dependent function. The Bernoulli integral (7) coincides with the Cauchy integral (8) in the steady-state case as $\partial_t = -c\partial_{X_1}, \partial_{x_1} = \partial_{X_1}$, and $\int \boldsymbol{v} \cdot \mathrm{d}\boldsymbol{r} = \varphi_v$. Finally, when the two-dimensionality is supported by the flow geometry or by other effects (say, the stratification), the problem (3–4) may be considered in the Stokes approximation when inertia of the fluid is negligible, i.e., $E_B = (p/\rho) + \varphi_b + \lambda c X_1$ and ω_3 grad $\Psi_3 = 0$. Then friction is always potential because the momentum equation (3) itself transforms into the Cauchy–Riemann conditions (5) for the vector potential $\nu\omega_3 = \lambda\,\Psi_3$ and the scalar one $\varphi_r = -E_B$, and the Bernoulli integral for steady-state creeping flows takes the form

$$\frac{p}{\rho} + \varphi_b + \lambda c X_1 + \lambda \int_{R_0}^{R} \boldsymbol{V} \cdot \mathrm{d}\boldsymbol{R} + \nu \int_{R_0}^{R} (\partial_{x_2}\omega_3\,\mathrm{d}X_1 - \partial_{X_1}\omega_3\,\mathrm{d}x_2) = E_0 \tag{9}$$

which coincides with the linear form of Eq. (7) for $\omega_3 = \omega_3(\Psi_3)$. So, in the general case of potential friction the kinematic problem is determined by a set of equations for the stream function and a harmonic function h,

$$J(\psi_3, h) = 0, \qquad \nu\Delta\psi_3 - \lambda\psi_3 = h, \qquad \Delta h = 0 \tag{10,11,12}$$

and solving of the dynamic problem reduces to calculating integrals of Eq. (7) or (9). In the case of free-surface waves, the dynamic boundary condition of pressure constancy on the free surface combined with Eq. (7) gives the dynamic equation for the wave shape.

III. Poiseuille–Couette Flows with the Linear Braking

To obtain admissible far-upstream conditions for the solitary-wave problem, one should solve the one-dimensional problem [Eqs. (7), (10–12)]. Equation (10) is satisfied identically *via* a parametric dependence $\Psi_3 = \Psi_3(x_2)$, $\omega_3 = \omega_3(x_2)$, and $h = h(x_2)$. The solution of [Eq. (12)] is a linear $h = -(\omega_0 + \omega_1 x_2)$, and the kinematic problem reduces to [Eq. (11)]. If the dimensionless variables of the laboratory frame are adopted as $x = x_1/L$, $y = x_2/h$, $u = v_1/U$, $\Psi = \Psi_3/Uh$, $\omega = \omega_3 h/U$, and $P = p/\rho U^2$, the considered problem becomes

$$\begin{gathered} \mathrm{d}_y^2\psi - Ha_\lambda\psi = -\Omega_0 - \Omega_1 y \\ \mathrm{d}_y\psi = u_1 \quad \text{for } y = y_1, \quad \mathrm{d}_y\psi = u_2 \quad \text{for } y = y_2, \quad \psi = 0 \quad \text{for } y = 0 \end{gathered} \tag{13}$$

where nonslip boundary conditions are presumed on the walls $y = y_1$ and $y = y_2$ moving with velocities $u = u_1$ and $u = u_2$, respectively. Its solution is characterized by the vorticity parameter $\Omega_1 = \omega_1 h/U$ and the Hartmann number $Ha_\lambda = \lambda h^2/\nu$

$$\frac{\psi}{2}\sqrt{Ha_\lambda} = \frac{\Omega_1}{Ha_\lambda}\left[\mu - \frac{\operatorname{sh}\mu\operatorname{ch}(\mu-\mu_2-\mu_1)}{\operatorname{ch}(\mu_2-\mu_1)}\right] + [u_2\operatorname{sh}(\mu-2\mu_1) - u_1\operatorname{sh}(\mu-2\mu_2)]\frac{\operatorname{sh}\mu}{\operatorname{sh}2(\mu_2-\mu_1)} \tag{14}$$

where $\mu = y\sqrt{Ha_\lambda}/2$, $\mu_1 = y_1\sqrt{Ha_\lambda}/2$, and $\mu_2 = y_2\sqrt{Ha_\lambda}/2$. The parameter $\Omega_0 = \omega_0 h/U$ and the Reynolds number with respect to the linear braking $Re_\lambda = Re_\nu/Ha_\lambda$ are excluded from the final results, and the pressure [Eq. (7)] $P = -\Omega_1 x/(Re_\nu\sqrt{\beta})$ depends on the Reynolds number $Re_\nu = Uh/\nu$ and the dispersion $\beta = h^2/L^2$ ($P = 0$ for $x = 0$). The solution clarifies an idea of potential friction, which makes possible steady-state viscous vortex flows. Under the conditions for steady motion[6] (§ 165) extended in the case of potential friction [Eq. (5)], being vortex ones in the general case the friction terms are equilibrated by the potential term of pressure (as it takes place in the case given) or by the other potential terms of the Bernoulli integral (7).

The velocity profile is a linear superposition of the Poiseuille flow modified by the linear braking ($u_2 = u_1 = 0$) given in Fig. 1a and the modified Couette flow ($\Omega_1 = 0$) given in Fig. 1b, which reduce to the well-known solutions for enough small Ha_λ:

$$\begin{aligned} u &= \frac{\Omega_1}{Ha_\lambda}\left[1 - \frac{\operatorname{ch}(2\mu-\mu_2-\mu_1)}{\operatorname{ch}(\mu_2-\mu_1)}\right] + \frac{u_2\operatorname{sh}2(\mu-\mu_1) - u_1\operatorname{sh}2(\mu-\mu_2)}{\operatorname{sh}2(\mu_2-\mu_1)} \\ &\to \frac{\Omega_1}{2}(y_2-y)(y-y_1) + (u_2-u_1)\frac{y-y_1}{y_2-y_1} + u_1 \quad \text{for } Ha_\lambda \to 0 \end{aligned} \tag{15}$$

An example of the modified Poiseuille–Couette flow [Eq. (15)] is given in Fig. 1c for different Ha_λ and $\Omega_1 = 2$, $u_1 = 0$, and $u_2 = 1$. When Ha_λ is large enough, the flow splits into the core and the Shercliff layers[12] of the length scale are inversely proportional to $\sqrt{Ha_\lambda}$, whereas the velocity scale does not depend on Ha_λ for the moving wall $y = y_2$ and does depend on it strongly for the motionless wall $y = y_1$. Being the implicit one in the general case [Eq. (11)], the dependence between ω and Ψ,

$$\omega + Ha_\lambda\psi - \Omega_1 y + \frac{\Omega_1}{\sqrt{Ha_\lambda}}\frac{\operatorname{sh}(\mu_2+\mu_1)}{\operatorname{ch}(\mu_2+\mu_1)} + \sqrt{Ha_\lambda}\,\frac{u_2\operatorname{ch}2\mu_1 - u_1\operatorname{ch}2\mu_2}{\operatorname{sh}2(\mu_2-\mu_1)} \tag{16}$$

changes in a nontrivial manner with respect to Ha_λ even in this simplest case of one-dimensional flows. The dependence is explicit and linear for the Couette flow and finite values of Ha_λ; then the vorticity tends to the constant

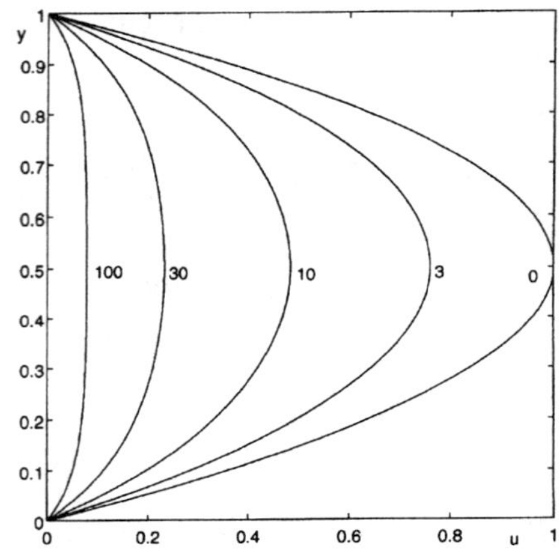

Fig. 1a Velocity profiles of the Poiseuille flow modified by the linear braking for different Hartmann numbers.

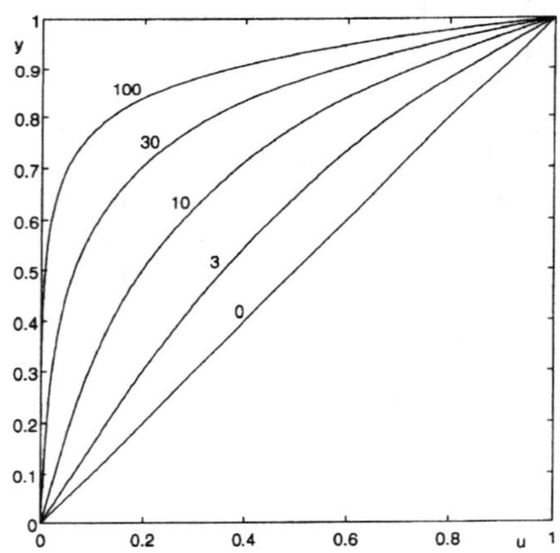

Fig. 1b Velocity profiles of the Couette flow modified by the linear braking for different Hartmann numbers.

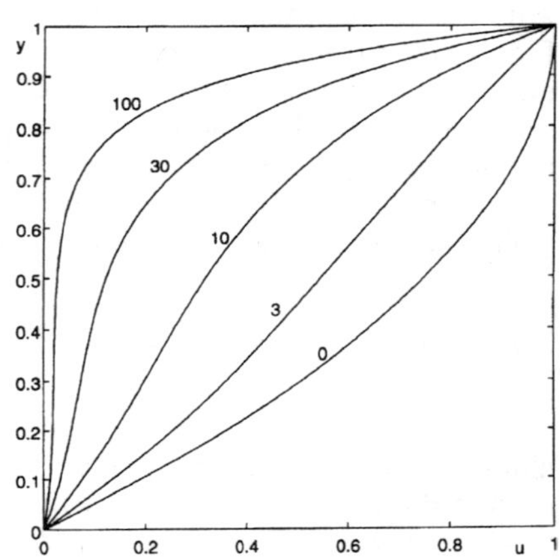

Fig. 1c Velocity profiles of the Poiseuille–Couette flow modified by the linear braking for different Hartmann numbers.

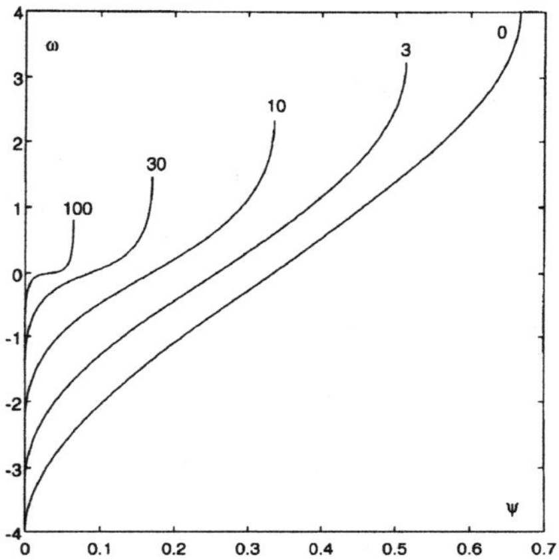

Fig. 2 Dependence of the vorticity on the stream function for the Poiseuille flow modified by the linear braking.

value x 5 2(u2 $- u_1)/(y_2 - y_1)$ and does not depend on the stream function in the limit of small Ha_λ. In the case of the Poiseuille flow ($\Omega_1 = 8$), the dependence given in Fig. 2 for different Ha_λ is implicit for finite values of Ha_λ, and if $Ha_\lambda \to 0$, the stream function takes a form of the cubic polynomial with respect to the vorticity $\Psi = -\frac{1}{2}\Omega_1\eta\left(\frac{1}{3}\eta^2 - y_s\eta + y_1y_2\right)$, $\eta = y_s + \omega/\Omega_1$, and $y_s = (y_2 + y_1)/2$. This example shows that the function $\omega = \omega_0$ sech $(\Psi - \Psi_0)$ met in different kinds of flows[5] may be fairly well approximated by Eq. (16) arising in the case of potential friction.

To estimate the validity boundaries of the linear-braking model [Eqs. (13) and (14)], the flow [Eq. (15)] was compared with the exact solution of Shercliff[13] for the rectangular channel $|x_2| < S$, $|x_3| < H$, which is written in the form admitting for two limit transitions to slot channels,

$$u = u_s\sum_{n=1}^{\infty}\frac{2(-1)^{n+1}}{k^3}\left[1 - C\frac{\mathrm{ch}(\mu_{2n-1}z)}{\mathrm{ch}(\mu_{2n-1}h_s)}\right] + s\frac{\mathrm{sh}(\mu_{2n-1}z)}{\mathrm{sh}(\mu_{2n-1}h_s)}\cos(kl_s y)$$

$$k = \pi\left(n - \frac{1}{2}\right), \qquad C = \mathrm{ch}(Ha\,h_s/2)\,\mathrm{ch}(Ha\,z/2) \tag{17}$$

$$S = \mathrm{sh}(Ha\,h_s/2)\,\mathrm{sh}(Ha\,z/2), \qquad \mu_{2n-1}^2 = (Ha/2)^2 + (kl_s)^2$$

where l_s, u_s, and h_s are the scale factors for the length, the velocity, and the Hartmann number, respectively. When $Ha = 0$, the flow [Eq. (17)] gives the Poiseuille flows $u = (1 - z^2)/2$ and $u = (1 - y^2)/2$ for infinite slot channels that are perpendicular and parallel to the external magnetic field and that imply the Hartmann (H scale) and Shercliff (S scale) characteristic systems of scales in each of the limits under consideration, correspondingly,

H scale:

$$z \equiv \frac{z}{H}, \quad y \equiv \frac{y}{H}, \quad u \equiv u / |\mathrm{d}_x p| \frac{H^2}{\nu\rho}$$

$$Ha = BH\sqrt{\sigma / \nu\rho}, \quad u_s = \frac{1}{s^2}, \quad l_s = s, \quad h_s = 1 \tag{18}$$

S scale:

$$z \equiv \frac{z}{S}, \quad y \equiv \frac{y}{S}, \quad u \equiv u / |\mathrm{d}_x p| \frac{S^2}{\nu\rho}$$

$$Ha = BS\sqrt{\sigma / \nu\rho}, \quad u_s = 1, \quad l_s = 1, \quad h_s = s$$

where $s = H/S$. The potential-friction flow [Eq. (15)] reduces to

$$u = \left[1 - \mathrm{ch}\left(\sqrt{Ha}\, y\right) / \mathrm{ch}\left(\sqrt{Ha}\right)\right] / Ha \tag{19}$$

where h equals to H or S for the H scale and S scale, respectively. Presenting a direct consequence of the linear braking model,[14] the flow of [Eq. (19)] may be termed the Sommréia–Moreau flow, and it differs from the Hartmann flow[12] $u = [1 - \mathrm{ch}(Ha\, z)/\mathrm{ch}(Ha)]/Ha$ th Ha mainly by the well-known characteristic dependence on the Hartmann number. The two-dimensionality of the Shercliff flow [Eq. (17)] is governed both by the braking electromagnetic force and the channel geometry. In the case of the slot channel perpendicular to the field ($s = 5$), the amplitude of Eq. (19), for $Ha = 0$, greatly differs from the Poiseuille profile in the slot channel. But increases in Ha rather quickly disrupt this discrepancy and validity of Eq. (19) in the core and are confirmed up to inaccuracy less than 3% for $Ha = 5$ (Figs. 3a and 3b). Even in the case of the square channel ($H = S = 1$), the flow of Eq. (19) fairly predicts (up to 3% $Ha = 5$) the velocity amplitude although the two-dimensionality is not expressed well. The situation changes qualitatively in the case of the slot channel parallel to the field when the channel geometry implies the two-dimensional Poiseuille flow coinciding exactly with Eq. (19) for $Ha = 0$. Then the growing Ha results in three-dimensional flow as

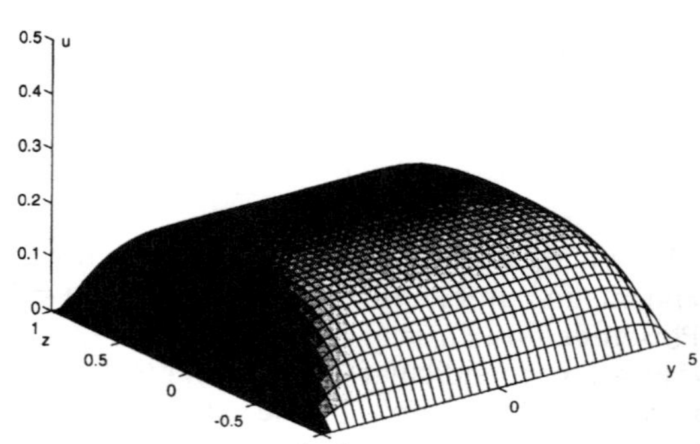

Fig. 3a Velocity profiles of the Shercliff flow for *Ha* = 5.

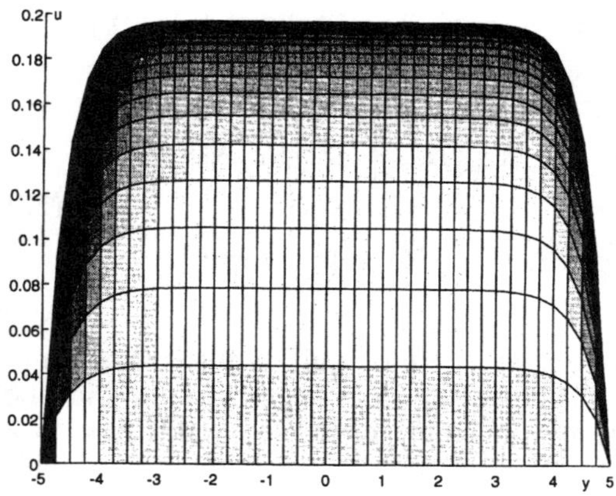

Fig. 3b The Shercliff layers of the considered flow.

the induction currents weakly interact with the magnetic field in the central domain ($z \approx 0$) of the slot channel, and the approximation (19) is valid for enough small Hartmann numbers (for $s = 1/5$, $Ha = 1/5$, inaccuracy of the core velocity is less then 4%).

IV. Solitary Vortex Couples

In the case of two-dimensional flows, we consider primarily the linear problem (9), (11), and (12) that is also applicable to large-scale geophysical flows because atmosphere behaves with respect to such flows rather like a weakly supercritical system than a highly developed turbulent one.[7] Transforming the problem to the polar coordinates $r = \sqrt{X_1^2 + x_2^2}\,/R, \theta$ and introducing dimensionless variables with the scales of length, velocity, stream function, and vorticity R, U, UR, and U/R, respectively, we get the problem

$$\partial_r^2 \psi + r^{-1}\partial_r \psi + r^{-2}\partial_\theta^2 \psi - Ha_\lambda \psi = h, \qquad \partial_r^2 h + r^{-1}\partial_r h + r^{-2}\partial_\theta^2 h = 0 \qquad (20)$$

where $Ha_\lambda = \lambda R^2/\nu$. Considering propagation of solitary vortex couples in the core of basic flow (19), we assume that the Hartmann number is large enough to consider the basic flow as a uniform one far upstream with respect to the vortex couple: ($r \to \infty$) $\Psi \to -r \sin\theta$. Then the general coupled solution may be written as

$$\psi = \left[-r + E/r - DK_1\left(\sqrt{Ha_\lambda} r\right)/\sqrt{Ha_\lambda}\right]\sin\theta, \quad h = Ha_\lambda (r - E/r)\sin\theta, \quad r > 1$$

$$\psi = -\left[Fr + CI_1\left(\sqrt{Ha_\lambda} r\right)/\sqrt{Ha_\lambda}\right]\sin\theta, \quad h = Ha_\lambda Fr \sin\theta, \quad r < 1 \qquad (21)$$

where I_1, K_1 are the modified Bessel functions of the first and second kinds, respectively. Parameters C, D follow from continuity conditions for normal and tangential components of the velocity (i.e., for Ψ and $\partial_r \Psi$) on the matching circle $r = 1$:

$$C=\sqrt{Ha_\lambda}\left\{(1-F)\left[2K_1\left(\sqrt{Ha_\lambda}\right)+\sqrt{Ha_\lambda}K_0\left(\sqrt{Ha_\lambda}\right)\right]-E\sqrt{Ha_\lambda}K_0\left(\sqrt{Ha_\lambda}\right)\right\}$$
$$D=\sqrt{Ha_\lambda}\left\{(1-F)\left[2I_1\left(\sqrt{Ha_\lambda}\right)-\sqrt{Ha_\lambda}I_0\left(\sqrt{Ha_\lambda}\right)\right]+E\sqrt{Ha_\lambda}I_0\left(\sqrt{Ha_\lambda}\right)\right\} \tag{22}$$

As continuity of the vorticity $\omega = -Ha_\lambda \Psi - h$ is equivalent to the continuity of h, it implies that $1 - E = F$. Integrating Eq. (9) we maintain that the pressure drop is uniform inside the matching circle and tends also to a constant value $Ha_\lambda (1 - S)$ for upstream with respect to the vortex couple

$$P=Ha_\lambda(F-S)r\cos\theta, \quad r<1, \quad P=Ha_\lambda\left[(1-S)+E/r^2\right]r\cos\theta, \quad r>1 \tag{23}$$

where the characteristic value of the pressure is taken as $\rho\nu U/R$ and $S = c/U$ is the dimensionless celerity of the vortex couple propagation. Continuity of the pressure yields the last equation for the arbitrary constants of Eq. (21) with the following solution: $E = S/2$ and $F = 1 - S/2$. Being the fundamental datum that maintains the motion, the constant pressure gradient of the basic flow [Eq. (19)] should coincide with that far upstream with respect to the solitary vortex couple: $\text{grad}\, p = \lambda\rho\,(U - c)$. Added by the kinematic relation $U = v_1 - c$ and the dimensional form of Eq. (19) for v_1, it enables one to calculate the last dynamic parameters of the vortex couple

$$c=-\frac{h^2}{\nu\rho}\frac{1}{Ha_f}\left[1-\frac{1}{2\,\text{ch}\left(\sqrt{Ha_f}\right)}\right]\text{grad}\,p, \quad U=\frac{h^2}{\nu\rho}\frac{1}{2Ha_f\,\text{ch}\left(\sqrt{Ha_f}\right)}\text{grad}\,p \tag{24}$$

and $S=2\,\text{ch}\left(\sqrt{Ha_f}\right)-1$, where $Ha_f = \lambda\, h^2/\nu$ is the Hartmann number of the basic flow.

Thus, we get a species of viscous solitary vortex couples traveling at the permanent speed c without changing the couple flow pattern in the opposite direction to the pressure gradient (Fig. 4a). The vortex couple propagates faster than the basic flow as the couple drag is smaller than the drag of the basic flow and being the potential one [Eq. (5)] the friction force is compensated by the external pressure gradient. Linearity conditions (V_1, $v_2 \ll c$) imply that Eq. (24) may be used for enough high Hartmann numbers of the basic flow ($S \gg 1$). In fact (see Fig. 5a), this condition is in agreement with the applicability condition of the basic flow obtained in the previous section. Influence of viscosity and braking yields a dependence between the radius of the vortex couple r_c defined as the solution of the equation $\Psi = 0$ and its dimensionless velocity S.

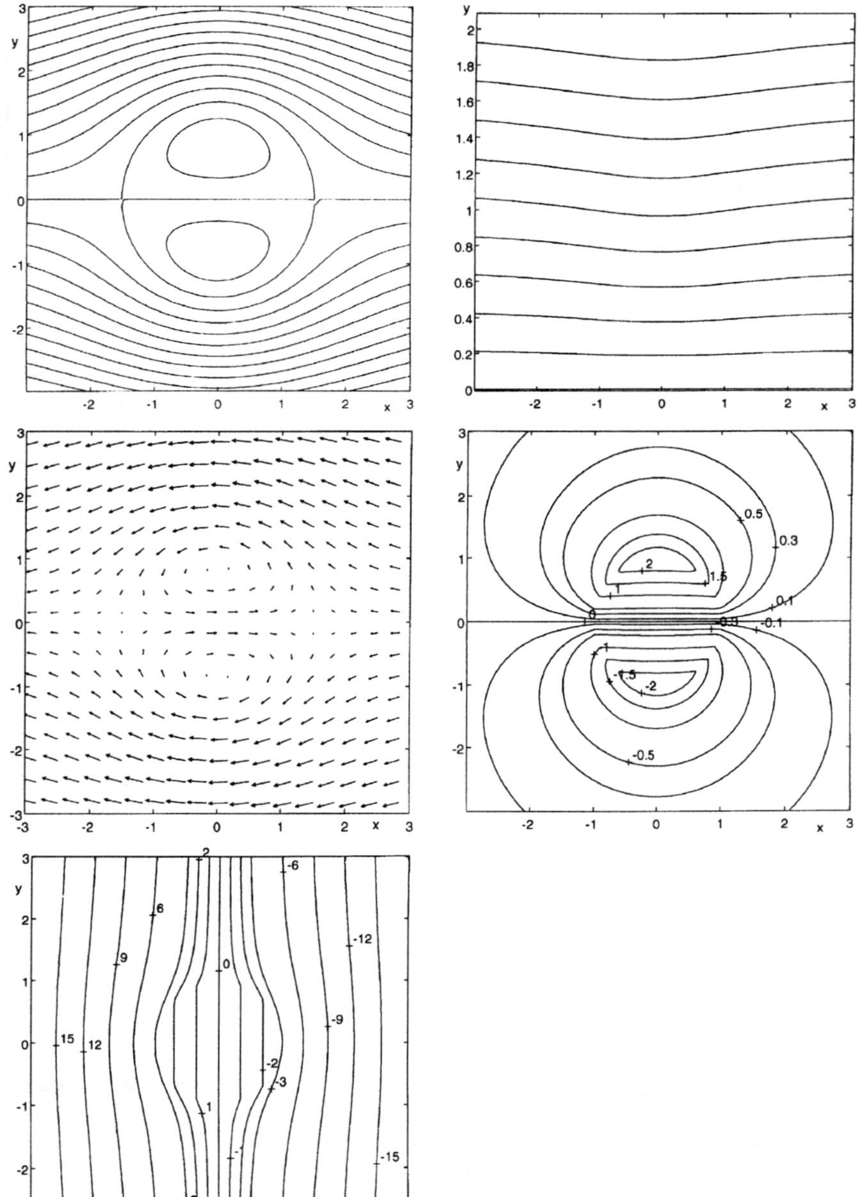

Fig. 4 The solitary vortex couple for $Ha_\lambda = 0.7$, $Ha_f = 5.7$, and $S = 10$: a) streamlines, values of $\Psi \in [-2, 2]$ with the step 0.2; b) streamlines in the laboratory reference frame, values of $\Psi \in [0, 18]$ with the step 2; c) velocity profiles; d) vorticity; and e) pressure.

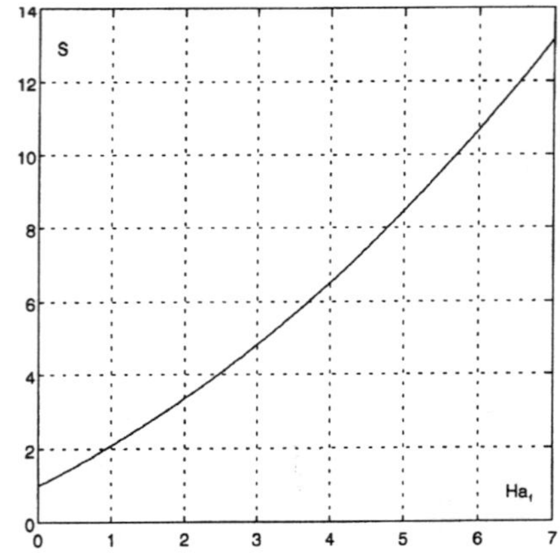

Fig. 5a Dimensionless velocity of viscous wavy-vortex couples versus the Hartmann number of the basic flow.

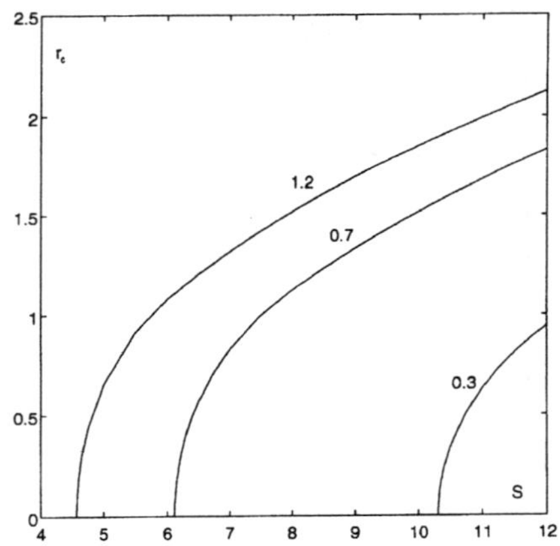

Fig. 5b The radius of the vortex couple r_c versus its speed of motion S.

$$
\begin{aligned}
&0.5 - K_1\left(\sqrt{Ha_\lambda}\right) I_1\left(\sqrt{Ha_\lambda}\, r_c\right) \Big/ r_c = 1/S, \qquad r_c < 1 \\
&\left[0.5 r_c^{-1} - I_1\left(\sqrt{Ha_\lambda}\right) K_1\left(\sqrt{Ha_\lambda}\, r_c\right)\right] \Big/ r_c = 1/S, \qquad r_c > 1
\end{aligned}
\tag{25}
$$

that is given in Fig. 5b; r_c increases versus S as the wave amplitude increases with the wave speed in the case of nonlinear inviscid free-surface solitary waves. In the laboratory frame the vortex couple takes the form of a concave wavy disturbance propagating with the speed c (Fig. 4b).

As regards kinematics, the viscous vortex couple differs from the inviscid Lamb vortex couple[6] by less intensive and more large-scaled circulation of the fluid (Fig. 4b) connected with the unbounded vorticity field, which is restricted by the vortex couple radius in the inviscid case (Figs. 4d and 6a). The principal difference appears in the case of pressure (Figs. 4e and 6b). Being a symmetric one in the inviscid case, the pressure distribution becomes asymmetric with the constant pressure drop inside the matching

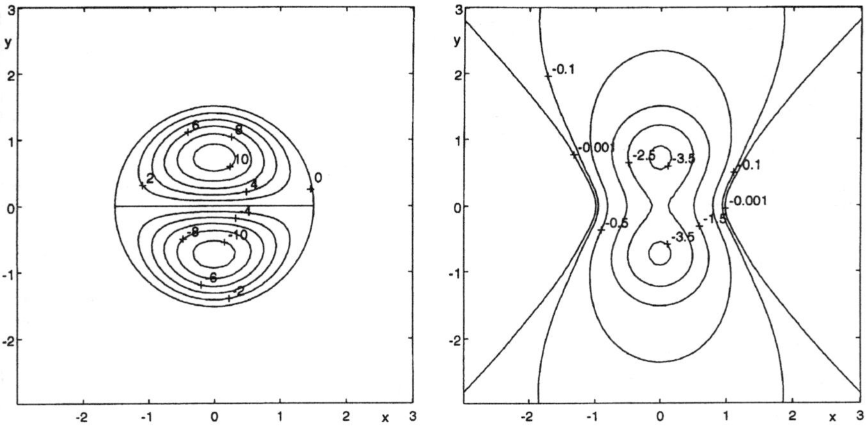

Fig. 6 The Lamb vortex couple for r_c = 1.5153: a) vorticity and b) pressure.

circle and far away from it. As such a pressure distribution enables the steady-state propagation of the vortex couple, the last one may be treated as a hydrodynamic carrier of the Bernoulli integral presenting the phase energy conservation law similar to the inviscid case of nonlinear two-level solitary waves.[15] Laboratory experiments on propagation of Lamb vortex couples in still water[4] have shown that the influence of the viscous friction and the linear braking results in damping of the inertial couples in absence of the external pressure gradient.

V. Coupled Solitary Waves

When the Hartmann number of the vortex couple, Ha_λ becomes small enough, its radius vanishes in the admissible interval of Ha_f (Figs. 5a and 5b), and the vortex couple transforms into a solitary coupled wave given in Figs. 7a and 7b. As vorticity and pressure of the coupled solitary waves are described by Eqs. (21–23) for appropriate values of parameters, the vorticity and pressure distributions of the coupled solitary waves differ from that of the solitary vortex couples by intensity. But the flow pattern differs qualitatively, as the flow only decelerates inside the coupled solitary wave and the flow separation does not occur. On the contrary to the free-surface inviscid solitary waves that are intrinsically nonlinear ones, the internal viscous coupled solitary waves are possible even in the linear limit because dependence of the wave amplitude on the celerity is provided by the viscous friction and the linear braking that keep the balance between the force produced by the constant external pressure gradient and the wave drag depending on its amplitude and celerity. The propagation velocity [Eq. (24)] is determined by the basic flow through Ha_f and grad p, while the velocity profile is controlled both by the basic flow via Ha_f and the wave shape via Ha_λ. The inverse transition from the coupled solitary wave to the solitary vortex couple, which takes place for the large enough wave amplitude, is well-known in the

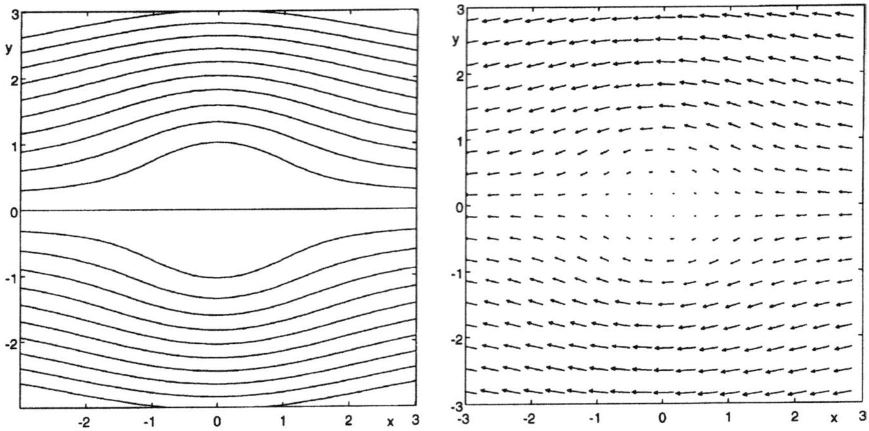

Fig. 7 The coupled solitary wave for $Ha_\lambda = 0.3$, $Ha_f = 5.7$, and $S = 10$: a) streamlines, values of $\Psi \in [-2, 2]$ with the step 0.2 and b) velocity profiles.

hydrodynamics of stratified flows.[16] Presumably, the separation of flow inside finite-amplitude internal solitary waves, which corresponds to a transformation of wavy disturbances into vortex ones, describes a transition of the basic flow into the turbulent regime.

Finally, to take into account nonlinear effects, we consider an inertial model of coupled solitary waves in the case of the linear dependence between the vorticity and the stream function $\omega = -Ha_\lambda \Psi$, when the problem corresponds to the Lamb vortex couple[6] (§ 165) with a complex-valued k ($\omega = k^2\Psi$). It is assumed that the flow is irrotational outside the matching circle and according to Eq. (5), the braking potential coincides with the velocity potential up to the multiplier λ. Inside the matching circle the vorticity transfer equation is satisfied identically because of the linear dependence, and the friction potential vanishes. The problem considered

$$
\begin{aligned}
&\partial_r^2\Psi + r^{-1}\partial_r\Psi + r^{-2}\partial_\theta^2\Psi - Ha_\lambda\Psi = 0, \qquad r<1 \\
&\partial_r^2\Psi + r^{-1}\partial_r\Psi + r^{-2}\partial_\theta^2\Psi = 0, \qquad r>1
\end{aligned}
\tag{26}
$$

has the general coupled solution

$$
\Psi = -AI_1\left(\sqrt{Ha_\lambda}\, r\right)\sin\theta, \quad r<1, \qquad \Psi = (-r + B/r)\sin\theta, \quad r>1 \tag{27}
$$

where parameters are chosen in such a way that the flow is uniform far upstream ($\Psi \to -r \sin\theta$ for $r \to \infty$) and the continuity conditions are satisfied for the velocity profile $r = 1$,

$$
A = 2\Big/\left[\sqrt{Ha_\lambda}\, I_0\left(\sqrt{Ha_\lambda}\right)\right], \quad B - 1 = -2I_1\left(\sqrt{Ha_\lambda}\right)\Big/\left[\sqrt{Ha_\lambda}\, I_0\left(\sqrt{Ha_\lambda}\right)\right] \tag{28}
$$

The pressure may be derived by Eqs. (7) and (8) outside and inside the matching circle, respectively,

$$P = -Ha_\lambda Sr\cos\theta + Re_v(Ha_\lambda \Psi^2 - V_x^2 - v_y^2)/2, \quad r<1$$
$$P = Ha_\lambda[(1-S) + B/r^2]r\cos\theta + Re_v(1 - V_x^2 - v_y^2)/2, \quad r>1 \tag{29}$$

where $Re_v = U R/v$ is the Reynolds number. The velocity profile follows from Eq. (26)

$$V_x = -1 + Br^{-2}\cos 2\theta, \quad v_y = Br^{-2}\sin 2\theta, \quad r>1 \tag{30}$$

$$V_x = -A\sqrt{Ha_\lambda}\left[I_0(\mu)\sin^2\theta + \mu^{-1}I_1(\mu)\cos 2\theta\right]$$

$$v_y = A\sqrt{Ha_\lambda}\left[I_0(\mu)/2 - \mu^{-1}I_1(\mu)\right]\sin 2\theta, \quad r<1$$

where $\mu = \sqrt{Ha_\lambda}\, r$. As far from the nonlinear solitary wave the pressure coincides with that for the linear solitary wave, the celerity is described also by Eq. (24) without restrictions of the creeping flow.

The inertial waves qualitatively differ from the linear ones by finite jumps of the vorticity and the pressure on the matching circle. Arising finite jumps of vorticity, pressure, or their gradients (as for the linear coupled solitary waves, for example) are a characteristic feature of all of the family of vortex-couple solutions verified by several laboratory and numerical experiments. Mathematically, existence of the jumps is explained by absence of cylindrical functions vanishing both at infinity and origin and, physically, correlates with turbulent perturbations of pressure and velocity taking place in the basic flow at enough large Reynolds numbers. The flow pattern of the

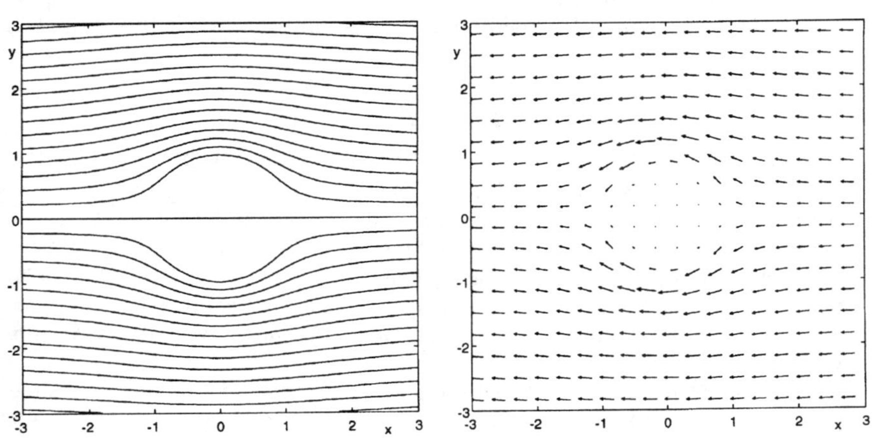

Fig. 8 The inertial coupled solitary wave for $Ha_\lambda = 70$ and $Ha_f = 100$: a) streamlines, values of $\Psi \in [-2.8, 2.8]$ with the step 0.2 and b) velocity profiles.

inertial coupled solitary wave is given in Figs. 8a and 8b for enough high Hartmann numbers, which clarify the dynamic wave structure. Similar to the case of channel flows considered in Section II, the flow may be split into the core of a curved cylindrical form with the quasiuniform vanishing velocity and the Shercliff layer of the scale inversely proportional to $\sqrt{Ha_\lambda}$, and the inertial coupled solitary wave may be treated dynamically as a curved cylinder in the case of high Hartmann numbers.

VI. Conclusion

The obtained Poiseuille–Couette flow modified by the linear braking provides a one-dimensional example of the potential-friction flows when the viscous and braking terms are equilibrated by the potential dynamic and pressure terms. Comparisons of this flow with the Shercliff flow enable one to determine more precisely the applicability frames of the quasi-two-dimensional model [Eq. (1)].

The viscid solitary vortex couple and coupled solitary wave give an example of two-dimensional steady-state potential-friction flows driven by a constant pressure drop that compensates the drag of wavy and vortex couples and provides constant propagation velocities and permanent flow structures. Influence of viscosity and braking results in the dependence of characteristic dimensions of the solitary vortex couple and the coupled solitary wave on the propagation velocity even for linear solutions. Such internal waves and vortices may be responsible for the well-known intermittency of the channel MHD flows.

Acknowledgment

The author would like to thank René Moreau for helpful discussions. Financial support of the Latvian Scientific Council and of the French government (for the stipend providing the possibility to finish this work) is gratefully acknowledged.

References

[1]Somméria, J., and Moreau, R., "Why, How, and When, MHD Turbulence Becomes Two-Dimensional," *Journal Fluid Mechanics*, Vol. 118, 1982, pp. 507–518.

[2]Pedlosky, J., *Geophysical Fluid Dynamics*, Springer–Verlag, New York, 1982.

[3]Thess, A., Somméria, J., and Jüttner, B., "Inertial Organization of a Two-Dimensional Turbulent Vortex Street," *Physics of Fluids*, Vol. 6, No. 7, 1994, pp. 2417–2429.

[4]Nguen, J.-M., and Somméria, J., "Experimental Characterization of Steady Two-Dimensional Vortex Couples," *Journal of Fluid Mechanics*, Vol. 192, 1988, pp. 175–192.

[5]Pastmanter, R. A., "On Long-Lived Vortices in 2-D Viscous Flows, Most Probable States of Inviscid 2-D Flows and a Soliton Equation," *Physics of Fluids*, Vol. 6, No. 3, 1994, pp. 1236–1241.

[6]Lamb, H., *Hydrodynamics*, Cambridge Univ. Press, 1932.

[7]Dolzansky, F. V., and Manin, D. Yu., "Free Shear Layers and Laboratory Simula-

tion of Atmospheric Circulation," *Izvestiya Akademii* and *Nauk SSSR, Fizika Atmosfery i Okeana*, Vol. 26, No. 12, 1990, pp. 1282–1288.

[8]Meleshko, V. V., and van Heijst, G. J. F., "On Chaplygin's Investigations of Two-Dimensional Vortex Structures in an Inviscid Fluid," *Journal of Fluid Mechanics*, Vol. 272, 1994, pp. 157–182.

[9]Miroshnikov, V. A., "Nonlinear Interaction of a Parabolic Sheared Current with a Free Surface," *Boundary Effects in Stratified and/or Rotating Fluids*, IPM, St. Petersburg, 1995, 113–115.

[10]Miroshnikov, V. A., "A Solitary Wave on the Surface of a Shear Stream in Crossed Electric and Magnetic Fields: Formation of a Single Vortex," *Magnetohydrodynamics*, Vol. 31, No. 1–2, 1995, pp. 169–185.

[11]Batchelor, G. K., "On Steady Laminar Flow with Closed Streamlines at Large Reynolds Number," *Journal of Fluid Mechanics*, Vol. 1, 1956, pp. 177–190.

[12]Moreau, R., *Magnetohydrodynamics*, Kluwer Academic, Dordrecht, The Netherlands, 1990.

[13]Shercliff, J. A., "Steady Motion of Conducting Fluids in Pipes Under Transverse Magnetic Field," *Proceedings of the Cambridge Philosophical Society*, Vol. 49, No. 1, 1953, pp. 136–144.

[14]Kliukin, A., private communication.

[15]Izrar, B., Lusseyran, F., and Miroshnikov, V., "Two-Level Solitary Waves as Generalized Solutions of the KdV Equation," *Physics of Fluids*, Vol. 7, No. 5, 1995, pp. 1056–1062.

[16]Davis, R. E., and Acrivos, A., "Solitary Internal Waves in Deep Water," *Journal of Fluid Mechanics*, Vol. 29, 1967, pp. 593–607.

IV. Astrophysics

Chapter 21

On the Driving Forces and Energy Sources of Planetary Magnetic Fields

Sh. Sh. Dolginov
POB 5250 Rehovot 76569 Israel

I. Introduction

In the past decades, different small-scale and large-scale flows in the liquid conductive core of the Earth have been studied. Initially weak magnetic fields can be amplified by these flows (theories of kinematic dynamo).

However, it remains necessary to discuss the problem of the driving force and energy sources of the dynamo mechanism. This problem was discussed by Gubbins and Masters[1] 15 years ago. They carefully analyzed the proposed hypothesis of the energy sources and driving forces for the geodynamo: 1) convection driving by heat released due to cooling of the liquid core consisting of the pure iron, and by heat released due to the core freezing (Ref. 2); 2) convection driving by gravitational differentiation in the core, considered as a binary alloy of heavy (iron) and light (sulfur) components. The iron supplements to the inner solid core. The light component rises upward due to archimedian force (Refs. 3–8) and 3) convection due to the Earth's core and mantle precession with different velocities under the action of external gravitational forces of the sun and the moon (Ref. 9).

Gubbins and Masters conclude, "Although it has not been possible to eliminate rigorously many of the suggestions that have been made for powering dynamos, gravitational energy release due to differentiation appears to be the most plausible." Indeed, the hypothesis of gravitational differentiation can explain the formation of the inner solid core, and gravitational energy converts to the magnetic field energy more effectively than the heat energy.

Within the frame of hypothesis of the principal role of the mechanism of the gravitational differentiation of the composition, it was difficult to explain the synchronization of the geotectonic and geomagnetic activity: long periods of continental stability, periods of state pole position, separated by apid pole wandering, tectonic activity, and magnetic reversals (see review by Anderson[10]). Such a correlation can be explained by the action of external forces on the convection in the core and in the Earth's crust, or by the action of thermal flux from the core into the mantle, changing the convection in the mantle, or by the action of both factors.

Gubbins and Masters suggested that the gravitationally powered model can be pursued only further in conjunction with the thermal evolution of the Earth as a whole.[1] A progress in the theory of a geodynamo based on the hypothesis of continuing Earth's cooling have been developed in the paper by Braginsky and Roberts.[11] Compositional convection may not dominate thermal convection as has been argued by Braginsky in his paper[4] and in the publications mentioned above.

A dynamo process based on dual (compositional and thermal) convection or a mechanism of "local turbulence" that ensures the turbulent transport of heavy and light components are described in Ref. 11 by a system of boundary conditions and equations. The authors[11] focused their paper on the problem of the driving mechanism for the Earth's dynamo. They make these efforts in order to isolate, as it is possible, the investigations from poorly known key parameters of the Earth's core: the cooling rate of the core, the density jump on the solid core boundary, the value of the kinematic viscosity in the core, and heat flow from the core to the mantle. In the framework of the geodynamo model,[1] it is difficult to explain the discovered small intensity magnetic field of Mars, some intensive magnetic field of Mercury with smaller size and slow rotation, and practical absence of the magnetic field for the Earth-like planet Venus.

In the equations governing convection in the Earth's core, presented in Ref. 11, the force $F_p = \rho\,[\Omega \times \omega] \times r$ is ignored. Here ω and Ω are velocities of angular rotation and precession of a mass of a unit volume ρ at the distance r from the center. This force was named Poincare force by Malkus.[9] Malkus connected the generation of the Earth's magnetic field with the precession of the Earth and the Poincare force as a driving force of the geodynamo. The results of the first laboratory experiments that tend to support this proposal have been presented in Ref. 12. But the theoretical analysis of the problem in the framework of geophysics leads to conclusions that the precession of the Earth can have only a small effect on geodynamo (Ref. 4 and 13), and even do not affect geodynamo at all (Ref. 14). Gubbins and Masters[1] hold the opinion that the hypothesis of a precession dynamo "suffers from not having an adequate theory to explain how it can generate magnetic fields, and even if experimental dynamo had worked, it would not be an easy matter to scale it to model the Earth." But it is not possible to agree with the following assertion of the authors[1]: "Much more work needs to be done on the magnetic coupling of the core and mantle and the subsequent evolution of the Earth-moon system before the precessional dynamo can be assessed or even understood."

The long discussion on the efficiency of different driving forces and energy sources for the geomagnetic field generation that was carried out in the framework of geophysics did not lead to a final solution because a criterion of efficiency parameters were used, which could not be tested directly.

The beginning of checking the possible role of precession in planetary magnetism was undertaken by the comparison of the magnetic fields of the Earth and Mars, i.e., planets having close angular velocities ω and precessional amplitudes α (Ref. 15). There is not one planet that would not add important proofs to this problem.

The important details of the precessional dynamo mechanism have been studied by laboratory experiments. A review of theoretical and experimental aspects of the investigation of flows in spinning and precessing limited regions, physical interpretation of the phenomena observed in the laboratory experiments, the application of these results to the planet, and the sources of energy for planetary dynamos is the content of Marcus.[16]

The subsequent development of experimental investigations related to the dynamo problem are reflected in the publications of Vanyo[17] and the book of abstract of reports presented at the Seminar "Earth's Deep Interior" (SEDI 94).

II. Toward a Realistic Scaling Law of Planetary Magnetism

Half a century separates us from a hypothesis put forth by P. Blackett and rejected by him: there is a direct proportionality between magnetic M and angular P moments of a rotating body (Ref. 18). Blacketts's hypothesis left a deep track in geophysics: 1) it demonstrated the perspective of laboratory experiments in geophysics; 2) it simulated magnetic experiments in deep mines, which show that the source of the geodynamo is in the liquid core; and 3) the relation M/P = const, as a scaling law as a sort of Bode's law in magnetism, was frequently utilized for prediction of the results of spacecraft experiments.

More physical meaning is contained in the scaling law based on the assumption that in the liquid core flows they realize a geothrophic balance (the balance of the Coriolis force and pressure gradient force) or a magnetostrophic balance (the balance of the Coriolis force and the Lorentz force). But the planets were not subordinated to these scaling laws (Refs. 19 and 20).

Malkus[12] proposed that by the magnetoturbulent flow in the core, Coriolis, Lorentz, and Poincare forces are balanced. Under such an assumption he estimated the flow velocity, the intensity of the toroidal and poloidal components of the geomagnetic field. Stevenson[21] applied these formulas to estimate the magnetic fields of other planets. He concluded that the dynamo is unlikely for Mercury, Venus, and Mars. The estimated intensity of the magnetic field of Jupiter and Saturn disagree with the observed intensity.

All of these failures cast doubts on the possibility of the existence of a formula that describes the intensity of the magnetic fields of all planets in comparison with a "base" field, the Earth's magnetic field. Stevenson writes,[22] "The field amplitudes of planetary dynamos are poorly understood.

Existing scale laws are naive because they deny the diversity of planets and poorly constrained because several laws perform equally well."

Another opinion is held by Parker.[23] He says that undertaking the dynamic of the convection in the core not knowing thoroughly the energy source and its intensity is very difficult. Therefore, it is not surprising, supposed Parker, that the magnetic field intensity of different planets are not connected by some simple relation of similarity, including angular velocity, core radius, and other values. In reality, knowledge of the energy source is a necessary but insufficient condition for an adequate scale law.

Malkus,[2] as was mentioned in the preceding, estimated parameters of convection and field components in the core, the power of the precession geodynamo reasoning from a formal and unknown assumption of equality of Poincare, Coriolis, and Lorentz forces in the core. In the framework of comparative planetology, it is not necessary to use unknown assumptions. Moreover, we can ignore the arguments in favor (the results of laboratory experiments[12]) and against (the daily variability of the Poincare force[4]) the role of the precession mechanism in magnetic field generation. The only serious assumption that we have in mind is that the dynamo process in all planets is similar to that in the Earth. Such an assumption is based on many fundamental facts:

1) There is the wonderful similarity in the ratios of dipole, quadrupole, and octupole harmonics of the Earth and Jupiter, planets which differ strongly in chemical composition, aggregate state and size (see Table 1).

2) The magnetic fields of a number of planets reveal similarity of the dipole axis tilt to the rotation axis and dipole polarity.

3) There is also a similarity in action of the external gravitational fields of the sun and the satellites of the planetary bodies: the axes of rotation of planets have different tilt to the normals to their orbital planes, and the orbits of their satellites are inclined to the planetary equational planes.

4) All planets have electrically conducting or half-conducting interiors (Ref. 24).

The method of construction of the precession dynamo model by full abstraction from poorly known details of the dynamo process is analogous to the method that is known as the "black box" method often applied with great success in physics (Ref. 25).

Table 1 Earth and planets have a similar structure of magnetic field

Planets	R_c, km	H_1, nT	H_2/H_1	H_3/H_1	$\Delta R/R$	i	$W \cdot 10^{-4}$ $k^4 \cdot cm^{-2}$	Field direction
Earth	3,480	30,500	0.13	0.09	0.067	11.5	0.046	S
Jupiter	64,340	419,200	0.20	0.15	0.11	10	5.44±0.43	N
Saturn	40,480	21,000	0.07	——	0.04	1	2.01±0.14	N
Uranus	14,700	22,840	0.7	?	0.3	60	0.042±0.047	N
Neptune	19,150	14,240	1.16	0.78	0.55	46.8	0.433	N

III. Precession Dynamo Model

Precession of the Earth occurs under the action of the gravitational fields of the sun and moon. Additional acceleration associated with precession $\omega = [\omega \times \Omega] = 2.25\ 10^{-16}\ s^{-2}$. The force experienced by the mass of volume ρ at the distance r from the rotational axis $F_p = \rho\ [\omega \times \Omega] \times r$ was named by Malkus as the Poincare force (ω and Ω are velocities of rotation and precession of the Earth, respectively). The velocity of precession Ω is proportional to dynamic flattening $f = (C-A)/c$, where C and A are the axial and equatorial moments of inertia. For the Earth f_e=3.27 10^{-3}. At the greater density and smaller sizes of the Earth's core, the dynamical flattening of the core is $f_c = 3/4f_e$. The core has to precess with less rate. At the distinctions between accelerations of material points of the core and mantle $\Delta\,\omega = 10^{-16}$ s^{-2} during the period of precession $T = 25{,}800$ years, the velocity of the core respective to the mantle may reach 156 m/s, but such great velocities were not determined by the geophysical methods on the surface of the Earth.

Consequently, there are mechanisms of coupling between the mantle and the core that allow for the existence of the torques of the core rotation $L_c = I_c\,\omega_c$ and mantle rotation $L_m = I_m\,\omega_m$ in magnitude ($\omega_c\ F, \omega_m$) and direction (existence of the angle ε between L_c and L_m).

Various estimations of coupling mechanisms in the rotating and precessing core-mantle system lead to significantly different notions about the character of flows inside the liquid core and opposite evaluations of efficiency of the precession in generation of the magnetic field. The contradiction appeared in the answers to two major questions:

1) Can the precession mechanism provide the energy necessary for generating and maintaining a geomagnetic field?

2) Can the external gravitational forces stimulate motions of the liquid required by kinematic models of the geodynamo (see Ref. 13 and also Ref. 26)?

The difficulties in answering these questions will be overcome with the help of laboratory experiments and the model of precessional dynamo in the framework of comparative planetology.

From the main equation of magnetohydrodynamics,

$$\frac{\delta B}{\delta t} = \nabla^2 B \eta_m + \mathrm{rot}[V \times B] \tag{1}$$

where η_m is the magnetic viscosity and V is the velocity of liquid motion, so that at $V = 0$ the field is dumped out in time $\tau = L^2/\eta_m$ and $\eta_m = C^{-2}/4\,\pi\,\sigma$, where C is the speed of light, σ is the electric conductivity, $L = R_c$, is the size of the magnetically active area, and τ is the magnetic capacity of the core. Assuming the dominating role of Poincare force F_p in the generation mechanism, it is essential to accept that $B \propto \tau\ F_p$,

$$B \propto R_c^3 \rho \omega \Omega \sin \alpha \,/\, \eta_m \tag{2}$$

Magnetic fields of a number of planets reveal surprising similarity of the dipole axis tilt to the rotational axis in correlations of dipole, quadrupole, and octupole harmonics of their magnetic fields, showing dipole polarity (see Table 2). The principle of physics similarity theory indicates that if two physical phenomena are similar, then by given characteristics of one it is possible to obtain the characteristics of the other by simple recount which is analogous to the transition from one coordinate system to the other one.[27]

Based on the principle of similarity and the supposition of the driving force of the dynamo mechanism to be common for all planets, the assumption of the possibility of the determination of the dipole field of any planet H by the dipole field of the Earth H_E can be made using the formula:

$$H_n = H_E(Q_E^3 / Q_n^3)(R_c^3 \Omega \omega \sin \alpha \rho / \eta_m) \tag{3}$$

where $Q = R_n/R_c$ and R_n, R_c are the radii of a planet and core, which means that the magnitude is related to the corresponding one for the Earth.

$$\Omega = 1.5 GMf \cos \alpha / \omega r^3 \tag{4}$$

where G is the gravitational constant, M and r are the mass and distance to the disturbing body responsible for precession, and f is the dynamical flattening of the planet.

The action of forces appearing in magnetically active areas at the convection of substances and appearing magnetic fields (Coriolis force, Lorentz force) are supposed to be model-similar in "base" planet Earth and any other planet.

IV. Model of Precession Dynamo for the Planets of Earth's Group

Data on the magnetic and gravitational fields of the planets of Earth's group and of the moon have been obtained by the spacecrafts of the former USSR and USA from 1959 to 1989. Table 2 gives parameters for the Earth, Mercury, Venus, Mars and the moon. The data are based on the results of research which are briefly discussed below.

A. Mercury

The magnetic field of Mercury was studied during the first (1974) and third (1975) flights of the Mariner 10 spacecraft. The magnetosphere of Mercury was crossed at an angle ~110 deg from the subsolar point under the limitations of determining the solar wind parameters. A review of the magnetic data and the models of internal structure is given in Ref. 28.

Intensity of the dipole field of Mercury on the surface at the equator in various models is found to be in the range from 165 to 330 nT. By data of Mariner 10, the rotation period T and the amplitude of precession α were specified[29]: $T = 58{,}646 \pm 0.005$ days; $\alpha = 1$ deg with the uncertainty ellipse

Table 2 Parameters of the precession dynamo model for the moon and terrestrial group planets

Parameter	Earth	Mercury 1	Mercury 2	Venus	Mars	Moon
R_p	6371	2440	2440	6051	3390	1738
R_c	3480	1600	1800	2840	1290	360
Q	1831	1525	1353	2130	2620	4828
$\omega \cdot 10^{-5}$	7.27	0.123	0.123	0.0299	7.30	0.269
f	$3.27 \cdot 10^{-3}$	$2.13 \cdot 10^{-4}$	$1.34 \cdot 10^{-4}$	$(6 \div 2) \cdot 10^{-6}$	$5.24 \cdot 10^{-3}$	$3.1 \cdot 10^{-4}$
Ω, 11/yr	50.2	1149	722.7	20	7.4	14.3
α, °	23.5	1	1	2.6	25.2	1.5
ρ, cm^{-3}	11	8.5	8.0	9.6	(85)	6
σ, Ω^{-1} m^{-1}	σE	$2\,\sigma E$	$2\,\sigma E$	σE	$0.54\,\sigma E$	σE
H_{meas}, nT	30,900	165	165	1	35	0.2
H_{calc}, nT		165	165	0.2	35	0.2
F_p, dyn cm^{-3}	$8.61 \cdot 10^{-7}$	$5.2 \cdot 10^{-9}$	$3.46 \cdot 10^{-9}$	$6.63 \cdot 10^{-11}$	$3.89 \cdot 10^{-8}$	$3.45 \cdot 10^{-11}$

with $\Omega \times L_s$ of 2.5 deg and 6 deg; the value $f = (1.2 \pm 0.9)\ 10^{-4}$, which was determined by the value of the harmonic J_2 of the gravitational field of Mercury published in 1977 (Ref. 30); and $J_2 = (8 \pm 6)\ 10^{-5}$ (at the period of rotation $T = 58$ days $f = 3/2J_2$).

Radial strains that appear in high and low latitudes while Mercury is moving around the sun along the orbit with a large eccentricity[31] have to stimulate the flow inside the liquid core along the circles of latitude and longitude having turbulent character to promote the dynamo process. In the majority of models of internal structure of Mercury, the radius of the liquid core $R_c = 1800$ km.

Representation of the magnetic field of Mercury as the sum of the dipole and quadrupole or eccentric dipole of much the same intensity[32,33] allows for the estimation of the upper limit of the magnetically active area of Mercury by determining the distance r, at which the first g_1^o and second g_2^o harmonics of the field are equal by formula of the spectrum of the Gauss series:

$$H_p = (2n+1)^{-1/2}\left(\frac{R_p}{r}\right)^{n+2} \times \sum\left[\left(g_n^m\right)^2 + \left(h_n^m\right)^2\right]^{1/2} \tag{5}$$

According to Whang[32] $g_1^o = 165$ nT and $g_2^o = 117$ nT. The results of the calculations are shown in Table 3.

The distance $r = 1600$ km at $R_c/R_p = 0.66$ is close to the boundary of the core in the model of Mercury's interior,[28] wherein $R_c/R_p = 0.69$. The conductivity σ in the terrestrial planets decreases with distance from the sun. In particular, for Mercury 2 $\sigma_{Mc} = 2\ \sigma_E$ and for Venus $\sigma_v = \sigma_E$.

The values of the dipole field of Mercury are determined with a smaller error (~30%) than the values of the dynamic flattening of the planet (~70%). Hence given in Table 2 are the values of for which the calculated values of H coincide with the measured ones.

B. Venus

The data listed in Table 1 are based on the work of Zharkov[34] containing an overview of the models of the planet's interior, observational data, and parameters of the planet's form. The value of dynamic flattening of Venus is given according to Akim et al.[35] The magnetic condition of Venus is sur-

Table 3

r, km	$\Delta g_{1,2}$
1500	−183
1500	−96
1600	+3
1610	+50

veyed by the spacecraft Mariner 5,[36] Venera 4,[37] Venera 9 and Venera 10,[38] and Pioneer–Venus.[39,40] The upper limit of the possible field of Venus $H_v \leqslant 10^{-4} M_E$ is shown in Ref. 41.

C. Mars

An overview of the models of the interior of Mars and the results of the investigations of the planet's magnetic field are given in the work by Dolginov[42]:

1) A determination of the dipole magnetic moment of Mars from the Phobos 2 satellite data is reported by Dolginov and Zhuzgov[43] and from a set of data from the spacecraft Mars 2, 3, and 5 and Phobos 2 by Dolginov.[44]

2) It is customary to assume that the conductivity of the core of Venus $\sigma_v = \sigma_E$ and that of the core of Mars $\sigma_{Ma} \leq \sigma_E$.

The radius of the liquid core of Mars is assumed to be R_c = 1290 km according to a model of the planet's interior.[45] With such a value of R_c, agreement is attained with the observed value of the dipole field at $\sigma_{Ma} = 0.56\ \sigma_E$. The current value of the parameters H_{Ma}, R_c, and σ_{Ma} of Mars allow for their variation in a range 1–1.3 with the model similarity of the dipole fields of Earth and Mars remaining true.

D. Moon

The problems of the magnetic fields of the moon are discussed in Section VI of this paper. Listed in the bottom line of Table 2 are the values of the Poincare force per unit of volume of the liquid cores of the terrestrial planets. In particular, the value F_p for Venus explains (without any speculative assumptions on the possible differences of its interior from that of the Earth) why Venus features no noticeable magnetic field: $F_{PV} : F_{PE} = 10^{-4}$ because of the smallness of the dynamic flattening of the planet and the amplitude of precession under the action of the sun's gravitational field ($\alpha \leq 1$ deg). This information, in conjunction with the results of Table 2, allows one to infer that the magnetic conditions of the Earth, Mercury, Mars, and the contemporary moon can be described numerically through a model of the precession dynamo, where the engine of the planetary dynamos is the external gravitational force of the Sun and the nearest satellites.

The ratio between Poincare forces acting on the planets Venus F_v and Earth F_E is $F_v/F_e = 8 \cdot 10^{-5}$. Venus can have a magnetic field only at $H_v < 2$ nT.

V. Model of Precession Dynamo for Giant Planets

Analysis of fields of giant planets having complicated satellite systems reveals the role of the external gravitational fields for the dynamo. Modern models of the giant planets' internal structure, as long as there is data on their magnetic, gravitational, and thermal fields, are based on the experimental results of the Pioneer 10 and 11 and the Voyager 1 and 2 missions. Results of the magnetic field surveys near Jupiter and Saturn are given by Smith et al.[46–48] Results of the Voyager 1 and 2 investigations near Uranus and Neptune are given by Ness et al.[49,51] and Connerney et al.[50,52] Models of the internal structure of giant planets are given in Refs. 53–58.

Table 1 gives boundaries of magnetoactive regions of the Earth and giant planets R_c, values of the dipole field H_1, ratios of quadrupole (H_2) and octupole harmonics (H_3) to H_1, displacement of eccentric dipole ΔR, inclination of the dipole axis to the rotation axis i, thermal flux density W, and hemispheres with a northern magnetic pole. The similarity between the parameters of the field of the Earth and Jupiter is surprising for planets with such different sizes and chemical structures. Comparison of H_1 and W values show that there is no correlation between the dipole field intensity and thermal flux density.

Enormous sizes of gas–liquid planets with complicated satellite systems create such conditions that are necessary for magnetic field generation (supercritical Reynolds numbers are reached for rather low values of conductivity σ). Stevenson[22] determines magnetic viscosity within the limits of 10^5–10^6 cm^2 S^{-1} c^{-1} in the molecular hydrogen layers for the pressures of P = 1 mB and T = 5000 K [corresponding σ is 7.2 ($10^{14} - 10^{13}$) s^{-1}]. Laboratory investigations[60] of the species typical for Uranus and Neptune give σ of $1 - 3 \cdot 10^{13}$ s^{-1} for pressures $P < 0.5$ mB.

Parameters of the satellites of giant planets are given in Tables 4 and 5 according to Ref. 61. Magnetoactive region sizes R_c for Jupiter and Saturn are determined from internal models of the planets. For Uranus and Neptune R_c is determined by use of Gauss series spectrum formula.[5] As was mentioned, R_c equals the distance r at which $\Delta H_{1,2} = 0$. Errors in determination of H_2 for Uranus result in two possible values of R_c corresponding to models Uranus 1 and Uranus 2 in Table 6. Results of the calculations of R_c are given in Table 5.

As dipole fields H_1 are known for giant planets rather accurately, formulas (3) and (4) enable us to re-estimate conductivity values. Table 7 gives for Jupiter, Saturn, and Uranus values of σ for which measured and calculated fields coincide. These values are in agreement with Stevenson's results.[22] Laboratory values of σ (Ref. 59) have been used for Neptune field calculations. As dipole fields of Jupiter, Saturn, and Uranus are defined by gravitational fields of their satellite systems, Tables 4–6 give values of the magnetic field induced by each satellite separately according to Eqs. (3) and (4). Table

Table 4 Saturn's magnetic field in precession dynamo model: the fields generated by the sun and the satellites

Source of precession	M, 10^{23} gr	r, km	α, deg	Ω, ^{11}/yr	H_p, nT
Mimas	0.38	$185.52 \cdot 10^1$	1.53	0.22	495
Tethys	7.6	$294.60 \cdot 10^1$	1.09	11.04	15,282
Titan	1345.7	$1221.8 \cdot 10^1$	0.33	27.42	12,298
Sun	$1.989 \cdot 10^{33}$	$14267 \cdot 10^6$	26.44	0.22	7,695

Table 5

Neptune		Uranus 1		Uranus 2		Uranus 3	
r, km	$\Delta H_{1,2}$	r, km	$\Delta H_{1,2}$	r, km	$\Delta H_{1,2}$	r, km	$\Delta H_{1,2}$
20,000	664	14,000	3569	14,000	−1095	13,000	3330
19,500	297	13,900	2497	14,200	−619	12,800	2070
19,200	57	13,800	2136	14,300	−92	12,600	596
19,150	0	13,700	1638	14,350	141	12,550	209
19,100	−15	13,600	710	14,400	350	12,500	−200
19,000	−123	13,500	548	—	—	12,450	−619
18,500	−694	13,400	−70	—	—	—	—
18,000	−1317	13,300	−713	—	—	—	—

6 gives the fields generated in Jupiter under the action of gravitational forces of the sun and of the closest satellites.[60] The results of calculations enable us to conclude the following:

1) The observed magnetic field is a sum of the fields generated by each satellite and the sun. Let us recollect that in dynamo theory the observed field is a sum of the fields of separate cyclonic cells.

2) Precession is effective for magnetic field generation even for small amplitudes α since it is performed during millions of years in the same rhythm.

The magnetic field of Jupiter is generated in the internal molecular shell rather than in the metallic hydrogen region.

Table 4 gives corresponding parameters of Saturn's satellite system along with the results of magnetic field calculations according to Eqs. (3) and (4). One should take into account the great difference between the Saturn rotation axis inclination to the normal to ecliptic plane (α = 26.5) and to normals to the satellites' orbits (α_{sat}). A vector sum of the fields enables us to explain the coincidence of the dipole axis and rotation axis. Neglecting small differences in α_{sat}, we obtain the field generated by satellites H_{sat} =

Table 6 Jupiter's magnetic field in precession dynamo model: the fields generated by gravitational forces of the sun and the satellites

Sources of precession	M, 10^{25} gr	r, 10^6 km	α, deg	Ω, ^{11}/yr	H_p, nT
Io	8.94	421.6	0.04	2.86	46,593
Europa	4.80	610.9	0.47	0.381	2,709
Ganymede	14.823	1070	0.195	28.97	235,575
Callisto	10.766	1883	0.281	3.82	13,565
Sun	$1.989 \cdot 10^{33}$	$778.9 \cdot 10^6$	3.04	1.015	119,397

Note. ΣH_6 = 417,840.

Table 7 Parameters of the precession dynamo model for the Earth and giant planets

Parameter	Earth	Jupiter	Saturn	Uranus 1	Uranus 2	Neptune
R_p, km	6.171	71.492	60,268	25,000	25,000	24,765
R_c, km	3.480	643.10	40,980	13,800	14,700	19.150
Q	1841	1.111	1.471	1.855	1.741	1.293
ρ, g · cm^{-3}	10	0.82	0.97	3.4	3.1	1.9
$\omega \cdot 10^{-5}$ cm^{-1}	727	17.58	16.38	10.12	10.12	10.83
α~, °	235	Table 4	Table 5	satellite	satellite	21
f	$3.27 \cdot 10^3$	0.0651	0.094	0.0195	0.0195	0.0181
Ω, 11/yr	50.2	Table 4	Table 5	satellite	satellite	48.39
σ, s^{-1}	$2.7 \cdot 10^{15}$	$5.93 \cdot 10^{14}$	$2 \cdot 10^{14}$	$1.81 \cdot 10^{15}$	$1.81 \cdot 10^{15}$	$1.08 \cdot 10^{13}$
H_{meas}, nT	30,500	419,200	21,000	22,840	22,840	14,240
H_{calc}, nT		417,940	21,000	16,885	22,915	14,240

28,075 nT. Let us consider a triangle ABC (see Fig. 1) with the side B = 21,000 nT (measured field that is parallel to the rotation axis) and C = 28,075 nT (the field H_{sat}). The angle between B and C is α = 10 deg (angle of inclination of dipole axes to rotation axes for Earth and Jupiter). Side A opposite to angle α completes the triangle. Let us determine side A and angle β. B = 21,000 nT and is the measured field oriented along the rotation axis; C = 28,075 nT and is the field generated under the action of all satellites, and inclined at α = 10 deg (as the dipolar fields of the Earth and Jupiter); $\Delta = B \sin \alpha / \sin b$ $b = \text{arctg}\,[B \sin \alpha/(C - B \cos \alpha)]$; and Δ = 8189 nT, and this value has an error of 6% in comparison with the field generated

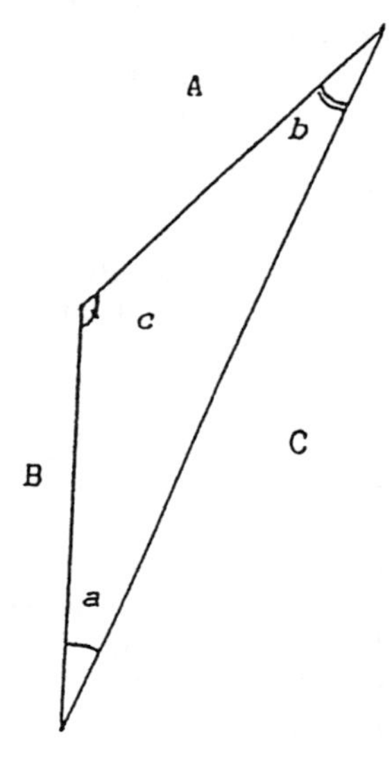

Fig. 1 Triangle ABC depicting directions of magnetic fields.

by the sun (see Table 4). $b = 26.5$ deg is in excellent agreement with the angle between Saturn's equator and its orbital plane.

The magnetic field of Saturn is generated in the inner molecular shell. The magnetic fields of the planets are results of the vectorial summation of the fields generated under the influence of the Sun and each one of the satellites. The boundaries of Uranus' and Neptune's magnetic active regions are calculated using the Gauss series formula, and they are situated at the distance r where the harmonics of H_1 and H_2 are equal.

Coincidence of the dipole axis of Saturn with the rotation axis and large angles i for Uranus and Neptune create difficulties for dynamo theories, e.g., small, but nonzero angles i are considered as a necessary condition for the magnetic field generation (one of the consequences of the Cowling theorem).

Apparently, two models are realized in the planetary dynamo process in order to satisfy conditions of the Cowling theorem. Indeed, as one can see from Tables 1, 2, 4, 6, and 7, magnetic fields of planets, which have dipole axes inclined to rotation axes, are generated under the action of the gravitational field of one source (the sun for Mercury and Mars) or under the action of the gravitational fields of satellites rotating at nearly complanar orbit (Uranus and Neptune). The Earth's magnetic field is generated under the action of two sources, the sun and the moon, but by coincident gravitational effects. Jupiter precesses under the action of the sun and Galilean satellites, but with small amplitudes. Only Saturn's magnetic field is generated under the action of two gravitational fields with a large angle between them. The possibility to explain the unique axisymmetric orientation of the Saturnian dipole as a result of the vector sum of two magnetic fields generated under the action of two nonparallel gravitational fields of different sources, undoubtedly, supports the model of the external gravitational fields as the driving force of planetary magnetic fields.

Table 5 shows that the magnetic field of Uranus is generated significantly deeper than the magnetic field of Neptune, at the depth where $\sigma U = 10^2 \sigma_N$. Small Uranian satellites can generate velocities necessary to reach supercritical Reynolds number only if $\sigma_U = 0.67\,\sigma_E$. The following questions arise:

1) Why is the magnetic field of Neptune smaller than that of Uranus, whereas the precession of Neptune produced by Triton is stronger than the precession of Uranus produced by its satellite system?

2) Why is the magnetic field of Neptune not generated at a larger depth in shells with higher conductivity?

Possibly the answer may be given by the general principle of thermodynamics that the system takes such states which minimize action of the external forces (le Chatelier principle[61]). Joule losses in the core are inversely proportional to the square of conductivity.[13] Generation of the Uranus field at larger depth may be one of the causes of negligible thermal flux of Uranus and significant thermal flux of Neptune.

The magnetic fields of Uranus and Neptune remind the structure of the geomagnetic field during the periods of inversion of polarity. In dynamo theory, the inversion is associated with a change of the structure of cyclonic cells in the core under the action of external and internal causes. The inclination of dipole may be defined by peculiarities of the dynamo process. For limited conductivity, the size of the cells should increase (in order to conserve conditions for the field generation), and, hence, their number N should decrease. Simple dependence of the inclination was proposed: $J \sim N^{-1/2}$ (Ref. 62).

Better conditions for the generation of quadrupole field exist for Reynolds numbers close to critical (which is possible for moderate conductivity). Geomagnetic dipole inclination reached 90 deg during inversions.

Planets of the Earth's group have northern magnetic poles in the southern hemispheres. Giant planets have opposite magnetic field polarity for the same orientation of the angular momentums. This is an unsolved enigma of the planetary magnetism. Thus, the assumption of the Uranus field being in the inversion stage looked very attractive. However, this assumption looks unlikely after revealing the same structure in the Neptune magnetic field since inversion is too rare an event to be observed simultaneously at two planets.

The intensity and unique topology of the Uranus and Neptune magnetic fields may be explained by the action of two effects of external gravitational fields: precession and tidal flows at very deep layers. These questions have been considered in Ref. 63 where the following scenario of Neptune field generation is described.

1) The lower boundary of tidal flows is located at the depth of 2000–3000 km from the Neptune surface. At this boundary the velocity of tidal flow is comparable with the velocity generated by precession ($V = r\,\Omega \sin \alpha/\eta f$). The upper boundary of the magnetoactive zone is located 2000–3000 km deeper. The conductivity in the region of tidal flows is insufficient to ensure supercritical Reynolds number, but it is sufficient for magnetohydrodynamic effect on the magnetic field spreading over this region. Processes in the region of zonal flows that may change the orientation of the field outside the region of generation are considered in Ref. 64 for Jupiter and Saturn. Latitudinal dependence of zonal flows in Uranus and Neptune is different from that for Jupiter and Saturn.

2) However, the deformation of the magnetic field by deep zonal flows cannot result in the appearance of quadrupole and octupole harmonics comparable with the dipole harmonics. Nevertheless, it is possible to assume that these deformations help processes resulting in co-ordinal changes of planetary field-like excurses and inversions. According to Levy,[65] deformation and inversion of the field starts with the appearance of cyclonic cells with opposite polarity in the equatorial region. Then the region of opposite polarity propagate toward higher latitudes. During this period the field becomes multidipole. Subsequent propagation toward high latitudes results in total field inversion. Levy's model of the field becomes unstable for the change of sign in 20–30% concluded that such a scenario is in agreement with the distribution of virtual magnetic poles of Earth's paleomagnetic field.[66]

3) Coincidence of two excurses (Uranus and Neptune) may be explained by synchronization well known in the solar system[67] and by the property of rotating bodies to gain the same rhythm even for very weak links between the objects.[68] Magnetic moments calculated by the formula (4) are shown in

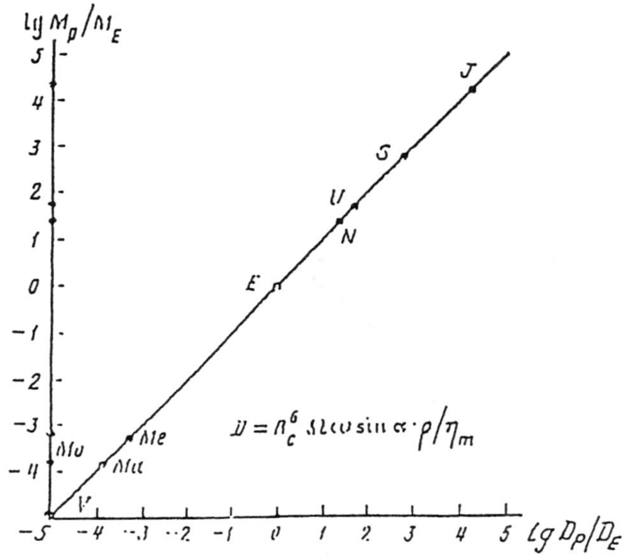

Fig. 2 The measured magnetic moments ratios (planets/Earth) versus their "dynamo number" ratios in the precessional dynamo model.

Fig. 2. Ordinates give the values $lg\ M_P/M_E$, and abcissas give $lg\ D_P/D_E$ for all investigated planets and the moon. D ($R^6\ \omega\Omega\rho$ sin α/η_n) is the dynamo-number for the model of precession dynamo.

Assumption on the dominating role of the external gravitational forces in the generation of planetary magnetic fields reveals strict regularity in the chaotic distribution of the dipole planetary field intensities. According to Parker,[23] the correct determination of the sources of energy for the dynamo process should result in a simple dependence of field intensity on few parameters. This problem has been solved by the preceding "numerical law" of precession dynamo within the framework of comparative planetology.

VI. Jupiter–Satellites System in the Model of Precession Dynamo

The research carried out by Pioneer 10 (1983), Pioneer 11 (1974) and Voyager 1 (1979), together with ground investigations, discovered amazing regularities in the material composition of Galilean satellites, the distribution of their average densities ρ, the synchronization of their own rotation periods $P_{\rm rot}$ and orbital rotation periods $P_{\rm orb}$ around Jupiter in orbits with eccentricities ε and average distances R. Even more remarkable are resonances nearing 2:1 in the average periods of orbital motions of Io and Europa ($n_1 - 2n_2 = 0.7395$ deg per 24 h). Absolutely the same resonance is discovered in orbital rotations of Europa and Ganymede. The heat flow of Io, relating to its mass M, exceeds 4 times the heat flow from Jupiter. Numercial values of the mentioned parameters are presented in Table 8,

Table 8 Parameters of Galileo satellites of Jupiter

	α, km	ρ, gr $\cdot$ cm^{-3}	M, gr $\cdot$ 10^{25}	R, km $\cdot$ 10^5	Σ	P_{orb}, days	P_{rot}, days	Wr, $S^{-1} \cdot 10^{-5}$	$H1$, erg $\cdot$ $^{-1} \cdot$ cm^{-2}
Io	1,815	3.55	8.24	4.22	0.0041	1.769	1.769	4.110	1500
Europa	1,569	3.04	4.80	6.71	0.0101	3.551	3.551	2.055	—
Ganymede	2,631	1.93	14.82	10.7	0.0015	7.155	7.155	1.027	—
Callisto	2,400	1.83	10.77	18.8	0.01	16.69	16.69	0.436	—
Jupiter	71,492	—	189,800	7780	0.484	43.33	0.413	17.58	4500

together with the average radius α, Galilean satellites angular velocities ω_i, and heat flow H_i.

All of these peculiarities are explained by the hypothesis of the Galilean satellites' evolution under the influence of massive Jupiter gravitational and heat fields and the influence of influx friction and heating treated in detail in the work of Hubbard.[69]

The dissipation of energy in the planet–satellite system in the framework of influx heating is proportional to E^2 and inversely proportional to R^6 depending on physical properties of the planet and satellite (Refs. 62 and 69). Hubbard[69] showed that the contemporary heat flow of Io and satellite volcanism cannot be explained either by preservation of accretional heat or by radioactive heating. It can be explained only by the continuing absorption of energy from Jupiter, by the mechanism of influx friction, and by the absorption of Jupiter rotational energy.

If Io were influenced only by Jupiter's gravitational field, the influx mechanisms would lead to Io's, synchronous rotation in circular orbit without dissipation of energy. This kind of Io motion is prevented by the gravitational influence of its nearest satellite Europa, which rotates with Io in the resonance mode approaching 2:1 [average velocities of Io (n_1)], and Europa (n_2) motion is determined by the equation $n_1 - 2n_2 = 0.7395$ deg/per 24 h. In turn, Europa and Ganymede rotate in the same resonance mode nearing 2:1. Jupiter's satellites' "forced" rotation into eccentric orbits, in contemporary age, is characterized by the values of E and R given in Table 8. Different scenarios of Galilean satellites' orbit evolution and energy dissipation in the system are considered in Ref. 69. Quite natural is the question on a link among processes in the Jupiter–satellites system, the mechanism of magnetic fields generation of Jupiter–Galilean satellites, and the magnetic field evolution of Jupiter itself.

A. On the Magnetic Fields of the Galilean Satellites

The assumptions about magnetic fields of the Jupiter satellites were based on the effects of Jupiter dekameter radiation modulation depending on satellites position in orbit[70] and on immediate results of measurements by Voyager.[71] It was assumed that the satellites possessed melted nuclei and dynamo fields generated by heat convection, and numerical values of the fields were predicted.[72] Dolginov[73] cast doubts on the efficiency of precession dynamo in the generation of supposed fields due to synchronization of their own and orbital rotations around Jupiter.

When the Galileo spacecraft was in the vicinity of Io, Europa, and Ganymede, its intensive influence on the Jupiter field was detected while only weak disturbance made by Callisto was registered. The research of gravitational fields performed with the help of the Galileo spacecraft allowed for creating models of inner design of all the satellites. The orbits of Io, Europa, and Ganymede are in the region of sufficiently intensive magnetic and plasma structures of Jupiter magnetosphere. This resulted in difficulties interpreting magnetic effects observed in the vicinity of Io and Europa.

Io Magnetic Field

The magnetic presence of Io was disclosed by a sharp decrease of the Jupiter field by 696 nT when in flight in the vicinity of Io at the height of 898 km. The Jupiter field, according to the 1996 model, was ~1835 nT. When the possible nature of the observed Io field was discussed, the following was considered: the induction in the Jupiter magnetic field, current systems arising when the ions and neutrals of the Io ionosphere interact with the plasma of the Jupiter magnetosphere, constant magnetizing of the layers containing magnetite at temperatures below that of Curie, and, lastly, the contemporary dynamo field. The best agreement with the observed effect is acquired when the source is presented as a dipole with the field intensity at the surface, near the equator ~1300 nT. The axis of the Io dipole is oriented antiparallel to the Jupiter dipole.[74–76]

Ganymede Magnetic Field

When the spacecraft flew near Ganymede, at the latitude of ~30 deg and at the height of 838 km, the Jupiter field decreased from ~420 to 118 nT. During the second flight at the latitude of ~79.8 deg and at the height of 264 km, the field of 1136 nT was measured. This indicated directly to a dipole character of the field and the proximity of the dipole axis to the satellite rotation axis. The effects under observation are explained by the action of the Ganymede dynamo field having the intensity of 750 nT at the surface near the equator, the axis of which is inclined to that of rotation by 10 deg (Refs. 77 and 78). The nature of the field was associated with constant magnetizing in the silicate mantle with the dynamo field in the ferro-nucleus or in a water shell.[77,79]

Europa Magnetic Fields

Galileo was in flight in the vicinity of Europa on December 19, 1996, and on February 20, 1997. So far, the results of the magnetic research have been published on the data of the 19.12.96 flight.

The magnetic influence of Europa was detected by disturbances in the Jupiter magnetic field, the intensity of which reaches 420 nT at the orbit of Europa. At the closest distance from Europe, the magnetometer revealed a sharp decrease of the Jupiter field by 50 nT (ΔH). One might suppose that the disturbance of such intensity may be related to various effects of the satellite interaction with the fields and plasma of the Jupiter magnetosphere. The authors of the experiment[79] treated in detail the possible influence of current systems associated with processes that were considered earlier, when the Io field was discussed. As distinct from the field that was observed in the vicinity of Io, the field ΔH is directed nearly perpendicular to the Jupiter field. This disturbance can be connected with Alfven waves generation; the intensity of those waves can reach 100 nT (Ref. 79). Nevertheless, space and time coincidences of ΔH effect and the nearest distance from the Europa surface makes it more probable that the field ΔH can be explained by an action of dipole, with the intensity of 120 nT at the equator surface, which is inclined at an angle of 135 deg to the rotation axis. Significant deviations of the dipole axis from the rotation axis may be linked to an action of higher harmonics of the field. They can be generated in the fields that are nearer to the surface. In Europa, this magnetoactive field can be an H_2O ocean with the depth of ~100–150 km, if its conductivity is sufficient. In contrast to Uranus and Neptune, where Reynolds numbers, sufficient for the field generation, are acquired at moderate conductivity but gigantic dimensions, in Europa such conditions can be achieved due to higher velocities of flows (see Section IV.C).

Nonmagnetic Callisto

When Galileo flew near Callisto at the height of ~1100 km and the latitude of ~13 deg from the Callisto equator, the magnetometer registered a disturbance in the Jupiter field of ~7 nT. As distinct from the disturbances registered near Io, Europa, and Ganymede, the Jupiter field, which has intensity at the Callisto orbit ~32 nT, was noticed to increase and not decrease. This disturbance is best explained by a disturbance in the current layer of the Jupiter magnetosphere when it interacts with Callisto, if the satellite has even low conductivity. The assumption of the reverse leads to the conclusion of the absence of Callisto's own field: extrapolation of ~7 nT to the surface by the dipole law leads to a field of ~17 nT on the equator, which is inclined to the axis of rotation at the angle of ~80 deg.[80] The absence of Callisto's own field is also established by plasma measurement data.[81] The absence of Callisto's field gives important information on the problem of the dynamo energy source and driving force. Table 9 presents the dipole nature of satellite fields. In this table, intensities at poles and dipole orientations are contrasted to the Jupiter field intensities B_0 in orbits of the satellites.[79]

The nature of Mars and Mercury's magnetic fields was proved by constant polarity of the observed fields when the interplanetary magnetic field polarity was changed. In the case of Galilean satellites, one cannot expect the same. The principle of physical similarity of magnetic fields generation, revealed in the solar system, may be effective. This possibility became more real owing to developing models of Galilean satellites' internal structure.

B. Models for Internal Structure of Galilean Satellites

Models for the internal structure of Galilean satellites that correspond to the results of their gravitational fields research are presented in the works of Anderson et al.[82–85] Following is the synopsis of the final results. During one-time and two-time flights near some of the satellites, data necessary to determine the coefficients J_2 and C_{22} to formulate gravitational potential (7) were obtained

$$V = \frac{GM}{r}\left[1 - \frac{1}{2}J_2\left(\frac{a}{r}\right)^2(3\sin^2\varphi - 1) + 3C_{22}\left(\frac{a}{r}\right)^2\cos^2\varphi\cos 2\lambda\right] \tag{6}$$

Table 9

Satellite	B_{surf}/B_0	$\cos(M\Omega)$	$\cos(MB_0)$
Io	~2600/1800	180°	5°
Europa	240/420	135°	65°
Ganymede	1500/120	170°	34°
Callisto	30/35	~80°	37°

$$C_{22} = (B-A)/4\ Ma^2 \tag{7}$$

$$J_2 = (C-A)/4\ Ma^2 \tag{8}$$

$C > B > A$ are the axial and equatorial moments of inertia, and J_2 is immediately linked to the dynamic compression f entering Eq. (4), which determines the velocity of precession:

$$f = \frac{3}{2}J_2 + \frac{1}{2}\frac{\omega^2 a^3}{M \cdot q} \tag{9}$$

$$C_{22} = \frac{3}{4}\alpha q \tag{10}$$

where q is the relation of centrifugal acceleration to the acceleration of gravity at the equator, $q \approx f$ (for the Earth $f = 3.2732 \times 10^{-3}$, $q = 3.4678 \times 10^{-3}$).

The degree of C/Ma^2 parameter deviation from the value of 0.4 determines the degree of mass differentiation in the satellite. Radial distribution of density is also characterized by differing from the value 0.5 of α parameter from Ref. 83. We limit ourselves by the parameters that determine dimensions r_c and nuclei mass m, dynamic compression of satellites f, and degrees of their differentiation. Other parameters in Eqs. (3) and (4) of the precession dynamo model are presented in Table 8.

So far, we do not know the precession amplitudes α, i.e., the angles between the rotation axis and normals to the planes of satellite orbits. By data obtained from the flights by Galileo, we give the values found in Table 10 to the models for the internal structure of the satellites[82–85] with $q = 0.7123 \cdot 10^{-3}$.

Table 10 Io

		Fe − FeS		Fe	
$C_{22} \cdot 10^{-3}$	C/Ma^2	r_c/a	m/M	r_c/a	m/M
0.532	0.371	0.572	0.279	0.397	0.143
0.559	0.378	0.517	0.202	0.357	0.105
0.586	0.386	0.447	0.130	0.310	0.068

Europa

When Galileo was twice in the vicinity of Europa, the second and third harmonics of the satellite gravitation potential were determined. Their average values are as follows:

$$J_2 = 0.215 \cdot 10^{-3} \quad \text{and} \quad 0.438 \cdot 10^{-3}$$

$$C_{22} = 0.365 \cdot 10^{-3} \quad \text{and} \quad 0.132 \cdot 10^{-3}$$

Average values of $J_3 = 0$ and $C_{33} = ? \cdot 10^6$ indicate a small deviation from hydrostatic equilibrium.

$$C/Ma^2 = 0.374 \quad \text{and} \quad 0.330$$

$$q = 4.97 \cdot 10^{-4} \qquad f = 577 \cdot 10^{-4}$$

In three-layer models one assumes that the existence of nuclei FeS at $r_c/a = 0.6, 0.45, 0.4$, and masses of $m/M = 47$–18%; Fe nucleus: $r_c/a = 0.4$–0.3 and $m/M = 20$–11%.

Ganymede

A three-layer model of the satellite from the core with the radius of 400–1300 km, silicate mantle and outside shell of H_2O ice with the thickness of ~800 km. The gravitational field is characterized by the following average values of coefficients: $J_2 = 0.126 \cdot 10^{-3}$ and $0.127 \cdot 10^{-3}$; $C_{22} = 0.378 \cdot 10^{-3}$ and $C_{22} = 0.383 \cdot 10^{-3}$; and $q = 1.903 \cdot 10^{-4}$. $C/Ma^2 = 0.3105$ is the least value of this parameter among other bodies of the solar system. Dynamic oblateness f is determined by Eq. (9):

$$f = 1.905 \cdot 10^{-4}$$

The following models of Ganymede nuclei are possible:

$$\text{Fe} - \text{FeS} \quad r_c/a = 0.2\text{–}0.5 \quad m/M = 2\text{–}33\%$$

$$\text{Fe} \quad r_c/a = 0.15\text{–}0.4 \quad m/M = 1.4\text{–}26\%$$

Callisto

From data obtained by ground observations and pictures of the surface from Voyager 1, the assumption was made that Callisto, as distinct from Ganymede and having nearly identical dimensions and density, kept a nondifferentiated composition of the mixture of ice and silicates. Gravitation research of Galileo resulted in numerical evidence of the nondifferentiated nature of the Callisto structure:

$$c/Ma^2 = 0.406 \quad \text{and} \quad 0.367 \qquad J_2 = (47.17 \pm 11.5) \cdot 10^{-6}$$

$$C_{22} = (14.3 \pm 3.2) \cdot 10^{-6} \qquad f = 7.05 \cdot 10^{-5}$$

For comparison here is the dynamic compression of the moon:

$$f = 3.1 \cdot 10^{-4}$$

C. Magnetic Fields of Galilean Satellites on the Model of Precession Dynamo

In the publications on magnetic experiments given in the preceding, the authors did not discuss the questions of energy sources and driving forces of dynamo fields. This problem was touched upon by Stevenson.[86] Having in mind that the process of substance differentiation in the depths of Saturn had come to an end long before and that gravitation energy that comes off during that process cannot be the source of dynamo process, Stevenson concluded, "Jupiter's satellite Ganymede has a magnetic field. Its cause could be an internal dynamo that was kick-started a billion years ago when the satellite was temporarily pushed into a more eccentric orbit."

The comparison of Io and Ganymede dipole fields in the framework of unreasonably applied Bode law[74,77] could not clarify the complicated problem. Note that the fields in the Jupiter–satellite system confirm the conclusion by Blacket[18] on the absence of any relation between the angular momentum P and magnetic moment M of rotating bodies: at similar polarities P of Jupiter and satellites, the polarities M of satellites are opposite to polarity M of Jupiter.

To survey the magnetic state of Galilean satellites in the framework of precession dynamo model, a number of stimulating factors exist:

1) The Jupiter–satellites system is similar, in model, to the solar system where the central body plays a determining part in the formation of the system.

2) In the system of Jupiter–satellites, the role of outside gravitational forces is the most significant as well as that of influx heating and synchronization mechanisms.

3) Rigorous synchronization is probably achieved in circular orbits that have no dissipation of energy. That is why one should expect that Galilean satellites rotating in eccentric orbits can precess with small amplitudes.

4) If the magnetic fields of Galilean satellites are generated by the mecha-

Table 11

Core Models	*H* - Jupiter		*H*-Earth	
Io	sin α	α°	sin α	α°
$r_c/a = 0.576\ r_c = 1045$	0.000108	0.006	0.0002	0.015
$r_c/a = 0.517\ r_c = 938$	0.000206	0.012	0.00026	0.012
$r_c/a = 0.447\ r_c = 811$	0.00053	0.03	0.00063	0.035
Europa				
$r_c/a = 0.6\ r_c = 940$	0.00012	0.007	0.00012	0.007
$r_c/a = 0.5\ r_c = 784$	0.00037	0.02	0.00037	0.02
$r_c/a = 045\ r_c = 706$	0.00069	0.04	0.00069	0.04
$r_c/a = 0.4\ r_c = 628$	0.0014	0.08	0.00135	0.08
Ganymede				
$r_c/a = 0.5\ r_c = 1317$	0.0069	0.4	0.0088	0.5
$r_c/a = 0.4\ r_c = 1050$	0.0268	1.5	0.0341	2.0
$r_c/a = 0.2\ r_c = 525$	sin α>1	——	sin α>1	——

nism of precession dynamo, this is mainly because of the influence of the gigantic gravitational field of Jupiter.

We do not know yet the angles α between the axes of satellite rotation and normals to the planes of their orbits around Jupiter. All other parameters in Eqs. (3) and (4) of the precession dynamo model are known. Also determined are precisions of the satellite internal structure models.

We will determine the values of angles α at which dipole fields of satellites, calculated by Eqs. (3) and (4), coincide with those determined by measurements. We use Jupiter and Earth as "basic" planets. Their magnetic moments differ by 10^5 times. Table 11 presents the results of calculations for the proposed models of satellite cores composed of FeS. Prior to the final results we intend to consider intermediate results of calculated parameters, determining the precession dynamo model for each of the satellites, the velocity Ω and the period T of precession, the velocity of generated flows V, and the Reynolds magnetic number R_m.

Io

$\Omega_{\mathrm{Io}} = 1.059 \cdot 10^{-7}$ s; $T_{\mathrm{Io}} = 1.08$ year. Compare these results to the results for the Earth: $\Omega_3 = 7.71 \cdot 10^{-12}$ s; $T_3 = 25{,}800$ years. These figures are difficult to compare. The unusual parameters of Io become less surprising if one takes into account that the period of Chandler fluctuations of the Earth $T_2 = 1.18$ year and the precession amplitude of Io is less than that of the Earth by about 580 times. The velocities generated by precession are determined by the equation $V = \Omega \sin \alpha\ r_c/f$ (13). Assuming that $r_c = 118$ km, $\Omega = 1.059 \cdot 10^{-7}$, $\sin \alpha = 0.0006$, and $f = 1.712 \cdot 10^{-3}$, we obtain $V \approx 3$ cm/s. The magnetic number of Reynolds is $R_m = r_c V/\eta$ (14). Assuming that $r_c = 811 \cdot 10^5$, $V = 3$ cm/s, $\eta = 2 \cdot 10^4$ cm^{2}s^{-1}, we obtain $R_m \approx 1.2 \cdot 10^4$.

Europa

$\Omega = 1.765 \cdot 10^{-8}$; $T = 11.3$ years; $V = 2.7$ cm/s; and $R_m = 8.5 \cdot 10^3$.

Ganymede

$\Omega = 2.815 \cdot 10^{-9}$; $T = 70$ years. Assuming that $r_c = 1050$ km, $\Omega = 2.815 \cdot 10^{-9}$, $\sin \alpha = 0.027$, and $f = 1.905 \cdot 10^{-4}$, we obtain $V \approx 42$ cm/s and $R_m \approx 2.1 \cdot 10^5$. Are these velocities feasible? The velocity of flows in the Earth's core, determined by Eq. (11), is $V_3 \approx 0.32$ cm/s. Besides, the flows generated by Eq. (11) are absolutely symmetric and cannot generate a magnetic field.[25] However, if in real conditions of planetary cores the flows acquire nonsymmetric character in the range of (0.5–1.0%), then R_m numbers are higher than critical, and field generation in Io, Europa, and Ganymede is quite feasible.

Callisto

Callisto does not seem to possess a melted core.[85] Because of the difference in orbits of Io and Callisto, in contemporary times the velocity of energy dissipation in Callisto is ~1300 times less than that in Io. More enigmatic is the heat and magnetic history of Ganymede.

Table 11 presents radii r_c corresponding to core models composed of FeS, $\sin \alpha$, and α, at which calculated values of dipole fields are equal to those determined by measurements at two "basic fields" differing from one another ($H_{\text{Jupiter}} = 417{,}900$ and $H_{\text{Earth}} = 30{,}500$). All of the parameters in Eqs. (3) and (4), relating to Jupiter and Earth, correspond to those in Tables 2, 6, 7, and 8. (Figures are kept in the given values for the convenience of calculation control.) The data in Table 11 and evaluation of parameters of dynamo process Ω, V, and R_m make it possible to conclude the following:

1) The agreement of angles α calculated by Eqs. (3) and (4), at gigantic difference of values in those equations relating to the Earth and Jupiter, indicate that the disclosed magnetic fields of Io, Europa, and Ganymede are similar in model to the fields of the Earth and Jupiter.

2) Dipole fields of these satellites are generated at small precession amplitudes because of great velocities of precession Ω under the action of the gigantic Jupiter gravitational field.

3) The conclusion in point 2 is also kept in case the conductivity in the cores of Galilean satellites is less than that of the Earth core σ_E by 2–3 times, which at calculations would lead to a corresponding increase of α value. (The possible decreases in conductivity in the Earth group planets with the increase of distance from the sun were mentioned in Section IV.)

4) The discovered increase of α angles with the augmentation of distance from Jupiter is probably connected with a relative growth of satellite resonance influence.

5) Under the Jupiter magnetic field influence, the dynamo process in the depth of satellites was, probably, more steady at the dynamo-field generation with the polarity reversible to that of the Jupiter field, at which the minimum of total magnetic energy is achieved.

6) The core of Ganymede with the radius $r_c = 0.2a$ cannot be the source of the field under observation.

D. Problems of the Jovian Magnetic Fields in the New Experimental Data

1) The model of a precession dynamo contains an assumption on the existence of synchronism in centennial variations of the magnetic fields of planets.[73] In the contemporary epoch, the Earth dipole field is diminishing. By measurement data from the Galileo spacecraft, an analytical model of the Jupiter field is being prepared, which will reflect the change of the Jupiter field during 22 years (model KK-96).

2) The agreement between the Jupiter dipole field values, calculated by Eqs. (3) and (4) and measured, was reached at the assumption that the conductivity of the Jupiter field generation is $\sigma_y = 5.93 \cdot 10^4\ s^{-1}$ (Tables 6 and 7). This led to a conclusion that the field of Jupiter is generated in the area of molecular, and not atomic, hydrogen, as was believed earlier. The same conclusion was reached by the results of experimental research reported in 1966 (Ref. 87).

3) Research from Galileo in the atmosphere of Jupiter at depths where pressure varies from 2 to 14 bar disclosed only an insignificant decrease of helium content in comparison with solar proportion.[88] This research indicates that the assumed, sedimentation of heavy helium is not the source of energy of the Jupiter magnetic field, but its driving force.

VII. Energy of the Precessional Geodynamo

The examination of the energy balance of the Earth's core invariably shows the existence of component $f\ [H \times E]$. dS, which is the flux of Pointing's vector that characterizes the external (relatively to the core's boundary) magnetic energy (Refs. 1, 5, 13, and 86). However, this component was not considered in the dependence of the core's parameters, and a specific dynamo mechanism in the noted publications. This restriction is eliminated by the analysis of the precessing dynamo model's formula[73]

$$H = KR_c^3 \rho \Omega \omega \sin \alpha / \eta_m$$

K is the proportionality coefficient. We ascribe to K the characteristics of the operator, which is dependent on the process, described in the MHD equations of induction, convection, and heat transfer. The detail characteristics of K are unknown to us, but it is possible to determine the dimensions of K:

$$K = \frac{\mathrm{cm^{-1/2} g^{-1/2} s^{-1} cm^2 s^{-1}}}{\mathrm{cm^3 s^{-2} g\, cm^{-3}}} = \mathrm{cm^{3/2} g^{-1/2}} = \frac{1}{|D|} \tag{11}$$

where D is the electric induction, $D = eE$, E is the intensity of the electric field, and e is the electrical permittivity. The dimensions $[e] = s^2\,cm^{-2} = [v^{-2}]$ are the dimensions of velocity

$$|HE| = \frac{R_c^3 \rho v^2 \Omega \omega \sin\alpha}{\eta_m} = \text{erg} \cdot \text{cm}^{-2}\text{s}^{-2} \tag{12}$$

and $v = \Omega \sin\alpha\, R_c/f \cdot \eta$ is the dynamic flattening of the core. According to Stewartson and Roberts,[26] V is the velocity stimulated by the precession in the liquid core. The preceding functions of the operator K can be described in the following formulas:

$$P = |HE| = \frac{R_c^3 \rho v^2 \Omega \omega \sin\alpha}{\eta + \nu + \chi} \tag{13}$$

where ν is the viscosity, χ is the thermal conductivity; η, ν, and χ have the same dimensions, $cm^2 s^{-1}$.

Considering the vector P flux to be anisotropic and changing it according to the formula $Z = Z_o \sin\varphi$ (φ is the geographical latitude), the following formula results:

$$\frac{dM_E}{dt} = \frac{2}{4\pi} \int_{\varphi=0}^{\pi/2} \int_{\lambda=0}^{2\pi} [E \times H] R_c^2 \sin^2\varphi \, d\varphi \, d\lambda = \frac{\pi R_c v^2 \rho \Omega \omega \sin\alpha}{2(\nu + \eta_m + \lambda)} (\text{erg} \cdot s^{-1}) \tag{14}$$

Let us estimate the numerical value of Eq. (12). According to Hide,[87] the sum of molecular and turbulent viscosity into the Earth's core was $\nu \leq 10^5$ cm s^{-1}. In the geodynamo model[17] a value $V \approx 10^3$ $cm^2 s^{-1}$ was adopted for the core. According to the results of the laboratory experiments,[16] the most important effect of the precession in the Earth's core is the appearance of the turbulence. The turbulent flux in liquid appears when the kinematic Reynolds number $R_K = R_c\, V/\nu \geq 100$ (Ref. 87). $V = 0.08$ cm s^{-1}; $\nu = 10^5$ cm^2 s^{-1}; and $R_k = 280$.

In the beginning of the dynamo process, the energy dissipation in the Earth's core is determined by the influence of the kinematic viscosity. The effect of the magnetic viscosity (ηm) increases with the magnetic field strengthening. Taking into account the generally accepted values of parameters included in formula (6) and the influence of η_m and ν, the energy altering rate $dM/dt \geq 3 \cdot 10^{17} \div 10^{19}$ erg $\cdot$ s^{-1}. These values can be compared with the experimentally seen changes of the magnetic field energy.

According to the spherical harmonic analysis of the geomagnetic field in the period between 1945–1965, the external energy of the dipolar field decreased with the rate of 5.1×10^{22} erg/year, and the summary energy of the multipolar field increased about 3.8×10^{22} erg/year. The common decrease of the energy of the magnetic field was 1.3×10^{22} erg/year = 4.5×10^{14} erg/s (Ref. 89).

Nowadays there are many versions explaining the decrease of the magnetic field energy. One of them is connected with the increase of the Earth's core energy dissipation while the multipolar harmonics are growing up.

VIII. Conclusion

The chaotic picture of the distribution of intensities of planetary dipole fields becomes regular when described by the "numerical law" of the model of precession dynamo. Some of the quantities necessary for this model are observed directly, whereas others can be determined from astronomical and geophysical observations and laboratory experiments. The physical sense of the formula describing the model of precession dynamo is quite clear. The formula determines the rate of change of external magnetic energy during the dynamo process. This quantity is determined from the geomagnetic observations on the surface and the data on the liquid core parameters. The latter also can be determined via magnetic and seismic data.

Nevertheless, we cannot determine the absolute values of dipole field via parameters of the precession dynamo model. This limitation was overcome by the revealed similarity between fields of different planets and by use of the principles of physical similarity. The Earth was chosen as a "base planet" for similarity studies.

Surprising accordance of dipole magnetic fields given by the equation of precession dynamo with reasonable parameters means that the external gravitational fields of the sun and the closest satellites are driving forces for dynamo processes. These fields cause precession and tidal effects, which result in turbulent convection of conductive fluid in magnetoactive regions.

Further studies of planets will give additional experimental data that may help to solve the problem of dynamo. So, one can expect that the Galileo spacecraft will reveal the secular variation of the Jovian magnetic field. The numerical value of secular variation will enable us to determine the upper boundary of the magnetoactive region in Jupiter by the use of the Hide method. The problem is whether it lies in the internal molecular sheath of Jupiter as it is determined here or in the region of metallic hydrogen, as was believed before. Also, it is important to know the sign of secular variation. Does the Jovian magnetic field decrease with time as the geomagnetic field does? Does the synchronization in this phenomenon exist as assumed by the model of the precession dynamo?

One can expect that the Mars-96 project will provide information on Martian paleomagnetism.[113]

The model of precession dynamo is open for numerical and experimental simulation. Obviously, efforts should be applied in order to reinvestigate processes in the system mantle-core.

Acknowledgment

The author thanks P. Israelevich for his assistance in preparing the manuscript.

References

[1]Gubbins, D., and Masters, T. G., "Driving Mechanisms for the Earth's Dynamo," *Advances in Geophysics,* Vol. 21, 1979, pp. 1–50.

[2]Verhoogen, J., "Thermal Regime of the Earth's Core," *Phys. and Planet. Inter.*, Vol. 7, 1973, pp. 47–53.

[3]Braginsky, S. I., "Kinematic Model of the Earth's Hydromagnetic Dynamo," *Geomag. Aeron.*, Vol. 4, No. 4, 1964, pp. 572–583.

[4]Braginsky, S. I., "Magnetohydrodynamics of the Earth's Core," *Geomag. Aeron.*, Vol. 4, No. 4, pp. 898–916.

[5]Gubbins, G., "Energetics of the Earth's Core," *Geophys.*, Vol. 43, 1977, pp. 453–464.

[6]Loper, D. E., "The Gravitationally Powered Dynamo," *Geophys. J. Royal Astron. Society*, Vol. 54, 1978, pp. 389–404.

[7]Gubbins, D., Masters, J. G., and Jacobs, J. A., "Thermal Evolution of the Earth's Core," *Geophys. J. Royal Astr. Soc.*, Vol. 59, 1979, pp. 57–98.

[8]Stevenson, D. J., Spohn, T., and Schubert, G., "Magnetism and Thermal Evolution of the Terrestrial Planets," *Icarus*, Vol. 54, No. 3, 1983, pp. 466–489.

[9]Malkus, W. V. R., "Precessional Torques as the Cause of Geomagnetism," *Journal of Geophysical Research,* Vol. 68, 1963, pp. 2871–2886.

[10]Anderson, D. L., "The Earth as a Planet: Paradigms and Paradoxes," *Science*, Vol. 223, No. 4634, 1984, pp. 347–355.

[11]Braginsky, S. I., and Roberts, P. H., "Equations Governing Convection in Earth's Core and in Geodynamo," *Geophys. Astrophys. Fluid Dynamics* (to be published).

[12]Malkus, W. V. R., "Precession of the Earth as the Cause of Geomagnetism," *Science*, 1968, Vol. 160, No. 3825, pp. 259–264.

[13]Rochester, M. G., Jecoos, J. A., Smylie, D. E., et al., "Can Precession Power the Geomagnetic Dynamo?" *Geophys. J. Royal Astron. Soc.*, Vol. 43, 1975, pp. 661–678.

[14]Loper, D. E., "Torque Balance and Energy Budget for Precessionally Driven Dynamos," *Phys. Earth and Planet. Inter.*, Vol. 11, 1975, pp. 53–60.

[15]Dolginov, Sh. Sh., "Magnetic Properties of Mars," Preprint No. 17, Moscow: IZMIRAN, 1973.

[16]Malkus, W. V. R., "Energy Sources for Planetary Dynamos," *Lectures on Solar and Planetary Dynamos*, edited by M. R. E. Proctor and A. D. Gibbert, Cambridge Univ. Press, Cambridge, England, UK, *University Press*, pp. 161–179.

[17]Vanyo, J. P., "A Geodynamo Powered by Luni-Solar Precession," *Geophysics Astrophysics Fluid Dynamics*, Vol. 59, 1991, pp. 200–214.

[18]Blacket, P. M. S., "A Negative Experiment Relating to Magnetism and Earth's Rotation," *Philosophical Transactions of the Royal Society of London, Series A*, Vol. 245, 1952, pp. 300–310.

[19]Curtis, S. A., and Ness, N. F., "Magnetostrophic Balance in Planetary Dynamos: Prediction for Neptune Magnetosphere," *Journal of Geophysical Research*, Vol. 91, No. A10, 1986, pp. 1003–1008.

[20]Buse, F. H., "Generation of Planetary Magnetism by Convection," *Phys. Earth and Planet. Inter.*, Vol. 12, 1976, pp. 350–358.

[21]Stevenson, D. J., "Planetary Magnetism," *Icarus*, Vol. 22, 1974, pp. 403–415.

[22]Stevenson, D. J., "Planetary Magnetic Fields," *Reports on Progress in Physics*, pp. 555–620.

[23]Parker, E. N., *Cosmical Magnetic Fields, Their Origin and Their Activity*, Georwdon Press, Oxford, England, UK, 1979.

[24]Zharkov, V. N., *The Earth's and Planets' Interiors*, Nauka, Moscow, 1983 (in Russian).

[25]Sedov, L. I., "Invention of Physical Models," *Ideas on Science of Past and Present*, Nauka, Moscow, 1973, pp. 63–67 (in Russian).

[26]Stewartson, K., and Roberts, P. H., "On the Motion of a Liquid in a Spheroidal Cavity of a Precessing Rigid Body," *Journal of Fluid Mechanics*, Vol. 17, 1963, pp. 1–20.

[27]Sedov, L. I., "Methods of Similarity and Dimensional Representation in Mechanics," Nauka, Moscow, 1965, p. 54.

[28]Ness, N. F., "Mercury: Magnetic Field and Interior," *Space Sci. Rev.*, Vol. 21, No. 5, 1978, pp. 527–553.

[29]Klassen, K. P., "Mercury's Rotation Axis and Period," *Icarus*, Vol. 28, No. 4, 1976, pp. 469–478.

[30]Esposito, P. B., Anderson, J. D., and Shapiro, I. F., "Experimental Determination of Mercury's Mass. *Sci. News,* Vol. 37, No. 24, p. 375.

[31]Burns, J. A., "Consequences of the Tidal Slowing of Mercury," *Icarus*, Vol. 23, 1976, pp. 453–458.

[32]Whang, Y. C., "Model Magnetosphere of Mercury," *Phys. Earth and Planet. Inter.*, Vol. 20, No. 2–4, 1979, pp. 218–230.

[33]Ng, K. H., and Beart, D., "Possible Displacement of Mercury's Dipole," *Journal of Geophysical Research*, Vol. 78, 1979, pp. 2115–2117.

[34]Zharkov, V., "Models of Internal Structure of Venus," *The Planet of Venus*, Nauka, Moscow, 1985, pp. 313–341.

[35]Akim, Z. L., Vlasova, Z. P., and Chuyko, N. V., "The Determination of Venus' Dynamic Flattening by the Trajectory Parameters of Its First Artificial Satellites," *Reports of the USSR Academy of Sciences*, Vol. 240, No. 3, 1978, pp. 556–559.

[36]Smith, E. J., Daris, L. J., Coleman, P. J., et al., "Magnetic Field Measurements Near Venus," 1965, Vol. 70, p. 1571.

[37]Dolginov, Sh. Sh., Yeroshenko, Ye. G., and Zhuzgov, L. I., "The Investigation of Magnetic Field from the Interplanetary Space Station Venus-4," *Space Research*, Vol. 6, No. 4, 1968, pp. 561–575.

[38]Dolginov, Sh. Sh., Zhuzgov, L. N., Sharova, V. A., et al., "The Magnetic Field and Magnetosphere of the Planet Venus," *Space Research*, Vol. 16, No. 6, 1978, pp. 827–863.

[39]Knudsen, W. G., Banks, P. M., and Miller, R. L., "A New Concept for Plasma Motion and Planetary Magnetic Field for Venus," *Journal of Geophysical Research*, Vol. 9, No. 1, 1982, pp. 165–168.

[40]Luhmann, J. G., and Russel, C. T., "Magnetic Fields in the Magnetospheric Holes of Venus: Evidence for an Intrinsic Field." *Journal of Geophysical Research*, Vol. 10, No. 1, 1983, p. 409.

[41]Phillips, J. I., and Russel, C. T., "Upper Limit on the Intrinsic Magnetic Field of Venus," *Journal of Geophysical Research*, Vol. 92, No. 3, 1987, pp. 2253–2263.

[42]Dolginov, Sh. Sh., "The Magnetic Field and Magnetosphere of the Planet Mars," *Space Research*, Vol. 29, No. 5, 1991, pp. 754–789.

[43]Dolginov, Sh. Sh., and Zhusgov, L. N., "The Magnetic Field and Magnetosphere of the Planet Mars," *Planet Space Sci.*, Vol. 39, No. 11, 1991, pp. 1493–1510.

[44]Dolginov, Sh. Sh., "Centennial Changes of the Magnetic Moment of the Planet Mars," *Space Research*, Vol. 30, No. 3, 1992, pp. 425–428.

[45]Binder, A., and Davis, D. R., "Internal Structure of Mars," *Phys. Earth and Planet. Inter.*, Vol. 7, 1973, pp. 477–485.

[46]Smith, E. J., Davis, L. J., Jones, D. E., et al., "The Planetary Magnetic Field and Magnetopause of Jupiter: Pioneer 10," *Journal of Geophysical Research*, Vol. 79, No. 25, 1974, pp. 3501–3513.

[47]Smith, E. J., Davis, L. J., Jones, D. E., et al., "Jupiter's Magnetic Field, Magnetosphere and Interaction with the Solar Wind: Pioneer 11," *Science*, 1975, Vol. 188, pp. 451–455.

[48]Smith, E. J., Davis, L. J., Jones, D. E., et al., "Saturn's Magnetic Field and Magnetosphere: Pioneer 11," *Science*, Vol. 207, No. 4429, 1980, pp. 407–410.

[49]Ness, N. F., Acuna, M., Bechanon, K. W., et al., "Magnetic at Uranus," *Science*, 1986, Vol. 233, pp. 85–89.

[50]Connerney, J. E. P., Acuna, M., and Ness, N. F., "The Magnetic Field of Uranus," *Journal of Geophysical Research,* Vol. 92, No. A13, 1987, pp. 15,329–15,336.

[51]Ness, N. F., Acuna, M. H., Burlaga, J. E. P., et al., "Magnetic Fields at Neptune," *Science*, Vol. 246, 1989, pp. 1473–1478.

[52]Connerney, J. I. P., Acuna, M. H., and Ness, N. F., "The Magnetic Field of Neptune," *Journal of Geophysical Research*, Vol. 96, Supplement, 1991, pp. 19,023–19,042.

[53]Anderson, J. D., Campbell, J. K., Jacobson, R. A., et al., "Radio Science with Voyager-2 at Uranus: Results on Masses and Densities of the Planet and Five Principal Satellites," *Journal of Geophysical Research*, Vol. 92, No. 13A, 1987, 14,877–14,883.

[54]Tyler, G. L., Sweetnam, D. N., Anderson, J. D., et al., "Voyager Radio Science Observations of Neptune," *Science*, Vol. 246, 1989, pp. 1456–1472.

[55]Pearl, J. C., and Conrath, B. S., "The Albedo, Effective Temperature and Energy Balance of Neptune as Determined from Voyager," *Journal of Geophysical Research*, Vol. 96, Supplement, 1991, pp. 18,921–18,930.

[56]Podola, M., Reinolds, R. T., and Young, R., "Post Voyager Comparisons of the Interiors of Uranus and Neptune," *Journal of Geophysical Research*, Vol. 17, No. 10, 1990, pp. 1737–1740.

[57]Hubbard, W. B., Nellis, W. J., Mitchell, A. C., et al., "Interior Structure of Neptune: Comparison with Uranus," *Science*, Vol. 253, No. 5020, 1991, pp. 648–651.

[58]Zharkov, V. N., "Internal Structure of Giant Planets," *Astronomical Journal*, Vol. 25, No. 6, 1981, pp. 627–649.

[59]Nellis, W. J., Hamilton, D. C., Holmes, N. C., et al., "The Nature of the Interior of Uranus Based on the Studies of Planetary Ices at High Dynamic Pressure," *Science*, Vol. 240, 1988, pp. 779–781.

[60]Rouskol, E. L., "Natural Satellites of Planets," *Summary of Science and Technology. Astronomy*, VINITI, Moscow, Vol. 23, 1986, p. 115.

[61]Landau, L. D., and Lifshits, E. M., *Statistical Physics. Part 1*, Nauka, Moscow, 1976, pp. 82–85.
[62]Kaula, W. M., "Thermal Earth's of Tidal Friction," *The Earth-Moon System*, edited by B. G. Marsden and A. G. W. Cameron, Plenum Press, New York, 1966, pp. 46–51.
[63]Dolginov, Sh. Sh., "Magnetic Fields of Uranus and Neptune: View from the Earth (Critical Survey)," *Geomagnetism and Aeronomy*, Vol. 33, No. 2, 1993, pp. 1–22.
[64]Kirk, R. L., Stevenson, D. J., "Hydromagnetic Constraints on Deep Zonal Flow in the Giant Planets," *Astrophysical Journal*, Vol. 316, 1987, pp. 836–846.
[65]Levy, E. H., "State of the Geomagnetic Fields and Its Reversals," *Astrophysical Journal*, Vol. 175, 1972, pp. 573–81.
[66]Holifman, K. A., "Polarity Transitions and the Geomagnetic Dynamo," *Science*, Vol. 196, No. 4296, 1977, pp. 1329–1334.
[67]Goldreich, P., "An Explanation of the Frequent Occurrence of Commensurable Mean Motions in the Solar System," *Monthly Notices of Royal Astronomical Soc.*, Vol. 130, No. 2–3, 1965, pp. 159–182.
[68]Blekhman, M. I., *Synchronization in Nature and Technology*, Nauka, Moscow, 1981.
[69]Hubbard, W. B., *Planetary Interiors*, Van Nustand Reynolds Co., 1984.
[70]Burns, J. A., "Jupiter's Decametric Radio Emission Belts of Its Galilean Satellites," *Science*, Vol. 159, 1968, pp. 971–972.
[71]Kivelson, M. G., Slavin, J. A., and Southwood, D. J., "Magnetospheres of the Galilean Satellites," *Science*, Vol. 205, 1979, pp. 905–908.
[72]Neubauer, F. M., "Possible Strengths of Dynamo Magnetic Fields of the Galilean Satellites and Titan," *Journal of Geophysical Research*, Vol. 5, No. 11, 1978, pp. 905–908.
[73]Dolginov, Sh. Sh., "The Precession of the Planets and Generation of Their Magnetic Fields," *Geomag. Aeron.*, Vol. 35, No. 1, 1995, pp. 1–29 (in Russian).
[74]Kivelson, M. G., Khurana, K. K., Walker, R. J., et al., "A Magnetic Signature at Io: Initial Report from the Galileo Magnetometer," *Science*, Vol. 273, 19.07.96, pp. 337–340.
[75]Kivelson, M. G., Khurana, K. K., Walker, R. J., et al., "Io's Interaction with the Plasma Torus: Galileo Magnetometer Report," *Science*, Vol. 274, 1996, pp. 396–398.
[76]Frank, L. A., Paterson, W. R., Ackerson, K. L., et al., "Plasma Observations at Io with the Galileo Spacecraft," *Science*, Vol. 274, 1996, pp. 394–395.
[77]Kivelson, M. G., Khurana, K. K., Russell, C. T., et al., "Discovery of Ganymede's Magnetic Field by the Galileo Spacecraft," *Nature*, Vol. 384, 1996, pp. 537–540.
[78]Gurnett, D. A., Kurth, W. S., Roux, A., et al., "Evidence for a Magnetosphere at Ganymede from Plasma-Wave Observations by the Galileo Spacecraft."
[79]Kivelson, M. G., Khurana, K. K., Russel, J. S., et al., "Europa's Magnetic Signature: Report from Galileo's Pass on 19.12.96," *Science*, Vol. 276, 1997, p. 1239.
[80]Khurana, K. K., Kivelson, M. G., Russell, C. T., et al., "Absence of an Internal Magnetic Field at Callisto," *Nature*, Vol. 387, 1997, p. 262.
[81]Gurnett, D. A., Kurth, W. S., Roux, A., et al., "Absence of a Magnetic-Field Signature in Plasma-Wave Observations at Callisto," *Nature*, Vol. 387, 1997, p. 261.
[82]Anderson, J. D., Sjorgen, W. L., and Schubert, G., "Galileo Gravity Results and the Internal Structure of Io," *Science*, Vol. 272, 1996, p. 709.

[83]Anderson, J. D., Lau, E. L., Sjorgen, W. L., et al., "Europa's Differentiated Internal Structure: Inferences from Two Galileo Encounters," *Science*, Vol. 276, 1997, p. 1236.

[84]Anderson, J. D., Lau, E. L., Sjorgen, W. L., et al., "Gravitational Constraints on the Internal Structure of Ganymede," *Nature*, Vol. 384, 1996, p. 541.

[85]Anderson, J. D., Lau, E. L., Sjorgen, W. L., et al., "Gravitational Evidence for an Undifferentiated Callisto," *Nature*, Vol. 387, 1997, p.264.

[86]"When Galileo Met Ganymede," *Nature*, Vol. 384, 1996, p. 511.

[87]Mellis, W. J., Weiz, S. T., and Mitchell, A. C., "Metallization and Electrical Conductivity of Hydrogen in Jupiter," *Science*, Vol. 273, 1996, pp. 936.

[88]Von Zahn, U., and Hunten, D. M., "The Helium Mass Fraction in Jupiter's Atmosphere," *Science*, Vol. 272, 1996, p. 849.

[89]Hubbard, W. B., Anderson, J. D., "Possible Flyby Measurements of Galilean Satellite Interior Structure," *Icarus*, Vol. 33, 1978, p. 336–341.

V. Experimental Methods and Data Analysis

Chapter 22

Characterization of Microfabricated Shear Stress Sensors

Eli Reshotko,* Tao Pan,† Daniel Hyman,‡ and Mehran Mehregany§
Case Western Reserve University, Cleveland, Ohio, 44106

I. Introduction

SHEAR stress measurement is of primary importance in aerodynamic and hydrodynamic design because it relates to a major part of the drag and performance characteristics. No universal solution exists for the accurate determination of wall shear stresses, but many indirect methods have been developed, including Preston tube, hot-wire or hot-film anemometry, and liquid surface techniques.[1] One disadvantage of these techniques is that many assumptions must be made about the fluid and its interactions with the sensor. A direct measurement of shear stress can be made with a floating element with a surface area parallel to the fluid flow. The shear forces cause the element to displace against a restoring mechanism, and this measurement can thus be directly related to the shear stress of the flow.

One important aspect of the work presented here is in the use of surface micromachining techniques, where conventional microfabrication techniques from the electronics industry have been optimized for miniature mechanical devices on the same substrate.[2] Moveable elements are photolithographically defined in thin films deposited on a silicon wafer substrate. They are freed when the so-called "sacrificial" layer beneath them is etched away. As they are fabricated out of surface films, these devices are intrinsically planar in nature and parallel to the substrate, both of which are requirements for floating-element shear-stress sensors.

Sensors can be passive or active. In a passive micromachined sensor, the floating element displaces in response to the flow, and this displacement is typically detected as a change in capacitance between two electrodes. A

*Department of Mechanical Aerospace Engineering.
†Department of Electrical Engineering and Applied Physics.
‡Department of Electrical Engineering and Applied Physics.
§Department of Electrical Engineering and Applied Physics.

surface micromachined, passive shear-stress sensor was previously reported by Schmidt et al.,[3,4] using a polyimide element with a tethered suspension system. One potential drawback of such a passive measuring system is that gaps around the floating element can vary with the shear stress, possibly resulting in spurious measurements. In an active sensor, an electrostatic mechanism is used to restore the floating element to its null displacement position, eliminating gap variance. The active design also enables the implementation of a "self-test" feature where the sensor response to a controlled input is examined to ensure proper sensor operation.

In our work heavily phosphorus-doped polysilicon is used as the structural material instead of polyimide. Polyimide absorbs water from the ambient in proportion to the relative humidity, contributing to instability and output drift. Polyimide also exhibits viscoelastic properties that cause drift. These effects are not present in polysilicon devices. Our polysilicon devices are also more than an order of magnitude smaller in area than the polyimide sensors previously reported.

II. Sensor Design

The shear-stress sensors developed at Case Western Reserve University (CWRU) are based on lateral resonance structures with folded beam suspensions as shown in Fig. 1, rather than tether or "fixed-beam" style suspension systems. Folded beam suspensions are more flexible than tether systems, allowing for higher sensor sensitivities.[5,6] After etching away the sacrificial layer in the surface micromachining process, the thin-film internal stresses play an important role in beam deflection characteristics. These residual stresses are relaxed in folded beam structures but not in tethers, resulting in more predictable sensor characteristics. Another advantage of the folded beam structure is that the deflection of the floating element is related to the shear force by a linear relationship:

$$\delta = \frac{F}{K}$$

where δ, F, and K are the deflection, shear force, and spring constant of the structure, respectively.

For the CWRU sensor design shown in Fig. 1, the floating plate overlaps the electrodes of the first polysilicon layer forming two parallel-plate capacitors. When shear force is applied by fluid flow over the surface, the floating plate deflects, changing the relative electrode overlap areas and, thus, capacitances. These two electrodes are driven by ac signals of the same magnitude, 180 deg out of phase, causing net charges to accumulate on the floating plate due to the unbalanced capacitances. The accumulated charge is fed to signal conditioning circuitry and converted into a voltage output.

CWRU has cooperated with Analog Devices to develop shear-stress sensors based on the existing ADXL line of automotive accelerometers. These active sensors have tether suspension systems and combine both the drive

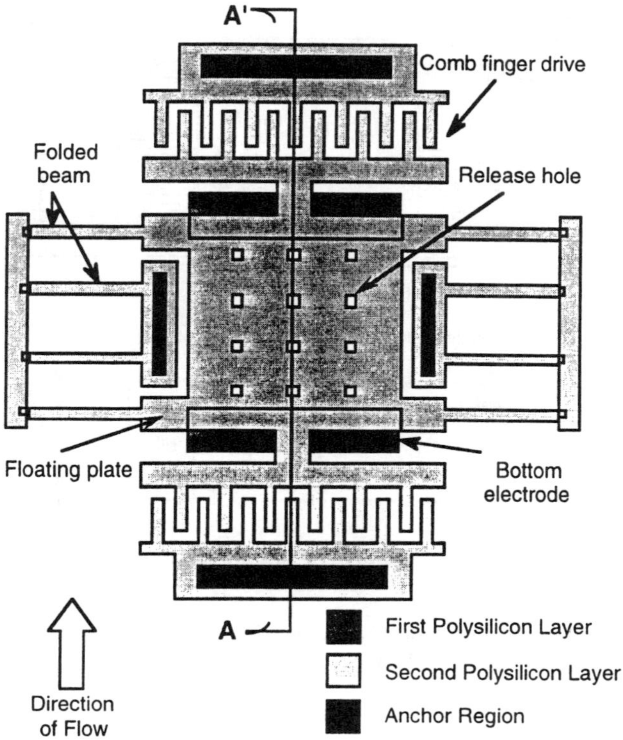

Fig. 1 Lateral resonant structure with folded beam suspension.

mechanism and the sense electrodes into the comb fingers on the sides of the floating plate (Fig. 2). When subjected to fluid flow, the shear force displaces the floating plate, causing the plate electrodes to move closer to one set of fixed electrodes and away from another. The two capacitances formed by the movable and fixed electrodes change accordingly, C_1 increasing and C_2 decreasing (Fig. 2). Charge accumulates on the plate due to the capacitance mismatch, measurable as plate voltage:

$$V = \frac{C_2 - C_1}{C_2 + C_1} V_i = \frac{\delta}{d} V_i$$

where δ is the deflection, d is the initial gap between comb fingers, and V_i is the magnitude of the driving signal.

On-chip circuitry provides active sensor operation as shown in Fig. 3 (Ref. 7). The plate voltage signal is modulated by an externally referenced oscillator and fed into a preamplifier. This signal is fed back to the floating plate as a bias signal. This bias incites electrostatic force to repel itself from the

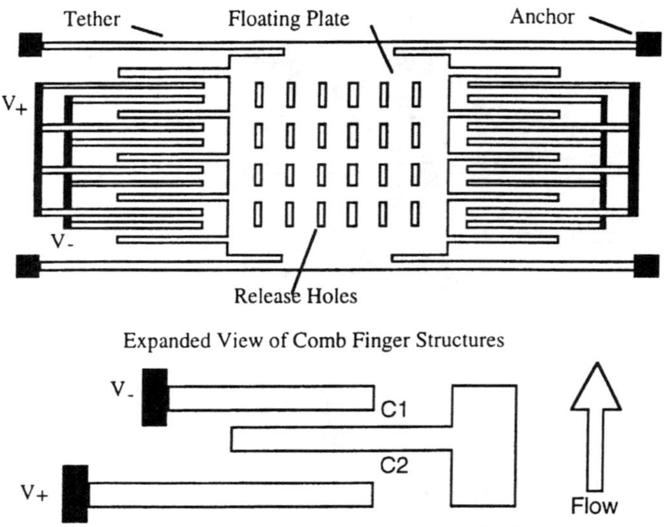

Fig. 2 CWRU/Analog Devices shear-stress sensor.

near electrode and toward the far electrode, causing it to move back to its zero displacement position. By working in this force-balanced mode, noise and distortion are reduced.

III. Sensor Fabrication

The CWRU sensor is fabricated in a four-mask surface micromachining process schematically illustrated in Fig. 4. The process starts with a bare silicon wafer with a layer of thermally grown silicon dioxide to release the stress between the substrate and the following layer. A silicon nitride layer is deposited to insulate devices from the substrate (Fig. 4a). The first polysilicon layer (0.5 μm) is then deposited and heavily doped to improve conductivity and also patterned to form a shield and bottom electrodes (Fig. 4b). A sacrificial oxide layer (2.2 μm) is deposited on top of the first polysilicon

Fig. 3 CWRU/Analog Devices sensor on-chip circuitry function diagram.

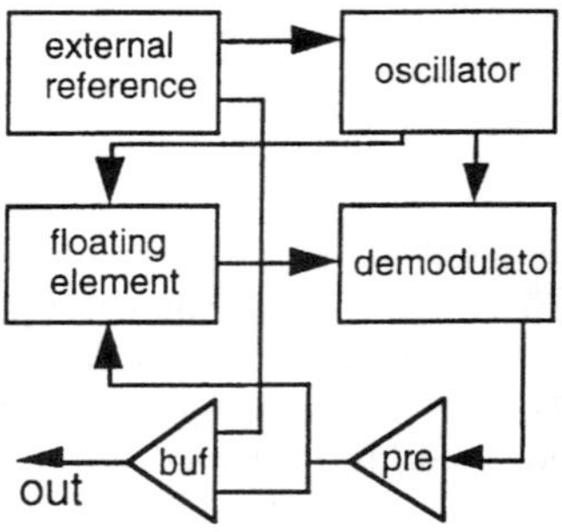

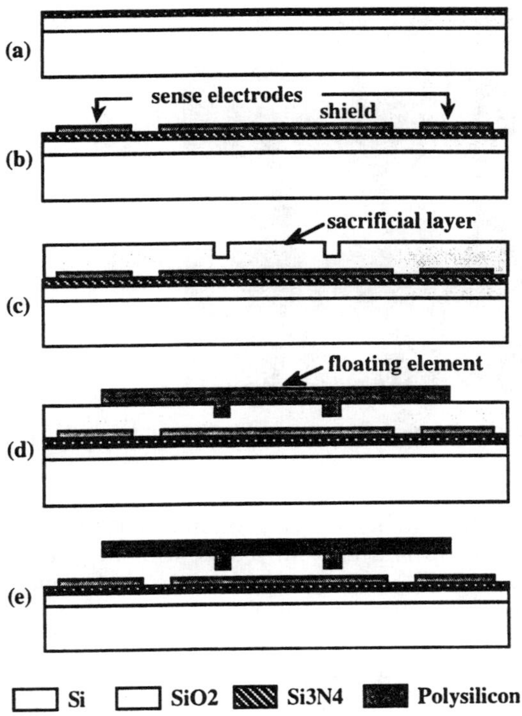

Fig. 4 CWRU shear-stress sensor process (cross section $A \div A$ in Fig. 1): a) grow oxide and deposit nitride, b) deposit and pattern first polysilicon layer, c) deposit and pattern oxide layer, d) deposit and pattern second polysilicon layer, and e) release device.

layer and bushing molds (0.5 μm deep) that form small bumps at the bottom of the second polysilicon layer to prevent the floating element from touching the lower electrodes (Fig. 4c). Anchor areas are also patterned where oxide is completely etched away (not shown in Fig. 4). The anchors provide mechanical support for the polysilicon structures and enable electrical connection between the two polysilicon layers. The second polysilicon layer (2.2 μm) is then deposited and heavily doped with phosphorus. The floating element and suspension are defined in Fig. 4d. The devices are then immersed in HF (49% wt.) to etch the sacrificial oxide, freeing the polysilicon floating elements (Fig. 4e). A SEM picture of the sensor is shown in Fig. 5. The CWRU/Analog Devices shear-stress sensor shown in Fig. 6 was fabricated by Analog Devices in its first iMEMs™ process.[8]

IV. Case Western Reserve University Sensor Calibration

The testing and calibration of sensors is conducted in a high aspect ratio continuum channel flow. The high aspect ratio (~100) is consistent with treating the flow as two dimensional. A sketch of our first test channel is shown in Fig. 7. A rotameter regulates gas flow from a nitrogen tank into the entrance port of the channel with atmospheric pressure at the exit. The sensor chip containing an array of sensors is inserted parallel to the fluid

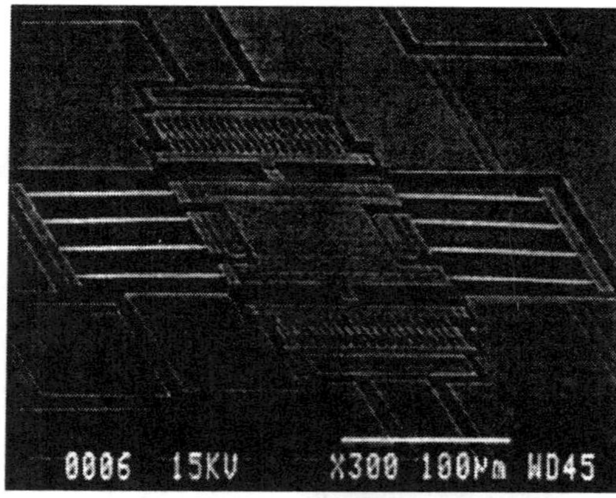

Fig. 5 SEM picture of CWRU shear-stress sensor with 100 μm beam.

flow near the exit. The channel width and height are defined by a brass shim between the machined aluminum top and bottom plates, with height selected in the range of 180 to 300 μm. The mean fee path at standard conditions is ~0.07 μm supporting the assumption of continuum flow.

For incompressible flow, using the momentum theorem of fluid mechanics, the shear stress on the channel wall is given by

$$\tau = -\frac{b}{2}\frac{\mathrm{d}p}{\mathrm{d}x}$$

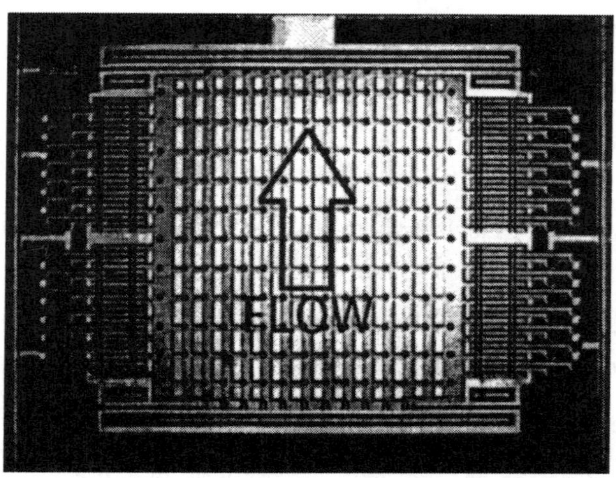

Fig. 6 CWRU/Analog Devices shear-stress sensor.

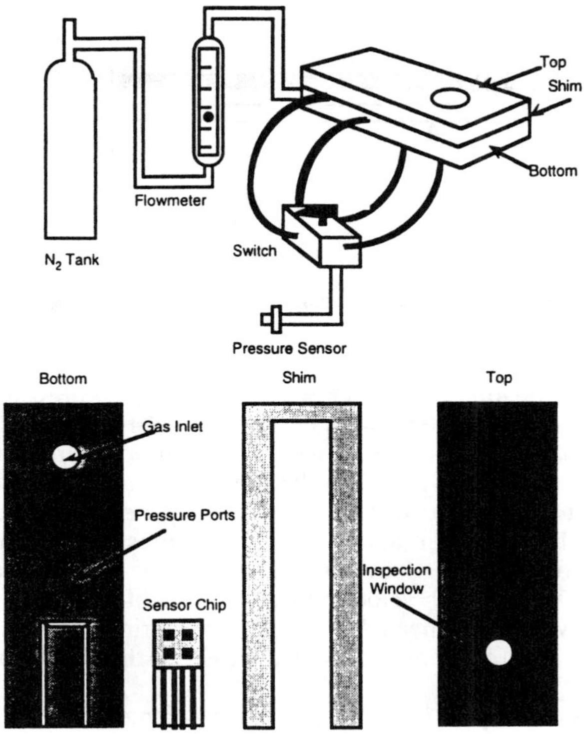

Fig. 7 Test setup for the calibration of microfabricated shear-stress sensors.

independent of the velocity profile. In this expression, b is the channel height, and $\mathrm{d}p/\mathrm{d}x$ is the channel pressure gradient. However because the pressure drop in the channel is a significant fraction of the inlet pressure, the channel flow must be treated compressibly. Again, using the momentum theorem and assuming an isothermal channel flow, the shear stress on the channel wall is

$$\tau = -\frac{b}{2}\frac{\mathrm{d}p}{\mathrm{d}x}(1-\gamma M^2)$$

where the pressure gradient, and the mean Mach number M are evaluated at the location of the shear-stress sensor. In this expression γ is the ratio of specific heats of the channel gas (air). The compressible flow in an isothermal channel is treated in some detail in Ref. 9. The pressure gradient is nonuniform.

There are four pressure taps in the channel routed through a manifold to a single commercial pressure transducer as shown in Fig. 8. To determine the pressure gradient at the sensor location, a second degree curve fit is made

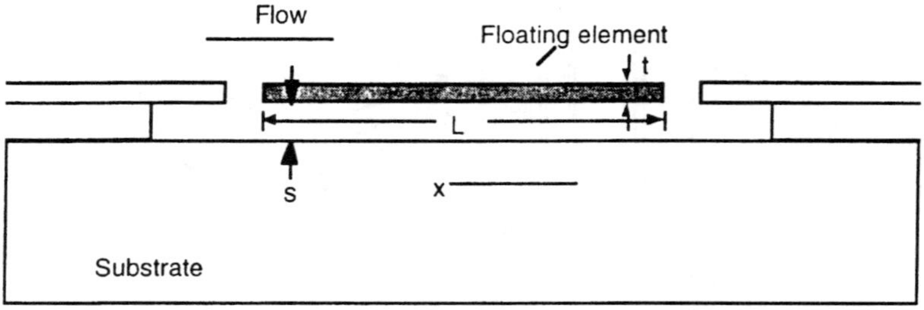

Fig. 8 Cross-sectional view of the gaps between the floating element and the surrounding area.

from the tap pressures. From $p(x) = ax^2 + bx + c$, the pressure gradient is found to be $p'(x) = 2ax + b$ where x is the distance from the sensor location to the edge of the sensor chip at which point atmospheric pressure is assumed. To determine the mean Mach number M, the velocity at the sensor location must first be calculated. This is done by measuring the volumetric flow rate at the rotameter, adjusting it for the pressure ratio between the rotameter and the sensor location and dividing by the channel cross-sectional area. Dividing this velocity by the speed of sound at the channel temperature yields the mean Mach number. During calibration, the exit Mach number must be kept small (<0.25) to keep the channel from "choking."

The effects of the gaps between the floating element and its surrounding solid surfaces must be considered in the calibration of the shear-stress sensor. The deflection of the sensing element or the restoring voltage in the case of an active sensor both depend on the total force exerted on all faces of the sensor. This includes also the pressure force on the fore and aft edge surfaces as well as the shear force on the bottom surface of the element shown in Fig. 8. Thus the force balance equation for the element becomes

$$F = \tau A - L\frac{\mathrm{d}p}{\mathrm{d}x}Wt - \frac{s}{2}\frac{\mathrm{d}p}{\mathrm{d}x}A$$

where A is the area of the floating element, L, W, and t are the length, width, and thickness of the floating element, respectively; and s is the height of the gap beneath the floating element.

The force per unit area can then be related to the channel shear stress as follows:

$$\frac{F}{A} = \tau\left[1 + \frac{2t + s/b}{1 - \gamma M^2}\right]$$

In a typical situation $t = s = 2\ \mu$m and $b = 200\ \mu$m, then because $(1 - \gamma M^2) \approx 1$, the value of F/A would exceed the shear stress by about 3%.

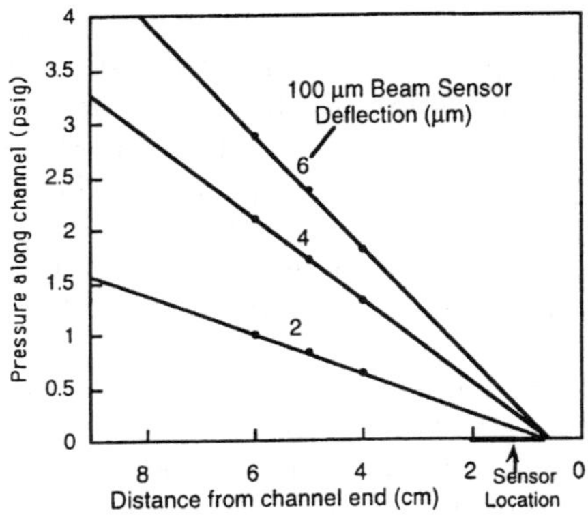

Fig. 9 Channel pressure graph used to interpolate the pressure gradient at the sensor location.

If the same sensor is used to measure shear stress in some other flow, then the shear stress is given by

$$\tau = \frac{F}{A} + \left[t + \frac{s}{2}\right]\frac{\mathrm{d}p}{\mathrm{d}x}$$

where the force F comes from the sensor calibration and the value of $(\mathrm{d}p/\mathrm{d}x)$ is at the sensor location for the flow being measured. Local pressure measurements are required for proper shear-stress evaluation.

A typical sensor response is shown in Fig. 10. Sensors with 100 μm beam lengths have sensitivities of 9 Pa per μm of floating element deflection. Early test results showed a "dead zone" for low flow rates where the sensor did not deflect, shown in Fig. 10. This was caused by stiction of the floating element to the substrate, introduced during the release process.

The last step in surface micromachining is the device release in which the SiO_2 sacrificial layer is removed. The usual procedure of releasing surface micromachined structures is to immerse the chip in hydrofluoric acid (49% wt.) to remove the SiO_2 surrounding the floating element. The devices are then rinsed thoroughly in DI water or methanol and dried. The surface tension of the receding liquid pulls the floating element down into contact with the substrate where they form a weak bond. This combination of bonding (sticking) and friction is called stiction; its adhesion mechanism has been investigated.[10,11] The main forces that play a role in stiction are thought to be (in order of strength) capillary, hydrogen bridging, electrostatic, and Van der Waals forces.

In the supercritical carbon dioxide (CO_2) drying process, the rinsing liquid is replaced by liquid CO_2, which is then heated to its supercritical state. The supercritical state of a fluid is above the critical temperature and pressure in which the fluid coexists in the gaseous and liquid states. The CO_2 is introduced in its liquid phase at a pressure greater than the critical pressure and heated into the supercritical region. The pressure of this supercritical fluid is then reduced, transforming into a gas without the traditional liquid/gas interface responsible for surface tension and capillary forces.[12]

The improvements brought by supercritical CO_2 drying process are demonstrated in both device yield and performance. Yield for 100 μm beam sensors has increased from approximately 25% to above 80%, whereas yield for 120 μm sensors increased from less than 5% to approximately 30%. In addition, traditionally dried devices show a certain latency in that they do not deflect with the flow until the shear forces exceeded the stiction force, as suggested in the deflection curve of Fig. 10. When released in the supercritical CO_2 drying process, the sensor deflection characteristic curve goes through zero.

All following sensors described have been released supercritically and do not demonstrate stiction. Figure 11 shows the reproducibility of sensor operation from sensor to sensor on the same die in the same channel mounting. Although the deflection curves are close to each other for the three different sensors with 100 μm beam lengths, they differ sufficiently to require separate calibration if 1% or better accuracy is desired.

The shear force F_{shear} is linearly related to displacement with a third-power relationship to the length of the beam L_b:

$$\frac{4EtW_b^3}{L_b^3}\delta = F_{\text{shear}}$$

where W_b is the width of the beam, E is Young's modulus of the doped polysilicon, and t is the thickness of the folded cantilever beam structure.

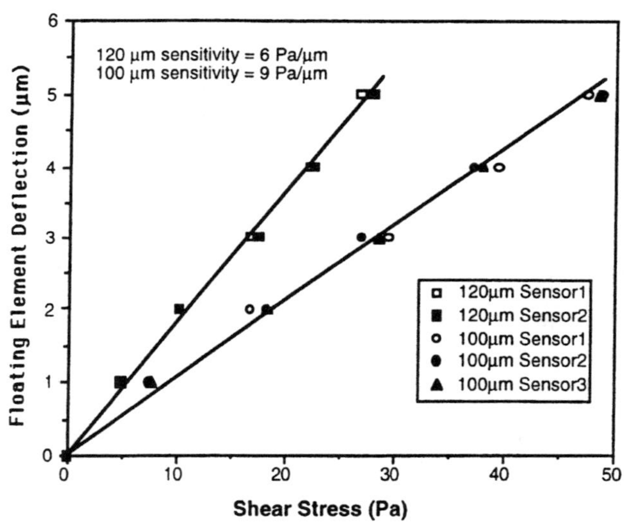

Fig. 10 Effects of stiction and the supercritical CO_2 release process on 100 μm beam sensors.

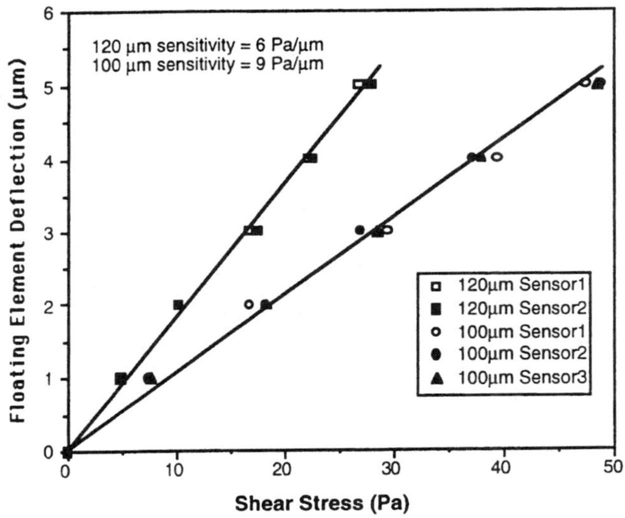

Fig. 11 Deflection curves for 100 μm and 120 μm beam length CWRU sensor.

The deflection characteristics of 120 μm beam devices are compared to 100 μm beam devices in Fig. 11. Sensitivity variation with beam length agrees with the third-power relationship predicted by the mechanical model with a 65% increase in flexibility for a 20% increase in beam length. Again the small spread between the deflection curves suggests the need for separate calibration of each sensor.

V. Case Western Reserve University/Analog Devices Sensor Testing

The CWRU/Analog Devices shear-stress sensor is supplied in a hermetic 10-pin TO-100 metal can package. In flow channel testing the can cap is removed, and the sensor is inserted into the inspection window. It is sealed with silicon gel to prevent leakage. Care is taken to mount the sensor surface flush with the surface of the bottom of the cover plate. Small inaccuracies of either the depth or angle of the sensor relative to the surface of the plate result in measurement error. In addition, if the sensor protrudes into the flow channel, its bonding wires can contact the bottom of the channel, shorting the device.

Figure 12 shows a schematic of the electrical connections to the sensor. The input voltage is a regulated + 5V. The external capacitors are 22 μF, 22 μF, and 10 μF. In device testing the self-test feature was used to ensure proper sensor operation both before and after data collection. The preamplified output V_{pr} was used, as the signals sent back to the plate were large enough to obtain accurate readings. The amplified output V_{out} was not used. The resistors in the preamplifier were laser trimmed at Analog Devices to increase sensitivities and match sensor-to-sensor output.

A typical response of a CWRU/Analog Devices shear-stress sensor is

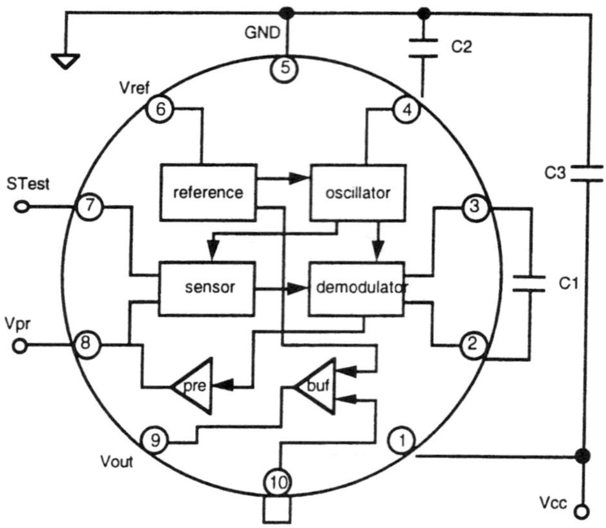

Fig. 12 External electrical connections to CWRU/Analog Devices shear-stress sensor.

shown in Fig. 13. In its linear regime, the sensor shows a sensitivity of 6 Pa per volt of output change. The sensor works in a force-balance mode so that the electrostatic force generated by the circuit counters the shear force. The output voltage is directly related to this generated force. At large shear stresses, the circuit cannot generate large enough forces to counter the shear forces, and the output saturates at this limit, as evident in Fig. 13.

One difficulty encountered in calibrating the CWRU/Analog Devices shear-stress sensors is their sensitivity to their angle of inclination in the flow channel. The imprecision of standard integrated circuit packages such as the TO-100 complicates the positioning of sensors parallel to the fluid flow.

Fig. 13 Typical response of CWRU/Analog Devices integrated shear-stress sensor.

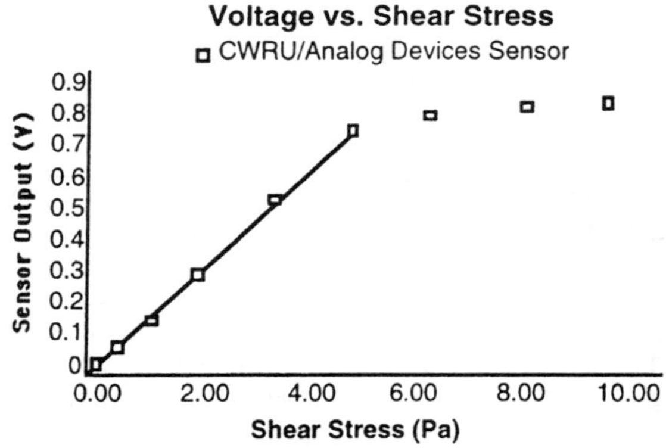

Slight misalignment on the order of a few degrees or several hundred microns results in sensor sensitivity variations of an order of magnitude. Saturation levels are also subject to variation. We did not have mounting problems with the CWRU sensors because the flat substrate itself was used as the package body.

VI. Sensor Redesign and Calibration

A number of improvements were performed in the redesign and calibration of the sensors to address accuracy and reproducibility concerns. First, the flow channel was redesigned to eliminate leakage and provide more accurate interpolation of the pressure gradient at the sensor. The CWRU sensors were fabricated with on-chip signal amplification circuitry to provide voltage response to shear stress without need for optical access to the device. Finally, the CWRU/Analog Devices sensors were repackaged for the new flow channel to eliminate fluctuations in sensor response and enable full sensor calibration of these devices.

The original flow channel was redesigned to address several problems encountered in early testing. A metal-glass seal was replaced with an o-ring type seal to eliminate leakage along the channel edges, a primary concern in the testing of both sets of sensors. The pressure taps in the new flow channel were now located both up- and downstream of the sensor location to provide a more accurate interpolation of the pressure gradient. In addition, isothermal flow conditions previously assumed in calculations were verified with three in-line thermocouples. Finally, to eliminate output variance with the CWRU/Analog Devices sensors, the sensor mounting region includes fine adjustment screws to enable flush sensor mounting to 25 microns.

The redesign of the CWRU sensors includes the monolithic integration of on-chip signal amplification circuitry. Because of the small size of the devices, capacitance changes were on the order of several femtofarads with very small signal changes on the floating plate. As a result of these small electrical signals and characteristics, device output was lost in parasitic capacitances associated with pad contact and wire bonding. Figure 14 shows the circuit schematic of the on-chip differential amplifier. The gate of M_1 is connected to the floating element with a diode necessary to bias the NMOS transistor. The gate of M_2 is the output node, isolating and amplifying the signal. Both M_1 and M_2 are depletion-mode NMOS transistors with channel

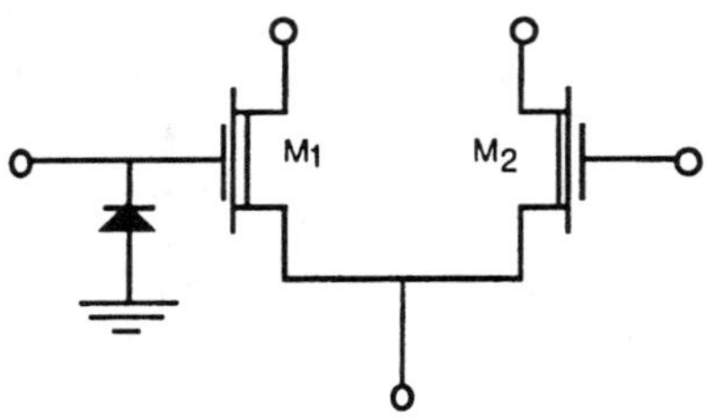

Fig. 14 On-chip differential pair amplifier.

widths of 120 μm and lengths of 12 μm. The voltage gain A_v of the differential pair is

$$A_v = g_m R_d$$

where g_m is the NMOS transconductance and R_d is the value of an off-chip resistor connected to the drains of the devices (not shown in Fig. 14). Typical gains are about 20.

The devices were driven with the same signal applied to both sets of electrodes. The expected output in such a case[13] is

$$V_o = A_1 V_G = A_1 \frac{\varepsilon_o t_{\mathrm{poly2}} L}{C_p d_o} 2V_d \sin \omega t \left[1/1 - \left(\frac{A}{d_o k} \tau \right)^2 \right]$$

where A_1 is the voltage gain of the differential amplifier, C_p is the parasitic capacitance at the gate of M_1, t_{poly2} is the thickness of the second polysilicon layer, L is the overlap between the stationary and movable comb fingers, and d_o is the initial gap between the comb fingers. We rewrite this as

$$V_o = \frac{a}{1 - b\tau^2} \sin \omega t$$

where a and b are constants determined for any driving voltage and frequency. Using an 8 kHz, 8 V peak-to-peak driving signal, the output voltage of the device was recorded with varying flow rates and the associated shear stress calculated for Fig. 15. The best curve fit ($R^2 = 0.962$) gives $a = 176.4$ and $b = 0.00054$.

The full calibration of the CWRU/Analog Devices sensors is enabled with the reproducible mounting scheme of the new package and flow channel. The reproducibility of the channel, device, and packaging is affirmed in Fig. 16. With corrections for the gage thickness correction factor and for fluid flow beneath the device, the sensitivity of the device is 1.03 V/Pa within the range from the dc offset of 0.2 V to 3.5 V. With the 200 mV offset, this corresponds to an actual output range of 0.5 V to 3.7 V, comparable to the published data for the Analog Devices ADXL50 accelerometer, claimed to operate linearly from 0.2 V to 3.8 V (Ref. 7).

The current apparatus and calibration process introduces a number of uncertainties in device measurement including limitations in instrument accuracy or operator-dependent uncertainties, only the greatest of which are described here. The uncertainty of the optical measurement of the channel height at the exact location of the sensor is calculated to be 20 μm because of the positioning of the sensor die flush to the channel wall and the depth of field of the measurement microscope. This 20 μm out of the channel height of 210, 250, or 400 μm corresponds to a fractional uncertainty of 9.5,

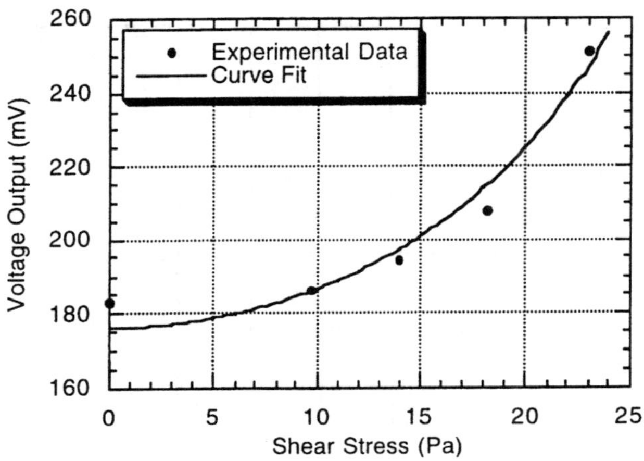

Fig. 15 Output of a second-generation horizontal-electrode CWRU shear-stress sensor with synchronous driving of sense electrodes.

8.0, or 5%, respectively. The curve fits for the pressure gradients vary up to 8.5, 5.5, and 0.5% for each fit made to the pressures in the trials with channel heights of 210, 250, and 400 μm, respectively. Based on the linearity specification for the pressure transducer, these curve fits can be as inaccurate as an additional 2.1% in the slope of the line at the location of the sensor. With these and additional, less significant, errors added in quadrature, the fractional uncertainty in the calculated value of shear stress in the calibration trials with channel heights of 210, 250, and 400 μm are 13.3, 10.3, and 5.4%, respectively. The data points in the linear regime of Fig. 16 are plotted and fitted for linear sensitivity in Fig. 17. A sensitivity of 1.02 ± 0.05 V/Pa is

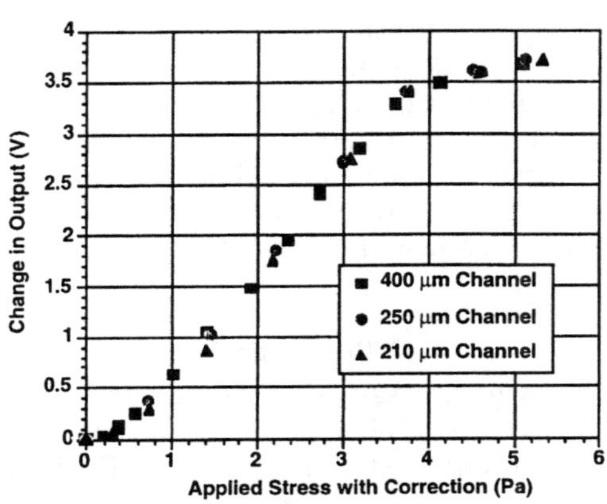

Fig. 16 CWRU/Analog Devices sensor calibration curves as corrected for gage forces and sensor under flow.

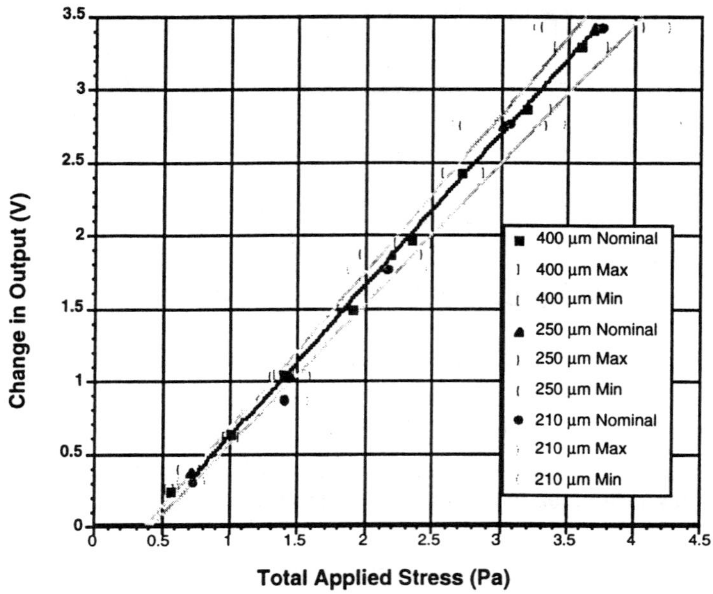

Fig. 17 Linear regime of calibration curves with nominal, maximum, and minimum sensor sensitivity curve fits corresponding to 1.02 ± 0.5 V/Pa.

obtained, corresponding to a 5% calibration accuracy. Note in Fig. 16 that the sensor characteristic is properly independent of channel height.

VII. Conclusion

A surface micromachined shear-stress sensor has been developed, and sensors of varying suspension beam lengths have been calibrated in a flow channel. The sensors responded linearly to shear force as designed with different shear-stress ranges related to the different stiffnesses of the folded beams. In collaboration with Analog Devices, an integrated shear-stress sensor was fabricated with voltage output linear to shear stress. A comprehensive calibration has been performed on these sensors to give a sensitivity of 1.02 V/Pa ± 5% over a range of 0.5–3.7 V.

Acknowledgment

This work was supported by NASA Lewis Research Center under Grant NAG3-1273. We thank Christopher A. Bang for the construction of the original testing channel and for the design of the floating element of the CWRU/Analog Devices integrated sensor. Technical assistance from Microfabrication Laboratory staff and Electronics Design Center staff at Case Western Reserve University is greatly appreciated.

References

[1]Winter, K. G., "An Outline of the Techniques Available for the Measurement of Skin Friction in Turbulent Boundary Layers," *Progress in Aerospace Sciences*, Vol. 18, 1977, pp. 1–57.

[2]Mehregany, M., Ko, W., Dewa, A., and Liu, C. C., *Introduction to Microelectromechanical Systems and the Multiuser MEMS Processes*, Notes from CWRU/MCNC MEMS Short Course, Aug. 1993.

[3]Schmidt, M., Howe, R., and Senturia, S., "A Micromachined Floating-Element Shear Sensor," *Transducers' 87* (Tokyo, Japan), 1987, pp. 383–386.

[4]Schmidt, M., Howe, R., Senturia, S., and Haritonidis, J., "Design and Calibration of a Micro-Fabricated Floating-Element Shear-Stress Sensor," *IEEE Transactions on Electron Devices*, Vol. 35, June 1988, pp. 750–757.

[5]Tang, W. C., Nguyen, T. C. H., and Howe, R. T., "Laterally Driven Polysilicon Resonant Micro-Structures," *Sensors and Actuators*, Vol. 20, 1989, pp. 25–32.

[6]Tang, W. C., Nguyen, T. C. H., Judy, M. W., and Howe, R. T., "Electrostatic-Comb Drive of Lateral Polysilicon Resonators," *Sensors and Actuators*, Vols. 21–23, 1990, pp. 328–331.

[7]Data sheet for ADXL50 monolithic accelerometer with signal conditioning, Analog Devices.

[8]Kuehnel, W., and Sherman, S., "A Surface Micromachined Silicon Accelerometer with On-chip Detection Circuitry," *Sensors and Actuators*, Vol. 45, 1994, pp. 7–16.

[9]Zucrow, M. J., and Hoffman, J. D., *Gas Dynamics*, Vol. I, Wiley, New York, 1976.

[10]Alley, R. L., Cuan, G. J., Howe, R. T., and Komvopoulos, K., "The Effect of Release-Etch Processing on Surface Microstructure Stiction," *Technical Digest, IEEE Solid-State Sensors and Actuators Workshop* (Hilton Head, SC), June 1992, pp. 202–207.

[11]Legtenberg, R., Elders, J., and Elwenspoek, M., "Stiction of Surface Micromachined Structures after Rinsing and Drying: Model and Investigation of Adhesion Mechanism," *Technical Digest, The 7th International Conference on Solid-State Sensors and Actuators* (Yokohama, Japan), 1993, pp. 198–201.

[12]Mulhern, G., Soane, D., and Howe, R., "Supercritical Carbon Dioxide Drying of Microstructures," *Technical Digest, The 7th International Conference on Solid-State Sensors and Actuators* (Yokohama, Japan), 1993, pp. 296–299.

[13]Pan, T., "Microfabricated Shear Stress Sensors," Ph.D. Thesis, Case Western Reserve Univ., Cleveland, OH, Jan. 1996.

Chapter 23

Mass Transfer Measurement at a Stirred Liquid-Liquid Interface by Means of an Electrochemical Method

Yves Fautrelle,* François Debray,† and Francis Dalard‡
Institut National Polytechnique de Grenoble, St. Martin d'Héres, 38400, France

Introduction

IN VARIOUS metallurgical processes the refining of the liquid metal is achieved by means of a mass transfer reaction that takes place at the interface between the liquid metal and a covering liquid (e.g., slag for liquid steel). For example sulphur is removed from liquid steel by a molten slag whose calcium reacts with the sulphur at the metal–slag interface. Thus, to determine the interfacial mass transfer coefficients as a function of the stirring conditions near the interface is of interest. In real situations, these experiments are of course very difficult to perform because of the hot environment. The present study in a cold mercury simulation consists of such a refining process.

Description of the Experiment

We use an electrochemical method to measure the mass transfer through the interface between the mercury and a covering aqueous electrolyte layer. The mercury is stirred either by gas bubbling injected through an immersed nozzle or by surface waves generated by low-frequency alternating electromagnetic forces. The facility is illustrated in Fig. 1.

The principle of the method consists in removing the amalgamated indium from the mercury by means of an aqueous sodium chloride electrolyte.[1] The indium has been chosen because it amalgamates very easily with mercury and is known to have very little effect on the value of the water–mercury

*EPM-MADYLAM, ENSHMG, BP 95.
†EPM-MADYLAM, ENSHMG, BP 95.
‡CREMGP, ENSEEG, BP 75.

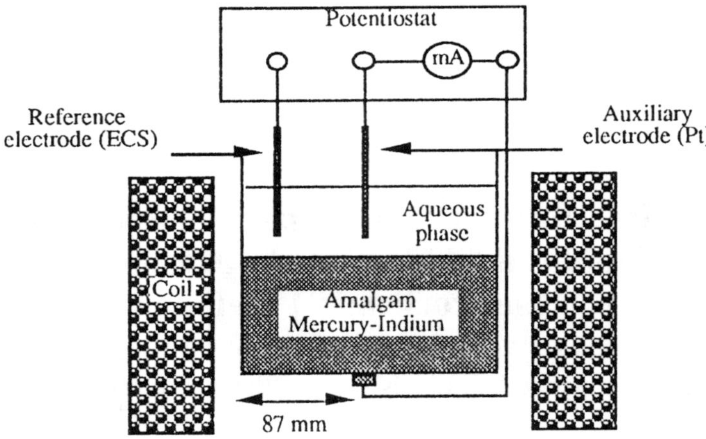

Fig. 1 Experimental setup.

interfacial tension.[2] Furthermore, its fast electron transfer kinetics allows it to reach limiting current conditions for a wide range of concentration and stirring conditions of the amalgam. Hence, we are able to simulate a metal-controlled reaction as it mostly occurs in the true metallurgical refining process.

Experimental Results

We use the electrochemical method to study two kinds of liquid metal stirring: gas bubbling in the liquid metal phase and interfacial stirring by means of electromagnetically driven surface waves (see Debray and Fautrelle[3] for a summary of this kind of surface excitation). We analyze the recorded Faraday current that contains both a slowly decaying mean part and superimposed fluctuations. The fluctuations are due to the bulk flow as well as the interface motion. The slowly decaying mean part is obtained only after a more drastic decay that can be shown to correspond to the time of establishment of the chemical diffusion layer on the mercury side.[4] In the case of strong stirring conditions, as we present here, this time is very short as compared with the duration of the experiments that are of the order of one minute. A signal sample is illustrated in Fig. 2 in the case where electromagnetically driven interfacial waves are generated with various amplitudes.

In the case of gas stirring, experimental results are illustrated in Fig. 3. The overall mass transfer coefficient k is found to increase with the gas flow rate approximately as $Q^{0.5}$. That tendency is consistent with data obtained in similar conditions with other experimental methods.[5] It also agrees with Levich–Davies semiempirical theory,[6] which predicts that k is proportional to $v^{3/2}$ with v being a typical pool velocity. Commonly admitted is that v is proportional to $Q^{1/3}$. As far as the electromagnetic stirring case is concerned,

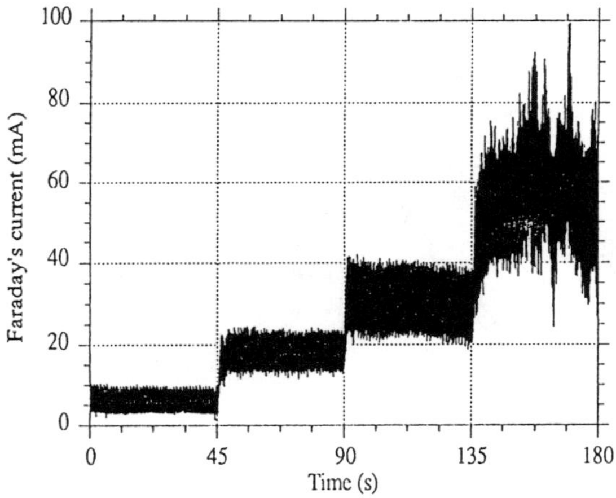

Fig. 2 Time evolution of the limiting current; the applied magnetic strength is increased every 45 s.

it has been possible to test the transfer efficiency of the interfacial waves as compared with the previous stirring condition. Figure 4 shows the evolution of the mass transfer coefficient with respect to the coil current electrical current I for various values of the magnetic field frequency.

We found that k evolves as $I^{2.1}$. In the latter case, the electromagnetic characteristic velocity obtained from local measurements[3] is found to be proportional to $I^{1.3}$. Accordingly, the latter result is also consistent with the Levich–Davies theory at least in the displayed case where the interfacial deformations were relatively well controlled.

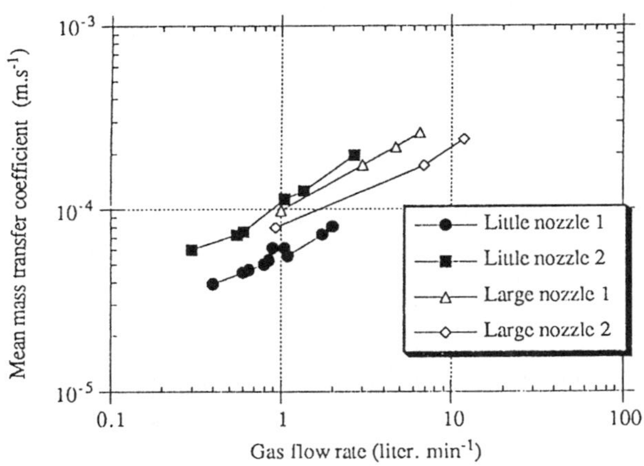

Fig. 3 Mean mass transfer coefficient versus the gas flow rate in the gas stirring conditions.

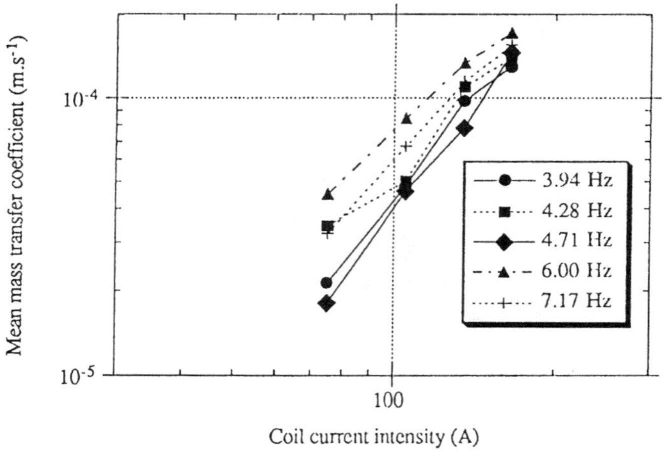

Fig. 4 Mean mass transfer coefficient versus the coil electrical current for various imposed interfacial wave frequencies.

The main advantage of using the electrochemical method as compared with more classical sampling techniques of oxido-reduction reaction at the interface is that it provides an instantaneous information on the mass transfer at the interface. Hence, we were able to analyze the interfacial dynamical behavior provided the oscillations of the waves do not exceed approximately 20 Hz. This is shown in Fig. 5 where the frequency spectrum of the signal shows the mode content of the interface oscillations. The observation must be made that the Faraday current fluctuations are related not only with the pulsating velocity but also to the change of the interfacial area. A clue of that effect can be observed in the spectrum of Fig. 5 where we may observe a significant enhancement of the pike at twice the frequency of the imposed wave motion equal to 3.6 Hz in this case.

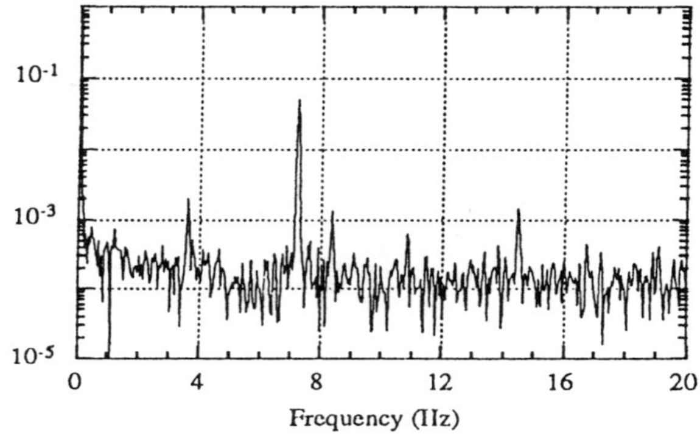

Fig. 5 Frequency spectrum of the Faraday current fluctuations for an applied wave frequency equal to 3.6 Hz.

Conclusion

The preceding experimental results show that the electrochemical method is now a powerful tool for measuring mass transfers at a liquid–liquid interface. In the case of the electromagnetically excited interfacial waves, the correlation between local height measurements that we performed and the overall measurement of the interfacial mass transfer permit us to understand the main mechanisms influencing the mass transfer at a moving liquid–liquid interface.

References

[1]Debray, F., Fautrelle, Y., and Dalard, F, "Measurement of Mass Transfer Across a Moving Mercury-Electrolyte Interface by an Electrochemical Method," *Experiments in Fluids*, Vol. 19, 1994, pp. 353–358.

[2]Wilkinson, M. C., "The Surface Properties of Mercury," *Chemical Reviews*, Vol. 72, No. 6, 1972, pp. 575–625.

[3]Debray, F., and Fautrelle, Y., "Free Surface Deformation Frequencies of an Electromagnetically Excited Mercury Layer," *Experiments in Fluids*, Vol. 16, 1994, pp. 316–322.

[4]Debray, F., Davoust, L., Fautrelle, Y., and Etay, J., "Mesure et Controle par Électrochimie D'une Convection D'Origine Solutale a Une Interface Entre Deux Liquides en Presence Ou Non D'un Champ Magnétique," *International Communications on Heat and Mass Transfers* (to be published).

[5]Porter, W. F., Richardson, F. D., and Subramanian, K. N., "Some Studies of Mass Transfer Across Interfaces Agitated by Bubbles," edited by A.W. D. Hills, Inst. of Mining and Metallurgy, London, 1972, pp. 79–111.

[6]Davies, J. T., *Turbulence Phenomena*, Academic, New York, 1972.

Chapter 24

Two-Phase Flow Measurement with an Electromagnetic Flowmeter

U. Opara* and I. Bajsić†
University of Ljubljana, Ljubljana, 1000, Slovenia

I. Introduction

TWO-PHASE gas-liquid flows are of common occurrence in power industry and chemical processes. Understanding the mechanics of these flows is therefore of great importance. Because two-phase flows are inhomogeneous in their structure and because the slip between phases is present in these flows, high-accuracy measurement is more difficult to execute than in single-phase flows. The main problem is to find a general theoretical solution, like a one-dimensional two-fluid model developed by Thibault et al.[1] According to Tanatugu et al.,[2] it is important to distinguish between weak electrolyte and liquid metal in gas-liquid flows. In electromagnetic flow measurement electric charges in the fluid are required to induce an electric field, so that various analytical models of the electromagnetic flowmeter, as reviewed by Baker,[3] refer only to conducting liquids.

The purpose of this paper is to present the possibility of a two-phase flow measurement with an induced voltage electromagnetic flowmeter, where one of the phases can be electrically insulating while the other one has to be electrically conducting. Because all electric charges are transferred just by the conducting phase, the basic flowmeter equation is modified; however, the principle of flowmeter operation stays the same: a movement of electric charges due to the fluid motion through the flowmeter magnetic field induces an electric field, and the difference of electric field potential on separate locations is detected by the electrodes as an induced voltage.

*M.Sc. Student, Faculty of Mechanical Engineering, Aškerčeva 6.
†Asst. Professor, Faculty of Mechanical Engineering, Aškerčeva 6.

II. Theory

The total two-phase volumetric flow $\dot{V}$ might be treated as a sum of the "gageable" phase volumetric flow $\dot{V}_1$ and "ungageable" phase volumetric flow $\dot{V}_2$:

$$\dot{V} = \dot{V}_1 + \dot{V}_2 \tag{1}$$

By applying volumetric quality β as a ratio of the ungageable phase volumetric flow to the total two-phase volumetric flow,

$$\beta = \frac{\dot{V}_2}{\dot{V}} \tag{2}$$

which can be, for each $|\beta| < 1$, expressed in the geometric series of the gageable phase volumetric flow:

$$\dot{V} = \dot{V}_1 + (\beta + \beta^2 + \cdots + \beta^n + \cdots)\dot{V}_1 \tag{3}$$

In this equation the first term is equal to the gageable phase volumetric flow, while the sum of other terms corresponds to the ungageable phase volumetric flow. The order of the ungageable phase contribution to the meter output depends on the flow regime in the meter tube.

A. Induced Voltage in Two-Phase Flow

The primary device of an induced voltage electromagnetic flowmeter contains the following elements: an electrically insulating meter tube; an electromagnet; and a pair of diametrically opposed point electrodes, each of which is mounted perpendicularly to the other. Considering axisymmetrical flow of an electrically homogeneous fluid without permittivity and conductivity gradients and assuming a uniform magnetic field by excluding the self-inductance and the displacement current from the Maxwell equations leads to the Shercliff[4] basic flowmeter equation that can be interpreted in various ways to produce a calibration factor, which in practice is usually determined by wet calibration:

$$q_{+-} = c(Rm, \mu, d) \cdot \frac{U}{i} \tag{4}$$

In this equation, the flowmeter output q_{+-} is proportional to the difference between two electric field potentials, known as induced voltage U, and inversely proportional to the magnetic flux density defined by coils current i where the calibration factor c depends on the magnetic Reynolds number Rm, fluids permeability μ, and pipe diameter d. According to Hewitt,[5] even in case of homogeneous flow of electrically heterogeneous fluid, the flowmeter output should be equal to the total volumetric flow:

$$\dddot{q}_{+-} = \dddot{V} \tag{5}$$

Bernier et al.,[6] however, obtained the following relation for the inhomogeneous flow of electrically heterogeneous fluid consisting of conducting and insulating phase:

$$\dddot{q}_{+-} = \frac{1}{1-\langle\dddot{\alpha}\rangle}\dddot{V}_1 \tag{6}$$

where the global void fraction $\langle\dddot{\alpha}\rangle$ is calculated as a sum of cylindrical bubbles to pipe cross-section ratios:

$$\langle\dddot{\alpha}\rangle = \frac{E\{\dddot{A}_2\}}{A} \tag{7}$$

Referring to the Bankoff and Armand model, mentioned later in this text, the value of volumetric quality becomes a constant if the global void fraction of a fully developed froth two-phase flow remains steady. With this in mind Eq. (6) can be expressed also as a function of volumetric quality and treated the same as Eq. (3). The flowmeter itself needs to be sensitive to the direction of flow by means of its sign (foward – positive, backward – negative), so that the averaging will give the right value even in the case of pulsating flow, while the time average (bar $^{-}$) should be equal to the ensemble average (bar $^{\cdots}$ or symbol $E\{\ \}$) of any random variable (two bars $^{\because\!\cdot}$).

B. Pressure Drop in Two-Phase Flow

On the basis of dimensional analysis, the pressure drop due to internal forces in fully developed turbulent single-phase flow can be calculated by using the Zlokarnick[7] expression:

$$p_{xo} = f(1/Re,\delta/d,\ell/d^5)\cdot\rho\cdot\dot{V}^2 \tag{8}$$

In this equation, the pressure drop p_{xo} due to internal forces is proportional to the total volumetric flow $\dot{V}$ and to the density ρ of incompressible fluid where the friction factor f depends on the inverse Reynolds number $1/Re$, relative wall roughness δ/d, and ratio ℓ/d^5 of the pipe length to its diameter to the power of five. Classifying terms of the Dobran[8] mixture momentum equation as pressure drops in two-phase flow enables us to express the pressure drop $\dddot{P}_{xo}$ due to internal forces as a sum of a static pressure drop $\dddot{P}_{xo,s}$ of internal flow and pressure drop $\dddot{P}_{xo,g}$ due to external fields:

$$\ddot{p}_{xo} = \ddot{p}_{xo,s} + \ddot{p}_{xo,g} \tag{9}$$

Here only the gravity force g is accepted to have a significant effect on a weak electrolyte. To evaluate these terms, the assumption is made that static pressure p has a flat profile along any cross-sectional plane perpendicular to the flow, whereas density ρ of each phase is treated in the same way also over a short length of pipe:

$$\ddot{p}_{xo} = (\ddot{p}_{\text{in}} - \ddot{p}_{\text{out}}) + [\langle\ddot{\alpha}\rangle\ddot{\rho}_2 + (1 - \langle\ddot{\alpha}\rangle)\ddot{\rho}_1]g\ell \tag{10}$$

Considering single-phase gas flow of negligible pressure drop due to internal forces and assuming two-phase gas-liquid flow of negligible pressure drop due to acceleration effects, all in the same vertical pipe, leads to the Ros[9] general emprical correlation:

$$\ddot{p}_{xo} = f(f_I f_{II} / f_{III}, \ell/d^5) \cdot \rho_1 \cdot \frac{1}{1 - \dddot{\beta}} \cdot \dddot{V}_1^2 \tag{11}$$

where incompressible liquid is considered as continuous phase and gas as a dispersed phase. The friction factor becomes modified, so that it also includes other components that contribute to the pressure drop due to internal forces, all involved in a product of f_I, f_{II}, and f_{III} friction factor. Equation (11) might be treated in the same manner as Eq. (3).

C. Induced Voltage and Pressure Drop in Two-Phase Flow

According to the previous statements, Eq. (6) and Eq. (11) can be approximated by using a general function:

$$\zeta = \sum_{i=0}^{n_\chi} c_i \chi^i \cdot \sum_{j=0}^{n_\upsilon} k_j \upsilon^j \tag{12}$$

in which all averaging factors and physical properties are included as constants c_i, k_j for individual cases. Variables χ, υ, ζ correspond to $\dddot{\beta}$, $\dddot{V}$, $\dddot{P}_{xo}$, $\ddot{q}_{+-}$ as shown in Table 1. Whereas orders of polynomial approximations n_χ and n_υ should be less or equal, too.

D. Void Fraction and Volumetric Quality in Two-Phase Flow

Hughmark's[10] empirical correlation for the Bankoff measure of the global void fraction to volumetric quality ratio

$$Z = Re_u^{1/6} Fr_u^{1/8} (1 - \dddot{\beta})^{-1/4} \tag{13}$$

Table 1 Coefficients of general function [Eq. (12)]

ζ	χ	υ
$\dddot{q}_{+-}$	$\dddot{\beta}$	$\dddot{V}$
p_{x_o}	$\dddot{\beta}$	$\dddot{V}$
$\dddot{q}_{+-}$	p_{x_o}	$\dddot{\beta}$
$\dddot{q}_{+-}$	p_{x_o}	$\dddot{V}$

by Re_u and Fr_u as Reynolds and Froude number at mixture superficial velocity, has a particular interval of Z between 10 and 20, where Bankoff K and Armand C factors for evaluating $\langle\ddot{\alpha}\rangle = K \cdot \dddot{\beta} = \dddot{\beta}/C$ can be taken as constants, that is, $K \approx 0.8$ and $C \approx 1.2$. Hence, in two-phase gas-liquid vertical downflow the Chisholm[11] modification of the Armand relation might be used:

$$\langle\ddot{\alpha}\rangle = \dddot{\beta} / [\dddot{\beta} + (1 - \dddot{\beta})^{1/2}] \tag{14}$$

E. Flow Regions and Stirring Effect in Two-Phase Flow

Ros[10] determined four significant dimensionless numbers, liquid velocity number Π_{u_1}, gas velocity number Π_{u2}, pipe diameter number Π_d, and liquid viscosity number Π_{η_1}, that generally describe two-phase gas-liquid vertical flow problems:

$$\Pi_{u_1} = \ddot{u}_1 \cdot (\rho_1 / g\varsigma)^{1/4} \tag{15a}$$

$$\Pi_{u_2} = \ddot{u}_2 \cdot (\rho_1 / g\varsigma)^{1/4} \tag{15b}$$

$$\Pi_d = d \cdot (\rho_1 g / \varsigma)^{1/2} \tag{15c}$$

$$\Pi_{\eta_1} = \eta_1 (g / \rho_1 \varsigma^3)^{1/4} \tag{15d}$$

In these equations $\ddot{u}_1$ represents the liquid superficial velocity, $\ddot{u}_2$ the gas superficial velocity, d the pipe diameter, η_1 the liquid viscosity, g the gravity acceleration, and ς the interfacial tension. The flow region where $1 < \Pi_{u_1} < 100$ and $1\ \Pi_{u2} < 100$ is treated as a froth flow, whereas for $\Pi_{u2} \leq 50 + \Pi_{u_1}$ liquid is considered as a continuous phase.

The influence of the flow pattern of intense velocity gradients perpendicular to the flow on flowmeter output was already studied by Haacke.[12] In practice this can be proven by applying a stirrer in the pipe as a transformer from froth to annular flow. The following set of six dimensionless numbers, absolute temperature number Π_T, atmospheric pressure number Π_{p_o}, volumetric quality number Π_β, volumetric flow number $\Pi_{\dot{V}}$, absolute pressure number Π_p, and flowmeter output number Π_{q+-}

$$\Pi_T = \dddot{T}_a / \dddot{T}_f \tag{16a}$$

$$\Pi_{po} = \ddot{p}_{o,a} / \ddot{p}_{0,f} \tag{16b}$$

$$\Pi_\beta = \dddot{\beta}_a / \dddot{\beta}_f \tag{16c}$$

$$\Pi_{\dot{V}} = \dddot{\dot{V}}_a / \dddot{\dot{V}}_f \tag{16d}$$

$$\Pi_p = \ddot{p}_a / \ddot{p}_f \tag{16e}$$

$$\Pi_{q_{+-}} = \dddot{q}_{+-,a} / \dddot{q}_{+-,f} \tag{16f}$$

characterizes the conditions between two experiments (that is two series of averaged variables), the one at which the stirrer was running (index a for annular flow of intense velocity gradients perpendicular to the flow) and the other at which the stirrer was not running (index f for froth flow of isotropic velocity gradients perpendicular to the flow). All of these numbers must be analyzed with the respect of variance, where any significant deviation from unity indicates repeatability error (stirrer effect).

III. Experiment

The two-phase flow measurement was limited to the system, where the conducting continuous phase was water (index 1) and the insulating dispersed phase was air (index 2). The measuring system is schematically shown in Fig. 1.

In the system water was circulating from the reservoir, through the pump, filter (dirt 50 μm), control valve, Coriolis mass flowmeter (range 300–6500 kg/h, accuracy $\pm$0.2% o.f.s. $\pm$1% o.r.), ball valve, and nonreturn valve to the Venturi tube.

Air was inserted into the system from the surroundings through the compressor, filter (dirt 0.01 μm, oil and vapor 0.01 mg/m^3), pressure regulator, thermal mass flowmeter (dual range 0–1.6 kg/h, 0–9.0 kg/h, accuracy $\pm$1.0% o.f.s. $\pm$0.5% o.r.), ball valve, and nonreturn valve to the Venturi tube.

Hereafter, the two-phase air-water volumetric downflow was measured by an electromagnetic flowmeter (slave meter), inlet and outlet gage pressure by the pressure gage (range 0–1 bar, class 0.1), global void fraction by the mechanically quick closing ball valves using a balance (range 0–2 kg, accuracy $\pm$0.1 g), inlet and outlet temperature by the resistance thermometer (Pt100, 1/3 B), and atmospheric pressure by the pressure calibrator (accuracy $\pm$0.5% o.r.).

Based on the values of mass flows, mean absolute temperature, and mean absolute pressure (master meters), considering dry air at the inlet and wet

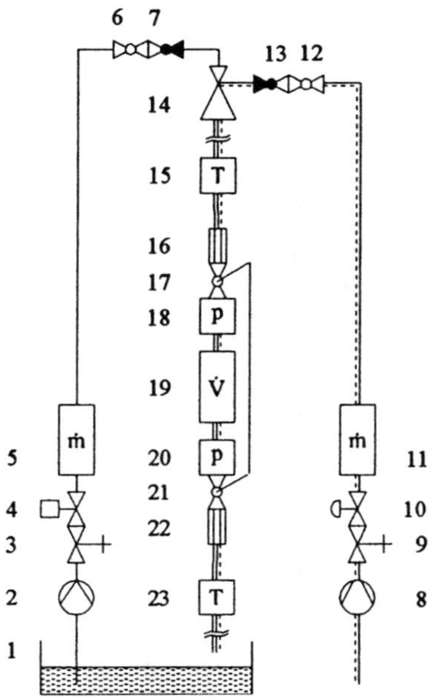

Fig. 1 Schematic of the measuring system reservoir 1; pump 2; filter 3; control valve 4; Coriolis mass flowmeter 5; compressor 8; filter 9; pressure regulator 10; thermal mass flowmeter 11; ball valve 6, 12; nonreturn valve 7, 13; venturi tube 14; resistance thermometer 15, 23; glass tube 16, 22; mechanically quick closing ball valve 17, 21; pressure gage 18, 20; and electromagnetic flowmeter 19.

air at the outlet of the Venturi tube and according to Refs. 13 and 14, volumetric flows of each phase were calculated. All meters were calibrated before the experiment and tested after it.

According to the continuity equation in the period of one measurement, the flow without interfacial transfer inside the electromagnetic flowmeter had to be steady, so that the rate of mass influx would be equal to the rate of mass efflux of each phase. A sample of 25 measurements was 100 s time averaged, long enough to meet the convergence of time and ensemble averages. Residual variance s_r

$$s_r = \sqrt{\frac{\sum_{i=1}^{n} (\zeta_{m,i} - \zeta_{e,i})^2}{n - (n_\chi + 1)(n_\upsilon + 1)}} \tag{17}$$

and relative deviation ε_r

$$\varepsilon_r = \frac{\zeta_m - \zeta_e}{\zeta_e} \tag{18}$$

between measured value (index m) and estimated value (index e) were calculated for individual approximation performed on this sample.

The two-phase flow section of the measuring system is shown in Fig. 2, where all dimensions are in millimeters. Piston valves were mounted because of safety reasons. The stirrer itself was used only to evaluate its effect

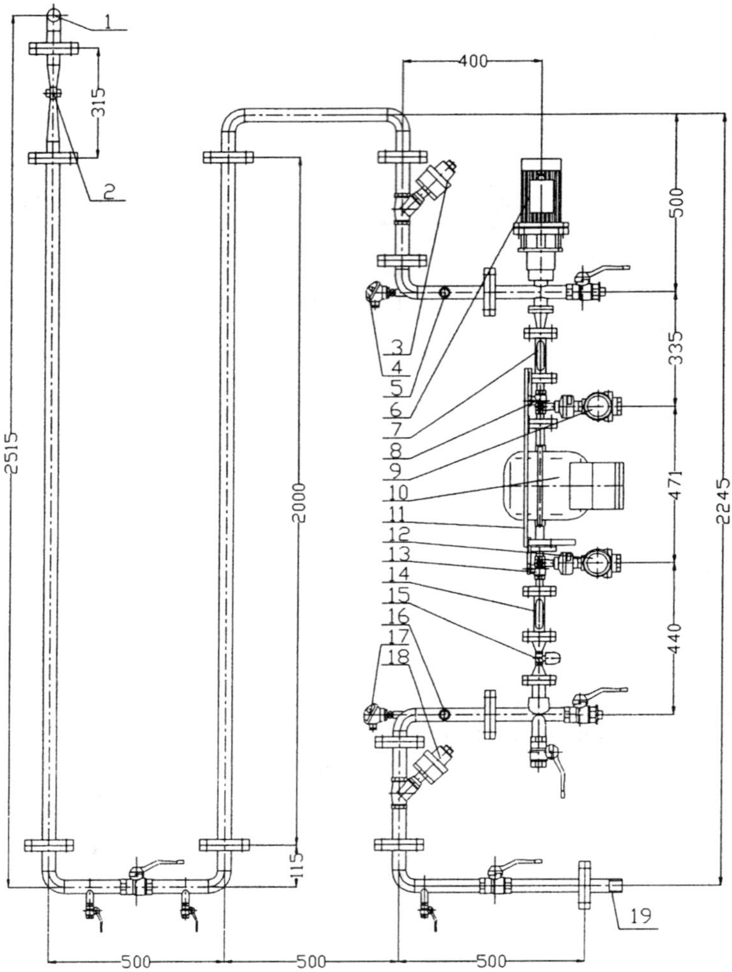

Fig. 2 Two-phase flow section of the measuring system water inlet 1; air inlet 2; piston valve 3, 18; resistance thermometer 4, 17; connection nipple 5, 16; stirrer 6; pneumatically quick closing ball valve 15; glass tube 7, 14; mechanically quick closing ball valve 8, 13; pressure gage 9, 12; electromagnetic flowmeter 10; hand lever 11; air-water outlet 19.

on the electromagnetic flowmeter output. A pneumatically quick closing ball valve was involved in the former investigation of the system dynamics. Glass tubes enabled the flow pattern to define visually. Pressure gage connections and release of mechanically quick closing ball valves were located at the same height to ensure the same scale of observation. The 7-m-long flexible tube went from the air-water outlet first 0.3 m down and then 2 m up to the reservoir.

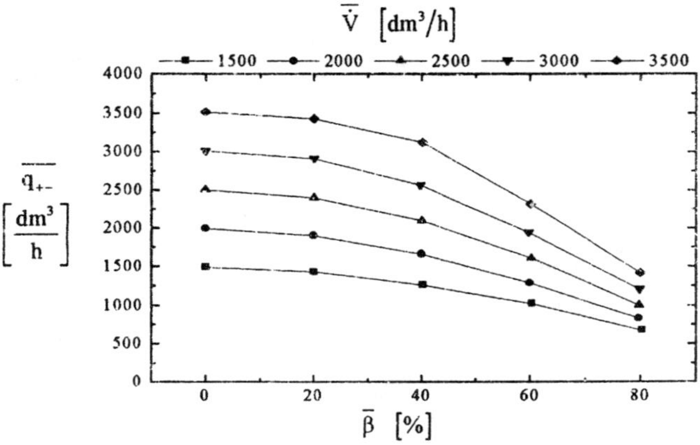

Fig. 3 The relation between $\overline{q_{+-}}$ and $\bar{\beta}$ at various $\bar{\dot{V}}$.

The Foxboro 1/2" flowtube 2800 loaded by the alternated current of 50 Hz was used, whereas the flowmeter transformer was developed in the laboratory by Smrečnik et al.[15] First tests, where the air was injected just before the fluid inlet into this electromagnetic flowmeter, were done by Opara et al.[16] in the air-water upflow on another test section.

A. Induced Voltage in Air-Water Flow

The experimental data shown in Figs. 3 and 4 presenting the relation $\overline{q_{+-}} = \overline{q_{+-}}(\bar{\beta}, \bar{\dot{V}})$ of Eq. (12) were approximated by the polynomial approximation of the orders $n_\beta = 2$ and $n_{\dot{V}} = 2$. The residual variance and relative deviation were calculated using Eqs. (17) and (18), respectively, and got values of $s_r = 27$ dm³/h and $\varepsilon_r \leq \pm 2.6\%$.

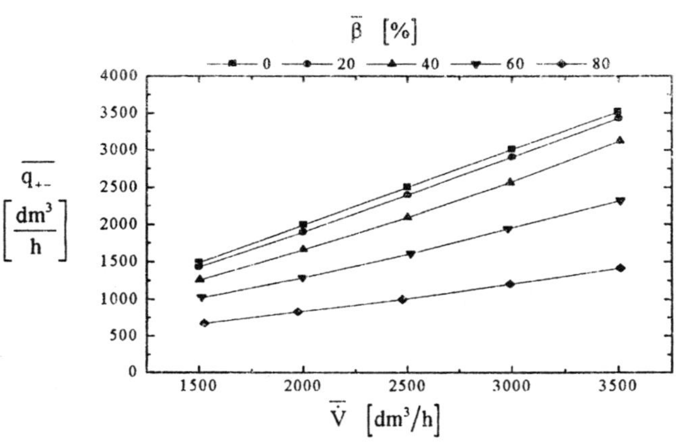

Fig. 4 The relation between $\overline{q_{+-}}$ and $\bar{\dot{V}}$ at various $\bar{\beta}$.

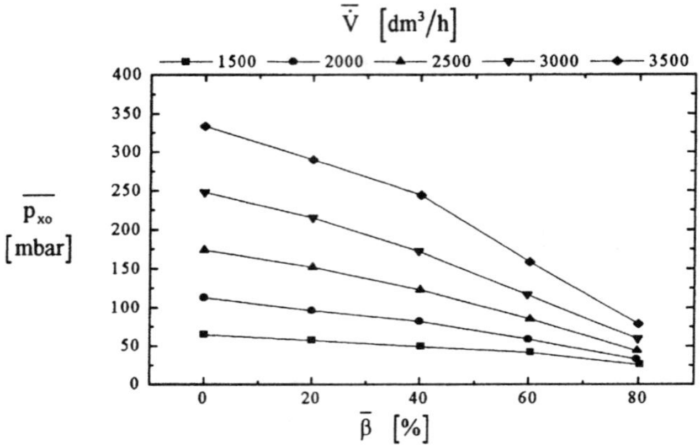

Fig. 5 The relation between $\overline{p_{xo}}$ and $\bar{\beta}$ at various $\bar{\dot{V}}$.

B. Pressure Drop in Air-Water Flow

The experimental data shown in Figs. 5 and 6 presenting the relation $\overline{p_{xo}} = \overline{p_{xo}}(\bar{\beta}, \bar{\dot{V}})$ of Eq. (12) were approximated by the polynomial approximation of the orders $n_{\beta} = 2$ and $n_{\dot{V}} = 2$. The residual variance and relative deviation were calculated using Eqs. (17) and (18), respectively, and got values of $s_r = 3$ mbar and $\varepsilon_r \leq \pm 4.0\%$.

C. Induced Voltage and Pressure Drop in Air-Water Flow

The experimental data shown in Figs. 7 and 8 presenting the relation $\overline{q_{+-}} = \overline{q_{+-}}(\overline{p_{xo}}, \bar{\dot{V}})$ of Eq. (12) were approximated by the polynomial

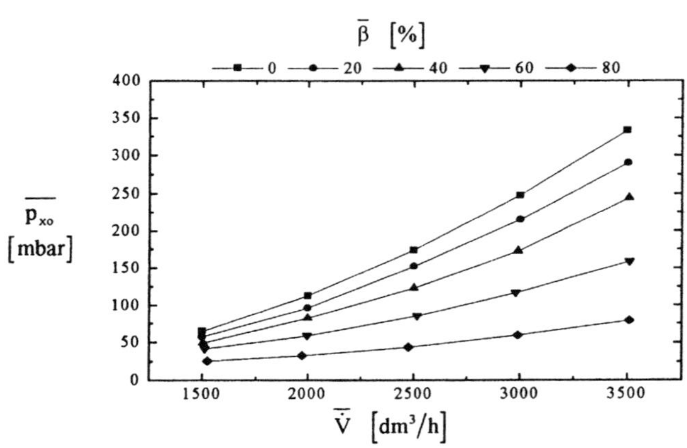

Fig. 6 The relation between $\overline{p_{xo}}$ and $\bar{\dot{V}}$ at various $\bar{\beta}$.

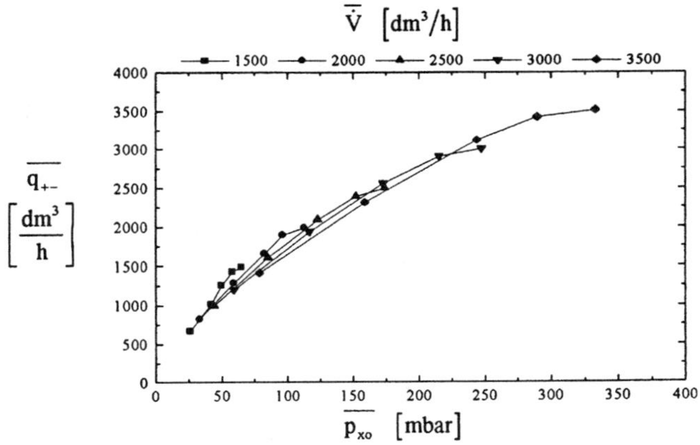

Fig. 7 The relation between $\overline{q_{+-}}$ and $\overline{p_{xo}}$ at various $\bar{\dot{V}}$.

approximation of the orders $n_{p_{xo}} = 2$ and $n_{\dot{V}} = 2$. The residual variance and relative deviation were calculated using Eqs. (17) and (18), respectively, and got values of $s_r = 38$ dm³/h and $\varepsilon_r \leq \pm 3.6\%$. Transition from homogeneous to inhomogeneous flow is obvious.

D. Void Fraction and Volumetric Quality in Air-Water Flow

Hughmark's Z plotted in Fig. 9 indicates that Bankoff's K and Armand's C might be taken as constants, that is, $K \approx 0.8$ and $C \approx 1.2$.

Hence, rather than experimental data, the Chisholm relation shown in Fig. 10 was used to predict the global void fraction $\langle\overline{\alpha}\rangle$ in the air-water downflow that was assumed to be fully developed.

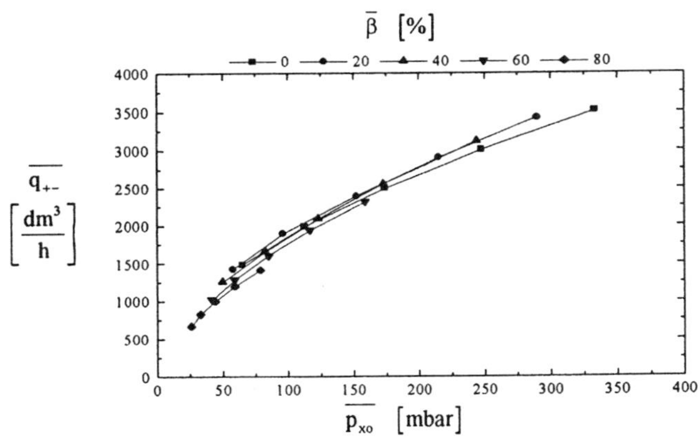

Fig. 8 The relation between $\overline{q_{+-}}$ and $\overline{p_{xo}}$ at various $\bar{\beta}$.

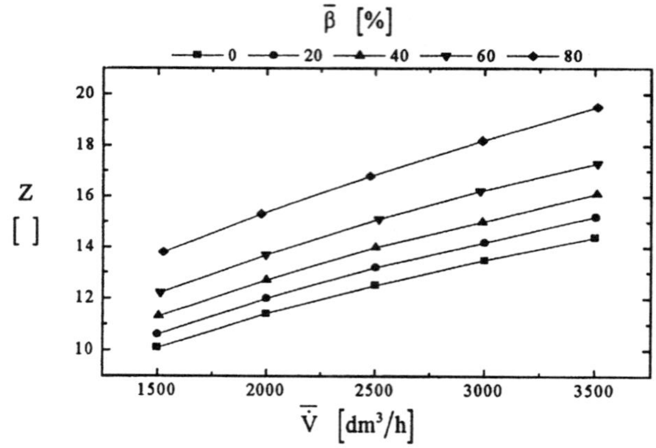

Fig. 9 Hughmark's Z.

E. Flow Region and Stirring Effect in Air-Water Flow

The liquid velocity number Π_{u_1} plotted against the gas velocity number Π_{u_2} in Fig. 11 indicates froth flow, as well as that the liquid might be treated as a continuous phase. The pipe diameter number and liquid viscosity number were constants in the entire region, that is, $\Pi_d = 4.77$ and $\Pi_{\eta_1} = 2.25 \cdot 10^{-3}$.

The flowmeter output number Π_{q+-} plotted against the absolute pressure number Π_p in Fig. 12 indicates stirrer effect because all other numbers, that is, absolute temperature number Π_T, atmospheric pressure number Π_{p_o}, volumetric quality number Π_β, and volumetric flow number $\Pi_{\dot{V}}$, have passed the $F_{0.99}$ test for the null hypothesis of equal means by the value $F = 3.77$ with 3 degrees of freedom in the numerator and 96 degrees of freedom in the denominator.

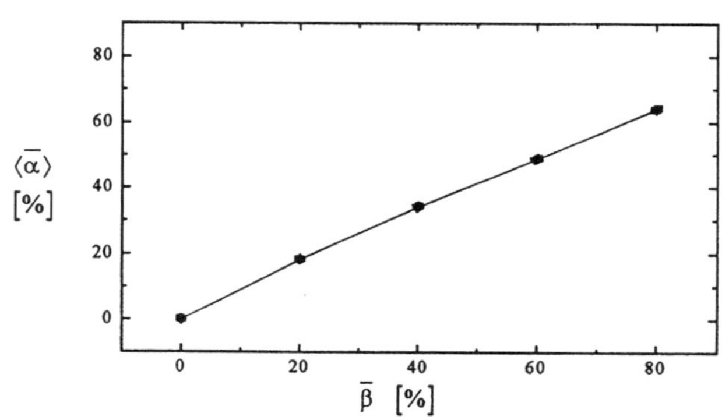

Fig. 10 Chisholm's $\langle\bar{\alpha}\rangle$.

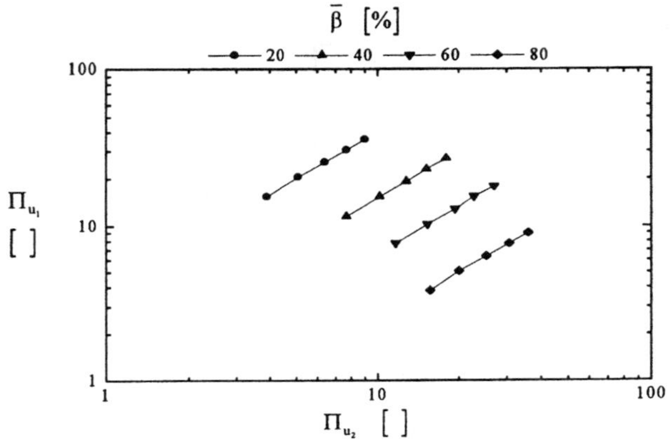

Fig. 11 Π_{u_1} **against** Π_{u_2} **indicates froth flow.**

Deviation of absolute pressure number is caused by the additional pressure gradient perpendicular to the flow when the stirrer is running, which may cause an error in calculating the phase densities. Even so, when this absolute pressure number approaches to unity, a significant deviation of flowmeter output number is apparent. This deviation due to the stirrer is caused by the annular flow pattern of intense velocity gradients perpendicular to the flow.

Let us write explicitly that the stirrer was running only in the case that is shown in Fig. 12. The schematic of the stirrer is shown in Fig. 13. Here, the pipe diameter before reduction was 32 mm and rotational speed of the stirrer 25 s^{-1}.

Flow patterns were also visually inspected using glass tubes of the sighthole length 90 mm and width 20 mm. An example at volumetric quality

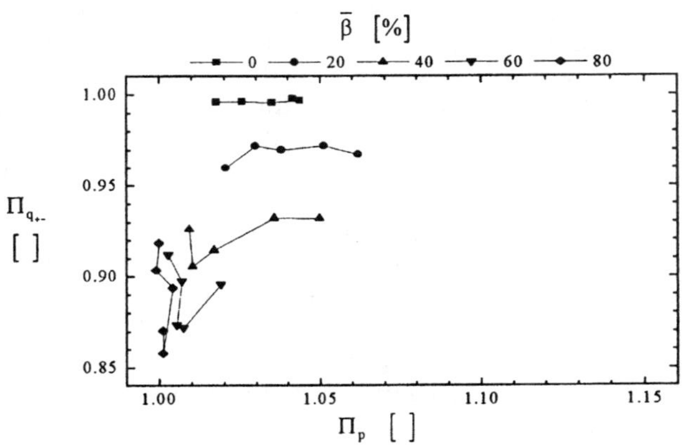

Fig. 12 Π_{q+-} **against** Π_p **indicates stirrer effect.**

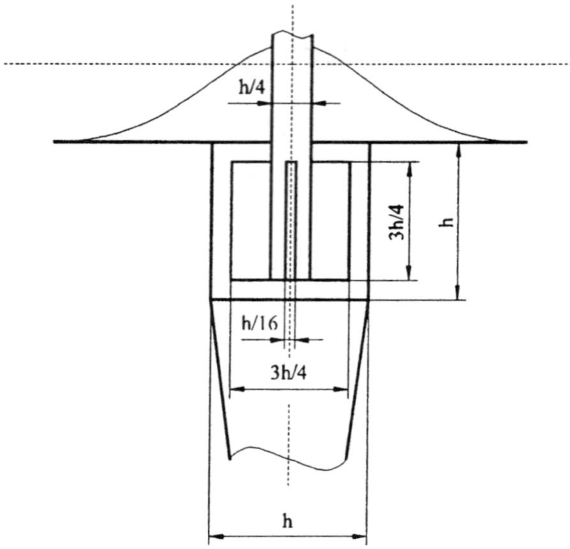

Fig. 13 The stirrer (h = 32 mm).

of 20% and total air-water volumetric flow of 1500 dm^3/h is shown in Fig. 14. Here, the snapshots were taken at the electromagnetic flowmeter inlet. The contrast between froth and annular flow is obvious. We emphasized that this contrast was not so significant at the electromagnetic flowmeter outlet.

IV. Conclusion

A simple approach, based on the assumption of gageable and ungageable phase, was applied to develop a set of three approximations that can be solved for each individual case of two-phase flow measurement with an electromagnetic flowmeter. According to this paper, the fully developed froth two-phase flow will ensure, same as fully developed turbulent single-phase flow, a regular operation of the electromagnetic flowmeter for the same reason of isotropic velocity gradients perpendicular to the flow. In practice this can be achieved by the pipe diameter reduction at a quoted range of minimum upstream and downstream lengths.

Although flowmeter output is physically rather a function of a global void fraction than of volumetric quality, the latter was inserted as a variable in all of these approximations with an intention to make them executable. The major on predicting the global void fraction instead of using experimental data can be found in an averaging problem of applied method data obtained by quick closing valves.

A question of measurement uncertainty and reliability remains open and is a subject of further study.

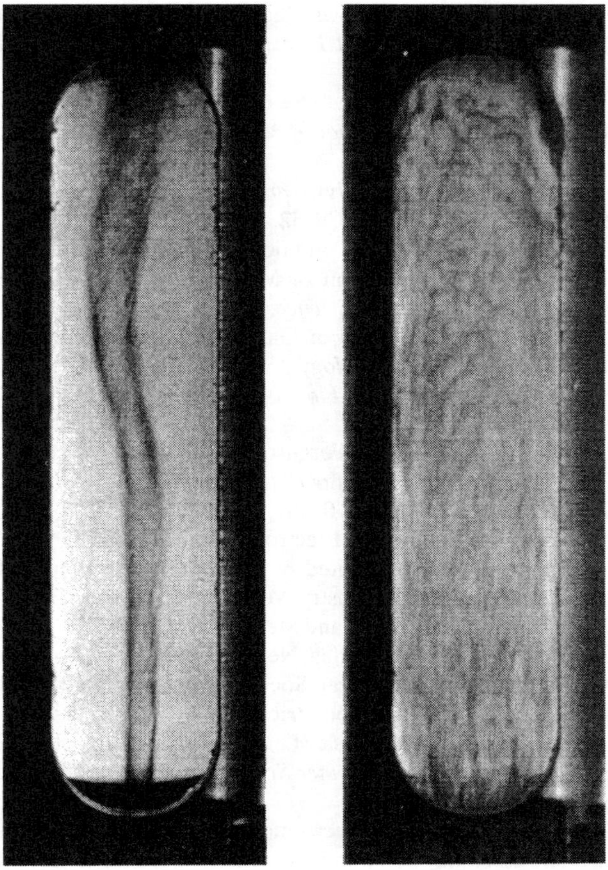

Fig. 14 Flow patterns at $\beta = 20\%$ and $\dot{V} = 1500\ dm^3/h$ (annular flow on the left is developed due to the stirrer).

References

[1]Thibault, J. P., and Seck, B., "Modelling of Magnetohydrodynamic Two-Phase Flow in Pipe," *Liquid Metal Magnetohydrodynamics*, edited by J. Lielpeteris and R. Moreau, Mechanics of Fluids and Transport Processes, Vol. 10, 1989, pp. 119–125.

[2]Tanatugu, N., Fujii-e, Y., and Suita, T., "Electrical Conductivity of Liquid Metal Two-Phase Mixture in Bubbly and Slug Flow Regime," *Journal of Nuclear Science and Technology*, Vol. 9, No. 12, 1972, pp. 753–755.

[3]Baker, R. C., "A Review of Recent Developments in Electromagnetic Flow Measurement," *Liquid-Metal Flows and Magnetohydrodynamics*, edited by H. Branover, P. S. Lykoudis, and A. Yakhot, Vol. 84, Progress in Astronautics and Aeronautics, AIAA, New York, 1983, pp. 225–259.

[4]Shercliff, J. A., *The Theory of Electromagnetic Flow-Measurement*, Cambridge Univ. Press, New York, 1962, pp. 22–27.

[5]Hewitt, G. F., "Oil/Water/Gas Flows Characteristics and Measurement," *Short Courses on Multiphase Flow and Heat Transfer*, Zurich, Switzerland, March 1992, p. 17B.

[6]Bernier, R. N., and Brennen, C. E., "Use of the Electromagnetic Flowmeter in a Two-Phase Flow," *International Journal of Multiphase Flow*, Vol. 9, No. 3, 1983, pp. 251–257.

[7]Zlokarnik, M., *Dimensional Analysis and Scale-up in Chemical Engineering*, Springer–Verlag, New York, 1991, pp. 29–32.

[8]Dobran, F., "On the Consistency Conditions of Averaging Operators in 2-Phase Flow Models and on the Formulation of Magnetohydrodynamic 2-Phase Flow," *International Journal of Engineering Science*, Vol. 19, No. 10, 1981, pp. 1353–1368.

[9]Ros, N. C. J., "Simultaneous Flow of Gas and Liquid as Encountered in Well Tubing," *Journal of Petroleum Technology*, Vol. 13, No. 10, 1961, pp. 1037–1049.

[10]Govier, G. W., and Aziz, K., *The Flow of Complex Mixtures in Pipes*, Krieger, Malabar, FL, 1987, pp. 335–362.

[11]Spedding, P. L., Hand, N. P., and Ferguson, M. E. G., "Prediction of Holdup in Two Phase Gas-Liquid Vertical Downflow," *International Symposium on Two-Phase Flow Modelling and Experimentation*, Rome, Italy, Oct. 1995, 1 (1995), pp. 191–196.

[12]Haacke, A. C., "Sensitivity of the Electromagnetic Flowmeter to Fluid Velocity Profile," *Flow Measuring Devices*, edited by R. E. Wendt, Jr., Flow—Its Measurement and Control in Science and Industry, Vol. 1, No. 2, 1974, pp. 735–743.

[13]Schmidt, E., "Properties of Water and Steam in SI-Units," *The 1967 IFC Formulation for Industrial Use*, Springer–Verlag, New York, 1979.

[14]ASHRAE, *Fundamentals*, American Society of Heating, Refrigerating and Air-Conditioning Engineers, 1993, Psychrometrics.

[15]Smrečnik, A., Fefer, D., and Bajsić, I., "The Electromagnetic Flowmeter (in Slovene)," *Electrotechnical and Computer Science Conference* (Portorož, Slovenia), Sept. 1996, Vol. A, pp. 495–498.

[16]Opara, U., and Bajsić, I., "The Electromagnetic Two-Phase Flowmeter (in Slovene)," *Electrotechnical and Computer Science Conference* (Portorož, Slovenia), Sept. 1995, Vol. A, pp. 353–356.

Chapter 25

Tau Equations: An Alternative to Buckingham's Pi Theorem

Paul S. Lykoudis*
Purdue University, West Lafayette, Indiana, 47907

I. Introduction

THE main weakness of Buckingham's Pi theorem is that one is obliged to know a priori the precise physical properties and parameters on which the answer to the problem posed depends. If this choice is not correct or incomplete, erroneous results are obtained.† Even so the nondimensional groupings that emerge through the Pi theorem are not interrelated to an equation. The researcher is called upon to find an arbitrary equation to fit data within the dimensionless numbers yielded by the Pi theorem. If the fitting is not good, other parameters need to be tried. It is important to know that at no time does the Pi theorem demand an explicit use of the conservation equations where the real physics of the problem reside. To these limitations, one should mention that the Pi theorem is not capable of determining nondimensional groupings in multiphase phenomena. But what is also true is that careful text writers caution their readers heavily on the pitfalls of the Pi theorem, and yet others disregard it completely.

The present work suggests an alternative. The paper is a combination of notions and ideas that were introduced in Lykoudis.[1,2] In the first paper all the nondimensional numbers were expressed in terms of characteristic times thus avoiding ad hoc definitions related to ratios of forces or energies with dangerous pitfalls. This concept has been helpful to scaling problems and other instances. In the second paper the method of average magnitude analysis was presented. This method is a mixture of the integral method and the order of magnitude analysis, a form of scale analysis.‡ The method

*School of Nuclear Engineering.

†An example that illustrates the use and limitations of the Pi theorem is the derivation of Einstein's $E \propto mc^2$ equation. An overzealous judgment is that if Einstein knew of the Pi theorem then his only accomplishment would have been the finding that the coefficient of proportionality is equal to 1.

‡Compare with Bejan's scale analysis.[3]

showed how the differential equations of conservation for laminar, steady-state, boundary-layer flows can be converted to a system of algebraic equations where the result is a sum of the order of magnitude of each term multiplied by a weight coefficient.

In the following section the method of the Tau equations is presented by using the method of average magnitude analysis in combination with the concept of characteristic times. In this union all of the terms in the equation of conservation of momentum, energy, and mass are converted to terms containing only characteristic times. The conservation equations in this form are called "the Tau equations." Depending on the nature of the problem, these algebraic equations are solved in terms of the unknown of the problem such as drag, via δ_M, heat transfer via δ_T, induced thermal velocity, etc. The result is a functional relationship that contains only ratios of characteristic times, where most can be identified with their regular names such as Reynolds, Rayleigh, Schmidt, etc. In Section III a number of problems are solved so that the reader can see how the Tau equations work.

II. Tau Equations

The appropriate nondimensional groupings and their functional interdependence that describe a transport phenomenon are solutions of the Tau equations.

A. Definitions

Equating the order of magnitude of each term in the equation of momentum, energy, and mass transfer with the inertia, heat, and mass convection terms, we define nondimensional characteristic times as follows:

$$t_c = \text{convective time} \propto L/U$$

$$t_f = \text{free fall time} \propto [L \,/\, g(\Delta\rho \,/\, \rho)]^{1/2}$$

$$t_p = \text{depressurization time} \propto L(\Delta p/\rho) \,/\, \rho^{-1/2}$$

$$t_{st} = \text{surface tension time} \propto (\rho L^3/\sigma)^{1/2}$$

$$t_M = \text{momentum diffusion time} \propto L^2 /\nu$$

$$t_H = \text{thermal diffusion time} \propto L^2 \,/\, \alpha$$

$$t_m = \text{mass diffusion time} \propto L^2 \,/\, \gamma$$

$$t_B = \text{electromagnetic rollover time} \propto \rho \,/\, \sigma_e B^2$$

$$t_{v.d} = \text{viscous dissipation time} \propto \nu \,/\, c_p \Delta T$$

$$t_q = \text{internal heat generation time} \propto \frac{\rho c_p \Delta T}{q'''}$$

From a combination of Maxwell's equations, we find

$$t_e = \text{magnetic flux diffusion time} \propto L^2/(1/\mu_e \cdot \sigma_e)$$

B. Equations

We write the equations of conservation of momentum, energy, and mass transfer in conventional notation including the terms of surface tension, ponderomotive force, viscous dissipation, and generation of heat:

$$\frac{\partial(u^2)}{\partial x} + \frac{\partial(uv)}{\partial y} = -\left(\frac{1}{\rho}\right)\frac{\partial p}{\partial x} + \frac{\sigma}{\rho R^2} + \nu\frac{\partial^2 u}{\partial y^2} + g\left(\frac{\Delta\rho}{\rho}\right) - \frac{\sigma_e B^2 u}{\rho} + F \tag{1a}$$

$$\frac{\partial(uT)}{\partial x} + \frac{\partial(vT)}{\partial y} = \alpha\frac{\partial^2 T}{\partial y^2} + \frac{\nu}{c_p \Delta T}\left(\frac{\partial u}{\partial y}\right)^2 + \frac{q'''}{\rho c_p \Delta T} \tag{2a}$$

$$\frac{\partial(uc)}{\partial x} + \frac{\partial(vc)}{\partial y} = \gamma\frac{\partial^2 c}{\partial y^2} \tag{3a}$$

By adding the order of magnitude of each term, we have

$$\frac{U^2}{L} = -\frac{1}{\rho}\frac{\Delta p}{L} + \frac{\sigma}{L^2} - \nu\frac{U}{\delta_M^2} + \rho g - \frac{\sigma_e B^2 U}{\rho} + F \tag{1}$$

$$U\frac{\Delta T}{L} = \alpha\frac{\Delta T}{\delta_T^2} + \frac{\nu}{c_p \Delta T}\left(\frac{U}{\delta_M}\right)^2 + \frac{q'''}{\rho c_p \Delta T} \tag{2}$$

$$U\frac{\Delta C}{L} = \gamma\frac{\Delta C}{\delta_m^2} \tag{3}$$

In the context of the method of average magnitude analysis, we have assumed in the preceding unit weight coefficients. For each of the preceding terms, substitute from the list of characteristic times the quantities Δ_p, σ , ν, g, σ_e, F, α, c_p δT, q''', γ to obtain

$$\tau_c^2 = \tau_p^2 + \tau_{s_t}^2 - \tau_M \tau_c \left(\frac{L}{\delta_M} \right)^2 + \tau_f^2 - \tau_B \tau_c + \tau_{\text{pot}}^2 \tag{4}$$

$$\tau_c = \tau_H \left(\frac{L}{\delta_T} \right)^2 + \frac{\tau_c^2}{\tau_{\text{v.d}}} \left(\frac{L}{\delta_M} \right)^2 + \tau_{q'''} \tag{5}$$

$$\tau_c = \tau_m \left(\frac{L}{\delta_m} \right)^2 \tag{6}$$

In the preceding equations the symbol τ has been chosen to define the inverse of the characteristic times $\tau = 1/t$. The quantity τ is proportional to the characteristic connective or diffusive velocities.*

To Eqs. (4) and (5) we have added the equation of mass transfer (6), which includes characteristic times leading to the Schmidt, Lewis, etc., numbers. One also could have introduced compressibility effects with the introduction of the acoustic velocity and Mach number. Another welcome generalization is to write the drift-flux equations of conservation equations for two-phase problems and generalize all of the above definitions.

One can now observe that in the Tau equation of conservation of momentum the square of the inverse of the convective time is equal to the sum of products of the inverse of two characteristic times formed as follows: For terms corresponding to nondissipative forces such as pressure, surface tension, gravitation, or forces derived from a potential, is the square of the inverse of the characteristic time of the term. For terms corresponding to dissipative forces such as viscous, electromagnetic, and the like, is the product of the inverse of the convective time and the characteristic time of the dissipative term. In the case of the Tau form of the energy and mass transfer equations, a similar statement can be made if these equations are multiplied by τ_c. The terms depending on viscosity, thermal conduction, and mass diffusivity are multiplied with the corresponding $(L/\delta)^2$. However, this ac-

*The author is aware that the Greek letter τ is often used to define time. It is too late to bring forth the argument that this was a poor choice since χ from the Greek $\chi\rho o\nu o\sigma$, (chronos as in chronometer) meaning time, would have been more appropriate. And it will be of little use to mention that the Greek word for velocity is $\tau\alpha\chi o\varsigma$ (tachos as in tachometer) or its modern version $\tau\alpha\chi\upsilon\tau\eta\sigma$, which makes τ appropriate. But if one writes tau with a hook at the bottom, and is willing to identify the leg and the hook with the letter *i*, then this *i* could stand for inverse. One could argue of course for the use of frequency for the inverse of time. But physically frequency makes little sense for the phenomena described here, whereas convective velocity and diffusive velocity are fitting descriptions.

tion is not necessary because when the final result is obtained in order to go from the characteristic Taus to their definitions, one needs to associate for each Tau the appropriate one for that term scale length. This step will be explained later.

With this understanding Eqs. (4–6) based on the same scale length are

$$\tau_c^2 = \tau_p^2 + \tau_\sigma^2 - \tau_M \tau_c + \tau_f^2 - \tau_B \tau_c + \tau_{\text{pot}}^2 \tag{7}$$

$$\tau_c = \tau_H + \frac{\tau_c^2}{\tau_{\text{v.d}}} + \tau_{q'''} \tag{8}$$

$$\tau_c = \tau_m \tag{9}$$

In some problems it is not obvious that there is a characteristic length to define characteristic times. Such is, for instance, the case of two-phase flow phenomena in which the wavelength of the Taylor instability is important. In such instances one has to use a separate physical argument to determine the characteristic length as it will be shown in one example later in this paper.

As already shown in Lykoudis,[1] nondimensional numbers are connected to the characteristic times of Eq. (1), or through their inverses τ as follows:

$$Re = \text{Reynolds number} = \frac{UL}{\nu} \propto \frac{\tau_c}{\tau_M}$$

$$M = \text{Hartmann number} = \frac{BL}{(\sigma_e / \mu)^{1/2}} \propto \left(\frac{\tau_B}{\tau_M}\right)^{1/2}$$

$$N = \text{magnetic interaction parameter} = \sigma_e B^2 L / \rho U \propto \frac{\tau_B}{\tau_c}$$

$$Pe = \text{Peclet number} = \frac{UL}{\alpha} \propto \left(\frac{\tau_c}{\tau_H}\right)$$

$$We = \text{Weber number} = \frac{\rho U^2 L}{\sigma} \propto \left(\frac{\tau_c}{\tau_{\text{st}}}\right)^2$$

$$N_\mu = \text{viscosity number} = We^{1/2} / Re \propto \frac{\tau_M}{\tau_{\text{st}}}$$

$$Ca = \text{capillarity number} \propto\ = We \,/\, Re \propto \frac{\tau_{\text{st}}^2}{\tau_c \tau_M}$$

$$Pr = \text{Prandtl number} = \frac{\nu}{\alpha} \propto \frac{\tau_M}{\tau_H}$$

$$Sc = \text{Schmidt number} = \frac{\nu}{\gamma} \propto \frac{\tau_M}{\tau_m}$$

$$Pr_m = \text{magnetic Prandtl number}\ = \sigma_e \mu_e \nu \propto \frac{\tau_M}{\tau_e}$$

$$Fr = \text{Froude number} = \frac{U}{[Lg(\Delta\rho \,/\, \rho)]^{1/2}} \propto \frac{\tau_c}{\tau_f}$$

$$Ly = \text{Lykoudis number} = \frac{\sigma_e B^2}{\rho.[Lg(\Delta\rho \,/\, \rho)]^{1/2}} \propto \frac{\tau_B}{\tau_f}$$

$$Gr = \text{Grashof number} = \frac{g(\Delta\rho \,/\, \rho)L^3}{\nu^2} \propto \left(\frac{\tau_f}{\tau_M}\right)^2$$

$$Eo = \text{Eotvos or bond number} = (\rho g L^2 \,/\, \sigma)^{1/2} \propto \frac{\tau_f}{\tau_{\text{st}}}$$

$$Ra = \text{Rayleigh number} = \frac{g(\Delta\rho \,/\, \rho)L^3}{\alpha\nu} \propto \left(\frac{\tau_f^2}{\tau_M \cdot \tau_H}\right)$$

$$Bo = \text{Boussinesq number} = \frac{g(\Delta\rho \,/\, \rho)L^3}{\alpha^2} \propto \left(\frac{\tau_f}{\tau_H}\right)$$

C. How the Tau Equations Are Used: Outline of Steps

Here is an outline of the steps that one can take to identify the appropriate nondimensional numbers and the governing equation of a given problem.

Step 1. In the Tau equations eliminate the terms that have no bearing on the physics of the problem. Assign the appropriate algebraic signs depending on the nature of the problem.

Step 2. For each term of the equations, decide what is the relevant length, such as the physical dimension of the geometry of the problem, boundary-layer thickness, Taylor instability length, etc.

Step 3. Solve the Tau equations for the unknowns of the problem.

Step 4. Nondimensionalize the answer and identify the nondimensional groupings.

Step 5. Insert the weight coefficients if they can be determined by the method of average magnitude analysis. If not, insert the minimum number of numerical constants to be determined from experiments or from asymptotic theoretical solutions or a mixture of the two.

Step 6. Use a statistical routine to find the values of the numerical constants for the best fit of the experimental data.

In the next section a number of problems will be solved to illustrate how these concepts are used in the application of the Tau equations.

III. Examples

A. Volumetric Laminar Flow in a Pipe

This is a simple example that serves to illustrate the method. Following steps 1 and 2, the Tau momentum equation is

$$\tau_c^2 = \tau_p^2 - \tau_M \tau_c \tag{10}$$

Assuming steady state the inertia term is zero (step 2). Here for all Taus the characteristic length is the radius R (step 3). We solve for the unknown τ_c (step 3) to find

$$\tau_c = \frac{\tau_p^2}{\tau_M} \tag{11}$$

From the definitions of the Taus, we readily deduce (step 4):

$$Q = R^3 \tau_c \propto R^3 \left(\frac{\tau_p^2}{\tau_M} \right) \propto \frac{R^4 (\mathrm{d}p / \mathrm{d}x)}{\mu} \tag{12}$$

From theory the coefficient of proportionality is $\pi/4$ (step 5).

B. Natural Convection over an Infinite Flat Plate

In this problem the characteristic length for the diffusion terms are the boundary-layer thicknesses and for the convective ones L. The Tau equations are (Steps 1 and 2)

$$\tau_c^2 = -\tau_M \tau_c \left(\frac{L}{\delta_M} \right)^2 + \tau_f^2 \tag{13}$$

$$\tau_c = \tau_H \left(\frac{L}{\delta_T} \right)^2 \tag{14}$$

In Eq. (13) the pressure term has been eliminated because from boundary-layer argument the term combines with the term τ_f. Here we have three unknowns, τ_c, δ_M, and δ_T. Combine Eqs. (13) and (14) to eliminate τ_c (Step 3):

$$\left(\frac{L}{\delta_T}\right)^4 + \left(\frac{\tau_M}{\tau_H}\right)\left(\frac{L}{\delta_M}\right)^4 = \left(\frac{\tau_f}{\tau_H}\right)^2 \tag{15}$$

As shown in Lykoudis,[2] it is legitimate to make $\delta_M = \delta_T$. Identifying L/δ_T with the unknown Nusselt number, we find (Steps 4 and 5)

$$Nu \propto \left(\frac{t_f^2}{t_H^2 + t_M t_H}\right)^{1/4} \propto \left[\frac{t_f^2 / t_M t_H}{1 + (t_H / t_M)}\right] \propto \left[\frac{t_f^2 / t_H^2}{1 + (t_M / t_H)}\right]^{1/4} \tag{16}$$

Using two constants A and B that are combinations of the weight coefficients which were made equal to one for convenience (step 5) and the definitions for the Rayleigh, Prandtl, and Boussinesq numbers, we obtain the familiar equations:

$$Nu \propto A\left(\frac{Ra}{1 + \dfrac{B}{Pr}}\right)^{1/4} \propto AB^{1/4}\left(\frac{Bo}{1 + Pr}\right)^{1/4} \tag{17}$$

Similar results can be obtained for the case of the infinite flat plate where better fractal correlations can be obtained in the fashion of Churchill and Ozoe.[4]

As another exercise, consider the same problem but in the presence of a transverse magnetic field. For very low Prandtl numbers and high magnetic fields, one finds that

$$Nu \propto \left(\frac{Ra}{Ly}\right)^{1/2} \propto \frac{\tau_f^2}{\tau_B \cdot \tau_H}$$

Clearly, by looking at this answer via the Taus one can readily see that viscosity plays no role in the answer as it is proper for liquid metals. The same is of course true if one looks at the answer via the numbers Ra and Ly, which are inversely proportional to V. This is a simple case, but in cases where combinations of more than one name-defined nondimensional number were raised to different powers on what the process on when the answer depends are not readily evident. For this reason it is best to establish the practice of using the combination of taus rather than the combination of the established nondimensional numbers with their names because one sees at a glance the processes involved and, also through the powers, their relative importance.* Such a practice, if adopted, will be most usefull in listing nondimensional parameters in scaling problems.

C. Nucleate Boiling in the Presence of a Magnetic Field

Here we will be concerned only with the determination of the nondimensional grouping that is relevant to the problem.

In the fashion of Lykoudis (1975), consider a spherical bubble that grows against an almost normal to the spherical growth magnetic field of B intensity. For a strong field the inertia forces are negligible, and the ponderomotive force is balanced by the pressure difference between the vapor pressure inside and the liquid pressure acting outside. Equation (4) gives

$$\tau_p^2 = \tau_B \tau_c \tag{18}$$

For this problem where there is a phase change inside a spherical shell of thickness δ growing with the square root of time, one can show that the energy equation (5) becomes

$$\tau_c = \tau_H Ja^2 \tag{19}$$

where

$$Ja = \text{Jakob number} = \frac{\rho_l c_p \Delta T}{\rho_v h_{fg}} \tag{20}$$

Elimination of τ_c from Eqs. (15) and (16) yields the grouping

$$\begin{aligned} \Lambda &= \text{Magnetic Interaction Boiling Parameter} \\ &= \frac{\tau_B \tau_H}{\tau_p^2} Ja^2 = \frac{\tau_B \tau_{Ja}}{\tau_p^2} \end{aligned} \tag{21}$$

To find Λ in terms of physical parameters, we need to replace the taus with their definition. If this is done, we obtain

---

*Adopting a shorthand notation for the combination of taus to indicate the nondimensional combination may be of interest. We propose the following: $Nu = N\,(f^2/B \cdot H)$. In this fashion the Nusselt number dependence on Re and Pr for a flat plate will be indicated as $Nu = Re^{0.8} \cdot Pr^{1/3} = N\,(c^{4/5} / M^{7/15} \cdot H^{1/3})$. Difficult to imagine is that after so many years of usage of named nondimensional numbers the practice will change to the one proposed here. The merit, of course, is obvious. Perhaps one of the readers will propose a better shorthand notion that could be adopted. Those undertaking scaling studies will be more interested.

$$\Lambda = \frac{\sigma_e B^2 \alpha}{\Delta p} Ja^2 = \frac{\sigma_e B^2 \alpha}{\gamma \Delta T} Ja^2 = \frac{\sigma_e B^2 \alpha}{(i_{fg}\rho_v / T_{\text{sat}})\Delta T} Ja^2 \tag{22}$$

In Lykoudis and Wagner (1975) the Λ parameter is successful in correlating experimental data. One should mention the inability of the Pi theorem to discern physical parameters pertaining to two different phases such as ρv and ρ_l as already mentioned in the Introduction.

D. Inception Criterion for Droplet Entrainment Through the Method of the Tau Equations

The Tau equations are now used to develop an inception criterion for droplet entrainment in two-phase concurrent film flow. This problem has been analyzed by Ishii and Grolmes.[5] The Ishii and Grolmes model was based on the roll-wave geometry as it develops at the interface of a thin liquid film flowing over a flat plate while a gas stream flows in the same direction above. Their argument states that "the entrainment becomes possible if the drag force from the high shear flow of gas acting on the wave crest exceeds the retaining force of the surface tension" (Ref. 5).

$$\frac{\rho_g v_r^2}{2} > \frac{\sigma}{\alpha} \tag{23}$$

In the application of this criterion, the viscous forces are included in a heuristic way by introducing a coefficient of proportionality defining the interfacial stresses, which are made proportional to the viscosity number N_μ. Their final result is given in terms of two equations applying at the low and high end of the viscosity number as shown in Fig. 1.

Here is a solution to this problem through the Tau equations resulting in one continuous equation in all ranges of N_μ. From the Tau momentum equation (4) we form the following inequality for inception of entrainment:

$$\tau_c^2 + \tau_M \tau_c > \tau_\sigma^2 \tag{24}$$

The length scale for all terms is the Taylor instability scale, which can be derived by the ad hoc momentum Tau equation:

$$\tau_f^2 = \tau_\sigma^2 \tag{25}$$

The solution for the unknown scale d is

$$d = \text{wavelength of Taylor instability} = \sqrt{\frac{\sigma}{g\Delta\rho}}$$

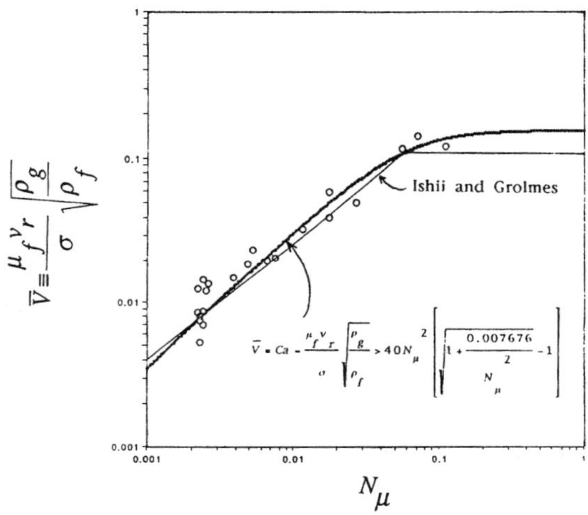

Fig. 1 Comparison of experimental data for inception of droplet entrainment with the Ishii and Grolmes two-line correlation and the continuous correlation of the present work derived through the Tau equations.

Solving Eq. (24) for the ratio $\tau_c / \tau_M \propto Re$, identifying the ratio $\tau_M / \tau_{st} \propto N_\mu$, and inserting the constants A and B as unknown weight coefficients to be determined from experiments, we find

$$Re_{df} > A\left[\sqrt{1+\frac{B}{N_\mu^2}}-1\right]$$

Here the Reynolds number is based on the Taylor instability wavelength and the fluid conditions.

In terms of the Capillarity number Ca_f,

$$Ca_f \equiv \frac{\mu_f v_f}{\sigma} > AN_\mu^2\left[\sqrt{1+\frac{B}{N_\mu^2}}-1\right] \tag{26}$$

For best fit of the experimental data, we determine the values $A = 40$ and $B = 0.007676$ with a correlation coefficient equal to 0.9751.

For small N_μ:

$$Ca_f > (A\sqrt{B})N_\mu$$

Meaning:

$$We_d = \frac{\rho_f v_f^2 d}{\sigma} > 3.504 \tag{27}$$

For high N_μ:

$$Ca_f > \left(\frac{AB}{2}\right)$$

Meaning:

$$Ca_f = \frac{\mu_f v_f}{\sigma} > 0.154 \tag{28}$$

In Fig. 1 the experimental data are compared with the two-line correlation of Ishii and Grolmes[5] and Eq. (26) of the present work.

V. Conclusions and Words of Caution

The Tau equations have been used to reveal the pertinent nondimensional combinations that govern a given laminar flow, mass and heat transfer problem, and also provide solutions to the equations of conservation in terms of unknowns such as heat and mass transfer rates, drag, critical velocities, etc. The method was illustrated in finding solutions to known problems such as natural convection over an infinite flat plate and bubble growth in nucleate boiling, but also in developing a new correlation for the critical velocity for droplet entrainment.

The fundamental thinking is based on collecting not a list of geometric, thermodynamic, and transport properties as is the case with the Pi theorem, but with a list of the taus that are important for a particular problem. On the other hand the Tau equations are not to be understood as a method that solves all imaginable problems. Its success depends on the modeling capability of the user to decide on physical and intuitive grounds as to what are the characteristic times and within them the characteristic lengths recognizing that each term in the equations of conservation does not necessarily admit of the same characteristic length. The method sometimes requires the ad hoc use of the Tau equations more than once, as was seen in the case of the droplet entrainment problem where a separate consideration of the momentum equation was needed to yield the appropriate characteristic length.

Because all the pertinent physical terms are involved in the Tau equations, the solutions obtained go smoothly from one asymptotic solution to the other. As an example in the problem of natural convection over an infinite flat plate, the equation obtained was continuous from the one extreme of low Prandtl numbers, when the Nusselt number depends on the Boussinesq number to the limit of high Prandtl number when the Nusselt number depends on the Rayleigh number. The same was true for the solution of the droplet entrainment problem where at the two extremes the critical velocity was found to depend on the Weber and Capillarity numbers.

References

[1]Lykoudis, P. S., "Nondimensional Numbers as Ratios of Characteristic Times," *International Journal of Heat and Mass Transfer*, Vol. 33, 1990, pp. 1568–1570.

[2]Lykoudis, P. S., "Introduction to the Method of Average Magnitude Analysis and Application to Natural Convection in Cavities," *Journal of Heat Transfer*, Vol. 117, 1995, pp. 604–610.

[3]Bejan, A., *Convection Heat Transfer*, 2nd ed., Wiley, New York, 1995.

[4]Churchill, S. W., and Ozoe, H., "Correlations for Laminar Forced Convection in Flow over an Isothermal Flat Plate and in Developing and Fully Developed Flow in an Isothermal Tube," *Journal of Heat Transfer*, Series C, Vol. 95, 1973, p. 78.

[5]Ishii, M., and Grolmes, M. A., "Inception Criteria for Droplet Entrainment in Two-Phase Concurrent Film Flow," *AIChE Journal*, Vol. 21, No. 2, 1975, pp. 308–318.

[6]Lykoudis, P. S., "Bubble Growth in the Presence of a Magnetic Field," *International Journal of Heat and Mass Transfer*, Vol. 19, 1976, pp. 1357–1362.

[7]Wagner, L., and Lykoudis, P. S., "Mercury Pool Boiling under the Influence of a Horizontal Magnetic Field," *International Journal of Heat and Mass Transfer*, Vol. 24, 1981, pp. 635–643.

VI. Energy Conversion

Chapter 26

Wall-Layer Microturbulence Phenomenological Model and a Semi-Markov Probability Predictive Model for Active Electromagnetic Control of Turbulent Boundary Layers

James C. S. Meng*

Naval Undersea Warfare Center, Newport, Rhode Island, 02841

I. Introduction

IN MOST underwater applications performance improvement is limited by the onset or presence of turbulence. Aside from some results that can be derived through dimensional reasoning, it is still not possible to solve from the basic equations a high-Reynolds-number turbulent flow having even the simplest boundary conditions. This inability to accurately predict the parameters of turbulent boundary layers severely limits technological advancement in vehicle dynamics, including the performance of underwater propulsors and sensors.

Over the past 20 years significant advances in flow visualization, instrumentation, and signal processing have provided a wealth of knowledge and insight into turbulent boundary flows. As has been pointed out by Cantwell,[1] there is a growing realization that turbulent flows are not totally random. The turbulence has been found to have a quasideterministic, repeatable pattern, as well as statistics that are characterized by a remarkable degree of order. The turbulence has been found to be composed of several identifiable events with unique characteristics; these events can be termed "microturbulent" events and their product can be termed "microturbulence." Progress in incorporating a structural model of microturbulent events into practical engineering methods, however, has been slow, and no one has yet explored how this ordered structure is to be connected with a truly predictive methodology. The investigation reported here attempts to make this connection and uses seawater electromagnetic turbulence control as a convenient prac-

*Division Newport.

tical application to test the methodology. Other turbulence control techniques could also benefit from this methodology. The overall intent is to present a new overview of the organized nature of turbulence and a road map for its control.

Turbulent boundary-layer drag is dominated by the turbulence production process, which in turn is characterized by identifiable, discrete microturbulent events that have been established by numerous investigators in the past. The discrete microturbulent events are summarized in Section II where the physical scales and sequential order of events is given in terms of the inner and outer scales of the boundary layer. The scaling relationships of these events, the structural model, and the effects of drag reduction on the microturbulence are discussed in Section III. The complete sequence of microturbulent events occurs on a moving frame and is conceptualized as the basis of the proposed probabilistic control process. Given natural evolution, i.e., no external influence, the transition probability from one event to the next is consistent with the limiting state probabilities, with little physics to reverse the transition. When a Lorentz pressure gradient is imposed, the transition probability is modified. As proposed here, the concept of electromagnetic control of turbulence using Lorentz pressure gradients is directed at affecting the frequency of occurrence of these microturbulent events in a manner that reduces the transition probability toward the mature end of discrete events, thereby interrupting the turbulence production cycle, reducing turbulence production and Reynolds stresses, and consequently reducing overall drag. A probabilistic model of the dependent microturbulent events is described in Section IV. Although there are global indicators of the Lorentz pressure gradient effects, to eventually provide real-time active control of the turbulent boundary layer one must find local indicators that instantaneously characterize the Lorentz effect. Several indicators based on the wall shear stress and wall pressure fluctuations are also described in Section IV. Section V presents conclusions and recommendations.

II. Wall-Layer Microturbulence Phenomenology

Although a large database on wall-layer microturbulence exists, the database has not been organized and presented in a comprehensive and coherent framework that lends itself to practical applications. This section offers a consolidated summary of microturbulence phenomenology with this goal in mind.

A. Near-Wall Microturbulence Phenomenology

Sequence of Microturbulent Events

As is well known, Kim et al.[2] found a quasiperiodic, three-stage sequence of events in the near-wall region of turbulent boundary layers and called the process *bursting*. This bursting is a continuous chain of events starting from a relatively quiescent wall flow; it involves the slow lifting of a low-speed streak, the formation of a relatively large, rapidly growing oscillatory motion, and relatively chaotic fluctuations called *breakup*. Kim et al.[2] also found that low-speed streak-lifting is triggered by large disturbances (implying away from the wall) already present in the flow and, hence, preceding the liftup. Corino and Brodkey[3] found an element of accelerated fluid farther away from the wall that occurred after the appearance of a low-speed wall streak. They called this phenomenon *sweep*, and they suggested that the interaction between sweep and low-speed streaks is fundamental to the ejection process. (Although it is recognized that streaks and near-wall longitudinal vortices are distinct dynamic elements, the two terms are used interchangeably in this article.)

Offen and Kline[4] concluded that the beginning and growth of a disturbance in the outer flow started upstream of and prior to the appearance of a low-speed streak near the wall, and they suggested that each liftup is associated with a sweep. They affirmed that sweep—now accepted and defined as high-momentum fluid that originates in the outer layer and moves toward the wall at a relatively shallow angle—is a disturbance that originates in the logarithmic region and is characterized by a mean motion toward the wall. They suggested that such disturbances are generated by the interaction of an earlier bursting farther upstream with the fluid motion in the logarithmic region. Offen and Kline[4] also gave a complete sequence of events: wall disturbances grow slowly and eventually lift up. Then the velocity field is perturbed in the region directly above the oscillating wall layer. Near the end of the burst's oscillatory growth, interaction between the burst fluid and the motion in the logarithmic region causes the formation of another large vortex-like structure; this structure then creates a large wallward moving disturbance in the outer flow, which is believed to be associated with another liftup process farther downstream, thus completing the quasiperiodic sequence.

Nakagawa and Nezu[5] focused on measuring the streamwise and vertical spatial characteristics of the ejection-sweep motions as a function of $yu_\tau/\nu \equiv y^+$, where y is the normal distance from the wall, u_τ is the friction velocity, and ν is the kinematic viscosity; they concluded that the spatial scale of the sweeps is larger than that of the ejections and, also, that ejections extend more widely downstream while the sweeps extend more widely upstream. Blackwelder and Eckelmann[6] used hot-film sensors to measure both streamwise and spanwise velocity gradients as a function of spanwise dimension, and they presented a detailed map of low-speed streaking, ejection, and sweep.

Clearly, the microturbulent events depend on each other and apparently follow a quasi-periodic sequence. This experimentally observed phenomenology of the microturbulent event sequence forms the basis for the Markov analysis to be presented later.

Head and Bandyopadhyay[7] demonstrated the effects of the Reynolds number (over the range of $500 < Re_\theta < 17{,}500$) on the wall turbulence structure. They made several major points and observations. First, at low Reynolds numbers, they found that the structure is composed of vortex loops or horseshoe vortices and, with increasing Reynolds number, these structures are stretched to become elongated hairpins or vortex pairs. Second, their unique upstream- and downstream-inclined visualization results showed that hairpin vortices, or stretched vortex loops or vortex pairs, are a major constituent of the turbulent boundary layer at all Reynolds numbers investigated and that these stretched vortex loops are substantially straight over a large proportion of their length and are inclined to the surface at a characteristic angle of 45 deg. Third, they found no evidence of large-scale coherent motions, other than a slow toppling or overturning motion, and they found that the large-scale structures appeared to be no more than random agglomerations of much narrower features inclined at something like 40 to 45 deg to the wall. Using the zero time delay correlations between the fluctuating wall pressure and streamwise velocities in the frame of reference moving with convection velocity, Snarski and Lueptow[8] conceptualized the large-scale boundary-layer structure as a primary spanwise-oriented vortex that rotates in the direction of the mean shear with its leading face at an angle of 18 deg to the wall and its trailing face at an angle of 45 deg. Conceivable in any flow visualization experiments is that the convergence of dye may likely reveal the 45-deg -trailing interface rather than the dye divergene leading face of 18 deg, thus leading to conclusion by Head and Bandyopadhyay.[7] Head and Bandyopadhyay[7] suggested that these structures may have their origin in longitudinal vortex motions very close to the wall. Their observations do not change the sequence of wall microturbulent events, although they hypothesize a different picture of that wall turbulence phenomenology, namely, that the wall turbulence is a relatively ordered structure of extended vortex loops.

As to the theoretical basis for the existence of the streaks, which are now accepted as the genesis of microturbulent events, a number of investigators have found consistent results. Bakewell and Lumley[9] obtained a space-time correlation function of the streamwise fluctuating velocities. Using the correlation data as input to an eigenfunction decomposition of the streamwise fluctuating velocity, they found a dominant large-scale structure of randomly distributed, counterrotating eddy pairs of elongated streamwise extent similar to the commonly observed low-speed streaks visualized by wall dye layers. Later, Asai and Nishioka[10] carried out a numerical simulation that explained the origin of the peak-valley wave structure near the wall. More recently, Sabry and Liu[11] applied a time-dependent quasi-two-dimensional formulation to simulate the formation of longitudinal vortices from initial Gortler vortices.

Geometry and Kinematics of Microturbulent Events

From their measurements of the space-time correlation of wall fluctuating velocity components, Kreplin and Eckelmann[12] found the spanwise center-to-center spacings $\Delta z^+ \approx 50$ and streamwise length $\Delta x^+ \approx 1200$ of counterrotating vortices. Using a variable-interval time averaging (VITA) process on data from an array of hot-wire anemometers, Gupta et al.[13] obtained detailed probability density functions (PDFs) of streak spacing over a range of Reynolds numbers and showed that the PDFs follow $\lambda^2 \exp[-(\alpha\lambda^2)]$, the Maxwell distribution, with an average $\lambda^+ \equiv \lambda U_\tau/\nu \approx 100$.

On wall-pressure fluctuations, Schewe[14] determined that the source of wall-pressure structures is located in the buffer region of the boundary layer, the same locale where microturbulent events are taking place. The convection velocity is approximately $12u_\tau$, and the pressure pulses of the same sign are separated at approximately $\Delta x^+ \approx 145$. Therefore, the distance between positive (sweep) and negative (ejection) pressure pulses is approximately $\Delta x^+ \cong 73$.

Kline et al.[15] concluded that the ejection of fluid away from the wall in the subsequent process is the central mechanism for energy, momentum, and vorticity transfer between the inner and outer layers; they presented PDFs for both the spanwise and the mean spanwise spacing with $\gamma^+ \cong 100$, and the breakups occurred in the region $10 < y^+ < 30$, with a breakup frequency F per unit span of $F^+ \equiv (Fu_\tau^3 / \nu^2) \approx 10^{-4}$ in the turbulent boundary layer over a flat plate with a zero pressure gradient. The characteristic frequency was measured to be $2\pi F^+ \lambda^+ = \omega^+ \equiv (\omega u_\tau^2 / \nu) \approx 0.06$, which is consistent with Black's[16] $\omega^+ \approx 0.056$ and Morrison et al.'s[17] $\omega^+ \approx 0.07$. These characteristic frequencies can be converted to duration $T^+ \equiv Tu_\tau^2 / \nu \approx 14$, which is comparable to the $T^+ = 20$ obtained by Kline et al.[15] Kim et al.[2] also presented histograms for the time interval between bursts T_B, i.e., the arrival time of the burst and derived $T_B{}^+ \approx 0.65\ Re_\theta{}^{0.73}$. Laufer and Narayanan[18] analyzed the mean period of the turbulent production mechanism in a boundary layer and found that $U_\infty T_B / \delta / \approx 5$ and is independent of Reynolds number over the range $Re_\theta = 10^3$ to 10^4. Rao et al.[19] verified Kim et al.'s $T_B{}^+ \approx 0.65\ Re_\theta{}^{0.73}$ relationship and found $T_B U_0/\delta^* \approx 32$ independent of the Reynolds number, where U_0 is the freestream velocity and δ^* is the displacement thickness. Bandyopadhyay[29] showed that $U_\infty T_B/\theta \approx 8.13A \times 10^{0.678H}$ and is dependent on the shape factor H, where A is an empirical constant. Falco[21] reasoned the identity of "typical eddies" with Offen and Kline's sweep, and he explained that the constancy of burst frequency across the outer region was a result of the uniform distribution of the average production of typical eddies across the outer region. Schewe[14] showed the pressure fluctuation frequency $\omega^+ \approx 0.52$, which is about an order of magnitude greater than what the shear-stress measurement data reveal. The pressure fluctuation histogram is nearly Gaussian, indicating that it is equally likely for ejection and sweep to take place, further substantiating the hypothesis that ejection and sweep are paired in microturbulent events. This information provides insights for practical applications. Schewe also determined that the pressure transducers of diameter $d^+ \equiv du_\tau/\nu \approx 20$ are required to resolve the pressure structures essential to turbulence. More recently, Löfdahl et al.[22] showed that there was a significant increase in pressure fluctuations with $d^+ < 20$.

When scaled on outer variables, spectra obtained for widely varying flow conditions collapse at low frequencies, suggesting that low-frequency spectral energy is associated with the large-scale structures in the outer portions of the boundary layer.[23] Similarly, when scaled on inner variables, these spectra collapse at high frequencies, suggesting energy at high frequencies is related to small-scale turbulent fluctuations in the near-wall region. Snarski and Lueptow[8] showed that within the boundary layer the spectra are broadband, whereas outside the boundary layer are more band-limited in character, representative of the organized irrotational motion of the turbulent/potential-flow interface. The presence of two characteristic frequencies near the wall can be traced to the two characteristic time scales: the time between events, similar to $T_B \approx 27\,\delta^* / U$ (the lower frequency $\omega\delta^* / U \approx 0.23$) and the event durations $\approx 5\,\delta^* / U_\infty$ (the higher frequency $\omega\delta^* / U_\infty \approx 1.2$). This bimodal spectral structure is found in fluctuation of streamwise velocity, pressure, and pressure-velocity. However from the more complete database, the energy content of the high frequency peak matches that due to the outer boundary layer large-scale motion rather than that from the individual ejection and sweep microturbulent events that correspond to durations on the order of $\approx 0.2\,\delta^* / U_\infty$. This valuable insight provides means to estimate the dynamic power contents of individual ejection and sweep events and those due to the large scale spanwise-oriented vortex motions and statistically periodic burst motion.

Snarski and Lueptow[8] showed that the wall-pressure measurements bring out the fact that there are groups of fluctuations: first is a large-scale low amplitude (low-frequency) that originates from the outer portions of the boundary layer to within the unsteady potential flow consistent with the wavy turbulent/potential flow interface outside the boundary layer; the second group are small-scale large amplitude (high-frequency) disturbances of the burst-sweep cycle. Therefore, they stated that the turbulent boundary layer contains organized coherent fluid motions that are random in occurrence in space and time but similar in character and in general terms has two types of coherent structures or organized motions: first is a quasicyclical ordered sequence of events in the near-wall region that is responsible for the majority of turbulence production in the boundary layer, i.e., the burst-sweep cycle; the second is a large-scale motion in the outer portions of the boundary layer that scales with the boundary layer thickness δ.

Brown and Thomas[24] used spatial correlations of wall shear stress and streamwise velocity as a function of normal distance away from the wall to determine an oblique angle of 18 deg away from the wall in the streamwise direction. Their analysis was based on separating the wall shear-stress experimental data into a slowly varying part due to the large-scale structure and a high frequency part due to the near-wall microturbulent bursting events. They suggested that wall shear-stress fluctuation is directly connected with the bursting phenomenon, and they gave an overall conceptual model of the wall shear stress during the entire cycle from sweep to breakup to lift to a new streak. Offen and Kline[4] showed that the sweeps are visible at $y^+ \approx 100$ and that they move at an angle of 6 deg relative to the plane of the wall. The 6 deg shallow angle is slightly greater than the 4.7 deg tangent angle of the $12u_\tau$ (horizontal) and u_τ (vertical) velocities well reported in the literature.

Using their space-time correlation of fluctuating velocity and normal gradients, Kreplin and Ecklemann[12] found the streamwide velocity component leading the ejected streak outward to be $\approx 13.8u_\tau$, which is essentially the mean velocity at $y^+ = 40$. Eckelmann[25] found that the sweep traveled forward with a nearly constant velocity almost corresponding to the friction velocity u_τ. Morrison et al.[12] measured the two-dimensional power spectra and found the characteristic near-wall streamwise and spanwise wavelengths to be $\lambda_x^+ \approx 630$ and $\lambda_z^+ \approx 135$, respectively. The tan $\beta \equiv \lambda_z^+ / \lambda_x^+ \approx 0.21$, or $\beta \approx$ 12 deg, can be considered to be a half-angle between one upstream sweep and two downstream ejections and vice versa. Bandyopadhyay[26] depicted the wall layer as an agglomeration of successively formed hairpin vortices arranged in a pattern with $\beta \approx 15.3$ deg. This overall geometry has an important implication for the control of microturbulent events. Murlis et al.[27] showed that the burst spacing increases as y/δ decreases, indicating that near the wall the burst spacing ranges from 3δ to 5δ for $Re_\theta \approx 1880$ to 4820, where δ is the boundary-layer thickness and θ is the momentum thickness.

Kim et al.[2] measured the fraction of turbulence production during the burst period and concluded that essentially all of the turbulence produced in the zone $y^+ < 100$ occurs during the burst period. Wallace et al.[28] measured the contributions to the Reynolds stress of the inward-moving accelerated elements of fluid and the outward-moving retarded elements of fluid and reported that they are about the same. Subramanian et al.[29] compared conditional sampling and averaging techniques in a turbulent boundary layer and found that none of the single-point detection techniques (i.e., VITA, HOLE, X-wire) was in good quantitative agreement with the rake detection technique although qualitative agreement did exist between single-point measurements and rake measurements. More interesting is the fact that during the ejection event at $y/\delta = 0.32$ the Reynolds stress is higher than the rms value, while during the sweep event Reynolds stress falls below the rms value. This fact can be very helpful in practical applications. From the preceding observations an overall conceptual schematic of the wall-layer microturbulent events can be depicted and is presented in Fig. 1.

To visualize in a Eulerian frame an event that is truly Lagrangian in nature, one must deal with the issue that definition of the sequence of events would depend on the frame of reference of observation. Based on reported experimental results, most of which are in a Eulerian frame, the microturbulent events observed in a fixed field view are the sweep, followed by oscillatory growth, and then ejection and low-speed streaks. In the Lagrangian frame, the order of sequence is reversed, i.e., ejection-growth/breakup-streaks-sweep. In a true physical sense, the Lagrangian sequence is the real sequence of events. Fundamental issues can be raised about this concept because the microturbulent events are by no means frozen in space and simply transported by advection. However, Nakagawa and Nezu[5] depicted a three-dimensional view of the microturbulent events occurring randomly in space and time with the coherent structure being convected downstream in frozen-turbulence structure, therefore allowing Eulerian observation of the convecting coherent structure. Schewe[14] also found nearly frozen structures that move with convection velocity $u_c = 12u_\tau$, also substantiating that the conceptual coherent structure repeats quasiperiodically.

In summary, selection of the event sequence depends on the application. For fixed point measurement and detection of microturbulent events, the

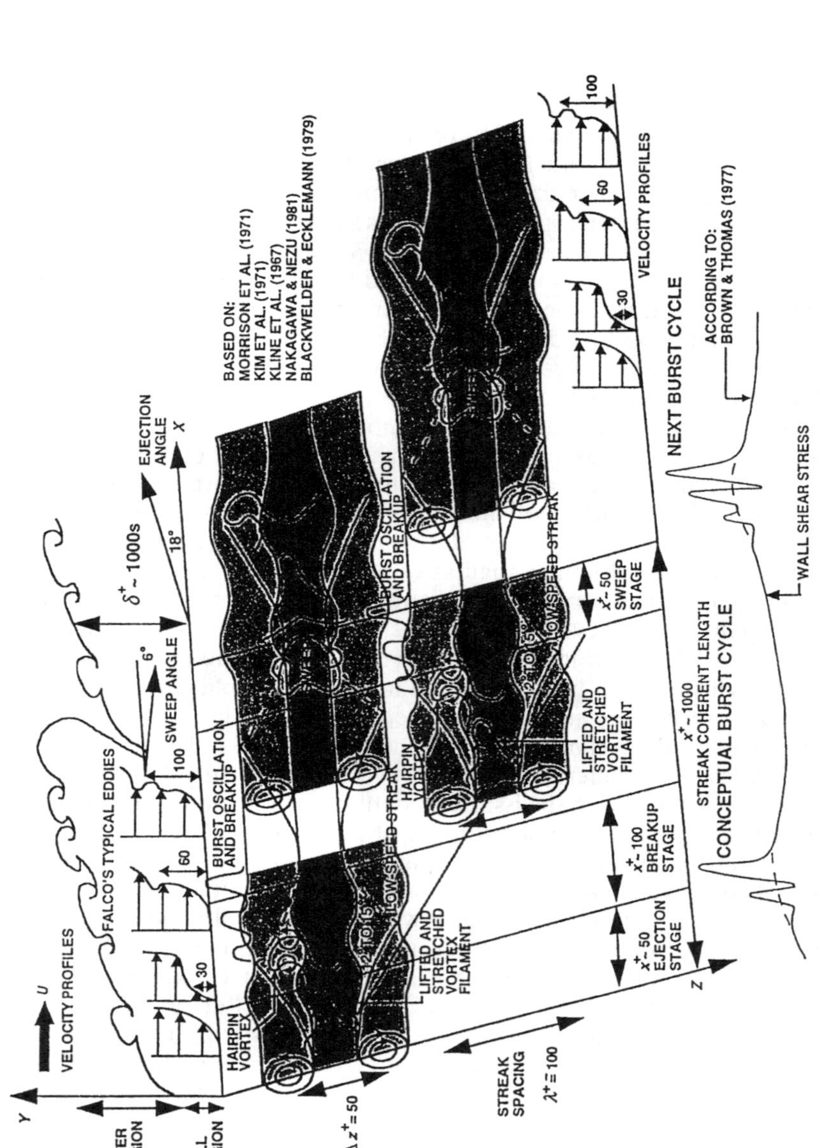

Fig. 1 Conceptual schematic of microturbulence phenomenology.

relevant signal sequence is the Eulerian frame of sequence, namely, sweep-breakup/oscillatory growth-ejection-streak-sweep. This is the sequence to be addressed in the later Markov process analysis (see Section 4).

B. Overview of Wall Turbulence Structural Models

Robinson[30] provided a comprehensive review of the coherent motions in the turbulent boundary layer. Several microturbulence structural models have been proposed; they can be categorized into the following three classes: kinematics, inviscid vortex dynamics, and vorticity layer dynamics with viscosity.

Kinematic Models

Two primary kinematic features of the near-wall layer flow, defined as flow within $y^+ < 100$, have been described by experimentalists as the streaks and the bursting events. The streaks delineate regions where the cross-stream motion converges and the streamwise velocity is in deficit relative to the local mean velocity, with the high-speed flows on both sides of the streaks. Kim and Moin[31] showed that the streak can result from the advection of a single streamwise vortex past a fluid marker in visualization experiments. During the burst event, the discrete ejection is quickly followed by a sweep in which high-speed fluid from upstream undercuts the erupting fluid and the local streamwise velocity exceeds the mean velocity. Ejections and sweeps are narrow-band events with small temporal and spatial (streamwise, spanwise, and normal distance from wall) scales. They are the fundamental means for regeneration and production of new turbulence, and it is clear that any methods of turbulence control must ultimately be based on the interruption of the cyclic events near the wall. Another striking feature of turbulent boundary layers are the strong shear layers protruding away from the wall into the flow, not considered as the near-wall flows, which are associated with high turbulence production. These shear layers have long lifetimes and impart high-amplitude pressure signatures at the wall, which is of particular interest for their detection in an active control algorithm.

Citing past studies of structures of turbulence that have been hampered by a limited number of sensors and limited field of view, flow visualization reveals only the consequences not the cause and the noncorrespondence of spatial and temporal measurements. Moin[32] illustrated via numerical simulations the errors of experimental measurements of near-wall turbulence intensities. More importantly, Moin showed that there are two kinds of vortices which can be identified in his numerical simulation results: those with tips above the legs (ejection) and others with the tip near the wall and upstream of the legs (sweep). His results also showed that there are more solitary vortices than counterrotating vortex pairs, which have a relatively short streamwise extend (100–200 wall units), long lifetimes, travel downstream without losing their identity, and appear to be the fundamental structures associated with regions of high turbulence production. Using a direct numerical simulation of a turbulent channel flow, Kim et al.[33] not only reproduced most of the experimentally established wall turbulence structure, but also showed the most interesting quantitative illustration that sweeps dominates in the near-wall region, i.e., $y^+ < 10$, whereas ejection dominates at $y^+ > 10$ through 100. This implies that when considering effective methodologies for skin-friction control those directly focused on inhibiting sweeps should be a primary candidate.

Kline and Robinson,[34,35] Robinson et al.,[36] and Robinson[30] have grouped the various experimentally observed coherent structures in the boundary layer into eight categories: low-speed streaks, ejections, sweeps, vortical structures (hairpin vortices with elongated trailing legs), near-wall shear layers, near-wall pressure pockets, δ-scale shear layers or "backs," and large-scale motions (horseshoe vortices that are as wide as their height and bulges and deep crevices in the intermittent interface).

Analyzing the near-wall layer flows by categorization of the kinematics is useful but does not yield any cause-and-effect insight into the flow dynamics; a model capable of reproducing the major dynamic features is desired. The following describes some near-wall turbulence structural dynamics models.

Vortical Structural Models

Summarizing four eras of turbulent boundary-layer structure, Robinson[30] compiled the extensive knowledge of vortices and vortical structures in terms of conceptual structural models. As is shown in Fig. 2, Theodorsen[37] proposed a vortex rollup model with vortical form that grows outward with heads inclined downstream at 45 deg that can be considered as the genesis of the near-wall structure model. Willmarth and Tu[38] proposed hairpin-shaped vorticity lines sloped downstream at about 10 deg from the wall, with the dominant element being trailing, nearly streamwise vorticity lines. Black[16] proposed outward-migrating horseshoe vortices shed from a near-wall instability, with multiple horseshoe elements in various stages of growth sharing a common frontlike trajectory in space. Offen and Kline[39] suggested that near-wall bursting processes are consistent with the passage of a horseshoe vortex, so that the feature structure is a lifted and stretched horseshoe-shaped vortex loop. Praturi and Brodkey[40] proposed a model based on the δ-scale shear-layer interface between high- and low-speed fluid upon which large-scale transverse vortices roll up, and outer region transverse vortices induce near-wall ejections and streamwise vortices, as well as bulges in the outer turbulent/potential interface and entrainment of new freestream fluid that appears as sweeps.

To relate a structural model to microturbulent events near the wall, the following discusses the preceding models from two basic physics perspectives.

Inviscid vortex dynamics models. Analyzing their numerical simulation results, Robinson and Kline[41] suggested that the low Reynolds number near-wall turbulence structure is dominated by tilted quasistreamwise vortices in the buffer region. The bursting process as seen from a stationary location is interpreted as the passage of a relatively long-lived, single quasi-

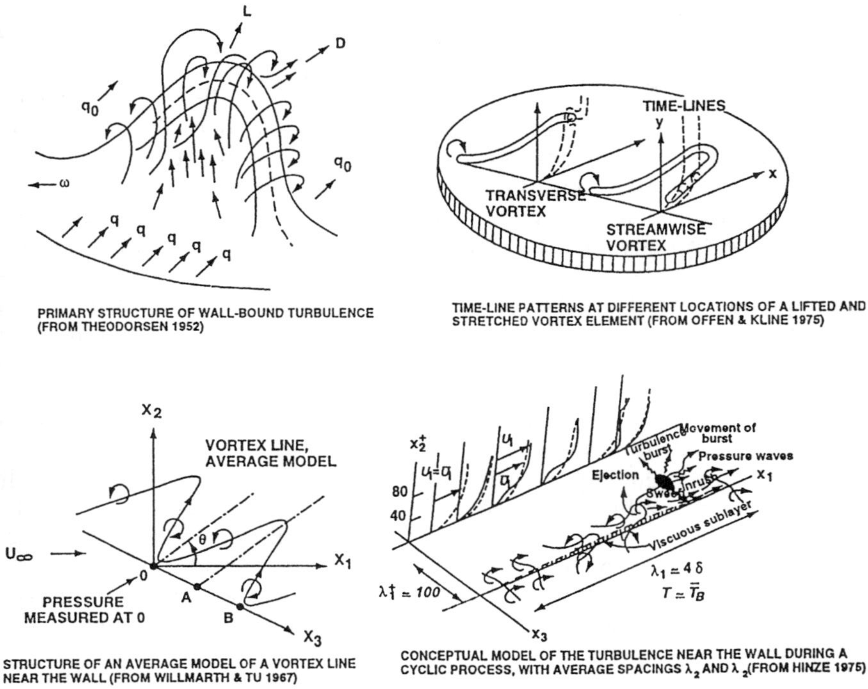

Fig. 2 Historical evolution of vortical structural models prior to 1980.

streamwise vortex, which ejects low-speed fluid away from the wall by vortex induction. Robinson and Kline[41] also concluded that the near-wall and outer portions of the flow structures are independent from each other although some of the events in the inner layers are "phase-locked to and triggered by outer region motions."

Perry and Chong[42] linked flow-visualization observations with the various quantitative measurements and scaling laws from anemometry studies by the concept of the Λ vortex. Tracing back to Townsend's[45] attached-eddy hypothesis, Perry and Chong forwarded the concept that the turbulent boundary layer has a granular structure with a colony of fine-scale potential flow Λ vortices. With the use of the potential flow Λ vortex concept, they formulated a unified theory of wall turbulence linking the mean velocity with the Reynolds shear stress, broadband turbulence, spectra, and even the entrainment processes. The irrotational Λ vortices are formed from wrapping of the sublayer vorticity sheets into vorticity rods. As the Λ vortices are stretched by the flow shear, the rods are brought closer together so that viscous diffusion will lead to vorticity cancellation, leading to the termination of an eddy. The mean spanwise separation of the legs of the Λ vortices follows the Kline scaling. Applying the Biot–Savart law, their numerical simulations illustrate how the observed flow pattern can be reproduced by moving Λ vortices. A systematic derivation was then given to show how a hierarchy of scales of eddies is necessary to extend the region of validity of the logarithmic law of the wall to arbitrary high y^+ values for increasing Reynolds numbers, thus completing the concept's validity of reproducing mean flow profile. Continuing the theme of hierarchies of eddies, broadband turbulence intensities and cross correlations, and spectra of turbulence were derived with favorable results compared with the experimental data.

Viscous vorticity layer dynamics models. Illustrating the time sequence of spanwise vorticity structure, Jimenez et al.[44] advanced a simple mechanism for the spontaneous ejection of vorticity and the associated low momentum fluid away from the wall and into the core. The mechanism is driven by the simple fact that the no-slip boundary condition at the wall must be satisfied during the stretching of the spanwise vorticity layer that leads to splitting and the regenerative formation of new shear layer. This hypothesis can be dubbed as "unsteady surface-layer separation" theory, which can also be named as the viscous vorticity layer dynamics model.

Smith[45] proposed a symmetric hairpin-shaped vortex model (Fig. 3); it explains the streak-bursting process, from oscillation to hairpin vortices to sweeps, in terms of interactions of multiple of Theodorsen's roll-up structure. Apparently unaware of Jimenez et al.'s[44] results, and producing essentially similar results of the theory of "unsteady surface-layer separation," Smith et al.[46] proposed a model simulating the dynamic physical processes in the

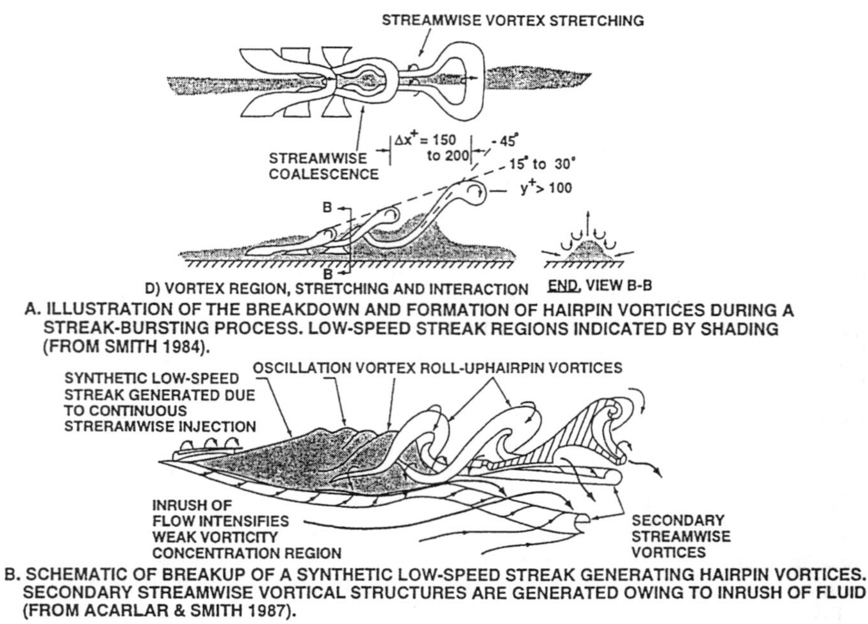

Fig. 3 Conceptual viscous symmetric vortical structural models.

near-wall region of a turbulent boundary layer at high Reynolds number. Their hypothesis is based on the concept that "hairpin vortex" is the basic flow structure in a turbulent boundary layer; all observed behavior of the near-wall flow can be explained in terms of how such inviscid vortices interact with one another, the background shear flow and the viscous flow near the surface.

Smith et al.[46] generalized their earlier symmetric hair-pin vortices to include asymmetric hairpin vortices and showed how the sublayer vorticity is transformed into streamwise vorticity. Summarizing the near-wall Reynolds stress production as result of the eruptive production of new vorticity ejection and a subsequent inrush sweep toward the wall, they illustrated how a two-dimensional vortex can produce a regenerative process of discrete eruption of the viscous flow near the wall and evolve into a sharply focused narrow-band outflow from the surface layer that culminates in the ejection of concentrated vorticity. Their criteria for such an occurrence are that the vortex is sufficiently strong and/or close to the surface for a sufficient period of time and viscosity is the catalyst in conjunction with the local adverse pressure gradient, so that it leads to the appearance of a recirculating eddy and a sharp "spike" of the displacement thickness. Using the Biot–Savart law, they simulated how a single and a group of three-dimensional inviscid hairpin vortices spread and grow in the streamwise and spanwise directions through interaction with the background shear flow. As the trailing legs move progressively toward the wall, more regenerative processes are initiated thus completing the turbulent development. Smith et al.'s[46] cause-and-effect dynamic sequence can be summarized as follows: any local concentration of vorticity above wall leads to formation of hairpin vortices in a shear flow passing of the hairpin vortices and induces streaks; legs of hairpin vortices move toward the wall that generates local adverse pressure gradient creating a local concentration of vorticity thus causing a local separation of wall layer—or an ejection, which evolves into new secondary hairpin vortices, thus completing the regeneration cycle.

In 1990, a conceptual model was proposed in which the streaks, streamwise vortices, near-wall shear layers and pockets of high-wall shear stress are spatially related. These various structural elements are regarded as part of a basic organized structure. In their model, near the wall, the streamwise velocity field comprises low- and high-speed streaks over which two attached eddies (or vortices) with opposite sense of rotation are arranged one above another. These attached eddies near the wall are quasistreamwise vortices. The vortices are stacked in the vertical direction rather than in the spanwise direction, and their configuration is more likely to be asymmetric. The ejection and sweep events can take place either side by side, in the spanwise direction, along a vortex, back to back in the streamwise direction across the single vortex at an oblique angle, or in between two vortices. Mechanisms leading to this configuration are believed to be interactions of the near-wall shear layer and the streamwise vortices with the wall that create new shear layers in the vicinity of the wall, which in turn create streamwise vortices, and the process repeats.

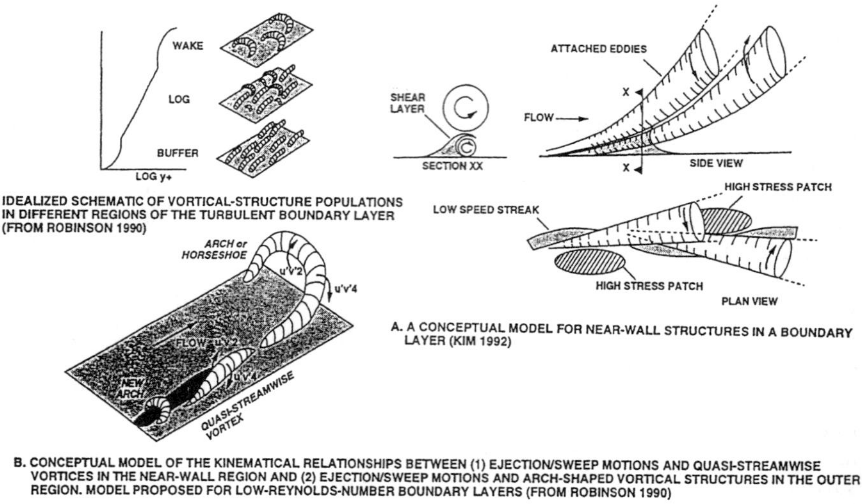

Fig. 4 Conceptual viscous asymmetric vortical structural models.

Further extended from Kim's[47] model (Fig. 4a) Robinson[48] proposed a hierarchical model of vortices (Fig. 4b); with quasistreamwise vortices they dominate the buffer region, and arch-like vortices in the wake region, and both arch-like and quasistreamwise vortices exist in the logarithmic region. Ejections and sweeps are consequences of streamwise vortices. No symmetry is required in Robinson's model. Robinson[30] then proposed that the wall region is populated by relatively strong quasistreamwise vortices with an upward tilt that increases with distance from the wall. These vortices are closely associated with both ejections and sweeps, and they are major contributors to the Reynolds shear stress in the near-wall region. Geometry of these vortices explain why wall-normal velocity fluctuations in the near-wall region are narrow and slightly elongated, and they tend to appear in side-by-side pairs. Buffer-layer vortices are an order of magnitude shorter than streaks in the sublayer. These vortices "drag" through the wall region leaving elongated trails of low-speed fluid from the upward-rotating sides of the vortices.

C. Scaling Relationships of Microturbulent Events

Table 1 presents a detailed comparison of the scales of wall turbulence. As can be seen, the microburst events has attracted the most attention from investigators because the largest part of the total Reynolds stress is produced during the short burst periods.[3] From this table one can compile and extract a number of important microturbulent event scales for later dynamic analysis. First are the dimensionless scales of microturbulent events (see Fig. 5). Note the reasonable agreement between the two curves for dimensionless distance between bursts (these are from Subramanian et al.[29] and Lu and Willmarth[49]). Both yield similar trends and are on the same order of magnitude as the well-known correlation length of the wall layer low-speed streaks. The dimensionless duration of ejection and bursting is $0.8\delta^*/U_\infty$, as given in Table 1 (Ref. 49). Similarly, the time period between bursts is given by Lu and Willmarth.[49] It is important to note that the burst period is about 70 times longer than the duration of ejection and bursting. Because of the microscales of the turbulent events, it is conceivably more difficult to inhibit each individual ejection and bursting over a small spatial scale and short temporal scale. Conceptually, the electromagnetic turbulence control (EMTC) pulse rate should be based on the burst period rather than on the individual ejection and burst durations.

Time Scales

From Corino and Brodkey,[3] the combined ejection-breakup duration can be calculated to be $0.8\delta^*/U_\infty$, which can be expressed in terms of inner scaling as $0.00136Re_x^{0.6}$ where Re_x is the streamwise length Reynolds number. Similarly, the sweep duration in terms of inner scaling can be expressed as $5.1 \times 10^{-4}Re_x^{0.6}$. The burst period is $30\delta^*/U_\infty$ or, in terms of inner scaling, $0.051Re_x^{0.6}$. Figure 6 illustrates the time scales in dimensional form. As flow speed increases for the same length x, the ejection-breakup and sweep durations decrease inversely as $Re_x^{-1.4}$, and the frequency increases as $Re_x^{1.4}$.

Length Scales

Dimensionless length scales have been provided by several investigators. As indicated in Table 1, the combined ejection and burst length scales can be obtained by multiplying the combined ejection and burst duration by the local streamwise velocity at $y^+ = 40$. The local velocity is calculated from the van Driest formula: $u^+ = 5.5 \log y^+ + 5.2 = 14$. Whereas one could question the Reynolds number dependence of these relationships, Purtell et al. established that, at least in the low-Reynolds number range, the law-of-the-wall and the extent of the logarithmic region are independent of the Reynolds number. Therefore, the combined ejection and breakup length is 11.2 $\delta^* u_\tau/U_\infty$ or, in terms of inner scaling, $0.0191Re_x^{0.6}$. Similarly, the burst's periodic length can be calculated to be $420\delta^* u_T/U$ or, in terms of inner scaling, 0.7148 $Re_x^{0.6}$. Lu and Willmarth[49] also gave the burst's periodic length to be 4.3δ, which translates, in terms of inner scaling, into 0.4327 $Re_x^{0.7}$, which is not very different from the expression above over the range of validity of $5 \times 10^5 < Re_x < 10^7$. These scales are illustrated in Fig. 7.

III. Drag Reduction Effects on Microturbulence Events

All of the preceding observed facts are fundamental to the thesis proposed here for the control of turbulence, viz., even though turbulent flow appears to be random and chaotic, the genesis of the turbulence is the near-wall microturbulent events, which have a remarkably quasiperiodic coherent structure, as well as deterministic, predictable statistics. By reduc-

Table 1 Scales of wall turbulence

Mean scales	Transitional flow regime	Turbulent wall region flow regime Kline's burst cycle[15]				Turbulent outer region flow regime	
	Spot	Streak	Ejection (streak-lifting)	Bursting (oscillatory growth and breakup)	Sweep	Typical eddy	Large-scale motion
Streamwise length, x^+	2 to 4δ	1 to 2δ (1200)[a]	0.1 to 0.2δ (20–40)	0.2 to 0.5δ (40–150) (120)[b]	0.1 to 0.3δ (20 to 90)	0.25δ (<200)	1.6 to 4δ
Spanwise length, z^+	2 to 4δ	(100) $\sigma \cong 0.4$	(5 to 50)[c] (10 to 30)[d]	(<100)	(<100)	(200)	0.5 to 1δ
Normal distance from wall, y^+	δ (500)	(0 to 30) (center @ 20–30)	(15 to 20)	(15 to 60)	(<100)	$\leq\delta$ (100) (edge of turbulent/ irrotational flow interface)	δ

Table 1 ***Continued***

Mean scales	Transitional flow regime	Turbulent wall region flow regime Kline's burst cycle[15]				Turbulent outer region flow regime	
	Spot	Streak	Ejection (streak-lifting)	Bursting (oscillatory growth and breakup)	Sweep	Typical eddy	Large-scale motion
Time duration, interval	Indefinite	$T^+ \cong 480$ up to 2500[e] $(F^+\lambda^+)^{-1} \cong 100$	$0.2\,\frac{\delta^*}{U_\infty}$ $T^+ \cong 20$	$0.6\,\frac{\delta^*}{U_\infty}, T^+ = 20 - 40$ $\left(32\frac{\delta^*}{U_\infty}, 6\frac{\delta}{U_\infty}\right)$ $\left(T_B\frac{u_\tau^2}{\nu} = 0.65\,Re_\theta^{0.73}\right)^f$	$0.3\,\frac{\delta^*}{U_\infty}$ $T_+ \cong 20$ $\left(40\frac{\delta^*}{U_\infty}\right)$	N.A.	$2.5\,\frac{\delta^*}{U_\infty}$
Burst rate per span, F	N.A.	N.A.	—	1.5×10^{-4} @ $\frac{dp}{dx} = 0$	—	—	—
Advection velocity in U_∞, u^+	0.5 to 0.8	N.A.	0.4 to 0.8 (12)	0.4 to 0.8 (12)	0.4 to 0.8 (12)	0.95	0.95
Vorticity strength	$\frac{U_\infty}{\delta^*}$ to $\frac{u_\tau^2}{\nu}$	$0.1\frac{u_\tau^2}{\nu}$	N.A.	$1.2\frac{u_\tau^2}{\nu}$	$\frac{u_\tau^2}{\nu}$	$0.4\frac{u_\tau^2}{\nu}$	$\frac{U_\infty}{\delta^*}$
References	64, 65	12,[a] 15, 56,[e] 66	29,[a] 3,[e] 15	2, 6, 12, 15, 19,[f] 27,[b] 49	2, 49	7, 21	21

N.A. = not applicable.

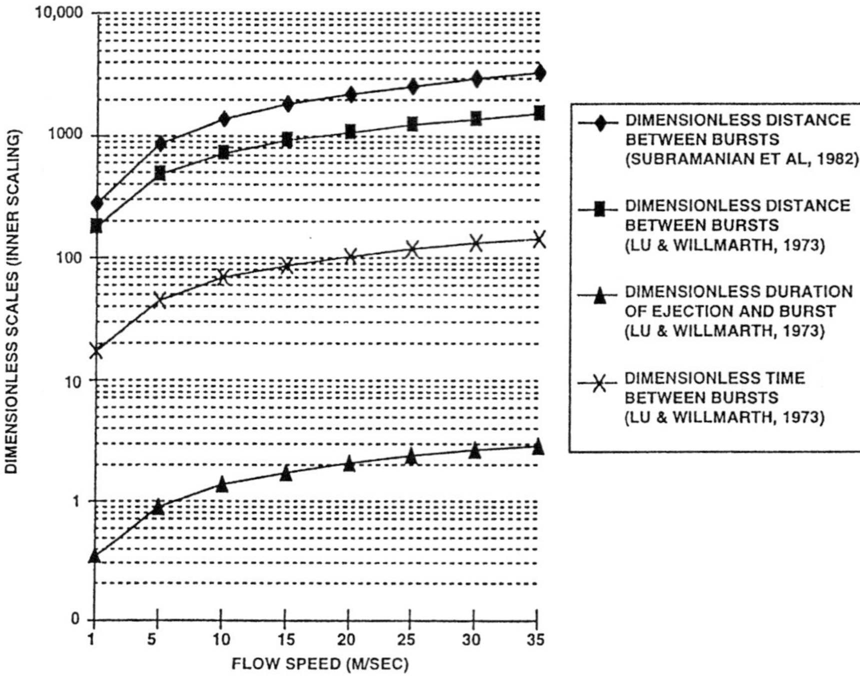

Fig. 5 Dimensionless time and length scales vs flow speed.

ing the clearly identifiable microturbulent events, such as bursting and sweep frequency of occurrence, one can expect to control and reduce turbulence production. However, as was pointed out by Robinson[30] that most statistical descriptions and models of turbulent boundary layer do not address the known existence of the quasiperiodic spatial and temporal characteristics, the intent here is to summarize effects on microturbulent events as a guide for practical applications and practical drag reduction approaches specifically focused on affecting near-wall structures.

Sirmalis[50] observed dyed turbulent boundary layers with low concentrations of polymer and found that the fine-scale turbulence was eliminated, leaving only coarse turbulence, and the boundary-layer thickness was thinned. His observations are consistent with the expected effects of polymer on microturbulent events, i.e., polymer reduces the frequency of the ejection, liftup, breakup, and sweep events. In fact, this thesis has been verified by Tiederman and Luchik,[51] who injected polymer into the sublayer and found increases in λ^+ and in the time between bursts T_B^+. Using wall pressure measurement, they also established that the increase in λ^+ and T_B^+ corresponds to drag reduction.

Kim[47] described a numerical experiment of active turbulence control by suppressing the sweep and ejection events associated with the streamwise vortices using blowing and suction. The results showed that the out-of-phase

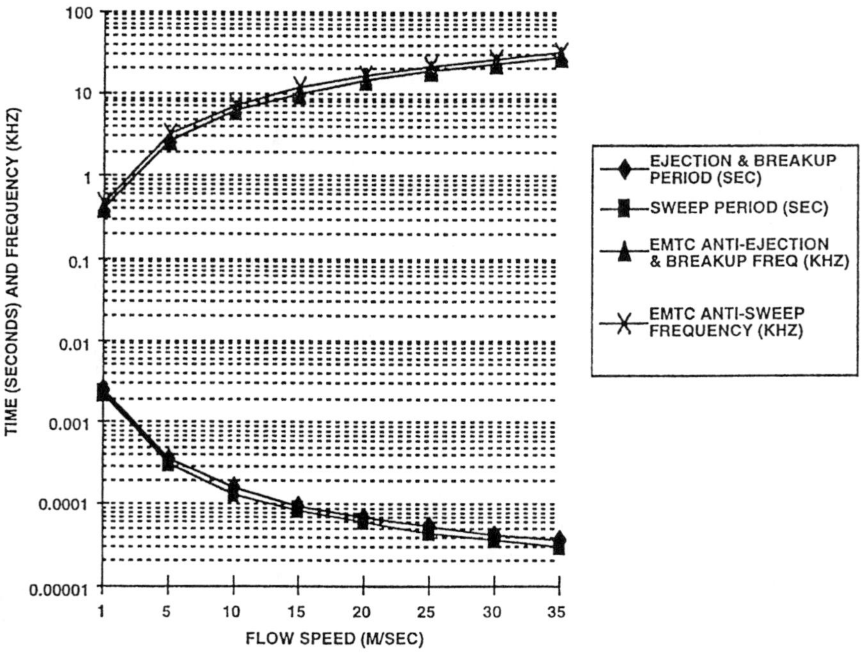

Fig. 6 Dimensional time scales vs flow speed.

blowing prevents the formation of high vorticity regions on the surface. The high vorticity regions are located not only off the surface, but also the strength of the vorticity is reduced, resulting in a reduction of the surface viscous drag. The structure of the wall-layer streaks has also been changed with their strength reduced considerably.

Using well-established numerical simulations, Choi et al.[52] showed significant drag reduction is achieved when the surface boundary condition is modified to suppress the dynamically significant coherent structures present in the wall region. Two key drag-reduction mechanisms were identified: first, deterring the sweep motion without modifying the primary streamwise vortices above the wall so that the high shear-rate regions on the wall are moved to the interior of the flow; and second, stabilizing and preventing lifting of the spanwise vorticity near the wall suppresses a source of new streamwise vortices above the wall. There is an apparent outward shift of turbulence statistics in the controlled flows that indicates a displaced virtual origin of the boundary layer and a thickened sublayer.

Choi et al.[52] found from their numerical simulations with active control that there is an upward shift in the log-law, namely the intercept of the log-law with $u^+ = y^+$ was increased from $y^+ \approx 10$ in the natural state to $y^+ \approx 15$. Peak locations of all turbulence intensities, production, and dissipation were all shifted away from the wall by the same amount. The viscous

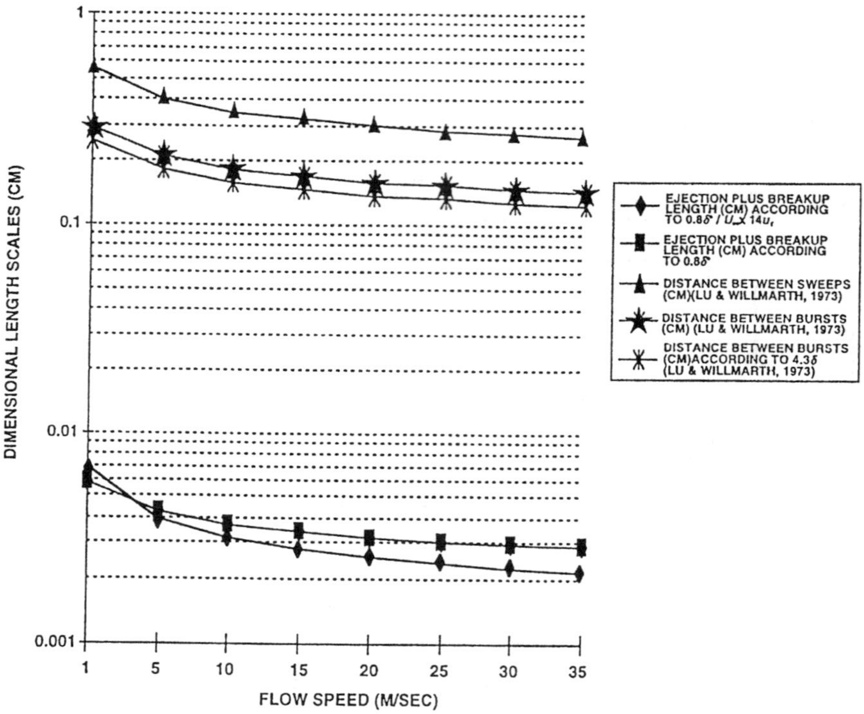

Fig. 7 Dimensional microturbulence length scales vs flow speed.

sublayer is thickened; the displacement thickness is increased while the momentum thickness is decreased, which is related to the skin friction. The streaky structures below $y^+ \approx 5$ were clearly diminished, and the streak spacing above $y^+ \approx 5$ was increased by the control. The local maximum of the streamwise vorticity fluctuations in the controlled flows are further away from the wall compared with that in the natural flows, suggesting that the sweeps are attenuated. Drag is reduced mainly by deterring the sweep motion without modifying the primary streamwise vortices above the wall, so that the high-shear-rate regions on the wall are moved to the interior of the flow ($y^+ \geq 5$). The active control changed the evolution of the wall vorticity layer by stabilizing and preventing lifting of the spanwise vorticity near the wall, thus weakening a source of new streamwise vortices above the wall. They also observed that the active control schemes do not alter the structure of the outer-wall turbulence but simply attenuate its strength and move the effective origin outward.

For passive control, Choi et al.[53] applied direct numerical simulations of turbulent flows over riblet surfaces and showed that riblets mitigate the Reynolds shear stress producing events by restricting the location of the streamwise vortices above the wetted surface such that only a limited area of the riblets is exposed to the sweeps and by impeding the spanwise cross-stream flows necessary to replace the near-wall fluid ejected away from wall during ejection. Effects on the mean velocity and turbulence statistics are limited to the inner region of the boundary layer. The observed global effects are upward shift in the log-law velocity profile, increased sublayer thickness, a displaced virtual origin of the wall, reduced momentum thickness, and increase in mean streak spacing. The upward shift of log-law velocity profile has been well observed in polymer flows. Through their simulations Kim et al.[33] found that the center of the streamwise vortex is located on average at $y^+ \approx 20$ with a diameter about $d^+ \approx 30$, so that the optimal spacing of the riblets is $s^+ \approx 20$ with included angle of 60 deg. A very interesting observation from Choi et al.'s[53] numerical simulation is the fact that sweep dominates Reynolds stress production near the wall while ejection dominates away from wall. This has an important implication that efficient wall skin-friction reduction methodologies should focus solely on reducing sweeps rather than both primary microturbulent events.

For active control, comparing the results of the active blowing and suction[52] with those of the unsteady blowing and suction,[54] Choi et al.[52] showed the former has a significant effect on turbulence statistics away from the wall whereas the latter has only marginal effects in the interior of the flow. The difference appears to be due to the use of a feedback control. Even in cases where the mass input at the wall is applied passively at the bursting frequency, useful interaction may not take place between control inputs and flow structure because of the randomness of turbulence structure in space and time.[55]

Choi et al.[52] applied numerical simulations to investigate a variety of strategies for active control of dynamically significant coherent structures to achieve skin-friction reduction. They found that wall pressure alone is not an adequate detector of the flow toward the wall or away from it. Surface shear stress correlates better with the normal velocity although the best indicator is the spanwise derivative on the wall of the normal gradient of spanwise velocity, which has little practical application potential.

IV. Markov Transition Probability as a Diagnostic Tool for Characterizing and Controlling Wall Microturbulent Events

A. Synopsis of Probabilistic Model Results of Wall-Layer Microturbulent Events

Microturbulence Probability Density Functions

Based on visualization results, Nakagawa and Nezu[5] stated that the probability of the occurrence of the streak spacing follows a log-normal distribution. Smith and Metzler,[56] employing hydrogen bubble wire flow visualization, confirmed this log-normal distribution of the streak spacing instead of the Maxwell distribution of Blackwelder and Eckelmann.[6] Smith and Metzler[56] stated that the streaks maintain their integrity and reinforce themselves for time periods up to an order of magnitude longer than the observed burst periods.

Kim et al.[2] obtained histograms of burst frequency versus y^+ and established that there is a close association with the turbulence production versus y^+ and that both distributions appear to be similar to a Rayleigh distribution. Rao et al.[19] suggested that the time between bursts follows a log-normal distribution. Bark[57] applied Rao et al.'s[19] results of the log-normal law of the time between bursts and assumed that the temporal and spatial structure is the same for all bursts with the strengths following a Poisson distribution.

Murlis et al.[27] obtained the conditional PDFs of turbulent zone lengths in the intermittent region; this intermittent turbulent zone length can be a qualitative representation of the burst's streamwise dimension. The peak PDF zone length $L^+ \approx 120$, which is half the length of a typical eddy, can be considered the most probable burst length and was found to be independent of Reynolds number. The burst length PDF appears, to the present author, to follow an Erlang distribution.

Eckelmann[25] presented the PDF of streamwise velocity, which is highly asymmetric near the wall. Mitchell and Hanratty,[57] who used an electrochemical wall shear stress meter in a pipe flow, obtained the PDF for the axial velocity over the entire velocity range and showed that it has a near-Gaussian distribution with a small skewness toward higher velocity. Bark's energy spectra of the streamwise velocity fluctuations were in agreement with measurements by Morrison et al.[17] Using a nine-sensor hot-wire vorticity meter, Vukoslavcevic et al.[58] measured the PDF of three vorticity fluctuations and showed that the spanwise component in the near-wall region, as close to the wall as $y^+ = 11.1$, is quite negatively skewed indicating that intense spanwise vorticity stretching predominates over vorticity compression there. Their results also show that the velocity and wall pressure spectra are very similar.

Sreenivasan and Antonia[59] presented the normalized wall shear stress PDF and showed that it is not Gaussian. They also compared the normalized power spectral density of wall shear stress versus frequency obtained by a number of measurements. Schewe[14] verified the histogram of wall-pressure fluctuation to be very nearly Gaussian, indicating that wall-pressure fluctuations encompass effects from both near-wall motions and large-scale motions away from the wall. From a statistical modeling point of view, this means that the wall-pressure PDF is the convolution of many PDFs of many preceding events and, therefore, should follow a Gaussian distribution. Blake[60] obtained the $p'^2/q \sim 3.4C_f$, where $q = 1/2\rho U_0^2$, over a range of Reynolds numbers from 8000 to 28,000; this relationship can also be used to calibrate the wall-pressure and shear-stress measurements.

Probabilistic Models of Microturbulent Events

Inspired by Kim et al.'s [2] results, Kovasznay[61] proposed a model of linear superposition of randomly distributed deterministic structures with a Poisson distribution of arrival statistics. Bark[57] derived a probabilistic model of the turbulent stresses based on a random system of bursts using an idealized model of the joint probability distribution in time and space of the occurrence of bursts. Unfortunately, no further development of similar models followed.

The known mathematical significance of the Rayleigh probability distribution, the ubiquitous log-normal distribution (see for example Aitchison and Brown[67]), the Poisson distribution, and the normal distribution can be applied to infer a possible functional relationship among the random variables. Such a relationship may offer additional insight into interpretation of the experimental data. It is well known that if two random variables x and y are normal and independent with zero mean and equal variance, then the random variable $z = \sqrt{(x^2 + y^2)}$ follows a Rayleigh distribution. If the random variable x follows a normal distribution, then the variable $z = e^x$ follows a log-normal distribution. A log-normal distribution also indicates that the random variable's variations are proportional to the variable itself. If two random variables are Poisson-distributed with parameters λ and μ, then their sum is also Poisson-distributed with parameters $\lambda + \mu$. If two random variables are normal-distributed, then their sum is also normal-distributed. By the central limit theorem the sum of multiple random variables of any distribution will nearly follow a normal distribution. It is useful to apply these facts to interpret the microturbulence PDFs, which is the focus of this section.

Table 2 summarizes the known PDFs of microturbulent random variables as an aid to inferring possible relationships among them. As can be seen, the wall shear stress, which is the sum of shear stresses in the streamwise and spanwise directions, follows a Rayleigh distribution. The low-speed streak spacing, the time between bursts, and the time between sweeps follow a log-normal distribution, whereas the burst length in the intermittent region follows an Erlang distribution. Random variables that follow a normal distribution are the axial velocity and its normal gradient, the spanwise velocity and its normal gradient, and the wall-pressure fluctuations. In the discussion that follows, several examples of experimental observations and probabilistic interpretations of the relationship among these random variables are presented to illustrate the insight a probabilistic model can offer.

Table 2 Known PDFs of microturbulent random variables

Generic PDF			
Rayleigh	Erlang	Log-normal	Gaussian (normal)
Frequency of burst vs y^+	Burst length in the intermittent region	Low-speed streak spacing	Axial velocity and normal gradient at $y^+ \cong 3$
Turbulence production vs y^+		Time between bursts	Spanwise velocity and normal gradient
Wall shear stress		Time between sweeps	Wall-pressure fluctuations

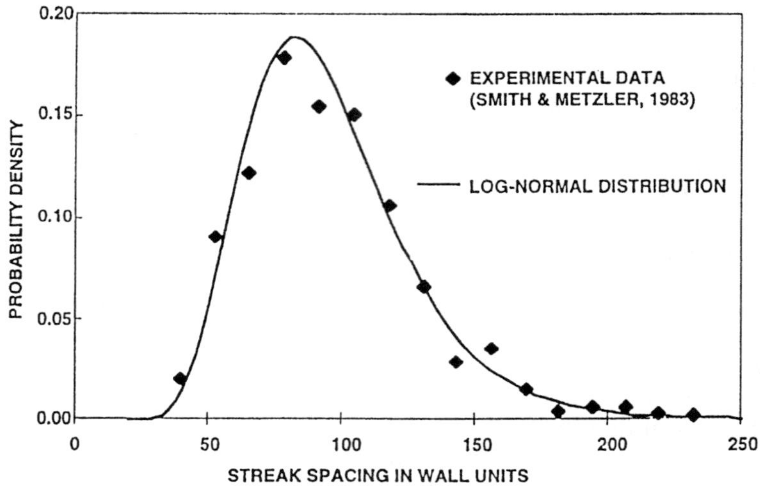

Fig. 8 Probability density function of streak spacing.

Figure 8 illustrates the fact that the low-speed streak spacing follows a log-normal distribution. The experimental data and the fact that the spacing follows a log-normal distribution were obtained by Smith and Metzler.[56] The mean streak spacing in terms of wall units is approximately 100. The probabilistic meaning of the streak spacing following a log-normal distribution can be traced to the fact that the spacing is approximately twice the cylindrical vortex core radius. The core radius is inversely proportional to the spanwise cellular motion velocity, which is proportional to the distance from wall. In other words, the fluctuation of low-speed streak spacing is proportional to the spacing itself; thus, its PDF should follow a log-normal distribution.

Figure 9 compares experimental data for the time interval between bursts; it shows, along with a log-normal distribution, data from Kim et al.[2] for $Re_\theta = 660$ and 1100 and data from Rao et al.[19] for $Re_\theta = 620$. Among several distributions tried—Erlang, Maxwell, Rayleigh and log-normal—the log-normal distribution provided the best fit to the data. Only the distribution best fitted to the Kim et al. data[2] with $Re_\theta = 1100$ is shown here; a better fit to the other two data sets can be obtained, but it was deemed not necessary for the purposes of this report. It can be reasoned that T_B, which describes the time interval between two arrivals, should follow an Erlang distribution with $n = 2$. The data clearly indicate that this is not the case implying that a scenario with bursts arriving perhaps from neighboring ailes of microturbulent events dominates the arrivals, rather than the conceptual scenario of an orderly advection of bursts from upstream.

Figure 10 compares the instantaneous streamwise velocity PDF obtained by Eckelmann[25] with a log-normal distribution. These data were analyzed and interpreted for their probabilistic meaning. Here the data have been normalized by the local mean streamwise velocity to allow comparison of velocities from two y^+ heights; the best fit was found to be a log-normal

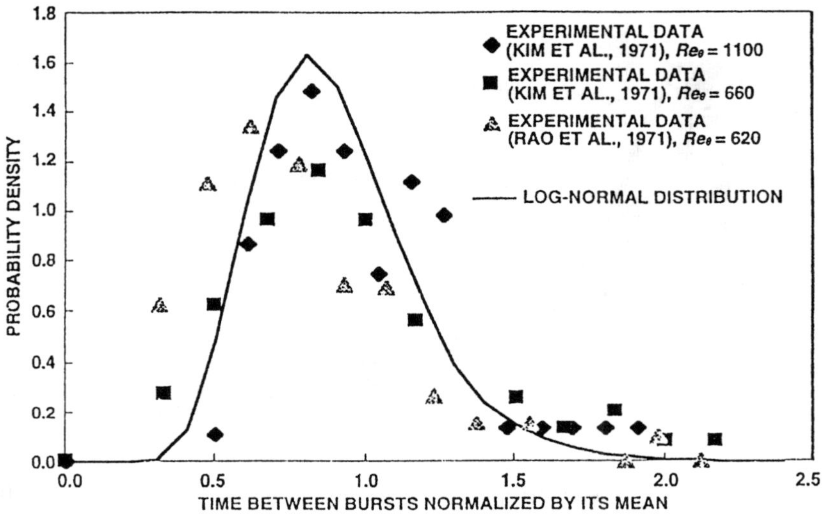

Fig. 9 Probability density function of time between bursts.

distribution. Because the mean streamwise velocity in the wall region is known to vary linearly with distance y^+, the streamwise velocity fluctuation also varies with the local streamwise velocity; thus, its PDF should follow a log-normal distribution.

Figure 11, from Murlis et al.,[27] shows experimental data for the burst length in the intermittent region along with an Erlang distribution. Because the burst length can be interpreted as the distance between ejection and

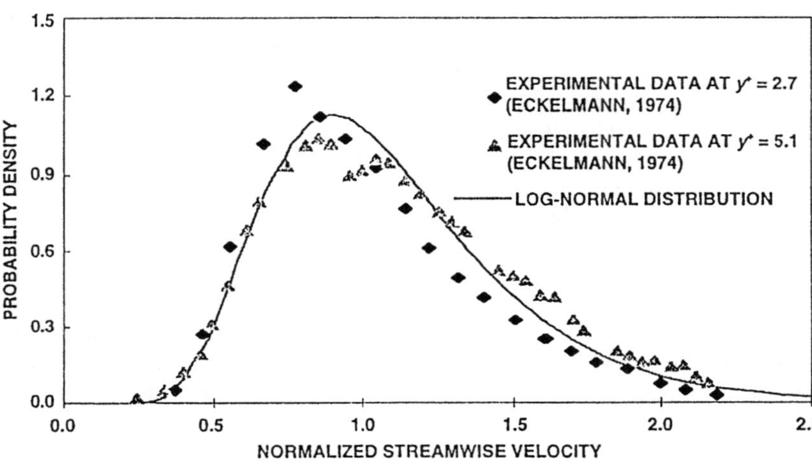

Fig. 10 Probability density function of instantaneous streamwise velocity normalized by local mean velocity.

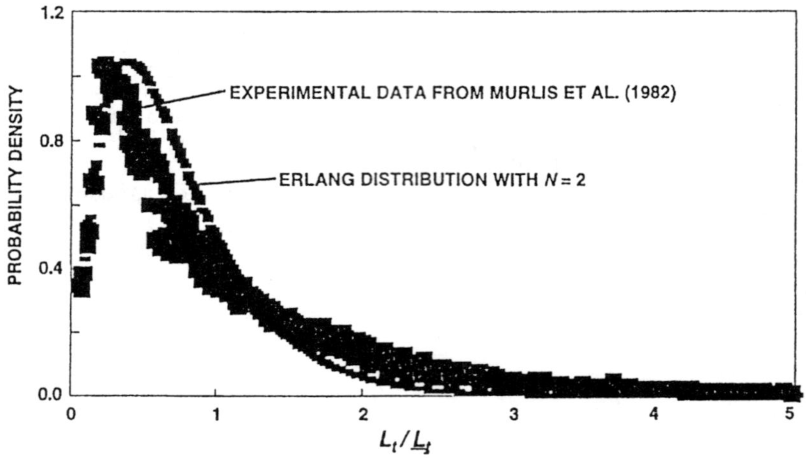

Fig. 11 Probability density function of burst length compared with Erlang distribution.

sweep, it is therefore the renewal process length from the arrival of ejection to the end of sweep. The arrival of ejection and sweep can be conceived as random variables following a Poisson process; thus, the burst length in the intermittent region should follow an Erlang distribution of order $n = 2$. As is shown in Fig. 11, the experimental data reasonably follow an Erlang distribution, illustrating the plausibility of a Poisson process for the ejection and sweep events, so that one can experimentally determine the arrival rate of each event that is near the wall and hence be able to predict the burst length in the outer (intermittent) region away from the wall.

In Fig. 12, experimental data for the wall shear stress (from Ref. 59) are shown together with a Rayleigh distribution. The data again reasonably conform to a Rayleigh distribution, although one can probably find a closer fit than that shown. Because, as has been demonstrated by Vukoslavcevic et al.,[5] the streamwise shear stress $\partial u/\partial y$ and the spanwise stress $\partial w/\partial z$ both follow nearly normal distributions and the wall shear stress is the square root of the sum of the squares of the streamwise and spanwise components, it is expected from probability theory that the wall shear stress should nearly follow a Rayleigh distribution as Fig. 12 shows. Antonia and Van Atta[63] showed that the Gaussianity of the velocity fluctuations is valid only as close to the wall as to $y/\delta \approx 0.1$ and that the spectrum of the Reynolds stress can be derived from the Gaussian assumption.

Figure 13 compares the experimental data of wall-pressure fluctuations obtained by Schewe[14] with a Gaussian distribution. The fit is exceedingly good, as is expected from the incompressibility of the fluid. The elliptic nature of the governing Poisson equation dictates that the wall-pressure fluctuations be a result of disturbances over a domain of influence significantly larger than the wall shear stress counterpart, so that it is a sum of many random variables; therefore, its PDF follows a Gaussian distribution.

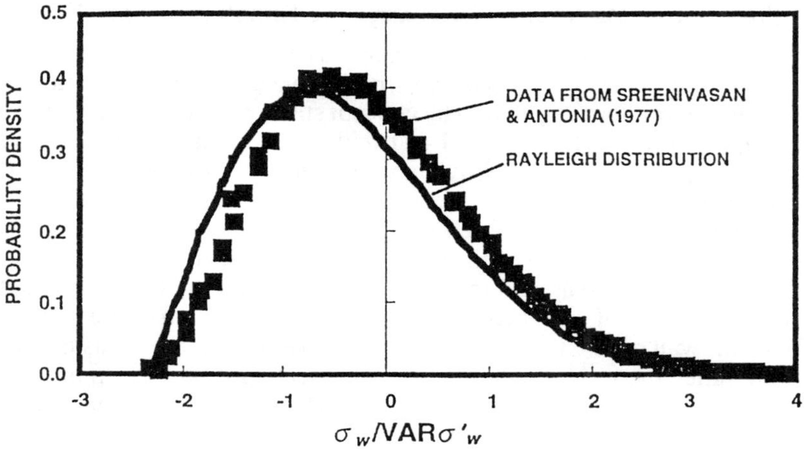

Fig. 12 Probability density function of wall shear stress compared with Rayleigh distribution.

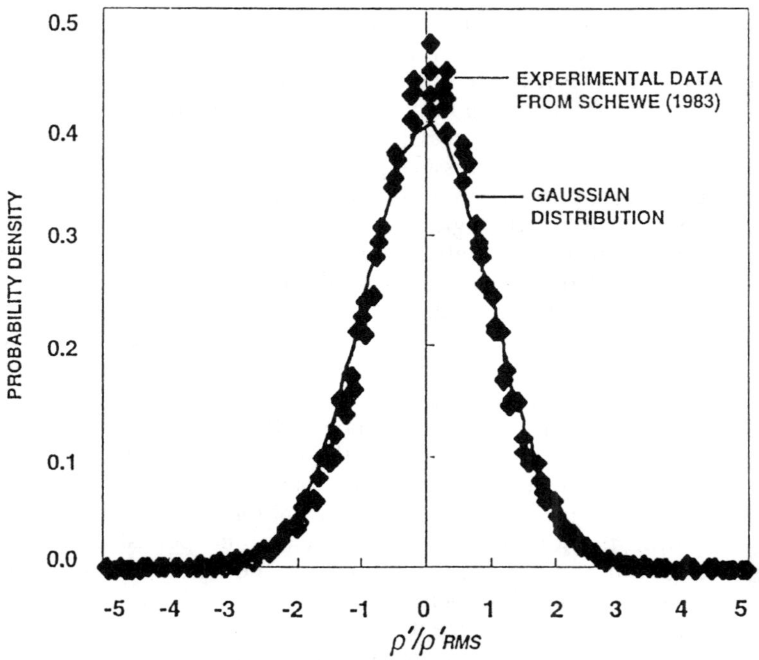

Fig. 13 Probability density function of wall-pressure fluctuation compared with Gaussian distribution.

B. Probabilistic Model and Electromagnetic Turbulence Control

Based on the preceding experimentally observed PDFs and the sequence of microturbulent events, one can construct a conceptual probabilistic model of dependent random variables. The concept starts with the low-speed streak spacing, which has a long-normal distribution; then, the liftups take place with a frequency PDF of Rayleigh distribution, followed by another Rayleigh distribution of the sweeps and a log-normal distribution of the mean time between bursts. Each microturbulent event contributes to the wall shear stress; the low-speed streak contributes at the low end with the liftup less than the mean and the sweeps greater than the mean, so that the wall shear stress PDF can be construed as the weighted sum of the individual PDFs. This wall shear stress PDF can be compared with the measured PDF to "fine tune" the weighting factors for practical applications. Similarly, the contributions from each microturbulent event to the wall-pressure fluctuations can be estimated in the same manner.

The weighting factors thus derived correlate to the percentages of time of each natural microturbulent state. When external electromagnetic influences are present, however, the weighting factors will vary because the electromagnetic influences will inhibit certain events (such as liftup and sweeps), thereby reducing turbulence production, and will promote other events (such as the low-speed streaks, which contribute to lower shear stress), thereby reducing the total drag over time.

The observed quasiperiodic nature of coherent microturbulent events that are taking place randomly in a spatial dimension near the wall can be applied to predict the next microturbulent event, thereby offering an opportunity to activate an opposing remote Lorentz pressure gradient to inhibit the turbulence production process. In other words, the turbulence control is based on applying Lorentz pressure gradients toward the wall during predicted liftup, Lorentz pressure gradients away from the wall during predicted sweep, and no pressure gradients during the low-speed streak. This is the fundamental basis of estimator-predictor feedback control of turbulence based on the known, dependent, random microturbulent event processes. By sampling the wall shear stress and pressure fluctuations and obtaining a real-time solution based on the Markov probabilistic model, the microturbulent state can be predicted, and the Lorentz pressure gradients applied, which should result in the total turbulent drag reduction.

In summary, the principles of electromagnetic turbulence control are to apply the Lorentz pressure gradient at the normal distance from the wall where the peak turbulence production occurs naturally. The objective is not to totally eliminate turbulence but only to inhibit turbulent ejection and sweep, allowing the low-speed streaks to exist and persist because they contribute to a more naturally stable mode of flow near the wall and are responsible for a low level of shear stress and therefore skin-friction drag.

Figure 14 illustrates the concept of applying Lorentz pressure gradient vectors to inhibit the naturally occurring microturbulent events. It shows the idealized Lorentz pressure gradient vectors inhibiting ejection and sweep simultaneously.

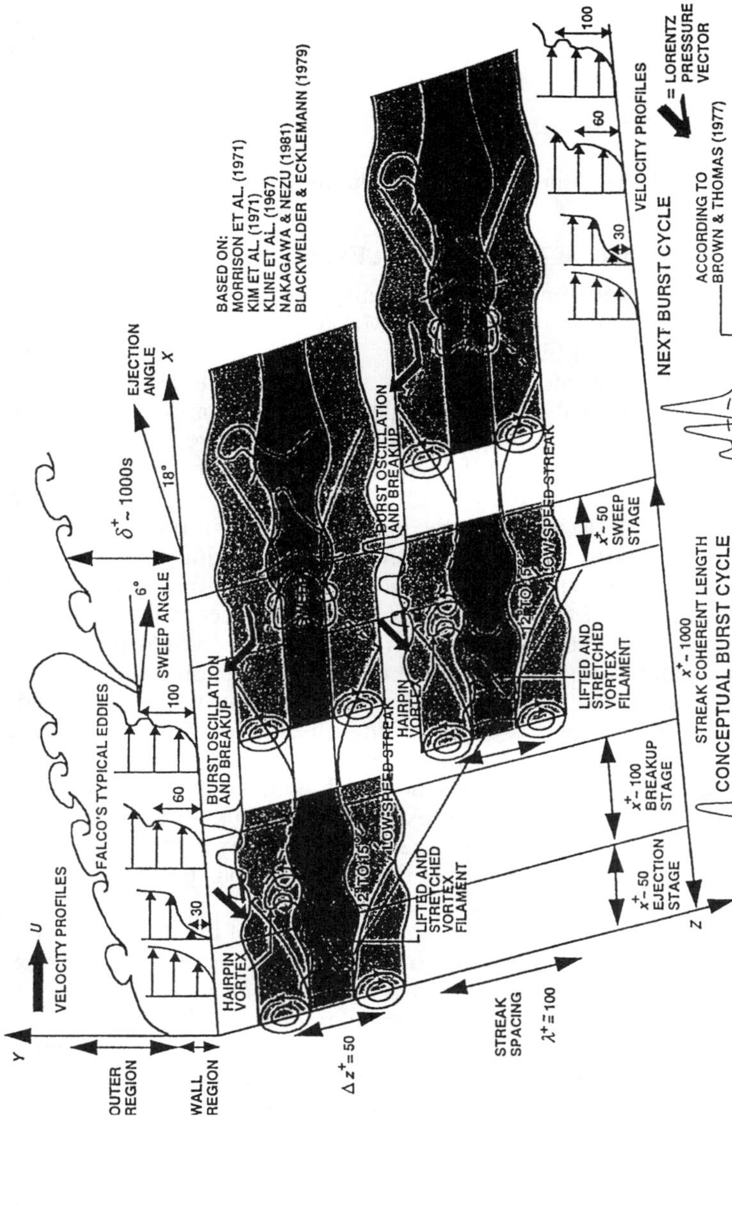

Fig. 14 Application of Lorentz pressure gradient vectors to inhibit microturbulent events.

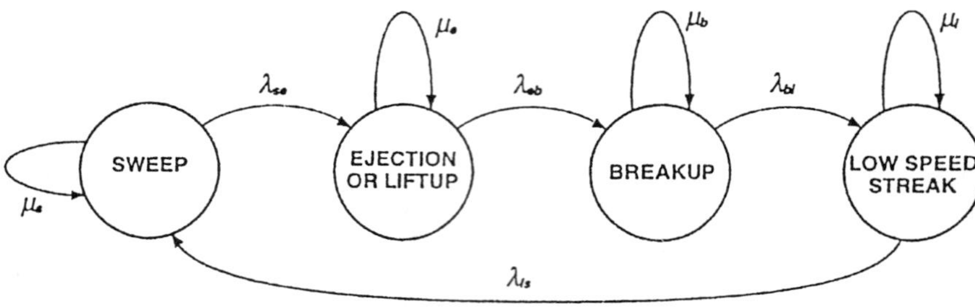

Fig. 15 Markov chain of stochastic processes.

C. Semi-Markov Chain of Microturbulent Events

The wall-layer microturbulent events at any particular location are mutually dependent and can be related to the Markov chain of stochastic processes, which is depicted in Fig. 15.

Note the similarity of this chain to the continuous birth-and-death chain well known in Markov analysis. It is a single-chain, quasicontinuous process. As shown in Fig. 15, the assumptions are that the entire event chain is characterized by the few identifiable discrete events, and the reverse transition probabilities from any state to the previous state are zero, except for the eventual low-speed streak state to sweep state. The transition probabilities and the distinctive states are characterized by the percentage of time resident in each state. One certainly can argue that not all low-speed streaks lead to liftup; although if the local Reynolds number exceeds the transition limit, it is certain that eventual transition will occur to the hairpin vortex liftup, which ultimately leads to the breakup state. One should further note that the turbulent events form a single-chain, quasicontinuous birth-and-death process. This assertion is based on the observation that no periodic states exist. It is important not to confuse Markov periodic states with the periodically occurring discrete states in a turbulent boundary layer. The former exist if, for every k number of experimental trials, a particular state emerges, which is certainly not the case in the wall layer microturbulent events. The latter are states that occur with predicable periodicity over certain spatial and temporal scales.

Several issues regarding the Markov representation of the discrete events in a turbulent boundary layer remain to be addressed. Among these issues are the satisfaction of the Markov process definition, i.e., that there is only a finite number of discrete states, the probability of any state depends only on the probability of the previous state in the time sequence, the existence of limiting states, and their corresponding limiting probabilities. Resolution of these issues must await future theoretical analyses.

Markov Transition Probabilities in the Absence of External Forces

The four microturbulent states are mutually dependent, and they occur according to a specific sequence. To describe the probabilistic model, one starts from the Chapman–Kolmogorov equation $p_i'(t) = -\mu_i p_i + \Sigma\lambda_{ji} p_j(t)$, where ′ represents the time derivative. This equation states that the rate of change of probability of any one state is determined by the regeneration rate of itself and the transition probabilities λ_{ji}, of other states multiplied by the probabilities of that state. Stated in terms of the four mutually exclusive microturbulent states—l for low-speed streak, e for ejection, b for breakup, and s for high-speed sweep—one can rewrite the preceding equation as follows:

$$p_e'(t) = -\mu_e p_e(t) + \Sigma\lambda_{je} p_j(t)$$

$$p_b'(t) = -\mu_b p_b(t) + \Sigma\lambda_{jb} p_j(t)$$

$$p_s'(t) = -\mu_s p_s(t) + \Sigma\lambda_{js} p_j(t)$$

where Σ is over all j's,

$$P_l(t) + P_e(t) + P_b(t) + P_s(t) = 1$$

If one assumes that the microturbulent states follow a continuously cyclic process, in other words the transition probabilities are nonzero only for those states in the observed sequence, i.e., $\lambda_{el} = \lambda_{es} = \lambda_{be} = \lambda_{bs} = \lambda_{sb} = \lambda_{be} = \lambda_{sl} = \lambda_{lb} = 0$, then the above equations become

$$p_e'(t) = -\mu_e p_e(t) + \lambda_{se} p_s(t)$$

$$p_b'(t) = -\mu_b p_b(t) + \lambda_{eb} p_e(t)$$

$$p_s'(t) = -\mu_s p_s(t) + \lambda_{ls} p_l(t)$$

$$p_l(t) + p_e(t) + p_b(t) + p_s(t) = 1$$

These equations can be solved analytically if all λ_{ij}s and μs are constants.

Markov Limiting Probabilities for Natural Turbulent Boundary Layers

All of the experimental measurements reported are based on long-duration averages that show the sequence of near-wall events: liftup and ejection, breakup, low-speed streak, and high-speed sweep toward the wall. The entire cycle then repeats itself. Therefore, one can assume that these states are mutually exclusive and exhaustively collective to describe any possible state of the turbulent boundary layer. Probabilistic analysis of results of any point diagnosis of local state in a turbulent boundary layer must, however, consider transition probabilities from all neighboring streaks and ensuing structures. This implies that even though the instantaneous point observations in an Eulerian frame of reference show random migration of neighboring streak's sequence of discrete events, such as the sequence of streak-to-vortex liftup then back to streak from a neighboring streak, the long-duration observation (i.e., over a time scale of $t \gg$ period of the break-up or burst cycle) shows only the above four discrete events, which occur periodically and indefinitely until flow conditions change. Therefore, probabilistically speaking, it suffices to say that after many statistical trials, i.e., as the number of observations approaches infinity, limiting probabilities (P_is) exist for each discrete event. In principle, one could then use the Chapman– Kolmogorov equation to derive the limiting state probabilities, provided that detailed measurement data exist over a very short time scale (a time scale shorter than a small fraction of the burst period) and provided that a reliable signal processing algorithm is available to record and detect all transition probabilities from neighboring states. This derivation would be very technically demanding although deserving investigation.

In the case of fully developed turbulent flows without externally applied unsteady forces, one expects that the stochastic process will asymptote to the limiting states, i.e., the time derivatives vanish, so that a stationary stochastic process can be established. This is mathematically possible if the regeneration rates and transition probabilities follow the known sequence and, therefore, the following conditions:

$$\mu_e + \lambda_{eb} = 1$$

$$\mu_b + \lambda_{bl} = 1$$

$$\mu_s + \lambda_{se} = 1$$

$$\mu_l + \lambda_{ls} = 1$$

With some rearrangement and substitution, the following solutions can be established in terms of the arrival rates μs:

$$p_l = 1 \Big/ \left(1 + \frac{1-\mu_l}{\mu_s} + \frac{(1-\mu_l)(1-\mu_s)}{\mu_s \mu_e} + \frac{(1-\mu_l)(1-\mu_s)(1-\mu_e)}{\mu_s \mu_e \mu_b}\right)$$

$$p_s = \frac{(1-\mu_l)}{\mu_s} p_l$$

$$p_e = \frac{(1-\mu_s)}{\mu_e} \frac{(1-\mu_l)}{\mu_s} p_l$$

$$p_b = \frac{(1-\mu_e)}{\mu_b} \frac{(1-\mu_s)}{\mu_e} \frac{(1-\mu_l)}{\mu_s} p_l$$

These rates can be determined experimentally from the mean time between arrivals. For practical applications, the arrival rates for ejection μ_e, breakup μ_b, and sweep μ_s can be assumed to be the same as the experimentally measured burst rate. The arrival rate for the low-speed streaks μ_l can be estimated by multiplying the burst rate by the ratio of streak coherence length to ejection and burst length ($\approx$12), or by the ratio of burst period to combined individual ejection-breakup-sweep duration ($\approx$30).

In summary, the measurement data, by definition, have already provided the information, based on the limiting states, that can help to determine the limiting probabilities. Specifically, one can use the experimentally observed mean time between occurrence of events to determine the birth rates of each event, and then apply the Chapman–Kolmogorov equation to calculate the limiting probabilities. Once the probabilities are determined, many statistical parameters can be obtained to aid practical implementation of EMTC. To illustrate the methodology, one can simplify the real-time calculation by assuming that the ejection and breakup states can be combined. This assumption then yields the following equations governing the limiting states condition, i.e., all derivatives vanish:

$$p_l = 1\Big/\left(1 + \frac{1-\mu_l}{\mu_s} + \frac{(1-\mu_l)(1-\mu_s)}{\mu_s \mu_e}\right)$$

$$p_s = \frac{(1-\mu_l)}{\mu_s}\, p_l$$

$$p_e = \frac{(1-\mu_s)}{\mu_e}\,\frac{(1-\mu_l)}{\mu_s}\, p_l$$

With this approach, only the arrival rates of the ejection/breakup and sweep states need to be measured from the experimental data.

Markov Transition Probabilities with EMTC Applied

In a natural turbulent boundary, the sequential order described in the phenomenological cycle is the only allowable mode of transition. Because all transition must occur with eventual certainty, one finds that the limiting states apply. Given that the frequency of occurrence of each state is proportional to the probability of each state, the experimentally obtained fractional times of ejection, breakup, sweep, and low-speed streaks serve as a guide to demark the wall-pressure fluctuation and wall shear stress histograms into four nonoverlapping regimes. The transition probabilities λ_{ij}s are related to the experimentally measured arrival rates.

When an external influence is activated, the transition probabilities no longer are certain, and the sequential order may not be preserved; in fact, some reverse transition is conceivable, although the natural tendencies must still be overwhelming. In this case, the full set of transition probabilities must be dealt with. For the unsteady case, the full set of state probability equations is as follows:

$$p_e'(t) = -\mu_e p_e(t) + (1 - \mu_s) p_s(t)$$

$$p_b'(t) = -\mu_b p_b(t) + (1 - \mu_e) p_e(t)$$

$$p_s'(t) = -\mu_s p_s(t) + (1 - \mu_l) p_l(t)$$

$$p_e(t) + p_b(t) + p_l(t) + p_s(t) = 1$$

To solve this set of equations, one must make some assumptions about the arrival rates. First, one can attempt to assume that the arrival rates are independent of time, as the experimental data indicate for the natural turbulent boundary layer. A question naturally arises: Does this assumption apply to the case where externally applied forces are present? Second, one assumes that the time-varying state probability does not depend on how much control is to be imposed. These assumptions are connected and are a function of the control methodology. If the control methodology is to completely eliminate turbulence, i.e., totally eliminate the microturbulent events, then the arrival rates would be zero, and the states representing each microturbulent event would not exist; the local flow would surely correspond to laminar flow conditions. Knowing that the microturbulent events represent the most stable flow configurations under turbulent conditions that nature can provide, the control methodology should be one that is in harmony with nature rather than one that tries to overcome nature. Therefore, the turbulence control methodology adopted here is to mitigate or minimize the occurrence of microturbulent events. Given that the control methodology is based on inhibiting turbulence rather than on totally eliminating it, the rate of change of state probabilities should be small compared with the limiting state probabilities, i.e., $\tilde{T}_B$ (d/dt) ln p_e, $\tilde{T}_B$ (d/dt) ln p_b, and $\tilde{T}_B$ (d/dt) ln p_s all should be much smaller than 1. Then, the first assumption (viz., that the arrival rates are not different from those in the natural state) is satisfied, and the second assumption is addressed by using only marginal control to maintain the turbulent state at a much lower level of turbulence production on resulting turbulence intensities.

Given the above assumptions, the differential equations with constant coefficients can be solved versus time. To illustrate the process, the preceding equations can be simplified by assuming that ejection is followed by breakup, so that only three states need to be considered, i.e.,

$$p_e'(t) = -\mu_e p_e(t) + (1 - \mu_s) p_s(t)$$

$$p_s'(t) = -\mu_s p_s(t) + (1 - \mu_l)[1 - p_e(t) - p_s(t)]$$

with the initial conditions $p_e(0), p_s(0), p_l(0)$ equal to the limiting state probabilities in natural turbulent boundary layers. The above equations can be represented by

$$\frac{\mathrm{d}}{\mathrm{d}_t}\begin{bmatrix} p_e(t) \\ p_s(t) \end{bmatrix} = \begin{bmatrix} -\mu_e & 1-\mu_s \\ -(1-\mu_l) & -(1-\mu_l)-\mu_s \end{bmatrix}\begin{bmatrix} p_e(t) \\ p_s(t) \end{bmatrix} + \begin{bmatrix} 0 \\ (1-\mu_l) \end{bmatrix}$$

This equation has the exponential solution $e^{\alpha t}$ where the αs are eigenvalues of the coefficient matrix

$$\begin{bmatrix} -\mu_e & 1-\mu_s \\ -(1-\mu_l) & -\mu_s-(1-\mu_l) \end{bmatrix}$$

The two α_1, α_2 are roots of the following quadratic equation:

$$\alpha^2 + (1+\mu_s+\mu_e-\mu_l)\alpha + 1 - \mu_l - \mu_s + \mu_e - \mu_e\mu_l + \mu_e\mu_s + \mu_l\mu_s = 0$$

Notice the sum of α_1 and α_2 is less than zero.

If one substitutes the following into the state probability differential equations,

$$\begin{bmatrix} p_e(t) \\ p_s(t) \end{bmatrix} = \begin{bmatrix} p_{eh}(t) \\ p_{sh}(t) \end{bmatrix} e^{\alpha l} + \begin{bmatrix} p_{en} \\ p_{sn} \end{bmatrix}$$

where p_{eh} and p_{sh} are the homogeneous parts of ordinary differential equations, one obtains the nonhomogeneous solution,

$$\begin{bmatrix} p_e(t) \\ p_s(t) \end{bmatrix} = \begin{bmatrix} p_{eh} \\ p_{sh} \end{bmatrix} e^{\alpha l} + \frac{(1-\mu_l)}{(1-\mu_l)(1-\mu_s)+\mu_e(1-\mu_l+\mu_s)}\begin{bmatrix} 1-\mu_s \\ \mu_e \end{bmatrix}$$

so that the complete solution becomes

$$\begin{bmatrix} p_e(t) \\ p_s(t) \end{bmatrix} = \frac{1}{2}\begin{bmatrix} p_e(o) - \dfrac{(1-\mu_l)(1-\mu_s)}{(1-\mu_l)(1-\mu_s)+\mu_e(1-\mu_l+\mu_s)} \\ p_s(o) - \dfrac{(1-\mu_l)\mu_e}{(1-\mu_l)(1-\mu_s)+\mu_e(1-\mu_l+\mu_s)} \end{bmatrix}(e^{\alpha_1 t}+e^{\alpha_2 t})$$

$$+ \frac{(1-\mu_l)}{(1-\mu_l)(1-\mu_s)+\mu_e(1-\mu_l+\mu_s)}\begin{bmatrix} 1-\mu_s \\ \mu_e \end{bmatrix}$$

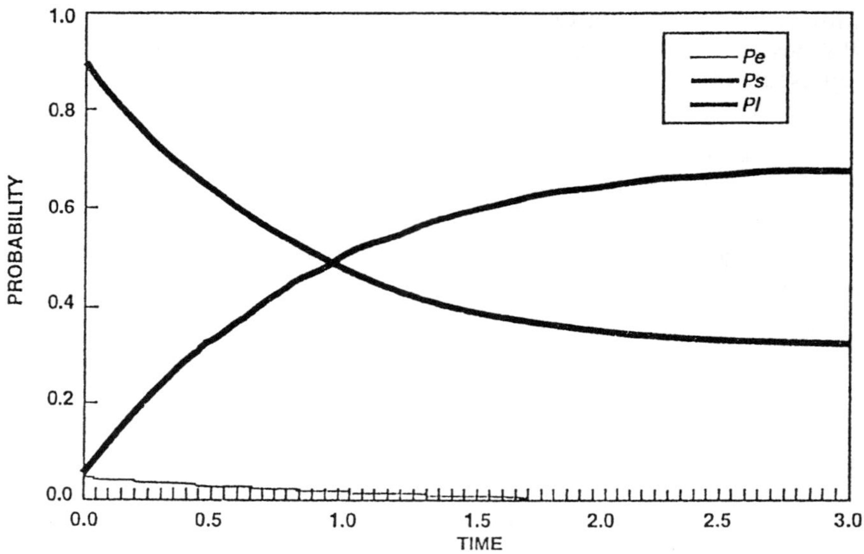

Fig. 16 Sample calculated results for discrete microturbulent Markov state probability vs time, illustrating crossover from low-speed streak to sweep.

and

$$p_l(t) = 1 - p_e(t) - p_s(t)$$

Note that α_1, α_2 are complex conjugates; p_e and p_s are always real numbers, which is consistent with the probability concept.

The consistency of these solutions can be demonstrated best by a simple example of numerical results. First of all, the values of the arrival rates μ_l, μ_e, and μ_s need to be determined from the experimental data. Using the time durations listed in Table 2 for ejection, breakup, and sweep, the μs can be evaluated to be $\mu_l \approx 31.5/32, \mu_e \approx 0.2/32$, and $\mu_s \approx 0.3/32$. The two αs can be found to be $\alpha_1 \approx 0.015 + 0.124i$ and $\alpha_2 \approx 0.015 - 0.124i$. From these values the probabilities of states e, s, and l can be calculated as a function of time. For a hypothetical case where $\mu_l = 0.37, \mu_e = 0.68$, and $\mu_s = 0.68$, the two roots are $0.87 + 0.47i$ and $-0.87 - 0.47i$. Figure 16 shows the crossover from low-speed streak to sweep.

Similarly, the full set of p_s, p_e, p_b, p_l equations can be solved; the analytic solutions are derived next:

$$\frac{\mathrm{d}}{\mathrm{d}t}\begin{bmatrix} p_s \\ p_e \\ p_b \end{bmatrix} = \begin{bmatrix} -(\mu_s + 1 - \mu_l) & -(1-\mu_l) & -(1-\mu_l) \\ 1-\mu_s & -\mu_e & 0 \\ 0 & 1-\mu_e & -\mu_b \end{bmatrix} \begin{bmatrix} p_s \\ p_e \\ p_b \end{bmatrix} + \begin{bmatrix} 1-\mu_l \\ 0 \\ 0 \end{bmatrix}$$

The exponent αs are eigenvalues of the matrix, i.e., solutions of the following equation

$$(\alpha + \mu_s + 1 - \mu_l)(\alpha + \mu_e)(\alpha + \mu_b)12(\alpha + \mu_b)(1-\mu_s)(1-\mu_l) \\ +(1-\mu_s)\,(1-\mu_e)(1-\mu_l) = 0$$

The total solution is therefore, as the simplified case shown before,

$$\begin{bmatrix} p_s \\ p_e \\ p_b \end{bmatrix} = \begin{bmatrix} p_{sh} \\ p_{eh} \\ p_{bh} \end{bmatrix} e^{\alpha l} + \begin{bmatrix} p_{sn} \\ p_{en} \\ p_{bn} \end{bmatrix}, \qquad p_l = 1 - p_s - p_e - p_b$$

where

$$\begin{bmatrix} p_{sn} \\ p_{en} \\ p_{bn} \end{bmatrix} = \frac{1-\mu_l}{\mu_b\mu_e(1-\mu_l+\mu_s)+\mu_b(1-\mu_l)(1-\mu_s)+\mu_e(1-\mu_l)(1-\mu_e)} \begin{bmatrix} \mu_b\mu_e \\ \mu_b(1-\mu_s) \\ (1-\mu_e)(1-\mu_s) \end{bmatrix}$$

D. Estimation of Instantaneous Markov States with External Forces Applied

Two familiar ways of presenting turbulent boundary-layer data are the wall shear stress and the wall-pressure probability histograms; both can be conceptually useful indicators for purposes of microturbulent event control. Shear stress is dominated by small-scale eddies near the wall, whereas the wall-pressure power density results from both large eddies away from the wall and small eddies near the wall. So, in a global sense, the wall shear stress fluctuations exist primarily as a consequence of the inner wall events, whereas the wall pressure fluctuations are due to both inner and outer wall events. The low end of the wall shear stress histogram represents the vortex liftup and ejection and breakup part of the cycle, whereas the high end of the histogram indicates the sweep part of the cycle. Similarly, the wall-pressure histogram provides another cross examination of the local microturbulent events where the shear stress and wall-pressure probes are collocated.

The spectra of both the shear stress and wall pressure provide different views of the effects of the same microturbulent events. The low wave number part of the wall-pressure spectrum characterizes the power density content of outer layer events, as does the high wave number end for near-wall inner layer events. The same holds true for the wall shear stress spectrum. Although there are many competing arguments for explaining the low wave number end of the wall-pressure spectrum, and although the high frequency and wave number end of the spectrum is controlled by the sensing frequency response, the objectives here deal only with the power density integrated under the spectrum, which is less sensitive to the details of the spectrum fine structure. For the shear-stress frequency spectrum, the same caveat is applicable.

The methodology for estimating the current state of the turbulent boundary layer in the presence of applied external forces can now be described, again using EMTC as an example. The PDFs of wall shear stress and wall-pressure fluctuation are shown in Fig. 17 and 18. They are as shown earlier (in Figs. 12 and 13) except that the PDFs are demarcated into several nonoverlapping regions. The first region is for the low-speed streaks, which generate the lowest level of shear-stress fluctuation, so that this region occupies the low end of the shear-stress PDF. Then comes ejection, which because of its induced counterflow has a local shear stress fluctuation that is above that of the no-ejection state but below the mean shear stress. For the breakup state, there is little or no specific measurement data describing how it affects the local shear-stress fluctuation. Intuitively, however, one would expect that the shear-stress fluctuation would be higher than the mean during breakup, so that it occupies the third region just to the right of the mean shear-stress value. During the sweep state, experimental evidence indicates increased shear-stress fluctuation because high-momentum fluid from the logarithmic region is imparted toward the wall; therefore, the sweep state is at the high end of the shear-stress PDF. With this relative ranking given, and with knowledge of the limiting probability of each state in a naturally occurring turbulent boundary layer (which equates to the areas under each region), one can then demarcate the PDF into nonoverlapping regions. The nonoverlapping comes from the fact that the four discrete states are mutually exclusive and exhaustively collective to describe the entire microturbulent cycle.

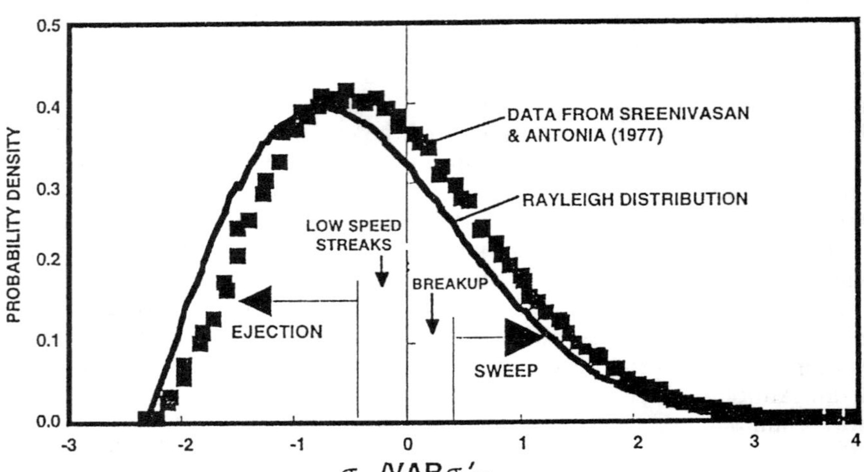

Fig. 17 Demarcation of microturbulent states on PDF of wall shear-stress fluctuation.

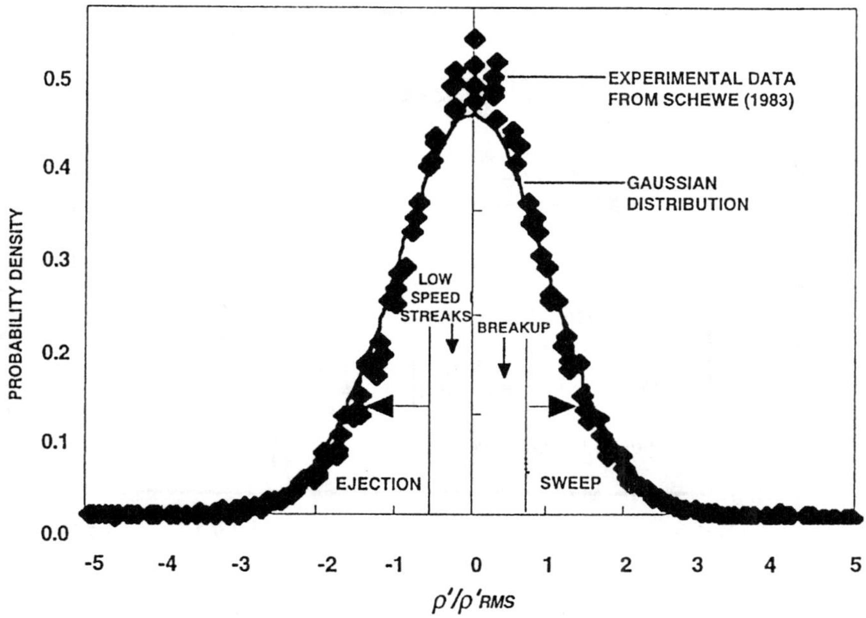

Fig. 18 Demarcation of microturbulent states on PDF of wall-pressure fluctuation.

Similarly, one can estimate the amount of pressure fluctuation power contained in ejection, sweep, and overall burst by examining the wall-pressure spectrum expressed in terms of inner variable dimensions. Figure 19 shows such an example. The wall-pressure spectra were obtained by Blake[60] and Schewe[14] both in terms of wall units, but Schewe's spectral levels were normalized by the spectral level at zero frequency and had to be adjusted for comparison. As can be seen, the two data sets agree. From the microturbulence phenomenology discussed earlier, the mean burst frequency $\omega^+ \approx 0.06$, and for the burst event frequencies measured by Schewe[14] $\omega^+ \approx 0.52$. These two frequencies demarcate the relative ranges of the wall events from events occurring away from the wall, so that the power contained in each category of events can be estimated by integrating over relevant ranges as is indicated in the figure.

E. Summary of Markov Transition Probability Methodology

As applied to characterizing and controlling wall-layer turbulence, the methodology of the Markov transition probability comprises two major steps: estimating and predicting. The estimator step involves the following. At time 0 before the active control is turned on, the flow characteristics are measured in real time by flush-mounted hot-film probes and miniature hydrophones. The wall shear stress and wall pressure fluctuation PDF histograms and spectra are generated by data processing. The histograms of the

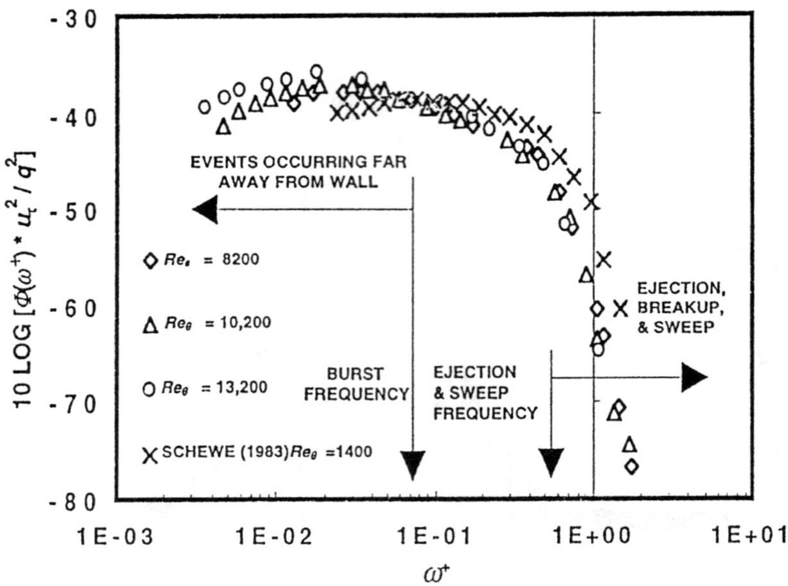

Fig. 19 Normalized smooth wall-pressure spectrum in wall units.

key parameters such as the time between bursts and the mean and variance of wall shear stress and wall-pressure fluctuation are determined, from which the dimensionless parameters are calculated. Then, the established wall turbulence phenomenology models are applied to estimate the mean time between bursts, burst duration, and spacing. The fractional time duration of each event as normalized by the average mean time between burst arrivals T_B is determined. The arrival rate of frequency of each microturbulent event is estimated from the fractional time duration. The limiting state probabilities are then determined from the arrival rate, as was derived in a previous section. These limiting state probabilities are stored. After several T_B periods, the limiting state probabilities are averaged and stored.

After several T_B periods, the current state is estimated by comparing the shear stress and wall-pressure fluctuations with the established PDFs. Then, solutions to the unsteady state probability equations are applied to estimate the state probability of the sensor location and determine the probability of current location within the next time step. The time step is limited by the lower bound of peak frequency $\omega^+ \approx 0.52$, the upper bound of $\omega^+ \approx 0.06$, and the desired accuracy. One rule of thumb is to resolve the highest frequency with six intermediate time steps, which translates into a stepping frequency of $\omega^+ \approx 3$. Selection of the time step must balance the computational load and the accuracy desired.

After the next state probabilities are obtained, a decision has to be made as to whether any external field is to be applied. This is the predictor step. For example, if the current state is low-speed streak and is moving toward sweep, then a normal Lorentz pressure gradient would be applied against the incoming high-momentum fluid motions. If the local state is currently ejection and is moving into breakup, then a decision would be made not to apply any Lorentz pressure gradient, thus letting the low-speed streaks re-establish themselves to maintain the low drag condition. If any external field is to be applied, the overall spatial pattern illustrated earlier in the wall-layer turbulence phenomenology would be applied over the entire physical dimension where the external field can be activated. In other words, by using the well-established phenomenology, which has been measured, visualized, published in the open literature, and thoroughly scrutinized over a Reynolds number range, only a finitely few sensors are needed to diagnose the Markov state at the local location and allow application of the external field globally with the known spatial pattern. This approach greatly reduces the burden of processing and control.

V. Summary and Conclusions

From a thorough overview of the published literature of wall-layer turbulent events, a conceptual phenomenology of microturbulence, including spatial and temporal patterns of discrete microturbulent events, has been established. An overall, end-to-end, conceptual pictorial representation of the microturbulent events has been constructed. The spatial and temporal patterns of the events invite the concept of controlling microturbulence in such a way that the turbulence-producing events are inhibited while the low-speed streaks, which are considered to be the most stable flow configuration naturally ubiquitous in many turbulent flow conditions beyond the wall turbulence, are preserved.

The repeatable, quasiperiodic, microturbulent events form the deterministic basis on which the Markov process is formulated. For a natural turbulent boundary layer with zero pressure gradient, the simplified Chapman–Kolmogorov equations for the transition probability are solved for the limiting probabilities. These limiting probabilities are related to the experimentally measured arrival rates of the ejection and sweep, which are Poisson-like processes. For a full four-state model (sweep, ejection, breakup, and low-speed streaks), four arrival rates are needed to completely describe the probability state solution; similarly, for a simplified three-state model (sweep, ejection, and low-speed streaks), three arrival rates are required.

With the presence of externally applied electromagnetic fields, the limiting probabilities no longer apply and the full, unsteady, Chapman–Kolmogorov state probability equations must be solved. The solutions to these linear ordinary differential equations are composed of homogeneous and nonhomogeneous parts. The homogeneous part is of the exponential solution form, with the exponents being eigenvalues of the transition probability matrix. The nonhomogeneous part results from the conservation condition of the state probabilities, i.e., they sum to one at all times. The nonhomogeneous solution can also be derived analytically in terms of the homogeneous solution. The initial conditions are satisfied by summing the three eigensolutions for the four states or the two eigensolutions for the simplified three-state model. The three-state simplification would be especially useful for high-speed applications where the spatial scales become the width of human hair and the temporal scales so short, into the acoustic frequency range, that the computational power required may exceed present capability.

Lastly, a specific method for EMTC implementation that is based upon both the local shear-stress and wall-pressure sensors is proposed. This method satisfies all of the known facts of microturbulence phenomenology. The time delay and active control steps are also described.

The next step in this work should be to conduct an experiment using the step-by-step approach described earlier. The experiment must include both quantitative shear-stress and wall-pressure measurements and flow visualizations with pulsed laser fluorescent dye to determine the effects of the externally applied fields on the naturally occurring turbulence. If turbulence is reduced, the shear-stress and wall-pressure fluctuations will be reduced, the coherence of the low-speed streaks will increase, and the burst rate will decrease. Fine tuning of the applied external field intensity and waveforms could result in a turbulence control solution that would have wide engineering applications and that would open a new chapter in the multidiscipline of fluid mechanics.

References

[1]Cantwell, B. J., "Organized Motion in Turbulent Flow," *Annual Reviews in Fluid Mechanics*, Vol. 13, 1981, pp. 457–515.

[2]Kim, H. T., Kline, S. J., and Reynolds, W. C., "The Production of Turbulence Near a Smooth Wall in a Turbulent Boundary Layer," *Journal of Fluid Mechanics*, Vol. 50, Pt. 1, 1971, pp. 133–160.

[3]Corino, E. R., and Brodkey, R. S., "A Visual Investigation of the Wall Region in Turbulent Flow," *Journal of Fluid Mechanics*, Vol. 37, 1969, pp. 1–30.

[4]Offen, G. R., and Kline, S. J., "Combined Dye-Streak and Hydrogen-Bubble Visual Observations of a Turbulent Boundary Layer," *Journal of Fluid Mechanics*, Vol. 62, Pt. 2, 1974, pp. 223–239.

[5]Nakagawa, H., and Nezu, I., "Structure of Space-Time Correlations of Bursting Phenomena in an Open-Channel Flow," *Journal of Fluid Mechanics*, Vol. 104, 1981, pp. 1–43.

[6]Blackwelder, R. F., and Eckelmann, H., "Streamwise Vortices Associated with the Bursting Phenomenon," *Journal of Fluid Mechanics*, Vol. 94, Pt. 3, 1979, pp. 577–594.

[7]Head, M. R., and Bandyopadhyay, P., "New Aspects of Turbulent Boundary-Layer Structure," *Journal of Fluid Mechanics*, Vol. 107, 1981, pp. 297–338.

[8]Snarski, S. R., and Lueptow, R. M., "Wall Pressure and Coherent Structures in a Turbulent Boundary Layer on a Cylinder in Axial Flow," *Journal of Fluid Mechanics*, Vol. 286, 1995, pp. 137–171.

[9]Bakewell, H. P., Jr., and Lumley, J. L., "Viscous Sublayer and Adjacent Wall Region in Turbulent Pipe Flow," *Physics of Fluids*, Vol. 10, No. 9, 1967, p. 188.

[10]Asai, M., and Nishioka, M., "Origin of the Peak-Valley Wave Structure Leading to Wall Turbulence," *Journal of Fluid Mechanics*, Vol. 208, 1989, pp. 1–23.

[11]Sabry, A. S., and Liu, J. T. C., "Longitudinal Vorticity Elements in Boundary Layers: Nonlinear Development from Initial Görtler Vortices as a Prototype Problem," *Journal of Fluid Mechanics*, Vol. 231, 1991, pp. 615–663.

[12]Kreplin, H.-P., and Eckelmann, H., "Propagation of Perturbations in the Viscous Sublayer and Adjacent Wall Region," *Journal of Fluid Mechanics*, Vol. 95, Pt. 2, 1979, pp. 305–322.

[13]Gupta, A. K., Laufer, J., and Kaplan, R. E., "Spatial Structure in the Viscous Sublayer," *Journal of Fluid Mechanics*, Vol. 50, Pt. 3, 1971, pp. 493–512.

[14]Schewe, G., "On the Structure and Resolution of Wall-Pressure Fluctuations Associated with Turbulent Boundary-Layer Flow," *Journal of Fluid Mechanics*, Vol. 134, 1983, pp. 311–328.

[15]Kline, S. J., Reynolds, W. C., Schraub, F. A., and Runstadler, P. W., "The Structure of Turbulent Boundary Layers," *Journal of Fluid Mechanics*, Vol. 30, Pt. 4, 1967, pp. 741–773.

[16]Black, T. J., "An Analytic Study of the Measured Wall Pressure Field Under Supersonic Turbulent Boundary Layers," NASA CR-888, 1968.

[17]Morrison, W. R. B., Bullock, K. J., and Kronauer, R. E., "Experimental Evidence of Waves in the Sublayer," *Journal of Fluid Mechanics*, Vol. 47, Pt. 4, 1971, pp. 639–656.

[18]Laufer, J., and Badri Narayanan, M. A., "Mean Period of the Turbulent Production Mechanism in a Boundary Layer," *Research Notes*, 1970, pp. 182–183.

[19]Rao, K. N., Narasimha, R., and Badri Narayanan, M. A., "The 'Bursting' Phenomenon in a Turbulent Boundary Layer," *Journal of Fluid Mechanics*, Vol. 48, Pt. 2, 1971, pp. 339–352.

[20]Bandyopadhyay, P. R., "Period Between Bursting in Turbulent Boundary Layers," *Physics of Fluids*, Vol. 25, No. 10, 1982, pp. 1751–1754.

[21]Falco, R. E., "Coherent Motions in the Outer Region of Turbulent Boundary Layers," *Physics of Fluids*, Vol. 20, 1977, p. S124.

[22]Löfdahl, L., Kälvesten, E., and Stemme, G., "Small Silicon Based Pressure Transducers for Measurements in Turbulent Boundary Layers," *Experiments in Fluids*, Vol. 17, 1994, pp. 24–31.

[23]Farabee, T. M., and Casarella, M. J., "Spectral Features of Wall Pressure Fluctuations Beneath Turbulent Boundary Layers," *Physics of Fluids A*, Vol. 3, No. 10, 1991.

[24]Brown, G. L., and Thomas, A. S. W., "Large Structure in a Turbulent Boundary Layer," *Physics of Fluids*, Vol. 20, No. 10, Pt. 11, 1977, pp. S243–S252.

[25]Eckelmann, H., "The Structure of the Viscous Sublayer and the Adjacent Wall Region in a Turbulent Channel Flow," *Journal of Fluid Mechanics*, Vol. 65, Pt. 3, 1974, pp. 439–459.

[26]Bandyopadhyay, P. R., "Comments on Reynolds Number Effects in Wall-Bounded Shear Layers," AIAA Paper 91-0231, 1991.

[27]Murlis, J., Tsai, H. M., and Bradshaw, P., "The Structure of Turbulent Boundary Layers at Low Reynolds Numbers," *Journal of Fluid Mechanics*, Vol. 122, 1982, pp. 13–56.

[28]Wallace, J. M., Eckelmann, H., and Brodkey, R. S., "The Wall Region in Turbulent Shear Flow," *Journal of Fluid Mechanics*, Vol. 54, 1972, p. 39.

[29]Subramanian, C. S., Rajagopalan, S., Antonia, R. A., and Chambers, A. J., "Comparison of Conditional Sampling and Averaging Techniques in a Turbulent Boundary Layer," *Journal of Fluid Mechanics*, Vol. 123, 1982, pp. 335–362.

[30]Robinson, S. K., "Coherent Motions in the Turbulent Boundary Layer," *Annual Reviews in Fluid Mechanics*, Vol. 23, 1991, pp. 601–639.

[31]Kim, J., and Moin, J. P., "Flow Structures Responsible for the Bursting Process," *Bulletin of American Physics Society*, Vol. 31, 1986, p. 1716.

[32]Moin, P., "Analysis of Turbulence Data Generated by Numerical Simulations," AIAA Paper 87-0194, Jan. 1987.

[33]Kim, J., Moin, J. P., and Moser, R., "Turbulent Statistics in Fully Developed Channel Flow at Low Reynolds Number," *Journal of Fluid Mechanics*, Vol. 177, 1987, pp. 133–166.

[34]Kline, S. J., and Robinson, S. K., "Quasi-Coherent Structures in the Turbulent Boundary Layer. Part 1: Status Report on a Community-wide Summary of the Data," *Near Wall Turbulence, Proceedings of the Zaric Memorial Conference*, edited by S. J. Kline, Hemisphere, New York, 1988, pp. 200–217.

[35]Kline, S. J., and Robinson, S. K., "Quasi-Coherent Structures in the Turbulent Boundary Layer. Part 1: Status Report on a Community-wide Summary of the Data," *Near Wall Turbulence, Proceedings of the Zaric Memorial Conference*, edited by S. J. Kline and N.H. Afgan, Hemisphere, New York, 1988, pp. 397–417.

[36]Robinson, S. K., Kline, S. J., and Spalart, P. R., "Quasi-Coherent Structures in the Turbulent Boundary Layer. Part 2: Verification and New Information from a Numerically Simulated Flat Plate," *Near Wall Turbulence, Proceedings of the Zaric Memorial Conference*, edited by S. J. Kline and N. H. Afgan, Hemisphere, New York, 1988, pp. 397–417.

[37]Theodorsen, T., "Mechanism of Turbulence," *Proceedings of Second Midwest Conference of Fluid Mechanics* (Columbus, Oh), 1952, pp. 1–18.

[38]Willmarth, W. W., and Tu, B. J., "Structure of Turbulence in the Boundary Layer Near the Wall," *Physics of Fluids,* Vol. 10, 1967, pp. S134–37.

[39]Offen, G. R., and Kline, S. J., "A Proposed Model of the Bursting Process in Turbulent Boundary Layers," *Journal of Fluid Mechanics*, Vol. 70, 1975, pp. 209–228.

[40]Praturi, A. K., and Brodkey, R. S., "A Stereoscopic Visual Study of Coherent Structures in Turbulent Shear Flow," *Journal of Fluid Mechanics*, Vol. 89, 1978, pp. 251–272.

[41]Robinson, S. K., and Kline, S. J., "Turbulent Boundary Layer Structure: Progress, Status and Challenges," *Structure of Turbulence and Drag Reduction*, edited by A. Gyr, Springer–Verlag, New York, 1990, pp. 3–22.

[42]Perry, A. E., and Chong, M. S., "On the Mechanism of Wall Turbulence," *Journal of Fluid Mechanics*, Vol. 119, 1982, pp. 173–217.

[43]Townsend, A. A., *The Structure of Turbulent Shear Flow*, 2nd ed., Cambridge Univ. Press, New York.

[44]Jimenez, J., Moin, P., Moser, R., and Keefe, L., "Ejection Mechanisms in the Sublayer of a Turbulent Channel," *Physics of Fluids*, Vol. 31, No. 6, 1988, pp. 1311–1313.

[45]Smith, C. R., "A Synthesized Model of the Near-Wall Behavior in Turbulent Boundary Layers," *Proceedings of the 8th Biennial Symposium on Turbulence*, edited by J. L. Zakin and G. Patterson, Univ. of Missouri–Rolla, Rolla, MO, 1984, p. 35.

[46]Smith, C. R., Walker, J. D. A., Haidari, A. H., and Sobrun, U., "On the Dynamics of Near-Wall Turbulence," Dept. of Mechanical Engineering and Mechanics, AFOSR Grant No. 89-0065, Lehigh Univ., Bethlehem, PA, April 1991.

[47]Kim, J. J., "Study of Turbulence Structure Through Numerical Simulations: The Perspective of Drag and Reduction," *AGARD Report 786, Special Course on Skin Friction Drag Reduction* (Brussels, Belgium), March 1992.

[48]Robinson, S. K., "Kinematics of Turbulent Boundary Layer Structure," Ph.D. Thesis, Stanford Univ., Stanford, CA, 1990; also NASA TM in press.

[49]Lu, S. S., and Willmarth, W. W., "Measurements of the Structure of the Reynolds Stress in a Turbulent Boundary Layer," *Journal of Fluid Mechanics*, Vol. 60, 1973, p. 481.

[50]Sirmalis, J. E., "A Study of the Drag Characteristics and Polymer Diffusion in the Boundary Layer of an Axisymmetric Body," Naval Underwater Systems Center, NUSC Technical Rept. 4860, Newport, RI, 1976.

[51]Tiederman, W. G., and Luchik, T. S., "Wall Layer Structure and Drag Reduction," School of Mechanical Engineering, Rept. PME-FM-82-2, Purdue Univ., Lafayette, IN, 1982.

[52]Choi, H., Moin, P., and Kim, J., "Active Turbulence Control for Drag Reduction in Wall-Bounded Flows," *Journal of Fluid Mechanics*, Vol. 262, 1994, pp. 75–110.

[53]Choi, H., Moin, P., and Kim, J., "Direct Numerical Simulation of Turbulent Flow Over Riblets," *Journal of Fluid Mechanics*, Vol. 255, 1993, pp. 503–539.

[54]Narasimha, R., "The Turbulence Problem; A Survey," *Journal of the Indian Institute of Science*, Vol. 64, No. A, 1983, p. 1.

[55]Bushnell, D. M., and McGinley, C. B., "Turbulence Control in Wall Flows," *Annual Reviews in Fluid Mechanics*, Vol. 21, 1989, pp 1–20.

[56]Smith, C. R., and Metzler, S. P., "The Characteristics of Low-Speed Streaks in the Near-Wall Region of a Turbulent Boundary Layer," *Journal of Fluid Mechanics*, Vol. 129, 1983, pp. 27–54.

[57]Bark, F. H., "On the Wave Structure of the Wall Region of a Turbulent Boundary Layer," *Journal of Fluid Mechanics*, Vol. 70, 1975, pp. 229–250.

[58]Vukoslavcevic, P., Wallace, J. M., and Balint, J.-L., "The Velocity and Vorticity Fields of a Turbulent Boundary Layer, Part 1. Simultaneous Measurement by Hot-Wire Anemometry," *Journal of Fluid Mechanics*, Vol. 228, 1991, pp. 25–51.

[59]Sreenivasan, K. R., and Antonia, R. A., "Properties of Wall Shear Stress Fluctuations in a Turbulent Duct Flow," *Journal of Applied Mechanics, Transactions of the ASME*, Sept. 1977, p. 389.

[60]Blake, W. K., "Turbulent Boundary-Layer Wall-Pressure Fluctuations on Smooth and Rough Walls," *Journal of Fluid Mechanics*, Vol. 44, Pt. 4, 1970, pp. 637–660.

[61]Kovasznay, L. S. G., "The Turbulent Boundary Layer," *Annual Reviews in Fluid Mechanics*, Vol. 2, 1975, pp. 95–112.

[62]Aitchison, J., and Brown, J. A. C., *The Lognormal Distribution*, Cambridge Univ. Press, New York.

[63]Antonia, R. A., and Van Atta, C. W., "Statistical Characteristics of Reynolds Stresses in a Turbulent Boundary Layer," *AIAA Journal*, Vol. 15, No. 1, 1977, pp. 71–75.

[64]Wygnanski, I., Sokolov, M., and Friedman, D., "On a Turbulent 'Spot' in a Laminar Boundary Layer," *Journal of Fluid Mechanics*, Vol. 78, Pt. 4, 1976, pp. 785–819.

[65]Cantwell, B., Coles, D., and Dimotakis, P., "Structure and Entrainment in the

Plane of Symmetry of a Turbulent Spot," *Journal of Fluid Mechanics*, Vol. 87, Pt. 4, 1978, pp. 641–672.

[66]Blackwelder, R. F., and Eckelmann, H., "The Spanwise Structure of the Bursting Phenomenon," *Structure and Mechanisms of Turbulence I, Lecture Notes in Physics, Vol. 75*, Springer–Verlag, New York, 1976.

[67]Acarlar, M. S., and Smith, C. R., "A Study of Hairpin Vortices in a Laminar Boundary Layer. Part 1. Hairpin Vortices Generated by a Hemisphere Protuberance," *Journal of Fluid Mechanics*, Vol. 175, 1987, pp. 1–141.

[68]Hinze, J. O., *Turbulence*, 2nd ed., McGraw–Hill, New York, 1975.

[69]Purtell, L. P., Klebanoff, P. S., and Buckley, F. T., "Turbulent Boundary Layers at Low-Reynolds Numbers," *Physics of Fluids*, Vol. 24, 1981, pp. 802–811.

Chapter 27

Numerical and Experimental Investigations of Flow Characteristics of Magnetohydrodynamic Propulsion Ducts

Choung Mook Lee* and Sang Joon Lee†
Pohang University of Science and Technology, Pohang, Korea
and
Chun Taek Kim‡
Korea Aerospace Research Institute, Taejon, Korea

I. Introduction

A MAGNETOHYDRODYNAMIC (MHD) ship propulsion offers several advantages over conventional mechanical propellers. The absence of a mechanical propeller system will reduce vibration in a ship and, thus, will reduce the mechanical noise. Also, there is less physical restriction on the speed of an MHD sea-water propulsion system.

The concept of MHD ship propulsion was conceived by Friauf[1] with an MHD duct-type propulsor. This was followed by a patent by Rice[2] as a similar idea of electromagnetic pump used for liquid metal. Way[3] of Westinghouse carried out experimental tests of MHD propulsion in July 1966 using a submersible model (EMS-1).

Japan Ship and Ocean Foundation[4] made a successful sea trial of the first super conducting experimental ship YAMATO-1. The ship had a displacement of 180 tons, length of 30 m, and internal-type superconducting magnets of 4 T. Although it was clear from the beginning that there was no economical advantage with a helium-cooled superconducting magnet, a great potential for the future feasibility for the MHD propulsion made the investigations justified.

---

*Professor, Department of Mechanical Engineering.
†Associate Professor, Department of Mechanical Engineering.
‡Research Engineer, Aeropropulsion Department.

The MHD propulsion efficiency is directly related to the intensity of the magnetic field, electric current density, duct shape, and inlet and outlet flow conditions. Numerous investigations on MHD duct flows were performed; however, experimental results of the velocity and turbulence distributions in a sea-water MHD thruster duct are scarce.

The objective of this study is to investigate the detailed flow characteristics in an MHD duct through numerical and experimental analyses.

In numerical analysis, a three-dimensional turbulent flow with nonuniform body force is investigated using a finite difference method (FDM). The numerical code developed under the present investigation is capable of treating three-dimensional flow in a duct of arbitrary cross sections under the influence of variable electric and magnetic fields.

To verify the numerical results, an experimental investigation of the flow characteristics in a rectangular duct of MHD thruster has been carried out. Velocity distributions inside the duct were measured by using a fiberoptic laser Doppler velocimetry (LDV) system and compared with the numerical results.

II. Theoretical Backgrounds

The system of equations describing the isothermal motion of an electrically conducting fluid under an applied magnetic field consists of the Navier–Stokes equations, Maxwell's equations, and Ohm's law for moving media. One can assume that the electrical conductivity is constant and isotropic.

By choosing a set of characteristic length L_c, velocity U_c, and magnetic field strength B_c, we can define the following nondimensional variables:

$$\begin{aligned} \boldsymbol{V}^* &= \frac{\boldsymbol{V}}{U_c}, \quad t^* = \frac{U_c t}{L_c}, \quad \nabla^* = L_c \nabla, \quad P^* = \frac{P}{\rho U_c^2} \\ \boldsymbol{B}^* &= \frac{\boldsymbol{B}}{B_c}, \quad \boldsymbol{E}^* = \frac{\boldsymbol{E}}{U_c B_c}, \quad \varphi = \frac{\varphi}{U_c B_c L_c}, \quad \boldsymbol{J}^* = \frac{\mu L_c \boldsymbol{J}}{B_c} \end{aligned} \tag{1}$$

where ρ is the fluid density, $\boldsymbol{V}$ the fluid velocity vector, P the pressure, μ the magnetic permeability, $\boldsymbol{J}$ the current density vector, $\boldsymbol{B}$ the magnetic field vector, φ the electric potential function $\boldsymbol{E} = -\nabla\varphi$, and t the time. The asterisk indicates a nondimensional variable and the bold-faced letters are vector quantities.

Now we can write the MHD equations in the following form, omitting the asterisks for the sake of simplicity:

$$\frac{\partial \boldsymbol{V}}{\partial t} + (\boldsymbol{V}\cdot\nabla)\boldsymbol{V} = -\nabla P + \frac{1}{Re}\nabla^2\boldsymbol{V} + Al(\boldsymbol{J}\times\boldsymbol{B}) \tag{2}$$

$$\boldsymbol{J} = Re_m\,(\boldsymbol{E} + \boldsymbol{V}\times\boldsymbol{B}) \tag{3}$$

$$E = -\nabla\varphi \tag{4}$$

$$\frac{\partial \boldsymbol{B}}{\partial t} + (\boldsymbol{V}\cdot\nabla)\boldsymbol{B} = (\boldsymbol{B}\cdot\nabla)\boldsymbol{V} + \frac{1}{Re_m}\nabla^2\boldsymbol{B} \tag{5}$$

$$\nabla^2 \varphi = \boldsymbol{B}\cdot\nabla\times\boldsymbol{V} - \boldsymbol{V}\cdot\nabla\times\boldsymbol{B} \tag{6}$$

The nondimensional parameters shown in the preceding equations and to be used subsequently are defined by

Reynolds number:

$$Re = \frac{U_c L_c}{\nu} \tag{7}$$

Magnetic Reynolds number:

$$Re_m = \frac{U_c L_c}{\nu_m} = \mu\sigma U_c L_c \tag{8}$$

Alfven number:

$$Al = \frac{B_c^2}{\mu\rho U_c^2} = \frac{N}{Re_m} \tag{9}$$

Interaction parameter (Stuart number):

$$N = \frac{\sigma B_c^2 L_c}{\rho U_c} = Al\cdot Re_m \tag{10}$$

where ν is the kinematic viscosity of the fluid, σ the electric conductivity, and $\nu_m = 1/\mu\sigma$ the magnetic diffusivity.

We substitute Eqs. (3) and (4) into Eq. (2) to find the final form of the MHD momentum equation as

$$\frac{\partial \boldsymbol{V}}{\partial t} + (\boldsymbol{V}\cdot\nabla)\boldsymbol{V} = -\nabla P + \frac{1}{Re}\nabla^2\boldsymbol{V} + N[(-\nabla\varphi + \boldsymbol{V}\times\boldsymbol{B})\times\boldsymbol{B}] \tag{11}$$

We should solve Eqs. (11), (5), and (6) with appropriate initial and boundary conditions to find the values of $\boldsymbol{V}$, $\boldsymbol{B}$, and φ in the fluid domain.

The right-handed Cartesian coordinate system will be employed. The main flow direction, the main magnetic field direction, and the main electric field direction are x, y, and z coordinates, respectively, in the duct.

The solution domain consists of the inlet duct, the test section, and the exit duct. And it is exactly the same as the experimental duct in size, as shown in Fig. 1. The test section has the dimension of 900 mm (L; length), 40 mm (H; height), and 70 mm (W; width) in x, y, and z directions, respectively.

In this study, only the steady-state case is considered. The following

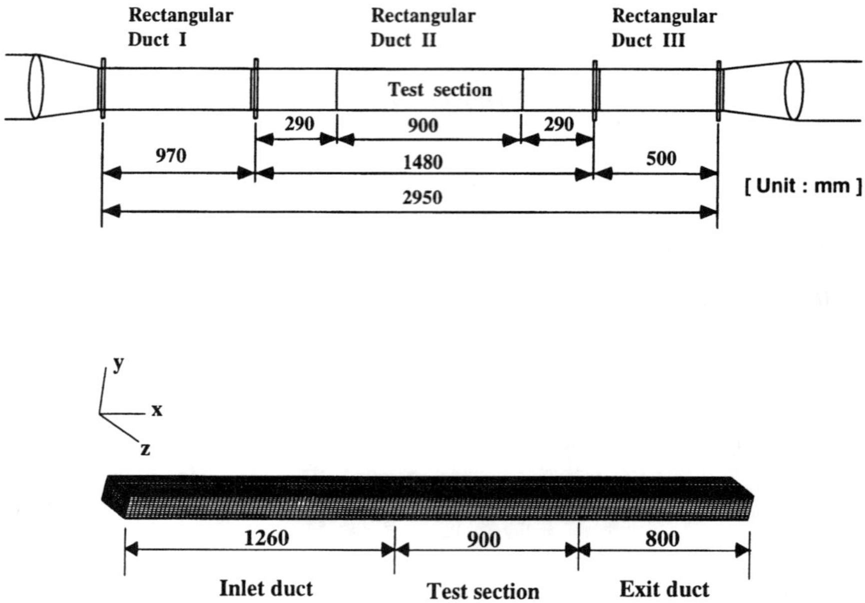

Fig. 1 Comparison of the solution domain between experimental apparatus and numerical grid.
a) Design of magnetohydrodynamic thruster duct
b) Grid of magnetohydrodynamic thruster duct used in numerical simulation

boundary conditions are applied to solve the equations: a uniform flow at the inlet, a fully developed flow at the exit, and no slip condition at the walls.

Figure 2 shows the applied magnetic field in magnetic field solution, which is constant in the region of the test section with the exponentially decaying to zero beyond the electrode edges at the boundary. The exponential decay of the applied magnetic field represents fringing at the magnetic field edges.

In the analysis of the electric field, φ is given on the electrodes, and on the nonconducting walls the following boundary conditions are given:

$$\frac{\partial \varphi}{\partial n} = 0 \tag{12}$$

where n is the unit normal vector at the duct wall boundary.

III. Numerical Solution

A computer code denoted as MHD-3D has been developed by adding the Lorentz force term as the body force, which is obtained by the magnetic field and electric field solver, to the computer code for solving the curvilinear Navier–Stokes equations in three-dimensional space (CNS-3D) (Ref. 5).

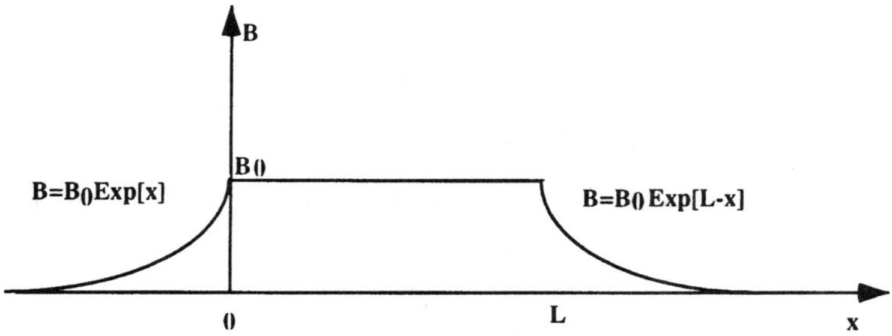

Fig. 2 Applied magnetic field.

For incompressible Newtonian fluid, the continuity, the conservative form of the momentum, magnetic field, and electric field equations have similar formulation and can be written as follows:

$$\boldsymbol{U}_t + \boldsymbol{E}_x + \boldsymbol{F}_y + \boldsymbol{G}_z = \boldsymbol{S} \tag{13}$$

where the subscripts mean partial derivatives with the respective variables, and

$$\boldsymbol{U} = \begin{Bmatrix} 1 \\ u \\ v \\ w \\ B1 \\ B2 \\ B3 \\ \varphi \end{Bmatrix} \qquad \boldsymbol{E} = \begin{Bmatrix} u \\ uu - vu_x \\ uv - vv_x \\ uw - vw_x \\ uB1 - v_m B1_x \\ uB2 - v_m B2_x \\ uB3 - v_m B3_x \\ \varphi_x \end{Bmatrix}$$

$$\boldsymbol{F} = \begin{Bmatrix} v \\ vu - vu_y \\ vv - vv_y \\ vw - vw_y \\ vB1 - v_m B1_y \\ vB2 - v_m B2_y \\ vB3 - v_m B3_y \\ \varphi_y \end{Bmatrix} \qquad \boldsymbol{G} = \begin{Bmatrix} w \\ wu - vu_z \\ wv - vv_z \\ ww - vw_z \\ wB1 - v_m B1_z \\ wB2 - v_m B2_z \\ wB3 - v_m B3_z \\ \varphi_z \end{Bmatrix}$$

$$
S=\left\{\begin{array}{c}
0 \\
(vu_x)_x+(vv_x)_y+(vw_x)_z-\frac{1}{\rho}P_x+\frac{\sigma}{\rho}[-\varphi_y+wB1)B3-(-\varphi_z-vB1)B2-(B3^2+B2^2)u] \\
(vu_y)_x+(vv_y)_y+(vw_y)_z-\frac{1}{\rho}P_y+\frac{\sigma}{\rho}[-\varphi_z+uB2)B1-(-\varphi_x-wB2)B3-(B1^2+B3^2)v] \\
(vu_z)_x+(vv_z)_y+(vw_z)_z-\frac{1}{\rho}P_z+\frac{\sigma}{\rho}[-\varphi_x+vB3)B2-(-\varphi_y-uB3)B1-(B2^2+B1^2)w] \\
B1u_x+B2u_y+B3u_z \\
B1v_x+B2v_y+B3v_z \\
B1w_x+B2w_y+B3w_z \\
(w_y-v_z)B1+(u_z-w_x)B2+(v_x-u_y)B3-(B3_y-B2_z)u-(B1_z-B3_x)v-(B2_x-B1_y)w
\end{array}\right\}
$$

where (u,v,w) is the velocity vector, $(B1,B2,B3)$ the magnetic field vector, and the subscripts denote partial derivatives. Equation (13) is transformed to a general curvilinear coordinate system (ξ,η,ζ), for numerical solution in the form of

$$
U_t+E_\xi\xi_x+E_\eta\eta_x+E_\zeta\zeta_x+F_\xi\xi_y+F_\eta\eta_y+F_\zeta\zeta_y+G_\xi\xi_z+G_\eta\eta_z+G_\zeta\zeta_z=S \tag{14}
$$

For turbulent flow computations, the present code has employed the standard k-ε turbulence model[6] to obtain the turbulent eddy viscosity. A theoretical analysis conducted by Hinatsu[7] on the effect of electromagnetic field on the turbulence shows that the effect is unnegligible. However, the analysis was based on the assumption that the constants involved in the equations for k and ε are not influenced by the presence of the electromagnetic field. Because of such an ambiguity and lack of available experimental evidence, the authors decided to use the k-ε model without considering the effect of the electromagnetic field.

Despite the negligence of the effect of the Lorentz force on the k-ε model, the agreement of the computed results of the mean velocities with the experimental results are judged to be reasonable, which, of course, may not be used for the justification of the negligence of the Lorentz force effect on the k-ε equations. To save the computational efforts, the widely used wall function approach[8] is employed to provide the near-wall boundary conditions for the momentum and the k-ε turbulence model.

The set of nonlinear partial differential equations governing the MHD flow is to be solved by means of finite difference approximations that are performed in the transformed domain. For treating the convection terms the hybrid scheme introduced by Patankar[9] is employed.

In this study, staggering grid systems are used. The velocity components u, v, and w are solved and stored at the grid nodes and the pressure P is evaluated at the corners of the control volume of u, v, and w. To estimate the pressure field, a pressure correction is derived approximately using SIM-

PLE-C algorithm.[10] A converging solution is obtained using the following criterion:

$$\text{Error} \equiv \frac{\left(|\nabla u|_{\max} + |\Delta v|_{\max} + |\Delta w|_{\max}\right)}{U} + \frac{|P|_{\max}}{\rho U^2} + \frac{\left(|\Delta B1|_{\max} + |\Delta B2|_{\max} + |\Delta B3|_{\max}\right)}{B} + \frac{|\Delta \varphi|_{\max}}{UBL} \leq 3.0 \times 10^{-4} \quad (15)$$

IV. Experimental Apparatus and Method

Figure 3 shows a schematic diagram of the experimental apparatus used in this study, consisting of a circulating sea-water loop, an MHD thruster duct, and a dipole electromagnet. The flow rate was monitored by a magnetic flow-meter and was controlled to meet the desired experimental conditions. The regulated sea water, which was taken from the Pohang bay, enters into the inlet duct. The water after the duct was recirculated to an inlet head tank by a pump through a reservoir tank. Joule heating of the electrodes increases temperature of the sea water in the loop with the lapse of time. A heat exchanger of 18 kW capacity was installed in the reservoir tank to maintain the sea-water temperature constant. During the experiment the temperature variation was sustained within ±1°C.

The coordinate system and the dimension of the MHD duct are shown in Fig. 4. The MHD duct was made of nonferrous substance such as acyl and PVC because metals can alter the magnetic field. The MHD thruster duct is 900 mm in length and has a rectangular cross section of 40 by 70 mm in

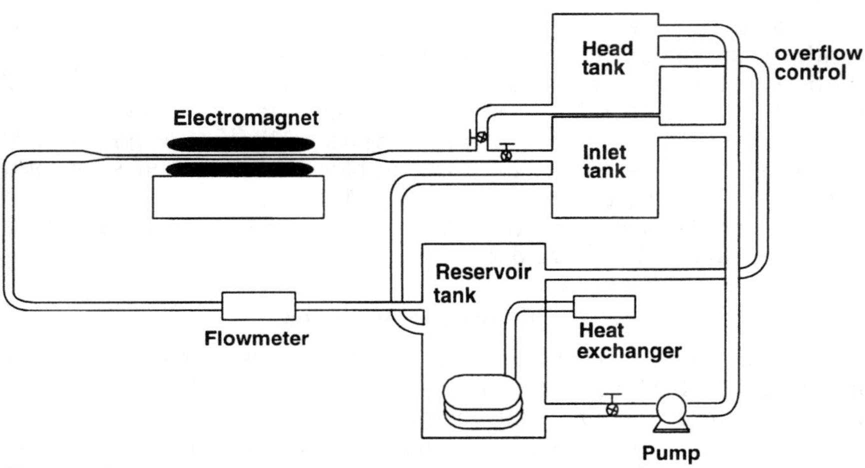

Fig. 3 Schematic diagram of the magnetohydrodynamic thruster experimental loop.

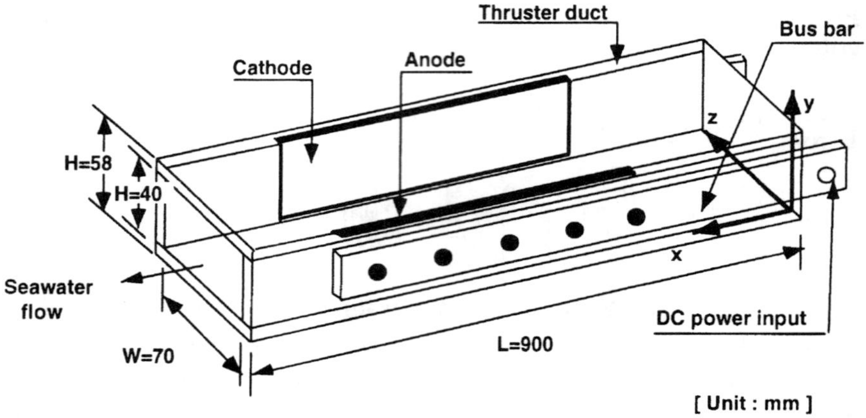

Fig. 4 The coordinate system and dimension of magnetohydrodynamic duct.

dimension. A 1.26-m-long rectangular inlet duct was installed in the upstream of the duct to make a stable flow at the MHD thruster inlet. The duct lies in the gap of 90 mm between two electromagnets and electrodes that are attached on both inner sides of the duct.

The dipole electromagnet was used to create a uniform magnetic field (within 0.01% variation) in the vertical direction of the duct. The maximum magnetic flux density attainable in the MHD duct was 1.2 T. The measured results confirmed the magnetic flux density being uniform throughout the space where the duct was to be installed.

Each electrode is connected with a bus bar for conducting electric current from the bus bar to the electrode.

A fiberoptic LDV system was used to measure the velocity distribution inside the duct along the streamwise direction as shown in Fig. 5. As seeding particles, aluminum powder was used. The fiberoptic LDV system consisted of a 4 W Ar-ion laser, optical lenses and mirrors, a fiberoptic cable, a photomultiplier, a frequency shifter, a counter-type signal processor, and an IBM PC. Because the side wall of the duct was blocked by electrodes, the laser beam was to be injected from the bottom. A 45 deg reflecting mirror was placed under the duct to reflect the laser beam at a 90 deg angle and form a measuring volume inside the duct.

V. Results and Discussions

In this analysis, working fluid is sea water whose property is as follows: density (ρ) = 1024 kg/m^3; kinematic viscosity (ν) = 1.044E-06 m^2/s; and electrical conductivity (σ) = 4.0 mho/m.

The grid system of $149 \times 15 \times 23$ size was used in the numerical computation. Figure 1b shows the three-dimensional representation of the grid system for the test section. Hydraulic diameter of this duct is D_h=area/wetted perimeter = 0.051 m.

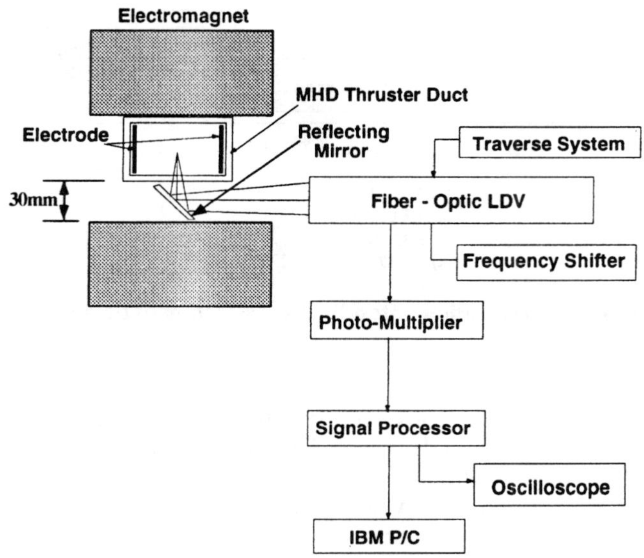

Fig. 5 Schematic of the fiberoptic laser Doppler velocimetry measurement system.

Numerical results were obtained for four different cases, and the flow conditions of each case are summarized in Table 1. Figure 6 represents the test lines used in the numerical calculation for selection of field variables.

Before the experiment, the uniformity of the magnetic field strength was confirmed and calculation of the x component Lorentz force in the duct was carried out; the results are shown in Fig. 7.

Uniform Lorentz force is found along the x direction of the duct except at the end region of the electrodes. The sharp peaks at the ends of the electrodes indicate an abrupt end effect of the electric potential. The end effects exist mainly because of the curvature of electric current vector $\boldsymbol{J}$ in the two edges of the electrodes.

Figure 8 shows pressure variations along the bottom plate of the MHD duct. The dashed and solid lines show numerical and experimental results,

Table 1 Summarization of the parameters for four cases

Parameter	Case 1	Case 2	Case 3	Case 4
Magnetic field, B	0.5 T	0.5 T	1.0 T	0
Electric current, I	100 A	50 A	50 A	0
Mean velocity, U	0.602 m/s	0.444 m/s	0.602 m/s	0.602 m/s
Reynolds num., Re	29,400	21,700	29,400	29,400
Hartmann num., Ha	1.56	1.56	3.12	0
Interaction par., N	8.27E-05	1.12E-04	3.31E-04	——

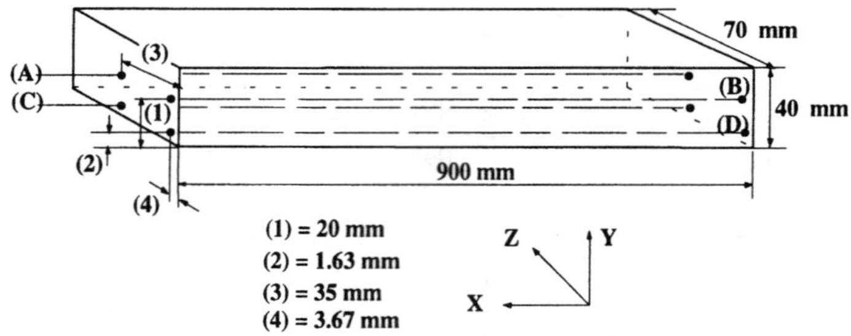

Fig. 6 Representation of test lines for selection of field variables.

respectively. Pressure increases linearly with the Lorentz force in the lengthwise direction of the MHD duct. Good agreement can be observed between the numerical and experimental results in the figure.

Figure 9 shows streamwise velocity distribution along the four test lines of the MHD duct. The solid and dashed lines represent numerical results of the cases 1 and 4 of Table 1, respectively. The quadrangles show corresponding experimental data. The numerical results reveal that the velocity profiles are not significantly affected by the Lorentz force. The numerical and experimental results agree well in the electrode region of the duct. The magnitudes of v and w velocities are comparatively small compared to u. The streamwise velocity distributions were measured for various magnetic field, electric current densities, and flow rates.

For velocity measurement initial flow was given to sweep out the bubbles in the duct using the head tank. The fiberoptic LDV probe was traversed horizontally (z axis) and vertically (y axis) with an interval of 5 mm at three downstream stations of x/L=0.0, 0.6, and 0.98. Thus, the streamwise

Fig. 7 *X*-directional Lorentz force distribution at B = 0.5 T, I = 100 A, and Re = 29,400.

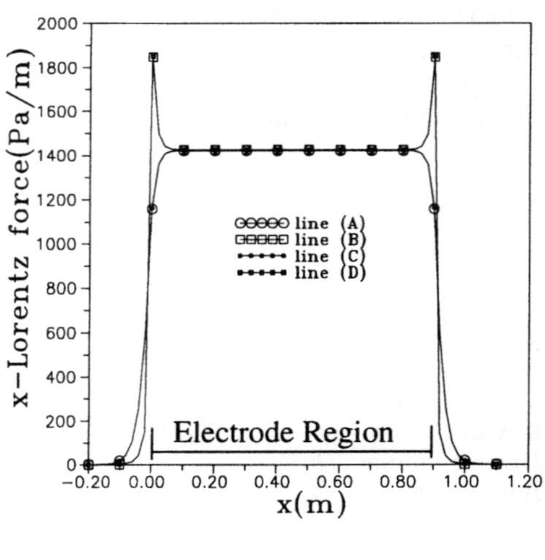

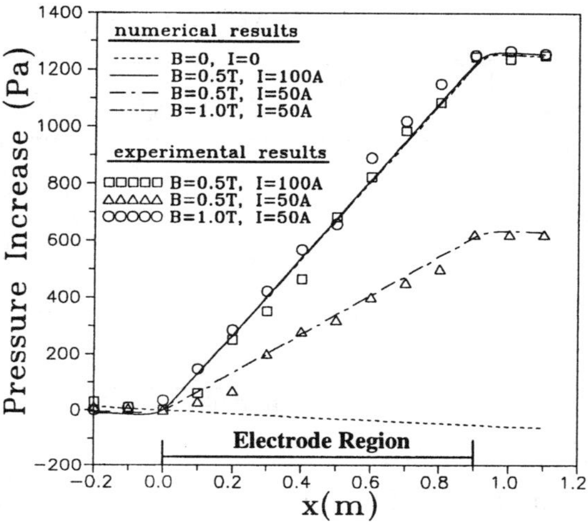

Fig. 8 Profiles of pressure variation at $y/H = 0$ and $z/W = 0.5$.

mean velocity and turbulent intensity profiles at each section were obtained.

The velocity measuring planes are shown in Fig. 10. Here the anode wall indicates $z/W = 0$ where W is the width of the duct that is 70 mm, and the cathode wall $z/W = 1$, the bottom wall $y/H = 0$, and the top wall $y/H = 1$, where H is the depth of the duct.

The computed u-component velocity profiles are compared with the experimental results with and without the Lorentz force in Fig. 11 for horizon-

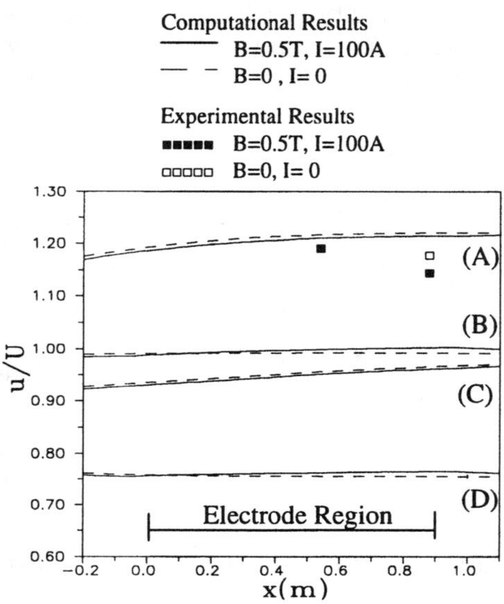

Fig. 9 Streamwise velocity distribution along x direction.

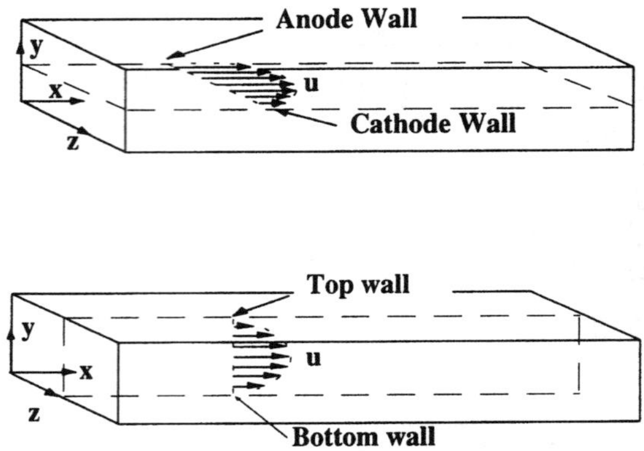

Fig. 10 Schematic view of velocity measuring planes.
a) Horizontal plane
b) Vertical plane

tal plane and in Fig. 12 for vertical plane. The computational results agree well with experimental results in the upstream region. As flow goes downstream, the two results in the zero Lorentz force case show symmetric velocity profiles, but the measured results with the presence of the Lorentz force show inflectional velocity profiles.

The agreement between the computed and the measured results in the presence of the Lorentz force seems to deteriorate for $x/L > 0.5$. The disagreement is caused by the shortcomings of the theory that ignores the presence of the bubbles in the field due to the electrolysis. The velocity profiles measured for different sets of the electric and magnet intensities showed similar results.

Figure 13 shows the mean velocity and turbulent intensity profiles measured along the horizontal and the vertical plane for the case 1. At $x/L = 0$, the velocity profiles are smooth, but the velocity distribution becomes irregular as the flow goes downstream. The velocity near the anode and upper surface increases at downstream and appears to accelerate as the flow goes

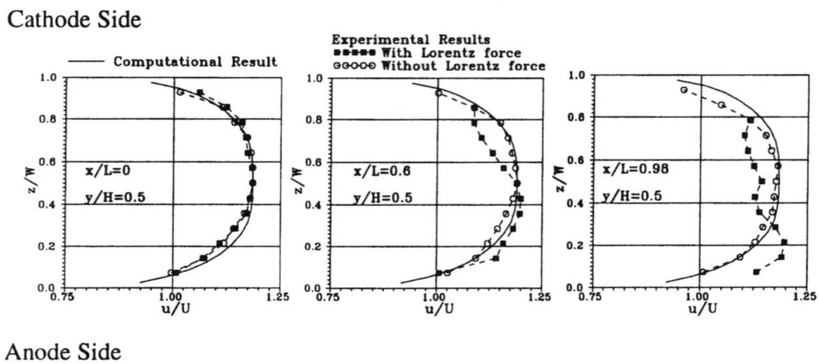

Fig. 11 The mean velocity distribution along horizontal line at $y/H = 0.5$ for $B = 0.5$ T, $I = 100$ A, and $Re = 29{,}400$.

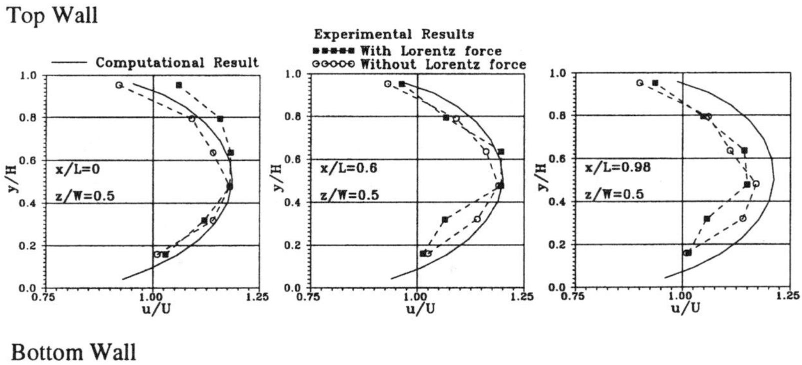

Fig. 12 The mean velocity distribution along the vertical line at $z/W = 0.5$ for $B = 0.5$ T, $I = 100$ A, and $Re = 29{,}400$.

downstream. The measured velocity profiles and turbulent intensities for the other cases, i.e., 2 through 4, exhibited similar trends.

From the experimental results, including those not presented in the paper, one could observe that the turbulence intensity decreases as the Lorentz force increases, whereas the change in the mean velocities with the magnitude of the Lorentz force are relatively small.

Electrical conductivity of sea water could be affected by the electrolytic bubbles at the surface of the electrodes. Bubbles were continuously formed and swept downward with the main stream while gradually rising toward the ceiling because of buoyancy. The velocity profiles are assumed to be attributed to the interaction between bubbles in sea water and magnetic field and spatial variation of the electrical conductivity due to the electrolytic bubbles. The turbulent intensity distribution has large values near the walls. This high turbulent intensity near the walls seems to be affected by the difference in the bubble distribution.

Figures 14–16 show the measured velocity profiles for various Lorentz forces at $x/L = 0$, 0.6, and 0.98, respectively. The solid line represents the velocity profile without the Lorentz force. The velocity profiles are smooth and symmetric and nearly the same at $x/L = 0$ for all conditions; however, they become irregular as the flow goes downstream. The experimental results of the horizontal-plane velocity for $B = 1.0$ T and $I = 50$ A are not presented in Fig. 15a because of the uncertainty of the results.

When the magnitude of the electric current is large, the average velocity in the duct increases, and a large quantity of bubbles are generated. The wavy cavities were observed occasionally at the upper surface of the exit duct. Thus, it can be postulated that under certain conditions the MHD duct flow could become unstable because of the electrolysis bubble and cavity phenomena.

The analysis of the instability in the MHD thruster duct is not investigated in the present work. Further investigations on bubbly flow with stability

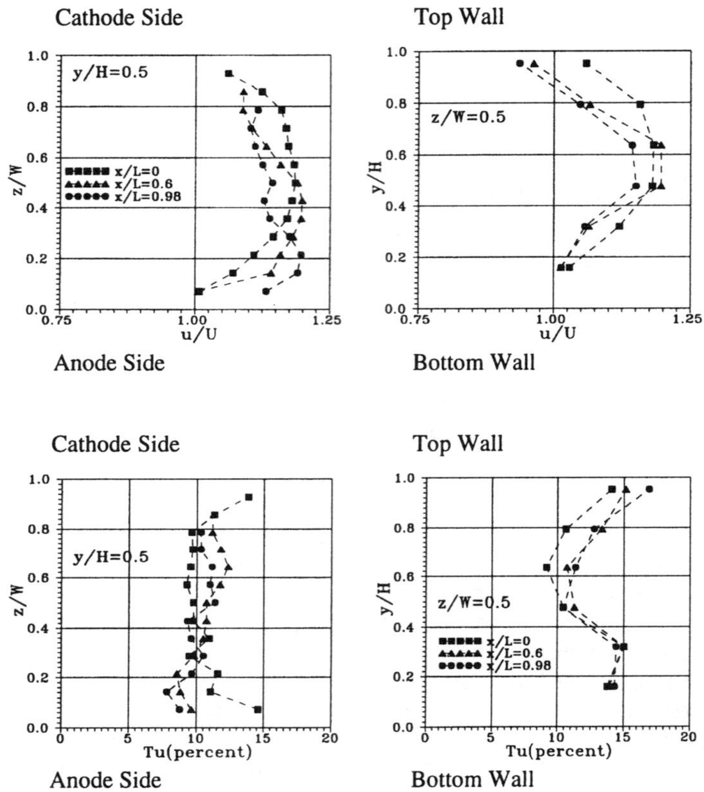

Fig. 13 Streamwise mean velocity and turbulent intensity profiles along the *x* direction at $B = 0.5$ T, $I = 100$ A, and $Re = 29{,}400$.
a) Horizontal plane velocity profiles
b) Vertical plane velocity profiles
c) Horizontal plane turbulent intensity
d) Vertical plane turbulent intensity

analysis will be necessary to evaluate the flow behaviors of an MHD duct flow more accurately.

VI. Concluding Remarks

The Lorentz force does not have much influence on the velocity profiles in a rectangular MHD duct in the region where the void fraction due to bubbles is small; however, when the bubble distribution density increases, the local Lorentz force density is affected and so the velocity distribution. The pressures along the MHD duct in the direction of the flow increases almost linearly proportional to the total Lorentz force.

The measured velocity profiles are in good agreement with the computational results in the upstream region of the MHD duct. The discrepancies in

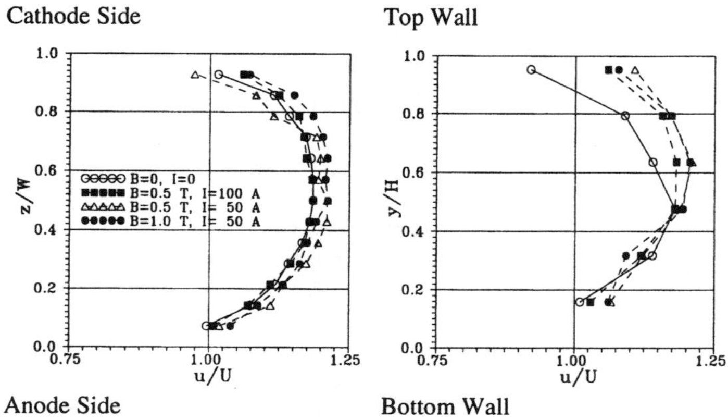

Fig. 14 Mean streamwise velocity profiles at $x/L = 0$.
a) Horizontal plane velocity at $y/H = 0.5$
b) Vertical plane velocity at $z/W = 0.5$

the aft region of the duct are assumed to be caused by the presence of the bubbles due to electrolysis. The flow inside the MHD duct was two-phase turbulent bubbly flow, and the flow characteristics seem to be affected by the nonuniform electric conductivity.

Further investigations of the bubble effects on the MHD duct flows are recommended in order to accurately predict the flow characteristics in an MHD propulsion duct.

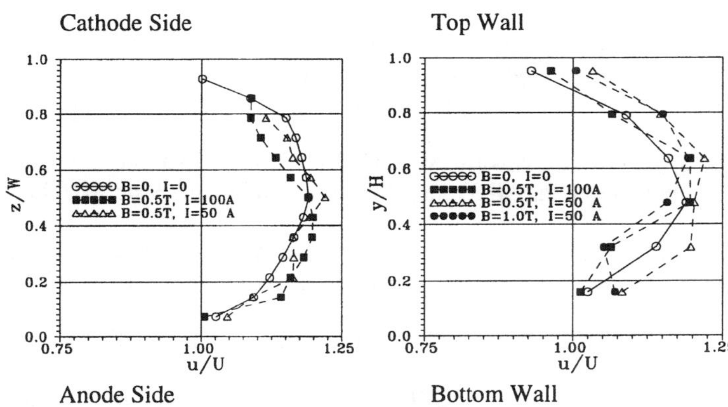

Fig. 15 Mean streamwise velocity profiles at $x/L = 0.6$.
a) Horizontal plane velocity at $y/H = 0.5$
b) Vertical plane velocity at $z/W = 0.5$

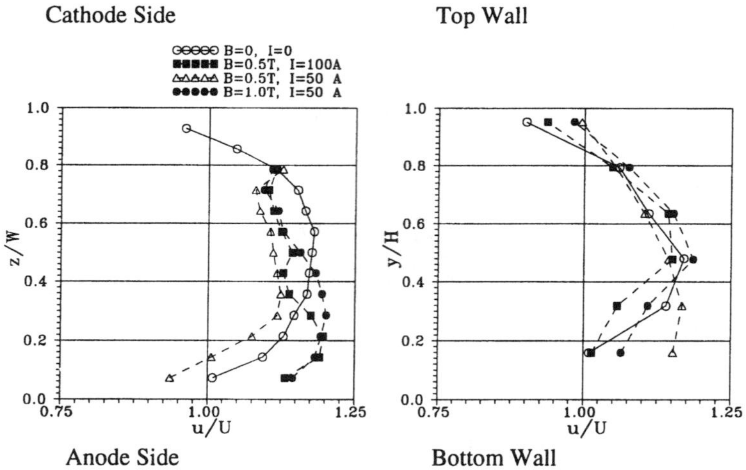

Fig. 16 Mean streamwise velocity profiles at $x/L = 0.98$.
a) Horizontal plane velocity at $y/H = 0.5$
b) Vertical plane velocity at $z/W = 0.5$

Acknowledgments

The authors acknowledge with gratitude the supports granted by Agency for Defense Development, Samsung Heavy Industries Company, and Advanced Fluids Engineering Research Center of Pohang University of Science and Technology. The assistance rendered by S. J. Kim, graduate student at Postech, in preparing the manuscript is deeply appreciated.

References

[1]Friauf, J. M., "Electromagnetic Ship Propulsion," *ASNE J.*, Feb. 1961, pp. 139–142.

[2]Rice, W. A., *U.S. Patent 2997013*, Aug. 22, 1961.

[3]Way, S., "Electromagnetic Propulsion for Cargo Submarines," *Journal of Hydronautics*, Vol. 2, 1968, pp. 49–57.

[4]Motora, S., Imaichi, K., Nakato, M., and Takezawa, S., "An Outline of the R&D Project on Superconducting MHD Ship Propulsion in Japan," *Proc. MHDS 91*, 1991, pp. 53–68.

[5]Chen, Y. S., "A Computer Code for Three-Dimensional Incompressible Flows Using Nonorthogonal Body-Fitted Coordinate Systems," NASA CR-178818, Mar. 1986.

[6]Launder, B. E., and Spalding, D. B., "Mathematical Models of Turbulence," Academic, London, 1972.

[7]Hinatsu, M., "k-ε Modeling of MHD Flow and Its Simulation of Channel Flow," *Journal of the Society of Naval Architecture—Japan*, Vol. 161, 1987, pp. 34–41.

[8]Launder, B. E., and Spalding, D. B., "The Numerical Computation of Turbulent Flows," *Comp. Meth. Appl. Mech. Engr.*, Vol. 3, 1974, pp. 269–289.

[9]Patankar, S. V., "Numerical Heat Transfer and Fluid Flow," McGraw–Hill, New York, 1980.

[10]Doormal, J. P., and Raithby, G. D., "Enhancements of the SIMPLE Method for Predicting Incompressible Fluid Flows," *Numerical Heat Transfer*, Vol. 7, 1984, pp. 147–163.

Chapter 28

Nonslip Creeping Electrolyte Flow Past a Moving Electromagnetic Cylinder

Victor Miroshnikov* and Andrejs Shishko†
Institute of Physics, Academy of Sciences, Salaspils LV-2169, Latvia

I. Introduction

THE recent activity in the scientific domain of the boundary-layer control by means of the electromagnetic (EM) forces[1,2] connected with the problem of magnetohydrodynamic (MHD) ship propulsion[3] has recently come under scientific scrutiny mainly in the flat geometry. In this first classical problem, periodic systems of current-supplying electrodes and permanent magnets form a quasiexponential profile of the EM force that terminates the boundary-layer thickness and operates its turbulization.[3,4] The present paper is devoted to the second classical problem connected with the EM control of the boundary-layer separation, which is expedient to study in the cylindrical geometry as the round cylinder is a bluff body with respect to the transverse flow.[5] More exactly, the suggestion to localize the EM forces within the boundary layer with the purpose of forming the necessary velocity and controlling the boundary-layer separation was made. The idea was borrowed from the theory and the practice of the inertial-circulation airfoils, where a distributed suction enables control of the boundary-layer separation, and the principal possibility of an EM separation control was demonstrated experimentally.[5]

The problem considered may be formulated with respect to distribution of an EM force needed to prevent the flow separation. In the current paper, the problem is solved exactly in the Stokes regime of a flow past the EM cylinder.

*currently ch. Manhattan College, Mathematics and Computer Science Department, Manhattan College Parkway, Riverdale, New York 10471.

II. Electromagnetic Cylinder

Similar to the case of a flat plane,[4] the EM system of a round cylinder-shaped body of radius R is constructed of magnetic and electric systems that are periodic in the axial x_3 direction to form a quasiexponential profile of the EM force. As the speed of the EM cylinder U is small enough ($Re \ll 1$) and the electrical conductivity of electrodes takes a finite value, induced currents are negligible compared with the applied ones, and induced magnetic fields are small compared with the external ones as well. In the reference frame of the moving body ($X_1 = x_1 - Ut$, x_2, x_3), the problem for electric current $\mathbf{j}$ and the magnetic field B may be described in the quasistationary approximation

$$\operatorname{div} \mathbf{j} = 0, \qquad \operatorname{rot} \mathbf{j} = 0, \qquad \operatorname{div} \mathbf{B} = 0, \qquad \operatorname{rot} \mathbf{B} = 0 \tag{1}$$

formulated for the electric ($\mathbf{j} = -\sigma \operatorname{grad} \Phi_E$) and magnetic ($\mathbf{B} = \operatorname{grad} \Phi_B$) potentials in the cylindrical frame of coordinates $\left[r = \sqrt{X_1^2 + x_2^2}, \theta = \operatorname{arctg}(x_2 / X_1), x_3\right]$

$$\frac{\partial^2 \Phi_i}{\partial r^2} + \frac{1}{r}\frac{\partial \Phi_i}{\partial r} + \frac{1}{r^2}\frac{\partial^2 \Phi_i}{\partial \theta^2} + \frac{\partial^2 \Phi_i}{\partial x_3^2} = 0, \qquad i = E, B \tag{2}$$

with the periodic boundary conditions describing disk-like permanent magnets and dipole-like current-supplying electrodes:

$$r = R, \qquad j_r = j_0 \cos(2\,\pi\, x_3/l)\sin\theta, \qquad B_r = B_0 \sin(2\,\pi\, x_3/l) \tag{3}$$

The three-dimensional solution of the problem (2–3) describing nonuniform magnetic and electric fields of the EM cylinder may be expressed using the modified Bessel's functions K_i of i order that vanishes exponentially at infinity in the cylindrical case:

$$\begin{gathered}
\frac{j_r}{j_0} = \frac{K_0(r_\beta) + K_2(r_\beta)}{K_0(\beta) + K_2(\beta)} \cos(z_\beta)\sin\theta \\
\frac{j_\theta}{j_0} = -\frac{-\mathrm{K}_1(r_\beta)}{r_\beta[K_0(\beta) + K_2(\beta)]} \cos(z_\beta)\cos\theta \\
\frac{j_z}{j_0} = -\frac{2\mathrm{K}_1(r_\beta)}{K_0(\beta) + K_2(\beta)} \sin(z_\beta)\sin\theta \\
\frac{B_r}{B_0} = \frac{K_1(r_\beta)}{K_1(\beta)} \sin(z_\beta), \qquad \frac{B_z}{B_0} = -\frac{K_0(r_\beta)}{K_1(\beta)} \cos(z_\beta)
\end{gathered} \tag{4}$$

where the following dimensionless coordinates are used: $r_\beta = \beta r/R$ and $z_\beta = \beta x_3/R$. Defined as the ratio of the cross-sectional perimeter to the period of the EM system parameter, $\beta = 2\pi R/l$ controls a spatial scale of the fields under consideration. If $\beta \gg 1$, the EM fields are localized in a narrow layer of scale $1/\beta$. When $\beta \approx 1$, the EM fields act at distances of R order. Small values of β correspond to the large-scale fields. Three-dimensional distributions of the monopole-like magnetic field and the dipole-like electric current possessing y axis of symmetry are given in Figs. 1 and 2, respectively. Intensities of both of the fields change signs periodically along the axial direction demonstrating that they are spatially rotating fields for a given value of $r_\beta = r_{\beta 0}$.

For the aim of methodical clarity, the total Lorentz force $\mathbf{j} \times \mathbf{B}$ may be expressed as a linear superposition of two-dimensional [Eq. (5)] and three-dimensional [Eq. (6)] parts,

$$
\begin{aligned}
\frac{f_{2r}}{j_0 B_0 f_0} &= \frac{K_1(r\beta)}{r_\beta} K_0(r_\beta)\cos\theta \\
\frac{f_{2\theta}}{j_0 B_0 f_0} &= \left[K_1^2(r_\beta) + \frac{K_0(r_\beta) + K_2(r_\beta)}{2} K_0(r_\beta)\right]\sin\theta
\end{aligned}
\tag{5}
$$

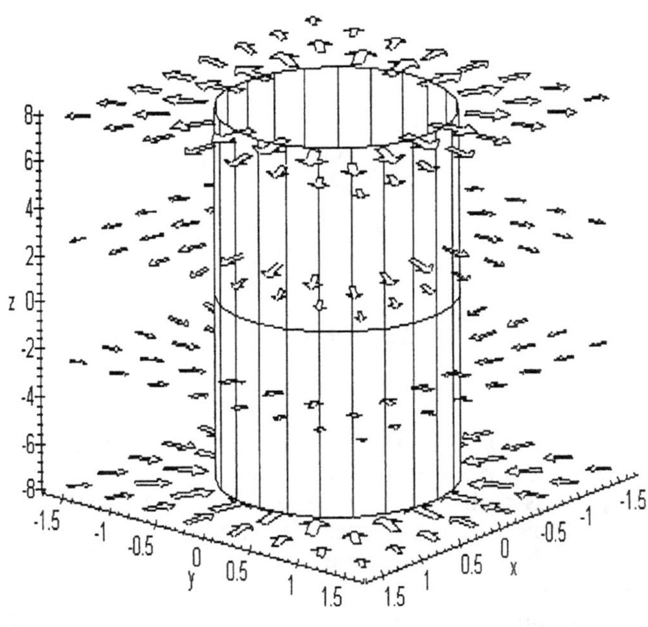

Fig. 1 Vectors of three-dimensional magnetic induction [Eq. (4)] are given on planes $z_\beta = \pm\pi/4, \pm\pi/2$ for $\beta = 0.2$.

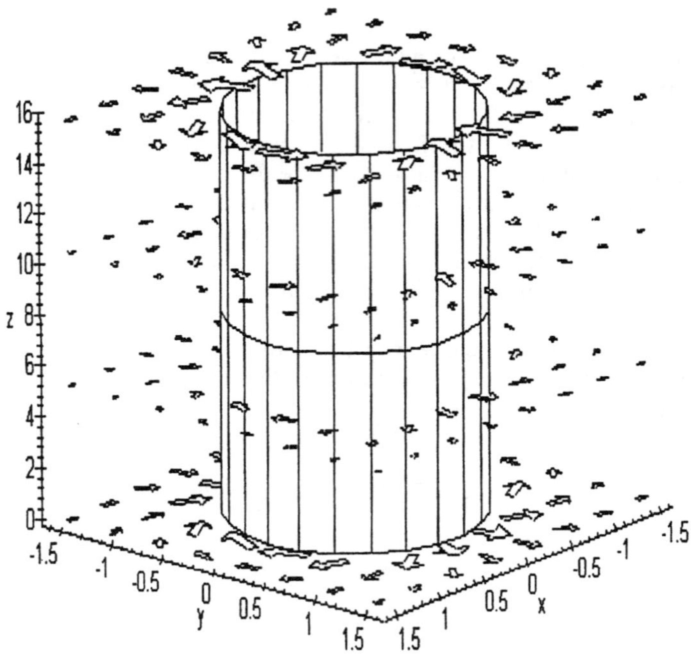

Fig. 2 Vectors of three-dimensional electric current [Eq. (4)] are given on planes $z_\beta = 0, \pi/3, 2\pi/3, \pi$ for $\beta = 0.2$.

$$\frac{f_{3r}}{j_0 B_0 f_0} = \frac{K_1(r_\beta)}{r_\beta} K_0(r_\beta)\cos(2z_\beta)\cos\theta$$

$$\frac{f_{3z}}{j_0 B_0 f_0} = \frac{K_1^2(r_\beta)}{r_\beta}\sin(2z_\beta)\cos\theta \tag{6}$$

$$\frac{f_{3\theta}}{j_0 B_0 f_0} = \left[-K_1^2(r_\beta) + \frac{K_0(r_\beta) + K_2(r_\beta)}{2} K_0(r_\beta)\right]\cos(2z_\beta)\sin\theta$$

where $f_0 = \{K_1(\beta)[K_0(\beta) + K_2(\beta)]\}^{-1}$. Characteristic properties of distribution of the Lorentz force are the following: primarily, its period is twice smaller than periods of the EM fields, and secondly, the spatial distribution of the force is nonlinear with respect to $K_i(r_\beta)$. In any cross section the Lorentz forcing field describes a dipole-like force field with the symmetry axis in the x direction, and the dipole field changes periodically in the axial direction. The two-dimensional Lorentz force (Fig. 3) repels the stream from the upstream stagnation point and presses to the downstream one. Because the two-dimensional force preserves the flow from the separation from the

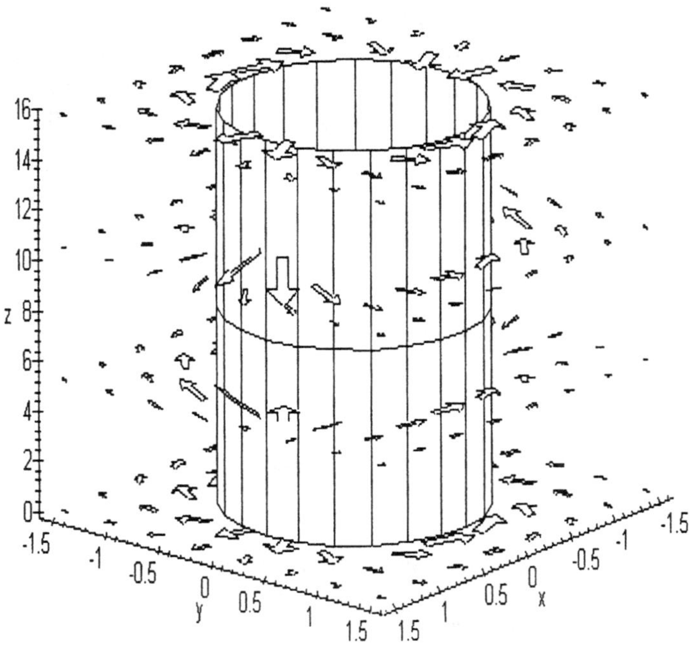

Fig. 3 Vectors of three-dimensional total Lorentz force described by Eqs. (5) and (6) are given on planes $z_\beta = 0, \pi/3, 2\,\pi/3, \pi$ for $\beta = 0.2$.

cylinder surface, its influence is always positive. The three-dimensional force changes its sign periodically with respect to z, and the influence of the total Lorentz force (Fig. 3) becomes positive or neutral ($z_\beta = \pi/2$).

In the case of the spherical geometry, the similar dipole EM force field was studied[6] for small Reynolds numbers [Re$=O(1)$] by the method of matching asymptotic expansions. A large variety of spatial distributions of the EM forcing was demonstrated, and it was shown that even two-dimensional flow past a spherical body may be generated for definite values of dimensionless parameters governing the problem. Using methods of the paper[6] in the cylindrical case, one can easily show that a steady creeping flow of an electrolyte may be formed using only a two-dimensional dipole force field. But it is impossible to generate such a dipole field by the EM fields [Eqs. (1) and (2)]. A simple analysis of the possible harmonics of Eq. (2) states that one should inevitably add the dipole field by a three-dimension force field with circulation disrupting possibilities for a two-dimensional nonslip fluid flow. This property of the considered problem is connected presumably with the well-known Stokes paradox that may be resolved only in three-dimensional space, and the EM fields [Eqs. (4)] resulting in the Lorentz forcing fields [Eqs. (5) and (6)] give the simple example of such a possibility.

III. Two-Dimensional Flow Past a Moving Electromagnetic Cylinder

As the geometry of the EM cylinder and the Lorentz force distribution (5) imply, the two-dimensionality of the flow under consideration, the velocity field of the fluid flow $\boldsymbol{v} = \upsilon_x \boldsymbol{i}_x + \upsilon_y \boldsymbol{i}_y = \upsilon_r \boldsymbol{i}_r + \upsilon_\theta \boldsymbol{i}_\theta$ in terms of the stream function

$$\boldsymbol{v} = \beta\,\mathrm{rot}_\beta(\psi \boldsymbol{i}_z), \quad \upsilon_x = \beta\frac{\partial\psi}{\partial y_\beta}, \quad \upsilon_y = -\beta\frac{\partial\psi}{\partial x_\beta}, \quad \upsilon_r = \frac{\beta}{r_\beta}\frac{\partial\psi}{\partial\theta}, \quad \upsilon_\theta = -\beta\frac{\partial\psi}{\partial r_\beta} \tag{7}$$

is described by the vorticity transport equation written in the frame moving with the cylinder

$$\frac{\partial\Delta_\beta\psi}{\partial x_\beta}\frac{\partial\psi}{\partial y_\beta} - \frac{\partial\Delta_\beta\psi}{\partial y_\beta}\frac{2\psi}{\partial x_\beta} - \frac{1}{Re}\Delta_\beta\Delta_\beta\psi = \frac{St}{\beta^3}\left(\frac{\partial f_{2x}}{\partial y_\beta} - \frac{\partial f_{2y}}{\partial x_\beta}\right) \tag{8}$$

where Δ_β is the Laplacian and the characteristic scales of the coordinates, the stream function, and the vorticity $\omega = -\beta^2\Delta_\beta\psi$ are R/β, UR, U/R, respectively. The Reynolds number $Re = UR/\nu$ and the electromagnetic Stuart number $St = j_0B_0R/\rho U^2$ defined with regard to the uniform velocity of the far-upstream flow ($\boldsymbol{v} = -\boldsymbol{i}_x$) represents the ratio of the inertial effects to viscous ones and the ratio of EM forces to inertial ones, correspondingly. Here $\boldsymbol{v}$ and ρ are the kinematic viscosity and the density of the electrolyte. To calculate the drag, necessarily one must evaluate the pressure from the steady two-dimensional Navier–Stokes equation:

$$\frac{1}{\beta^2}\,\mathrm{grad}_\beta\left(Eu\,p + \frac{|\boldsymbol{v}|^2}{2}\right) - \Delta_\beta\psi\cdot\mathrm{grad}_\beta\psi - \frac{1}{Re}\,\mathrm{rot}_\beta(\Delta_\beta\psi\boldsymbol{i}_z) = \frac{St}{\beta^3}\boldsymbol{f}_2 \tag{9}$$

forced electromagnetically by $\boldsymbol{f}_2 = f_{2r}\boldsymbol{i}_r + f_{2\theta}\boldsymbol{i}_\theta = \beta\,\mathrm{rot}_\beta\,(\psi_{2f}\,\boldsymbol{i}_z)$, where $Eu = p_0/(\rho U^2)$ is the Euler number and ψ_{2f} is the vector potential of the EM force [Eq. (5)]:

$$\psi_{2f} = f_0\frac{K_1(r_\beta)}{r_\beta}K_0(r_\beta), \quad f_{2r} = \frac{\beta}{r_\beta}\frac{\partial\psi_{2f}}{\partial\theta}, \quad f_{2\theta} = -\beta\frac{\partial\psi_{2f}}{\partial r_\beta} \tag{10}$$

In the case of a creeping flow ($Re \ll 1$), Eq. (8) reduces to the Laplace equation as the sum of the viscous friction and the EM force is a potential function:

$$\Delta_\beta(\beta^2\omega - Ha^2\psi_{2f}) = 0 \tag{11}$$

where $Ha^2 = ReSt = j_0 B_0 R^2/(\nu \rho U)$ is the electromagnetic Hartmann number. Then the kinematic problem for the stream function becomes

$$\frac{\partial^2 \psi}{\partial r_\beta^2} + \frac{1}{r_\beta}\frac{\partial \psi}{\partial r_\beta} + \frac{1}{r_\beta^2}\frac{\partial^2 \psi}{\partial \theta^2} = -Ha^2 f_0 \beta^{-3} K_1(r_\beta) K_0(r_\beta) \sin(\theta)$$

$$x \to \infty, \quad \psi \to -(r_\beta / \beta)\sin\theta, \quad \rho = 1, \quad \psi = 0, \quad \partial\psi / \partial r_\beta = 0 \tag{12}$$

The solution of problem (12) may be obtained only for the critical value of the EM Hartmann number when the balance between the viscous and the forcing terms exists. The larger the cylinder speed, the stronger EM fields are needed to form the nonslip flow pattern without separation points:

$$\psi = \frac{r_\beta \sin\theta}{\beta K_0^2(\beta)}\left[K_0^2(r_\beta) - K_0^2(\beta) - K_1^2(r_\beta) + \frac{\beta^2 K_1^2(\beta)}{r_\beta^2}\right], \quad Ha_c^2 = \frac{4\beta^2}{f_0 K_0^2(\beta)} \tag{13}$$

Compared with the dashed streamlines of the potential slip flow past a cylinder, streamlines of the two-dimensional nonslip velocity field are given in Fig. 4a, demonstrating formation of upstream and downstream wakes. In the case of both potential and rotational body force $f_2/\beta = -\nabla_\beta \varphi_{2f} + \mathrm{rot}_\beta(\psi_{2f}\, \boldsymbol{i}_z)$, the dynamic problem for the pressure distribution may be written as the Cauchy–Lagrange integral because Eq. (9) takes the form of the Cauchy–Riemann conditions,

$$\frac{\partial \Omega}{\partial x_\beta} = \frac{\partial P}{\partial y_\beta}, \quad \frac{\partial \Omega}{\partial y_\beta} = -\frac{\partial P}{\partial x_\beta}, \quad \Omega = \frac{\beta^2 \omega}{Re} - St\psi_{2f}, \quad P = Eu\, p + St\varphi_{2f} \tag{14}$$

and the integral becomes

$$Eu\, p + St\varphi_{2f} + \int\left[\frac{\partial}{\partial y_\beta}\left(\frac{\beta^2\omega}{Re} - St\psi_{2f}\right)\mathrm{d}x_\beta - \frac{\partial}{\partial x_\beta}\left(\frac{\beta^2\omega}{Re} - St\psi_{2f}\right)\mathrm{d}y_\beta\right] = C \tag{15}$$

where C is a constant. The nonlinear version of the integral for inviscid flows is given elsewhere.[7] Because the EM force is entirely vortical ($\omega = Ha^2\psi_{2f}/\beta^2$) in the case considered, the pressure is uniform.

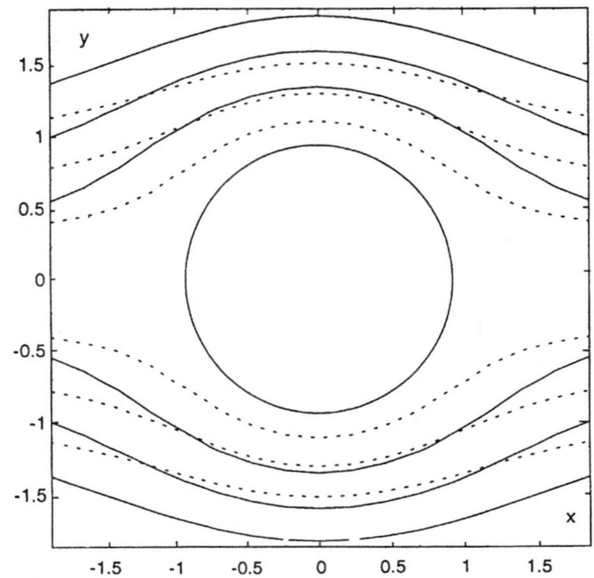

Fig. 4a Streamlines of the nonslip two-dimensional flow [Eq. (13)] compared with those of the slip potential flow (dotted lines).

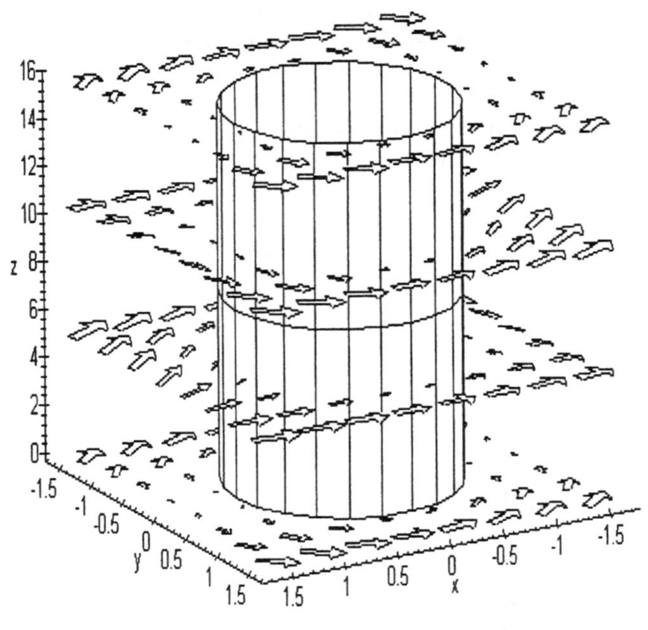

Fig. 4b Vectors of the total velocity described by Eqs. (7), (13), (19), and (21) are given on planes $z_\beta = 0, \pi/3, 2\,\pi/3, \pi$ for $\beta = 0.2$.

IV. Electromagnetic Forced Three-Dimensional Flow

The vector potential of three-dimensional Lorentz force [Eq. (6)] defined by equations $\boldsymbol{f}_3/j_0\, B_0 = \beta \operatorname{rot}_\beta \boldsymbol{A}_f$, div $_\beta \boldsymbol{A}_f = 0$ may be readily calculated, as it has only two components

$$A_{fr} = \frac{f_0}{2\beta}\left[K_0^2(r_\beta) - K_1^2(r_\beta) + \frac{K_1(r_\beta)}{r_\beta} K_0(r_\beta)\right]\sin(2z_\beta)\sin\theta$$

$$A_{f\theta} = -\frac{f_0}{2\beta}\frac{K_1(r_\beta)}{r_\beta} K_0(r_\beta)\sin(2z_\beta)\cos\theta \tag{16}$$

while the third one determines the two-dimensional problem since $A_{fz} = \psi_{2f}$. The transfer equation of the three-dimensional vorticity ($\boldsymbol{w} = \beta \operatorname{rot}_\beta \boldsymbol{v}$) again takes the form corresponding to the transfer equation (11) of two-dimensional vorticity

$$\Delta_\beta(\beta^2 \boldsymbol{w} - Ha_c^2 \boldsymbol{A}_f) = 0 \tag{17}$$

Equation (17) yields that the three-dimensional vorticity which may be obtained by the method of separable variables is generated by the EM force up to the vectorial harmonic function

$$w_r = \frac{2\beta}{K_0^2(\beta)}\left[\Omega_2 K_2(2r_\beta) - \Omega_0 K_0(2r_\beta) + K_0^2(r_\beta) - K_1^2(r_\beta) + \frac{K_1(r_\beta)}{r_\beta} K_0(r_\beta)\right]\sin(2z_\beta)\sin\theta$$

$$w\theta = \frac{2\beta}{-K_0^2(\beta)}\left[\Omega_2 K_2(2r_\beta) + \Omega_0 K_0(2r_\beta) + \frac{K_1(r_\beta)}{r_\beta} K_0(r_\beta)\right]\sin(2z_\beta)\cos\theta \tag{18}$$

$$w_z = \frac{2\beta}{K_0^2(\beta)}\Omega_1 K_1(2r_\beta)\cos(2z_\beta)\sin\theta$$

where Ω_0, Ω_1, Ω_2 are constants. Further solving may be implemented by the two-dimensional method used in the preceding section, but following direct methods[6] and keeping the cylindrical system of coordinates is much simpler. Written with the angular dependencies of the Lorentz force (6)

$$\upsilon_r = K_0^{-2}(\beta) V_r(r_\beta)\cos(2z_\beta)\cos\theta, \qquad \upsilon_\theta = K_0^{-2}(\beta) V_\theta(r_\beta)\cos(2z_\beta)\sin\theta$$

$$\upsilon_z = K_0^{-2}(\beta) V_z(r_\beta)\sin(2z_\beta)\cos\theta \tag{19}$$

the Navier–Stokes equations

$$\frac{d^2V_r}{dr_\beta^2}+\frac{1}{r_\beta}\frac{dV_r}{dr_\beta}-\left(\frac{2}{r_\beta^2}+4\right)V_r-\frac{2}{r_\beta^2}V_\theta+4K_0(r_\beta)\frac{K_1(r_\beta)}{r_\beta}=0$$

$$\frac{d^2V_\theta}{dr_\beta^2}+\frac{1}{r_\beta}\frac{dV_\theta}{dr_\beta}-\left(\frac{2}{r_\beta^2}+4\right)V_\theta-\frac{2}{r_\beta^2}V_r+4\left[K_0^2(r_\beta)-K_1^2(r_\beta)+K_0(r_\beta)\frac{K_1(r_\beta)}{r_\beta}\right]=0 \quad (20)$$

$$\frac{d^2V_z}{dr_\beta^2}+\frac{1}{r_\beta}\frac{dV_z}{dr_\beta}-\left(\frac{1}{r_\beta^2}+4\right)V_z+4\frac{K_1^2(r_\beta)}{r_\beta}=0$$

may be easily split by the symmetric $V_+ = V_r + V_\theta$ and asymmetric $V_- = V_r - V_\theta$ variables, and the velocity distribution takes the form

$$V_r=V_2K_2(2r_\beta)-K_1^2(r_\beta)+V_0K_0(2r_\beta)-K_0^2(r_\beta),\qquad V_2=K_1^2(\beta)/K_2(2\beta)$$

$$V_\theta=V_2K_2(2r_\beta)-K_1^2(r_\beta)-V_0K_0(2r_\beta)+K_0^2(r_\beta),\qquad V_0=K_0^2(\beta)/K_0(2\beta) \quad (21)$$

$$V_z=2\left[V_1K_1(2r_\beta)-K_0(r_\beta)K_1(r_\beta)\right],\qquad V_1=K_0(\beta)K_1(\beta)/K_1(2\beta)$$

When the corresponding parameters of vorticity [Eq. (18)] become $\Omega_2 = V_2 - V_1$, $\Omega_0 = V_0 - V_1$, and $\Omega_1 = V_0 - V_2$, then the stream function $\boldsymbol{A}$ may be readily obtained from the definition $\boldsymbol{v}$ = rot $\boldsymbol{A}$. Similar to the two-dimensional case, the pressure vanishes in the three-dimensional case, as well.

V. Discussion

The nonslip flow pattern of the three-dimensional flow [Eq. (21)] summarized with the two-dimensional flow [Eq. (10)] is given in Fig. 4b. The flow is formed by the interaction of three mechanisms: the viscous friction on the cylinder surface corresponding to the boundary condition $\partial\psi/\partial r = 0$, the inviscid deceleration and deflection related with the boundary condition $\psi = 0$, and the EM forcing described by inhomogeneous terms of Eqs. (12) and (20). Integration of the viscous stresses along the cylinder surface gives the drag force per unit length $F_x = 4\pi\nu\rho U\beta K_1(\beta)/K_0(\beta)$, which linearly depends on velocity of the EM cylinder.

One of the characteristic properties of the considered flow (Fig. 5) is a well-expressed minimum of the critical Hartmann number dependence versus β ($Ha_c(0.2) \approx 3.57$). To understand this phenomenon, surfaces of constant modulus of the total velocity are given in Figs. 6a and 6b for different values of β. When the EM force is concentrated in the boundary layer ($\beta \gg$

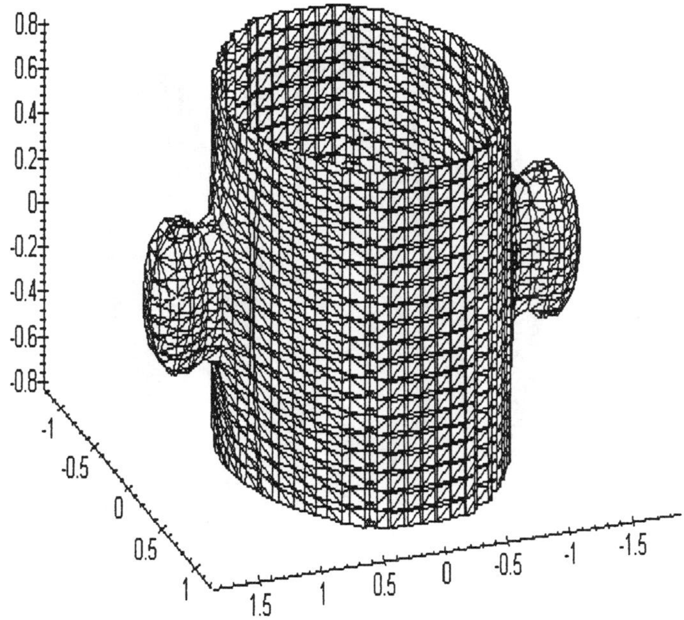

Fig. 5 The critical Hartmann number Ha_c versus β.

1), the high values of EM fields are necessary to decelerate the stream rapidly at small length scales of the EM forcing layer. The spatial scale of the EM force increases with the decrease of β, and it becomes possible to decelerate the stream using smaller values of the EM fields.

Let us describe an effective shape of the EM cylinder by a domain, where the far-upstream velocity diminishes by the order of value. Then the decrease of β corresponds to the formation of drop-shaped wave cutters upstream and downstream of the cylinder (Figs. 6a and 6b), and the EM cylinder effectively takes a form of the three-dimensional streamlined body. The minimal value of the EM Hartmann number provides the most streamlined effective shaping (Fig. 7). Here, we have a parallel effect of all the three deceleration mechanisms that have the same spatial scale by the order of value. For smaller values of β, the cutters separate from the cylinder (Fig. 6b), and the Hartmann number again increases as the deceleration mechanisms act consequently.

Acknowledgments

Financial support of the Latvian Scientific Council is gratefully acknowledged. One of the authors would like to thank the French government for the stipend providing the possibility to finish the paper.

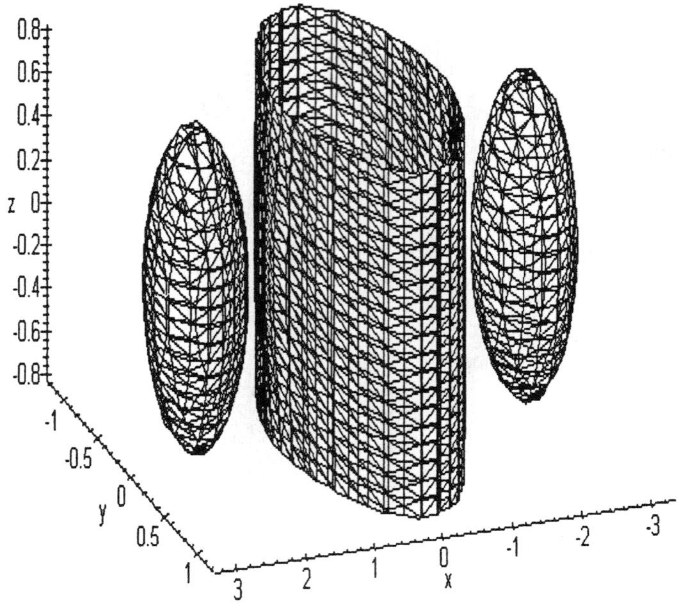

Fig. 6a Surface of constant modulus of the total velocity $|v| = 0.1$ for $Ha_c = 3.85$, and $\beta = 0.4$.

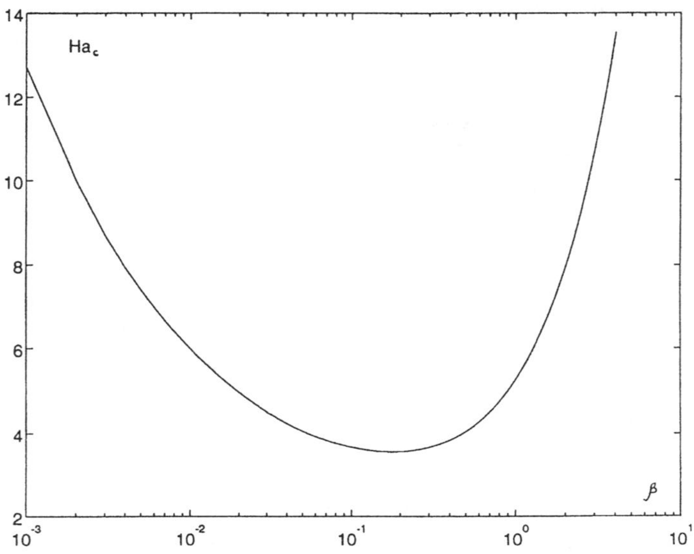

Fig. 6b Surface of constant modulus of the total velocity $|v| = 0.1$ for $Ha_c = 4.06$ and $\beta = 0.05$.

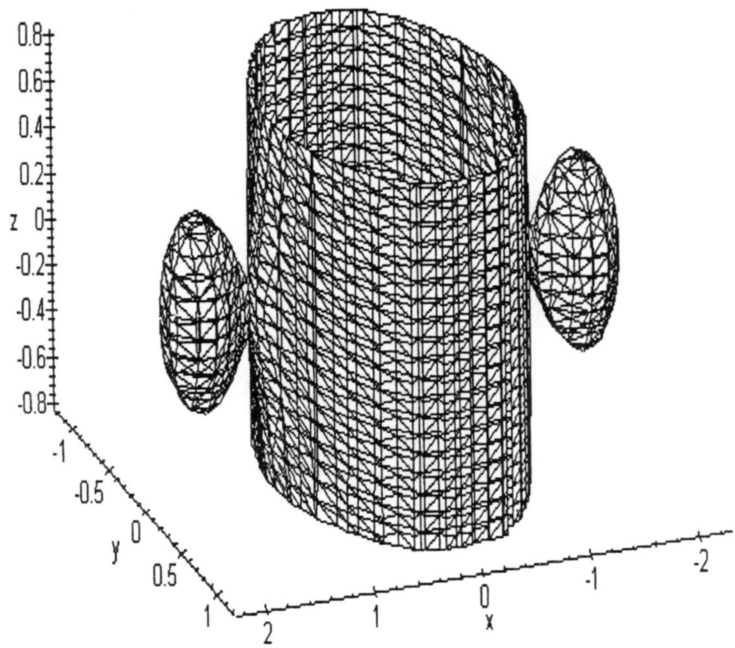

Fig. 7 Effective shape of the EM cylinder (|v| = 0.1) for Ha_c = 3.57 and β = 0.2.

References

[1]Gailitis, A., and Lielausis, O., "On a Possibility to Reduce the Hydrodynamic Resistance of a Plate in an Electrolyte," *Applied Magnetohydrodynamics*, Vol. 12, Inst. of Physics, Riga, 1961, pp. 143–146 (in Russian).

[2]Tsinober, A., and Stern, A. G., "Possibility of Increasing Flow Stability in a Boundary Layer Flow by Means of Crossed Electric and Magnetic Fields," *Magnetohydrodynamics*, Vol. 3, No. 2, 1967, pp. 152–154.

[3]Meng, J. C. S., Henock, C. W., and Hrubes, J. D., "Seawater Electromagnetohydrodynamics: A New Frontier," *Magnetohydrodynamics*, Vol. 30, No. 4, 1994, pp. 483–506.

[4]Henock, C., and Stage, J., "Experimental Investigation of a Salt Water Turbulent Boundary Layer Modified by an Applied Streamwise Magnetohydrodynamic Body Force," *Physics of Fluids*, Vol. 7, No. 6, 1995, pp. 1371–1383.

[5]Lielausis, O., Gailitis, A., and Dukure, R., "Boundary Layer Control by Means of Electromagnetic Forces," *Proceedings of International Conference on Energy Transfer in MHD Flows* (Cadarche, France), 1991, pp. 5–9.

[6]Miroshnikov, V. A., "Motion of a Sphere in Crossed Electric and Magnetic Fields at Small Reynolds Numbers," *Magnetohydrodynamics*, Vol. 17, No. 1, 1981, pp. 63–71.

[7]Miroshnikov, V., "A Solitary Wave on the Surface of a Shear Stream in Crossed Electric and Magnetic Fields: Formation of a Single Vortex," *Magnetohydrodynamics*, Vol. 31, Nos. 1–2, 1995, pp. 169–185.

Chapter 29

On Perspectives of Plasma Magnetohydrodynamic Installations

A. E. Sheindlin and E. B. Matveyeva
Institute for High Temperatures,
Izhorskaya St 13/9 127412 Moscow, Russia

I. The State of the Arts

AFTER impressive experiments in the late 1950s when the possibility of creating plasma magnetohydrodynamic (MHD) generators was demonstrated, the prospects of broad application of MHD installations in energy sector became the object of keen interest. In many countries and primarily in the United States and the Soviet Union, extensive research and development work started in order to have large-scale MHD installations for capacious power plants and for specific peak-load power installations for short-duration runs.

Quite significant was the work undertaken in the Soviet Union to develop large-scale experimental installation U-02 and pilot installation U-25 possessing all required parts of possible MHD power plants. Exclusively valuable results and extensive experience were accumulated in the development and operation of large-scale combined MHD installation. It gave impetus to building a high-capacious MHD power plant in the Soviet Union near Riazan although because of certain technical, political, and economic reasons the construction of the plant was not accomplished.

In the United States quite prominent was the work of AVCO-Everett Laboratory to develop large Mark MHD generators and some components of open-cycle MHD installations. Similar kinds of research carried out by Westinghouse, General Electric, and other U.S. research and industrial centers were rather promising. Scientific research and engineering developments undertaken in France, the United Kingdom, Poland, Italy, India, and other countries turned out to be very important and fruitful. Relatively large-scale experimental installations were developed, and serious theoretical studies were carried out.

The work was mainly concentrated around the task of developing open-cycle MHD, i.e., installations using plasma of combustion products of proper fuels. Some research was aimed at developing closed-cycle MHD at non-equilibrium plasma with the objective of using it in combination with high-temperature gas cooled nuclear reactors and in installations operating at fossil fuel with special high-temperature heat exchangers.

Why did such euphoric enthusiasm over plasma MHD energy conversion start to decline in the mid-1970s and reach the rather low level at present? The available explanation consists, in my opinion, of three items.

We shall briefly consider all three reasons. The main advantage of plasma MHD is determined by high thermal efficiency; assessed net coefficient was about 50–55%, and in perspective 60% efficiency was contemplated. In comparison with large, mainly steam turbine power installations of about 40% efficiency existing in the 1960s and 1970s, that was an extremely attractive feature of the MHD power conversion method. However, in the late 1970s progress in gas turbine technology resulted in the increase of gas turbine working temperatures. Application of proper technologies in stationary energy units made it possible for energy equipment producers to start manufacturing high-temperature gas turbines as topping components for steam turbine units. Such extensively applied now combined gas and steam units have over 50% efficiency rate, which in some cases can reach 55%. High-temperature gas and steam units operate in a thermal cycle similar to the MHD installation cycle, and MHD generator is a topping to the gas and steam section of the cycle. High "turbine" efficiency of gas turbines, their relative simplicity, and actual potential for mass production at moderate cost are decisive factors for gas and steam units taking over plasma MHD installations.

The second reason of the decreased interest to plasma MHD energy conversion lies in certain technical difficulties in practical implementation of the MHD conversion principle. We shall briefly dwell on those difficulties moving consecutively along the working fluid tract in the installation. That kind of analysis will be carried out for open-cycle MHD installations. We shall start with the combustion chamber and the system of ionizing seed injection. The combustion chamber should ensure proper fuel burning at the reached temperature of higher than 2500°C and the required presence of the ionizing seed for sufficient ionization. Provided the required gas dynamics for fuel and oxidizer mixing and the injected seed, fuel, oxidizer, and seed should be in the combustion chamber for a certain period of time, its volume and surface should correlate properly. Given the rather high temperature of ionized combustion products (of plasma) and corresponding optical features, it leads to noticeable heat loss in combustion chamber walls, and, therefore, to noticeable loss of the chamber efficiency that shall directly impact the efficiency of the installation as such. Now we shall consider a MHD generator comprising a MHD channel and a magnet system. Because of the relatively low electric conductivity of plasma in the channel even at extreme assumed magnetic induction of 5–6 T, the length of the MHD channel in the required superconducting magnet system reaches 15–20 m that leads to noticeable heat losses along with certain "electric" and gas-dynamic losses in the channel. Thus, "turbine" efficiency of the channel becomes considerably lower than that for the gas turbine. We should also take into account existing losses in the diffuser at the outlet of the MHD channel (before the inlet to steam boiler). In the long run it practically leads to equal thermal efficiency of gas and steam unit and MHD installation despite the fact that working fluid temperature in MHD installation is about 1000° C higher than in the gas and steam unit.

The third reason is tied up with the high cost of MHD installation that is primarily determined by a high cost of the superconducting magnet system and high-temperature air preheater.

The preceding statement makes it clear why the considered MHD installations lose in comparison with modern gas and steam units.

The preceding comparison was made for open-cycle MHD installations. The state of the art for closed-cycle MHD installations, though operating at considerably lower temperatures, is no better because of other problems inherent to these methods—problems of working fluid heating in high-temperature heat exchanger and losses at nonequilibrium ionization of the working fluid.

Thus, following the undertaken analysis we shall see whether the plasma MHD energy conversion method has any perspectives. We shall define the directions for research and development studies to increase considerably future application for plasma MHD energy conversion. We shall also give a promising example of plasma MHD installations.

II. Required Research and Development

The priority should be given to the potential increase of plasma conductivity in an MHD generator. Electric conductivity of the equilibrium plasma is determined by its parameters, i.e., temperature and pressure, as well as by molecular composition. At first glance little can be done here. However, the importance of increasing the conductivity is so great that any attempts in that direction should be encouraged, e.g., research aimed at using working fluid plasma with minimum water vapor content can turn out very promising. It is worth investigating some technological schemes of MHD installations with thermochemical conversion allowing efficient coal burning as the energy source in the MHD cycle. One such scheme will be given in the third section of the paper.

Speaking about closed-cycle MHD installations, one should closely study the suitability of some metal vapor, the conductivity of which can be very high at proper parameters.[1] For nonequilibrium plasma MHD installations, we should look into discovering different efficient nonequilibriums at minimum energy expenditure. Special attention should be paid to developing MHD channels with plasma of peculiarly high average efficient conductivity, which can be illustrated by the example based on developing a so-called MHD channel with gas and plasma interacting flow.[2]

One of the main reasons limiting the efficiency of plasma MHD generators is low conductivity of plasma, especially at low temperatures. Even easy-ionizing seeds of alkali metals and salts (K, Cs) do not increase the conductivity of equilibrium ionized gas at normal pressure and temperature lower than 2000 K and more than 10 S/m that predetermines unacceptable low density of electric power in the current, even at modern extreme values of magnet field induction and current velocity in the channel. This "lower" limitation together with the "upper" limitation to the temperature interval of the working fluid at 2800–3000 K will determine the extreme value of the efficiency for the total enthalpy conversion at the level of 20% or max 25%. This very matter combined with that inherent to the MHD-generator low isoenthropic efficiency deprives the MHD generator of a beneficial comparison with other high temperature energy generating units.

At present two ways of solving low-temperature limitation are considered. One of them, already mentioned, implies the use of nonequilibrium plasma as working fluid. The other, less advanced in the research method of broadening the working temperature interval in the MHD generator, is based on the inhomogeneous quasiperiodical flow of working fluid in the MHD-generator channel.

Periodical inhomogeneity of the flow is determined by several (normally 2–10) relatively high-temperature plasma current-carrying clots. Those clots periodically appear in the working fluid flow at the inlet to the channel due to, for example, pulse electric discharge or different grades of enriching the oxidizing agent by oxygen. Conductivity inhomogeneity (that can be quite significant in the case of nonconducting main stream) predetermines inhomogeneity of interaction with the magnetic field: conductive current-carrying clots are restrained in the outer magnetic field; nonconductive (or low-conductive) gas practically does not interact with the outer magnetic field directly, but it can perform beneficially pushing current-carrying clots through the magnetic field. The heat balance in a current-carrying clot usually ensures joule dissipation slightly surpassing expansion work, so that the temperature and, consecutively, the conductivity will not decrease along the channel. At the same time the primary gas pushing current-carrying clots within the channel gets cooled down to actually arbitrary temperatures. The current-carrying clot acts here as a gas turbine blade. Thus, given sufficient interaction of the main working fluid with the current-carrying clots, the temperature interval available in principle for conversion in such an MHD generator will considerably increase toward the low temperature range. By "effective" temperature of the working fluid, we mean the average temperature of one fragment of the working fluid stream that includes one clot and a corresponding main working fluid stream part. The amount of current-carrying clots typically makes up to 10–20%.

Principal difficulties of implementing this concept are caused first by the procedure of stable controlled clot generation at the inlet to the channel and second by the necessity to retain the clots during their stay in the channel, to have time and space alternative electric loading, and other typical to conductive MHD generators problems, e.g., near-electrode losses.

Experimental and analytical studies for MHD generators have recently been pursued by an international research group within the frames of a trilateral international project: "MHD Generation of Electric Power in Flows with Gas-Plasma Interaction" in Eindhoven Technological University (the Netherlands), in Bologna University (Italy), and in the Institute for High

Temperatures of the Russian Academy of Sciences (Russia). Series of experiments on MHD installations at Eindhoven Technological University helped to work out methods of stable clot generation on entering the channel by pulse electric discharge, demonstrated stable electric power generation in MHD generators with inhomogeneous streams of several tens kW, and presented effects of MHD interaction in inhomogeneous streams; thus, key elements of the physical concept were proved. Large-scale parametric simulations for different description levels undertaken in the University of Bologna confirmed preliminary assessments of achieving possible total enthalpy conversion coefficient of 30–40% at relatively high values of isoenthropic ratio. In the Institute for High Temperatures of the Russian Academy of Sciences, the physical concept of MHD generation in inhomogeneous flows with gas and plasma interaction was complemented by including physical processes and phenomena discovered and described during the processing and interpretation of the data of physical and numerical experiments.

Thus, we can state that principal parts of the physical concept of MHD energy generation in inhomogeneous streams with gas and plasma interaction have found experimental and analytical support although additional research and development are absolutely essential.

Other prospective investigations should be devoted to the efficient superconductive system of the plasma MHD generator. Along with consideration of optimal magnet systems, we should look into the possible transfer from traditional superconducting Nb- and Ti-based alloys to more promising materials. Structural development of superconducting cables for high-value currents is very important. Even the application of high-temperature superconducting materials, which are not, unfortunately, good for a strong current, could turn out worthwhile in retrospect.

Investigations on injection and extraction of ionizing seeds are required. In open-cycle MHD installations K_2CO_3 potassium seeding is used, as a rule. As mentioned in the preceding material, the time of complete ionization of the seed is extremely important. Therefore, to carry the investigation to obtain the appropriate dispersion of the seed and the minimum amount of the so-called "inert" substance ensuring the required dispersion of the seed before or right in the combustion chamber is reasonable. Extraction of the ionizing seed from the main installation loop is of special importance. To what degree the combustion products are purified exactly from the seed at the outlet and how the seed (potassium) could be used for purifying the combustion products from certain harmful components would be significant. Speaking about required plasma MHD installation investigations, we should not forget that being "high-temperature" these installations demand different efficient heat exchangers, both, of regenerating and recuperating types. The heat exchangers must operate at temperature values of 2000°C in different gas media, oxidizing and regenerating, as well.

Construction materials for heat exchangers and MHD channels need large-scale material investigations. We should state that regrettably the scale of the required research has been reduced recently, especially in the field of creating necessary composite materials and ceramics.

Considering MHD installations, we should not omit the problem of specific inversion of the acquired direct current electric power. All potential peculiar features of the required invertors with a multielement structure of altering electric potential determine complex and rather expensive construction. That part of plasma MHD installation has always demanded more extensive investigations because it is an essential part of the MHD installation control system as well.

Finally, elaboration of the advanced MHD installations needs a well-developed search for scheme solutions with an account of available scientific and technological achievements. The search process should be, in our opinion, focused on scheme development of plasma MHD installations operating on solid fuel. At present there is practically no competition with regard to steam and gas installations because the specifics of gas turbine operation lie in the imperative use of clean fuels, i.e., natural gas or liquefied fuel.

The third section will present an example of promising findings in developing coal-fired plasma MHD installation with thermochemical fuel conversion in the cycle.

III. Advanced Magnetohydrodynamic Systems with Chemical Regeneration of Heat

The effect of chemical regeneration of heat in the MHD cycle is ensured by endothermal reactions with the participation of primary fuel, which are facilitated by the presence of the physical heat of combustion products after the MHD channel. In particular, when fuel is mixed with combustion products, its carbon containing components interact at high temperature with water vapor and carbon dioxide in the combustion products. Conversion of this kind, yielding synthesis gas, can be accomplished by a scheme similar to that typical of MHD installation with heat supplied to the stoichiometric mixture of a portion of the combustion products with fuel in a reactor/heat exchanger beyond the MHD channel.[3] Combined with the preheating of the oxidizing agent to a high temperature, chemical regeneration of heat raises the efficiency of MHD power plants to more than 60% (Ref. 4). However, in the case of a gaseous fuel, such conversion is effective at a reactor temperature not lower than 1300 K, which engenders serious difficulties in developing an appropriate heat exchanger; in the case of coal, the reaction takes place at even higher temperatures.

An alternative conversion technique consists of mixing high-temperature combustion products leaving the MHD generator with primary fuel in a ratio that significantly exceeds the stoichiometric ratio and subsequent separation of the cooled converted mixture of gases. A ternary MHD cycle using this idea, which reduces to minimum technical problems associated with designing an efficient coal-fired MHD plant, is outlined in the following.[5]

Ternary Magnetohydrodynamic Cycle with Thermochemical Conversion

The block diagram is given in Fig. 1 (Ref. 6). Magnetohydrodynamic generator and gas turbine both working with synthesis gas produced in a

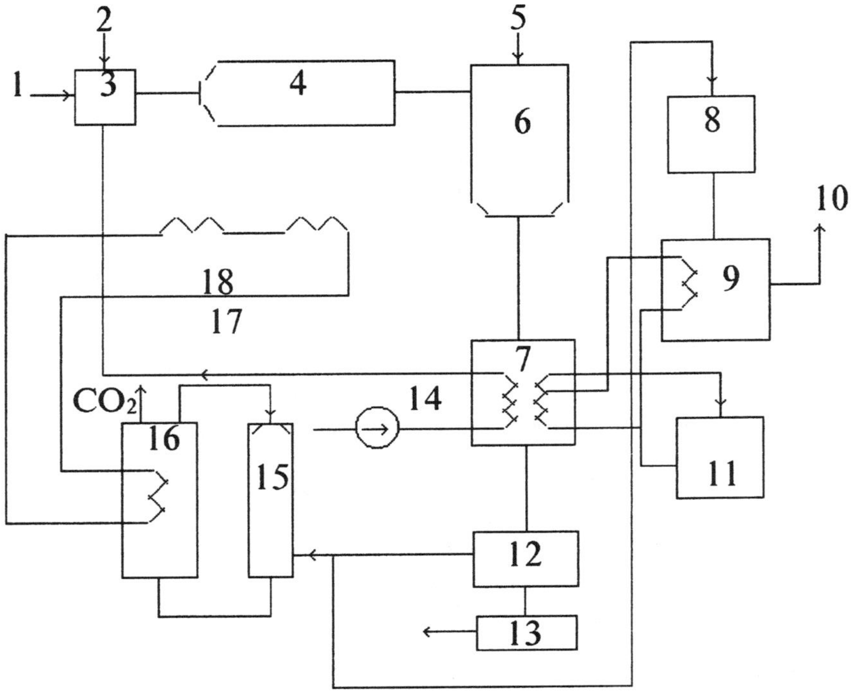

Fig. 1 Block diagram of a combined MHD power plant with thermochemical conversion: 1, seed; 2, oxygen; 3, combustion chamber; 4, MHD generator channel; 5, fuel; 6, thermochemical converter; 7, steam generator I with built-in synthesis-gas preheater; 8, gas turbine unit; 9, steam generator II; 10, exhaust of the gas/steam plant; 11, steam turbine unit; 12, filters; 13, seed extraction from slag; 14, synthesis-gas compressor; 15, CO_2 absorber; 16, desorber; 17, synthesis-gas; 18, cooling system of the MHD-generator.

thermochemical converter by mixing the combustion products that leave the MHD generator with primary solid (pulverized) fuel. Conversion takes place in the temperature range of 2500 to 1700 K. To exclude seed capture, slag is withdrawn from the high-temperature part of the flow path where synthesis gas is brought by a recirculation system from purifying filters. The synthesis gas proceeding to the gas turbine is obtained from the cooled converted gas mixture by condensation and separation of moisture and dust removal, and the gas proceeding to the MHD generator has carbon dioxide removed (with the use of one of conventional methods, e.g., carbonate purification).[7] The heat derived from cooling the gases leaving the thermochemical converter and the gas turbine is used in the steam-turbine cycle. The steam and gas unit can be arranged by a known scheme of steam injection into a gas turbine as shown in Fig. 2 or by using a separate steam turbine. Both variants have almost equal resulting thermal efficiency. The low-grade heat of cooling the MHD flow path elements is used for gas

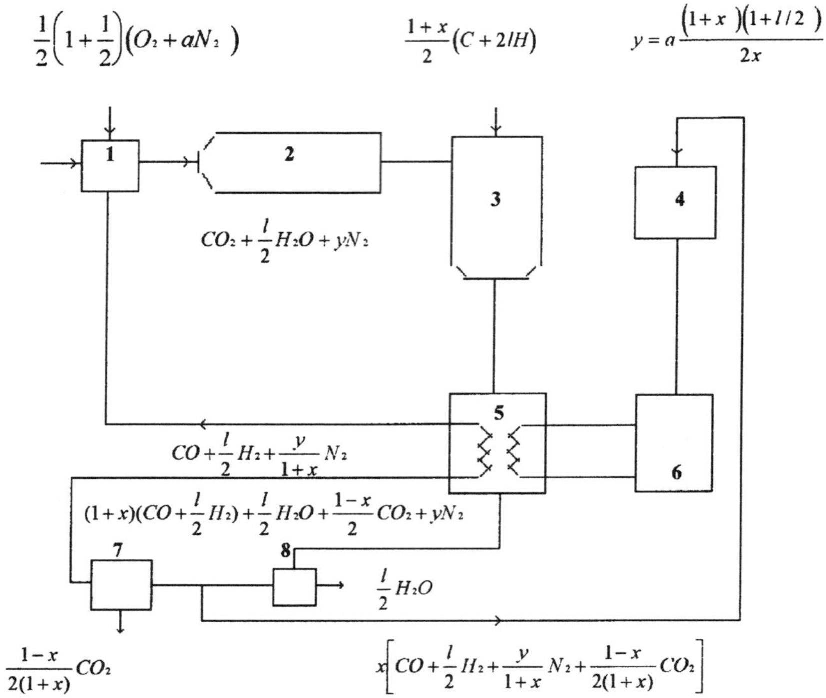

Fig. 2 Material balance of the main components of the working media in a thermochemical conversion/MHD power plant; 1, combustion chamber; 2, MHD generator; 3, thermochemical converter; 4, gas turbine unit; 5, heat exchanger/steam generator; 6, steam turbine unit; 7, gas separator; 8, H_2O-condenser/dust cleaner. 2*l* is the H/C atomic ratio in primary fuel, and *x* is the fraction of synthesis gas consumed outside the MHD cycle.

cleaning. Pure oxygen is used as the oxidizing agent in the combustion chamber of the MHD generator.

Scheme of Working Media Streams

In the considered type of thermochemical conversion/MHD power plant, the main variable parameter of material balance is x fraction of synthesis gas consumed in a gas turbine related to the flow rate through the MHD generator. The composition of synthesis gas is influenced by the converter-outlet temperature, which determines mole relation H_2/C. At temperatures higher than 1500 K, this relation is lower than the relative content of those components in the primary dried coal and shifts to H_2 increase at the temperature falling at the converter outlet. As the chosen value of this temperature is close to the indicated border, we can give nearly exact approximating expressions of relative mole expenditures in the material balance scheme in Fig. 2. Circulation and accumulation of nitrogen in the

MHD generator loop exclude operation at $x = 0$ and demand the use of oxygen as an oxidizing agent. On the other hand, the high specific enthalpy of the working fluid in the MHD generator (~7 MJ/kg as compared to 4.5–5 MJ/kg in traditional MHD power plant schemes) and high operating temperatures in the thermochemical converter are synonymous to the highly intensive working processes, that is, very efficient MHD energy conversion and/or relatively small dimensions and cost of the equipment.

Thermal Efficiency

The efficiency of MHD power plants with thermochemical conversion (TCC-MHD) depends mainly on two decisive parameters, namely, the enthalpy conversion coefficient in the MHD generator η_N and the relative out-of-MHD-generator consumption of synthesis gas x. The temperature value of ~2000 K at the outlet of the MHD channel is considered a limiting factor for η_N; below that temperature insufficient conductivity will not provide efficient power generation. The η_N values of 0.25–0.35 can be taken as good at the initial plasma enthalpy ~7 MJ/kg in MHD installation in comparison with η_N of 0.2 at 4.5–5 MJ/kg for conventional MHD power plants. The range of x values is limited by 0.15 from below because of the necessary extraction of nitrogen accumulated in the MHD loop and by 0.6 from above, which is the value of the boundary temperature at the converter outlet (~1700 K as shown below). Preheating of the initial media for the MHD generator and steam-gas unit in the heat exchanger after the converter reaches 1100 K. The MHD channel with ceramics-enforced walls is considered at temperature 2200 K. Heat losses in channel walls make up to 10% of the heat converter to electricity; in the combustion chamber and nozzle, losses make up to 3% of the inlet thermal capacity. The losses involved in a dc/ac conversion amount to 2%. The MHD channel characteristics were calculated by the method of design development for the commercial Ryazan MHD power plant; the range of possible parameter values is given in Fig. 3. The efficiency of the channel diffuser equals here to 0.5. Losses for the working media compression are accounted in the supposition that three-stage compressors with intermediate cooling are used. Oxygen is produced by an air-separating unit with a power consumption of 0.4 kW/m.[3]

The use of a modern high-temperature gas turbine with synthesis gas makes the efficiency of the gas-steam section equal to 0.5. We have considered already two equivalent variants of implementing a gas-steam unit: the scheme of steam injection to gas turbine and the use of separate gas turbine. The steam is generated by the heat of combustion products at the gas turbine outlet (Fig. 1) and to a less amount by the heat of conversion products that were not all used for regenerating heating of synthesis gas in the MHD generator loop after thermochemical converter (heat exchanger, Fig. 1). When the mentioned steam flows mix, they are heated here additionally because of the high temperature of the source.

To simplify the technological scheme of MHD installation, the heat derived from cooling high-temperature elements of MHD loop is not used in the steam cycle but is applied for the technological needs of the gas separa-

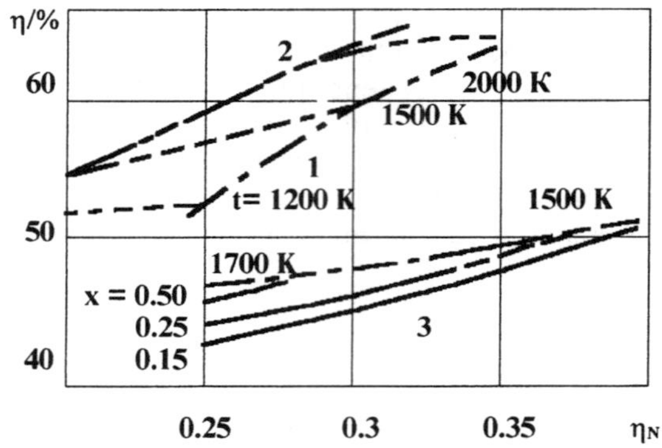

Fig. 3 Efficiency of an MHD installation with thermochemical conversion as a function of η_N for the MHD generator: 1, cycle with methane cracking; 2, cycle with steam conversion of methane [*t* is the temperature of regenerative-air preheat (Ref. 4)]; 3, combined cycle (simplified) for MHD installation thermochemical conversion by solid fuel injection at different fractions *x* of synthesis gas used in the gas/steam plant and maximum temperatures at converted outlet of 1700 K and 1500 K.

tion system (at temperature 120–140°C). The heat up to 11 K kal/mole is estimated for CO_2 separation from the converted mixture. Coal is dried by low-potential heat as well.

Detailed calculations were carried for different temperatures behind the converter and primary fuel compositions with an account of radiation in the converter that slightly increase the heat portion going to the steam section; dependence of the efficiency on the capacity level was also investigated. The calculations of thermal efficiency of the TCC-MHD plant for typical Kuznetsk grades of coal with an MHD generator capacity of 400 MW and higher are given in Fig. 3 as certain parametric dependencies. Note that the calculations have not shown serious efficiency differences in using lean and highly volatile coals, where as for natural gas the corresponding values of efficiency are ~1% lower. Figure 3 also shows the possible performance of this cycle at lowering the temperature at the outlet of the converter to 1500 K. The gained rise of the efficiency demands from the converter and the MHD generator more complexity (to increase η_N). On the other hand, even drastic decrease of the conversion coefficient in the MHD generator does not lead to a significant drop of thermal efficiency of the cycle, if the *x* value of synthesis gas consumed in the gas-steam section is correspondingly increased. the extreme value of *x* is confined to the minimum permissible temperature in the converter in accordance with the thermal balance conditions.

Nevertheless, despite comparable advantages of this type of MHD installation its efficiency is considerably lower than the efficiency attained by using in the cycle not only fuel conversion but also high-temperature regenerating heating of the oxidizing agent, see Ref. 4. Figure 3 presents data on the MHD installation efficiency with thermochemical conversion for the cycle with methane cracking (curve 1) and cycle with steam conversion of methane with regenerative-air preheating up to 1200 to 2000 K (curve 2). Prospective development of open-cycle MHD power plants should obviously follow that principle.

The preceding features put the priority on developing the proposed coal-fired MHD power plant and achieving efficiency similar to that of gas-fired gas/steam plants. This goal is quite attainable with less complicated MHD topping and environmental requirements observed as well. Thus, realistically we can develop rather simple, efficient, and flexible coal-fired MHD plants with an MHD generator operating with synthesis gas at real values of η_N and large consumption of synthesis gas in a gas turbine unit.

Specific Features of the Gas Dynamic Flow Tract of Magnetohydrodynamic Generator

The feature that distinguishes the MHD flow tract in the TCC-MHD power plant we are speaking about is the use of combustion products of a clean fuel, such as synthesis gas, regardless of the type of primary fuel used. This feature will ensure simpler and experimentally tested solutions to technical problems involved in the development of components for the MHD generator flow tract.

We shall note first that, because of thermochemical regeneration, the thermal power at the inlet to the MHD generator is higher than the thermal power of the corresponding amount of primary fuel. In the example considered here, the ratio is 1.52 at $x = 0.6$ and $\eta_n N = 0.25$. From the diagram in Fig. 1, we can see that the mass flow rate of the high-enthalpy working fluid of the MHD generator is also higher than the corresponding mass flow rate of combustion products of the primary fuel if both are burned with the same oxidizing agent, i.e., specific flow rate of the oxidizing agent is substantially reduced. Instead of the reaction of direct oxidation of carbon as the main fuel component

$$C + O_2 = CO_2$$

the reaction that takes place in the MHD combuster is that of final oxidation of synthesis gas:

$$CO + 1/2O_2 = CO_2$$

For those reasons the use of oxygen as an oxidizing agent in the thermochemical converter of the MHD cycle does not reduce the efficiency of MHD plants, as it usually does. The use of oxygen eliminates the need for high-temperature ceramic preheaters of the oxidizing agent in the MHD cycle. The cost of a suitable oxygen plant is less than a quarter of the cost of preheaters required for conventional MHD power plants of the same capacity and with autonomous preheating of the oxidizing to 1700 K.

Owing to a reduction of the relative content of ballast nitrogen, the specific enthalpy of the working fluid increases slightly with the parameter x of the relative fuel flow rate outside the MHD cycle by approximately 5% with x varying from 0.15 to 0.6 at $\eta_N = 0.25$, reaching 7 MJ/kg. At a calculated thermal power capacity of a combined TCC-MHD power plant of 1000 MW(t), the flow rate of the combustion products in the MHD generator at $x = 0.6$ is close to 150 kg/s, and the pressure is ~30 bar. The adiabatic temperature of the flame is 3300 K. At combustion chamber dimensions of $D^{\sim} = 0.9$ m and $L = 1.2$ m, the total heat losses at the cold wall will be 30 MW(t) (~3%). Note, however, that the MHD generator power capacity in this case will be below a reasonable optimal figure. Therefore, for a plant with an optimum power capacity of the MHD generator above ~500 MW(e), as well as in operation at other values of x, relative losses in the combustion chamber will be less than 3% even at cold walls. Because the MHD generator works with a clean fuel, the walls can be protected by spraying a layer of zirconia dioxide, for example, by periodically injecting it as powder in the wall region. The specific heat fluxes will then decrease depending on the initial enthalpy to 2.5–3.5 MW/m^2 or by a third. We envisage that the walls will be simply water cooled, and the cooling water will have the parameters for ensuring only the process needs of a gas separation system (up to 140°C).

The high specific enthalpy of the working fluid not only at the inlet to the MHD generator but also at the outlet (~4.5 MJ/kg), even at conversion coefficients as high as 0.35, ensures in an optimized supersonic flow a plasma conductivity at the outlet of the order of several S/m and determines effective MHD interaction in the channel with a superconducting magnet system with a "traditional" level of magnetic induction of 5–6 T. Because, as was in the preceding material, the development of large superconducting magnet systems with high magnetic induction can present fairly serious difficulties and will require substantial capital costs, consideration was given to MHD operation at lower magnetic induction levels of 2–4 T. Preliminary optimization of the flow was performed, in particular, for a linear MHD generator with a typical ratio between average cross section and length of the order 1:10, with the potential of longitudinal Hall field not higher than 2.5 kV/m, Faraday current density up to 1 A/cm^2, average wall temperature of 2200 K, and heat fluxes in the working section of the channel up to approximately 1 MW/m^2. Calculations show that the optimum Mach number in the flow significantly exceeds 2 at the inlet and tends to drop to approximately 1 at the outlet, whereas electrical efficiency is 0.82–0.85. (An increase of the enthalpy conversion coefficient is accompanied by an increase of electrical efficiency.) Magnetic induction required at $\eta_N = 0.25$ is 3 T at atmospheric pressure at the outlet of the channel and 2 T in the exhaust version at a pressure of 0.5 bar. The respective values of the stagnation pressure in the flow at the channel outlet are 1.3 and 0.7 bar.

Low magnetic induction values in the channels of the TCC-MHD plants have significant advantages for superconducting magnet systems: the permissible current density in the superconductor can be substantially increased (by a factor of 5 when the field strength is reduced from 6 to 2 T), and the electromagnetic forces in the winding are reduced. Steel can be used in the first place to close, by a magnetic circuit, the induction flux outside the cryostat fitted with winding of the magnet for shielding the field and for improving the layout of the equipment in the engine room of the MHD generator, and, thus, allowing reduced exposure of electrical devices and personnel to the magnetic field, and in the second place steel can be used to reduce the required number of ampere-turns of the winding of the superconducting magnet system and the forces in it. The reduction of ampere-turns of the winding plays an important role in making use of superconducting magnet systems that have a steel core with poles. When $B = 2$ T, for instance, the number of ampere-turns for the winding of the superconducting magnet system of an MHD generator with a capacity of 200–400 MW is a third to half that of the iron version. The available estimates suggest that the total cost of a superconducting magnet system, which is the most expensive and capital-intensive element of the MHD generators for combined TCC-MHD plants at an output capacity of 500–300 MW at $B = 2$ T (exhaust version), can be of an order of magnitude less than the cost of a superconducting magnet system with $B = 6$ T, typical of traditional MHD power plants using natural fuel combustion products with the same output capacity and the same length of the working section of the linear MHD channel. At $B = 3$–3.5 T the total cost of the superconducting magnet system would amount to only 20–25% of its cost at $B = 6$ T, and the total output capacity of the TCC-MHD plant would be 1000 MW, compared with the output capacity of a traditional MHD power plant of only 500–700 MW.

Thus, combined-cycle power-generating plants incorporating an MHD stage can operate at a higher heat-to-electric energy conversion efficiency than any other known plants. However, realization of this basic advantage requires higher quality components, particularly high-temperature regenerative heat exchangers capable of working with seeded combustion products of an ash-bearing fuel. Work in this field should be aimed both at improving thermal schemes and at developing more advanced equipment.

The use of intracycle thermochemical conversion of coal in MHD plants provides an example of how experience acquired with the development and construction of gas-fired MHD plants can be used in developing coal-fired MHD power stations. This approach seems to be promising because it not only enables us to avoid many difficulties inherent to coal-fired installations but also facilitates the development of an efficient MHD generator by raising the parameters of the working fluid and the degree of regeneration in the MHD cycle.

The technological scheme of power installations of this type could be rather simple and at the same time meet the highest environmental protection standards. Thus, the considered coal-fired MHD power plants are obviously competitive with gas turbine units on solid fuel that demand pregasification of coal.

IV. Conclusions

Gas turbine units that facilitated development of highly efficient gas-steam plants on clean fuel, i.e., natural gas and liquified fuel, have certain advantages in comparison with the plasma MHD installation of the first generation. To develop competitive promising plasma MHD installation of the next generation, which would be, ahead of all, coal-fired, we need extensive research and development programs. Those programs should be aimed primarily at the following:

1) There should be a *high degree of heat regeneration in the topping cycle.* Both variants of regeneration, high-temperature heat exchangers and/or chemical coal conversion, should be upgraded to the required parameters.

2) *High efficiency of energy conversion in the topping cycle* should exist. Two parameters, i.e., enthalpy conversion coefficient and inherent efficiency of the MHD generator (nozzle and diffuser included), should be simultaneously very high. The value of efficient plasma conductivity is fairly significant here.

3) There should also be *minimum heat losses in the topping cycle*, which refers to all parts of the gasdynamic tract.

4) *Maximum efficiency of the bottoming cycle* should exist. All possible thermal scheme configurations with gas-steam units and steam plants should be optimal.

The data available at present prove that the preceding goals are quite attainable. However, total implementation of the required MHD programs will demand an impressive amount of time and material resources. In that respect correct arrangement of efforts in research and development plays a key role. One of the strategic goals will be launching the research program and developing an integrated pilot installation. The scale of the installation should be considered very thoroughly. Minimum scale should ensure that positive working effects prevail over dissipative losses in the cycle components of a demo installation. Such extensive programs could hardly be implemented by one country. A solution could be found in launching an international project. Available technical support for such a project is quite sufficient.

References

[1]Biberman, L. M., Likalter, A. A., and Yakubov, I. T., "MHD-Generator with Nonideal Plasma of Alkali Metal Vapours," Vol. 20, No. 3, 1982, pp. 565–578.

[2]Bityurin, V., "5 New Concept. MHD Electrical Power Generation: The Status Report," edited by F. Negrini.

[3]Nosach, V. G., Kozllyuk, V. N., and Marchenko, G. V., "A Method for Utilizing the Heat in Gases Leaving the MHD Generator," Sovient Patent 303 344 (Avtorskoe Svidetelstvo) Byulleten Informatsii 10, 1987 (in Russian).

[4]Shumiatskii, B. Ya., Sokolskii, A. G., Pereletov, I. I., Khokhlov, L. K., Rogachev, A. P., and Koptev, V. M., "The Effect of Methods of Attaining High Temperatures on Thermal Efficiency of Open-Cycle MHD Installations," *Teplofizika Vysokikh Temperatur*, Vol. 7, No. 4, 1969, pp. 754–465.

[5]Carasse, J. M. P., "Open MHD Cycle and Method of Operation of MHD Cycle," U.S. Patent 3467 842, 1969.

[6]Sheindlin, A. E., and Kovbasyuk, V. I., "Perspectives in Energy," "Open-Cycle Magnetohydrodynamic Energy Conversion: State of the Art Prospects for the Future," Vol. 1, 1991, pp. 179–192.

[7]Leitis, I. L., *The Theory and Practice of Chemical Power Technology*, Khimya, Moscow, 1988 (in Russian).

Chapter 30

Nuclear Driven Closed-Circle Magnetohydrodynamic Generator on Self-Sustained Thermal Convective Flows

V. V. Glaskov* and O. A. Sinkevich†
Moscow Power Engineering Institute, Moscow, 111250, Russia

I. Introduction

IN THE works of Refs. 1 and 2, the steady-state medium movements along the closed contour of a varied cross section in the absence of the external mass force field were examined both analytically and numerically. Investigation was carried out for the arbitrary equation of state of the moving medium. The general view of the investigated model loop is shown in Fig. 1. The variation of the cross-section diameter and pressure distribution along the examined model contour is given in Fig. 2. It was shown that in case of the medium being heated in the widest duct cross section and cooled in the narrowest one, the medium movement along the contour can become self-maintained. We called these steady-state flows self-sustained thermal (ST) flows. The dimensionless gas velocity, temperature, and density distributions along the contour, characteristic for ST flows, are represented in Fig. 3. The scales for temperature and density are their consecutive values in cross section 0-0 and for the velocity is the local speed of the sound there.

The closed contour of the varied duct cross section with the heater and cooler is in fact the heat machine with a thermodynamic cycle of a working medium similar to that of the common heat machine. The thermodynamic cycle of the working medium is similar to the well-known Cannot cycle with the heat supplied under a pressure higher than that from which it is removed. As such a cycle is realized under the condition of subsonic motion

*Assistant Professor, Heat Physics Department, Krasnokazarmennaya 14.

†Professor, Head of Plasma Research Group, Heat Physics Department, Krasnokazarmennaya.

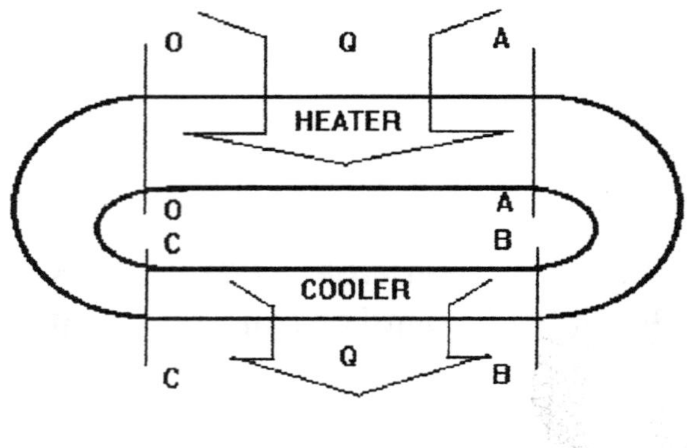

Fig. 1 The model loop investigated in Refs. 1 and 2.

of the medium in a closed loop of variable cross section, then a higher pressure corresponds to a larger cross section and to a deceleration of the flow whereas a lower pressure corresponds to a smaller cross section and to an acceleration of the flow. (In supersonic flows, in accordance with the condition of inversion of action,[3] the regularity is inverse.)

The main difference of a ST machine from the traditional one is the following: in a traditional heat machine released energy is accumulated in the flywheel (i.e., in piston combustion engine), but in an ST heat machine the work performed at the stage of adiabatic expansion of the working medium, in the narrowing part of the loop, is accumulated in the kinetic energy of the flow itself. The much similar cycle as in the ST machine is realized in well-known ramjet engines. The proposed scheme is in fact the ramjet engine whose working medium cycle is really closed. (In theoretical analysis the ramjet thermodynamic cycle is closed conditionally and corresponds to the scheme depicted in Fig. 4.)

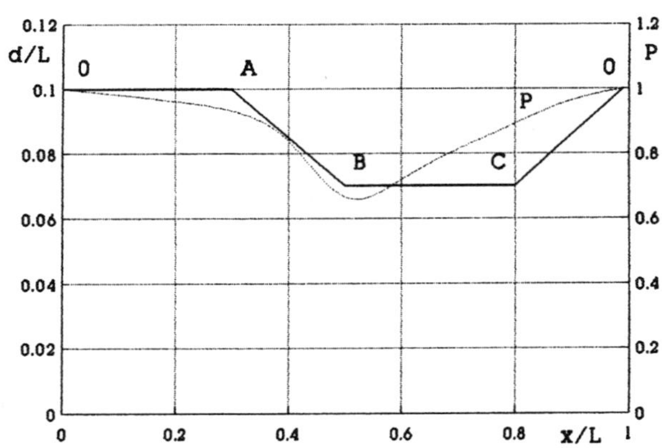

Fig. 2 Cross-section diameter (solid line) and pressure distribution (dashed line) along the contour.

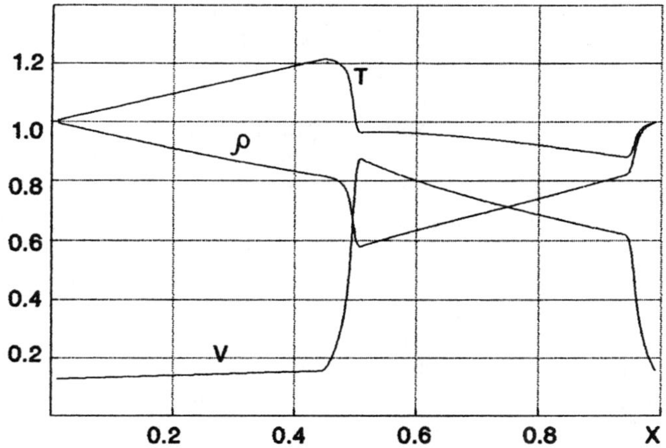

Fig. 3 Dimensionless gas velocity, temperature, and density along the loop.

The much similar ideas about the possibility of the very existence of self-sustaining heat-convection flow were expressed earlier (for example, the remark about "heat resistance" during the motion of gas in a heated pipe[3] and energy cycle mentioned in Ref. 4), but they did not receive the treatment they deserved. Obviously, the general conclusion regarding the existence of self-maintaining heat-convection flow is in no way associated with the fact whether or not the selected closed loop is ideal.

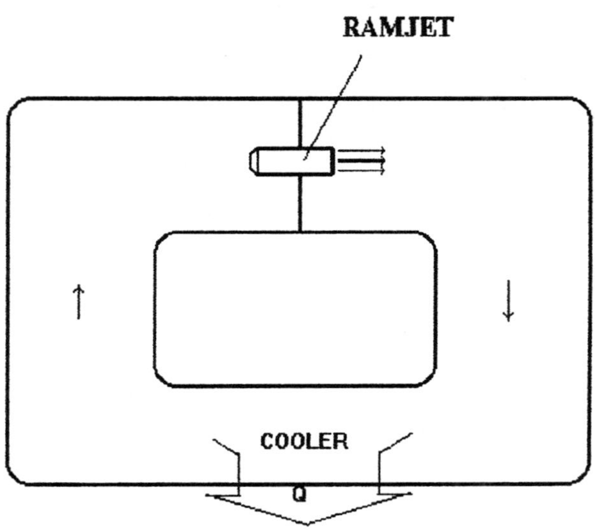

Fig. 4 Ramjet in closed contour.

In the works of Refs. 1 and 2, we examined the case when the total work performed in the thermodynamic cycle was spent to compensate dissipation losses because of the motion of gas in the loop corresponding to the heat convection engine functioning in the idle mode. However, self-sustained heat-convection flow has the ability to perform useful work as well; anyway, as far as the theory is concerned, there are no obstacles preventing it. The results obtained indicate the possibility of attaining fairly high velocities of flow in a loop. Taking into account this fact, the possibility to use this facility to develop a closed-cycle magnetohydrodynamic (MHD) generator[5–7] exists. The general view of the investigated model MHD loop for ST flow is shown in Fig. 5. Such an MHD generator, with the motion of a conducting working medium being supported by self-sustaining heat-convection flows, possesses the indisputable advantage of not having any moving parts. A similar heat-convection MHD generator, using a mixture of noble gases with the vapor of alkali metals and coupled with an advanced gas-cooled nuclear reactor, can form a united power unit with possible applications on the Earth and in space power systems.

Of particular interest are the prospects of utilizing pressurized gaseous nuclear fuel (UF_6) as a working medium in the contour encased in beryllium and carbon sheaths where the conditions for neutron breeding are fulfilled.[8] As the major part of the medium is concentrated in the widest part of the contour, the major heat release also will be concentrated there, thus making it possible to combine in one single module a gaseous nuclear reactor and an MHD channel. Gaseous fuel usage allows continuous removal of high radioactive waste from the contour during the process of its exploitation thus preventing the threat of environmental radioactive pollution even in the case of total destruction of the reactor. The preliminary results on this item are published in Ref. 6. The schematic view of such a module is depicted in Fig. 6.

In this work we deal with conventional Faraday type MHD generators[4] using self-sustaining heat-convection flows. It was shown that their heat efficiency may reach some 10–15% at relatively low flow Mach numbers in

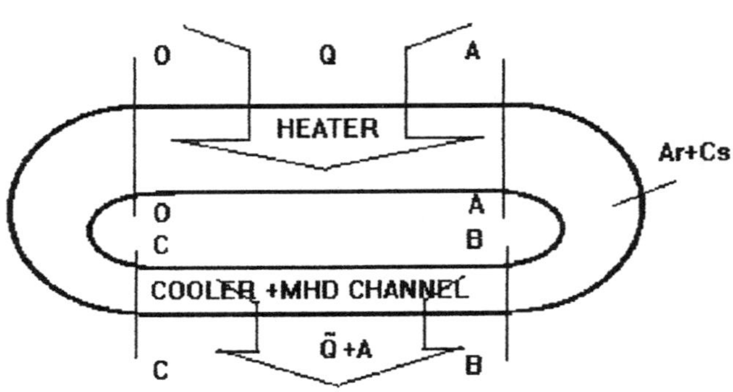

Fig. 5 Model loop with MHD channel investigated in Refs. 5–7.

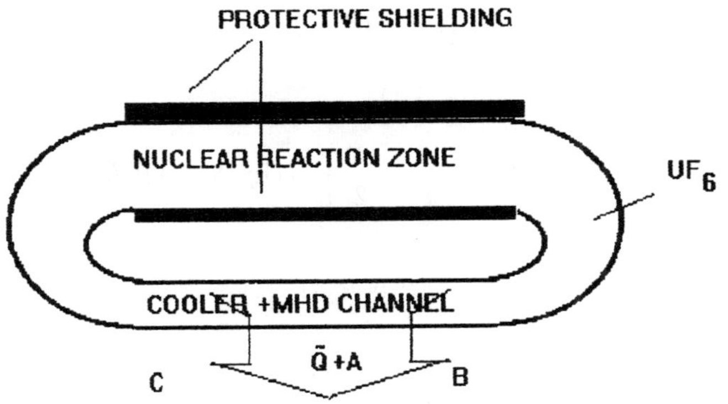

Fig. 6 Gas nuclear reactor with MHD channel in one module.

the loop. The possibility exists that for self-sustained heat, convection flows with the Mach numbers close to unity or for supersonic flows (there are some arguments in favor of the possible existence of supersonic self-sustaining heat-convection flows or mixed super- and subsonic flows in specially profiled loop), the heat efficiency can reach rather high values (at least 40%). Using as the base the solution obtained in Ref. 7, to generalize the results to other schemes of MHD generators (Hall generators, loop generators, and so on) is not difficult.

II. Model of Heat-Convection Flows in Closed Loop

Now we consider the motion of a medium in a closed loop of variable cross section in the presence of heat transfer with the walls. The profile of the loop is shown by a dashed line in Fig. 2, where D is the dimensionless hydraulic diameter of a channel normalized to the full length L of the loop along its median line and X is the dimensionless longitudinal coordinate along the median line. Let the OA, AB, BC, and CO segments designate the parts of the loop where the medium is heated (a heater with a constant cross section), the converging nozzle where the medium is accelerated, the constant cross-section part where the medium is cooled (cooler), and the widening diffuser section of the loop where the medium is decelerated, respectively. Using a well-known hydraulic model, we can average all parameters of the medium with respect to the cross section of the loop. For the averaged parameters of the medium and for the stationary motion while ignoring the heat flux along the direction of the motion and centripetal acceleration in closed loop, the following conventional one-dimensional system of equations can be obtained:

$$\frac{\partial}{\partial x}(\rho VS) = 0 \tag{1}$$

$$\rho V \frac{\partial}{\partial x} V = -\frac{\partial P}{\partial x} - \frac{\xi \rho V^2}{2d} - jB \tag{2}$$

$$\rho V \frac{\partial}{\partial x}\left(C_v T + \frac{P}{\rho} - \frac{V^2}{2} \right) = (\Pi q / S) - kjVB \tag{3}$$

$$P = R_\mu \rho T \tag{4}$$

$$j = \frac{VB}{(\sigma_g^{-1} + \sigma_l^{-1})} = (1-k)\sigma_g VB \tag{5}$$

Here x is the longitudinal coordinate along the central line of the loop; ρ, V, T, and P are the density, velocity, temperature, and gas pressure in the loop, respectively; j is the density of the current passing through the gas from one wall of the MHD channel to the opposite one; B is the intensity of the transverse magnetic field in the channel; σ_g is the electrical conductivity of gas; σ_l is the electrical conductivity of the external load per unit length of the channel ($\sigma_l = 0$ in the case of a broken external circuit, and $\sigma_l \to \infty$ in the case of a short circuit of the MHD generator); $k = 1/(1 + \sigma_l/\sigma_g)$ is the coefficient of MHD generator load ($k = 0$ in the case of a short circuit and $k = 1$ in the case of a broken external circuit[4]); d is the current value of the hydraulic diameter of the channel (in the channel of rectangular cross section, $d = 2ab/(a+b)$, where a and b are the sides of the rectangle); S is the value of the current cross section of the channel; Π is the value of the current perimeter of the channel; q is the density of heat flux from the channel surface; and R_μ is the universal gas constant.

III. Condition for Existence of Stationary Heat-Convection Flow in a Closed Loop

In a closed loop for the stationary flow, the following periodical conditions must be valid:

$$\rho(0) = \rho(L), \qquad V(0) = V(L), \qquad T(0) = T(L) \tag{6}$$

At first glance a contradiction emerges from three conditions (6) for the existence of the stationary state of system (1–5), because formally there is only one free external parameter at the defined geometry of the channel and equation of state for the working medium. This one free external parameter is the density of the heat flux from the wall q. However, under stationary condition, the flow rate through the cross section of the loop remains constant ($\rho VS = G =$ const), and the heat flux applied to the medium in the heater is equal in value to the heat flux removed in the cooler plus useful work performed by the medium. So we have one more condition:

$$C_v T(0) + \frac{P(0)}{\rho(0)} + \frac{V(0)^2}{2} = C_v T(L) + \frac{P(L)}{\rho(L)} + \frac{V(L)^2}{2}$$

Therefore, only one (any) of conditions (6) is actually independent, and the formulation of the problem is correct.

In a further investigation of an MHD generator on self-sustaining heat convection flow, we consider only the case when the MHD generator channel coincides with the cooled section of the loop. By designating $F = jB/\rho C_s$, $\Lambda = \Pi q/G + Fk$ (here C_s is the sound velocity), and assuming that F is constant along the length of the MHD generator channel in order to find stationary solution of system (1–4), we write this system in the form:

$$\rho VS = G = \text{const} \tag{7}$$

$$\frac{\partial \ln V}{\partial x}\left(1 - \frac{V^2}{\gamma R_\mu T}\right) = -\frac{\partial \ln S}{\partial x} + \frac{\Lambda}{C_p T} + \frac{\xi V^2}{2R_\mu T d} + \frac{F}{R_\mu T} \tag{8}$$

$$\frac{\partial}{\partial x}\left(C_p T + \frac{V^2}{2}\right) = \Lambda \tag{9}$$

Leaving in these equations only the terms of the second order of the flow Mach number ($M^2 = v^2/C_p^2$), we rewrite these equations in the form:

$$\frac{\partial T}{\partial x} = \frac{\Delta}{c_v} - \frac{\partial}{\partial x}\frac{V^2}{2C_P} \tag{10}$$

$$\frac{\partial \ln V}{\partial x} = -\frac{\partial l \ln S}{\partial x}\left(1 + \frac{V^2}{C_s^2}\right) + \frac{\Lambda}{C_p T}\left(1 + \frac{V^2}{C_s^2}\right) + \frac{\xi V^2}{2R_\mu T d} + \frac{F}{R_\mu T} \tag{11}$$

Assuming that $F/\Lambda \ll 1$, we obtain a zero-order approximation for the temperature and velocity of gas in the loop. From Eqs. (10) and (11) we have

$$
\begin{aligned}
&T^{\#}_{OA} = T_0(1+z), \quad z + x/l_q \\
&l_q^{-1} = \frac{\Lambda}{C_P T_0}, \quad x \in (OA) \\
&V_{OA} = V_0(1+z) \\
&T^{\#}_{AB} = T_0(1+l_1), \quad l = \int_O^A \mathrm{d}z, \quad x \in (AB) \\
&V_{AB} = V_0(1 + l_1 - y), \quad y = z - \int_O^B \mathrm{d}z, \quad x \in (BC) \\
&V_{BC} = V_0(1 + l_1 - y)\frac{d_0^2}{d_2^2} \\
&T^{\#}_{CO} = T_0, \quad x \in (CO) \\
&V_{CO} = V_0 \frac{d_0^2}{d^2}
\end{aligned}
\tag{12}
$$

Here the subscripts OA, AB, DC, and CO indicate the value of the given variable in the corresponding interval of the loop, whereas subscripts 0 and 2 indicate the value of the variables at the beginning of the heated section and at the beginning of the cooled section of the loop, respectively.

By rewriting Eq. (10)

$$
T = T^{\#} + (V_0^2 - V^2)/2\,C_p
$$

we can rewrite the expression (11) for velocity in the form

$$
\begin{aligned}
\frac{\partial \ln V}{\partial x} = &-\frac{\partial \ln S}{\partial x} + \frac{\Lambda}{C_p T^{\#}}\left(1 - \frac{V_0^2}{2C_p T^{\#}}\right) \\
&+ \frac{V^2}{\gamma R_\mu T^{\#}}\left(\frac{\Lambda(y+1)}{2C_P T^{\#}} - \frac{\partial \ln S}{\partial x} + \frac{\gamma \xi}{2d}\right) + \frac{F}{R_\mu T^{\#}}
\end{aligned}
\tag{13}
$$

For stationary solution the following two conditions must be fulfilled:

$$\oint\left(\frac{\partial T}{\partial x}\right)\mathrm{d}x = 0, \quad \oint\left(\frac{\partial \ln V}{\partial x}\right)\mathrm{d}x = 0 \tag{14}$$

As mentioned in the preceding statements, only one of conditions (14) for stationary motion in the loop is really independent. For further analysis the condition for the velocity of gas flow is selected as independent. This condition can be reduced with an accuracy to the fourth-order terms with respect to the Mach number in the form:

$$\oint\left[\frac{V^2}{T^\#}\left(\frac{\gamma\xi}{2d} + \frac{\Lambda(\gamma+1)}{2C_pT^\#} - \frac{\partial \ln S}{\partial x}\right) + \frac{F}{R_\mu T^\#}\right]\mathrm{d}x = 0 \tag{15}$$

For the velocity v and temperature T, the initial approximation (12) is employed.

By integrating Eq. (15) for the case when the hydraulic resistance of the loop is constant and segments $[OA] = [BC] = L_1$ and $[AB] = [CO] = L_2$, we can derive the following expressions relating the parameters of the self-sustaining heat-convection flow including the MHD effects:

$$\frac{\Lambda V_0^2}{2C_pT_0} = \frac{(F/A_2) + (\xi V_0^2/2d_0)}{(A_1/A_2) - (\xi L_1/2d_0)} \tag{16a}$$

or in the following form,

$$\frac{Nu}{RePr}\cdot\frac{T_h - T_0}{T_0} = \frac{(1-k)R_{\sigma 0} + \xi A_2}{aA_2[k(1-k)R_{\sigma 0} + 1]} \tag{16b}$$

Here $R_{\sigma 0} = 2\,\sigma_g B^2 d_0/\rho_0 V_0$ is the MHD parameter of magnetohydrodynamic interaction defined by velocity V_0, T_h is the character temperature in the heater, Nu is the Nusselt number, Re is the Reynolds number defined by the velocity v_0, Pr is the Prandt] number,

$$\alpha = \left(\frac{A_1}{A_2} - \frac{\xi L_1}{2d_0}\right), \quad A_1 = [(d_0/d_2)^4 - 1]$$

$$A_2 = 1 + (d_0/d_2)^5 + \frac{2}{L_1}\int_A^B [d_0/d(x)]^5\,\mathrm{d}x$$

Now let us determine the thermal efficiency of the MHD generator on self-sustaining heat-convection flow as the ratio of total useful work produced by one kilogram of the working medium along the full length of the MHD channel to the total amount of the heat received by this quantity of medium in the heater. As follows from Eq. (3) when the parameter F is constant along the MHD generator channel, the expression for the efficiency has a very simple form:

$$\eta = kF/\Lambda$$

Using the expression for Λ from Eq. (16), we can obtain the following expression for the thermal efficiency:

$$\eta = kM^2 \frac{(\gamma - 1) f[(A_1/A_2) - (\xi L_l/2d_0)]}{2(f/A_2 + 1)} \tag{17}$$

Here,

$$f = 2d_0F/\xi V_0^2 = 2V_2B^2d_0/\xi\rho V_0^2(\sigma_g^{-1} + \sigma_i^{-1})$$

As clearly seen from Eq. (17), the value of thermal efficiency of the MHD generator on self-sustaining heat-convection flow is substantially dependent on two parameters: 1) the Mach number of the flow in the loop, defined preliminarily by the geometry of the loop and by the intensity of the heat flux from the walls of the loop; and 2) the parameter f, dependent on the gas conductivity and the value of the magnetic field in the channel. The thermal efficiency dependence on f is monotonously increasing, and at $f \rightarrow$ the thermal efficiency reaches its maximum:

$$\eta_{\text{max}} \approx kM_{\text{max}}^2 \frac{(\gamma - 1)}{2}[1 - (d_2/d_0)^4] \tag{18}$$

Here $M_{\text{max}}^2 = M_0^2(d_0/d_2)^4$ corresponds approximately to the maximum Mach number of the subsonic self-sustaining heat-convection flow in the loop. The obtained formula is valid for the Mach number up to M = 0.6–0.7, which corresponds to the efficiency of the MHD generator of about 15%. The heat efficiency of the investigated model loop with MHD generator via parameter $y = (1 - k)\, R_\sigma/\xi A_2 \sim B$ is given in Fig. 7. Here $R_\sigma = 2\,\sigma_g d_0 B_0^2/\rho_0 V_a$ is the parameter of the MHD interaction, and V_{a} is the flow velocity at the beginning of the MHD section of the loop. The dependence of the heat efficiency MHD generator on ST flow via maximum Mach number in the closed contour is shown in Fig. 8.

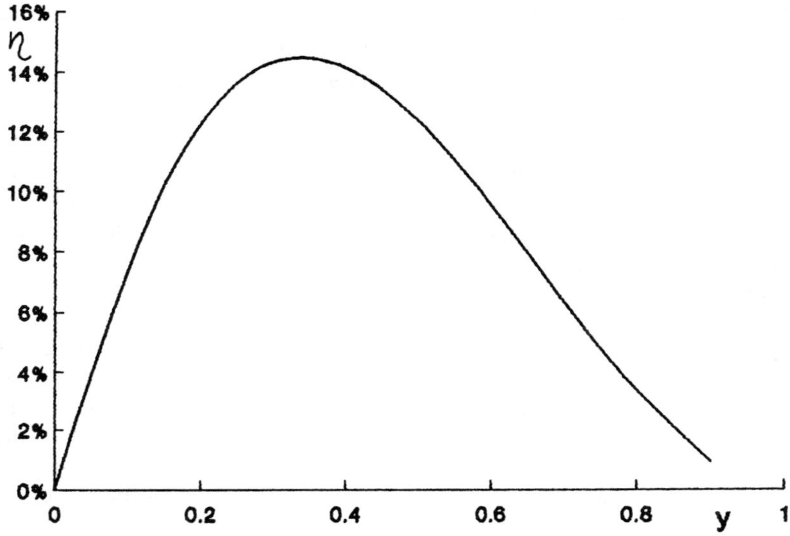

Fig. 7 Heat efficiency of MHD channel via parameter $y \sim B$.

We also can express the MHD efficiency on self-sustaining heat-convection flow in the form:

$$\eta = 8k(1-k)R_\sigma\left[1-(1-k)\frac{R_\sigma}{\xi A_2}\right]^2 St^2(\gamma-1)(\alpha/\xi)^3 \tag{19}$$

Here $St = q\,/\,\rho_o c_p T_0 \sqrt{\gamma R_\mu T_0}$ is the modified Stanton number.

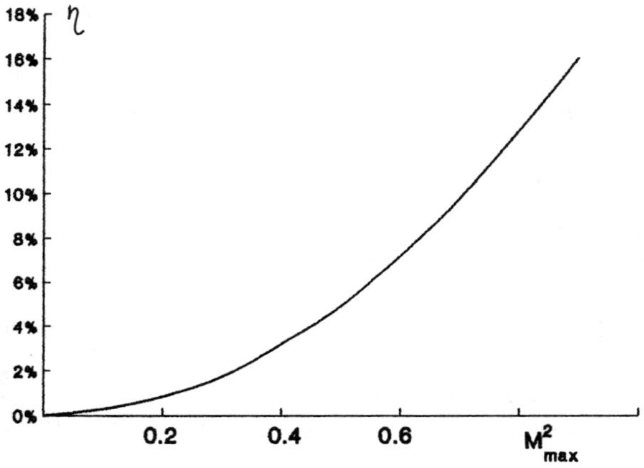

Fig. 8 Heat efficiency of MHD channel via maximum Mach number in the loop.

IV. Magnetohydrodynamic Generator Combined with an Advanced Gas-Cooled Nuclear Reactor

The proposed MHD-generator scheme with an advanced gas-cooled reactor as a heat source can result in a new generation of powerful compact and reliable space energy systems. The general view of the investigated module loop combining the nuclear reactor with the MHD generator is shown in Fig. 6. As a working medium, the noble gas mixtures with alkali metal vapors could be used, or, possibly, nonequilibrium plasma generated by some kind of electric discharge or irradiation. The use of pressurized gaseous nuclear fuel (UF_6) [8] as a working medium, in the contour encased in beryllium and carbon sheaths where the conditions for neutron breeding are fulfilled, can solve most of the problems common for closed contour MHD generators to ease heat transfer in the contour and maintain acceptable levels of electric conductivity in the MHD channel. The resulting united power unit can have unique qualities: gaseous fuel usage allows continuous removal of high radioactive waste from the contour during the process of its exploitation thus preventing environmental radioactive pollution even in the case of total destruction of the reactor. The reactor, realizing the proposed scheme will be absolutely safe, simply cannot pose the threat of radioactive pollution as the volume of radioactive waste in the contour will be reduced to a minimum.

In our laboratory the codes were developed for calculating MHD generators utilizing nonequilibrium plasmas as the working medium. In Ref. 9 MHD generators using nonequilibrium plasmas of noble gases seeded with alkali metals could operate with high efficiency in strong magnetic fields, and several technical projects for suppressing and regulating the development of the ionization instability in closed-cycle MHD generators also were proposed. The special numerical code for computing the plasma turbulence properties (the turbulent electrical and thermal conductivity, the turbulent Hall parameter, and so on) was developed.[10,11]

Theoretically, there could be a closed-circle MHD generator on self-sustaining heat-convection supersonic flow in the loop (in that case the cross section of the cooler has to be more than that of the heater). The heat efficiency of such MHD can be very high. The resulting power generating installation can have unique qualities and reliability, being capable for use on ground and in space for both types, ac and dc, of electrical energy generation.

References

[1]Glaskov, V. V., "Self-Maintained Thermal Convection Gas Flows in the Absence of External Mass Forces Field," *Proceedings of the 31st Eurotherm Conference* (Bochum, Germany), 1993, pp. 51–65.

[2]Glaskov, V. V., and Sinkevich, O. A., "Self-Sustaining Heat-Convection Flows in the Absence of Mass Forces," *Fluid Dynamics,* Vol. 29, No. 2, 1994, pp. 166–170.

[3]Abramovich, N. G., *Applied Gas Dynamics*, Nauka, Moscow, 1969.

[4]Rosa, R. J., *Magnetohydrodynamic Energy Conversion*, McGraw–Hill, New York, 1968.

[5]Glaskov, V. V., and Sinkevich, O. A., "Closed-Cycle MHD Generator with Self-Sustaining Heat-Convection Flows," *High Temperature*, Vol. 32, No. 2, 1994, pp. 271–274.

[6]Glaskov, V. V., and Sinkevich, O. A., "Nuclear Driven Closed Circle Magnetohydrodynamic Generator on Self-Maintained Thermal Convection Flow," *Proceedings of the Second Conference on Energy Transfer in Magnetohydrodynamic Flow* (Aussois, France), Vol. 2, 1994, pp. 661–670.

[7]Glaskov, V. V., "Self-Sustaining Thermoconvection Flows: Theory and Possible Applications," Ph.D. Thesis, Technical Sciences, Moscow Power Engineering Inst., Moscow, 1994.

[8]Thorn, K., and Schwenk, F. C., "Gaseous Fuel Reactor Systems for Aerospace Applications," *Journal of Energy*, Vol. 1, No. 5, 1977, pp. 267–276.

[9]Sinkevich, O. A., "Ionization Phenomena in Low-Temperature Plasma with Strong Magnetic Fields," Doctorate Dissertation, Physics and Math Department, Inst. of High Temperature, Academy of Sciences of the USSR, Moscow, 1978.

[10]Sinkevich, O. A., and Chuklova, V. T., "The Non-Linear Interaction of Waves During the Development of Ionization Instability in Non-Equilibrium Magnetized Plasma," *High Temperature*, Vol. 31, No. 3, 1993, pp. 306–312.

[11]Artemov, V. I., Levitan, Yu. S., and Sinkevich, O. A., *Instabilities and Turbulence in Low-Temperature Plasmas*, Publishing House of Moscow Power Engineering Inst., Moscow, 1994.

Chapter 31

Numerical Computation of Current Density Vector and Lorentz Force in a Magnetohydrodynamic Disk Generator

Alberto Geri* and Giuseppe Maria Veca†
Università di Roma "La Sapienza," Roma, 00184, Italy
and
Alessandro Salvini‡
Università di Roma Tre, Roma, 00146, Italy

Nomenclature

α	= thermic diffusion coefficient, m^2/s
$\boldsymbol{B}$	= magnetic field flux density vector, T
B_k	= component of magnetic field flux density vector $(k=r,\theta,z)$, T
C_e	= average velocity of thermal agitation
c_p	= specific heat at constant pressure, J/K
$D_{x,y}$	= geometrical distance between two, m
$\boldsymbol{d}_{x,y}$	= unit vectors of oriented segments of Fig. 2c
$\boldsymbol{E}$	= electrical field, V/m
E_{Fk}	= ideal independent Faraday's voltage source $(k=r,\theta,z)$, V
E_{Hk}	= ideal controlled Hall's voltage source $(k=r,\theta,z)$, V
H	= length along the z axis of the sc coil cross section, m
h	= length along the z axis of each partition of the coil, m
h_t	= total enthalpy of the gas, J
I_{scm}	= dc current flowing into sc coils, A
$\boldsymbol{J}$	= current density vector, A/m^2
L	= length along the r axis of the sc coil cross section, m

*Researcher, Dipartimento di Ingegneria Elettrica, Via Eudossiana 18.
†Full Professor, Dipartimento di Ingegneria Elettrica, Via Eudossiana 18.
‡Researcher, Dipartimento di Ingegneria Elettronica, Via Della Vasca Navale 84.

l = length along the r axis of each partition of the coil, m
m_e = electron mass
n = atomic and molecular density
n_H, n_L = number of subdivisions of magnet cross section
P = generic node of the equivalent electrical network
p = pressure, N/m^2
Q = cross-sectional dimension from which the motion quantity transmission occurs
e = electron electric charge
R_k = resistances of equivalent electrical network (k=r,θ,z), Ω
r = radius (first coordinate of cylindrical system), m
T = temperature, K
t = time, s
U = internal energy of the gas, J
V = electric voltage, V
$\boldsymbol{v}$ = plasma velocity vector, m/s
z = height (third coordinate of cylindrical system), m
β = Hall parameter
δ = mass density of the fluid, kg/m^3
ε = dielectric permittivity, F/m
θ = azimuth (second coordinate of cylindrical system), rad
ϑ = segmentation angle of the sc coil, rad
μ_o = magnetic permeability, H/m
μ_v = dynamic viscosity of fluid, Ns/m^2
σ = electrical plasma conductivity, S/m
τ = average time between two electron collisions
ω = cyclotron frequency

I. Introduction

IN THESE last years the technical and scientific interest on the magnetohydrodynamic (MHD) disk generators (see Fig. 1) has been growing. The disk generator is a valid alternative as regards that linear because of the encumbrance reduction, the easiest construction of the sc magnet, the best opportunity of connection to inverters groups of ac lines, and the possibility of a high enthalpy extraction. The channel is composed of two facing disks, and the excitation magnetic field is generated by means of an sc magnet having annular shape (Helmholtz coil). The ionized gas flows into the channel with a high radial velocity that creates with the magnetic flux density a distribution of electromotive forces (emfs) due to Faraday's induction. Thus, a strong Faraday current begins to flow, being the plasma self-short-circuited. Furthermore, jointly with Faraday's emfs, a second type of electromotive force arises because of Hall's effect. This phenomenon, due to the interaction between the current density and magnetic field, allows recovery of an electrical power by connecting an external load to annular electrodes placed on the disk channel walls (Fig. 2). The evaluation of the three-dimensional current density distribution within the plasma can be pointed out by

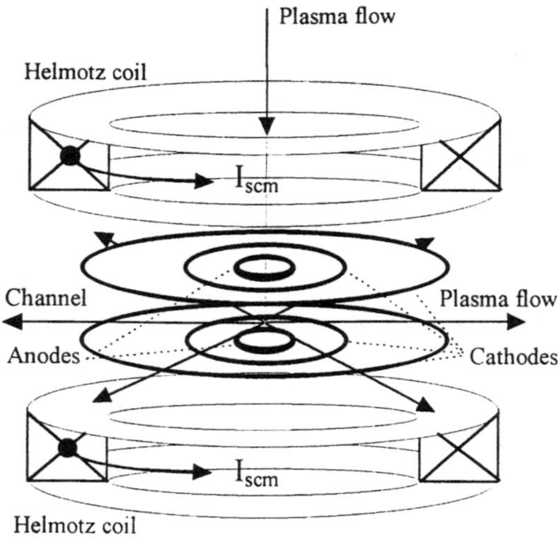

Fig. 1 Sketch of an MHD disk generator, a three-dimensional view.

solving the fundamental set of differential equations governing the magnetohydrodynamic flows:[12]

The equation of energy conservation:

$$\delta\frac{\partial}{\partial t}\left(\frac{v^2}{2}+U+\frac{\varepsilon E^2}{2}+\frac{B^2}{2\mu_0}\right)+\delta\boldsymbol{v}\cdot\nabla\left(\frac{v^2}{2}+U\right)=-\nabla\cdot(\boldsymbol{v}p)-\nabla\cdot\left(\frac{\boldsymbol{E}\times\boldsymbol{B}}{\mu_0}\right) \tag{1}$$

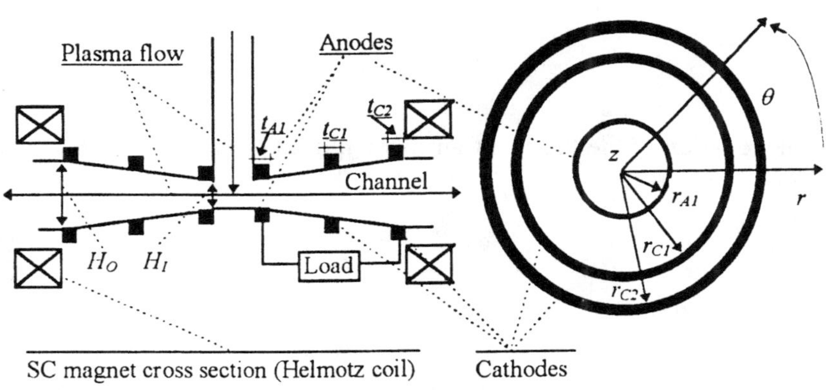

Fig. 2 Sketch of an MHD disk generator: a) lateral view and b) frontal view. Numerical values: r_{A1} = 0.090 m, r_{C1} = 0.250 m, r_{C2} = 0.380 m, t_{A1} = 0.070 m, t_{C1} = 0.020 m, t_{C2} = 0.040 m, H_I = 0.0124 m, and H_O = 0.0280 m.

The conservation of momentum:

$$\delta\left(\frac{\partial}{\partial t}+\boldsymbol{v}\cdot\nabla\right)\boldsymbol{v}=\boldsymbol{j}\times\boldsymbol{B}-\nabla\left(p-\frac{1}{3}\mu_v\nabla\cdot\boldsymbol{v}\right)+\mu_v\nabla^2\boldsymbol{v} \tag{2}$$

The equation of continuity:

$$\left(\frac{\partial}{\partial t}+\boldsymbol{v}\cdot\nabla\right)\delta=\delta\nabla\cdot\boldsymbol{v} \tag{3}$$

The equation of the generalized Ohm law:

$$\boldsymbol{j}=(\boldsymbol{E}_L+\boldsymbol{v}\times\boldsymbol{B})-\frac{\beta}{B}(\boldsymbol{j}\times\boldsymbol{B}) \tag{4}$$

The equation of total enthalpy of plasma:

$$\mathrm{d}h_t = c_p\,\mathrm{d}T \tag{5}$$

The equation of state:

$$\mathrm{d}p - R(\mathrm{d}\delta + \mathrm{d}T) = 0 \tag{6}$$

The relation of Mach number:

$$M=\frac{v}{\sqrt{kRT}} \tag{7}$$

The equation of the thermal transmission from plasma to external channel walls:

$$\alpha\nabla^2 T = -\nabla T \tag{8}$$

The Hall parameter:

$$\beta=\omega\tau=\frac{eB}{m_e n Q c_e} \tag{9}$$

This set of differential equations can be integrated with great difficulty: their solution requires a longer computation time and deep computing resources. For this reason three-dimensional models are not encouraged, and therefore some simplifying hypotheses are frequently introduced [e.g., the $(B_r r,\theta,z)$ and $B_\theta(r,\theta,z)$ components of the magnetic flux density are neglected, whereas the $B_z(r,\theta,z)$ component is often assumed to be uniform]. However, in many technical cases, it is interesting to have a three-dimensional model of the MHD device for a correct evaluation of its performance. An alternative to the direct integration of differential MHD equations is provided by equivalent circuital models that can be advantageously used to simulate the. For this reason the authors have extended their circuital model, already applied and validated to linear gnerators,[3–7] to MHD disk generators. The circuital approach can significantly reduce the calculation cost and can make available a greater amount of information with the minimum computational resources (time and memory). Nevertheless, even if this method allows the investigation of the macroscopic global performance of the device, the method is unable to describe the microscopic behavior of the plasma physics and fails when detailed analysis of the plasma core behavior is required. (In this case, it is necessary to take into account the time dependence of all physical plasma parameters.) To get a believable estimation of the main electrical parameters (such as plasma conductivity, Hall parameter, electromotive forces due to both Faraday and Hall effects), the design configuration and the measured values of the Fuji facilities experiments (FUGJI-1 DSIK-F3a)[8,9] have been assumed as input data for the circuital model.

II. Magnetohydrodynamic Disk Generator Modeling

A. Magnetic Field Computation

The sc magnet is a Helmholtz coil, and, in the usual cases, twin magnets are used to produce the excitation field (see Figs. 1 and 2). The three-dimensional map of the magnetic flux density due to the sc coils has been determined by means of a calculation code developed by the authors. It is based on a traditional method that allows the evaluation of the magnetic flux density components (B_r, B_θ, and B_z) at every point inside and outside the plasma. Initially, the procedure replaces each magnet coil (defined by a restricted number of information, i.e., shape, dimensions, and position of each coil) with a filiform representation consisting of several elementary turns (a suitable choice of their number $n_H \cdot n_L$ makes the magnetic field computation sufficiently accurate) arranged in order to approximate the effective shape of the coil (see Fig. 3a). Then, the code determines all straight segments in which the elementary turns can be divided (see Fig. 3b). Finally, the total magnetic field at each assigned point (e.g., at the center of gravity of each elementary partition of the plasma) is computed superimposing the contribution of each segment (see Fig. 3c) that can be evaluated by means of the well-known integral formulation given by

$$\boldsymbol{B} = \frac{I_{\text{scm}} \cdot 10^{-7}}{D_{MP}} \left[\frac{D_{Ai,M}}{D_{Ai,P}} + \frac{D_{A(i+1),M}}{D_{A(i+1),P}} \right] [\boldsymbol{d}_{Ai,A(i+1)} \times \boldsymbol{d}_{P,M}] \tag{10}$$

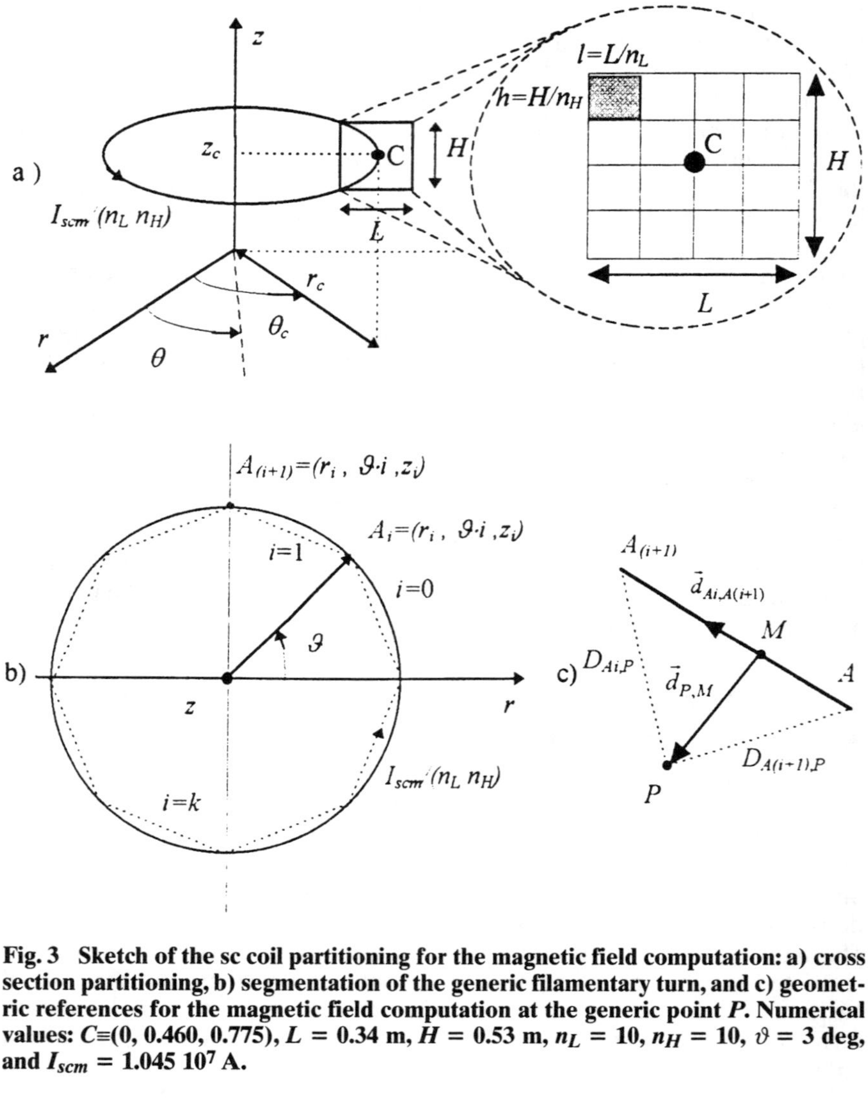

Fig. 3 Sketch of the sc coil partitioning for the magnetic field computation: a) cross section partitioning, b) segmentation of the generic filamentary turn, and c) geometric references for the magnetic field computation at the generic point *P*. Numerical values: *C*≡(0, 0.460, 0.775), L = 0.34 m, H = 0.53 m, n_L = 10, n_H = 10, ϑ = 3 deg, and I_{scm} = 1.045 10^7 A.

B. Equivalent Electrical Network

The plasma is modeled by means of an equivalent electrical lumped parameter network in stationary conditions. The ionized gas has been ideally partitioned into a generic number of small volumes (see Fig. 4). At each center of gravity [indicated with $P(\theta,r,z)$ in Fig. 4], an electrical node has been associated. Six electrical branches, having the same directions of the reference system, start from this node. Each branch is composed by a resistance R, an ideal current controlled voltage source E_H, and an ideal inde-

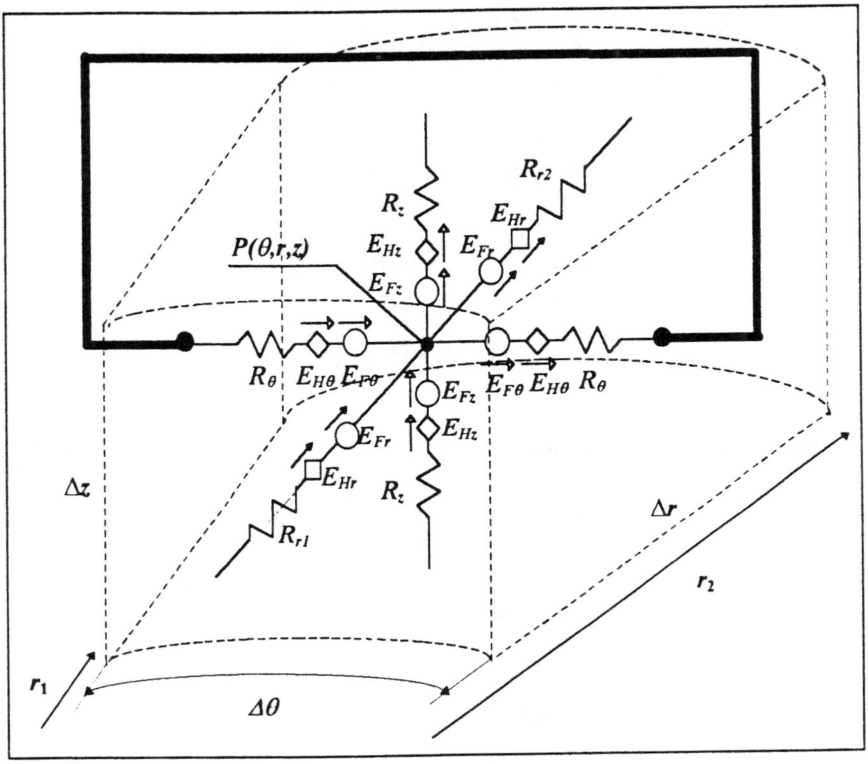

Fig. 4 Three-dimensional equivalent electrical network of each elementary volume. Numerical values: $\Delta r = 0.009$ m, $\Delta z = 0.002$ m, and $\Delta\,\theta = 1$ deg.

pendent voltage source E_F. The resistive bipole R takes into account all electric macroscopic characteristics of the plasma. These values are deduced from the duct geometry and from the distribution of the plasma conductivity. In correspondence with each branch of the generic point P, we have (see Fig. 4)

$$R_{r1}(P) = \frac{1}{\sigma(P)\Delta h \Delta\theta}\,\ell n\left(\frac{r_P}{r_1}\right) \qquad R_{r2}(P) = \frac{1}{\sigma(P)\Delta h \Delta\theta}\,\ell n\left(\frac{r_2}{r_p}\right)$$

$$R_\theta(P) = \frac{1}{2\sigma(P)}\,\frac{n^2\Delta\theta r}{\Delta h \Delta l} \tag{11}$$

$$R_z(P) = \frac{1}{2\sigma(P)}\,\frac{\Delta h}{\Delta\theta\Delta l}$$

The independent ideal voltage source E_F takes into account the Faraday electromotive force due to the external product between the plasma velocity and the excitation magnetic field. They are given by

$$E_{Fr}(P) = [\boldsymbol{v}(P) \times \boldsymbol{B}(P)]_r \frac{(r_2 - r_1)\Delta\theta}{2}$$

$$E_{F\theta}(P) = [\boldsymbol{v}(P) \times \boldsymbol{B}(P)]_\theta \frac{n^2 \Delta\theta r_P}{2\Delta h \Delta l} \tag{12}$$

$$E_{Fz}(P) = [\boldsymbol{v}(P) \times \boldsymbol{B}(P)]_\theta \frac{\Delta h}{2\Delta\theta \Delta r \Delta l}$$

The current controlled ideal voltage source E_H takes into account the Hall effect, coming from the external product between the current density distribution within the plasma and the excitation magnetic field, as follows:

$$E_{Hr}(P) = -\frac{\beta(P)}{\sigma(P)|\boldsymbol{B}(P)|}[\boldsymbol{J}(P) \times \boldsymbol{B}(P)]_r \frac{(r_2 - r_1)\Delta\theta}{2}$$

$$E_{H\theta}(P) = -\frac{\beta(P)}{\sigma(P)|\boldsymbol{B}(P)|}[\boldsymbol{J}(P) \times \boldsymbol{B}(P)]_\theta \frac{n^2 \Delta\theta r_P}{2\Delta h \Delta l} \tag{13}$$

$$E_{Hz}(P) = -\frac{\beta(P)}{\sigma(P)|\boldsymbol{B}(P)|}[\boldsymbol{J}(P) \times \boldsymbol{B}(P)]_\theta \frac{\Delta h}{2\Delta\theta r_P \Delta l}$$

All of these bipoles allow the simulation of the device in steady-state conditions for the assigned value of the load resistance R_L and of the dc current flowing into sc coils I_{scm}. All physical quantities (such as pressure, temperature, velocity, and so on) must be assumed as input data (i.e., these values must be evaluated apart from the simulation). The model exploits the electric and geometric symmetry of the device. For this reason the authors have modeled only a slice of the plasma having an angular amplitude $\Delta\theta$. To take into account the boundary conditions (along the lateral surface of the slice, i.e., on the plane $\theta = 0$ and $\theta = \Delta\theta$), we have connected all branches placed along the θ axis in short circuit (see Fig. 4). In addition, the value of the load resistance has been modified in such a way to consider that only a small slice of the plasma has been modeled ($R'_L = 2\pi R_L/\Delta\theta$).

C. Computational Algorithm

The equivalent electrical network of the device has been solved by means of the nodal analysis. The presence of controlled generators has been taken into account with a suitable iterative method. This procedure is based on the initial evaluation of the Farady emfs [which only depend on the assigned velocity vector distribution within the plasma and from the computed three-dimensional magnetic field map as shown in Eqs. (12)]. Then, solving the equivalent electrical network (assuming as excitation the computed Farady emfs), the possibility to estimate the current density distribution inside the plasma exists. These currents are used for the evaluation of the Hall emfs of first attempt by means of Eqs. (13). Thus, the network is solved again using as excitation both the Hall emfs (just computed) and the Faraday emfs (previously computed). The up-to-date current density distribution is then used to evaluate the Hall emfs of second attempt. The iterative procedure stops when the analysis reaches the fixed degree of accuracy.

III. Numerical Simulation

The authors have applied their model to reproduce the results of the Fuji facilities experiments (FUJI-1 DSIK-F3a).[8,9] The design configuration of the device and the conditions for the experiment have been assumed as input data for the circuital model (see Fig. 5). The sc coils (see Fig. 3) have been chosen in a such way as to generate within the plasma a magnetic field having the same B_z values corresponding with the inlet and the outlet of the channel (see Fig. 6). In Fig. 7 are summarized the results of simulations executed by means of the circuital model. In this figure are also plotted the calculated values (using an analytical approach) and the measured data as taken from the analysis in Shoida's papers.[8,9]

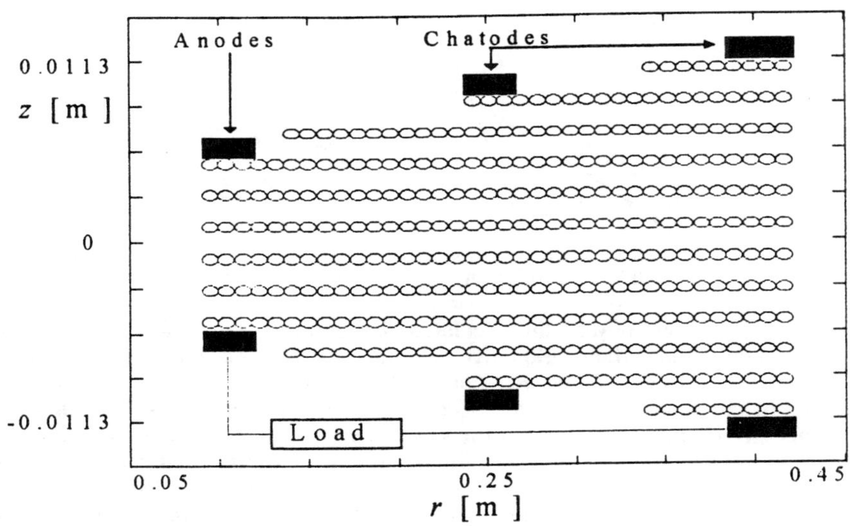

Fig. 5 Partitioning of the plasma slice.

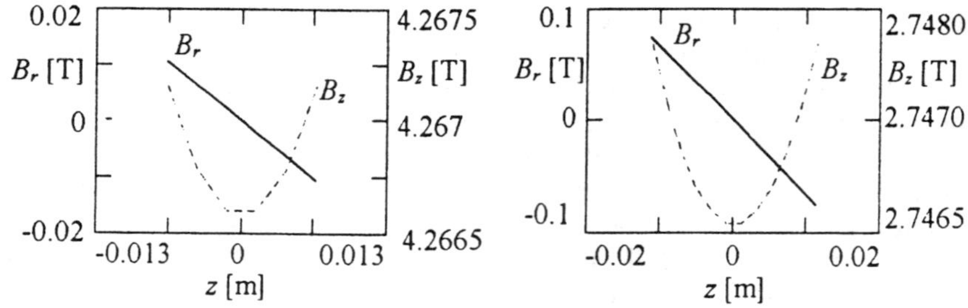

Fig. 6 Evolution of the magnetic flux densities within the ionized gas.

IV. Conclusions and Remarks

In this paper the authors have described a circuital model able to simulate MHD disk generators. The results of the simulations executed seem to confirm that the proposed model is suitable to perform the investigation of MHD disk generators. In fact, the curves obtained using this model (matching with a sufficient degree of accuracy both the experimental data and the numerical computations based on more sophisticated analytical approaches) confirm the reliability of the simulations executed. These preliminary results encourage the authors to continue the development of their model with the aim of improving its accuracy. At the same time they are

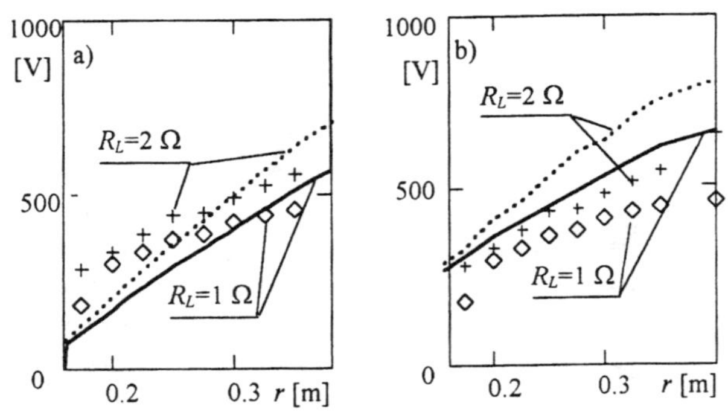

Fig. 7 Comparison between the calculated distribution of the Hall potential along the radial direction with experimental and numerical values as taken from Shioda's papers[8,9] for two kind of load resistance (R_L = 1 Ω and R_L = 2 Ω). The curves plotted by continuous lines are the calculation values whereas scattered points are the measured values. The result of the simulations executed by circuital model are drawn in a) while Shioda's analyses are drawn in b). Experimental conditions and numerical values: working gas, Ar+Cs; stagnation gas temperature, 1850 K; thermal input, 2.6 MW; stagnation gas pressure; 0.46 MPa; seed fraction, ≈ 2 10^{-4}; Hall parameter, 4 ÷ 7; gas velocity, 500 m/s; electrical conductivity, 250 S/m; and swirl, 0.02 ÷ 0.2.

working to make the procedure more independent from the several experimental data that are used as input data in the actual version of the code (i.e., to make it able to perform the simulation without preliminary experimental investigations). All of these efforts are justified by the observation that the circuital model can operate an efficient and fast evaluation of the global electrical performance of the device (e.g., to evaluate the total output electric power), which are those of interest from an engineering point of view.

Acknowledgment

The authors want to thank S. Shioda and his collaborators of the Technological Institute of Tokyo for the support kindly given in bibliographical research regarding data from experimental tests in the Fuji facilities.

References

[1]Rosa, R. J., *Magnetohydrodynamic Energy Conversion*, Hemisphere, Washington, DC, 1987.

[2]Messerle, H. K., *Magnetohydrodynamic Electrical Power Generation*, Unesco Energy Engineering Series, Wiley, New York, 1995.

[3]Geri, A., Salvini, A., and Veca, G. M., "A 3D model of a MHD Faraday Linear Generator," *Conference Proceedings of 32nd SEAM* (Pittsburgh, PA), June 1994.

[4]Geri, A., Salvini, A., and Veca, G. M., "Magnet for MHD Linear Generator," *IEEE Transactions on Magnetics*, Vol. 30, No. 4, July 1994, pp. 1847–1850.

[5]Geri, A., Salvini, A., and Veca, G. M., "A 3D Model for MHD Hall Linear Generator," *Conference Proceedings of 2nd ETMF Pamir* (Aussois, France), Vol. 2, 1994, pp. 759–768.

[6]Geri, A., Salvini, A., and Veca, G. M., "Comparison Between Numerical Integration and 3D Circuital Approach for Analysing the Performance of MHD Linear Generators," *Conference Proceedings of 33rd SEAM* (Tullahoma, TN), 1995, pp. IV3-1–IV3-8.

[7]Geri, A., Salvini, A., and Veca, G. M., "MHD Linear Generator Modelling," *IEEE Transactions on Magnetics*, Vol. 5, No. 2, June 1995, pp. 465–468.

[8]Tsunoda, K., Harada, N., Kabashima, S., and Shioda, S., "Numerical Simulation Studies on FUGJ-1 Experiments," *Energy Conversion Management*, Vol. 34, No. 6, 1993, pp. 447–456.

[9]Tsunoda, K., Saito, Y., Yamasaki, H., Harada, N., Yoshikawa, K., Kabashima, S., and Shioda, S., "Supersonic Flow Behavior under Strong Lorentz Force," *JSME International Journal*, Series B, Vol. 37, No. 4, 1994, pp. 752–459.

Chapter 32

Self-Consistent Approach to Modeling of High-Current Arc Spots on Electrodes of a Magnetohydrodynamic Generator

M.S. Benilov*
Universidade da Madeira, Funchal, 9000, Portugal

I. Introduction

THE electric current can flow to electrodes of a magnetohydrodynamic (MHD) generator in a diffuse regime, being distributed over the electrode surface more or less uniformly or in regimes with arc spots. The regimes with spots are usually divided into a micro-arc regime and a high-current arc regime. The micro-arc regime is characterized by many small fast-varying spots; in the high-current arc regime the current is localized in one or several slowly varying spots.

From the point of view of practice, the most desirable is the diffuse regime in which the rate of the erosion of the electrodes is the lowest. This regime may be realized if the average temperature of the electrode surface is high enough (say, of order of 1500 K or higher for anodes and of order of 2000 K or higher for cathodes) or in special conditions (e.g., Ref. 1). From the point of view of a theoretical description, the diffuse regime is simpler than the regimes with arc spots because the regime may be treated in the framework of a one-dimensional model (all parameters vary only in the direction normal to the electrode surface; variations in the directions along the surface are much slower and may be neglected). At present, a well-developed theory exists for this regime (Ref. 2 and references therein) that does not rely on empirical parameters described in the framework of a single-model phenomenon in a wide range of conditions, i.e., for both cathode and anode regimes, with and without a magnetic field, and which agrees with the available experimental data.

Unfortunately, to maintain the electrode surface at the high temperature necessary for the diffuse regime on the cathode is not technologically simple

*Departámento de Física, Largo do Município.

and in most cases cathodes of MHD generators operate in the micro-arc regime with the occasional appearance of high-current arc spots. The high-current spots are, however, highly undesirable because they result in a rapid destruction of the construction of the MHD channel. Up to now this has been one of the major obstacles hindering a practical utilization of stationary MHD generators.

A choice of the design of the electrodes of an MHD generator and of their electrical loading preventing appearance of high-current arc spots is of crucial importance. This choice could be facilitated by a self-consistent theoretical model of high-current cathode arc spots in the conditions of MHD generators. Unfortunately, such a model is still absent up to now in spite of many years of investigations; available models (see, e.g., references in Ref. 3) are either semiempirical or based on rather arbitrary theoretical suppositions, usually involving some or other implementations of the "principle" of minimum voltage.

On the other hand, considerable progress has been achieved recently (e.g., Ref. 4 and references therein) in the theory of cathode spots in high-pressure arc discharges for typical conditions, e.g., for arc welding or plasma torches. In particular self-consistent calculation models have been developed. The aim of this paper is to apply this theory to the conditions of an MHD generator. It is found that the specific features of the MHD plasma (presence of an alkali seed and a high velocity of the flow) were found not to affect substantially the cathode spots and the calculation, models of cathode spots in high-pressure arc discharges mentioned in the preceding paragraph may be directly applied to the conditions of an MHD generator.

II. Model of Cathode Spots in High-Pressure Arc Discharges

A brief review of the state of the art in the theory of cathode spots in high-pressure arc discharges is given in Ref. 4. Two approaches have been developed in the literature: one based on the numerical modeling of the entire system arc cathode, including the arc column, the near-cathode plasma layer and the cathode body (Refs. 5 and 6 and references therein); and a semianalytical approach[4] combining a model of the near-cathode plasma layer with a model of the heat transfer in the cathode body. Note that the calculation model,[4] in contrast to previous investigations carried out in the framework of a similar semianalytical approach, enables one to determine all parameters of spots, including its radius, in a self-consistent way, without using empirical parameters or arbitrary theoretical suppositions. In what follows, the model[4] is used, which is simple and treats the physics of the near-cathode plasma layer in detail.

The calculation procedure suggested in Ref. 4 is as follows. The plasma is divided into the arc column, which is assumed to be in local thermodynamic equilibrium, and the near-cathode plasma layer, in which deviations from the equilibrium are localized. The estimates show that the contribution of the heat flux from the arc column to the energy balance of the near-cathode layer is not decisive. Therefore, the effect of the arc column on the near-cathode plasma layer and on the spot itself may be neglected. Furthermore, calculation of the near-cathode plasma layer and of the structure of the spot itself (i.e., of the distribution of parameters along the electrode surface) may be carried out separately. More specifically, the procedure is as follows.

As a first step, one calculates in the one-dimensional approximation characteristics of the near-cathode layer in function of the local surface temperature T_w and of the near-cathode voltage drop U. In particular, the density of the heat flux from the plasma to the cathode surface $q(T_w,U)$ and of the density of the electric current from the plasma to the cathode surface $j(T_w,U)$ should be found. These characteristics do not depend on the total arc current or on the shape of the cathode, and to calculate them only once for every given combination of the cathode material, the plasma gas, and its pressure is sufficient.

After this, one can calculate the temperature distribution inside the cathode body and on the surface. This amounts to solving inside the cathode the (multidimensional) heat conduction and current continuity equations with the boundary conditions supplied by the dependencies $q(T_w,U)$ and $j(T_w,U)$. (If the Joule heat production in the body of the electrode is unessential, which is usually the case, the current continuity equation may be excluded from consideration.) After this problem has been solved, one can substitute the distribution of the surface temperature obtained as a part of the solution into the dependencies found at the previous step, thus determining distributions of parameters of the near-cathode layer along the cathode surface. Integrating the current density surface distribution, one finds the arc current corresponding to the considered value of parameter U. This step can be repeated several times with different values of U until a desired value of the arc current is reached.

The first step described in the preceding paragraphs can be made by means of the model of the near-cathode plasma layer developed in Ref. 4. In the framework of this model, the layer as a whole is subdivided into three zones: a space-charge sheath, an ionization layer, and a layer of thermal relaxation.

The space-charge sheath is a region where deviations from quasineutrality are localized. This layer is of the thickness of the order of several hundredths of a micrometer adjacent to the cathode surface. The space-charge sheath is collision-free, and currents of different species of charged particles are constant across the sheath. A major fraction of the electric current in the sheath is transported by the electrons emitted by the cathode surface; however, a contribution of the ions coming from the quasineutral plasma is also substantial. There also may be an appreciable current of the electrons "counter-diffusing" from the plasma to the cathode.

The ionization layer is a (quasineutral) region adjacent to the space-charge sheath in which deviations from ionization equilibrium are localized. Its thickness is of the order of 10 μm. The fraction of the ion current at the inner boundary of the ionization layer equals that in the sheath; as the distance from the cathode surface increases, the ion current falls down to a very small value at the outer boundary. Because the flux of the ions leaving the ionization layer for the sheath is much higher than the flux of the ions entering the ionization layer from the layer of thermal relaxation, ionization must prevail over recombination in the ionization layer. Losses of the electron energy in elastic collisions with heavy particles are negligible in the ionization layer.

The (major) part of the near-cathode layer that is external with respect to the ionization layer represents the layer of thermal relaxation. Thickness of this layer is of the order of 100 μm. In the layer of thermal relaxation, both quasineutrality and ionization equilibrium hold, whereas temperatures of the heavy particles and of the electrons are different. The main charge carriers are the electrons.

Of these three zones the most important are the ionization layer, in which the ion flux tq the cathode is generated, and the space-charge sheath, in which the ions and the emitted electrons are accelerated. The layer of thermal relaxation is of minor importance. Accordingly, the space-charge sheath and the ionization layer are accounted for in the model,[4] and the layer of thermal relaxation is disregarded. The sheath is described analytically. The results of the numerical solution[7] of the multifluid equations are used to describe the ionization layer. The boundary condition for the charged particle density at the inner boundary of the ionization layer is supplied by the Bohm criterion.

The second step mentioned in the preceding paragraphs was considered in Refs. 8 and 9. In particular, an approximate asymptotic theory of cathode spots has been developed for a case when the function q is localized in a narrow range of the surface temperatures, which, according to the calculations, is the case for the conditions of the high-pressure arc spots. Simple asymptotic formulas have been obtained for integral parameters of spots on a planar cathode (such as the spot radius, the integral current per spot, and the integral heat flux inside the cathode body removed by heat conduction) for the case when the transversal dimension is much larger than the spot radius.

Detailed calculation results obtained by means of this model for various combinations plasma/cathode for atmospheric-pressure arcs are reported in Ref. 4. The results agree with the measurements of the spot temperature reported recently in the literature.

III. Application to Conditions of a Magnetohydrodynamic Generator

Characteristic features distinguishing the plasma in MHD channels from, e.g., the plasma in welding arcs or in plasma torches, is the presence of an alkali seed (usually 1–2%) and a high velocity of the plasma flow (of the order of 1 km/s). Consider first the effect of the high velocity of the plasma flow.

The flowfield in an MHD channel may be divided into two zones: the core, in which the velocity distribution is more or less uniform and the viscosity and heat conductivity effects are inessential, and the near-wall viscous boundary layer. Thickness of the boundary layer in full-scale MHD channels is of the order of 1–10 cm. The boundary layer itself may be subdivided into the turbulent core and the laminar sublayer. Thickness of the latter is of the order of 100 μm.

As was mentioned in the preceding section, the two zones playing a decisive role in the near-cathode plasma layer are the ionization layer and the space-charge sheath, and their total thickness is of the order of 10 μm. They are positioned at the bottom of the laminar sublayer. Assuming that the flow velocity at the edge of the sublayer is about half of the velocity in the inviscid core and that inside the sublayer the velocity decreases approximately linearly to zero at the cathode surface, one finds that the flow velocity is of the order of 50 m/s in the ionization layer and practically zero in the sheath. Because the characteristic diffusion velocity of the ions in the ionization layer is comparable to the thermal velocity, which is of the order of 1 km/s, one can conclude that the convective effects in the near-cathode plasma layer are not decisive. Thus, the high velocity of the plasma flow in MHD channels does not affect substantially the cathode spots.

This means that the model described in the preceding section may be directly applied to cathode spots in MHD channels. The next question is whether the alkali seed should be taken into account. The results[4] indicate that the electron temperature in the near-cathode layer is quite high, and the ionization degree is close to unity even in the case of inert gases. In such a situation, the presence of a small amount of an alkali seed cannot affect substantially the physics of spots (in contrast to that of the diffuse regime, where the effect of an alkali seed is quite substantial).

Thus, presence of an alkali seed may in the first approximation be disregarded. For example, if one is interested in the calculation of cathode spots in a closed-cycle MHD generator working with a potassium-seeded argon, one can use the results given by the model[4] for pure argon. Parameters of cathode spots obtained in such a way are shown in Table 1. The work function of the electrode material is taken equal to 4.55 eV, which corresponds to tungsten; the pressure is 1 atm. The input parameter in Table 1 is the voltage drop in the near-cathode layer U whereas all of the other parameters of spots, including the current per spot, are results of evaluation. To emphasize, however, one can always choose U in such a way that I will be equal to any given value. In particular, values of U in the calculations presented in Table 1 have been chosen in such a way that I varies between approximately 40 A and 300 A. One can see that the electron temperature is indeed quite high. The plasma is close to full ionization with a large fraction of the multiply charged ions.

IV. Conclusions

The estimates show that thickness of the near-cathode plasma layer in which the processes crucial for maintenance of the spot are localized is much smaller than the thickness of the laminar sublayer in MHD channels. Therefore the high velocity of the plasma flow in MHD channels has a minor effect on the characteristics of high-current cathode spots. The plasma in the near-cathode layer is nearly fully ionized; therefore, presence of the alkali seed also does not produce a major effect.

Table 1 Parameters of cathode spots

U, V	I, A	T_*, K	r_*, mm	$\langle j \rangle$, Am^{-2}	Q, W	T_{e*}, K	Z_*	j_{i*}/j_*	j_{e*}/j_*
22.0	295	4380	054	32×10^8	972	41,312	272	023	008
22.5	267	4370	052	32×10^8	928	41,851	276	023	007
23.0	234	4350	049	31×10^8	878	41,904	276	024	007
23.5	211	4335	047	31×10^8	839	42,155	278	025	006
24.0	188	4320	045	30×10^8	800	42,365	279	025	006
24.5	168	4300	043	29×10^8	766	42,267	278	026	005
25.0	152	4290	041	29×10^8	734	42,677	280	027	005
25.5	137	4275	040	28×10^8	704	42,766	281	027	004
26.0	123	4260	038	27×10^8	675	42,816	281	028	004
26.5	112	4245	037	26×10^8	651	42,814	281	029	003
27.0	99	4230	035	25×10^8	623	42,766	281	030	003
27.5	92	4220	034	25×10^8	604	43,029	282	030	003
28.0	82	4205	033	24×10^8	581	42,897	281	031	002
28.5	75	4195	032	24×10^8	562	43,084	282	031	002
29.0	68	4180	031	23×10^8	542	42,866	281	032	002
29.5	63	4170	030	22×10^8	527	42,983	282	033	002
30.0	56	4160	029	21×10^8	508	43,079	282	033	002
30.5	52	4150	028	21×10^8	494	43,145	283	034	001
31.0	47	4140	027	20×10^8	479	43,178	283	034	001
31.5	43	4130	027	19×10^8	465	43,179	283	035	001
32.0	40	4120	026	19×10^8	453	43,152	283	036	001

U, the near-cathode voltage drop; I, current per spot; T_*, characteristic surface temperature inside the spot; r_*, the spot radius; $\langle j \rangle$, the average current density in the spot; Q, the integral heat flux into the cathode body; T_{e*}, characteristic electron temperature in the near-cathode layer; Z_*, the average ion charge in the layer; j_{i*}/j_*, fraction of the ion current; and j_{e*}/j_*, fraction of the current of counterdiffusing electrons.

In general, the specific features distinguishing the plasma in MHD channels from, e.g., the plasma in welding arcs or in plasma torches, do not affect appreciably the cathode spots. Hence, the theory developed previously for calculation of characteristics of cathode spots in slowly moving nonseeded plasmas may be applied also to the conditions of MHD generators. An example of the calculation is given.

Acknowledgment

The work has been performed within the activities of the project *Creation, Diagnostics, and Modelling of Reactive Plasmas* of the program Praxis XXI.

References

[1]Svensson, R., Engvall, K., Holmlid, L., and Braun, J., "High Emissivity Electrodes for MHD Channels," *Proceedings of the XI International Conference on MHD Electrical Power Generation* (Beijing, China), Vol. 1, International Academic Publishers, Beijing, 1992, pp. 248–252.

[2]Benilov, M. S., Bochkarev, G. G., and Rogov, B. V., "Modeling of Diffuse Current Transfer in a Near-Electrode Layer of the High-Pressure Molecular Plasma," *IEEE Transactions Plasma Sciences*, Vol. 23, No. 4, 1995, pp. 742–749.

[3]Rosa, R. J., Krueger, C. H., and Shioda, S., "Plasmas in MHD Power Generation," *IEEE Transactions on Plasma Sciences*, Vol. 19, No. 6, 1991, pp. 1180–1190.

[4]Benilov, M. S., and Marotta, A., "A Model of the Cathode Region of Atmospheric Pressure Arcs," *Journal of Physics D: Applied Physics*, Vol. 28, No. 9, 1995, pp. 1869–1882.

[5]Kaddani, A., Simonin, O., and Delalondre, C., "Numerical Investigation of the Cathode Region of Electric Arcs," *Proceedings of the 12th International Symposium on Plasma Chemistry* (Minneapolis, MN), edited by J. V. Heberlein, D. W. Ernie, and J. T. Roberts, Vol. III, Univ. of Minnesota, Minneapolis, MN, 1995, pp. 1443–1448.

[6]Lowke, J. J., Morrow, R., and Haider, J., "A Simplified Unified Theory of Arcs and Their Electrodes," *Proceedings of the 12th International Symposium on Plasma Chemistry* (Minneapolis, MN), edited by J. V. Heberlein, D. W. Ernie, and J. T. Roberts, Vol. III, Univ. of Minnesota, Minneapolis, MN, 1995 pp. 1449–1454.

[7]Benilov, M. S., "The Ion Flux from a Thermal Plasma to a Surface," *Journal of Physics D: Applied Physics*, Vol. 28, No. 2, 1995, pp. 286–294.

[8]Benilov, M. S., "Nonlinear Heat Structures and Arc-Discharge Electrode Spots," *Physical Review*, Vol. E 48, No. 1, 1993, pp. 506–515.

[9]Benilov, M. S., "A Self-Consistent Analytical Model of Arc Spots on Electrodes," *IEEE Transactions on Plasma Sciences*, Vol. 22, No. 1, 1994, pp. 73–77.

Chapter 33

Two- and Three-Component Alkali-Metal Alloys with Controlled Properties as Advanced Working Fluids for LMMHD Apparatus

Dmitry N. Kagan,* Galina A. Krechetova,† and Evald E. Shpilrain‡
Institute for High Temperatures of the Russian Academy of Sciences, Moscow, 127412, Russia

Introduction

THE programs for the elaboration and investigation of thermophysical properties of working fluids and coolants with controlled characteristics for power and technology installations (first for energy-conversion installations by the LMMHD method, fast-neutron nuclear reactors, solar power sources) were developed in the Institute for High Temperatures of the Russian Academy of Sciences. These working fluids and coolants are based on two- and three-component alloys of alkali and alkali-Earth metals and have the working range of a liquid phase from 200 to 400 K up to 1500 to 1800 K.

These alloys, reserving all of the advantages of pure components, acquire additional merits: the possibility of regulation of thermo-, electro-, and nuclear-physical properties by varying of the component fractions; the maximum wide range of working temperatures of the liquid phase having both high-temperature and low-temperature fluids working simultaneously; and the maximum effectiveness over all this range concerning hydrodynamic and heat exchange characteristics.

For example, the alloys corresponding to special points of phase diagrams (eutectics), which have extraordinarily low temperatures of crystallization ($t \approx -80°C$), can be obtained with the help of certain regulations of component fractions. This temperature is the minimum one for all of the metal liquids known in the world, and the working fluid remains in liquid state at

*Head of Laboratory and Professor, IVTAN, Izhorskaya 13/19.
†Senior Researcher, IVTAN, Izhorskaya 13/19.
‡Head of Division and Professor, IVTAN, Izhorskaya 13/19.

any Earth temperature and for most real conditions on a large orbital station.

The purpose of the study was to obtain the correct thermodynamic description of binary and ternary alkali liquid-metal systems as a basis for working fluids and coolants of the new kind in the area of compositions and temperatures $0 \le x_i \le 1$ and $200 \le T \le 1500$K. This means building the set of characteristic functions (potentials), which in coordinates P, T, x_i are Gibbs energies of the formation of the systems (partial and integral functions, absolute and excessive values), and thus the activities and coefficients of activity of components: $\Delta\overline{G}_i$, ΔG, $\Delta\overline{G}_i^*$, ΔG^*, α_i, and γ_i. Then all of the derivatives, i.e., the rest thermodynamic functions that are enthalpy and entropy of formation and excessive heat capacity ($\Delta\overline{H}_i$, ΔH, $\Delta\overline{S}_i$, ΔS, $\Delta\overline{S}_i^*$, ΔS^*, and ΔC_p), can be obtained.

The total algorithm of the method, experimental equipment, and the data in the mass were presented in Ref. 1. Now the results of fine analysis of the above thermophysical properties in dependence on the temperature and concentration are submitted.

Method and Experiment

Direct measurement of the change of partial Gibbs energy of components at formation of liquid-metal alloy in wide temperature range, which would be performed by the EMF-method,[2,3] is extraordinarily difficult because of the absence of steady high-temperature solid electrolytes with needed ionic composition used as a membrane of concentration cells. Determining partial pressure and activity of components with the help of the experimental study of absorption spectra by the atomic absorption method[4,5] is limited by temperatures because of the influence of the radiation contribution of corresponding lines. Effusion methods at high temperatures (with either optic- or mass-spectroscopic or chemical determining of flow composition) are not suitable for alkali-metal systems because of the high saturation pressures and technological difficulties of operating with high intensity atomic flows of alkali metals leaving an effusion cell for the area of installation.

Instead of a direct measurement of the Gibbs energy of alloy formation $\Delta\overline{G}_i$ or the activity of components α_i at high temperatures, the method is based on the calculation of these functions in the broad area of temperatures and concentrations by the integration of differential equations of chemical thermodynamics:

$$[\partial \ln \alpha_i / \partial(1/T)]_{p,x_i} = \Delta\overline{H}_i / R \quad \text{or} \quad [\partial(\Delta\overline{G}_i / T) / \partial(1/T)_{p,x_i} = \Delta\overline{H}_i \tag{1}$$

The solutions of these equations have a kind

$$\ln \alpha_i(x_i, T)]_{p,x_i} = \ln \alpha_i(x_i, T_i) - R^{-1} \int_{T_1}^{T} \Delta\overline{H}_i(x_i, T) T^{-2}\, dT \tag{2}$$

$$\Delta\overline{G}_i(x_i,T)=\Delta\overline{G}_i(x_i,T_1)(T/T_1)-\int_{T_1}^{T}\Delta\overline{H}_i(x_i,T)T^{-2}\,\mathrm{d}T \tag{3}$$

where the underintegral function $\Delta\overline{H}_i\ (X_i,T)$, the enthalpy of formation in all areas of named parameters of the state and the boundary conditions $\alpha_i = f(X_i,T_1 = \text{const})$ or $\Delta\overline{G}_i = RT\ \ell n\ \alpha_i = f(X_i,T_1 = \text{const})$ of the concentration dependencies of activities or Gibbs energies at one (not high) reference temperature $T_1 = 400$ K, are determined by experiment.

The proposed scheme does not require any special assumptions, can be provided with reliable input experimental data, and as a result can be permitted to obtain an inwardly agreed upon thermodynamic description of the studied systems.

For the solution of both experimental tasks, two groups of installations were constructed. The first group includes a complex of calorimetric apparatus with a level of sensitivity permitting the determination of excessive thermodynamic functions of liquid-metal systems.[6] The second group is based on the determination of activities of components via partial pressure of saturated vapor obtained by the measurement of the intensivity of their atomic flows with the help of the effusion method using electron-beam bombardment.[7]

Results

In accordance with the algorithm of the method, both the enthalpy of formation (at all the concentrations and temperatures) and the activity of components (at all the concentrations and one temperature $T_1 = 400$ K) were measured for binary and ternary systems: Cs-Na, K-Na, Cs-K, Cs-K-Na, Ba-Cs, Be-Ba, Be-Sr, Be-Ca, Be-Mg, Ca-Sr, Ca-Ba, Sr-Ba, Mg-Ba, Mg-Ca, and so CsF-Cs and Li-H (lithium hydride) at parameters just named. Then the activity (and Gibbs energy) at all of the concentrations and temperatures was calculated [Eqs. (2) and (3); therefore, the rest of the functions were obtained with the help of standard thermodynamic relations].[1] The results for the binary alkali-metal systems are demonstrated in Fig. 1 (enthalpies of formation) and Fig. 2 (activities).

Description of the obtained experimental data with the help of an adequate model theory permits us to give an interpretation for them and to ground approximation equations, which would be sufficiently sensitive to translate the probable peculiarities of thermodynamic properties behavior. As a rule the model theories do not provide a reliable description for all thermodynamic functions simultaneously and all the more their derivatives, in particular for the area of the maximum ordering of liquid-metal systems.[8] Therefore the following method of approximation of characteristic functions (Gibbs energies of formation ΔG) seems to be rather effective[9]: contributions of enthalpy term ΔH (experimentally obtained enthalpy of formation) and entropy term $T\Delta S$ (obtained as a difference between experimental values of enthalpy of formation and Gibbs energy of formation) are

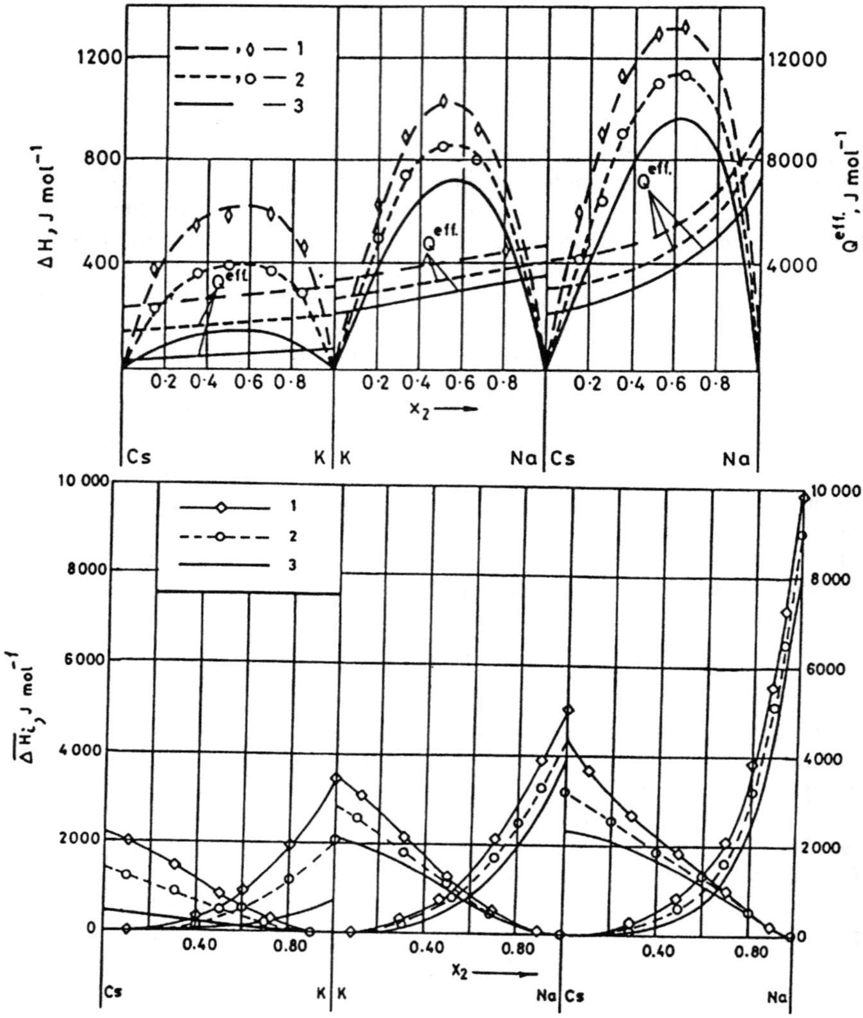

Fig. 1 Integral (ΔH) and partial ($\Delta \overline{H}_i$) enthalpy of formation of binary alkali-metal alloys up to 1200 K. $Q^{\text{eff}} = \Delta H / x_1 x_2$ is the effective parameter of interaction, J/mol; x_2 is the mole fraction of component 2. 1,2,3—this work: 1, 1200 K; 2, 800 K; and 3, 400 K.

being described separately. Each of them is being approximated by the model theory to be adequate named to this term of the Gibbs energy of formation. It gives the possibility of a reliable description for both contributions to the Gibbs energy of formation and, as a result, for this function itself.

The entropy term must take into account both the big size factor of Hume–Rothery[10] $\xi = V_{\text{atomic}_i} / V_{\text{atomic}_j}$ in alloy (e.g., for Cs-Na $\xi = 3$) and the

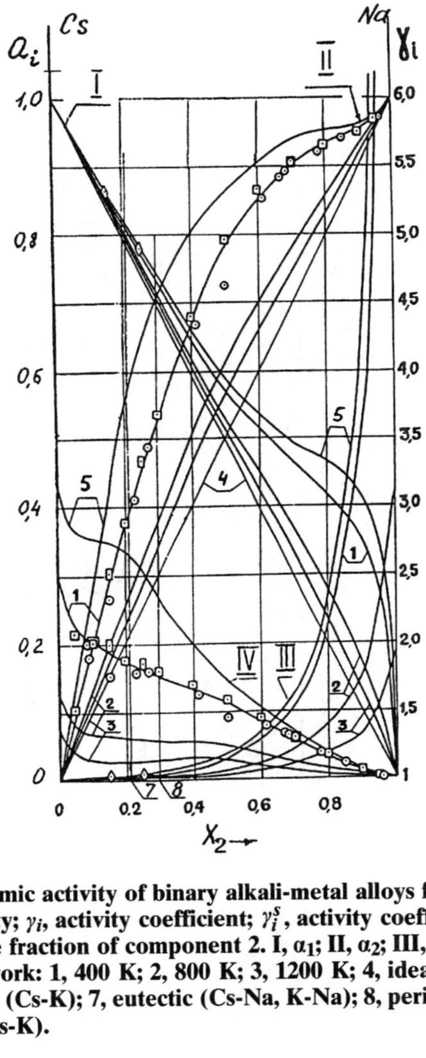

Fig. 2 Thermodynamic activity of binary alkali-metal alloys from liquidus lines up to 1200 K. α_i, activity; γ_i, activity coefficient; γ_i^s, activity coefficient on solidus line (Cs-K); and x_2, mole fraction of component 2. I, α_1; II, α_2; III, $\gamma_1 = \alpha_1/x_1$; and IV, $\gamma_2 = {}_2/x_2$. 1,2,3—this work: 1, 400 K; 2, 800 K; 3, 1200 K; 4, ideal solution; 5, liquidus lines; 6, solidus lines (Cs-K); 7, eutectic (Cs-Na, K-Na); 8, peritectic (Cs-Na, K-Na); and 9, azeotropic (Cs-K).

a) Cs-Na

b) K-Na

c) Cs-K

specificity of the liquid state. To answer both these demands, the entropy of formation was constructed as a sum of lattice one from the Flory–Huggins theory ΔS_{FH} (the first approximation as a basis) and liquid-state correction by a model of hard-sphere liquid ($\Delta S_{\text{hs}} - \Delta S_{\text{FH}}$) (Ref. 9):

$$\Delta S_{\text{FH}} = -R\sum_{i=1}^{2} x_i \,\ell n\, \varphi_i = -R\{x_1 \,\ell n\, [\xi x_i/(x_2 + \xi x_1)] + x_2 \,\ell n\, [x_2/(x_2 + \xi x_1)]\} \qquad (4)$$

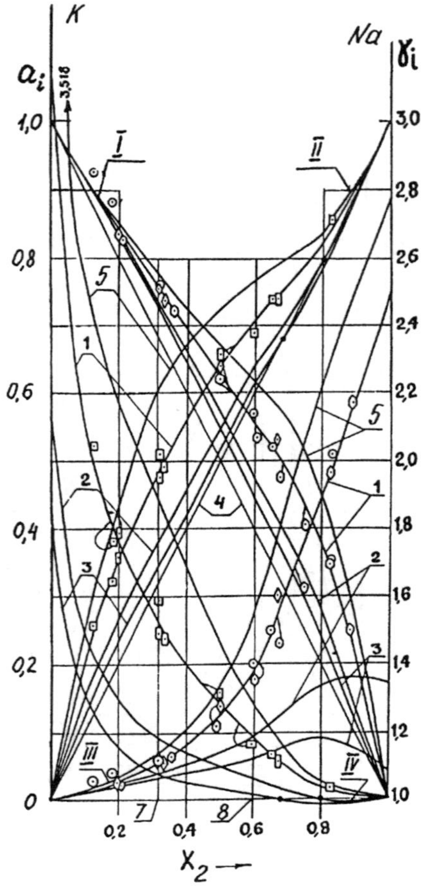

Fig. 2 (continued)

$$\Delta S_{\text{hs}} - \Delta S_{\text{FH}} = \Delta S_{\text{v}} + \Delta S_{\psi} + \Delta S_{r} \tag{5}$$

Here x_i, φ_i are the mole and volume fractions of components; $\Delta S_v = R\,\ell n\,(V/V_{id})$, ΔS_ψ, ΔS_r are the terms connected respectively with the excessive mole volume of alloy, alteration of packing density of spherical ions at alloy formation, and the difference between diameters $(2r_i)$ of hard spheres. All of the terms of expression [Eq. (5)] depend on these diameters, and little concentration dependence $r_i = f(x_i)$ (no more then 3% over all the concentration range $0 \le x_i \le 1$) was needed to fit the approximated parameter of the theory for the best description of the experimental values $T\Delta S = \Delta H_{\text{exp}} - \Delta G_{\text{exp}}$.

In spite of the fact that both ΔG and ΔH are evidently nonsymmetrical functions of concentration, the entropy term is close to the symmetrical one.

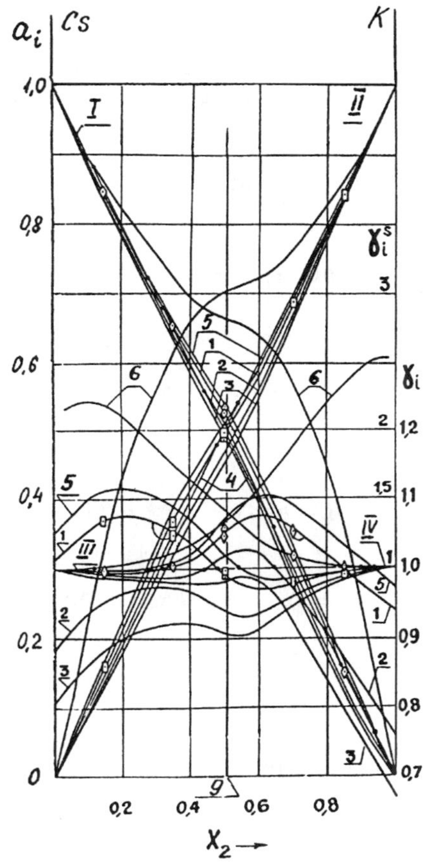

Fig. 2 (continued)

Analysis of expressions (4) and (5) shows that the reason is partial mutual elimination of contributions in expression (5). Therefore the main role in entropy term belongs to the lattice Flory–Huggins contribution, which is not so far from ideal. Such character of entropy in studied systems proves that the ground of nonideality and corresponding behavior of the Gibbs energy is the enthalpy term in a large measure but not the entropy one.

According to the preceding method, enthalpy of formation is the under-integral function in Eqs. (2) and (3) for the determination of the Gibbs energy of formation and the independent item for the approximation of it. The quasichemical model proposed by Guggenheim[11] and developed for these liquid-metal alloys[8] taking into account the alteration of the structure of short-range order comparatively with chaotic distribution as well as concentration and temperature dependences of energies of all kinds of pair interparticle interactions (e.g., three kinds for binary systems) is used for the description of the enthalpy of formation.

Short-range order parameter η is connected with excessive energies of interparticle interaction $\Delta\varepsilon_{ii}$, $\Delta\varepsilon_{ij}$ by the equations of quasichemical equilibrium (6) and (7):

$$(B-1)(1+\eta)^2 x_1 x_2 - B\eta = 0 \tag{6}$$

$$\begin{aligned} 1 - \eta x_2 / x_1 &= \exp(-\Delta\varepsilon_{11} / kT) \\ 1 - \eta x_1 / x_2 &= \exp(-\Delta\varepsilon_{22} / kT) \\ 1 + \eta &= \exp(-\Delta\varepsilon_{12} / kT) \end{aligned} \tag{7}$$

where

$$B = \exp(-2\Delta\varepsilon^{**} / kT) \tag{8}$$

$$\Delta\varepsilon^{**} = \Delta\varepsilon_{12} - 0{,}5(\Delta\varepsilon_{11} + \Delta\varepsilon_{22}) \tag{9}$$

$$\Delta\varepsilon_{12} = \Delta\varepsilon_{12} - (\varepsilon_{21}^0 x_1 + \varepsilon_{12}^0 x_2), \quad \Delta\varepsilon_{ii} = \varepsilon_{ii} - \varepsilon_{ii}^0 \tag{10}$$

The ε_{ii}^0 and ε_{ij}^0 correspond to pure components and infinitely dilute the solution respectively.

The effective parameter of interaction $Q^{\text{eff}} = \Delta H / x_1 x_2$ (where ΔH is the integral enthalpy of formation obtained by the experiment) depends on x_i and η (Ref. 8):

$$Q^{\text{eff}} = Q_0 + Q_\eta + Q_x \tag{11}$$

where terms

$$Q_0 = \Delta\overline{H}_2^0 x_1 + \Delta\overline{H}_1^0 x_2 \tag{12}$$

$$Q_\eta = (\Delta\overline{H}_2^0 x_1 + \Delta\overline{H}_1^0 x_2)\eta \tag{13}$$

$$\begin{aligned} Q_x = -0{,}5zRT[&(x_1 / x_2 - \eta)\ln(1 - \eta x_2 / x_1) \\ &+ (x_2 / x_1 - \eta)\ln(1 - \eta x_1 / x_2) + 2(1 + \eta)\ln(1 + \eta)] \end{aligned} \tag{14}$$

(z is the coordination number and $\Delta\overline{H}_i^0$ is the partial enthalpy of formation in infinitely dilute solution) are responsible for the subregular contribution

(12), the alteration of the structure of short-range order (13), and the energies of interparticle interaction (14) at the formation of solution. Having found by experiment the values of enthalpy of formation, one can calculate the short-range order parameter η for each value of enthalpy with the help of Eqs. (11–14) and, therefore, parameters of model B and $\Delta\varepsilon^{**}$ by Eqs. (6–10).

In contrast to functions ΔH, Q^{eff}, η, and $\Delta\varepsilon_{ij}$, function $\Delta\varepsilon^{**} = f(x_i)$ turned out to be sufficiently linearized. By reason of this fact the approximation over all the concentrations was fulfilled namely for these functions. A high precision of this operation can be achieved even at a limited set of concentrations. Therefore this rather small varying value was chosen as a fitting parameter of the theory at approximation of the enthalpy of formation. Then all of the parameters of the theory as a function of concentration are being recreated in consecutive order: η by Eq. (6), $\Delta\varepsilon_{ij}$ by Eq. (7), and approximating functions for Q^{eff} or ΔH by Eqs. (11–14).

The long-wave limit of concentration correlation function $S_{cc}(0)$ (Ref. 9):

$$S_{cc}(0) = N\langle(\Delta x_i)^2\rangle = RT(\partial^2\Delta G/\partial x_i)^{-1}_{p,T} = RT[(\partial\Delta\overline{G}_i/\partial x_i)/(1-x_i)]^{-1}_{p,T}$$
$$= [(\partial \ln a_i/\partial x_i)/(1-x_i)]^{-1}_{p,T} \tag{15}$$

was computed for all of the areas of temperatures and concentrations on the basis of the preceding analytical description of the Gibbs energy of formation (Fig. 3). These results were compared with those available at $T \approx 400$ K literature data of difraction experiment [structure factor $S(0)$][12] as well as some thermodynamic experiments[2,3] and calculations.[9] There was an accord-

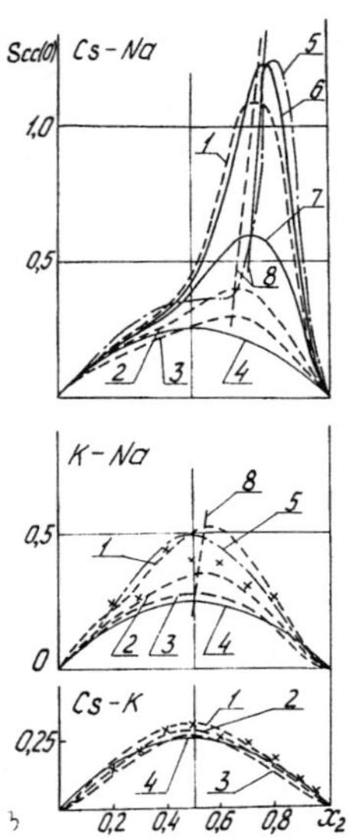

Fig. 3 Concentration correlation function of binary alkali-metal systems from 400 to 1200 K. 1,2,3—this work: 1, 400 K; 2, 800 K; 3, 1200 K; 4, ideal solution; 5, 383 K (Ref. 2); 6, 383 K (Ref. 3); 7, 473 K [3]; 8, extremum lines; x, 383 K (Ref. 9); and X_2, mole fraction of component 2.

ance concerning both the magnitude and coordinate of maximum of function S_{cc} (0) ($0{,}7 \leq x_i \leq 0{,}8$ for system Cs-Na).

Discussion and Conclusions

Behavior of concentration correlation function $S_{cc}(0)$ corroborates the existence of concentration long-range fluctuations in some liquid-metal systems, i.e., micro-nonhomogeneous (claster) structure of the liquid alloy and therefore decreasing the thermodynamic stability of such a system as a homogeneous liquid phase. This fact correlates with the character of the concentration and temperature dependencies of a short-range order parameter and the energies of the interparticle interaction[8] as well as with their phase diagrams, e.g., for Cs-Na system liquidus-line in the named concentration range is practically horizontal, i.e., increasing the concentration fluctuations pointed out the tendency to phase stratification.[23]

The important peculiarity in the behavior of thermodynamic functions takes place: inversion (change of sign) in deviation from one of the thermodynamic activity coefficients γ_i (or deviation from zero of the excessive Gibbs energy of formation $\Delta \overline{G}_i^* = RT \ln \gamma_i$). As consequence of this phenomenon, the inversion of excessive entropy $\Delta \overline{S}_i^*$ is discovered; such an effect is also displayed in the integral excessive functions ΔG^* and ΔS^*. This phenomenon is analogous to the inversion of the compressibility factor for one-component liquids in the area of negative values of the second virial coefficient ($T < T_{\text{Boyle}}$).

Analysis of the expressions

$$\Delta \overline{S}_i^* = \Delta \overline{H}_i / T - R \ln \gamma_i, \qquad \Delta S^* = \Delta H / T - \Delta H / T - \Delta G^* / T \tag{16}$$

reveals that in liquid-metal systems under the test the different thermodynamic classes of solutions exist depending on temperature, concentration, and size factor. (It must be emphasized that in studied systems $\Delta H > 0$, $\Delta H_i > 0$, see Fig. 1.)

1) The area where the solution is like the athermal one when enthalpy term is less essential is

$$\ln \gamma_i > 0, \quad \Delta \overline{G}_i^* > 0, \quad \Delta \overline{H}_i \ll RT \ln \gamma_i, \quad \Delta \overline{S}_i^* < 0$$

or

$$\Delta G^* > 0 \quad \Delta H \ll \Delta G^*, \quad \Delta S^* < 0 \tag{17}$$

$$\Delta \overline{S}_i^* \approx -R \ln \gamma_i, \quad \Delta \overline{S}_i \approx -R \ln \alpha_1, \quad \text{or} \quad \Delta S^* \approx -\Delta G^* / T \tag{18}$$

In this case only the description of a non-ideal system entropy may be practically carried out by a simple changing concentration for the activity.

2) The area of regular solutions (at more high temperatures) where Hildebrand's condition ($\Delta \bar{S}_i^* = 0$ or $\Delta S^* = 0$) (Ref. 13) corresponding to

$$\Delta \bar{H}_i = RT \ln \gamma_i = \Delta \bar{G}_i^* \quad \text{or} \quad \Delta H = \Delta G^* \tag{19}$$

is realized. It is the only case in thermodynamics when the activity of the component or the excessive Gibbs energy of formation can be determined in calorimetric experiment. Discovering this area and correctly stating its boundaries have principal methodic significance for the experimental investigation of thermodynamic properties of the liquid-metal systems and the building of their phase equilibrium diagrams.

3) If the temperature increases, the area appears where the solution may be named Raoult's solution:

$$\ln \gamma_i = 0 \quad \Delta \bar{G}_i^* = 0 \quad \Delta \bar{S}_i^* > 0 \quad \text{or} \quad \Delta G^* = 0 \quad \Delta S^* > 0 \tag{20}$$

$$\Delta \bar{S}_i^* = \Delta \bar{H}_i / T \quad \text{or} \quad \Delta S^* = \Delta H / T \tag{21}$$

In this case saturated vapor pressure is equal to the ideal one, and excessive entropy is defined by the heat effect of formation only. Apparently it is the reason that a given class of solutions is the only one in which the entropy is being reproduced satisfactorily enough by lattice model theories (quasichemical, Flory–Huggins, etc.). In common cases these theories have the well-known limitation of reproducing the behavior of formation enthalpy but not entropy.

4) At maximum temperatures both terms in expressions (16) are of the same signs (no compensation) and essential:

$$\ln \gamma_i < 0 \quad \Delta \bar{G}_i^* < 0 \quad \Delta \bar{S}_i^* > 0 \quad \Delta G^* < 0 \quad \Delta S^* > 0 \tag{22}$$

i.e., the common case of real solutions in mutual liquid alloys of alkali metals.

The analysis of the effective parameter of interaction Q^{eff} as function of concentration and temperature gives the possibility of determining the quantity of dependence on size factor (Cs-Na, $\xi = 3,0$; K-Na, $\xi = 1,9$; Cs-K, $\xi = 1,5$) and thus the ranges of existence of the second necessary condition (simultaneous with Hildebrand's condition) of the regular solutions in liquid alkali metal alloys, symmetry of thermodynamic functions.

Acknowledgment

The authors would like to acknowledge support for this work by the Russian Foundation for Basic Research (RFBR).

References

[1]Bystrov, P. I., Kagan, D. N., Krechetova, G. A., and Shpilrain, E. E., *Liquid-Metal Coolants for Heat Pipes and Power Plants*, Hemisphere, New York, 1990.

[2]Ichikawa, K., Granstaff, S. H., and Thompson, J. C., "Chemical Potentials and Related Thermodynamics in Liquid NaCs Alloys," *Journal of Chemical Physics*, Vol. 61, No. 10, 1974, pp. 4059–4062; also Vol. 59, No. 4, 1973, pp. 1680–1683; also *Journal of Physics F*, Vol. 4, No. 1, 1974, pp. L9–10.

[3]Neale, F. E., and Cusack, N. E., "Thermodynamic Properties of Liquid NaCs Alloys," *Journal of Physics F: Metal Physics*, Vol. 12, No. 12, 1982, pp. 2839–2850; also *Journal of Non-cristalline Solids*, Vol. 61/62, Pt. 1, 1984, pp. 169–174.

[4]Cafasso, F. A., Khanda, V. M., and Feder, H. M., "Thermodynamic Properties and Ordering in Liquid Na-K Alloys," *Advances in Physics*, Vol. 16, No. 63, 1967, pp. 535–543.

[5]Ciurylo, J., and Rozwadowski, M., "Saturated Vapour Pressure of Cs over KCs Solution," *Acta Physics Polonica*, Vol. A51, No. 4, 1977, pp. 583–588.

[6]Kagan, D. N., and Krechetova, G. A., "Experimental Methods for Measuring of Thermodynamic (Caloric) Properties of Alkali Metals," *Handbook of Thermodynamic and Transport Properties of Alkali Metals*, edited by R. V. Ohse, Blackwell Scientific Publishing Ltd., Oxford, 1985, Chap. 6.3.1, pp. 359–410.

[7]Kagan, D. N., "Measurement of Activity of Components in Liquid Alloys of Alkali Metals by Effusion Method on Electronbeam Installation," *Thermophysics of High Temperatures*, Vol. 26, No. 2, 1988, pp. 360–369; also No. 3, pp. 478–491.

[8]Kagan, D. N., and Krechetova, G. A., "Characteristics of Interparticle Interaction in Liquid Binary Systems of Alkali Metals, Dependence on Temperatures," *Thermophysics of High Temperatures*, Vol. 19, No. 2, 1981, pp. 432–435.

[9]Visser, E. G., Alblas, B. P., De Hosson, J. T., and Van der Lugt, W., "Thermodynamic Calculations for the Liquid Systems Cs-Na, K-Na, Pb-Li," *Journal of Physics F: Metal Physics*, Vol. 10, No. 8, 1980, pp. 1681–1692; also *Physica*, Vol. 114B, No. 1, 1982, pp. 59–66.

[10]Hume-Rothery, W., and Raynor, G. V., *The Structure of Metals and Alloys*, 3rd ed., Inst. of Metal Monograph, London, 1954.

[11]Guggenheim, E. A., *Mixtures*, Oxford Univ. Press., Oxford, England, 1952.

[12]Huijben, M. J., Alblas, B. P., Van der Lugt, W., et al., "Structure Properties of the Liquid Systems CsNa, KNa, CsK by Ray and Neutron Difraction Experiments," *Acta Crystallographica*, Vol. A35 1979, pp. 431–436; also *Journal of Physics F: Metal Physics*, Vol. 7, 1977 pp. L119–121; also Vol. 10, 1980, pp. 531–539; also *Physica*, Vol. 97B+C 1979, pp. 338–350; also Vol. 106B, 1981, pp. 22–29; Vol. 101B, 1980, pp. 177–184.

[13]Hildebrand, J. H., and Scott, R. L., *Regular Solutions*, Prentice–Hall, New York, 1962; also *Solubility of Non-Electrolytes*, Dover, New York, 1964.

Chapter 34

Inductive-Conductive Magnetohydrodynamic Pumps for Liquid Metals

Arkady B. Kapusta,* Alexandra I. Likhachova,†
and Boris M. Mikhailovich‡
Panbas State Academy of Civil Engineering and Architecture, Donetsk Region, 339023, Ukraine
and
Boris M. Mikhailovich‡
Donetsk State Technical University, Donetsk, 34000, Ukraine

I. Introduction

IN PRESENT time different conductive and inductive technological magnetohydrodynamic (MHD) devices, used for action on conducting mediums, are known. Synthesis of opportunities and advantages of both methods led to the creation of inductive-conductive MHD pumps (MHD pumps of a double-feeding type).

Such inductive-conductive MHD pumps are intended for metallurgical and foundry manufacture (transference and dosing of metals and alloys) and nuclear power (moving of heat-carriers in contours of primary cooling nuclear reactors). The pumps can be executed with plane linear or screw channel in a kind of set of electromagnetic modules, located around a channel with corresponding symmetry.

Each module consists of an electromagnet and a lowering transformer, the secondary winding of which includes a part of a working zone of a channel. The set $3N$ of modules feeds from a three-phase alternating current circuit (N is the quantity of modules in a phase). Thus, the offered pumps represent a combination $3N$ of single-phase conductive pumps of an alternating current and inductive pump. They have advanced reliability and easy servicing.

*Head of Department, 2, Derzhavina str., 23 Makeyevka.
†Ph.D. Student, 2, Derzhavina str., 23 Makeyevka.
‡Associate Professor, Department of Power Engineering, 58, Artem str.

As far as devices, using the interaction of a rotating magnetic field with a conducting transfering medium is the most perspective and effective for the solution of technological problems;[1] new devices with inductive-conductive MHD pumps with screw channels are elaborated.[2]

II. Formulation of the Electrodynamic Problem

At the elaboration of pumps of such type, in particular, there is the problem of the rational arrangement of modules. This problem is solved with the help of the following mathematical model, taking into account the screw symmetry of the MHD phenomena, essentially appearing in screw channels with arbitrary corners of a course.

The electrodynamic processes are considered in a screw channel of an inductive-conductive pump with electrically conducting walls, filled by conducting liquid. The conductivity of the liquid is close to the conductivity of the walls of a channel. The channel is placed in the inductor with explicity poles, exciting a magnetic field and running on a screw line along a channel. The explicity poles of inductor, arranged at the external and internal cylindrical walls and forming a channel, are executed from an ideal ferromagnetic ($\mu \to \infty$). Besides that the conductive electric field, running in the same direction, is made to face the walls of a channel by electrodes.

Thus, in the problem the effects are considered, arising at the interaction of a rotating magnetic field with fields of density of currents. One of them is induced by a primary rotating magnetic field, and the second is moved in a conductive way. The magnetic field of the conductive current is not taken into account.

For the screw channel of a small width $\rho\delta$ (Fig. 1), such a problem in the screw coordinate system [the screw coordinates (ρ, θ, ζ) are connected with

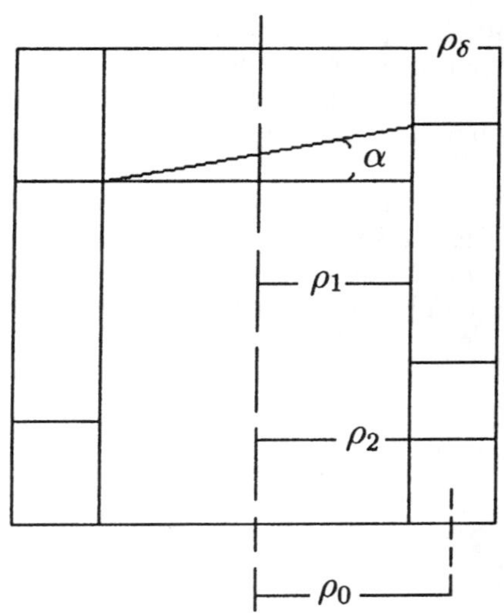

Fig. 1 Schematic diagram of a screw channel.

the cylindrical coordinates (ρ, φ, z) by the relations[1,3] $r = \rho, \varphi = (\theta \cos \alpha - \zeta \sin \alpha)/\rho$, and $z = (\theta \sin \alpha + \theta \cos \alpha)]$ is described by the following dimensionless equations for the ζ component of vector potential of a magnetic induction a_ζ:

$$\frac{\partial^2 a_\zeta}{\partial \rho^2} + \frac{\partial a_\zeta}{\partial \rho} + \beta a_\zeta + \delta_\theta^2 \frac{\partial^2 a_\zeta}{\partial \theta^2} = \overline{\omega}\left(\frac{1}{2\pi}\frac{\partial a_\zeta}{\partial \tau} + \upsilon_\theta \delta_\theta \frac{\partial a_\zeta}{\partial \theta}\right) \tag{1}$$

with boundary conditions on poles of magnets

$$\left.\frac{\partial a_\zeta}{\partial \theta}\right|_{\rho=\rho_l, \theta_{n-1} \le \theta \le \theta_n} = C_l \exp\left[2\pi i\left(\tau - \frac{n-1}{3}\right)\right], \quad (l = 1,2; n = 1,2,3) \tag{2}$$

and scalar electric potential Φ

$$\delta_\theta^2 \frac{\partial^2 \Phi}{\partial \theta^2} + \delta_\zeta^2 \frac{\partial^2 \Phi}{\partial \zeta^2} = 0 \tag{3}$$

with boundary conditions

$$\delta_\zeta \left.\frac{\partial \Phi}{\partial \zeta}\right|_{\zeta=\pm 1} = \exp\left[2\pi i\left(\tau - \frac{n-1}{3} + \psi\right)\right], \quad (n = 1,2,3) \tag{4}$$

where $\beta = (2 - 3 \sin^2 \alpha)\sin^2 \alpha$, $\delta_\theta = 1/(2\pi \cos \alpha)$ and $\delta_\zeta = 1/(2\pi \sin \alpha)$ are the relative dimensions of the area lengthwise of coordinates θ, ζ; α is the corner of a course of a screw line; $H_2 = \sqrt{\rho_l^2 \cos^2 \alpha + \rho_0^2 \sin^2 \alpha}/\rho_0$ is the Lame factor; ρ_1, ρ_2 are the coordinates of internal and external channel surfaces; $\rho_0 = (\rho_1 + \rho_2)/2$; $\omega = \mu_0 \sigma \rho_0^2 \omega$ is the relative frequency of fields; μ_0 is the magnetic constant; σ is the conductivity of a liquid; v_θ is the velocity of a liquid; τ is the moving time; ψ is the phase lag between the magnetic and electric fields; θ_{n-1}, θ_n are the θ coordinates of the nth pole of the magnet; and $C_l = H_2/\delta\theta$, $(l = 1, 2)$.

III. Problem Decision

We search the decision of problems (1) and (2) in the form

$$a_\zeta = a(\rho,\theta) \cdot e^{2\pi i \tau} \tag{5}$$

Substitution of Eq. (5) into Eq. (1) gives for $a = a(\rho,\theta)$

$$\frac{\partial^2 a}{\partial \rho^2} + \frac{\partial a}{\partial \rho} + \beta a + \delta_\theta^2 \frac{\partial^2 a}{\partial \theta^2} = \overline{\omega}\left(ia + \upsilon_\theta \delta_\theta \frac{\partial a}{\partial \theta} \right) \tag{6}$$

Representing a in

$$a(\rho,\theta) = \sum_k a_k(\rho) \cdot e^{-ik\theta} \tag{7}$$

and executing the formal procedure of the Galerkin method, we receive for $a_k(\rho)$ the following system of equations

$$\frac{\mathrm{d}^2 a_k}{\mathrm{d}\rho^2} + \frac{\mathrm{d}a_\mathrm{k}}{\mathrm{d}\rho} - {}_k^2 a_k = 0 \tag{8}$$

where ${}_k^2 = \delta_\theta^2 k^2 + i\overline{\omega}\delta_\theta k S_k - \beta$ and $S_k = (1/\delta_\theta k - \upsilon_\theta)$—sliding, with boundary conditions

$$a_k\big|_{\rho = p_i} = \frac{C_l}{k^2}\left(e^{ik\theta_n} - e^{ik\theta_{n-1}}\right) \cdot e^{-2\pi i(n-1)/3}, \qquad (l = 1, 2; n = 1, 2, 3) \tag{9}$$

The decision of problems (8) and (9) has the form

$$a_k(\rho) = D_1 e^{\gamma 1k\rho} + D_2 e^{\gamma 2k\rho} \tag{10}$$

where $D_1 = Q(C_1 e^{\gamma 2k\rho 2} - C_2 e^{\gamma 2k\rho 1}) / E_k, D_2 = Q(C_2 e^{\gamma 1k\rho 1} - C_1 e^{\gamma 1k\rho 2}) / E_k, Q = (e^{ik\theta_n} - e^{ik\theta_{n-1}}) \cdot e^{-2\pi i(n-1)/3}, E_k = k^2(e^{\gamma 1k\rho 1 + \gamma 2k\rho 2} - e^{\gamma 1k\rho 2 + \gamma 2k\rho 1})$, and γ_{1k}, γ_{2k} are the roots of characteristic equations of a system (8).

Thus, we receive the solution for the ζ component of vector potential of a magnetic induction

$$a_\zeta = \sum_k \left(D_1 e^{\gamma 1k\rho} + D_2 e^{\gamma 2k\rho}\right) \cdot e^{i(2\pi\tau - k\theta)} \tag{11}$$

where we define the ζ component of density of induced current

$$j_{\zeta i} = -i\overline{\omega} S_k \sum_k \left(D_1 e^{\gamma 1k\rho} + D_2 e^{\gamma 2k\rho}\right) \cdot e^{i(2\pi\tau - k\theta)} \tag{12}$$

and the ρ component of magnetic induction in the cavity of a channel

$$b_\rho = \frac{-ik}{H_2} \sum_k \left(D_1 e^{\gamma 1k\rho} + D_2 e^{\gamma 2k\rho}\right) \cdot e^{i(2\pi\tau - k\theta)} \tag{13}$$

For decision of problems (3) and (4) for scalar electric potential, we search it in a form

$$\Phi = \sum_k \Phi_k(\zeta) \cdot e^{-ik\theta} \tag{14}$$

Substitution of Eq. (14) into Eq. (3) and fulfillment of the formal procedure of the Galerkin method reduce to the system of equations

$$\frac{d^2\Phi_k}{d\zeta^2} - \nu_k^2 \Phi_k = 0 \tag{15}$$

where $\nu_k^2 = k^2(\delta_\theta/\delta_\zeta)^2$ with boundary conditions

$$\left.\frac{d\Phi_k}{d\zeta}\right|_{\zeta=\pm1} = -P_k \cdot \exp\left[2\pi i\left[\tau - (n-1)/3 + \psi\right]\right] \tag{16}$$

where $P_k = \frac{1}{i\delta_\zeta k}(e^{ik\theta_n^*} - e^{ik\theta_{n-1}^*})$ and $\theta_{n-1}^*, \theta_n^*$ are the coordinates of the nth conductive electrode.

The decision of the system of equations (15) is searched in a form

$$\Phi_k = G_1 ch\nu_k\zeta + G_2 sh\nu_k\zeta \tag{17}$$

With the accounting of boundary conditions (16)

$$G_1 = 0, \quad G_2 = -\frac{P_k}{\nu_k ch\nu_k},$$

$$\Phi_k = -\frac{P_k}{\nu_k}\frac{sh\nu_k\zeta}{ch\nu_k} \tag{18}$$

Substituting Eq. (18) into Eq. (14), we shall record finally for scalar electric potential

$$\Phi = -\sum_k \frac{P_k}{\nu_k}\frac{sh\nu_k\zeta}{ch\nu_k}\exp\left(i\left\{2\pi\left[\tau - (n-1)/3 + \psi\right] - k\theta\right\}\right) \tag{19}$$

where we define the density of a conductive entered current

$$j_{\zeta c} = \delta_\zeta \sum_k \frac{P_k ch v_k \zeta}{ch v_k} \exp\left(i\left\{2\pi\left[\tau - (n-1)/3 + \psi\right] - k\theta\right\}\right) \tag{20}$$

The received relations (12) and (20) permit one to calculate the resulting significance of the density of the electric current

$$j_\zeta = j_{\zeta i} + j_{\zeta c} \tag{21}$$

and with the accounting of Eq. (13) into receiving a density of the electromagnetic forces in a screw channel of a MHD pump,

$$f_\theta = \Re j_\zeta \cdot \Re b_\rho \tag{22}$$

IV. Results and Applications

The typical distribution of the density of the electromagnetic forces along the θ coordinate on a length of polar division θ_p ($\theta_p = 1$) is shown on Fig. 2; τ average for the same variant is shown on Fig. 3 (maximum of the τ average force is chosen as scale; $\overline{\omega} = 3$, $\alpha = 5$, $\rho_0 = 1$, $\rho_\delta = 0, 1$, $\psi = 0$, and $\upsilon = 0.1$). Obtained data completely correspond to the physical representations about the electromagnetic processes in the screw channel and made use of analysis

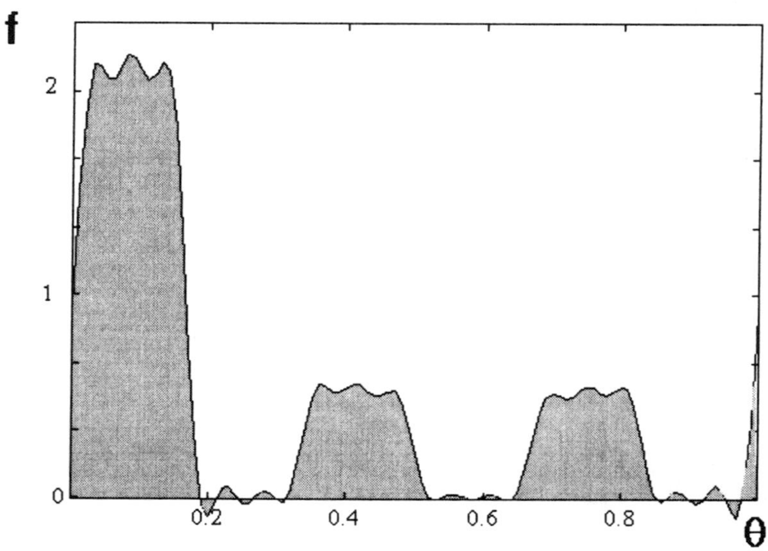

Fig. 2 Distribution of density of electromagnetic forces along the pole pitch (single turn of a screw channel).

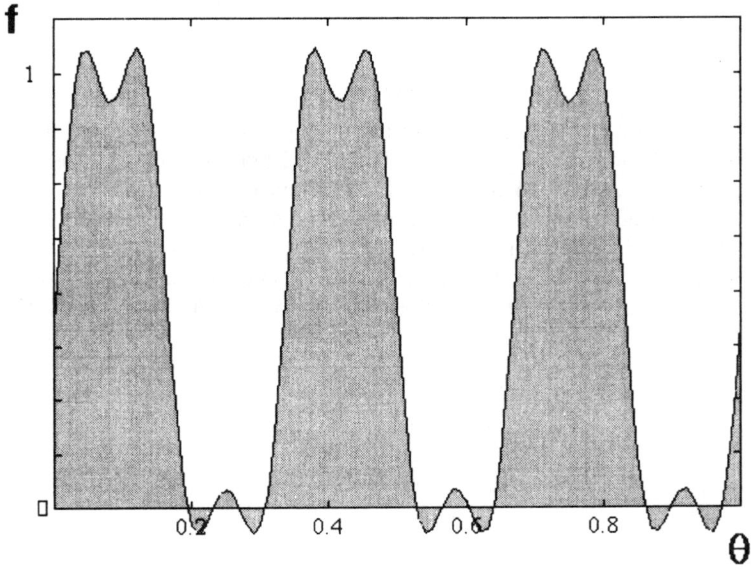

Fig. 3 Time-averaged distribution of the density of electromagnetic forces.

of the optimal placement of the magnetic poles and conducting electrodes and at the choice of the quantity of the pump's modules.

Thus, the received results are used at the advanced development of the inductive-conductive MHD pump with a screw channel.

On Fig. 4 a design of such a pump is shown. On frame 1 a screw channel 2 is placed, around which three identical modules are symmetrically located. Each of them consists of a transformer 3 and a magnet 4. The MHD pump is intended for a transfer liquid tin and lead with productivity up to 1 liter

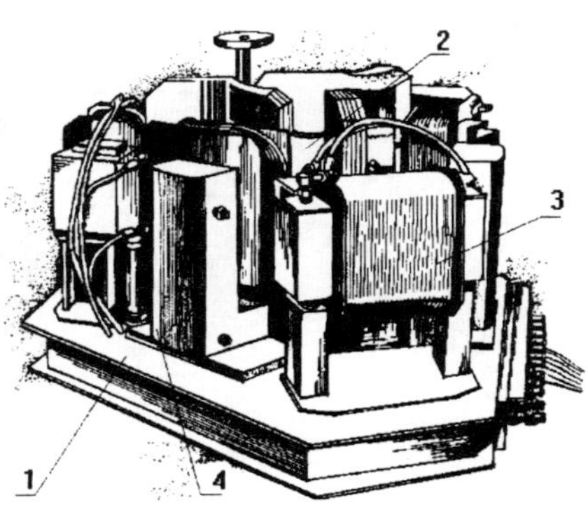

Fig. 4 Inductive-conductive pump.

per second at a pressure of about 0.15 MPa in an interval of temperatures up to 500°C.

References

[1]Reymal, L. R., Kapusta, A. B., and Likhachova, A. I., "Magnetohydrodynamic Processes in Devices with Curvilinear Channels," Energoatomizdat, Moscow, 1986.

[2]Kapusta, A. B., Likhachova, A. I., and Mikhailovich, B. M., "Elaboration of Spiral Inductive-Conductive Pump," *Proceedings of the 14th International Riga Conference on MHD*, Inst. of Physics, Riga, Latvia, 1995, p. 65.

[3]Zibold, A. F., Kapusta, A. B., Likhachova, A. I., and Reymal, L. R., "3-D Flow in Screw Channel of Inductive Pump," *Magnitnaya Gidrodinamika*, No. 2, 1986, pp. 121–124.

Chapter 35

Direct Numerical Modeling of Kinematic Dynamo Effect: Application to the French Fast Breeder Reactor Phenix Core's Geometry

F. Plunian,* A. Alemany,† and Ph. Marty‡
Laboratoire des Ecoulements Géophysiques et Industriels, Grenoble, 38041, France
and
Ph. Massé§
Laboratoire de Mécanique et de Génie des Procédés, ENSPG, Grenoble, 38026, France

Nomenclature

$\boldsymbol{A}$ = vector potential, v/s/m^{-1}
$\boldsymbol{B}$ = magnetic induction, T
$\boldsymbol{E}$ = electric field, V/m^{-1}
$\boldsymbol{H}$ = magnetic field strength, A/m^{-1}
Q = flow rate, m^3/s^{-1}
R_m = magnetic Reynolds number
$\boldsymbol{V}$ = velocity field, m/s^{-1}
μ = magnetic permeability, H/m^{-1}
μ_0 = magnetic permeability of free space, H/m^{-1}
σ = electrical conductivity, Ω^{-1}/m^{-1}
φ = electric potential, V

I. Introduction

DYNAMO effect describes the conversion of kinetic energy into magnetic energy by the motion of an electrical conductor medium. This

*BP53X.
†BP 53X.
‡BP 53X.
§Institut des Sciences Nucléaires, 53 ruedes martyrs.

phenomenon permits us to explain the existence of a magnetic field in most of the astrophysical objects, like the sun or the Earth. Without such a mechanism, Earth's magnetic field would have disappeared a long time ago. Besides, the Earth's magnetic field's polarity reversals observed by the paleontologists can find consistent justifications in these mechanisms. They result from the coupling of several physics disciplines, such as thermodynamics (for Earth, the primary source is from the thermic origin), electromagnetism, and fluid mechanics.

In the case of the Earth, the dynamo effect is desirable (Earth magnetic shield); in revanche it is not the case in the large industrial realizations containing liquid sodium.[1] Thus, the fast breeder reactor (FBR) authorities have requested the support of researchers to realize the measurements inside the reactors. Even if some results have shown a semidynamo mechanism correlated to thermoelectrical currents in the first loop of a FBR,[2] no dynamo effect has actually been observed.[3,4] However, one part of the reactor has never been the object of direct electromagnetic measurements: the core of the reactor, which contains the plutonium and the uranium fuel.

We have been requested to carry out a general program concerning the analysis of the different possible sources of the magnetic field inside the first sodium loop of the French reactor Phenix. Several analytical results have already been found.[5,6] The present paper deals with new results obtained by direct numerical simulation in order to predict the kinematic dynamo effect inside the core of the reactor.

The core of a FBR seems to be favorable to the dynamo effect for several reasons (part II): 1) the flow inside the core possesses a strong helicity; 2) the magnetic Reynolds number at the scale of the core is larger than 10; 3) some ferromagnetic elements inside the core may disturb the phenomena because of their high magnetic permeability ($10^3\mu_0$); and 4) the core is surrounded by a ferromagnetic crown that may act in favor of the dynamo effect.[5]

The complexity of the flow frontiers, the extended spatial scale of the flow geometry (three decades), an approximated knowledge of the turbulent flow in the core, and the heterogeneousness of the magnetic properties of the materials are the main difficulties to face. The time evolution of magnetic induction B is given by

$$\partial \boldsymbol{B} / \partial t = \mathrm{rot}(\boldsymbol{V} \times \boldsymbol{B}) + R_m^{-1} \nabla^2 \boldsymbol{B} \tag{1}$$

Dynamo effect is reached unless $B(t) \to 0$ when $t \to \infty$.

The nonsymmetric distribution of the ferromagnetic elements within the core leads to a heavy modeling (three-dimensional and unstationary) of Eq. (1) by the finite element method (Part III). Two cases have been simulated. The first one (Part IV) is treating the case of a homogeneous core (without any ferromagnetic elements). The second one (Part V) is treating the case of a heterogeneous core (with 10% of ferromagnetic elements).

II. Hypothesis

The following hypotheses are concerning the FBR Phenix.

A. Fast Breeder Reactor Geometry

The reactor is composed of two sodium loops (Fig. 1). The primary loop is contained in a tank (1) with the sodium motion being forced by three pumps (2). The sodium is heated inside the core (4) and gives its heat to the secondary loop (6) through three primary heat exchangers (5). The sodium motion of the secondary loop (6) is forced by three pumps (7). It gives its heat to water loops (9) through secondary exchangers (8).

B. Fast Breeder Reactor Core Geometry

The FBR's core is composed of three main parts (Fig. 2): the lateral neutronic protection made of ferromagnetic materials, the outer core where the motion of sodium is slow, and the inner core that has the following characteristics: mean axial velocity of the sodium in the inner core, $V_0 = 5$ m/s $-$ 1; mean flow rate, $Q = 3.7$ m^{-3}/s^{-1}; efficient radius of the core, $(Q/\pi V_0)^{0.5} = 0.5$ m; mean electrical conductivity (temperature between 400 and 560°C), $\sigma = 4.1\ 10^{6}\Omega^{-1}\text{m}^{-1}$; and magnetic permeability of nonferromagnetic elements, $\mu_0 = 4\pi\ 10^{-7}\text{H m}^{-1}$.

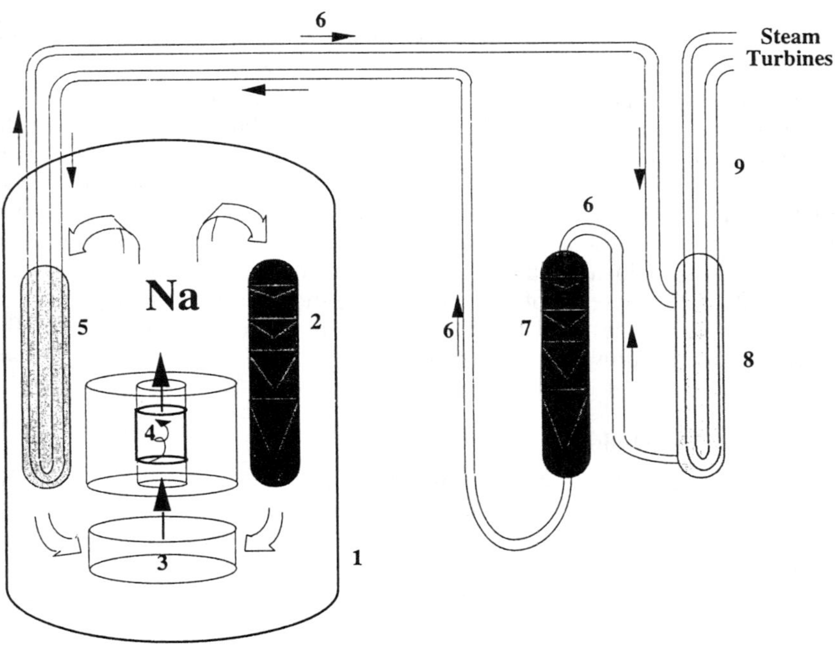

Fig. 1 Principal scheme of a compact type fast breeder reactor.

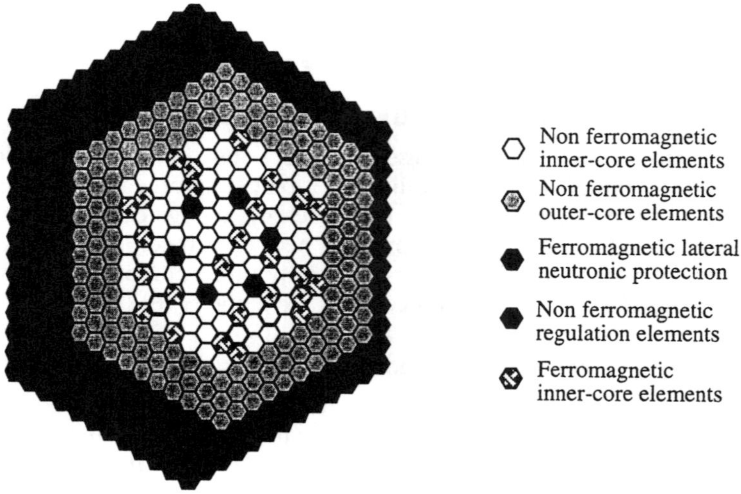

Fig. 2 Horizontal section of the Phenix reactor.

These values lead to the following magnetic Reynolds number:

$$R_m = \sigma\mu_0 V_0 (Q/\pi V_0)^{0.5} = 13 \tag{2}$$

which is high enough to generate some eventual dynamo effect.

The inner core is composed of about a hundred cylindrical elements of the hexagonal section. It is then obvious that the magnetic Reynolds number at the scale of each element should be lower than one and then unable to generate dynamo effect. However, the sodium motion in each element is guided by 217 helical space wires for optimal thermal exchange reasons. This action leads to a local azimuthal component of the velocity field with a double effect: the increase of the magnetic Reynolds number to a value close to one (for the same axial flowrate), and a nonzero helicity value of the motion, which is known to act in favor of a dynamo.[7]

C. Core's Flow Model

One of the difficulties of the numerical modeling of the kinematic dynamo comes from the necessity to introduce a sufficiently simple model of the motion in each hexagonal element, keeping nonetheless the characteristics essential to the appearance of the phenomena: magnetic Reynolds number and helicity. The simplified model of the velocity field that is presented in the following is based on a previous study offering experimental and numerical results.[8] The mean flow rate and the mean helicity in each element have been computed on the basis of these results and reported in

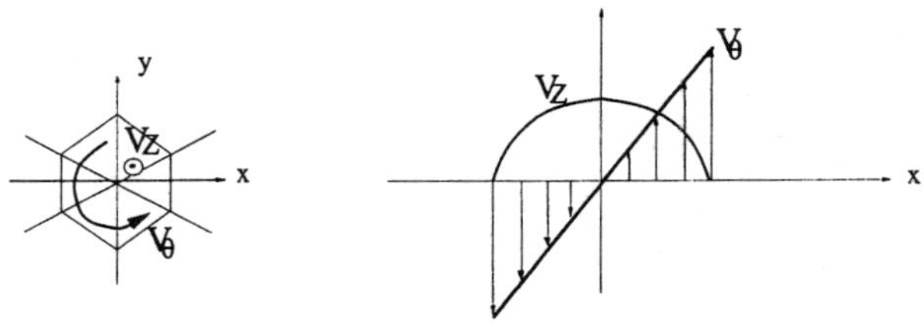

Fig. 3 Velocity field in one hexagonal element (horizontal section).

the following velocity model (Fig. 3). In each hexagonal element of symmetry axis $r = r_0$, the velocity field is defined in Cartesian coordinates by

$$V_x = -(y - y_0)\omega \tag{3a}$$

$$V_y = +(x - x_0)\omega \tag{3b}$$

$$V_z = 2V_0\{1 - [(x - x_0)^2 + (y - y_0)^2]/R_2\} \tag{3c}$$

with $\omega = 100$ rad/s^{-1} and $V_0 = 1.8$ m/s^{-1}.

D. Electromagnetic Properties

The walls of 10% of the inner core elements are made with ferromagnetic materials (Fig. 2). Their magnetic permeability is 10^3 times larger than the rest of the core. A computation based on magnetostatic assumptions[9] leads to a magnetic interaction rate between the inside and the outside of each ferromagnetic element lower than 4%.* Thus the ferromagnetic wall is actually acting as a magnetic screening strongly limiting the interactions between the inner element and the outer element. As the local magnetic Reynolds number is too low to lead to magnetic self-excitation (around 1), the role of the sodium motion in each ferromagnetic element is considered as insignificant for dynamo action. Therefore, each one of them is replaced by a full ferromagnetic hexagonal cylinder with no velocity inside (Fig. 4).

The lateral neutronic protection (Fig. 2) is made of ferromagnetic material. Its presence is modeled by a homogeneous Neuman frontier condition on potential vector A. This implies a magnetic field direction normal to the frontier.

---

*Let us consider, for example, an outer current density creating an outer magnetic field. Then, a part of this magnetic field can penetrate into the wall and also into the inner element. If the magnetic permeability of the wall is equal to 10^3, then the intensity of the inner element penetrating the magnetic field is lower than 4% of the outer magnetic field.

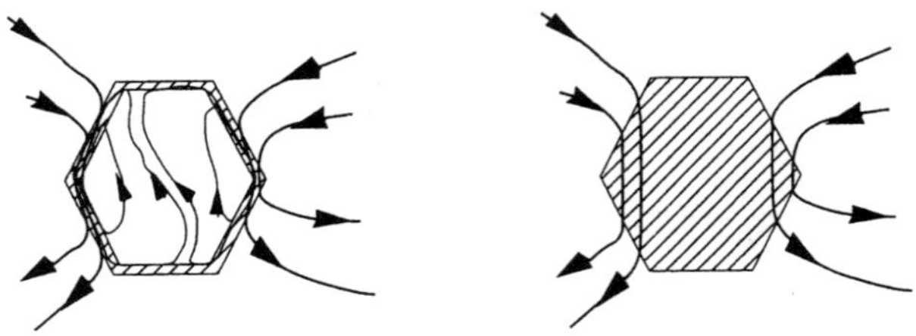

Fig. 4 Modeling of one hexagonal element with its ferromagnetic wall. The latter is acting as a magnetic screening between the inside and the outside of the element.

III. Numerical Model

A. (*A*, *W*) Modeling

Because of the complexity of the geometry, the number of nodes is very large (60,000). To be solved, the time-dependent linear system must not be ill-conditioned or nonsymmetric. Vector and electric potentials $\boldsymbol{A}$ and φ are found to be optimal unknowns on the spatial continuity point of view.[10] They verify the following system equivalent to Eq. (1):

$$\begin{cases} \operatorname{curl}\left(\dfrac{1}{\mu}\operatorname{curl}\boldsymbol{A}\right)=\sigma\left(-\dfrac{\partial \boldsymbol{A}}{\partial t}-\operatorname{grad}\varphi+\boldsymbol{V}\times\operatorname{curl}\boldsymbol{A}\right) \\ \operatorname{div}\sigma\left(-\dfrac{\partial \boldsymbol{A}}{\partial t}-\operatorname{grad}\varphi+\boldsymbol{V}\times\operatorname{curl}\boldsymbol{A}\right)=0 \\ \operatorname{div}\boldsymbol{A}=0 \end{cases} \tag{4}$$

the post-processing equations being given by

$$\boldsymbol{E}=-\operatorname{grad}\varphi-\frac{\partial \boldsymbol{A}}{\partial t},\quad \boldsymbol{j}=\sigma\left(-\operatorname{grad}\varphi-\frac{\partial \boldsymbol{A}}{\partial t}+\boldsymbol{V}\times\operatorname{curl}\boldsymbol{A}\right),$$
$$\boldsymbol{B}=\operatorname{curl}\boldsymbol{A},\quad \boldsymbol{H}=\frac{\operatorname{curl}\boldsymbol{A}}{\mu} \tag{5}$$

A classic finite element Galerkin projection of the first-order time-dependent finite difference scheme of Eq. (4) involves a nonsymmetric ill-conditioned system. The nonsymmetric part of the time-dependent matrix can be symmetrized introducing a new unknown W defined by[10]

$$\varphi=\frac{\partial W}{\partial t} \tag{6}$$

The ill-conditioned system given by the term curl(μ^{-1} curl A) can be mixed with the gage equation div A by a penalty method to obtain a Laplacian formulation. The weak form of Eq. (4) is then derived on the vectorial space of projection functions α_i, and scalar functions β_i:

$$\begin{cases} \iiint_{(\Omega)} \left\{ \frac{1}{\mu} \text{curl}\, \alpha_i \cdot \text{curl}\, \boldsymbol{A} + (\text{penalty}) \text{div}\, \alpha_i \, \text{div}\, \boldsymbol{A} \right. \\ \quad \left. + \sigma \alpha_i \cdot \left(\frac{\partial(\boldsymbol{A} + \text{grad}\, W)}{\partial t} - \boldsymbol{V} \times \text{curl}\, \boldsymbol{A} \right) \right\} \text{d}V \\ \quad - (\text{penalty}) \oiint_{(\text{I})+(\Gamma)} \alpha_i \cdot \boldsymbol{n}\, \text{div}\, \boldsymbol{A}\, \text{dS} + \oiint_{(\text{I})+(\Gamma)} \alpha_i \cdot (\boldsymbol{n} \times \boldsymbol{H})\, \text{d}S = 0 \\ \iiint_{(\Omega)} \sigma\, \text{grad}(\beta_i) \cdot \left(\frac{\partial(\boldsymbol{A} + \text{grad}\, W)}{\partial t} - \boldsymbol{V} \times \text{curl}\, \boldsymbol{A} \right) \text{d}V \\ \quad + \oiint_{(\text{I})+(\Gamma)} \beta_i \sigma \left(\frac{\partial(\boldsymbol{A} + \text{grad}\, W)}{\partial t} + \boldsymbol{V} \times \text{curl}\, \boldsymbol{A} \right) \cdot \boldsymbol{n} \text{d}S = 0 \end{cases} \tag{7}$$

The convective terms $\boldsymbol{V} \times$ curl $\boldsymbol{A}$, in reason of their dependence on V, cannot lead to a symmetric matrix. Then they must be rejected to the second member of the equations using an iterative process between them and the rest of the system. The resolution takes the following form:

$$(\Delta t M + L) X_i^{n+1} = L X^n + \Delta t F\left(X_{i-1}^{n+1}\right) \tag{8}$$

where $X_i^{n+1} = (A_x, A_z, W)$ at the time step $(n + 1)\Delta t$ and at the ith iteration and X^n these values at the preceding time step. In this expression, $(\Delta t\ M + L)$ is symmetric and well-conditioned, and F is the quadri-vectorial function corresponding to the convenctive terms.

This formulation has been described with the Flux–Expert* generator package[11] with second order three dimension isoparametric curvilinear finite elements, and the linear systems are solved by incomplete Cholesky conjugate gradients iterative algorithm (see Ref. 12 for more details).

B. Stability

To describe time dependence, an implicit first-order finite differences algorithm is used. As the implicit method is unconditionally stable, there is no stability constraint on time-step size.

To treat a spatial derivative of the convective terms, a central difference scheme is used rather than an upwind scheme. The reason for this choice is

---

*FLUX-EXPERT is a registered trademark from DT21 S.A., 8 chemin dees Pèles 38240 Meylan, France.

to avoid adding numerical attenuation to the physical instability when evidence is searched. Therefore the following relation must be fullfilled on each mesh element of the domain:

$$(R_m)_{\text{mesh element}} < 2 \tag{9}$$

where $R_m = \sigma\mu VL$ with V and L being the characteristic velocity and dimension of the mesh element.

C. Precision

The time step is adjusted so that it is smaller in each mesh element than the convection and diffusion characteristic times. Indeed, these mechanisms are essential for the dynamo effect. As $(\Delta t)_{\text{diff}} = \sigma\mu(\Delta x)^2$ and $(\Delta t)_{\text{conv}} = \Delta_x V^{-1}$,

$$\begin{cases} \text{if } \left(R_m\right)_{\text{mesh}} < 1, \text{ then } \Delta t < \left(R_m\right)_{\text{mesh}} \Delta x V^{-1} \\ \text{if } \left(R_m\right)_{\text{mesh}} > 1, \text{ then } \Delta t < \Delta x V^{-1} \end{cases} \tag{10}$$

At last, mesh size must be small enough to represent the largest variations of the unknowns without numerical rebound.

D. Preliminary Tests

Some preliminary tests showing good accuracy with previous analytical results have been done. A simulation of the 1988 Gailitis et al. experiment has been done for a magnetic Reynolds number equal to 19. The simulation was self-excited in conformity with the extrapolation of the experimental results.[13]

IV. Homogeneous Case

Several simulations have been done for a different number of hexagonal elements, from 1 to 61. For each one of them, the whole domain is homogeneous in terms of magnetic permeability and electrical conductivity without taking care of the ferromagnetic lateral neutronic protection. Outside the hexagonal elements the velocity field is equal to zero. Inside each element the velocity field is defined by Eqs. (3a–3c). The characteristics of the simulations are summarized in Table 1.

The main goal of these simulations is to evaluate the number of elements (each one containing a screw motion) necessary to obtain dynamo effect. This point is interesting from a fundamental point of view as an intermediate case between a single and an infinity of screw motions. Indeed, the single screw motion has been shown to self-excite above a critical magnetic Reynolds number of 17 for a body motion and 20 for a Couette–Poiseuille profile.[14,15] In the infinite case there always exists magnetic axial wave numbers that are self-excited depending on the magnetic Reynolds number.[6,16]

Table 1 Simulation parameters of the homogeneous and heterogeneous cases

Element radius, m	Domain radius, m	$\sigma, \Omega^{-1}m^{-1}$	μ_0, H m^{-1}	V_z, m s^{-1}	ω_{max}, rad/s^{-1}	Δt, s
0.06	2	4.10^6	$4\pi 10^{-7}$	3.6	100	0.009

Another issue of these simulations is the evaluation of the critical size of an FBR core above which self-excitation may occur. The French Superphenix reactor's core, for example, is composed of 364 hexagonal elements (instead of 103 for Phenix).

A. Magnetic Growth Rate in Function of the Number of Elements

On Fig. 5 the time evolution of magnetic energy is sketched for different numbers of elements. In each case the magnetic energy is decreasing, so that the dynamo effect is not reached. Nevertheless, the decrease is less strong when the number of elements is larger. Therefore one can expect that for a sufficiently large number of elements, the dynamo can be reached. An extrapolation of the results is shown in Fig. 6 displaying the number of elements in the function of the growth rate. A zero value of the latter is obtained for a number of elements varying between 300 and 900 depending on the extrapolation method (linear or quadratic) that is used. To avoid this extrapolation lack of precision, the best idea would be to raise the number

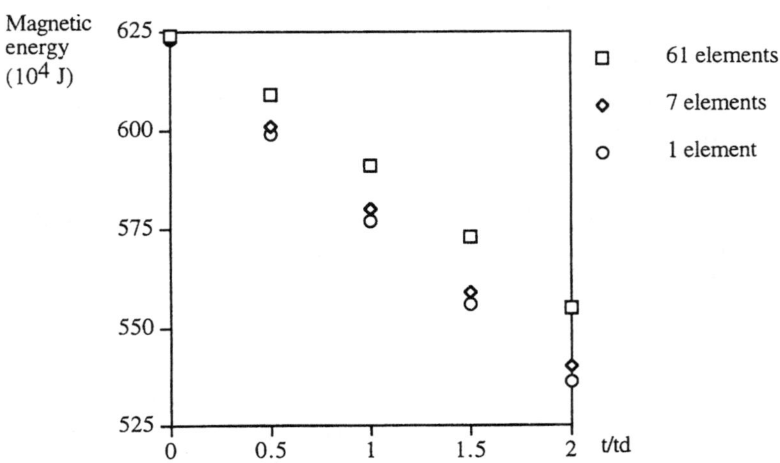

Fig. 5 Magnetic energy in function of adimensional time t/t_d for different numbers of elements. The diffusion time t_d is calculated on the basis of one element radius R: $t_d = \sigma\mu_0 R^2 = 0.09$ s.

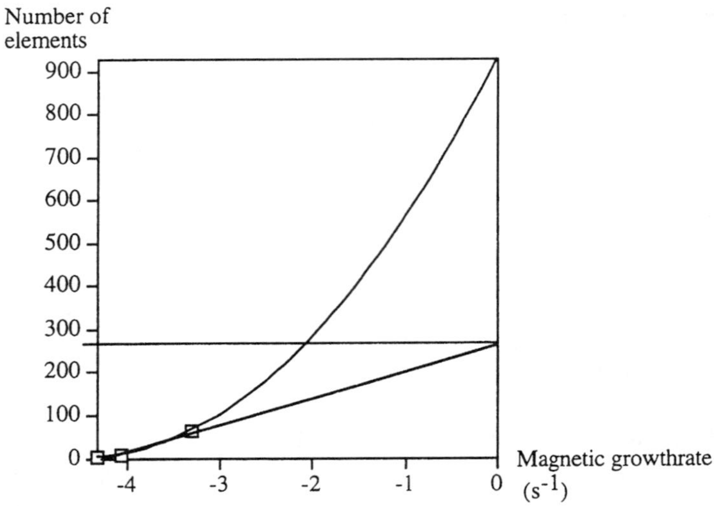

Fig. 6 Extrapolation of the minimum number of elements needed to obtain a positive growth rate of magnetic energy E. The latter is evaluated by the Lyapunov exponent: $t^{-1}\ell n E$.

of elements within the direct simulation. This is difficult, however, because of the computer's limitations. In the 61 elements case, for example, computation on a Cray C94 Vectorial computer lasted about 10 for each time step. Nevertheless, these results show that the FBR Phenix is subcritical in the case of homogeneous core.

B. Simulation of 61 Elements

At the initial time ($t = 0$), $B_z = 0$ and a nonzero horizontal initial magnetic field is given with a variation scale equal to the core's size. As the motion scale is equal to one element's size, one could expect that the vertical component B_z induced by the motion would also be at the scale of one element's size. Figure 7 shows that it is actually the case. But, a large scale B_z is also taking place (Fig. 7), probably resulting from an inverse cascade generation process.[17]

V. Heterogeneous Case

On the basis of the assumptions summarized in Section II, a realistic simulation of the Phenix FBR's core has been done. All of the different regions (Fig. 2) have been represented. However, because of the limited computer machine capacity, only 91 elements have been modeled instead of 127 in reality. To keep the preconditionment of the resolution system healthy the permeability of the ferromagnetic elements is taken equal to $10^2\mu_0$ instead of $10^3\mu_0$. This action can affect the result in the way of an underesti-

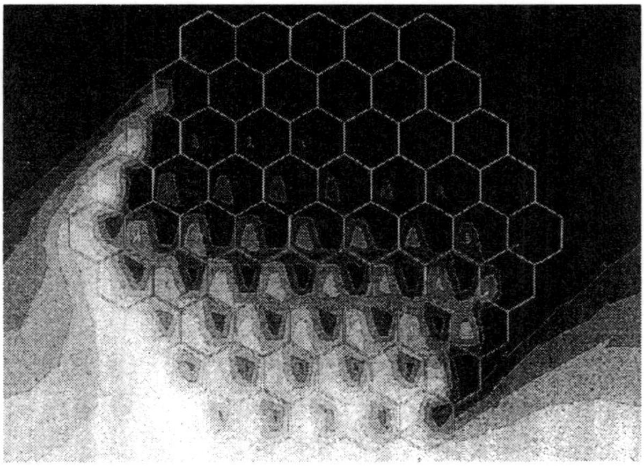

Fig. 7 Isovalues of the magnetic field axial component B_z in a horizontal section. From dark to light, negative to positive values of B_z. B_z is present at the scale of each hexagonal element and also at the scale of the core.

mation of the dynamo effect probability. The other characteristics of the simulation are identical to the previous homogeneous case (Table 1).

A. Growth Rate Within Different Regions of the Core

On Fig. 8, the time evolution of magnetic energy is sketched for different regions of spatial integration of the magnetic energy density $B^2/s2\mu$. In any nonferromagnetic region the magnetic energy is decreasing along the time. Then, the dynamo effect is not obtained at the scale of the FBR's core. Nevertheless, inside the ferromagnetic elements the magnetic energy is strongly increasing. The corresponding growth rate is very strong (around 40 s^{-1}) compared to the decrease in the other regions ($-6\ s^{-1}$), possibly leading quickly to a large value of the magnetic field.

B. Magnetic Field Repartition

As expected by the magnetostatic theory (Section II.D), the magnetic field lines are concentrated on the periphery of the ferromagnetic elements (Fig. 9). Its intensity is growing inside the ferromagnetic elements and decreasing outside.

VI. Discussion

Maybe the most important question concerning the preceding results is to know if the growth of the magnetic energy within the ferromagnetic elements comes from a real dynamo effect. The question is difficult to answer because even if the computation time is twice the characteristic diffusion time inside a nonferromagnetic element, then it is nonetheless only a little

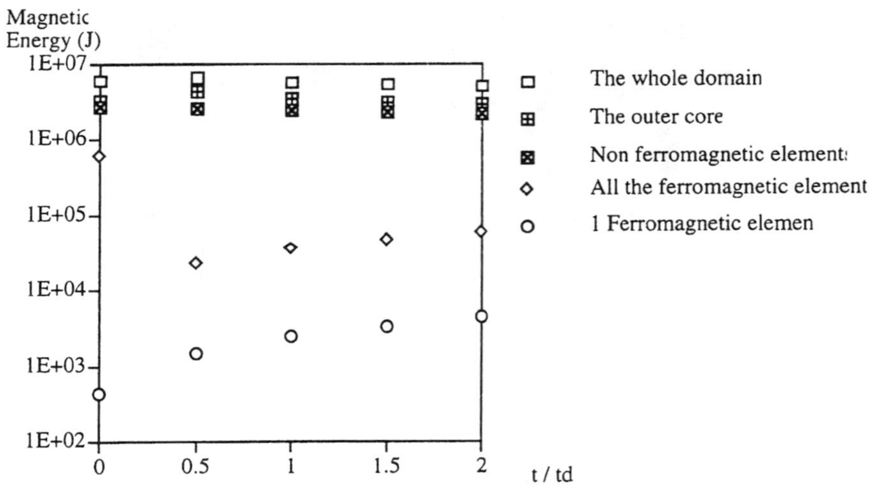

Fig. 8 Magnetic energy in function of adimensional time t/t_d in the presence of ferromagnetic elements. The different curves are corresponding to different regions of the domain on which the magnetic energy has been computed. The diffusion time t_d is calculated on the basis of one nonferromagnetic element $t_d = \sigma\mu_0 R^2 = 0.09$ s.

fraction of the characteristic diffusion time of a ferromagnetic element. Thus, seemingly if one could increase the number of time steps at least by a factor of 100, then possibly one could know if the phenomena are only the result of a convective transient state or if the magnetic field will be increas-

Fig. 9 Horizontal magnetic field in a horizontal section around a ferromagnetic hexagonal element.

ing in all of the domains by a dynamo effect. A definitive answer cannot be given with the present model as the computation time needed for four time steps is already very large (40H of Vectorial Cray computer). However, the opportunity of doing such computations must be possible on parallel and high parallel Cray computers. The present model is now being reprogrammed to be able to run on such computers.

For the FBR authorities the previous results are yet very important. Indeed, even if the magnetic field increase is only a convective transient state, then the increase can still occur during the functioning of the reactor. This sudden increase is probably dependent on the initial magnetic conditions. Suppose that at a certain time the initial conditions are favorable, then the strong amplification of the magnetic field will be accompanied by a growth of the electromagnetic forces. The latter can then lead to eventual large strains on the element's structure, which is not desirable.

Acknowledgment

F. P. was supported by the fast breeder reactor Phenix under Contract 5061 3A099450.

References

[1]Bevir, M. K., "Possibility of Electromagnetic Self-Excitation in Liquid Metal Flows in Fast Reactors," *Journal of the British Nuclear Engineering Society*, Vol. 12. No. 4, 1973, pp. 455–458.

[2]Kirko, G. E., "Phénomènes Électromagnétiques Dans les Grandes Masses de Métaux Liquides," trad. CEA No. 2.722.1, 1984.

[3]Garnier, J., "Synthèse et Critique des Travaux de l'Académie des Sciences Soviètique Concernant les Effects Thermoelectromagnetohydrodynamiques Dans les Surgénérateurs," Tech. Note NR, 1986.

[4]Alemany, A., and Marty, Ph., "L'étude des Perturbations Magnétiques au Voisinage des Pompes Secondaires de Superphénix," 1rst rapport INPG-LEGI/EDF-CLI, 1995.

[5]Marty, Ph., Ajakh, A., and Thess, A., "Magnetic Fields in Fast Breeder Reactors: New Results on Thermoelectricity and Dynamo Effect," *Magnetohydrodynamics*, Vol. 30, No. 4, 1994.

[6]Plunian, F., Alemany, A., and Marty, P., "Influence of Magnetohydrodynamic Parameters on Electromagnetic Self-Excitation in the Core of a Fast Breeder Reactor," *Magnetohydrodynamics* (to be published).

[7]Léorat, J., Pouquet, A., and Frisch, U., "Fully Developed MHD Turbulence Near Critical Magnetic Reynolds Number," *Journal of Fluid Mechanics*, Vol. 104, 1981, pp. 419–443.

[8]Lafay, J., Menant, B., and Barroil, J., "Local Pressure Measurements and Peripherical Flow Visualization in a Water 19-Rod Bundle Compared with FLICA II B Calculations: Influenced of Helical Wire-Wrap Spacer System," ASME, Aug. 1975.

[9]Durand, E., *Magnétostatique*, Masson and Cie, 1968.

[10]Biro, O., and Kurt, P., "On the Use of the Magnetic Vector Potential in the Finite

Element Analysis of Three-Dimensional Eddy Currents," *IEEE Transactions on Mag.*, Vol. 25, No. 4, 1989.

[11]Massé, Ph., "Modeling of Continuous Media Methodology and Computer-Aided Design of Finite Element Programs," *IEEE Transactions on Mag.*, Vol. 20, No. 5, 1984, pp. 1885–1890.

[12]Plunian, F., and Massé, Ph., "An Optimal Preconditioned Scheme for FE Modelization of Kinematic MHD Dynamo Effect," *Proceedings of Ecomas96* (to be published).

[13]Plunian, F., Alemany, A., Marty, Ph., and Massé, Ph., "A Finite Element Modeling of the 1988 Gailitis et al MHD Dynamo Experiment," *Magnetohydrodynamics* (submitted for publication).

[14]Ponomarenko, Yu. B., "On the Theory of Hydromagnetic Dynamo," *Zh. Prikl. Mekh. Tekhn. Fiz.*, USSR, Vol. 6, 1973, pp. 47–51.

[15]Gailitis, A., Karasev, B. G., Kirillov, I. G., and Lielausis, O., and Ogorodnikov, A. P., "The Helical MHD Dynamo," *Proc IUTAM*, Riga, 1988.

[16]Roberts, G. O., "Dynamo Action of Fluid Motions with Two-Dimensionnal Periodicity," *Philosophical Transactions of the Royal Society of London*, Vol. A271 1972, p. 411.

[17]Frisch, U., Pouquet, A., Léorat, J., and Mazure, A., "Possibility of an Inverse Cascade of Magnetic Helicity in MHD Turbulence," *Journal of Fluid Mechanics*, Vol. 68, Pt. 4, 1975, pp. 769–778.

Chapter 36

Influence of Magnetic Field Inhomogeneity on the Operation of Magnetohydrodynamic Devices

A.V. Zagorskii*
Institute of Theoretical and Applied Mechanics, Novosibirsk, 630090, Russia

I. Introduction

WE WILL consider the operation of a magnetohydrodynamic device of the type of plasma driven electromagnetic rail accelerator, a railgun (Fig. 1), in which the acceleration is a result of the interaction between the current through the plasma and the magnetic field. The magnetic field can be created by the current-carrying rails as well as by a special outer source. A great number of works are devoted to the investigation of such accelerators.[1-5] The state of railgun plasma armature (PA) and the acceleration dynamics of the projectile are determined by the ratio of the accelerating force to the momentum losses connected with the turbulent friction and entrainment of a mass eroded off the channel walls. At the beginning of the process, the Lorentz force exceeds losses quite significantly. This result leads to a relatively small PA length (about several channel calibers) and a high acceleration rate. But as velocity and PA mass increase, the losses become comparable with the pushing force. As a result, acceleration stops, and the PA length drastically increases.

The system of equations describing the railgun operation in a quasi-one-dimensional approach is as follows:

$$\frac{\partial \rho}{\partial t} + \frac{\partial \rho u}{\partial x} = q_e$$

$$\rho \frac{\partial u}{\partial t} + \rho u \frac{\partial u}{\partial x} = -\frac{\partial p}{\partial x} + F_m - f_t - u q_e$$

---

*Senior Research Worker, Laboratory of Fast-Process Physics.

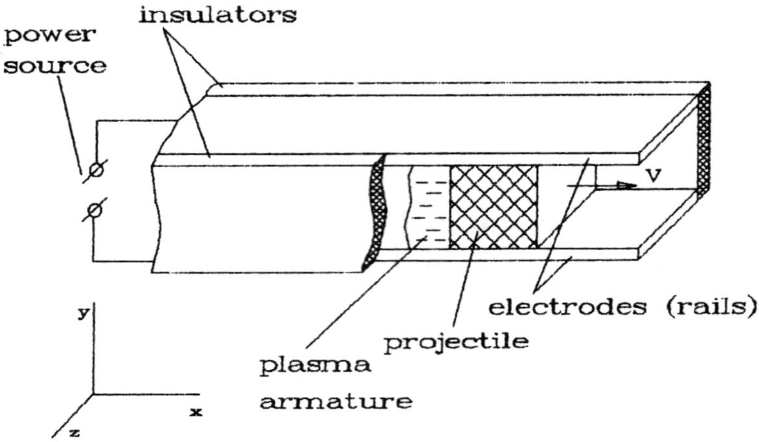

Fig. 1 Principal railgun layout.

$$\rho\frac{\partial e}{\partial t}+\rho u\frac{\partial e}{\partial x}=-p\frac{\partial u}{\partial x}+Q_j+q_e\left(\frac{u^2}{2}-e\right)+Q_t+Q_r+Q$$

$$\frac{\mathrm{d}V}{\mathrm{d}t}=p_1S/M_p$$

In this system F_m is the Lorentz force, f_t is the term describing the momentum losses caused by turbulent friction, Q_t is the corresponding source of heat, Q_r describes the radiation energy losses, q_e is the intensity of the erosion mass source, M_p is the projectile mass, p_1 is the pressure at the projectile back surface, V is the projectile velocity, and S is the squared channel cross section.

The expression for the electromagnetic force is

$$F_m=-\overline{\boldsymbol{B}\times j}=\overline{B_z j_y}-\overline{B_y j_z}$$

Writing the magnetic field and current density in the form of the sums of their averaged values and variations and taking into account that $\overline{B}_y=0, \overline{j}_z$ we obtain

$$F_m=F_0=\overline{f_d}$$

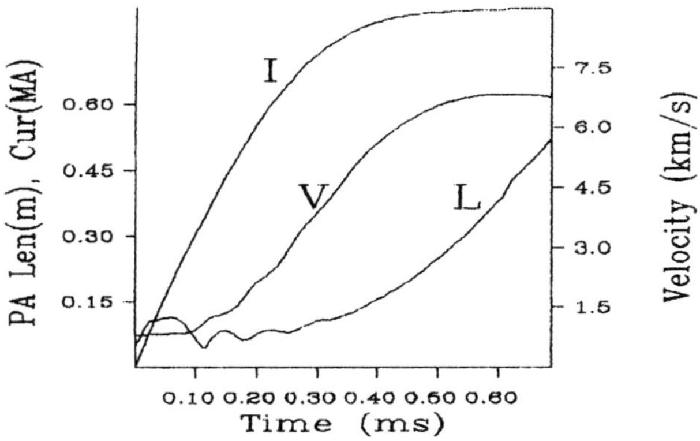

Fig. 2 Typical integral characteristics of the acceleration.

where

$$F_0 = \overline{B_z j_y}$$

$$\overline{f_d} = \overline{B_y j_z} - \overline{\Delta B_z \Delta j_y} \qquad (1)$$

and where averaging is being made over the channel cross section. In all papers devoted to the one-dimensional simulation of PA dynamics, the assumption was made that the current density is uniformly distributed over a channel cross section and only its y component was taken into account, i.e., assuming that $F_m = F_0$. Typical temporal dependences of the total current, PA length, and projectile velocity calculated using model Ref. 4 for the acceleration of the body of 3.8 g mass in the railgun fed from the capacitor energy store are represented in Fig. 2. Numerical calculations[4,5] and qualitative analysis[6] show that during launching PA oscillates near the state of force equilibrium. The equilibrium PA length increases with velocity. The present paper considers the magnetic field inhomogeneity effects on the PA state and launching dynamics.

II. Interaction of the Moving Plasma Armature with the Magnetic Field

The magnitude of the magnetic field nonuniformity depends on a particular technical realization of its source. The magnetic field distribution in the space between two parallel cooper rails of the square cross section 1 × 1 cm (Fig. 3) is shown in Fig. 4. The transient problem of the electromagnetic field propagation in this system has been solved using the method from Ref. 7. The results presented correspond to the instant 10 μs after the current appears in the rails. As the thickness of the skin layer grows, the field

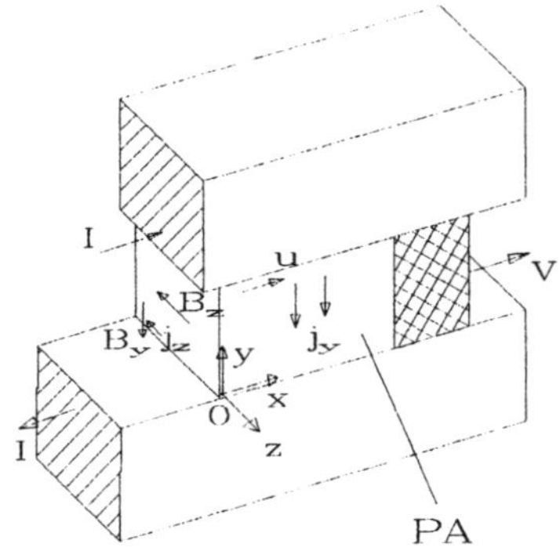

Fig. 3 Fields and currents in a railgun PA.

inhomogeneity increases. In the near-electrode area the deviation of the field from its mean value is rather significant.

The nonuniformity of the distribution of the current's y component and appearance of the j_z component are connected with the electromagnetic interaction of the moving conductive gas and the magnetic field. Inhomogeneity of the B_z causes distortion of j_y. Field component B_y induces a z component of current density. If this takes place near the conductive wall (electrode surface), the induced currents are closed through it. Magnetohy-

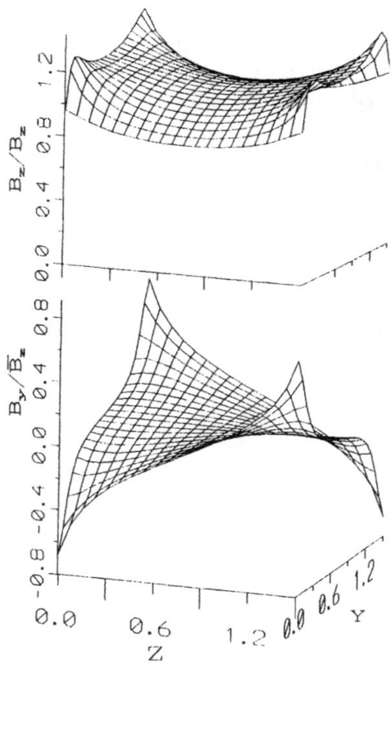

Fig. 4 Magnetic field distribution over the channel cross section.

drodynamic flows in the magnetic field perpendicular to the conducive wall are described in Ref. 8. In such flows the averaged values of second-order summands in Eq. (1) differ from zero, i.e., the decelerating force $\overline{f_d}$ appears.

The problem of the current density distribution in a channel cross section has been solved. The uniformity of the conductivity and gas velocity profiles in the cross section was assumed. The current density were determined from the Ohm's law:

$$\boldsymbol{j} = \sigma\,(\boldsymbol{u} \times \boldsymbol{B} - \nabla\varphi)$$

For the electrostatic potential φ the equation

$$\Delta\varphi = 0$$

with the boundary conditions $\partial\varphi/\partial z = -uB_y$ on the isolating channel walls and $\varphi_1 = \varphi_{el}, \varphi_2 = 0$ on the electrode surface was solved. The ratio f_d to F_0 for the field configuration, presented in Fig. 4, is shown in Fig. 5.

III. Influence of the Field Inhomogeneity on the Plasma Armature State

Let us consider the PA dynamic neglecting the turbulent friction and erosion. Because the values of ΔB_z and B_x are proportional to the mean field and f_d is proportional to the plasma conductivity and velocity, present the mean value of the deceleration force in the form of

$$f_d = k\sigma u\overline{B_z^2}$$

The momentum equation in the projectile-connected coordinate frame is

$$\rho\frac{\partial u}{\partial t} + \rho u\frac{\partial u}{\partial x} = -\frac{\partial p}{\partial x} + \overline{B_z j_y} - k\sigma(u+V)\overline{B_z^2} - a\rho \tag{2}$$

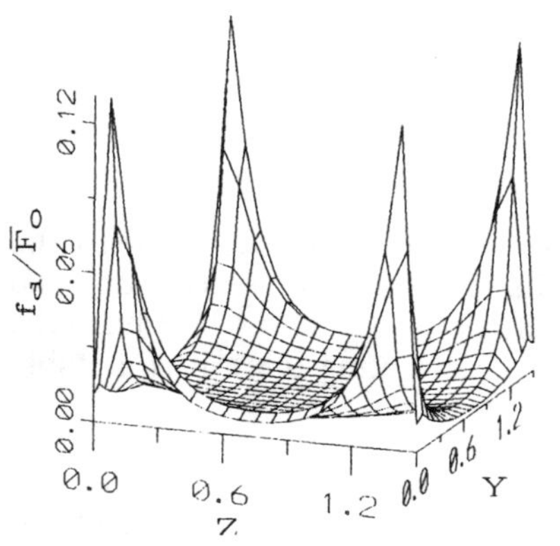

Fig. 5 Distribution of the deceleration electromagnetic force.

where a is the projectile acceleration. Integrating Eq. (2) along the conductive region taking into account the boundary conditions ($\rho_0 = 0$, $u_1 = 0$, subscripts 0 and 1 correspond to the parameters at the free boundary of PA and at the projectile surface) and multiplying by the cross section square S, we obtain

$$z_1 M \frac{du_0}{dt} = -p_1 S + z_2 \overline{B_{z0}} Ib - z_3 k\sigma V \overline{B_{z0}^2} Sl - z_4 \sigma \overline{B_{z0}^2} Slu_0 - aM$$

where l is the PA length, M is its mass, I is the total current, and b is the electrode separation. Coefficients $z_1 - z_4$ depend on the velocity, density, magnetic field, and current density distributions along the PA. Assume them to be independent on time. Assuming plasma to be isothermal and writing the state equation in the form of $p = c^2\rho$, we obtain $p_1 = z_5 c^2 M/Sl$. Taking into account that $a = p_1 S/M_p$ and $dl/dt = -u_0$, the momentum equation is written in the following form:

$$\frac{d^2 l}{dt^2} + Hl\frac{dl}{dt} = \frac{A}{l} + Cl - 2D \tag{3}$$

where $A = z_5 c^2(1 + M/M_p)/z1$, $C = z_3 kS\sigma \overline{B_{z0}^2} V/z_1 M$, $D = z_2 \overline{B_{z0}} Ib/2z_1 M$, and $H = z_4 Sk\sigma \overline{B_{z0}^2}/z_1 M$.

If $CA \leq D^2$, Eq. (3) has two stationary states:

$$l^*_{1,2} = D/C \pm \sqrt{(D/C)^2 - A/C}$$

If $CA > D^2$, the force equilibrium is impossible.

For singular numbers of Eq. (3) linearized near its stationary station, we have

$$\lambda_{1,2} = -Hl^*/2 \pm \sqrt{(Hl^*/2)^2 + 4(C - A/l^{*2})}$$

It is possible to show that the first singular point corresponding to the greater equilibrium PA length is of "saddle" type, and the second one is a stable node if the system is close to the critical state $AC = D^2$ (magnetic decelerating force is comparable with the accelerating one); it is a stable focus in the opposite case.

The analysis performed suggests the following:

1) If $AC > D^2$, PA has no stationary state and expends limitlessly. Respectively, losses increase limitlessly, and acceleration becomes impossible at

such a regime. This limits the reachable velocity and mass of the accelerated plasma:

$$VM \leq \frac{z_2^2}{z_3 z_5} \frac{I^2}{k\sigma c^2(1 + M/M_p)}$$

2) If the PA length exceeds critical value during launching because of some reason or PA is not compact at the beginning of the process the plasma region becomes unstable and decays. Indirect support for this comes from experiments in Ref. 9 where discharge stratification was observed when the PA passes through the area of great inhomogeneity of the magnetic field.

3) If the PA state is near-critical, oscillations do not appear in it.

IV. Influence of Field Nonuniformity on the Electrode Erosion

The character of the magnetic field distribution has a great effect on an energy balance of the near-electrode area. Induced currents cause the additional plasma heating. As a result, the heat flux to the electrode surface increases and intensifies its destruction.

Estimations show that during the contact time (about 10 μs at a velocity of 5 km/s) the electrode can be heated up to the temperature of melting; the point of boiling is usually not reached. Therefore the quantity of eroded mass involved into the acceleration is determined by the mechanisms of destruction of a liquid metal film on the electrode surface.

At the surface of separation "melting/plasma," there appears the Kelvin–Helmholtz instability. For the perturbation wave lengths that are of interest for us (of the order of the film thickness 1 . . . 10 μm) and the plasma parameters typical of the macroparticles launching (plasma density is about 10 kg/m^3), the characteristic time of its development is $\sqrt{\rho_{liq}/\rho_{pl}}\,\delta/2\pi V^*$ where δ is the film thickness and V^* is the character for our problem gas velocity in the boundary layer, which determines the pressure variation over the disturbed surface. Reference 10 shows that for the turbulent flow this is the velocity corresponding to the distance from the surface equal to the disturbance wave length. Estimated boundary-layer thickness in our case is 10 . . . 100 μm. Respectively, character velocity is 0.1 . . . 1 of the mean-mass one, and the time of disturbance development is about 1 . . . 10 μs and is comparable with the contact time.

The second possible mechanism of film destruction is the MHD instability of it. Figure 4 shows that the magnetic field magnitude B_z decreases from the electrode surface to the channel symmetry plane. This makes possible the development of the interchange instability with the time of disturbance growth $\sqrt{\mu\rho_{liq}\delta/B_z/\partial B_z/\partial y}$, which is about 1 . . . 10 μs for the typical field magnitudes (up to 30 T).

Thus, the joint heat and mechanical action of the nonuniformity of the magnetic field can intensify the processes of electrode destruction and involve the eroded mass into acceleration.

V. Results of Two-Dimensional Simulation of Plasma Armature Dynamics

The flow is considered in a two-dimensional approach in the xOy plane. Model Ref. 4 generalized to a two-dimensional approach is used. The following system of equations of gas dynamics with the source terms (the denotations used are traditional) is solved:

$$\frac{\partial \rho}{\partial t} + \nabla \cdot (\rho \boldsymbol{v}) = q_e$$

$$\rho \frac{\partial \boldsymbol{v}}{\partial t} + \rho(\boldsymbol{v} \cdot \nabla)\boldsymbol{v} = -\nabla p + \overline{B_z j_y} - f_d - f_t - \boldsymbol{v} q_e$$

$$\rho \frac{\partial e}{\partial t} + \rho \boldsymbol{v} \cdot \nabla e = -p \nabla \cdot \boldsymbol{v} + Q_j + q_e \left(\frac{\boldsymbol{v}^2}{2} - e \right) + q_d + q_t + Q_r + Q$$

In this system f_t and f_d are terms describing the momentum losses because of turbulent friction and magnetic field inhomogeneity; q_t and q_d are corresponding sources of heat; Q_τ and Q_{tur} are the summands describing the radiation and turbulent transfer of energy; and q_e is the intensity of the erosion mass source. We supposed that x, the component of the current, while solving the Maxwell equations may be neglected. Then having averaged the electromagnetic magnitudes along z for $\overline{j_y}$, we have the integral-differential equation:

$$\varepsilon - b \left[\int_{x_l}^{x_r} \frac{\partial \overline{j_y}}{\partial t} G(x,x')\,\mathrm{d}x' + \overline{j_{y_r}} G_r \overline{u_r} - \overline{j_{yl}} G_l \overline{u_l} \right] - \overline{uBb}$$

$$= L_m c \left[\int_{x_l}^{x_r} \frac{\partial \overline{j_y}}{\partial t}\,\mathrm{d}x' + \overline{j_{y_r} u_r} - \overline{j_{yl}}\,\overline{u_l} \right] + R_m c \int_{x_l}^{x_r} \overline{j_y}\,\mathrm{d}x'$$

$$+ 2\gamma \left(x_l \int_{x_l}^{x_r} \overline{j_y}\,\mathrm{d}x' + \int_{x_l}^{x} \mathrm{d}x \int_{x}^{x_r} \overline{j_y}\,\mathrm{d}x' \right) / S_I + \overline{j_y} b / \overline{\sigma}$$

where γ is the specific resistance of the electrode material; S_I is the area occupied by current in rails; c is the electrode width; x_t and x_τ are coordinates of the left- and right-hand boundaries of the plasma region; ε is the emf of the power source; L_m, R_m are the initial inductance and resistance of the railgun; G is the function depending on the channel geometry; and $\overline{u}$, $\overline{\sigma}$, and $\overline{B}$ are the y-averaged magnitudes of velocity, conductivity, and magnetic field, respectively. The magnetic field is calculated by direct integration according to Biot–Savart's law. Copper is assumed to be a plasma-forming material. Thermodynamic parameters of plasma were calculated by the method represented in Ref. 11; the values of conduction were taken from the tables of Ref. 12. Radiation processes were taken into account in the approximation of radiative heat conduction. The required coefficients were calculated by the method described in Ref. 13. The assumption has been made that the plasma radiates from a surface as an absolutely black body. The plasma viscosity was calculated by the formula obtained from the elementary kinetic theory: $\eta \approx 2 \times 10^{-13} T\sigma$.

The problem of heat distribution in copper electrodes was solved. The assumption has been made that the electrodes are heated from the surface by the plasma radiation and cooled at the expense of heat removal inside. The Joule heating of the electrodes was not taken into account.

The regime of 3.8-g projectile acceleration in a 3-m long channel having a square cross section with a calibre of 1.5 cm was considered. The capacity of a bank of capacitors is equal to 0.35 F; the initial induction and resistance are 1.2 μH and 0.47 mΩ, respectively. The initial velocity is 0.6 km/s. The initial mass of plasma was taken at 0.16 g. In the test of Ref. 14, the output velocity was 6.4 km/s.

The deceleration force was presented in the form of

$$f_d(x, y) = k_d(y) u \sigma B_z^2(x, y)$$

where $k_d(y)$ is calculated according to the obtained distributions of the field components taking into account the presence of dielectric walls as was described in the preceding paragraphs. In the calculations the dependence $k_d(y)$ was approximated by the expression

$$k_d(y) = 0.1 \cdot [(y - b/2)/(b/2)]^6$$

An expression for the additional heat release has the following form:

$$q_d = j_z^2(x, y)/\sigma = k_d u^2 \sigma B_z^2(x, y)$$

The assumption was made that the whole melted metal penetrates into the near-electrode layer.

Figure 6 shows temporal distributions of the velocity, mass, and length of PA with and without taking into account the field nonuniformity effects

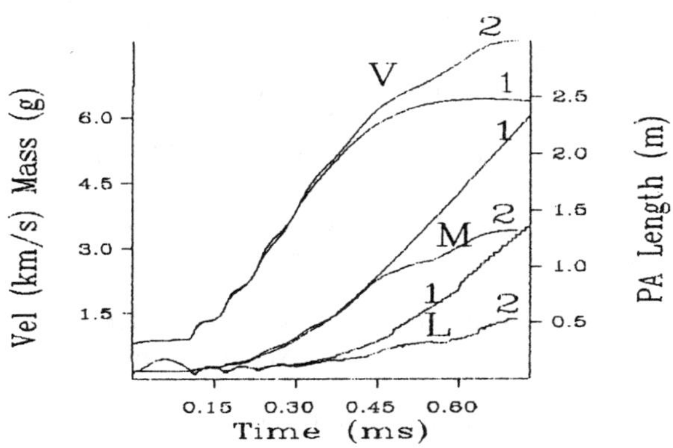

Fig. 6 Influence of the magnetic field inhomogeneity on the railgun characteristics; 1, uniform field and 2, nonuniform field.

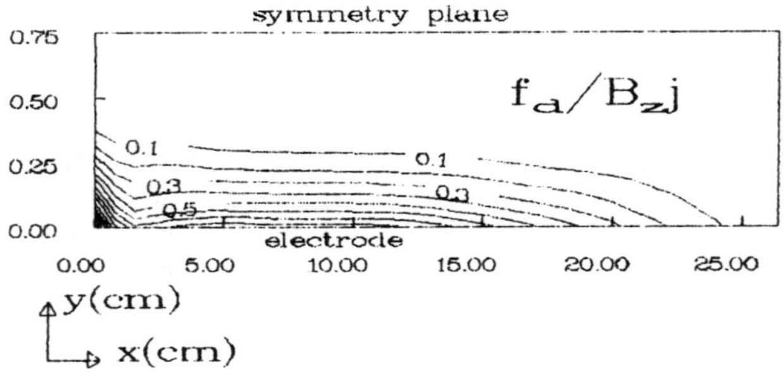

Fig. 7 Distribution of the deceleration force in a xOy plane.

(curves 1 and 2, respectively). After the velocity of 4.5 km/s, these effects lead to a drastic increase in PA length, intensification of erosion, and as result decrease the final velocity significantly. For example, the spatial distribution of the ratio of the deceleration force to the pushing one f_d/B_zj_y at the instant when projectile velocity reaches 5.5 km/s is represented in Fig. 7.

VI. Conclusions

Performed analysis of the influence of the magnetic field distribution over the channel cross section on the railgun operation suggests that the field inhomogeneity causes additional limitations of the outlet characteristics of such devices. It can initiate the decay of plasma armature and intensifies the erosion increasing of accelerated mass.

References

[1]Proceedings of the 5th Symposium on Electro-Magnetic Launch Technology, *IEEE Transactions on Magnetics*, Vol. 27, No. 1, 1991.

[2]Proceedings of the 6th Symposium on Electro-Magnetic Launch Technology, *IEEE Transactions on Magnetics*, Vol. 29, No. 1, 1993.

[3]Proceedings of the 7th Symposium on Electro-Magnetic Launch Technology, *IEEE Transactions on Magnetics*, Vol. 31, No. 1, 1995.

[4]Zagorskii, A. V., and Katsnel'son, S. S., "Dynamics of Plasma Armature in a Rail Accelerator," *High Temperatures*, Vol. 29, No. 3.

[5]Rolader, G. E., Batteh, J. H., Powell, J. D., and Desai, P. V., "Transient Modelling of Railgun Plasma Armatures," *IEEE Transactions on Magnetics*, Vol. 25, No. 1, 1989, pp. 489–494.

[6]Zagorskii, A. V., "The Qualitative Analysis of Plasma Armature Dynamics in a Rail Launcher Channel," *Isvestiya SO AN SSSR, Sib. fiz.-tekhn. Zhurn.*, No. 6, 1991, pp. 52–56 (Russian).

[7]Zagorskii, A. V., "On the Calculation of Electro-Magnetic Fields in Conductors," *Izv. SO AN SSSR, ser. tech. nauk*, No. 1, 1990, pp. 103–106 (Russian).

[8]Branover, H. H., "Turbulent Magnetohydrodynamic Flows in Pipes," Zinatne Publ., Riga, 1967, pp. 21–39 (Russian).

[9]Calvin, H. A., Virostek, S. P., and Anderson, J. J., "A Triggering Mechanism for Secondary Arcs," *IEEE Transactions on Magnetics*, Vol. 29, No. 1, 1993, pp. 757–762.

[10]Jackson, P. S., and Hunt, J. G. R., "Turbulent Wind Flow over a Low Hill," *Quart. Jour. Roi. Met. Soc.*, Vol. 101, No. 430, 1975, pp. 929–955.

[11]Koval'skaya, G. A., "Composition and Thermophysical Properties of Non-Ideal Copper Plasma," *Russ. Journ. of Theor. and Appl. Mech.*, Vol. 2, No. 3, 1992, pp. 249–263.

[12]Ermakov, V.V., and Kalitkin, N. N., "Tables of Electrical Conductivity and Electron Heat Conductivity of Dense Plasma," Acad. Sci. USSR, Inst. Applied Mathematics, Moscow, Dep. 2813-78, 1978 (Russian).

[13]Zeldovich, Ya. B., and Raizer, Yu. P., "Physics of Shock Waves and High Temperature Hydrodynamic Phenomena," Academic, New York, 1963.

[14]Zheleznyi, V. B., Zagorskii, A.V., Kudryavtsev, A.V., Plekhanov, A.V., and Katsnel'son, S. S., "Theoretical and Experimental Modelling of Railgun Operation," *Appl. Mech. and Techn. Phys.*, Vol. 34, No. 2, 1993, pp. 32–35.

Chapter 37

Study of the Magnetohydrodynamic Turbulent Flow in the Electromagnetic Valve

S. Frandoş*
Iprolam Research Institute of Bucharest
O. Peşteanu†
University Transilvania of Braşov
and
V. Fireţeanu‡
University Politechnica of Bucharest

I. Introduction

THE electromagnetic liquid flow control is a new application of the MHD, in which the flow of an electric conductive liquid through a channel takes place in an electromagnetic field. The development of this technique is in connection with the flow control of the molten metal in continuous casting processes.

The electromagnetic valve uses a poly-phased inductor; the generated field of which has the character of a progressive wave. The electromagnetic forces in the channel exert a complex action on the liquid spurt: a breaking or an accelerating effect, accompanied by a compressing one.

This work deals with the MHD flow through a cylindrical electromagnetic valve by the numerical modeling of the electromagnetic and hydrodynamic phenomena. The FEM method is considered for the computing of the electromagnetic forces, as well as an approach of the flow study on the basis of the Navier–Stokes equation. The k-ε model of the turbulent flow is associated. The steady state of the flow is determined to be the final solution of a transient regime. Numerical results are presented for the two kinds of tasks of the valve: the breaking and accelerating of the flow. Also, with the aid of

*Electrical Engineering Dept., 313 Splaiul Independentei 77206, Bucharest, Romania

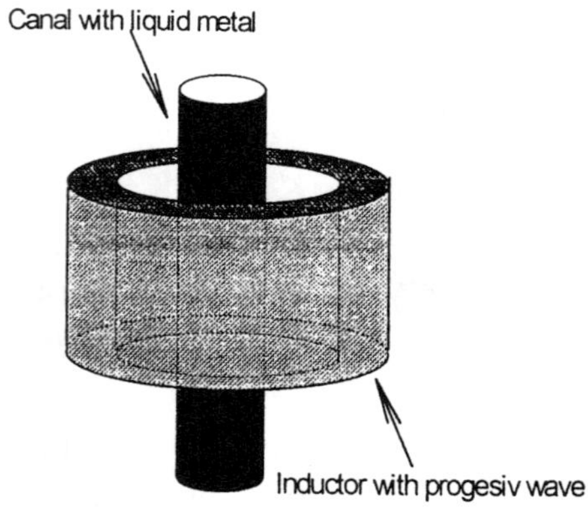

Fig. 1 Electromagnetic valve model.

the authors' dedicated programs VALVELMG and VALVMHD, the optima values of the frequency and of the inductor height are determined.

II. Electromagnetic Force Computation in the Liquid Metal

Fig. 1 presents a general view of an induction electromagnetic valve with cylindrical symmetry. The computing domain of the electromagnetic field and of the associated dynamic effects, Fig. 2, is represented by an axial section of the valve. This is delimited by the *AD* symmetry axis and by the *ABCD* contour to which the electromagnetic field magnitude vanished. The cylindrical coordinate system $[r, \varphi, z]$ is used. The subdomain $AA'D'D$ represents the liquid metal with the characteristics $\sigma \neq 0$ and $\mu = \mu_0$. *EFGH* is the source subdomain composed from the coils of a three-phase linear

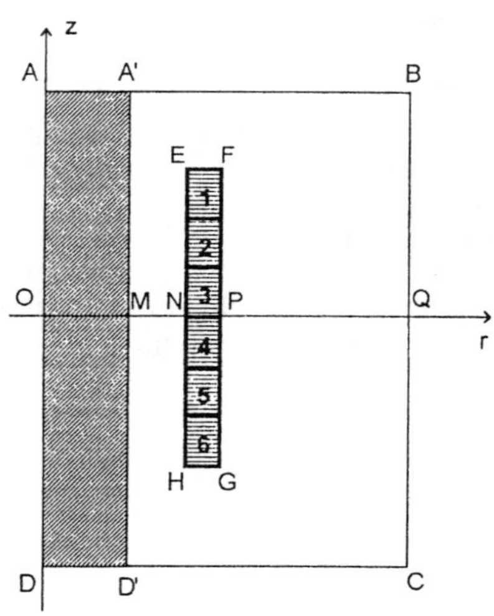

Fig. 2 Electromagnetic field computing domain.

inductor. The material electromagnetic properties in this subdomain are $\sigma = 0$ and $\mu = \mu_0$. The electromagnetic effects are provided by applying electric currents through the six coils. Thus currents have the direction of φ coordinate, $\boldsymbol{J}_s$ $[0, \boldsymbol{J}_s, 0]$ and the expressions:

$$J_{si} = \sqrt{2} \cdot J \cdot \cos\left(\omega t + i\frac{2\pi}{6}\right) \tag{1}$$

where $i = 1, \ldots, 6$ is the coil's identifier, accordingly with Fig. 2. The rest of the computing domain has the properties $\sigma = 0$ and $\mu = \mu_0$.

The electromagnetic field model is that of the quasistationary field with eddy currents in the metalic subdomains. The vector of the complex magnetic potential $\boldsymbol{A}$ introduced by the relation $\boldsymbol{B} = \text{rot}\,\boldsymbol{A}$ satisfies the general equation:

$$\text{rot rot}\,\boldsymbol{A} = \mu \boldsymbol{J}_s - \mathrm{j}\omega\mu\sigma\,\text{rot}\,\boldsymbol{A} \tag{2}$$

Accordingly with the associated two-dimensional model of the problem, the vector of the magnetic potential is oriented in the same direction with the current density after φ coordinate $\boldsymbol{A}$ $[0, \boldsymbol{A}(r, z), 0]$. The unknown potential $\boldsymbol{A}(r,z)$ is determined by the finite-element method. Introducing the auxiliary unknown $\boldsymbol{U}(r,z) = r\boldsymbol{A}(r,z)$, the associated function of Eq. (2) has the expression:

$$F = 2\pi \iint_{ABCD} \left\{ \frac{1}{2r}\left[\left(\frac{\partial U}{\partial r}\right)^2 + \left(\frac{\partial U}{\partial z}\right)^2\right] - \mu_0 J_s U + \frac{j\omega\mu_0\sigma}{2r} U^2 \right\} \mathrm{d}z\,\mathrm{d}r$$

$$+ \int_{ABCD} \frac{U}{r} \cdot \frac{\mathrm{d}}{\mathrm{d}n}\left(\frac{U}{r}\right) \mathrm{d}s \tag{3}$$

In any point of the computing domain limits $ABCD$ (Fig. 2), the condition $A = 0$ is considered; therefore, the last term in Eq. (3) is null.

The volume density of the electromagnetic force is performed with the relation:

$$\boldsymbol{f} = \text{Real}\,\{-\boldsymbol{j}\omega\sigma\boldsymbol{A} \times \text{rot}\,\boldsymbol{A}\} \tag{4}$$

For the following geometrical dimensions $OM = 40$ mm, $ON = 80$ mm, $OP = 80 + 36$ mm, $NE = NH = 3 \times 54$ mm (Fig. 2), the electric conductivity of the liquid metal (steel) $\sigma = 7 \cdot 10^6\ \Omega^{-1}\text{m}^{-1}$, the frequency $f = 700$ Hz, the current density in the coils $J = 10\ A/\text{mm}^2$, represented in Figs. 3–5 are the bidimensional fields of the magnetic flux density $B(r,z)$ and of the two components of the electromagnetic force $f_z(r,z)$ and $f_r(r,z)$. These solutions

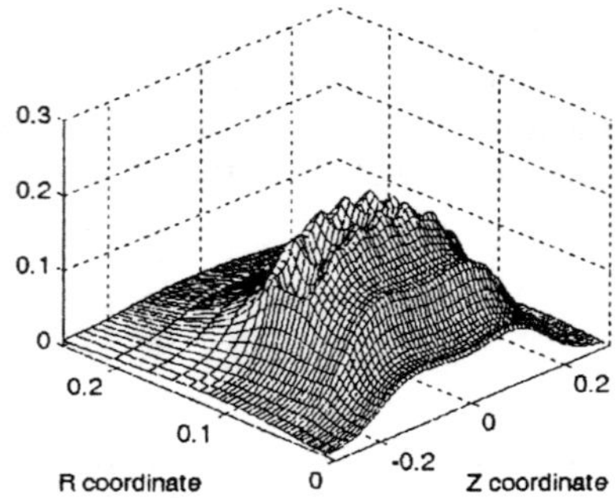

Fig. 3 Magnetic flux density, $B(r,z)$ [T].

were obtained with the dedicated program VALVELMG elaborated by the authors.

For various frequencies the values of the total axial force, achieved either with VALVELMG or with the professional software FLUX2D, are presented in Table 1.

III. Magnetohydrodynamic Flow Model of the Electromagnetic Valve

The study of the liquid-metal flow through the channel valve has the purpose of determining the velocity and the pressure fields in $AA'D'D$ subdomain (Fig. 2) on the liquid actions' gravity forces, electromagnetic forces, and viscosity forces. For cylindrical symmetry reasons, the velocity in any point of the flow domain is considered to have only two components $\boldsymbol{v}[v_r,0,v_z]$. In the integration of the velocity field equations, the $k - \varepsilon$ turbulence model is associated.

A. Mathematical Model

The liquid flow phenomena are described by the Navier–Stokes equation and completed with the continuity equation for incompressible liquids. The

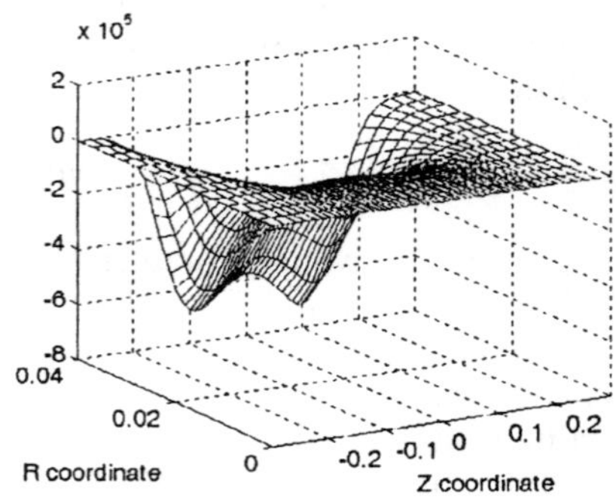

Fig. 4 Radial electromagnetic force, f_z (r, z) [N/m^3].

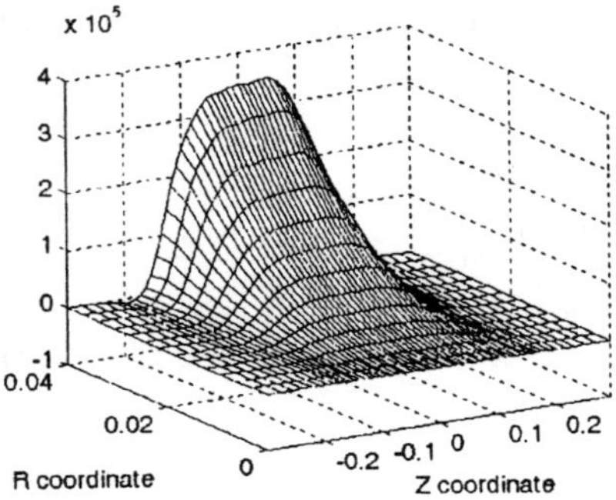

Fig. 5 Axial electromagnetic force, f_r (r, z) [N/m^3].

projections of the Navier–Stokes equation on the axis Or and Oz (Fig. 2) have the expressions:

Oz:

$$\rho \cdot \left[\frac{\partial v_z}{\partial t} + \frac{\partial (v_z \cdot v_z)}{\partial z} + \frac{1}{r} \cdot \frac{\partial (r \cdot v_r \cdot v_z)}{\partial r} \right] = -\frac{\partial p}{\partial z} + f_z + \frac{\partial}{\partial z}\left(2 \cdot \eta_{\text{ef}} \cdot \frac{\partial v_z}{\partial z} \right)$$

$$+ \frac{1}{r} \cdot \frac{\partial}{\partial r}\left[r \cdot \eta_{\text{ef}} \cdot \left(\frac{\partial v_z}{\partial r} + \frac{\partial v_r}{\partial z} \right) \right] \tag{5}$$

Or:

$$\rho \cdot \left[\frac{\partial v_r}{\partial t} + \frac{\partial (v_z \cdot v_r)}{\partial z} + \frac{1}{r} \cdot \frac{\partial (r \cdot v_r \cdot v_r)}{\partial r} \right] = -\frac{\partial p}{\partial r} + f_r + \frac{\partial}{\partial z}\left[2 \cdot \eta_{\text{ef}} \left(\frac{\partial v_z}{\partial r} + \frac{\partial v_r}{\partial z} \right) \right]$$

$$+ \frac{1}{r} \cdot \frac{\partial}{\partial r}\left(r \cdot \eta_{\text{ef}} \cdot \frac{\partial v_r}{\partial r} \right) - 2\eta_{\text{ef}} \frac{v_r}{r^2} \tag{6}$$

where ρ is the liquid density, p is the local pressure, and η_{ef} is the local effective viscosity. The effective viscosity yields like the sum $\eta_{\text{ef}} = \eta + \eta_t$ of the dynamic viscosity η and the turbulent viscosity η_t.

Table 1 Axial force values

Frequency, H_z	Axial force, N	Axial Force, N
300	56.7	55.2
500	78.6	77.4
700	88.5	87.4
800	90.5	89.6
900	91.2	90.5
1000	91	90.4
——	FLUX2D	VALVELMG

The $k - \varepsilon$ model for the turbulent viscosity computation[1] is composed of the equations

$$\rho \cdot \left[\frac{\partial k}{\partial t} + \frac{\partial}{\partial z}(v_z \cdot k) + \frac{1}{r} \cdot \frac{\partial}{\partial r}(r \cdot v_r \cdot k) \right] = \frac{\partial}{\partial z}\left(\frac{\eta_t}{\sigma_k} \cdot \frac{\partial k}{\partial z} \right)$$

$$+ \frac{1}{r} \cdot \frac{\partial}{\partial r}\left(\frac{\eta_t}{\sigma_k} \cdot r \cdot \frac{\partial k}{\partial r} \right) + P - \varepsilon \tag{7}$$

$$\rho \cdot \left[\frac{\partial \varepsilon}{\partial t} + \frac{\partial}{\partial z}(v_z \cdot \varepsilon) + \frac{1}{r} \cdot \frac{\partial}{\partial r}(r \cdot v_r \cdot \varepsilon) \right] = \frac{\partial}{\partial z}\left(\frac{\eta_t}{\sigma_\varepsilon} \cdot \frac{\partial \varepsilon}{\partial z} \right)$$

$$+ \frac{1}{r} \cdot \frac{\partial}{\partial r}\left(\frac{\eta_t}{\sigma_\varepsilon} \cdot r \cdot \frac{\partial \varepsilon}{\partial r} \right) + \frac{\varepsilon}{k} \cdot (c_1 \cdot P - c_2 \cdot \varepsilon) \tag{8}$$

In Eqs. (7) and (8)

$$P = \eta_t \cdot \left\{ 2 \cdot \left[\left(\frac{\partial v_z}{\partial z} \right)^2 + \left(\frac{\partial v_r}{\partial r} \right)^2 + \left(\frac{v_r}{r} \right)^2 \right] + \left(\frac{\partial v_z}{\partial r} + \frac{\partial v_r}{\partial z} \right)^2 \right\} \tag{9}$$

is the generating rate of kinetic energy k, and ε is the dissipation rate of k. Turbulent viscosity η_t is achieved with the formula:

$$\eta_t = \rho \cdot c_\mu \cdot \frac{k^2}{\varepsilon} \tag{10}$$

In the relations (7–10) $\sigma_k = 1$, $\sigma_\varepsilon = 1{,}3$, $c_\mu = 0{,}09$, $c_1 = 1{,}44$, and $c_2 = 1.92$ are the model constants.

Because the $k - \varepsilon$ model equations are not available close to the channel walls where the velocity tends to vanish, in the Navier–Stokes equations the viscosity efforts are replaced with the friction efforts:

$$\tau_w = \rho \cdot U\tau \cdot \frac{v_z}{|v_z|} \tag{11}$$

where v_z is the axial velocity and the factor U_τ has the expression:

$$U_\tau = C_r \cdot \frac{|v_z| \cdot c_\mu^{1/4} \cdot k^{1/2}}{\frac{1}{\chi} \cdot \ell n\left(\rho \cdot [9 \cdot (R - r) \cdot c_\mu^{1/4} \cdot k^{1/2}/\eta]\right)} \tag{12}$$

In the relation (12) R is the inner channel radial dimension, r is the current coordinate on Or axis, $\chi = 0{,}4$ is the von Kármán's constant,[1] and C_r is a coefficient that depends on the rough wall.

In the points from the vicinity of the wall, the condition $\partial k/\partial r = 0$ is considered, and the following dependence between k and ε is used:

$$\varepsilon = \frac{c_\mu^{3/4} \cdot k^{3/2}}{\chi \cdot (R - r)} \tag{13}$$

The continuity equation for the incompressible liquids div $\boldsymbol{v} = 0$ has the expression:

$$\frac{\partial v_r}{\partial r} + \frac{v_r}{r} + \frac{\partial v_z}{\partial z} = 0 \tag{14}$$

B. Numerical Integration of the Magnetohydrodynamic Flow Equations

The numerical integration of Eqs. (5–7) is performed by the finite-difference method. Because of the system nonlinearity and as a result of time dependence of the unknowns v_r, v_z, ε, and k, a time iterative algorithm is used to find the steady-state solutions. To ensure the solution stability,[2] the limitation of the time step δt is required:

$$\delta t \leq 1 \Big/ \frac{2 \cdot \eta_{\text{ef}}}{\rho} \cdot \left(\frac{1}{h_r^2} + \frac{1}{h_z^2} \right) + \frac{|v_r|_{\max}}{h_r} + \frac{|v_z|_{\max}}{h_z} \tag{15}$$

where h_r and h_z are the minimum values of the discrete step spaces on the Or and Oz axis, respectively, and $|v_r|_{\max}$ and $|v_z|_{\max}$ are the absolute values of the minimum local velocities.

At the beginning of the iteration, the velocity and pressure components are set to zero in all computing domains. The turbulent viscosity is initialized with a nonzero value, but with at least one size order below the dynamic viscosity value.

On the frontier $A'D'$ (Fig. 2) the following conditions are imposed:

$$v_r = 0, \quad v_z = 0, \quad \frac{\partial k}{\partial r} = 0, \quad \text{and} \quad \varepsilon = \frac{c_\mu^{3/4} \cdot k^{3/2}}{\chi \cdot (R - r)} \tag{16}$$

On the AD channel axis there are used homogeneous Neumann conditions. For the AA' bounder the radial velocity is considered zero, and for all other unknowns the homogeneous Neumann conditions are imposed.

On the lower bounder DD' there are the special conditions. Taking into account that the turbulence once stimulated by the electromagnetic forces

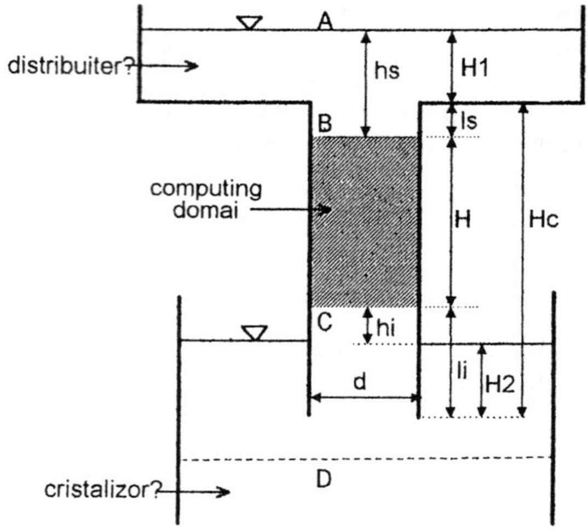

Fig. 6 Ensemble structure of the electromagnetic flow-control system needed in continuous casting.

could extend beyond the computing domain, the conditions on this limit must preserve the motion of the liquid that passes through. Thus, considered on this limit at each time-step iteration are the velocity values that are obtained with the aid of a Taylor series, which is developed along the *Oz* axis. The terms are retained up to the second derivatives. For the k and ε unknowns the homogenous Neumann conditions are used.

At each iteration the velocity and the pressure fields are modified to correct the continuity equation that must be more strictly respected during the iteration process. The pressure field supports a second modification to fulfill the limit conditions described in the following.

In Fig. 6 the ensemble structure of the electromagnetic flow-control system needed in the process of continuous casting is sketched.

At levels B and C (Fig. 6) the pressure means p_i' and p_s' must respect Bernoulli's equation:

between A and B,

$$\rho \times g \times h_s = p_s' + \rho \frac{v2}{2} + \rho \frac{v2}{2}\left(\lambda_s \frac{l_s}{d} + \xi_s\right) \tag{17}$$

between C and D,

$$p_i' + \rho \cdot g \cdot h_i + \rho \frac{v2}{2} = \rho \frac{v2}{2}\left(\lambda_i \frac{l_i}{d} + \xi_i\right) + \rho \cdot g \cdot h_d \tag{18}$$

In relations (17) and (18) v is the mean velocity through the canal, ρ is the liquid density, λ is Darcy's coefficient for linear pressure losses, and ξ is the coefficient of the local pressure loses at the sudden change of the transversal section of the flow. Geometrical dimensions used in these relations are noted on Fig. 6. Null means velocities are considered at the levels A and D of Fig. 6. The following is obtained:

$$p'_s = \rho \cdot g \cdot hs - \rho \frac{v^2}{2}\left(\lambda \frac{ls}{d} + \xi_s + l\right) \quad p'_i = \rho \cdot g \cdot hi + \rho \frac{v^2}{2}\left(\lambda \frac{ls}{d} + \xi_s - 1\right) \tag{19}$$

Let $\Delta p_s = p'_s - p_s$ and $\Delta p_i = p'_i - p_i$ be the deviations of the mean values p_s and p_i from the computed values p'_s and p'_i. Taking into account that the relations (19) must characterize the means pressure at level B and C of Fig. 6, linear variation is considered between the differences Δp_s at the superior and Δp_i at the inferior limits of the computing domain.

At each iteration step the turbulent viscosity is computed. The numerical integration procedure of the flow equations is performed until a sufficiently small error between two consecutive iteration steps is obtained, or until the maximum acceleration in the computing domain satisfies, with a good approximation, the zero value in comparison with the gravity acceleration. In these conditions the transient regime is considered finished, and the solutions correspond to the stationary regime of the turbulent MHD flow.

C. Numerical Application

Based on the mathematical model and on the numerical integration procedure presented above, the program VALVMHD was developed. Running after the VALVELMG program and with its results (electromagnetic force in $AA'D'D$ subdomain), for this program are necessary in addition to some geometrical parameters. These parameters are particularized below for the application starting in section II.A. See Fig. 6: $H_1 = 100$ mm, $H_2 = 100$ mm, $H_c = 1000$ mm; l_i ,l_s, and H are automatically generating according to the inductor height h_i, so that $H = 3h_i$, $l_i = l_s$, and $H + l_i + l_s = H_c$. Also $\rho = 7200$ kg/m³ and $\eta = 0.048$ Ns/m² for molten steel are considered.

For the case in which the progressive wave of the electromagnetic valve acts is opposite sense with the flow, i.e., the valve work in the breaking regime, presented in Fig. 7 are: the field of the axial velocity component (Fig. 7a), the field of the radial velocity component (Fig. 7b), the field of the turbulent viscosity (Fig. 7c), and the field of the pressure (Fig. 7d).

The breaking effect is characterized by the reduction of the axial velocity near the channel wall in front of the inductor. In this area the electromagnetic forces are greatest (Fig. 7a). As a result, the spurt is obtruded, and the flow will be forced through the center of the channel. The radial velocity field emphasized the compressed effect because of the radial electromagnetic force. This effect consists of a stimulated motion of liquid toward the canal axis at the entrance of the liquid in the electromagnetic field. The liquid motion from the axis toward the wall is a consequence of the continuity equation. Because of the significant gradient of the axial and radial velocity component fields in the front of the inductor, the turbulent viscosity is increased, thus contributing to the enhancement of the breaking effect.

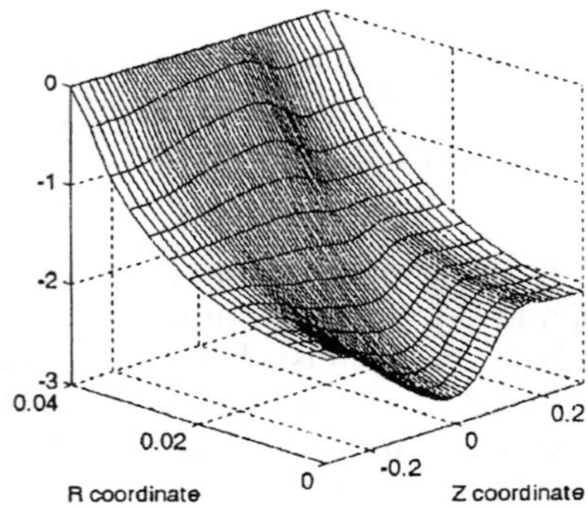

Fig. 7a The axial velocity, v_z (r, z) [m/s].

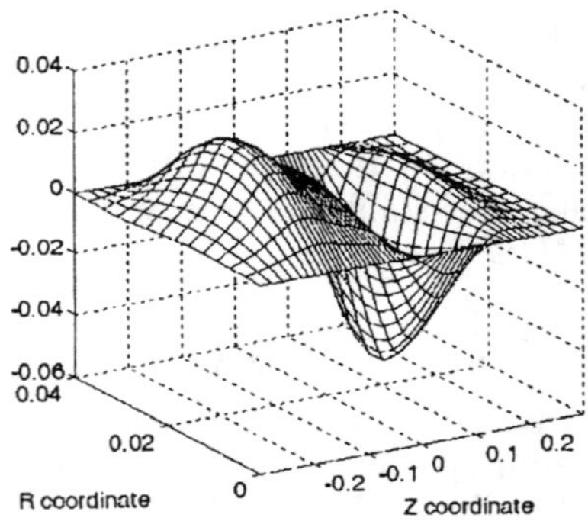

Fig. 7b The velocity, v_r (r, z) [m/s].

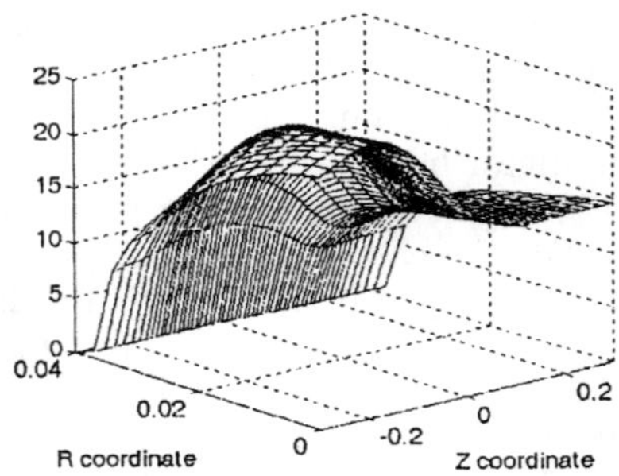

Fig. 7c The turbulent viscosity, η ± (r, z) [Ns/m^2].

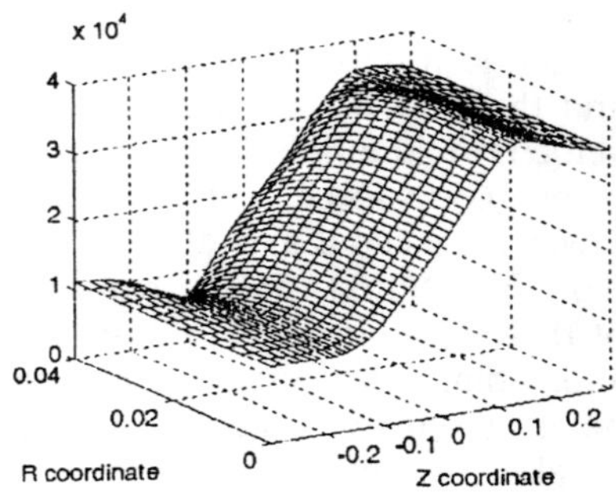

Fig. 7d The pressure field, P − (r, z) [N/m^2].

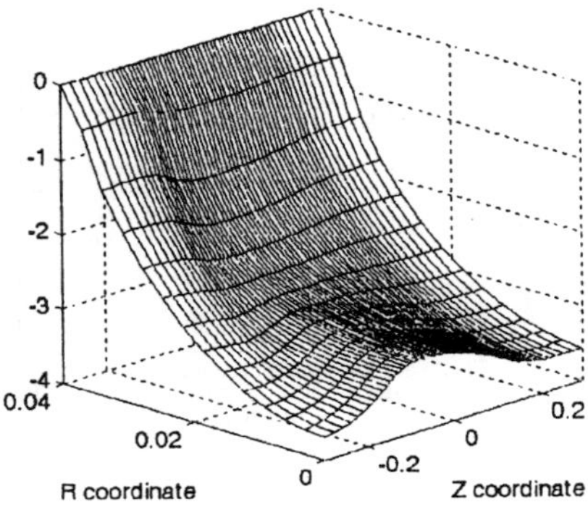

Fig. 8a The axial velocity, v_z (r, z) [m/s].

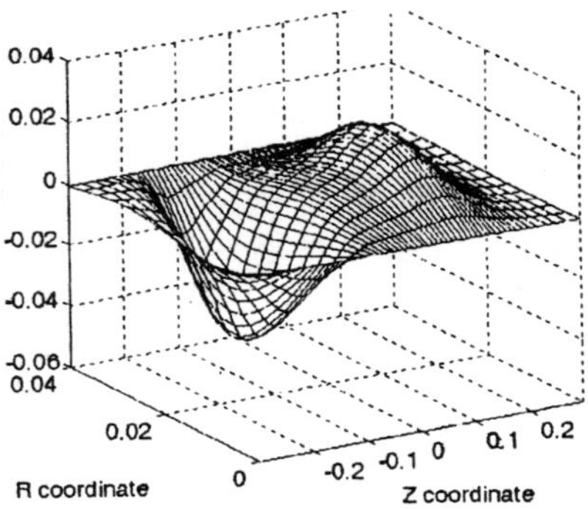

Fig. 8b The radial velocity, v_r (r, z) [m/s].

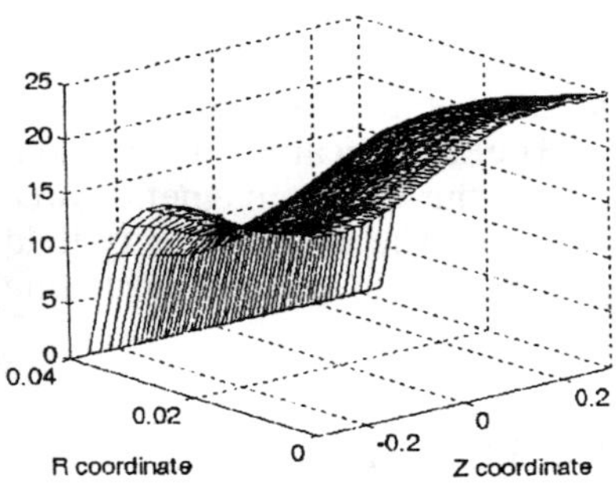

Fig. 8c The turbulent viscosity field, η ± (r, z) [Ns/m^2].

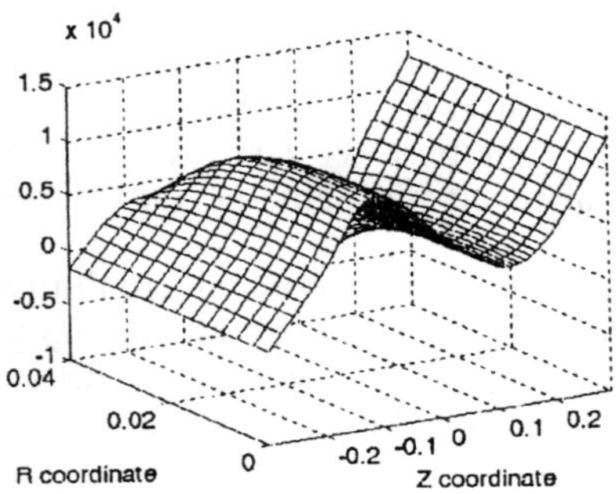

Fig. 8d The pressure field, P − (r, z) [N/m^2].

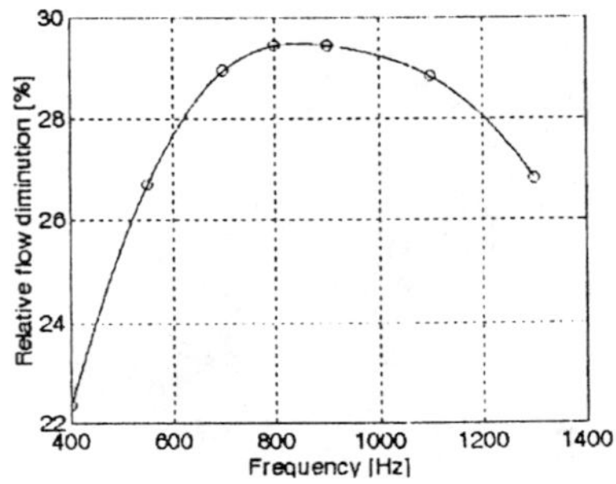

Fig. 9 Relative flow diminution

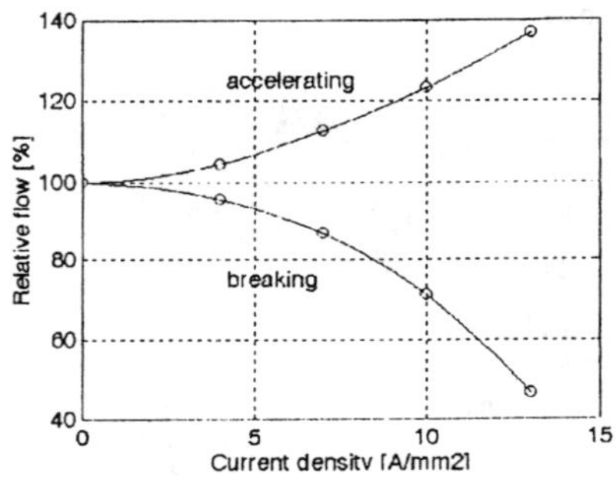

Fig. 10 Flow control capability

In the regime of the liquid acceleration, the velocity, the viscosity, and the pressure fields are presented in Fig. 8. The amplification of the axial velocity near to the wall is observed because the electromagnetic force acts in the same sense with the liquid global flow. The radial velocity field has changed because it is needed to supply with liquid the entrance zone near to the wall.

IV. Numerical Results Obtained with VALVELMG and VALVMHD

A. Optimum Values of the Frequency Supply and of the Inductor Height

In Fig. 9 the relative flow diminution depending on the frequency supply in the case of the breaking effect is represented. It is considered like reference the flow in the case of a null electromagnetic force (free flow). The optima frequence is about 800 Hz. A similar result is obtained even for the accelerating regime.

For the optimum frequency the variation of the flow control capability depending on the inductor height was studied. The flow-control capability is defined as the ratio between the accelerating and breaking flow difference and the evidence for an optimum value for the inductor height at about 360 mm.

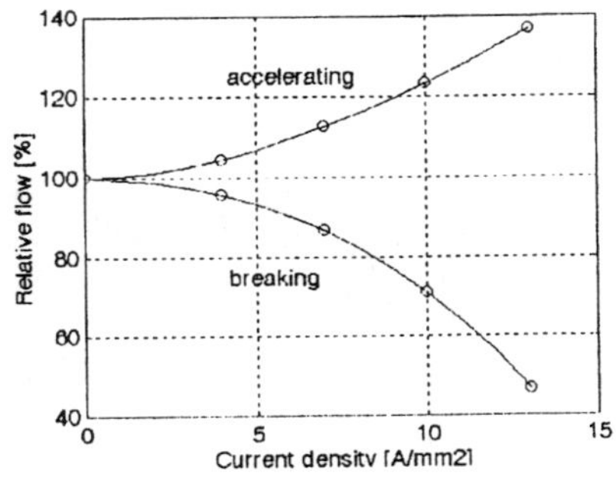

Fig. 11 Relative flow vs. current density

B. Control Flow Characteristics

Using the optimum values of the frequency and of the inductor height the relative flow depending on the current density for the two regimes of the valve was calculated. Figure 11 shows the relative limits of the control flow for a given current density.

C. Considering the Inner Surface State of the Valve Flow Canal

The $k - \varepsilon$ computational model for the turbulent viscosity take into account the friction effort τ_w, which is calculated with the relation (9). The friction effort depends on the rough wall through the C_r coefficient. In Fig. 12 is represented the velocity v_z component field for the case in which $\tau_{w2} = 0{,}1 \cdot \tau_{w1}$ where τ_{w1} is the friction effort with which the solutions from Fig. 7 were obtained. The observation was made that for τ_{w2} the v_z components are greater and the shape of the velocity field in the canal represents a more "flattened" parabolic shape than in Fig. 7a. The turbulent viscosity (Fig. 13) is smaller in the case of the wall with lower rough because the gradient of the velocity v_z component in direction Or is smaller. This situation offers the possibility to adjust the friction coefficient C_r according to the experimental results.

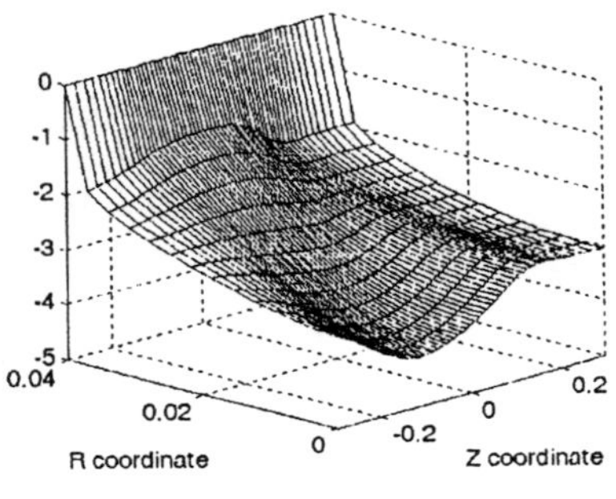

Fig. 12 The axial velocity, v_z (r, z) [m/s].

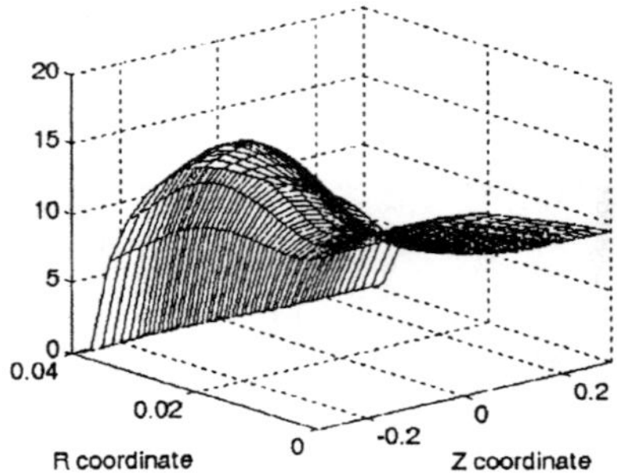

Fig. 13 The turbulent viscosity, $\eta \pm (r, z)$ [Ns/m^2].

References

[1]Peşteanu, O., *Simulation of Turbulent Molten Metal Flow in Electromagnetic Field*, VDI Verlag GmbH, Dusseldorf, Germany, 1994.

[2]Racoveanu, N., Dodescuşi, Gh., Mincu, I., *Metode Numerice Pentru Ecuaţii cu Derivate Parţiale de Tip Parabolic,* Editura Tehnică, Bucharest, 1977.

[3]El-Kaddah, N., and Szekely, J., "The Turbulent Recirculating Flow Field in a Coreless Induction Furnace, a Comparison of Theoretical Predictions with Measurements," *Journal of Fluid Mechanics,* Vol. 133, 1983, pp. 37–46.

VII. Solidification

Chapter 38

Effects of Thermoelectric Magnetohydrodynamics on the Solidification of a CuAg Alloy

P. Lehmann*
DTA/CEREM/DEM/SES, Grenoble, 38054, France
R. Moreau†
EPM-Madylam (CNRS-INP Grenoble), St. Martin d'Hères, 38402, France
D. Camel‡
DTA/CEREM/DEM/SES, Grenoble, 38054, France
and
R. Bolcato§
EPM-Madylam (CNRS-INP Grenoble), St Martin d'Hères, 38402, France

I. Introduction

THE effect of a steady-state magnetic field on solidification is not always well understood. When the interface between the solid and the melt is planar (Bridgman crystal growth), recent studies[1,2] have shown that the magnetic field damps the natural thermal convection in the liquid bulk. A measurement of the level of convection is given by the effective partition coefficient k_{eff} ($k_{eff} = C_S/C_0$, C_S and C_0 being respectively the concentration in the formed solid and in the liquid bulk): k_{eff} tends to 1 when the regime is purely diffusive and decreases to the value of the thermodynamic partition coefficient k under pure convective mass transport. The measured variations of k_{eff} with the magnetic field agrees with theory.

When the solid-melt interface becomes dendritic, magnetohydrodynamic (MHD) damping is still efficient in the bulk liquid. But unexpected experimental behaviors have been observed, depending on the composition

*Commissariat à l'Energie Atomique, Centre d'Etudes Nucléaires de Grenoble, cedex 9.
†ENSHMG BP 95.
‡Commissariat à l'Energie Atomique, Centre d'Etudes Nucleéaires de Grenoble, cedex 9.
§ENSHMG BP 95.

of the alloy and the experimental conditions. Youdelis et al.[3] applied a 3.4 T transverse field on a directionally solidifying A1 4% mass Cu alloy. The value of the effective partition coefficient decreased with the field just as if the magnetic field enhanced (and no more braked) mass transport in the bulk liquid. More recently, Tewari et al.[4] solidified a tin-lead alloy vertically under solutally destabilizing conditions. In the dendritic regime, neither primary spacing between the dendrites nor effective partition coefficient were affected by the field. In the cellular regime, zones depleted from cells could be observed on the metallographic images in a cross section. The last experiments were made by Alboussière et al.[5] and Laskar.[6] They suppose that these effects are due to the interaction between MHD and thermoelectric effects. Their results agree qualitatively with this assumption, but no quantitative comparison can be made with the analytical and numerical models.

The purpose of this paper is to present experimental results obtained in a configuration that allows one to check the available models. We first briefly review the thermoelectric effects occurring in a dendritic network. We then compare the motion because of the thermoelectric effects with MHD braking in the dendritic mushy zone. Our experimental device is presented, the experimental results are discussed, and we are shown that we are now able to predict to the leading order the effects of the magnetic field on natural convection in the dendritic network.

II. Thermoelectric Effects Around a Dendrite

In this part, we focus on the Seebeck effect, which is essentially known for its application to thermocouples. The other thermoelectric effects (Peltier, Thomson) have been shown to have negligible effects on solidification.[6] If two media of different thermoelectric powers (η_S,η_l) are connected together and if a difference of temperature ΔT is applied between the two junctions, then a voltage $\Delta V = (\eta_s - \eta_l)\,\Delta T$ is measured (Fig. 1a). This property can also be expressed by the modified Ohm's law derived from thermodynamics of irreversible processes[7]:

$$\frac{\boldsymbol{j}}{\sigma} = -\nabla V - \eta \nabla T \tag{1}$$

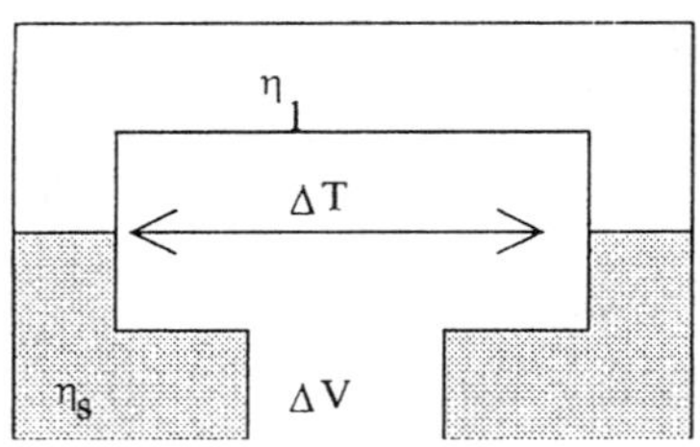

Fig. 1a A thermocouple.

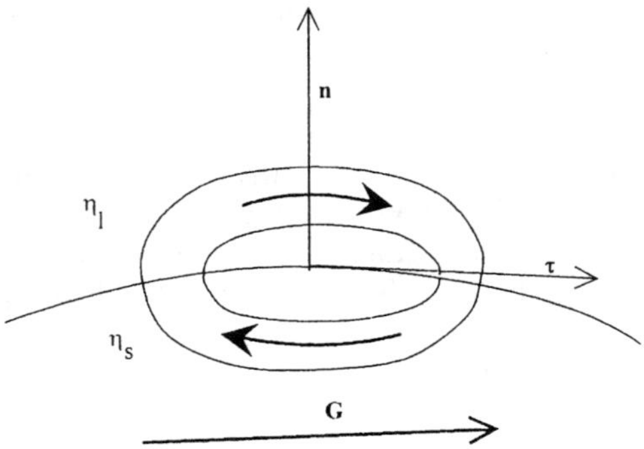

Fig. 1b An elementary portion of the solid liquid interface.

V being the electric potential, σ the electrical conductivity, and $\boldsymbol{j}$ the current density. A voltage may thus appear even if no current circulates. Consider now an elementary part of a solid-liquid interface submitted to a temperature gradient (Fig. 1b). Provided the solid and liquid phases show a difference of thermoelectric power, each part of the solid-melt interface can be looked upon as an elementary thermocouple. But, this thermocouple is now a shortcut. So, the previous measured voltage now generates a current, and each portion of the interface behaves as an elementary current generator. Calculating the circulation of Eq. (1) around this elementary contour yields[8]:

$$\left(\frac{\boldsymbol{j}_s}{\sigma_s} - \frac{\boldsymbol{j}_l}{\sigma_l}\right)\cdot\tau = -(\eta_s - \eta_l)\cdot\nabla T\cdot\tau \tag{2}$$

where τ is the tangent vector to the interface, and s and l subscripts stand for solid and liquid. The curl of Eq. (1) can be written

$$\nabla \times \frac{\boldsymbol{j}}{\sigma} = \nabla T \times \nabla\eta \tag{3}$$

If each media is assumed to be homogeneous, then the thermoelectric power depends only on the temperature; $\nabla \times \boldsymbol{j}/\sigma$ is zero, and the current derives from a potential. If the thermoelectric power depends weakly on the temperature, which is usually the case in the considered range of temperature (around 100 K), this potential is expressed as

$$\Phi = V + \eta T \tag{4}$$

The divergence of the current must remain zero, so that the thermoelectric problem reduces to solve a Laplace equation

$$\Delta\Phi = 0 \tag{5}$$

with the boundary condition (2) and the continuity condition of the component of the current normal to the interface.

Around an idealized dendrite, several solutions have been found both analytically[9] or numerically.[6] They have shown that the currents circulate in the same direction as the temperature gradient. No current crosses the solid-melt interface except at the tip and at the bottom of the structure. An elementary solution exists where Φ is a linear function of z (uniform currents).[5] Consider two media, infinite along the z direction, alternating along the x direction and submitted to a temperature gradient along the z direction. The current in the liquid phase is given by

$$\boldsymbol{J}_{\mathrm{TE}} = \frac{\sigma_s \sigma_l f_s(\eta_s - \eta_l)}{\sigma_s f_s + \sigma_l f_l} \nabla T \approx \frac{\sigma_s \sigma_l^2}{(\sigma_s - \sigma_l)^2} f_s(\eta_s - \eta_l) \nabla T \tag{6}$$

where f_s and f_l are respectively the solid and liquid volume fractions. This solution has been shown to be a good approximation for the value of the currents along a cell or a dendrite by comparison with numerical calculations[6] and with experimental voltage measurements on a solidifying interface.[9] The order of magnitude of the currents in our solidifying AgCu alloy is about $10^5\ \mathrm{A \cdot m^{-2}}$ (Table 1). We will use this approximation in the following discussion.

III. Thermoelectric Magnetohydrodynamic and Magnetohydrodynamic Damping

If now a magnetic field $\boldsymbol{B}$ is applied on the solidifying solid-melt interface, a Lorentz force $\boldsymbol{j}_{\mathrm{TE}} \times \boldsymbol{B}$ appears in the dendritic network. This force induces a new movement in the mushy zone, which has to be compared to the corresponding MHD braking effect. Laskar[6] calculated numerically the thermoelectric magnetohydrodynamic (TEMHD) motion around a single idealized dendrite, i.e., at the micro-scale. But experimental observations as average liquid fraction or eutectic zones depleted from dendrites (freckles) are made at the scale of several dendrites. So our purpose here is to get a formulation at an averaged meso-scale. Considering the dendritic network as a mushy zone leads to Darcy's law:

$$\frac{\mu \boldsymbol{U}}{K} = -\nabla p - +(\boldsymbol{j}_{\mathrm{TE}} + \boldsymbol{j}_{\mathrm{MHD}}) \times \boldsymbol{B} \tag{7}$$

p being the pressure, U the average velocity in the mush, μ the viscosity, and $\boldsymbol{j}_{\mathrm{MHD}}$ the MHD braking current. Both the electric and hydrodynamic equations are linear, so that each current can be calculated separately. An order of magnitude of the MHD currents is roughly $\boldsymbol{j}_{\mathrm{MHD}} = \sigma_l B U$. The velocity[10] in the mushy zone is about 10^{-5} m/s^{-1}, so that the MHD current would be three orders of magnitude smaller than the thermoelectric current. Thus,

Table 1 Numerical data

ρ_0, kg/m^{-3}	β_o	$\sigma, \Omega^{-1}\mathrm{m}^{-1}$	$\sigma_1, \Omega^{-1}\mathrm{m}^{-1}$	$\eta_l - \eta_s$, VK^{-1}	μ, Pa/s	M_l, K (fraction mass)
8620	0.15	$12.3 \cdot 10^6$	$510 \cdot 10^6$	$7.5 \cdot 10^{-6}$	$4.22 \cdot 10^{-3}$	410

Table 1 (cont.)

C_0 (mass fraction)	k	G, k/m^{-1}	H, mm	l, mm	D, m^2/s^{-1}	Γ, k/m
0.6	0.09	$1.38 \cdot 10^4$	6	2	$3.3 \cdot 10^{-9}$	$1.4 \cdot 10^{-7}$

MHD effects would be definitively negligible at this scale, whereas they are shown to be significant at the micro-scale (see Laskar[6]).

Indeed, even with a moderate magnetic field, the Hartmann layer may become smaller (25 μm at 1 T) than the primary dendritic spacing λ (around 200 μm). However, this Hartmann layer is still thicker than the small scales as secondary or tertiary arms (Fig. 2a). So the interdendritic flow will be controlled by the Hartmann layer. The only possibility to introduce this effect into the meso-scale model is to take into account both viscous friction and the MHD braking at the scale of the pore. This is done by introducing a new value of the permeability. Let us consider an elementary interdendritic channel comparable to a duct flow as sketched on Fig. 2b. The forces on a portion of fluid $S_l \cdot dx$ are such that equilibrium is written:

$$-\mathrm{d}P \cdot S_l - 2 \cdot \tau \cdot h \cdot \mathrm{d}x + \langle j_{\mathrm{MHD}} \rangle B \cdot S_l \cdot \mathrm{d}x = 0 \tag{8}$$

where τ, the friction due to the Hartmann layer δ, and the current density j are given by

$$\tau \approx \mu \frac{\langle u \rangle}{\delta}, \qquad \delta = \left(\sqrt{\frac{\sigma_l}{\mu}} B \right)^{-1} \tag{9}$$

$$\boldsymbol{j} = \sigma\,(\boldsymbol{E} + <\boldsymbol{u}> \times \boldsymbol{B}) \tag{10}$$

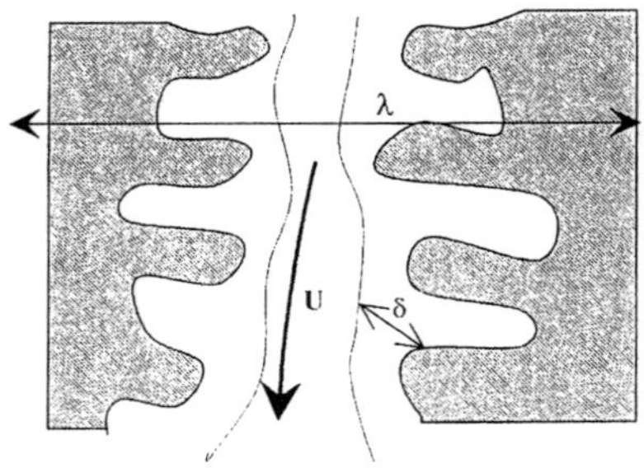

Fig. 2a The Hartmann layer δ in an interdendritic channel.

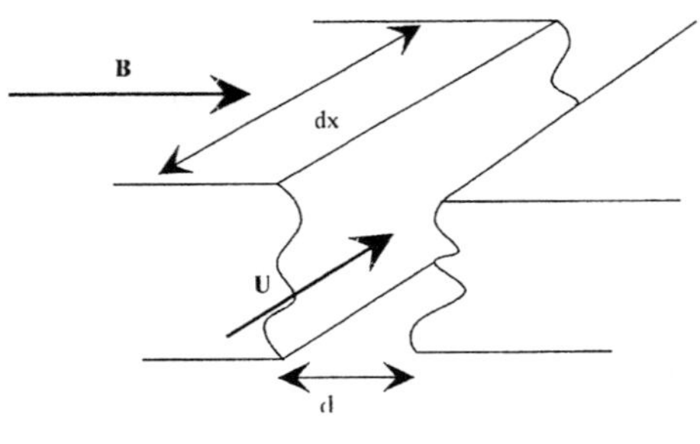

Fig. 2b This channel looked upon as an elementary duct flow.

where $<u>$ is the local average velocity that is related to Darcy's mesoscopic velocity by $U = f_l<u>$. To verify the zero divergence condition of the current, we suppose that the current loops close locally so that $f_s j_s + f_l j_l = 0$. If we define a Hartmann number based on the effective size d of the liquid channel, $Ha = d\delta$ with $f_l = d^2/\lambda^2$, straightforward calculations lead to

$$\nabla p = -\mu \frac{\boldsymbol{U}}{\lambda^2 f_l^2}[2Ha + (1-\bar{\sigma})Ha^2], \qquad \bar{\sigma} = \frac{f_l\sigma_l}{f_l\sigma_l + f_s\sigma_s} \tag{11}$$

If we now compare this equation to Eq. (7), the effect of MHD on a porous media is clearly not to create currents at a large scale but to change the value of the permeability. A compilation of permeability measurements on dendritic mushy zones[11] has shown that the permeability for a flow parallel to the primary arms can be approximated by a Hagen–Poiseuille expression: $K_0 = 3.75\ 10^{-4}\lambda^2\ f_l^2$. We thus write the modified permeability

$$K_{\mathrm{MHD}} = \frac{K_0}{2Ha + (1-\bar{\sigma})Ha^2} \tag{12}$$

Even for moderate values of Ha, the permeability can be reduced by an order of magnitude. This regime will only be reached if the Hartmann layer becomes smaller than the interdendritic liquid size. In the case of our AgCu alloy, MHD effects may begin to appear around 0.5 T. Under this value we will only have to deal with a pure thermoelectric Lorentz force. This result agrees with the numerical calculations of Laskar.[5]

IV. Thermoelectric Magnetohydrodynamic Forces Compared to Buoyancy Forces

To demonstrate the thermoelectric effect, Alboussière et al.[5] and Laskar[6] solidified directionally a Bi 40 at % Sn and a Cu 55% mass Ag alloy (according to the phase diagram, the solute is the silver; the eutectic concentration being at 71.9% mass). They used a vertical solutally stabilizing configuration with pulling rates sufficiently high to avoid possible thermal residual convection (60 μm/s). The magnetic field (0.6 T) was applied transverse to the solidification direction to generate a movement perpendicular to the solidification direction. On metallographic images, a radial macrosegregation can be observed, and an eutectic zone depleted from dendrites appears on one side of the sample. But it must be pointed out that no correlation can be made with the predicted intensity of the force; moreover, the orientation of the sample versus magnetic field is not easily marked. So, the direction of the flow cannot be controlled.

To avoid this problem, the Laplace force necessarily had to be compared to another well-known force. So we decided to solidify the alloy horizontally with lower magnetic fields. In this case, both thermal and solutal buoyancy forces may be taken into account and the Boussinesq approximation

$$\rho = \rho_0[1 + \beta_T(T - T_0) + \beta_C(C - C_0)] \tag{13}$$

where ρ is the density; and β_T and β_C, are the thermal and solutal expansion coefficients, are valid. Under the assumption of local thermodynamic equilibrium,[12] the temperature is related to the concentration by the liquidus slope m_l ($T = T_0 + m_l C$) in the mushy zone: β_T is about 10^{-4} and β_C/m_l about 10^{-2}. So thermal convection can be neglected. Thus, the equation of the motion written in low magnetic field approximation is

$$\frac{\mu \boldsymbol{U}}{K} = -\nabla p - \rho_0 \beta_C \frac{G}{ml} x \cdot \boldsymbol{g} + \boldsymbol{j}_{\mathrm{TE}} \times \boldsymbol{B} \tag{14}$$

x being the solidification axis (x is zero at the liquid mush interface), G the temperature gradient (supposed uniform), and m_l the liquidus slope. The transverse magnetic field is applied in such a way that the buoyancy forces (downward) can be opposed by the thermoelectric effect (upward). At higher magnetic fields we expect to inverse the buoyancy driven movement.

V. Experimental Device and Results

We are using a Cu 60% mass Ag, which has both a good electrical conductivity and a high thermoelectric power (Table 1). The samples were solidified in a melting zone furnace with a 6-cm-long liquid zone (Fig. 3). The first reason for this choice was the perfectly horizontal configuration of the experiment. Moreover, after an initial transient about the length of the liquid zone, a permanent regime is reached, and the formed solid is always at the concentration C_0 of the initial rod, the liquid being at C_0/k_{eff} [k_{eff} has been shown to be between 1 and 0.85 (Ref. 10)]. The fact a steady state is always reached facilitates the comparison between the samples.

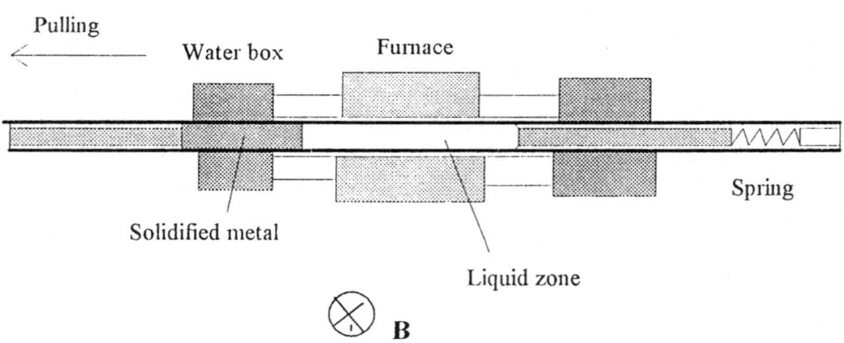

Fig. 3 Experimental device.

The alloy was first prepared from 5N materials in a 5.8 ± 0.05 mm ID quartz crucible and then placed in 6.0 ± 0.05 mm ID crucible to allow free movement of the sample. A compensation device with a spring ensured that the liquid alloy filled perfectly the crucible to avoid any free surface convection phenomena (Marangoni). Two quartz capillaries were embedded in one of the sample: thermocouples were placed inside to measure the thermal gradients. In the mushy zone, it was found to be 138 K/cm ±8 K/cm. The complete liquid zone was placed in the magnetic field. It must be pointed out that with a horizontal transverse magnetic field, the MHD braking in the liquid is negligible.[13]* Magnetic fields up to 0.6 T were applied.

The metallographic images are presented on Fig. 4. On Fig. 4a, at 0 T, the radial segregation is important, and an eutectic zone characteristic from natural convection can be seen at the bottom. On Fig. 4b, the value of the magnetic field is 0.03 T: the structure of the dendritic network is distorted with large eutectic zones in the middle. A longitudinal metallographic image shows that no steady state is reached and that the movement must be swirling within the mushy zone. This is the case when the thermoelectric and buoyancy forces cannot balance each other. At 0.05 T the primary spacing is drastically modified. This change is characteristic of a very low convection level.[10] The primary spacing in the central part of the section is measured around 650 μm whereas the theoretical value under a pure diffusive regime given by the Trivedi formula (Appendix A) is found at 700 μm. This value of the field is the critical value where the two forces equilibrate. At 0.2 T (Fig. 4d) the convection movement is reversed, and an eutectic zone appears at the top of the sample. Up to 0.6 T, the eutectic zone is still larger at the top of the sample which means that a new flow regime controlled by $\boldsymbol{j}_{\text{TE}} \times \boldsymbol{B}$ remains established with an upward vertical velocity component. We have plotted the variation of the primary spacing measured in the central part of the sample with the magnetic field (Fig. 4e). It can be noticed that excepted around the critical field, it remains nearly at the same value. But at higher fields, it increases slowly again, which shows MHD braking begins to become efficient.

VI. Critical Value of the Magnetic Field

An estimate of the critical value of the magnetic field can be derived from an order-of-magnitude analysis. To gain some insight into the nature of the flow, we solved Eq. (14) supposing that the permeability is constant in the mushy zone. This can seem a drastic assumption as the permeability is known to change by several orders of magnitude through the dendritic network (theoretically from zero in the solid to infinity in the liquid). However, it has been shown[10] that the movement is restricted to a zone where the permeability is not too low; elsewhere, the fluid can be assumed to be at rest. We can thus suppose the permeability is uniform, but that the extent of the flow is limited to a part of the mushy zone. Moreover, the statement can be made that the permeability brakes equally the buoyancy movement and the thermoelectric movement, so that such a modeling seems well adapted to the comparison

*If the movement was perfectly two-dimensional, it would be zero.

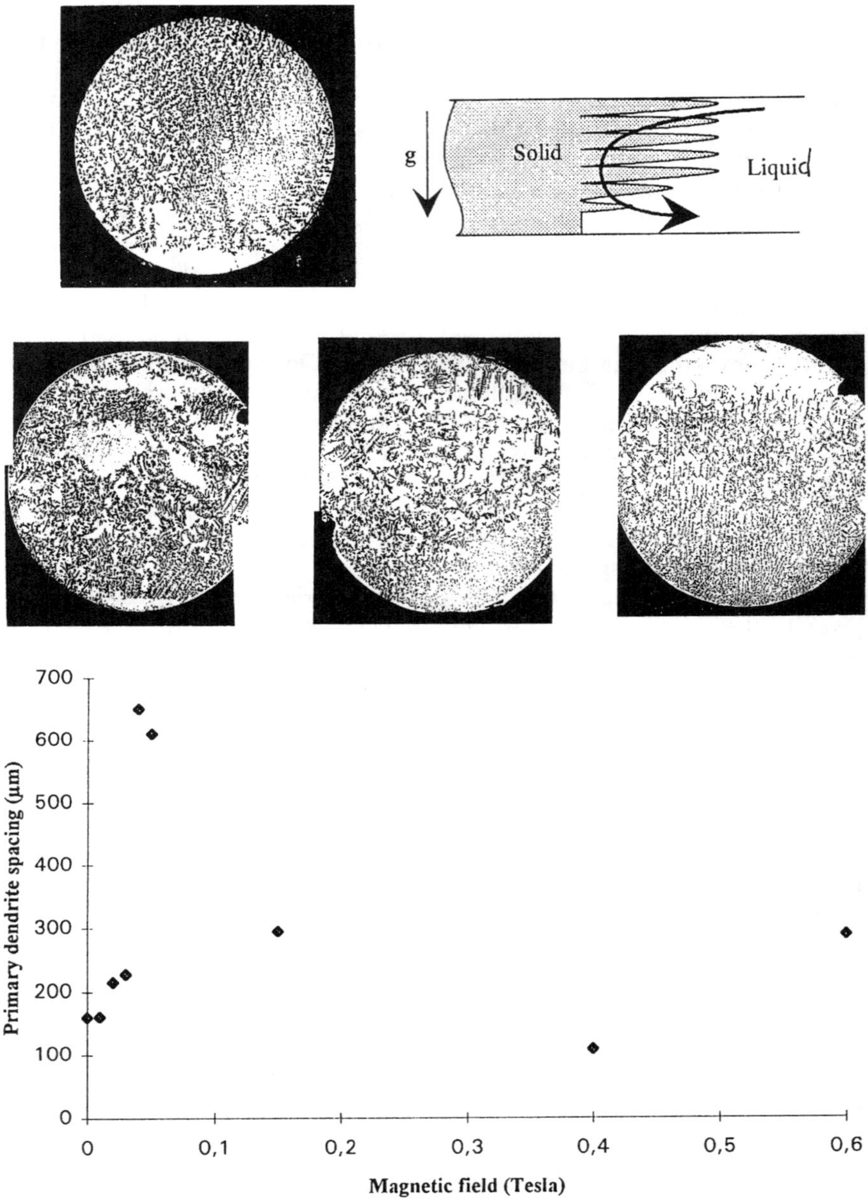

Fig. 4 Experimental results.

a) No field—the flow is shown on the left (downward)

b) B = 0.03T—a distorted morphology of the mushy zone

c) B = 0.5T—buoyancy and Lorentz forces equilibrate each other

d) B = 0.2T—the flow is reversed (upward)

e) Variations of the primary spacing with the magnetic field

between the two forces. In our AgCu alloy, quenched interfaces showed the liquid fraction was around 0.9 even in the bottom of the dendrites. So we considered the flow to circulate in the whole mushy zone.

Two calculations with different assumptions have been made. In the first one we suppose the thermoelectric current to be uniform over the mushy zone and zero outside as done in Ref. 8 with cells. If we introduce the stream function Ψ, the curl of Eq. (14) yields (Fig. 5a):

$$\Delta\Psi = \frac{K}{\mu}\left[\rho_0\beta_C g\frac{G}{m_l} - j_{\mathrm{TE}}B\delta(x)\right] \tag{15}$$

where $\delta(x)$ is the Dirac function as the curl of the constant thermoelectric Lorentz force is zero except at the liquid-mush boundary. The boundary conditions at the solid-mush boundary are obvious no-normal flow conditions. Let us now consider the liquid-mush interface. In the liquid bulk no forces exist. Because of the very low velocities, the movement is purely viscous and has no other role than allowing the continuity equation to be verified. The pressure gradient in the mushy zone is about the order of magnitude of $\mu U/K$ whereas in the liquid bulk it is $\mu U/H^2$ with H being the height of the crucible. The fraction K/H^2 is about 10^{-5}. We must then assume that at the liquid-mush boundary, the pressure gradient is zero, i.e., $p = 0$. The solution is found as an expansion of orthogonal functions:

$$\Psi = \frac{K\rho_0\beta_C gGH^2}{\mu m_l}\left[\frac{1}{2}z(z-1) + \frac{4}{\pi^3}\Omega_1(x,z) - F\frac{l^2}{H^2}\cdot\left(x + \frac{8}{\pi^2}\Omega_2(x,z)\right)\right] \tag{16}$$

where the z dimension is scaled by the width H and the x dimension by the depth l (the functions Ω_1 and Ω_2 are given in Appendix B). The important dimensionless parameter that evaluates the ratio between the thermoelectric and buoyancy forces within the mushy zone is F:

$$F = \frac{\sigma_s\sigma_l(\eta_s - \eta_l)\langle f_s\rangle m_l B}{(\sigma_s\langle f_s\rangle + \sigma_l\langle f_l\rangle)\rho_0\beta_C g\cdot l} \tag{17}$$

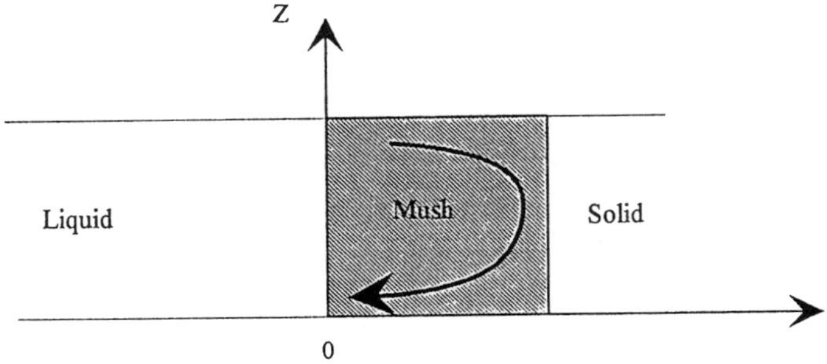

Fig. 5a The mushy zone.

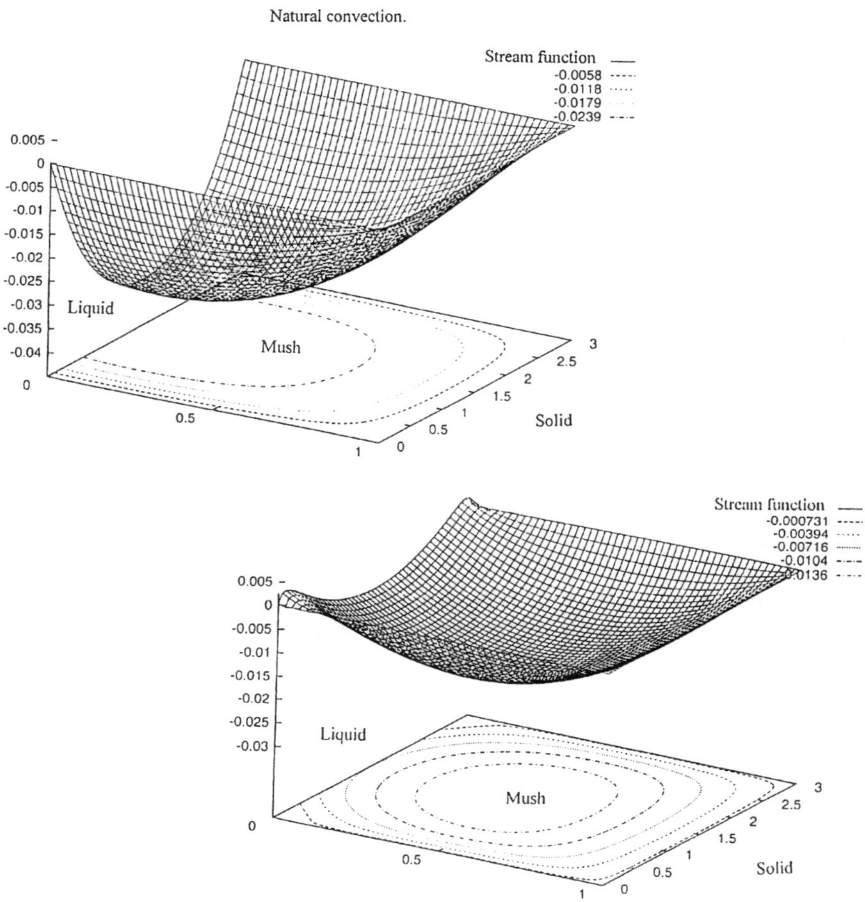

Fig. 5b The streamlines and the stream function for different values of *F*.

We have plotted the Ψ function and the streamlines for several values of F in the case of our AgCu alloy (Fig. 5b). At $F = 0.3$ a swirl appears in the mushy zone, which is consistent with the metallographic image (Fig. 4b). At $F = 0.5$ equilibrium is approximately reached. This model is interesting because it shows the experimental flow patterns. But the critical field is found at 4 mT instead of 40 mT. So we conclude that a constant value of the thermoelectric current overestimates the real effect.

Let us now assume that the value of the thermoelectric current depends on the solid fraction. Actually, the solid fraction is modified by the intensity of the convection. But, near the equilibrium, the solute transport regime is little affected by convection: the typical convection velocity is negligible compared to the pulling rate. So to the leading order, we suppose the liquid fraction follows a diffusive solute distribution law.[12]

$$f_l = \left(1 + \frac{Gx}{m_l C_0}\right)^{-1/1-k} \Rightarrow f_s \approx \frac{1}{1-k} \cdot \frac{Gx}{m_l C_0} \tag{18}$$

Under this assumption, the thermoelectric Lorentz force depends to the leading order on the x position as well as the buoyancy forces, and the equilibrium gives

$$\frac{\sigma_s \sigma_l^2 G(\eta_s - \eta_l) B}{(\sigma_s - \sigma_l)^2 (1-k) C_0 \rho_0 \beta_C g} \tag{19}$$

The predicted value of the critical magnetic field is now about 15 mT, which is consistent with the experimental value. But the situation observed in Fig. 4b cannot anymore be reached. It is clear experimentally that Lorentz and buoyancy forces never equilibrate completely. Only a complete (but quite complex) modeling of the problem taking into account the coupling between the permeability and the flow could give more accurate solutions.

VII. Conclusion

A local current density exists within the dendritic network because of the Seebeck effect. The value of these currents has been theoretically estimated. When a magnetic field is applied, a Lorentz force is created in the mushy zone. This force can brake buoyancy driven convection or even reverse the movement. However, MHD braking has to be taken into account as soon as the Hartmann layer becomes significantly smaller than the primary spacing: MHD effects are then modeled by a modification of permeability.

The value of the magnetic field necessary to equilibrate buoyancy and thermoelectric Lorentz forces has been assessed under two assumptions. It is shown that the variation of the thermoelectric current with the solid fraction has to be taken into account. But some simple calculations agree with the experimental observations. From a more practical point of view, TEMHD may used to reduce radial segregation or perhaps to enhance it: the solute could then be pushed in a part of the sample, and the material would be purified.

Appendix A

The Trivedi formula for interdendritic spacing under pure diffusive mass transport is

$$\lambda = 6.5\Gamma^{0.25} D^{0.25} [m_l C_0(1-k)]^{0.25} G^{-0.5} R^{-0.25}$$

with D being the diffusion coefficient of the solute, Γ the capillary parameter, and R the pulling rate.

Appendix B

The orthogonal functions $\Omega_1(x,z)$ and $\Omega_2(x,z)$ are

$$\Omega_1(x,z) = \sum_{n=0}^{\infty} \frac{\cosh\left[(2n+1)\frac{\pi x}{e}\right]\sin[(2n+1)\pi z]}{(2n+1)^3\cosh\left[(2n+1)\frac{\pi}{e}\right]}$$

$$\Omega_2(x,z) = \sum_{n=0}^{\infty} \frac{\cosh\left[(2n+1)\frac{\pi e}{2}\left(z-\frac{1}{2}\right)\right]\cos\left[(2n+1)\frac{\pi x}{2}\right]}{(2n+1)^2\cosh\left[(2n+1)\frac{\pi \varepsilon}{4}\right]}$$

$$e = \frac{H}{l}$$

Acknowledgments

The present work was conducted within the frame of the GRAMME agreement between the CNES and the CEA. The authors are very grateful to Dr. Alboussière for his advice and for the time he spent on fruitful discussions on the topic.

References

[1]Neubrand, A. C., Ph.D. Thesis, INPG, France, 1995.

[2]Alboussière, T., Garandet, J. P., and Moreau, R., *Journal of Fluid Mechanics,* Vol. 253, 1993, pp. 545–563.

[3]Youdelis, W. V., and Dorward, D. C., *Canadian Journal of Physics,* Vol. 44, 1966, pp. 139–150.

[4]Tewari, S. N., Shah, R., and Song, H., *Met. Transactions A,* Vol. 25A, 1994, pp. 1535–1544.

[5]Moreau, R., and Camel, D., *C. R. Academy of Science Paris,* Ser. II, 1991, pp. 479–489.

[6]Laskar, O., Ph.D. Thesis, INPG, France, 1994.

[7]Callen, H. B., *Physical Review,* Vol. 13, No. 11, 1948, pp. 1349.

[8]Shercliff, J. A., *Journal of Fluid Mechanics,* Vol. 91, No. 2, 1979, pp. 231–251.

[9]Favier, J. J., Lehmann, P., Drevet, B., Garandet, J. P., Camel, D., and Coriell, S., *Lecture Notes in Physics,* Vol. 464, 1996, pp. 77–94.

[10]Dupouy, M. D., Camel, D., and Favier, J. J., *Acta Metallurgica,* Vol. 37, 1989, pp. 1143–1157.

[11]Poirer, D. R., *Met. Transactions B,* Vol. 18B, 1987, pp. 245–255.

[12]Flemings, M. C., *Solidification Processing,* McGraw–Hill, New York, 1974, pp. 77–83.
[13]Moreau, R., *Magnetohydrodynamics,* Kluwer Academic, Norwell, MA, 1990, pp. 111–136.

Chapter 39

Overcooling as a Possible Cause of Thermoelectric Currents and Magnetic Fields in the Case of Metal Solidification

A. Miķelsons, O. Lielausis, J. Kļaviņš, and J. Šmits
Institute of Physics, Latvian Academy of Sciences, Latvia

IT IS well known that thermoelectric effects take place in electric circuits where the conductors are made of different materials and the temperatures of contacts T_1 and T_2 are different (Fig. 1). In this case emf $E = \alpha(T_1 - T_2)$ arises, where α is the thermal power coefficient. But in a physical sense thermoemf must arise under much more general conditions when in a definite domain the values of the thermoelectric power and of the temperature are not constant and additionally the requirement is fulfilled that grad α is directed orthogonal to grad T. Even in a given material, reasons for α const can be very different—different stages of mechanical tensions, irradiation, etc. At least, the transfer from one physical state to another, in the case under consideration, from the liquid phase to the solid one. It means, at solidification. It is right to expect an essential value of grad α in such a situation, directed perpendicular to the front of the phase change. But, if we are remaining in the frame of the theory of ideal solidification, there are no reasons for thermoemf because along the front of solidification grad $T = 0$.

One can easily confute this argument if the ability of molten metals to overcooling before solidification is taken into account. Under definite conditions depending on the cooling rate, the deepness of overcooling may reach up to 30% of the melting temperature. Because the process is of a very complicated physical character, there is reason to expect that the degree of overcooling will never be constant over the whole domain under consideration which means that grad T can be oriented in different unexpected directions.

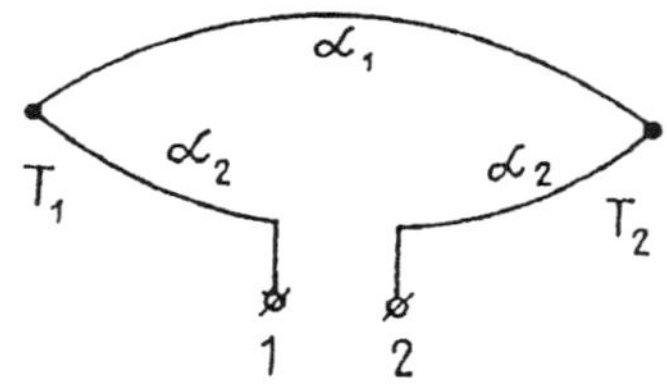

Fig. 1 Thermoelectric circuit

For instance, let us imagine that the surface of a liquid metal is cooled by a water jet. Directly under the jet, the metal must be overcooled remarkably more deeply compared with the surrounding region. Such an experiment has been carried out with molten gallium. Gallium (melting point 29.1°C) has been melted in a cylindrical crucible made of stainless steel with enamel inside the surface to avoid electrical contacts between the crucible and the molten metal. The gallium temperature in the crucible at the start of the experiment was in all cases equal to 30°C (Fig. 2a). A magnetic field sensor was placed outside the crucible in the middle of its height 2 mm from the molten metal surface. The crucible diameter was equal to 60 mm. The gallium temperature was measured in two points: at the crucible bottom (T_2) and near the cooling jet (T_1).

The procedure of the experiment was simple. At the starting moment a definite amount of water at temperature 10°C was poured onto the surface of the molten metal. To realize a more intense cooling in a localized region, a syringe was installed in the occupied by the water volume. By means of discharging this syringe, a jet could be directed toward the surface of the liquid metal. Then the syringe was filled again, and the procedure of discharging was repeated. Corresponding changes of temperature were fixed by the mentioned thermocouples placed at the bottom of the vessel and close to the cooling jet. Measurements in both these points give evidence of an essential overcooling. Let us remember that the temperature of ideal solidification of Ga equals 29.1°C. These 6–8 discharges can be seen on Fig. 2c, too (in the form of small spikes on the general curve). During these discharges, the mean temperature of the water reached 15°C, and the whole amount of the covering liquid was changed on a new portion with the same initial temperature 10°C. The large-scale periodic fluctuations on the curve in Fig. 2c correspond to these procedures of a general replacement of the covering water. Figure 2c shows the dependence of the measured magnetic field on time. A little solid islet formed after the first ($t \sim$ 0–80 s) and second ($t \sim$ 100–180 s) cooling series, but it melted during the exchange of water before the third series. During the third and further impulsive cooling series, a stable, well-observable islet was formed on the gallium surface. Starting with this third series (at $t > 200$ s), sharply increasing curve was recorded by the magnetic field sensor.

The second experiment described already in Ref. 1 was even more simple. To increase the intensity of thermal effects instead of gallium, metallic Bi (melting point 271°C) was used. Specimens of this material were exposed to

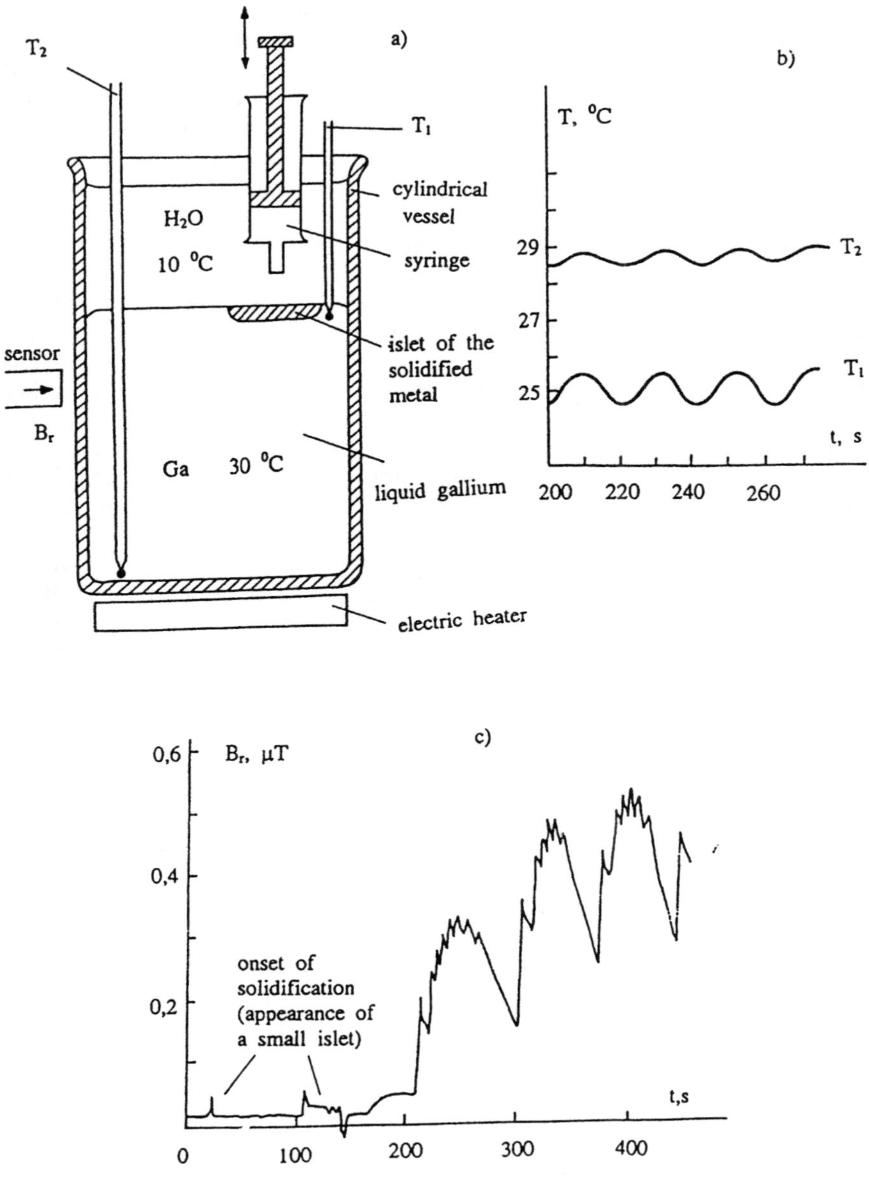

Fig. 2 Experimental facility on measured values

falling $d \sim 3$ mm water drops, evaporating in approximately a second and dissipating approximately 100 cal of energy each. When the temperature of Bi was remarkable above the melting point, the surface remaining liquid and the magnetometer did not fix any field. If the temperature was lowered close to the melting point, each drop caused the formation of a solid islet (diameter ~1 cm, thickness ~0.2 cm). A sharp pike in the field measurements with the length of a few seconds was fixed (Fig. 3). At the same time the islet remained not melted for approximately 10–15 s. This occurrence can be explained because the thermoelectric effects must be more intense in situations when the nonhomogeneity of temperature fields is more expressed.

An attempt was made to demonstrate the role of overcooling in the process of generation of the thermoelectric currents not only by physical, but also by numerical, simulation. As mentioned, the traditional approach to the problem of solidification cannot be used as a base for a corresponding theoretical treatment because along the front of crystallization the temperature is postulated constant and the necessary temperature gradients are absent. At the same time there are no doubts today that solidification of real metals can be connected with a remarkable overcooling. It is enough to allow for a definite overcooling, and the simplest calculating schemes can deliver conditions for thermocurrents. Figure 4 corresponds to the following conditions on the surface of an initially fully liquid metal with the melting point at 271°C and an allowed constant degree of overcooling 30°C. Starting with the moment $t = 0$ inside a circle of radius $r = 5$ mm, the temperature of the surface is abruptly lowered and kept constant at 10°C. A practically flat islet starts to grow at $t = 2$ s (under considerations in Fig. 4) reaching a thickness of 0.36 mm.

The resulting curve represents the gradient of temperature in the radial direction in dependence on r. A line along the "bottom" of the islet at $z = 0.36$ mm is chosen as the most characteristic. The conclusion could be drawn that in the region where the temperature jumps from 240 to 270°C the radial

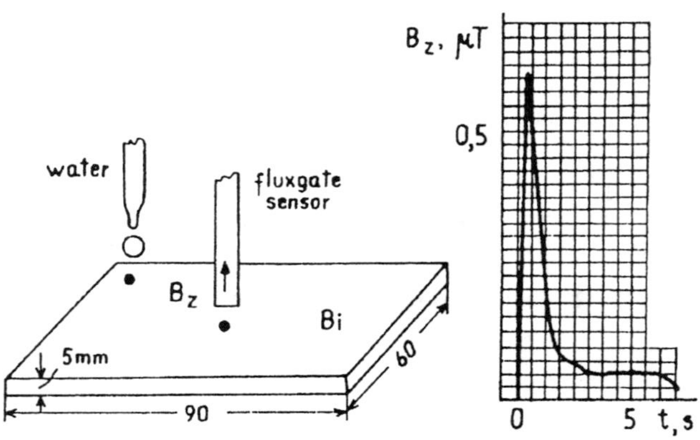

Fig. 3 Magnetic field measured values

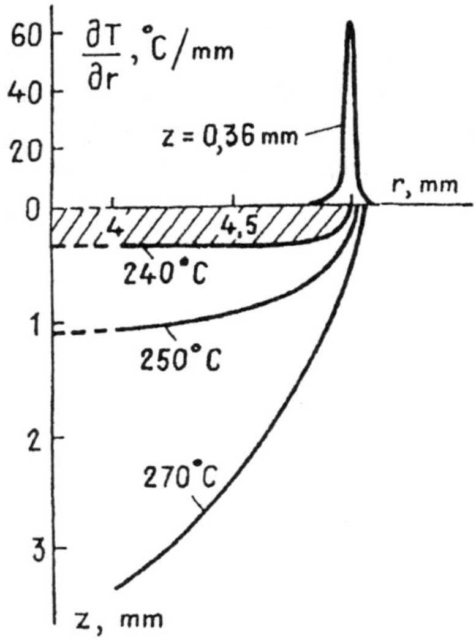

Fig. 4 Temperature gradient values

temperature gradient reaches values of the order of 50°C/mm. The model is very arbitrary. For example, we need values of temperature gradients not along a line in the liquid phase, but along the liquid-solid contact surface. Without essential difficulties the model could be improved to deliver more realistic temperature fields. But here we are meeting a principal contradiction. If we are thinking about some estimates for currents and corresponding magnetic fields, then practically we cannot apply ourselves to the task even with the roughest estimates for the temperature fields. Simply, nothing is known about the second component of the task—values of gradients of thermoelectric power. The high mechanical tensions inside the body of the islet should be taken into account when the question arises: What are the features of the transfer zone from the liquid to the solid state in the case of overcooling? We must conclude that in the case under consideration the possibilities of numerical simulation are limited. It means that at the moment the idea about the role of overcooling can be applied mainly to a phenomenological description of the process. But at least, it removes the mentioned at the beginning principle contradiction, if the task is considered in the traditional statement.

The presented references show that thermoelectric phenomena will remain in the field of interest of Riga researchers for a rather long time.

References

[1]Miķelsons, A., Lielausis, O., Kļaviņš, J., and Šmits, J., "Thermocurrents and Magnetic Fields at Intense Local Surface Cooling of Molten and Solid Metals," *The 14th International Riga Conference on MHD, Book of Abstracts*, 1995, p. 69.

[2]Vītola, V., Kārkliņš, J., Miķelsons, A., and Sjomins, S., "The Study of Thermal e.m.f.

in Nonhomogeneous Media," *Proceedings of the Latvian Academy of Sciences*, Phys.-Techn. series, 1978, Vol. 2, pp. 63–68 (in Russian).

[3]Lielausis, O., Miķelsons, A., Shcherbinin, E., and Gelfgat, Yu., "Electric Currents in Molten Metals and Their Interaction with a Magnetic Field," *Proceedings of the Metallurgical Application of Magnetohydrodynamics Symposium*, Cambridge, 1982, p. 234.

Chapter 40

Effects of Oscillatory Electromagnetic Force Fields on Microstructure of Solidifying Aluminum Alloys

Charles Vivès*

Université d'Avignon, Laboratoire de Magnétohydrodynamique, Avignon, F84000, France

Nomenclature

At = ampere-turn
a = mold width
$\mathbf{B}$ = amplitude of the induced magnetic field
$\mathbf{Eo}$ = rms of the electromotive field generated by u and Bo
$\mathbf{E}$ = amplitude of the externally imposed electric field
$\mathbf{e}$ = amplitude of the alternating fluid motion
$\mathbf{F}$ = electromagnetic body force
fs = solid fraction
$\mathbf{g}$ = gravity
h = mold height
h' = depth
I = amplitude of the electric current intensity
i = variable electric current intensity
$\mathbf{j}$ = conduction current density
$\mathbf{jd}$ = displacement current density
L = mold length
P = amplitude of the oscillatory pressure
t = time
$\mathbf{u}$ = variable velocity
U = amplitude of the variable velocity
γ = acceleration
ε = electric permittivity
μ = magnetic permeability
ρ = density
σ = electric conductivity of the alloy

*Professor, 33 rue Louis Pasteur.

φ = phase angle
ω = current angular frequency

Introduction

OVER the two last decades, owing to their impact on industrial casting processes, an increasing interest has been shown in fundamental and applied investigations bearing on metal solidification either in the presence of free convection or when various dynamic treatments, generating forced convection, are applied in the melt during freezing.

A number of examples can be found in the literature where external forces have been applied to induce fluid flow during solidification to refine grain size. These methods include rotation of the mold, mechanical or electromagnetic stirring of the melt, and rheocasting.[1–7] Under these conditions, grain structures of castings and ingots change from columnar-dendritic to equiaxed dendritic or globular when they are solidified in the presence of a sufficiently intense forced convection, which generally promotes both the evacuation of superheat and the homogenization of the melt temperature. The phenomena occuring during these various treatments are currently rather well understood.

Moreover, several investigators have found that mechanical vibrations of both sonic and ultrasonic character, when applied during the solidification of metals and alloys, modify conventionally obtained macrostructures and microstructures.[8,9] The most commonly observed effect is the suppression of undesirable dendritic and columnar zones and the development of a fine-grained equiaxed structure. In fact, the effects produced when high intensity sonic or ultrasonic waves are propagated through molten metals can be listed under three main headings: grain refinement, dispersive effects, and degassing, resulting in reduced porosity. There appear to be two distinct views regarding the mechanism, which may be explained by the cavitation effects and by the influence of the fluid flow phenomena.

Cavitation is the term used to describe the formation of bubbles, or cavities, in liquid. These cavities may be filled with air, or vapor, or may be almost empty; they can be produced in liquids by the passage of sonic or ultrasonic waves, provided they are of suitable frequency and intensity. Because of the oscillation of the medium, regions of compression and rarefaction are formed. In the rarefaction regions a “negative pressure” (tension) may exist, and air or vapor bubbles can be formed. In most liquid metals a considerable amount of gas is present in the form of very small bubbles, which are, in most circumstances, seeded from pre-existing gas pockets.[10] Liquid vapor may also evaporate into the partial void produced by the sudden expansion of the undissolved gas bubbles. The usefulness of cavitation in processes such as cleaning, dispersion, and grain refinement is largely due to the very high pressure produced locally on the collapse of the cavities. During this collapse time, the walls of the bubble are forced inwards until they impinge on the small nuclei of gas or vapor, contained in the cavity, which is, at the time, severely compressed. The demonstration has been made that the pressure in the bubbles immediately prior to their final collapse can attain several thousand atmospheres.[9,10] Thus, when the bubbles finally disappear, extremely powerful shock waves occur, which are responsible for most of the phenomena brought about by cavitation. In particular, during metals and alloys solidification, the forces associated with cavitation result in the dislocation of growing crystals. This splitting up of crystals effectively produces many more nuclei around which new crystals can form. Thus, in this manner, the crystals never grow beyond a certain size.

The other view is that vibrations also give rise to considerable agitation of the melt and result in the newly formed nuclei being distributed throughout the solidifying pool, so that the crystallization takes place uniformly inside the entire volume. Moreover, the vibrations have much of the same effects as turbulence, dispersing small crystals so that more of them grow, with the result that the grain size is reduced.

The sonic or ultrasonic irradiation of molten metals is mainly carried out with magnetostrictive, or piezoelectric, transducers (Fig. 1). Coupling rods made of quartz, graphite, and various ceramic materials have been used to communicate vibrations to a molten metal, and these materials are attached to the transducer by special cements. However, such a technique presents several disavantages. The oscillating rods are very rapidly dissolved when they are immersed into molten aluminum alloys, and this circumstance provokes an undesirable contamination of the metal. Moreover, the intensity of cavitation is greatest near the transducer or the coupling rod face; thus, the use of such a system is principally justified for the treatment of metal mixtures on a small scale. Furthermore, in view of both the cost and the bulkiness of the equipment, the production of large aluminum alloy ingots by continuous casting seems unrealistic. Therefore, the transmission of sonic or ultrasonic vibra-

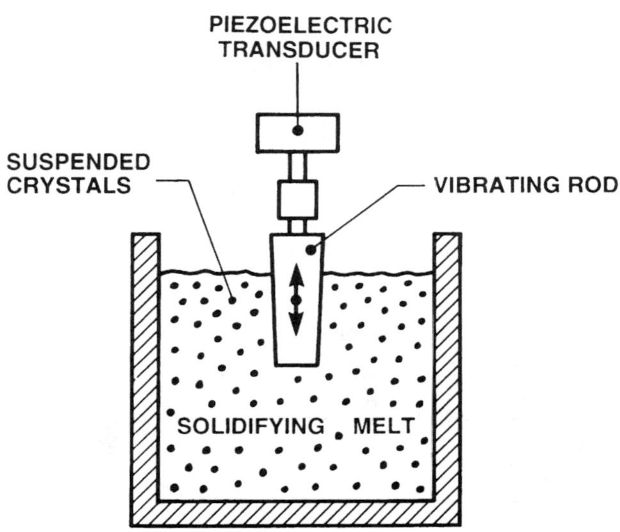

Fig. 1 Schematic sketch of a system for a mechanoacoustic irradiation of melts.

tions to the solidifying melt is not an easy task and largely determines the potential application for improving the microstructures of cast materials.

The main objective of this investigation is to study, very probably for the first time, the effects of forced vibrations, produced by electromagnetic force patterns, during the freezing of representative aluminum alloys. The working principle of an electromagnetic device, able to set solidifying aluminum alloy melts in vibration, is described. The method, which consists of the simultaneous application of an oscillatory electric field and a steady magnetic field, presents the following properties: 1) both the main features of the electromagnetically induced alloy flows, and the amplitude of the pulsating electromagnetic pressure, are well known in each point of the bath, and 2) are characterized by a fairly uniform intensity throughout the whole melt volume. Hence, this technique will lend itself relatively easily to an accurate investigation of the influence of both fluid flow and cavitation on the grain refinement. From the standpoint of a fundamental study, this is a major advantage with respect to the conventional mechanoacoustic processes where fluid flow and cavitation can be very intense near the transducer face, but decrease rapidly with distance from the vibrating rod.

Despite the fact that the working principle of this device may seem very simple, the magnetohydrodynamic phenomena occuring during these tests are rather complex and will not be thoroughly examined here. Consequently, our purpose is only restricted to the presentation of the main characteristics of the fluid flow and electromagnetic phenomena acting efficiently on the solidification processes. Representative examples of macro- and microstructure of electromagnetically irradiated aluminum alloy samples will be exhibited solely in an attempt to follow the evolution of both the size and shape of the grains during solidification with the oscillating flow intensities and the cavitation conditions.

Working Principle

The principle consists of the simultaneous application of a stationary magnetic field $\boldsymbol{B_o}$ and of a sinusoidal electric current of frequency N and of intensity $i = I \sin \omega t$, during the course of solidification of an alloy contained in a parallelepipedic tank. The width, height, and length of the vessel will be represented by a, h, and L, respectively. In our experiment, the externally imposed horizontal electric and magnetic fields are perpendicular (Figs. 2 and 3). The

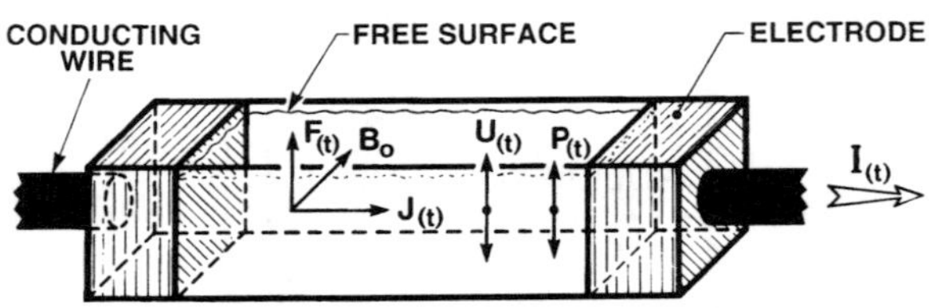

Fig. 2 Principle of production of forced electromagnetic vibration in solidifying melt.

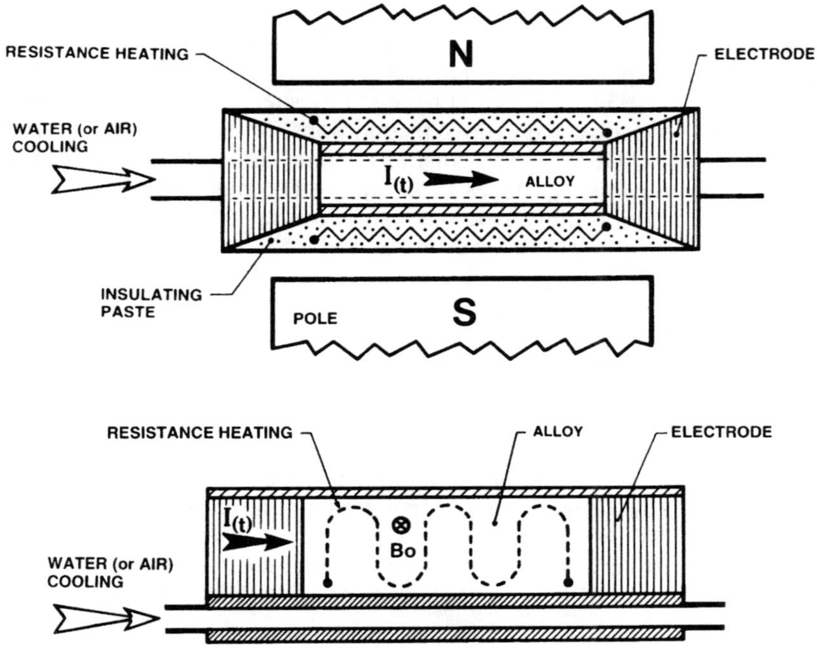

Fig. 3 Schematic sketch of the experimental device.
a) View from above
b) Front view

alternating current density $\boldsymbol{j} = \boldsymbol{J} \sin \omega t$ induced a periodic magnetic field $\boldsymbol{b} = \boldsymbol{B} \sin(\omega t + \varphi)$ of the same frequency, according to the Maxwell equation $\mu o \boldsymbol{j} =$ **Curl** $\boldsymbol{b}$ [on account of the low frequency of the electromagnetic parameters, the displacement current $\boldsymbol{j}d = \varepsilon o\,(\mathrm{d}E/\mathrm{d}t)$ is negligible here[11]]. From the hydrodynamic standpoint, the melt flow may be considered as the superposition of three motions: two alternating flows of N and $2N$ frequency, respectively, and an unstable recirculating flow that may be resolved into a steady component and a random component.

The externally imposed stationary magnetic field $\boldsymbol{Bo}$ interacts with the electric current of density $\boldsymbol{j}$ and gives rise to a vertically directed periodic electromagnetic force $\boldsymbol{j} \times \boldsymbol{Bo}$ of frequency N, which in turn forces the solidifying melt into vibration. The resultant vertical electromagnetic force applied to the whole melt volume is given by $F = BoIL \sin \omega t$ and generates an electromagnetic vibrating pressure $P = (BoIL/a)\sin \omega t$. In the ideal case, where the solidifying alloy is not in contact with any material wall, the whole liquid, or semisolid, mass is set in uniform vibration under the action of the $\boldsymbol{j} \times \boldsymbol{Bo}$ electromagnetic force (namely, the alloy particles vibrate in phase, and the melt behaves as a rigid body). The result is that the main peak parameter values, given by very elementary calculations, are the following:

displacement $e = JBo/\rho\omega^2$), velocity $u = JBo/\rho\omega$, and acceleration $\gamma = JBo/\rho$, where ρ is the melt density. Actually, velocity measurements will show that, under our experimental conditions, the real values of these parameters are reduced about 30%. because of the strong shear stresses caused by the presence of the vertical walls (no-slip conditions), which promote an energy dissipation provoked by the viscous effect and induce a phase difference between the individual particles in motion.[16] The high intensity of turbulence is also an important cause of energy losses. It should be noted that, owing to the presence of the lower horizontal wall of the mold, which is not sufficiently elastic, the negative cycle part of both the displacement and the velocity of an alloy particle cannot exist here.

On the other hand the $\boldsymbol{j} \times \boldsymbol{b}$ force consists of an oscillating component of frequency $2N$ and of a time-independent component. Though the time-smoothed component is mainly irrotational (pinching force), the strength of its rotational part is sufficient enough to engender relatively unstable recirculating flows characterized by rather slow local velocities on the order of a few cm/s^{-1}, whereas the periodic component has a tendency to set the melt vibrating. An order of magnitude calculation readily shows that when $\boldsymbol{Bo}$ is greater than 0.2 T, these two last forces are weak with regard to $\boldsymbol{j} \times \boldsymbol{Bo}$. For example, the ratio of $\boldsymbol{j} \times \boldsymbol{Bo}$ (N frequency) to the oscillating part of $\boldsymbol{j} \times \boldsymbol{b}$ ($2N$ frequency) is on the order of 30 when the strength Bo reaches 0.7 T.

Experimental Apparatus and Procedure

The crucible, whose internal dimensions were 2 × 4 × 12 cm, was made of stainless steel. To prevent spillout of the solidifying melt, particularly as it is vibrated at full electromagnetic power, the pool was hermetically closed by an upper horizontal wall (2 × 12 cm) made of stainless steel and covered by a ceramic insulator. So as to impede untimely alloy freezing, an electric resistance, sunk in an insulating paste, was placed inside the gap formed by the double casing constituting the vertical walls (4 × 12 cm) perpendicular to the magnetic field $\boldsymbol{Bo}$ (Fig. 3). The lower horizontal wall might be internally cooled, either by air or water. Hence, the combination of these adjustable heating and cooling systems enabled regulation, practically at will, of the heat extraction rate and, consequently, of the solidification time duration.

An alternating voltage of 50 Hz frequency was applied between the stainless steel electrodes of 2 × 4 cm cross section permitting the passage of a horizontal sinusoidal current through the solidifying alloy to take place. The current intensity I was modulated at will from 0 to 3500 A (peak value) by electric equipment that included an autotransformer, followed by an intensity transformer. A horizontal, stationary, and nearly uniform magnetic field $\boldsymbol{Bo}$, perpendicular to the current density $\boldsymbol{j}$ and likely to vary within the range 0–0.7 T, was imposed by means of an electromagnet between two cylindrical 16-cm-diam poles (Figs. 2 and 3). This device enabled achieving relatively high amplitudes $\boldsymbol{J}$ of the electric current density, the electromagnetic body force $\boldsymbol{j} \times \boldsymbol{Bo}$, and the electromagnetic pressure P. As an example, for $I = 3500$ A and $Bo = 0.7$ T, the corresponding values of the main periodic parameters were $\boldsymbol{J} = 4.38\ 10^6$ A/m^{-2}, $\boldsymbol{j} \times \boldsymbol{Bo} = 3.06\ 10^6$ N/m^{-3} and $P =$ 122,500 Pa (i.e., 1.21 bar). In the case of the simultaneous application of an magnetic field strength $\boldsymbol{Bo} = 0.7$ T and of an electric current intensity amplitude $I = 3300$ A, the ratio of the peak value of the vibrating electromagnetic force to the gravity force applied to the melt is on the order of 130!

Various aluminum alloys were first melted using an electric furnace and then poured into the preheated mold in the presence of the externally imposed alternating current i, which contributed efficiently to the melt heating. Depending on the experiments, the vessel was filled either partially, or completely. When a melt superheat of 10°C above the liquidus temperature was reached, the resistance heating was cut off so that first the molten metal and then the liquid-solid mixture could cool relatively slowly (cooling rate on the order of 4°C/nm^{-1}), while the magnetic field $\boldsymbol{Bo}$ was applied simultaneously. At the completion of the fluid consistency controlled from velocity measurements, the water cooling was started to stop the crystal growth and to solidify the remaining liquid phase rather rapidly.

Experiments, consisting in simultaneous measurements of local velocity $\boldsymbol{u}$ and temperature, were carried out during the freezing of the vibrating liquid-solid mixture. To this end, a 0.25-nm-diam thermocouple was fastened to a potential sensor[20] connected to a microprocessor voltmeter. This specific velocity measurements technique has already been described.[12] Accordingly, with this arrangement the possibility existed to simultaneously follow, from recordings, the evolution of the local velocity with the temperature (or the solid fraction) within the solidying mixture.

Representative cases of solidification were examined through the experimentation of aluminum alloys characterized by either wide, or narrow, freezing range. The experimental base materials were an A 356 aluminum alloy (Al, 92.2%; Si, 7.2%; Fe, 0.15%; Mg, 0.34%; Ti, 0.13%), a 2024 aluminum alloy (Al, 92.2%; Cu, 4.2%; Mg, 1.6%; Si, 0.5%; Fe, 0.5%; Mn, 0.7%; Ti, 0.15%; other impurities, 0.15%), and a 1050 aluminum alloy (Al, 99.5% and traces of Si, Fe, Cu, Mn, Mg, Zn and Ti, liquidus: 664°C, solidus: 635°C). Several samples were extracted from each irradiated alloy ingot and then were revealed by means of specific etchants.

Experimental Results and Discussion

In these experiments the pressure in the solidifying liquid metal is the sum of static and oscillating terms (Fig. 4). Actually, the metallostatic pressure pgh', at depth h', and the fluctuating dynamic pressure $1/2\ \rho u^2$ can be neglected here. Accordingly, if the electromagnetic pressure amplitude is greater than the atmospheric pressure, then the pressure in the liquid will be negative (tension) for part of each period [Fig. 4(d)].

On the other hand velocity measurements showed that the intensity of the fluctuating flow is relatively intense (Fig. 5). When the electromagnetic pressure is optimal, the shear rate du/da reaches 3.10^2 s^{-1}; namely, a value comparable to the magnitude order obtained in the rheocasting operation.[6,7] Moreover, the trend of the crest velocity-fraction solid curves is quite similar to that already plotted during the solidification of thixotropic alloy slur-

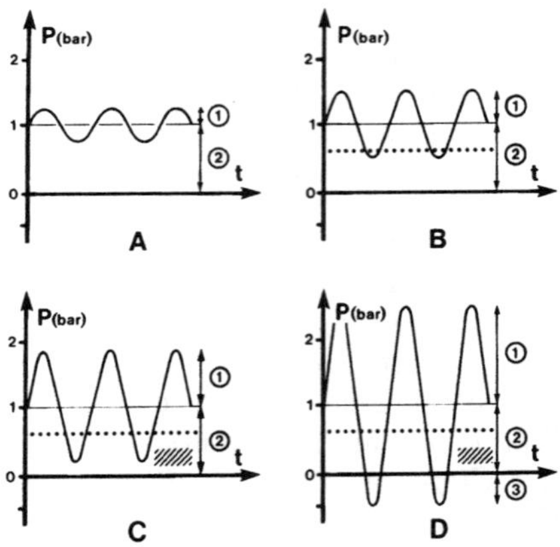

Fig. 4 Influence of the vibrating electromagnetic pressure amplitude on the grain refinement: a) coarse refinement (P = 0.25 bar), b) refinement due to shear rate effects, (P = 0.52 bar), c) simultaneous influence of shear rate and cavitation (P = 0.84 bar) and d) predominance of cavitation effects producing a fine microstructure (P = 1.40 bar); 1) electromagnetic pressure amplitude, 2) static pressure, and 3) tension generated in liquid (negative pressure); and the dotted line corresponds to the onset of a significant grain refinement provoked by the oscillating flow and the hachured area to the critical threshold zone of cavitation.

ries.[6,7] It appears that during the irradiation of the semisolid alloy the apparent viscosity remains practically unchanged up to a solid fraction on the order of 0.5, whereas the velocity decreases dramatically when the solid fraction exceeds 0.55.

The production of cavitation in liquid metals strongly depends on the most volatile undissolved gas content of the liquid, and it has been established that, in the case of aluminum melts, the hydrogen content is the controlling factor.[9–13] The solubility of hydrogen in aluminum is dependent on both the partial pressure of gas and the melt temperature. At a constant

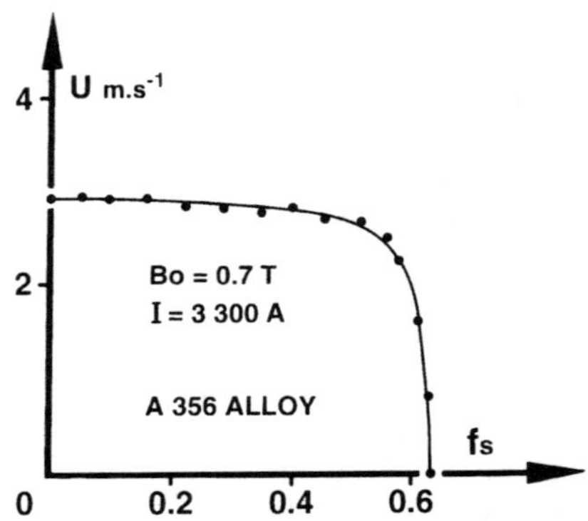

Fig. 5 Example of evolution of the oscillating velocity peak with the solid fraction.

temperature, the equilibrium concentration of gas in solution is proportional to the square root of the partial pressure.[13] For example, at a melt temperature of 650°C, the hydrogen level is on the order of 0.3 ppm and the corresponding equilibrium pressure is 0.29 bar.[13]

Cavitation takes place with full efficiency during the negative pressure part of a cycle, or a series of cycles (Fig. 4d), and nucleation follows either as a result of the modification in equilibrium temperature caused by the pressure change during the collapse of the cavitation bubbles or as a result of cooling the surface of the bubbles by evaporation during its growth. Under this circumstance, cavitation may occur in several points inside the liquid and on the walls mold.

The micrograph, presented in Fig. 6a, exhibits typical columnar-dendritic structures and corresponds to the as-cast A356 aluminum alloy. Figures 6b–6f relate to the alloy solidification in the presence of increasing amplitudes of forced electromagnetic vibrations. Examination of Fig. 6b, which is connected with the electromagnetic pressure peak displayed in Fig. 4a, namely 0.25 bar, reveals the presence of both coarse dendritic fragments and big clusters. The fluid flow intensity here (velocity peak on the order of 60 cm/s^{-1}) appears to be sufficient to prevent the formation of an interlocked

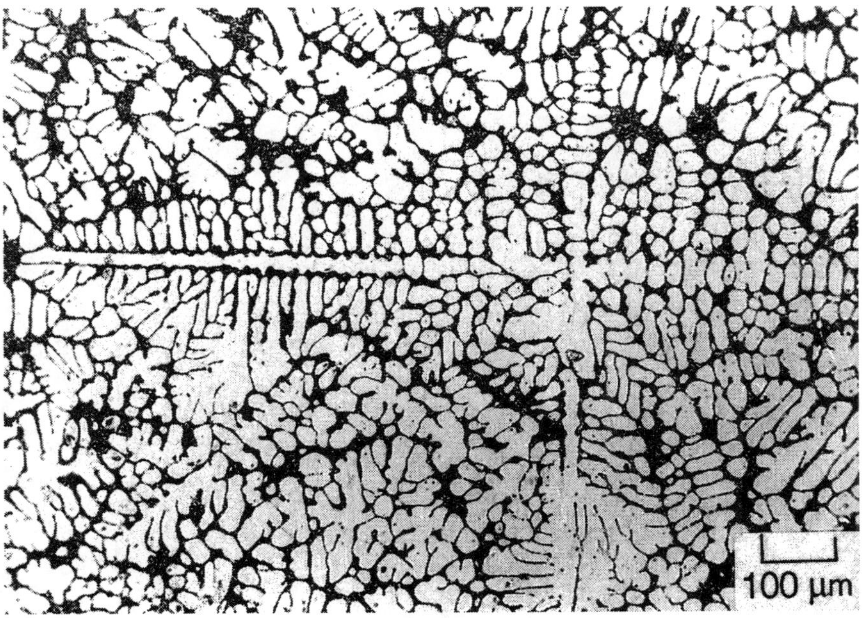

Fig. 6 Microstructures of an A 356 aluminum alloy: a) as-cast material; b)–f) electromagnetically irradiated material where b) *P* = 0.3 bar, c) *P* = 0.52 bar, d) *P* = 0.35 bar (applied during 10 nm), followed by *P* = 1.16 bar (applied during 15 s), e) *P* = 1.16 bar (applied during 12 nm), f) *P* = 0.35 bar (applied during 8 nm), followed by *P* = 1.16 bar (applied during 2 nm).

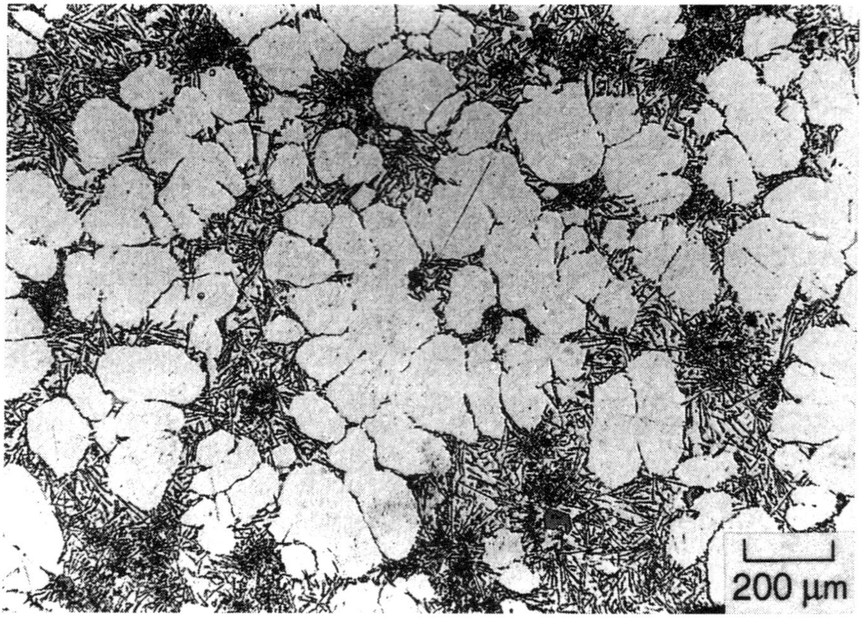

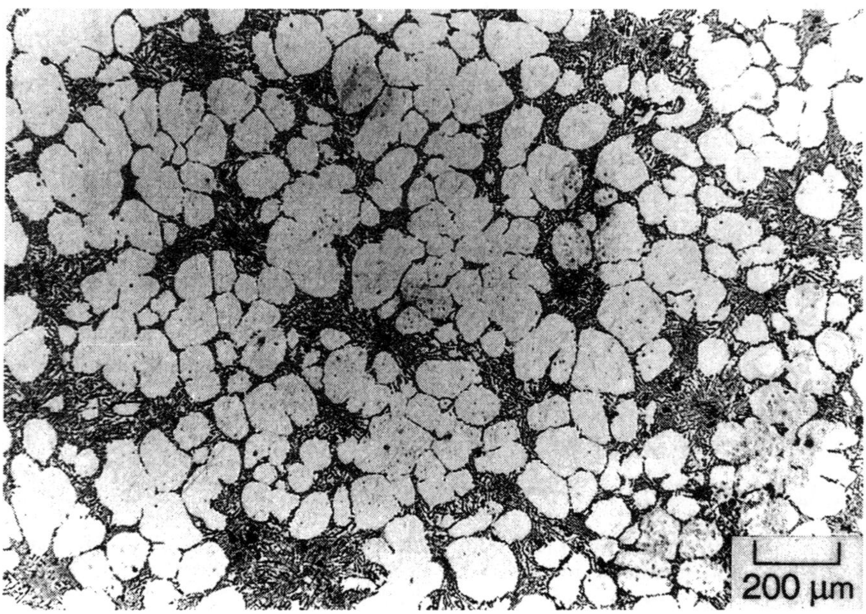

Fig. 6 (cont.)

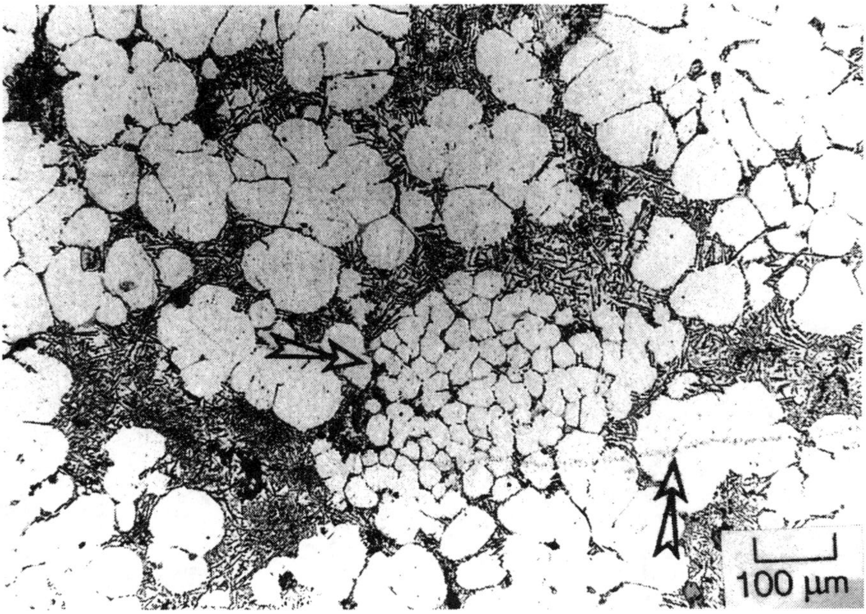

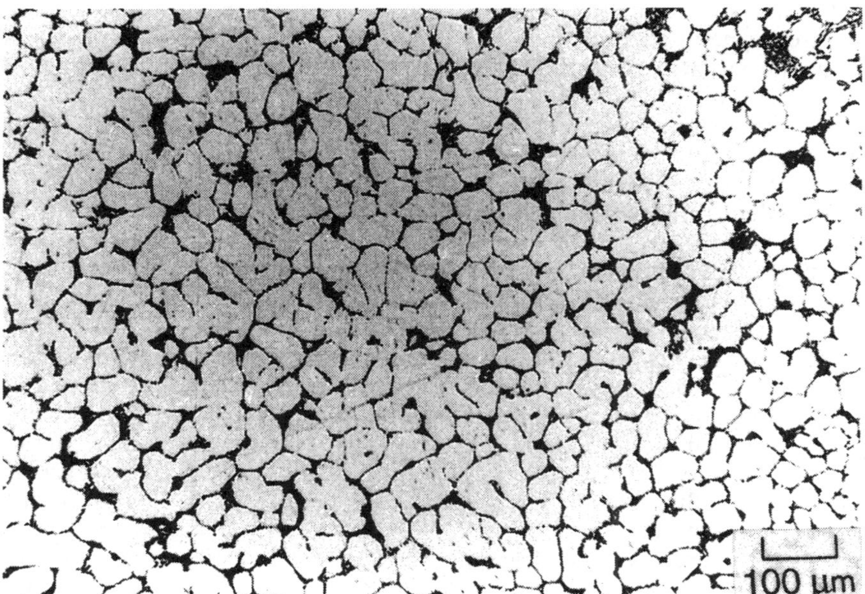

Fig. 6 (cont.)

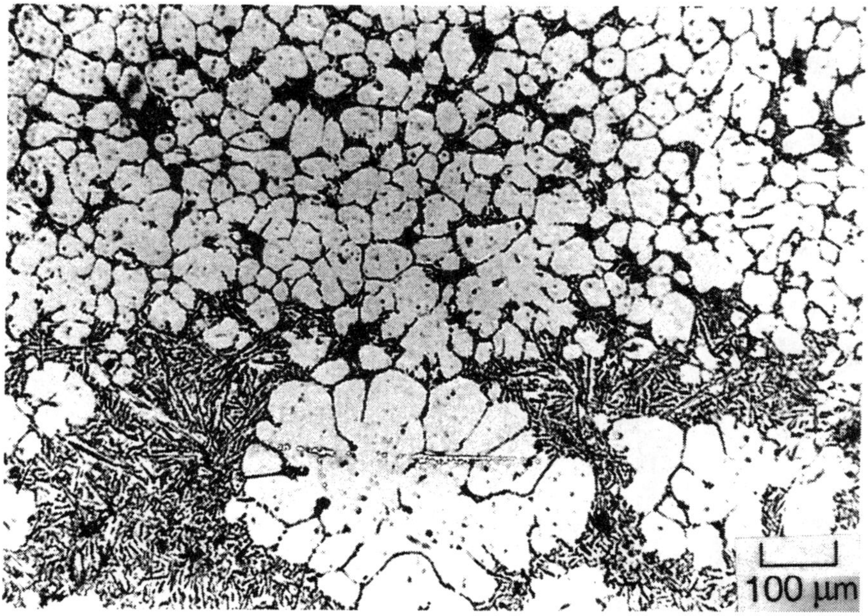

Fig. 6 (cont.)

network of dendritic shape. Examination of Fig. 6c, which corresponds to the electromagnetic pressure crest conveyed in Fig. 4b, shows that the electromagnetic vibrations can refine the structure (average grain diameter of about 120 μm) even in the absence of cavitation. This observation can be interpreted by the extent of the hydrodynamic effects (velocity peak upper than 1 m/s^{-1}), which are quite similar to those encountered in the rheocasting techniques.[5–7] Because of the intense oscillating flow and the high degree of turbulence, the whole volume of the semisolid mixture is practically isothermal. Consequently, the crystallization proceeds simultaneously throughout the undercooled melt around a number of suspended nuclei. This shearing process breaks dendritic particles and deforms them into a spheroidal shape (Fig. 6d).[57]

Figure 6d relates to the following test: first an electromagnetic pressure amplitude of 0.35 bar was applied to the solidifying alloy for 10 min from the liquidus temperature with a cooling rate of 3.5°C/nm^{-1}. Next, a pressure peak of 1.16 bar was imposed suddenly through a rapid augmentation of the current intensity for a relatively short time (15 s), and then the semisolid alloy was rapidly cooled by water with a cooling rate of 100°C/nm^{-1}. In this experiment, the first operation corresponds to grain refinement, consisting mainly of the presence of an agglomeration of nondendritic crystals and due only to hydrodynamic effects in the absence of cavitation phenomena, whereas the second operation is executed under the cavitation conditions. The presence, indicated by arrows, of an area of about 600 μm mean diameter composed of much smaller grains, and of a crack propagating along a distance of approximately 1 mm, can be seen. This observation is understood by the disintegration of a cluster, which gives rise to the formation of about 500 crystals of 40 μm average diameter, probably under the action of a single cavitation event producing impressive mechanical damage over a microscopic region. Identical repetitive tests have confirmed this observation. In such a case where the irradiation time is rather short, cavitation in the solidifying two-phase mixture does not take place uniformly, and systematic trials showed that the regions of intense cavitation are quite random from one experimental run to the next.

The microstructure displayed in Fig. 6e has been obtained from a trial performed in the presence of electromagnetic vibrations related to the cavitation conditions. Negative pressure peaks of 1.16 bar were transmitted, 50 times per second, to the solidifying melt for 12 min, from the liquidus and with a cooling rate of 4°C/nm^{-1}. Examination of Fig. 6 reveals that the grains are much finer (50 μm mean grain size) and less globular than those produced by hydrodynamic effect in the absence of cavitation (Fig. 6c). Moreover, the cavitation effects result in the total disappearance of the numerous large clusters observed in Figs. 6b–6d. Similar results concerning a significant reduction of the grain size, under the action of the cavitation phenomena, were obtained on a 2024 aluminum alloy. The macrographs depicted in Fig. 7 reveal the dramatic reduction of the macrograin size of an electromagneti-

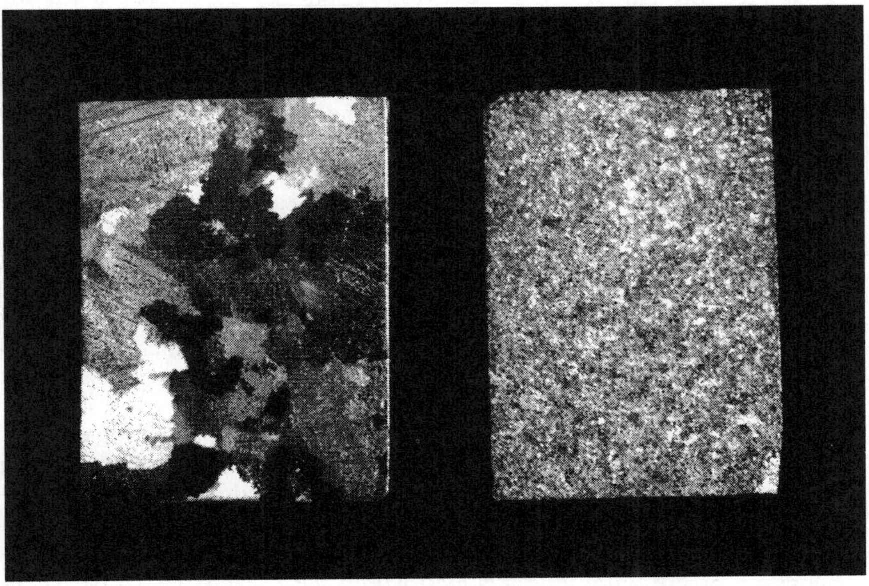

Fig. 7 Macrostructures of a 1050 aluminum alloy conventionally cast (left-hand side) and electromagnetically irradiated during 12 nm, *P* = 1.16 bar (right-hand side); the width of the samples is 22 mm.

cally irradiated 1050 aluminum alloy, which is characterized by a relatively narrow solidification interval. Contrary to the case of mechanically vibrated melts,[14] the electromagnetically irradiated semisolid alloy mixtures appear to exhibit uniform grain refinement across the entire section because, in this experiment, the electromagnetic forces that induce oscillating motion act uniformly throughout the semisolid melt. Moreover, one should note that the cavitation effects, which are exclusively provoked by the electromagnetic pressure fluctuations, are likely to be effective even after the arrest of the two-phase mixture motion, provided that gas bubbles remain inside the melt and upon the crucible walls.

Among the possible mechanisms related to the presence of sonic or ultrasonic vibrations, the cavitation phenomenon is well known for the erosion provoked on the solid walls of hydraulic facilities. Here the fluctuating electromagnetic body forces can momentarily reduce the pressure on the wall to such a low, or even negative, value that transient hydrogen-filled cavities form on the wall. When the cavities collapse, the impact of the liquid metal with the interface is likely to produce very high pressure over a small local region. The emitted shock wave may have an amplitude of up to 1 GPa (10 k bar).[9,10,13]

Figure 8 shows the severe erosion damage undergone by the stainless-steel electrode surfaces, which were initially well-surfaced and very smooth. In these experiments the electrodes were subjected for only about 25 min to an electromagnetic pressure amplitude of 1.16 bar. The presence of eroding cavities, brought about by electromagnetically produced cavitation, provides striking evidence of the powerful forces acting during this phenomenon.

When a solid piece of metal is exposed to cavitation, owing to the pitting action, the metal is gradually eroded. The time-integrated erosive effects of many such events can be readily observed in Fig. 8. Moreover, evident from the erosion patterns, is the fact that cavitation takes place more favorably at certain points than at others. In particular, in Fig. 8a this effect results in the presence of a number of small pinholes and of a crater of about 7 mm in diameter. These hollows constitute gas pockets, which behave as reserves of gas bubbles or voids, and are conducive to cavitation. When the mold is used several times, some pits caused by single cavitation events increase and become bigger excavations with the result that the gas is increasingly entrapped during the mold filling from one trial to the next; consequently, cavitation is effectively enhanced.

Therefore, because of the increase both in number and size of the hollows on the electrode surfaces, the cavitation threshold is evolving during the irradiation time. Accordingly, in this experimental apparatus the cavitation does not occur at a precise critical threshold, but within a zone that was delimited through a series of trials and corresponding to an electromagnetic pressure amplitude included within the range 0.65–0.8 bar (Figs. 4c and 4d).

Figure 6f presents a horizontal slice of an A 356 alloy located in the neighborhood of an electrode. Two distinct and well-separated zones can be

Fig. 8 Examples of cavitation erosion damage caused on electrode surface (irradiation time: 25 nm, P = 1.16 bar).

readily seen: a first region, which is the nearest from an electrode, characterized by the presence of small grains of about 50 μm mean diameter; and a second region distinguished by the existence of large clusters suspended in the solidified liquid matrix. The first area relates to intense cavitation effect provoked both by increasing nucleation and by the disruption of particles or conglomerates. Methodical trials, corresponding to increasing irradiation time, showed that the first zone has a tendency to develop from the electrode walls, constituting substantial reserves of gas bubbles, to the vessel center. According to several authors,[12–14] in a concentrated mass of bubbles the collapse of one may initiate the collapse of a neighbor, and the combined shocks from this "cloud cavitation" can act at a relatively great distance. This was particularly observed when cavitation occurred near a solid surface.[14]

To summarize, when the cavitation conditions are reached, the grain refinement arises from two main mechanisms consisting in shock wave impacts, either starting from the electrode wall (Fig 6f) or arising from various locations of the melt bulk, as already shown in Fig. 6d. In our experiments, fine and homogeneous microstructure was achieved throughout the vibrated ingots after irradiation times on the order of 10 min and corresponding to 30,000 negative pressure cycles.

Conclusion

A new magnetohydrodynamic method of transmitting forced vibrations to solidifying aluminum alloy melts has been developed. The working principle of this device lies in the simultaneous application of crossed oscillating electric fields and stationary magnetic fields in a parallelepipedic crucible containing an alloy in the course of solidification.

From the viewpoint of a fundamental investigation, this arrangement presents two main advantages with respect to the conventional mechanoacoustic processes of liquid irradiation: the peak values of the vibrating velocity and the electromagnetic pressure are relatively well known at each point of the melt and are very nearly homogeneous on the horizontal planes of the bath. These properties have allowed us to establish, after methodical tests, that the cavitation threshold depends on the surface state of the internal vessel walls and that the cavitation effects increase with the electromagnetic pressure peak. Moreover, two distinct causes of grain refinement, represented by fluid flow and cavitation phenomena, have been revealed. In the absence of cavitation, and for a sufficient intensity of the oscillating flow, the columnar-dendritic crystallization is replaced by a microstructure characterized by the formation of the agglomeration of globular particles. On the other hand, when an alloy is solidified in the presence of well-developed cavitation situations, a very fine and homogeneous microstructure has been observed throughout the irradiated ingot, whatever the freezing range width; moreover, the clusters totally disappear, and the grains are smaller and less rounded than those produced above the cavitation threshold.

The ability for industrial application (continuous casting and permanent mold casting) of a noncontact method of electromagnetic irradiation of solidifying alloy melts is currently considered in our laboratory. This work investigates the irradiation of solidifying alloys by ultrasonic magnetoacoustic waves vibrating at the fundamental resonance frequency.

References

[1]Tsavaras, A., and Brody, H., *Journal of Metallurgy*, Vol. 1, 1984, p. 31.

[2]Vivès, Ch., and Perry, Ch., *International Journal of Heat Mass Transfer*, Vol. 29, 1986, p. 21.

[3]Vivès, Ch., *Metallurgical Transactions*, Vol. 20B, 1989, p. 623.

[4]Vivès, Ch., *Metallurgical Transactions*, Vol. 20B, 1989, p. 631.

[5]Flemings, M. C., *Metallurgical Transactions B*, Vol. 22B, 1991, p. 260.

[6]Vivès, Ch., *Metallurgical Transactions*, Vol. 23B, 1992, p. 189.

[7]Vivès, Ch., *Metallurgical Transactions*, Vol. 24B, 1993, p. 493.

[8]Campbell, J., *Int. Met. Review*, Vol. 2, 1981, p. 71.

[9]Abramov, O., *Ultrasound in Liquid and Solid Metals*, CRC Press, Boca Raton, Fl, 1994, p. 289.

[10]Leighton, T., *The Acoustic Bubble*, Academic, London, 1994, p. 531.

[11]Shercliff, J., *A Textbook of Magnetohydrodynamics*, Pergamon, Oxford, England, 1986, p. 44.

[12]Vivès, Ch., *International Journal of Heat Mass Transfer*, Vol. 33, 1990, p. 2585.

[13]Lillicrap, D., *Electrowärme International*, Vol. 44, p. 116.

[14]Szekely, J., *Fluid Flow Phenomena in Metals Processing*, Academic, New York, 1979, p. 305.

VIII. Incompressible Flow

Chapter 41

Role of Ekman Pumping in Confined Magnetohydrodynamic Flows

P. A. Davidson,* D. Kinnear,† and D. Short‡
University of Cambridge, CB2 1PZ, Cambridge, England, United Kingdom

I. Introduction

MAGNETIC fields are frequently used to heat, pump, or stir liquid metals, particularly in casting or welding processes. The flow that results from the Lorentz forces can have a profound influence on the metallurgical structure of the resulting material, particularly as this motion controls the transport of heat, impurities, and crystal fragments. It is important, therefore, that we understand the general structure and scaling of these flows. This understanding is the aim of our paper.

We focus on confined flows, where at least part of the Lorentz force tends to induce swirl. Consequently, we exclude unconfined flows, levitated masses, and flows without swirl. Our starting point is Davidson's[1] theoretical description of axisymmetric, swirling flow and the subsequent extension of this work by Davidson et al.[2] Our primary purpose is to highlight the crucial role played by Ekman pumping in allowing such flows to achieve a steady state. We consider axisymmetric flows with swirl, nonaxisymmetric flows with swirl, and finally flows where the dominant force does not produce swirl, but rather produces poloidal motion in the r-z plane. The last of these topics constitutes a direct extension of the previous studies by Davidson[1] and Davidson et al.[2]

*Department of Engineering, Trumpington Street.
†Department of Engineering, Trumpington Street.
‡Department of Engineering, Trumpington Street.

II. Energy Constraints and Ekman Pumping

Consider a confined turbulent flow driven by the rotational Lorentz force $\boldsymbol{F}$. The flow is assumed to be steady-on-average and is governed by the time averaged Navier–Stokes equation in the form

$$\frac{\mathrm{D}\boldsymbol{u}}{\mathrm{D}t}=\boldsymbol{u}\cdot\nabla\boldsymbol{u}=-\nabla\left(\frac{p}{\rho}\right)+\frac{\partial}{\partial x_i}\left[\frac{\tau_{ij}}{\rho}\right]+\boldsymbol{F},\quad(\nabla\cdot\boldsymbol{u}=0) \tag{1}$$

Locally, the Lorentz form is balanced by the fluid inertia or else by gradients in pressure and turbulent shear stresses (Reynolds stresses). However, the shear stresses play a critical role in establishing global equilibrium. This is evident if we integrate Eq. (1) around a closed streamline to give

$$\oint_c[\boldsymbol{F}\cdot\boldsymbol{u}]\,\mathrm{d}t=\oint_c\left[\frac{\tau_{ij}}{\rho}\frac{\partial u_i}{\partial x_j}\right]\mathrm{d}t-\oint_c\left[\frac{\partial}{\partial x_j}\left(\frac{\tau_{ij}u_i}{\rho}\right)\right]\mathrm{d}t \tag{2}$$

(Here t is the time used in a Lagrangian sense.) This is an energy equation. It states that all of the energy acquired by a fluid particle as it passes through a region forced by $\boldsymbol{F}$ must be subsequently destroyed if the flow is to remain steady. The left-hand side represents the work done by $\boldsymbol{F}$ on a fluid particle as it passes once around a closed streamline. The two terms on the right-hand side represent the destruction and cross-stream diffusion of mechanical energy. The last term redistributes the energy of the mean flow but does not alter the global energy balance because $\boldsymbol{u}$ is zero on the boundary. Evidently, if the flow is to achieve a steady state, then each forced streamline must pass through a region of significant shear. However, such regions are generally confined to thin boundary layers or internal shear layers. A critical question is, therefore; How does the flow arrange itself to ensure that all forced streamlines are flushed through shear layers where they can dump their excess energy? We shall see that in many cases the answer lies in Ekman pumping.

Davidson[1] was the first to suggest that Ekman pumping is critical to confined, forced, swirling flows. Ekman pumping provides an ideal candidate for the entrainment of the core flow by the boundary layers. This is best illustrated by the textbook problem of "spin-down" of tea in a teacup. Here the tea is initially in a state of (almost) inviscid rotation, and we examine the process by which it comes to rest. Within the core of the flow, the centrifugal force is balanced by a radial pressure gradient, and this gradient is also imposed throughout the boundary layer on the base of the teacup. However, the swirl in this boundary layer is diminished through viscous drag, and there is a corresponding reduction in the centrifugal force. The result is a local imbalance between the imposed pressure gradient and the centrifugal acceleration, causing a radial inflow at the base of the teacup. The tea flows toward the axis, eventually drifting up and out of the boundary layer. Con-

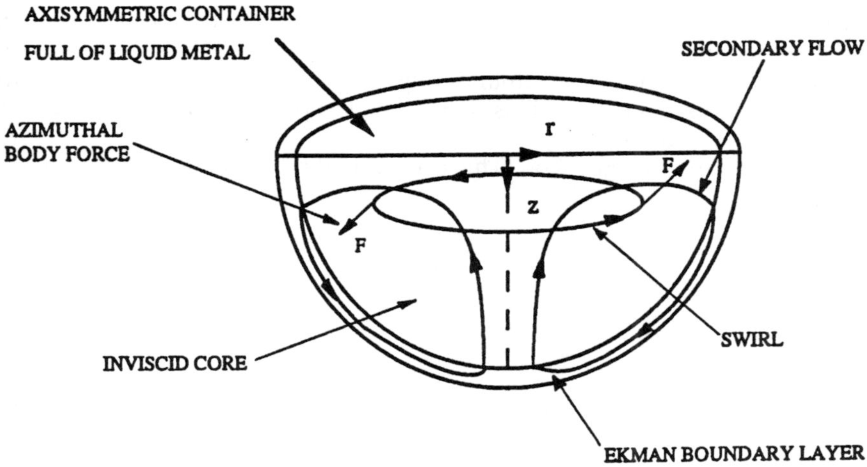

Fig. 1 Swirl induced by an azimuthal body force.

tinuity then requires that the boundary layer is replenished via the side walls, so that a complete recirculation is established. This is Ekman pumping. It dominates the "spin-down" process because, as each fluid element passes through the bottom boundary layer, it gives up a significant fraction of its kinetic energy. The liquid finally comes to rest when all of the contents of the teacup have been flushed through the Ekman boundary layer.

Although "spin-down" is a transient phenomenon, an analogous process can occur in forced, steady flows consisting of one or more vortices, each driven by a region of magnetic forcing. All that is required is a solid boundary adjacent to each vortex. The fluid particles are then cycled back and forth between an inviscid core region where they collect energy, and an Ekman boundary layer, where their energy is removed. The simplest configuration to analyze is that of a forced swirl in an axisymmetric container (Fig. 1). The swirl is driven by an azimuthal Lorentz force, which could arise from a rotating magnetic field or else from crossed, static, magnetic, and electric fields. This is the basis of Davidson's model,[1] which we now briefly review.

III. Forced Swirl in an Axisymmetric Cavity

In this section we use cylindrical polar coordinates (r, θ, z). Consider an axisymmetric flow driven by an azimuthal Lorentz force. This is shown schematically in Fig. 1. Such a flow is typical of that encountered in the rotary magnetic stirring of a nonferrous alloy in continuous casting. We might anticipate that Ekman layers form on the cavity walls. This leads, via Ekman pumping, to a secondary, poloidal flow in the r-z plane. Therefore, to divide the flow into two regions is natural: an inviscid core flow, where the fluid is spun up, and turbulent boundary layers, where energy is extracted from the mean flow. Now all of the streamlines are closed in the (r-z) plane, and so we may apply Eq. (1) to the r-z projection of each streamline. Consequently, all of the streamlines must pass through the Ekman layer to release their excess energy. This simple fact has profound consequences for the structure of the flow.

Consider first the core flow. This is governed by the inviscid Euler equations:

$$\boldsymbol{u} \cdot \nabla\Gamma = rF \tag{3}$$

$$\boldsymbol{u} \cdot \nabla(\omega_\theta / r) = \frac{\partial}{\partial z}\left(\frac{\Gamma^2}{r^4}\right) \tag{4}$$

Here Γ is the angular momentum $\omega_\theta\, r$, which increases monotonically in the core as the fluid is spun up. The second equation governs the vorticity of the secondary, poloidal motion (u_r, o, u_z.). Now let u_b and u_c be characteristic values of the recirculation in the boundary layer and in the core. Then Eq. (4) tells us that u_b is of the same order as $u\theta$. Also, because all of the streamlines are flushed through the boundary layer, continuity of mass requires that u_b and u_c are related by

$$u_c \sim u_b\,(\delta/R) \sim u_\theta\,(\delta/R) \tag{5}$$

where δ is the boundary-layer thickness and R is the maximum radius of the cavity. The fact that the recirculation in the core is weak has profound implications for the primary swirl flow. From Eq. (5) we have

$$\boldsymbol{u} \cdot \nabla(\omega_\theta/r) \sim \Gamma^2\delta^2/R^7$$

and following from Eq. (4) the interior swirl flow is independent of z to second order in δ.

$$\Gamma_c = \Gamma_c(r)[1 + O(\delta/R)^2] \tag{6}$$

This states that, no matter what the distribution of the body force, the core angular momentum will be independent of z. We can now determine the core recirculation from Eq. (3).

$$u_r = rF / \Gamma_c'(r)$$

Because all of the fluid is recycled via the boundary layer, we are allowed to evaluate the mass flux in the Ekman layer (see Ref. 1).

$$\dot{q}(r_s) = 2\pi r_s \int_0^{\delta} u_b \, \mathrm{d}n = 2\pi r_s \int_0^{z_s} u_r \, \mathrm{d}z = \frac{2\pi}{\Gamma_c'(r_s)} \int_0^{z_s} r_s^2 F \, \mathrm{d}z = \frac{1}{\Gamma_c'(r_s)} \frac{\mathrm{d}T}{\mathrm{d}r_s} \tag{7}$$

Here (r_s, z_s) are the coordinates of the boundary, n is the distance measured normal to the wall, and T is the total magnetic torque applied to the fluid between $r = 0$ and $r = r_s$. The core equations contain no additional information, and so we must now turn to the turbulent boundary layer. If we integrate the azimuthal equations of motion across the boundary layer, we obtain,

$$\frac{\mathrm{d}T}{\mathrm{d}s} - \frac{\mathrm{d}}{\mathrm{d}s}\left\{ \chi \frac{\Gamma_c}{\Gamma_c'(r_s)} \frac{\mathrm{d}T}{\mathrm{d}r_s} \right\} = \frac{2\pi r_s^2 \tau_0}{\rho}$$

where τ_0 is the wall shear stress, s is the curvilinear coordinate measured along the wall from the base of the cavity, and χ is a dimensionless coefficient of order one defined by

$$\int_0^{\delta} [\Gamma_c - \Gamma] u_b \, \mathrm{d}n = \chi \Gamma_c \frac{\dot{q}}{2\pi r_s}$$

Substituting from the boundary-layer mass flux using Eq. (7) and introducing the skin-friction coefficient c_f

$$c_f = -2\overline{u_i' u_j'} / u^2$$

our momentum integral reduces to

$$\frac{\mathrm{d}T}{\mathrm{d}s} - \frac{\mathrm{d}}{\mathrm{d}s}\left\{ \chi \frac{\Gamma_c}{\Gamma_c'(s)} \frac{\mathrm{d}T}{\mathrm{d}s} \right\} = \pi c_f \Gamma_c^2 \tag{8}$$

If the distribution of F is known, this can be integrated to give the core angular momentum distribution. Davidson et al.[2] have shown that Eq. (8) is insensitive to the value of χ. A value of 1/6 would be typical of a turbulent boundary layer. The skin-friction coefficient, on the other hand, has a strong influence on the magnitude of the swirl, reflecting the fact that the global magnetic torque must be balanced by shear stresses on the boundary. One estimate of c_f is

$$c_f \approx 0.052\,[\Gamma_c(R)/\nu]^{-1/5} \tag{9}$$

based on the universal law-of-the-wall.

Davidson et al.[2] have compared the predictions of Eqs. (6) and (8) with numeral and laboratory experiments. They found that the angular momentum distribution is indeed independent of depth, irrespective of the distribution of the applied body force. Also, Eq. (8) gives a reasonable estimate of the radial variation in swirl. There was some indication, however, that Eq. (9) represents an underestimate of the skin-friction coefficient. A typical result is shown in Fig. 2. Here the flow has been computed in a 45 deg cone using a finite-difference code. The flow is turbulent, and the turbulent shear stresses are estimated using the k-ε model. The applied body force is concentrated in the upper region of the cone and decays with depth according to

$$F = \frac{1}{2}\Omega_f^2 r \exp[-9z/L], \qquad \Omega_f = 1.462 \text{ rad/s}$$

Figure 2 shows the contours of Stokes' stream function and swirl velocity. Evidently, despite the rapid decay of F with z, the swirl is independent of depth. Moreover, all of the streamlines do indeed pass through the boundary layer. A comparison between theory and computation is shown in Fig. 2c. Theory overestimates the swirl by around 15%, probably because Eq. (9) underestimates the friction.

IV. Ekman Pumping in Nonaxisymmetric Flows

Ekman pumping must occur whenever there is an imbalance between the centrifugal force outside a boundary layer and the centrifugal force inside a boundary layer. Consequently, we should see Ekman pumping whenever the axis of a vortex intersects a solid boundary. Symmetry is not required. This being the case, we might expect Ekman pumping to exist in a wide range of confined, magnetohydrodynamic (MHD) flows. One simple example of three-dimensional Ekman pumping is shown in Fig. 3. Here, liquid metal fills an open-topped rectangular container, and a horizontal body force is applied near the surface of the fluid. Using the coordinate system shown, the body force per unit mass has the form

$$\boldsymbol{F} = F_o \exp\,[-5(1 - y/H)]\,\sin(\pi x/H)\hat{\boldsymbol{e}}_x$$

The force decays exponentially with depth and drags the fluid across the surface of the trough causing a horizontal, columnar vortex to form. The trough is 5 × 5 × 15 cm, and the Reynolds number is 400. The flow is laminar and has been computed using a finite-element code. Despite the fact that the

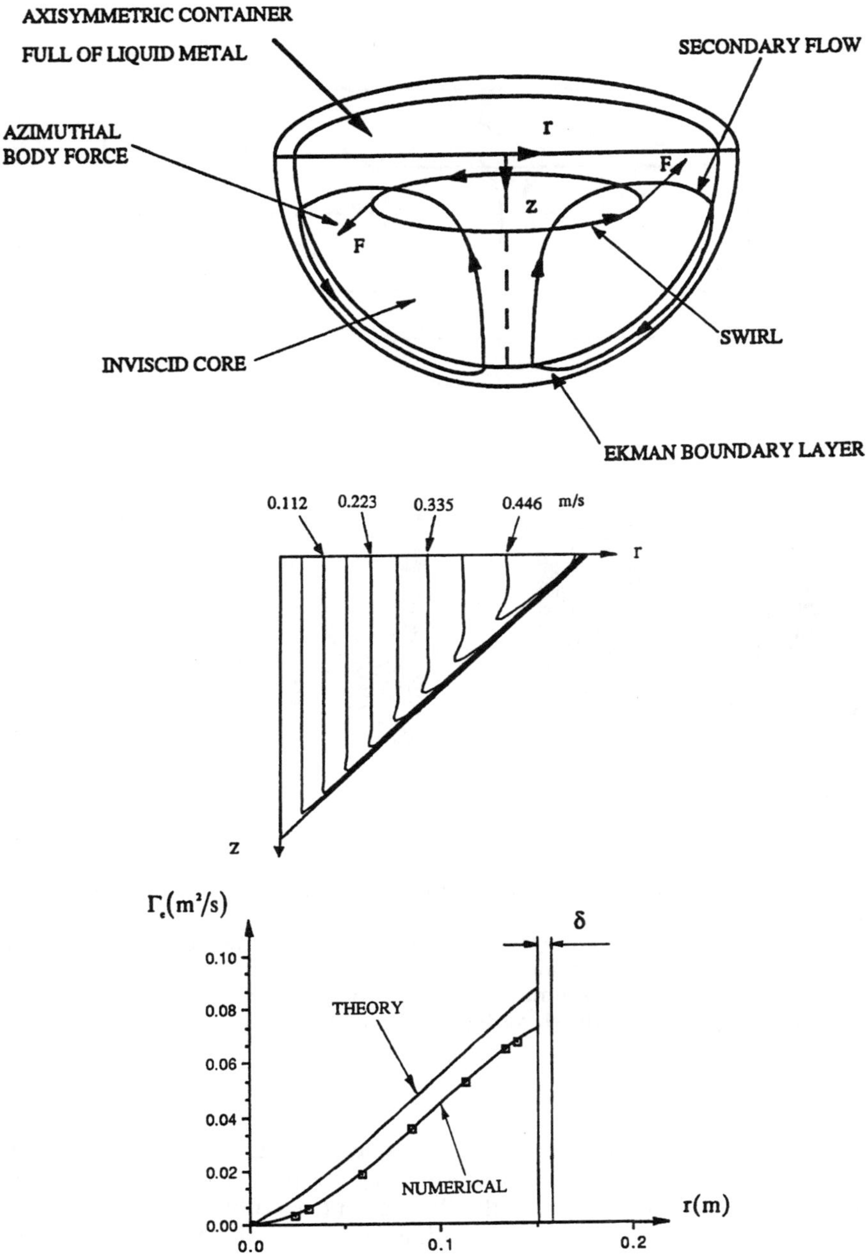

Fig. 2 Computed flow in a cone.
a) Streamlines (I = 0.092 Nm, R = 0.158m)
b) Contours of constant swirl velocity (m/s) (I = 0.092 Nm, R = 0.158 m)
c) A comparison between theory and numerical computation

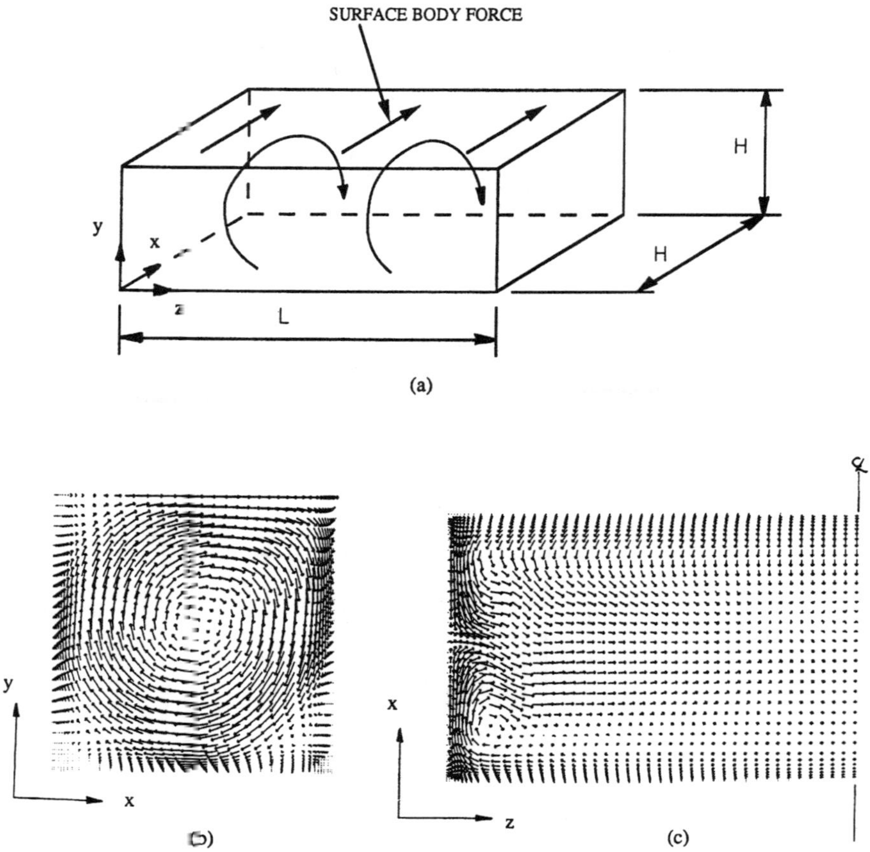

Fig. 3 Computations of a columnar vortex driven in a trough ($\nu = 1.25 \times 10^{-5}$ m²/s).

a) A tangential Lorentz body force is applied near the surface of an open-topped rectangular box containing liquid metal

b) A columnar vortex is formed in the x-y plane (z = 7.5 cm)

c) Ekman pumping is clearly visible in the horizontal plane (y = 2.75 cm)

Lorentz force is concentrated near the surface, the fluid inertia is such that the vortex is roughly symmetrical. More significantly, Ekman pumping is clearly visible in Fig. 3c. Now, if the flow is controlled by Ekman pumping, then the magnitude of the u_θ can be estimated from the balance between the Coriolis force and the body force in the core. Let $u\theta$ be the swirl velocity and u_r be the weak core recirculation. Then,

$$u_r \, u_\theta / H \sim F_o$$

where u_r is of order δ/H times the velocity in the Ekman layer. It follows that

$$u_\theta^2 \sim F_o H^2/\delta$$

Substituting for the laminar Ekman layer thickness gives Davidson's[1] estimate of u_θ:

$$u_\theta \sim (F_o H)^{1/2} \left[\frac{(F_o H)^{1/2} H}{\nu} \right]^{1/3}$$

Davidson et al.[2] computed the flow for a range of viscosities and found the velocity scales as

$$u_\theta \sim \nu^{-0.338}$$

which is consistent with our model of Ekman pumping.

V. Role of Ekman Pumping in Stabilizing Poloidal Vortices

In this final section we focus on flows where the Lorentz force is largely, but not exclusively, poloidal. That is, in cylindrical polar coordinates the dominant force is $(\boldsymbol{F}_r, 0, \boldsymbol{F}_z)$. However, we also allow for a small but finite azimuthal force F_θ. Such a situation is typical of welding or electric-arc remelting of ingots. In both cases, the current spreads radially outward from a source at the center of the surface. This current $(J_r, 0, J_z)$ gives rise to an azimuthal magnetic field $(0, B_\theta, 0)$ and so to a poloidal force. However, often in these processes there is an external, vertical magnetic field B_z, which arises from remote inductors. The interaction of B_z with J_r gives rise to a second, azimuthal force, typically somewhat smaller in magnitude than the poloidal force.

The inviscid core equations for this flow are

$$\boldsymbol{u} \cdot \nabla(\omega_\theta / r) = \frac{\partial}{\partial z}\left(\frac{\Gamma_c^{\,2}}{r^4}\right) + \frac{G}{r} \tag{10}$$

$$\boldsymbol{u} \cdot \nabla \Gamma_c = r F_\theta \tag{11}$$

where G is the azimuthal component of the curl of $\boldsymbol{F}_p$. With swirl present, Ekman pumping will ensure that at least some of the core flow is entrained through the boundary layers. If a significant fraction of the core flow is flushed through the Ekman layers, then continuity requires that the left-hand side of Eq. (10) is small. Let us make this assumption and check later to see if the solution is self-consistent. Under these conditions the poloidal

Lorentz force is balanced by the centripetal acceleration, and Eq. (10) reduces to

$$\frac{\partial \Gamma_c^2}{\partial z} = -r^3 G \tag{12}$$

However, the global torque balance requires

$$\pi c_f \oint \Gamma_c^2 \, \mathrm{d}s = \int r F_\theta \, \mathrm{d}V \tag{13}$$

where s is a curvilinear coordinate following the boundary. From this second equation we deduce

$$\Gamma_c \sim c_f^{-1/2} L_g \left(F_\theta L_g\right)^{1/2} \tag{14}$$

where L_g is a typical geometric length scale.

Let us now check that this swirl is indeed large enough to counterbalance the poloidal Lorentz force. This force balance requires

$$\Gamma_c \geq L_g (F_p L_g)^{1/2}$$

so our solution is self-consistent, provided

$$F_\theta \gg c_f F_p \sim 0.01 F_p \tag{15}$$

Now some, but by no means all, laboratory experiments and industrial processes could satisfy Eq. (15). Consider, for example, an induction furnace. The simplest form of furnace consists of a cylindrical vessel around which is placed a concentric, high-frequency induction coil. The magnetic field and associated Lorentz force is, to a large extent, poloidal. However, the geometry can never be strictly axisymmetric because the turns in the induction coil are helical rather than circular and because the current in the coil must be applied by remote inductors. In general, then, a small azimuthal force will always be present. To satisfy Eq. (15), however, requires that $B_\theta \gg 0.1 B_p$, which does represent a significant departure from symmetry. Arguably, we would not expect to see the swirl dominated flow of Eq. (12) in induction furnaces.

An alternative example is the radial spreading of direct current from a point electrode placed on the surface of a liquid-metal pool. If the current flow is poloidal, then its self-induced magnetic field B_θ is azimuthal. The Lorentz force is then strictly poloidal. However, often stray fields will be present, arising from remote inductors, or even from the Earth's own magnetic field. If the stray field has a vertical component that greatly exceeds 0.01 B_θ, then the azimuthal Lorentz force will satisfy Eq. (15). Unlike the induction furnace; we require only 1% departure from symmetry, which is a modest requirement.

That Eq. (15) is satisfied in at least some experiments seems plausible. In such a situation we must decide whether the fluid adopts a poloidal-dominated flow or the swirl-dominated flow of Eq. (14). In practice this may depend on the precise geometry of the force distribution and of the container. However, there are three ways in which we may help establish the preferred configuration. The first is to determine which flow is energetically more favorable. If the poloidal flow is preferred, then there is no need to look further. The second is to determine whether the more obvious candidate, the poloidal vortex, is linearly stable. If it is not, and the swirl flow has a lower energy, then Ekman pumping begins to look promising. The third and most important task is to determine whether Eq. (12) represents a flow that is physically realizable.

The first of these issues is readily resolved. Provided the container has an aspect ratio of order one, the global kinetic energy or our two candidate flows is

$$E_p \sim c_f^{-1} F_p L_g^4, \qquad E_\theta \sim c_f^{-1} F_\theta L_g^4$$

Evidently, for a small but finite azimuthal force, say $F_\theta \sim 0.1F_p$, the swirl-dominated flow is energetically more favorable. The question of stability is, however, slightly more difficult to resolve. The rate of transfer of kinetic energy from a poloidal flow to an azimuthal disturbance is

$$\dot{E}_\theta(t) = -\int u_r \left(\frac{u_\theta^2}{r}\right) \mathrm{d}V = -\int \frac{\Psi}{r^4} \frac{\partial \Gamma^2}{\partial_z} \mathrm{d}V \tag{16}$$

It would appear that, if a poloidal flow converges radially in precisely those regions where the swirl is largest, then it is likely to be unstable. This has been confirmed in liquid-metal pools that are fed with a direct current. Two examples are shown in Fig. 4. In the first case, shown in Fig. 4a, current is fed into the pool from above. The interaction of this current with its own magnetic field generates a poloidal force. This, in turn, induces a flow that converges at the surface. Conversely, when the current is fed upward through an insulated wire to the meniscus and then spreads back down through the melt, the flow diverges at the surface. (See Fig. 4b.) Both of these cases are discussed by Moreau.[3] Now the current $\boldsymbol{J}$ is largest at the surface, so that any azimuthal Lorentz force generated by a stray magnetic field will be concentrated in this region. Consequently, in Fig. 4 we expect the case a) will be unstable while case b) is stable. This is precisely what is found in practice. A detailed analysis of case a) shows that a purely poloidal flow cannot exist for all but the lowest of Reynolds numbers, and this has been confirmed by experiment.[3] Case b), on the other hand, is found to be stable, both in theory and in practice.

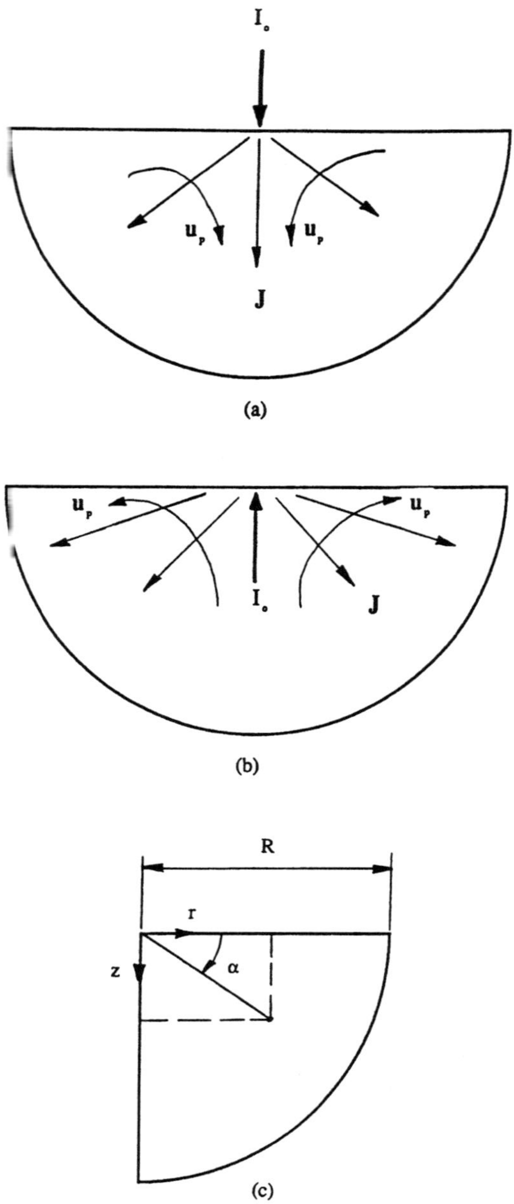

Fig. 4 Flow induced in a hemispherical pool by the injection of current. Case a) is unstable for all but the lowest of currents. Case b) is stable for all values of current.

Let us now address the third and most difficult issue of whether Eqs. (11) and (12) represent a physically realizable flow. To focus the discussion, we shall consider the weld-pool problem shown in Fig. 4. Let I_o be the current fed to the surface, B_θ be the self-field induced by I_o, and $B_o\,\hat{\boldsymbol{e}}_z$ be a much weaker stray field induced either by remote inductors or else by the Earth's magnetic field. It is convenient to introduce a stream function for $\boldsymbol{J}$, defined through

$$\boldsymbol{J} = \nabla \times [(\varphi/r)\hat{\boldsymbol{e}}_\theta]$$

Ampère's law shows that φ is related to the self-induced field B_θ by

$$B_\theta = \mu\varphi/r$$

while continuity of current requires

$$\nabla_*^2(\varphi) = \frac{\partial^2\varphi}{\partial z^2} + r\frac{\partial}{\partial r}\left(\frac{1}{r}\frac{\partial\varphi}{\partial r}\right) = 0$$

It is readily confirmed that $\boldsymbol{F}_p$, $\boldsymbol{F}_\theta$, and G are determined by φ as follows:

$$\boldsymbol{F}_p = -(\mu/2\rho r^2)\nabla\varphi^2 \tag{17}$$

$$rF_\theta = -\frac{B_o}{\rho}\frac{\partial\varphi}{\partial z} \tag{18}$$

$$G = -\frac{\mu}{\rho r^3}\frac{\partial\varphi^2}{\partial z} \tag{19}$$

If Ekman pumping is dominant, then we must satisfy Eqs. (11) and (12), which reduce to

$$\Gamma_c^2(r,z) = (\mu/\rho)\varphi^2 + f(r) \tag{20}$$

$$u\cdot\nabla\Gamma_c = -\frac{B_o}{\rho}\frac{\partial\varphi}{\partial z} \tag{21}$$

Here, $f(r)$ is an arbitrary function of r, which may be expressed in terms of Γ_c and φ at the surface. This leads to

$$\Gamma_c^2(r,z) = \Gamma_c^2(r,0) + (\mu/\rho)\left[\varphi^2(r,z) - \varphi^2(r,0)\right]$$

Now for case a), where the current spreads uniformly away from the point source, the stream function for $\boldsymbol{J}$ is

$$\varphi_A = (I_o/2\pi)[1 - \sin \alpha]$$

where α is the angle measured down from the horizontal surface. For case b), we must add to this the stream function associated with an infinitely long wire carrying a current of $-I_o$. The resulting stream function is

$$\varphi_B = -(I_o/2\pi)\sin \alpha$$

The angular momentum distributions for our two geometries are then,

$$\Gamma_c^2(r,z) = \Gamma_c^2(r,0) - (\mu/\rho)(I_o/2\pi)^2 \sin \alpha(2 - \sin \alpha) \quad [\text{case a}] \; (22)$$

$$\Gamma_c^2(r,z) = \Gamma_c^2(r,0) + (\mu/\rho)(I_o/2\pi)^2 \sin^2 \alpha \quad [\text{case b}] \; (23)$$

Now on the z axis we have $\Gamma_c = 0$, except perhaps at the singular point (0,0). If this were not so, then the kinetic energy of the swirl would be infinite. These equations follow:

$$0 = \Gamma_c^2(0,0) - (\mu/\rho)(I_o/2\pi)^2 \quad (\text{case a})$$

$$0 = \Gamma_c^2(0,0) + (\mu/\rho)(I_o/2\pi)^2 \quad (\text{case b})$$

The second of these is clearly contradictory, and we conclude that Ekman pumping cannot be realized in this configuration. This is consistent with the observed stability of the poloidal vortex in case b). However, case a) appears to give a consistent solution provided Γ_c is singular at (0,0). This local singularity in Γ does not lead to an infinite kinetic energy, and so may be tolerated.

In case a) the global torque balance (7.12) requires

$$\pi c_f \int \Gamma_c^2 \, ds = T_o = B_o I_o R^2/3\rho \quad (24)$$

This allows us to determine the (weighted) average value of the surface angular momentum $\Gamma_c(r,0)$. Substituting Eq. (22) into Eq. (24) gives

$$\int_0^R \frac{\Gamma_c^2(r,0)}{(R^2 - r^2)^{1/2}} \, dr = \frac{\mu}{\rho}\left(\frac{I_o}{2\pi}\right)^2\left[2 - \frac{\pi}{4}\right] + \frac{B_o I_o R}{3\pi\rho c_f} \quad (25)$$

For our solution to be consistent, $\Gamma_c^2(r,0)$ must exceed the second term on the right of Eq. (22). From Eq. (25) we can see that this is satisfied, provided

$$B_o \gg c_f B_\theta \sim 0.01 B_\theta \tag{26}$$

which is the same criterion as Eq. (15). In this case the contours of constant Γc are almost vertical.

It remains to ensure that our swirl flow can satisfy Eq. (21). For the point electrode problem this takes the form

$$u \cdot \nabla\Gamma_c = \frac{B_o I_o}{2\pi\rho} \frac{r^2}{\left(r^2 + z^2\right)^{3/2}} > 0 \tag{27}$$

Because the contours of Γ_c are almost vertical, we merely required that the recirculation in the core points radially outward. This is entirely consistent with Ekman pumping as indicated in Fig. 1. We conclude therefore that a swirl dominated flow is physically realizable in case a), though not in case b).

The idea that a fluid which is subjected to a strong poloidal force should, under favorable conditions, degenerate into a swirling flow is counter to intuition. We must turn to the experimental evidence to support this hypothesis. Moreau[3] reports a striking example of this kind of spontaneous rotation. When a current is fed into a weld pool from above, the pool spontaneously rotates for all but the lowest currents (15 amps). Then the current is fed from below; however, no such rotation occurs. That swirl can dominate the velocity field even when the Lorentz forces are largely poloidal is indicative of the effectiveness of Ekman pumping in dissipating energy, thus providing a stable, low-energy state.

References

[1]Davidson, P. A., "Swirling Flow in an Axisymmetric Cavity of Arbitary Profile, Driven by a Rotating Magnetic Field," *Journal of Fluid Mechanics*, Vol. 209, 1992, pp. 35–55.

[2]Davidson, P. A., Short, D. J., and Kinnear, D., "The Role of Ekman Pumping in Confined, Electromagnetically-Driven Flows," *European Journal of Mech.*, Vol. 14, No. 6, 1995, pp. 795–821.

[3]Moreau, R., *Magnetohydrodynamics*, Kluwer, Dordrecht, The Netherlands, 1990.

Chapter 42

Numerical Simulation of Flow Driven by a Rotating Magnetic Field in a Rotating Cylindrical Vessel

I. Grants* and Yu. M. Gelfgat†
Institute of Physics, Latvian Academy of Sciences, Riga, Latvia

Introduction

A ROTATING magnetic field (RMF) is widely used in many technological processes, for instance, in metallurgy, semiconductor crystal growth, etc., in order to homogenize melts and reinforce heat/mass transfer. Axial variation of melt angular velocity near the solidification interface or crucible bottom causes disbalance of radial pressure gradient and centrifugal force. It forces the melt to circulate in the meridional plane, the so-called Ekman pumping. The meridional flow is the most important factor determining heat/mass transfer in the melt and, as a result, the parameters of technological process product. Direction and pattern of the meridional flow are fixed toward the axis in the Ekman boundary layer and, due to incompressibility, in the direction from the solidification front[1,2] that may turn out not to be optimal. Priede[1] using an analytical steady self-similar boundary-layer type solution showed that imposed rotation of a crucible itself could change a direction and pattern of the meridional flow. Besides, it was shown that the imposed rotation suppressed both the differential (to a vessel) angular velocity of the core region and characteristic magnitude of the meridional flow.[1] Existence of another branch of a possible steady-state solution for counter-rotation of a vessel and the absence of a solution for a certain range of parameters were shown as well.

The present paper deals with the following problems for further theoretical investigation of the imposed rotation effect on a flow induced by RMF:

1) Finding out of a numerical solution, including a two-dimensional one, of the problem, to prove earlier obtained results. The numerical simulation

*Researcher, Laboratory of MHD Technology.
†Professor, Head of Laboratory of MHD Technology.

allows us to find a solution, which is not of a boundary-layer type and not self-similar (i.e., essentially two-dimensional).

2) Checking of the existence of a second solution and finding out the conditions to reach it; the numerical simulation allows us to trace evolution to a steady solution from initial ones. Different initial conditions or conditions in "physical infinity" can result in different steady states because different self-similar ones were found there.

3) Investigation of an essentially two-dimensional flow in the range of parameters, where no self-similar solution exists. Actually, a flow in this range can be either unsteady or not self-similar, or both. The numerical simulation can answer this question.

4) Investigation of action of a steady magnetic field (SMF) on a flow in a given range of parameters. SMF can be useful to obtain a steady self-similar flow in the region, where it does not exist, in order to get a greater variety of sufficiently smooth flow patterns.

Model

A stretch of the investigated theoretical system is shown in Fig. 1. A rotating magnetic field vector $\boldsymbol{B}_0$ = const rotates around the vessel axis with a constant angular velocity ω_0. Because the fluid is electrically conducting, RMF induces oscillating currents, which interact with RMF, so giving rise to oscillating emf with a frequency $2\omega_0$ and averaged value in the direction of RMF. These volumetric forces make the fluid rotate with an azimuthal velocity $v_\alpha = r\Omega$, where Ω is the angular velocity of a fluid. Besides, the vessel itself rotates with an angular velocity Ω_w. When a fluid in the volume of a vessel rotates with a different angular velocity than that of the vessel, the meridional flow with velocity $\boldsymbol{v} = (v_r, v_z)$ appears. A theoretical model is based on the following assumptions: 1) frequency of RMF rotation is high enough to neglect an oscillating part of the emf; 2) frequency of RMF rotation is too low to cause an essential skin effect; 3) induction of RMF is too small to drive flow with angular velocity comparable to that of RMF

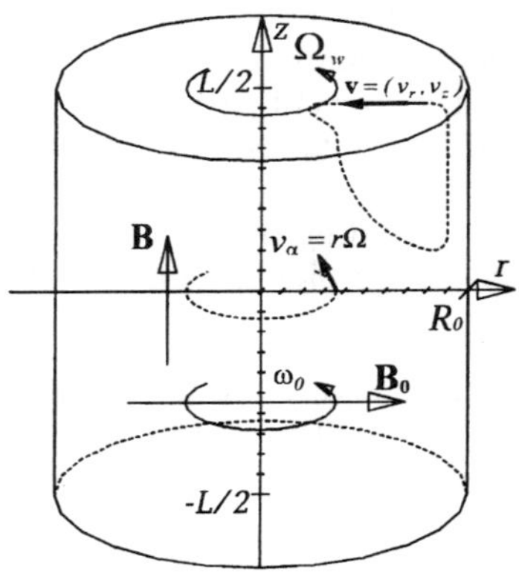

Fig. 1 Sketch of the model. Ω_w is the angular velocity of vessel rotation, B_0 is the induction, ω_0 is the angular velocity of a rotating magnetic field, and B is the induction of a steady magnetic field.

rotation; where the flow is considered to be incompressible, axially symmetric, and laminar but inertia dominated.

These assumptions turn out to be applicable for common crystal growth conditions. On the other hand, they permit separation of electromotive and hydrodynamic problems and provide reception of two-dimensional or even one-dimensional (self-similar) sets of equations.

A nondimensionalized* axisymmetric Navier–Stokes set of equations† in an angular velocity, meridional vorticity, and stream function formulation is obtained in a form[2] of

$$r^2 \frac{\partial \Omega}{\partial t} = \nabla\left[r^2(\nabla\Omega - \boldsymbol{v}\Omega)\right] + r^2 Te \cdot k(r,z) - \mathrm{H}a^2 r \frac{\partial \varphi}{\partial z} \tag{1}$$

$$r^{-2} \frac{\partial W}{\partial t} = \nabla\left[r^{-2}(\nabla W - \boldsymbol{v}W)\right] + \frac{\partial \Omega^2}{\partial z} - \frac{1}{r} Ha^2 \frac{\partial v_r}{\partial z} \tag{2}$$

$$\boldsymbol{v} = -\frac{1}{\rho}\boldsymbol{e}_\alpha \times \nabla\psi \tag{3}$$

$$Wr^{-2} = \nabla(r^{-2}\nabla\psi) \tag{4}$$

$$\nabla(r^{-2}\nabla\varphi) = -\frac{\partial\Omega}{\partial z} \tag{5}$$

Here $W(r,z)$ is the momentum of the meridional vorticity; $\Omega(r,z)$ is the angular velocity; $\psi(r,z)$ is the Stokes stream function of a meridional flow $\boldsymbol{v}(r,z)$; $k(r,z)$ is the distribution of electromotive nondimensionalized torque of magnitude Te due to RMF; $Te = 0.5\sigma\omega_0 B_0^2 L^4 \rho^{-1} \nu^{-2}$ is the electromagnetic Taylor number; $\varphi(r,z)$ is the stream function of electric currents due to SMF; $Ha = \sigma^{1/2}(\rho\nu)^{1/2} BL$ is the Hartmann number characterizing electromagnetic braking due to SMF; and σ, ρ, ν denote conductivity, density, and kinematic viscosity of liquid, respectively.

Imposed rotation of a vessel is encountered through boundary conditions for angular velocity:

$$\Omega|_S = \Omega_w$$

*Height of vessel L, diffusion time of vorticity L^2/ν, characteristic current densities $\sigma\nu B_0/L$ and $\sigma\nu B/L$ are used as length, time, RMF, and SMF induced current scales, respectively.

†The Maxwell's equations and Ohm's law are used to define $k(r,z)$ and obtain Eq. (5) (Refs. 2 and 3).

where Ω_w is the nondimensional angular velocity of a vessel and S is a vessel surface.

One-dimensional corresponding sets of equations can be obtained[4] using von Kármán similarity variables*:

$$v\alpha = \Omega(z,t); \quad v_z = H(z,t); \quad v_r = rf(z,t) = -0.5rH'(z,t) \tag{6}$$

Substituting of Eq. (6) into the Navier–Stokes set of equations yields the following one-dimensional set:

$$\frac{\partial \Omega}{\partial t} + H\Omega' - H'\Omega = \Omega'' + Te - Ha^2\left(\Omega - p\overline{\Omega}\right) \tag{7}$$

$$\frac{\partial H''}{\partial t} + HH''' + 2\left(\Omega^2\right)' = H^{IV} - Ha^2 H'' \tag{8}$$

Here p denotes a parameter characterizing wall conductivity: $p = 0$ corresponds to well-conducting walls and $p = 1$ to insulating ones; and $\overline{\Omega}$ is the average value of Ω over the height z.

Boundary conditions for Eqs. (7) and (8) are $\Omega(\pm 1/2) = \Omega_w, H(\pm 1/2) = H'(\pm 1/2) = 0$, where $z \pm 1/2$ corresponds to end walls.

We continue investigations done by means of boundary-layer approach[1] via direct numerical simulation of Eqs. (7) and (8) and corresponding two-dimensional set (Eqs. [1–5]). Basically, a flow in a zero SMF ($Ha = 0$) is studied. The action of SMF is examined for a definite range of vessel rotation rate in a separate section only.

Comparison of Analytical and Numerical Solutions

First of all, the results of the numerical solution of the self-similar problem were compared to those obtained by Priede.[1] Under an inertia dominated regime the parameter, which characterized the imposed rotation effect and determined influenced flow patterns, was the so-called interaction parameter[1]:

$$P = \Omega_w Te^{-2/3}$$

defining a ratio of an angular velocity of vessel rotation and a characteristic angular velocity of a flow due to RMF in a nonrotating vessel. Figures 2 and 3 show characteristic velocity profiles in a boundary layer at end walls for different—small, medium, and rapid—co- ($P > 0$) and counter-rotation ($P < 0$) of the vessel. In both cases increasing of the imposed rotation reduces flow velocities as boundary-layer thickness except for the case of small counter-rotation. In the range of $-0.07 < P < 0$, these characteristics in-

*$f(z) = -0.5H'(z)$ follows from flow continuity.

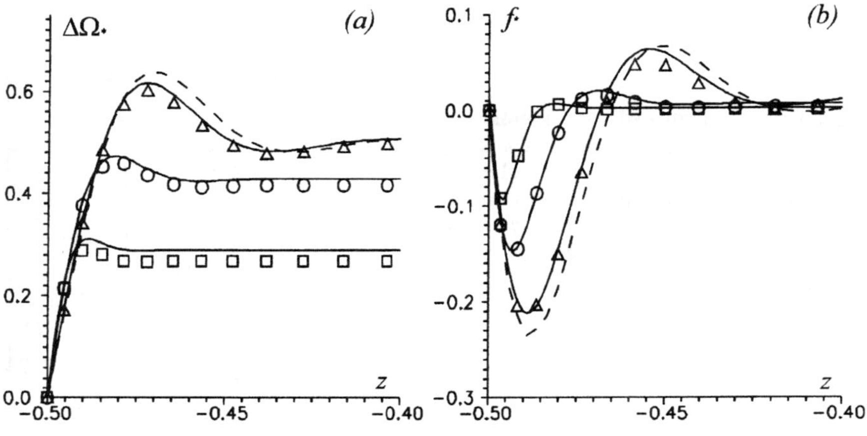

Fig. 2 Velocity distributions at the end wall under co-rotation of the vessel: a) differential angular ($\Delta\Omega^* = (\Omega - \Omega_w)Te^{-2/3}$) and b) radial ($f^* = f\,Te^{-2/3}$). Numerical solutions at $P = 0.06$ (Δ), $P = 0.6$ (○), and $P = 2.5$ (□) are shown by dots. Curves represent the results of an analytical solution at corresponding P. A dashed curve depicts results for nonrotating vessel ($P = 0$).

crease with the increasing of vessel rotation. Further increasing of counter-rotation will result in a lack of numerical solution. In the range of $-1.5 < P < -0.07$, numerical solution diverges together with unlimited growth of velocity rates. This range, where no self-similar solution can be obtained, turned out to be wider than the one predicted analytically $-1.5 < P < -0.3$, and no branch of steady solution 2 (Fig. 4) was obtained numerically. This branch was discussed[1] to be unstable and so failure to find them numerically

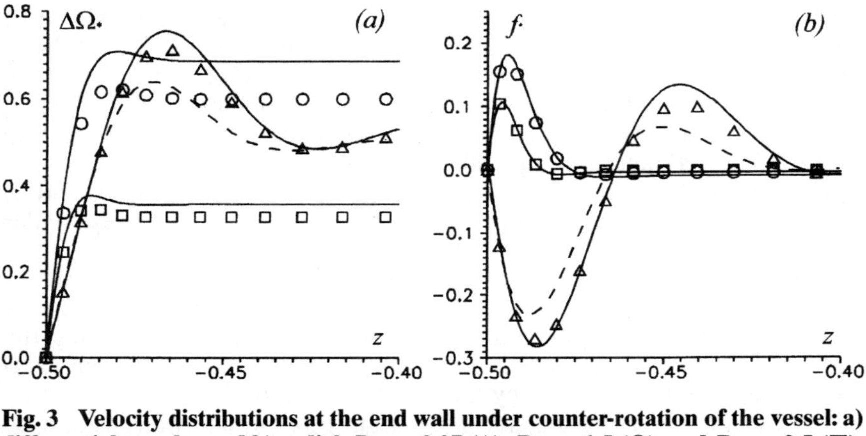

Fig. 3 Velocity distributions at the end wall under counter-rotation of the vessel: a) differential angular and b) radial. $P = -0.07$ (Δ), $P = -1.5$ (○), and $P = -2.5$ (□).

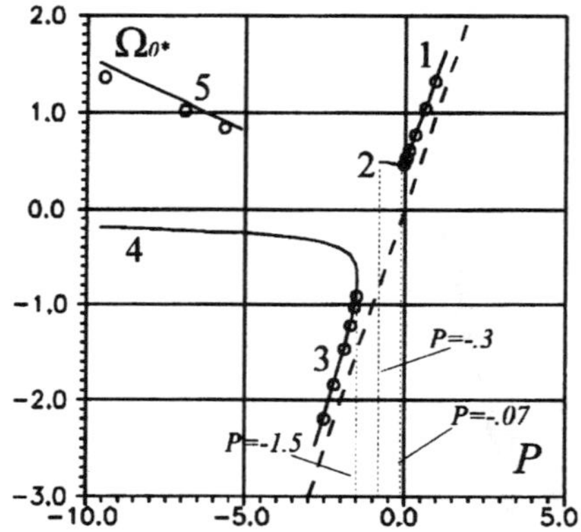

Fig. 4 Dependence of core angular velocity ($\Omega_0{}^* = \Omega_0 Te^{-2/3}$, where Ω_0 is the angular velocity of core) on angular velocity of the vessel rotation. (○) numerical results: continuous curves are the analytical solution and the dashed line is the angular velocity of vessel.

cannot be treated as a disagreement of results. The lowest value of this range ($P = -1.5$) fits well in the analytical results. In Fig. 3 we clearly observe that at $P < -1.5$ the direction of a meridional velocity changes. This feature of the imposed rotation application is mentioned to be of great importance. So, the conclusion can be made that the direct numerical simulation performed confirms the analytical results for branches 1–3 (Fig. 4).

Second Solution for a Vessel with Counter-Rotation

The analytical results predicted the existence of a second solution (branch 4, Fig. 4) at counter-rotation of a vessel ($P < -1.5$). The numerical simulation with a number of different initial conditions did not prove it. Regardless of the initial state, a steady solution of branch 3 (Fig. 4) was always obtained. Further increase of counter-rotation rate led to a previously unknown steady state solution (branch 5, Fig. 4) at a rapid counter-rotation of the vessel ($P < -5.5$). This solution develops in an initially nonrotating or corotating RMF layer and has the following properties (Fig. 5ab): 1) direction of the core rotation is opposite to that of the vessel; 2) existence of two pairs of meridional flow loops diverging as in the core and as at end walls; 3) flow velocities are much larger than in the case of a "normal" solution (branch 3, Fig. 4), and, consequently, an angular momentum flux through a rigid wall due to a shear is much larger than in the normal case; and 4) increasing of the vessel rotation causes an increase of the core angular velocity in an opposite direction (Fig. 4).

Actually, we note here that a self-similar solution does not account for phenomena occurring outside the core, i.e., angular momentum can be driven from or to physical infinity by radial convection. Considering the third property mentioned in the preceding material, clearly such a self-similar flow should be supported by additional forcing in the physical infinity; otherwise, a much larger angular momentum flux through walls cannot be provided. Obviously, such a solution with increasing shear due to the im-

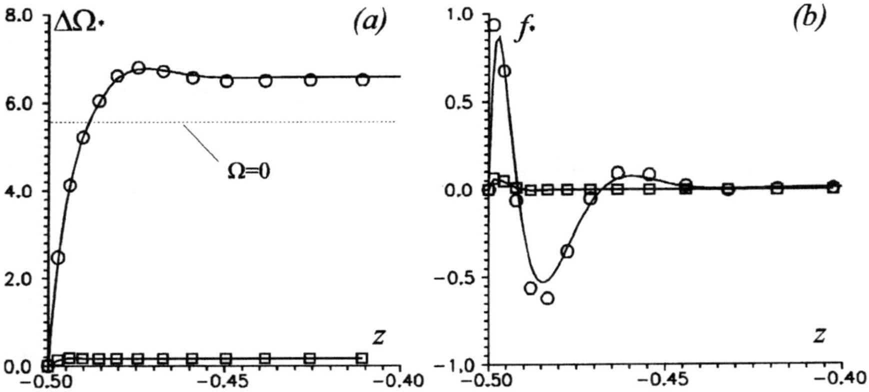

Fig. 5 Velocity distributions in the boundary layer at end walls, corresponding to "normal" (□) and second (○) solutions: a) differential angular and b) radial. Dots are the numerical solution and curves are the analytical ones.

posed counter-rotation increase requires extra forcing in the region at the vessel's side walls.

Figures 6 and 7 show comparison of the numerical and analytical solutions presented in the form of dependencies defining the analytical solution.[1] The h_∞ characterizes the magnitude and direction of the core meridional circulation, but $R = 0.5(\Omega_0 + \Omega_w)/(\Omega_0 - \Omega_w)$ is the inverse Rosby number, an intermediate parameter representing an inverse value of a relative differential core rotation. Numbers in these figures denote branches corresponding to those in Fig. 4. A "new" solution (branch 5) was obtained by the analytical method[1] as well.

Why this branch was missed while the analytical method was applied was seen. The domain of R (which is a main variable of the analytical solution), where a new branch is defined, is quite narrow: $(-0.3667 < R < -0.3620)$,

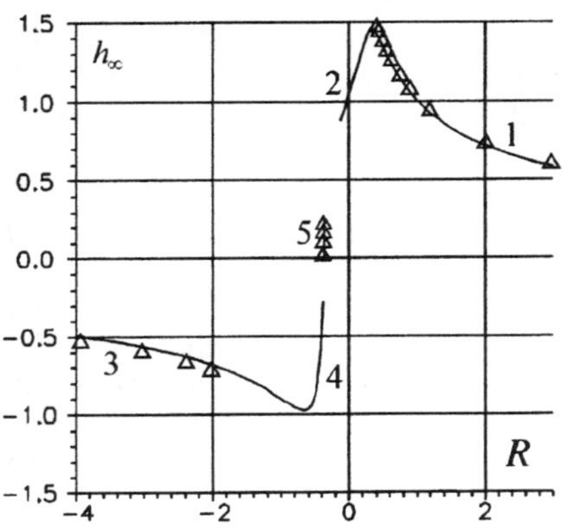

Fig. 6 Core meridional circulation as a function of the inverse Rosby number.

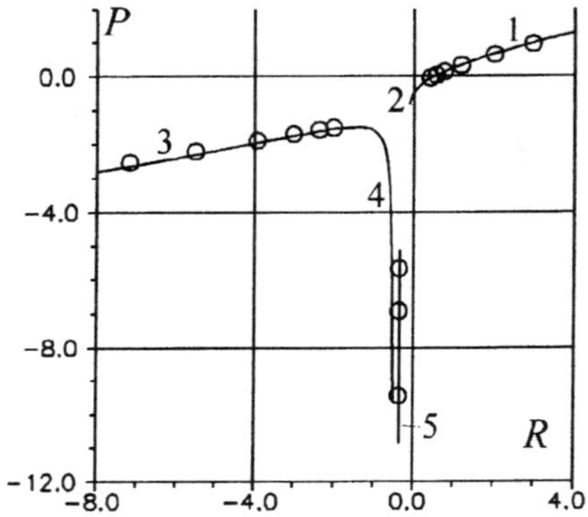

Fig. 7 Dependence linking the inverse Rosby number and angular velocity of the vessel.

and the analytical solution here is quite sensitive to the initial approximation.

This fact illustrates that despite the property common to analytical solutions, which supplies more general information, there can occur situations where some features of the same equations may be missed by an analytical method and found by a numerical one. Two-dimensional numerical solutions were found also corresponding to this branch. As noted, was possible with additional forcing at side walls only. Without a solution corresponding to a normal one (Fig. 4) is always obtained.

Investigation of Essentially Two-Dimensional Flows

The analytical and numerical simulation shows that there does not exist any steady solution of the self-similar problem [Eqs. (7) and (8)] in a certain range of parameters. So, what actually happens to a flow in this range? Is it the absence either of a steady or a self-similar solution? Besides, the question on the self-similar result application to a more real radially bounded system has not been answered yet.

The two-dimensional problem (1–5) was solved numerically using a finite-difference code derived from employing the control volume method. An exponential type approximation obtained by an integral identity technique was used for both $r^2\Omega$ and $r^{-2}W$. Integration over time was performed applying the Peaceman–Rachford implicit alternating direction time-stepping scheme.[2,3]

Two-dimensional calculations confirmed the principal results of the self-similar solution in a range where it existed. In this range a two-dimensional solution with electromotive torque $Te \cdot k(r,z)$ turned out to fit well to a self-similar one with $Te' = Te \cdot k(r,z)$ (where the bar sign denotes averaging over the whole volume). Actually, at counter-rotation slight change of the angular velocity in the core in radial direction (Fig. 8d) was observed. Nevertheless, these deviations are not essential. Obviously, they are determined by slight disbalance of an angular momentum driven from and to infinity in a corresponding self-similar solution.

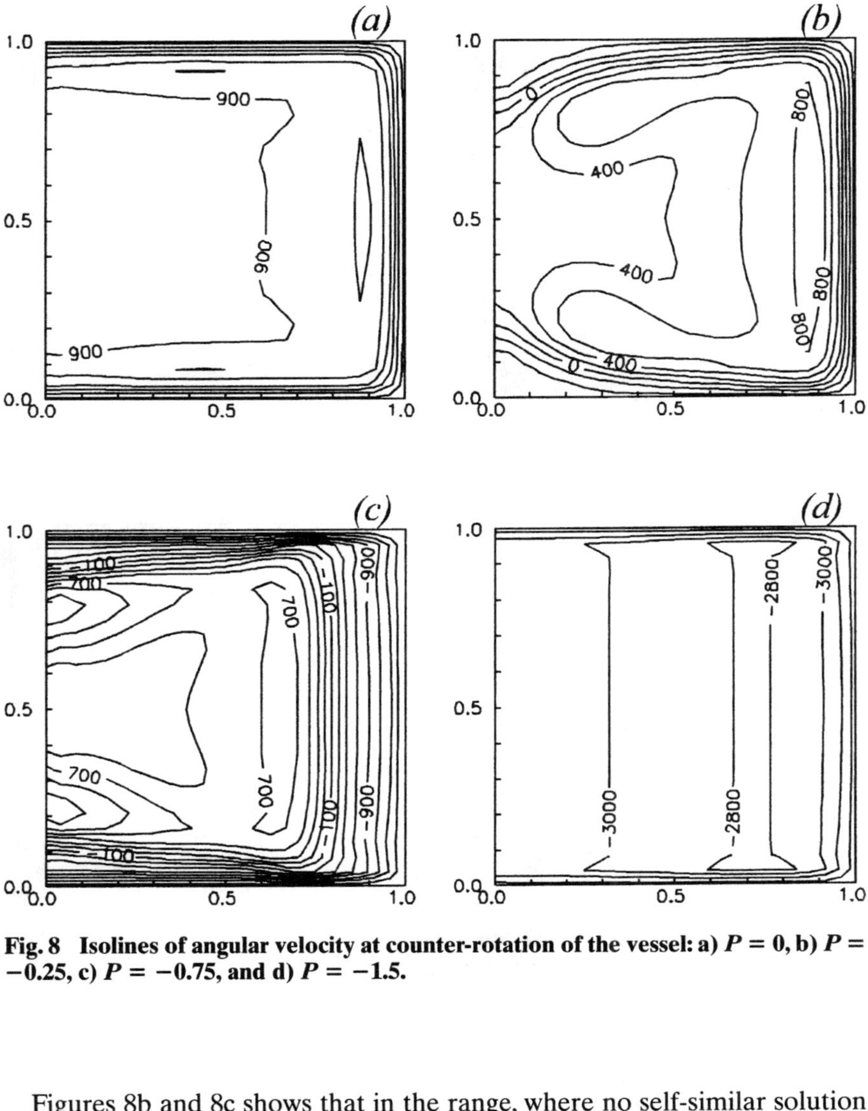

Fig. 8 Isolines of angular velocity at counter-rotation of the vessel: a) $P = 0$, b) $P = -0.25$, c) $P = -0.75$, and d) $P = -1.5$.

Figures 8b and 8c shows that in the range, where no self-similar solution exists, the core flow with a strong viscous angular momentum flux in the radial direction appears as well as boundary layers with varying thickness. Nevertheless, the flow remains steady. Figure 9c illustrates an intermediate stage of the meridional flow direction change when two pairs of meridional flow loops appear. It results in a very extended boundary layer at a side wall (Fig. 8c). Figures 8c and 8d show a seemingly paradox feature of the vessel counter-rotation to reduce viscous stress at a side wall. Actually, when one keeps in mind that the meridional flow direction also changes near the side wall, the preceding phenomenon becomes obvious.

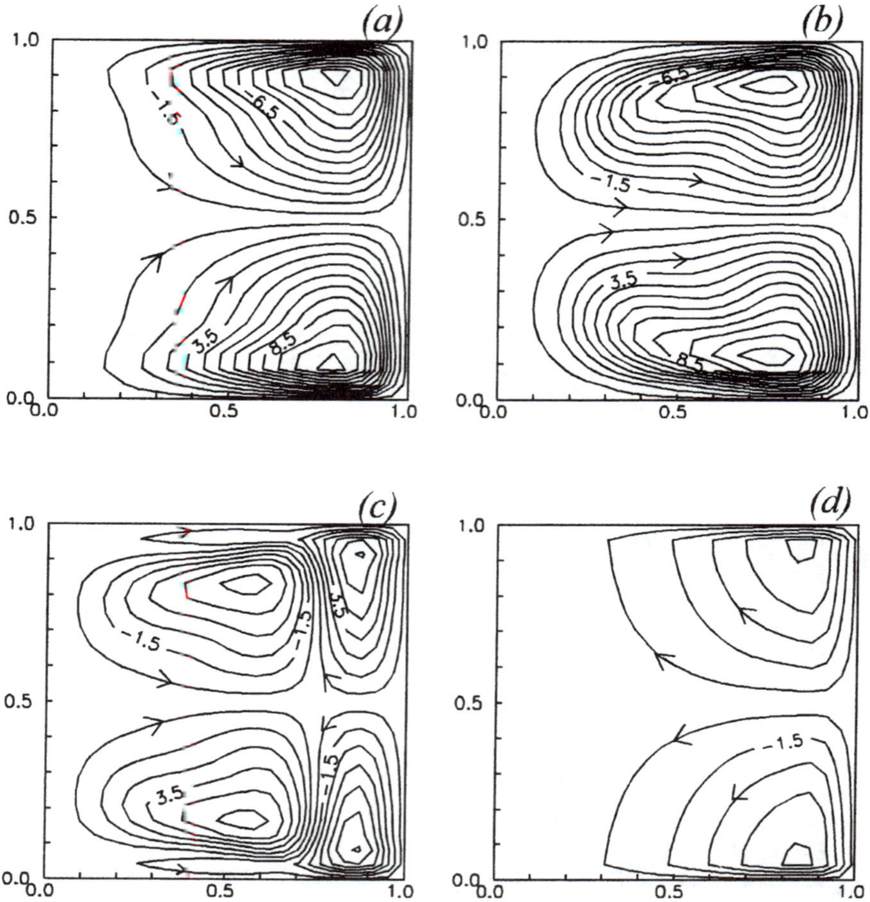

Fig. 9 Streamlines of meridional flow at counter-rotation of the vessel: a) $P = 0$, b) $P = -0.25$, c) $P = -0.75$, and d) $P = -1.5$.

Action of Steady Magnetic Field on Flow

The preceding considerations state that there exists a range of vessel counter-rotation angular velocity ($-1.5 < P < -0.07$) where it is impossible to obtain a self-similar flow. Instead, in real radially bounded systems an essentially two-dimensional flow with boundary layers of varying thickness and flows with a more complicated angular velocity distribution appears. The occurrence of flows with extra meridional flow loops is possible. So there may appear a variety of probably useful flow patterns with undesirable flow inhomogeneity. A SMF can be applied to obtain a self-similar (more homogeneous, respectively) flow at any velocity of counter-rotation.

Induction necessary to make em due to SMF comparable to the inertial ones in the core and boundary layers is characterized by the Hartmann number $Ha \sim Te^{1/3}$ in the case of insulating end walls and $Ha \sim Te^{1/4}$ in the case of well-conducting ones.[4] Numerical simulation showed that applying the SMF with an induction of this magnitude allows one to obtain a steady solution of self-similar set at any rate of counter-rotation. A characteristic feature of the solutions found in this regime was the appearance of another couple of secondary flow loops, making meridional flow to circulate in the opposite direction at the end walls (as it occurs at $P < -1.5$). Changing of P allows changing of the ratio of characteristic sizes and velocities of these vortices. Corresponding two-dimensional investigations fit well to the self-similar ones meaning that SMF actually smoothes the azimuthal flow patterns and at the same time allows us to obtain a variety of double coupled meridional flow patterns of the type in Fig. 9c.

Conclusions

The numerical simulation of the imposed rotation on the flow induced by RMF confirmed the principal results of the analytical solution: 1) reduction of flow velocity magnitude and boundary-layer thickness applying extra rotation of the vessel and 2) possibility to change direction of the meridional flow.

The numerical self-similar solution was found to fit well to the corresponding analytical one. It showed that the range, where no self-similar solution existed, was wider than what followed from the analytical investigations. In the range where the analytical solution exists but should be unstable[1] $-0.3 < P < -0.07$, a corresponding numerical solution diverges.

The analytical solution at counter-rotation predicted the existence of two steady solutions. Only one of them, for which the flow linearizes and tends to a von Kármán boundary-layer-type flow, when the counter-rotation velocity increases, is proved numerically. The second solution is unreachable because of a variety of initial conditions. Instead, another steady solution for rapid counter-rotation with a number of interesting properties is first found numerically and later analytically. This solution keeps the core angular velocity in the direction of RMF ignoring fast counter-rotation of the vessel. Such a flow has a much larger integral viscous shear torque than an electromotive torque because of RMF, that is, why such a flow should be provided by additional forcing in the RMF direction outside the core.

The two-dimensional numerical solution confirmed principal results of the self-similar one and allowed investigation of the flow in the range of vessel rotation rate values where the self-similar solution did not allow it. The discovery was made that in that range the thickness of boundary layers varied along end walls, and they could even fly apart from rigid walls. In this range the flow with an extra couple of meridional flow loops can be generated. The change of the meridional flow direction is accompanied by considerable increasing of boundary-layer thickness at a side wall.

A variety of flow patterns can be obtained in the range where no self-similar solution exists. A characteristic feature of these patterns is the flow inhomogeneity. Studies of the self-similar solution show that SMF with $Ha \sim Te^{1/3}$ in the case of insulated and $Ha \sim Te^{1/4}$ in the case of well-conducting end walls makes a self-similar flow possible at any velocity of the vessel rotation. The corresponding two-dimensional simulation confirmed that the SMF allowed us to obtain a variety of self-similar smooth flow patterns in the preceding-mentioned range of counter-rotation.

The obtained results permit us to conclude that the imposed rotation of the vessel is a powerful means to control flow patterns induced by the RMF, thereby providing opportunities to choose optimal ones for every particular technological process.

References

[1]Priede, J., "Effect of Imposed Rotation on a Swirling Fluid Flow Driven by a Rotating Magnetic Field," *Proceedings of the Second International Conference on Energy Transfer in MHD Flows* (Aussois, France), Vol. 1, 1994, pp. 87–96.

[2]Gelfgat, J., Priede, J., and Sorkin, M. Z., "Numerical Simulation of a MHD Flow Induced by a Rotating Magnetic Field in a Cylindrical Container of Finite Length," *Proceedings of the International Conference on Energy Transfer in MHD Flows* (Cadarache, France), 1991, pp. 181–186.

[3]Priede, J., Thesis, 1993.

[4]Grants, I., Priede, J., and Gelfgat, J. M., "MHD Flow in Flat Layer in a Rotating and Steady Magnetic Fields," *Magnetohydrodynamics Journal*, Vol. 31, No. ©, 1995, pp. 141–150.

Chapter 43

Clarification on the Theory of the Electromagnetic Striction

R. Moreau,* P. Colpo,† and A. Gagnoud‡
Laboratory E.P.M.-MADYLAM, St.Martin d'Héres, 38400, France

I. Introduction

SOME metallurgical applications use an electromagnetic nozzle to control the flow rate and the radius of liquid metal jets. To characterize such a process, a theoretical model is useful to predict the striction phenomenon and to size the experimental device (power supply, coil size), which should predict how the jet free surface depends on parameters such as the coil current and the height of liquid metal above the striction. The explanation initially proposed by Garnier[1] to interpret his experiments on electromagnetic striction, founded on the Bernoulli relation including the magnetic pressure, predicts that the jet should separate from the wall below the coil. As a matter of fact, the experiments have demonstrated that the separation takes place above the coil. In this paper a new model of the flow is proposed aiming at clarifying the understanding of this phenomenon and at predicting both the position of the separation and the free surface shape downstream of the separation. The new ingredients of this model are 1) the internal overpressure due to the curvature of the streamlines and 2) the contact angle of the separation due to the surface tension.

II. Presentation of the Analytical Model

A. Former Model

Consider a liquid metal jet flowing into a coil where a high frequency current is circulating (Fig. 1a). A magnetic pressure $B_o^2/2\mu$ is created into the liquid metal jet where B_o is the local value of the magnetic field at the

*ENSHMG BP 95, 1340 rue de la Piscine.
†ENSHMG BP 95, 1340 rue de la Piscine.
‡ENSHMG BP 95, 1340 rue de la Piscine.

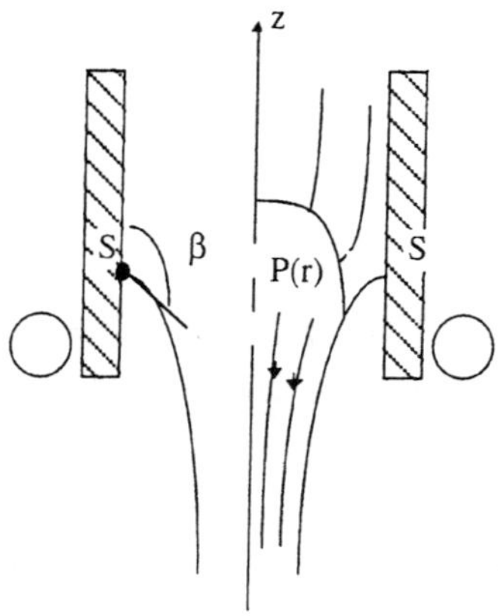

a

Fig. 1a Streamlines, pressure variation across the radius, contact angle β.

surface. To model the phenomenon, the formulation expressed by Garnier[1] is based on Bernoulli and mass conservation:

$$\frac{\rho U_o^2}{2} + \frac{B_o^2}{2\mu} = \text{cst} \tag{1}$$

$$U_o R_o^2 = \text{cst} \tag{2}$$

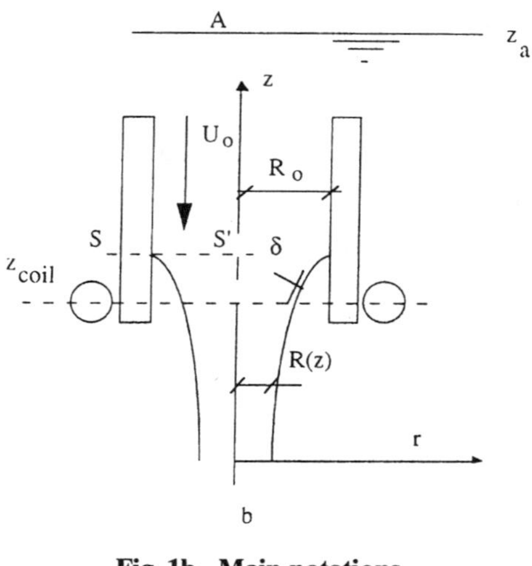

b

Fig. 1b Main notations.

which assume that the streamlines are quasiparallel to the jet axis. Then, the striction formula deduced from the Eq. (1) and (2) is

$$\frac{R_\infty}{R_o} = \left(1 + \frac{B_o^2}{\mu \rho U_o^2}\right)^{-1/4} \tag{3}$$

This model conjectures that, where the magnetic pressure is decreasing, the kinetic energy increases, and, by continuity, the jet cross section decreases. Therefore, this model cannot predict the right position of the separation from the wall that is located above the coil where the magnetic pressure increases. Nevertheless, the calculated striction ratio (3) does not significantly disagree with the measurements.

B. New Model

The main idea of our new model is to write a better momentum budget for each slice of thickness dz taking into account not only the magnetic overpressure but also the internal overpressure due to the curvature of the streamlines. To do this, an approximation of the velocity distribution is necessary from which the pressure distribution across the jet $P(r)$ is derived. As the inertial contribution introduces the first derivatives $R'(z)$ and $R''(z)$ of the unknown expression $R(z)$, the momentum balance must be a differential equation to which appropriate boundary conditions have to be associated. One of them expresses the role of surface tension at the separation point S, which controls the contact angle (the same as in steady conditions, typically $\beta = 135$ deg because S is a stagnation point). It imposes the slope $R'(z_s)$ taken as unity in the following.

This model is built with a few other simplifying assumptions: the frequency is so high that the electromagnetic skin depth δ is much smaller than R (therefore, the magnetic overpressure is $B_o^2/2\mu$, as in Garnier); the capillary overpressure is neglected because R is much larger than the capillary length scale; the influence of gravity within the short length over which striction takes place is neglected (but the weight of liquid metal above the striction is not neglected); and the modeled expressions of the velocity components are derived from a stream function to guarantee the mass conservation.

Above the striction (section SS' on Fig. 1b) $R(z) = R_o$, the streamlines are parallel, and the velocity components are approximated (Q denotes the flow rate):

$$U_r = 0; \quad U_z = U_o = -\frac{Q}{\pi R_o^2} \tag{4}$$

From the ordinary Bernoulli relation, the pressure at S' (on the axis) is

$$P_{s'} = \rho g (z_a - z_s) - \frac{\rho U_o^2}{2} \tag{5}$$

Below the striction where $R(z)$ is an unknown function, the stream function is taken as

$$\psi = \frac{Qr^2}{2\pi R_o^2} \tag{6}$$

so that the velocity components have the following expressions:

$$U_r = \frac{1}{r}\frac{\partial \psi}{\partial z} = -\frac{QR_r'}{\pi R^3} \tag{7}$$

$$U_z = -\frac{1}{r}\frac{\partial \psi}{\partial r} = -\frac{Q}{\pi R^2} \tag{8}$$

Substituting them into the radial component of the motion equation gives

$$\frac{\partial P}{\partial r} = (\boldsymbol{j} \wedge \boldsymbol{B})r - \frac{\rho Q^2 r}{\pi}\left(\frac{R''}{R^5} - \frac{2R'^2}{R^6}\right) \tag{9}$$

From our assumption on δ, the integration of the first term on the second member is just $B_o^2/2\mu$. Then, after integration of Eq. (9), we get the pressure distribution within the jet:

$$P(r,z) = \frac{B_o^2}{2\mu} + \frac{\rho Q^2}{2\pi^2}\left[\frac{R''}{R^3} - \frac{2R'^2}{R^4}\right]\left[1 - \frac{r^2}{R^2}\right] \tag{10}$$

The second term in this expression (10), because of inertia, exhibits the main originality of this model. It is illustrated by the quantity P_i on Fig. 2.

Expression (10) must guarantee the pressure continuity in S'. So identifying $P(o, z_s)$ with P_s', we get

$$R_o R_s'' = 1 + \frac{2g(z_a - z_s)}{U_o^2} - \frac{B_o^2(S)}{\mu \rho U_o^2} \tag{11}$$

This is quite an important relation because it gives the value of R_s'' (a needed boundary condition) and it explicitly introduces the height $z_a - z_s$ of liquid metal above the striction (an important parameter from the experiments).

Finally, having expressions for velocity components and pressure in terms of the unknown function $R(z)$, to derive the ordinary differential equation for $R(z)$ from the momentum balance applied to a slice of fluid of thickness dz (see Fig. 3) is elementary. The general expression is

$$\oiint \rho\, U_z(\boldsymbol{U}.\boldsymbol{n})\, \mathrm{d}S = -\mathrm{d}Fp + \mathrm{d}Fm \sin\theta \tag{12}$$

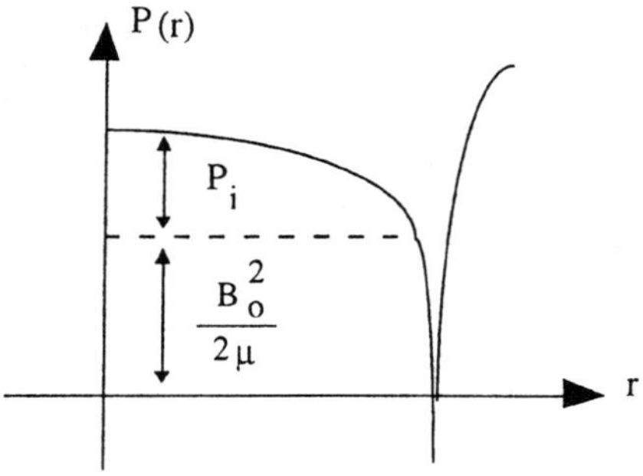

Fig. 2 Pressure variation across the jet.

where

$$Fp = \int_o^R p(r,z) 2\pi r \, \mathrm{d}r, \qquad \mathrm{d}Fm \sin\theta = \frac{B_o^2(s)}{2\mu} 2\pi R \, \mathrm{d}R$$

It is straightforward to get

$$-\frac{R'^2}{2R^4} + \frac{R''}{2R^3} + \frac{1}{R^4} + \frac{B_o^2}{\mu\rho U_o^2} \frac{1}{R_o^4} = \mathrm{cte} \tag{13}$$

where the two first terms represent the influence of the streamlines' curvature and the other terms representing the Bernoulli form.

The constant in the second member is determined far below the coil when $R = R_\infty$, $R' = R'' = 0$, and $B_o = 0$. The constant's value is therefore $1/R_\infty^4$. Relation (13) may also be applied in section z_s where R_o, R_s', and R_s'' are known to yield the new striction formula

$$\left(\frac{R_o}{R_\infty}\right)^4 = 1 + \frac{B_o^2}{2\mu\rho U_o^2} + \frac{g(z_a - z_s)}{U_o^2} \tag{14}$$

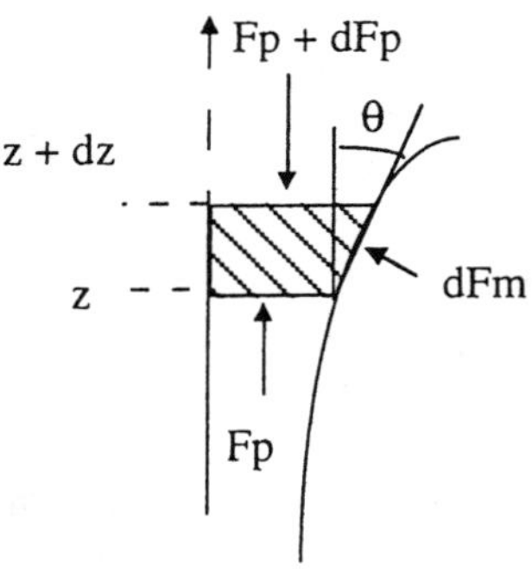

Fig. 3 Domain to which the momentum theorem is applied.

which exhibits two main differences with Garnier's previous formula (3): it explicitly includes the height of liquid metal above the striction $z_a - z_s$, and the effective magnetic pressure is only half of $B_o^2/2\mu$.

Now in differential equation (13) as well as in the boundary condition (11), the velocity U_o appears, which depends on parameters such as the height $z_a - z_s$ or the electric current I passing in the coil. This velocity may be related to the asymptotic value U and R by the relation:

$$U_o = U_\infty \frac{R_\infty^2}{R_o^2}; \qquad U_\infty = \sqrt{2g(z_a - z_\infty)} \tag{15}$$

where z_∞ is the height where the radius reaches its limit R. Again, noticeably the height of liquid is taken into account.

III. Numerical Procedure and Results

The algorithm is presented in Fig. 4. Essentially, the algorithm contains two main loops: one to iterate on $R(z)$ until convergence on the distribution of the magnetic field $B_o(z)$ and the other to iterate on the ordinate z_s of the separation point. In the first one, the solution of Eq. (13) is based on the Runge–Kutta 4 method and determines R_S'' in order that $R_\infty' = R_\infty'' = 0$. At each profile $R(z)$ is associated with a distribution of the magnetic field $B_o(z)$ computed with the code Ophelie[2] available in our laboratory. When the convergence on the couple $R(z), B_o(z)$ is reached, the numerical approximation of R_s'' is compared with the exact value given by Eq. (11), and the ordinate z_s is changed until a good agreement is obtained.

Different jet free surfaces have been computed for some combinations of the two main parameters I and $z_a - z_{\text{coil}}$. A few examples are shown in Fig. 5, which demonstrates that the predicted position of the separation is above the coil and yields a global shape of the free surface quite similar to Garnier's observations.

On a more practical point of view, Figs. 6–8 show how the coil current and the height of liquid metal above the coil influence the two main results: the striction ratio R_∞/R_o and the fluid velocity U_o (in the nozzle above the striction). Noticeably, an increase of the coil current by 50% (from 6500 A to 10,000 A) only reduces the confined radius R_∞ by 7%. This exhibits a limitation of this technique, which may be interpreted as follows. When I (or the power supply) is increased, z_s also increases, and, as consequence, the liquid metal free surface goes farther from the coil, giving more space for the magnetic flux. Therefore, the value of B_o increases much less than the parameter I.

IV. Concluding Remarks

As expected, the introduction of the streamlines curvature and the contact angle at the separation from the wall allow the prediction of jet free surfaces quite in agreement with experiments. Let us add a few remarks:

1) When we try to impose values of R_s' much smaller than unity (smooth contact angles), the computed solution never goes to a finite and positive

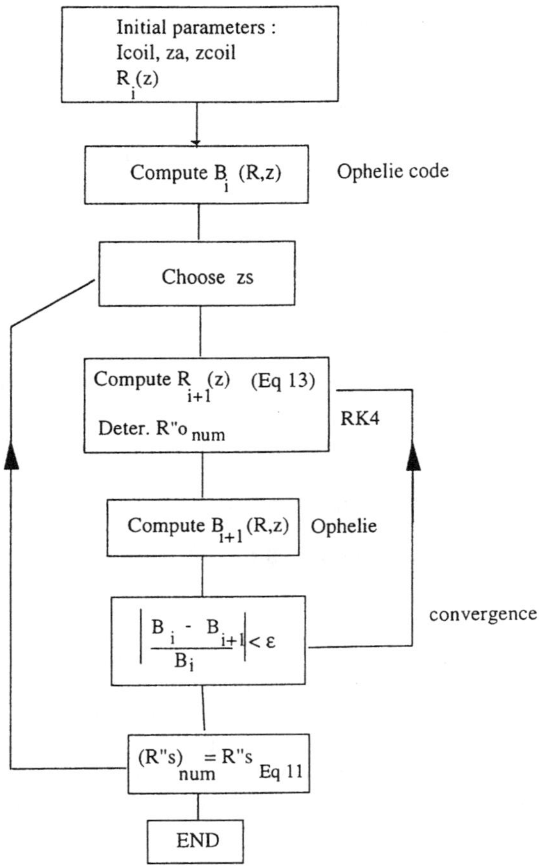

Fig. 4 Resolution algorithm.

Fig. 5 Computed and experimental metal liquid jet profiles.

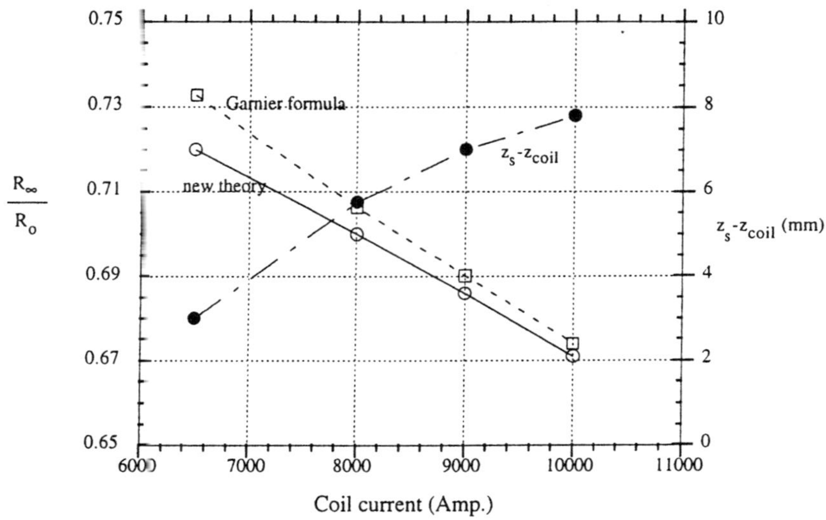

Fig. 6 Striction ratio and separation height vs the coil current ($z_a - z_c = 0.205$ m).

value of R_∞. The high surface tension of liquid metals therefore appears as a necessary condition for the existence of the electromagnetic striction.

2) The striction formula (14) involves a magnetic pressure reduced to half of the value expected from Garnier's formula.

3) No head losses at all are involved in this model. This seems justified because the electromagnetic head losses due to the stirring effect involve

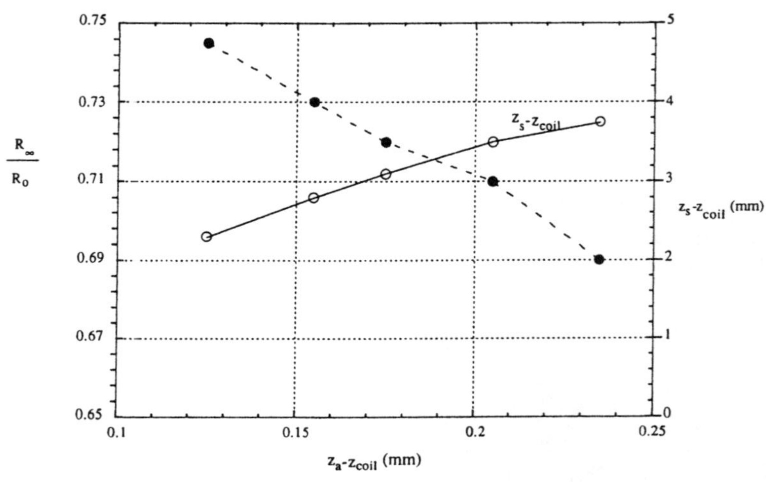

Fig. 7 Striction ratio and separation height vs the liquid height (*I*coil = 6500 A).

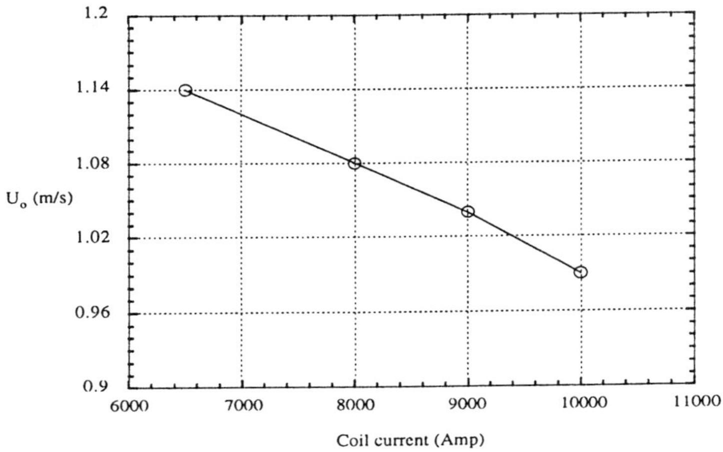

Fig. 8 Liquid metal jet velocity vs coil current.

velocities much smaller than U_o or $B_0^2/2\mu\rho$ (see for instance, Moreau[3], p. 228). The corresponding kinetic energy is therefore negligible by comparison with the magnetic energy.

References

[1]Garnier, M., "Une Analyse des Possibilités de Contröle Électromagnetique des Surfaces Libres des Métaux Fondus," Thèse d'Etat, USM, INP Grenoble, France.

[2]Gagnoud, A., and Leclercq, I., "Electromagnetic Modelling of Induction Melting in Cold Crucibles', *IEEE Transactions on Magnetics*, Vol. 24, No. 1, 1992, pp 573–575.

[3]Moreau, R., *Magnetohydrodynamics*, Kluwer, Dordrecht, The Netherlands, 1990.

Chapter 44

Physical Nature of Reduction Cell Instabilities

R. I. Lindsay* and P. A. Davidson†
University of Cambridge, CB2 1PZ, Cambridge, England, United Kingdom

I. Introduction

ALUMINUM is produced by the electrolysis of aluminum oxide in Hall–Heroult cells. These consist of a thin layer of electrolyte, known as cryolite, about 5-cm thick floating on a pool of liquid aluminium about 20-cm deep. The two fluids are bounded below and bounded laterally by a carbon cathode block. A carbon anode block acts as an incomplete top boundary, allowing the cryolite to rise up about 25 cm in the 10-cm-wide channels that surround the block. The whole cell is typically of dimensions 8×4 m in plan (Fig. 1).

Current is passed through the anode into the cryolite layer where the reduction process takes place producing the denser molten aluminum, which makes its way to the bottom of the cell, and oxygen, which reacts with the carbon to give carbon dioxide, gradually burning away the anode. The gas escapes through the open channels of the cell. The anode must be continuously lowered and eventually replaced. The liquid aluminum is regularly drawn from the bottom of the cell.

A common problem in aluminum reduction cells is the appearance of gravity waves at the cryolite-aluminum interface, which, if their amplitude becomes too large, can cause the aluminum to rise and short circuit with the anode. This is a form of interfacial instability and is because of the electromagnetic body force. There is some evidence that this instability is triggered by the cryolite layer being too thin. It acts to limit the efficiency of the cell by requiring a thicker layer of electrically resistant material. The period of the disturbance is typically that of gravitational modes, i.e., about 30 s, the growth rates are very slow, on the order of hours, and the wavelengths are 1 m or more. Very small changes to the background magnetic field B can bring about the instability.

*Department of Engineering, Trumpington Street.
†Department of Engineering, Trumpington Street.

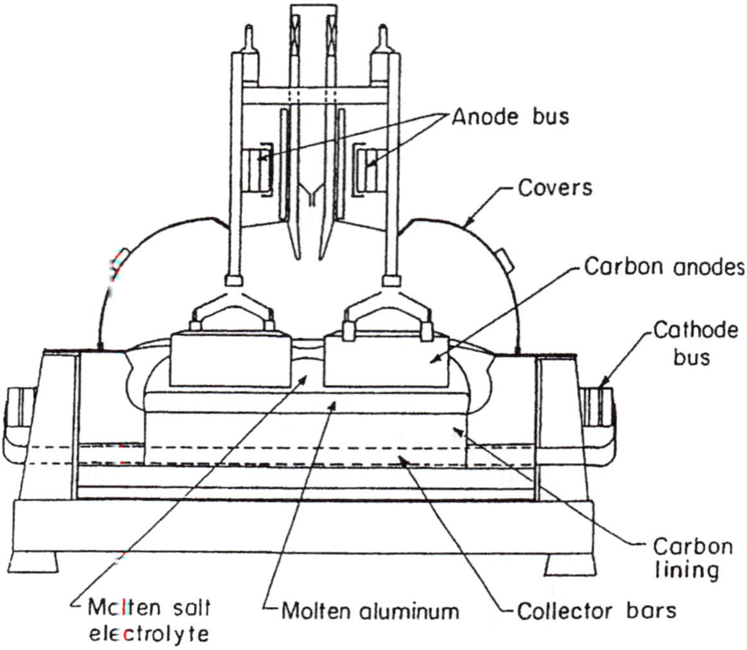

Fig. 1 A typical aluminum reduction cell.

One of the first theoretical studies of the instability is by Urata,[1] who highlights the importance of the horizontal gradients of the vertical background magnetic field B_z on the onset of instability. Moreau and Ziegler[2] use a perturbation analysis to show the role of the horizontal component of the current. This work has been extended by Pigny and Moreau.[3] Sneyd[4] focused his study on the effect of the horizontal gradients of the horizontal background magnetic field B_x but later showed that a much stronger instability arises because of the vertical gradients.[5]

Davidson and Boivin[6] introduced a global energy expression, extended by Davidson,[7] which not only confirms the importance of the horizontal gradients of B_z to unstable interfacial waves but also provides a physical mechanism. This mechanism explains the results of Sneyd.[45]

The earlier models assume that the cell is of finite depth but infinite transverse dimensions. They describe the perturbed interface as a traveling wave and neglect the effects of the lateral cell boundaries and channels. This leads to a dominant mechanism in which there is a flux of mechanical energy from the far field, which is induced by the action of the wave pressure on the transverse velocity component. These transverse velocity components are themselves induced by the electromagnetic body force.

More recent work by Sneyd and Wang[8] models the cell as a finite rectangular tank. This leads to a description of the interface as a general combination of gravity-wave modes, i.e., standing waves. The instability is shown to occur via mode interactions. If two modes have similar frequencies, they may resonate, the electromagnetic perturbation force due to one mode feeding energy into another, and become unstable at a low current threshold. The stability criteria then depends only on B_z.

However, both theories idealize the cell configuration in that they both ignore the open channels at the top perimeter of the cell. This is important as the channels could influence the form of the interfacial disturbance. (The channels were investigated by Moreau and Evans,[9] but only in the limits of very wide or very narrow channels on the cryolite layer and only in the context of mean flow.) This raises the question of whether the open channels affect the way in which traveling waves reflect at the cell boundary to produce standing waves. In this paper we discuss one possible mode of interaction between the channels and the finite rectangular cell, namely that of simple harmonic sloshing. We show that this sloshing has no appreciable effect on the standing-wave frequencies: the corresponding natural frequencies differ too much to allow for interaction. Because sloshing only represents one of many ways in which the channels may play a part, the effect of the channels is still an open question. However, these preliminary results seem to support the use of the standing-wave model.

II. Traveling-Wave Model

In this section, we review briefly the earlier traveling-wave models of cell stability. If we begin by neglecting the finite transverse dimensions of the cell, as in Ref. 2, then the perturbed interface may be assumed to take the form of a traveling wave η:

$$z_s = \eta\,(x,t) = \eta_0 \exp\,[st - ikx] \quad \text{with} \quad s = \sigma + i\omega_n \tag{1}$$

where σ gives the growth rate and ω_n the natural frequencies. Figure 2 gives the cell geometry.

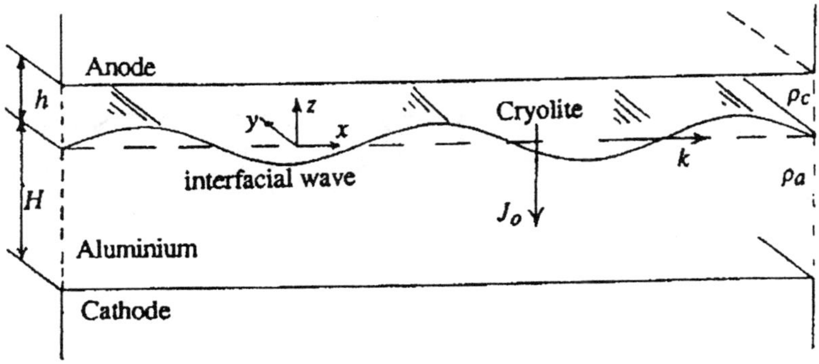

Fig. 2 The cell geometry.

We assume that the characteristic time scale for the wave is much greater than the diffusion time for the magnetic field, as is observed experimentally. The conductivities of the materials are ranked as follows:

$$\sigma_{\text{aluminum}} \gg \sigma_{\text{carbon}} \gg \sigma_{\text{cryolite}} \tag{2}$$

This means that the current is refracted toward the vertical as it enters the cryolite and in the limit of small σ_{cryolite} is taken to be purely vertical $\boldsymbol{J} \simeq J_z$. The disturbance of the interface causes a spatial variation in the current field. We would expect the normal current to increase at the wave crests and decrease in the valleys, resulting in a spatially periodic current flow in the aluminum. Because the resistivity is much lower in the aluminum, these current loops close within the aluminium layer (Fig. 3). In the long-wavelength limit, the current is considered to be almost purely horizontal in the aluminium, i.e., $\boldsymbol{J} \simeq J_x$.

We may show that the work done by the first-order perturbation in the magnetic body force δF is given by[7]

$$\delta \boldsymbol{F} \cdot \boldsymbol{u} = \nabla \cdot \left[\frac{(\boldsymbol{b} \cdot \boldsymbol{u})\boldsymbol{B} + (\boldsymbol{B} \cdot \boldsymbol{u})\boldsymbol{b}}{\mu} \right] - \nabla \cdot \left(\frac{\boldsymbol{B} \cdot \boldsymbol{b}}{\mu} \boldsymbol{u} \right) - 2\tau_{ij} s_{ij} \tag{3}$$

where $\boldsymbol{u}$ is the velocity field; $\boldsymbol{B}$ and $\boldsymbol{b}$ are the unperturbed and perturbed magnetic fields, respectively; μ is the permeability; τ_{ij} is the perturbed Maxwell shear stress; and s_{ij} is the strain-rate tensor for the motion. The three terms on the right-hand side correspond to magnetic shear work on the boundary, magnetic pressure work on the boundary, and the conversion of energy by the Maxwell shear stresses.[7]

By using a linear drag term, assuming that the spatial gradients in the background magnetic field $\boldsymbol{B}$ are small in comparison with the interfacial wavelength, and also restricting the investigation to marginally un-

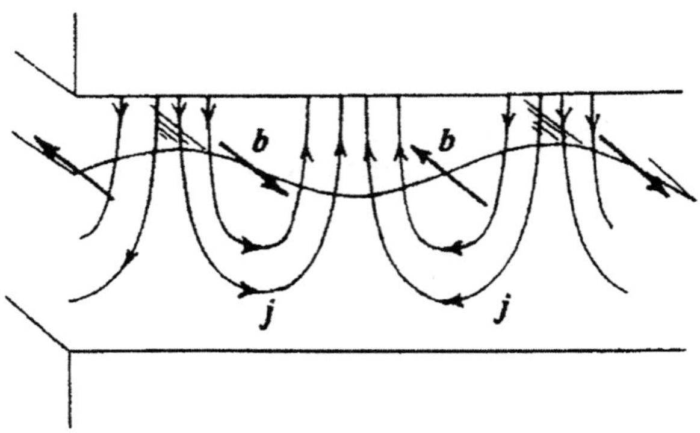

Fig. 3 The perturbation in the current and magnetic fields.

stable waves, we may show that Eq. (3) leads to the integrated energy equation[7]:

$$\frac{\partial E_w}{\partial t} + \frac{\partial}{\partial x}\int k_w U_{0x}\,\mathrm{d}z + \frac{\partial}{\partial y}\int k_w U_{0y}\,\mathrm{d}z + U_{0x}\frac{\partial}{\partial x}\left(\frac{1}{2}\Delta\rho g\eta^2\right)$$

$$= -\rho\int u_i u_j S_{0ij}\,\mathrm{d}z - \rho\kappa\int \boldsymbol{u}^2\,\mathrm{d}z - \frac{\partial}{\partial x}\int\left(p + B_y b_y/\mu\right)u_x\,\mathrm{d}z$$

$$-\int p_g\frac{\partial u_y}{\partial y}\,\mathrm{d}z + \frac{1}{\mu}\int\frac{\partial B_x}{\partial y}b_y u_{gx}\,\mathrm{d}z + \frac{1}{\mu}\int\frac{\partial B_z}{\partial_y}b_y u_{gz}\,\mathrm{d}z \tag{4}$$

where E_w is the vertically integrated mechanical wave energy, k_w is the local kinetic energy, U_0 is the unperturbed velocity field, ρ is the fluid density, κ is the friction coefficient, and g is the acceleration due to gravity. The last three terms on the left-hand side of the energy equation correspond to the redistribution of the wave energy because of the mean flow and do not represent potential sources of instability. The first term on the right-hand side, when averaged, represents the interaction of the wave stress with the strain-rate tensor S_{0sij} and is purely mechanical in nature. The second term is purely dissipative and cannot lead to a magnetohydrodynamic instability. The third term, the wave energy flux with augmented pressure, averages to zero over a wavelength.

The final three terms, however, play a more important role. The first can lead to a transfer of energy when the wave pressure p_g and the lateral expansion and contraction of the wave, represented by the strain $\partial u_y/\partial y$, are in-phase. The second and third term give the work done by the horizontal and vertical Maxwell stresses, respectively.

By only allowing for linear gradients in the magnetic field and ignoring the effects of the background motion U_0, as well as that of mechanical dissipation κ, Davidson[7] has shown that, in the case of long wavelengths, the dominant source of energy is the work done by p_g on the transverse motion u_y. The growth rate for such waves is given by

$$\sigma = \frac{J_0}{2\bar{\rho}\omega_n}\frac{\partial B_z}{\partial y}\frac{1}{(kh)(kH)} \tag{5}$$

where J_0 is the background current, $\bar{\rho}$ is the average density, and h and H represent the undisturbed depth of the cryolite and aluminium layers, respectively.

The general criteria for stability is a low value of ΔB_z. Any finite value of $\partial B_z/\partial y$ in Eq. (5) would lead to a nonzero growth rate, but in practice there is a gradual competition between ΔB_z and friction. The threshold value is determined by viscous dissipation.

Curiously, the source of energy for the instability is a flux of mechanical energy from the far field rather than a conversion of magnetic to mechanical energy as would be the case for work done by the Maxwell shear stresses. If the fluids were assumed to be contained within a finite cell, there would be no source for this mechanical energy; however, there would still be a source of instability via the Maxwell shear stress τ_{zy}. The model fails to predict the source of this far-field energy and what happens to the waves as they approach the cell boundaries.

III. Standing-Wave Model

A. Interacting Modes

In these models, the aluminum reduction cell is assumed to be a finite rectangular tank. The disturbed interface takes the form of standing waves, with the general term a combination of gravity-wave modes:

$$\eta(x,y,t) = \sum_{m,n=0}^{\infty} \hat{\eta}_{mn}(t) \cos\left(\frac{m\pi x}{L_x}\right) \cos\left(\frac{n\pi y}{L_y}\right) \tag{6}$$

with L_x and L_y representing the horizontal dimensions of the box. This analysis is only valid for a closed box, as formulated by Sneyd and Wang[8] and others.

The amplitudes of the gravity modes are weakly coupled by the electromagnetic force. Associated with each wave mode is a corresponding disturbance in current $\boldsymbol{J}_{mn}$, which has the same wavelength as η_{mn}. However, the expression for the magnetic body force $\boldsymbol{J}_{mn} \times \boldsymbol{B}$ involves multiple wavelengths and has a different orientation from the mn wave. This provides a means of coupling different modes. This occurs for two reasons: first, the magnetic field $\boldsymbol{B}$ is spatially varying, and second, $\boldsymbol{B}$ has a component in the z direction $\boldsymbol{B}_z$, which when coupled with the predominantly transverse $\boldsymbol{J}_{mn}$ produces a force perpendicular to the initial mode. Thus, energy can be transferred from one wave to another via the magnetic field $\boldsymbol{B}$.

The general stability criteria depends primarily on the magnitude of the vertical component of the magnetic field $\boldsymbol{B}_z$. The system can become unstable at a very low threshold value, but only after two natural frequencies collide on the real axis, as in Fig. 4.

Destabilization occurs through the resonant interaction of two modes. The electromagnetic force feeds energy from one mode into the other, so that a pair of adjacent frequencies can become unstable. This effect is more likely to occur when the two natural frequencies of the modes are similar. This is because the natural frequencies of the modes must move in a continuous manner in the complex plane as the magnetic field is increased. Modes can only go unstable in complex conjugate pairs, requiring that two modes first approach and then coincide on the real axis.

A useful means of understanding the resonant interaction is through the use of the following simple mechanical analogy.

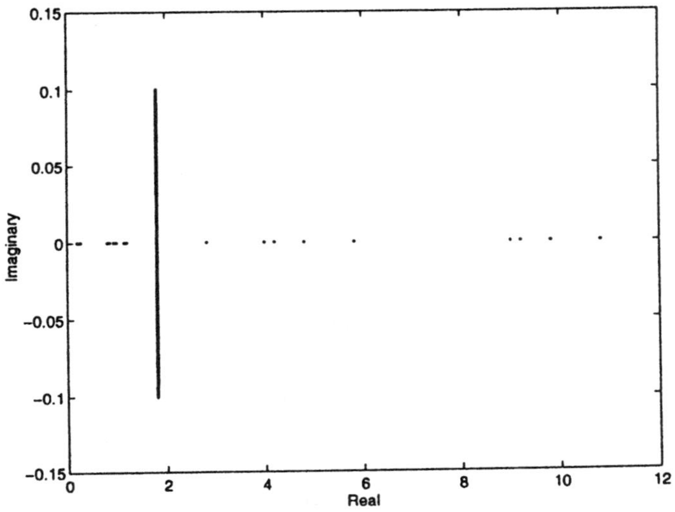

Fig. 4 The normal frequencies of a particular aluminum reduction cell, showing the resonant interaction between two modes. Two frequencies approach each other along the real axis, collide, and then go unstable.

B. Simple Mechanical Analogy of Standing-Wave Mode Interactions

The essential physical mechanism underlying the mode interactions in the standing-wave model can be captured by considering a compound pendulum with two degrees of freedom. We imagine a highly conducting solid plate of thickness H and horizontal dimensions L_x and L_y corresponding to the aluminum layer. The plate is suspended at a distance h_0 from a solid boundary, and the only movements allowed are the angular displacements θ_x and θ_y around the x and y axes, respectively. The area between the plate and ceiling is taken to be filled with a poorly conducting medium corresponding to the cryolite layer. (See Fig. 5.)

We begin by assuming that there exists a background current $\boldsymbol{J}_o$, which is purely vertical, and a background magnetic field $\boldsymbol{B}_o$. The transverse components of $\boldsymbol{B}_o$, due to the vertical current, are assumed to be much smaller than the vertical component. The current is tapped at the center of the plate in the equilibrium configuration I_0. The time scale of the movement is so much larger than that of the current, so that $\boldsymbol{J}$ and $\boldsymbol{B}$ are taken to always be fully diffused. We make the additional assumptions that

$$\sigma' \ll \sigma \tag{7}$$

so that the resistance in the system is dominated by the "electrolyte" layer, and

$$\rho' \ll \rho \tag{8}$$

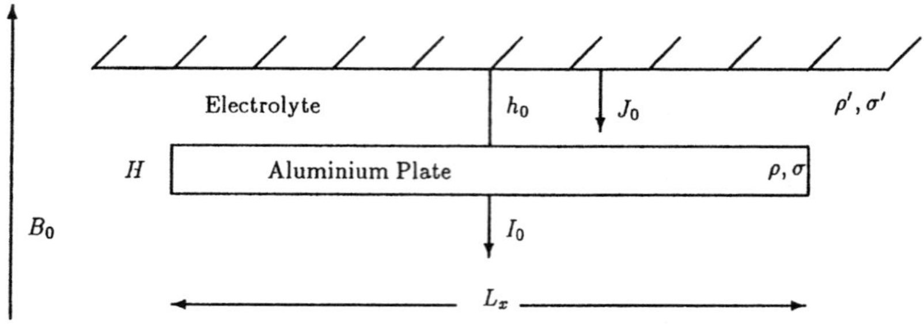

Fig. 5 The compound pendulum.

so that the intermediate medium has negligible inertia. Finally, we assume that

$$L_x, L_y \ll h_0, H \tag{9}$$

Using the above conditions, we can solve for the electric potential Φ:

$$\Phi = V_0 \frac{z}{h(x, y)} \tag{10}$$

where V_0 corresponds to the potential at the top boundary and h is the local separation of the plate and the top boundary:

$$h\,(x, y) = h_0 + \theta_y x - \theta_x y \tag{11}$$

From Eqs. (10) and (11) we can solve for the current $\boldsymbol{J}$, in the electrolyte:

$$\boldsymbol{J} = -J_0 \left(1 + \frac{\theta_x}{h_0} y - \frac{\theta_y}{h_0} x \right) \hat{e}_z = \boldsymbol{J}_0 + \boldsymbol{j} \tag{12}$$

Because the aluminum plate is such a good conductor, any current that reaches the plate flows horizontally within the plate. The unperturbed current $\boldsymbol{J}_0$ is tapped off centrally, but the perturbed current $\boldsymbol{j}$ recirculates in closed loops (just as in the traveling-wave model for the aluminum reduction cell). We can calculate the perturbation in the current moving along the plate in each direction $\delta \boldsymbol{I}$ by integrating the expression for $\boldsymbol{j}$, the perturbed current:

$$\delta I_x = \frac{-J_0 L_y \theta_y}{2h_0} \left[x^2 - \left(\frac{L_x}{2} \right)^2 \right] \tag{13}$$

$$\delta I_y = \frac{J_0 L_x \theta_x}{2h_0}\left[y^2 - \left(\frac{L_y}{2}\right)^2\right] \tag{14}$$

This, in turn, gives an expression for the perturbed magnetic body force $\delta\boldsymbol{F}$:

$$\delta F_x = -\frac{J_0 L_x \theta_x B_0 L_y^3}{12h_0} \tag{15}$$

$$\delta Fy = -\frac{J_0 L_y \theta_y B_0 L_x^3}{12h_0} \tag{16}$$

By taking moments about the pivot point, we can derive two equations that describe the motion of the plate:

$$\left(\frac{\ddot{\theta}_x}{L_x^2}\right) + \frac{12gh_0}{L_y^2}\left(\frac{\theta_x}{L_x^2}\right) = -\frac{J_0 B_0}{\rho H}\left(\frac{\theta_y}{L_y^2}\right) \tag{17}$$

$$\left(\frac{\ddot{\theta} y}{L_y^2}\right) + \frac{12gh_0}{L_x^2}\left(\frac{\theta_y}{L_y^2}\right) = +\frac{J_0 B_0}{\rho H}\left(\frac{\theta_x}{L_x^2}\right) \tag{18}$$

The skew symmetry of the coupled system [Eqs. (17) and (18)] ensures a route to instability. The pendulum being disturbed by an angle θ_y around the y axis brings about a flow of current δI_x in the x direction. This results in a force $-\delta F_y$ in the y direction that in turn causes the pendulum to swing by an angle θ_x. This disturbance brings about a flow of current δI_y that gives a force $-\delta F_x$, which then acts to amplify the initial disturbance θ_y.

The frequency of the system ω is given by

$$\omega^2 = \frac{1}{2}\left(\omega_x^2 + \omega_y^2\right) \pm \sqrt{\left(\omega_x^2 + \omega_y^2\right)^2 - \left(\frac{2J_0 B_0}{\rho H}\right)^2} \tag{19}$$

where $\omega_x^2 = 12gh_0/L_y^2$ and $\omega_y^2 = 12gh_0/L_x^2$. The system goes unstable when

$$\frac{J_0 B_0}{6\rho g h H} \geq \frac{\left|L_x^2 - L_y^2\right|}{L_x^2 L_y^2} \tag{20}$$

Thus, we have that the stability criteria depends on the vertical component of the background magnetic field.

The compound pendulum shares many properties with the closed-box problem. The current is essentially vertical J_z in the upper layer and horizontal $J_{x,y}$ in the bottom layer. In both cases, the wavelength of the mechanical disturbance of the bottom layer is the same as that of the disturbance in the current $\boldsymbol{J}$. However, because of coupling with the vertical magnetic field $\boldsymbol{B}_o$, the force resulting from this disturbance acts perpendicularly to the initial disturbance. Thus, both systems allow the transfer of energy between modes, and in both systems the amount of energy transferred depends on B_z.

IV. Cell Boundaries

The closed-tank model introduced by Sneyd and Wang[8] looks like an improvement on the infinite-box model that went before it. But the question remains whether it is physically realistic. In practice, the aluminum reduction cell is not completely closed but instead features open channels around the top. These open channels may affect the way in which traveling-wave disturbances reflect from the side boundaries to produce standing waves, as they provide another means of energy transfer.

In an attempt to incorporate the effect of the cryolite channels on the hydrodynamics of the cell, we now consider a two-dimensional cell configuration as shown in Fig. 6. The bulk of the fluid is contained in a box of width L_x and height $h + H$, where h and H are the depths of the unperturbed cryolite and aluminium layers, respectively. Some of the cryolite rises to a height l in channels of width d on either side of the anode.

We investigate the case of simple sloshing. Along the imagined boundary between the rectangular main cell and the channels, i.e., at $z = h$, the cryolite has the following vertical velocity v_z:

$$v_z = \begin{cases} U \sin \omega t & \text{if} \quad x \in (0, d) \\ 0 & \text{if} \quad x \in (d, L_x - d) \\ -U \sin \omega t & \text{if} \quad x \in (L_x - d, L_x) \end{cases} \tag{21}$$

where U is the amplitude of the oscillation. Instead of using the boundary condition of zero normal velocity along the top of the cell, as is done for the closed-box calculations, we use the condition (21), expressed as a Fourier series. We can then solve Laplace's equation for the velocity potential of the cryolite ξ_c, as a function of the undetermined frequency of the system ω.

From the resulting expression for the velocity potential, we derive a formula for the perturbed pressure p_c on the upper boundary, which takes into account the effects of the interfacial disturbance. Integrating this equation for the pressure under each channel must yield the same value as integrating the expression for the pressure perturbation obtained by considering only the changes in hydrostatic and dynamic pressure because of the movement of the fluid in each channel, i.e., to first order,

$$p_c = \rho_c U \left(l\omega - \frac{g}{\omega} \right) \cos \omega t \qquad \text{for} \qquad x \in (0, d) \text{ and } z = h \tag{22}$$

with ρ_c the density of the cryolite.

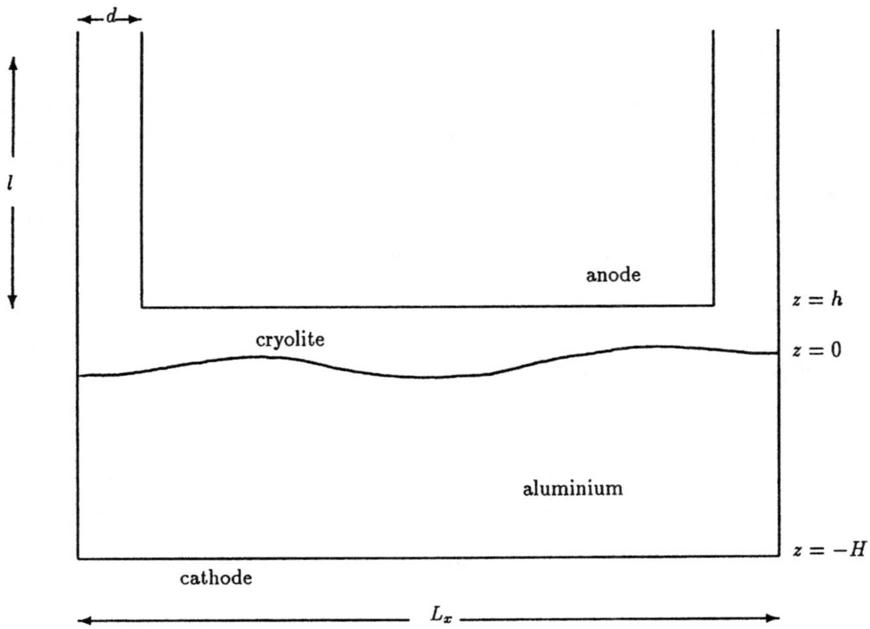

Fig. 6 The cell configuration including channels.

Matching the pressure integrals and rearranging give us an infinite order polynomial in ω, which must be truncated and the roots solved numerically to find the new characteristic frequencies of the coupled system. By taking the limit $d/L_x \to 0$, we recover the natural frequencies of the closed-box problem as expected. In the limit $\rho_a \to \rho_c$, we obtain the "pure" sloshing frequencies.

The so-called sloshing frequencies are very different from the natural frequencies associated with standing waves in a closed box. Thus, the two systems do not interact appreciably. Rather than modifying the standing-wave frequencies, allowing a simple sloshing mode simply adds a much higher frequency mode to the existing natural frequencies. Because high frequency responses are more vulnerable to viscous damping and are not observed in practice, it appears that the channels do not effect the dynamics of the cell. However, we should note that although sloshing is the most obvious way in which the channels could affect the system it is not the only way in which they could play a role.

V. Conclusions

We conclude that there are quite distinct physical differences in the traveling-wave and standing-wave instabilities. This leads to quite different stability criteria and different explanations of the physical mechanisms involved. Although one might expect to be able to recover the results of one model by taking an appropriate limit, this is quite clearly not the case.

In the case of the finite rectangular cell problem, we see that instability arises through the transfer of energy between modes. The introduction of the compound pendulum analogy makes the mechanism clear. The transfer is made possible through the coupling of wave modes in the expression for the magnetic body force. Because the perturbed current ***j*** has the same wave properties as the interface disturbance, the coupling of wave modes is thus crucially dependent on the value of the vertical component of the background magnetic field B_z.

The results of Sneyd and Wang[8] seem to be vindicated by the fact that the frequencies of the closed-box system are not much changed by the channels at least not in cases considered here.

References

[1]Urata, N., "Magnetics and Metal Pad Instability," *Light Metals*, 1985, pp. 581–589.

[2]Moreau, R. J., and Ziegler, D., "Stability of Aluminium Cells: A New Approach," *Light Metals*, 1985, pp. 359–364.

[3]Pigny, S., and Moreau, R., "Stability of Fluid Interfaces Carrying an Electric Current in the Presence of a Magnetic Field," *European Journal of Mech. B*, Vol. 11, 1991, pp. 1–20.

[4]Sneyd, A. D., "Stability of Fluid Layers Carrying a Normal Electric Current," *Journal of Fluid Mechanics*, Vol. 156, 1985, pp. 223–236.

[5]Sneyd, A. D., "Interfacial Instabilities in Aluminum Reduction Cells," *Journal of Fluid Mechanics*, Vol. 236, 1992, p. 111.

[6]Davidson, P. A., and Boivin, R. F. "Hydrodynamics of Aluminum Reduction Cells," *Light Metals,* 1991, pp. 1199–1211.

[7]Davidson, P. A., "An Energy Analysis of Unstable, Aluminum Reduction Cells," *European Journal of Mech. B.*, Vol. 13, 1945, pp. 15–32.

[8]Sneyd, A. D., and Wang, A., "Interfacial Instability Due to MHD Mode Coupling in Aluminum Reduction Cells," *Journal of Fluid Mechanics*, Vol. 263, 1994, pp. 343–359.

[9]Moreau, R., and Evans, J. W., "An Analysis of the Hydrodynamics of Aluminum Reduction Cells," *Journal of the Electrochemical Society*, Vol. 131, 1984, pp. 2251–2259.

Chapter 45

On Hydrodynamics, Heat and Mass Transfer Control in a Cylindrical Vessel by Rotating Magnetic Fields of Different Types

Yu. Gelfgat,* J. Krūmiņš,† and A. Abricka‡
Institute of Physics, Latvian Academy of Sciences, Riga, Latvia

Introduction

APPLICATION of rotating magnetic fields (RMF) for the solution of different technological problems requires widening of their action on characteristics of appearing flows in order to purposefully control heat/mass transfer in liquid and at phase interfaces. The present investigation discusses such possibilities for several nonstandard variations of RMF action on a liquid: with an asymmetric position of an RMF inductor and a cylindrical liquid volume of finite dimensions along the axis and with an action of two counter-rotating magnetic fields on a liquid volume. Here we consider distribution of electromagnetic forces in a fluid and their influence on velocity patterns, temperature distribution, and shape of the solidification front for different values of the Teylor, Grashof, and Prandtl numbers.

The preceding problems were investigated under definite conditions for single crystal growth with a temperature profile given along the cylinder axis. Two variations of the problems are considered:

1) A completely liquid or partially solidified volume is under the action of one RMF inductor when the ratio of their diameters is different as well as their positions with regard to the symmetry center of the volume either with or without electrically conducting edges and walls of a vessel (Fig. 1a); and

2) The volume is under the action of a field induced by two inductors of counter-rotating magnetic fields with a changed relation between electromagnetic forces generated by the both inductors (Fig. 1b).

---

*Professor, Head of Laboratory of MHD Technology.
†Laboratory of MHD Technology.
‡Researcher, Laboratory of MHD Technology.

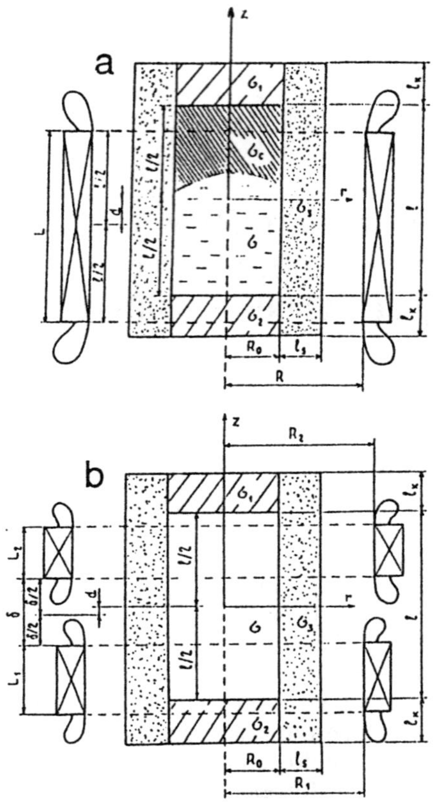

Fig. 1 Computational models: a) a liquid volume under the action of a field generated by one RMF inductor and b) a liquid volume under the action of a field generated by two RMF inductors.

The walls and edges of the vessel were assumed to be nonconducting.

Computation of Electromagnetic Force and Field Distribution

An original technique for calculation of an electromagnetic field of an RMF inductor at its idle operation[1,2] was worked out to account for the RMF inductor finite dimensions. This technique changes an actual current distribution in an inductor for an equal system of closed current loops covering an entire active surface of the inductor. Dimensions of these current loops were assumed to be infinitesimal.

The analysis shows that current strength in the loops is

$$I_k = \frac{A_o R}{p} \cdot e^{\omega t - \alpha p}$$

where A_o is the amplitude of the inductor linear current loading, R is its radius, p is a number of pole pairs, and α is the azimuthal angle.

From electrodynamics we know that a closed current loop has a magnetic dipole moment, which in this case is

$$\mathrm{d}\boldsymbol{M} = I_k \cdot \mathrm{d}S \cdot \boldsymbol{e}_r$$

and in void this dipole moment generates a field; its vector potential can be found due to the equation:

$$\mathrm{d}\boldsymbol{A} = \frac{\mu_o}{4\pi} \cdot \frac{\mathrm{d}\boldsymbol{M} \times \boldsymbol{V}}{V^3}$$

where $\boldsymbol{V}$ is the vector from a loop location to an observation point, but $\mathrm{d}S$ is the loop square.

Integrating $\mathrm{d}\boldsymbol{A}$ across the entire inductor surface, we get a resulting vector potential in the point of observation:

$$\boldsymbol{A} = \int_0^{2\pi} \int_{-L/2}^{L/2} \mathrm{d}\boldsymbol{A}$$

where L is the inductor length. We define an electromagnetic field of its idle operation:

$$\boldsymbol{H}^o = \frac{1}{\mu_o} \operatorname{rot} \boldsymbol{A}; \qquad \boldsymbol{E}^o = -\frac{\partial \boldsymbol{A}}{\partial t} = -i\omega \boldsymbol{A}$$

which in this case is presented by analytical formulas.

Two goals are achieved in such a way:

1) Together with the magnetic field distribution, we get that of an electric field as well, that is impossible, if the Biot–Savart law is used in calculations; and

2) The action of currents in frontal parts of the inductor winding is accounted for at the same time.

Let us suppose that in a known field of idle operation $\boldsymbol{H}^o, \boldsymbol{E}^o$, a conducting body of finite dimensions separated into zones with different electrical conductivities, is introduced. A noninductive approximation is the only restraint here, i.e., an electrical conductivity of a body should be small enough; in order a magnetic field of currents $\boldsymbol{H}^1$ induced in the body was small, too, comparing to $\boldsymbol{H}^o$, i.e., an actual magnetic field $\boldsymbol{H} \approx \boldsymbol{H}^o$.

An electric field, otherwise, will change considerably, and an actual electric field $\boldsymbol{E}$ should be presented as $\boldsymbol{E} = \boldsymbol{E}^o + \boldsymbol{E}^1$ where $\boldsymbol{E}^1$ is the field violation (distortion) caused by the presence of the body.

Because of the noninductive approximation rot $\boldsymbol{E}^1 = 0$, a scalar potential $\boldsymbol{U}$ can be introduced, which satisfies the Laplace equation $\Delta U = 0$ and $\boldsymbol{E}^1 = -\text{grad}\,\boldsymbol{U}$.

The conditions for the electric field transition over the boundary of two media can serve as boundary conditions: tangential components of the field and normal components of current should be continuous there, i.e.,

$$E_{t1} = E_{t2}; \qquad \sigma_1 E_{n1} = \sigma_2 E_{n2}$$

where σ_1 and σ_2 are the electrical conductivities of medium 1 and medium 2.

Having solved the equation $\Delta U = 0$, we find $\boldsymbol{E}$ distribution, and according to

$$\boldsymbol{f} = \mu_o \sigma \boldsymbol{E} \times \boldsymbol{H}$$

we find the distribution of electromagnetic force $\boldsymbol{f}$.

In such an approach, when the field of idle operation and field violation are found separately, a position for a conducting body is not defined before hand. Under such circumstances there exists an opportunity to consider variations for the RMF inductor and liquid volume relative position and their other characteristics. We should note that because an electromagnetic field possesses additivity features, the proposed computational technique permits us to consider resulting fields appearing under the action of several, two in particular, RMF inductors.

Calculation of Motion Characteristics of Liquid and Temperature Distribution

In a common case a motion of liquid is characterized by velocity fields $\boldsymbol{v} = v_r(r,z); v_\varphi(r,z); V_z(r,z)$ and pressure $p = p(r,z)$.

New variables can be introduced instead of v_r, v_φ, v_z, and p: Ω is the angular velocity of azimuthal rotation; and W is the moment of velocity rotor and ψ function of Stocks flow, which are defined as

$$\Omega = \frac{v_\varphi}{r}; \qquad W = r\left(\frac{\partial v_r}{\partial z} - \frac{\partial v_z}{\partial r}\right)$$

$$\frac{1}{r}\frac{\partial \psi}{\partial r} = -v_z; \qquad \frac{1}{r}\frac{\partial \psi}{\partial z} = v_r$$

Then, the system of equations turns into

$$\frac{\partial \Omega}{\partial t} = -\frac{1}{r^2}\,\text{div}\left(\text{r}^2 \boldsymbol{I}\right) + \frac{1}{\rho r}\left[\boldsymbol{J} \times \boldsymbol{B}_o\right]_\varphi$$

$$\frac{\partial W}{\partial t} = -r^2 \operatorname{div}\left(\frac{1}{r^2}\boldsymbol{J}\right) + \frac{r}{\rho}\left[\operatorname{rot}(\boldsymbol{j} \times \boldsymbol{B}_o)\right]_\varphi + r^2 \frac{\partial \Omega^2}{\partial z}$$

$$\boldsymbol{I} = -\nu \operatorname{grad} \Omega + \boldsymbol{v}\Omega$$

$$\boldsymbol{J} = -\nu \operatorname{grad} W + \boldsymbol{v}W$$

Referring to a dimensionless form of equations, the following parameters are chosen as characteristic: R_o is the radius of a vessel, $v_o = \nu/R_o$ is the velocity, and $t_o = R_o^2/\nu$ is the time. In a dimensionless form the system of equations turns into

$$r^2 \frac{\partial \Omega}{\partial t} = \operatorname{div}\left[r^2 \operatorname{grad} \Omega - \boldsymbol{v}\Omega\right] + F_\varphi$$

$$r^{-2} \frac{\partial W}{\partial t} = \operatorname{div}\left[r^{-2} \operatorname{grad} W - \boldsymbol{v}W\right] + \frac{\partial \Omega^2}{\partial z} - \frac{Gr}{r}\frac{\partial T}{\partial r}$$

$$\operatorname{div}\left[r^{-2} \operatorname{grad} \Psi\right] = r^{-2}W$$

The system of hydrodynamic equations is supplied with equations of heat/mass transfer:

$$\frac{\partial T}{\partial t} = \operatorname{div}\left[\frac{1}{Pr} \operatorname{grad} T - \boldsymbol{v}T\right]$$

Boundary conditions are presented as

$$\Psi|_w = 0; \quad \Omega|_w = 0; \quad \left.\frac{\partial \Omega}{\partial z}\right|_{f.s} = 0; \quad W|_{f.s.} = 0$$

$$\left.\frac{\partial \Omega}{\partial r}\right|_{r=0} = 0; \quad W|_{r=0} = 0$$

Here

$$F_\varphi = \frac{R_o^2}{\nu^2} \cdot \frac{r}{\rho} \cdot f_\varphi, \quad Pr = \frac{c_p \rho \nu}{\lambda}$$

$$Gr = \frac{\beta g \Delta T R_o^3}{\nu^2}; \quad Te = \frac{\mu_o^2 \sigma \omega R_o^4 A_o^2}{2\rho \nu^2}$$

where f_φ is the density of the φ component of an electromagnetic force acting in the point discussed; $\rho, \nu, c_p, \beta, \sigma$ are, correspondingly, the density, kinematic viscosity, heat capacity, coefficient of volumetric expansion, and electrical conductivity of a liquid. The obtained system of equations was solved by the method of finite differences. To solve grid equations, the method of changing directions was applied.

In equations for hydrodynamics and thermal conductivity, a motion in the solid phase is halted by introducing a large viscosity coefficient. A convective heat transfer is accounted for in the equation for thermal conductivity.[3]

Release of latent solidification heat in a temperature range is considered by the method proposed in Ref. 4. A δ function is introduced in the area under discussion, which permits us to insert a softened and effective heat capacity in the equation for thermal conductivity.

Influence of Design Parameters of the "Liquid Cylinder—Rotating Magnetic Field Inductor" System on Value and Distribution of Electromagnetic Forces

Electrodynamic and hydrodynamic phenomena occurring in a cylindrical volume under the action of RMF depend on the following parameters: an inductor linear current loading A, a number of pole pairs p, a ratio of diameters of a liquid cylinder and inductor R/R_o, a ratio of lengths of a cylinder and inductor l/R, a ratio between length and diameter of a liquid cylinder l/R_o, a shift of the liquid cylinder center along the z axis with regard to the center of an inductor d/R_o, thicknesses of hard electrically conducting edges and side walls l_k/R_o and l_s/R_o, and temperature distribution on a side surface of a liquid volume and cylinder edges.

Magnetic field intensity determines the value of an electromagnetic force, which is proportional to the square of the field intensity. Field intensity can be changed by various methods, the simplest of which is the variation of current strength in windings (coils) of an inductor. The rate of an electromagnetic force depends also on design parameters of a device: it decreases if a gap between a liquid cylinder and inductor enlarges, if a number of pole pairs of a RMF inductor increases, and if a vessel has electrically conducting side walls. Electrical conductivity of hard edges of a vessel with a fluid also affects the rate of the force if their electrical conductivity is larger than that of the fluid. For example, if their electrical conductivity is five times larger than that of a fluid, the electromagnetic force increases up to two times.

To study the action of the preceding parameters on hydrodynamic and heat characteristics, two components of a fluid motion are considered: an azimuthal flow, which is primary due to the fact that an electromagnetic force applied has the azimuthal component only, and a secondary, meridional flow generated by the action of centrifugal and buyoancy forces. The calculations considered are isolines of electromagnetic force f = const, isolines of angular velocity Ω = const or isolines of azimuthal velocity v_φ = const, streamlines Ψ = const, and isolines of temperature T = const.

Examples of characteristic computational results will be discussed in the following. Meanwhile, we should note that in an isothermal case when thermogravity convection determined by a temperature drop is ignored, a hydrodynamic pattern of azimuthal and meridional flows occurring does not practically depend on a manner of the electromagnetic force distribution if an RMF inductor is located symmetrically with regard to a liquid volume. As a matter of fact, variations of ratio of geometrical dimensions, numbers of pole pairs, electrical conductivities confining a liquid volume, and other factors are equivalent to a corresponding change of a linear current loading in an RMF inductor.

The only factor affecting a hydrodynamic pattern of flows appearing in a fluid is the symmetry factor of the RMF inductor location with regard to a plane perpendicular to an axis of a liquid volume and dividing it into halves. We assume here that an RMF inductor and liquid volume have one common axis.

Distribution of Electromagnetic Forces, Velocities, and Temperatures in an Asymmetrically Located Rotating Magnetic Field Inductor

Let us analyze an interaction between a distribution character and rate of an electromagnetic force and characteristics of a conducting fluid motion for different distances between central planes of a liquid volume and inductor. Let us first consider an isothermal flow (without thermogravity convection) in a cylindrical vessel whose edges and side walls have different electrical conductivities with various ratios of diameters of the inductor and liquid volume.

Figure 2 illustrates the results of a numerical simulation of the problem for the case of a symmetric location of an inductor and liquid volume (d = 0). As found out,[5] at a symmetric distribution of electromagnetic forces a pattern of the azimuthal velocity comprises a central core, near which a fluid rotates with an approximately constant angular velocity and boundary layers, where the velocity quickly comes to zero. At the same time a maximal calculated velocity is at about 0.7–0.8 of the radius length from the center. Meridional flows manifest themselves in two vortices located symmetrically with regard to the symmetry plane of the system under discussion. The preceding-mentioned conceptual velocity patterns remain the same, even if the electromagnetic force distribution changes by, for example, variations in a number of pole pairs, or geometrical ratio in the system, or electrical conductivities of hard walls of a vessel. Only absolute values of azimuthal and meridional velocities change.

Changes in the preceding velocity patterns appear if the electromagnetic force distribution in a fluid turns asymmetric. Here how such a situation is achieved is insignificant. The simplest way to achieve the goal is to shift an inductor along a common axis with regard to a liquid volume (d is dimension). Figures 3 and 4 show the computational results for the cases with d = 0.5 and d = 1.

The diagrams in Figures 3 and 4 show that a character of velocity patterns changes crucially. Meridional flows,[6] the most interesting from a practical point of view, differ greatly near upper and lower edges of a cylindrical volume. The rate of difference because of flow types and strength depends

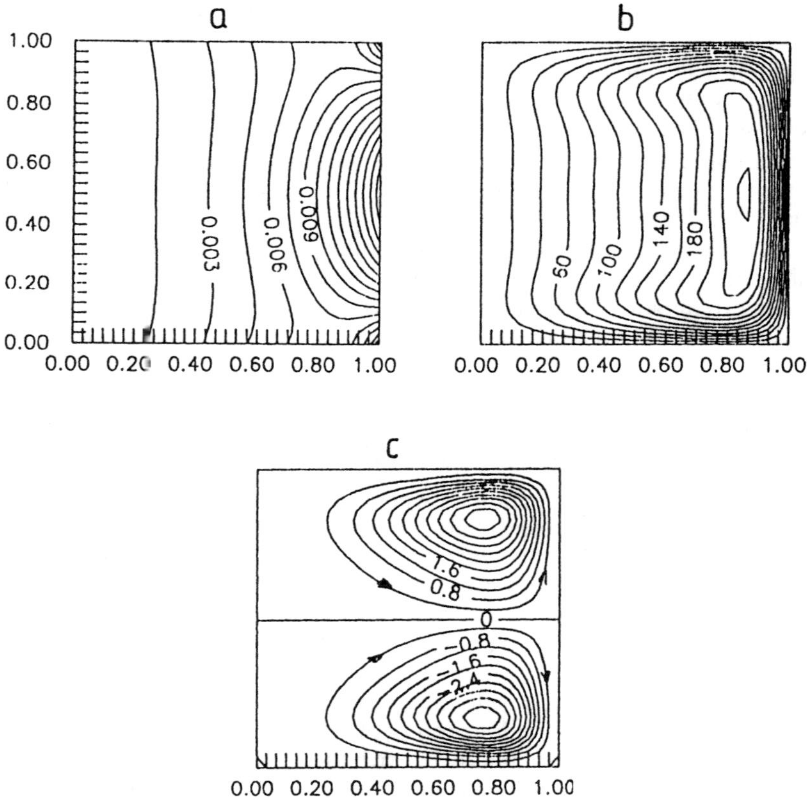

Fig. 2 Distribution of a) an electromagnetic force, b) azimuthal velocity, and c) streamlines of meridional flow for the case of a symmetrical position of an RMF inductor with $d/R_o = 0$, $l/R_o = 1$, $R/R_o = 2$, $p = 1$, and $Te = 7 \cdot 10^5$.

on d. In other words, in some growth technologies the possibility exists to purposefully control heat/mass transfer characteristics, for example, varying the level of axial symmetry at melting and solidification fronts.

Let us analyze now a nonisothermal flow and discuss the action of thermogravity convection on velocity patterns determined by the RMF action. To investigate the problem under action of all factors on the temperature distribution and shape of the solidification front in a solidifying sample seems wise. That is why the following phenomena and conditions were taken into account in the numerical simuation: solidification heat of a solidifying melt, different variations of boundary conditions on the walls of cylinders, free convective flows caused by a given temperature profile, different thermal conductivity values for solid and liquid phases, and different velocities of motion of a vessel with a melt along a given temperature profile.

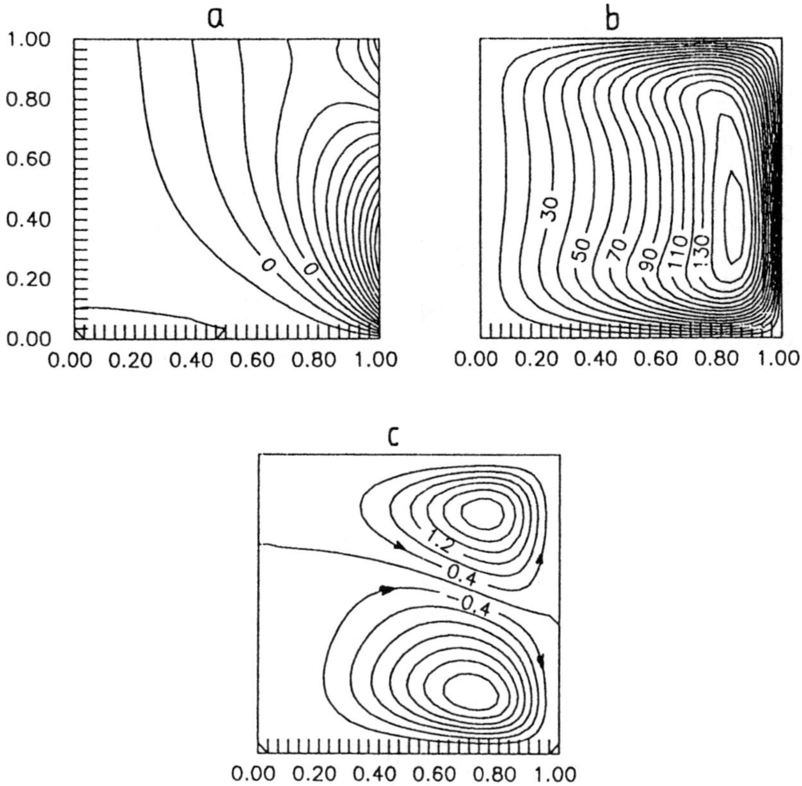

Fig. 3 The same as in Fig. 2 but with d/R_o = 0.5.

The calculations were performed with the following values of parameters: Gr = 0–10^6; Te = 0–5.10^5; pulling velocity of a vessel with a melt v = 0–300 mm/h; and the relation of thermal conductivities of liquid and solid phases λ_l/gik$_s$ = 1–10. Physical constants of a melt and solidifying sample were similar to those of GaAs.

The results of numerical simulations evidence that without RMF and with $v = 0, Gr = 0, \lambda_l = \lambda_s$, and equal temperature gradients in solid and liquid zones $[(dT/dz)_s = (dT/dz)_l]$, the temperature distribution in a cylindrical vessel causes a flat shape of the solidification front. A similar result is obtained if a weak free convective flow appears in the case discussed ($Gr \approx 10^5$), but not taking into account solidification heat in the calculations. Under conditions close to real (with account for all factors) but without the RMF action on a melt, a solidification front line becomes concave (Fig. 5). Curving of the interface increases with enlarging of the Gr number as well as velocity v.

Application of an RMF generated, for example, by a symmetrically located RMF inductor (Fig. 1a) causes a slight flattening of the solidification

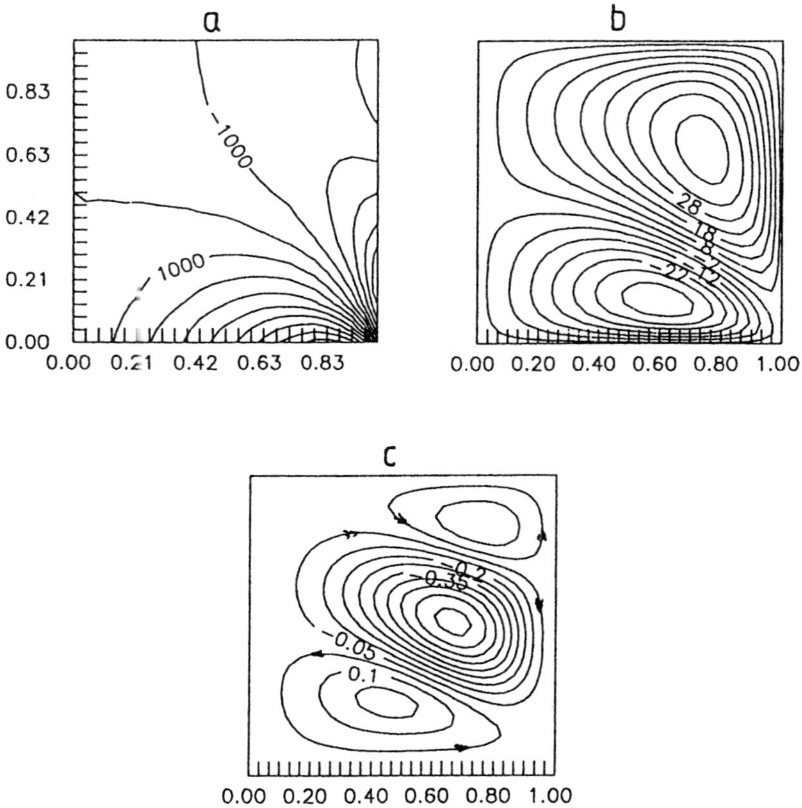

Fig. 4 The same as in Fig. 2 but with d/R_o = 1.0.

front line (Fig. 6) under conditions of strict quantitative correspondence of strengths of convective and meridional flows under the RMF action. If the correspondence was violated (Fig. 7), the interface greatly bends, even tending to an M-shape of the solidification front. Thermogravity convection together with a meridional flow induced by RMF will significantly change a resulting velocity pattern of the latter (compare Figs. 6 and 7). In this case in a fluid, there appear two stable meridional vortices with broken symmetry.

We must emphasize that for every situation (temperature drop, wall conductivity, pulling velocity, etc.) there exists an optimal value of RMF ensuring a decreasing of the interface curving. It is illustrated by Fig. 8, where one finds a pattern of meridional flows and temperature distribution under conditions similar to those of Fig. 7, but with an enlarged Grashof number $Gr = 5 \cdot 10^5$. In this case the shape of the solidification front tends to flatten.

A series of special calculations showed that application of an RMF inductor asymmetrically positioned along the vessel's axis provided a much wider range for the solidification front shape control.

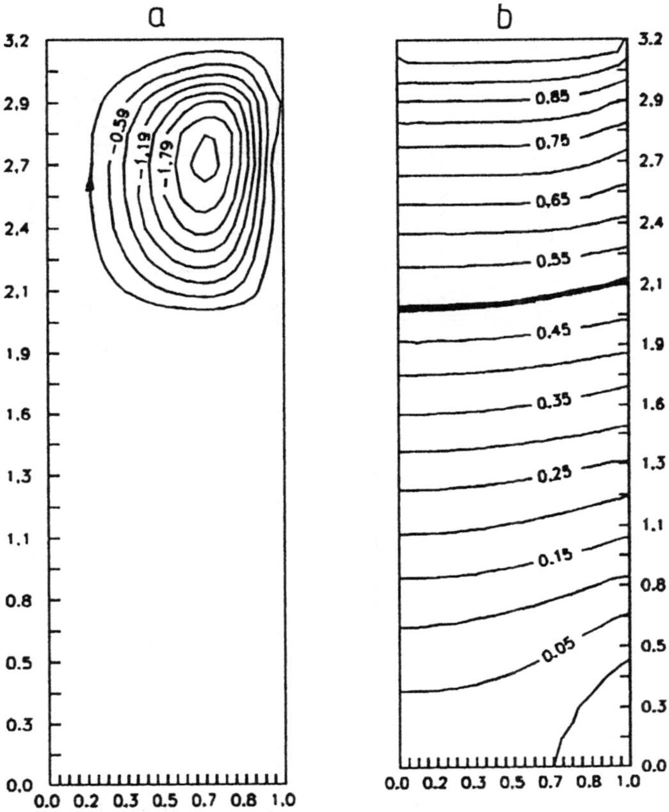

Fig. 5 Streamlines of a) a meridional flow and b) isotherms at $d = 0$, $Gr = 10^5$, and $Te = 0$.

In conclusion we should stress that the obtained calculation results for the velocity distribution in RMF and the field action on the shape of the solidification front are in a good agreement with experimental results. The experiments conducted quantitatively and qualitatively prove the possibility to use the proposed technique for the calculation of main characteristics of a technological process.

Distribution of Electromagnetic Forces, Velocities, and Temperatures with Two Counter-Rotating Magnetic Fields

Hydrodynamic and heat characteristics are most effectively controlled by the application of a superimposition of two counter-rotating magnetic fields (Fig. 1b). In this case varying linear current loading in the RMF inductors and their position along a vessel with a fluid, it is possible to obtain one- and

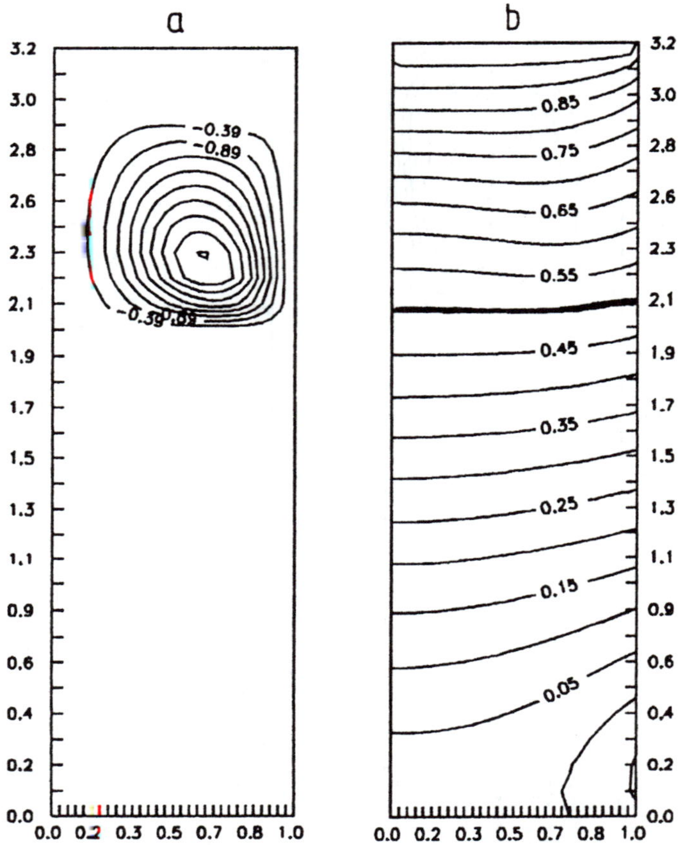

Fig. 6 Streamlines of a) a meridional flow and b) isotherms at d = 0, $Gr = 10^5$, and $Te = 10^5$.

multivortex velocity patterns and, correspondingly, required conditions for heat/mass transfer.

The preceding statement can be proved by computational results for such magnetohydrodynamic (MHD) action on a fluid. First, let us consider an isothermal case without thermoconvective flows ($Gr = 0$) when a fluid is affected in opposite directions by electromagnetic forces induced by two inductors placed at a distance d from each other. Figure 9 presents the results of a numerical simulation for such a case with $Te_1 = Te_2$. Here, the force and velocity distributions are strictly symmetric. A distinguishing feature is the appearance of an expressed shear layer between two meridional vortices.

With the variation of a Te_1/Te_2 relation in RMF inductors, the situation sharply changes. For example, if $Te_2/Te_1 = 10$ but with $(Te_1 + Te_2)/2 = 5 \cdot 10^5$, as in the former case, a pattern of meridional flow loses its symmetry

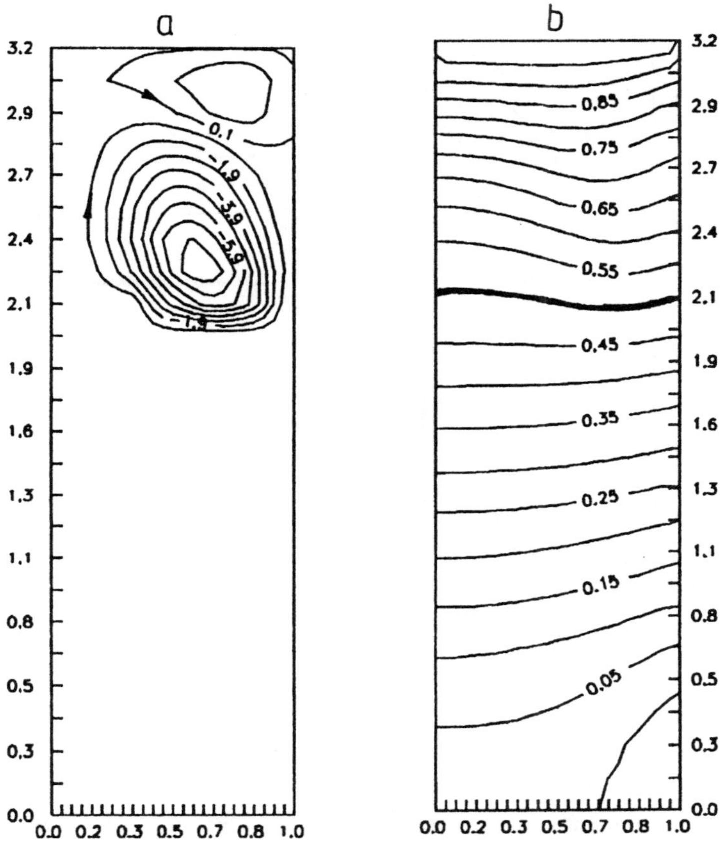

Fig. 7 Streamlines of a) a meridional flow and b) isotherms at d = 0, Gr = 10^5, and Te = 3.5. 10^5.

(Fig. 10). A meridional flow caused by Te_2 forces out a similar vortex caused by Te_1, and the strengths of these flows become different at upper and lower edges of a cylindrical volume.

In a nonisothermal case when a fluid is affected as much by electromagnetic as buyoancy forces and additional thermoconvective flows occur, a pattern of the velocity distribution differs greatly. For example, with $Te_2/Te_1 = 1(Te_1 = Te_2 = 5 \cdot 10^5)$ and $Gr = 10^6$ in a fluid instead of symmetric velocity patterns (Fig. 10) in the central zone of a volume, there appears one meridional vortex and practically no motion near edges (Fig. 11).

With asymmetric current loading in the inductors, for example, $Te_2/Te_1 = 10(Te_1 = 5 \cdot 10^5)$ and $Gr = 10^6$, the meridional velocities can again arrange two-vortex asymmetric patterns even with an opposite rotational direction near the edges of a cylindrical volume (Fig. 12). Patterns of the temperature distribution in a fluid change correspondingly (Figs. 11 and 12).

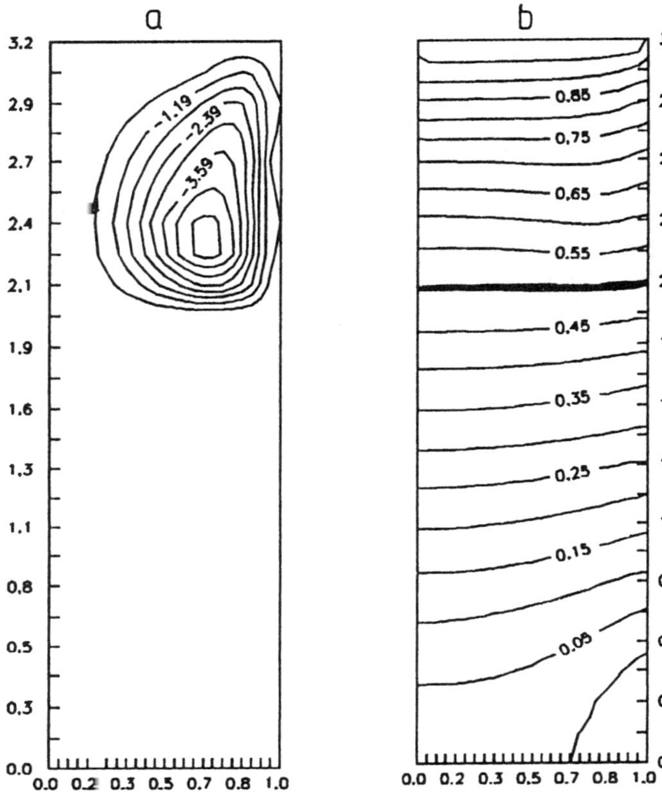

Fig. 8 Streamlines of a) a meridional flow and b) isotherms at d = 0, $Gr = 5 \cdot 10^5$, and $Te = 3.5.\ 10^5$.

The preceding examples of the obtained computational results prove the possibility to control hydrodynamics and heat/mass transfer in a fluid by two counter-rotating magnetic fields for a wide range of parameters. But the data of the numerical simulations performed must be proved experimentally. That is why to estimate their correctness, special experiments on measuring distribution of azimuthal velocities in a cylindrical ampoule with InGaSn melt were carried out.

An ampoule with the melt was placed coaxially in bores of two RMF inductors with opposite rotational directions similar to the diagram in Fig. 1b.

The results of measuring are depicted in Fig. 13 in comparison to corresponding calculations for the conditions of the experiment.

From the figure the idea follows that the data for the experiment and calculation are in a good agreement that is proved by the preceding discussion.

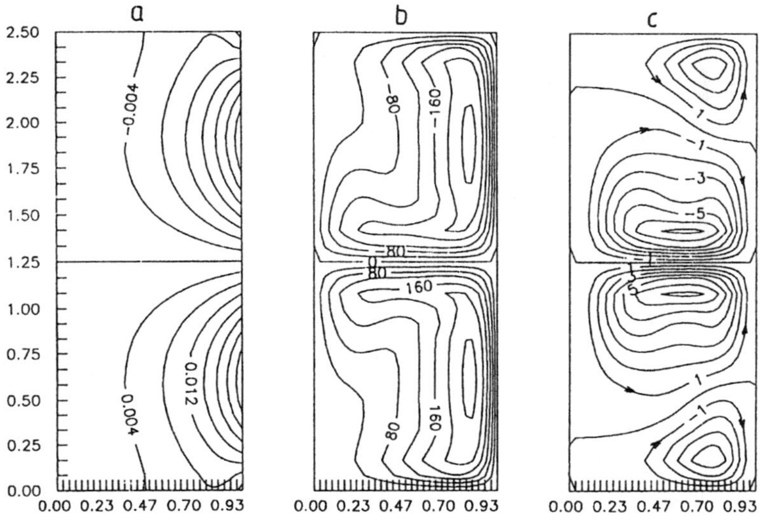

Fig. 9 Distribution of a) an electromagetic force, b) azimuthal velocity, and c) streamlines of meridional flow for the case of two RMF inductors with rotation in opposite directions and with $l/R_o = 2.5$, $L_1/R_o = L_2/R_o = 1$, $R_1/R_o = R_2/R_o = 2$, $p_1 = p_2 = 1$, $\delta/R_o = 0.5$, $d/R_o = 0$, $Te_1 = Te_2 = 5.10^5$, and $Gr = 0$.

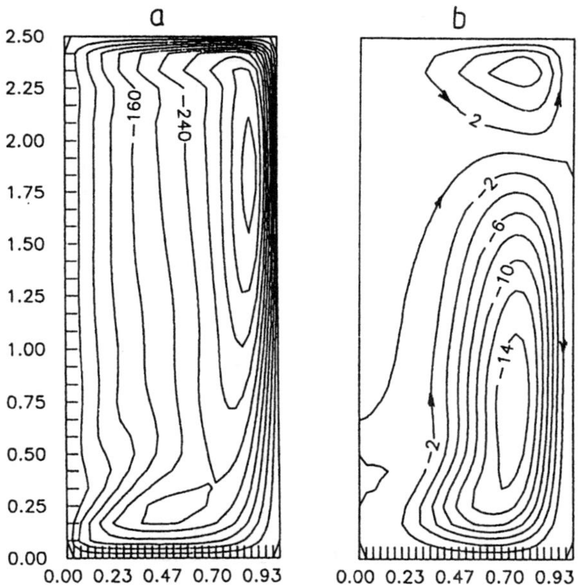

Fig. 10 Distribution of a) azimuthal velocity and b) streamlines of a meridional flow as in Fig. 9, but with $Te_2/Te_1 = 10$ and $Te_1 + Te_2/2 = 5 \cdot 10^5$.

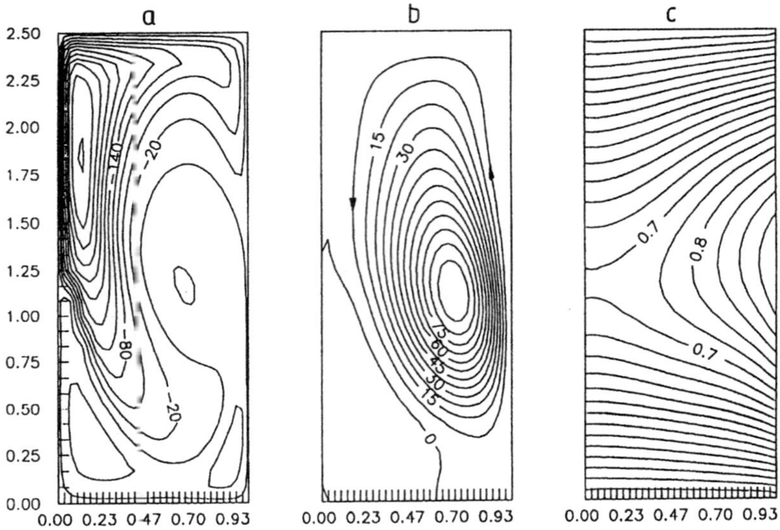

Fig. 11 Comparison of computational and experimental values for azimuthal velocity. Parameter values in calculation and experiment: $l/R_o = 5, L_1/R_o = L_2/R_o = 6.07$, $R_1/R_o = R_2/R_o = 12.14, p_1 = p_2 = 1, \delta/R_o = 2.86, d/R_o = 0, R_o = 14$ mm, $Te_1 = Te_2 = 2.7 \cdot 10^6$, and $Gr = 0$ (in calculations).

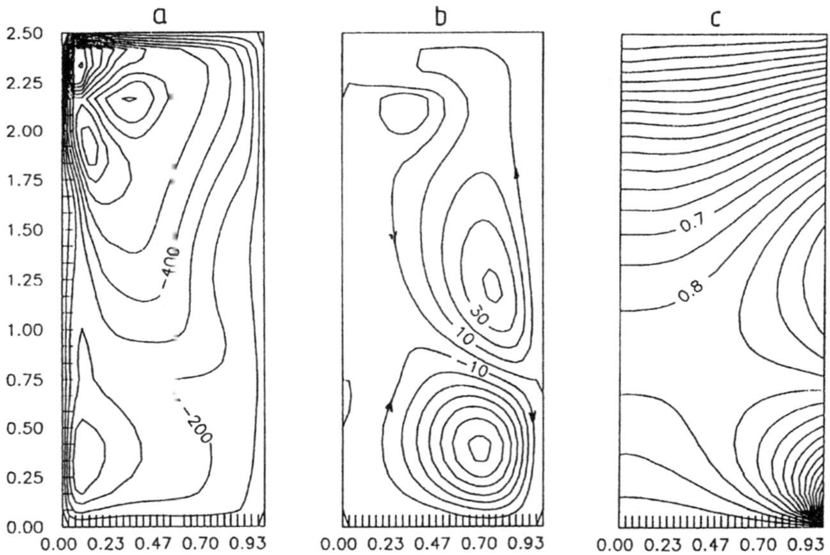

Fig. 12 Distribution of a) azimuthal velocity, b) streamlines of a meridional flow, and c) temperature in the case similar to that in Fig. 9 for a nonisothermal flow at $Gr = 10^6$.

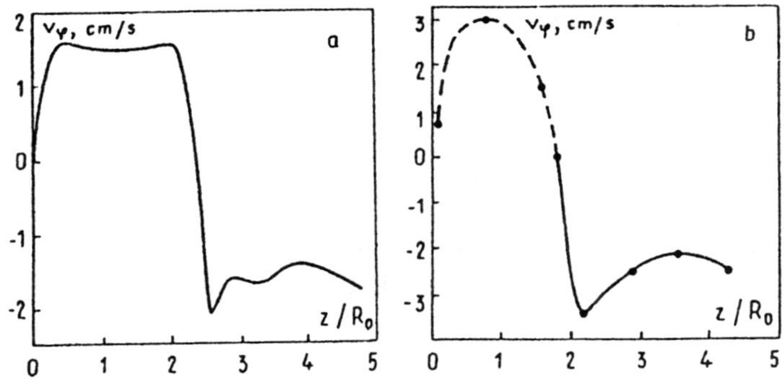

Fig. 13 The same as in Fig. 12, but with $Te_2/Te_1 = 10$, $Te_1 + Te_2/2 = 5 \cdot 10^5$, and $Gr = 10^6$.

Conclusions

1) A mathematical model for the RMF calculation is proposed, which allows for the accounting of finite dimensions of an inductor and liquid volume confined by walls with different electrical conductivities as well as asymmetry of their position with regard to each other.

2) Motion of an RMF inductor along the axis of a cylindrical vessel permits us to significantly widen the range of the velocity pattern characteristics and heat/mass transfer control in the liquid.

3) In calculations thermoconvective flows in an initial melt should be obligatorily accounted for to define the optimal characteristics of RMF for definite technological processes.

4) Superimposition of two counter-rotating magnetic fields along the axis of an ampoule with a melt are an effective means to affect hydrodynamic and heat/mass transfer characteristics in liquid.

5) The proposed technique for the calculation of electromagnetic fields, forces, hydrodynamic and heat characteristics in a melt allows us to define the required parameters of an RMF ensuring optimal technological conditions for the solution of different practical problems.

References

[1]Krūmiņš, J., and Priede, J., "Calculation of Electromagnetic Field and Force Distribution in an Inductionless Approximation Taking into Account Finite Dimensions of an Inductor and Field of End Parts of Winding Coils," *Magnitnaya Gidrodinamika*, Vol. 30, No. 2, 1994, pp. 259–265 (in Russian).

[2]Krūmiņš, J., "Calculation of Electromagnetic Force in a Rotating Magnetic Field in the Noninductive Approximation with Account for Finite Dimensions of a Device and Field of Frontal Parts of Winding," *Latvian Journal of Physics and Technical Sciences*, No. 5, 1995, pp. 65–72.

[3]Bulygin, L. L., "Calculation of a Stephani Two-Phase Problem for Electromagnetic Convection by the Method of Variational Inequality," *Modelling of Physical*

Processes in Continuous Media, edited by Latvian State Univ., 1985, pp. 3–25 (in Russian).

[4]Samarsky, A. A., and Moiseenko, B. D., "Efficient Procedure of a Throughout Counting for a Stephani Multi-Dimensional Problem," *Journal of Computational Mathematics and Mathematical Physics*, No. 5, 1965, pp. 816–827 (in Russian).

[5]Gorbachev, L. P., Nikitin, N. V., and Ystinov, A. L., "On MHD Rotation of an Electrically Conducting Fluid in a Cylindrical Vessel of Finite Dimensions," *Magnitnaya Gidrodinamika*, No. 4, 1974, pp. 32–42 (in Russian).

[6]Gelfgat, Yu. M., and Priede, J., "MHD Flows in a Rotating Magnetic Field," *Magnitnaya Gidrodinamika*, No. 1, 1995, pp. 214–230 (in Russian).

Chapter 46

Numerical Three-Dimensional Stability Analysis of the Magnetohydrodynamic Cylinder Wake Flow

V. Shatrov,* G. Mutschke,† and G. Gerbeth‡
Research Center Rossendorf, Inc., Dresden, 01314, Germany

Introduction

RECENTLY there has been considerable progress in understanding the transition of the hydrodynamic cylinder flow to three-dimensionality on the way toward turbulence.[1–4] At a Reynolds number of about 188.5, the quasiperiodic two-dimensional flow creating the well-known von Kármán vortex street becomes absolutely unstable to periodic three-dimensional disturbances with a wavelength of about 4 diam.[1]

In the present paper we extend this stability problem to the magnetohydrodynamic (MHD) case, i.e., we consider the flow of an electrically conducting fluid around an insulating circular cylinder exposed to a constant external magnetic field. One motivation is simply to investigate particular MHD features of open flow instabilities in a model problem best understood in hydrodynamics. Besides the usual *Re*, the MHD cylinder flow is characterized by a second parameter $N \sim B_0^2/v_\infty$ and offers, depending on the direction of the applied magnetic field, a large variety of flow configurations. Earlier work was devoted to the two-dimensional stability problem, i.e., the calculation of the neutral stability curve in the (*N*, *Re*)-parameter plane separating steady and unsteady (quasiperiodic) regimes on the two-dimensional flow in the plane perpendicular to the cylinder axis.[5] Early experimental results from the 1960s are summarized in Ref. 6.

In this paper we are restricted to a geometry where the magnetic field is aligned with the oncoming flow, for a sketch see Fig. 1. For this particular case Hunt supposed that there might exist three-dimensional instabilities in parameter regions where the flow is two-dimensional stable.[7] The basis for his conclusion is the interesting point that the Squire theorem, well known

*P.O. Box 510119; on leave from ITAM, Novosibirsk, Russia.
†P.O. Box 510119.
‡P.O. Box 51019.

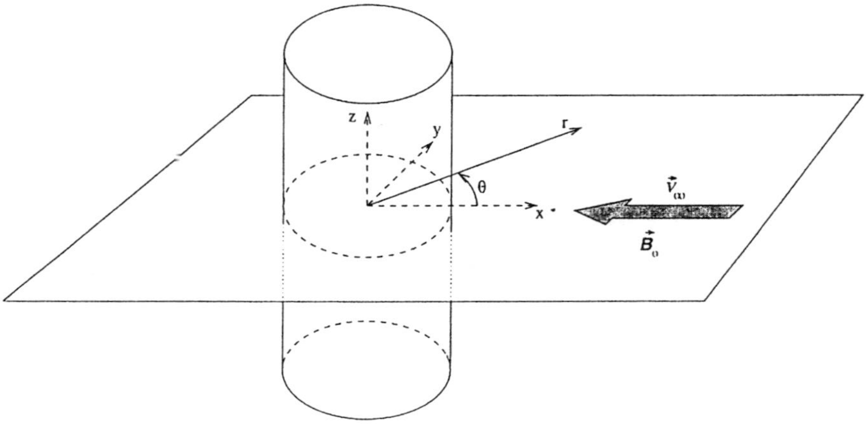

Fig. 1 Sketch of the geometry.

from the local stability analysis of parallel flows,[8] is no longer valid in the MHD case. This means that for constant interaction parameter N and increasing Re, three-dimensional instabilities might occur before two-dimensional flow becomes unstable because the external magnetic field influences two- and three-dimensional instabilities in different ways.

In the following we numerically perform a linear three-dimensional stability analysis of the two-dimensional flow. By monitoring the time evolution of the integral energy of the three-dimensional disturbances in the computational domain, we can decide whether the flow is globally unstable or stable.

Problem Definition

We consider the flow of an electrically conducting, incompressible viscous liquid around an insulating circular cylinder in an external uniform magnetic field B_0 aligned with the oncoming flow: $\boldsymbol{B}_0 \parallel \boldsymbol{v}_\infty$. We restrict to the inductionless approximation, i.e., $Rm = \mu\sigma v\, d \ll 1$. Scaling the length by the cylinder radius a = d/2, the velocity by v, the pressure by ρv_∞^2, the time by a/v_∞, the magnetic field by B_0, the electric field by $v_\infty B_0$, and the electrical current density by $\sigma v_\infty B_0$ respectively, yields the following nondimensional governing equations:

$$\frac{\partial \boldsymbol{v}}{\partial t} + (\boldsymbol{v} \cdot \nabla)\boldsymbol{v} = -\nabla p + \frac{2}{Re}\Delta \boldsymbol{v} + \frac{N}{2}\boldsymbol{j} \times \boldsymbol{B}_0 \tag{1}$$

$$\text{div}\ \boldsymbol{v} = 0 \tag{2}$$

$$\Delta\varphi = \text{div}\ (\boldsymbol{v} - \boldsymbol{B}_0) \tag{3}$$

with

$$j = E + v - B_0, \qquad E = -\nabla\varphi, \qquad B_0 = \pm e_x \tag{4}$$

where e_x is a unit vector in x direction. The two characteristic parameters are $Re = dv_\infty/\upsilon$ as the Reynolds number and $N = \sigma\beta_0^2 d/\rho v_\infty$ as the interaction parameter ($N = Ha^2/Re$).

Now we perform a linear three-dimensional stability analysis of a basic two-dimensional flow (steady or periodic) by assuming small three-dimensional disturbances v' to the two-dimensional flow v^0:

$$v = v^0 + v', \qquad |v'| \ll |v^0| \tag{5}$$

The electric potential is only due to three-dimensional perturbations, i.e., $\varphi = \varphi'$, because there is no electric field in the basic two-dimensional case. Linearization of Eq. (1) with respect to v' allows us to look for solutions of the form

$$v' = v'' \cdot e^{\lambda t}, \qquad \varphi' = \varphi'' \cdot e^{\lambda t} \tag{6}$$

Supposing periodicity in the cylinder spanwise direction z characterized by a wave number k, each mode of the three-dimensional disturbances can be investigated separately, and the problem reduces to an essentially two-dimensional one:

$$\left(v_r'', v_\theta'', v_z''\right)^T = \left[v_r^1 \cos(kz), v_\theta^1 \cos(kz), v_z^1 \sin(kz)\right]^T \tag{7}$$

$$\varphi'' = \varphi^1 \sin(kz) \tag{8}$$

Numerical Method

The two-dimensional flow is determined using an established stream function vorticity finite difference method [$v^0 = -\mathrm{curl}(\Psi e_z)$, $\omega_z^0 = \Delta\Psi$] similar to Ref. 9; details are described in Ref. 5. The linear three-dimensional problem was treated in a $v - \omega$ finite difference formulation similar to Ref. 10. The total accuracy is second order in space and time, whereby the spatial discretization of all elliptic equations including the Poisson equation for the electric potential was done with fourth-order accuracy. We used an exponential polar grid $r_i \sim e^{\lambda i}$, $r_{\max} = 50$ equidistant in θ direction with up to 256×256 points. All boundary conditions were implemented in second order, as shown in Table 1.

The numerical procedure used to investigate the three-dimensional stability was the following: At some initial time $t = 0$, a three-dimensional vorticity disturbance

$$\left(w_r^1, w_\theta^1, w_z^1\right)^T = \left(0, a_0 \frac{k}{r^3} \sin^2\frac{\theta}{2}, \frac{a_0}{2r^4} \sin\theta\right)^T \tag{9}$$

Table 1 Boundary conditions

	Two-dimensional	Three-dimensional
$r = 1$	$v^0 = 0$	$v = 0, \frac{\partial \varphi'}{\partial r} = 0$
$r = r_{max}$	$\Psi^0 = (r - \frac{1}{r}) \sin \theta$	$v' = 0, \varphi' = 0$
Outflow angle	$\frac{\partial \Psi^0}{\partial x} = 0, \frac{\partial \omega^0}{\partial x} = 0$	$\frac{\partial v'}{\partial x} = 0, \frac{\partial \varphi'}{\partial x} = 0$

of amplitude $a_0 \sim 10^{-2}$ was introduced to a developed two-dimensional flow being either steady or quasiperiodic. Typically, after some transient time of the order of $t_0 \sim r_{max}$ the flow entered the expected linear regime where the three-dimensional disturbance will either increase or be damped exponentially in time. To determine the global three-dimensional stability of the flow, we monitored the total three-dimensional energy:

$$E_{3d}(t) = \iiint^{2\pi/k_z} \mathrm{d}x\, \mathrm{d}y\, \mathrm{d}z \frac{\boldsymbol{v}^2}{2} \tag{10}$$

(integrated over the computational domain) in time. In the following the slope of the logarithm of this quantity in the linear regime is called the *temporal growth rate* (TGR). Figure 2 shows an example of the temporal growth rate obtained by this method at $Re = 200$, $N = 0.4$, and $k = 0.8$. In general, our global linear stability analysis has the advantage that it detects only absolute instabilities and excludes those of a purely convective nature.

To validate our numerical approach, we performed intensive tests. The main error in the TGR was due to the finite spatial grid spacing; therefore, dense grids had to be applied. Figure 3 shows a Richardson extrapolation for the TGR with up to 256×256 points reducing the error to less than 5%. Further, we compared our results for the purely hydrodynamic problem at $Re = 200$ with data published recently by Henderson[1] and found an excellent agreement (see Fig. 4).

Results

Figure 5 shows the neutral two-dimensional stability curve in the (Re, N) parameter plane for an aligned magnetic field. Above the curve any two-dimensional disturbance will be damped out in time, and the flow finally will become steady, whereas below the curve the magnetic field is too weak and can only slightly modify the persisting Kármán vortex street.

Hunt[7] performed analytical work on the relation between two- and three-dimensional instabilities but was limited to a local analysis of parallel flows. Transferring his results to the present case results in a tangent to our two-dimensional stability curve beginning at the origin (see Fig. 5). The parallel

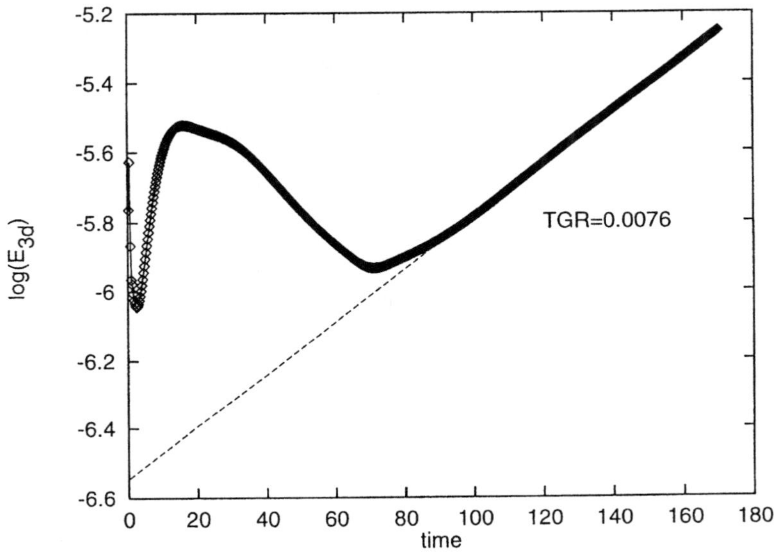

Fig. 2 Typical behavior of the logarithm of the global three-dimensional energy E_{3d} in time (Re = 200, N = 0.4, k = 0.8). After a transient time the flow enters a linear regime, and the temporal growth rate TGR can be detected.

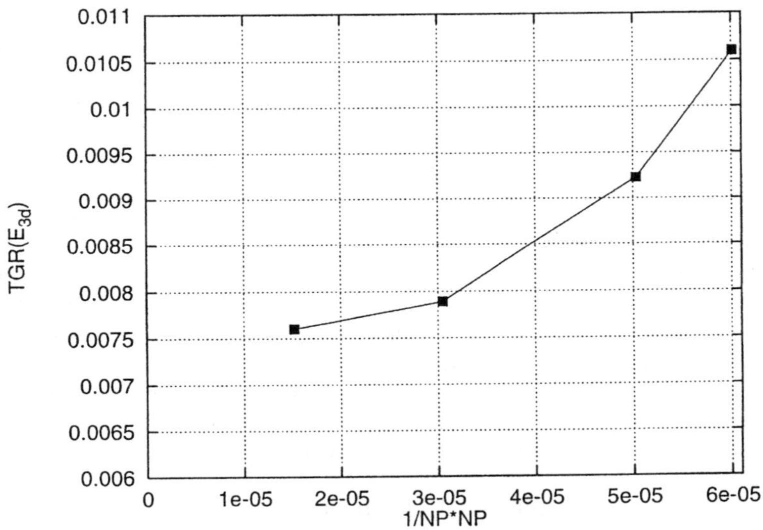

Fig. 3 Richardson extrapolation for the temporal growth rate of E_{3d} at Re = 200, N = 0.4, and k = 0.8. NP denotes the number of grid points in both radial and azimuthal directions.

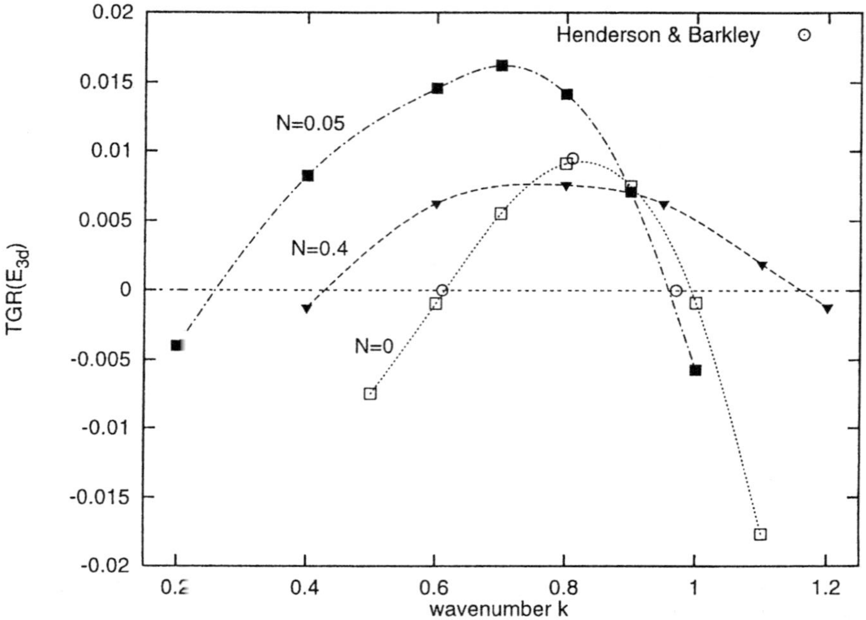

Fig. 4 Temporal growth rates of the global three-dimensional energy at *Re* = 200 for different *N*. The results from Henderson[2] are in good agreement with the case *N* = 0.

flow analysis would now distinguish between three different regions in the parameter space

Region I: Above the tangent and above the two-dimensional stability curve before their point of contact, the flow is stable to both two- and three-dimensional disturbances, i.e., steady and two-dimensional.

Region II: Below the two-dimensional stability curve the flow is primarily two-dimensionally unstable; three-dimensional instability will occur at higher *Re*.

Region III: Between the two-dimensional stability curve and the tangent beyond their point of contact, the flow is two-dimensionally stable but three-dimensionally unstable. This clearly shows that the Squire theorem is no longer valid in the MHD case.

Of course, the MHD cylinder wake is neither a parallel flow, nor does local instability always lead to global instability; therefore, the results of Hunt[2] cannot be applied straightforwardly and give only a first qualitative hint.

First results of our numerical global linear three-dimensional stability analysis are shown in Figs. 4 and 6–9. Figure 4 shows three curves of temporal growth rates at $Re = 200$ and different interaction parameters. A weak magnetic field of $N = 0.05$ widens the range of unstable wave numbers compared to the hydrodynamic case $N = 0$. The most unstable wave number

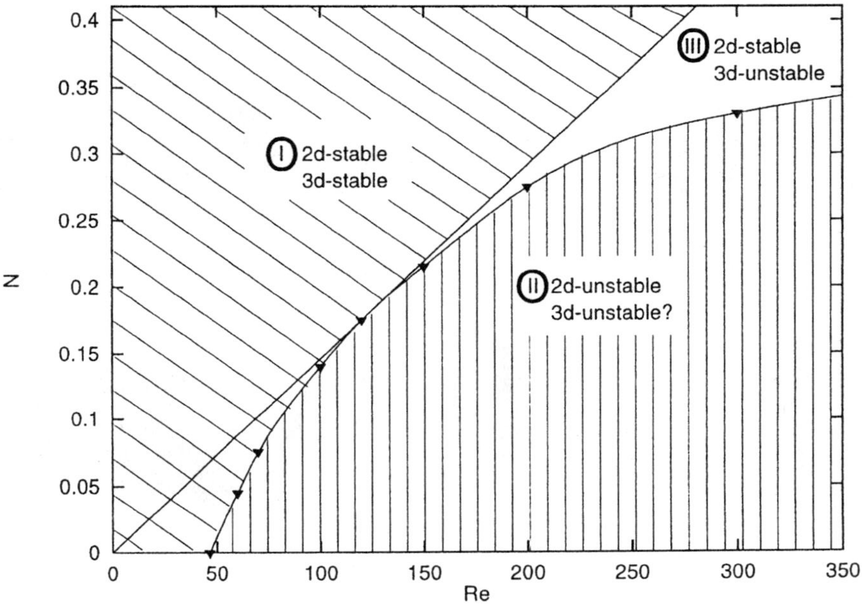

Fig. 5 Neutral two-dimensional stability curve for an aligned magnetic field (-▽-) and its tangent defining three different stability regions if Hunt's parallel flow analysis would be applied.

is slightly decreased, whereas the TGR at this wave number is larger than the maximum TGR for $N = 0$. Thus, such a weak magnetic field seems to amplify the three-dimensional instability of the flow.

The larger value of $N = 0.4$ shows a broader range of unstable wave numbers than in the hydrodynamic case. The maximum TGR value is smaller than for $N = 0$, i.e., here the magnetic field has a damping influence on the three-dimensional instability. One might suppose that for larger magnetic fields this instability can be suppressed completely. However, at $N = 0.4$ the flow is still three-dimensionally, whereas the corresponding two-dimensional flow is already steady, as can be seen from Fig. 5. The critical value of neutral two-dimensional stability at $Re = 200$ is about $N_c^{2d} \approx 0.27$, so three-dimensional instability above the two-dimensional threshold is clearly detected.

To characterize the three-dimensional instabilities, we plotted in Figs. 6–8 isolines of the two-dimensional stream function Ψ^0 the two-dimensional vorticity ω_z^0, and the local three-dimensional energy density

$$e_{3d} = \frac{\boldsymbol{v}^2}{2} \tag{11}$$

for a wave number $k = 0.8$ at a time step arbitrarily chosen inside the linear regime for all three values of N. At first sight the differences between Fig. 6

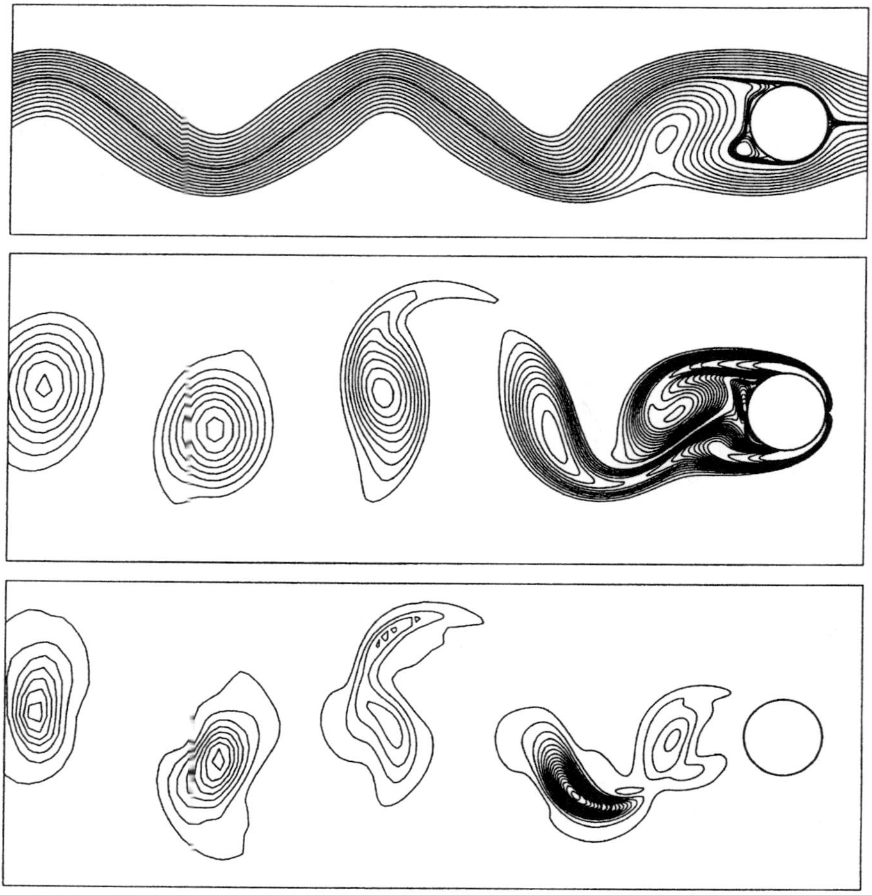

Fig. 6 Snapshot of Ψ^0, ω^0, and e_{3d} (from top to bottom) at Re = 200, N = 0, and k = 0.8 in the linear regime.

and 7 seem to be small. One can further see that the maxima of e_{3d} for both unsteady and steady flows are closely arranged with the shedded two-dimensional vortices and the steady two-dimensional recirculation eddies, respectively.

Figure 9 shows the three-dimensional energy density on the downstream axis. For $N = 0.05$ the local variation of e_{3d} resembles the vortex street oscillations of the two-dimensional flow, whereas for $N = 0.4$ the steady two-dimensional flow results in a smooth shape of e_{3d}. In both cases the absolute maximum of e_{3d} is located at about 3 diam downstream as in the hydrodynamic case: for $r \gg 1$ the three-dimensional energy density decreases exponentially.

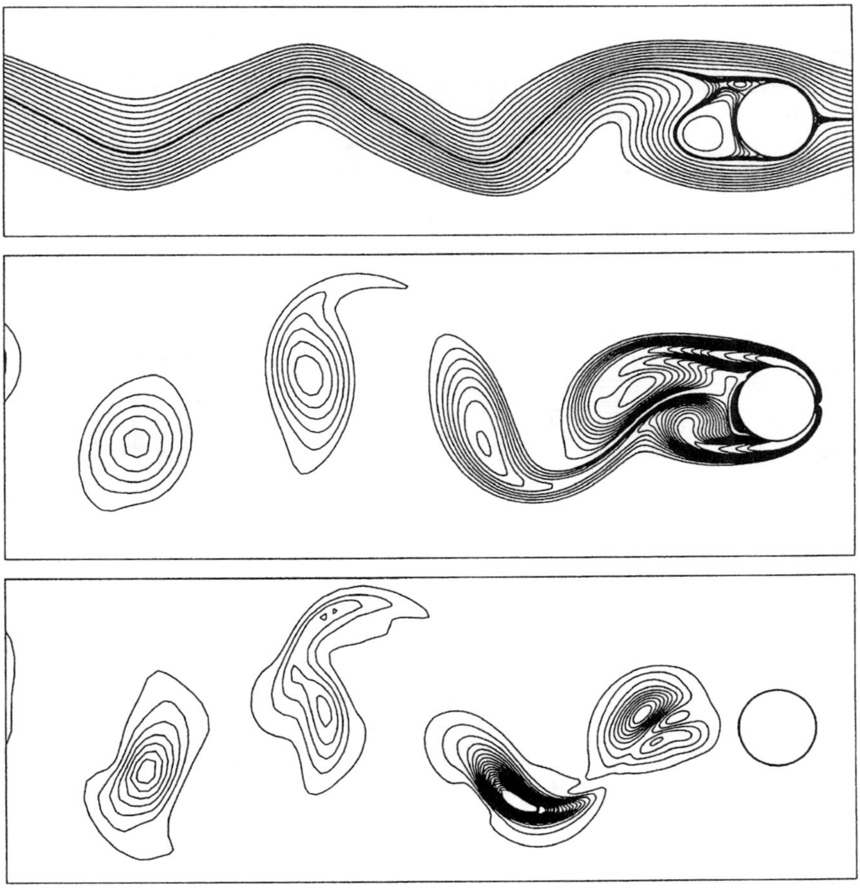

Fig. 7 Snapshot of Ψ^0, ω^0, and e_{3d} (from top to bottom) at $Re = 200$, $N = 0.05$, and $k = 0.8$ in the linear regime.

Summary and Outlook

We presented first results concerning the influence of an aligned magnetic field on the relation between two- and three-dimensional instabilities in the MHD cylinder flow. The computational effort to determine temporal growth rates of three-dimensional instabilities as in Fig. 4 is large (~ 5h CRAY-YMP for one TGR value) and makes a systematic investigation of the (Re, N, k) parameter space for all three possible directions of the magnetic field difficult.

By monitoring the time evolution of the integral energy of the three-dimensional disturbances; we detected global three-dimensional instability in a parameter region where the flow is two-dimensionally stable, i.e., steady.

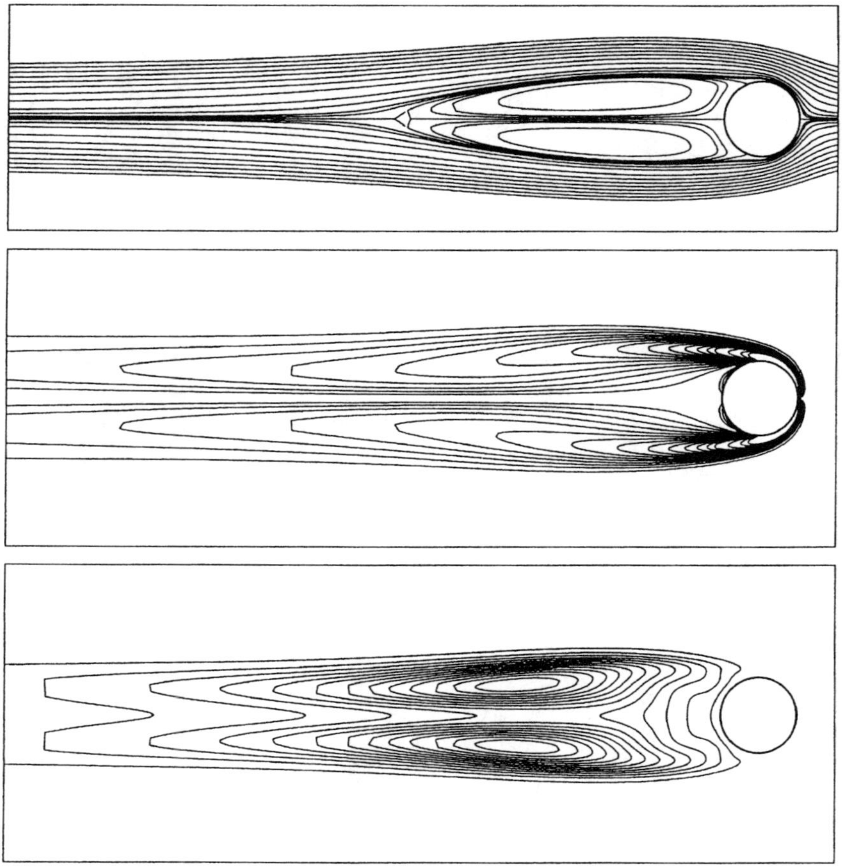

Fig. 8 Snapshot of Ψ^0, ω^0, and e_{3d} (from top to bottom) at Re = 200, N = 0.4, and k = 0.8 in the linear regime.

This is qualitatively in line with the result of Hunt,[7] notwithstanding the fact that his work was restricted to a local parallel flow stability analysis.

Three-dimensional instability of a two-dimensional flow is unique for the cylinder problem. Interesting to observe is the fact that even this standard hydrodynamic problem of the flow around a cylinder represents one example where the influence of an applied magnetic field on instabilities is not always a damping one and also strongly depends on the character of the instability. To verify these results experimentally would be highly desirable. Recent experiments[11] show a slight amplification of low-frequency fluctuations because of an aligned magnetic field and support the widening of the unstable wave number region found here.

Detailed explanations for the action of the aligned magnetic field on three-dimensional instabilities have to be left for future work. The matter

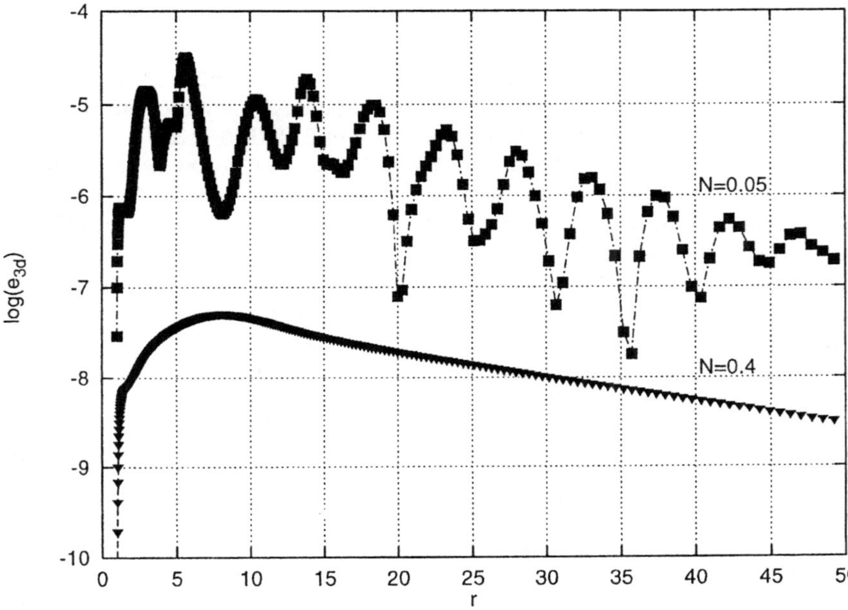

Fig. 9 Logarithm of the three-dimensional energy density e_{3d} versus radius r downstream at $y = 0$ for $Re = 200$ and $k = 0.8$.

seems to be more complex than in the hydrodynamic case because varying the magnetic field changes the two-dimensional flow simultaneously. In the future we will extend this work to larger *Ha* to determine whether the magnetic field finally damps out the three-dimensional disturbances, i.e., if a stability curve exists above which the flow is stable to all instabilities. We will further investigate the other two directions of the magnetic field. For a transverse field it is known that not only the vortex street is damped but also the steady recirculation eddies vanish for increasing *Ha*.[5] Therefore, one might expect that three-dimensional disturbances are more strongly damped here than in the aligned field case. For B_0 aligned with the cylinder axis, we expect three-dimensional disturbances to be strongly suppressed because here the Lorentz force does not act on the aligned vorticity.

Acknowledgments

Financial support under grant No. Ge 682/3-4 as well as grants for several visits of V. Shatrov at Research Center Rossendorf from "Deutsche Forschungsgemeinschaft" are gratefully acknowledged. The work is supported by the European INTAS program under grant No. 94-1504.

References

[1]Henderson, R. D., and Barkley, D., "Secondary Instability in the Wake of a Circular Cylinder," *Physics of Fluids*, Vol. 8, 1996, p. 1683; also Barkley, D., and Henderson, R. D., "Three-Dimensional Floquet Stability Analysis of the Wake of a Circular Cylinder," *Journal of Fluid Mechanics*, Vol. 322, 1996, p. 2.

[2]Williamson, C. H. K., "Oblique and Parallel Modes of Vortex Shedding in the Wake of a Circular Cylinder at Low Reynolds Numbers," *Journal of Fluid Mechanics*, Vol. 206, 1989, p. 38; also Williamson, C. H. K., "The Natural and Forced Formation of Spot-Like Vortex Dislocations in the Transition of a Wake," *Journal of Fluid Mechanics*, Vol. 243, 1992, p. 393; also Williamson, C. H. K., and Prasad, A., "A New Mechanism for Oblique Wave Resonance in the 'Natural' Far Wake," *Journal of Fluid Mechanics*, Vol. 256, 1993, p. 269.

[3]Karniadakis, G. E., and Triantafyllou, G. S., "Three-Dimensional Dynamics and Transition to Turbulence in the Wake of Bluff Bodies," *Journal of Fluid Mechanics*, Vol. 238, 1992, p. 1; also Tomboulides, A. G., Karniadakis, G. E., and Triantafyllou, G. S., "A Mechanism of Period Doubling in Free Shear Flows," *Physics of Fluids A*, Vol. 4, 1992, p. 1329.

[4]Noack, B. R., Koenig, M., and Eckelmann, H., "Three-Dimensional Stability Analysis of the Periodic Flow Around a Circular Cylinder," *Physics of Fluids A*, Vol. 5, 1993, p. 1279; also Noack, B. R., and Eckelmann, H., "A Low-Dimensional Galerkin Method for the Three-Dimensional Flow Around a Circular Cylinder," *Physics of Fluids A*, Vol. 5, 1994, p. 124; also Noack, B. R., and Eckelmann, H., "A Global Stability Analysis of the Steady and Periodic Cylinder Wake," *Journal of Fluid Mechanics*, Vol. 270, 1994, p. 297; also Zhang, H.-Q., Frey, U., and Noack, B. R., "On the Transition of the Cylinder Wake," *Physics of Fluids*, Vol. 7, 1995, p. 779.

[5]Shatrov, V., Mutschke, G., and Gerbeth, G., "Numerical Simulation of the Two-Dimensional MHD-Flow Around a Circular Cylinder," *Proceedings of the International Conference on Energy Transfer in Magneto HydroDynamic Flows PAMIR* (Aussois, France), Vol. 2, 1994, pp. 745–756.

[6]Tsinober, A. B., *MHD Flow Around Obstacles*, Zinatne, Riga, Latvia, 1970 (in Russian).

[7]Hunt, J. C. R., "On the Stability of Parallel Flows with Parallel Magnetic Fields," *Proceedings of the Royal Society of London*, Series *A*, Vol. 293, 1966, pp. 342–358.

[8]Squire, H. B., "On the Stability for Three-Dimensional Disturbances of Viscous Fluid Flow Between Parallel Walls," *Proceedings of the Royal Society of Love Series A*, Vol. 142, 1933, p. 62.

[9]Phuoc Loc, Ta., *Journal of Fluid Mechanics*, Vol. 100, 1980, p. 111.

[10]Roache, P. R., *Computational Fluid Mechanics*, Hermosa Publishers, Albuquerque, NM, 1976.

[11]Alemany, A., Gerbeth, G., Mutschke, G., Pilaud, A., and Weier, T., "On the Stability of the MHD Flow Around a Cylinder in an Aligned Magnetic Field," *PAMIR Conference* (Aussois, France), 1994.

Chapter 47

Two-Dimensional Calculation for Fluid Flow in Magnetic Centrifugal Casting

Y. Yang,* W. Zhang,† Q. Liu,‡ and Z. Hu§
Institute of Metal Research, Chinese Academy of Sciences, Shenyang, 110015, China

Nomenclature

$\boldsymbol{B}$	= magnetic flux density, T
d_w	= distance to the inner solid wall of the mold, m
$\boldsymbol{f_b}$	= body force per unit volume, N m^{-3}
$\boldsymbol{f_c}$	= centrifugal force per unit volume, N m^{-3}
$\boldsymbol{f_e}$	= electromagnetic force per unit volume, N m^{-3}
$\boldsymbol{f_g}$	= gravity, N m^{-3}
f_{int}	= electromagnetic force on the interface, N m^{-3}
$f(r, \theta)$	= z component of curl ($\boldsymbol{f_b}$)
$\boldsymbol{j}$	= inductive current density, A m^{-2}
n	= rotation rate of the casting mold, rpm
p	= pressure, Pa
$\hat{\boldsymbol{r}}$	= radial unit vector, m
r	= radial coordinate, m
r_0, r_i	= radii for the outer and internal surfaces, m
r_{int}	= radius for the interface, m
u, v	= radial and tangential components of velocity vector V, m s^{-1}
u_f, v_f	= radial and tangential components for flow velocity, m s^{-1}
$\boldsymbol{V}$	= velocity vector of fluid, m s^{-1}
z	= coordinate, m
Δ	= Laplacian
δ_r, δ_θ	= mesh steps
ε	= relative error for convergence
η	= kinematic viscosity, m^2 s^{-1}

*Professor, State Key Laboratory of RSA, 72 Wenhua Rd.
†State Key Laboratory of RSA, 72 Wenhua Rd.
‡State Key Laboratory of RSA, 72 Wenhua Rd.

$\hat{\theta}$ = tangential unit vector, m
θ = angular coordinate
μ = dynamic viscosity, N s m^{-2}
ρ = density of melt, kg m^{-3}
σ = electric conductivity, Ω^{-1} m^{-1}
Φ = electric potential, V
ψ = scalar form of stream function, m^2 s
Ω = angular velocity, s^{-1}
ω = scalar form of vorticity, s^{-1}
∇ = Hamilton operator in cylindrical coordinate system

I. Introduction

ELECTROMAGNETIC metallurgy is a challenging technology in material processing. Well documented is the fact that the structures and properties of metals are effectively improved by applying an electric or magnetic field to metallurgical processes. Because the melt flow in the vicinity of the solidification front plays an important role in affecting nucleation, crystal growth, as-cast structures, segregation, and the like, numerous techniques have been developed to control fluid flow in solidification processing. Among these techniques, electromagnetic stirring has proven to be one of the most effective and economical methods.[1–3] Within decades of its introduction, the application of electromagnetic stirring to continuous casting processes was so successful that it was applied widely in industry and brought about beneficial effects in the metallurgical quality of finished products.

Fluid flow driven by an electromagnetic force is of great theoretical and practical interest, and the magnetically stirred flow of liquid during phase transformation has been investigated by several authors. Tewari,[4] Kobayashi,[5] and Vives[6] with their co-workers have experimentally investigated the influence of electromagnetic stirring on the as-cast structures and segregation under various solidification conditions. Mihelcic and Wingerath[7] and Hall and Walker[8] have numerically simulated given laminar melt flow in a magnetic Czochralski process. Spitzer[9] presented a turbulent flow model for continuous casting with the stirring in the presence of a rotating magnetic field.

The present work is focused on the electromagnetically forced flow in electromagnetic centrifugal casting (EMCC) that has been developed in our previous research[10] and shows many advantages in the fabrication of pipes to get a further understanding of the effects of fluid flow on solidification of metals under the condition of the coexistence of an electromagnetic force and a centrifugal force.

II. Mathematical Model

The model is based on a horizontal centrifugal casting system in a transverse magnetic field as shown schematically in Fig. 1, which illustrates the formation of a melt pipe via the rotation of the mold at a speed Ω in a

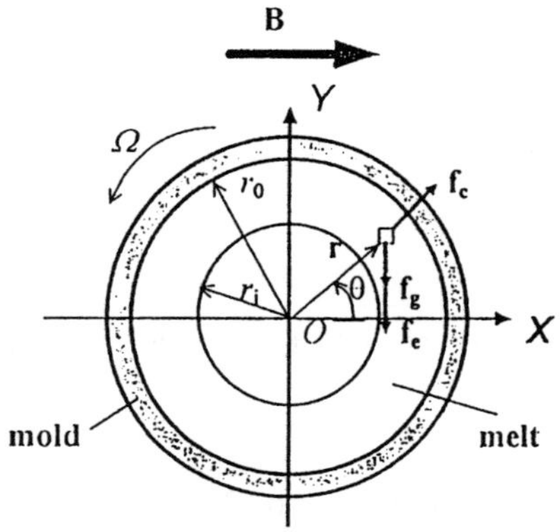

Fig. 1 The geometry of EMCC system and the body forces on melt.

magnetic field with a strength B. After the melt pipe is formed, the melt still rotates together with the mold in the presence of the magnetic field, so that an electromagnetic force is created in the melt. To simplify the computation of flow and force fields in the melt, the melt is considered to be incompressible, and the fluid flow in the infinitely long mold is laminar and in steady state. Only the two-dimensional case is taken into account. Moreover, the transverse magnetic field imposed on the rotating melt is assumed to be stationary and uniform. And, an induced magnetic field caused by induced current through the melt pipe is ignored.

The governing equations of fluid motion are the Navier–Stokes equations [Eq. (1)] and the continuity equation [Eq. (2)]:

$$\rho\,(\boldsymbol{V}\cdot\nabla)\,\boldsymbol{V} - \mu\nabla^2\boldsymbol{V} + \nabla p = \boldsymbol{f}_b \tag{1}$$

$$\nabla\cdot\boldsymbol{V} = 0 \tag{2}$$

where $\boldsymbol{f}_b$ denotes the volumetric body force including the electromagnetic force (Lorentz force) $\boldsymbol{f}_e$, centrifugal force $\boldsymbol{f}_c$, and gravity $\boldsymbol{f}_g$ per unit volume. According to electromagnetic theory, the Lorentz force can be written as

$$\boldsymbol{f}_e = \boldsymbol{j}\times\boldsymbol{B} \tag{3}$$

where $\boldsymbol{j}$ is the inductive current density in the melt and is given by Ohm's law:

$$\boldsymbol{j} = \sigma\,(-\nabla\Phi + \boldsymbol{V}\times\boldsymbol{B}) \tag{4}$$

and

$$\nabla\cdot\boldsymbol{j} = 0 \tag{5}$$

Here, to simplify the computation, the term of electric potential Φ is neglected. Thus the Lorentz force may be rewritten as

$$f_e = (\sigma \boldsymbol{V} \times \boldsymbol{B}) \times \boldsymbol{B} \tag{6a}$$

The vector form of $\boldsymbol{f_e}$ in cylindrical coordinates is

$$\boldsymbol{f_e} = -\sigma B^2\left(u \sin^2\theta + \frac{v}{2}\sin 2\theta\right)\hat{\boldsymbol{r}} - \sigma B^2\left(\frac{u}{2}\sin 2\theta + v\cos^2\theta\right)\hat{\boldsymbol{\theta}} \tag{6b}$$

where $\hat{\boldsymbol{r}}$ and $\hat{\boldsymbol{\theta}}$ are the unit vectors.

It is convenient for computational purposes to rewrite the Navier–Stokes equations in terms of the stream function ψ and vorticity ω. The scalar form of the transformed equation may be expressed as

$$u\frac{\partial\omega}{\partial r} + \frac{v}{r}\frac{\partial\omega}{\partial\theta} - \eta\left(\frac{\partial^2\omega}{\partial r^2} + \frac{1}{r}\frac{\partial\omega}{\partial r} + \frac{1}{r^2}\frac{\partial^2\omega}{\partial\theta^2}\right) = f(r,\theta) \tag{7}$$

$$\nabla^2\psi + \omega = 0 \tag{8}$$

where $f(r, \theta)$ designates the z component of $(\nabla \times \boldsymbol{f_b})$, η denotes the kinematic viscosity, and the velocity components are determined by

$$u = \frac{1}{r}\frac{\partial\psi}{\partial\theta} \quad \text{and} \quad v = -\frac{\partial\psi}{\partial r} \tag{9a,9b}$$

On the interface of the mold and the melt, the no-slip boundary condition must be satisfied:

$$\psi = 0, \quad \frac{\partial\psi}{\partial r} = -\Omega r_0 \quad \text{at} \quad r = r_0 \tag{10}$$

On the inner surface of the melt pipe where a free surface is present, the no-shear condition yields

$$\frac{\partial^2\psi}{\partial r^2} - \frac{1}{r}\frac{\partial\psi}{\partial r} - \frac{1}{r^2}\frac{\partial^2\psi}{\partial\theta^2} = 0 \quad \text{at} \quad r = r_i \tag{11}$$

We noticed that, on one hand, the melt rotates with the mold in the magnetic field at the velocity v; but, on the other hand, the melt will flow in the mold under the action of the electromagnetic force. Here, we use a term *flow velocity* to describe the relative flow of the fluid in the rotating coordinate system. And, we are interested in the flow velocity of the fluid because it has great influences on the solidification of metals. To discuss the problem more conveniently, one can divide the forces into radial and tangential components, both of which are responsible for radial and tangential flow of the melt. Hence, the flow of the fluid can be considered as the combination of the tangential and radial components corresponding to the force. According to the Eq. (6b), evidently, the tangential flow of the fluid is in the reverse direction of mold rotation.

Let u_f and v_f denote the radial and tangential components of the flow velocity in the rotating coordinate system, then

$$u_f = u \quad \text{and} \quad v_f = v - \Omega r \tag{12}$$

By applying the divergence to Eq. (1), the viscous term on the left-hand side of Eq. (1) vanishes. We have a Poisson equation

$$\Delta p = -\rho \nabla \cdot (\boldsymbol{V} \cdot \nabla)\boldsymbol{V} + \nabla \cdot \boldsymbol{f_b} \tag{13}$$

with the boundary condition

$$p = 0 \quad \text{at} \quad r = r_i \tag{14}$$

where

$$\Delta = \frac{\partial^2}{\partial r^2} + \frac{1}{r}\frac{\partial}{\partial r} + \frac{1}{r^2}\frac{\partial^2}{\partial \theta^2}$$

III. Computational Results

The numerical approach to determining the flowfield was based on a finite difference approximation of Eqs. (7–14) with the domain discretized using a uniform grid. To avoid the numerical instability associated with a large Reynolds number, the second-order upwind scheme was applied to the advective terms following Peyret and Taylor's suggestions.[11] Numerical experiments showed that results did not depend on the mesh size provided that the size was small enough. The equations were solved by a successive-over-relaxation SOR method. The iteration procedure was stopped when convergence was obtained. A relative criterion for convergence was used that was defined as the ratio of the maximum difference between the results of two neighboring iterations to the maximum value of the first of the two neighboring iterations. After convergence was obtained, the maximum residuals of the difference equations for Eqs. (7), (8), and (13) were of order 10^{-3}. After calculations of the velocity and pressure fields were performed, the body force imposed on the melt was computed.

The computations were carried out for an Al melt at 700°C. The rotation rate of the mold (n) was 1000 to 3000 rpm, and the magnetic flux density (B) was 0.2 to 0.4 Ta. The other data for computation are listed in Table 1.

Table 1 Data used in computations

Dimensions of the melt pipe	
Outer radius	r_0 = 40 mm
Inner radius	r_i = 25 mm
Controlling parameters	
Rotation rate	n = 1000–3000 rpm
Magnetic flux density	B = 0.2–0.4 T
Physical properties of the melt	
Density	ρ = 2350 kg m^{-3}
Dynamic viscosity	μ = 1.148 × 10^{-3} N s m^{-2}
Kinematic viscosity	η = 4.885 × 10^{-7} m^2/s^{-1}
Electrical conductivity	σ = 4.0274 × 10^6 Ω^{-1} m^{-1}
Discretizations	
Radial mesh	δ_r = 0.375 mm
Azimuthal mesh	δ_θ = 1.8 deg
Relative error of convergence	ε ≤ 10^{-4}

A. Flow Velocity

Figure 2 shows the radial and azimuthal distributions of the tangential component (v_f) of the flow velocity. On the inner wall of the mold, the component v_f remains zero due to the no-slip condition, but it abruptly increases to a maximum value over a very short distance from the solid wall (d_w), and then decreases with an increase of the distance from the wall. Near the free surface it begins to increase slightly. Along the azimuthal direction, the distributions of flow velocity are characterized by fluctuations that exhibit a sine-like behavior with respect to the angular coordinate, but the oscillating frequency remains twice that of the mold rotation.

The distribution of the velocity component is determined by the body forces, especially the Lorentz force acting on the fluid. As the melt rotates together with the mold, the Lorentz force in the bulk fluid is proportional to the rotational velocity that depends upon the distance to the mold wall and the azimuthal coordinate. Consequently, the fluid on the inner surface of the mold should bear the strongest electromagnetic force, and the force is reduced gradually with the increase of d_w. The no-slip boundary condition requires that the fluid has no relative motion to the surface of the wall. But as for the fluid near the wall, the tangential component of the Lorentz force on it is so great that flow there is promptly accelerated, and this results in a sharp increase in the velocity distribution versus d_w. The fluid on the free surface satisfies the no-shear condition, under which the viscous resistance to the flow is smaller than that in the bulk fluid, and this accordingly leads to the slight increase in the flow velocity near the free surface.

The rotation rate (n) and magnetic flux density (B) can greatly affect the velocity components. By increasing the rotation rate n from 1000 to 3000 rpm, v_f increases dramatically; in particular, the maximum value of the velocity near the solid wall seems to be proportional to the rotation rate. The

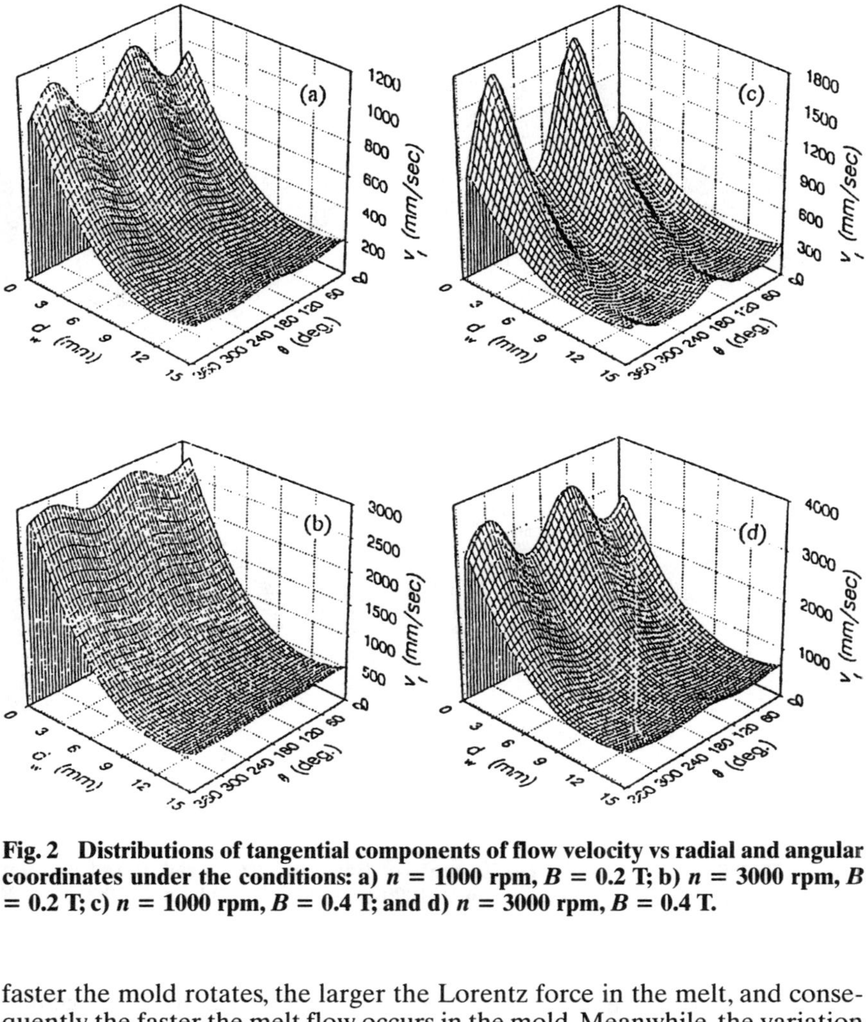

Fig. 2 Distributions of tangential components of flow velocity vs radial and angular coordinates under the conditions: a) n = 1000 rpm, B = 0.2 T; b) n = 3000 rpm, B = 0.2 T; c) n = 1000 rpm, B = 0.4 T; and d) n = 3000 rpm, B = 0.4 T.

faster the mold rotates, the larger the Lorentz force in the melt, and consequently the faster the melt flow occurs in the mold. Meanwhile, the variation of magnetic flux density can bring about a remarkable change in v_f, but it does not influence the flow near the internal surface as effectively as the rotation velocity. When B is increased without changing n, the maximum values of the flow velocities are accordingly raised to some extent as are the amplitudes of angle-dependent fluctuations. In other words, the intensification of magnetic field has a slight influence on the tangential flow near the free surface of melt but greatly enhances the azimuthal fluctuation near the solid wall.

Figure 3 shows the distributions of the radial component of the flow velocity (u_f) vs spatial coordinates and the controlling parameters. The u_f component remains zero on the interface of mold/melt and gradually increases with an increase in d_w. There are also angle-dependent oscillations

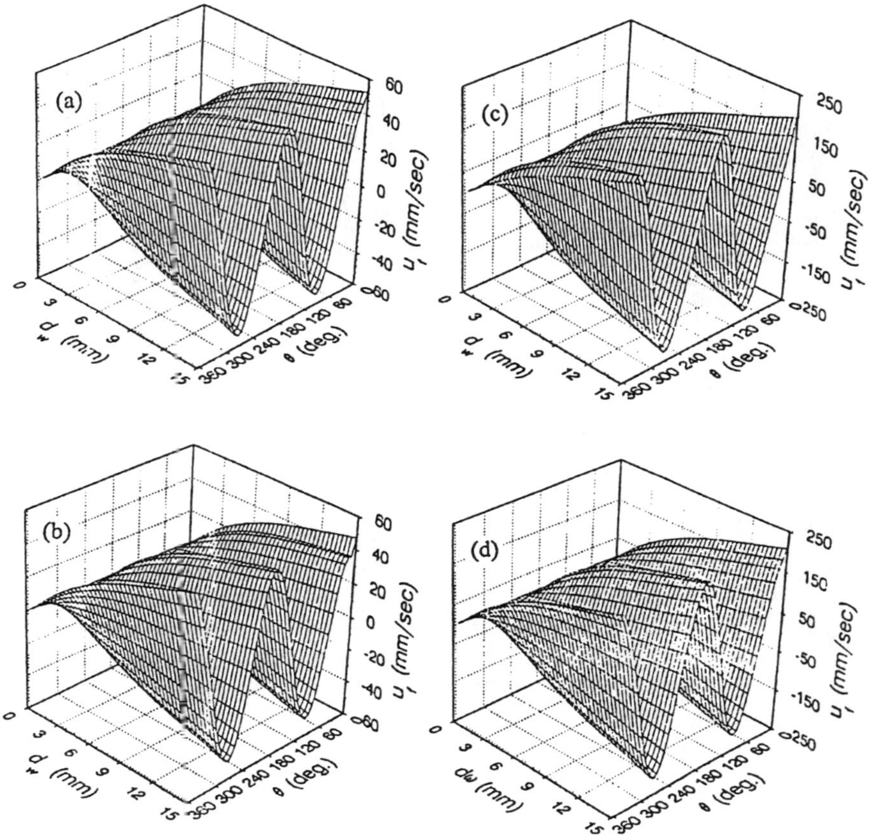

Fig. 3 Distributions of radial components of flow velocity vs radial and angular coordinates under the conditions: a) n = 1000 rpm, B = 0.2 T; b) n = 3000 rpm, B = 0.2 T; c) n = 1000 rpm, B = 0.4 T; and d) n = 3000 rpm, B = 0.4 T.

in the u_f component, and they have the same frequency as in the tangential component. Unlike the tangential component, the u_f component is not dramatically influenced by the rotation rate of the mold, but it shows a close dependence on the magnetic field if B varies in the range 0.2 to 0.4 T. The computational results show that the u_f component has an approximate linear relation to B. For instance, when B is 0.2, 0.3, and 0.4 T, the maximum value of the u_f component is 55.8, 132.5, and 210.4 mm/s,$^{-1}$ respectively. The cases for n = 2000 and 3000 rpm are identical to the above.

B. Body Forces

The relationship between the fluid motion and the body forces is complex. The forces cause fluid to flow, and the change of the flow velocity must result in variation in the body forces. As for a steady-state flow, the forces and the

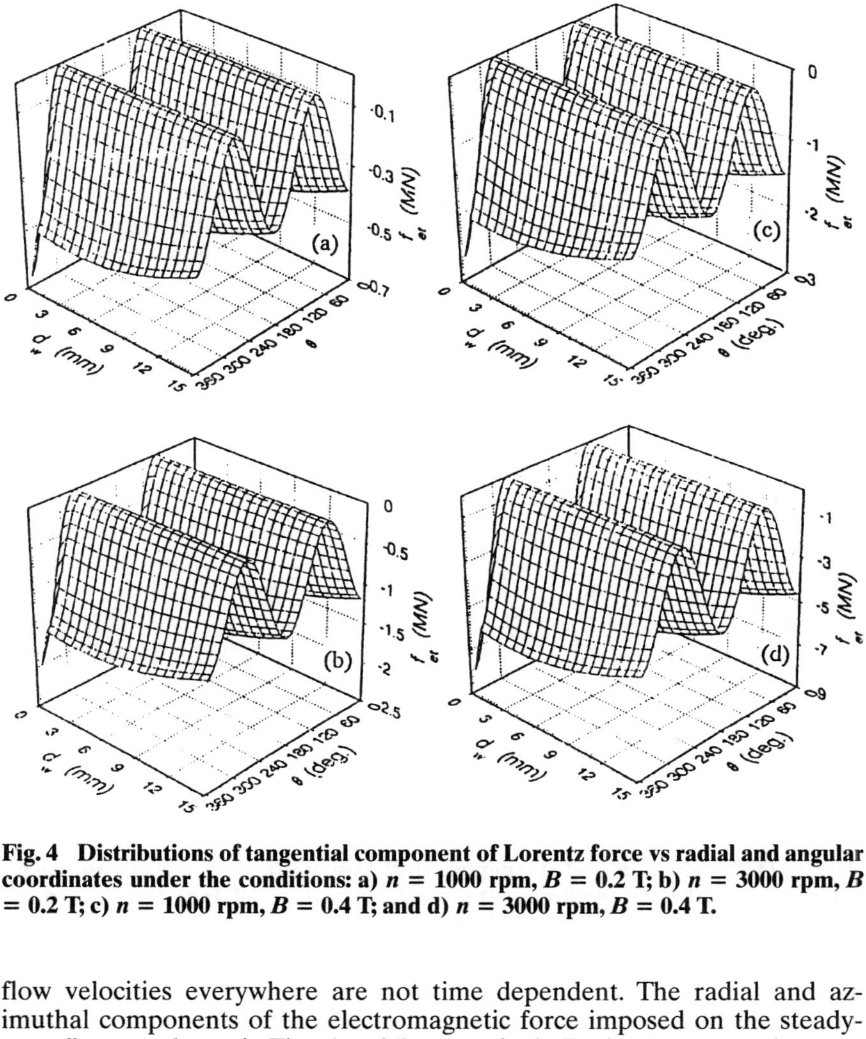

Fig. 4 Distributions of tangential component of Lorentz force vs radial and angular coordinates under the conditions: a) n = 1000 rpm, B = 0.2 T; b) n = 3000 rpm, B = 0.2 T; c) n = 1000 rpm, B = 0.4 T; and d) n = 3000 rpm, B = 0.4 T.

flow velocities everywhere are not time dependent. The radial and azimuthal components of the electromagnetic force imposed on the steady-state flow are shown in Figs. 4 and 5, respectively. Both components have no marked changes in radial distributions, except for the abrupt decrease of the radial component adjacent to the solid wall.

And, both components are found to fluctuate as sine and cosine functions of the angular coordinate, respectively. As shown in Figs. 4 and 5, any increase in the magnetic flux density or rotation rate of the mold may cause an increase in the force. If n is increased from 1000 to 3000 rpm or B from 0.2 to 0.4 T, the maximum value for the radial component of f_e on the bulk liquid is increased by approximately four times and for the tangential component three to four times.

Centrifugal forces under the given conditions are shown in Fig. 6 in which a) and b) indicate those in magnetic-field-free processes that are angle-in-

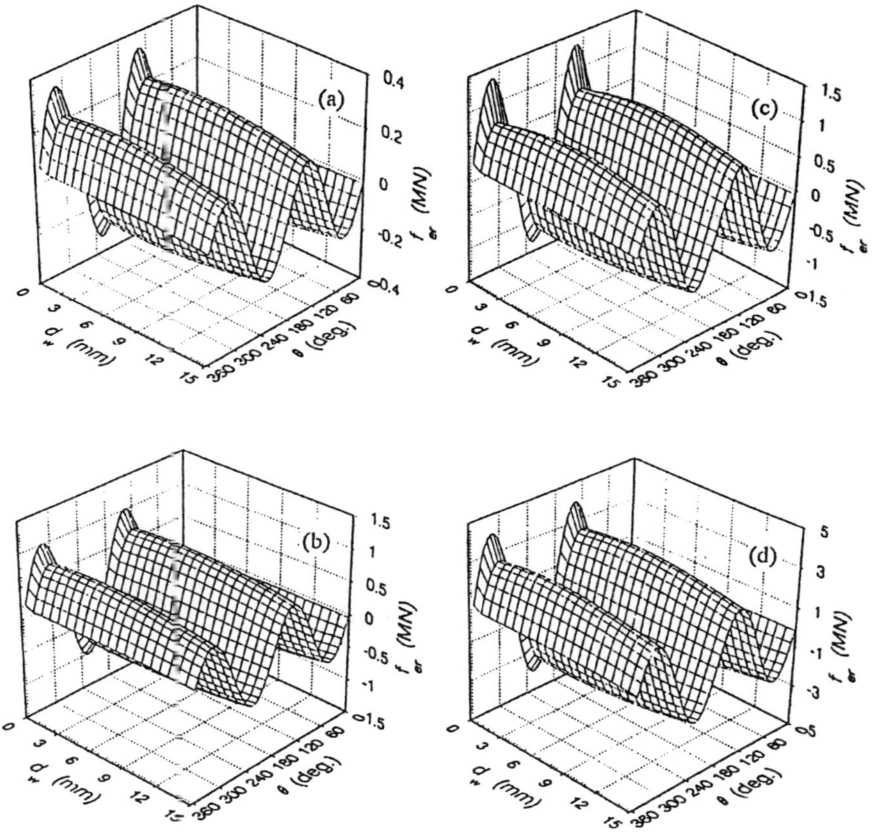

Fig. 5 Distributions of radial component of Lorentz force vs radial and angular coordinates under the conditions: a) n = 1000 rpm, B = 0.2 T; b) n = 3000 rpm, B = 0.2 T; c) n = 1000 rpm, B = 0.4 T; and d) n = 3000 rpm, B = 0.4 T.

dependent; and c) through f) describe the influence of magnetic field and rotation speed of the mold. As for the cases in the presence of a magnetic field, the centrifugal force near the solid wall decreases. This results from the decrease in the tangential component of the flow velocity. Because the value of the centrifugal force is proportional to the square of the rotation speed of the local fluid, there must be a decrease in the centrifugal force even though the decrease of the flow velocity is small. If the mold rotates faster, it is easily understandable that a stronger centrifugal force field should be produced. However, the intensification of the imposed magnetic field leads to an amplification in the scope of the fluctuations without increasing the average level of the force.

We should note that the gravity per unit volume f_g vanishes when taking the curl as well as the divergence of the momentum equation; namely, the

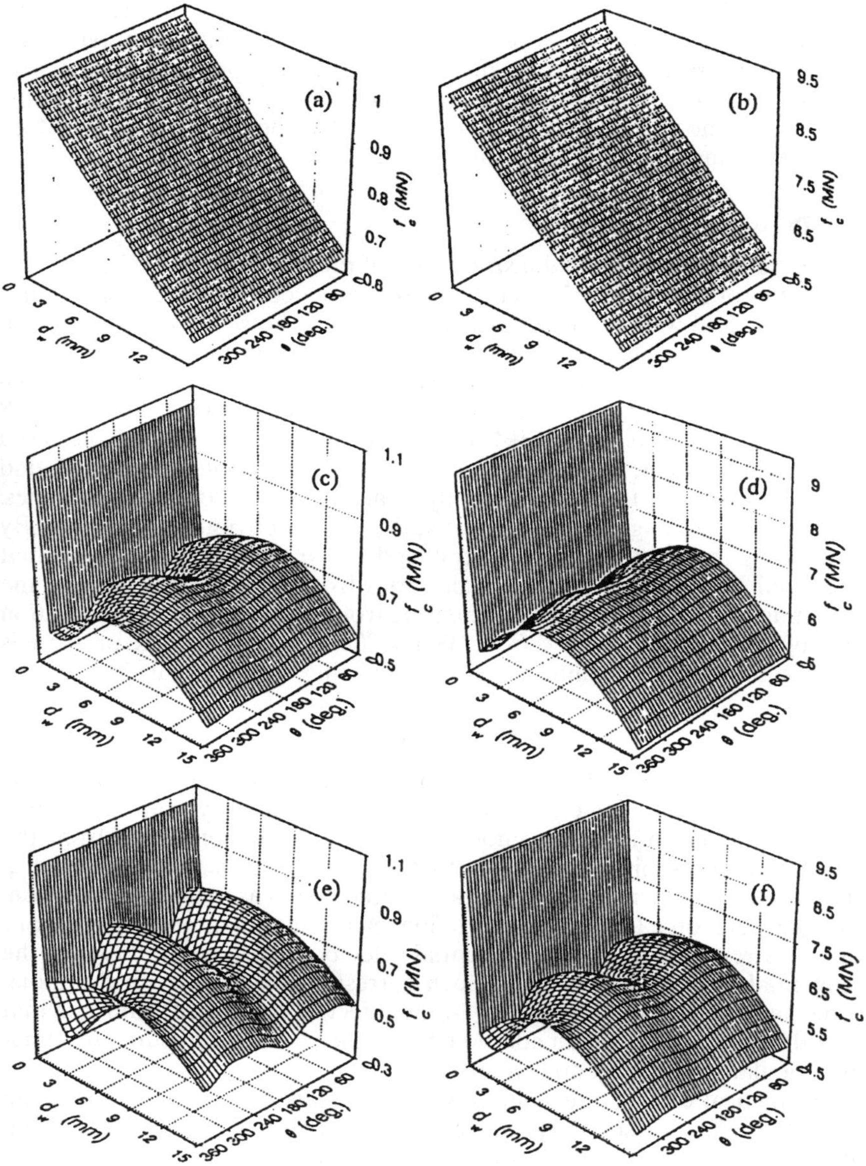

Fig. 6 Distributions of the centrifugal force vs radial and angular coordinates under the conditions: a) n = 1000 rpm, B = 0 T; b) n = 3000 rpm, B = 0 T; c) n = 1000 rpm, B = 0.2 T; d) n = 3000 rpm, B = 0.2 T; e) n = 1000 rpm, B = 0.4 T; and f) n = 3000 rpm, B = 0.4 T.

body force terms in the stream function and vorticity equations [Eqs. (7) and (8)] and the pressure equation [Eq. (13)] do not include the gravity. This implies that the gravity has no effect on the flow and pressure. As a matter of fact, the gravity leads to a lowering of the center-of-gravity position of the melt pipe in horizontal centrifugal casting, and, without regard to this effect, the gravity-induced flow is also neglected.

C. Pressure

The radial and azimuthal distributions of pressure are shown in Fig. 7. In the cases for $B = 0$, Figs. 7a and 7b, the pressure decreases with the distance from the mold wall d_w increasing. As for EMCC, however, the pressure on the solid wall reaches the highest value, whereas on the free surface the pressure remains zero as the preliminary assumption, and its azimuthal distribution also has the cyclic characteristics. If the magnetic flux density B is increased from 0.2 up to 0.4 T with the rotation velocity of the mold remaining constant, the average pressure within the whole domain is found to decrease except that there are only small changes in some certain zones. Meanwhile, the angle-dependent fluctuations are obviously enhanced. By comparing the cases for $n = 1000$ rpm and $B = 0.2$–0.4 T, one may find that the maximum pressure on the wall varies from 8.5 to 10.5 kPa with the minimum pressure on the wall decreasing from 7.5 to 1.8 kPa. If the rotation rate of the mold is increased from 1000 to 3000 rpm, the average pressure is increased with the maximum pressure increasing by more than 20 kPa.

IV. Discussion

From the preceding results was found that all of the distributions of the flow velocity, body forces, and pressure have oscillatory features though there are some differences among them. The oscillation arises from the complex relationship between the forces and fluid flow. As shown in Eq. (6b), both components of the force are characterized as periodic fluctuation. The periodic force leads the fluid to flow in a similar manner. The positions of the extreme values for the tangential velocity correspond to those for the tangential force but there is no such correspondence between the radial force and radial flow velocity due to the effect of the centrifugal force that is in the radial direction. In addition, fluctuations of the pressure are somehow similar to the tangential velocity.

In practical casting, the fluid flow near the liquid/solid interface has an important effect on solidification. Some interesting phenomena were found in the previous work[10] on electromagnetic centrifugal casting, one of which was that electromagnetic stirring promoted the formation of equiaxed grains in solidification, and the columnar-equiaxed transition was related to the break-off of dendrites. According to the computational results, the maximum tangential velocity is in general located near the interface. The tangential component of the electromagnetic force on the interface is

$$f_{\text{int}} = \sigma B^2 \Omega r_{\text{int}} \cos^2 \theta \tag{15}$$

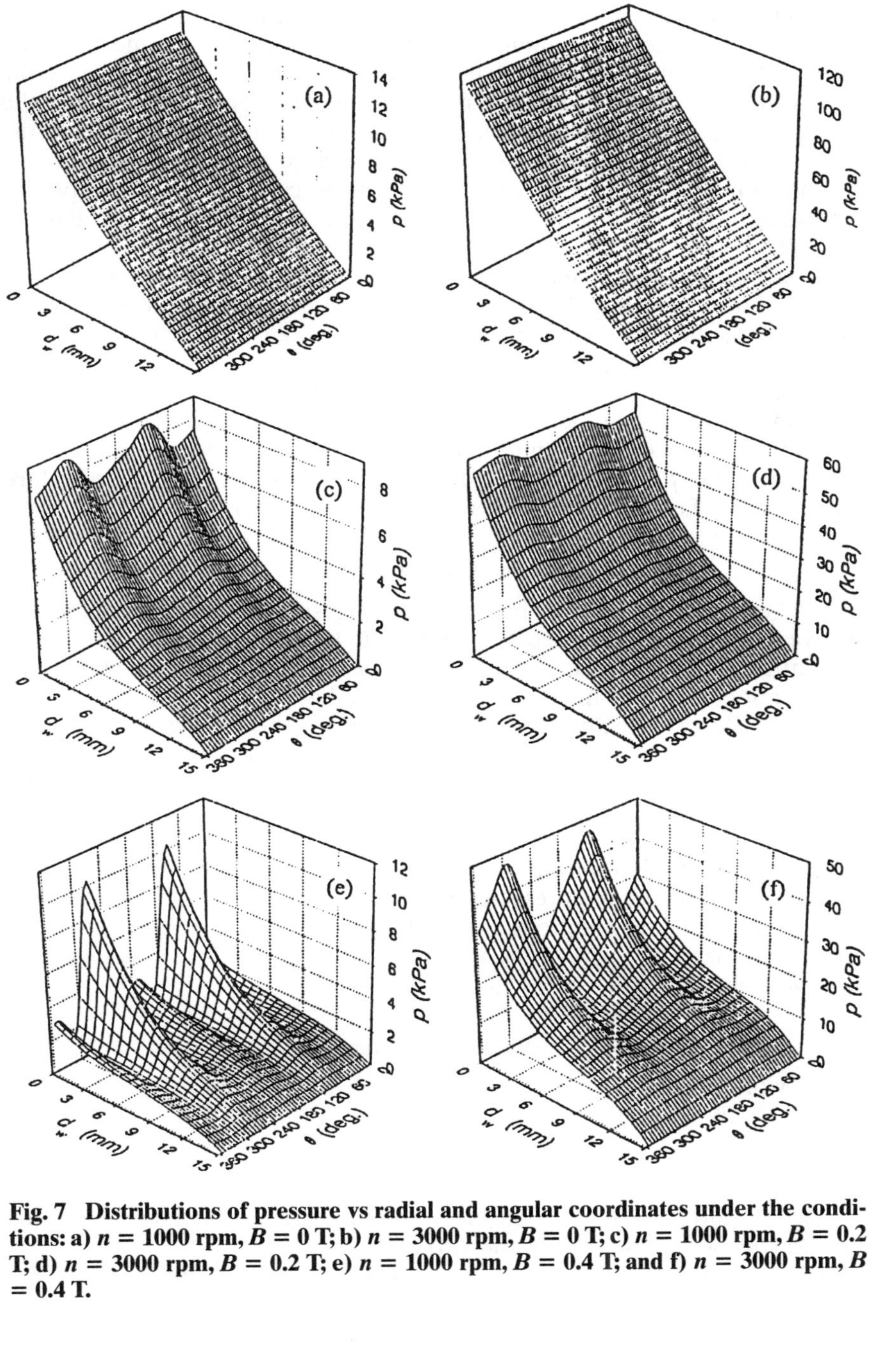

Fig. 7 Distributions of pressure vs radial and angular coordinates under the conditions: a) n = 1000 rpm, B = 0 T; b) n = 3000 rpm, B = 0 T; c) n = 1000 rpm, B = 0.2 T; d) n = 3000 rpm, B = 0.2 T; e) n = 1000 rpm, B = 0.4 T; and f) n = 3000 rpm, B = 0.4 T.

When the total stress is high enough to exceed the fracture strength of the columnar dendrite, the columnar dendrites may be broken, and the fragments of the dendrites are probably taken away from the interface by the radial flow and then become the nuclei of equiaxed grains. The higher the rotation rate of the mold or the magnetic flux density, the stronger the electromagnetic force on the solid/liquid interface and the more equiaxed grains form. Therefore, one can control the as-cast structure of pipes by adjusting the controlling parameters, rotation rate, or magnetic flux density in EMCC processes.

Pressure is one of the most important factors in controlling metallurgical quality during centrifugal casting, and pressure decrease resulting from the adoption of the magnetic field will probably make it difficult for molten metal to obtain perfect shapes and structures in casting. In accordance with this point of view, although a high intensity of magnetic field is effective to force melt flow in solidification, it is not appropriate for practical casting if the rotation of the fluid cannot compensate for the decrease in pressure. Therefore, this problem requires that the mold in the EMCC procedure should rotate faster than in conventional centrifugal casting.

V. Conclusions

1) The tangential flow of the melt is strongest near the solid wall and is weakened with the increase of the distance to the solid wall. The radial flow, in contrast, is increased with the increase of the distance to the wall. Both the rotation rate and magnetic flux density have great influence on the tangential flow, but the rotation rate has little effect on the radial flow.

2) Body forces in steady state are dependent on the controlling parameters. Any change of the rotation rate or magnetic intensity leads to an alteration of the body forces.

3) The pressure field is also affected greatly by the controlling parameters: the faster the mold rotates, or the lower the intensity of magnetic field employed, the higher the pressure level reaches.

4) In azimuthal distributions, flow velocity, pressure, and body forces are all characterized by fluctuations, the amplitudes of which are influenced by the controlling parameters.

Acknowledgment

The work was supported by the National Natural Science Foundation of China (No. 59431010).

References

[1]Asai, S., "Birth and Recent Activities of Electromagnetic Processing of Materials," *ISIJ Int.*, Vol. 29, No. 12, 1989, pp. 981–992.

[2]Kolesnickenko, A. F., "Electromagnetic Processes in Liquid Material in the USSR and East European Countries," *ISIJ Int.*, Vol. 30, No. 1, 1990, pp. 8–26.

[3]Garnier, M., "Electromagnetic Processing of Liquid Materials in Europe," *ISIJ Int.*, Vol. 30, No. 1, 1990, pp. 1–7.

[4]Tewari, N., "Effect of Magnetic Field on the Microstructure and Macrosegregation in Directionally Solidified Pb-Sn Alloys," *Met. Mat. Trans.*, Vol. A25, 1994, pp.1535–1541.

[5]Kobayashi, S., Ishimura, S., Yoshihara, M., and Sugitani, Y., "Factors Equiaxed Zone Generation in Electromagnetic Stirring," *Trans. ISIJ*, Vol. 28, 1988, pp. 939–944.

[6]Vives, C., and Perry, C., "Effect of Electromagnetic Stirring during the Controlled Solidification of Tin," *Journal of Heat Mass Transfer*, Vol. 29, No. 1, 1986, pp. 21–33.

[7]Mihelcic, M., and Wingerath, K., "Three Dimensional Simulations of the Czochralski Bulk Flow in a Stationary Transverse Field and in a Vertical Magnetic Field: Effects on the Asymmetry of the Flow and Temperature Distribution in the Si Melt," *Journal of Crystal Growth*, Vol. 82, 1987, pp. 318–326.

[8]Hall, M. C., and Walker, J. S., "Heat and Mass Transfer through a Liquid Metal in an Infinitely Long, Rotating Cylinder with a Uniform, Transverse Magnetic Field," *Journal of Heat Mass Transfer*, Vol. 36, No. 14, 1993, pp. 3509–3514.

[9]Spitzer, K. H., Dubke, M., and Schwerdtfeger, K., "Rotational Electromagnetic Stirring in Continuous Casting of Round Strands," *Met. Trans.*, Vol. 17B, 1989, pp. 119–131.

[10]Yang, Y., Liu, Q., Jiao, Y., and Hu, Z., "Application of Steady Magnetic Field for Refining Solidification Structure and Enhancing Mechanical Properties of 25Cr-20Ni-Fe-C Alloy in Centrifugal Casting," *ISIJ Int.*, Vol. 35, No. 4, 1995, pp. 389–392.

[11]Peyret, R., and Taylor, T. D., *Computational Methods for Fluid Flow*, Springer–Verlag, New York, 1983, pp. 182–199.

Chapter 48

Bifurcations in a Magnetofluid with Helical Forcing

N. Sheehafer* and F. Feudel‡
Max-Planck-Gruppe Nichtlineare Dynamik, Universität Potsdam, Potsdam, D-14415, Germany
and
B. Galanti‡
Weizmann Institute of Science, 76100, Rehovot, Israel

I. Introduction

A PROMINENT subject in magnetohydrodynamics (MHD) is the dynamo problem,[1] namely, the question under which conditions of electrically conducting fluids act as self-excited dynamos. "Self-excited" here means that no electric or magnetic fields are externally applied. The majority of studies in this field has been kinematic. Kinematic dynamo theory studies the conditions under which a prescribed velocity field can amplify, or at least prevent from decaying, some seed magnetic field, completely disregarding the equations governing the motion of the fluid. The most successful branch of kinematic dynamo theory is the theory of the turbulent dynamo,[2,3] which has supplied evidence that the presence of kinetic and magnetic helicities is favorable for a dynamo effect. With $\boldsymbol{v}$, $\boldsymbol{B}$, and $\boldsymbol{A}$ denoting fluid velocity, magnetic field, and a magnetic vector potential, the densities per unit volume of kinetic and magnetic helicity are defined by

$$H_K = \boldsymbol{v} \cdot \operatorname{curl} \boldsymbol{v}, \qquad H_M = \boldsymbol{A} \cdot \boldsymbol{B} \tag{1}$$

*PF 601553.
‡PF 601553.
†Department of Chemical Physics.

Simple examples of strongly helical flows are provided by the so-called *ABC* flows,

$$\boldsymbol{v} = \boldsymbol{v}_{ABC} = (A \sin z + C \cos y, B \sin x + A \cos z, C \sin y + B \cos x) \tag{2}$$

where A, B, and C denote constant coefficients. By satisfying curl $\boldsymbol{v} = \boldsymbol{v}$, they have the Beltrami property curl $\boldsymbol{v} \times \boldsymbol{v} = \boldsymbol{0}$, and in general (if $ABC \neq 0$) there are domains in the flow where the streamlines are chaotic. For these reasons, they have received much interest,[4–6] notably as candidates for fast dynamos (kinematic dynamos for which the growth rate of the magnetic field remains bounded away from zero as the magnetic diffusivity tends to zero). The *ABC* flows are steady solutions of the incompressible Euler equations. They are also steady solutions of the incompressible Navier–Stokes equations [Eq. (4) with the magnetic field dropped] if an external body force

$$\boldsymbol{f} = -\nu\mu\boldsymbol{v}_{ABC} = \nu\boldsymbol{v}_{ABC} \tag{3}$$

called *ABC* forcing in the following, just compensating for viscous losses is applied. They are then stable below and unstable above a certain critical strength of the forcing or critical Reynolds number, respectively.[7,8] Imposing this kind of forcing, Galanti et al.[9] studied the complete system of the incompressible MHD equations. Numerically simulating the system, they found that at some critical value of the Reynolds number a stable *ABC* flow without a magnetic field loses stability to a time-periodic state with a magnetic field. In the present paper we continue and extend the study of Galanti et al.[9] by systematically applying numerical methods of bifurcation analysis. In particular, we study the sequence of bifurcations that occur if the strength of the *ABC* forcing is raised. Furthermore, by applying a generalized *ABC* forcing of varying degree of helicity, we investigate the influence of the degree of helicity on the transition from an always existing stable stationary solution to time-dependent states and the role of the helicity for a dynamo effect.

In Section II we cast the three-dimensional MHD equations into spectral form and explain their truncation as well as the external forcing used. Then in Section III we describe the sequence of bifurcations obtained when applying the pure *ABC* forcing, whereas in Section IV a bifurcation diagram for the case with the generalized *ABC* forcing is presented. Section V, finally, contains a brief conclusion.

II. Basic Equations, Truncation, and Forcing

We start from the equations for a nonrelativistic, incompressible, electrically conducting fluid with constant material properties,[10]

$$\frac{\partial \boldsymbol{v}}{\partial t} + (\boldsymbol{v}\cdot\nabla)\boldsymbol{v} = \nu\Delta\boldsymbol{v} - \text{grad } p - \frac{1}{2}\text{grad } \boldsymbol{B}^2 + (\boldsymbol{B}\cdot\nabla)\boldsymbol{B} + \boldsymbol{f} \tag{4}$$

$$\frac{\partial \boldsymbol{B}}{\partial t}+(\boldsymbol{v}\cdot\nabla)\boldsymbol{B}=\eta\Delta\boldsymbol{B}+(\boldsymbol{B}\cdot\nabla)\boldsymbol{v} \tag{5}$$

$$\operatorname{div}\boldsymbol{v}=0, \qquad \operatorname{div}\boldsymbol{B}=0 \tag{6}$$

where the density has been set equal to unity, p is the thermal pressure, ν the kinematic viscosity, η the magnetic diffusivity, and $\boldsymbol{f}$ an external body force.

We apply periodic boundary conditions on a cube of side length 2π, which is equivalent to considering the motion on the torus $T^3 = [0,2\pi] \times [0,2\pi] \times [0,2\pi]$. The mean values of $\boldsymbol{v}$ and $\boldsymbol{B}$, and consequently also of $\boldsymbol{f}$, are assumed to vanish,

$$\int_{T^3}\boldsymbol{v}\,\mathrm{d}^3\boldsymbol{x}=\boldsymbol{0}, \qquad \int_{T^3}\boldsymbol{B}\,\mathrm{d}^3\boldsymbol{x}=\boldsymbol{0}, \qquad \int_{T^3}\boldsymbol{f}\,\mathrm{d}^3\boldsymbol{x}=\boldsymbol{0} \tag{7}$$

The periodicity assumption implies that the Fourier representations of $\boldsymbol{v}$, $\boldsymbol{B}$, p, and $\boldsymbol{f}$,

$$\boldsymbol{v}(\boldsymbol{x})=\sum_{\substack{\boldsymbol{k}\in\mathbb{Z}^3\\ \boldsymbol{k}\neq 0}}\boldsymbol{v}_{\boldsymbol{k}}\exp(i\boldsymbol{k}\cdot\boldsymbol{x}) \tag{8}$$

$$\boldsymbol{B}(\boldsymbol{x})=\sum_{\substack{\boldsymbol{k}\in\mathbb{Z}^3\\ \boldsymbol{k}\neq 0}}\boldsymbol{B}_{\boldsymbol{k}}\exp(i\boldsymbol{k}\cdot\boldsymbol{x}) \tag{9}$$

$$p(\boldsymbol{x})=\sum_{\boldsymbol{k}\in\mathbb{Z}^3}p_{\boldsymbol{k}}\exp(i\boldsymbol{k}\cdot\boldsymbol{x}) \tag{10}$$

$$\boldsymbol{f}(\boldsymbol{x})=\sum_{\substack{\boldsymbol{k}\in\mathbb{Z}^3\\ \boldsymbol{k}\neq 0}}\boldsymbol{f}_{\boldsymbol{k}}\exp(i\boldsymbol{k}\cdot\boldsymbol{x}) \tag{11}$$

can be differentiated term by term with respect to the spatial coordinates. In Fourier space Eq. (6) takes the form

$$\boldsymbol{v}_{\boldsymbol{k}}\cdot\boldsymbol{k}=0, \qquad \boldsymbol{B}_{\boldsymbol{k}}\cdot\boldsymbol{k}=0 \tag{12}$$

and is automatically satisfied if we write

$$\boldsymbol{v}_{\boldsymbol{k}}=v_{\boldsymbol{k}}^{(1)}\boldsymbol{e}_{\boldsymbol{k}}^{(1)}+v_{\boldsymbol{k}}^{(2)}\boldsymbol{e}_{\boldsymbol{k}}^{(2)}, \qquad \boldsymbol{B}_{\boldsymbol{k}}=B_{\boldsymbol{k}}^{(1)}\boldsymbol{e}_{\boldsymbol{k}}^{(1)}+B_{\boldsymbol{k}}^{(2)}\boldsymbol{e}_{\boldsymbol{k}}^{(2)} \qquad \text{for } k\neq\boldsymbol{0} \tag{13}$$

with (real) "polarization" unit vectors $e_k^{(1)}, e_k^{(2)}$ perpendicular to k,

$$e_k^{(i)} \cdot k = 0, \quad e_k^{(1)} \cdot e_k^{(2)} = 0, \quad e_k^{(i)} \cdot e_k^{(i)} = 1, \quad e_{-k}^{(i)} = e_k^{(i)} \quad \text{for } i = 1,2 \tag{14}$$

The last condition in Eq. (14) ensures that

$$v_{-k}^{(i)} = v_k^{*(i)}, \qquad B_{-k}^{(i)} = B_k^{*(i)} \tag{15}$$

for real $\boldsymbol{v}(\boldsymbol{x})$ and $\boldsymbol{B}(\boldsymbol{x})$ (an asterisk indicates the complex conjugate). By using these representations for $\boldsymbol{v}_k$ and $\boldsymbol{B}_k$, we furthermore easily get rid of both the thermal, grad p, and magnetic, grad $\boldsymbol{B}^2/2$, pressure terms in Eq. (4) and arrive at the following infinite-dimensional system of ordinary differential equations (ODE):

$$\frac{\mathrm{d}v_k^{(j)}}{\mathrm{d}t} = -\nu k^2 v_k^{(j)} - i \sum_{\substack{p \in \mathbb{Z}^3 \\ p \neq 0, k}} \sum_{\alpha,\beta=1}^{2} \left[e_p^{(\alpha)} \cdot e_k^{(j)}\right]\left[e_{k-p}^{(\beta)} \cdot k\right]\left[v_p^{(\alpha)} v_{k-p}^{(\beta)} - B_p^{(\alpha)} B_{k-p}^{(\beta)}\right] + f_k^j \tag{16}$$

$$\frac{\mathrm{d}B_k^{(j)}}{\mathrm{d}t} = -\eta k^2 B_k^{(j)} - i \sum_{\substack{p \in \mathbb{Z}^3 \\ p \neq 0, k}} \sum_{\alpha,\beta=1}^{2} \left[e_p^{(\alpha)} \cdot e_k^{(j)}\right]\left[e_{k-p}^{(\beta)} \cdot k\right]\left[B_p^{(\alpha)} v_{k-p}^{(\beta)} - v_p^{(\alpha)} B_{k-p}^{(\beta)}\right] \tag{17}$$

The $f_k^{(j)}$ on the right of Eq. (16) is defined by

$$f_k^{(j)} = f_k \cdot e_k^{(j)} \qquad j = 1,2 \tag{18}$$

In our numerical calculations, we have used an isotropic truncation in wave-number space, following Lee,[11,12] who segmented k space into successive shells $n^2 - n < k^2 \leq n^2 + n, n = 1,2, \ldots$. In most of our calculations, we have taken into account three shells corresponding to 89 k vectors, which amounts to studying a system of 712 ODEs. But partially, to test the influence of the degree of truncation, up to nine shells were included into the computations, corresponding to 1847 k vectors and 14,776 ODEs, respectively. We did not apply a Fourier–Weierstrass method,[13] in which, in order to keep high Reynolds number turbulence feasible, k space is thinned out by taking into account only k vectors in selected shells, disregarding shells in between. By contrast, we have taken into account all wave vectors in a full sphere. In real space this means that the effect of small spatial scales, corresponding to wave numbers outside the sphere, is neglected.

Because of the periodic boundary conditions, the total energy flow through the boundary of our cube vanishes, so that in order to compensate for viscous and ohmic losses, and thus to allow for nontrivial time-asymptotic states, some kind of external forcing has to be applied. We have used the ABC forcing, given by Eqs. (3) and (2), with

$$A = B = C = f \tag{19}$$

For this forcing the MHD equations are equivariant with respect to a discrete symmetry group that contains 24 elements and is isomorphic to the octahedral group O (the rotation group of the cube).[5,14] A number of bifurcations observed in the system are related to these symmetries.

Besides the pure ABC forcing, we have also applied a generalized ABC forcing given by

$$\boldsymbol{f} = \nu\left[(1-\lambda)\boldsymbol{v}_{ABC} + \lambda \boldsymbol{v}_{\bar{A}BC}\right] \tag{20}$$

where

$$\boldsymbol{v}_{\bar{A}BC} = (A\cos z + C\sin y, B\cos x + A\sin z, C\cos y + B\sin x) \tag{21}$$

and λ is a parameter varying between 0 and 0.5. The $\boldsymbol{v}_{\bar{A}BC}$ satisfies curl $\boldsymbol{v}_{\bar{A}BC} = -\boldsymbol{v}_{\bar{A}BC}$, and for $\lambda = 0.5$ its addition in the forcing term "kills" the helicity on average in the volume, whereas $\lambda = 0$ corresponds to the original ABC forcing.

Kinetic and magnetic Reynolds numbers R and Rm may be defined by

$$R = \frac{f}{\nu}, \qquad Rm = \frac{f}{\eta} \tag{22}$$

We have been restricting ourselves to the case $\nu = \eta$ (magnetic Prandtl number equal to unity). R and λ have been our bifurcation parameters.

III. Bifurcation Sequence for Pure ABC Forcing

In this section we describe the bifurcations observed in the case of the pure ABC forcing ($\lambda = 0$). For weak forcing (small R), there exists a stable stationary solution, namely the ABC flow [given by Eq. (2)] with a vanishing magnetic field, and all system trajectories are attracted by this solution, which is symmetric with respect to all of the 24 transformations of the symmetry group O. Raising R, we have traced the steady-solution branch. In each step of the tracing to detect bifurcation points the eigenvalues of the Jacobian matrix have been calculated. The steady state loses stability in a Hopf bifurcation, leading to a periodic solution with a magnetic field as the only time-asymptotic state.

Table 1 summarizes the values of the Reynolds number at which the Hopf bifurcation occurred for different truncations. Discernible in Table 1 is a

Table 1 Reynolds number at which the Hopf bifurcation was observed for different truncations in wave-number space.

Number of shells in $\boldsymbol{k}$ space	Number of modes (equations)	Critical R for Hopf bifurcation
2	40 (320)	4.6
3	89 (712)	5.7
4	194 (1,552)	12.7
5	369 (2,952)	8.2
6	594 (4,752)	<12.5
7	895 (7,160)	<11.0
9	1847 (14,776)	<12.0

dependence of the critical Reynolds number on the degree of truncation with a tendency to higher values for weaker truncation. The latter may be because of an increase of the energy dissipation with an increasing number of modes (as a consequence of the presence of smaller scales).

Because of a very long run time of the program calculating the eigenvalues of the Jacobian matrix, for the cases with more than five shells in $\boldsymbol{k}$ space only upper bounds for the critical Reynolds number are given obtained from simulations of single-system trajectories (i.e., simulations of the system for selected initial conditions).

In the remainder of this section, as well as in the next section, we present results obtained by applying the (simple) three-shell truncation. In Fig. 1 the instability of the ABC flow is demonstrated for a Reynolds number only

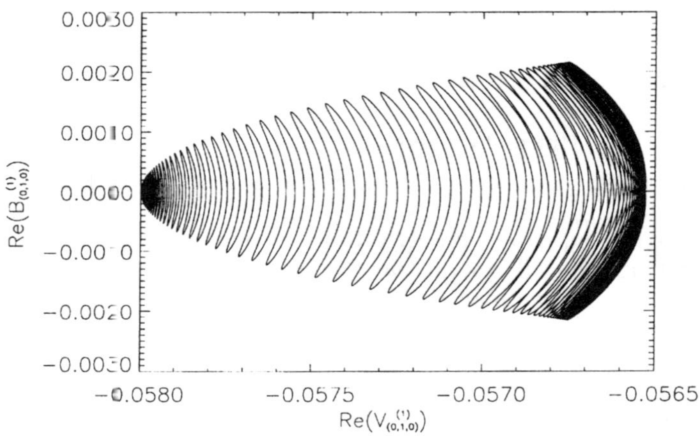

Fig. 1 Example of a trajectory demonstrating the instability of the primary steady state after the Hopf bifurcation for R = 5.8.

slightly above the Hopf value. How a trajectory escapes the originally stable fixed point (to the left in the figure; initial conditions very close to this point had been chosen) and approaches a periodic orbit (to the right in the figure) can be seen.

The Hopf bifurcation is nondegenerate. Just one pair of complex conjugate eigenvalues crosses the imaginary axis, the derivative of their real part with respect to R being different from zero. Consequently, just one periodic solution originates, which is still symmetric with respect to the full group O. But for $R = 7.7$ three new stable periodic solutions appear, which can be transformed into each other by certain elements of O. The three new solutions do not bifurcate from the stable symmetric cycle, which coexists with them. We suppose that they are generated in (due to the symmetry simultaneous) saddle-node (turning-point) bifurcations.

For $R = 11.55$ also the symmetric branch loses part of its symmetry. The cycle bifurcates into four new periodic solutions, which can be be transformed into one another by elements of O. Thus now altogether seven stable periodic solutions coexist.

Both branches, the four-solutions one and the three-solutions one, lead finally to chaotic states if R is increased further. Here we describe the transition for the three-solutions branch. In a second Hopf bifurcation at $R = 16$, a torus solution, as shown in Fig. 2, is generated (more precisely, three stable tori, namely from each of the three periodic solutions of the considered branch just one). At $R = 20$ the torus decays to a chaotic attractor, depicted in Fig. 3. The appearance of chaos has been verified by calculating the five largest Lyapunov exponents. For this calculation we have used an algorithm by Shimada and Nagashima.[15] Figure 4 shows the Lyapunov exponents in dependence on the integration time. A good convergence is discernible, as well as that at least the largest exponent is positive.

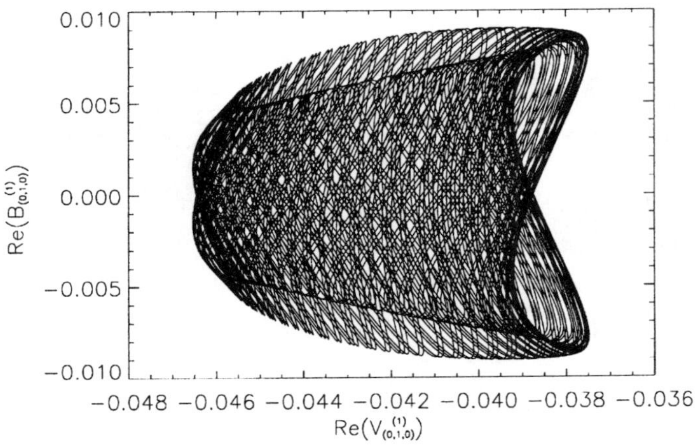

Fig. 2 Torus solution for R = 16.

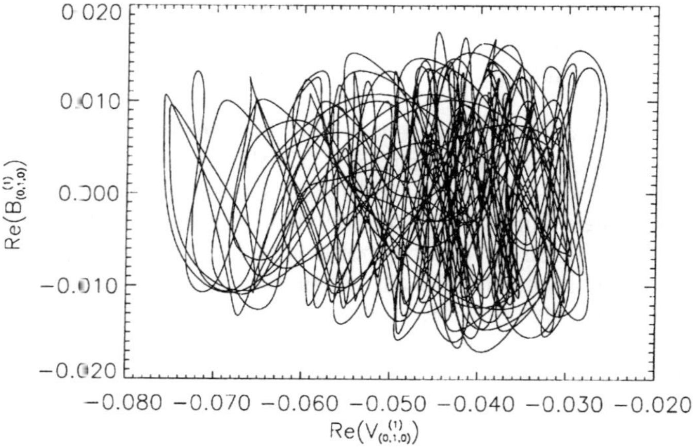

Fig. 3 Trajectory on chaotic attractor for $R = 20$.

After the first Hopf bifurcation the magnetic energy increases to a value of about 10% of the total energy. A similar ratio between magnetic and kinetic energies has been observed in three-dimensional MHD simulation experiments.[16] The volume-averaged magnetic helicity [cf. Eq. (1)] is negative, thus opposite in sign to the kinetic helicity, as also found by Galanti et al.[9]

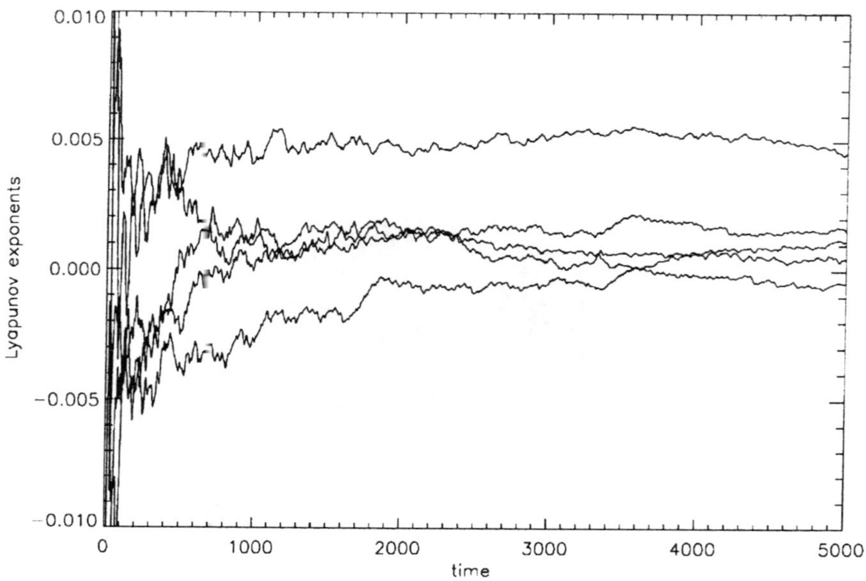

Fig. 4 The five largest Lyapunov exponents versus integration time for $R = 20$.

Above the critical value for the Hopf bifurcation, we did not find any interval in Reynolds number where no dynamo action occurs. Such a gap in dynamo action was found by Galloway and Frisch[6] in a kinematic study of the induction equation with prescribed stationary velocity field (*ABC* flow). Similar gaps were also observed in alpha-effect mean-field studies.

IV. Bifurcation Diagram for Generalized *ABC* Forcing

Now we present the bifurcation diagram obtained when applying the generalized *ABC* forcing, given by Eq. (20). For weak forcing (small R), there always exists a stable stationary, globally attracting solution (which coincides with the the original *ABC* flow only in the special case of $\lambda = 0$). Keeping fixed λ and raising R, we have traced the steady-solution branch. For $\lambda < 0.4$ the steady state loses stability in a Hopf bifurcation, but at $\lambda = 0.4$ the type of the first bifurcation, as well as the character of the time-asymptotic states after this bifurcation, change. Whereas for $\lambda < 0.4$ a magnetic periodic state is the (only) new attractor new nonmagnetic states emerge for λ between 0.4 and 0.5.

The original steady state has been traced also in the region beyond the first bifurcation where it is unstable and where it undergoes secondary bifurcations. The locations of primary and secondary bifurcations in the parameter plane are shown in Fig. 5. Thick solid and dashed lines, respectively, indicate the primary bifurcation of the original steady state. For $0.4 < \lambda < 0.49$ the Hopf bifurcation is preceded by a bifurcation in which two real eigenvalues of the Jacobian matrix become positive; these two eigenvalues are equal already before the bifurcation because of one of the symmetries in the system. The bifurcation results in three coexisting new stable stationary solutions, which can be transformed into each other by certain elements of the symmetry group of the problem (which is a subgroup of the symmetry group O in the case of the pure *ABC* forcing). For $\lambda = 0.4$ this bifurcation occurs simultaneously with the Hopf bifurcation.

At $\lambda = 0.49$, showing up as a kind of cusp in Fig. 5, the primary bifurcation changes its character again, namely from the real bifurcation observed for $\lambda < 0.49$ to another Hopf bifurcation for $\lambda > 0.49$. This Hopf bifurcation leads to a nonmagnetic periodic state. At the cusp-like transition point the imaginary part of the pair of complex-conjugate eigenvalues responsible for the Hopf bifurcation to the right is zero, and the complex-conjugate pair coincides with the real pair responsible for the bifurcation to the left.

The new stationary and periodic solutions bifurcating from the original stationary one for $0.4 < \lambda < 0.5$ are stable only over very small intervals of the bifurcation parameter R and lose their stability directly to nonmagnetic chaotic states. The presence of chaos has again been verified by calculating, for selected values of the bifurcation parameters, the largest Lyapunov exponents.

The chaotic attractors found here are strongly suggestive of Shilnikov-type homoclinic chaos. In a three-dimensional flow, this type of chaos is

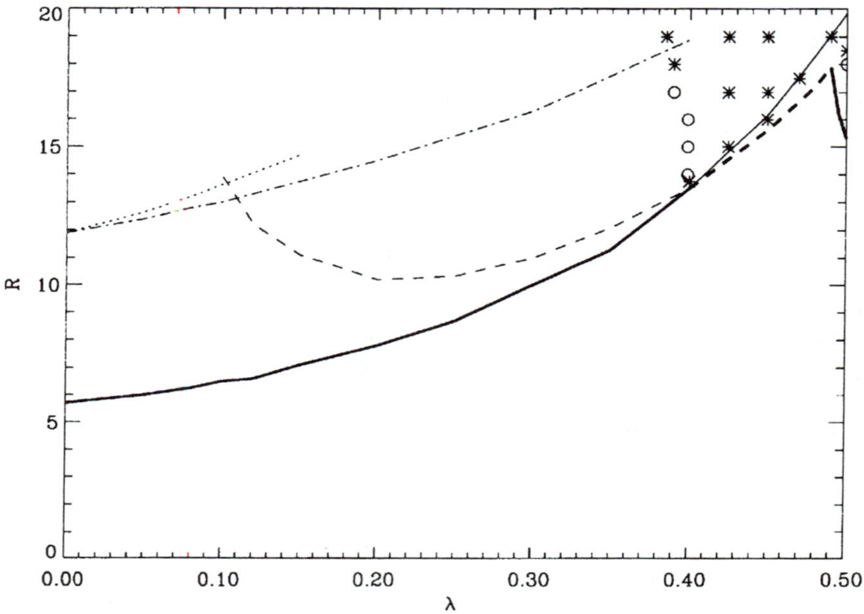

Fig. 5 Locations of primary and secondary bifurcations of the original stationary solution in the λ-R plane. Solid line and dashed-dotted line: a single pair of complex conjugate eigenvalues crosses the imaginary axis; dashed line: two real eigenvalues pass through zero; dotted line: two pairs of complex conjugate eigenvalues cross the imaginary axis. Asterisks indicate points at which, by means of simulations, nonmagnetic chaotic (Shilnikov-like) time-asymptotic states have been found, whereas circles correspond to magnetic periodic attractors.

related to the existence of a homoclinic trajectory to a saddle point with a real eigenvalue and a pair of complex-conjugate eigenvalues, the real eigenvalue having a larger magnitude than the real part of the complex eigenvalues (cf. Ref. 17, Sec. 6.5). The trajectory shown in Fig. 6 seems to be very close to such a homoclinic orbit. It approaches the unstable fixed point along a path apparently close to a one-dimensional stable manifold, spirals outward (thereby seemingly close to a two-dimensional unstable manifold), approaches the fixed point again, and so on. In our case the phenomenon seems to be connected with the degenerate bifurcation at $\lambda = 0.4$ (crossing of solid and dashed lines in Fig. 5). In problems with two parameters, the occurence of homoclinic orbits in the vicinity of such points of degeneracy is a generic phenomenon (see, e.g., Ref. 18).

In Fig. 5 asterisks indicate points at which nonmagnetic Shilnikov-like chaotic attractors have been found. The nonmagnetic chaotic domain extends also to λ values less than 0.4, thereby causing a disappearance of the dynamo effect already present at smaller Reynolds numbers.

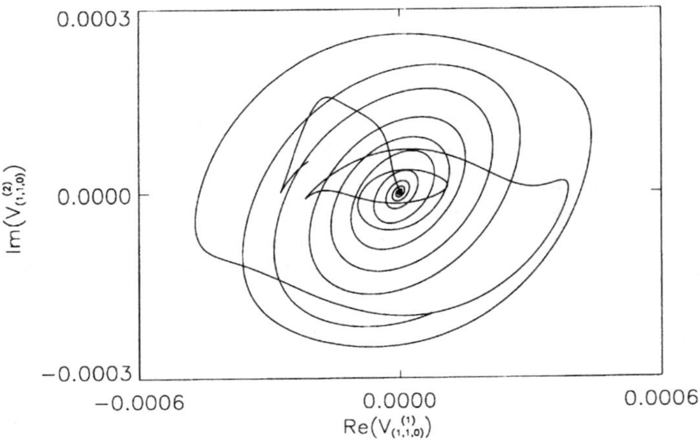

Fig. 6 Projection of a trajectory onto the plane spanned by two of the velocity components for $\lambda = 0.45$ and $R = 15.7$.

V. Conclusion

The MHD equations with pure *ABC* forcing (i.e., forcing with maximum helicity) show a generic dynamo effect that occurs via a Hopf bifurcation. All magnetic states are time-dependent. No gaps in dynamo action are observed for Reynolds numbers above the Hopf value. Such gaps, known from kinematic studies with stationary velocity field (pure *ABC* flow), are probably a result of the arresting of the velocity field. The Hopf bifurcation is followed by further bifurcations leading to chaos. A number of the observed bifurcations are related to the symmetries in the system. The final transition to chaos seems to follow the standard Ruelle–Takens–Newhouse scenario (decay of torus).

The transition from an always existing stable stationary solution to time-dependent states is decisively influenced by the degree of helicity in the forcing. Only if the helicity exceeds a certain threshold value will a Hopf bifurcation lead to a magnetic periodic state (i.e., to a dynamo effect). For helicities below the threshold value, the transition is more complex, but always the ensuing time-dependent states, including chaotic ones, are nonmagnetic. The character of the chaos found here is different from that for pure *ABC* forcing. We note that the absence of a dynamo effect refers to Reynolds numbers not too much above the first bifurcation value; the possibility exists that at sufficiently large Reynolds numbers a dynamo effect occurs even if the helicity is below the threshold value.

For helicities only slightly above the threshold value, the dynamo effect may again disappear for further increased Reynolds number, but this is an effect different from the gaps in dynamo action observed in kinematic studies with prescribed (pure) *ABC* flow.

References

[1]Roberts, P. H., and Soward, A. M., "Dynamo Theory," *Annual Review of Fluid Mechanics*, Vol. 24, 1992, pp. 459–512.

[2]Moffatt, H. K., *Magnetic Field Generation in Electrically Conducting Fluids*, Cambridge Univ. Press, Cambridge, England, UK, 1978.

[3]Krause, F., and Rädler, K.-H., *Mean-Field Magnetohydrodynamics and Dynamo Theory*, Akademie-Verlag, Berlin, 1980.

[4]Arnold, V. I., and Korkina, E. I., "The Growth of a Magnetic Field in a Three-Dimensional Steady Incompressible Flow," *Vest. Mosk. Univ. Mat. Mekh.*, Vol. 3, 1983, pp. 43–46 (in Russian).

[5]Dombre, T., Frisch, U., Greene, J. M., Hénon, M., Mehr, A., and Soward, A. M., "Chaotic Streamlines in the ABC Flows," *Journal of Fluid Mechanics*, Vol. 167, 1986, pp. 353–391.

[6]Galloway, D., and Frisch, U., "Dynamo Action in a Family of Flows With Chaotic Streamlines," *Geophys. Astrophys. Fluid Dyn.*, Vol. 36, 1986, pp. 53–83.

[7]Galloway, D., and Frisch, U., "A Note on the Stability of a Family of Space-Periodic Beltrami Flows," *Journal of Fluid Mechanics*, Vol. 180, 1987, pp. 557–564.

[8]Podvigina, O., and Pouquet, A., "On the Non-Linear Stability of the 1:1:1 ABC Flow," *Physica D*, Vol. 75, 1994, pp. 471–508.

[9]Galanti, B., Sulem, P. L., and Pouquet, A., "Linear and Non-Linear Dynamos Associated With ABC Flows," *Geophys. Astrophys. Fluid Dyn.*, Vol. 66, 1992, pp. 183–208.

[10]Roberts, P. H. *An Introduction to Magnetohydrodynamics*, Longmans, London, 1967.

[11]Lee, J., "Steady-State Simulation of 2D Homogeneous Turbulence," *Physica D*, Vol. 37, 1989, pp. 417–422.

[12]Lee, J., "Topology of Trajectories of the 2D Navier-Stokes Equations," *Chaos*, Vol. 2, 1992, pp. 537–563.

[13]Eggers, J., and Grossmann, S., "Does Deterministic Chaos in Fully Developed Turbulence Imply Intermittency?" *Physics of Fluids A*, Vol. 3, 1991, pp. 1958–1968.

[14]Arnold, V. I., "On the Evolution of a Magnetic Field Under the Influence of Advection and Diffusion" *Some Problems of Modern Analysis*, edited by V. M. Tikhomirov, Moscow State Univ., Moscow, 1984, pp. 8–21 (in Russian).

[15]Shimada, I., and Nagashima, T., "A Numercial Approach to Ergodic Problem of Dissipative Dynamical Systems," *Progress of Theoretical Physics*, 1979, pp. 1605–1616.

[16]Meneguzzi, M., Frisch, U., and Pouquet, A., "Helical and Non-Helical Turbulent Dynamos," *Physical Review Letters*, Vol. 47, 1981, pp. 1060–1064.

[17]Guckenheimer, J., and Holmes, P., *Nonlinear Oscillations, Dynamical Systems, and Bifurcations of Vector Fields,* Springer–Verlag, New York, 1983.

[18]Nicolis, G., and Gaspard, P., "Bifurcations, Chaos and Self-Organization in Reaction-Diffusion Systems," *Continuations and Bifurcations: Numerical Techniques and Applications*, edited by D. Roose, B. de Dier, and A. Spence, Kluwer, Dordrecht, The Netherlands, 1990, pp. 43–70.

Chapter 49

Experimental Results on Magnetohydrodynamic Free Convection in a Horizontal Cylinder

L. Davous* and R. Moreau†
Laboratory EPM-MADYLAM, St. Martin d'Hères, 38400, France
A. C. Neubrand‡
DTA/CEREM/DEM/SES, Grenoble, 38054, France
and
R. Bolcato§
Laboratory EPM-MADYLAM, St. Martin d'Hères, 38400, France

I. Introduction

MAGNETIC fields are known as efficient tools to damp the convection flow that occurs within the liquid pool of crystal growth furnaces. Here, this paper presents a review on the recent experimental results about the magnetohydrodynamic (MHD) free convective flow within the horizontal Bridgman configuration in which the liquid pool behaves as a differentially heated horizontal cylindrical enclosure. For the sake of clearness, the involved configuration is precisely defined from a geometrical viewpoint on Fig. 1.

The literature is relatively abundant in papers dealing with common natural convection into enclosures. But as soon as a uniform magnetic field is applied to a confined free convective flow, papers about this subject become harder to find. Nevertheless, Garandet et al.[1] determine a two-dimensional analytical formulation of the velocity field in the vertical longitudinal midplane (XZ). The authors show that far from the ends the axial

*ENSHMG BP 95, 1340 rue de la Piscine.
†ENSHMG BP 95, 1340 rue de la Piscine.
‡Commissariat à l'Energie Atomique, Centre d'Etudes Nucléaires de Grenoble, 17, Avenue des Martyrs, cedex 09.
§ENSHMG BP95, B40 rue de la Piscine.

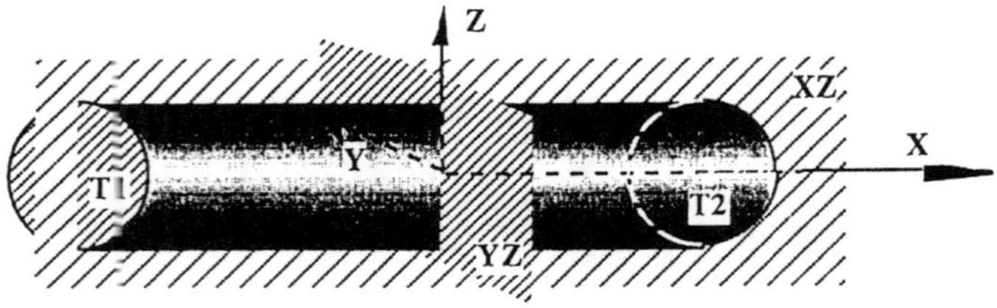

Fig. 1 Sketch of the studied configuration: the aspect ratio L/R_o is 20.

component velocity field scales as Gr/Ha^2 where the Grashof number Gr and the Hartmann number Ha are the two relevant nondimensional numbers that characterize the velocity when the flow is self-established. Besides, in the end-regions, Garandet et al.[1] demonstrate that the vertical component of the velocity field scales as $Gr/Ha^{3/2}$ within these recirculating end layers (of typical thickness $1/Ha^{1/2}$).

Assuming Ha much larger than one, Alboussière et al.[2] still confirm that within the middle part of the enclosure (whereas the flow is self-established), a MHD organization occurs with a core and Hartmann layers. The authors demonstrate that when the cross section is symmetric with respect to the horizontal midplane these Hartmann layers are electrically inactive. This last noticeable result entails that only a very small fraction of the current is passing through the Hartmann layers.

Cowley[3] using an expansion in powers of the effective Rayleigh number Ra, defined as ratio $Ra = PrGr/Ha^2$ where Pr is the Prandtl number shows that the weak temperature variations (only some few degrees) present in the cross section are able to drive four vortices in each quadrant of the central cross section of the cylindrical cavity. The experimental results are in full agreement with this analytical prediction. The Rayleigh number Ra is the nondimensional number that guides the MHD reorganization due to the influence of the magnetic field on this confined free convective flow.

II. Description of the Experiment

Because the Magnetic Stabilization of Convection and Turbulence (MASCOT) experimental setup has already been described in another paper,[4] only the main features about the experiment are gathered in this part. The experimental device, sketched on Fig. 2a, involves a cylindrical mercury-filled cell (radius R_o = 20 mm, length L = 400 mm) made of glass and equipped with 55 thermocouples all distributed on 11 half circles of the wall. Two water-cooled copper disks are located at the ends of the cylindrical cell whose temperatures T_1 (cold end wall) and T_2 (hot end wall) are controlled at a precision of ±0.01 K. A special device, a thick thermally insulating coat surrounded by an external copper cylinder, is located all around the cell between the two copper end walls to get an

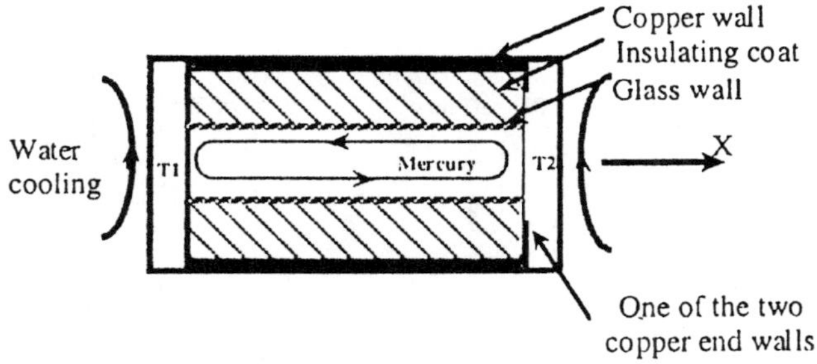

Fig. 2a Overview of the MASCOT experimental setup, including the following properties: a) 55 thermocouples located at the wall; b) rotation around the X axis; c) aspect ratio $L/R_o = 20$; and d) a movable probe inside the cell.

adiabatic cylindrical wall. Thus, when the fluid is at rest, the thermal pattern is purely conductive, and the related temperature gradient G_o is uniform $[G_o = (T_2 - T_1) / L]$. The whole cell may be rotated around its X axis in such a way that every probe allows a temperature to be obtained at any point on a circle. The whole setup is placed horizontally between the two poles of a magnet that is able to provide a uniform vertical magnetic field B_o (0 to 0.6 T). A movable probe, also sketched on Fig. 2b and built especially to detect temperature oscillations whose typical amplitude is a few hundred K, is able to rotate around and to translate along the X axis of the cylinder directly inside mercury.

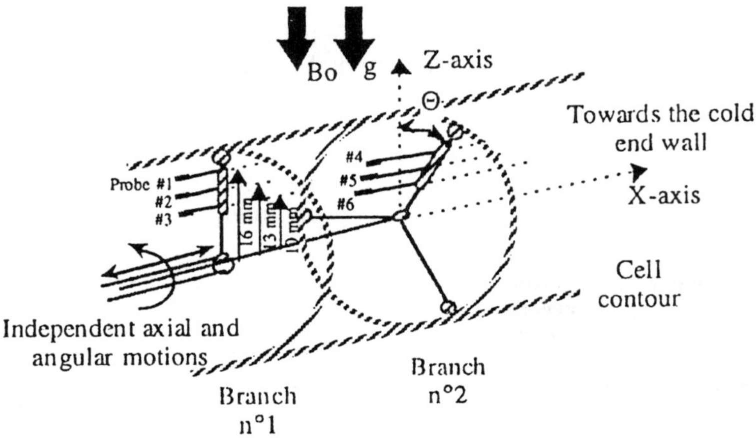

Fig. 2b The movable probe is composed of two dynamically independent branches; both are equipped with three probes.

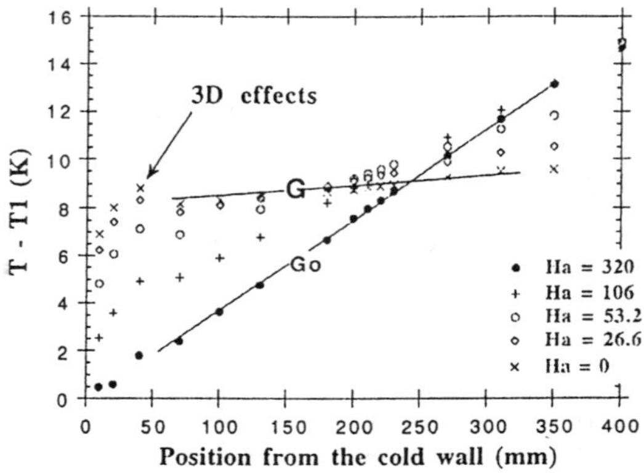

Fig. 2c Basic principle of the measurement of heat transfer at the end wall.

III. Experimental Results

A. When the Applied Magnetic Field B_o Is Uniform and the Flow Is Steady

Because of buoyancy, some fluid goes up in the hot end-regions while it goes down in the cold end-regions. These buoyancy forces partially balance the pressure forces and, thus, because of the pressure variations, some hot fluid travels near the top generatrix while some cold fluid travels near the bottom generatrix from the cold wall toward the hot one. As a consequence, the experimental temperature distributions exhibit a centro-symmetric feature. Let us denote x, the abscissa along a generatrix at the surface of the cylindrical fluid domain that varies from 0 (cold end wall) to 400 mm (hot end wall), and $T_{\alpha^\circ}(x)$, the x dependence of the temperature on the generatrix located at the angular position $\alpha°$ from the horizontal midplane (XY) of the cell. On Fig. 3 are plotted the experimental curves $T_{180^\circ}(x)$. A first point to

Fig. 3 Temperature distribution $T_{45^\circ}(x)$ along the generatrix located at 45 deg to the horizontal midplane (XY) vs the Hartmann number.

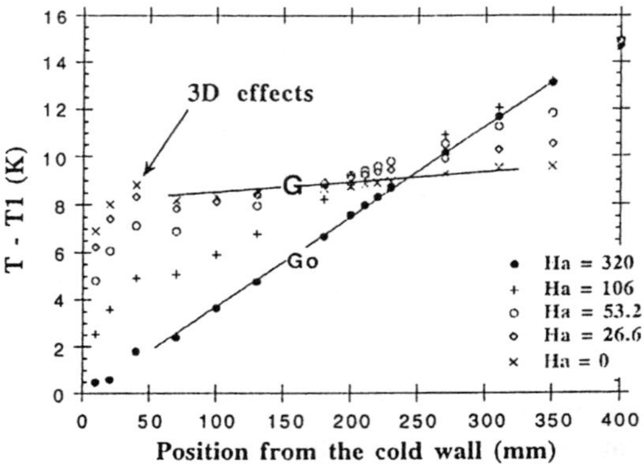

emphasize is the uniform linear temperature gradient G in the middle part of the cell, which strongly varies with the applied magnetic field (see Fig. 4). The value of G, or the ratio G/Go, reveals itself to be an adequate indicator of the convection level. When Ha is larger than 150, any difference between G and Go, thus between the temperature distribution and a conduction pattern, disappears. A second feature to notice is the presence of a temperature jump that takes place on a length of about 7 diam from the end wall on the $T_{45°}(x)$ plot. Such an amazing pattern can be explained by the three-dimensional end-regions flow. Let us consider two fluid particles that travel near the top generatrix on both sides of the vertical symmetry plane. Rather than going straight to the cold end, they are descending symmetrically and increasing their distance from the vertical longitudinal midplane. That this phenomenon is quite conspicuous on the horizontal generatrix located at $\alpha = 0°$ or 45° has been checked experimentally. This involves a significant local increase of the temperature above the value corresponding to an extrapolation of the linear axial dependence.

On Fig. 5 is represented the temperature profile $T_{X=200}(\alpha°)$ around the central circle [at the location $x = 200$ mm by analogy with the notation $T_{45°}(x)$]. For large Hartmann numbers, accordingly with the former experimental results, the temperature is uniform within the cross section in agreement with a pure conduction pattern. For some moderate Hartmann number values, and as predicted by Cowley,[3] we find that the curve $T_{X=200}(\alpha°)$ is well-fitted by a sinus function whose amplitude characterizes the maximum temperature variation across the cell cross section. Surprisingly, this amplitude first increases when Ha increases from 0 to 26.6, and then decreases to scale as Ra when Ha increases again to 320. This last tendency is the explicit sign of the damping effect of the magnetic field. We consider that the first increase of the temperature variation within the cross section has two complementary explanations. A first explanation of this character is the good ability of the magnetic field to efficiently stabilize the convection flow for Ha of the order of a few units: the effective viscosity is then reduced to the molecular level, and the velocity has to be higher than in a turbulent

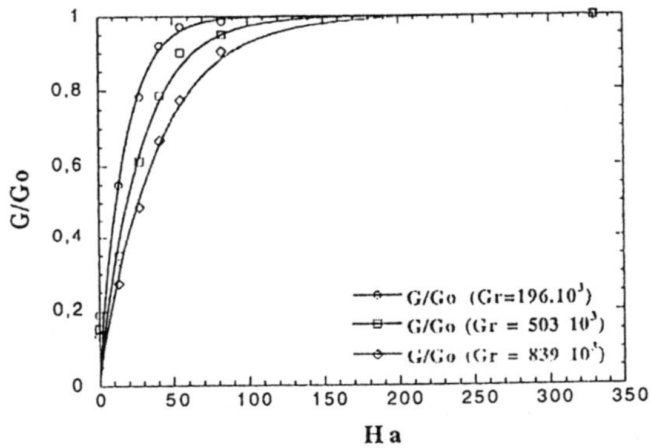

Fig. 4 Ratio between the actual axial temperature gradient G and its value G_o corresponding to pure conduction vs both Ha and Gr.

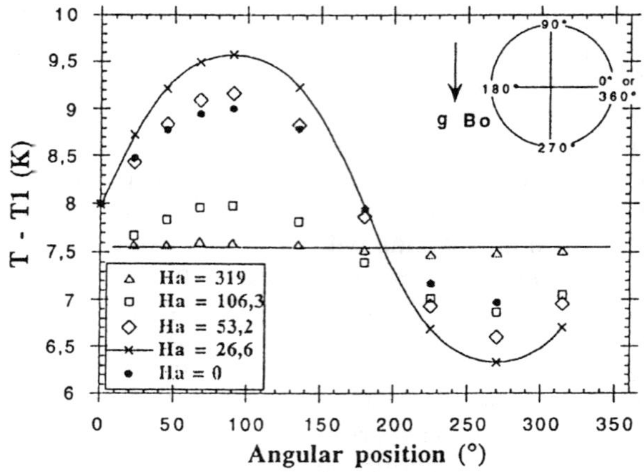

Fig. 5 Temperature distribution around the central circle located at $x = 200$ mm for $Gr = 839 \times 10^3$ and vs the Hartmann number.

regime. A second explanation, necessary in the range $Ha > 10$, is that the increase of the temperature variation across the section is due to the MHD reorganization of the flow mentioned in previous papers.[3,4]

Cowley[3] also predicts the presence of four vortices in each quadrant of the middle cross section (where the flow is supposed to be self-established). Accordingly, on Fig. 6, the typical deformation of the temperature distributions around the circles located at 100 mm from the central cross section demonstrates that these four vortices induce a centro-symmetric diverging flow both in the upper and in the lower part of the cylinder. Such results

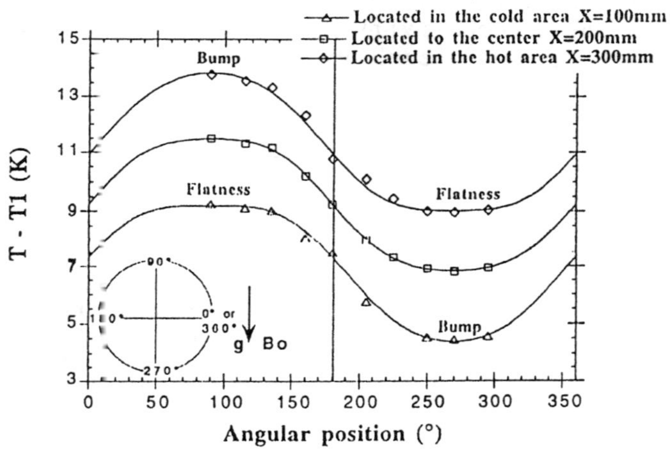

Fig. 6 Typical deformation of temperature distributions demonstrating the presence of a centro-symmetric diverging flow for $Ha = 26{,}6$ and $Gr = 10^6$ ($Ra = 35$).

have only been observed for Rayleigh numbers larger than 10. On the contrary, when $Ra \sim 0.1$ to 1, the previous deformation of the temperature profiles vanishes, and the exact opposite deformation (also centro-symmetric) arises, which is more and more obvious for the temperature distributions around circles located nearer and nearer the end walls (Fig. 7). This action may be explained by the presence of diffusive side layers along the end walls, whose thickness varies in the vertical direction. Because of the influence of these diffusive side layers, temperature is no longer an even function with respect to Z. A same phenomenon has also been observed on any cross section located out of the center of the cylinder by the way of numerical simulations realized for $Ra \ll 1$ (Davoust et al.[5]). The quadri-polar symmetry (four transversal vortices) stays only in the areas where the transversal symmetry of temperature distribution is still valid. Finally, this new MHD organization, which takes place inside the cylinder, is clearly depending on the value of the Rayleigh number, and the velocity field is mainly marked by the asymptotic law $\boldsymbol{U_c} \sim Gr/Ha^2$. This last remarkable result may be enforced by estimating the Nusselt number.

Lykoudis[6] has been the first to predict a Nusselt number Nu for MHD heat transfer in the configuration of a convective boundary layer flow along a hot vertical plate in the presence of a horizontal transverse magnetic field. The author finds that Nu decrases proportionally to the Lykoudis number $Ly = Ha^2/Gr^{1/2}$. Several decades later, Ozoe et al.[7] studied experimentally the MHD convective flow of gallium within a differentially heated cube. They found, in this geometry, that heat transfer was well-described by the parameter Λ which scales as the nondimensional number $Ha/Gr^{1/3}$. The

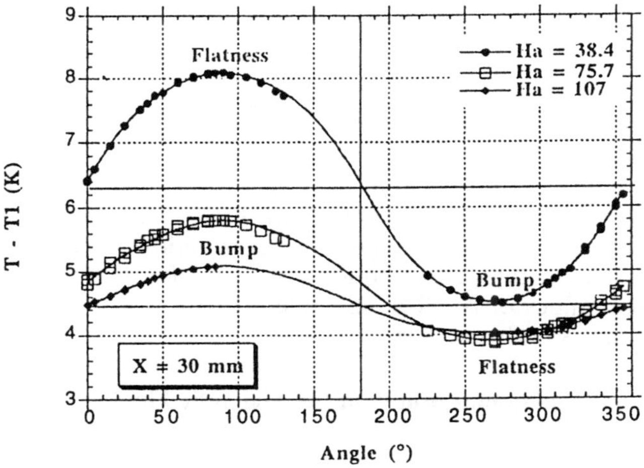

Fig. 7 The deformation of temperature distributions typical of four transversal vortices disappears: a new and opposite deformation, also centro-symmetric, arises because of the diffusivity of thermal side layers, as *Ra* scales from 13 down to 1.8.

parameter Λ was chosen to be referenced to the nonmagnetic situation and calculated by the ratio:

$$\Lambda = [Nu - 1]/[Nu_{(Ha=0)} - 1]$$

This choice seems to us to be not really appropriate because the classical sense of the Nusselt number implies that Nu is the ratio between the actual heat flux (which takes into account both convection and conduction) and the reference heat flux related to a pure conduction (obtained when Ha is maximum). In this work, Nu has been estimated two distinct ways.

First, direct measurements of heat transfer at the end walls have given an estimation Nu_1 using the following procedure: because a given flow rate of water was used to control accurately temperatures T_1 and T_2 at the end walls at a given convection level (fixed value of Ra), some temperature elevation (a few hundred K) could be associated, whose physical meaning was nothing but the conservation of total heat flux Φ_{tot} across the cylinder (Fig. 2c). This is to be compared with a second estimation of the Nusselt number Nu_2 calculated from the conservation of heat flux along any cross section of the cylinder:

$$\varphi_{tot} = \iint_{center} \left[-K\nabla T + \rho C_p \boldsymbol{u} T \right] ds$$

The notations K, ρ, and C_p designate respectively the thermal conductivity, the density, and the heat capacity of mercury. Then, according to this last relation, from the temperature distribution around the central circle and from the value of the axial temperature gradient G, it is possible to calculate Nu_2. The agreement between Nu_1 and Nu_2 demonstrates the relevant law that describes the evolution of Nu scales as

$$Nu = G/G_o \left[1 + 7/24 * Ra^2 \right]$$

This theoretical expression of Nu has been established assuming that the asymptotic theory of Cowley[3] is still true even if Ra is large, provided that the amplitude of the central temperature distribution δT is taken from the experimental temperature distribution. An overall figure (Fig. 8) gathers the three former values of Nu. The agreement between experiments and the asymptotic law is fairly good when $Ra < 1$.

B. When a Slight Nonuniformity of the Magnetic Field b Is Superposed to B_o

The influence of a slight nonuniformity of the imposed magnetic field has been experimentally examined for the case of the fully established flow in the midpart of the cylinder. Considering a weak perturbation b of the imposed magnetic field $\boldsymbol{B_o}$ in an asymptotic approximation ($Ha \gg 1$),

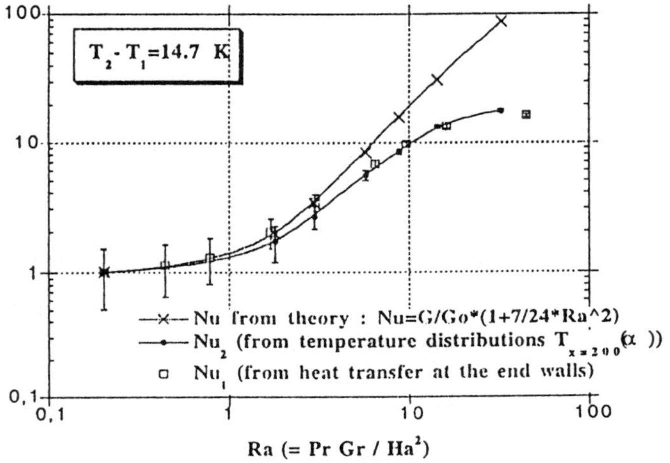

Fig. 8 Two experimental determinations of the Nusselt number *Nu* are compared with the analytical pronostic of Cowley[3] (*Nu* scales as Ra^2 when $Ra < 1$).

Neubrand et al.[8] carried out a first-order perturbation analysis. Because of the interaction between $\boldsymbol{b}$ and the nonperturbed part of the velocity field $\boldsymbol{U_c}$, additional electric current densities are created within the core of the flow. The resulting perturbation of the velocity field $\boldsymbol{u_c}$ scales either as $\lambda.Gr/Ha$ or as $\lambda.Gr/Ha^2$ (λ is the ratio $\boldsymbol{b}/\boldsymbol{B_o}$) depending on whether the Hartmann layers are electrically active or not. Thus, with perfectly electrically conducting walls, the velocity disturbance $\boldsymbol{u_c}$ is always found to be of the order of $\lambda.Gr/Ha^2$ whatever the involved classes of nonuniformity (see Neubrand et al.[8]). But as soon as the walls are electrically insulating and the magnetic field nonuniformity $\boldsymbol{b}$ is of the first class (see Fig. 9a), the electric lines of the additional current density may recycle through the Hartmann layers and $\boldsymbol{u_c}$ scales as $\lambda.Gr/Ha$.

An experimental checking of this pronostic has been led with MASCOT. On Figs. 9b and 9c the predicted analytical shape of the velocity perturbation $\boldsymbol{u_c}$ and the experimental temperature distributions around the central circle $T_{X=200}$ (α) and $[T + t]_{X=200}$ (α) for the case $Ha = 116.5$ are presented.

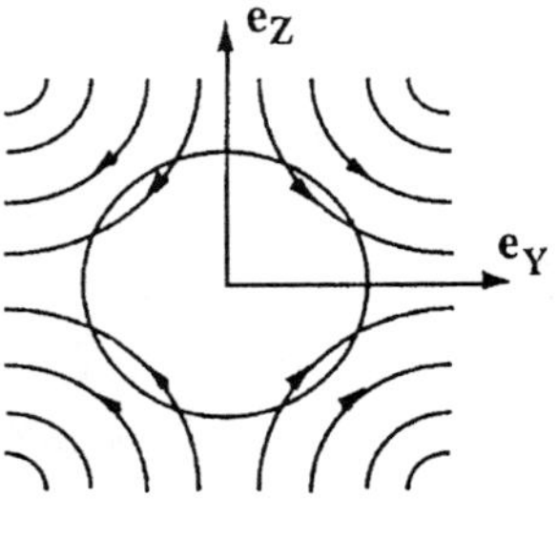

Fig. 9a Nonuniformity of the first class: $b = \lambda(Ye_Z + Ze_Y)$.

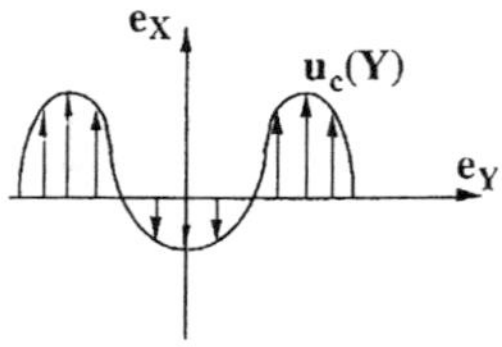

Fig. 9b The axial perturbed part of the velocity field: $u_c(Y)$.

By the notation $[T + t]_{X=200}$ (α), we designate the experimental temperature distribution around the central circle in the presence of the magnetic nonuniformity. Then, $t_{X=200}$ (α) stands as the temperature perturbation distribution around the central circle related to the magnetic field nonuniformity b.

The qualitative agreement between Neubrand et al.'s theory and the experiment is the presence of the two small bumps quite conspicuous on the $[T + t]_{X=200}$ (α) profile for $\alpha \sim 90°$ or $\alpha \sim 270°$ to the horizontal midplane (XY). This is explained by the shape of the axial $\boldsymbol{u_c}$ profile versus the Yaxis. In the fully established region the axial velocity $\boldsymbol{U_c}$ is either positive in the lower part of the cylinder, like $\boldsymbol{u_c}$ at $\alpha \sim 270°$, or negative in the upper part, unlike $\boldsymbol{u_c}$ at $(\alpha \sim 90°$. This is why the perturbed temperature distribution $[T + t]_{X=200}$ $(\alpha$ exhibits hotter temperatures at $\alpha \sim 90°$ or at $\alpha \sim 270°$ than the $T_{X=200}(\alpha)$ profile. Finally, to validate the asymptotic law $\boldsymbol{u_c} \sim \lambda.Gr/Ha$ from the temperature perturbation $t_{X=200}$ (α) deduced from the $[T + t]_{X=200}$ (α) profile, the amplitude δt of $t_{X=200}$ (α) is compared with the amplitude δT of the $T_{X=200}$ (α) profile. These are displayed on Fig. 10 for a few Hartmann numbers. As expected, δT scales as $PrGr/Ha^2$, and δt scales as $PrGr/Ha$.

C. When the Flow Is Unsteady

This third part is devoted to the experimental investigation of the stability of the laminar MHD convective flow within a cylinder. Here, again, many results are already available on nonmagnetic situations in the literature. Let us summarize the main results from the experiments carried out by Pratte et al.[9] about the oscillatory behavior of the convective flow of mercury enclosed in shallow rectangular boxes submitted to some given horizontal temperature gradients. We demonstrated that the finite width of the chamber is of great importance the values of the critical Grashof numbers Gr_c for

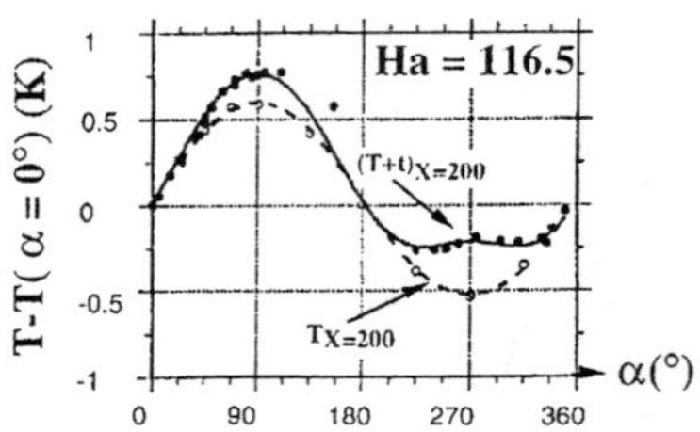

Fig. 9c Temperature distributions around the central circle (x = 200 mm), $T_{x=200}(\alpha)$, and $[T + t]_{x=200}$ (α), either with a uniform applied magnetic field B_o or with a slightly perturbed applied magnetic field $b + B_o$.

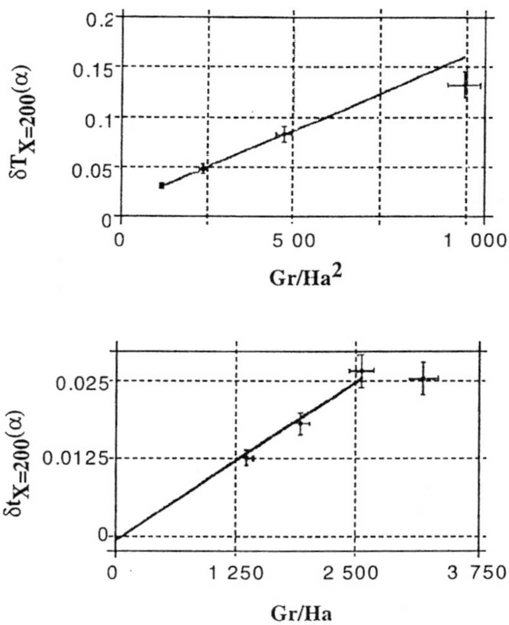

Fig. 10 Amplitudes $\delta T_{X=200}$ (α) and $\delta t_{X=200}$ (α) of the temperature distributions with a uniform magnetic field B_o and a slightly nonuniform magnetic field $B_o + b$ suggesting that in the last case Hartmann layers become electrically active.

which first thermal oscillations appear. The aspect ratio (length by width) entails relevant effects on the topology of the instabilities that present the characteristic to be either three-dimensional longitudinal waves (*X*-axis oscillating rolls) or two-dimensional transversal waves (*Y*-axis oscillating rolls), and the three-dimensional oscillating waves are the first to appear. Hurle et al.[10] have experimented with thermal convection within a differentially heated parallepipedic cavity under the influence of a magnetic field. The resulting damping of temperature fluctuations in molten gallium ($Pr \ll 1$) is confirmed. This stabilization by the application of a magnetic field is found to reveal hysteretic behavior. But in this experience, the upper surface of the melt is supposed to be sufficiently thermally insulating via a thin film of silicon oil. Thermo-capillar effects are hence neglected. No more information is delivered about the nature of these thermal oscillations, and hysteresis is just mentioned.

Using the MASCOT experiment some features about the hysteretic transition between the laminar and turbulent flows are now presented. First, as predicted by the linear theories developed by Bojarevics[11] and Kaddeche,[12] critical Hartmann numbers Ha^* for which oscillations disappear are only of the order of 10 when $Gr \sim 10^5$ to 10^6. Because no differences on the values of Ha^* were detected, we preferred to use the more sensitive movable probe

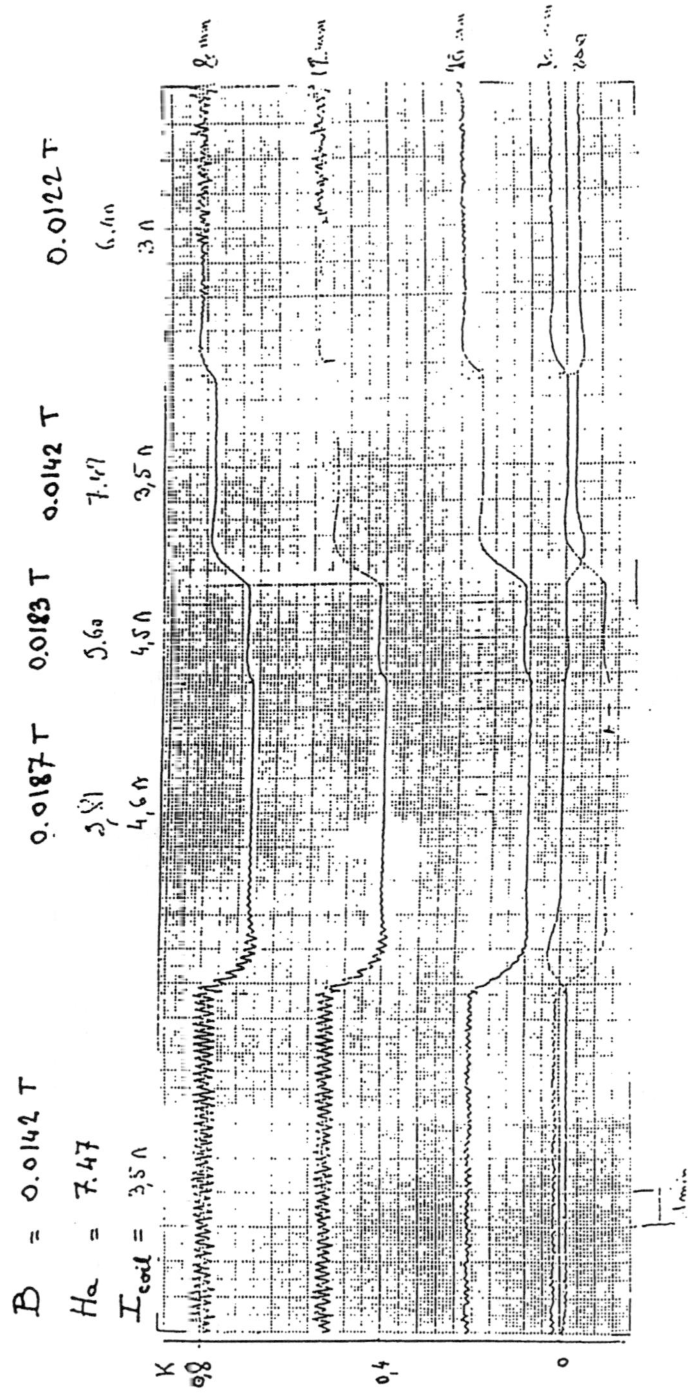

Fig. 11a Hysteresis between values of the Hartmann numbers Ha_1 and Ha_2 related to the destabilization of the laminar flow and the stabilization of the turbulent flow and suggesting a subcritical bifurcation.

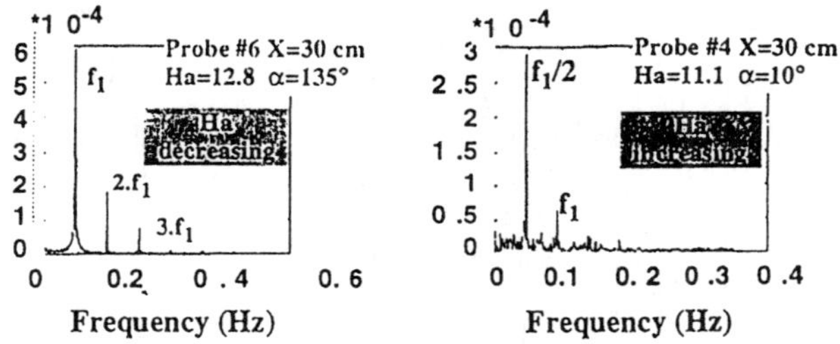

Fig. 11b Spectral contents of thermal oscillations when *Ha* decreases (left-hand side) and when *Ha* increases (right-hand side). The difference between the fundamental frequencies (f_1 in the first case and $f_1/2$ in the second case) suggests also a subcritical bifurcation.

directly put within the core (described on Fig. 2c) rather than the probes located at the wall whose sensitivity is limited by the diffusivity of temperature through the Hartmann layers. Moreover, the movable probe is made of two dynamically independent branches, and each one of them is equipped with three thermocouples. Because of this specific design, we were able to check that the velocity field intensity was too small (maximum 1 cm/s) to let the movable generate some dynamical disturbances. Thanks to a very accurate control of the temperatures T_1 and T_2 at the end walls, some thermal oscillations, whose amplitudes are of the order of a few hundredth K, were easily put in evidence.

Hysteresis on the laminar-turbulent transition is demonstrated both with the amplitude (see for example Fig. 11a) and with the spectral content of the oscillations for a temperature difference: $T_2 - T_1 = 15$ K. As illustrated by Fig. 11b, when *Ha* decreases down to Ha_1^* a monoperiodic thermal oscillation whose fundamental frequency is $f_1 \sim 0.087$ Hz and with at least two harmonics first appears. But when *Ha* increases up to Ha_2^*, oscillations whose fundamental frequency f_2 is half the former frequency f_1 arise from a pre-existing turbulent signal. As a consequence, hysteresis is obvious by the differences between both the critical Hartmann numbers and the fundamental frequencies of the oscillations. This result also suggests that the transition is subcritical: only a weakly nonlinear analysis would explain these experimental results well. Finally, the shape of the thermal oscillations is also dependent on the location of the probe within the cylinder as shown by Fig. 12. A possible explanation is the arising of a spatio-temporal chaos because the aspect ratio of the cell is large.

The stabilization of the laminar flow has been realized for other Grashoff numbers. The critical Hartmann numbers Ha_2^* for which any thermal oscillations are no longer detected from the broadband noise are related to electronics scales as $Gr^{1/2}$. This last result is self-consistent with Hurle et al.[10] who found that the critical temperature difference ΔT, for which any thermal disturbances vanish, follows a scaling law $\Delta T \sim Ha^2$.

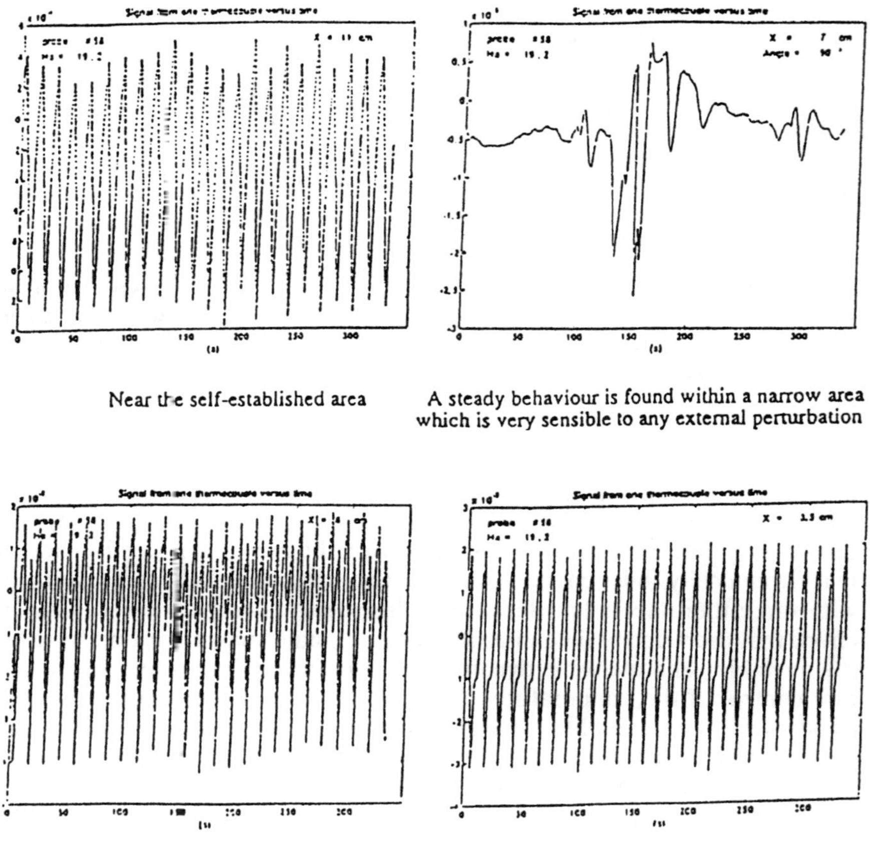

Near the self-established area

A steady behaviour is found within a narrow area which is very sensible to any external perturbation

3D recirculating area

Near the thermal boundary layer

Fig. 12 The shape of the thermal oscillations is strongly dependent on the location of the probe within the cylinder.

IV. Conclusion

A recent experimental investigation of the MHD free convection inside a differentially heated horizontal mercury-filled cylinder whose end walls are regulated to the temperatures T_1 and T_2 ($T_1 < T_2$) has been presented. By measuring temperatures both at the wall and within mercury, the influence of a uniform vertical magnetic field $\boldsymbol{B_o}$ either on the steady or the unsteady regimes has been studied. Because of buoyancy, upward to the hot end and downward to the cold end, a primary monocellular cell is created; the resulting axial velocity field $\boldsymbol{U_c}$ in the fully self-established part of the enclosure (central area) scales as Gr/Ha^2 when Ha is larger than 150. As predicted by Cowley,[3] the typical deformation of the temperature distributions around the central circle proves that a secondary transversal flow, involving four vortices in each quadrant of the central cross section, is present. In the end regions, the temperature distribution along a generatrix is explained well by a three-dimensional recirculation of the flow. The strong ability of the magnetic field to damp out the convective flow is demonstrated both by the decreasing of the amplitude of the central temperature distribution and the decreasing of the Nusselt number.

As soon as Ha increases up to a few units, first, thermal oscillations, whose typical frequencies are of the order of 0.01 Hz, arise from chaos, and second, the flow becomes laminar: the effective viscosity is reduced to the molecular level. This turbulent-laminar transition of the flow is subcritical.

Finally, the influence of a slight nonuniformity $\boldsymbol{b}$ of the imposed magnetic field is also examined. As predicted by Neubrand et al.,[8] the perturbed temperature distribution around the central cross section demonstrates that the perturbed velocity field $\boldsymbol{u_c}$ may scale as Gr/Ha if the magnetic disturbance $\boldsymbol{b}$ owns the convenient symmetry.

References

[1]Garandet, J. P., Alboussière, T., and Moreau, R., "Bouyancy Driven Convection in a Rectangular Enclosure with a Transverse Magnetic Field," *International Journal of Heat Mass Transfer*, Vol. 35, 1992, pp. 741–748.

[2]Alboussière, T., Grandet, J. P., and Moreau, R., "Bouyancy Driven Convection with an Uniform Magnetic Field," Part I: Asymptotic Analysis ($Ha \gg 1$), *Journal of Fluid Mechanics*, Vol. 253, 1993, pp. 545–563.

[3]Cowley, M., "The Effect of a Buoyancy-Driven Magnetohydrodynamic Flow on the Temperature Distribution in a Horizontal Cylinder,"*Proceeding of the Second International Conference on Energy Transfer in Magnetohydrodynamic Flows* (Aussois, France), 1994.

[4]Davoust, L., Moureau, R., Bolcato, R., Alboussière, T., and Neubrand, A. C., "Influence of a Vertical Magnetic Field on Convection in the Horizontal Bridgman Crystal Growth Configuration," *Proceedings of the Second International Conference on Energy Transfer in Magnetohydrodynamic Flows* (Aussois, France), 1994.

[5]Davoust, L., Bertrand, F., Moreau, R., Tanguy, P. A., and Bolcato, R, R., "Recent Results on MHD Damped Convection in the Horizontal Bridgman Configuration," *MAHYD 95, 14th International RIGA Conference on Magnetohydrodynamic* (Jurmala, Latvia), 1995.

[6]Lykoudis, P. S., "Natural Convection of an Electrically Conducting Fluid in the Presence of a Magnetic Field," *International Journal of Heat Mass Transfer*, Vol. 5, 1961, pp. 23–24.

[7]Ozoe, H., and Okada, K., "The Effect of the Direction of the External Magnetic Field on the Three-Dimensional Natural Convection in a Cubical Enclosure," I*nternational Journal of Heat Mass Transfer*, Vol. 32, 1989, pp. 1939–1954.

[8]Neubrand, A. C., "Convection Naturelle et Ségré en Solidification Bridgman Sous Champ Magnétique," Thèse CENG-CEREM/DEM, Inst. National Polytechnique de Grenoble, Grenoble, France, 1995.

[9]Pratte, J. M., and Hart, J. E., "Endwall Driven, Low Prandtl Number Convection

in a Shallow Rectangular Cavity," *Journal of Crystal Growth*, Vol. 102, 1990, pp. 54–68.

[10]Hurle, D. T. J., Jakeman, E., and Johnson, C. P., "Convective Temperature Oscillations in Molten Gallium," *Journal of Fluid Mechanics,* Vol. 64, 1974, pp. 565–576.

[11]Bojarevics, V., "Buoyancy Driven Flow and Its Stability in a Horizontal Rectangular Channel with an Arbitrary Oriented Transversal Magnetic Field," *Proceeding of the Second International Conference on Energy Transfer in Magnetohydrodynamic Flows* (Aussois, France), 1994.

[12]Kaddeche, S., "Magnétohydrodynamique et Ségrégation en Configuration de Bridgman," Thèse LMFA, Ecole Centrale de Lyon, Lyon, France, 1995.

Chapter 50

Free Mode Determination on an Electromagnetically Shaped Free Surface

S. Bourdrez,* J. Etay,† and Y. Fautrelle‡
Institut National Polytechnique de Grenoble, St. Martin d'Hères, 38402, France

Introduction

WHEN a liquid metal pool is submitted to a low-frequency electromagnetic field, the alternating part of the electromagnetic forces may excite free surface waves (see for example Galpin and Fautrelle[1]). Similar effects may be obtained in induction heating processes (e.g., the cold crucible melting) when the high-frequency applied electromagnetic field contains a low-frequency modulation. The origin of those waves may be explained by resonance phenomena as soon as the frequency of the modulation is of the order of the eigenfrequencies of the free oscillation modes. Thus, to determine those frequencies is of interest. We have focused our attention on the determination of the free oscillation modes of an electromagnetically shaped free surface.

Governing Equations

We are concerned with the case illustrated in Fig. 1. We consider a liquid metal pool located in coil supplied with high-frequency alternating electric currents. The magnetic field is large enough so that in the magnetostatic approximation the free surface is deformed under the action of the mean part of the Lorentz forces. If $z_S(r)$ denotes the free surface static deformation, then z_S satisfies the equation

$$z_S(r) = \frac{\gamma}{\rho g}\left[K(r) - K(0)\right] - \frac{< B_S^2(r) >}{2\mu\rho g} \tag{1}$$

*EPM-MADYLAM, ENSHMG, BP 95.
†EPM-MADYLAM, ENSHMG, BP 95.
‡EPM-MADYLAM, ENSHMG, BP 95.

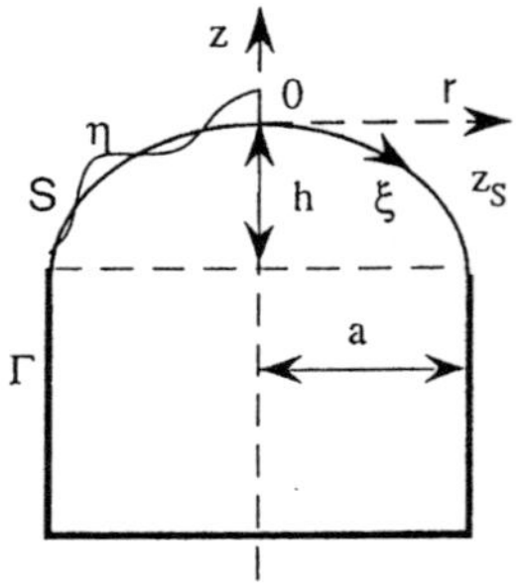

Fig. 1 Scheme of the geometry.

where $K(r)$ is the total curvature of the free surface, $B_S(r)$ the magnetic field distribution along the free surface, γ the surface tension, g the gravity, μ the magnetic permeability, and ρ the density. The symbol $< >$ denotes the time-average operator.

Let η be a small perturbation of the free surface normal to it. In the inviscid approximation, the corresponding fluid motion perturbation is governed by the Euler linearized equations:

$$\rho \frac{\partial \boldsymbol{u}}{\partial t} = -\nabla p + \boldsymbol{f} \tag{2}$$

$$\nabla \cdot \boldsymbol{u} = 0 \tag{3}$$

where $\boldsymbol{u}$, $\boldsymbol{f}$, and p respectively denote the velocity field, the electromagnetic force perturbation, and the pressure. The linearized boundary conditions on both the velocity and pressure fields are the following:

$$\boldsymbol{u} \cdot \boldsymbol{n} = 0 \text{ on the solid boundaries } \Gamma \tag{4a}$$

$$\frac{\partial \eta}{\partial t} = \boldsymbol{u} \cdot \boldsymbol{n} \text{ on the free surface S} \tag{4b}$$

$$p = \rho g \eta \cos \beta - \gamma \nabla_S^2 \eta \tag{4c}$$

where ∇_S^2 is the Laplacian operator along the surface and β the angle between the gravity and the normal.

To solve Eqs. (2) and (3) with the boundary conditions (4), we have used the inner product method proposed by Sneyd and Wang.[2] To that purpose, we first define normalized curvilinear coordinates (ξ, ζ), ξ being the curvilinear abscissa along the free surface and ζ the coordinate normal to it. Then, we solve the eigenvalue problem for the functions $\hat{E}\ (\xi, \theta)$ such that

$$\nabla_S^2 E - \lambda^2 E = 0, \quad \text{with } \hat{E} = E(\xi)\cos(n\theta), \tag{5}$$

and the boundary conditions

$$\frac{\partial E}{A\xi} = 0 \quad \text{for} \quad \xi = 0, \quad 1 \text{ corresponding to } r = 0, a \tag{6}$$

with a being the radius of the pool and λ a nondimensional eigenvalue.

Solution of Eqs. (5) and (6) yields an infinite number of eigenfunctions E_m, the so-called wave functions corresponding to each eigenvalue λ_m. System (5) and (6) has been solved numerically by means of a first-order finite-element method. In the case of a plane surface, E is identical to the Bessel function of the first kind $J_n(\lambda_m r)$.

According to the Sturm–Liouville theorem, the functions E_m form a complete set of orthogonal functions, and for a given n η may be expanded as follows:

$$\eta_n = \sum_{m=0}^{+\infty} \eta_m(t) E_m(\xi) \cos(n\theta) \tag{7}$$

We then introduce the functions φ_m such that

$$\nabla^2 \varphi_m = 0 \tag{8}$$

with the following boundary conditions

$$\varphi_m = E_m \text{ on the free surface } S \tag{9a}$$

$$\frac{\partial \varphi_m}{\partial n} = 0 \text{ on the solid boundaries } \Gamma \tag{9b}$$

The problem (8) and (9) have been solved by means of a finite-element method with linear interpolation functions.

Once the functions E_m and φ_m are determined, they may be used to derive the governing equation for η. Multiplying Eq. (2) by φ_m, integrating over the whole domain V, then transforming the volume integrals yield

$$\rho \int_S \varphi_m \frac{\partial^2 \eta}{\partial t^2} \, \mathrm{d}s + \int_S \frac{\partial \varphi_m}{\partial n} \left(\rho g \eta \cos\beta - \gamma \nabla_S^2 \eta\right) \mathrm{d}s = \int_V \nabla \varphi_m \cdot \boldsymbol{f} \, \mathrm{d}v \tag{10}$$

Introducing the expansion (7) in Eq. (10), using the orthogonality properties of the functions φ_m, and applying the boundary conditions (4) yield the differential equations governing each mode $\eta_m(t)$ of the free surface perturbation, namely,

$$\rho \frac{\mathrm{d}^2 \eta_m}{\mathrm{d}t^2} + (\rho g \alpha_{ml} - \gamma \beta_{ml}) \eta_l = \frac{1}{|E_m^2|} \int_V \nabla \varphi_m \cdot \boldsymbol{f} \, \mathrm{d}v \tag{11}$$

where α_{ml} and β_{ml} are infinite matrices whose expressions are

$$\alpha_{ml} = \int_S \cos\beta \left(\frac{\partial \varphi_m}{\partial n}\right)_S E_l \, \mathrm{d}s \quad \text{and} \quad \beta_{ml} = \int_S \left(\frac{\partial \varphi_m}{\partial n}\right)_S E_l \, \mathrm{d}s$$

The normalization factor in the RHS of Eq. (11) is defined as follows:

$$|E_m^2| = \int_0^1 E_m^2 \sqrt{g_S}\, ds$$

$\sqrt{g_S}$ being the Jacobian value of the transformation on the free surface.

Worthwhile to observe is the fact that in the case of plane free surface the matrices α_{ml} and β_{ml} are diagonal. In the general case because of the curvature of the surface, the matrices are nondiagonal, and the free modes are coupled.

The RHS of Eq. (11) involves the Lorentz force perturbation. The determination of that term requires the determination of the perturbation of the magnetic field because of the free surface perturbation. In the present paper, we shall ignore the RHS of Eq. (11), and we shall restrict our analysis to the study of the effect of the free surface curvature on the eigenfrequencies. The validity of that assumption will be discussed in the next section.

Results and Discussions

When the RHS of Eq. (11) is neglected, the equation reduces to a generalized pendulum equation. The solutions of Eq. (11) are of exponential form, namely,

$$\eta_1 = \eta_{10} \exp(is_1 t) \tag{12}$$

The eigenfrequencies simply are the roots of the following determinant:

$$\det\left[g\alpha_{ml} - \frac{\gamma}{\rho}\beta_{ml} - s^2\delta_{ml} \right] = 0 \tag{13}$$

δ_{ml} being the Kronecker symbol.

In practice, the expansion (7) is truncated, and we only keep a finite number of modes, say N. Thus, the dimension of the matrices is $N*N$. To have a good accuracy, we must ensure that the meshes of the grid used to calculate numerically the functions E_m and φ_m are consistent with the highest order mode retained in the truncation.

Equation (13) has been solved numerically. We have tested various given dome shapes. The physical values used in the calculations correspond to those of the mercury. The free surface static deformation has been approximated by the following family of curves:

$$z_S(r) = h\left(\frac{r}{a}\right)^4 \tag{14}$$

where h is the height of the static dome as shown in Fig. 1.

Figure 2 shows the effect of the truncation level N on the eigenfrequencies. The values of the eigenfrequencies of the lowest mode converge quite rapidly as N increases (e.g., for $m = 1$ or 2). As for the highest modes, the value of N must obviously be increased to obtain an acceptable accuracy.

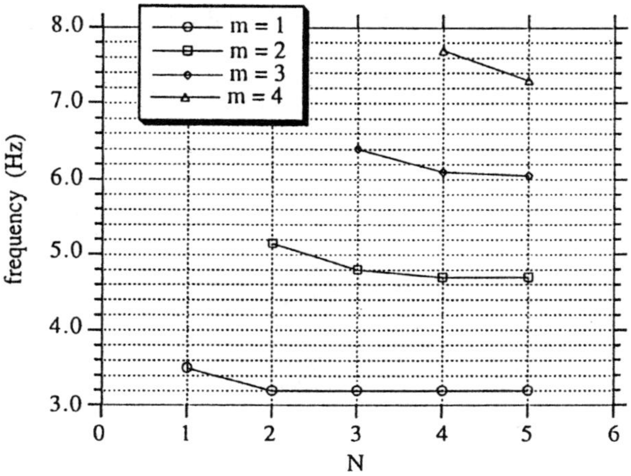

Fig. 2 Influence of the truncation level on the values of the eigenfrequencies: n = 0 (axisymmetric modes) and various m, dome height h = 3 cm, pod radius a = 3 cm, and total pool height $h_{\pm}$ = 7 cm.

The eigenfrequencies decrease as the height h of the dome increases. That result is illustrated both in Fig. 3 and in Table 1. The physical explanation may be related to two different mechanisms:

1) Because of the curvature of the free surface, its actual area is larger than in the flat surface case, and the effective radius of the pool is larger.

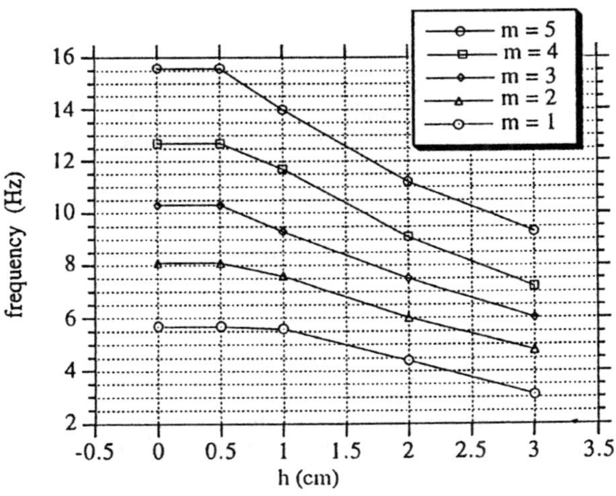

Fig. 3 Eigenfrequencies vs the dome height for n = 0 (axisymmetric modes) and various m, dome height h = 3 cm, pod radius a = 3 cm, and total pod height $h_{\pm}$ = 7 cm.

Table 1 Eigenfrequencies of the free surface modes for various values of n and m

	$n = 0$	$n = 1$	$n = 2$	$n = 3$
$m = 0$	—	3.91	5.08	6.03
	—	3.71	4.72	5.75
	—	2.48	3.18	4.31
$m = 1$	5.68	7.01	7.97	8.95
	5.55	6.51	7.23	8.10
	3.22	3.85	4.48	5.11
$m = 2$	8.06	9.28	10.40	11.52
	7.53	8.47	9.34	10.27
	4.72	5.30	5.86	6.49
$m = 3$	10.33	11.72	12.94	14.15
	9.39	10.58	11.74	12.79
	6.03	6.74	7.37	8.03
$m = 4$	12.78	14.22	15.59	16.89
	11.75	12.96	14.22	15.64
	7.33	8.24	8.98	9.74
$m = 5$	15.64	16.92	18.39	19.53
	14.07	15.49	16.63	18.09
	9.30	10.04	10.94	11.70

Top value: $h = 0$; middle value: $h = 1$ cm; and bottom value: $h = 3$ cm.

2) The overall effect of the gravity is less important because the free surface is inclined with respect to the vertical axis, especially near the crucible. Both effects tend to reduce the values of the eigenfrequencies.

It is of interest to have an estimate of the order of magnitude of the Lorentz force perturbations that have been neglected in Eq. (11). Garnier and Moreau[3] have investigated the stability of a plane free surface under the influence of an alternating magnetic field. They found that the Lorentz forces lead to the appearance of a typical eigenfrequency f_B whose order of magnitude was

$$f_B = O\left(\frac{B_0}{a(\mu\rho)^{1/2}}\right)$$

B_0 being a typical value of the magnetic field strength. That value must be compared with the typical frequency of the gravity waves f_g that have been presently considered and whose order of magnitude is

$$f_g = O\left(\left(\frac{\lambda g}{a}\right)^{1/2}\right)$$

Comparison of f_B and f_g leads to the following ratio:

$$\frac{f_B}{f_g} = O\left(\frac{B_0}{(a\lambda\mu\rho g)^{1/2}}\right) \tag{15}$$

Then, we may introduce the free surface height h, which can be estimated from Eq. (1). The ratio (15) may be rewritten as follows:

$$\frac{f_B}{f_g} = O\left(\left(\frac{h}{a\lambda}\right)^{1/2}\right) \tag{16}$$

The latter estimate shows that our results are strictly valid only for small static deformation or high-order modes. In the present case the numerical value of the above ratio is at most equal to approximately 0.4 for $h = 3$ cm and the lowest mode. This indicates that the present calculation yields quite good approximations of the eigenvalues of the free oscillation modes of the surface.

References

[1]Galpin, J. M., and Fautrelle, Y. R., "Liquid Metal Flows Induced by Low Frequency Alternating Fields," *Journal of Fluid Mechanics*, Vol. 239, 1992, pp. 383–408.

[2]Sneyd, A. D., and Wang, A., "Interfacial Instability due to MHD Mode Coupling in Aluminum Reduction Cells," *Journal of Fluid Mechanics*, Vol. 263, 1994, pp. 343–359.

[3]Garnier, M., and Moreau, R., "Effect of Finite Conductivity on the Inviscid Stability of an Interface Submitted to a High Frequency Magnetic Field," *Journal of Fluid Mechanics*, Vol. 127, 1983, pp. 365–377.

Chapter 51

On the Magnetohydrodynamic Flow Around a Nonconducting Spherical Droplet in Crossed Electric and Magnetic Fields

Ashish D. Patel* and Nagy El-Kaddah†
University of Alabama, Tuscaloosa, AL, 35487

Introduction

THE use of electromagnetic forces for the separation of particles from conducting liquids is used in a number of applications. These range from relatively mature applications such as selective sorting of minerals[1] to emerging applications such as purification of molten metals from non-metallic inclusions.[2] The force field in these applications is generated either by passing a uniform dc through the liquid via electrodes and crossing it with an external steady magnetic field,[1] or by induction from a time-varying magnetic field.[2] In such a force field suspended particles with electrical conductivity different from that of the fluid experience forces, causing them to migrate in the fluid. The net force on the particle is essentially the sum of the electromagnetic force in the particle and the normal forces on the surface of the particle resulting from the applied force field in the fluid.

Although the applied force field in electromagnetic separation is unidirectional, the problem of determining the net force on the particle is quite complex because of the distortion of the initially uniform electric current in the liquid around the particle as shown in Fig. 1. The local distortion of the field induces additional electromagnetic forces in the vicinity of the particle, altering the magnetic pressure distribution around the particle and hence the expulsion force. Furthermore, it also induces a flow around the particle, which may affect the migration velocity of a particle, a key parameter in evaluating the performance of electromagnetic separation processes.

In contrast to the extensive experimental and theoretical work, which has been published on the determination of the electromagnetic expulsion force,[1,3,4] the fluid flow problem has received very little attention. The flow

*Department of Metallurgical and Materials Engineering.
†Department of Metallurgical and Materials Engineering.

Fig. 1 Sketch illustrating the local distortion of the electric field around a nonconducting particle.

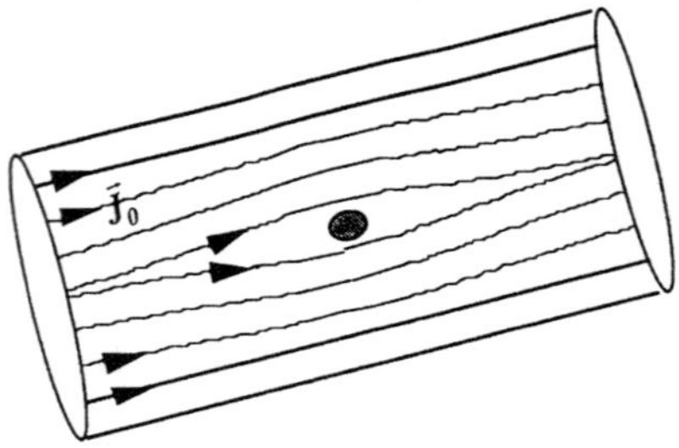

around submerged particles in crossed electric and magnetic fields has been studied for particles with simple geometry, namely, a sphere and a cylinder. The flow around an infinitely long cylinder in an applied axial force field has been investigated by Ricou,[5] Marty and Alemany,[6] and more recently by Gerbeth et al.[7,8] For such a system the flow is one dimensional along the axis of the cylinder. Ricou[5] presented an analytical solution for the flow around a nonconducting cylinder at the limit of a low Hartmann number. Marty and Alemany[6] extended his analysis to conducting cylinders and presented analytical solutions for the flow at low and high Hartmann numbers. They also measured the velocity field around the cylinder at high Hartmann numbers, and their analytical solution was found to be in good agreement with measurements.

The flow around a solid sphere in crossed uniform electric and magnetic fields was investigated analytically by Leenoy and Kolin[9] in the early 1950s. They examined the flow resulting from the interaction between the disturbed current and the applied magnetic field at low Reynolds numbers. Their solution showed that the induced flow is two-dimensional, characterized by four recirculating loops in θ and φ directions with no radial velocity component. Gil[10] and Miroshinkov[11] have extended their analysis to high Reynolds numbers by including the inertia term in the fluid flow equations. In these studies, the inertia term was linearized using the Oseen approximation.[12] Their solutions showed that the flowfield is essentially that obtained by Leenov and Kolin,[9] but with a relatively smaller radial velocity component.

In a recent paper by the authors,[3] that the interaction between the disturbed magnetic field and the applied current also generates electromagnetic forces around the sphere was shown. These forces, which have been neglected in previous studies, are rotational in nature and will act to induce a flow in the radial direction. The purpose of this study is to examine the role played by the principal electromagnetic forces on the nature of the flow in and around a nonconducting liquid droplet and to investigate the effect of viscosity of the droplet on the velocity field.

Electromagnetic Body Force

Let us consider a nonconducting spherical particle of radius R suspended in a conducting fluid. Let us further consider that a uniform dc $\boldsymbol{J}_o$ passes

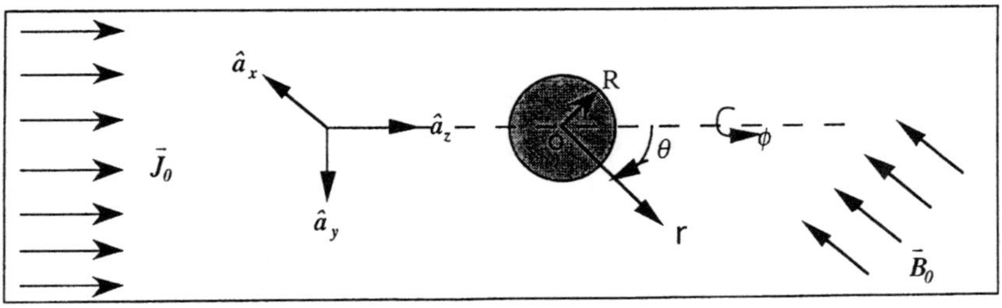

Fig. 2 Sketch illustrating the system and particle coordinates.

through the liquid and is crossed by a uniform steady magnetic field $\boldsymbol{B}_o$ as shown in Fig. 2. Because the electric current cannot pass through the particle, the initial uniform current distribution in the liquid will be distorted in the vicinity of the particle. The current distribution in the liquid may, therefore, be considered as the superposition of $\boldsymbol{J}_o$ and a disturbed current $\boldsymbol{j}'$:

$$\boldsymbol{J} = \boldsymbol{J}_o + \boldsymbol{j}' \tag{1}$$

Similarly, the magnetic flux density at a point in the fluid can be represented as the sum of the applied magnetic field $\boldsymbol{B}_o$ and the magnetic field associated with the electric current:

$$\boldsymbol{B} = \boldsymbol{B}_o + \boldsymbol{b}_o + \boldsymbol{b}' \tag{2}$$

where $\boldsymbol{b}_o$ and $\boldsymbol{b}'$ are the magnetic field associated with $\boldsymbol{J}_o$ and $\boldsymbol{j}'$, respectively.

By using these definitions of the current and the magnetic field, the Lorentz force in the fluid may be expressed as

$$\boldsymbol{F}_{\text{em}} = [\boldsymbol{J}_o \times (\boldsymbol{B}_o + \boldsymbol{b}_o) + \boldsymbol{j}' \times (\boldsymbol{B}_o + \boldsymbol{b}_o) + (\boldsymbol{J}_o \times \boldsymbol{b}') + (\boldsymbol{j}' \times \boldsymbol{b}')] \tag{3}$$

This expression permits the specification of individual forces generated in the fluid as a result of the interaction with the field and with the particle. The force term $\boldsymbol{j}' \times (\boldsymbol{B}_o + \boldsymbol{b}_o)$ describes the interaction between the disturbed current and the applied magnetic field, whereas the last two terms in the left-hand side of Eq. (3) represent the interaction between the current and the magnetic field associated with the disturbed current.

The problem of determining the electromagnetic force is now reduced to the evaluation of the disturbed current in the system. In particle coordinates (r, θ, φ), see Fig. 2, the current distribution around the sphere is axially symmetrical and has two components. If we define the following dimensionless group

$$\boldsymbol{r}^* = \frac{r}{R}, \quad \boldsymbol{j}'^* = \frac{\boldsymbol{j}'}{J_o}, \quad \boldsymbol{b}'^* = \frac{\boldsymbol{b}'}{\mu_o J_o R_o}, \quad \boldsymbol{b}_o^* = \frac{\boldsymbol{b}_o}{\mu_o J_o R_o}, \quad \boldsymbol{F}^* = \frac{\boldsymbol{F}}{J_o B_o} \tag{4}$$

then the dimensionless disturbed current and the associated magnetic field in the fluid are

$$j_r'^* = -\frac{1}{r^{*3}}\cos\theta, \quad j_\theta'^* = -\frac{1}{2r^{*3}}\sin\theta, \quad b_\varphi'^* = -\frac{1}{2r^{*2}}\sin\theta, \quad b_{o\varphi}^* = \frac{1}{2}r^*\sin\theta \tag{5}$$

From Eq. (3) to (5), the four individual force components normalized with respect to the applied force may be expressed as

$$\boldsymbol{F}_o^* = \boldsymbol{a}_z \times (\boldsymbol{a}_x + Y\boldsymbol{b}_o^*) = \nabla^*\left(r^*\sin\theta\sin\varphi - \frac{Y}{4}r^{*2}\sin^2\theta\right) \tag{6}$$

$$\begin{aligned}\boldsymbol{F}_1^* &= \boldsymbol{j}'^* \times (\boldsymbol{a}_x + Y\boldsymbol{b}_o^*)\\ &= \left[\nabla^*\left(-\frac{1}{4r^{*2}}\sin\theta\sin\varphi + \frac{Y}{4r^*}\sin\theta\right) + \nabla^* \times \left(\frac{3}{8}\frac{\sin 2\theta\cos\varphi}{r^{*2}}\right)\boldsymbol{a}_r\right]\end{aligned} \tag{7}$$

$$\boldsymbol{F}_2^* = Y(\boldsymbol{a}_z \times \boldsymbol{b}'^*) = Y\left\{\nabla^*\left[\frac{1}{2r^*}\left(\frac{1}{2}\sin^2\theta - 1\right)\right] + \nabla^* \times \left[-\frac{1}{8r^*}\sin 2\theta\boldsymbol{a}_\varphi\right]\right\} \tag{8}$$

$$\boldsymbol{F}_3^* = Y(\boldsymbol{j}'^* \times \boldsymbol{b}'^*) = Y\left[\nabla^*\left(-\frac{1}{8r^{*4}}\cos^2\theta\right) + \nabla^* \times \left(-\frac{1}{8r^{*4}}\sin 2\theta\boldsymbol{a}_\varphi\right)\right] \tag{9}$$

Equations (6–9) show that the applied force field $\boldsymbol{F}_o$ is irrotational, as expected, and the disturbed field produces stirring forces in the fluid. Inspection of these equations reveals that there are two characteristic rotational force fields, one in θ and φ directions and the other in r and θ directions, of different dependence on the applied electric and magnetic fields and on the size of the particle. $\boldsymbol{F}_1$ is of an $O(J\,B)$, whereas $\boldsymbol{F}_2$ and $\boldsymbol{F}_3$ are of an $O(\mu_o J^2\,R)$, and for a given applied force field their relative magnitude can be described by the dimensionless group Y $(\mu_o J_o R/B_o)$.

Low Reynolds Number Solution of the Velocity Field

In the previous section three distinct forces $\boldsymbol{F}_1$, $\boldsymbol{F}_2$, and $\boldsymbol{F}_3$ were identified as the driving force for the flow around spherical particles in cross electric and magnetic fields. In this section we shall examine the nature of the induced flows by these forces in and around a spherical droplet with arbitrary viscosity at low Reynolds numbers ($Re < 1$). Because of the complexity of the force field, as a first step to solve the fluid flow problem the interaction between the velocity and the magnetic fields was ignored.

The governing equations describing the steady flow in the liquid and the particle are

$$-\nabla p^{(o)} + \mu_f{}^{(o)}\nabla^2 \boldsymbol{u}^{(o)} + \boldsymbol{F}_{\text{em}} = 0 \quad \text{and} \quad \nabla \cdot \boldsymbol{u}^{(o)} = 0 \tag{10}$$

$$-\nabla p^{(i)} + \mu_f{}^{(i)}\nabla^2 \boldsymbol{u}^{(i)} = 0 \quad \text{and} \quad \nabla \cdot \boldsymbol{u}^{(i)} = 0 \tag{11}$$

where p is the pressure, μ_f is the viscosity, u is the velocity of the fluid, and F_{em} is the electromagnetic body force. The superscripts o and i refer to the exterior and interior regions, respectively, of the sphere.

Using the following dimensionless groups

$$\boldsymbol{u}^{*(i,o)} = \frac{\boldsymbol{u}^{(i,o)}}{U}, \quad p^{*(i,o)} = \frac{p^{(i,o)}R}{\mu^{(i,o)}U}, \quad U = \frac{\mu_o J_o^2 R^3}{\mu_f{}^{(o)}}, \quad \lambda = \frac{\mu_f{}^{(i)}}{\mu_f{}^{(o)}}$$

Eqs. (10) and (11) take the following general form:

$$-\nabla^* p^{*(i,o)} + \nabla^{*2}\boldsymbol{u}^{*(i,o)} + \frac{1}{Y}\boldsymbol{F}^*_{\text{em}} = 0 \quad \text{with} \quad \nabla^* \cdot \boldsymbol{u}^{*(i,o)} = 0 \tag{12}$$

where

$$\boldsymbol{F}^*_{\text{em}} = \begin{cases} \boldsymbol{F}_1^* + \boldsymbol{F}_2^* + \boldsymbol{F}_3^* & \text{for } r^* > 1 \\ 0 & \text{for } r^* < 1 \end{cases} \tag{13}$$

Where , $\boldsymbol{F}_1^*$, $\boldsymbol{F}_2^*$, and $\boldsymbol{F}_3^*$ are the electromagnetic forces given by Eqs. (7–9).

The boundary conditions required for solving these equation are zero radial velocity at the surface of the sphere, the continuity of tangential velocity, and shear stresses at the surface of the droplet, zero velocity gradient at the axis of symmetry, and zero velocity at infinity.

Because Eq. (12) is linear, the velocity field is the sum of the flows generated by each force component of the force field:

$$\boldsymbol{u}^{*(i,o)} = \boldsymbol{u}_1^{*(i,o)} + \boldsymbol{u}_2^{*(i,o)} + \boldsymbol{u}_3^{*(i,o)} \tag{14}$$

The inner and outer velocity fields generated by the force $\boldsymbol{F}_1$, which drives the flow in θ and φ directions, are

$$\boldsymbol{u}_1^{*(i)} = -\frac{1}{Y}\frac{r^{*2}}{4(\lambda+4)}\left(\cos\theta \sin\varphi \boldsymbol{a}_\theta + \cos 2\theta \cos\varphi \boldsymbol{a}_\varphi\right) \tag{15}$$

and

$$\boldsymbol{u}_1^{*(o)} = -\frac{1}{Y8r^*}\left(1 - \frac{1}{r^{*2}}\frac{\lambda+2}{\lambda+4}\right)(\cos\theta\sin\varphi\boldsymbol{a}_\theta + \cos 2\theta\cos\varphi\boldsymbol{a}_\varphi) \quad (16)$$

Equations (15) and (16) show that the flow is two-dimensional in θ and φ directions. Note that the magnitude of the velocity is directly proportional to $(1/Y)$.

The velocity fields produced by $\boldsymbol{F}_2$ are given by

$$\boldsymbol{u}_2^{*(i)} = -(A_2 r^{*3} + B_2 r^*)(2 - 3\sin^2\theta)\boldsymbol{a}_r + (5A_2 r^{*3} + 3B_2 r^*)\sin\theta\cos\theta\boldsymbol{a}_\theta \quad (17)$$

and

$$\boldsymbol{u}_2^{*(o)} = -\left(\frac{C_2}{r^{*2}} + \frac{D_2}{r^{*4}} + \frac{1}{16}\right)(2 - 3\sin^2\theta)\boldsymbol{a}_r - \left(\frac{2D_2}{r^{*4}} - \frac{1}{8}\right)\sin\theta\cos\theta\boldsymbol{a}_\theta \quad (18)$$

where the constants A_2, B_2, C_2, and D_2 are given by

$$A_2 = \frac{1}{20(\lambda+1)}, \quad B_2 = \frac{-1}{20(\lambda+1)}, \quad C_2 = -\frac{1}{40}\left(\frac{3+5\lambda}{1+\lambda}\right), \quad D_2 = \frac{1}{80}\left(\frac{1+5\lambda}{1+\lambda}\right)$$

Here it should be noted that the solution of Eq. (12) for $\boldsymbol{F}_2$, which is of $O(1/r^2)$, satisfies the boundary conditions at the surface of the sphere, but not far away from the sphere. Equation (18) is therefore only applicable close to the surface of the droplet.

The velocity fields produced by $\boldsymbol{F}_3$ are given by

$$\boldsymbol{u}_3^{*(i)} = -(A_3 r^{*3} + B_3 r^*)(2 - 3\sin^2\theta)\boldsymbol{a}_r + (5A_3 r^{*3} + 3B_3 r^*)\sin\theta\cos\theta\boldsymbol{a}_\theta \quad (19)$$

and

$$\boldsymbol{u}_3^{*(o)} = -\left(\frac{C_3}{r^{*2}} + \frac{D_3}{r^{*4}} + \frac{1}{16r^{*3}}\right)(2 - 3\sin^2\theta)\boldsymbol{a}_r - \left(\frac{2D_3}{r^{*4}} + \frac{1}{16r^{*3}}\right)\sin\theta\cos\theta\boldsymbol{a}_\theta \quad (20)$$

and the constants A_3, B_3, C_3, and D_3 are given by

$$A_3 = \frac{-1}{160(\lambda+1)}, \quad B_3 = \frac{1}{160(\lambda+1)}, \quad C_3 = -\frac{1}{160}\left(\frac{6+5\lambda}{1+\lambda}\right), \quad D_3 = \frac{-1}{160}\left(\frac{4+5\lambda}{1+\lambda}\right)$$

Equations (17–20) show that the flow driven by $\boldsymbol{F}_2$ and $\boldsymbol{F}_3$ is in r and θ directions. We should note that the velocity fields $\boldsymbol{u}_2^{*(i,o)}$ and $\boldsymbol{u}_3^{*(i,o)}$ are independent of the parameter Y.

General Features of the Flow Around the Particle

The preceding analysis shows that the flow around spherical particles is essentially a superposition of two characteristic velocity fields: one in θ and φ directions $\boldsymbol{u}_1$, and the other $(\boldsymbol{u}_2+\boldsymbol{u}_3)$ in r and θ directions. These two velocity fields have a different dependences on the applied fields. The θ and φ velocity field depends on the strength of the applied force field, whereas the flow in r and θ directions depends only on the strength of the applied current. In the following we shall examine the main features of these flows, as well as the effects of the induced flow inside the droplet and the ratio of the applied current to magnetic field Y, on the velocities of these two characteristic flowfields.

Figures 3 and 4 show typical stereographic projections of velocity streamlines in the plane normal to the direction of the magnetic field for the interior and exterior flows of θ and φ characteristic velocity field $\boldsymbol{u}_1$. These figures reveal that the rotational flow in θ and φ directions is characterized by four recirculating loops, two in front and two in the back of the projected plane. Figures 5 and 6 show the flow pattern resulting from $\boldsymbol{u}_2$ and $\boldsymbol{u}_3$, which

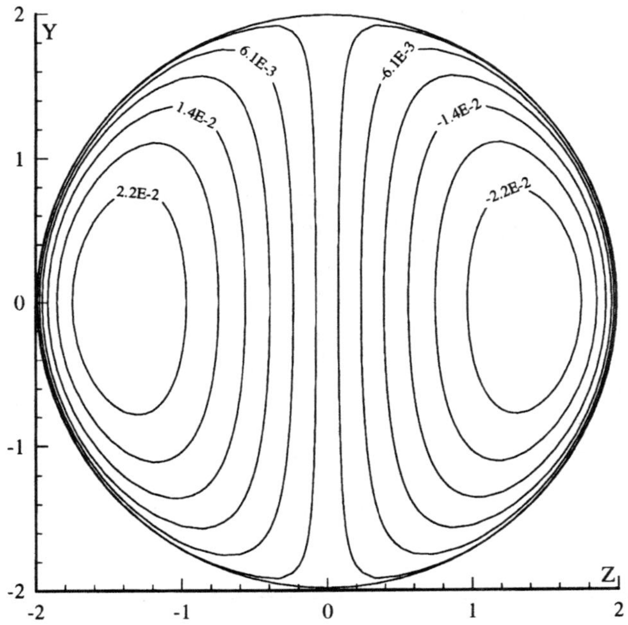

Fig. 3 Streamlines of the flow outside the sphere, generated by $F_1^*\pi$ at $r^* = 2$; $Y = 1$ and $\lambda = 1$.

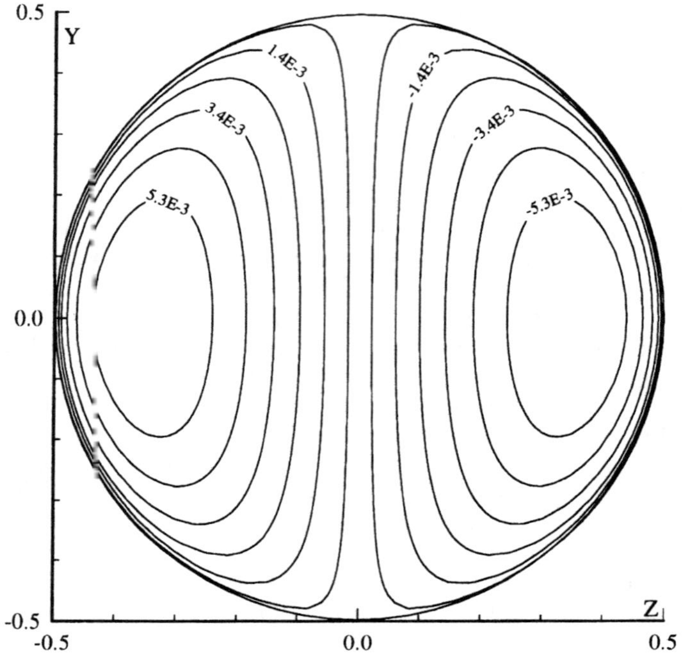

Fig. 4 Streamlines of the flow inside the sphere, generated by $F_1^* \pi$ at $r^* = 0.5$; $Y = 1$ and $\lambda = 1$.

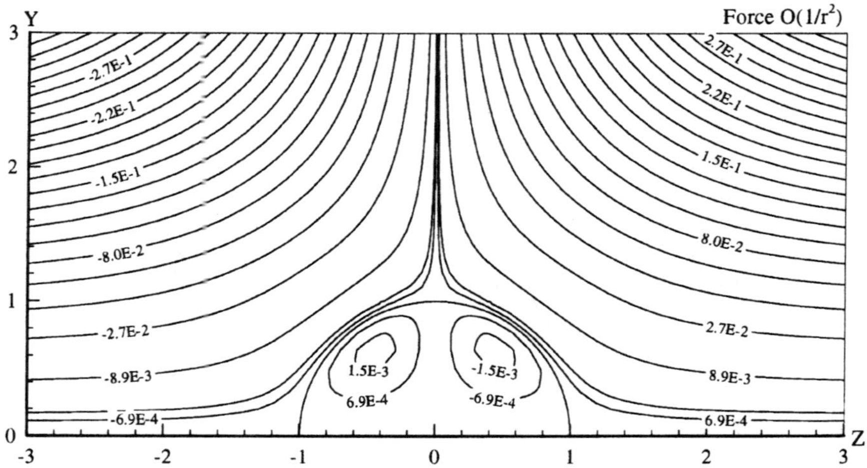

Fig. 5 Streamlines of the flow inside and outside the sphere, generated by F_2^* for $\lambda = 1$.

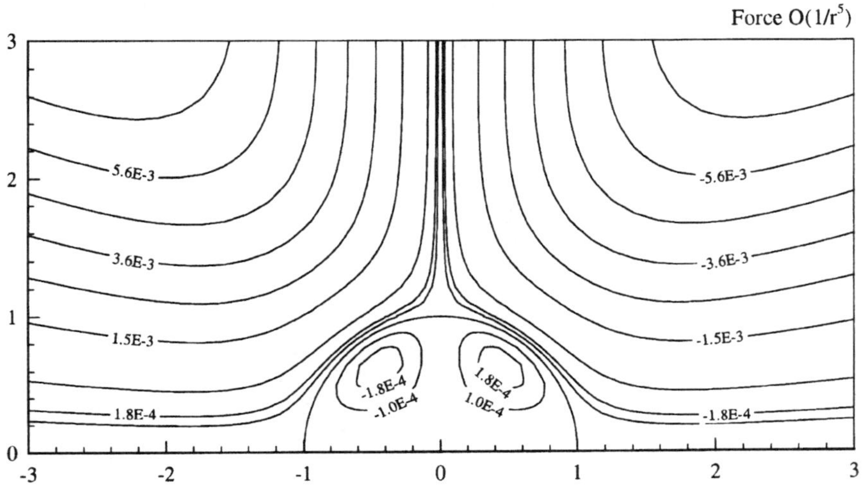

Fig. 6 Streamlines of the flow inside and outside the sphere, generated by F_3^* for $\lambda = 1$.

constitute the r and θ characteristic flow, respectively. These figures show that the fluid outside the sphere is driven either toward the surface of the sphere, Fig. 5, or away from the sphere, Fig. 6, along the axis of symmetry, which is the direction of the applied current. These figures also show that the flow inside the sphere is a recirculating one with four loops, two above and two below the equator. From the values of the stream functions, the r and θ characteristic flow is dominated by the $\boldsymbol{u}_2$ velocity field.

The effects of the magnetic field and the induced flow in the droplet on the θ and φ characteristic flows are shown in Figs. 7 and 8. Figure 7 shows the variation of $\boldsymbol{u}_1$ with r at three different values of particle to fluid viscosity λ, 0.1, 1, and 10. This figure clearly shows that the flow inside the droplet affects the velocity profile in the surrounding fluid. As seen in this figure, decreasing λ increases the mean velocity around the particle and brings the maximum velocity closer to the surface of the particle, which is the source of the vorticity of the force field. The difference in the maximum velocity of the almost inviscid droplet ($\lambda = 0.1$) and that of the solid particle ($\lambda = 10$) is not very large, about 20%. Interesting to see is that λ does not affect the secondary flow region around the droplet. For $Y = 1$ the flow domain is about three times the radius of the particle. Figure 8 shows the variation $\boldsymbol{u}_1$ with r at different values of Y. This figure shows that Y has a strong influence on both the velocity and the secondary flow region. These findings reflect the direct dependence of the velocity on $1/Y$, Eqs. (15) and (16).

The effect of λ on $\boldsymbol{u}_2$ and $\boldsymbol{u}_3$ velocity fields are shown in Figs. 9 and 10. Figure 9 shows that λ modifies the $\boldsymbol{u}_2$ velocity profile in the vicinity of the particle. Because of the limitation of the viscous flow solution for $\boldsymbol{F}_2$ discussed in the preceding section, the effect of λ on the $\boldsymbol{u}_2$ flow region around

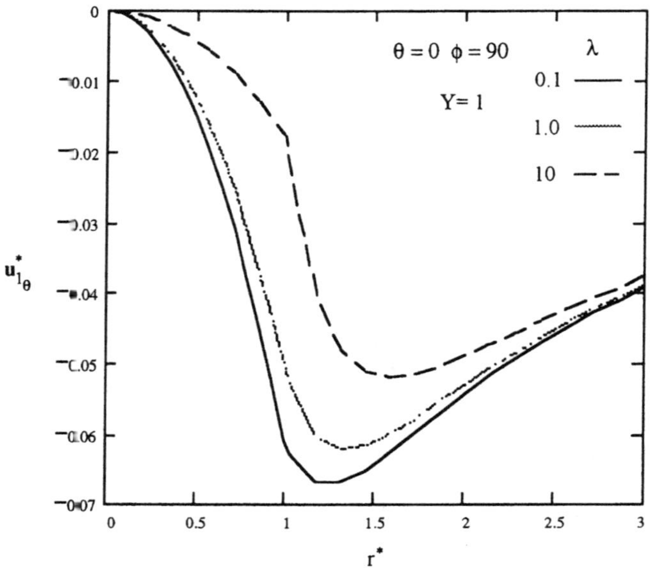

Fig. 7 Effect of the dimensionless viscosity λ on the tangential velocity $u^*_{1_\theta}$, generated by $F_1^*\pi$ for $Y = 1$ and $\theta = 0°$, $\varphi = 90°$.

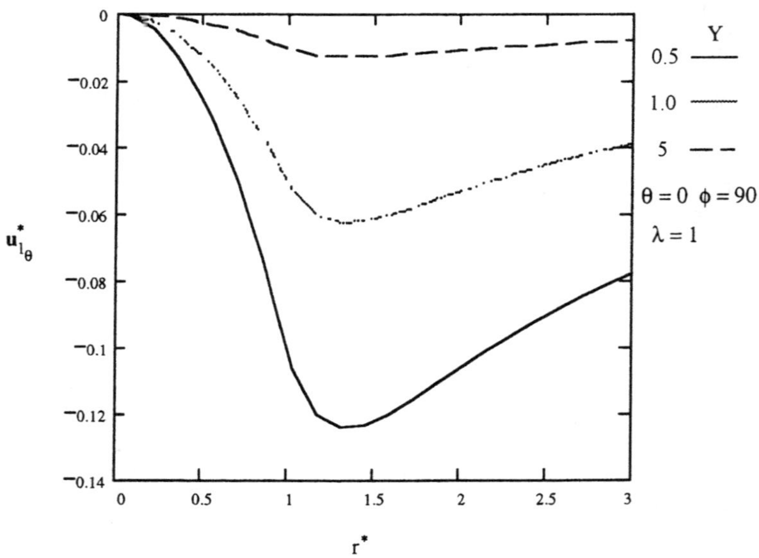

Fig. 8 Effect of the ratio of the applied current to the applied magnetic field Y on the tangential velocity $u^*_{1_\theta}$, for $\lambda = 1$ and $\theta = 0°$, $\varphi = 90°$.

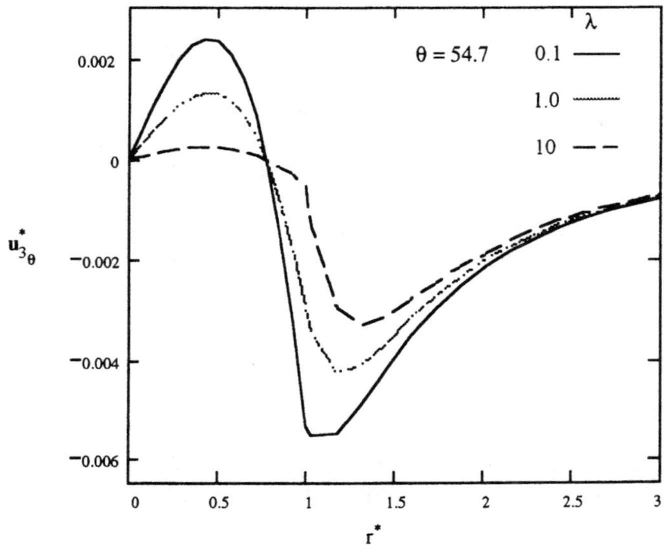

Fig. 9 Effect of the dimensionless viscosity λ on the tangential velocity $u^*_{3_\theta}$, generated by F^*_3 for $\theta = 54.7°$.

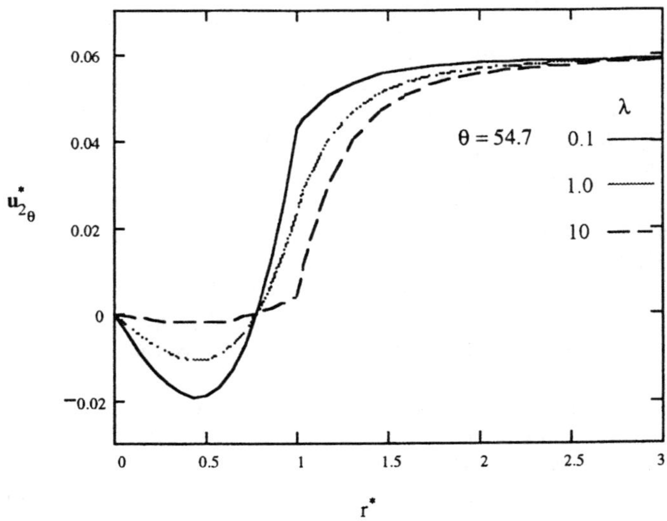

Fig. 10 Effect of the dimensionless viscosity λ on the tangential velocity $u^*_{2_\theta}$, generated by F^*_2 for $\theta = 54.7°$.

the sphere cannot be deduced. Figure 10 shows that decreasing λ intensifies the flow around the particle and brings the maximum velocity closer to the surface of the particle.

Finally, the overall flow structure around the particle depends on the ratio of the applied electric and magnetic fields as well as the size of the particle. If the applied magnetic field is stronger than the magnetic field associated with the applied current, i.e., $Y < 1$, the flow around the particle is predominantly in θ and φ directions. Increasing the current density at the same value of the applied force and/or increasing the particle size intensifies the r and θ flow. When Y is much larger than 1, the flow reduces to an axially symmetrical one.

Acknowledgments

Support for this work is provided by the U.S. Department of Energy, DOE cooperative agreement # DE FC 07 931D13232. Ashish D. Patel would like to thank the University of Alabama TriCampus Materials Science Program for providing financial support.

References

[1]Andres, U., *Magnetohydrodynamic and Magnetohydrostatic Methods of Separation*, Wiley, New York, 1976.

[2]El-Kaddah, N. Patel, A. D., and Natrajan, T. T., "The Electromagnetic Filtration of Molten Aluminum Using an Induced-Current Separator," *Journal of Metals*, Vol. 47, No. 5, 1995, pp. 46–49.

[3]Patel, A. D., and El-Kaddah, N., "On the Theory of Electromagnetic Separation in Alternating Electromagnetic Fields," *International Symposium on Electromagnetic Processing of Materials*, Nagoya, ISIJ, 1994, pp. 115–120.

[4]Andreev, A. V., and Andres, U. Ts., "Influence of Shape, Orientation, and Conductivity of Bodies on Their Electromagnetic Expulsion from Conducting Liquids," *Zh. Prikl. Mekh. I Tekn. Fiz.*, No. 2, 1964, pp. 140–143.

[5]Ricou, R., "Perturbations des Parametres Electriques et Mecaniques Dues al' Introduction d' un Porte-Sonde, Au Sein d' un Fluid Electroconducteur, en Presence d' un Champ de Force Electromagnetique," These D' Universite, Univ. D' Aix Marseille III, France, 1975, pp. 3–28.

[6]Marty, P., and Alemany, A., "Theoretical and Experimental Aspects of Electromagnetic Separation," *Metallurgical Applications of Magnetohydrodynamics*, edited by H. K. Moffatt and M. R. E. Proctor, The Metals Society, London, 1982, pp. 245–259.

[7]Gerbeth, G., Thess, A., and Marty, P., "Theoretical Study of the MHD Flow Around a Cylinder in Crossed Electric and Magnetic Field," *European Journal of Mechanics*, Vol. 9B, No. 3, 1990, pp. 239–257.

[8]Thess, A., Gerbeth, G., and Marty, P., "Electromagnetically Driven Flow Around a Cylinder," *Sixth Beer-Sheva International Seminar On Magnetohydrodynamic Flows and Turbulence* (Beer-Sheva, Israel), 1990, pp. 535–550.

[9]Leenov, D., and Kolin, A., "Theory of Electromagnetophoresis: I. Magnetohydro-

dynamic Forces Experienced by Spherical and Symmetrically Oriented Cylindrical Particles," *Journal Of Chemical Physics*, Vol. 22, No. 4, 1954, pp. 683–688.

[10]Gil, B. B., "Investigation of Electromagnetic Expulsion of Solid Particles under Different MHD Separation Regimes," *Inst. Goryuchikh Iskopaemykh, Ministerstvo Ugol'noi Promyshlennosti (MUP) SSSR*, Moscow, 1969.

[11]Miroshnikov, V. A., "Motion of a Sphere in Crossed Electric and Magnetic Fields at Small Reynolds Numbers," *Mag. Gidrodin.*, No. 1, 1980, pp. 70–78.

[12]Happel, J., and Brenner, H., *Low Reynolds Number Hydrodynamics*, Martinus–Nijhoff, Dordrecht, The Netherlands 1983.

Chapter 52

Force Effects in Conductive Fluid Caused by the Action of Rotating Magnetic Field with In-Phase Modulated Frequency

V. P. Shamota* and Ok. A. Ponomarenko†
Donbas State Academy of Civil Engineering and Architecture, Donetsk Region, 339023, Ukraine

I. Formulation of the Problem

IN THE present work is investigated the electrodynamic phenomenon arising in conducting liquids, filling infinitely long cylindrical vessels of circular section on an external surface infinitely thin of a dielectric wall of which is distributed infinitely thin a current layer, exciting rotated magnetic field (RMF). The other space is filled in ideal ferromagnetic ($\mu \to \infty$).

Magnetohydrodynamics is the phenomena that in such model are described in a cylindrical system of coordinates alternative r, φ, z by the nondimensional equation for the z component of vectorial potential of a magnetic induction ($\boldsymbol{b}$ = rot$\boldsymbol{a}$)

$$\varpi_0 \left\{ \frac{1}{2\pi} \frac{\partial a_z}{\partial r} + \frac{\upsilon}{r} \frac{\partial a_z}{\partial \varphi} \right\} = \Delta a_z \tag{1}$$

where $\varpi_0 = \mu_0 \sigma \omega_0 R_0^2$ is the relative frequency basic RMF, $\Delta = (\partial^2/\partial r^2) + (1/r)(\partial/\partial r) + 1/r^2(\partial^2/\partial\varphi^2)$, $\tau = t^*/T_0, T_0 = 2\pi/\omega_0, v = v^*/\omega R_0, a_z = a_z^*/A_0, r = r^*/R_0$, $A_0 = \mu_0 NJR_0$, NJ is the linear current of the load of inductor, υ is the average velocity of stationary motion of liquid, and asterisks label the dimension variables.

*Assistant Professor, 2, Derzhavina str., 23 Makeyevka.
†Ph.D. Student, 2, Derzhavina str., 23 Makeyevka.

II. Amplitude Modulation

The amplitude modulated RMF is set by the following boundary condition.

$$\left.\frac{\partial a_z}{\partial r_\tau}\right|_{r=1} = -\left[1+\varepsilon\cos\left(2\pi\tilde{\omega}_1\tau - m\varphi\right)\right]e^{i(2\pi r-p\varphi)} \tag{2}$$

where ε is the dimensionless depth of modulation of amplitude, p is the number of pairs of inductor poles (wave-number basic RMF), and m is the wave number of a wave of modulation, $\tilde{\omega}_1) = \omega_1/\omega_0$. If $m = 0$, we name the modulation synchronous; if $m \neq 0$, we call it in-phase.

Taking into account that $\cos x = (e^{ix} + e^{-ix})/2$, we shall write Eq. (2) as

$$\left.\frac{\partial a_z}{\partial \tau}\right|_{r=1} = e^{i\alpha_1} - \frac{\varepsilon}{2}\left(e^{i\alpha_2} + e^{i\alpha_3}\right) \tag{3}$$

where $\alpha_1 = 2\pi\tau - p\varphi$, $\alpha_2 = 2\pi(1 + \tilde{\omega}_1))\tau - (m + p)\varphi$, and $\alpha_3 = 2\pi\,(1 - \tilde{\omega}_1))\tau + (m - p)\varphi$.

Expression (3) superposition of three waves is basic with frequency 2π and phase speed $2\pi/p$, high-frequency direct with the frequency $2\pi(1 + \tilde{\omega}_1))$ and phase speed $2\pi(1 + \tilde{\omega}_1))/(m + p)$, and low-frequency reverse with the frequency $2\pi(1 - \tilde{\omega}_1))$ and phase speed $2\pi(1 - \tilde{\omega}_1))/(m - p)$. At $m = p$ a reverse wave degenerates into the oscillation. If $m = \tilde{\omega}_1)p$, the phase and group speeds of all three waves are identical.

The solution of problems (1) and (3) are searched in a kind of superposition of three waves:

$$a_z = \sum_{n=1}^{3} a_n(\tau)e^{i\alpha n} \tag{4}$$

Substituting Eq. (4) into Eqs. (1) and (3) and grouping the members with identical value of parameter n, receive three problems for $a_\pi(\pi)$

$$i\omega_0 v_n S_n a_n = L_n a_n = 1, 2, 3 \tag{5}$$

where $v_1 = p$, $v_2 = m + p$, $v_3 = m - p$, $S_1 = 1/p - \upsilon/r$, $S_2 = [(1 + \tilde{\omega}_1))/(m + p)] - v/r$, $S_3 = [(1 + \tilde{\omega}_1))/m - p)] - (\upsilon/r)$ is the sliding, $L_n = (\partial^2/\partial r^2) + (1/r\,(\partial/\partial r) - v/r^2)$.

Boundary conditions to problem (5) are

$$\left.\frac{\partial a_1}{\partial \tau}\right|_{r=1} = -1 \tag{6}$$

$$\left.\frac{\partial a_2}{\partial \tau}\right|_{r=1} = \left.\frac{\partial a_3}{\partial \tau}\right|_{r=1} = \frac{\varepsilon}{2} \tag{7}$$

The solution of problems (5–7) has

$$a_n(\tau) = C_n \cdot J_{\nu n}(k_n \tau) \tag{8}$$

where $k_n = i\sqrt{i\varpi_0 \nu_n S_n}$, $C_n = -Q_n / J'_{\nu_n}(k_n)$, $Q_1 = -1$, $Q_2 = Q_3 = -\varepsilon/2$, $J'_{\nu_n}(k_n)$ $= dJ_{\nu_n}(k_n r)/dr\Big|_{r=1}$

The radial component of a magnetic induction is

$$b_r = \frac{1}{r}\frac{\partial a_z}{\partial \varphi} \tag{9}$$

The axial component of density of a current is

$$j_z = -\varpi_0\left(\frac{1}{2\pi}\frac{\partial a_z}{\partial \tau} + \frac{\upsilon}{r}\frac{\partial a_z}{\partial \varphi}\right) \tag{10}$$

And the azimuthal component of density electromagnetic forces is

$$f_\varphi = \Re b_r \cdot \Re j_z = \frac{Q_0}{\tau}\sum_{n'j=1}^{3}(a_{n1}\sin\alpha_n + \alpha_{n2}\cos\alpha_n)(a_{j1}\sin\alpha_j + a_{j2}\cos\alpha_j)\nu_n S_j \tag{11}$$

where a_{n1}, a_{j1} are real and a_{n2}, a_{j2} are imaginary parts of a_n (). The expression (11) represents a superposition of a stationary field and waves of sixth types: direct waves with the frequency 4π; phase speed $2\pi/p$ and amplitudes; proportional ε^0 $(n = j)$ and $\varepsilon^2(n = 2; J = 3)$; direct and reverse waves with the frequency $2\pi\varpi_1$); phase speed $2\pi\varpi_1)/m$ and amplitudes ~ ε $(n = 1; j = 3$, direct and $n = 1; j = 2$, reverse); direct with the frequency $2\pi(2 + \varpi_1)$); phase speed $2\pi(2 + \varpi_1))/(m + 2p)$ and amplitude ~ ε and reverse with the frequency $2\pi(2 - \varpi_1)$); and phase speed $2\pi(2 - \varpi_1))/(m - 2p)$ and amplitude ~ ε. At cases $m = p$ and $m = 2p$, the reverse waves of vectorial potential and density of forces degenerate into the oscillations.

III. Frequency Modulation

In the case of frequency modulation of currents and exciting RMF, the boundary condition (2) has

$$\left.\frac{\partial a_z}{\partial \tau}\right|_{\tau=1} = -e^{i[2\pi r\omega(\tau) - p\varphi]}, \tag{12}$$

where $\omega(\tau) = 1 + \varepsilon\cos(2\pi\tilde{\omega}_1)\tau + m\varphi)$ and ε is the nondimensional deviation of frequency.

We shall record Eq. (12) as

$$\left.\frac{\partial a_z}{\partial \tau}\right|_{r=1} = -A(\tau,\varphi)\cdot e^{i(2\pi\tau - p\varphi)}, \tag{13}$$

where $A(\tau,\varphi) = \exp 2\pi i e\tau\cos(2\pi\overline{\omega}_1\tau - m\varphi)$.

While $\varepsilon\tau \ll 1/2\pi$, the expression (13) is resulted in a kind (3).

Really

$$A(\tau,\varphi) = \cos\left[2\pi\varepsilon\tau\cos(2\pi\tilde{\omega}_l\tau - m\varphi)\right] + i\sin\left[2\pi\varepsilon\tau\cos(2\pi\tilde{\omega}_l\tau - m\varphi)\right]$$

As far as $0 \le |\cos(2\pi\tilde{\omega}_1)\tau - m\varphi)| \le 1$, $\cos 2\pi\varepsilon\tau + \sin 2\pi\varepsilon\tau \le 1$. Owing to a small of $2\pi\varepsilon\tau$ $A(\tau,\varphi) \simeq 1 + i2\pi\varepsilon\tau/\cos(2\pi\tilde{\omega}_1)\tau - m\varphi)$, Eq. (12) will be recorded in a kind similar to Eq. (3):

$$\left.\frac{\partial a_z}{\partial \tau}\right|_{r=1} = -\exp\left[i(2\pi\tau - p\varphi)\right] - \pi\varepsilon\tau\left(\exp\left\{i\left[2\pi(1+\tilde{\omega}_1)\tau - (m+p)\varphi + \pi/2\right]\right\}\right.$$
$$\left. + \exp\left\{i\left[2\pi(1-\tilde{\omega}_1)\tau + (m-p)\varphi + \pi/2\right]\right\}\right)$$

Solution of problems (1) and (14) in this case are similar to the solutions of problems (1) and (3), and the physical effects, described by these problems, are identical. The existence of high- and low-frequency oscillations in a rotating liquid metal reduce to the intensification of many processes in casting technology.[4]

References

[1]Kapusta, A. B. and Scheluchin, E. M., "To Account of Electromagnetic Volumetric Forces in a Liquid Metal, Placed in a Rotated Magnetic Field," *Tehnicheskaya Electromagnitnaya Gidrodinamika*, Donetsk, Ukraine, 1968, pp. 225–232.

[2]Dalberg, E., "On the Action of a Rotating Magnetic Field on a Conducting Liquid," Studsvik, AE-447, Nicöping, Sweden, 1972.

[3]Pierce, J., Almost *All about Waves,* MIT Press, Cambridge, MA 1974.

[4]Herrmann, E. "Handbuch des StranggieBens," Aluminium–Verlag GmbH, Düsseldorf, Germany, 1960.

Chapter 53

Using the "External Friction" Model for Describing Quasilaminar and Turbulent Flows Caused by a Rotating Magnetic Field

A. B. Kapusta* and V. P. Shamota†
Donbas State Academy of Civil Engineering and Architecture, Donetsk Region, 339023, Ukraine

I. Introduction

THE devices with the rotating magnetic field are used for stirring and mixing of liquid metals and electrically conducting melts in metallurgy, foundry, and crystal growth.[1] One major application is continuous steel casting. So the optimization of this device's design is of great interest. It needs a simple method for the average angular velocity of fluid motion calculation.

The "external friction" model is a simple and effective method of the description of average quasilaminar and turbulent rotary flows, produced by a rotating magnetic field in a cylindrical vessel. The main idea of this method consists in the approximation of the Reynolds stress tensor by a vector, the components of which are proportional to average flow velocity. Nonlinear hydrodynamic effects are thus taken into account with the help of the external friction factor λ which in an accepted model depends on the average significance of flow vorticity.

The one-dimensional laminar rotary flow of a liquid at an increase of its velocity, as is known, goes in turbulent through some of the quasilaminar stages. First the quasilaminar structures are the Taylor vortices.[2] Then arises a wave disturbance, extending in an azimuthal direction along these vortices. Further there are two and more harmonics of running waves that, in a final total, result to complete turbulization of rotary flow. But even in a completely turbulent flow, the process of transfer of energy from a main rotary flow to large-scale vortical flows and further to more small-scale is saved.[3]

*Head of Department, 2, Derzhavina str., 23 Makeyevka.
†Assistant Professor, 2, Derzhavina str., 23 Makeyevka.

These features of rotary flow in the external friction model are taken into account with the help of the universal empirical parameter ε, the whole significances of which correspond to quasilaminar flows and flows with advanced turbulence and the half-whole significance of which corresponds to flows with harmonic by-wave structures.

II. Formulation of the Problem

With the use of the external friction model are studied nonlaminar flow viscous incompressible conductive liquids in cylindrical vessels of a circular section with infinitely thin dielectric walls. The vessel is placed in the ideal inductor with one pair of poles, which produce the rotating magnetic field. Under the ideal inductor is understood the device with a smooth surface of its side, on which is in regular intervals distributed an infinitely thin layer of an electric current, exciting the first harmonic of a rotating magnetic field. The inductor is executed from a ideal ferromagnetic ($\mu \to \infty$). In a noninductive approximation such a flow is described in a cylindrical system of coordinates r, ρ, and z; the axis z coincides with an axis of the vessel's symmetry by the following nondimensional equation.

$$\frac{d^2 \upsilon}{dr^2} + \frac{1}{r}\frac{d\upsilon}{dr} - \frac{\upsilon}{r^2} - \beta^2 \upsilon = -Ha_a^2 r \tag{1}$$

with boundary conditions

$$\upsilon\big|_{r=1} = 0; \qquad \upsilon\big|_{r=0} < \infty \tag{2}$$

where υ is the average on coordinate z (for quasilaminar flows) and on the group (for turbulent flows) azimuthal nondimensional velocity $\beta^2 = \lambda + Ha_a^2$; $\lambda = 0{,}0165 \exp(11{,}6{,}\varepsilon) \cdot (Re_\omega < \Omega >)^{1-\varepsilon}/f(\delta)$ is the external friction coefficient; $< \Omega >= \frac{1}{2}\int_0^1 \left|\mathrm{rot}_z \vec{\upsilon}\right| r dr$ is the half average on the vortical lead section of absolute significance of a vortex flow; ε is the unique empirical parameter completely determining the structure of flow; $\delta = Z_0/R_0$ is the attitude of half of a vessel's height to its radius; $Re_\omega = \omega R_0/\nu$ is the Reynolds number based on the rotating magnetic field velocity; and $Ha_a = Ha/\sqrt{2}$ is the acting significance of a Hartmann number $Ha = B_0 R_0 \sqrt{\sigma/\eta}$, $f(\delta) = \exp[Ath(\ln\delta/AC)]$, where $A = 11.23947$, $C = 0.98584$. The function $f(\delta)$ satisfies. the following requirements: $f(\delta) \approx \delta$ for $\delta \leq 10$ and $f(\delta) \to \exp A$ for $\delta \to \infty$.

Number A is defined as a result of theoretical research of the instability of the laminar rotary current, excited by the rotating magnetic field in an infinitely long cylindrical vessel in relation to the stationary disturbances of the Taylor vortices type.

III. Velocity Profile

The solution of problems (1) and (2) is

$$v = \frac{Ha_a^2}{\beta^2}\left[r - \frac{I_1(\beta r)}{I_1(\beta)}\right] \tag{3}$$

where I_1 is the modified Bessel function.

For small significances β and Ha_a ($\beta < 1, Ha_a < 1$),

$$v \simeq \frac{Ha_a^2 r(1-r^2)}{8+\beta^2} \tag{4}$$

For large significances β and Ha_a ($\beta \gg 1, Ha_a \gg 1$),

$$v \simeq \frac{Ha_a^2}{\beta^2}\left(r - \frac{e^{\beta(r-1)}}{\sqrt{r}}\right) \tag{5}$$

The expressions (3–5) are the generalization of the formulas, describing the one-dimensional laminar rotary flow[4] at $\lambda(<\Omega>\varepsilon) \to 0$ and $\beta \to Ha_a$. The velocity profile of the turbulent flow, which was calculated under the formula (3) as it is visible from Fig. 1, is in good conformity with the experiment.[5]

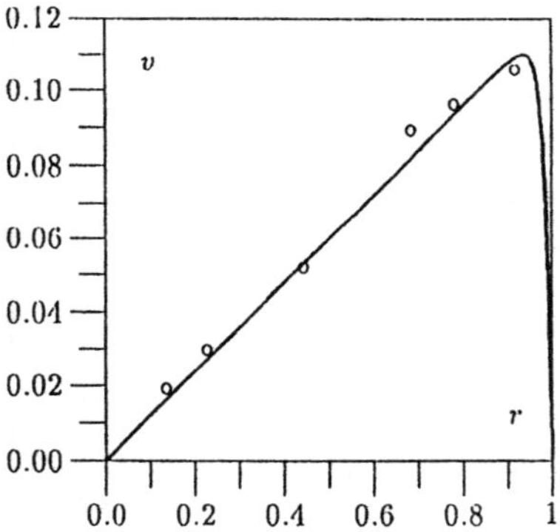

Fig. 1 The velocity profile for Ha_a = 22.2, Re_ω = 7.375.10^6, δ = 4, and ε = 0.

IV. Average Angular Velocity

From Eq. (3) is easily received the algebraic equation for the definition of an average vorticity $<\Omega>$:

$$<\Omega>^{2-\varepsilon} + \frac{Ha_a^2 f(\delta)(<\Omega>-1)}{0.0165\exp(11.6\varepsilon)Re_\omega^{1-\varepsilon}} = 0 \tag{6}$$

We shall note that the laminar flow has a maximum vorticity, and the average vorticity does not depend on Ha_a, Re_ω, and $f(\delta)$.

Depending on the significance of the $Re_\omega <\Omega>$ number, the external friction factor can be growing or decreasing the function of parameter ε. The border between these modes is determined from a condition

$$\frac{\partial\lambda}{\partial\varepsilon} = 0 \tag{7}$$

This condition corresponds to the significance of number $Re\omega <\Omega> = 109.098$.

The given model correctly described the various steps of transition of the kinetic energy of a primary large-scale rotary current, produced by electromagnetic field forces, to more small-scale movements (ordered or chaotic); necessary is that the external friction coefficient decreased with the increase of ε, i.e., the inequality $Re_\omega < 109.098$ was executed. At $Re_\omega < 109.098$ the external friction coefficient grows with the increase of ε, and that within the framework of a given model means the transfer of energy from small-scale structures to large-scale to currents with negative viscosity.

From Eq. (6) it is possible to express Ha_a and then to substitute in a received expression found in the critical significance of a $Re_\omega <\Omega>$

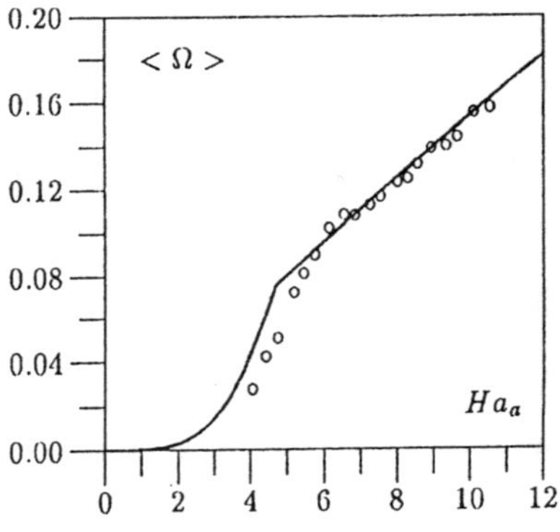

Fig. 2 The dependence of average angular velocity of the Hartmann number for δ = 6.67 and Re_ω = 1.45 × 10⁶.

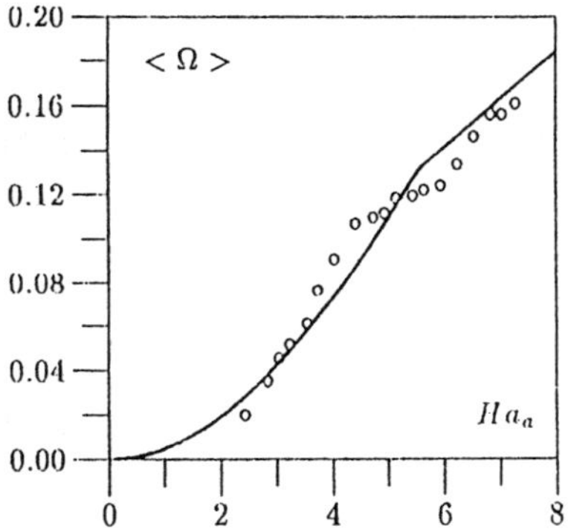

Fig. 3 The dependence of average angular velocity of the Hartmann number for δ = 8.82 and Re_ω = 8.29 × 10^5.

number. As a result, the simple formula for finding a border of the section of two various modes of the current curve is

$$Ha_a = \sqrt{\frac{0.0165\exp(23.2)}{f(\delta)(Re_\omega - 109.098)}} \tag{8}$$

Thus $Ha_a = f(Re_\omega)$ as received is, as it is visible from the comparison with the experiments, the border of transition from the laminar-vortical structure of the current to the advanced turbulent flow.

The point of transition on Figs. 2 and 3 form $\varepsilon = 3/2$ to $\varepsilon = 0$ is found from Eq. (8).

V. Friction Force Moment

In the external friction model we supposed that

$$\lambda \upsilon = \left[\frac{1}{r}\frac{\partial\left(r\overline{u'\upsilon'}\right)}{\partial r} + \frac{1}{r^2}\frac{\partial\left(r\overline{\upsilon'^2}\right)}{\partial\varphi} + \frac{1}{\delta r}\frac{\partial\left(r\overline{\upsilon'\omega'}\right)}{\partial z}\right]Re_\omega \tag{9}$$

Leaving only tangential stress and assuming that

$$\frac{\partial\left(r\overline{u'\upsilon'}\right)}{\partial r} \simeq \frac{\partial\left(r\overline{u'\omega'}\right)}{\partial z} \tag{10}$$

we have

$$\lambda \upsilon \simeq \frac{1}{r}\frac{\partial\left(r\overline{u'\upsilon'}\right)}{\partial r}\left(1 + \frac{1}{\delta}\right)Re_\omega \tag{11}$$

If to enter the Reynolds stress of friction τ_R as

$$\tau_R = Re_{\omega}\overline{u'v'} \tag{12}$$

then

$$\lambda v = \frac{1}{r}\frac{\partial(r\tau_R)}{\partial r}\left(1+\frac{1}{\delta}\right) \tag{13}$$

and

$$\tau_R = \frac{\lambda}{(1+1/\delta)r}\int rv\,\mathrm{d}r \tag{14}$$

The complete moment of friction forces on a cylindrical vessel's wall thus will appear equal:

$$M = \frac{\lambda}{(1+1/\delta)}\int_0^1 rv\,\mathrm{d}r \tag{15}$$

On Figs. 4 and 5 the dependence of a moment of friction of a liquid about the vessel's wall from the size of a magnetic field and comparison received curve with experimental data is submitted. The point of transition from one mode to the other was from Eq. (8).

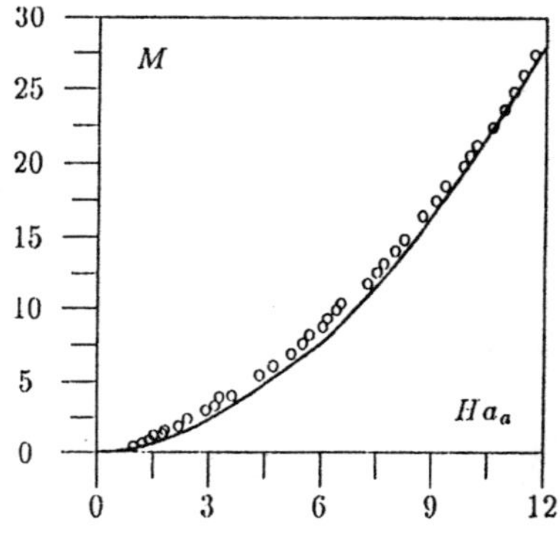

Fig. 4 Moment of friction forces at $\overline{\omega}$ = 0.212, δ = 7.4, and Re_{ω} = 8.29 × 10^5.

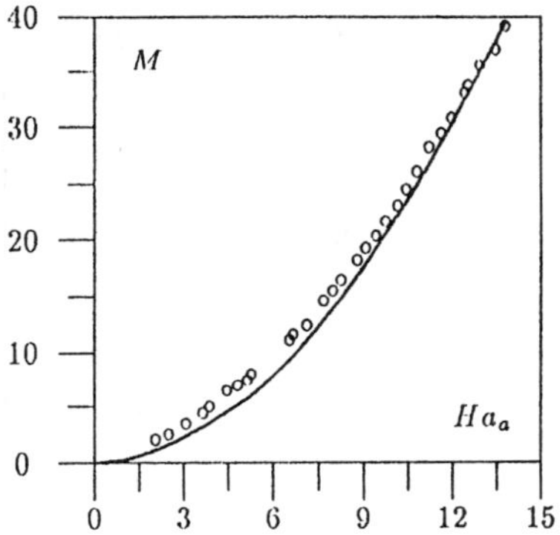

Fig. 5 Moment of friction forces at $\overline{\omega}$ = 0.212, δ = 5.4, and Re_ω = 1.514 × 10^6.

VI. Stability of Flow

The power valuation of stability of a laminar one-dimensional flow, excited by the magnetic field received in Ref. 6, can be essentially improved and is physically justified with the help of the known theorem about the change of kinetic energy of a mechanical system, which can be recorded as

$$\int_0^1 \upsilon^2 r\,\mathrm{d}r = \frac{2N_c}{Re_\omega}\left\{Ha_a^2\int_0^1\left(1-\frac{v}{r}\right)r^3\,\mathrm{d}r - \int_0^1\left|\frac{\partial \upsilon}{\partial r}-\frac{\upsilon}{r}\right| r\,\mathrm{d}r\right\} \tag{16}$$

Change of the kinetic energy of the laminar flow at the acceleration of a conducting liquid from a condition of rest up to the velocity v equally to the algebraic sum of the electromagnetic forces work and forces of friction. Here N_c is the significance average on the radius of the number of revolutions with which there is instability. The condition (16) is executed for any $N > 0$. Therefore, for the definition of the neutral curve to know N_c is necessary, or coordinate through one point on the neutral curve, with the help of which N_c can be determined. Then with the help of Eq. (16) all curves of neutral stability (Fig. 6) are restored. It has appeared that the curve thus received by the neutral stability precisely coincides with the curve calculated by the method of small disturbances.[7]

Successful use of Eq. (16) for the definition of a border of stability of the laminar current, excited by a rotated magnetic field, made us apply a similar procedure for the description of structural transitions between modes of current with the different significances ε, or the border of transition from laminar, vortical modes of the current to the advanced turbulent current. In this case the equation of any appropriate border has a kind of equation

$$\int_0^1 \upsilon_l^2 r\,\mathrm{d}r = \frac{2N_c}{Re_\omega}\left\{Ha_a^2\int_0^1\left(1-\frac{\upsilon_c}{r}\right)r^3\,\mathrm{d}r - \int_0^1\left|\frac{\partial \upsilon_c}{\partial r}-\frac{\upsilon_c}{r}\right| r\,\mathrm{d}r - \frac{\lambda}{1+1/\delta}\int_0^1 r\upsilon_\varepsilon \mathrm{d}r\right\} \tag{17}$$

where υ_l is the laminar flow velocity and $\upsilon\varepsilon$ is the average velocity of quasilaminar or of turbulent flow.

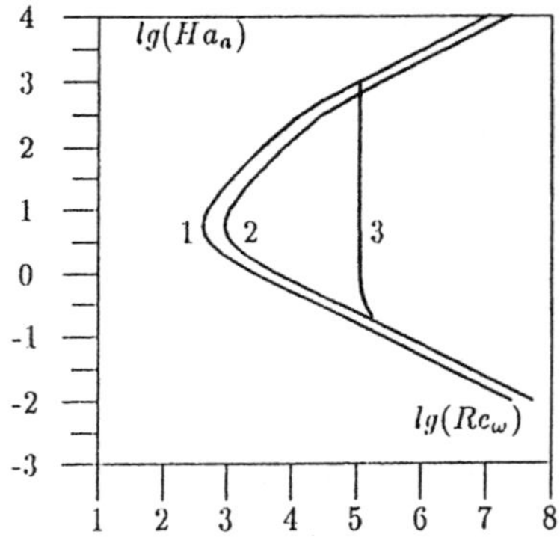

Fig. 6 The neutral stability curves for $\delta \to \infty$, 1 is the laminar flow stability boundary, 2 is the genuine turbulence boundary, and 3 is the curve that satisfies the condition $2\lambda/2\varepsilon = 0$.

The second member in the right part of Eq. (17) describes the work of forces of molecular friction, and the third describes the work of forces of turbulent friction, i.e., work of Reynolds stress. As well as in Eq. (16), the main problem consists of the definition N_c, a finding for which is offered the use of condition (8). Assuming that from the condition (8) it is possible to determine the coordinates of points, which lie on the appropriate neutral curve; it is possible to calculate N_c for each transition and to calculate it with the help of Eq. (17) of appropriate curve neutral stability.

The results of accounts are shown in Fig. 6. For any fixed significance Rc_ω at an increase of the Hartmann number, the laminar flow goes in quasilaminar ($\varepsilon > 0$). Then at a further increase of the Hartmann number, the quasilaminar mode of the flow is replaced by a mode with $\varepsilon = 0$, which corresponds to geniune turbulence. Such sequence of the shift of modes is confirmed experimentally. The area between curve 2 and vertical branch curve 3, in visible, should relate to flows with negative viscosity.

References

[1]Salk, M., Gelfgat, Y. M., et al., "CdTe Crystal Growth in the Soviet Facility ZONA 4," *Microgravity Science Technology*, Vol. 2, 1993, pp. 88–90.

[2]Taylor, G., "Stability of a Viscous Liquid Contained Between Two Rotating Cylinders," *Proceedings of the Royal Society Landa, Series A*, Vol. 223, 1923.

[3]Joseph, D., "Stability of Fluid Motions," Springer–Verlag, Berlin, 1976.

[4]Kapusta, A. B., "The Motion of Conducting Fluid by the Action of Rotating Magnetic Field," *Magnitnaya Gidrodinamika*, No. 3, 1968, pp. 71–76.

[5]Kapusta, A. B., Dremov, V. V., and Doronin, V. I., "The Experimental Investigation of Mercury Motion in the Rotating Magnetic Field," *Technicheskaya Electromagnitnaya Gidrodinamika*, Donetsk, Ukraine, USSR, 1970, pp. 187–191.

[6]Kapusta, A. B., "On the Stability of Azimuthal Flow Caused by the Rotating Magnetic Field," *Tehnicheskaya Electromagnitnaya Gidrodinamika*, Donetsk, Ukraine, USSR, 1968, pp. 166–173.

[7]Kapusta, A. B., and Zibold, A. F., "The Effect of Rotating Magnetic Field Symmetry on the Stability of Axisymmetric Flow," *Magnitnaya Gidrodinamika*, No. 3, 1981, pp. 134–136.

Chapter 54

Numerical Simulation of Two-Dimensional Unsteady Magnetohydrodynamic Duct Flows

L. Leboucher
Forschungszentrum für Technik und Umwelt, Karlsruhe, 76021, Germany

Introduction

Generalities

FULL numerical simulations of inertial flows are often limited to the steady regime and are always limited to very low Hartmann numbers[1,2] ($Ha \sim 50$). The Hartmann number is related to the ratio between the electromagnetic and the viscous forces. It is therefore a measure of the magnetic field strength for a given fluid in a duct of a given scale.

Magnetohydrodynamic (MHD) boundary layers can be much thinner than pure hydrodynamic layers. A boundary layer perpendicular to the magnetic field has its thickness scaled with Ha^{-1}. The thickness of a boundary layer parallel to the magnetic field scales with $Ha^{-1/2}$. Therefore, numerical simulations are limited by the resolution of the boundary layers at high Hartmann numbers.

Governing Equations

The flow of an electrically conducting fluid under the influence of an external imposed magnetic field is governed by the laws of conservation of mass, electric charge, and momentum, and by Ohm's law:

$$\nabla \cdot \boldsymbol{v} = 0 \tag{1}$$

$$\nabla \cdot \boldsymbol{j} = 0 \tag{2}$$

$$\frac{1}{N}\frac{\partial \boldsymbol{v}}{\partial t} + (\boldsymbol{v}\cdot\nabla)\boldsymbol{v} = -\nabla p + \boldsymbol{j}\times\hat{\boldsymbol{y}} + \frac{1}{Ha^2}\nabla^2\boldsymbol{v} \tag{3}$$

$$\boldsymbol{j} = -\nabla\Phi + \boldsymbol{v}\times\hat{\boldsymbol{y}} \tag{4}$$

The duct has its axis in the x direction and has a rectangular cross section in the (y,z) plane. The magnetic field $\boldsymbol{B} = B\hat{\boldsymbol{y}}$ is constant and aligned with the y direction. The induced field is neglected. The dimensionless variables are scaled as follows. The length scale a is the half-width of the duct in the y direction. The scale of the velocity $\boldsymbol{v}$ is the mean velocity v_0. The dimensionless unit of time is the length scale divided by the mean velocity. The current density $\boldsymbol{j}$ is scaled with the product of the electrical conductivity σ of the fluid, the mean velocity, and the strength of the magnetic field B. The pressure is scaled with the mass density ρ of the fluid times the square of the mean velocity times the interaction parameter. The dimensionless magnetic field is scaled with its own strength B, so that its dimensionless representation is the unit vector $\hat{\boldsymbol{y}}$. There are two dimensionless parameters in this problem. The interaction parameter $N = \sigma B^2 a/(\rho v_0)$ characterizes the ratio between the electromagnetic and the inertia forces, and the Hartmann number squared is defined as $Ha^2 = \sigma B^2 a^2/\mu$.

The electrically conducting walls are modeled using the thin wall condition[3] and are treated numerically in a similar way to the iterative method used by Sterl.[1] The Hartmann layers are modeled[4] rather than computed numerically.

Finite Difference Scheme

Discretization and Integration Method

The set of MHD equations (1–4) is solved with the fractional step method[5] for the pressure term whereas all other terms are computed explicitly with a second-order Adams–Bashforth scheme. The spatial discretization is based on a staggered grid as in the finite volume method and uses second-order central differences for all terms except for the advection term that uses a second-order upwind scheme.[6]

Numerical Stability

The standard finite volume discretization, where scalars are computed at the cell center and vectors at the edges of the cell, is found to give a numerically unstable finite difference scheme. This is characterized by strong oscillations at the corners of the duct and occurs for partially or perfectly electrically conducting walls, and for high Hartmann numbers (strong magnetic field) or coarse grid spacing in the direction perpendicular to the magnetic field and to the velocity. Note that this instability is independent of the grid spacing within the Hartmann layers, i.e., the grid spacing in the direction parallel to the magnetic field. The reason for the instability is that the scheme is not monotone.[7]

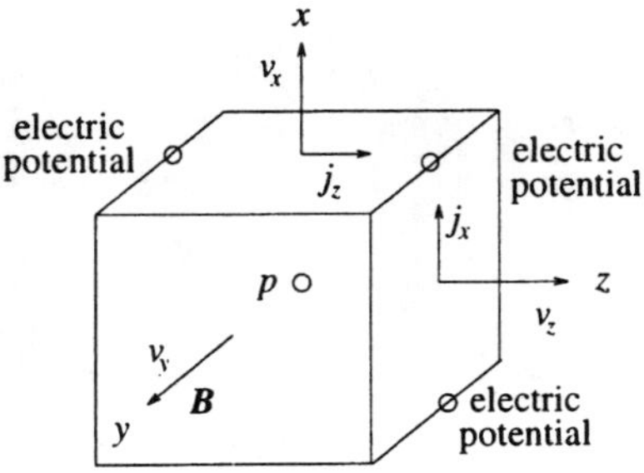

Fig. 1 Fully staggered grid.

A monotone and numerically stable finite difference scheme is found by discretizing the governing equations on a fully staggered grid.[8] In such a grid not only are the scalar quantities are shifted from the vector quantities, but the electromagnetic quantities are shifted from the fluid mechanics quantities (Fig. 1). The stability conditions of this scheme are the same as those of ordinary computational fluid dynamics, except for the diffusion limit on the time step. Because the Lorentz force is computed explicitly, it inevitably gives a stronger limitation on the time step compared to computational fluid dynamics without magnetic field.

Simulations

Introduction

To create a shear layer within a flow through an insulating duct, the walls perpendicular to the magnetic field can be supplied with a pair of electrically conducting strips (Fig. 2). This idea was originally proposed by Kolesnikov.[9] Such a device recently has been modeled analytically and numerically by Bühler[10] as a two-dimensional flow between two infinite plates and was also investigated experimentally as a duct flow.[11]

Unstable Flow in a Duct with Conducting Strips

The code was run for the same parameters as those of an experiment performed by Debray.[11] The parameters are the following: $Ha = 300$, $N = 29$, $Re = 3108$, the ratio of the distance between the two plates to the strip

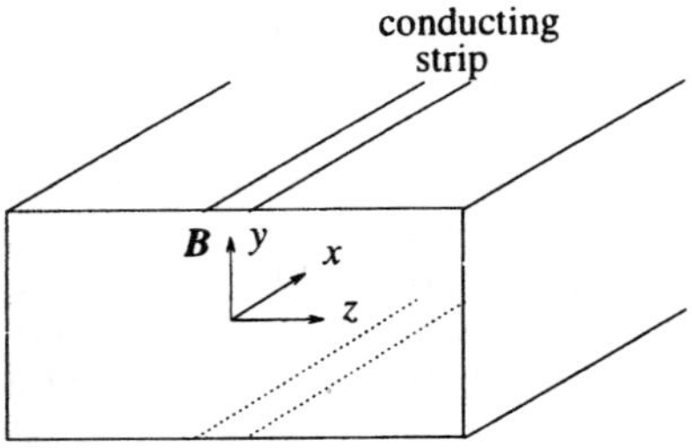

Fig. 2 Duct with electrically conducting strips.

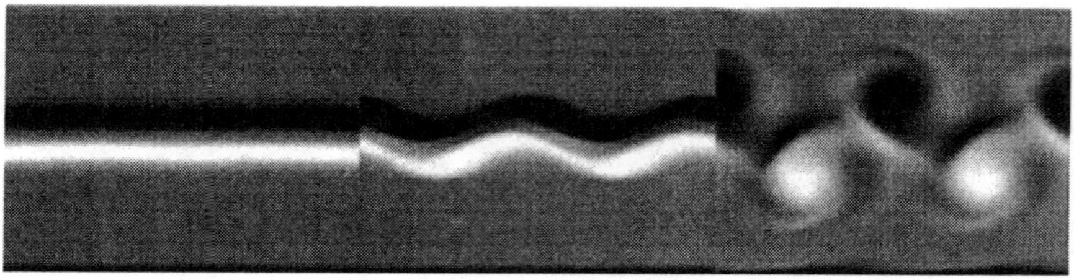

Fig. 3 Unstable MHD shear layer in a duct. Evaluation of the vorticity in the plane ($Y = 0$).

width = 6, and all walls insulating, the size of the grid in x, y, z directions = $27 \times 16 \times 128$ cells. The computational domain is periodic in the direction of the flow and has a wavelength of 2.79, corresponding to the wavelength that was found experimentally, whereas the width of the duct in the direction perpendicular to the magnetic field is 4. The width in the direction parallel to the magnetic field is 2, and the width of the strips is 0.33. The simulation is initialized with a small perturbation on the velocity. This perturbation is rapidly amplified as large eddies around the strip (Fig. 3).

Conclusion

A full staggered grid allows explicit central finite differences to simulate unsteady MHD duct flows. The highest Hartmann number that can be reached is not limited by numerical instabilities, as is the case when using an ordinary stagered grid, but by the computer power only.

Acknowledgment

This work was done with the helpful advice of L. Bühler about the physics of MHD duct flows and with the experience of C. Günther in numerical analysis. R. Mößer has dealt with the various aspects of the mainframe computer's operating system, and the author is grateful to the people of the Forschungszentrum Karlsruhe.

References

[1]Sterl, A., "Numerical Simulation of Liquid-Metal MHD Flows in Rectangular Ducts," *Journal of Fluid Mechanics*, Vol. 216, 1990, pp. 161–191.

[2]Lenhart, L., "Magnetohydrodynamik in Rechteckgeometrien. Volle Numerische Simulation," Forschungszentrum für Technik and Umwelt, FZKA-Rep. No. 5317, ISSN 0303-4003, Germany, April 1994.

[3]Walker, J. S., "Magnetohydrodynamic Flow in Rectangular Ducts with Thin Conducting Walls," *Journal de Mécanique*, Vol. 20, 1981, pp. 79–112.

[4]Bühler, L., "Magnetohydrodynamische Strömungen Flüussiger Metalle in Allgemeinen Dreidimensionalen Geometrien unter der Einwirkung Starker, Local Variabler Magnetfelder," FZKA-Rep. No. 5095, ISSN 0303-4003, Forschungszentrum für Technik und Umwelt, Karlsruhe, Germany, Feb. 1993.

[5]Blair Perot, J., "An Analysis of the Fractional Step Method," *Journal of Computational Physics*, Vol. 108, 1993, pp. 51–58.

[6]Günther, C., "Conservative Versions of the Locally Exact Consistent Upwind Scheme of Second Order," *International Journal for Numerical Methods in Engineering*, Vol. 34, No. 3, 1992, pp. 793–804.

[7]Hirsh, Ch., *Numerical Computation of Internal and External Flows*, Vol. 2, Wiley, 1990, pp. 525–526.

[8]Shimomura, Y., "Large Eddy Simulation of Magnetohydrodynamic Turbulent Channel Flows under a Uniform Magnetic Field," *Physics of Fluids A*, Vol. 3, No. 12, 1991, pp. 3098–3106.

[9]Kolesnikov, Y. B., "Two-Dimensional Turbulent Flow in a Channel with Inhomogeneous Electrical Conductivity of the Wall," *Magnetohydrodynamics*, Vol. 8, 1972, pp. 70–74.

[10]Bühler, L., "Instabilities in 2D MHD Flows," *Proceedings of the 2nd International Conference on Energy Transfer in Magneto Hydro Dynamic Flows* (Aussois, France), 1994, pp. 463–472.

[11]Debray, F., FZKA-Rep. No. 5723, ISSN 0303-4003, 1997, in press.

Chapter 55

Comparison of Analytical Solutions for Magnetohydrodynamic Flow Development in the Entrance Region of a Flat Channel

Savely Khosid*
Technion, Haifa 32000, Israel

I. Introduction

LAMINAR magnetohydrodynamic- (MHD) flow development in the hydrodynamic entrance region of ducts with various cross sections has been extensively investigated in the last few decades. The inlet velocity profile is usually different from the fully developed profile and such an entrance region exists where development of flow occurs. The whole working region of an MHD device may sometimes be labeled as an entrance region because the velocity distribution here is changed continuously and does not reach the fully developed velocity profile.[1]

Usually investigations of the "entrance problem" involve some form of Prandtl's boundary-layer approximation and are based on the boundary-layer equations. Various approximate methods have been proposed to solve these equations. Five different approaches have been used for the analysis of the entry problem: 1) the time-dependent method, 2) series expansions, 3) integral methods, 4) linearization of the inertia terms, and 5) a numerical finite-difference solution.

The time-dependent approach was proposed by Shercliff.[2] He considered a development of velocity profiles with time instead of axial distance. The axial position is then calculated as the time multiplied by the mean flow velocity. As noted by Yang and Yu,[3] this method is not precise and does not agree with the flow in the actual steady-state entrance problem.

The second method was developed by Schlichting[4] for a nonmagnetic flow and was applied to the magnetic case by Roidt and Cess.[5] In this

*Doctoral Student, Department of Applied Mathematics.

method the flowfield in the entrance region is divided into two zones: flow near the inlet (upstream zone) and a downstream zone that approaches the fully developed flow. In the upstream zone the flow near the walls is assumed to be analogous to boundary-layer flow over a flat plate (with the Blasius series solution). The central core is assumed to be inviscid. The velocity profile in the downstream zone is expressed as the fully developed profile, plus a deviation velocity that approaches zero with increasing axial position. The solution for the whole entrance region is then obtained by matching the two solutions. This method is too cumbersome for use. Moreover, Van Dyke[6] noted that the central core cannot really be uniform and showed that Schlichting's method is valid only for conditions far downstream. Because of the preceding reasons, the first and the second approaches are not taken into consideration.

The third approach is the momentum integral boundary-layer analysis. The first solution for nonmagnetic flow in ducts was presented by Schiller.[7] The cross section of the duct was considered to be composed of two regions: a boundary layer developing near the wall and an inviscid fluid core. Schiller proposed a parabolic velocity distribution within the boundary layer and applied the Bernoulli equation in the core to determine the pressure distribution in the axial direction. Other researchers reworked Schiller's solution using a higher order of polynomial representation of velocity within the boundary layer,[8] or taking into account the energy loss because of viscous dissipation.[9–11] These integral methods were then widely used for MHD flows using a simple parabolic[12] or a four-order Pohlhausen profile.[13] This approach is very attractive because of its simplicity and capability of analytically treating certain related problems (such as heat transfer and diffusion). However, this solution contains certain drawbacks that were pointed out by Moffatt.[14] Indeed, the polynomial velocity distribution in the boundary layer asymptotically leads to the Pouiseuille fully developed parabolic profile. However, for the magnetic case the Hartmann profile must finally be reached, which is impossible for polynomial representation. To overcome this problem, Moffatt introduced the Hartmann-like velocity distribution and showed that such a distribution leads to more accurate results and provides solutions without discontinuity, in contrast to those with the polynomial velocity profile (see Ref. 15). Review and validation of integral methods for MHD channel flows were carried out by Heywood and Moffatt[15] and by Vatagin et al.[16]

The fourth method involves a linearization of the inertia terms. Two different approaches have been proposed here. The first approach was developed by Langhaar.[17] The inertia terms were replaced by a term proportional to the axial velocity, and the quasisimilar velocity profiles were obtained. Langhaar then applied the momentum equation evaluated at the centerline and a momentum integral equation to determine the unknown proportional parameter.

Targ, on the other hand, replaced all nonlinear terms by the derivative of the longitudinal velocity and then integrated the linearized momentum equation to obtain the pressure gradient (see Ref. 18 for discussion). Targ's solution provides a velocity profile that develops too slowly near the entrance. Sparrow et al.[18] generalized Targ's linearization by introducing an unknown weighting factor into a constant slug-flow approximation and then used the integral forms of momentum, energy, and linearized momentum equations to determine the velocity profile. This method has been widely applied to the MHD entrance flow problem.[19–21]

The fifth approach involves a reduction of the governing equations to finite-difference form with subsequent numerical solution. Shohet,[22] Shohet et al.,[23] Hwang et al.,[24] and Yang and Yu[3] all used this method to solve the entrance problem under different conditions. A comparison of the results shows that each of the mentioned numerical methods is accurate and may be used to validate an analytical solution. In the present study for a flat channel case we choose the method by Hwang et al.,[24] which is simple and may easily be adapted to various conditions (for an annular channel the method by Shohet[22] is appropriate).

We must note that the velocity profiles obtained by each of the analytical methods mentioned in the preceding paragraphs differ from one another only by a small percentage. This is of little significance for calculations of heat transfer in a channel. However the solutions to other problems such as flow stability and flow transition are very sensitive to a velocity profile. Therefore, to know when each method gives the best results within the parameters of flow, magnetic field, and channel shape is desirable. To compare the results of different publications is difficult, however, because of the variety of criteria used for the conclusion that a fully developed flow is established. It might be the velocity at the centerline of a channel, boundary-layer thickness, wall shear stress, etc. Note that in available reviews (see Refs. 15 and 16) the different solutions were not reduced to unified criteria. Moreover, some methods (for example, Langhaar-like linearization) were not included in the analysis.

The purpose of this work is to validate the available analytical solutions by comparison with numerical solutions and provide the reader with a set of available solutions. To compare different solutions, we analyze them from one point of view and under unified criteria. Such analysis is performed in the present paper for the developing MHD flow in the entrance region of a flat channel with a uniform transverse magnetic field.

In the present analysis, the system for symbolic calculations *Mathematica*[25] was widely used. This modern tool is very convenient in the rearrangements of discussed analytical solutions, and long closed-form final expressions can readily be obtained.

A. Criteria for Validation of Analytical Solutions

To validate various approximate solutions for entry MHD channel flow and to show that these methods are acceptable, the criteria of such validation must be determined. The list of the most general criteria was given by Heywood and Moffatt[15] and includes the following items:

1) The approximate solution must agree with experimental data and/or the exact analytical or numerical solution to the problem.

2) Where no exact solution to the general problem exists, the approximate solution must agree with exact solutions for any limiting cases that may exist.

3) The approximate solutions using several different, but physically reasonable, velocity distributions should show reasonable agreement with one another.

To compare particular approximate solutions, these general criteria must be presented in the more specific form, and the parameters of the comparison will be determined as follows:

1) The length of the developing region L_{95}, L_{98}, and L_{99}, where the subscript shows the percentage of the velocity at the centerline relative to the fully developed velocity;

2) Deviation of velocity profiles from the exact numerical profiles in some cross sections of the channel;

3) Deviation of the centerline velocity along the channel from its exact shape;

4) When close to the channel entrance, the approximate results should be asymptotic to the exact edge solution (i.e., Heywood and Moffatt[15] suggested employing the Blasius solution here, but the last solution is not valid close to the channel entrance, as noted by Schlichting[4]);

5) Ratio of friction factor to its calculated value;

6) Incremental pressure drop K and its final value K_∞;

7) In the limiting case (when $M \to 0$), all of the preceding criteria ($1 \div 6$) must be valid for ordinary hydrodynamic equivalent flow.

In what follows, these criteria are applied to the flat channel entry problem.

II. Formulation of the Flat-Channel Problem

The geometry of the MHD parallel-plate channel is shown in Fig. 1. The height of the channel is 2α and taken to be much smaller than the width of the channel d. The channel must be long compared with the development length and height. In this figure the fluid enters the channel on the left. The origin of the coordinate system is placed at the center of the channel entrance. The variable y increases in an upward direction, and the variable z forms a right-handed coordinate system with x and y axes. The velocity profile develops toward the direction of increasing x.

The external magnetic field is applied in the positive y direction from the bottom to the top of the channel.

Two perfectly conducting plates are placed in planes parallel to the $x = y$ plane at distances of $(d/2)$ and $(-d/2)$ from the origin. These plates enclose the outside of the channel, and, if d is sufficiently large, they will not affect the velocity development at the middle of the channel, as was illustrated by Hughes and Young.[26]

The whole description of the MHD-flow development problem is complicated, and a number of assumptions have been made in order to simplify the problem and to be consistent with the physical geometry of the problem. We consider the incompressible viscous flow of an electrically conducting fluid.

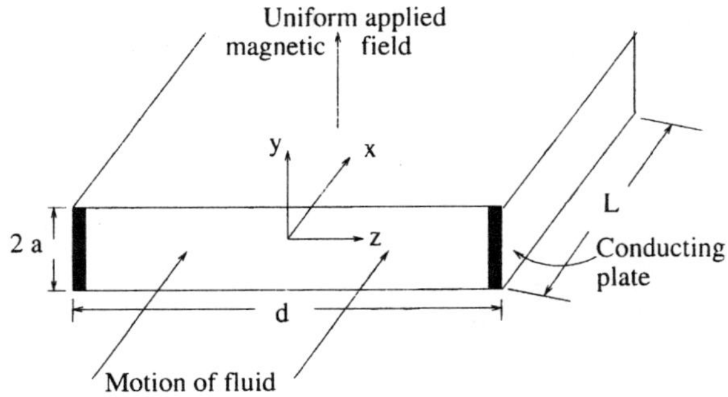

Fig. 1 Geometry of the parallel-plate channel.

The following assumptions are made in the ensuing analysis (see Refs. 1, 5, 13, 24, and 26): 1) the flow is laminar and steady-state; 2) all fluid properties (i.e., mass density ρ, specific heat C_p, thermal conductivity k, kinematic viscosity ν, magnetic permeability μ_e, and electrical conductivity σ) are constant; 3) the hydrodynamic Reynolds number is large; 4) the usual Prandtl boundary-layer assumptions apply; 5) all quantities are constant in the z direction, i.e., the flow is two-dimensional; 6) the effect of gravitational force is negligible; 7) the displacement currents are negligible because the rapid oscillations do not exist and the flow velocity is relatively small; 8) the magnetic Reynolds number in terms of the entry length is small, i.e., the induced magnetic field is negligible in comparison with the applied field $B0$ (which is assumed to be constant and equal) to B_y; 9) the Hall and ion-slip effects are considered negligible; 10) the electric field term E_y measured across the insulated channel walls is zero; 11) the electric field strength in the z direction E_z is constant (and equal to E_0); and 12) the fluid is electrically neutral, i.e., no surplus electrical charge distribution is present in the fluid.

Bearing in mind the preceding assumptions, the equations governing the problem may be written in the following form (for details see analysis performed by Roidt and Cess[5]):

$$\frac{\partial u}{\partial x}+\frac{\partial v}{\partial y}=0 \tag{1}$$

$$u\frac{\partial u}{\partial x}+v\frac{\partial u}{\partial y}=-\frac{1}{\rho}\frac{\mathrm{d}p}{\mathrm{d}x}+\nu\frac{\partial^2 u}{\partial y^2}-\frac{\sigma B_0}{\rho}(E_0+uB_0), \tag{2}$$

where u and v are the velocity components in the x and y directions, respectively, and p is the pressure (which is assumed constant across the channel width).

To decrease the number of parameters and for the sake of uniformity, the governing equations can be nondimensionalized using the following set of dimensionless variables and parameters based upon the characteristic half-width of the channel a and the mean velocity $\overline{u}$:

$$u^* = \frac{u}{\overline{u}}, \quad v^* = \frac{av}{\nu}, \quad x^* = \frac{x}{aRe} = \frac{x\nu}{a^2\overline{u}}, \quad y^* = \frac{y}{a}$$
$$p^* = \frac{p - p_0}{\rho\overline{u}^2}, \quad M = B_0 a\sqrt{\frac{\sigma}{\rho\nu}}, \quad e = -\frac{E_0}{B_0\overline{u}} \tag{3}$$

Then Eqs. (1) and (2) become, respectively,

$$\frac{\partial u}{\partial x} + \frac{\partial v}{\partial y} = 0 \tag{4}$$

$$u\frac{\partial u}{\partial x} + v\frac{\partial u}{\partial y} = -\frac{\mathrm{d}p}{\mathrm{d}x} + \frac{\partial^2 u}{\partial y^2} + M^2(e - u) \tag{5}$$

where, for convenience, the asterisks have been dropped with the understanding that all quantities are now nondimensional.

The preceding equations must be solved subject to appropriate boundary and initial conditions. There are the usual no-slip boundary conditions at the channel walls

$$u(x, \pm 1) = v(x, \pm 1) = 0 \tag{6}$$

and the inlet velocity profile

$$u(0, y) = u_0(y) \tag{7}$$

which is assumed symmetric. Note that some methods of solution for Eqs. (4) and (5) are able to treat the uniform inlet velocity profile only. These and other analytical methods will be discussed in the next section of the paper.

III. Review of Analytical Methods for a Solution of the Parallel-Plate Channel Flow Problem

As mentioned in the Introduction, the time-dependent approach by Shercliff[2] and the series-matching procedure developed by Schlichting[4] and applied to the magnetic case by Roidt and Cess[5] are not included in this analysis because of their restrictions and complexity. Other methods are listed below.

A. Integral Method

First we consider the integral boundary-layer analysis in which it is assumed that the flow consists of a developing boundary layer together with a uniform core. The velocity of the flow's central core may be determined as

$$u_c = \frac{1}{1 + \frac{1 - \sin\text{h}\, M/M}{\cos\text{h}\, M - 1}\delta} \tag{8}$$

Inside the developing boundary layer the most appropriate velocity distribution is the Hartmann-like velocity profile introduced by Moffatt[14]

$$\frac{u}{u_c} = \frac{\cos\text{h}\, M - \cos\text{h}\, M\left(1 - \frac{y}{\delta}\right)}{\cos\text{h}\, M - 1}, \qquad \text{for } 1 - \delta \le y \le 1 \tag{9}$$

The solution presented by Eqs. (8) and (9) becomes complete when the relation between the downstream distance x and the boundary-layer thickness δ is determined. This relation may be obtained in two ways. In the first approach (for example, see Refs. 13, 14, and 27) the integral forms of the continuity and momentum equations are used. When this approach is applied to the MHD flow, the pressure drop in the momentum equation is replaced by its value in the inviscid core of the flow (Bernoulli equation)

$$-\frac{\mathrm{d}p}{\mathrm{d}x} = u_c \frac{\mathrm{d}u_c}{\mathrm{d}x} + M^2(u_c - e) \tag{10}$$

Then, integrating Eq. (5) over the channel cross section, one obtains the following relation for the variation of the boundary-layer thickness δ with the downstream distance x:

$$\frac{\mathrm{d}x}{\mathrm{d}\delta} = \frac{\frac{\mathrm{d}}{\mathrm{d}\delta}\left(\int_0^1 u^2\,\mathrm{d}y - \frac{u_c^2}{2}\right)}{M^2(u_c - 1) - \left(\frac{\partial u}{\partial y}\right)_{y=0}} \tag{11}$$

The final closed-form expression for $x(\delta)$ that completes the first integral solution easily may be obtained from Eqs. (8), (9), and (11) with the aid of *Mathematica*.[25]

The integral boundary-layer solution for the MHD flow problem described in the preceding is adequate for the certain values of the Hartmann number M (see Refs. 14, 15, and 27). This solution is also correct in the vicinity of the inlet where the central core remains unchanged, and the boundary layer is very similar to that developing over a flat plate (Blasius solution). However, for the limiting case of small values of M, the above solution leads to the Schiller solution,[7] which is incorrect when far away from the channel entrance. This error was pointed out by Campbell and Slattery[9] and is because of viscous dissipation in the central core, which was not taken into account by Schiller and other investigators using the integral momentum equation.[11]

Fargie and Martin[11] have shown that the methods that neglect viscous dissipation predict the entrance length L_{99} for nonmagnetic initially uniform flow as a half of its exact value, obtained as a result derived from a sound numerical solution of the basic Navier–Stokes equations. Those methods that take the viscous dissipation into account predict much more accurate results for the entrance length.

In light of the preceding material, we present here a new integral method that takes into consideration the effect of viscous dissipation.

The analysis is patterned after a method developed by Bhatti and Savery[10] for nonmagnetic flow, in which the mechanical energy equation is used. The pressure gradient is determined at the demarcation line between the boundary layer and the central core. Because of its relative simplicity, this method is more attractive than that derived by Campbell and Slattery,[9] where the macroscopic mass balance, momentum balance, and mechanical energy equations are utilized simultaneously. In the present analysis the mechanical energy equation is obtained by multiplying the momentum equation (5) by the velocity u and integrating over the cross section. This yields

$$\frac{\mathrm{d}}{\mathrm{d}x}\int_0^1 \frac{u^3}{2}\,\mathrm{d}y = \left(M^2 e - \frac{\mathrm{d}p}{\mathrm{d}x}\right) - \int_0^1 \left(\frac{\partial u}{\partial y}\right)^2 \mathrm{d}y - M^2 \int_0^1 u^2\,\mathrm{d}y \tag{12}$$

The first term on the RHS of Eq. (12) is determined at the edge of the boundary layer, where $y = 1 - \delta$ and $u = u_{\mathrm{bound}}$ and leads to the following expression

$$M^2 e - \frac{\mathrm{d}p}{\mathrm{d}x} = -\frac{\partial^2 u_{\mathrm{bound}}}{\partial y^2} + M^2 u_{\mathrm{bound}} \tag{13}$$

Substituting Eq. (13) into Eq. (12) and applying the Hartmann-like velocity distribution (9), the variation of the thickness of boundary layer δ with the axial distance x is obtained, completing the solution.

B. Linearized Analysis

This approach includes linearization of the nonlinear inertia terms on the LHS of the momentum equation.[5] Two types of linearization are usually used. The first was suggested by Langhaar[17] for the nonmagnetic case and then derived by Sue Min-Lin[28] for the MHD channel flow. The inertia terms on the LHS of the momentum equation (5) are replaced by

$$u\frac{\partial u}{\partial x} + v\frac{\partial u}{\partial y} = (\gamma^2 - M^2)u \tag{14}$$

where the function $\gamma(x) > M$ and the solution of the linearized equation is obtained in the form

$$u = \frac{\cosh\gamma - \cosh(\gamma y)}{\cosh\gamma - \sinh\gamma/\gamma} \tag{15}$$

To fully specify the velocity distribution, to determine the relation between γ and x is necessary. To do this, the momentum equation is evaluated at the centerline of the channel (for $y = 0$) and also integrated over the cross section of the channel.

Eliminating the pressure gradient, one obtains the variation of x with γ as

$$\frac{\mathrm{d}x}{\mathrm{d}\gamma} = \frac{\dfrac{\mathrm{d}}{\mathrm{d}y}\left(\displaystyle\int_0^1 u^2\,\mathrm{d}y - \frac{u_0^2}{2}\right)}{M^2(u_0 - 1) + \left(\dfrac{\partial u}{\partial y}\right)_{y=1} - \left(\dfrac{\partial^2 u}{\partial y^2}\right)_0} \tag{16}$$

One should note that all velocity profiles [Eq. (15)] are Hartmann-like profiles with various values of parameter γ. Finally the Hartmann fully developed velocity distribution is achieved. The pressure drop is determined on the basis of the mechanical energy equation (12) (see Appendix A).

The second and most frequently used approach is based on the analysis by Sparrow et al.,[18] where the following linearization was introduced:

$$u\frac{\partial u}{\partial x} + v\frac{\partial u}{\partial y} + \frac{\mathrm{d}p}{\mathrm{d}x} = \varepsilon\frac{\partial u}{\partial x} - F(x) \tag{17}$$

where $\varepsilon(x)$ and $F(x)$ are yet unknown functions to be determined later. Substituting Eq. (17) into Eq. (5) and integrating over the channel cross section, the function $F(x)$ is eliminated, and the following linear equation is obtained[20]

$$\frac{\partial u}{\partial \overline{x}} = \frac{\partial^2 u}{\partial y^2} - \left(\frac{\partial u}{\partial y}\right)_{y=1} + M^2(1 - u) \tag{18}$$

where the stretched axial coordinate $\overline{x}$ is related to the physical coordinate x through a stretching factor $\varepsilon(x)$ by

$$x = \int_0^{\overline{x}} \varepsilon\,\mathrm{d}\overline{x} \tag{19}$$

The solution for the velocity u satisfying the boundary conditions (6) and the symmetric initial conditions (7) is as follows:

$$u(\bar{x}, y) = u_f + \sum_{1}^{\infty} A_i \left(\frac{\cos b_i y}{\cos b_i} - 1 \right) e^{-(b_i^2 + M^2)\bar{x}} \tag{20}$$

where the fully developed profile u_f is the Hartmann solution

$$u_f = \frac{\cosh M - \cos \mathrm{h}(My)}{\cosh M - \sin \mathrm{h}\, M/M} \tag{21}$$

Coefficients b_i are the roots of the equations

$$b_i = \tan b_i \tag{22}$$

Coefficients A_i depend on the inlet velocity profile as follows:

$$A_i = \frac{2}{b_i^2} \int_0^1 \left[u_0(y) - u_f \right] \left(\frac{\cos b_i y}{\cos b_i} - 1 \right) \mathrm{d}y \tag{23}$$

and for the uniform inlet velocity profile

$$A_i = \frac{2}{\left(b_i^2 + M^2 \right)} \tag{24}$$

The stretching factor $\varepsilon(\bar{x})$ may be written in the simplified form derived by Wiginton and Wendt[29]

$$\varepsilon(\bar{x}) = \frac{\dfrac{\mathrm{d}}{\mathrm{d}\bar{x}} \displaystyle\int_0^1 u^3 \, \mathrm{d}y}{\dfrac{\mathrm{d}}{\mathrm{d}\bar{x}} \displaystyle\int_0^1 u^2 \, \mathrm{d}y} - 2 \tag{25}$$

The pressure drop along the channel between the entrance section and any downstream section is found by integrating Eq. (5) over the cross section and along the length of the channel. The result is

$$p_0 - p = M^2 \left(\frac{\cosh M}{\cosh M - \sinh M / M} - e \right) x + K(M, x) \tag{26}$$

where $K(M, x)$ is the incremental pressure drop because the flow development depends on the inlet velocity profile (see Appendix B). Note that this approach is very attractive because it enables one to consider the nonuniform inlet velocity profile and to obtain the closed-form analytical solution for the coupled thermal-convection problem.[20,21]

C. Finite-Difference Solutions

As shown by Setayesh and Sahai,[30] in the case of constant transport properties the simple finite-difference schemes used by Shohet et al.[23] and Hwang et al.[24] are appropriate. A comparison of these two numerical schemes carried out in the framework of the present paper shows that the method used by Hwang et al.[24] is slightly more accurate than that employed by Shohet et al.[23] For this reason the numerical method by Hwang et al.[24] is chosen to validate the approximate analytical solutions. The governing equations are written in implicit finite-difference form, and calculations advance column by column along the mesh network until the flow becomes fully developed.

IV. Results and Discussion

A comparison of the different analytical solutions is performed by analysis of the variations in the centerline velocity and the wall friction factor along the channel. Initial velocity profile at the channel entry is assumed uniform. We must note that among the discussed analytical methods the linearization procedure by Chen and Chen[20] is only able to treat something other than uniform initial velocity profile (such as a frequently used parabolic profile, for example). The application of the integral boundary-layer method to the nonuniform initial velocity distribution case does not appear to be justified because the boundary layer does not occur due to the absence of discontinuity of boundary conditions here. Hence, the physical meaning of the boundary-layer thickness δ in this case (see Ref. 12) remains unclear.

Figure 2 shows the variation of the centerline velocity along the channel for three different values of the Hartmann number M (such as $M = 1, 4$, and 10). These quantities are commonly used for a comparison of results. In this case we do not consider extremely large values of the Hartmann number M because of its complexity. The entry region becomes very short, the pressure distribution in the channel cross section is not uniform, and the second derivatives cannot be neglected in the governing equations, which means that the full Navier–Stokes equations must be discussed and the numerical solution of simplified governing equations (4 and 5) may be inadequate in this case.

As one can see in Fig. 2, all analytical methods predict faster flow development than the numerical calculations. Figure 2a confirms the conclusion that if the developing region is relatively long, then the second integral method (which takes into account the viscous dissipation) is better than the first integral methods, which are completely unsuitable in the far-field. However, in the near field close to the inlet the simplest first integral method provides the most appropriate results, whereas the second integral method fails. At the initial stage of flow development, the central core remains undisturbed and does not affect the developing boundary layer. As a result,

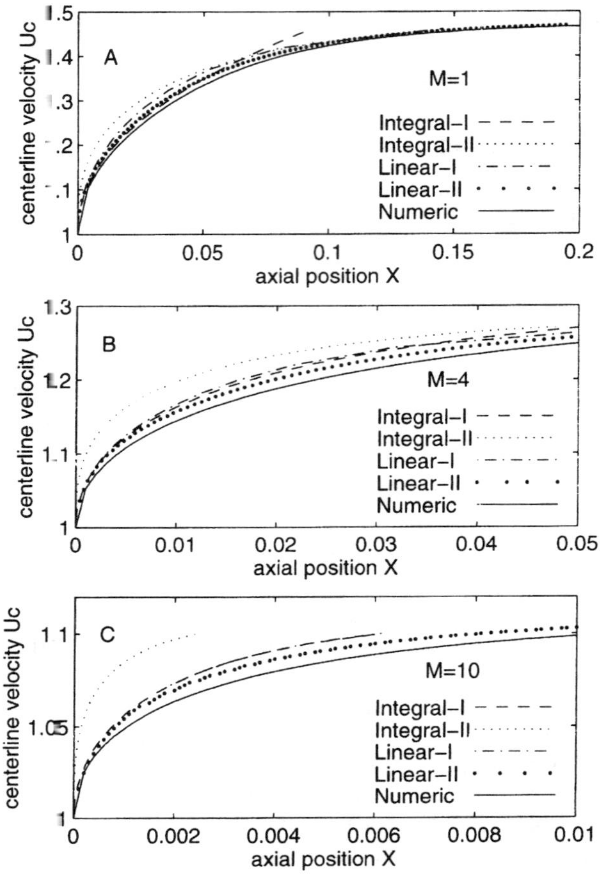

Fig. 2 Centerline velocity variation along the channel: *A*, weak magnetic field; *B*, intermediate magnetic field; and *C*, strong magnetic field.

the first-type linearization[28] that includes the immediate acceleration of the central core also fails. The second-type linearization[20] is appropriate in the entire developing region for small values of the Hartmann number M.

When the Hartmann number M increases (see Figs. 2b and 2c), the entry region decreases (approximately as $1/M^2$), and the second integral method becomes unacceptable. At the same time the first-type linearization,[28] which consists of the family of Hartmann-like quasisimilar profiles, provides the same results as the first integral method. However, these two methods predict a very short length of the entry region. The second-type linearization is better than other methods, although it is also flawed in that it predicts relatively fast flow development.

Another important parameter of MHD-flow development is the friction factor at the channel wall. Figure 3 presents the variation of the friction

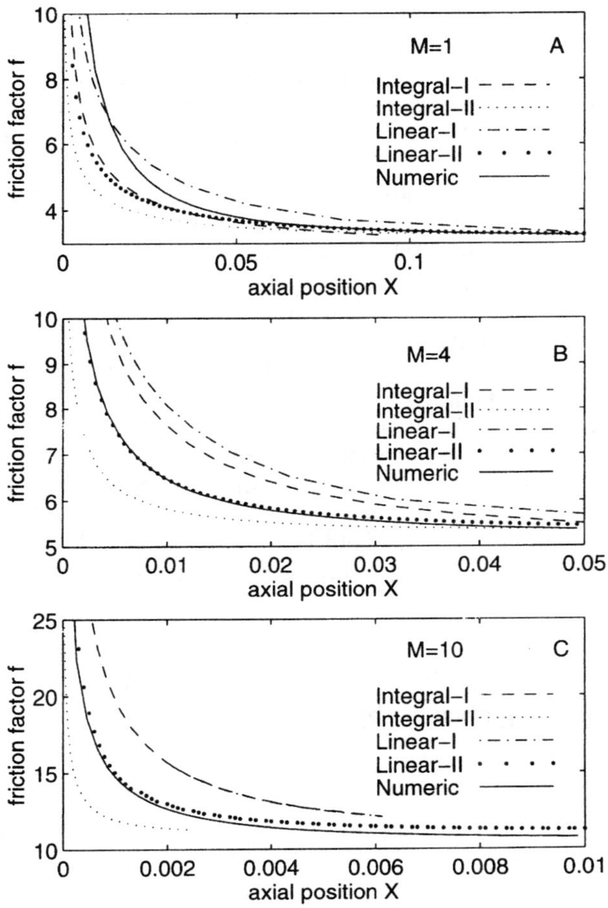

Fig. 3 Friction factor variation along the channel: *A*, weak magnetic field; *B*, intermediate magnetic field; and *C*, strong magnetic field.

factor along the channel for various values of the Hartmann number M. When the magnetic field is weak (see Fig. 3a), the first-type linearization agrees with the results of numerical calculations in the vicinity of the channel entry. The results of the first integral method and the second-type linearization are similar, but their predictions lie below the numerical curve. In the far field the second-type linearized method results match those of the numerical calculations, whereas the curve of the first integral method runs away. The predictions of the second integral solution are far from the reality in the near field, but very accurate in the far field, whereas the accuracy of the first-type linearization is low.

Figure 3b shows that for intermediate values of M the second-type linearization method by Chen and Chen[20] is excellent in each zone of the flow whereas all other methods fail. However, in the far field the results obtained using the second integral approach are very close to the numerical curve.

For large values of the Hartmann number M, all methods fail, apart from the second-type linearization that agrees with the numerical solution in the near field (see Fig. 3c). However, in the farfield the accuracy of this method is low.

The present graphical results make it clear that the second-type linearization method developed by Chen and Chen[20] appears to be the best available analytical solution to the MHD-flow development in a channel. The additional advantages of this method may be used to treat the heat transfer problem and to account for the nonuniform initial velocity distribution without any modification of the method (see Ref. 21). However, this method predicts very low local friction at the channel wall, near the inlet, in the presence of a weak magnetic field.

The first integral boundary-layer method that uses the velocity distribution proposed by Moffatt[14] is appropriate for predicting velocity close to the channel entry in the presence of weak and intermediate magnetic fields. All other methods are less accurate and may be used for qualitative predictions only.

When the applied magnetic field becomes stronger, the accuracy of all analytical solutions is reduced, and these solutions cannot be substituted for the numerical solution.

Appendix A: The Pressure Drop

The pressure drop obtained by Langhaar's procedure [see Eqs. (12) and (15)]:

$$p_0 - p = \int_0^1 \frac{u^3 - 1}{2}\,\mathrm{d}y - (M^2 e)x + \int_0^x \left\{ \int_0^1 \left(\frac{\partial u}{\partial y} \right)^2 \mathrm{d}y + M^2 \int_0^1 u^2\,\mathrm{d}y \right\} \mathrm{d}x$$

Appendix B: The Incremental Pressure Drop

The incremental pressure drop for the case of uniform inlet velocity distribution [see also Eqs. (19), (22), and (25)]

$$K(M,x) = \frac{M^2 + M\cosh(M)\sinh(M) - 2\sinh(M)^2}{2\left(M\cosh(M) - \sinh(M)\right)^2}$$

$$+ 2\sum_1^\infty \frac{b_i^2}{\left(b_i^2 + M^2\right)^2} e^{-\left(b_i^2+M^2\right)^{2\bar{x}}} \left(e^{-\left(b_i^2+M^2\right)^{2\bar{x}}} - 2 \right)$$

$$+ 2\sum_1^\infty \frac{b_i^2}{\left(b_i^2 + M^2\right)^2} \int_0^{\bar{x}} e^{-\left(b_i^2+M^2\right)^{2\bar{x}}} \varepsilon\,\mathrm{d}\bar{x}$$

References

[1]Branover, H., *Magnetohydrodynamic Flow in Ducts*, Wiley, New York, 1978.

[2]Shercliff, J. A., "Entry of Conducting and Non-Conducting Fluids in Pipes," *Cambridge Philosophical Society*, Vol. 52, No. 3, 1956, pp. 573–583.

[3]Yang, H. K., and Yu, C. P., "Combined Forced and Free Convection MHD Channel Flow in Entrance Region," *International Journal of Heat and Mass Transfer*, Vol. 17, 1974, pp. 681–691.

[4]Schlichting, H., "Laminare Kanaleinlaufstromung," *Z. Angew, Math. Mech.*, Vol. 14, 1934, pp. 368–373.

[5]Roidt, M., and Cess, R. D., "An Approximate Analysis of Laminar MHD Flow in the Entrance Region of a Flat Duct," *Journal of Applied Mechanics*, Vol. 84, Mar. 1962, pp. 171–176.

[6]Van Dyke, M., "Entry Flow in a Channel," *Journal of Fluid Mechanics*, Vol. 44, 1970, pp. 813–823.

[7]Schiller, L., "Die Entwicklung der Laminaren Geschwindigkeitsverteilung und ihre Bedeutung fur Zahigkeitsmessungen," *Z. Angew. Math. Mech.*, Vol. 2, 1922, pp. 96–106.

[8]Siegel, R., "The Effect of Heating on Boundary Layer Transition for Liquid Flow in a Tube," D.Sc. Thesis, Massachusetts Inst. of Technology, 1953.

[9]Campbell, W. D., and Slattery, J. C., "Flow in the Entrance of a Tube," *Journal of Basic Engineering*, Vol. 85, Mar. 1963, pp. 41–46.

[10]Bhatti, M. S., and Savery, C. W., "Heat Transfer in the Entrance Region of a Straight Channel: Laminar Flow with Uniform Wall Heat Flux," *Journal of Heat Transfer*, Vol. 99, 1977, pp. 142–144.

[11]Fargie, D., and Martin, B. W., "Developing Laminar Flow in a Pipe of Circular Cross-Section," *Proceedings of the Royal Society of London, Series A*, Vol. 321, 1971, pp. 461–476.

[12]Maciulaitis, A., and Loeffler, A. L., Jr., "A Theoretical Investigation of MHD Channel Entrance Flow," *AIAA Journal*, Vol. 2, No. 12, 1964, pp. 2100–2103.

[13]Dhanak, A. M., "Heat Transfer in Magnetohydrodynamic Flow in an Entrance Region," *Journal of Heat Transfer*, Vol. 87, May 1965, pp. 231–236.

[14]Moffatt, W. C., "Analysis of MHD Channel Entrance Flows Using the Momentum Integral Method," *AIAA Journal,* Vol. 2, No. 8, 1964, pp. 1495–1497.

[15]Heywood, J. B., and Moffatt, W. C., "Validity of Integral Methods in MHD Boundary-Layer Analyses," *AIAA Journal,* Vol. 3, No. 8, 1965, pp. 1565–1567.

[16]Vatagin, A., Lubimov, G., and Regirer, S., *MHD Channel Flows*, 1970, pp. 274–352 (in Russian).

[17]Langhaàr, H. L., "Steady Flow in the Transition Length of a Straight Tube," *Journal of Applied Mechanics*, Vol. 9, 1942, pp. 55–58.

[18]Sparrow, E. M., Lin, S. H., and Lundgren, T. S., "Flow Development in the Hydrodynamic Entrance Region of Tubes and Ducts," *Physics of Fluids*, Vol. 7, No. 3, 1964, pp. 338–347.

[19]Snyder, W. T., "Magnetohydrodynamic Flow in the Entrance Region of a Parallel-Plate Channel," *AIAA Journal*, Vol. 3, No. 10, 1965, pp. 1833–1838.

[20]Chen, T. S., and Chen, G. L., "Magnetohydrodynamic Channel Flow with an Arbitrary Inlet Velocity Profile," *Physics of Fluids*, Vol. 15, No. 8, 1972, pp. 1531–1534.

[21]Yu, C. P., and Hendrix, C. E., "Linearized Analysis of Magnetohydrodynamic Entrance Flow in a Vertical Channel with Combined Thermal Convection," *Applied Scientific Research*, Vol. 33, Nov. 1977, pp. 369–384.

[22]Shohet, J. L., "Velocity and Temperature Profiles for Laminar Magnetohydrodynamic Flow in the Entrance Region of an Annular Channel," *Physics of Fluids*, Vol. 5, No. 8, 1962, pp 879–884.

[23]Shohet, J. L., Osterle, J. F., and Young, F. J., "Velocity and Temperature Profiles for Laminar Magnetohydrodynamic Flow in the Entrance Region of a Plane Channel," *Physics of Fluids*, Vol. 5, No. 5, 1962, pp. 545–549.

[24]Hwang, C. L., Li, K. C., and Fan, L. T., "Magnetohydrodynamic Channel Entrance Flow with Parabolic Velocity at the Entry," *Physics of Fluids*, Vol. 9, No. 6, 1966, pp. 1134–1140.

[25]Wolfram, S., *Mathematica*, Addison–Wesley, Reading, MA, 1991.

[26]Hughes, W. F., and Young, F. J., *The Electromagnetodynamics of Fluids*, R. E. Krieger Publishing Co., Malabar, FL, 1989, Chap. 7.

[27]Tan, C. W., "Laminar MHD Channel Entrance Flows," *AIAA Journal*, Vol. 3, No. 7, 1965, pp. 1369–1371.

[28]Min-Lin, S. "Laminar MHD Flow in the Entrance Region of a Flat Channel," *Acta Mechanica Sinica*, Vol. 6, No. 2, 1963, pp. 168–170.

[29]Wiginton, C. L., and Wendt, R. L., "Flow in the Entrance Region of Ducts," *Physics of Fluids*, Vol. 12, No. 4, 1969, pp. 465–466.

[30]Setayesh, A., and Sahai, V., "Heat Transfer in Developing Magnetohydrodynamic Pouiseuille Flow and Variable Transport Properties," *International Journal of Heat and Mass Transfer*, Vol. 33, No. 8, 1990, pp. 1711–1720.

Chapter 56

Oscillatory Convection in a Cylindrical Cavity and Stability of Magnetohydrodynamic Hadley Flows

H. Ben Hadid,* S. Kaddeche,† and D. Henry*
Ecole Centrale de Lyon/Université Claude Bernard-Lyon 1
Ecully Cedex, 69131, France

I. Introduction

SOLIDIFICATION of metallic alloys or semiconductors is often accompanied by oscillatory flows that affect the crystal quality by creating undesirable striations. The origin of such oscillations for moderate Prandtl number fluids is still not well understood. A number of experimental studies have been done to identify the conditions for the appearance of such oscillations.[1–4] Hurle et al.[1] have shown that when the horizontal imposed temperature gradient ∇T is increased beyond a critical value ∇T_c thermal oscillations appear spontaneously even in the absence of solidification. So investigating hydrodynamic and/or thermal instabilities without solidification seems reasonable to aid in understanding the appearance of such oscillations.

Some theoretical works[5–9] have been performed in this field. For example, Gill[5] has developped a theory based on a normal mode analysis to investigate the conditions for the marginal oscillations to appear in a cavity of infinite cross section. This theory provides results comparing within a factor of two with the numerical results found by Hart[6,7] for the case of

---

*Senior Researchers, Laboratoire de Mécanique des Fluides et d'Acoustique UMR CNRS 5509.

†Doctoral Student, Laboratoire de Mécanique des Fluides et d'Acoustique UMR CNRS 5509 ECL, BP 163.

rigid conducting boundaries cavity. Smith and Davis[8] have investigated the stability of thermocapillary flows generated by a horizontal temperature gradient imposed along a thin liquid layer and found that the unsteady hydrothermal waves involve a new mechanism of instability deriving its energy from the horizontal thermal gradients. Laure and Roux[9] generalized the stability study as a function of the Prandtl number for different boundary conditions.

A possibility to suppress these oscillations can be obtained by the use of a magnetic field. Hurle et al.[1] have investigated the effect of a constant and uniform magnetic field on the threshold conditions and have shown that the critical Rayleigh number increases linearly with the square of the Hartmann number. Despite this potentiality, only few theoretical and numerical works have been devoted to the control of the instabilities by a magnetic field. In the case of thermocapillary flows, we can mention the recent work of Priede et al.[10]

In this paper we first study the convective flows appearing in long confined cavities corresponding to the horizontal Bridgman configuration, and particularly in the cylinders. We present the characteristics of the stationary flows and different types of oscillatory flows that emerge when increasing the Grashof number. We focus on the symmetry properties of such flows. Concerning the action of a magnetic field, we follow the treatment of Laure and Roux[9] to study the evolution of the instability onset as a function of the applied magnetic field for a rigid cavity of infinite cross section. We consider the case of a constant magnetic field either vertical or transverse.

II. Mathematical Model

We consider a cavity filled with an electrically conducting liquid metal submitted to a horizontal temperature gradient that generates a laminar convective motion.

The basic equations for the melt flow submitted to a constant magnetic field are the Navier–Stokes equations including the Lorentz force $\boldsymbol{J} \times \boldsymbol{B}$ and the Boussinesq approximation for buoyancy because the metallic liquid is considered incompressible. The three-dimensional conservation equations of momentum and energy are made dimensionless using H, H^2/ν, $U_{\text{ref}} = Gr\nu/H$, ∇TH, $\boldsymbol{B}_0$, and $\sigma_e U_{\text{ref}} \boldsymbol{B}_0$ as scales for length, time, velocity, temperature, magnetic flux density, and electric current density, respectively (H is the diameter of the cylindrical cavity for the direct numerical simulations or the height of the fluid layer for the linear stability study, ν the kinematic viscosity, σ_e the electric conductivity of the melt, and ∇T the diffusive temperature gradient). Using these reference quantities and the frame of reference given in Fig. 1, the governing equations in the melt may be written as

$$\nabla \cdot \boldsymbol{v} = 0 \tag{1}$$

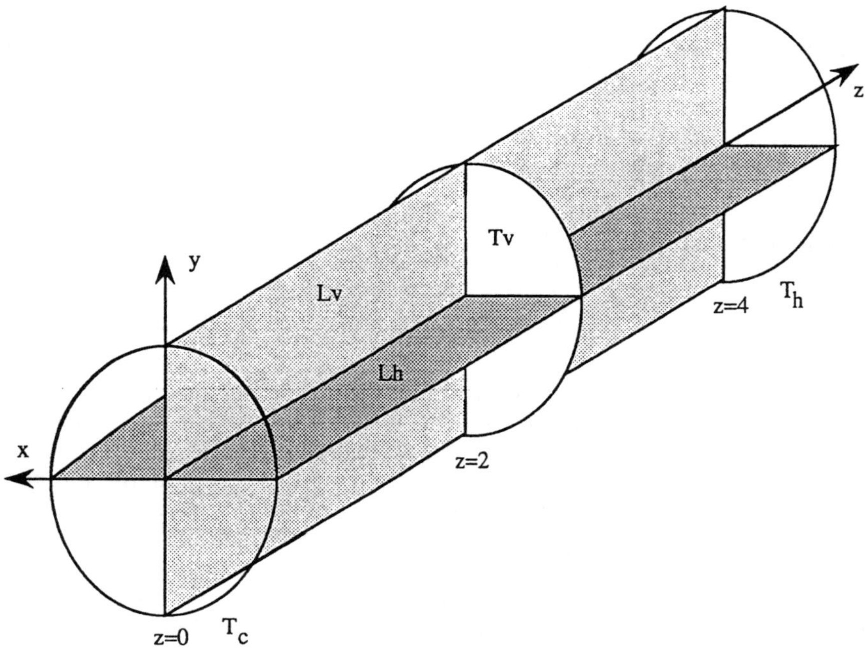

Fig. 1 Geometry of the cylindrical cavity and of the frame of reference.

$$\frac{\partial \boldsymbol{v}}{\partial t} + Gr(\boldsymbol{v} \cdot \nabla)\boldsymbol{v} = -\nabla p + \nabla^2 \boldsymbol{v} + \theta \boldsymbol{e}_y + Ha^2 \boldsymbol{J} \times \boldsymbol{e}_{B_0} \tag{2}$$

$$\frac{\partial \theta}{\partial t} + Gr(\boldsymbol{v} \cdot \nabla \theta) = \frac{1}{Pr} \nabla^2 \theta \tag{3}$$

where $\boldsymbol{v}$ is the dimensionless velocity field [$\boldsymbol{v}$ = (u,v,w)], p denotes the dimensionless static pressure, $\boldsymbol{e}_{B_0}$ 2 the unit vector in the direction of $\boldsymbol{B}_0$, and θ is the dimensionless temperature defined as $\theta = (T - T_m)/(\nabla TH)$ where T_m is the mean temperature in the cavity. The magnetic flux density vector $\boldsymbol{B}$ in the Lorentz force is set equal to $\boldsymbol{B}_0$ (the applied field) as the induced field $\boldsymbol{b}$ is small ($Re_m \ll 1$, where Re_m is the magnetic Reynolds number). The dimensionless electric current density J is given by Ohm's law for a moving fluid:

$$\boldsymbol{J} = \boldsymbol{E} + \boldsymbol{v} \times \boldsymbol{e}_{B_0} \tag{4}$$

where $\boldsymbol{E}$ is the dimensionless electric field. Here, because $Re_m \ll 1$, the electric field can be written as the gradient of an electric potential ($\boldsymbol{E} = -\nabla\varphi$). The equation of conservation for the electric current density gives

$$\nabla \cdot \boldsymbol{J} = 0 \tag{5}$$

which, combined with Eq. (4), leads to

$$\nabla^2\varphi = \nabla \cdot \left(\boldsymbol{v} \times \boldsymbol{e}_{\boldsymbol{B_0}}\right) \tag{6}$$

The boundary conditions for the velocity are the usual no-slip conditions $u = v = w = 0$ at all of the walls. For the electric current density $\boldsymbol{J}$, insulated conditions $\boldsymbol{J} \cdot \boldsymbol{n} = 0$ are adopted at all of the boundaries.

The dimensionless parameters appearing in Eqs. (2) and (3) are the Grashof number $Gr = g\beta\nabla TH^4/\nu^2$, the Prandtl number $Pr = \nu/\kappa$, and the Hartmann number $Ha = \boldsymbol{B_0}H\sqrt{\sigma_e/\rho\nu}$, where β is the coefficient of volumetric expansion and κ the thermal diffusivity.

The numerical simulations are performed in a cylindrical cavity of aspect ratio $A = L/H$ with isothermal endwalls and insulated lateral walls (i.e., $\nabla T \cdot \boldsymbol{n} = 0$) (Fig. 1). The Prandtl number is set equal to 0.026, a value corresponding to experiments with mercury.[11] The results presented here have been obtained without magnetic field ($Ha = 0$). The spatial discretization of the governing equations (1–6) is based on an isoparametric spectral element method,[12] which is a high-order weighted residual technique. The cylindrical geometry is divided in five subdomains; in each one $9 \times 9 \times Nz$ (following x, y, and z, respectively) Gauss–Lobatto–Legendre collocation points were used. The number of collocation points following the z direction (the axis of the cylinder) is increased from 35 to 51 when the aspect ratio is changed from 4 to 10. The time discretization of Eqs. (2) and (3) employs a high-order time-differencing scheme combined with a splitting algorithm.[13]

The action of the magnetic field is studied by a linear analysis in normal modes in the simplified case of monodimensional Hadley-type flows. The different instabilities are characterized as a function of the Hartmann number Ha, the Prandtl number Pr, and the thermal properties (conducting or insulating) of the horizontal boundaries.

III. Numerical Simulation of Convection in Circular Cylinders

The flow is studied without a magnetic field for Grashof numbers up to 5×10^5 and two aspect ratios $A = 4$ and $A = 10$. Transition to time-dependent flows is obtained when the Grashof number is increased beyond some critical value that is found to depend on the aspect ratio.

We need now to define different planes inside the circular cylinder used to characterize the results (Fig. 1): the (Lv) plane (longitudinal vertical

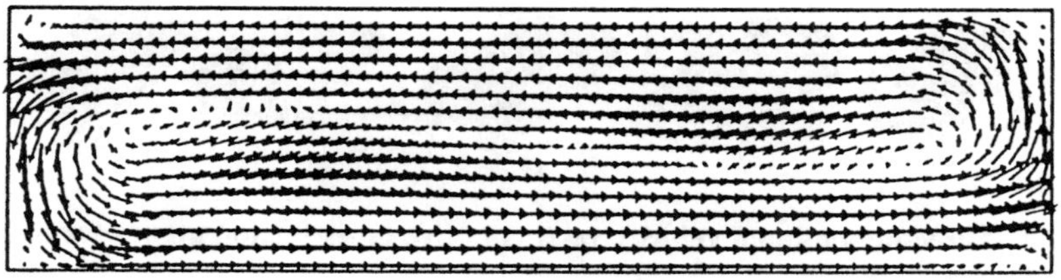

Fig. 2 Velocity vectors in the (*Lv*) plane for a steady situation: $A = 4$ and $Gr = 1.5 \times 10^5$.

middle plane, main plane of circulation), the (*Lh*) plane (longitudinal horizontal middle plane), and the (*Tv*) plane (transverse vertical middle plane). The original symmetries in such a confined cavity are a reflection symmetry (*Sp*) with respect to the (*Lv*) plane and a reflection symmetry (*Sa*) with respect to the (*Th*) line (transverse horizontal centerline). The combination of these two symmetries also gives a symmetry (*Sc*) with respect to the center point of the cavity. In the steady situations these symmetries are found to remain valid.

Steady-State Flows

For the steady-state regime the buoyancy flow in the cylindrical cavity corresponds to two horizontal counterflows moving from the hot wall above and from the cold wall below. A progressive increase of the Grashof number produces progressive modifications in the flow structure. In the (*Lv*) plane (Fig. 2) the main feature is the onset of secondary vortices in the end regions inside the large convective loop. These vortices, already visible for $Gr = 5 \times 10^4$, are fully developed for $Gr = 1.5 \times 10^5$ where they clearly affect the circulation in the (*Lh*) plane (Fig. 3). Small recirculations appear also in the vicinity of the bottom corner of the hot wall and at the top corner of the cold wall (Fig. 2). Further modifications corresponding to the development of the three-dimensional character of the flow are visible in Fig. 4 where fully three-dimensional velocity vectors field at five cross-section planes is given. In this figure clearly the flow is

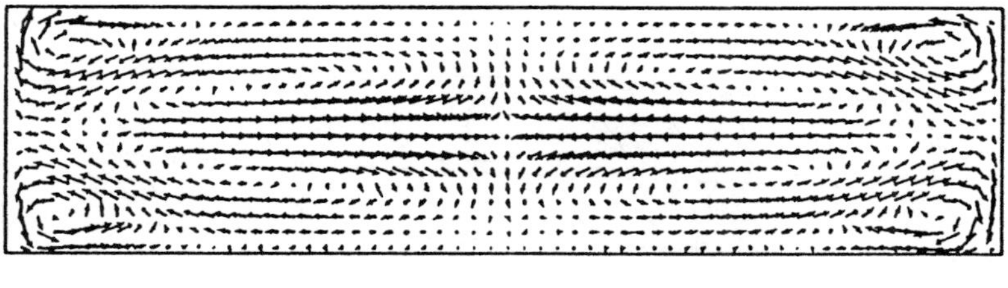

Fig. 3 Velocity vectors projection in the (*Lh*) plane for a steady situation: $A = 4$ and $Gr = 1.5 \times 10^5$.

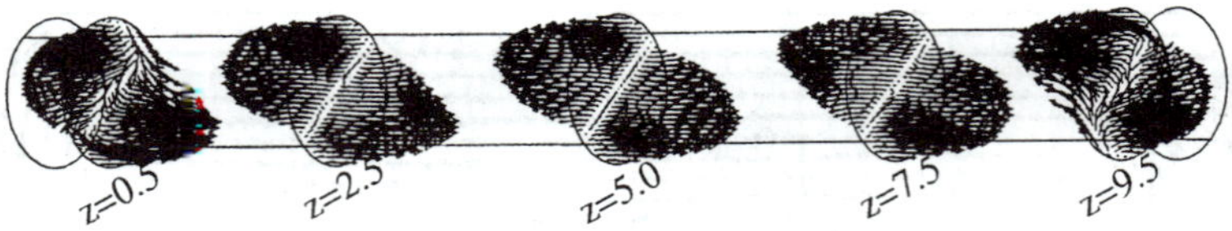

Fig. 4 Three-dimensional velocity vectors in five cross-section planes for a steady situation: $A = 10$ and $Gr = 3 \times 10^5$.

rather concentrated near the (Lv) plane when it leaves the end walls and then spreads out over the whole width.

From Fig. 5 where the isotherm lines are presented for three horizontal planes, it is clear that the temperature field is affected by the convective flow. However, despite the large Gr numbers the isotherms are slightly deformed because of the fact that the fluid considered is a good thermal conductor ($Pr = 0.026$). In contrast to the case $A = 4$ where the effect of the end recirculations concerns the whole length of the cavity for $A = 10$, the central region is not affected by these recirculations. In fact, Fig. 5 clearly shows that in the central region the isotherm lines are nearly invariant indicating that the flow is well-established.

Oscillatory Flows

When the Grashof number is increased beyond a certain critical value Gr_c, the convective flow in the cavity becomes oscillatory. We have then simulated different situations beyond the thresholds and tried to characterize them.

Different types of instabilities are found depending on the aspect ratio of the cavity. For long cavities ($A = 10$) the oscillatory flow seems to be initiated in the central region (Fig. 6) where longitudinally invariant velocity profiles are obtained in the steady situation. These perturbations, which could be a consequence of the shear stresses existing in these zones, propagate then in the whole cavity (Figs. 6b and 6c). They lead to the breaking of the main symmetries, the symmetry Sp with respect to the main plane of circulation and Sa with respect to the center transverse axis. Only the symmetry Sc with respect

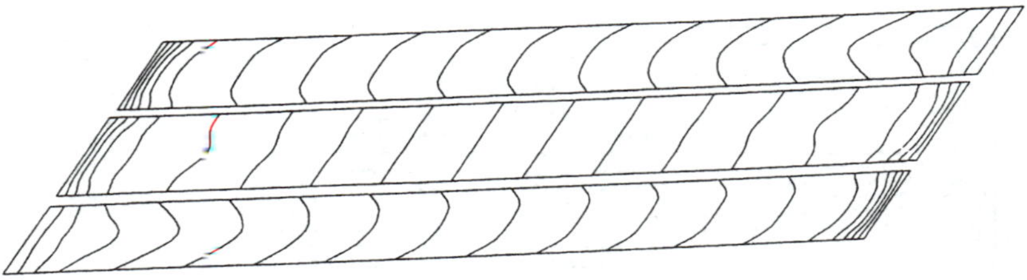

Fig. 5 Isotherm lines in three horizontal planes inside the cavity for a steady situation: $A = 10$ and $Gr = 3 \times 10^5$.

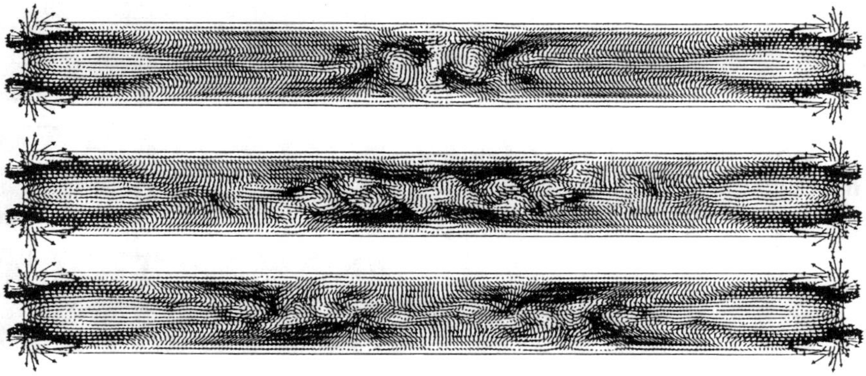

Fig. 6 Velocity vectors projection in the (Lh) plane at three different times for an oscillatory situation: $A = 10$ and $Gr = 5 \times 10^5$.

to the center point of the cavity is maintained. This instability that affects the direction transverse to the flow (breaking of *Sp*) is rather of a three-dimensional type, but, as it seems to concern a thin layer in the core region, it does not disturb the global structure of the flow, which remains unicellular (Fig. 7).

For shorter investigated cavities ($A = 3$ and 4), another type of instability appears first (Figs. 8–10). This instability seems to be connected to the recirculation flows existing in the end parts of the cylinder. These recirculations extend up to the core region where they interact together creating a stagnation zone. The instability of this stagnation zone seems to be for smaller cylinders the reason for the onset of oscillatory flows. This instability is rather of a two-dimensional type, affecting mainly the longitudinal direction. The instability breaks then the symmetries with respect to the center of the cavity, the *Sa* and *Sc* symmetries, but keeps the *Sp* symmetry, which affects the transverse direction (Figs. 9 and 10).

The experimental results of Hurle et al.[1] have shown a decrease of the critical Grashof number as the aspect ratio is increased from 2 to 5. Here, in contrast, we can note that at $Gr = 2 \times 10^5$ the flow for shorter cavities is already time dependent whereas it is still steady for $A = 10$. This behavior could be connected to the fact that the thresholds correspond in fact to two different types of instability.

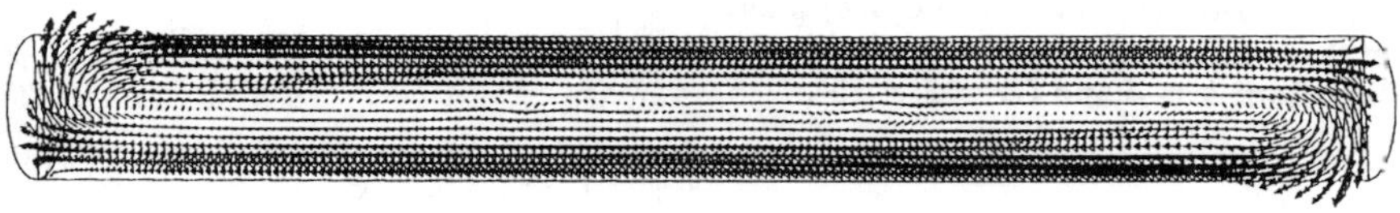

Fig. 7 Velocity vectors in the (Lv) plane at a given time for an oscillatory situation: $A = 10$ and $Gr = 5 \times 10^5$.

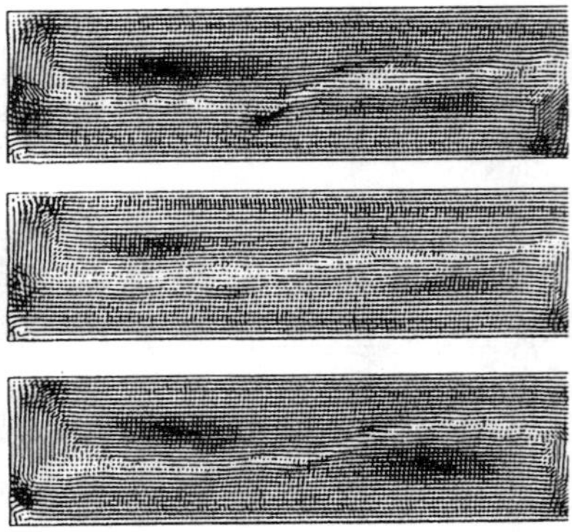

Fig. 8 Velocity vectors in the (Lv) plane at three different times for an oscillatory situation: $A = 4$ and $Gr = 5 \times 10^5$.

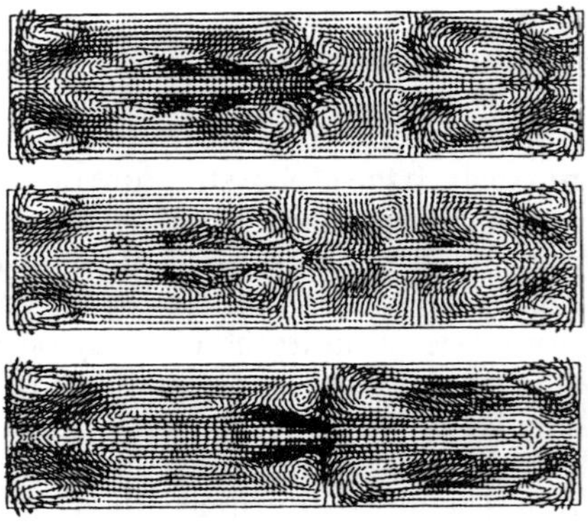

Fig. 9 Velocity vectors projection in the (Lh) plane at three different times for an oscillatory situation: $A = 4$ and $Gr = 5 \times 10^5$.

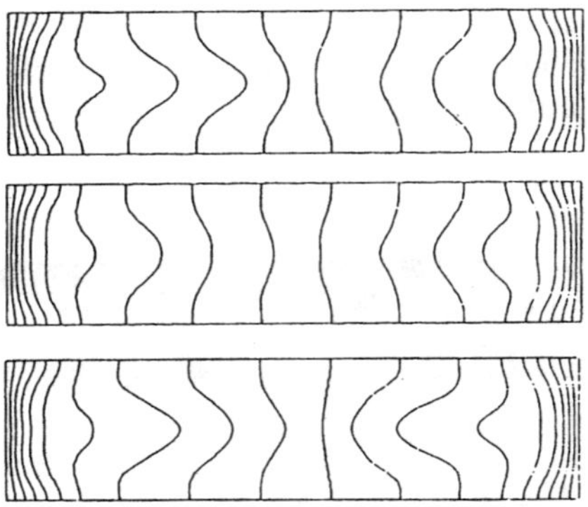

Fig. 10 Isotherm lines in the (Lh) plane at three different times for an oscillatory situation: $A = 4$ and $Gr = 5 \times 10^5$.

IV. Action of a Uniform Magnetic Field on the Stability of Hadley Circulation

We consider in the following a fluid layer bounded with rigid horizontal walls and subject to an external constant magnetic field. In such a system, owing to the applied horizontal temperature gradient, a parallel flow sets up. In the presence of a constant vertical magnetic field, the basic flow is given by the following analytical expressions for y between -0.5 and 0.5:

$$u_0(y) = \frac{1}{Ha^2}\left[\frac{\sinh(Hay)}{2\sinh(Ha/2)} - y\right] \tag{7}$$

Thermally insulating walls:

$$T_0(x,y) = \frac{x}{2} + \frac{GrPr}{Ha^2}\left[\frac{\sinh(Hay)}{2Ha^2\sinh(Ha/2)} - \frac{y^3}{6} + y\left(\frac{1}{8} - \frac{\cosh(Ha/2)}{2Ha\sinh(Ha/2)}\right)\right] \tag{8}$$

Thermally conducting walls:

$$T_0(x,y) = \frac{x}{2} + \frac{GrPr}{Ha^2}\left[\frac{\sinh(Hay)}{2Ha^2\sinh(Ha/2)} - \frac{y^3}{6} + y\left(\frac{1}{24} - \frac{1}{Ha^2}\right)\right] \tag{9}$$

These expressions show a strong decrease of the flow intensity with the Hartmann number and a velocity profile that becomes progressively linear in y. In the presence of a transverse magnetic field, the basic flow is the one obtained without a magnetic field because, due to the electric potential effect, the electric current and consequently the Lorentz force vanish in the whole cavity.

To analyze the linear stability of the basic state ψ_0, the governing equations are linearized with respect to small perturbations ψ': $/\psi = \psi_0 + \psi'$, where ψ represents all the variables $(\boldsymbol{v}, p, \theta, \phi)$. The perturbations can be written in normal modes of the form: $\psi'(x, y, z) = \psi'(y) \exp[i(hz + kx) + \omega t]$, which leads to an eigenvalue problem. The normal mode analysis consists for fixed values of Ha and Pr in the determination of the critical values of the Grashof number Gr and of the wavelengths h and k for which the basic flow loses its stability. The numerical procedure is based on a Tau method that uses Chebyshev polynomials as trial functions.[9]

The results of Laure and Roux[9] in the absence of a magnetic field show that two types of convective instabilities develop in such systems: two-dimensional modes ($k = 0$) that consist in the development of marginal cells with their axis perpendicular to the basic flow and three-dimensional modes ($h = 0$) for which the marginal cells have their axis parallel to the basic flow.

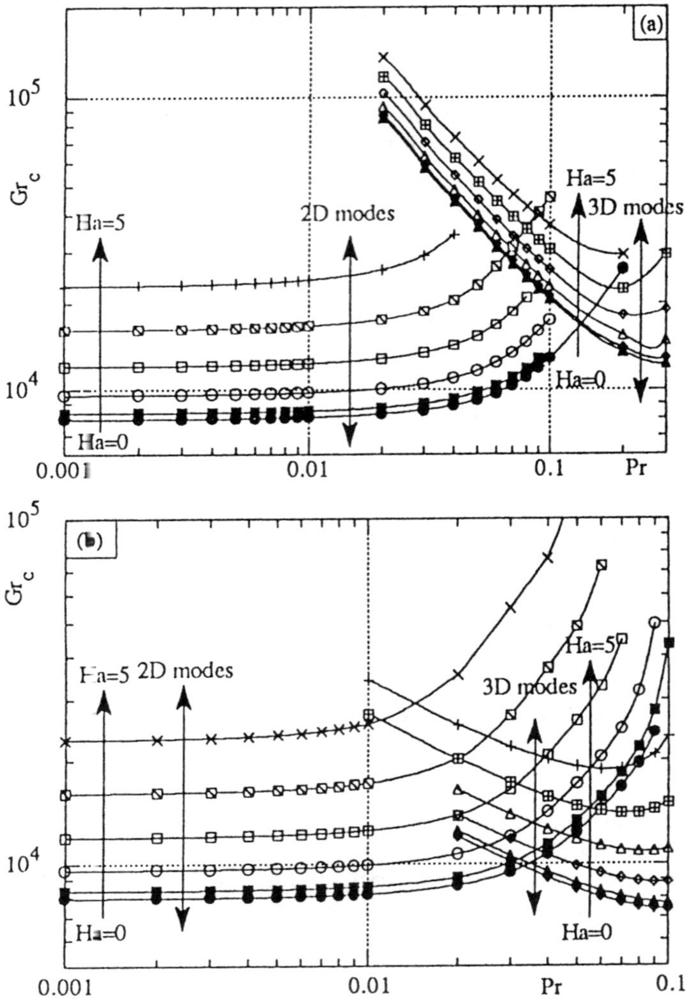

Fig. 11 Instability thresholds Gr_c versus Pr for Hadley flows submitted to a vertical magnetic field: a) thermally conducting and b) insulated horizontal boundaries.

In general, the two-dimensional modes prevail for low Prandtl numbers whereas the three-dimensional modes are dominant for moderate Prandtl numbers. In what follows we then define Pr_T as the Prandtl number at which the transition between two- and three-dimensional modes occurs.

Vertical Magnetic Field

The results are presented in Figs. 11a and 11b through the evolution of the critical Grashof number Gr_c as a function of the Prandtl number Pr for different values of the Hartmann number. Two types of thermal boundary conditions are considered either conducting (Fig. 11a) or insulated (Fig. 11b).

For the steady two-dimensional instability the critical Grashof number Gr_c is seen to be almost constant for low Prandtl number fluids ($10^{-3} \leq Pr \leq 10^{-2}$), then begins to increase with the Prandtl number for all Ha. The behavior of the neutral curves for the three-dimensional oscillatory modes is similar for all of the investigated Hartmann numbers. We find that the critical Grashof number Gr_c first decreases with Pr, then reaches a minimum before beginning to increase under the action of the stabilizing effect of the vertical temperature gradient.

Figures 12a and 12b clearly show that for the two-dimensional instabilities Gr_c increases exponentially with Ha^2 for both thermally insulating and conducting horizontal walls: $Gr_c \sim \exp(Ha^2/25)$. Note that for this two-dimensional instability the induced electric potential φ vanishes in the whole cavity. Thus, the action of the Lorentz force is reduced to the damping factor $Ha^2(\boldsymbol{v} \times \boldsymbol{e}_y) \times \boldsymbol{e}_y$. Moreover, the length of the marginal cells increases with the Hartmann number. Concerning the three-dimensional instability, Gr_c varies linearly with Ha^2. Furthermore, when Ha is increased the marginal cells become narrower, and the frequency of oscillations increases.

We can also note from Figs. 11a and 11b that Pr_T decreases rapidly when the Hartmann number is increased. In practice, this means that in laboratory experiments with a magnetic field one could expect to observe longitudinal instabilities instead of transverse ones, even for low Prandtl number fluids. Furthermore, the difference in the trend of the two types of instabilities with respect to Ha may be attributed to the fact that the three-dimensional instabilities occur under nonvanishing electric potential φ whereas this electric potential vanishes for the two-dimensional instability case. Indeed, the induced electric potential φ attenuates the stabilizing effect of the applied magnetic field by reducing the magnitude of the Lorentz force.

Transverse Magnetic Field

The stabilizing effect of a transverse magnetic field is found less significant than that of a vertical one, both on two- and three-dimensional instabilities. Moreover a transverse magnetic field changes the nature of the two-dimensional instabilities that become oscillatory. The corresponding oscillation frequencies first increase significantly with Ha, then reach a constant value weakly depending on Pr for sufficiently high Ha ($Ha \geq 50$). A similar feature is observed for the three-dimensional modes, but in this case the constant value reached by the frequencies depends on Pr.

In this case asymptotic laws for high Ha can be drawn. The critical Grashof numbers Gr_c are displayed in Fig. 13a for the two-dimensional modes and in Fig. 14a for the three-dimensional modes, the wave numbers h and k in Figs. 13b and 14b. We can notice, for $Ha \leq 5$, that the values of Gr_c, h, and k are almost constant, independent of Ha. Then for sufficiently

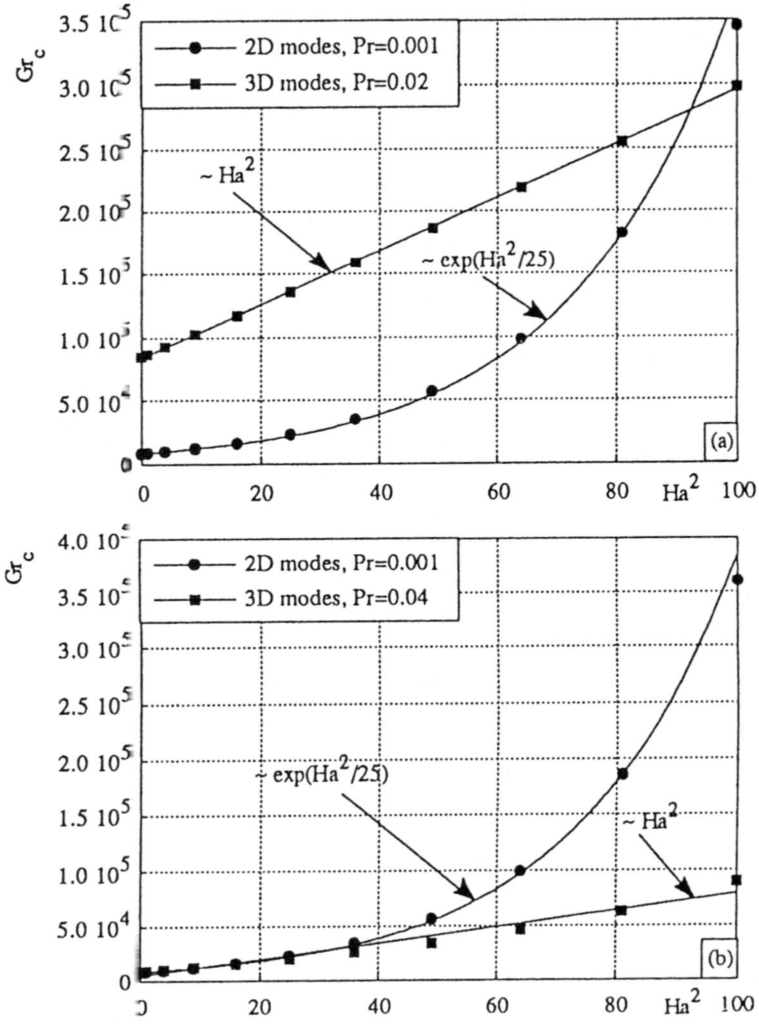

Fig. 12 Instability thresholds Gr_c versus Ha for Hadley flows submitted to a vertical magnetic field: a) thermally conducting and b) insulated horizontal boundaries.

high Ha ($Ha \geq 50$), Gr_c is seen to increase almost linearly with Ha whatever the nature of the considered instability, while the wave numbers h and k decrease as Ha^{-1}. The dependence with Pr of the two-dimensional modes is very weak whereas the one for the three-dimensional modes corresponds to Pr^{-1} in the range $0.001 \leq Pr \leq 0.01$. We can also note that, contrary to the case of a vertical magnetic field, the size of the marginal cells for the three-dimensional instabilities increases with Ha.

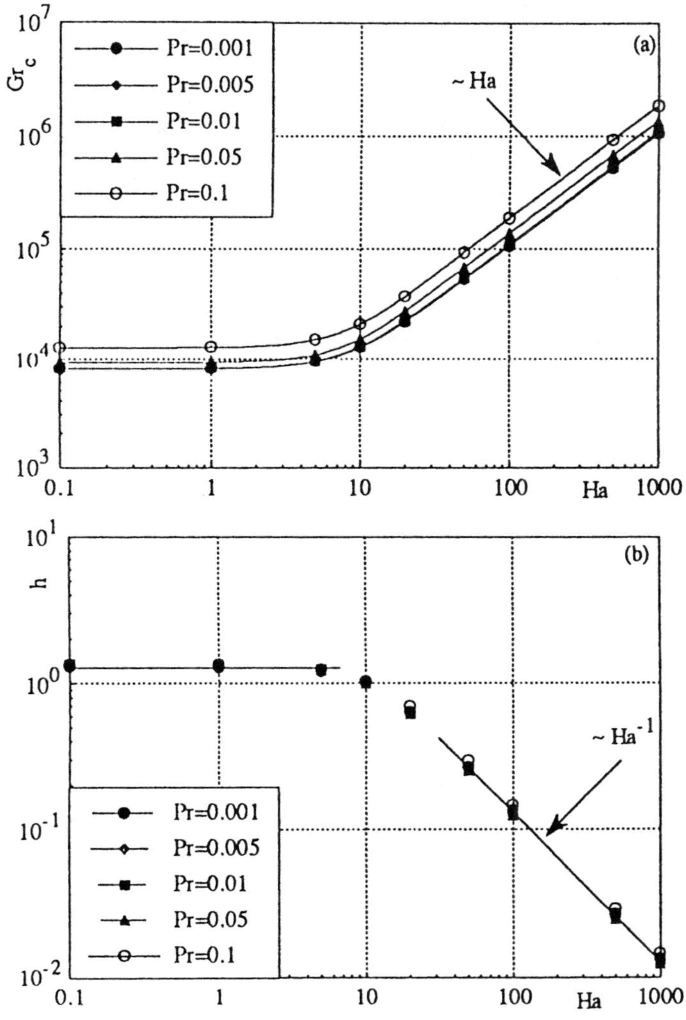

Fig. 13 Instability thresholds Gr_c and wavelengths h versus Ha for the two-dimensional modes in Hadley flows submitted to a transverse magnetic field (thermally conducting horizontal boundaries).

V. Conclusion

This study on oscillatory convection has first shown that different types of instabilities could be the cause for the onset of oscillatory flows. In the case of cylindrical cavities, a first instability that is obtained for large aspect ratios seems to correspond to a three-dimensional shear-stress instability initiated in the central part of the cavity. A second type of instability is obtained for

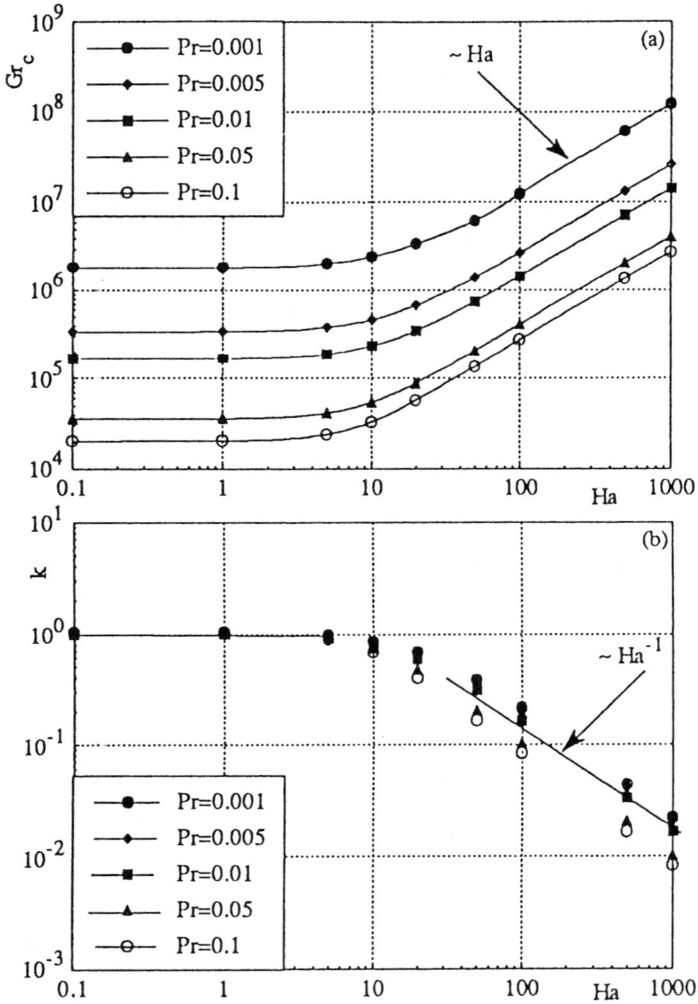

Fig. 14 Instability thresholds Gr_c and wavelengths k versus Ha for the three-dimensional modes in Hadley flows submitted to a transverse magnetic field (thermally conducting horizontal boundaries).

smaller aspect ratios and seems to correspond to the instability of the stagnation zone generated in the center of the cavity by the end recirculation flows. Each type of oscillatory flow breaks some particular symmetries among those existing originally in such cavities and were kept in the stationary situations.

The action of a magnetic field on Hadley circulations is found to stabilize the flows by increasing the instability thresholds. For a uniform vertical magnetic field the two-dimensional steady instabilities that prevail for low *Pr* are strongly stabilized with Gr_c varying exponentially with Ha^2, and the length of the marginal cells is increased. The three-dimensional oscillatory instabilities that prevail initially only for larger *Pr* become progressively the dominant ones even for low *Pr* as *Ha* is increased. For these instabilities the thresholds vary as Ha^2 and the marginal cells decrease in length. The stabilizing effect obtained with a transverse magnetic field is less efficient than that with a vertical one. The critical Grashof number that is little affected for small *Ha* varies then proportionally to *Ha*. The two- and three-dimensional modes are both oscillatory and correspond to marginal cells increasing in length with *Ha*.

Acknowledgments

This work is supported by the Centre National des Etudes Spatiales. The calculations were carried out on a Cray YMP C98 computer with the support of the CNRS through the Institut du Dévelopment et des Ressources en Informatique Scientifique and on an SP2 parallel computer with the support of the CNUSC. We wish to thank R. Moreau and Alboussière from MADYLAM and J. P. Garandet from CEN Grenoble for helpful discussions of this work.

References

[1]Hurle, D. T. J., Jakeman, E., and Johnson, C. P., "Convective Temperature Oscillations in Molten Gallium," *Journal of Fluid Mechanics*, Vol. 64, 1974, p. 565.

[2]Pimputkar, S. M., and Ostrach, S., "Convective Effects in Crystals Grown from Melts," *Journal of Crystal Growth*, Vol. 55, 1981, p. 614.

[3]Hurle, D. T. J., *Proceedings of the International Conference on Crystal Growth*, Pergamon, New York, 1966, p. 659.

[4]Jakeman, E., and Hurle, D. T. J., *Rev. Phys. in Tech.*, Vol. 3, 1972.

[5]Gill, A. E., "A Theory of Thermal Oscillations in Liquid Metals," *Journal of Fluid Mechanics*, Vol. 64, 1974, p. 577.

[6]Hart, J. E., "A Note on Stability of Low Prandtl Number Hadley Circulation," *Journal of Fluid Mechanics*, Vol. 132, 1983, p. 271.

[7]Hart, J. E., "Stability of Thin Non-Rotating Hadley Circulations," *Journal of Atmospheric Sciences*, Vol. 29, 1972, p. 687.

[8]Smith, M. K., and Davis, S. H., "Instabilities of Dynamic Thermocapillary Liquid Layers. Part I: Convective Instabilities," *Journal of Fluid Mechanics,* Vol. 132, 1983, p. 119.

[9]Laure, P., and Roux, B., "Synthèse des Résultats Obtenus par L'Étude de Stabilité des Mouvements de Convection Dans Une Cavité Horizontale de Grande Extension," *Comptes Pendus Hebdomadaires des Seances de l'Academie des Sciences*, Vol. 305, 1987, p. 1137.

[10]Priede, J., Thess, A., and Gerbeth, G., "Thermocapillary Instabilities in Liquid Metals: Hartmann-Number Versus Prandtl-Number," *Second International Conference on Energy Transfer in Magneto Hydro Dynamic Flows* (Aussois, France), Vol. 2, 1994, pp. 571–580.

[11]Moreau, R. Private communication.

[12]Patera, A. T., "A Spectral Element Method for Fluid Dynamics: Laminar Flow in a Channel Expansion," *Journal of Computational Physics*, Vol. 54, 1984, pp. 468–488.

[13]Karniadakis, G. E., Israeli, M., and Orszag, S. A., "High-Order Splitting Methods for the Incompressible Navier–Stokes Equations," *Journal of Computational Physics*, Vol. 97, 1991, pp. 414–443.

Chapter 57

Global Simulation of Bulk Crystal Growth: Investigation of Turbulent Convection under the Influence of an Axisymmetric Magnetic Field

R. Assaker,* N. Van den Bogaert,† and F. Dupret‡
CESAME, Louvain-la-Neuve, B-1348, Belgium

I. Introduction

MOST semiconductor single crystals used in the electronics industry, such as silicon, are grown using the Czochralski technique.[1] Poly-crystalline material is introduced into the crucible inside the furnace chamber and heated by means of a graphite resistor until full liquefaction is attained. A single crystal seed is then dipped into the melt. The pulling begins when a meniscus is formed between the seed and the molten material, and a cylindrical crystal is thence grown from the melt. To preserve the process axisymmetry, the crystal and crucible are rotated with given angular velocities.

Several highly coupled physical phenomena take place in this process, such as heat transfer and fluid flow. Heat exchanges in the furnace are because of radiation between the different constituents, conduction inside the solid and liquid parts, and convection in the molten semiconductor and the ambient gas. The melt flow is induced by buoyancy forces, surface tension gradient, and forces due to crystal and crucible rotations. Other driving forces, such as concentration gradients, also can be present but will not be discussed here. The temperature in the entire furnace and the flow in the melt depend on the process operating parameters, which are mainly the pull rate, the heater power, and the crystal and crucible rotation rates. This dependence is how-

*Research Engineer and Ph.D. Student, Unité de Mécanique Appliquée, Université Catholique de Louvain, 4–6, au. G. Lemaitre.

†Doctor Research Engineer, Unité de Mécanique Appliquée, Université Catholique de Louvain, 4–6, av. G. Lemaitre.

‡Professor, Unité de Mécanique Appliquée, Université Catholique de Louvain, 4–6, av. G. Lemaitre.

ever highly nonlinear and difficult to predict and is in particular a hard task to guess exactly the influence of a parameter variation upon the final crystal quality. This is where numerical simulation can play a helpful role.[2,3]

Silicon crystals of a 300 mm diam can now be grown using modern Czochralski pullers. With such a large crystal size and melt volume, new difficulties and challenges for crystal growers arise from the physics of the fluid flow. The latter gets indeed turbulent and can induce striations in the crystal because of temperature fluctuations. Oxygen concentration in the crystal, which is of the first importance for product quality, is also governed by the melt flow. To damp out turbulence and to control the flow inside the crucible, one can use external Lorentz forces, as generated by the interaction of an imposed magnetic field with the flow of an electrically conducting fluid.[4,5] Axial, horizontal, or cusp magnetic fields are used in the crystal growth industry.[6]

The objective of this paper is to present global simulation results predicting the temperature field in an entire Czochralski silicon furnace, together with the velocity field in the melt.[7] We will concentrate on studying the influence of turbulent convection upon the melt-crystal interface shape and upon the temperature profile along the crucible wall and the melt-crystal axis, both in the presence and the absence of a cusp magnetic field. A comparison between turbulent and laminar convection under the influence of a cusp magnetic field will also be carried out. The Prandlt mixing length model[8] is used to represent turbulent effects.

The equations governing the melt and crystal behavior are described in the second section of this paper, together with the numerical method used to solve the problem. In the third section, simulation results are discussed while the most important conclusions of these numerical experiments are stressed in the last section.

II. Governing Equations and Numerical Method

Both laminar and turbulent flow models are considered in the present paper. In the turbulent case, any physical quantity is in principle defined by means of Gibbs averaging (i.e., by repeating the experiment until convergence of the average is reached). In what follows, the same symbols $\boldsymbol{v}, p, T, \boldsymbol{J}$ will denote the velocity, pressure, temperature, and electric current density fields in the case of a laminar flow, and their averaged counterparts in the case of a turbulent flow. Introducing the quasisteady assumption and using the Boussinesq approximation, the mass, momentum, and energy equations are written as

$$\nabla \cdot \boldsymbol{v} = 0 \tag{1}$$

$$\rho(\boldsymbol{v} \cdot \nabla)\boldsymbol{v} = -\nabla p + \nabla \cdot [(\mu + \mu_t)(\nabla \boldsymbol{v} + \nabla \boldsymbol{v}^T)] + \rho g(1 - \beta(T - T_0)) + J \times B \tag{2}$$

$$\rho c(v \cdot \nabla)T = \nabla \cdot [(k + k_t)\nabla T] \tag{3}$$

where ρ, μ, β, c, and k denote the fluid specific mass, dynamic viscosity, thermal expansion coefficient, specific heat, and thermal conductivity, whereas $\boldsymbol{B}$ and $\boldsymbol{g}$ stand for the magnetic field and the acceleration of gravity, and T_0 is a reference temperature. The additional viscosity μ_t and thermal conductivity k_t are introduced in the turbulent case only to account for the extra terms appearing when the products $\boldsymbol{vv}$ and $\boldsymbol{v}T$ are averaged.

From the classical magnetohydrodynamic approximation the electric current density is related to the magnetic field, the velocity, and the electric field $\boldsymbol{E}$ by Ohm's law:

$$\boldsymbol{J} = \sigma\,(\boldsymbol{E} + \boldsymbol{v} \times \boldsymbol{B}) \tag{4}$$

where σ stands for the electric conductivity. Because of the low value of the magnetic Reynolds number, the magnetic field is assumed to be rigid, and the effect of the induced magnetic field is neglected. When the magnetic field is direct, the electric field is irrotational and derives from a potential φ,

$$\boldsymbol{E} = -\nabla\varphi \tag{5}$$

from which the equation governing φ is

$$\nabla \cdot [\sigma(\nabla\varphi - \boldsymbol{v} \times \boldsymbol{B})] = 0 \tag{6}$$

since the electric current is divergence free because the displacement current is negligible. Any cusp magnetic field can be obtained easily by superposing appropriate pairs of Helmholtz coils, which are symmetrically disposed with respect to the melt surface and in which the electric current flows in opposite directions. The analytical solution for one coil of radius R and current I is given by the formula

$$\begin{cases} B_r(r,z) = \dfrac{\mu_0 Ikz}{4\pi r\sqrt{Rr}}\left[\dfrac{R^2+r^2+z^2}{(R-r)^2+z^2}E(k^2) - K(k^2)\right] \\[2ex] B_z(r,z) = \dfrac{\mu_0 Ikz}{4\pi r\sqrt{Rr}}\left[\dfrac{R^2-r^2-z^2}{(R-r)^2+z^2}E(k^2) - K(k^2)\right] \end{cases} \tag{7}$$

where μ_0 is the magnetic permeability of vacuum and k^2 is defined by the relation

$$k^2 = \frac{4Rr}{(R+r)^2+z^2} \tag{8}$$

and where $K(k^2)$ and $E(k^2)$ are the elliptic integrals

$$\begin{cases} K(k^2) = \int_0^{\pi/2} \frac{d\theta}{\sqrt{1 - k^2 \sin^2 \theta}} \\ E(k^2) = \int_0^{\pi/2} \sqrt{1 - k^2 \sin^2 \theta}\, d\theta \end{cases} \tag{9}$$

A single pair of coils is used in the present paper.

Our objective here is to investigate the effect of a cusp magnetic field on turbulence in Czochralski growth by means of the Prandlt mixing length model.[5,8] This latter consists simply in expressing the turbulent viscosity and thermal conductivity in the following form:

$$\mu_t = \rho l_m^2 \dot{\gamma} \tag{10}$$

$$k_t = \rho c l_m^2 \dot{\gamma} \tag{11}$$

where $\dot{\gamma}$ is the shear rate

$$\dot{\gamma}^2 = \frac{1}{2}\left(\nabla \boldsymbol{v} + \nabla \boldsymbol{v}^T\right) : \left(\nabla \boldsymbol{v} + \nabla \boldsymbol{v}^T\right) \tag{12}$$

while l_m is the mixing length. The latter is related to the distance d to the flow domain boundary by the relation

$$l_m = C_1 d \tag{13}$$

where $C_1 = 0.4$ is the von Kármán constant. We must emphasize however that turbulent effects are likely to be anisotropic when a magnetic field is applied. As the Prandlt model is based on introducing an isotropic turbulent viscosity μ_t and an isotropic thermal conductivity k_t, nonisotropic effects cannot be captured by this approach. Moreover, a complete damping of turbulence is not possible when the crystal and/or the crucible are rotating because the shear rate (and thus μ_t and k_t) remain higher than a given minimum, whatever the intensity of the magnetic field is. As a consequence, results should be considered cautiously and compared to experiments before definite conclusions could be drawn. Nevertheless, the advantage of our approach is to provide a single model for both laminar and turbulent flows and to allow for a coupling with global heat transfer calculations in view of the model simplicity.

The distance d to the boundary must be evaluated, having in mind that its role is significant mainly in the shear layers that develop near the crucible walls and along the melt-atmosphere and melt-crystal interfaces. To this aim, we solve for d in the following equation:

$$\Delta\Delta d + \alpha d = 0 \text{ in } \Omega \tag{14}$$

with the boundary conditions

$$\begin{cases} \dfrac{\partial d}{\partial n} = 1 \\ d = 0 \end{cases} \quad \text{on} \quad \partial\Omega \tag{15}$$

in such a way that a smooth result is obtained, while d is [from conditions (15)] very close to the expected values near the domain boundary $\partial\Omega$. The parameter α is normally vanishing, but a different value can be selected to improve the result in the core of the melt. Note that two-dimensional (and not axisymmetric) calculations are carried out to obtain d from Eqs. (14) and (15).

Appropriate boundary conditions for $\boldsymbol{v}$, T, and φ must be introduced to close the system. No slip is assumed along the crucible wall and the solidification front. Free slip is assumed along the melt-atmosphere interface. The electric potential is evaluated in the crystal and the melt,[9] whereas a vanishing outgoing electric current density $\boldsymbol{J} \cdot \boldsymbol{n}$ is imposed along the crystal-melt boundary. Temperature calculations are coupled to the global heat transfer in the overall furnace according to the theory developed by Dupret et al.[7] Basically, the furnace is subdivided into macro-elements, at the interfaces of which temperature continuity and balance of heat flux are imposed. Macro-elements are either transparent radiative enclosures (with the assumption of diffuse gray radiation), solid bodies (in which the energy equation is a balance between heat diffusion and power supply), thin shells, and finally the crystal and melt domain, together with their interface along which solidification heat is released during growth (while temperature remains constant).

Our numerical scheme is based on using a finite-element method. Radiative macro-elements are covered by one-dimensional meshes, whereas conductive macro-elements are covered by two- or one-dimensional meshes—in the case of thin components. The melt-crystal is a particular macro-element whose interface position is unknown, whereas its temperature is imposed to be the thermodynamic melting temperature. Discretization of the equations describing each macro-element is performed using the Galerkin method and provides a linear algebraic system relating nodal heat fluxes and temperatures (or fourth powers of temperatures) along the macro-element boundaries. The global heat transfer discrete system is finally obtained by assembling all of the equations arising from the different macro-elements. This system is solved by a succession of local and global steps,[10] which are performed until convergence of the solid-liquid interface position and tri-junction temperature are achieved. The problem we solve is inverse because the heater power is unknown, while the crystal radius is imposed.

During each local step, a reduced problem is solved on the domain occupied by the crystal and the melt. Thermal boundary conditions are obtained from the previous global step. This local problem consists in solving simultaneously the energy, mass, and momentum equations, together with the equations related to the electric potential in case a magnetic field is considered. Unknowns are the nodal temperatures, velocities, pressures, electric potentials, and nodal axial positions of the solid-liquid interface. To solve this highly nonlinear system, decoupling is carried out between, on the one hand, calculation of the temperature field together with the interface position and, on the other hand, calculation of the velocity field together with the electric potential. These two subproblems are alternatively solved (using Newton iterations) until convergence is obtained. The partial decoupling of the local problem allows for performing calculations within a reasonable computation time, whereas the coupling between velocities and electric potential remains necessary to maintain a good convergence. In a previous paper[5] calculations of velocities and electric potential were performed iteratively, which led to significant difficulties to reach convergence. The distance to the boundary, which is needed for solving the momentum equations, must be calculated as related to the changing solid-liquid interface position. Without affecting the convergence, this evaluation is performed only once for the current local step, and not for each substep. Finally, to respect the balance between equations and unknowns, the melting temperature is imposed on all interface nodes, except the tri-junction, because the axial position of this latter is fixed by the imposed crystal shape. The linear system relating heat fluxes and temperatures along the melt/crystal boundary is obtained by using the calculated velocity field and interface position as parameters in the energy equation.

During each global step, the heat transfer problem is solved on the overall furnace by assuming that all matrices and vectors characterizing each macro-element are known from the previous local step. Again, Newton iterations are used to solve the global system, which is also highly nonlinear. During each global step, the tri-junction is imposed to be at the melting temperature, whereas the heater power is an unknown. Suitable boundary conditions required to solve the subsequent local step are selected on the basis of this global solution.

III. Results and Discussions

The first objective of the examples we investigate here is to show the influence of melt convection upon the temperature field and the shape of the solidification interface with or without the effect of a cusp magnetic field. The second objective is to compare laminar and turbulent flow calculations in the presence of the magnetic field. Four quasisteady global simulations are performed. In the first case melt convection is neglected, which means that only temperature field and melt-crystal interface position are computed. In the next two simulations melt convection is taken into account using the mixing-length turbulence model in the presence and in the absence of a cusp magnetic field. In the final simulation we compute melt convection in the presence of a cusp magnetic field using the classical laminar flow model.

All simulations discussed in this paragraph are global, and results have been obtained using the preceding method. The geometry of the Czochralski silicon furnace is shown in Fig. 1a together with the finite-element mesh used

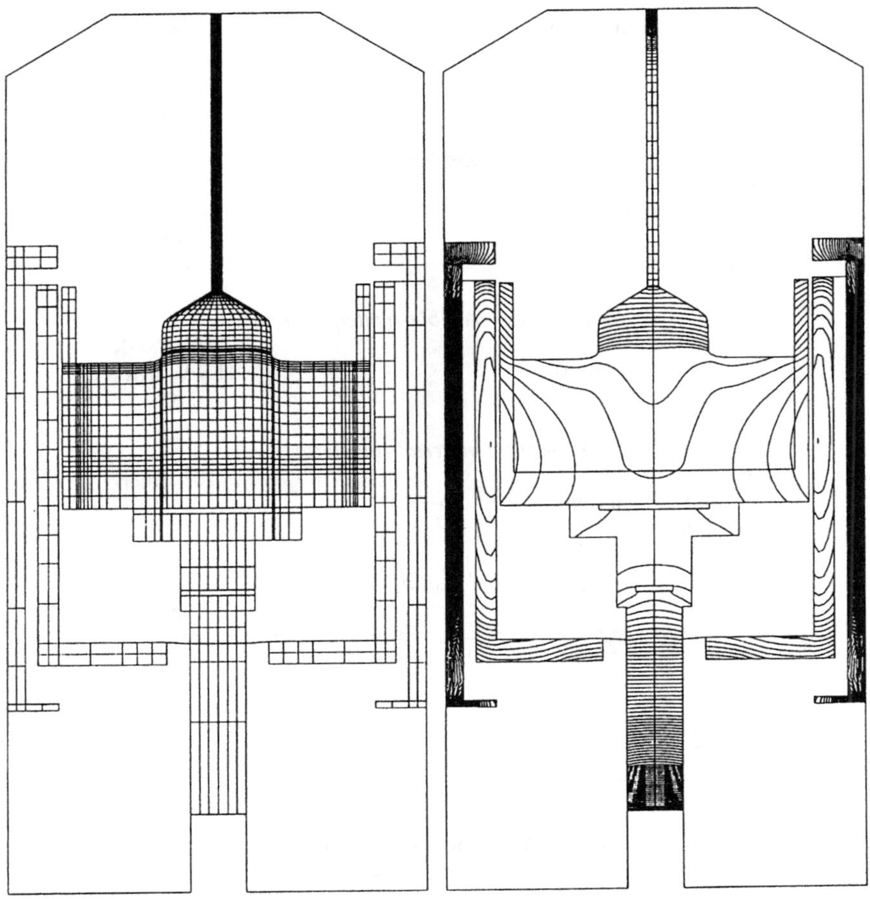

(a) (b)

Fig. 1 Obtained when taking melt convection into account without the effect of a magnetic field are a) finite-element mesh covering the furnace and b) global temperature field by steps of 20 K.

Table 1 Physical properties of the different constituents of the furnace

Liquid Silicon	
Density	2530 Kg/m^3
Heat capacity	2.39 10^6 J/m^3K
Dynamic viscosity	7 10^{-4} kg/ms
Coefficient of thermal expansion	1.4 10^{-5} K^{-1}
Latent heat of fusion	4.14 10^9 J/m^3
Melting temperature	1685 K
Thermal conductivity	42.9 W/mK
Emissivity	0.3
Electrical conductivity	12.3 10^5 S/m
Liquid-solid wetting angle	11 deg
Solid Silicon	
Heat capacity	2.39 J/m^3K
Thermal conductivity	21.6 W/mK
Emissivity	0.507
Electrical conductivity	4.3 10^4 S/m
Solid bodies (crucible, heater, pedestal)	
Heat capacity	3.3 10^6 J/m^3K
Thermal conductivity	41.9 W/mK
Emissivity	0.81
Insulating material (insulator)	
Heat capacity	3.3 10^5 J/m^3K
Thermal conductivity	0.1 W/mK
Emissivity	0.3
Steel (external wall of the furnace)	
Emissivity	0.3

Table 2 Parameters of the simulation

Crystal rotational rate	1.57 rad/s (15 rpm)
Crucible rotational rate	−0.53 rad/s (−5 rpm)
Cusp magnetic field	
(B_r)max	0.3 T
(B_z)max	0.6 T
Pulling rate	1.67 10^{-5} m/s (6 cm/h)
Coolant temperature on the wall	300 K
Crystal diameter	0.1 m

for the calculations. The physical properties of the materials used in the furnace and the simulation parameters are given in Tables 1 and 2. Figure 1b depicts the temperature field in the entire furnace when melt convection is taken into account.

In what follows we will focus on the melt-crystal macro-element. Hence, calculation results will be shown for one half of the melt and the crystal only.

A. Influence of Melt Convection on the Temperature Field and Interface Shape

The temperature field in the melt and the crystal is shown in Figs. 2a–2c for the cases when melt convection is neglected, when the flow is computed in the absence of a magnetic field, and when it is under the influence of a cusp magnetic field, respectively. In Fig. 2a heat is transferred by conduction only. The number of isotherms in the melt is eight. Temperatures are quite high with a maximum of 1850 K. The solidification interface is concave, whereas the power dissipated by the resistor to grow this 10-cm-diam crystal is 110 kW. When introducing turbulent melt convection (Fig. 2b), the number of isotherms in the melt goes down to five and isolines are obviously deformed by the flow, whereas the maximum temperature is 1789 K and the heater power is 99 kW. This general decrease is caused by heat convection in the flow and by the diffusive effect of turbulence. The maximum turbulent conductivity computed in the melt is 101 W/mK, which is about two times the actual conductivity of liquid silicon (42.9 W/mK). The interface position on the axis of symmetry is located much below its previous position, and the interface shape is slightly different with a typical inflection point.

When a cusp magnetic field is applied (with maximum radial and axial components of 0.3 and 0.6 T in the melt), seven isotherms are present in the liquid phase, and the maximum temperature is 1828 K. Heat transfer efficiency is lower than in the case without a magnetic field. This is because of flow breaking and turbulence damping by the field. The maximum stream function is 2.55 cm^3/s, instead of 4.9 cm^3/s in the absence of the magnetic field. The maximum turbulent conductivity decreases from 101 W/mK to 51 W/mK when the magnetic field is applied, whereas heater power needed to grow the crystal is 106 kW.

Temperature profiles along the melt-crystal axis and the bottom and vertical wall of the crucible are shown in Figs. 3a and 3b, respectively. Three cases are considered, namely the situations where conduction only is present in the melt, where turbulent convection is taken into account and where a cusp magnetic field acts while a turbulent flow model is used. Although temperature gradient in the crystal is almost the same in these three cases (Fig. 3a), the temperature in the crystal is slightly lower at any given distance from the seed when melt convection is computed in the absence of a magnetic field. This shows that the axial temperature profile is simply translated vertically in such a way that the interface remains at the melting temperature. On the other hand, the temperature profile in the melt is completely different in each case. When turbulent convection is considered, the temperature is relatively low, and the profile of convection is more or less flat.

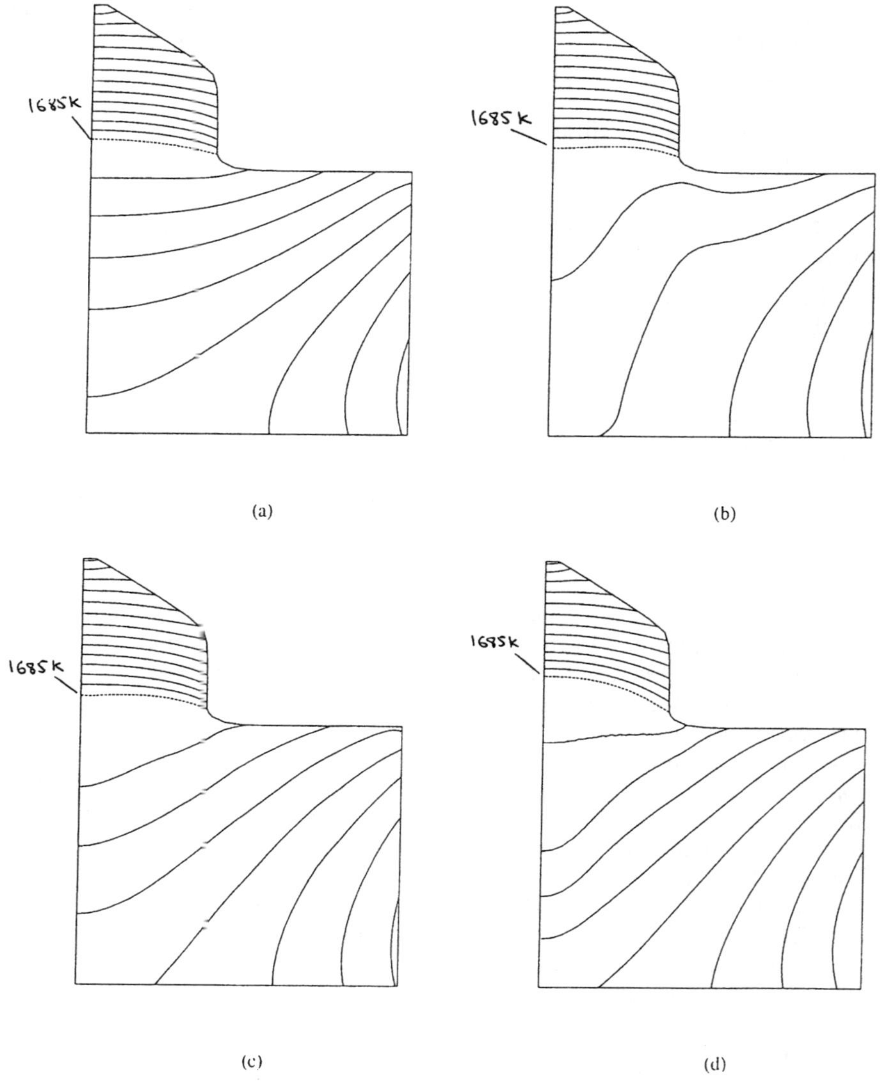

Fig. 2 Temperature field in the melt and crystal by steps of 20 K: a) when melt convection is neglected, b) when melt convection is calculated using a turbulent model in the absence of a magnetic field, c) when melt convection is calculated using a turbulent model under the influence of a cusp magnetic field, and d) when melt convection is calculated using a laminar model under the influence of a cusp magnetic field.

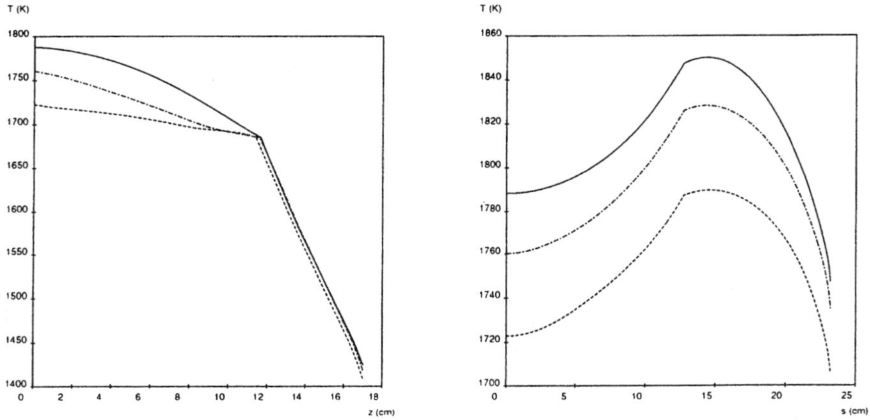

(a) (b)

Fig. 3 Temperature profiles a) along the axis of symmetry from the bottom of the melt to the top of the crystal and b) along the crucible bottom and vertical wall. Solid, dashed, and dash-dotted lines refer to neglecting convection, calculating convection with a turbulent model in the absence of a magnetic field, and calculating convection with a turbulent model under the influence of a cusp magnetic field.

When convection is neglected, the temperature gradient in the melt is important, and the temperatures reached are 5% higher than in the previous case. The cusp case is an intermediate one. Let us emphasize that temperature is an important parameter along the crucible-melt interface because of its influence on crucible erosion and oxygen dissolution into the melt. Figure 3b shows that this temperature is 60° higher when melt convection is neglected than when the flow is computed. Applying a cusp magnetic field increases the temperature along the crucible by about 35°.

B. Influence of a Cusp Magnetic Field on the Melt Flow

The azimuthal and meridional flow is generated by buoyancy forces and by crystal and crucible counter-rotations. Marangoni convection is neglected. The crystal is rotating at 15 rpm and the crucible at -5 rpm.

When no magnetic field acts, the meridional flow pattern exhibits two main vortices rotating in the counterclockwise direction. One of these vortices is located under the crystal whereas the other one is adjacent to the crucible vertical wall (Fig. 5a). The maximum value of the stream function in this case is 4.9 cm^3/s. Two small clockwise rotating eddies, one order of magnitude smaller than the main vortices, are also present. One of them, because of crystal rotation, is located between the solidification interface and the main vortex. The other one is located between the two principal vortices while being too small to be shown. The azimuthal flow exhibits two

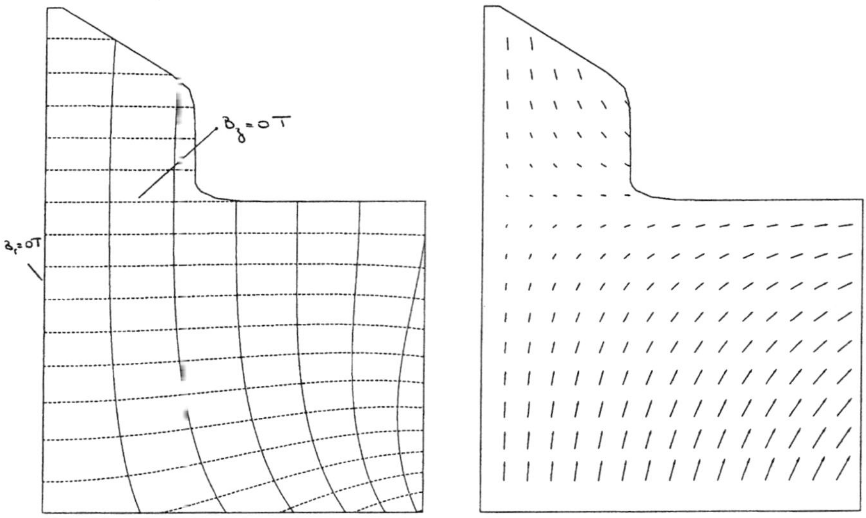

(a) (b)

Fig. 4 Imposed cusp magnetic field; a) radial (solid lines) and axial (dashed lines) components by steps of 0.05 T and b) field vectors. The radial and axial components vanish on the axis of symmetry and at the melt surface, respectively.

regions (Fig. 5b). Near the melt-crystal interface the flow is rotating in the same direction as the crystal, while the fluid rotates in the same direction as the crucible anywhere else. Near the bottom of the crucible, a sharp change in the azimuthal velocity occurs, and a boundary layer develops to let the flow match the crucible velocity.

When a cusp magnetic field is applied (Fig. 4), the meridional flow is a superposition of a big counterclockwise rotating eddy and a small clockwise rotating vortex located near the melt-crystal interface (Fig. 6a). The clockwise vortex is not shown in the figure for it is one order of magnitude smaller than the other one. The maximum stream function reached in the presence of a magnetic field is 2.4 cm^3/s, which is about one-half of the value obtained without a magnetic field (4.9 cm^3/s). The azimuthal motion shows that the flow undergoes a rigid body rotation that is almost sticking to the crucible (Fig. 6b).

C. Comparison Between Turbulent and Laminar Models

As explained in Section II, turbulence modeling should be considered cautiously when a magnetic field is applied. The difficulty arises from the anisotropic character of the magnetic field. To compare turbulent and laminar calculations in the presence of a magnetic field, we have attempted to decrease the turbulent contributions in the flow and heat equations (i.e., the

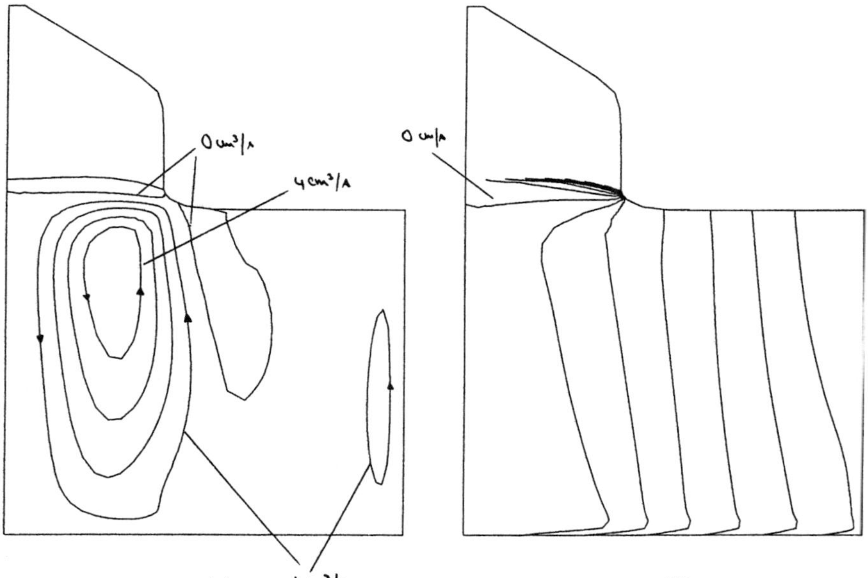

(a) (b)

Fig. 5 As calculated by means of a turbulent model in the absence of a magnetic field are a) the stream function in the melt by steps of 1 cm³/s and b) the azimuthal velocity by steps of 1 cm/s.

turbulent viscosity μ_t and thermal conductivity k_t) by letting the value of the coefficient C_1 in Eq. (13) tend to zero. Convergence is achieved down to $C_1 = 0.15$ as long as the actual dynamic viscosity of liquid silicon is used (7.0 10^{-4} Kg/ms). In this case the turbulent contribution is low but not nil. The lack of convergence for lower values of C_1 is essentially because of the coarseness of the finite-element mesh used. To get a laminar solution with the same mesh, we have thus increased the dynamic viscosity to 7.0 10^{-3} Kg/ms. Comparison with turbulent calculations using a dynamic viscosity of 7.0 10^{-4} Kg/ms is still possible because, in this case, the turbulent contribution to the total viscosity is 5.2 10^{-2} Kg/ms, which is about 100 times larger than the actual viscosity.

The temperature field obtained using the laminar convection model in the presence of a magnetic field is shown in Fig. 2d. The maximum temperature reached in the melt is 1858 K to be compared with 1828 K in the turbulent case (Fig. 2c). The general profile is similar in these two cases. The melt-crystal interface is however much more concave when convection is computed using the laminar model. The meridional and azimuthal melt flows computed using the laminar model are shown in Figs. 7a and 7b, respectively. The azimuthal velocity is the same as in the turbulent case (Fig. 6b and 7b). The meridional flow is still a superposition of two main vortices. Whereas counterclockwise vortex has the same shape as in the turbulent case (Fig. 6a), the

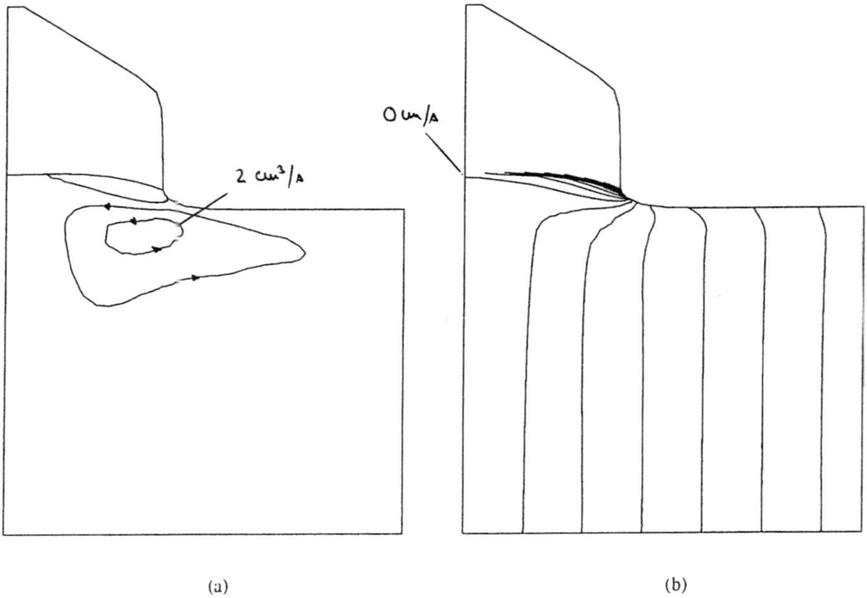

Fig. 6 As calculated by means of a turbulent model in the presence of a cusp magnetic field are a) the stream function in the melt by steps of 1 cm³/s and b) the azimuthal velocity by steps of 1 cm/s.

maximum stream function is 3.2 cm³/s instead of 2.55 cm³/s in the turbulent case. This 20% increase is because of the decrease of the total viscosity induced by the absence of turbulence. In turbulent calculations the total viscosity was indeed 5.47 10^{-2} Kg/ms, whereas in the laminar case it is only 7.0 10^{-3} Kg/ms. As a conclusion, the important difference between laminar and turbulent models in the presence of a cusp magnetic field concerns mainly the clockwise rotating eddy located beneath the crystal interface in which the maximum value of the stream function is -4.0 cm³/s instead of -0.8 cm³/s. Further comparisons are necessary to draw definite conclusions. We should note that laminar calculations give nonsmooth contour lines (Fig. 7), which means that a more refined finite-element mesh should be used in that case.

IV. Conclusions

We have developed a quasisteady model to predict the global heat transfer, the shape of the solidification front, and the melt flow in a Czochralski silicon furnace. The mixing length model is used to account for turbulent effects in the flow whereas the influence of a configurated magnetic field is easily modeled.

The question is posed of whether the laminar or the turbulent flow model must be used when the magnetic field acts while the dimensionless flow

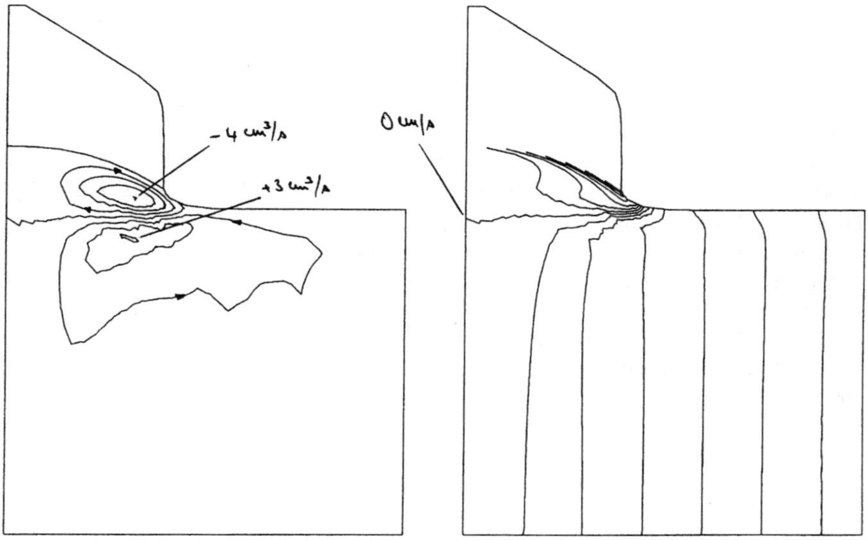

(a) (b)

Fig. 7 As calculated by means of a laminar model in the presence of a cusp magnetic field are a) the stream function in the melt by steps of 1 cm³/s and b) the azimuthal velocity by steps of 1 cm/s.

numbers are high. Although results are quite similar in both cases in most of the flow domain, significant discrepancies are exhibited beneath the solidification front because of the important difference in viscosity between both models. As turbulence is likely to be anisotropic when a magnetic field acts, the exact solution is not known, which means that no definite answer can be given today without experimental comparisons.

Acknowledgments

This paper presents research results of the Belgian Programme on Interuniversity Poles of Attraction initiated by the Belgian State Prime Minister's Office for Science, Technology, and Culture. The scientific responsibility rests with its authors.

References

[1]Hurle, D. T. J., *Handbook of Crystal Growth*, North-Holland, Amsterdam, 1994.

[2]Dupret, F., and Van den Bogaert, N., "Modelling Bridgman and Czochralski Growth," *Handbook of Crystal Growth*, Vol. 2b, edited by D. T. J. Hurle 1994, Ch. 15.

[3]Brown, R. A., *AIChE Journal*, Vol. 34, 1988, pp. 881.

[4]Moreau, R., *Magnetohydrodynamics*, edited by Kluwer Academic, Norwell, MA, 1990.

[5]Assaker, R., Van den Bogaert, N., and Dupret, F., "Dynamic Global Simulation of Bulk Crystal Growth under the Influence of an Axisymmetric Magnetic Field," *Magnetohydrodynamics,* to be published.

[6]Series, R. W., and Hurle, D. T. J., "The Use of Magnetic Fields in Semiconductor Crystal Growth," *Journal of Crystal Growth*, Vol. 113, 1991, pp. 305–328.

[7]Dupret, F., Nicodème, P., Ryckmans, Y., Wouters, P., and Crochet, M. J., "Global Modelling of Heat Transfer in Crystal Growth Furnace," *International Journal of Heat Mass Transfer*, Vol. 33, No. 9, 1990, pp. 1849–1871.

[8]Schlichting, H., *Boundary Layer Theory*, McGraw–Hill, New York, 1986.

[9]Hjellming, L. N., and Walker, J. S., "Melt Motion in Czochralski Crystal Puller with an Axial Magnetic Field: Isothermal Motion," *International Journal of Fluid Mechanics*, Vol. 164, 1986, pp. 237–273.

[10]Ryckmans, Y., Nicodème, P., and Dupret, F., *Journal of Crystal Growth*, Vol. 99, 1990, pp. 702.

Chapter 58

Nonlinear Waves with Electromagnetic Interaction in Aluminum Electrolysis Cells

V. Bojarevics*
University of Greenwich, London, SE18 6PF, England, United Kingdom

Introduction

AN ALUMINUM electrolytic cell contains two overlying fluid layers of small density difference and large horizontal extent relative to their depth. The interface waves are similar to stratified sea layers, but the situation is complicated by the passage of the electric current exceeding in modern cells a magnitude of 200,000 A. The interface waves change the electric current distribution in accordance to depth variation of poorly conducting electrolyte that produces high-density horizontal currents in the relatively thin and well-conducting liquid aluminum. The wave development depends on the magnetic field distribution created by the current supplying busbars and by the current in the cell. The interface stability problem is of great practical importance because the electrolytic aluminum production is a major electrical energy consumer and is related to the environmental pollution rate.

Urata et al.[1] were the first to introduce a coupled wave equation to describe the interface evolution in electrolysis cells. In their model the electric current is coupled to the interface deformation. In Ref. 2 a systematic perturbation expansion is developed for the fluid dynamics and electric current problems that permitted the reduction of the three-dimensional problem to a two-dimensional one for the leading-order expansion terms.

The procedure is more generally known as "shallow water approximation," which can be extended for the case of weakly nonlinear and dispersive waves. A generalized KdV equation was derived by Hofman[3] for small finite amplitude gravity waves in the presence of a magnetic field. Oshima and Yamane[4] applied numerical analysis of the generalized Boussinesq equations to investigate the unidirectional propagation of a solitary wave and a

*Senior Research Fellow, School of Computing and Mathematics, Wellington St.

hydraulic bore and obtained qualitative agreement to their experimental results. The Boussinesq formulation permits the generalization of the problem for nonunidirectionally propagating waves, accounting for side walls and for a two-fluid layer interface, e.g., Renouard et al.[5] found a good correspondence to the experimental results for resonantly interacting waves in a channel mounted on a rotating platform.

In the present work shallow-layer generalized Boussinesq equations are derived for wave motions depending on two horizontal variables x,y for two layers and with the electromagnetic interaction that makes the flow velocities rotational. Thus the velocity potential alone will not be adequate for the flow representation. The mathematical problem is put in "weak" formulation, and coupled equations in Fourier space are derived. The solution method is a discrete time stepping where the full nonlinear problem is linearized within each small time step and the resulting eigenvalue/eigenvector problem solved numerically by high accuracy LAPACK routines. The procedure can be considered as an extension of the linear stability problem. The simulated wave behavior will be validated against the one-dimensional finite amplitude standing wave analytical solutions by Tadjbakhsh and Keller.[6] Different magnetically modified wave patterns will be presented graphically.

Statement of the Problem

An idealized electrolysis cell can be represented by three layers shown in Fig. 1 where the top layer (3) is solid carbon anode, (2) is the liquid electrolyte of density $\rho_2 = 2.1 \cdot 10^3$ kg/m^3, and (3) is molten aluminum - $\rho_1 = 2.3 \cdot 10^3$ kg/m^3. The horizontal dimensions L_x and L_y are assumed to be much larger than the typical depth H, and the interface wave typical amplitude A is assumed to be small relative to the depth. Thus two small parameters of the problem are

$$\delta = \frac{H}{L} \ll 1$$

$$\varepsilon = \frac{A}{H} \ll 1$$

and they are assumed to be related as $\varepsilon = O(\delta^2)$.

With the purpose to derive weakly nonlinear shallow layer approximation Boussinesq equations, we will need to estimate the terms in the full equations of motion. Therefore, nondimensional variables are introduced using the following scales: L for coordinates x,y; $\varepsilon\sqrt{gH}$ for velocity $\boldsymbol{v}$, $L/\sqrt{gH}$ for time t, $\rho_1\, gH$ for pressure p, and IB_o/L^2 for electromagnetic force $\boldsymbol{f}$ (B_o is typical magnetic field magnitude and I the total electric current). According to the small depth assumption, a stretched nondimensional coordinate $\overline{z} = z/(L\delta)$ is defined, and the nondimensional interface deformation is represented as $\overline{h}_o = \varepsilon\, \varsigma\, (x, y, t)$. With these definitions the nondimensional fluid

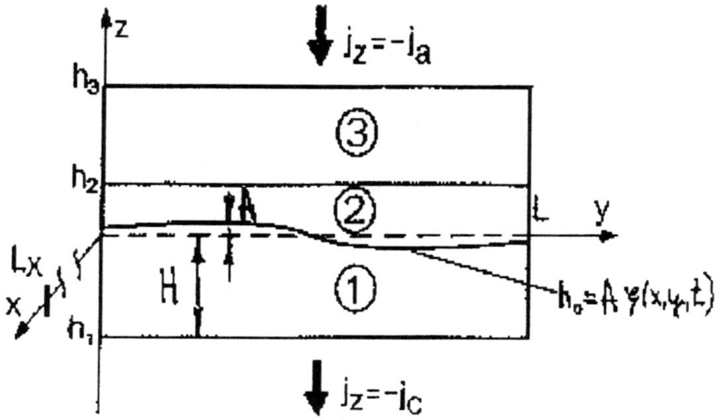

Fig. 1 Idealized electrolysis cell.

flow equations, continuity, horizontal momentum, and vertical momentum transport, are, respectively,

$$\partial_k \boldsymbol{u}_k + \delta^{-1}\partial_{\bar{z}} w = 0 \tag{1}$$

$$\partial_t u_j + \varepsilon\partial_k\left(u_k u_j\right) + \varepsilon\delta^{-1}\partial_{\bar{z}}\left(u_j w\right) = -\varepsilon^{-1}\partial_j p + Re^{-1}\left(\varepsilon\delta^{-2}\partial_{\bar{z}\bar{z}} u_j + \varepsilon\partial_{kk} u_j\right) + Ef_j \tag{2}$$

$$\partial_t + \varepsilon\partial_k\left(u_k w\right) + \varepsilon\delta^{-1}\partial_{\bar{z}} w^2 =$$

$$-\varepsilon^{-1}\delta^{-1}\partial_{\bar{z}} p + Re^{-1}\left(\varepsilon\delta^{-2}\partial_{\bar{z}\bar{z}} w + \varepsilon\delta_{kk} w\right) + Ef_z - \varepsilon^{-1}\delta^{-1} \tag{3}$$

where the Einstein summation convention is assumed over the repeating j and k indexes (equal to 1 or 2, respectively for x, y coordinates) used, and the nondimensional governing parameters are the Reynolds number, $Re = L\varepsilon\sqrt{gH}/\nu$, and electromagnetic interaction parameter, $E = IB_{o/}L^2\rho g\varepsilon\delta$. Formally, the Boussinesq equations can be derived if assuming $Re^{-1}\varepsilon\delta^{-2} = O(1)$ and $E = O(1)$, and the leading terms of velocity expansion in the small depth parameter are given by

$$\boldsymbol{u}(x,y,z,t) = \boldsymbol{u}_o(x,y,t) + \delta\boldsymbol{u}_\delta(x,y,z,t) + o(\delta) \tag{4}$$

The continuity equation (1) shows that the vertical velocity expansion starts with the second term only:

$$w(x,y,z,t) = \delta w_\delta + o(\delta) \tag{5}$$

Depth-averaged quantities are introduced according to the definition:

$$\hat{u}_j(x,y,t) = \frac{1}{\bar{h}_i - \bar{h}_o} \int_{\bar{h}_o}^{\bar{h}_i} u_j \, \mathrm{d}\bar{z} \tag{6}$$

The same depth-averaging procedure is formally applied to the fluid flow equations. With the kinematic boundary condition at the interface

$$w(h_o) = \delta \partial_t \varsigma + \varepsilon\delta u_k(h_o) \partial_k \varsigma \tag{7}$$

the depth-averaged continuity equation for each of the two fluid layers with the variable depths $h_i(x,y,t)$ is

$$\partial_t \varsigma = \partial_j (\bar{h}_i - \varepsilon\varsigma) \hat{u}_j, \tag{8}$$

which is accurate to all orders in ε and δ.

The depth-averaged horizontal momentum equations for each of the fluid layers are

$$\partial_t \hat{u}_j + \varepsilon \hat{u}_{ko} \partial_k \hat{u}_{jo} + O(\varepsilon\delta) = -\varepsilon^{-1} \partial_j P - \partial_j \varsigma + \delta^2 \frac{\bar{h}_i^2}{3} \partial_{tjk} \hat{u}_{ko} + O(\varepsilon\delta^2) - \mu \hat{u}_j + E\hat{f}_j \tag{9}$$

where the continuity of the pressure at the interface is satisfied by introducing $P(x,y,t)$, the pressure at the interface common to both layers. The linear friction law is introduced in Eq. (9):

$$-\mu \hat{u}_j = \frac{1}{\bar{h}_i - \bar{h}_o} \mathrm{Re}^{-1} \varepsilon \delta^{-2} \partial_{\bar{z}} u_j \Big|_{\bar{h}_o}^{\bar{h}_i} \tag{10}$$

which was used previously for shallow water identifications both for sea tide simulations[7] and aluminum electrolysis problems.[8] The appropriate boundary conditions to solve Eq. (8) and (9) are zero normal velocity at the side walls:

$$u_n = 0 \tag{11}$$

The momentum (9) and continuity (8) equations for the two fluid layers can be combined in one nonlinear wave equation for the interface:

$$\left\langle \frac{\rho}{\bar{h}} \right\rangle \partial_{tt}\varsigma - \delta^2 \frac{1}{3} \langle \rho \bar{h} \rangle \partial_{ttjj}\varsigma + \left\langle \frac{\mu\rho}{\bar{h}} \right\rangle \partial_t \varsigma + \langle \rho \rangle \partial_{jj}\varsigma$$

$$= \left\langle \rho E \partial_j \hat{f}_j \right\rangle - \varepsilon \left\langle \frac{\rho}{\bar{h}} \partial_{tj}(\varsigma u_{jo}) + \rho \partial_j (u_{ko} \partial_k u_{jo}) + \frac{\mu\rho}{\bar{h}} \partial_j (\varsigma u_{jo}) \right\rangle \tag{12}$$

where $\langle F \rangle = F_1 - F_2$ denotes a difference of quantity in the two layers. Note that $\hat{u}_0 = u_0$ according to Eq. (4), and Eq. (12) contains only the leading-order velocities that can be determined from linear fluid flow equations. For this purpose these velocities are expressed as

$$u_{xo} = -\partial_y \Psi + \partial_x \chi$$

$$u_{yo} = -\partial_x \Psi + \partial_y \chi \tag{13}$$

$$(\mathrm{curl}\, \boldsymbol{u}_o)_z = \partial_{kk} \Psi$$

then

$$\mathrm{div}_2 \boldsymbol{u}_o = \partial_{kk\chi}$$

and we need to solve the following additional linear equations:

Curl of linear momentum equation (in layer i):

$$\partial_{tkk}\Psi = \mu_i \partial_{kk}\Psi + E_i\, \mathrm{curl}_z \hat{f}_i \tag{14}$$

Linear continuity equation:

$$\partial_t \varsigma = \bar{h}_i \partial_{kk} \chi \tag{15}$$

The coupled electromagnetic force is expressed similarly as in Ref. 2. The horizontal electromagnetic force appearing in Eqs. (12) and (14) is given by

$$\hat{\boldsymbol{f}}_j = -\frac{\sigma_1}{\sigma_2}\varepsilon\delta\nabla\Phi \times \boldsymbol{B}_{oz} = O(1) \tag{16}$$

The equation for the electric potential $\Phi(x,y,t)$ in the aluminum is[2]

$$s^{-1}\bar{h}_1\bar{h}_2\partial_{kk}\Phi + \Phi - \frac{\bar{h}_1}{\bar{h}_3 - \bar{h}_2}\Phi = j_o\varsigma \tag{17}$$

where $s = (\sigma_2/\sigma_1)\delta^{-2} = O(1)$, electrical conductivities for liquid aluminum are $\sigma_1 = 3.3 \cdot 10^6$, and for liquid electrolyte are $\sigma_2 = 2 \cdot 10^2\,\Omega^{-1}\,\mathrm{m}^{-1}$. The nonconducting side walls of the electrolysis cell mean the boundary condition:

$$\partial_n\Phi = 0 \tag{18}$$

Solution and Results

The coupled equations of the problem (11–18) are put in a weak form by integrating against sufficiently regular functions over the cell and accounting for the boundary conditions. The solutions of the problem can be expressed in the form:

$$\varsigma(x,y,t) = \sum_{k,j} e^{\omega_j t}\varsigma_{kj}\cos k_x x \cos k_y y, \tag{19}$$

where the wave vector $\boldsymbol{k} = \left(\frac{\pi m}{L_x}, \frac{\pi n}{L_y}\right)$ $m, n = 0, 1, 2, \ldots$, and similar expressions for other unknowns Ψ, χ, Φ.

The problem is homogenous and results in a nonlinear eigenvalue problem for ω_j. For the linear problem $\varepsilon = 0$, solving the homogenous problem for the eigenvalues and eigenvectors is possible. Then the evolution problem with initial conditions is solved by a double summing up of all eigenvectors for all of the respective eigenvalues. The same method can be applied to the nonlinear problem when $\varepsilon \neq 0$, performing a linearization over a small time interval Δt and computing the linear velocity components from the previous time step. The end of the previous time step gives new initial conditions for the new linearized problem. At each time step the computed eigenvectors need to be normalized according to these initial conditions. Accurate routines for these linear algebra tasks are supplied in the LAPACK program package. Typical runs were made with 16*16 modes in the horizontal directions, and the time steps up to 2.5 were accessible.

Significant efforts were done to validate the problem final formulation and the numerical simulation program against the previously published results. The simple analytical case for the two-mode (0,1) and (1,0), interaction when

$$\varsigma = \varsigma_{1,0} \cos \frac{\pi}{L_x} x + \varsigma_{0,1} \cos \frac{\pi}{L_y} y, \tag{20}$$

permits the recovery of the previous linear theory[2] ($\varepsilon = 0, \delta = 0, \mu = 0$) result for the critical uniform vertical magnetic field magnitude:

$$B_{0\mathrm{cr}} = \frac{(\rho_1 - \rho_2) g \pi^4 H^2}{16 I} \left| \frac{L_x}{L_y} - \frac{L_y}{L_x} \right|$$

In particular, the square cell case $L_x = L_y$ is always unstable if the friction coefficient $\mu = 0$. This is the simplest case of the so-called rotating wave instability. The present theory permits us to show that the instability holds even for the dispersive case with finite $\delta \neq 0$. However, for

$$\mu \geq \frac{8 I B_0}{(\rho_1 - \rho_2) g \pi^4 H^2}$$

this simplest case becomes stable.

Other checks were made against the analytical results[6] for a pure hydrodynamic standing wave in the weakly nonlinear case. If started from a (1,0) one-dimensional gravity wave (Fig. 2), the evolving wave is losing its cosine

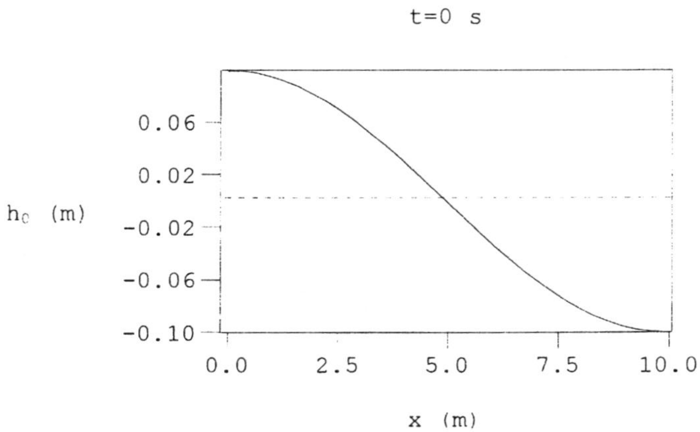

Fig. 2 Wave profile at $t = 0$.

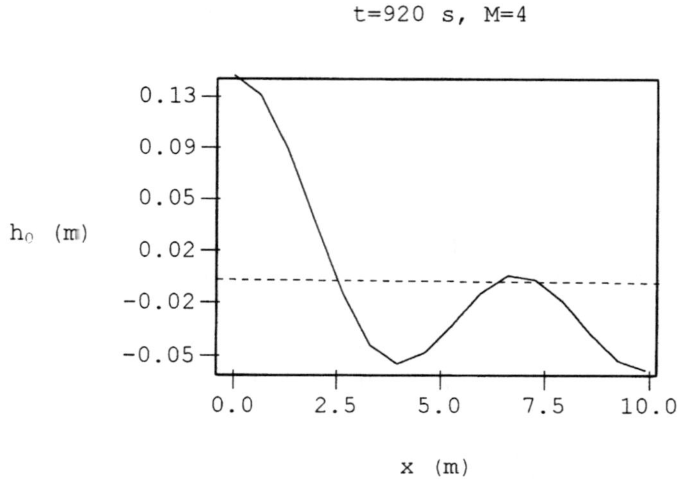

Fig. 3 Wave profile at t = 920 s, m = 9.

profile in space and time, and finally the profile's maximum elevation is not equal to the minimum depression (Fig. 3). The present simulation profile is the same as the analytical result[6] for $m = 4$. The present numerical method permits us to increase number of modes; thus Fig. 4 shows the profile for $m = 8$, which is different from the four-mode case. The result was checked

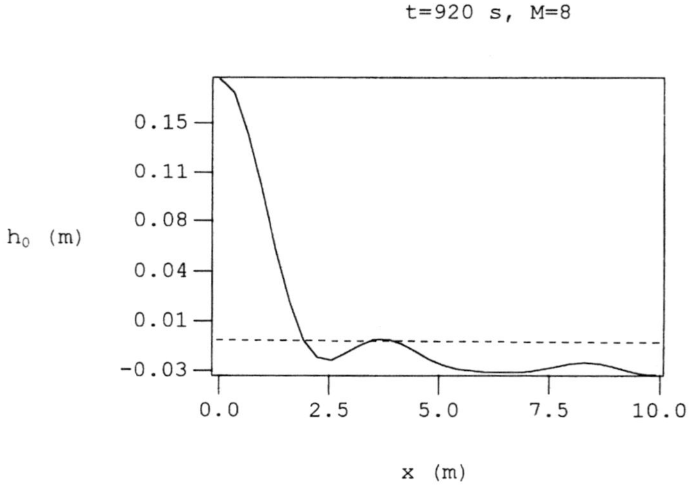

Fig. 4 Wave profile at t = 920 s, m = 8.

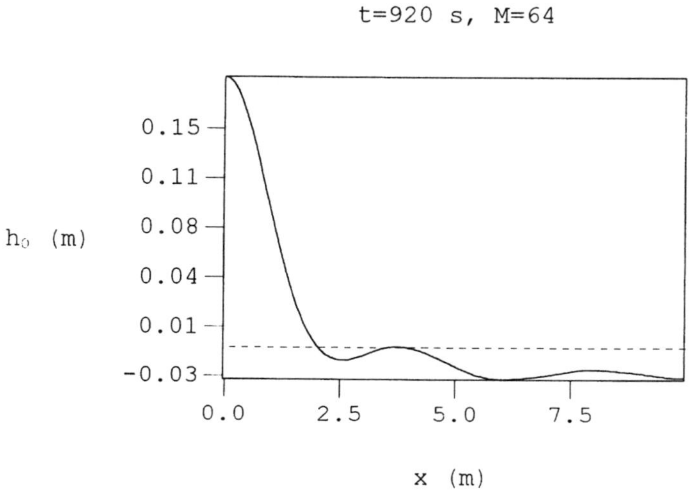

Fig. 5 Wave profile at t = 920 s, m = 64.

as increasing still more the number of modes; Fig. 5 shows a computed profile for $m = 64$, which is the same as for $m = 8$.

A hydrodynamic solitary wave was attempted to simulate numerically starting with a step-like perturbation. We found that for the considered two-fluid particular case the dispersive effects were much stronger expressed than the nonlinear. Therefore a single-fluid case ($\rho_2 = 0$) was considered for the following test exercise. Figure 6 shows the initial pulse resulting after the Fourier representation was applied to the step-like perturbation. This pulse travels the channel length ($L = 10$ m) many times with

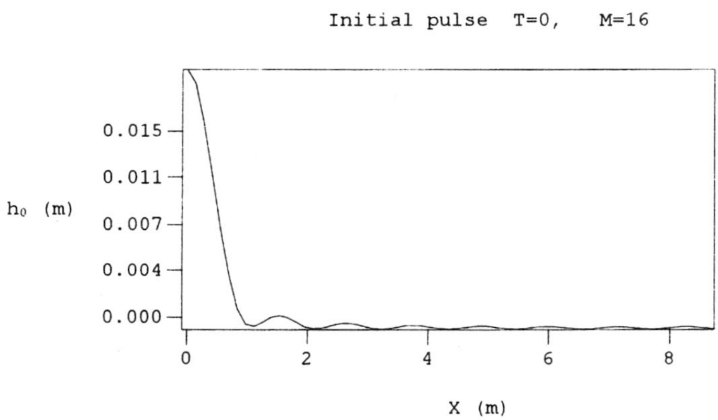

Fig. 6 Wave profile, initial pulse

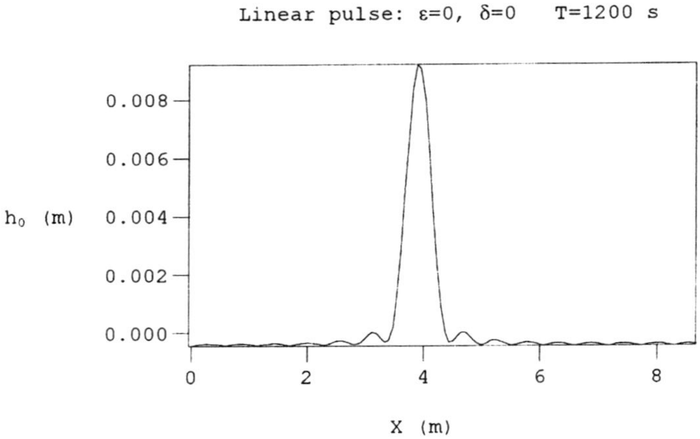

Fig. 7 Wave profile, linear pulse.

reflections, the typical period being approximately 30 s. Figure 7 shows the profile at 1200 s. The nonlinear wave without dispersion ($\delta = 0$) shows a steep front profile (Fig. 8). Yet a pure dispersive pulse exhibits a typical dispersive tail shown in Fig. 9. Finally the dispersive nonlinear wave is shown in Fig. 10.

In the case of an applied uniform vertical magnetic field, the numerically simulated solitary wave initially starts to bounce from one side to another (Fig. 11) and at longer times establishes a quasistationary pattern that is totally different from the nonmagnetic case. For a typical electrolysis cell situation this is shown in Fig. 12. Remarkably, this result is not dependent on the initial conditions. Starting from a depression perturbation, the initial

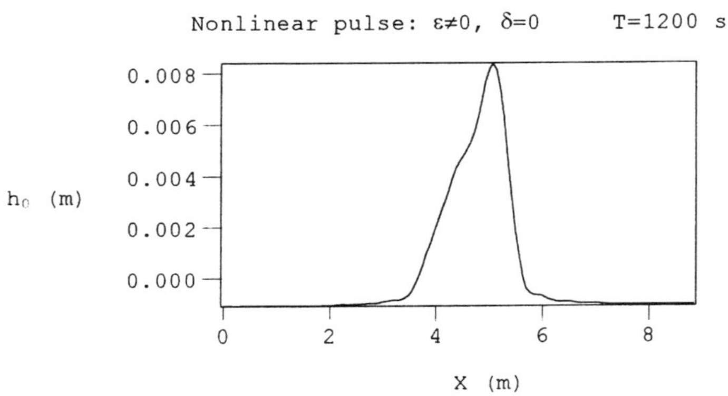

Fig. 8 Wave profile, nonlinear pulse.

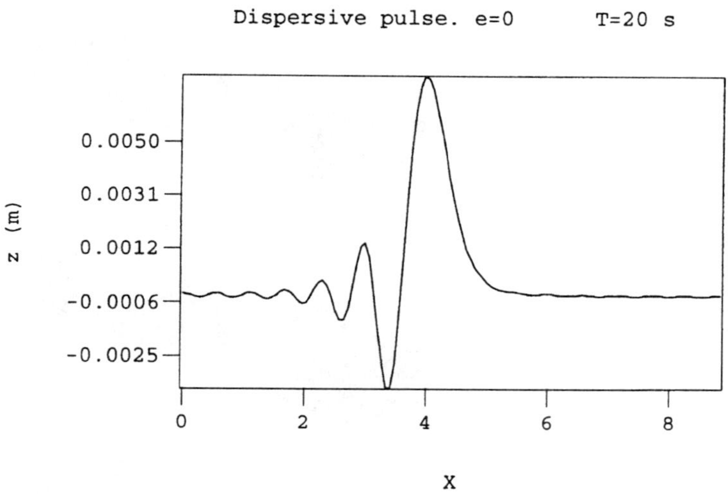

Fig. 9 Wave profile, dispersive pulse.

traveling wave is shown in Fig. 13, yet the long-time behavior, shown in Fig. 14, is quite similar to the previous case (compare Fig. 12). However, we need to note that the particular wave patterns evolving from the electromagnetic interaction depend on the cell geometrical and material parameters and on the magnetic field distribution that is affected by the current supply and removal at the cell.

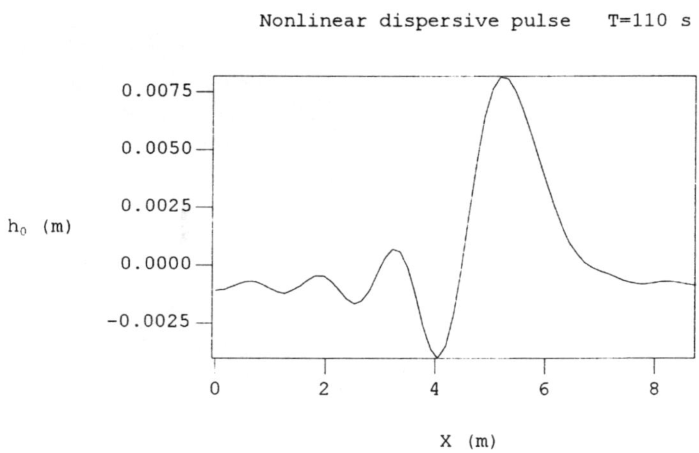

Fig. 10 Wave profile, nonlinear dispersive pulse.

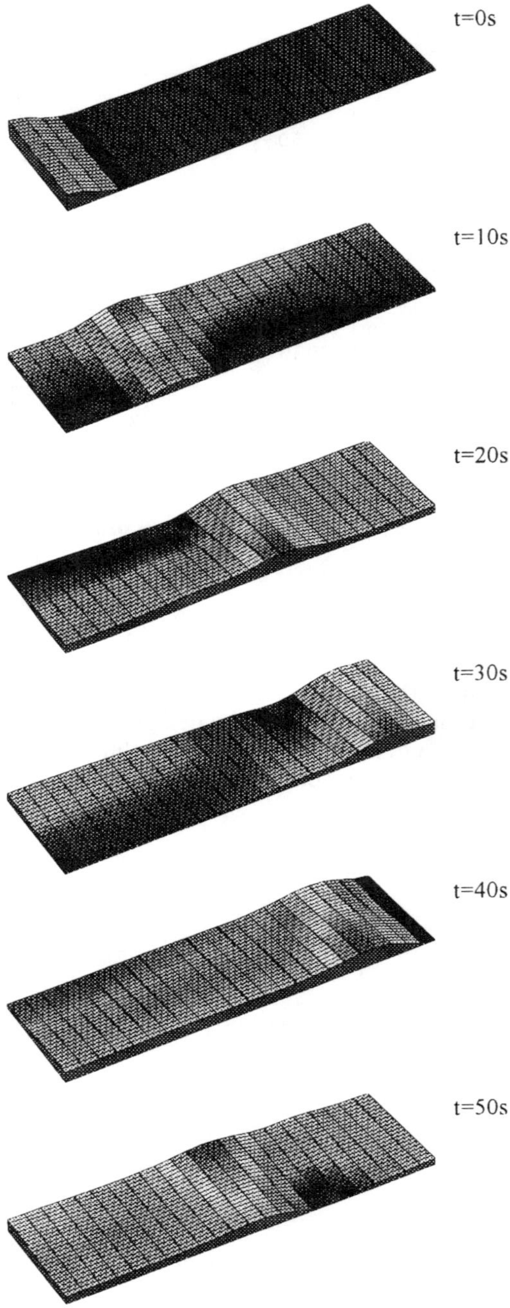

Fig. 11 Initial propagation of the step perturbation: $Lx = 9.8$ m, $Ly = 3.2$ m, $Bz = 0.003$ T, and $I = 160{,}000$ A.

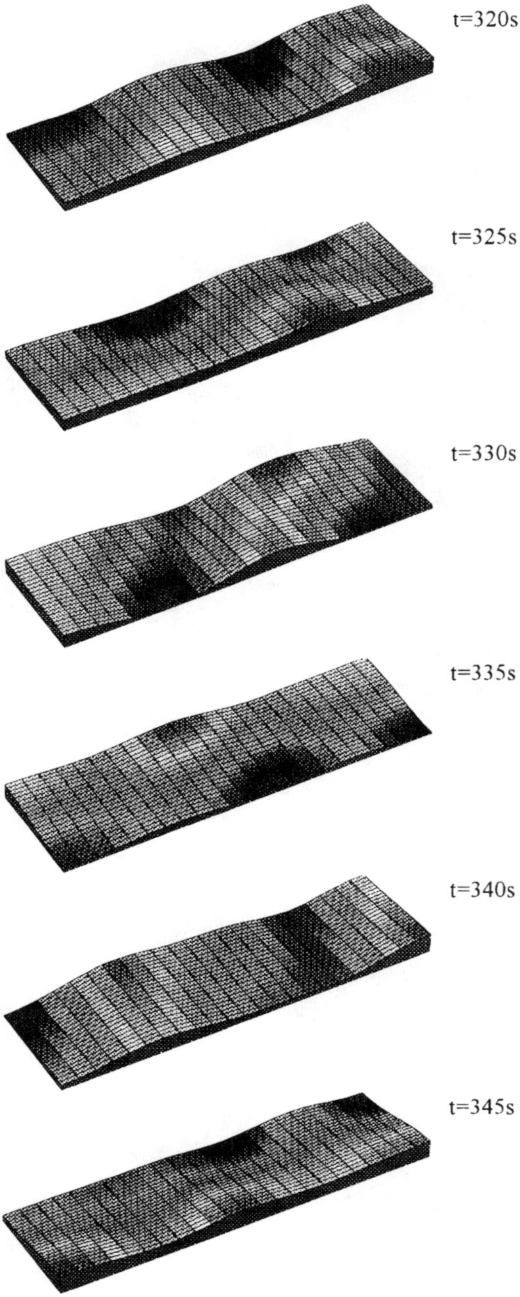

Fig. 12 Established oscillations (from initial step perturbation): *Lx* = 9.8 m, *Ly* = 3.2 m, *Bz* = 0.003 T, and *I* = 160,000 A.

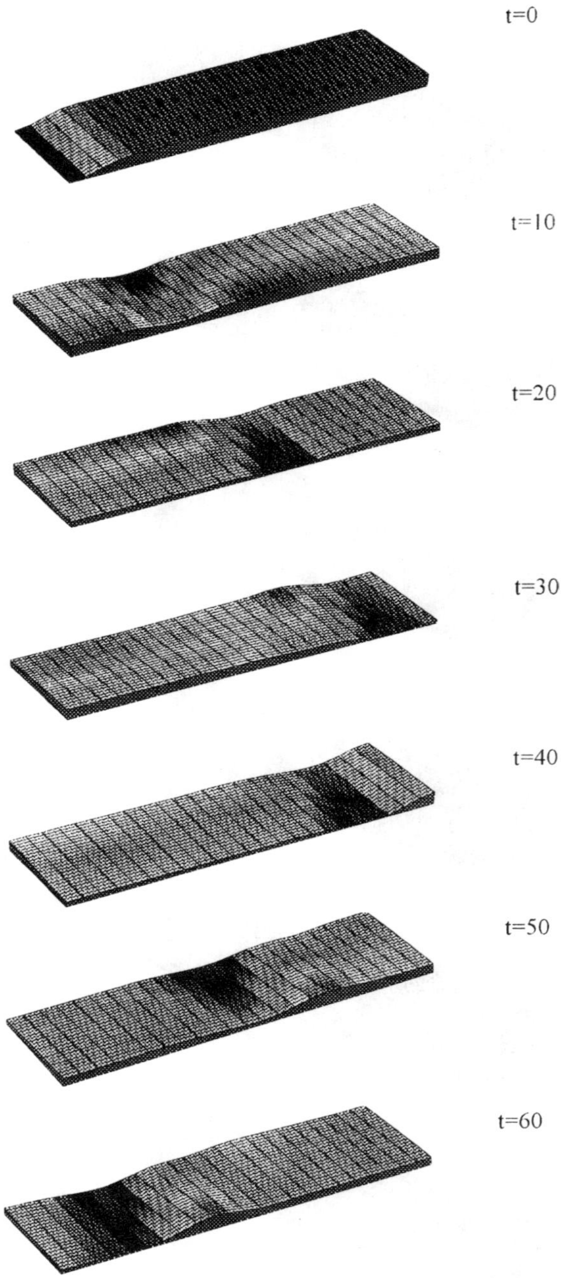

Fig. 13 Initial propagation of the depression at two fluid interfaces: Lx = 9.8 m, Ly = 3.2 m, Bz = 0.003 T, and I = 160,000 A.

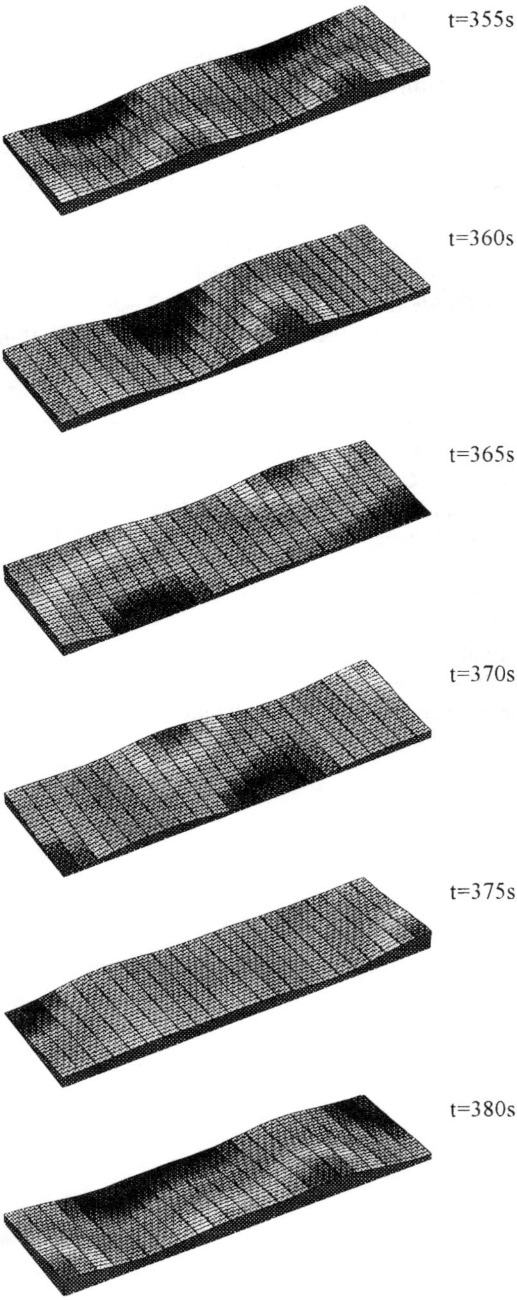

Fig. 14 Established oscillations (from initial depression perturbation): Lx = 9.8 m, Ly = 3.2 m, Bz = 0.003 T, and I = 160,000 A.

Acknowledgment

The research described in this publication was made possible in part by Grant N LBF000 8 from the International Science Foundation.

References

[1]Urata, N., Mori, K., and Ikeuchi, H., "Behavior of Bath and Molten Metal in Aluminium Electrolytic Cell," *Keikinzoku*, Vol. 26, No. 11, 1976, pp. 573–600.

[2]Bojarevics, V., and Romerio, M. V., "Long Waves Instability of Liquid Metal-Electrolyte Interface in Aluminium Electrolysis Cells: A Generalization of Sele's Criterion," *European Journal of Mecha B/Fluids*, Vol. 13, No. 1, 1994, pp. 33–56.

[3]Hofman, M., "Nonlinear Wave on the Free Surface of an Electrically Conducting Liquid," *Wave Motion*, Vol. 5, April 1983, pp. 115–124.

[4]Oshima, S., and Yamane, R., "Nonlinear Wave of Liquid Metals under a Transverse Magnetic Field," *Liquid Metal Flows: Magnetohydrodynamics and Applications,* Vol. 111, Progress in Astronautics and Aeronautics, *edited by H. Branover, M. Mond, and Y. Unger, AIAA, Washington, DC, 1987, pp. 100–118.*

[5]Renouard, D. P., Tomasson, G. G., and Melville, W. K., "An Experimental and Numerical Study of Nonlinear Internal Waves," *Physics of Fluids, A*, Vol. 5, No. 6, 1993, pp. 1401–1411.

[6]Tadjbakhsh, I. and Keller, J. B., "Standing Surface Waves of Finite Amplitude," *Journal of Fluid Mechanics*, Vol. 8, 1960, pp. 442–451.

[7]Heemink, A. W., "Two Dimensional Shallow Water Flow Identification," *Applied Mathematical Modelling*, Vol. 12, 1988, pp. 109–118.

[8]Moreau, R., and Ewans, J. W., "An Analysis of the Hydrodynamics of Aluminium Reduction Cells," *Journal of Electrochemical Society*," Vol. 131, No. 10, pp. 2251–2259.

Chapter 59

Transfer Processes in the Manufacturing of Single Silicon Crystals and High-Temperature Alloys by Increasing Flow Intensity

A. Mühlbauer,* A. Jakovics,† J. Virbulis,‡ and E. Westphal‡
University of Hanover, Hanover, Germany

I. Introduction

INTRODUCTION of new or improvement of existing melting processes is characterized by the growing use of an electromagnetic (EM) field. Both of these concern the growth of single silicon crystals with large diameters (Fig. 1a) and the melting of industrial used high-temperature alloys like titanium-aluminides (TiAl) within the IFCC (Fig. 1b). The following features are characteristic for these processes:

1) The melting of conductive material is caused by the induced high-frequency currents. Corresponding to the typical data shown in Table 1, the penetration depth of the EM field normalized on the radius r_0 of the melt is very small:

$$\hat{\delta} = \frac{\delta}{r_0} = \sqrt{\frac{1}{\pi f \sigma \mu_0 r_0^2}} = \sqrt{\frac{2}{\hat{\omega}}} \ll 1 \tag{1}$$

where $\hat{\omega}$ is the dimensionless frequency of the EM field. As a result of that, not only the density of heat sources

$$q = j2/2\sigma \tag{2}$$

*Professor, Institute of Electroheat.
†Guest Professor, Institute of Electroheat.
‡Research Associate, Institute of Electroheat.

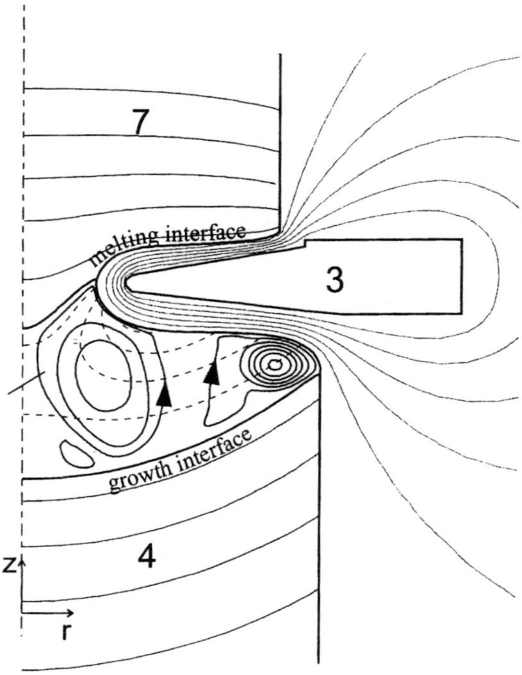

Fig. 1 Scheme of the FZ process a) and the IFCC b) with depiction of melt flow and EM lines of force: 1, melt; 2, cold crucible; 3, inductor; 4, skull or single crystal; 5, bottom of crucible; 6, slitted bottom; 7, feed rod; and ○, points of measurement.

but also the density of EM forces

$$f^E(r,z) = \left(f_r^E, 0, f_z^E\right) = \Re\left[j; B^*\right]/2 \tag{3}$$

exponentially decreases with the coordinate l perpendicular to the surface of the charge ($\sim e^{-2l/\delta}$). So there are quasi-no-direct effects of the EM field outside a thin boundary layer with thickness δ.

2) Because of very high power densities (2) at the surface, necessary for melting these materials, a strong deformation of the melt surface cannot be avoided. The shape of the free melt surface is governed by the balance of hydrostatic and EM pressure along the surface S (Ref. 1):

$$(h_m - z)\rho g = p^E\big|_S = \boldsymbol{B}^2/4\mu_0\big|_S \tag{4}$$

where h_m is the height of the supported melt area and the $\boldsymbol{B}$ is the magnetic induction. In both cases the EM supported melt is grounded on a solid phase of the same material, which is called skull in the IFCC process. The shape of the crystallization interface is governed by the temperature distribution,

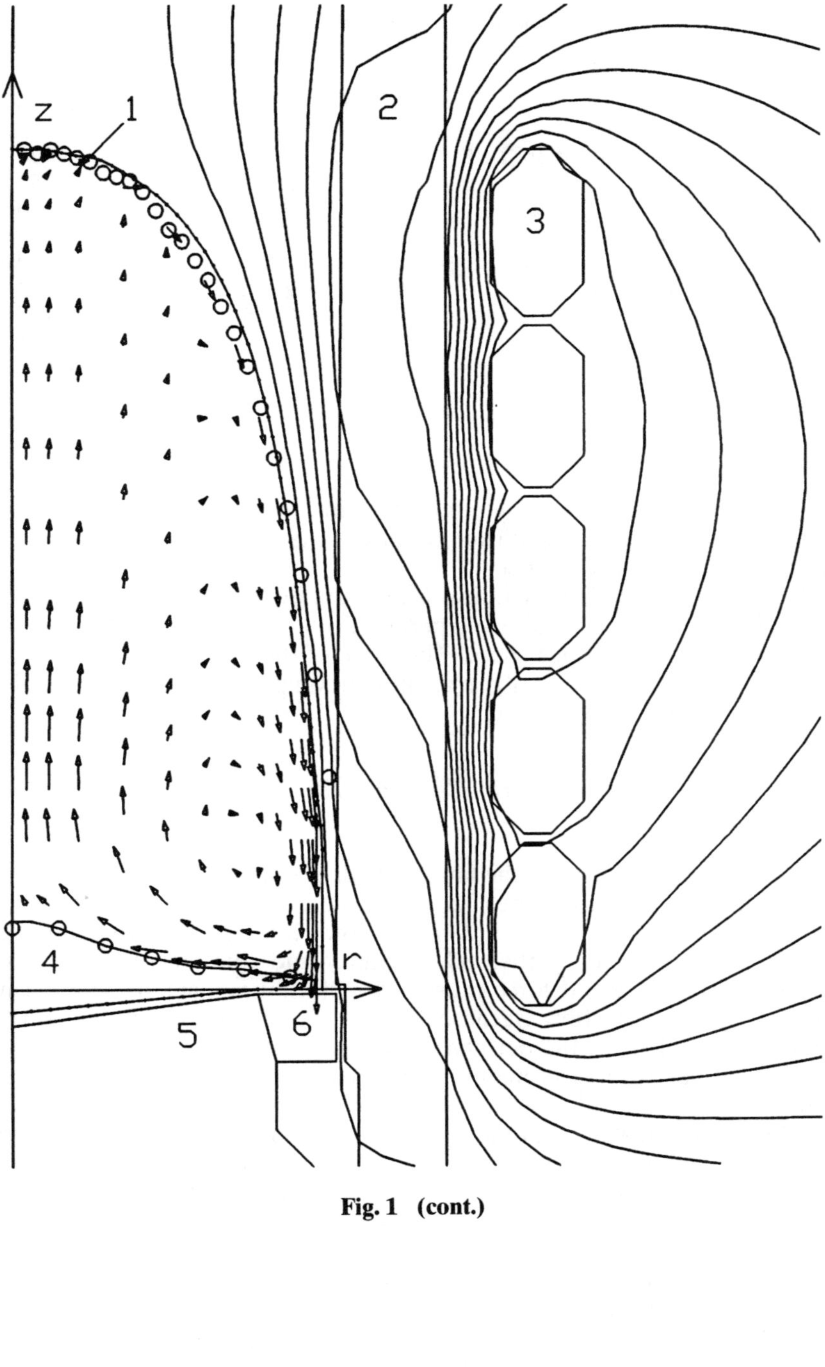

Fig. 1 (cont.)

Table 1 Characteristic parameters for the floating zone (FZ) and IFCC processes

Parameter, unit	Silicon, FZ process	TiAl, IFCC process
Temperature of crystallization, T_0, K	1685	1770
Density, ρ, kg/m^3	2530	3750
Electrical conductivity, σ, $1/\Omega \cdot$ m	$1.2 \cdot 10^6$	$1.0 \cdot 10^6$
Kinematic viscosity, ν_L, m^2/s	$3.4 \cdot 10^{-7}$	$2.9 \cdot 10^{-7}$
Frequency of the EM field, f, Hz	$2.8 \cdot 10^6$	$4–10 \cdot 10^3$
Characteristic current, I_0, A	800	3000
Characteristic radius, r_0, m	0.05–0.075	0.07
Dimensionless frequency, $\hat{\omega}$	$6.6–14.8 \cdot 10^4$	154–385
Normalized penetration depth, δ/r_0	0.0055	0.07–0.11

which is strongly affected by convection inside the melt. Caused by the effects of support and crystallization, various shapes of the melt surface with concave and convex areas occur.

3) The mean turbulent melt flow is initiated by the rotational component of the EM force which is directed perpendicular to the surface and changes its value along the surface shape. The drive of the flow is described by the rotor of this force that has only an azimuthal component and is acting mainly within the penetration depth (1):

$$\left(\text{rot } f^E\right)_\alpha \approx \partial f_n^E / \partial\tau \tag{5}$$

where n is the normal and τ the tangential direction along the surface of the melt. Within the floating zone (FZ) process the strength of the current and the penetration depth of the EM field is low (Table 1); therefore, the melt flow is also governed by thermal buoyancy and rotation of the crystal.[2] Inside the IFCC the EM force caused by melt flow is much more intensive and is dominating the flow. Because of this sort of drive, the melt flow is very intensive especially in the peripheral areas and at the surface of the melt, which is proved by measurements of the melt flow in conventional coreless induction furnaces.[3]

All this goes to show that the resulting melt flow depends very much on the shape of the melt area. The melt flow also determines the following technologically desired properties for the melting process:

1) It is important for the balance of temperature inside the melt so that partial overheating can be avoided and the shape of the crystalization interface can be regulated. Within the FZ process thermal stress, which can cause fractures, is reduced.

2) The melt flow leads to a better homogenization of the melt components, e.g., with regard to the distribution of doping elements inside the

silicon, so that local variations of the material properties are reduced. This raises the material quality for further processing and application.

3) Also it governed the stability of the free surface of the melt, one of the main process parameters, which can limit the productivity of the melting process.

Growth of the supported melt mass is connected to an increase of the magnetical induction on the melt surface [Eq. (4)]. On one hand, this leads to intensification of buoyancy and EM-caused melt flow, which benefits the desired homogenization inside the melt. On the other hand, this produces some unwanted effects like nonstationary melt flow pattern and instability of the free melt surface. This type of nonstationary melt flow occurs in industrial FZ processes.[2] With an increasing diameter of the crystal, also Reynolds numbers ($Re = v_0 r_0 / v_L > 10^3$) are reached that are typical for the development of a turbulent flow. In difference within the IFCC melt flow caused by buoyancy is secondary, but already in small devices the EM-caused melt flow shows a distinct turbulent pattern.[3] This is conditioned by the characteristic dependence of the EM-caused melt flow upon frequency and inductor current

$$\boldsymbol{v}_0 \sim I_0 \Big|_{\hat{\omega}=\text{const}}; \qquad \boldsymbol{v}_0 \sim \hat{\omega}^{0{,}25} \Big|_{I_0=\text{const}} \tag{6}$$

In the IFCC the current is higher and the frequency is lower (Table 1). The average intensity of turbulent flow fluctuation v' and dynamic pressure p_T that occurs on the free melt surface can be approximated by

$$\sqrt{< v'v' >} \sim \boldsymbol{v}_0\,; \qquad p_T \sim v_0^2 \tag{7}$$

The melt flow within the IFCC is expected to show turbulent-caused perturbations (instabilities) on the free surface [Eq. (6)].

For the modeling of the nonstationary laminar and turbulent melt flow pattern within an axisymmetric area with any shape of surface taking into account the rotation ($v_\alpha \neq 0$), a method based on the k-ε; turbulence model[4] is developed.

II. Description of the Mathematical Model and Calculation Method

The basis for the numerical simulation of the axisymmetrical flow in the charge while using natural hydrodynamic variables is formed by the following equations:

1) Equations of motion with the variable effective viscosity $v_E = v_L + v_T$ and the velocity components $\boldsymbol{v}(r, z, t) = (v = v_r, w = v_\alpha, u = v_z)$:

$$\begin{gathered}\frac{\partial u}{\partial t} + \frac{1}{r}\frac{\partial (ruv)}{\partial r} + \frac{\partial u^2}{\partial z} \\ = -\frac{1}{\rho}\frac{\partial p}{\partial z} + \frac{f_z^E + f^{\text{Th}}}{\rho} + 2\frac{\partial}{\partial z}\left(v_E \frac{\partial u}{\partial z}\right) + \frac{1}{r}\frac{\partial}{\partial r}\left[r v_E \left(\frac{\partial v}{\partial z} + \frac{\partial u}{\partial r}\right)\right]\end{gathered} \tag{8}$$

$$\frac{\partial w}{\partial t}+\frac{vw}{r}+v\frac{\partial w}{\partial r}+u\frac{\partial w}{\partial z}=\frac{\partial}{\partial z}\left(\nu_E\frac{\partial w}{\partial z}\right)+\frac{1}{r^2}\frac{\partial}{\partial r}\left[r\nu_E\left(r\frac{\partial w}{\partial r}-w\right)\right] \tag{9}$$

$$\frac{\partial v}{\partial t}+\frac{1}{r}\frac{\partial(rv^2)}{\partial r}+\frac{\partial(uv)}{\partial z}$$
$$=-\frac{1}{\rho}\frac{\partial p}{\partial r}+\frac{f_r^E}{\rho}+\frac{2}{r}\frac{\partial}{\partial r}\left(r\nu_E\frac{\partial v}{\partial r}\right)+\frac{\partial}{\partial z}\left[\nu_E\left(\frac{\partial v}{\partial z}+\frac{\partial u}{\partial r}\right)\right]-2\frac{\nu_E v}{r^2} \tag{10}$$

where $f^{Th} = -\rho g\beta_{\mathrm{Th}}(T-T_0)\boldsymbol{e}_z$ is the buoyancy force.

2) The incompressibility condition.

3) Transport equations for the specific turbulence energy k and the dissipation rate ε of this energy by using the RNG extension of the k-ε turbulence model[5]:

$$\frac{1}{r}\frac{\partial(rv)}{\partial r}+\frac{\partial u}{\partial z}=0 \tag{11}$$

$$\frac{\partial k}{\partial t}+\frac{1}{r}\frac{\partial(rvk)}{\partial r}+\frac{\partial(uk)}{\partial z}$$
$$=q-C_k\varepsilon+\frac{\partial}{\partial z}\left[\left(\frac{\nu_T}{\sigma^k}+\nu_L\right)\frac{\partial k}{\partial z}\right]+\frac{1}{r}\frac{\partial}{\partial r}\left[r\left(\frac{\nu_T}{\sigma^k}+\nu_L\right)\frac{\partial k}{\partial r}\right] \tag{12}$$

$$\frac{\partial\varepsilon}{\partial t}+\frac{1}{r}\frac{\partial(rv\varepsilon)}{\partial r}+\frac{\partial(u\varepsilon)}{\partial z}=\left[C_{1\varepsilon}q-\left(C_{2\varepsilon}+\frac{C_\nu\eta^3(1-\eta/\eta_0)}{1+\beta\eta^3}\right)\varepsilon\right]\frac{\varepsilon}{k}$$
$$+\frac{\partial}{\partial z}\left[\left(\frac{\nu_T}{\sigma^\varepsilon}+\nu_L\right)\frac{\partial\varepsilon}{\partial z}\right]+\frac{1}{r}\frac{\partial}{\partial r}\left[r\left(\frac{\nu_T}{\sigma^\varepsilon}+\nu_L\right)\frac{\partial\varepsilon}{\partial r}\right] \tag{13}$$

with

$$q=\nu_T\left\{2\left[\left(\frac{\partial u}{\partial z}\right)^2+\left(\frac{\partial v}{\partial r}\right)^2+\left(\frac{v}{r}\right)^2\right]+\left(\frac{\partial u}{\partial r}+\frac{\partial v}{\partial z}\right)^2+\left(\frac{\partial w}{\partial z}\right)^2+\left[r\frac{\partial}{\partial r}\left(\frac{w}{r}\right)\right]^2\right\} \tag{14}$$

as a function that describes the generation of the turbulent energy and

$$\eta=\sqrt{q/(C_\nu\varepsilon)} \tag{15}$$

While working with the k-ε model, the following values for parameters are used in the transport equations:

$$C_\nu = 0.084; \quad C_{1\varepsilon} = 1.42; \quad C_{2\varepsilon} = 1.68; \quad C_k = 1.0$$
$$\sigma k = 0.72; \quad \sigma\varepsilon = 0.72; \quad \beta = 0.012; \quad \eta_0 = 4.38 \tag{16}$$

The kinematic turbulent viscosity is calculated by

$$\nu_T = C_\nu k^2/\varepsilon \tag{17}$$

The traditional boundary conditions for the normal and tangential velocity components and characteristics of turbulence are used on the axis of symmetry and free surface of the liquid:

$$v_n = 0; \quad \frac{\partial v_\tau}{\partial n} = 0; \quad \frac{\partial k}{\partial n} = \frac{\partial \varepsilon}{\partial n} = 0; \quad w\big|_{r=0} = 0 \tag{18}$$

The boundary conditions for the tangential velocity components are defined within the melt close to the solid wall. There it depends on the normalized distance between the point of the examined flow and the wall. Outside this viscous sublayer the logarithmic wall function for the turbulent flow is used.

The distribution of the vector potential of the magnetic field with high frequencies is calculated by using the boundary element method.[6] The solution of Eqs. (8–17) is achieved on the basis of the modification of the SIMPLE procedure.[7] The velocity components v_α, v_r, and v_z, as well as scalar values (p, k, ε, v_E), are determined on mutually shifted differential grids. The not-equidistant grid is fitted to the shape of the free surface, which is iteratively computed from the balance of hydrostatic and EM pressure at the surface [Eq. (4)] (Ref. 8). For the analyzed processes the melted areas show a relatively simple shape that is marked by the free melt surface and the crystallization interface. This has the advantage to use a large-meshed grid without much loss of accuracy, which leads to relatively low needs of computing power, e.g., for the discretization of the melt area within the IFCC, only 1200 elements are used, so that computation is possible on a Pentium-PC. Within the modeling of the FZ process, not only the melt flow pattern but also the temperature distribution and the growth interface are computed. For the discretization the finite-element method[2] with 3000 elements is used, and the computing is done on a Sun workstation.

III. Discussion of the Results

A. Floating Zone Process

Besides the pulling rate and the inductor current, the rotation velocity of the crystal is a parameter that can be easily varied for the influencing of the FZ process. In the FZ system shown in Fig. 1a, the melt flow was investigated depending on the rotation velocity up to 20 rpm that corresponds to 0.1 m/s in the outlet area of the crystal.

For conditions corresponding to the industrial production (Table 1), the buoyancy-caused melt flow dominates over the EM-caused flow within the meridonal plane. The intensity of EM forces in the FZ system is relatively small in comparison with the EM force in IFCC because the current in coil is 2–3 times lower and the thickness of the skin layer is about 20 times thinner. Although the shape of the melt in the FZ system and in IFCC is different, the relations (5) and (6) are valid in both processes.

The maximal meridional flow velocities are reached at small rotation rates, and they correspond to the laminar flow area ($Re < 4000$). However, the flow is unstable, and periodic oscillations of the intensity and the extent of vortices take place (Fig. 2b). The buoyancy-caused melt flow in the

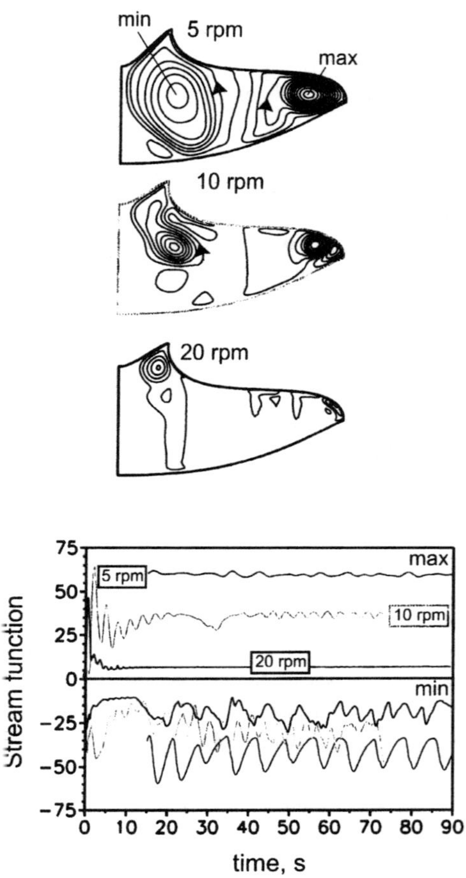

Fig. 2 Characteristic distribution of the stream function a) and of its oscillations within a FZ process by rotation velocities 0, 5, 10, and 20 rpm.

peripheral areas can be suppressed almost completely by increasing the rotation velocity of the crystal (Fig. 2a). The velocity oscillations are also suppressed and tend to zero. The inner vortex decreases and is shifted to the upper side. Thus the melt velocity and its oscillations at the growth interface are relatively small. On the one hand, this decreases the intensity of striations because the dopants are built in different layers of single crystal more regularly. On the other hand, because of the insufficient melt mixing the radial dopant distribution can be more irregular.

The low-frequency oscillation of the melt flow velocity in the central vortex has a period T_P that is smaller than the period T_R of the crystal rotation (for $T_R = 12$ s is $T_P \approx 8$ s and for $T_R = 6$ s is $T_P \approx 4$ s). Both periods T_P and T_R are approximately proportional $T_P \sim T_R$. The significant oscillations of the outer vortex at 10 rpm are caused by the strong interaction between the buoyancy and centrifugal forces (Fig. 2a). At a low rotation rate (5 rpm) and a high rotation rate (20 rpm) when the intensity of buoyancy and centrifugal forces differ to a high extent, the oscillations in the outer vortex are insignificant.

The rotation velocity influences the temperature field because of the melt convection and therefore the shape of the growth interface is also influenced. Figure 3 shows that with the increasing of the rotation rate the deflection of the growth interface increases. This leads to unwanted larger thermal stress in the single crystal. Although the increasing of the rotation rate reduces the flow oscillations at the growth interface, it also increases the

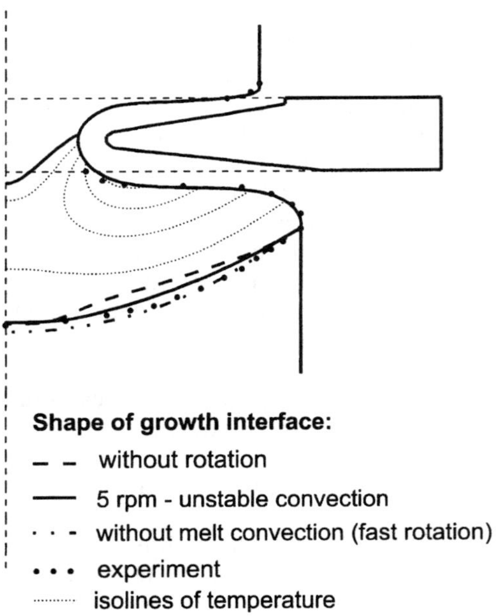

Fig. 3 Effect of rotation on the shape of the growth interface in the FZ process and the isolines of temperature in the melt.

deflection of growth interface and reduces melt mixing and, therefore, is not the general solution for the technical FZ applications.

B. Induction Furnace with a Cold Crucible

Main parameters for the IFCC process are the dimensionless frequency $\hat{\omega}$ and the electrical power, or rather the inductor current I_0. These parameters govern not only the shape of the melt area by influencing the EM supporting of the melt and the development of the skull but also the transportation of heat within the melt. Though not always possible to fit the construction of the IFCC perfectly to all its tasks, these parameters can be changed easily and used for the controlling of the IFCC process. For an IFCC shown in Fig. 1b, numerical investigations are made assuming an aluminum melt with a mass m_{Al} =4.5 kg and a density ρ_{Al} = 2370 kg/m^3 where the current (I_0 = 1500–4000 A) and the frequency ($\hat{\omega}$ = 120–1200) are varied.

The penetration depth of the EM field in the melt and the copper crucible is very small within the used frequency spectrum [Eq. (1)]. Therefore the hydrostatic approximated shape of the flow area changes only a little with the variation of the frequency if the current is left constant. Figures 4a and 4c show that for I_0 = 3000 A the change of the height of the supported melt is lower than 2%. The decrease of the characteristic flow velocity v_0, which is the maximum velocity on the axis of symmetry, with a growing frequency is a little bit lower as expected from Eq. (6) $v_0 \sim \hat{\omega}^{-0.2}$. For the limiting case $\hat{\omega} \rightarrow \infty$, the EM forces become potential (rot f^E = 0), and, therefore, the intensity of the flow becomes zero ($v_0 \rightarrow 0$).

According to Eq. (7), a decrease of the intensity of the turbulent flow oscillations $\sqrt{<v'v'>}$ on the free melt surface can be expected if the frequency is raised. Direct interactions between the EM field and the turbulent flow oscillations can be neglected because of the low penetration depth. The dynamic pressure caused by this pulsation can be expressed as $p_T \sim <v'v'>$. Therefore a relatively quick decrease of the pressure of the turbulent pulsations at the free surface of the melt is following from Eq. (6) $p_T \sim \hat{\omega}^{-1/2}$ if the current I_0 is held constant. So a calm-down of the free surface of the melt with regard to the short-lived turbulent oscillations can be expected. Also the development of inertia waves and low frequency oscillations of the flow that are connected to the rotation of the melt within the meridional plane[3] lose significance because the average velocities decrease. However, the increase of the frequency does not directly affect the development of surface waves.[1]

For the increase of the electrical power, three regions with essential differences in the characteristic of the melt flow can be noticed:

1) The melt is partly supported by the EM field. This case is mostly connected with the existing skull on the bottom and the cooled wall of the crucible.

The maximum of the force normal to the surface of the melt is reached in the middle area of the melt height (Fig. 5a). Without a skull the force density remains on a relatively high level even in the zone where the melt is in

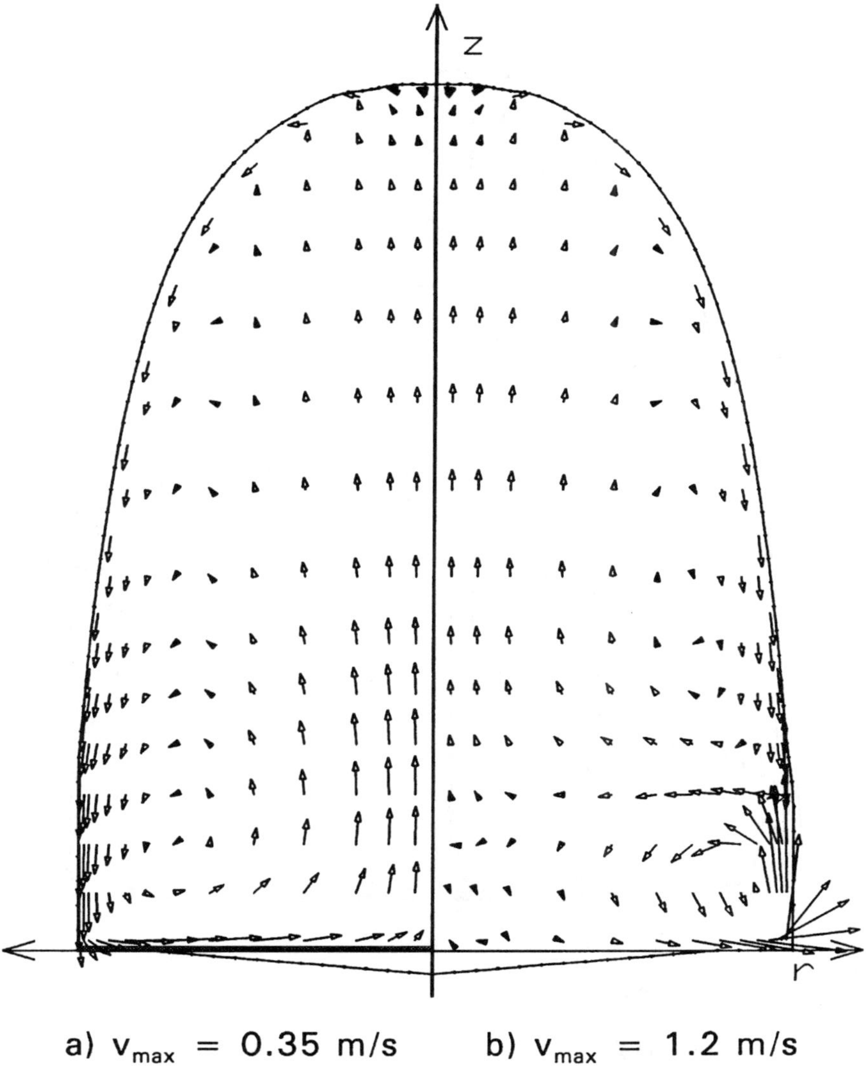

Fig. 4 Shape of the free surface and the distribution of velocity of the melt flow in an IFCC with r_0 = 0,7 for different conditions: a) v_{max} = 0.35 m/s, I_0 = 3000 A, $\hat{\omega}$ = 1200, with skull at the bottom; b) v_{max} = 1.2 m/s, I_0 = 3000 A, $\hat{\omega}$ = 1200, without skull at the bottom; c) v_{max} = 0.45 m/s, h_m = 0.187 m, I_0 = 3000 A, $\hat{\omega}$ = 120; and d) v_{max} = 0.45 m/s, h_m = 0.138 m, I_0 = 1500 A, $\hat{\omega}$ = 1200, with partly supported melt by a wall.

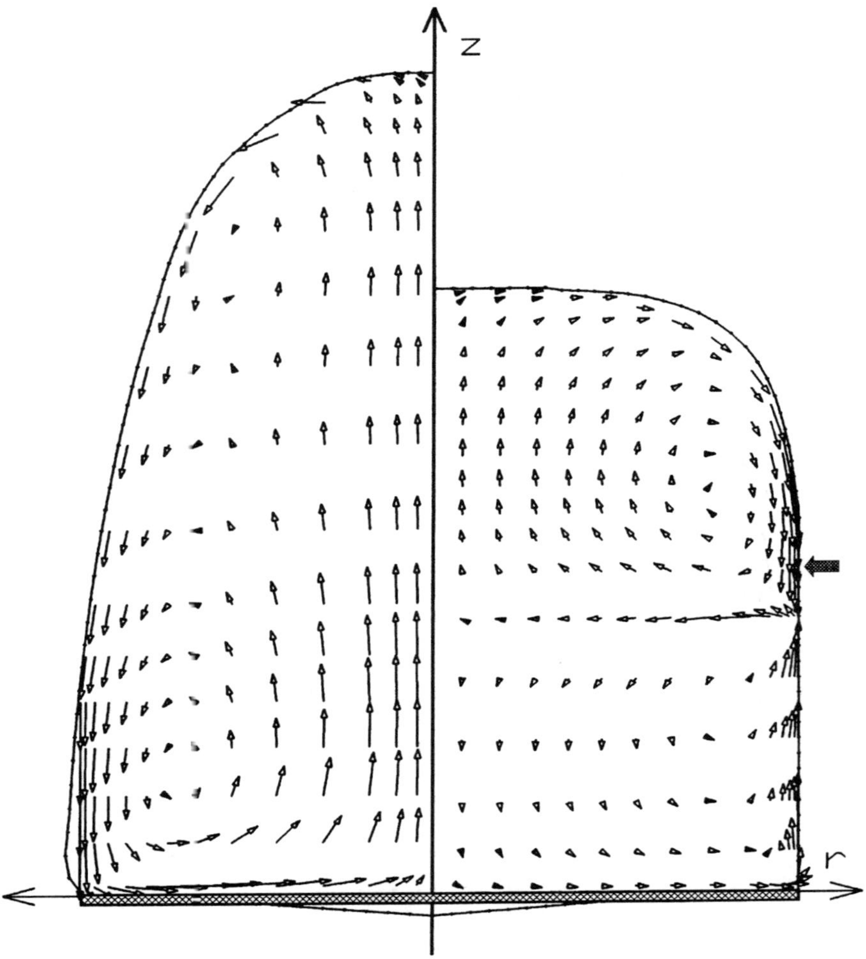

c) $v_{max} = 0.45$ m/s; $h_m = 0.187$ m

d) $v_{max} = 0.45$ m/s; $h_m = 0.138$ m

Fig. 4 (cont.)

contact with the wall. Caused by relatively low value of the rotor of force in the bottom area ($f_r^E \approx$ const), a double loop that is characteristic for conventional coreless induction furnaces occurs in the meridional plane (Fig. 4d). The velocity of flow in the upper loop at the free surface of the melt is always directed downward. With growing EM support of the melt, the upper loop increases downward, and the lower loop is suppressed within the vertical direction.

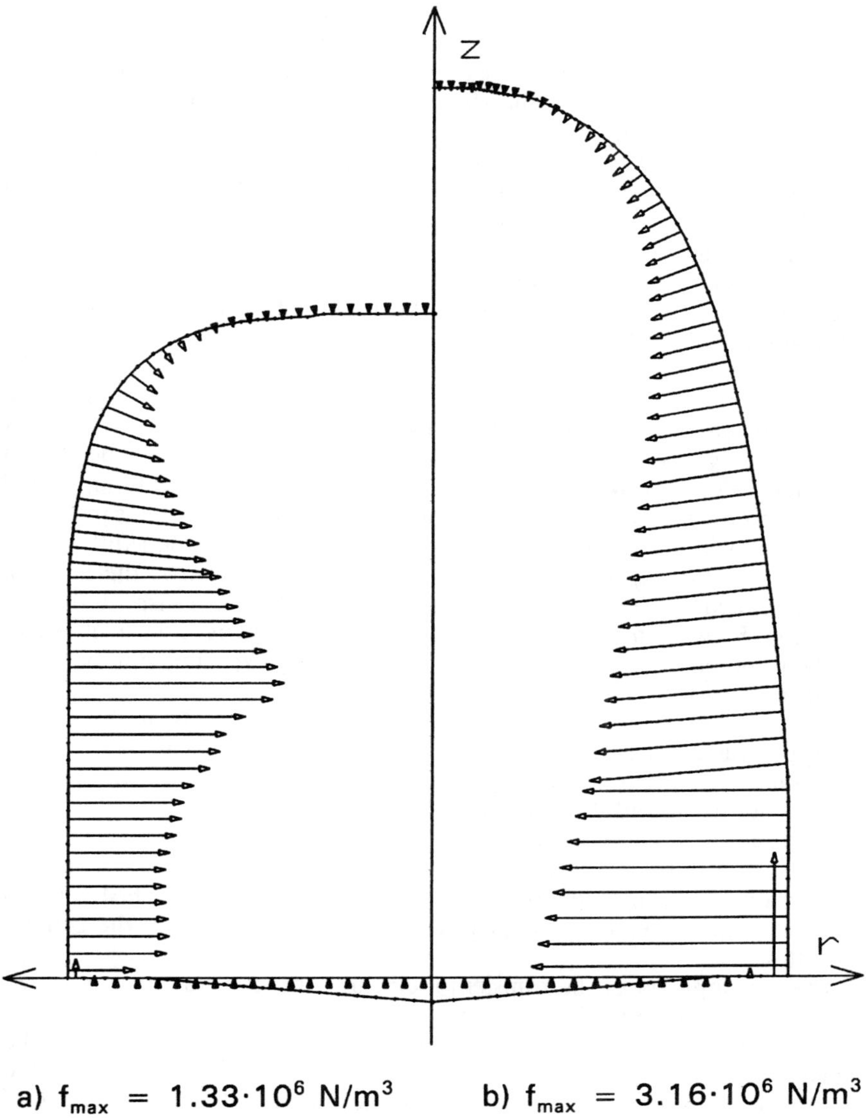

Fig. 5 Distribution of the density of the EM force at the melt surface in an IFCC with $\hat{\omega} = 1200$: a) $f_{max} = 1.33 \cdot 10^6$ N/m^3, $I_0 = 1500$ A, $h_m = 0.138$ m and b) $f_{max} = 3.16 \cdot 10^6$ N/m^3, $I_0 = 3000$ A, $h_m = 0.184$ m.

The development of a skull in the areas where the melt is in contact to the crucible is mainly affecting the lower loop. The absolute intensity of the melt flow can even grow caused by locally increasing values of the rotor of force [Eq. (5)]. Because the shape of the melt flow area is essentially changing with the increase of the inductor current and accordingly the growing of the EM support, there is no proportionality [Eq. (6)] between the characteristic velocity and the inductor current. Therefore intensity of the flow with partly supported melt is similar to the case the melt fully supported (Fig 4a,4d).

2) The melt is fully supported by the EM force. Only on the bottom of the crucible a skull remains that causes the melt in this area to be shielded against the EM field (Fig. 4a).

According to Eq. (4), the maximum of the magnetic induction and the corresponding EM force density, which is directed mainly radially, is reached on the bottom of the supported melt near the skull. Therefore the melt flow is directed downward along the whole free surface and toward the axis of symmetry along the skull (Figs. 4a and 4c). In contrast to a partly supported melt (Fig. 4d), only one loop occurs. The maximum intensity of this melt flow is reached within the lower area of the free surface near the skull. Corresponding to Eq. (2), this area of the free surface also shows the greatest density of heat sources, and with the melt flow this heat is transported along the skull. So we can expect that the skull is melted more intensively especially in the peripheral area, which is confirmed by experimental measurement of the shape of the skull (Fig. 1b).[8]

We must point out that the EM field has different effects on the kinetic energy of the average melt flow within different melt surface areas:

In the upper areas of the melt surface, the flow is directed toward the surface and is slowed down by the EM force $(\boldsymbol{v} \cdot \boldsymbol{f}^E) < 0$. This reduces the kinetic energy of the total flow.

In the lower area where the melt flow is directed along the skull, the EM force shows the same direction $(\boldsymbol{v} \cdot \boldsymbol{f}^E) > 0$ and because the EM force is much stronger here, this area governs the rotation direction of the general melt flow.

The development of the transient melt flow from the state of rest shows that in the beginning of the process the melt is accelerated very quickly. With the increase of dissipation, the characteristic average intensity of the quasistationary state melt flow is reached soon. The general flow within the whole melt is mainly governed by inertia, and the development of this flow is slower than one order of magnitude. Therefore the resulting melt flow pattern is strongly connected to the shape of the skull and the free surface in the lower areas where the effect of the EM force is maximum:

A strong diversion of the intensive flow along the surface is caused geometrically by the small radius of the melt surface.

Because the force distribution within the melt depends on the development of the skull, the rotor of force [Eq. (5)] can change its sign, and the moment of force can change its direction.

3) If the induced power is increased further [Eq. (2)], the whole skull is melted especially in the peripheral areas, and the shielding of the melt against the EM field disappears.

In this case the melt on the slitted bottom of the crucible (6 in Fig. 1b) is affected with additional EM forces that are mainly directed upward (Fig 5b). Therefore the rotor of force [Eq. (5)] changes its sign and reaches great negative values in the area of transition between the slitted bottom and the free melt surface. As a result, an additional, very intensive loop occurs that is located in this peripheral area near the bottom (Fig. 4b). A comparison of the flow intensities with I_0 = const shows that within the upper loop the maximum flow intensity roughly stays the same whether there is a skull or not; in contrast the maximum flow velocity within the additional lower loop is several times higher. This causes the construction elements of the IFCC to be extremely stressed from the intensive turbulent flow that occurs in this area. Therefore such extreme conditions cannot be recommended for industrial applications.

IV. Conclusion

A numerical calculation method using a model for nonstationary laminar flow, and the RNG extension of the k-ε model for medium turbulent flow pattern within any axisymmetric area was developed. The model also takes into account azimuthal velocities. All of this makes the program system useful for many tasks in designing industrial equipment.

Investigations on the nonstationary melt flow within the FZ process with different rotation velocities have shown that with increasing rotation velocity the meridional flow is strongly stabilized and suppressed. Also the shape of the growth interface becomes more curved. Therefore a steady and fast rotation does not fit all technical requirements.

Within the investigations on the turbulent melt flow in an IFCC, an essential connection between intensity and structure of the flow and the support of the melt and the development of the skull that depends on the inductor current was found out. With increasing frequency of the current, an essential calm-down with regard to some disturbance caused by the melt flow at the free surface is expected. The melt flow significantly differs from laws known for furnaces with a ceramic crucible. As long as the skull on the bottom is not melted, the maximum flow intensity can stay nearly constant with an increasing inductor current. But a further increase of the induced power leads to critical melt flow pattern within the IFCC that may not be useful for industrial applications.

References

[1]Hegewaldt, F., Buligins, L., and Jakowitsch, A., "Transiente Badkuppenausbildung beim Einschalten eines Induktionstiegelofens," *Electrowärme International B,* Vol. 51, No. 1, 1993, pp. 28–43.

[2]Mühlbauer, A., Muiznieks, A., Virbulis, J., Lüdge, A., and Riemann, H., "Interface, Shape, Heat Transfer and Fluid Flow in the Floating Zone Growth of Large Silicon-

Crystals with Needle-Eye Technique," *Journal of Crystal Growth*, Vol. 151, 1995, pp. 66–79.

[3]Baake, E., Mühlbauer, A., Jakovich, A., and Nacke, B., "Experimental and Numerical Investigations of the Turbulent Material Exchange in the Melt of Induction Crucible Furnaces," *International Symposium on Electromagnetic Processing of Materials*, Nagoya, 1994, pp. 529–535.

[4]Jakovich, A., and Mühlbauer, A., "Use of k-ε Turbulence Model for Calculation of Flows in Coreless Induction Furnaces," *Progress in Turbulence Research*, Vol. 162, Progress in Astronautics and Aeronautics, AIAA, Washington, DC, 1994, pp. 323–336.

[5]Yakhot, V., Orszag, S. A., Thangam, S., Gatski, T. B., and Speziale, T. B., "Development of Turbulence Models for Shear Flows by a Double Expansion Technique," *Physics of Fluids A*, Vol. 4, 1992, pp. 1510–1520.

[6]Mühlbauer, A., Muiznieks, A., and Jakowitsch, A., "Modellierung des Elektromagnetischen Feldes in Induktionsöfen Mit Kaltem Tiegel," *Electrowärme International B*, Vol. 49, No. 3, 1991, pp. 130–141.

[7]Patankar, S., *Numerical Heat Transfer and Fluid Flow*, Hemisphere, New York, 1980.

[8]Westphal, E., Mühlbauer, A., Muiznieks, A., and Virbulis, J., "Study of the Heat Fluxes in an Induction Furnace with Cold Crucible," *Advances in Engineering Heat Transfer*, Southampton, 1995, pp. 471–480.

Chapter 60

Flow Structure in Special Channel Having Step and Obstacle

F. Tazioukov* and F. Garifoullin†
Institute of Mechanics, Kazan, 420111, Russia

Nomenclature

C = nondimensional solute
D = solute diffusivity
h = half-width of the channel
U_0 = velocity at the channel inlet
ν = kinematic viscosity
Ψ = stream function
ω = vortex function

I. Introduction

THE formation of a polarized gel-layer on a membrane surface limits its working period.[1] This phenomenon, often referred to as concentration polarization, is highly flux limiting and so needs intense investigation.

This investigation is an attempt at numerical simulation of the flow within the channel between two parallel, flat ultrafiltration membranes, as shown in Fig. 1, and a computer visualization of flow pattern.

The step and obstacle (1 and 2 in Fig. 1) are the flow turbulizators and serve as the reduction of the polarized gel layer, forming on the membrane surfaces and so for the membrane's self-purification.

II. Mathematical Formulation

Let us consider as a hydrodynamical model of the investigated process; the case in which Newtonian homogeneous fluid-containing macromolecular soluted flows within the channel, like that shown in Fig. 1. The following

*2131 Lobachevsky Str., P.O. Box 559.
†2131 Lobachevsky Str., P.O. Box 559.

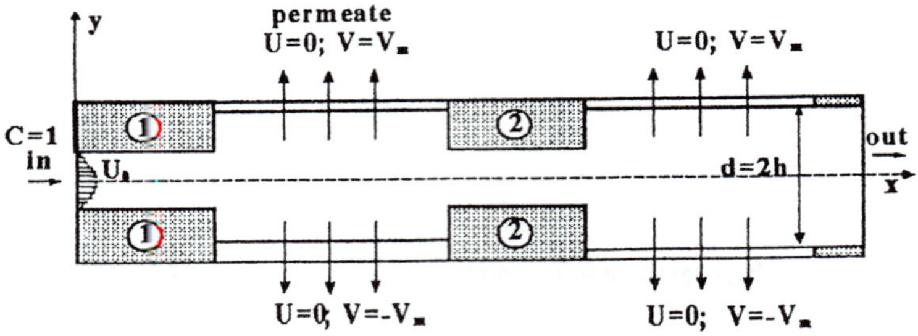

Fig. 1 Model under study.

suggestions were used for our model: a) flow is two-dimensional and symmetrical about the midplane of the channel; b) natural convection effects are assumed to be negligible; and c) product water is assumed to be permeable in both the upper and the lower membranes.

Under those conditions, in the absence of gravitation forces, our model can be described by the following systems of equations:

$$\frac{\partial \omega}{\partial t} + \frac{\partial \psi}{\partial y} \cdot \frac{\partial \omega}{\partial x} - \frac{\partial \psi}{\partial x} \cdot \frac{\partial \omega}{\partial y} = \frac{1}{Re} \cdot \nabla^2 \omega \tag{1}$$

$$\omega = -\nabla^2 \psi \tag{2}$$

$$\frac{\partial C}{\partial t} + \frac{\partial \psi}{\partial y} \cdot \frac{\partial C}{\partial x} - \frac{\partial \psi}{\partial x} \cdot \frac{\partial C}{\partial y} = \frac{1}{Pe} \nabla^2 C \tag{3}$$

where $Re = U_0 h/\nu$; $Pe = U_0 h/D$.

Appropriate boundary conditions for the diffusion equation are

$$C(0, y) = 1 \tag{4}$$

$$\left. \frac{\partial C}{\partial y} \right|_{y=0} = 0 \tag{5}$$

$$\frac{\partial C}{\partial y} = 0 \qquad \text{on the solid walls} \tag{6}$$

$$\frac{\partial C}{\partial y} = Pe_w \cdot C \qquad \text{on the membrane} \quad (7)$$

where

$$Re_w = \frac{V_m h}{\nu}; \qquad Pe = \frac{V_m h}{D}$$

The equations of motion and diffusion in the unsteady two-dimensional setup are solved by the finite-element method with computer visualization assuming a no-slip velocity condition on the membrane is surface.

Equation (4) is the condition for constant permeation flux, and Eq. (7) is the so-called gel polarization model.[2]

III. Results and Discussion

Some results of our numerical simulation are presented in Figs. 2 (lines of concentration), 3 (vortex lines), and 4 (stream function) for next values of the basic parameters $Re = 100$; $Re_W = 0{,}2$; and $Pe_W = 0{,}3$.

On these figures the stream function, vortex line, and concentration in different moments of the process are represented ($\nabla t = 0.0005$, $N_1 = 1000$ iterations, N_2=9000 iterations).

As can be seen from Figs. 2–4, the setting of the obstacles is very useful for the prolongation of the working period of the membrane surface, as far as vortex, formatting here, influences on macromoleculars transport from the concentrated polarized layer to the bulk flow. Thus, the net effect of

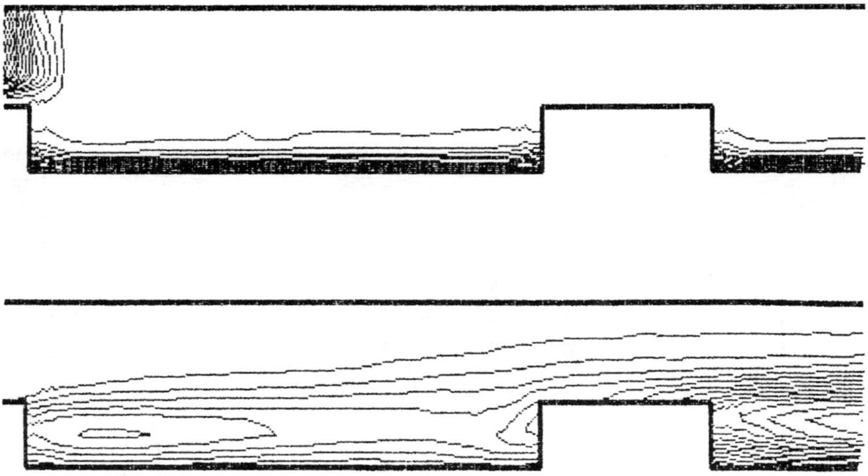

Fig. 2 Lines of concentration: upper-N_1 iterations and lower-N_2 iterations.

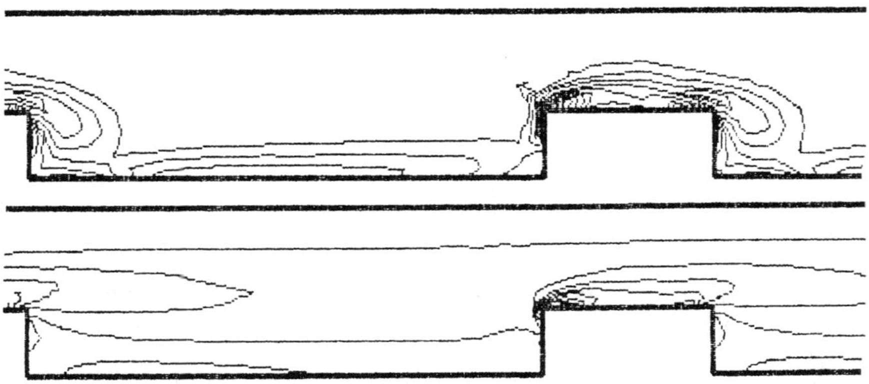

Fig. 3 Lines of vortex: upper-N_1 iterations and lower-N_2 iterations.

using these obstacles is the reduction of polarization and so increases the flux rates through the membrane.

References

[1]Porter, M. C., 'Concentration Polarization with Membrane Ultrafiltration," *Industrial and Engineering Chemistry Product Research Development*, Vol. 11, 1972, pp. 234–248.

[2]Singh, R., and Laurence, R., "Influence of Slip Velocity at a Membrane Surface on Ultrafiltration Performance—I. Channel Flow System," *International Journal of Heat and Mass Transfer,* Vol. 22, 1979, pp. 721–29.

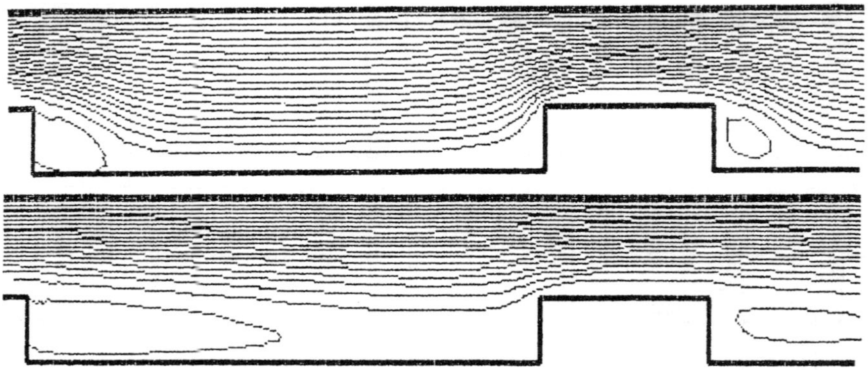

Fig. 4 Lines of stream function: upper-N_1 iterations and lower-N_2 iteration

Chapter 61

Dispersion Influence in the Plug-Flow Electrochemical Reactor Design

M. Olea* and L. Oniciu†
University "Babes-Bolyai" Cluj, Romania

I. Introduction

MEASUREMENT of the residence time distribution (RTD) is carried out by a tracer technique, where an inert substance (KCl) with some measurable physical property (the electrical conductivity here) is introduced into the fluid at the vessel inlet.

We used a pulse tracer signal. The pulse test stimulus consists of a near instantaneous introduction of a small amount of tracer at $t = 0$. The response of this tracer at the vessel exit indicates the type, or the characteristics, of flow in the vessel.

The usual raw data obtained from a tracer test are a set of tracer concentrations at known time intervals.[1]

II. Experimental Data

We have built an experimental reactor that has had the same dimensions as the industrial one. It was a rectangular channel with the following dimensions: width, $l = 0.05$ m, length, $L = 0.970$ m, and height, $h = 7*10^{-3}$ m.

Because the tracer injection has been at the reactor entrance and the detection has been in two points—one somewhere within the vessel and one at the vessel exit—a "closed-open" system (or vessel) has been considered.

The tracer detection has been done in two points because the tracer injection was not the ideal one.

We studied two system flows: a laminar flow and a turbulent one in our short electrode. It was a short electrode because its length was $L \leq 10^{-4} d_e ReSc$ (Ref. 2).

Table 1 presents the experimental results for the laminar flow: $Re = 1306$, $u = 0.107$ ms^{-1}, and $Q_{v0} = 1.25*10^{-5}$ m^{3}s^{-1}. Table 2 presents the experimen-

*Lecturer, 11 Arany Janos St., Faculty of Chemistry, 3400 Cluj-Napoca, Romania.
†Professor, 11 Arany Janos St., Faculty of Chemistry, 3400 Cluj-Napoca, Romania.

Table 1 RTD measurements for laminar flow

t, s	λ_1*10,^{3} s	λ_2*10,^{3} s	D_e/uL	Pe
3	4.9	0.1	——	——
6	0.8	0.12	——	——
9	0.35	1.651	——	——
12	0.25	5	0.042	23.81
15	0.15	3.6	——	——
18	0.1	2.2	——	——
21	——	1.3	——	——
24	——	0.1	——	——

tal results for the turbulent flow: $Re = 2733$, $u = 0.224$ ms^{-1}, and $Q_{v0} = 2.65*10^{-5}$ m^{3}s^{-1}.

For a "closed-open" system, the variance is[3]

$$\Delta\sigma^2 = \frac{\Delta\sigma^2(t)}{\bar{t}^2} = 2\frac{D_e}{uL} + \left(\frac{D_e}{uL}\right)^2\left[\exp\left(-\frac{2uL}{D_e}\right) + 4\exp\left(\frac{-uL}{D_e}\right)\right] \tag{1}$$

$$\Delta\sigma^2(t) = \sigma^2(t_2) - \sigma^2(t_1) \tag{2}$$

$$\sigma^2(t) = \frac{\sum_{i=1}^{N} t^2\lambda(t)\Delta t}{\sum_{i=1}^{N}\lambda(t)\Delta t} - \bar{t}^2 \tag{3}$$

Table 2 RTD measurements for turbulent flow

t, s	λ_1*10,^{3} s	λ_1*10,^{3} s	D_e/uL	Pe
3	14.5	0.09	——	——
6	2	0.95	——	——
9	0.6	14.82	——	——
12	0.58	3.91	0.014	71.43
15	0.2	3.38	——	——
18	0.1	2.15	——	——
21	0.09	1.53	——	——
24	——	0.1	——	——

$$\bar{t} = \frac{\sum_{i=1}^{N} t\lambda(t)\Delta t}{\sum_{i=1}^{N} \lambda(t)\Delta t} \tag{4}$$

Because Δt = constant, the equations become

$$\sigma^2(t) = \frac{\sum_{i=1}^{N} t^2\lambda(t)}{\sum_{i=1}^{N} \lambda(t)} - \bar{t}^2 \tag{5}$$

$$\bar{t} = \frac{\sum_{i=1}^{N} t\lambda(t)}{\sum_{i=1}^{N} \lambda(t)} \tag{6}$$

where Q_{v0} is the volumetric flow rate, m^3 s^{-1}; D_e the dispersion number, m^2 s^{-1}; u the mean velocity, m s^{-1}; $\bar{t}$ the average residence time, s; λ the electrical conductivity, s; Re the Reynold's number; Sc the Schmidt's number; Pe the Peclet's number; Bo the Bodenstein's number; d_e the hydraulic mean diameter, m; τ the residence time, s; R the reaction rate per unit volume, mol m^{-3} s^{-1}; C the concentration, mol m^{-3}; and x the dimension of reactor, m.

Solving Eq. (1) (by a numerical method), we obtained the value of Peclet number (Pe = 23.81 for the laminar flow and, respectively, Pe = 71.43 for the turbulent one).

For small amounts of dispersion (e.g., turbulent flow), we can use the following approximation, too:

$$\Delta\sigma^2 = \frac{2}{Pe} \tag{7}$$

Then,

$$Pe = \frac{2\bar{t}^2}{\Delta\sigma^2(t)} \tag{8}$$

and now Pe = 70.47.

This compares reasonably well to the value given by Eq. (1).

This value of Pe corresponds to a low value of dispersion (turbulent flow). The value of Pe for the laminar flow corresponds to a mean value of dispersion. Then, we tried to verify the correlation of *Pe* (or *Bo*) versus *Re* and *Sc* by our experimental data.

So, for the laminar flow

$$Bo = \frac{192}{ReSc} \tag{9}$$

and for the turbulent one,

$$Bo = \frac{0.28}{C_f^{1/2}} \tag{10}$$

where C_f = friction factor.[1]

$$C_f = 0.079\ Re^{-1/4} \tag{11}$$

Here, $Sc = 1070$ and then

$$Bo_{\text{laminar}} = 1.8*10^{-4}$$

$$Bo_{\text{turbulent}} = 2.678$$

On the other hand, from Bodenstein's number

$$Bo = \frac{ud_e}{D_e} \tag{12}$$

we have obtained:

$$De_{\text{laminar}} = 7.27\ \text{m}^2\ \text{s}^{-1}$$

$$De_{\text{turbulent}} = 1.02*10^{-3}\ \text{m}^{-2}\ \text{s}^{-1}$$

From experimental data

$$De_{\text{laminar}} = 4.3*10^{-3}\ \text{m}^2\ \text{s}^{-1}$$

$$De_{\text{turbulent}} = 1.45*10^{-3}\ \text{m}^2\ \text{s}^{-1}$$

Therefore, the experimental value of the dispersion number for the turbulent flow is very close to the calculated one.

But, for the laminar flow there is a dramatic change in the dispersion number from the measurements to the calculus. In this case we must use the following equation to calculate the friction factor (coefficient):

$$C_f = \frac{96}{Re} \tag{13}$$

In this case $C_f = 0.0735$, $Bo_{\text{laminar}} = 1.04$, and $D_e = 14*10^{-3}$ m^2 s^{-1} (a reasonable value; more acceptable).

III. Conclusions

In conclusion, we can say that the turbulent flow in our rectangular channel (filter-press type reactor) is very nearly of an ideal one: a plug flow because the dispersion number has a low value.

In such a case we can write an analytical model to describe the behavior of the electrochemical reactor without having consideration for the dispersion.

But, if the flow is a laminar one, we must consider the influence of dispersion in the reactor modeling. In this case it must solve the following equation for the material balance:

$$\frac{1}{Pe}\frac{\mathrm{d}^2C}{\mathrm{d}Z^2} - \frac{\mathrm{d}C}{\mathrm{d}Z} - \tau R = 0 \tag{14}$$

where $Z = x/L$.

References

[1]Scott, K., *Electrochemical Reaction Engineering*, Academic, London, 1991, Chap. 5.

[2]Rousar, I., Micka, K., and Kimla, A., *Electrochemical Engineering*, Vol. 2, Academia Praha, 1986, Chap. 22.

[3]Bischoff, K. B., *Chemical Engineering Science*, Vol. 2, 1960, pp. 226–270.

IX. Plasma MHD

Chapter 62

Rotating Plasma in the Closed Magnetic Trap

K. B. Abramava and G. A. Galechyan
Joffe Physical–Technical Institute, St. Petersburg, Russia
M. L. Lev
Technion—Israel Institute of Technology, Israel
A. A. Semenov, E. M. Sklyarovar, and A. V. Voronin
Joffe Physical–Technical Institute, St. Petersburg, Russia

I. Introduction

THE study of plasma confinement and heating in closed magnetic traps with toroidal field such as tokamaks or stellarators have been actively and successfully proceeded in the world. The closed systems with a poloidal field such as levitron, spherator, and the Tornado discussed here have been studied much less even though they have certain advantages in comparison with the toroidal field systems.[1–3] Their principal merit is the stable field structure and potentiality to confine a plasma with high plasma-to-magnetic field pressure ratio β.

The less activity in the studying of the systems with a poloidal field can be attributed to difficulties in their practical realization. The problem is that the conductors carrying a current that generates the poloidal field should reside in the "interior" of a plasma. They cannot (or almost cannot) be fixed rigidly. This paper demonstrates that the designing difficulty can be gotten over.

The Tornado is a closed magnetic trap with a nearly poloidal field.[4,5] B. Lehnert suggested the adequate Tornado method of plasma heating with a rotating plasma technique.[6] This method uses the drift motion of charged particles in a cross electric and magnetic field. The drift motion not only accumulates energy associated with a whole plasma rotation around the system axis but, in the case of collisional plasma, disperses energy into the plasma temperature.

Because the magnetic field of the trap is closed, there are no energy losses along magnetic lines, which makes obtaining a high rotation velocity essentially easier. As a result, a high temperature can be achieved because of the relationship between the plasma temperature and velocity. One more advantage of the rotating plasma heating is plasma purification that occurs because of centrifugal effect.

On the Tornado-650 (Ref. 7) the Alfven's velocity limit[8] has been exceeded by an order of magnitude, and the velocity of 3×10^7 cm/s was reached. There were 10^{13} cm^{-1} in plasma density, 1 keV of plasma rotation energy, and 200 eV of ion temperature on that device. These parameters were obtained under the magnetic field in a barrier of 0.15 T and a plasma volume of 0.8 m^3. The total stored energy in a pulse was 100 kJ.

These results allowed the planning of the Tornado-X installation with the 1-T magnetic barrier field. The concentration of 10^{14} cm^{-3}, ion energy of 8 keV, and ion temperature of 1.5 keV are expected with it.

The Tornado-X installation has been pushed recently, and the first results on plasma investigation have been obtained.

II. Conception of a Tornado Magnetic Field

First in Ref. 9 a toroidal system, imagined as the interior of a superconducting sphere with the poles connected by a thin diametric conductor, has been considered. Because of the superconductivity of the sphere, the toroidal magnetic field generated by currents flowing through the conductor and over the sphere is closed. The surface $r = r_0$, where r_0 is the superconducting sphere radius, forms a magnetic separatrix. Further still, the field would remain closed even if the diametric conductor is elongated and coiled as though to lay on a sphere with $r_{in} < r_0$, forming a spherical helix. But in this case the spiral conductor winds almost in azimuthal planes, and the poloidal field component predominates.

Reference 10 shows that the superconducting sphere is not necessary if an additional current, which is geometrically analogous to the first, flows through a conductor placed on a sphere with radius $r_{out} = r_0^2/r_{in}$, and the current value is $I_{out} = -I_{in}\sqrt{(r_{in}/r_{out})}$ where I_{in} is the current through inner helix, I_{out} is the current through outer helix, and the minus means that the currents flow in the opposite directions. If the preceding conditions are met, the magnetic field inside the radius r_0 is the same as in the case when the superconducting sphere is placed at $r = r_0$.

Another essential feature of the field should be noted. There is a space with a lower magnetic field strength around the central region. Experiments with models and numerical calculations have shown that the central space of a lower magnetic strength is surrounded with a magnetic barrier that is adjacent to the spherical separatrix. This system of two spherical spiral currents generating the closed magnetic field with spherical separatrix and magnetic barrier was named "tornado."

III. Magnetic Field Calculation

The model calculated (Fig. 1) consists of two spherical helices and two straight jumpers that connect the poles in pairs. The latter are required to support the internal helix mechanically as well as to supply the current to it. They lay along the system axis z, and the current through them generates azimuthal components of magnetic fields B_φ only. All of the conductors are assumed to be infinitely thin.

The field lines were derived by the numerical integration of a system of differential equations:

$$d\eta/B_\eta = \eta \, d\theta/B_\varphi = \eta \cos\theta \, d\varphi/B_\theta = ds/B$$

where ds is the differential of the arc length of a field line, whereas B is the field strength. The system was integrated by the fourth-order Runge–Kutta method.

Calculation of sufficiently long field-line sections allows one to imagine the structure of magnetic surfaces in the Tornado.

To make a classification, the magnetic field line may be named, by convention, as "individual" and "common." The individual lines always encircle a single turn of the inner helix (Fig. 2). In the pole region an individual line escapes into space between the helices and turns over a sphere adjacent to the separatrix inside. Repeating such passes forms a closed toroidal surface that is spherical near the separatrix and slips near the pole as a "stocking" around the inner helix conductor. A layer of such surfaces closes the trap "outside" as well as "inside" from the inner helix, and both "outside" and "inside" are external for plasma.

A common line can encircle more than one turn of the inner spiral (Fig. 3). Winding, the common line drifts through angle φ. In the course of the drift, the line jumps suddenly and changes the number of conductor turns encircled by one. The second jump in a symmetric point returns it. These lines form the toroidal surface with two holes where the helical conductor, submerged in individual line surfaces, enters and comes out (Fig. 4). There are surfaces that envelope one-two, three-four, five-six and seven-eight conductors.

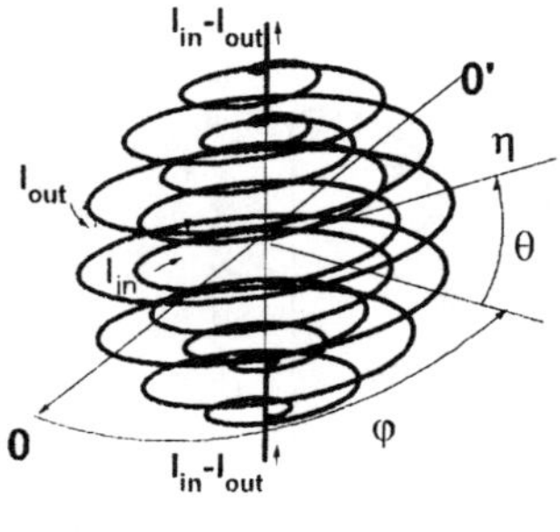

Fig. 1 Scheme of two spiral Tornado traps.

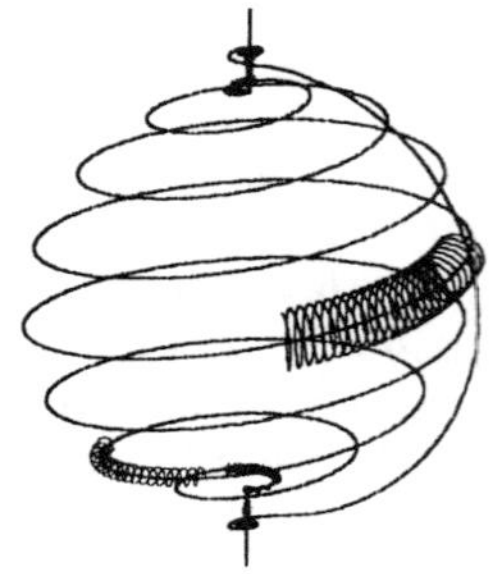

Fig. 2 Fragments of an individual magnetic line

IV. Construction of the Tornado-X

The Tornado-X magnetic trap consists of two similar spherical helices that are coiled with equal pitch in angle θ [r_{in} = 0.34 m, r_{out} = 0.48 m, $\theta = (1/2\ n)\ \varphi$, and $n = 8$ is the number of turns]. The 50-mm stainless steel helices, vacuum chamber, and electrical scheme are shown in Fig. 5. Final projection, decision of material, and manufacturing of helices had been carried out by scientific, engineering, and technical staff of the Division of Fusion Plasma Physics, Alfven Laboratory in Stockholm.

The electrical current is supplied to the trap by a coaxial cable connected through the coaxial inlet in the chamber top with the inner and outer spiral. The lower poles of the helices are electrically connected. Thus the spirals are connected in a series to the current source, but the current's ratio in spirals can be varied by the connection of an additional resistance parallel to either of the helices.

The magnetic field is generated by the pulse current that is set up at the discharge of the capacity energy storage with the possible initial energy of 425 kJ and voltage up to 5 kV. The energy storage consists of two alike capacity bank units that start to discharge one after another with 1-ms delay, which allows the setting up of the quasistationary magnetic field. The nonuniformity of the flat-topped pulse is 20%.

The trap is installed in the vacuum chamber with a volume of 2.5 m^3. The chamber is designed as a cylinder with two ellipsoidal doors closing its butt-ends. The cylinder diameter and distance, center to center, between doors are 1.5 m. The chamber walls 10-mm thick are made from stainless steel. They are

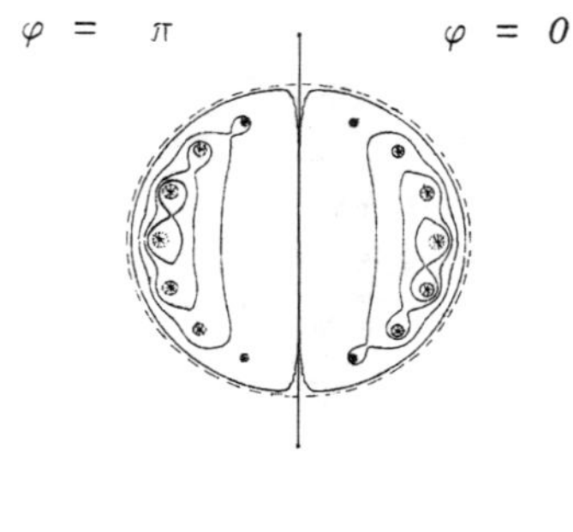

Fig. 3 The intersection traces of magnetic surfaces formed by "common" lines and the $\varphi = 0$, $\varphi = \pi$ plane. The dashed line is a separatrix intersection.

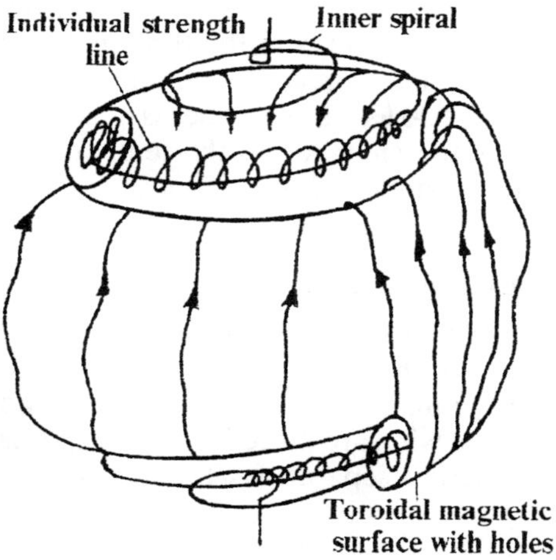

Fig. 4 An example of the toroidal magnetic surfaces with two holes formed by "common" magnetic lines.

remote from the spirals, so that the magnetic field decreases 10 times at this distance; this excludes the disturbance of the trap magnetic field from the current induced in the wall during the pulse. Coaxial rods withdrawn from the trap poles are fixed in the Teflon isolators that are securely locked at the top and bottom flange of the vacuum chamber. The mechanical overloads at the trap poles can be up to 10 T during the current pulse. Therefore four circle frames are designed along the cylindrical walls to protect the chamber. The

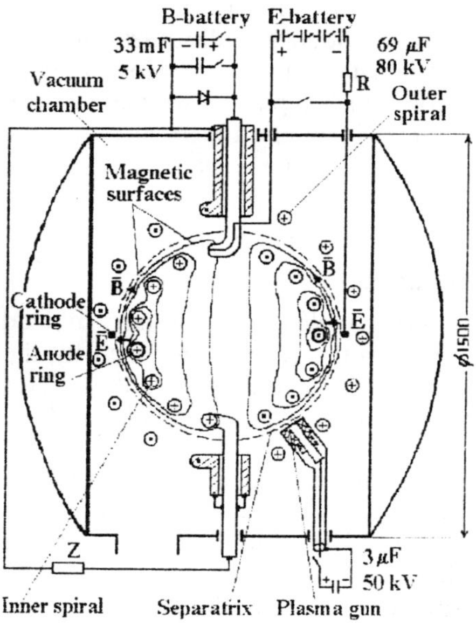

Fig. 5 Scheme of the Tornado-X experiment.

seals fabricated of thermotolerant vacuum rubber are used in chamber design. Pre-evacuation is proceeded by the pump unit with the throughput of 150 l/s, and high vacuum pumping is performed by several titanium absorption pumps with the total throughput of 11,000 l/s. The chamber is heated by 10 quartz halogen lamps, installed inside the chamber, to the temperature more then 100°C. The final pressure in the chamber is 5×10^{-7} torr.

V. Rotation Plasma Technique

The principal plasma heating (and also its freeing from heavy impurities) is affected by plasma rotation in the crossed magnetic and electric field. To apply the electric field transversely to the magnetic one, several extended electrodes, made from 3-mm stainless steel wire, are placed between spirals around the whole 2π arc outside the spherical separatrix. These electrodes serves as a cathode.

To eliminate extraneous discharges, the spirals are isolated with Teflon film. The central turn of the inner spiral is coated over the Teflon film with molybdenum foil. This molybdenum foil serves as an anode. To convey voltage to the foil, a thin wire is laid along the turns of the inner spiral and comes out through the upper pole. Thus the electrical field is directed transversely to the trap magnetic field.

The voltage on the electrodes is provided by the 80-kV capacitor bank with the energy of 200 kJ. The capacitor bank consists of four sections charged up to 20 kV, which are switched into a series at the moment when they connect to the electrodes. This arrangement gives a safe and secure operation. Duration of the electrical pulse can be as long as 2 ms. The current through the plasma is up to 100 A. The current limitation is set up to preclude the electrode sputtering and associated impurity entering into the hydrogen plasma.

A rotating plasma is heated effectively provided its full ionization. In a partially ionized plasma there is an essential channel of energy losses owing to the charge exchange process. In the Tornado-X a plasma with concentration above 10^{12} cm^{-3} would not be fully ionized with the 100-A current pulse only. A special plasma source is required supplying a plasma when the electrical field is applied.

VI. Plasma Injector

The pulse erosion-type injector is used to provide the preliminary ionized plasma. As a working material, there is polyethylene decomposed by the electrical discharge into ionized hydrogen (67%) and carbon (33%). Later when the rotation starts, the heavy carbon impurities are removed out of the trap owing to the centrifugal effect.

The electrical discharge develops over the surface of a polyethylene spacer, which is fixed at the face of two coaxial electrodes. The electrodes are thin-walled, of stainless steel, tubes 40 and 20 mm in diameter. The face of the injector is located near the spherical separatrix. The power is supplied to the injector from the 4-kJ energy storage with the initial voltage on the capacity bank up to 50 kV. The plasma concentration in the trap is regulated by this initial voltage and can be as much as 10^{14} cm^{-3}.

VII. Experiments Without Plasma Rotation

The first experiments on Tornado-X have been carried out with the purpose of investigating the confining features of the trap. The trap was filled in by a plasma for interval 150 μs, the filling finishing when the magnetic field came to the quasistationary value.

The plasma decay was studied. The electron concentration in the equatorial plane, averaged over the trap diameter, was measured with the HCN-laser interferometer at the 337-μm wavelength. The charge particle confinement time was obtained as dependent on the magnetic field strength, on currents-through-helices ratio $\boldsymbol{m} = \boldsymbol{I}_{\text{out}}/\boldsymbol{I}_{\text{in}}$, and on the injector current value.

From Fig. 6a we can see that the maximum confinement time is at the current's ratio $\boldsymbol{m}$ = 1. The theoretical value provided the existence of the

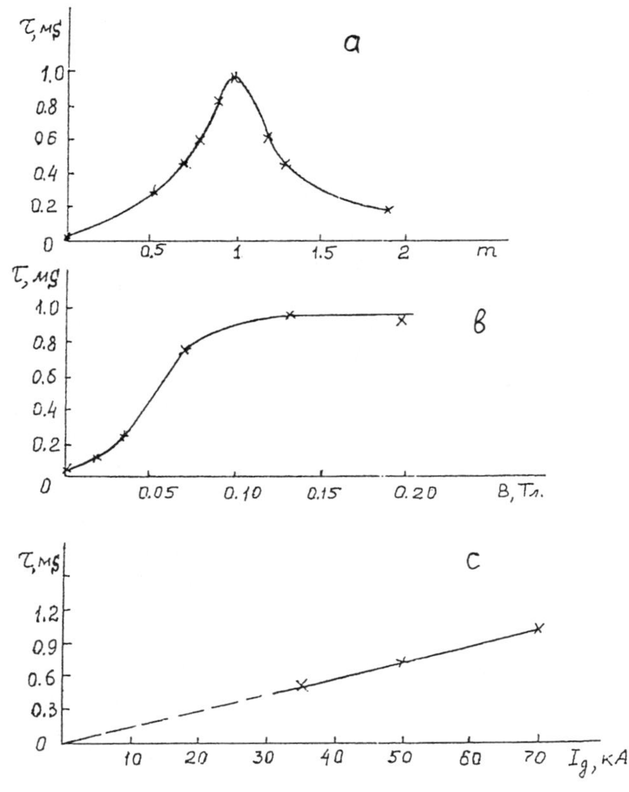

Fig. 6 Confinement times dependencies: a) on the current-through-helices ration *m*, b) on the magnetic field, and c) on the injector current.

separatrix, and the closed magnetic field is $\boldsymbol{m} = 0.85$. The discrepancy is possibly caused by the nonuniformity of the current distribution over the spiral conductors cross section. In the theory the conductors are assumed to be infinitely thin.

With the increasing of the magnetic field was found that the confinement time grows quadratically and amounts up to 950 μs at a magnetic field value of $B = 0.13$ T (Fig. 6b). Further increasing of the magnetic field does not affect on confinement time, as it seems, because of the mechanism of electron–ion volume recombination. Figure 6c shows that the confinement time can be extended with the higher injector current. Most likely the point is that the plasma obtained by the erosion-type injector is not fully ionized. For further experiments to obtain the 100% ionization and to suppress the recombination process during the magnetic field pulse, a considerable increase of the injection current is necessary.

The, integral-over-discharge-interval, photographs of the volume occupied by the plasma, made in its own radiation, shows that a plasma glow is rather uniform inside the separatrix and sharply decreases at the separatrix profile.

Thus these experiments suggest that the Tornado-X magnetic trap is closed. The plasma losses are governed by the classical diffusion law. The charged particle confinement time is at maximum when the current ratio in helices is near the theoretical value.

VIII. Experiments with Plasma Rotation

The following experiments on the plasma rotation in the crossed magnetic and electric field have been carried out with the purpose of plasma heating.

In this case before filling in the trap with plasma, the electric field was applied between the anode and cathode. When the plasma enters into the gap, the discharge ignites, and the current flows through the plasma, limited with a resister $\boldsymbol{R}$. Interaction of the current and magnetic field should lead to the plasma rotation. The energy supplied converts to the rotation energy. The voltage on the plasma builds up, and the current, on the contrary, decreases. At the end of the magnetic field quasistationary period, the cathode-anode gap is shorted by a low-resistance shunt circuit.

Figure 7 presents the time dependencies of the magnetic field, electrical characteristics, plasma density, and $\boldsymbol{H}_\beta$ -line radiation from the plasma. The magnetic field reaches its quasistationary value 0.2 T in the barrier at the moment 550 μs from the start and holds about the same until 1750 μs.

The electrical field is switched on at 100 μs before the plasma injector starts to operate The plasma enters into the trap for the 150-μs interval beginning from the moment of 550 μs. At this time the voltage over the gap drops to the value of several hundred volts according to the plasma conductivity. But in following 100 μs, the voltage increases up to 5 kV. This voltage rise appears to be reasonably accounted for in the plasma rotation actuating.

The duration of the high-voltage period varies somewhat in different discharges, but to the end of the quasistationary magnetic field interval the

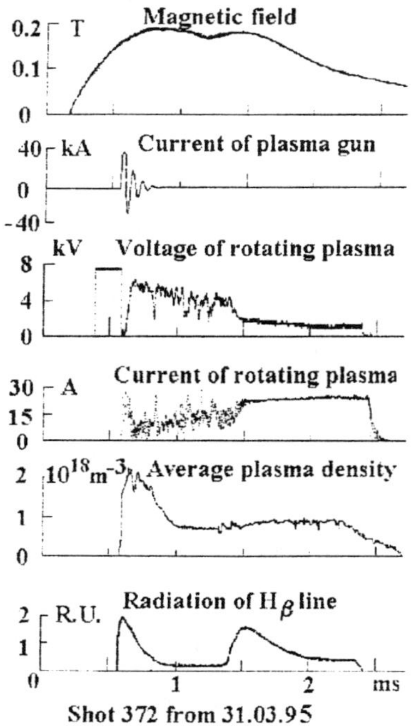

Fig. 7 Measured parameters time dependencies.

voltage becomes less than 1 kV always. At the moment of 2400 μs, the voltage pulse is released by the shorting switch circuit.

The plasma current may run to 30 A at a minimum voltage during a transition period and reduces to 10 A at 5 kV during quasistationary rotation. The instantaneous power, supplied during the discharge, amounts to 100 kWt at the transition period when the plasma is ionized and accelerated, but then it settles at 50kWt that can be associated with a constant rotation velocity.

The average electron concentration comes up to 2×10^{12} cm^{-3} at the transition period and decreases slowly to 10^{12} cm^{-3} at the end of the rotation process. The $\boldsymbol{H_\beta}$-line radiation lasts during the transition period but then disappears and rises again when rotation ends. Such a behavior of the radiation suggests that the whole neutral component is ionized during the transition period.

The simple evaluation based on magnetic and electrical measurements gives the rotation velocity 5×10^7 cm/s. So it seems that supercritical velocity is reached in the Tornado-X. The extrapolation of the Tornado-650 data allows the ion temperature to be found of 0.5 keV. Accurate measurements of temperature, rotation velocity, and energy as well as spectral analysis are the objectives for further experiments on Tornado-X.

Not only a thermonuclear fusion aspect of this system should be considered. The use of the system as a neutron source on a smaller scale was suggested in Ref. 11. The strong centrifugal effect that was shown in plasma cleaning can be used for the light atoms or isotopes separation.[12]

References.

[1]Lehnert, B., *Journal of Nuclear Energy, Part C*, Vol. 1, No. 1, 1959, p. 40.

[2]Freemam, R., Jonson, L., Okabayshi, M., Pacher, G., Schmidt, J., and Yoshikawa, S., *Fourth European Conference of Controlled Fusion and Plasma Physics* (Rome, Italy), 1970.

[3]Morozov, A. I., *Plasma Physics,* Vol. 18, No. 3, 1992, p. 305 (in Russian).

[4]Galechyan, G. A., and Peregood, B. P., *Soviet Physics Technical Physics*, Vol. 14, 1970, p. 1272.

[5]Vorobiev, G. M., Galechyan, G. A., and Peregood, B. P., *Third European Conference on Controlled Fusion and Plasma Physics*, Utrecht, 1969.

[6]Lehnert, B., *Physica Scripta*, Vol. 12, 1975, p. 166.

[7]Peregood, B. P., et al., "Termoyadernyj Sintez," *Voprosy Atomnoj Nauky i Techniky*, Vol. 3, No. 16, 1984, p. 46.

[8]Alfven, H., *On the Origin of the Solar System*, Claredon Press, Oxford, England, UK, 1954.

[9]Skornyakov, G. V., *Soviet Physics—Technical Physics*, 7, 1962, p. 187.

[10]Peregood, B. P., and Semenov, A. A., *Soviet Physics—Technical Physics*, Vol. 16, 1972, p. 1824.

[11]Lehnert, B., *Nuclear Instruments and Methods*, Vol. 129, 1975, p. 27.

[12]Abramova, K. B., Lev, M. L., and Voronin, A. V., Rept. to the Research Inst. of Chromatography, Moscow, on contract No. 175/117, Ioffe Physical Technical Inst. (in Russian).

Chapter 63

Turbulent Plasma Heating in Satellite Alfven Resonances in Devices with Bumpy Magnetic Field

I. O. Girka* and V. I. Lapshin†
Kharkiv State University, Kharkiv, 310077, Ukraine

I. Why Do Satellite Alfven Resonances Arise?

REGIONS of local resonances play an important role when electromagnetic waves propagate in nonuniform plasma.[1–5] Even weak effects (thermal motion of particles, finite inertia of electrons, weak nonlinearity, and dissipation) can significantly affect the conversion of these waves into small-scale oscillations and provide for their efficacious absorption[4,6–11] in these regions.

The external steady magnetic field is often bumpy $\boldsymbol{B} = B_r\boldsymbol{e}_r + B_z\boldsymbol{e}_z$ in laboratory plasmas:

$$B_r = B_0(\varepsilon'/k_b)\sin(k_b z), \qquad B_z = B_0[1 + \varepsilon\cos(k_b z)] \tag{1}$$

here $\varepsilon' \equiv \partial\varepsilon/\partial r$ and $k_b = 2\pi/L$ where L is the axial period of $\boldsymbol{B}$. The nonuniformity of $\boldsymbol{B}$ is supposed to be weak, $|\varepsilon| \ll 1$. Note that Eqs. (1) do not provide the condition for automatic fulfillment of the fundamental equation div $\boldsymbol{B} = 0$. Substituting Eqs. (1) into this fundamental equation, we obtain the relation that determines the dependence of the small parameter ε upon the radial coordinate r. This relation has the form of the Bessel equation. That is why ε is proportional to the modified Bessel function of the zeroth order, which means that one may assume ε to be the constant if the period of axial inhomogeneity is large as compared with the radius of the chamber. In the opposite case when the period of axial inhomogeneity is small, then the value ε decreases approximately exponentially with going away from the

*Associated Professor, Deputy Dean of the Department of Physics and Technologies, Svobody sq., 4.
†Professor, General Director of the National Scientific Center, Kharkiv Institute of Physics and Technologies, Svobody sq., 4.

plasma periphery. Inhomogeneity of the steady magnetic field is essential in the narrow layer k_b^{-1} near the metal wall in this case.

When the Alfven wave (AW) $\propto \exp(i\psi), \psi = k_z z + m\vartheta - \omega t$ is excited by the antenna and weak periodic inhomogeneity (1) of B results in the arising of satellite harmonics in AW electric and magnetic fields,

$$E^{\sim}_{r,\vartheta} = \left[E^{(0)}_{r,\vartheta}(r) + E^{(+)}_{r,\vartheta}e^{ik_b z} + E^{(-)}_{r,\vartheta}e^{-ik_b z}\right]e^{i\psi} \tag{2}$$

Rapid growth of these satellite harmonics and their conversion into small-scale kinetic AW occur in the vicinity of satellite Alfven resonances (SAR) $r = r_A^{(\pm)}$, in which

$$\varepsilon_1^{(0)} \approx \frac{c^2}{\omega^2}(k_z \pm k_b)^2 \tag{3}$$

Here the component $\varepsilon_1^{(0)}$ of the cold collisionless plasma tensor ε_{ik} is

$$\varepsilon_1^{(0)}(r) \equiv \varepsilon_1(r,z) = 1 - \sum_i \frac{\omega_{\text{pi}}^2(r)}{\omega^2 - \omega_{\text{ci}}^{(0)2}} \tag{4}$$

ω_{pi} and ω_{ci} are plasma and cyclotron frequencies of ions, $\omega_{\text{ci}}^{(0)} = eB_0/(m_i c)$.

II. Method for Ion-Cyclotron Turbulence Description

The reversed effect of kinetic parametric ion-cyclotron plasma instabilities on pumping AW, which excites these instabilities, is studied here.

The maximum growth rate γ of short ion-cyclotron oscillations ($k_r\rho_{\text{Li}} \gg 1$) at the frequency

$$\omega = n\omega_{\text{ci}}^{(0)}\left(1 + \frac{1}{\sqrt{2\pi}k_r\rho_{\text{Li}}}\right)$$

is equal[12] to

$$\gamma \sim n\omega_{\text{ci}}\frac{T_i u}{T_e v_{\text{Ti}}} \tag{5}$$

where k_r is the transverse wave number of short ion cyclotron waves excited by the pumping wave (these waves form the spectrum of weak turbulence), $\rho_{\text{Li}} = V_{\text{Ti}}/\omega_{\text{ci}}$ is the ion Larmor radius, $V_{\text{Ti}} = \sqrt{T_i m_i}$ is the ion thermal velocity, n is the integer, T_i and T_e are the temperatures of ions and electrons, and $u = cE^{\sim} / B_0$ is the relative velocity of ions and electrons in the field of the pumping wave. The level of energy density for these oscillations (the level of noise) under conditions of saturation for parametric kinetic ion cyclotron instability is of the order[13]

$$\frac{W}{n_i T_i} \sim \frac{T_i}{T_e}\left(\frac{u}{v_{\text{Ti}}}\right)^4 \tag{6}$$

when the saturation is caused by induced scattering of cyclotron waves on free ions.

At the nonlinear stage of cyclotron oscillations evolution, ions scatter on turbulent pulses. They say that the arising of the effective frequency of scattering is the effective frequency of collisions ν_{eff}, which leads, in turn, to losses of pumping wave energy with the damping rate γ_{eff}. Under the conditions of saturation for parametric instability,[14]

$$\gamma_{\text{eff}} W_e = \gamma W \tag{7}$$

where W_e is the energy density for pumping wave in plasma,[15]

$$W_e = \frac{1}{16\pi\omega} \frac{\partial(\omega^2 \varepsilon_{ij})}{\partial\omega} E^*_{oi} E_{0j} \tag{8}$$

If γ_{eff} is known, one can find ν_{eff} and use it instead of ν_{col} (ν_{col} is the frequency of couple Coulomb collisions) in studying the propagation of electromagnetic waves with the frequency $\omega < \omega_{\text{ci}} : \gamma_{\text{col}} \Rightarrow \gamma_{\text{eff}}$.

III. Alfven Waves in Satellite Alfven Resonances in the Linear Approximation

Amplitudes $B^{\pm}(r)$ of the nearest satellite harmonics of axial magnetic field $B_z^{\sim}$ are small as compared with those of the main harmonic $|B^{\pm}| \sim |\varepsilon B^{(0)}$ everywhere except in the SAR vicinity. The following equation for the description of amplitudes $B^{\pm}(r)$ in resonance regions is obtained from the Maxwell equations:

$$\frac{1}{r}\frac{\partial}{\partial r} \frac{r}{k_{\pm}^2[\varepsilon_1^{(0)} - (N_z \pm N_b)^2]} \cdot \Bigg\{ \varepsilon_1^{(0)} - (N_z \pm N_b)^2 + i\varepsilon_1^{(c)}$$

$$+ \left[\varepsilon_T - \varepsilon_b + \frac{\varepsilon_1^{(0)}}{\varepsilon_3} \right] \frac{c^2}{\omega^2} \frac{\partial^2}{\partial r^2} \Bigg\} \frac{\partial B^{(\pm)}}{\partial r} + q(k_z \pm k_b, r) B^{(\pm)} + \hat{G}^{(\pm)} B^{(0)} = 0 \tag{9}$$

here

$$k_{\pm}^2 \equiv k_{\perp}^2(k_z \pm k_b, r), \quad k_{\perp}^2 = (\omega/c)^2 [\varepsilon_1^{(0)} - N_z^2](1 - \mu^2)$$

$$\mu = \varepsilon_2^{(0)} / [\varepsilon_1^{(0)} - N_z^2], \quad \varepsilon_2^{(0)}(r) \equiv \varepsilon_2(r,z) = -\sum_i \frac{\omega_{\text{pi}}^2(r)\omega}{\left[\omega^2 - \omega_{\text{ci}}^{(0)2}\right]\omega_{\text{ci}}^{(0)}} \tag{10}$$

$$q(k_z, r) = 1 - \frac{m^2}{k_{\perp}^2 r^2} + \frac{m}{r}\frac{\partial}{\partial r}\left(\frac{\mu}{k_{\perp}^2}\right)$$

The value $\varepsilon_1^{(c)}$ provides an account of particle collisions,[16]

$$\varepsilon_1^{(c)} = \sum_{a,b} \frac{\omega_{pa}^2 \nu_{ab}}{\omega\left[\omega^2 - \omega_{ca}^{(0)2}\right]^2} \left(\frac{\omega^2 + \omega_{ca}^{(0)2}}{\omega^2 - \omega_{ca}^{(0)2}} - \frac{e_b m_a}{e_a m_b} \frac{\omega^2 + \omega_{ca}^{(0)} \omega_{cb}^{(0)}}{\omega^2 - \omega_{ca}^{(0)2}} \right) \tag{11}$$

Here e_a is a charge of particles of the a sort and ν_{ab} is the frequency of the collisions between particles of sort a and particles of sort b. The value ε_T in Eq. (9) provides an account of finite ion Larmor radius,[17]

$$\varepsilon_T = \sum_i \frac{3\omega_{\text{pi}}^2(r)\omega^2 v_{\text{Ti}}^2(r)}{\left[\omega^2 + \omega_{\text{ci}}^{(0)2}\right]\left[\omega^2 + 4\omega_{\text{ci}}^{(0)2}\right]c^2} \tag{12}$$

Electrons' inertia is taken into account in Eq. (9) in the component ε_3. As long as the nonuniformity of the steady magnetic field is weak, one can use the expression for ε_3 known from the zero approach ($\varepsilon \equiv O$).

Expressions for $\varepsilon_1^{(c)}$, ε_T, and ε_3 have the same form as in the case of the main Alfven resonance (AR) unlike the other two. First, we do not write the analytical expression for the differential operator $\hat{G}^{(\pm)}$ because it is too cumbersome[18]; it is the small value of the first order $|\hat{G}^{(\pm)} B^{(0)}| \propto |\varepsilon B^{(0)}$. Second, the value ε_b, caused by the curvature of the magnetic surfaces in the resonant region, affects on the width $\Delta r \sim |k_1|^{-1}$ of SAR,

$$\varepsilon_b = \frac{\mu^2}{1-\mu^2} \frac{\varepsilon'^2}{4k_b^3}(k_b \pm 2k_z)\Big|_{r=r_A^{(\pm)}} \tag{13}$$

$$k_1 = \left[-\frac{\omega^2}{c^2} \frac{\partial \varepsilon_1^{(0)}}{\partial r} \frac{\varepsilon_3}{(\varepsilon_T - \varepsilon_b)\varepsilon_3 + \varepsilon_1^{(0)}} \right]^{1/3} \Bigg|_{r_A^{(\pm)}} = |k_1| e^{i\varsigma} \tag{14}$$

We suppose plasma parameters and amplitudes of the main harmonics of the pumping wave to vary weakly in the vicinity of SAR, i.e., we solve Eq. (9) using the method of "the narrow slab." In this case in the linear with respect to the pumping-wave amplitude approach, Eq. (9) can be transformed into a nonuniform Airy equation. Then the solution $\partial B^{(\pm)}/\partial r$ of Eq. (9) is expressed through the Airy function $u_0(\xi)$,

$$\frac{\partial B^{(\pm)}}{\partial r} = \frac{\omega}{ic} \varepsilon_2^{(0)2} k_1 \left[\frac{\partial e_1^{(0)}}{\partial r} \right]^{-1} A^{(\pm)} u_0(\xi) \tag{15}$$

$$u_0(\xi) = s \int_0^\infty \exp\left[is\left(t\xi + t^3/3\right)\right] \mathrm{d}t \tag{16}$$

Here the value $A^{(\pm)}$ varies weakly in the SAR regions,[18]

$$A^{(\pm)} \sim E_{\vartheta}^{(\pm)} - \frac{icm}{r\varepsilon_2^{(0)}\omega} B^{(\pm)} \sim E_{\vartheta}^{(0)} \tag{17}$$

The value S defines the direction of propagation for the small-scale kinetic AW,

$$s = \operatorname{sgn}\left\{\left[\partial\varepsilon_1^{(0)} / \partial r\right]_{r_A^{(\pm)}} \operatorname{Re}(k_1)\right\} \tag{18}$$

The argument of Airy function ξ is equal

$$\xi = k_1\left[r - r_A^{(\pm)}\right] - i\eta, \quad \eta = -k_1\varepsilon_1^{(c)}\left[\partial\varepsilon_1^{(0)} / \partial r\right]_{r_A^{(\pm)}}^{-1} \tag{19}$$

The asymptotic of $u_0(\xi)$ for large arguments $Re(\xi) \to -\infty$ has the following form:

$$u_0(\xi) \to \frac{i}{\xi} + \left(\frac{\pi^2}{-\xi}\right)^{1/4} \exp\left\{-is\left[\frac{2}{3}(-\xi)^{3/2} + \frac{\pi}{4}\right]\right\} \tag{20}$$

The second addendum in Eq. (20) corresponds to small scale kinetic AW. We suppose this wave to damp strongly in the narrow vicinity of SAR because of the Landau mechanism. Slow $[\propto(-\xi)^{-1/4}]$ decreasing of the kinetic wave amplitude results in the following fact. The region, where nonlinear effects can be essential, is nonsymmetrical with respect to $r = r_A^{(\pm)}$ because the Airy function $U_0(\xi)$ has its maximum when $\xi = -1,8$.

The following value of RF power absorbed at the unit length of the plasma cylinder in the SAR region was obtained in linear with respect to field amplitude approach,

$$P_r^{(\pm)} = \pi r_A^{(\pm)} Re\left\{\int\left[j_r^* E_r^{(\pm)}\right] \mathrm{d}r\right\} \tag{21}$$

$$P_r^{(\pm)} = \frac{\pi\omega}{4}\left\{r\left[\partial\varepsilon_1^{(0)} / \partial r\right]^{-1} \varepsilon_2^{(0)2}\right\}_{r=r_A^{(\pm)}} |A^{(\pm)}|^2 \tag{22}$$

RF power $P^{(\pm)}$ absorbed in the SAR is proportional to the small parameter square $P^{(\pm)} \propto \varepsilon^2$ when calculated in a linear approach. Nevertheless, the

SAR can contribute sufficiently to plasma heating under perfect conditions.[18]

IV. Effect of Striction and Turbulence on Alfven-Wave Field Distribution in Satellite Alfven Resonance

If the pumping wave is powerful, then nonlinear effects become essential. Here we take into account the effect of striction nonlinearity because it is designative for the fast wave processes ($V_{ph} \gg V_{Ti}$; V_{ph} is wave-phase velocity). [Nonlinearity at the second harmonic plays the main role for slow wave processes,[19] $V_{ph} \lesssim V_{Ti}$ (Ref. 19).]

Striction changes plasma density $n(r) \Rightarrow n_{NL}$,

$$n_{\rm NL} = n(r) \cdot \exp(-U{\sim}/T) \tag{23}$$

Potential energy of particles[20] in the pumping-wave field has the following form in the case of a bumpy magnetic field:

$$U^{\sim} = U^{(0)} + U^{(2)} \tag{24}$$

here

$$U^{(0)} = \sum_{\alpha} \{e_\alpha^2 n_\alpha / [4m_\alpha n_e(\omega^2 - \omega_{c\alpha}^2)]\}$$
$$\times \left\{ \left|E_r^{(0)}\right|^2 + \left|E_\vartheta^{(0)}\right|^2 + i(\omega_{c\alpha}/\omega)\left[E_r^{(0)*}E_\vartheta^{(0)} - E_r^{(0)}E_\vartheta^{(0)*}\right]\right\} \tag{25}$$

the addendum $U^{(2)} \propto \varepsilon^2$ varies in the SAR rapidly,

$$U^{(2)} = \sum_{\alpha} \left[e_\alpha^2 n_\alpha / (4m_\alpha n_e)\right]\left[\left|E_r^{(\pm)}\right|^2 / (\omega^2 - \omega_{c\alpha}^2) + \left|E_z^{(\pm)}\right|^2 / \omega^2\right] \tag{26}$$

The field $E_z^{(0)}$ is neglectfully small for the AW. Nevertheless, amplitudes $E_z^{(\pm)}$ of its satellite harmonics resonantly grow in the SAR vicinity, and they do it even more rapidly, $\propto \left[\varepsilon_1^{(0)} - (N_z \pm N_b)^2\right]^{-2}$, than $E_r^{(\pm)} \propto \left[\varepsilon_1^{(0)} - (N_z \mp N_b)^2\right]^{-1}$. That is why we can neglect the second addendum in Eq. (26) only if

$$\left|k_z k_1 c^2 / (\omega^2 \varepsilon_3)\right| \leq 1 \tag{27}$$

As long as the striction is weak ($U^{\sim} \ll T$), it is enough to account for the change $n(r) \Rightarrow n_{NL}$ in Eq. (9) only for the addendum $\left[\varepsilon_1^{(0)} - (N_z \pm N_b)^2\right]$, which is small in the SAR region,

$$\varepsilon_1^{(0)} - \left(N_z \pm N_b\right)^2$$
$$\Rightarrow \left[\partial\varepsilon_1^{(0)} / \partial r\right]_{\big|r_A^{(\pm)}} \left[r - r_A^{(\pm)}\right] + \left[\varepsilon_1^{(0)} - 1\right](-U^{\sim} / T) + o(U^{\sim 2} / T^2) \tag{28}$$

As long as the amplitudes of the main harmonics of the AW fields vary weakly in the SAR vicinity, then we account for the striction potential $U^{(0)}$ [Eq. (25)] results only in the moving of the SAR region away from the plasma axis for the small distance:

$$\delta r = \left[\varepsilon_1^{(0)} - 1\right]\left[U^{(0)} / T\right]\left[\partial\varepsilon_1^{(0)} / \partial r\right]_{r_A^{(\pm)}}^{-1} \tag{29}$$

Striction nonlinearity affects the field distribution in the SAR region if

$$\left[\varepsilon_1^{(0)} - 1\right]\left[U^{(2)} / T\right] > \left|(1 / k_1)\left[\partial\varepsilon_1^{(0)} / \partial r\right]_{\big|r_A^{(\pm)}}\right| \tag{30}$$

If inequalities (27) and (30) are true, then from Eq.(9), we obtain that $E_r^{(\pm)}$ is equal in the SAR:

$$\left|E_r^{(\pm)}\right| \sim \left\{A^{(\pm)} E_0^2 \varepsilon_2^{(0)} / \left[\varepsilon_1^{(0)} - 1\right]\right\}^{1/3} \tag{31}$$

here

$$\frac{1}{E_0^2} = -\sum_{\alpha} \frac{e_\alpha^2 n_\alpha}{4 m_\alpha n_e T(\omega^2 - \omega_{c\alpha}^2)} \tag{32}$$

From Eq. (31) we note that if the axial inhomogeneity of the external magnetic field is not weak,

$$\varepsilon \gg \left[\Omega E_\vartheta^{(0)} / E_0\right]^{2/3} \tag{33}$$

then amplitude (31) of the satellite harmonic is greater in the SAR than that of the main harmonic. The width Δr_{NL} of the region, in which striction is sufficient, is of the following order in accordance to Eq. (31):

$$\Delta r_{\mathrm{NL}} \sim \left[\varepsilon_1^{(0)} - 1\right]\left[\partial\varepsilon_1^{(0)} / \partial r\right]_{r_A^{(\pm)}}^{-1}\left[\varepsilon\omega E_\vartheta^{(0)} / (\omega_{\mathrm{ci}} E_0)\right]^{2/3} \tag{34}$$

If the Coulomb collisions weakly affect a propagation of AW $\nu_{\mathrm{eff}} \gg \nu_{\mathrm{col}}$, then turbulence can have an influence on the stationary distribution of the fields of the pumping electromagnetic wave in the SAR. Because amplitudes of satellite harmonics appear to be large compared with those of the pumping wave, then effective imaginary dissipative addendum caused by the turbulence

$$\varepsilon_1^{(0)} + i\varepsilon_1^{(c)} \Rightarrow \varepsilon_1^{(0)} + i\varepsilon_{1\mathrm{eff}} \tag{35}$$

becomes greater inside the SAR than outside of it,

$$\varepsilon_{1\mathrm{eff}} = \frac{16\pi n T_i}{B_0^2}\left(\frac{c\left|E_r^{(\pm)}\right|}{v_{\mathrm{Ti}} B_0}\right)^3 \left(\frac{c}{v_{Ti}}\right)^2 \left(1 + \frac{k_z^2 c^2}{\omega_{\mathrm{pi}}^2}\right) \tag{36}$$

Power W_T, absorbed at the unit of the plasma cylinder length in the SAR layer because of the effective dissipation – ion cyclotron turbulence, equals

$$W_T \sim (\omega/4) r_A^{(\pm)} \Delta r_{\mathrm{NL}} \varepsilon_{1\mathrm{eff}} \left|E_r^{(\pm)}\right|^2 \tag{37}$$

$$W_T \sim \frac{4\pi\omega n T_i r_A^{(\pm)}}{B_0^2} \left\{\frac{\partial \ln\left[\varepsilon_1^{(0)}\right]}{\partial r}\right\}^{-1} \left[\frac{\varepsilon\left|E_\vartheta^{(0)}\right|\omega E_0^2}{B_0^3 \omega_{\mathrm{ci}}}\right]$$

$$\times \left(\frac{c}{v_{\mathrm{Ti}}}\right)^5 \left(1 + \frac{c^2 k_z^2}{\omega_{\mathrm{pi}}^2}\right) \left[\frac{\varepsilon\omega E_\vartheta^{(0)}}{\omega_{\mathrm{ci}} E_0}\right]^{4/3} E_0^2 \tag{38}$$

From Eq. (38) we note that the power $W_T \propto \left[\varepsilon E_\vartheta^{(0)}\right]^{7/3}$ unlike that absorbed in the case of the weak pumping wave $P_r^{(\pm)} \propto \left[\varepsilon E_\vartheta^{(0)}\right]^2$, which is explained in the following way. Although the amplitude of the satellite harmonic in Eq. (37) increases with the increasing of the pumping-wave amplitude $E_\vartheta^{(0)}$ much more slowly, $E_r^{(\pm)} \propto \left[\varepsilon E_\vartheta^{(0)}\right]^{1/3}$, than in the linear approach [Eq. (15)], nevertheless, power W^T [Eq. (37)] increases with the increasing of $E_\vartheta^{(0)}$ also because of the increasing of the turbulent dissipation $\varepsilon_{1\text{eff}} \propto (\varepsilon E_\vartheta^{(0)})$ and because of the broadening of Δr_{NL}, where $\Delta r_{NL} \propto \left[\varepsilon E_\vartheta^{(0)}\right]^{2/3}$.

As long as the amplitude of the satellite harmonic is greater in the SAR than that of the main harmonic in the devices with significant [Eq. (33)] axial inhomogeneity of B, then the density of the power absorbed in the SAR is greater than one absorbed because of turbulent dissipation outside of the SAR.

V. Conclusion

There are regions of satellite Alfven resonances [Eq. (3)] in which sufficient plasma heating can take place in the devices with a bumpy magnetic field [Eq. (1)] in addition to that observed in the main local Alfven resonance. In the linear at the respect to amplitudes of the satellite harmonics approach, RF power $P_r^{(\pm)}$ [Eq. (22)] absorbing at the unit length of the plasma cylinder in the SAR regions because of ion-cyclotron turbulence is proportional to the small parameter ε [Eq. (1)] square. Scattering of the plasma particles on the turbulent pulsations is responsible for the pumping Alfven wave absorption in the SAR rather than a couple Coulomb collisions if the pumping wave is powerful enough, $\nu_{\text{eff}} \gg \nu_{ab}$. This condition is true in the SAR for the value of the pumping-wave amplitude, which is much greater than that necessary for the main AR.

Striction results, first of all, in the moving of the SAR regions away from the axis of the plasma cylinder for the distance δr. The expression (29) for δr is similar to that known for the main AR. Second, striction affects the distribution of satellite harmonics amplitudes in the SAR regions if the condition (30) is valid. Inequality (30) is more strict than that taking place in the main AR. Turbulent plasma heating W^T [Eq. (38)] can be even more sufficient in the narrow layer ($r - r_A^{(\pm)} < \Delta r_{\text{NL}}$) in the SAR region rather than outside of the SAR.

Acknowledgment

This work is supported by the Science and Technology Center in Ukraine, project No. 253.

References

[1]Uberoi, C., "Alfven Waves in Inhomogeneous Magnetic Fields," *Physics of Fluids*, Vol. 15, 1972, pp. 1673–1675.

[2]Chen, L., and Hasagawa, A., “Plasma Heating by Spatial Resonance of Alfven Wave,” *Physics of Fluids*, Vol. 17, No. 7, 1974, pp. 1399–1403.

[3]Tataronis, J. A., “Energy Absorption in the Continuous Spectrum of Ideal MHD,” *Journal of Plasma Physics*, Vol. 13, 1975, pp. 87–92.

[4]“High Frequency Plasma Heating,” *Materials of All Union Simposium*, Inst. of Applied Physics of USSR Academy of Sciences, Gorky, Russia, 1983 (in Russian).

[5]Golant, V. E., and Fedorov, V. I., *High Frequency Methods of Plasma Heating in Toroidal Fusion Devices*, Energoatomizdat, Moscow, 1986 (in Russian).

[6]Gildenburg, V. B., “About Nonlinear Effects in Inhomogeneous Plasma,” *Zhurnal Experimentalnoi i Teoreticheskoi Fiziki*, Vol. 46, No. 6, 1964, pp. 2156–2163 (in Russian).

[7]Golant, V. E., and Pilija, A. D., “Waves’ Linear Transformation and Absorption in Plasma,” *Uspehi Fizicheskih Nauk*, Vol. 104, No. 3, 1971, pp. 413–457 (in Russian).

[8]Litvak, A. G., “Dynamic Nonlinear Electromagnetic Phenomena in Plasma,” *Problems of Plasma Theory*, Vol. 10, Atomizdat, Moscow, 1980, pp. 164–242 (in Russian).

[9]Stix, T., and Swanson, D., “Waves Propagation and Transformation in Inhomogeneous Plasma,” *Basics of Plasma Physics*, Vol. 1, Energoatomizdat, Moscow, 1983, pp. 333–364 (in Russia).

[10] Timofeev, A. V., “To the Theory of Alfven Oscillations of Inhomogeneous Plasma,” *Problems of Plasma Theory*, Vol. 17, Energoatomizdat, Moscow, 1989, pp. 157–244 (in Russian).

[11]Lapshin, V. I., Stepanov, K. N., and Shtrasser, V. O., “Excitation of Nonlinear Kinetic Wave in the Region of Local Alfven Resonance,” *Fizika Plazmi*, Vol. 18, No. 5, 1992, pp. 660–663 (in Russian).

[12]Kitsenko, A. B., Panchenko, V. I., and Stepanov, K. N., “Parametric Excitation of Ion Acoustic Plasma Oscillations in Alternate Electric and Steady Magnetic Fields,” *Zhurnal Tekhnicheskoi Fiziki*, Vol. 43, No. 7, 1973, pp. 1437–1444 (in Russian).

[13]Mikhailenko, V. S., and Stepanov, K. N., “Theory of Weak Parametric Plasma Turbulence,” *Zhurnal Experimentalnoi i Tekhnicheskoi Fiziki*, Vol. 87, No. 1(7), 1984, pp. 161–176 (in Russian).

[14]Silin, V. P., *Parametric Effect of Powerful Radiation on Plasma*, Nauka, Moscow, 1973 (in Russian).

[15]Landau, L. D., and Lifshits, Ye. M., *Theoretical Physics. Vol. 8. Continuous Media Electrodynamics*, Nauka, Moscow, 1982 (in Russian).

[16]Akhiezer, A. I., Lapshin, V. I., and Stepanov, K. N., “To the Theory of Magnetohydrodynamic Waves Damping in High Temperature Plasma,” *Zhurnal Experimentalnoi i Teoreticheskoi Fiziki*, Vol. 70, No. 1, 1976, pp. 81–91 (in Russian).

[17]Akhiezer, A. I., et al., *Plasma Electrodynamics*, Vol. 1, Pergamon, Oxford, England, UK, 1975.

[18]Girka, I. A., Lapshin, V. I., and Stepanov, K. N., “Alfven Wave Heating of Plasma in Devices with Bumpy Magnetic Field,” *Contributed Papers of the 20th EPS Conference on Controlled Fusion and Plasma Physics* (Lisbon, Portugal), July 1993, Pt. 3, pp. III-877–III-880.

[19]Davydova, T. A., and Lashkin, V. M., “Modulation Instability and Solitons Appearance at Ion-Ion Resonance,” *Fizika Plazmi*, Vol. 16, No. 8, 1990, pp. 907–915 (in Russian).

[20]Milenko, V. P., Sidorov, V. P., and Tsypin, V. S., "Nonlinear Transformation of Alfven Waves in a Hot Plasma," *Contributed Papers of the 17th EPS Conference on Controlled Fusion and Plasma Heating* (Amsterdam), June 1990, Vol. 14B, Pt. 3, pp. 1145–1148.

[21]Klima, R., "The Drifts and Hydrodynamics of Particles in a Field with a High - Frequency Component," *Czechoslovak Journal of Physics*, Vol. B18, 1968, pp. 1280–1291.

Chapter 64

Magnetohydrodynamic Flows with Temperature, Current Layers

Savelii S. Katsnel'son*
Institute of Theoretical and Applied Mechanics, Novosibirsk, 630090, Russia

I. Introduction

FOR a series of technological processes, high-enthalpy gas flows consisting of narrow high-temperature zones are necessary. In particular, to use such flows in the magnetohydrodynamic (MHD) transformations of heat energy into electric (the MHD generators) seems rather risky.[1,2] It would allow one to increase the effective conductivity of a working body without an essential increase in the heat content of the main flow. Having provided a rather high velocity of flow, one may reduce the time of heat contact in such a zone with the given section of the channel surface so that it will not be destroyed. With this the high-temperature layer from which the main current runs will be a piston transforming the flow enthalpy into the useful work. The main difficulty of the realization of such an idea is connected with the organization and stability of such a flow.

The practical possibility of creating the stratified flow with the local current-carrying zones appeared after the discovery of a physical phenomenon—the TC-layer effect (temperature, current).[3] The essence of the phenomenon is the following. In the plasma where the electric current runs, the local increase of temperature results in the conductivity increase, the growth of current density in this place, and an additional release of heat at the expense of the Joule dissipation. Under definite conditions the temperature increase results in the development of heat instability there. Further, while achieving a dynamic equilibrium between heat supply and the energy losses, there appear the stable structures—the TC layers in the flow. The characteristic peculiarity of the TC layer is the current localization in the definite mass of the gas and its force effects on the remaining flow.

*Head of Research Group, Laboratory of Fast-Process Physics.

The first calculating, theoretical investigations of the TC-layer effect may be found in Refs. 4 and 5. The main experimental and theoretical works on the organization of the flow with the TC layer were performed at the Institute of Theoretical and Applied Mechanics SD RAS.

II. Experimental Installation and the Peculiarities of a Flow in the Disk Channel of the Installation

The principal scheme of the experimental installation with the disk MHD channel is represented in Fig. 1. The radial channel (Fig. 1a) is formed by two disks, 1, made of organic glass with the diameter of 350 mm with changeable distance between the disks (the width of the MHD channel). Into the walls of the channel, the working winding, 2, providing the inductive collection of electric power is mounted, and the holes for the pressure transducers and the Rogowsky coils are placed as well as the optic windows for the IR diagnostics of spectral measurements and the photographing the plasma flow.

The outer magnetic field is created by an ironless electromagnet of constant current performed by the Helmholtz scheme, 3. The temperature perturbance in the flow was created by a pulse inductive discharge. The inductor turns, 4, and a pair from each side is placed on the radius of 55 mm. The scheme of the plasma source was determined by the choice of the working body. In the case of gaseous plasma-forming matter, a coaxial electrode accelerator was used, and for obtaining the alkali metal plasma an electric arc generator of the plasma with evaporating electrodes was used. The conductive collection of electric energy was performed with the help of an electrode insert (Fig. 1b). For diagnostics purposes the insert with segmented electrodes was used.

The experimental investigations at the given installation were performed by V. P. Fomichev and G. A. Pozdnyakov. The experiments were conducted

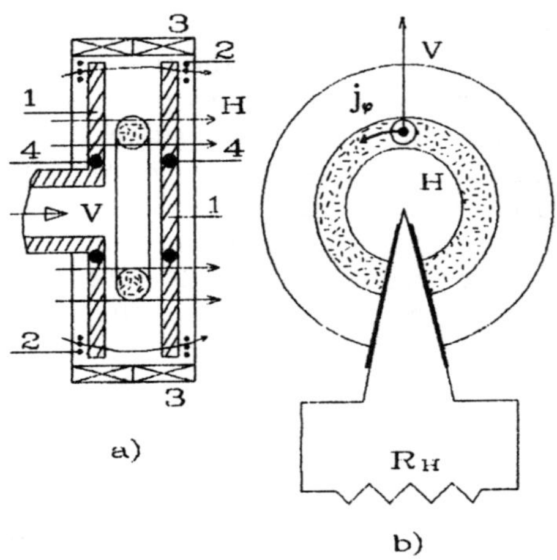

Fig. 1 The principal layout of the experimental facility: a) radial channel and b) electrode insert.

on air, argon, carbon dioxide gas, and sodium plasma. Excluding the latter, 2% of hydrogen was added to the gas for diagnostics purposes to measure the electron density on the H_β line.

The investigation of the flow in the channel without a magnetic field has shown that the plasma extending in the channel has an axisymmetric character. On the frame filming the region of intensive lighting in the central part of the channel (the region of flow deceleration near the end walls) is observed. While moving the plasma through the channel and connecting with it, extending and cooling the intensity of the lighting rapidly decreases such as it is not always registered. The pressure measurements have shown that near the entrance part the flow is strongly inhomogeneous. However, beginning from the radius equaling approximately 60 mm, a quasistationary characteristic for the supersonic flow distribution of the pressure is maintained.

To reveal the peculiarities of the flow structure in the experimental channel, calculations were performed. The system of gas dynamic equations is written in a weakly divergent form and has the following view:

$$\frac{\partial f}{\partial t} + \frac{\partial F(f)}{\partial z} + \frac{\partial G(f)}{\partial r} + H(f,r) = 0$$

$$f = \begin{vmatrix} \rho \\ \rho u \\ \rho v \\ e \end{vmatrix}, \quad F = \begin{vmatrix} \rho u \\ \rho u^2 + p \\ \rho uv \\ (e-p)u \end{vmatrix}, \quad G = \begin{vmatrix} \rho v \\ \rho uv \\ \rho v^2 + p \\ (e+p)v \end{vmatrix}, \quad H = \frac{1}{r}\begin{vmatrix} \rho v \\ \rho uv \\ \rho v^2 \\ (e+p)v \end{vmatrix}$$

where $e = \rho u^2/2 + \rho v^2/2 + p/(\gamma - 1)$.

The parameters of the falling shock wave and the channel sizes used in the calculation have been taken from the experiment in which the diameter of a straight cylindrical tunnel was equal to 55 mm, and the width of the radial channel was 24 mm. The shock wave with the Mach number $M = 6.3$ was created in argon at the initial pressure 0.03 bar.

The results of the calculations of the character of the flow at the time moment when the reflected shock wave stops are represented in Fig. 2. Further the process develops mainly in the radial direction. As we see behind the reflected shock wave in the zone of the symmetry axis, a stagnation zone of increased pressure appears in the upper rarefaction wave in which a turn of the flow in the radial direction takes place. The velocity in this direction is close to the gas velocity behind the falling shock wave, and in the radial direction the shock wave that joins the diffracted wave with time is formed.

A complex structure of the flow appears in the axis direction and is connected with the fact that the reflected shock wave interacts with the rarefaction wave passing from the channel angle. This occurence results in the curvature of the shock wave and the appearance of an additional gradi-

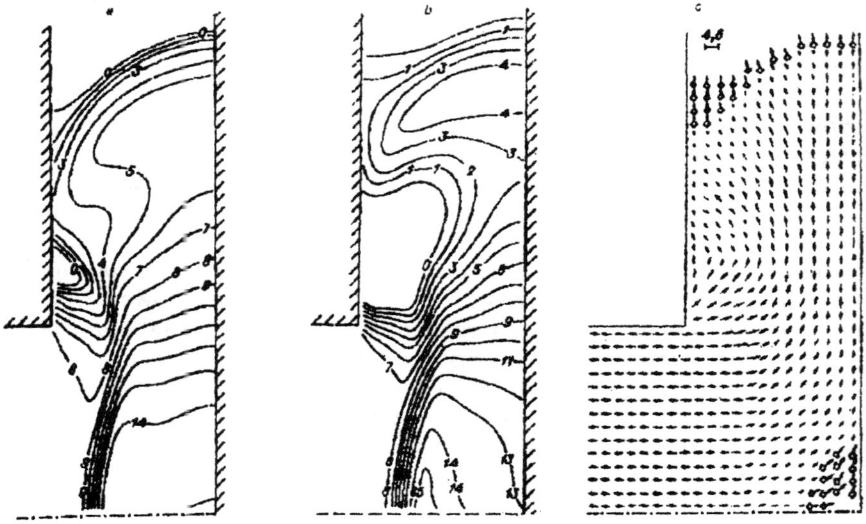

Fig. 2 Distribution of gas parameters in the radial channel: a) pressure field p = 2 bar (0), 10 (5), 80 (14); b) density field ρ =1 (0), 2.5 (5), 4.5 (10), 6.5 (15); and c) velocity field.

ent of density (contact discontinuity) behind it (isohorus has the bend point in this place).

While moving the flow in the radial direction (on the same radii in the experiment), the flow close to the one-dimensional one is maintained.

III. Spontaneous Formation of the Temperature, Current Layer in the Plasma Flow

At small intensities of the magnetic field, the experimental picture of the flow practically differs little from that described in the preceding section. However, when the magnetic pressure becomes comparable with the static pressure in plasma, the flow character changes sharply. So, at the magnetic field $B > 0.3$ T at some distance behind the weakly illuminating first front, a brightly illuminating ring that begins to decelerate effectively appears. From this brightly illuminating zone along the direction to the channel center, the shock wave runs. This picture is observed on the exposure photography and on the photosweep of the process; the graphic reflection of which is represented in Fig. 3.

The electric measurements have shown that the zone of maximum density of the electric current coincides with the position of the intensively illuminating ring; whereas, while moving the ring along the channel, the spatial separation of the maximums of pressure and of current density takes place. The results of the measurement of the electron density (Fig. 4) testify about stopping the plasma cooling in the current ring that is formed.

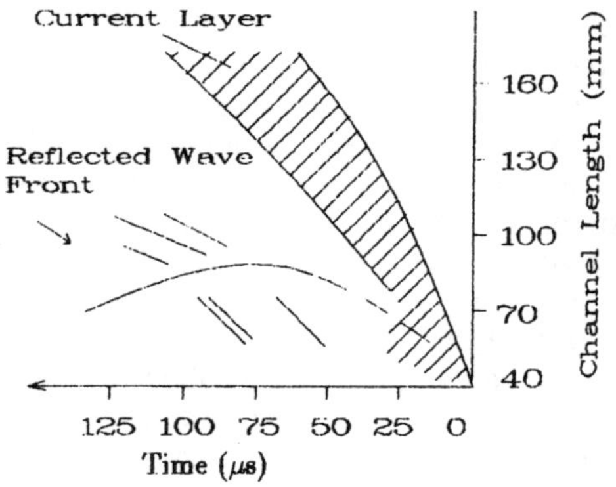

Fig. 3 Photosweep of the process: Ar + 2% H_2, P_0 = 2.5 · 10^{-4} bar, and B_0 = 0.32 T.

While analyzing the results, we supposed that the first reason for the TC-layer appearance might be the skinning of the magnetic field. To verify this supposition, the calculations in the approximation of one-dimensional MHD equations were performed.

Let the supersonic source be in the radial cylindrical channel on the radius τ_0 = 0.5 m. The gas from this source runs to the side of the large radii. Perpendicular to the flow velocity, the outer magnetic field homogeneous on space and changeable with time is applied. As an initial flow, the stationary flow of argon with the parameters on the source T_0 = 9000 K, u_0 = 2.5 km/s,

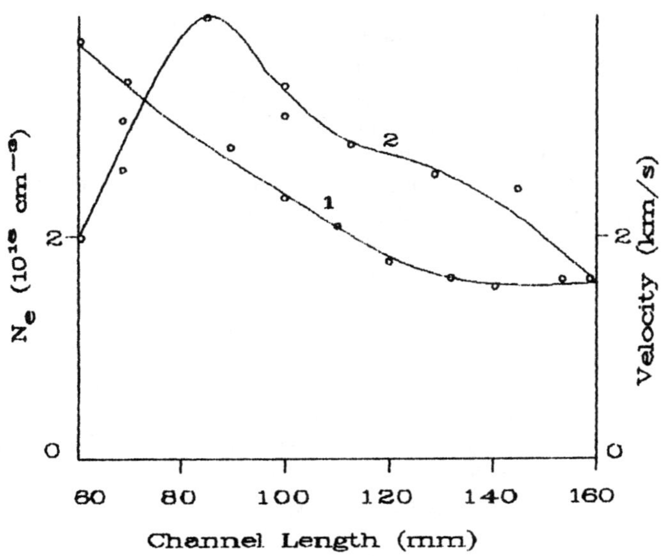

Fig. 4 Electron density in *T* layer and *T*-layer velocity vs radius: Ar + 2% H_2, P_0 = 2.6 · 10^{-4} bar, and B_0 = 0.32 T. Curve 1 is the velocity, and 2 is the electron density.

$p_0 = 3.76$ bar is considered. The law of the external magnetic field change is given as follows: during the first 100μs, the magnetic induction increases linearly from 0 to 1.5 T and then remains as the constant one.

The data on the calculation of the coefficient of heat conductivity and the conductance of argon were taken from Refs. 6–8. The account of radiation was performed in the approximation of the optically thin layer; the degrees of blackness were taken from Ref. 9. As we know, if a supersonic flow in an extending channel exists, then the gas velocity along the channel length increases, and its density and temperature decrease. In our case, the temperature from 9000 K at the input decreases almost up to 3000 K at the output of the channel such that the gas there is not ionized practically. The magnetic field changeable with time is skinned inside the channel in the place where the level of conductivity is rather high. In the region of the skin layer where mainly all of the electric current is concentrated, the heating of gas at the expense of the Joule dissipation takes place.

As the calculations showed (Fig. 5), the rapid increase of temperature in the region of the skin layer results in the formation of the TC layer in this place and the root reconstruction of the flow structure. As a result of the intensive heat supply, the pressure rapidly increases in the TC layer, and by the moment of time ~60 μs, a separation of the region of increased pressure with the formation of shock waves is observed. The shock wave running down the flow turns out to be in the region of a weakly ionized gas and ionizes the gas additionally.

As applied to the concrete experimental installation where the intensity of the external magnetic field in the radial channel increases with the radius, the following problem was solved. The shock wave moving with the Mach number 5 along the mixture of helium with 3% of Cs at $p_0 = 0.028$ bar and $T_0 = 300$ K runs from the straight tunnel into the disk channel. The radius of the straight tunnel and the width of the crack h of the radial channel are equal to 27.5 cm and 24 cm, respectively. The external magnetic field in the disk channel is distributed by the law $B_e = B_0 \cdot [0.4 + 0.037 \cdot (r/h)]$.

While solving the initial system of equations on the basis of the estimates, the following simplifying suppositions were made:

1) Let us neglect the contribution of heat transfer at the expense of radiational and conductive heat exchange in total energetic balance.

2) Let us neglect the Hall current as compared with the main one.

3) Let us consider the problem in the approximation of small magnetic numbers of Reynolds, i.e., we neglect the inductive magnetic field.

Because of supposition 2), the problem will have the axial symmetry: V (v, 0, u); $j(0, j_\varphi, 0)$; and E $(0, E_\varphi, 0)$. In the approximation of the small magnetic numbers of Reynolds and the constant external magnetic field B_b $(0, 0, B_0)$ from the equation $(1/r)(\partial/\partial r)rE_\varphi = 0$ and the boundary condition $E_\varphi(0, t) = 0$, it follows that $E_\varphi = 0$. In this case the Ohm law has the following form: $j_\varphi = -\sigma v B_0$.

Because helium has a high potential of ionization ($I_{\mathrm{He}} = 24.46$) in the temperature range under consideration, Helium is not ionized practically, and one may use the Sakh equation for finding the degree of ionization $\alpha =$

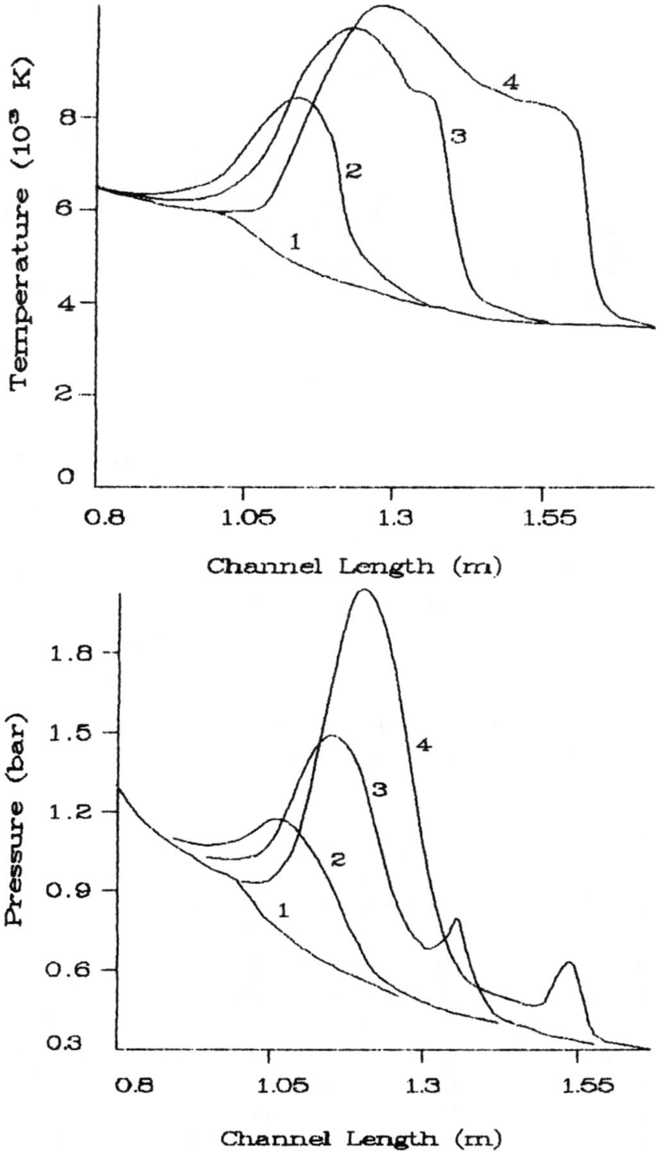

Fig. 5 Temperature and pressure distributions in the Argon flow under the condition of the spontaneous *T*-layer appearance: $t = 30$ μs (curve 1), 60 (2), 90 (3), and 120 μs (4).

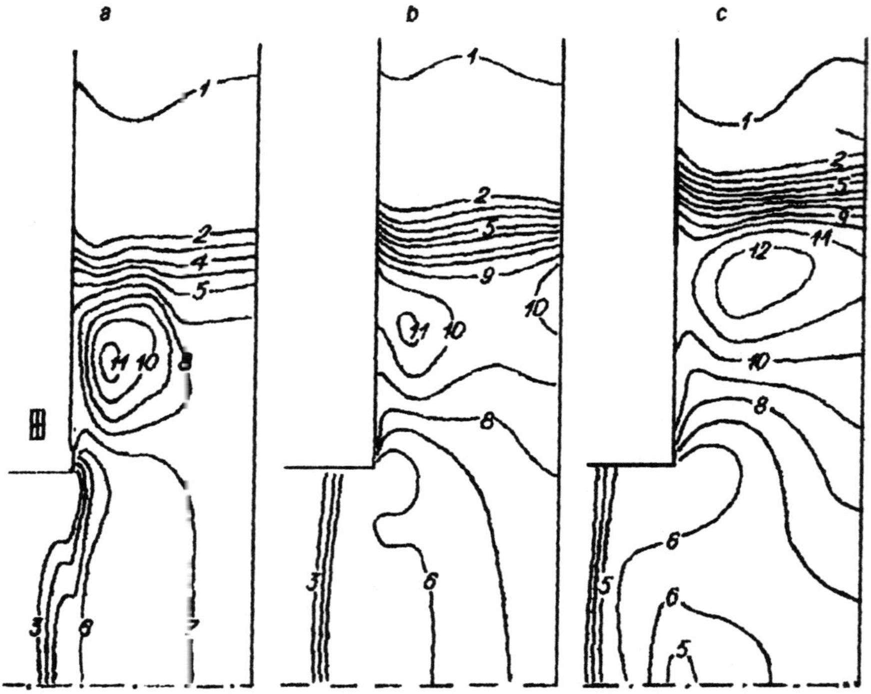

Fig. 6 Temperature fields in the MHD channel at different instants: a) $t = 240\ \mu s$; b) $t = 288\ \mu s$; and c) $t = 336\ \mu s$. $T/T_0 = 1$ (1); 3.5 (5); and 6 (10).

$n_e/(n_{ics} + n_{cs})$. The state equation may be written in the form $p = [1 + (1 + \alpha) \cdot v/(1 - v)]\ n_{He}kT$ where at small concentrations of seeding v transforms into the equation of state for the main gas—helium.

The results of the calculation for $B_0 = 0.25$ T in the form of the field of isothermal lines for the various moments of time are represented in Fig. 6. The picture of a flow shown in Fig. 6a corresponds to the instant when the part of the shock wave reflected from the end wall enters the cylindrical shock tunnel. The upper part of the reflected shock wave having been reflected for the second time from the first wall of the channel creates the zone of increased temperatures. By the moment of time $t = 288\ \mu s$, this zone occupies already all of the width of the disk channel (Fig. 6b) and keeps on moving in the radial direction in the increasing magnetic field. The further interaction of the formed high-temperature ring with the magnetic field results in its additional heating and the increase of conductance, which is clearly seen from Fig. 6c obtained at $t = 336\ \mu s$. Note that the temperature distribution in the high-temperature zone with time becomes more homogeneous. Thus, if at the initial stage of the appearance of the TC layer the flow in the current zone has the characteristic two-dimensional structure, then the further it approaches the one-dimensional structure.

IV. Generation of the Temperature, Current Layer in the Flow

For the proper organization of the flow with current-carrying layers, to be able to generate the TC layer in a proper place is necessary and may be done by way of creating a local pertubation of conductance. The latter may be obtained at the expense of the heating of some mass of a gas by the local addition of easily ionized seedings.

For the first time the possibility of the generation of the TC layer on the temperature perturbation in the flow of weakly conducting gas is shown in Ref. 4. In this work only the initial stage of the TC-layer development was studied because the absence of heat conductance and radiation losses of energy resulted with time in an unlimited growth of temperature in the perturbation zone. Later the analogous problem was considered in Ref. 10 where the real physical properties of the medium were taken into account, which allowed the observation of the dynamics of the TC-layer development.

Under the conditions close to the experimental ones, the numerical experiment was conducted. The aim of this experiment was the elucidation of the character of the flow and the determination of the conditions under which the initial perturbation of temperature (conductance) develops into the TC layer.

The stationary supersonic flow of argon in the radial channel of a constant cross section was considered. Perpendicular to the flow velocity, the external magnetic field was applied. To determine the conditions of "distribution" of a perturbation and the development into the TC layer, the initial velocity of the flow $u_0 = 1.5 \ldots 12$ km/s, the initial temperature of the pertubation $T_p = 6500 \ldots 10{,}400$ K, the width of the perturbation zone $\delta = 0.05 \ldots 8$ cm, and the magnetic induction of the external field $B_0 = 0.7 \ldots 2$ T were varied. The initial temperature of the flow and the pressure were given; $T_0 = 6500$ K and $p_0 = 1$ bar, respectively. At these parameters the main flow does not interact practically with the magnetic field. Isochoric perturbation of temperature was created at the initial moment of time, near the source. For the regimes when the perturbation was "continued," the character of its development and the flow picture differ negligibly. In Fig. 7 the distributions of velocity, those of pressure, temperature, and electric current density in the flow are represented for the set of parameters $u_0 = 6$ km/s, $T_p = 8500$ K, $\delta = 1$ cm and $B_0 = 2$ T. At the initial stage ($t = 5 \cdot 10^{-2}$ μs) a skinning of the magnetic field at the boundaries of perturbation because of its movement occurs, and the characteristic "two-hill" profile of the current is formed. Further, the current profile becomes close to the "parabolic" one with the maximum inside the zone. In the region of this maximum, the largest deceleration of the plasma takes place. From this place the compression waves that strengthen the initial shock wave (at the expense of the discontinuousity break decay) pass upstream. The rarefaction waves passing downstream catch and weaken the initial shock wave.

The general regularities of the development of temperature perturbation into the TC layer obtained as a result of numerical experiment may be noted.

1) After a rather long period of time, the maximum temperature in the TC layer $T_{\max}$ asymptotically approaches ~13,000 K independent of the initial

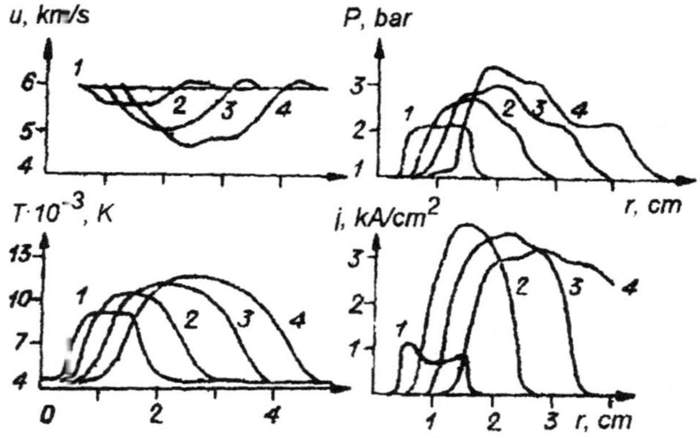

Fig. 7 Dynamics of TC-layer development under the conditions of model experiment: t = 0.05 μs (curve 1); 1 (2); 2 (3); and 3 μs (4).

parameters and is a consequence of maintaining an equilibrium between the Joule dissipation and the losses for radiation.

2) The rate of the T_{max} growth with time decreases while increasing the initial temperature in perturbation.

3) At the given T_{dis} and u_0 the rate of the T_{max} growth with time decreases with the decrease of B_0.

4) The greater the initial temperature of perturbation T_{dis} is, the less the curves $T_{max}(t)$ differ from each other at different u_0 and B_0.

In the experiments on the initiation of the TC layer on a local temperature perturbation, the regimes with spontaneous appearance of the TC layer were chosen. Further, increasing the initial pressure in the installation the velocity and temperature of the flow were reduced such that it practically stopped to interacting with the magnetic field. Then the heating in the ring zone with the cross section about 2 × 2 cm² by inductive discharge during approximately 2 μs was carried out. For this time the energy about 4 J was introduced into the plasma. As a result, in this zone the electron concentrations, pressure, and temperature increased, which provided rather intensive interaction of the flow with the magnetic field.

The moment of heating was chosen in such a way that the perturbation zone appeared behind the front of the shock wave caused by the collision of the plasma mass with the end wall and by the turning of the flow in the radial direction.

As a result of the heating, the conditions providing a rather intensive interaction of the flow with the magnetic field appear. While moving along the channel, the heated ring zone was not only cooled but also heated.

In Fig. 8 the results of the similar experiment in argon are represented. On the filming the formation of the high-temperature current layer and that of the reflected shock wave appearing because of its deceleration in the magnetic field are seen. While moving along the channel, the TC layer is behind the front of the wave.

Fig. 8 Frame photorecording of the movement of a heated zone through the channel: Ar + 2% H_2; $B_0 = 4.8$ T; and $P_0 = 1.5 \cdot 10^{-2}$ bar ($125 \cdot 10^3$ frames per second).

The results of the experiments with the use of the medium He + 2% H_2 as a working body showed that an artificial generation of the TC layer is possible. With this possibility the picture of the development of the temperature perturbation into the TC layer is more demonstrative than in argon and air, and this effect is observed even at lower initial pressures. However, the attempts to obtain the TC layer in pure helium were not successful ones. The rapid photorecording of the process showed that in the zone of heating the development of radial instabilities of the flow takes place.

The investigations confirming the possibility of the creation of local high-temperature zones in the plasma flow of alkali metal were carried out.

In Fig. 9 the results of the calculation of the dynamics of the development of the TC layer from the originally given isochoric perturbation of the

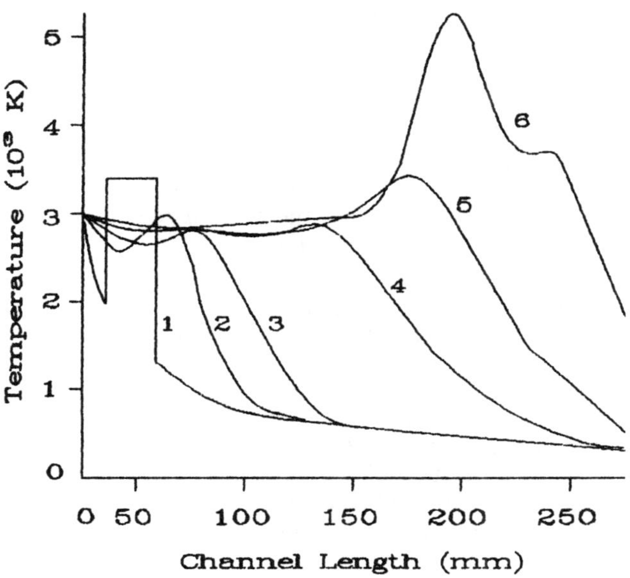

Fig. 9 The temperature disturbance development in the radial channel. Sodium plasma: $M_0 = 1.1$, $P_0 = 1.5$ bar, $T_0 = 3000$ K, $T_{div} = 3500$ K, and $B =$ T. $t = 0$ (curve 1), 10 (2), 20 (3), 50 (4), 70 (5), and 90 μs (6).

Fig. 10 Photosweep of the temperature disturbance development in a sodium plasma: $B_0 = 0.64$ T.

temperature in Na flow under the conditions corresponding to the experimental installation are represented. At the initial stage can be seen that a decomposition of the region of increased pressure with the temperature decrease takes place. The increase of the perturbation and the TC-layer development takes place only from 70 μs. Opposite of the noble gases in the given case, the current zone (because of the potential of the Na ionization) is more longitudinal.

In Fig. 10 the photosweep of the process of the TC-layer generation in the Na plasma flow is represented. That the current region is decelerated and the reflected shock wave appears is seen.

V. Influence of Nonequilibrium Processes on the Temperature, Current-Layer Development

A. Statement of the Problem

Let us consider the helium flow with inhomogeneously distributed seedings of alkali metal in the plane channel of the constant cross section (Fig. 11). The investigation of the process of the interaction of electroconductive gas with the magnetic field will be performed in the frames of two-temperature magnetic hydrodynamics taking into account the ionizing kinetics. The system of initial equations is

$$\frac{\partial \rho}{\partial t} + \frac{\partial}{\partial x}(\rho u) = 0$$

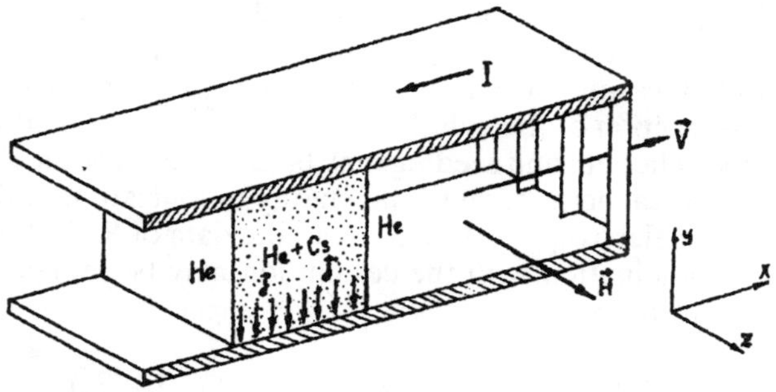

Fig. 11 Scheme of the flow in a MHD channel.

$$\rho\frac{\partial u}{\partial t}+\rho u\frac{\partial u}{\partial x}+\frac{\partial p}{\partial x}=j\cdot B$$

$$\rho\frac{\partial \varepsilon}{\partial t}+\rho u\frac{\partial \varepsilon}{\partial x}=\frac{\partial}{\partial x}\left(\lambda\frac{\partial T}{\partial x}\right)-p\frac{\partial u}{\partial x}-Q_{\text{rad}}+æ_{ea}(T_e-T)$$

$$\frac{\partial}{\partial t}\left[n_e(3kT_e/2+I)\right]+\frac{\partial}{\partial x}\left[n_e u(5kT_e/2+I)\right]=j^2/\sigma-æ_{ea}(T_e-T)$$

$$\frac{\partial n_e}{\partial t}+\frac{\partial}{\partial x}(n_e u)=\nu n_a n_e-\beta n_e^3$$

$$p=\frac{R}{\mu}\rho T$$

$$\frac{\partial B}{\partial x}=-j$$

$$j=\sigma(E-uB)$$

$$æ_{ea}=3km_e(8kT_e/\pi m_e)^{1/2}n_e(Q_{\text{He}}^e n_{\text{He}}/m_{\text{He}}+Q_{\text{Cs}}^e n_{\text{Cs}}/m_{\text{Cs}}+Q_i^e n_e/m_i$$

$$\sigma=\frac{n_e e^2}{m_e\sqrt{3kT_e/m_e}\cdot\left(n_n<Q_n^e>+n_e Q_i^e\right)}$$

where ε is the inner energy of the gas; I is the potential of the ionization of seeding; $\nu(T_e)$ is the function of ionization; $\beta(T_e)$ is the recombination function; $n_n = \Sigma_s r_s$ is a number of neutral particles in the unit of the volume; $Q^e_{He}, Q^e_{Cs}, Q^e_i$ are the cross sections of the collision of the electron with the atoms of helium, cesium, and ions, respectively; and $< Q^e_n > = \Sigma_s n_s Q^e_s / n_n$ is the mean cross section of the collision of an electron with neutral particles.

The mixture of helium and seeding may be considered as an ideal gas with the constant heat capacity because the energy of ionization in this model is subscribed to the electron gas. According to the data of Ref. 11, the function of electron recombination with the cesium ion may be represented by the approximation formula

$$\beta(T_e) = 1.11 \cdot 10^{-22}/T_e \cdot \exp(10.17 \cdot 10^3/T_e)$$

with an error of 5%. The ionization function $\nu(T_e)$ is determined from the condition of detailed equilibrium $\nu = \beta K$ where $K = 4.85 \cdot 10^{15}\ \Sigma_+/\Sigma_0 \cdot T_e^{3/2} \times \exp(-I/kT_e)$; for cesium $I = 3.98$ eV, and the statistic sums of the ion and atom are $\Sigma_+ = 1$ and $\Sigma_0 = 2$.

The cross sections of the collision are equal to $Q^e_{He} = 5.9 \cdot 10^{-16}\text{cm}^2$, $Q^e_{Cs} = 3.85 \cdot 10^{-14}$ cm^2, respectively,

$$Q^e_i = \pi e^4 (1 + \omega/2)^2 / \left[4k^2 T_e^2 (1 + \omega)^2\right] \cdot \ln(1 + 2/\omega)$$

where

$$\omega = 2e^2/kT_e \cdot [4ke^2/k \cdot (n_e/T_e + n_i/T_i)]^{1/2}$$

The estimates performed showed that at the characteristic parameters of the problem ($T_* \sim 5000$ K, $t_* \sim 10^{-4}$ s, $l^* \sim 5$ cm, and $\rho_* \sim 10^{-4}$ cm^3) the contribution of the heat conductivity and of radiation into the energetic balance may be neglected.

B. Results of the Calculations

The helium flow with the initial parameters $p_0 = 3.2$ bar, $u_0 = 3.3$ km/s, and $T_0 = 2040$ K in the channel shown in Fig. 11 is considered. The external magnetic field is B_0=const=2 T. A seeding is introduced into the flow locally. The initial width of the zone with the seeding is 4 cm The seeding concentration is $z = 1.5\%$. The dynamics of the process are represented in Fig. 12. The electron temperature instantly (for $t \sim 0.1\ \mu$s) increases for 500 K. Further, the smooth increase of the temperature takes place and by the thirtieth μs achieves ~3800 K. The ionization equilibrium is maintained starting from the time moment $t = 2\ \mu$s. The electron density increases smoothly from the initial equilibrium value $\approx 10^{14}$ ($T_e = T = 2040$ K) up to $2 \cdot 10^{16}$.

The temperature of heavy particles increases under the influence of two factors: adiabatic heating in the comparison wave formed by a decelerating electrodynamic force and the energy transfer at the collisions with "hot"

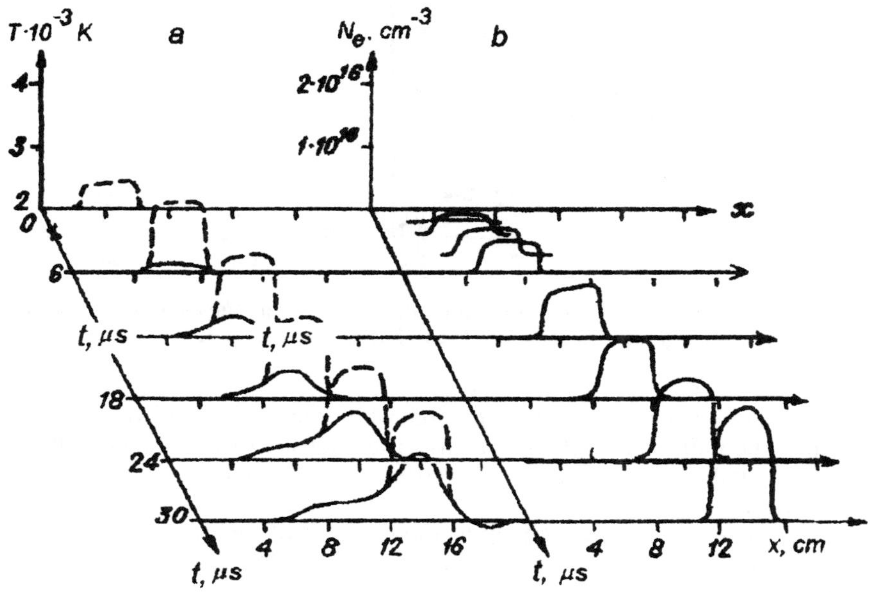

Fig. 12 Dynamics of the current layer: a) temperature of electrons and heavy particles and b) electron density. Asterisk means the moment of reaching ionization equilibrium - - - T_e, — T_i.

electrons. In the graphic of the temperature distribution (see Fig. 12) for the time moment $t = 30\ \mu s$, the shock wave, passed away from the current layer, and the rarefaction wave in which the temperature has decreased lower than the initial level are seen.

During the calculations, the Hall parameter $\omega_e\tau_e$ was determined. Its distribution (first homogeneous by the current zone) to the end of the interaction is reconstructed in accordance with the density distribution. In the region of maximum influence (the maximum of current density), the parameter $\omega_e\tau_e \sim 0.4$. In the rarefaction wave the parameter increases, having achieved the values ~1 at the right boundary of the conducting zone. However, the current density, and hence, the deceleration force here are small.

To elucidate the role of the intensity of the magnetic field, the variants with two values of B_0—1.0, 1.5, and 2.0 T—were calculated. The discovery was made that the time for maintaining the ionization equilibrium strongly depends on B_0. Thus, in the regime with $B_0 = 1.0$ T the equilibrium does not take place up to the output of the current zone from the MHD channel. At $B_0 = 1.5$ T it is achieved by the seventh μs, and for $B_0 = 2.0$ T by the second μs. In Fig. 13 the distribution of pressure and temperature in the flow by the moment $t = 30\ \mu s$ for the regimes from $B_0 = 2.0$ T and $B_0 = 1.5$ T are represented. This result demonstrates a property of the effect of the TC layer

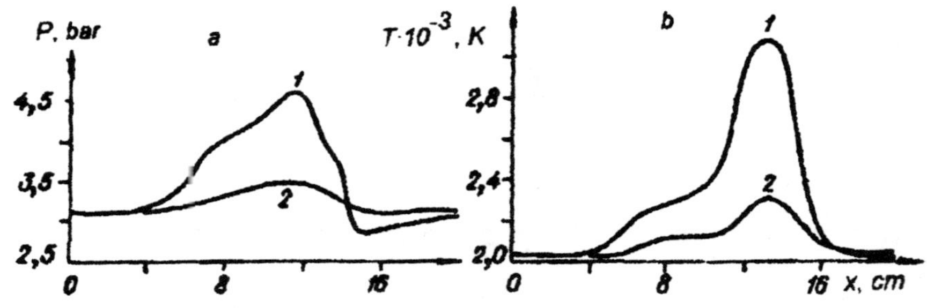

Fig. 13 Distribution of the pressure and temperature along the channel at instant t = 30 μs:1 is the regime with B_0 = 0.2 T, and 2 is the B_0 = 0.15 T.

as an essentially nonlinear phenomenon, when relatively small variations of the initial conditions (in the given case, the intensity of the magnetic field) result in qualitative changes, i.e., the formation of a high-temperature zone in the flow.

As the calculations showed, up to the moment of the maintaining of the ionization equilibrium in the plasma the gas flow does not interact practically with the magnetic field, and there is no increase of temperature in the zone with seeding. In the given case one may introduce three characteristic periods of time determining the process of interaction: t_p is the time of maintaining the ionization equilibrium; t_T is the time of the TC-layer development; and $t_* = l_*/u_*$ is the characteristic gasdynamic time. The necessary condition for the realization of the TC-layer development in the flow of inhomogeneous and nonequilibrium plasma is the fulfillment of the conditions $t_p < t_T < t_*$.

VI. Temperature, Current Layer in the Molecular Gas Flow

The investigations on the TC-layer generation in the flows of complex molecular gases and, in the first turn, the products of burning are of definite interest. At the present moment in this direction there are no results. We investigated the regimes of spontaneous formation and generation of the TC layer in the flow air plasma. However, the installation parameters were such that in the MHD channel gas was completely dissociated, and the results differ from the experiments with the noble gases.

The effort to generate the TC layer in the flow of the plasma obtained from the carbonic gas was made. However, under the conditions of the operation of the given installation even at the maximum possible energy input into the heating zone and the intensity of the external magnetic field 6.4 T, we failed to obtain the stable and unambiguous picture of TC-layer development. In Fig. 14 frame photorecording of the plasma movement in the channel is represented. We see that there is no formation of the true ring zone. Inhomogeneities appear on the radius as on the angle. The spirelike zones of increased brightness testify about the presence of essential Hall currents directed along the radius and comparable with the main current.

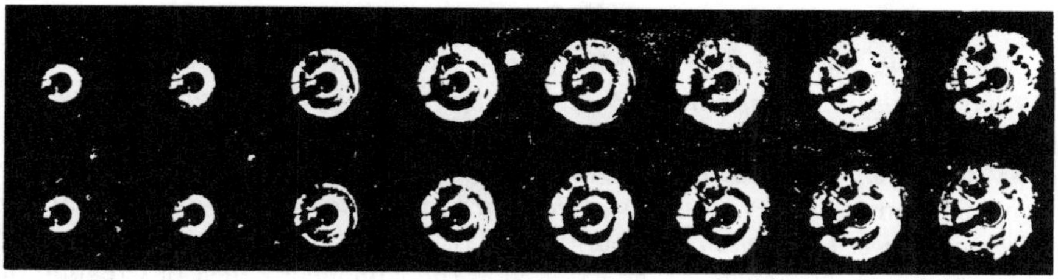

Fig. 14 Frame photorecording of the movement of a heated zone through the channel CO_2; $B_0 = 0.64$ T; and $P_0 = 4 \cdot 10^{-3}$ bar.

The possibility of generating the TC layer in the combustion products flow in the channel of a constant cross section was studied theoretically. As these products, dissociating water vapors were considered. The initial perturbation of the conductivity was given by the local introduction of the seeding with the mole concentration 1.5%. With this, we supposed that at the given temperatures T_{dis} Cesium would be in a free state; besides, the energy losses for the radiation and heat conductivity were not taken into account.

The errors introduced into the calculation by these suppositions have a general tendency to increase the temperature value in the interaction zone. However, the calculation showed that the temperature in the current layer increases for some inconsiderable value as compared with the total temperature background, which is connected with the fact that the dissociating gases have a very large heat inertion (heat capacity of water vapors is 15 . . . 20 times more than the heat capacity of a one-atomic gas), and because of this, the error in the energy balance does not result in a large error in the temperature determination. The calculation showed that with time, in the zone of conductivity, temperature and current density do not increase and there is no increase in the perturbation.

VII. Conclusion

The plasma flow interacting with the magnetic field has undergone a series of MHD instabilities. The development of the latter destabilizes the flow and finally decreases the degree of its interaction with the field. The TC-layer phenomenon testifies about the fact that because of the nonlinear processes in the plasma the instability development under definite conditions may result in the reconstruction of the flow and its stabilizing at some new level. For the first time it was obtained as a result of a numerical experiment and then was investigated experimentally.

The main results here were obtained for the plasma flows of one-atomic gases. The results represented in Section VI testify only about the necessity to continue the investigations on the organization of the flows with the TC layer in the products of burning.

References

[1]Ricateu, P., and Zettwoog, P., "MHG Conversion in Inhomogeneous Flow," *Proceedings of the 4th Symposium on Engineering Aspects of Magnetohydrodynamics* (Berkeley, CA), 1963.

[2]Milewski, J., "On the Striated Flow in the Induction Synchronous Magnetohydrodynamic Generator," *Proceedings of the Third International Symposium on MHD Electrical Power Generation* (Salzburg, Austria), 1966.

[3]Tikhonov, A. N., Samarskii, A. A., et. al., "The Nonlinear Effect of Creation of Self-Supported High-Temperature Gas Layer at Transient Magnetohydrodynamic Processes," *Soviet Physics—Doklady*, Vol. 173, No. 4, 1967, pp. 808–811.

[4]Degtyarev, L. M., Zaklyazminskij, L. A., Kurdyumov, C. P., et al., "Development of the Finite Local Disturbances of Electrical Conductivity in the Flow of a Weakly Conducting Gas in the Magnetic Field at Arbitrary Magnetic Reynolds Numbers," *Teplofizica Vysokih Temperatur*, Vol. 7, No. 3, 1969, pp. 471–478.

[5]Tikhonov, A. N., Samarskii, A. A., et. al., "TC-Layer Effect in Magnetohydrodynamics," Preprint IAM AS USSR, Moscow, 1969.

[6]Neumann, W., "Uber Radialen Temperaturverlauf in Stationaren und Impulsmodulirten Argon-Hochtemperaturborgen," *Beitraus der Plasma Physics*, Vol. 2, No. 2, 1962, pp. 80–115.

[7]Spitzer, L., and Harm, R., "Transport Phenomena in Completely Ionized Gas," *Physical Review*, Vol. 89, No. 5, 1953, pp. 977–981.

[8]Nighan, W. L., "Electrical Conductivity of Partly Ionized Noble Gases," *Physics of Fluids*, Vol. 12, No. 1, 1969, pp. 162–171.

[9]Katsnel'son, S. S., and Koval'skaya, G. A., *Thermophysical and Optical Properties of Argon Plasma*, Nauka, Novosibirsk, Russia, 1985.

[10]Katsnel'son, S. S., "The Solving of Some Transient Problems of Magnetohydrodynamics," Ph. D. Disc. Dept. of Physical–Mathematical Sciences, Novosibirsk, Russia, 1975.

[11]Biberman, L. M., Vorob'yov, V. S., and Yakubov, I. T., "Non-Equilibrium Low-Temperature Plasma. IV. Functions of Ionization and Recombination," *Teplofizika Vysokih Temperaur*, Vol. 7, No. 4, 1969, pp. 593–603.

Chapter 65

Free Magnetohydrodynamic Flow of the Low-Density Vacuum-Arc Cathodic Jet

M. Keidar,* I. Beilis,† R. L. Boxman,‡ and S. Goldsmith§
Tel Aviv University, Tel Aviv, 69978, Israel

I. Introduction

THE low-density part of the cathodic plasma jet is very important in the technological applications of the cathodic arc as a source of a high-energy metal ion flux, which can be used to produce thin-film coatings with advantageous properties for many applications[1] and can be used as an ion source for accelerators.[2] Plasma-jet characteristics in the low-density region were studied experimentally.[3–6] These discoveries were made: a significant kinetic energy is associated with the directed ion motion, the ion-current density falls off with axial distance from the cathode, and the ion current is about 10% of the total arc current.

A popular configuration for the vacuum-arc plasma source uses a ring anode. The ion-current distribution generated on the cathode and passing through the anode in the presence of a magnetic field was studied by Cohen et al.[7] with a multielement probe. The measured ion-current density distribution without a magnetic field was approximately a cosinusoidal function of the angle with respect to the normal to the cathode surface, whereas with an axial magnetic field the ion-current distribution became peaked in the axial direction.

A theoretical study of vacuum-arc plasma jets has been concentrated on the plasma jet formation and magnetic collimation effects. Beilis et al.[8,9] and Hantzsche[10] studied the plasma jet without the presence of an axial magnetic field for the various plasma density regions and reviewed other works in this field. In the low-density part of the plasma jet, an axial magnetic field can have a very significant effect on the jet expansion. The magnetic field

*Electrical Discharge and Plasma Laboratory, P. O. Box 39040.
†Electrical Discharge and Plasma Laboratory, P. O. Box 39040.
‡Electrical Discharge and Plasma Laboratory, P. O. Box 39040.
§Electrical Discharge and Plasma Laboratory, P. O. Box 39040.

influence on the low-current-density quasineutral plasma beam was analyzed by Morozov.[11] A static electric field that can exist in a low-density plasma if both an external magnetic field and electron drift are present together was shown. When the electron cyclotron radius is small in comparison to the radius of the system containing the plasma, the electrons are confined. Charge separation generates an electric field that confines the ions whose cyclotron radius may be larger than the system radius.

Boxman[12] studied the current-carrying plasma flow in the interelectrode gap between two disc electrodes. The fluid flow was analyzed assuming a uniform current density and a mass flow constriction caused by the magnetic pinch force that was obtained near the anode. A constriction in the current flow, assuming a uniform mass distribution, was also calculated caused by the Hall effect. The influence of the self-magnetic field on the current-carrying plasma flow was investigated by Izraeli et al.[13] and Sholhet et al.[14] for the vacuum arc and by Gonzales et al.[15] and Chen and Pfender[16] for the atmospheric arc. A constriction of the current flow at the anode region was obtained for some conditions.

One of the main difficulties in the jet expansion problem is the formulation of the physical conditions at the plasma-vacuum boundary. The free boundary conditions at the interface between the plasma and the vacuum in the case of large current density, where the self-magnetic field plays a significant role, was discussed by Huysmans et al.[17] The plasma boundary was defined by the condition that the magnetic field lines cannot point out of the plasma (i.e., the normal component of the magnetic field at the plasma boundary must equal zero). The discontinuity in the tangential component of the magnetic field across the boundary determines the surface current. The boundary condition that the normal component of the plasma velocity to the fixed plasma boundary must be zero was used by Kress and Riedel.[18]

In the present work the problem of the plasma-jet expansion in an axial magnetic field will be analyzed taking into account the self-magnetic field. The jet has a defined initial radius, which is smaller than the inner radius of a ring anode, and has no contact with any other material wall, resulting in a free expand into the vacuum.

II. Model Formulation

Let us discuss a plasma jet with characteristic radius R_0 (Fig. 1) at some distance from the cathode surface where the jets from the individual cathode spots in the case of the multi-cathode-spot vacuum arc are averaged, and, therefore, all plasma parameters will be assumed to be uniform. This plasma is sufficiently rarefield, so that inelastic collisions (ionization, recombination, etc.) do not influence the plasma density. In addition we considered that the electron and ion temperatures are different and that they do not change substantially during expansion. Under these conditions the electric field in the interelectrode gap does not influence considerably the expansion of the plasma jet, but may play the dominant role in the current collection at the anode. The anode does not influence the flowfield and only acts as a current collector. The imposed axial magnetic field has only an axial

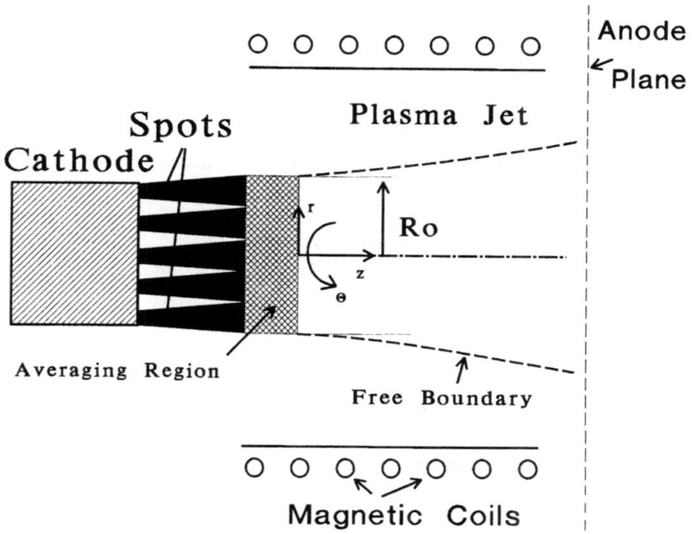

Fig. 1 Sketch of the vacuum-arc geometry and model of the vacuum-arc plasma expansion.

component. The geometric position of the free boundary will be determined as part of a self-consistent solution.

The steady-state plasma flow in the vacuum arc is characterized by a directed velocity V of about 10^4 m/s, which is larger than the thermal velocity.[6] The plasma jet of the low-current vacuum arc ($I_{arc} \sim$ 100–300 A) has a plasma density in the rarefield part of the cathode region[9] of about 10^{20} m^{-3} and a characteristic radius of about a few centimeters. Taking into account that the electron temperature (T_e) is about 1–2 eV, we find that the Debye length L_D is 10^{-6} m, and the mean free path for electrons L_c is 10^{-3} m. The magnetic diffusion time τ_m is about 10^{-7} s. This problem is characterized by a time scale of $\tau_f = L/V$, where L is the interelectrode region length. For $L \sim$ 0.1 m τ_f is about 10^{-5} s.

The current-carrying plasma flow in the magnetic field will be studied with the aid of the following assumptions based on the preceding estimations and previous calculations[9]:

1) The plasma is fully ionized and consists of two species of charged particles.

2) During the expansion, the temperature of each species is not substantially changed and thus the process may be considered as isothermal.

3) The range of the plasma parameters at the starting plane and in the large part of the expansion region, of a characteristic length L, is such that the mean free path for electron-ion collisions (L_c) is much smaller than the L. Thus, the hydrodynamic approximation may be used to describe the flow of the plasma in the magnetic field.

4) The electron component is not inertial and thus $m_e(\boldsymbol{V}_e \cdot \nabla)\boldsymbol{V}_e$ may be neglected in the conservation of momentum equation.

5) The plasma is quasineutral, i. e., $|N_e - Z_i N_i|/N_e \ll 1$, where N_e and N_i are the electron and ion densities, respectively, and Z_i is the ionicity.

Let us define the electric current density $\boldsymbol{j} = eN(\boldsymbol{V}_i - \boldsymbol{V}_e)$ and the plasma conductivity $\sigma = (e^2 N/m_e \nu_{ei})$ where ν_{ei} is the electron-ion collision frequency[19]

$$\nu_{\mathrm{ei}} = \frac{(\ell n\, \Lambda/10) Z_i N_e [\mathrm{m}^{-3}]}{3 \cdot 10^{10} T_e^{3/2} [\mathrm{eV}]}$$

and where T_e is the electron temperature and $\ell n \Lambda$ is the Coulomb logarithm. In addition we use the condition $\nu_{ei} N_e = \nu_{ie} N_i$. The ions and electrons are assumed to be ideal gases with partial pressures $P_\alpha = kT_\alpha N_\alpha$, where $\alpha = e, i$, and k is Boltzmann's constant. The steady-state behavior of this current-carrying plasma may be described by the following set of equations:

$$m_i(\boldsymbol{V}_i \cdot \nabla)\boldsymbol{V}_i = -k(Z_i T_e + T_i)\nabla \ell n\, N + \frac{\boldsymbol{j} \times \boldsymbol{B}}{N} \tag{1}$$

$$\boldsymbol{j} = \sigma\left[\boldsymbol{E} + \frac{kT_e}{e}\nabla \ell n\, N - \frac{\boldsymbol{j} \times \boldsymbol{B}}{eN} + (\boldsymbol{v}_i \times \boldsymbol{B})\right] \tag{2}$$

$$\nabla \cdot (\boldsymbol{V}_i N) = 0 \tag{3}$$

$$\nabla \cdot \boldsymbol{j} = 0 \tag{4}$$

where $\boldsymbol{V}_i$ and $\boldsymbol{V}_e$ are ion and electron velocities and $\boldsymbol{E}$ is the electric field.

The situation in which the plasma parameters are such that the electrons are magnetized and the ions are unmagnetized will be analyzed, i.e., where $\rho_i \gg L \ll \rho_e$, where ρ_e, ρ_i are the electron and ion Larmor radii, respectively. Assuming cylindrical symmetry, Eqs. (1–4) may be written in cylindrical component form in the r-z plane (Fig. 1):

$$V_z \frac{\partial V_r}{\partial z} = -V_r \frac{\partial V_r}{\partial r} - C_s 2 \frac{\partial \ell n\, N_r}{\partial r} + \frac{j_\theta B_z}{N m_i} - \frac{j_z B_\theta}{N m_i} \tag{5}$$

$$V_z \frac{\partial V_z}{\partial z} = -V_r \frac{\partial V_z}{\partial r} + \frac{j_r B_\theta}{N m_i} - \frac{1}{1 - \dfrac{V_z^2}{c_s^2}}$$

$$\cdot \left(V_z \frac{\partial V_r}{\partial r} - \frac{j_r B_\theta}{N m_i} + V_r V_z \frac{\partial \ell n\, N}{\partial r} + \frac{V_r V_z}{r} - V_r \frac{\partial V_z}{\partial r}\right) \tag{6}$$

$$\left(C_s^2 - V_z^2\right)\frac{\partial N}{\partial z} = NV_z\frac{\partial V_r}{\partial r} + \frac{j_r B_\theta}{m_i} + V_r V_z\frac{\partial N}{\partial r} - NV_r\frac{\partial V_z}{\partial r} + \frac{NV_r V_z}{r} \quad (7)$$

$$j_z = \sigma\left(E_z + \frac{kT_e}{e}\cdot\frac{\partial \ln N}{\partial z} - \frac{j_r B_\theta}{eN}\right) \quad (8)$$

$$j_r = \frac{\sigma}{1+\beta_e^2}\left(E_r + \frac{kT_e}{e}\cdot\frac{\partial \ln N}{\partial r} + \frac{j_z B_\theta}{eN}\right) \quad (9)$$

$$j_\theta = j_r\beta_e \quad (10)$$

$$B_\theta = \frac{\mu_0}{r}\int_0^r J_z\zeta\, \mathrm{d}\zeta \quad (11)$$

where V_z and V_r are the axial and radial components of the ion velocity; j_r, j_z, j_θ are the radial, axial, and azimuthal components of the current density; μ_0 is the magnetic permeability of vacuum; $C_s{}^2 = k(Z_i T_e + T_i)/m_i$; and $\beta_e = \omega_e/\nu_{ei}$ where $\omega_e = eB/m_e$. In the momentum conservation equation for the ions, a term containing B is omitted because the ions are unmagnetized.

The set of equations (8–11) and Eq. (4) will be used to express the equation for the electric field in explicit form. Taking into account that the Coulomb logarithm $\ln\Lambda$ has a weak dependence on the density N, we consider that $\nu_{ei} = \mathrm{const}\cdot N$. An explicit expression for the potential distribution φ may be obtained from Eq. (14) using the definition $\boldsymbol{E} = -\nabla\varphi$.

In this work the current-carrying plasma flow in the magnetic field is analyzed using the free-jet expansion approach. We assume that the arc runs to an annular anode and that the characteristic size of the cathodic plasma (e.g., the cathode radius or the cathode spot radius) is smaller than the inner radius of the anode; thus, a free jet flow occurs. We further assume that the anode does not influence the flowfield but only plays the role of a current collector.

III. Boundary Conditions

Mathematically, the free boundary problem consists of a system of partial differential equations to be satisfied within the bounded domain together with the necessary boundary conditions. One part of the problem interface, the free boundary, is unknown and must be determined as part of the solution. A common method for the solution of the free boundary problem is the trial-free boundary method.[20,21]

Generally, to solve the system of equations (5–11) knowing the geometric position of the plasma boundary where two kinds of the boundary conditions exist is necessary: 1) hydrodynamic conditions for the plasma density and velocity and 2) electrostatic conditions for the electric potential and electric field.

A. Starting Plane

In the vacuum arc a plasma is generated in the cathode spot. This small region is very dense, so that gas pressure is more than the magnetic pressure from the self-magnetic field. A quasi-one-dimensional model of such a plasma expansion shows that a conical supersonic jet is formed.[9] A self-consistent expansion model predicts a conical angle of about 45 deg. This means that the radial component of the plasma-jet velocity may be significant at large axial distances (about 100 initial cathode spot radii). We will start with our calculation in this region. A uniform distribution of the plasma, gasdynamic, and electrical parameters is assumed at the starting plane. The ratio of the radial to the axial velocities is introduced as a parameter.

B. Free Boundary

In this section the boundary conditions at the free plasma boundary will be described briefly. The model of the free plasma-jet boundary calculation and boundary condition were described in detail in Ref. 22.

If the plasma is in a stable steady state, the shape of the boundary is described by a function $\alpha(z)$, which is the angle of the jet boundary (see Fig. 2). The main problem then is to develop an equation for this function. Let us introduce the spatial coordinate n in the direction normal to the boundary and the coordinate τ in the direction tangential to the boundary. The plasma velocity direction is tangential to the boundary, and therefore the normal component V_n is equal to zero (point A in Fig. 2). At the point C the normal n_1 and tangential τ_1 coordinates are transferred and turned relative to the normal and tangential coordinate point A. The angle between the tangential axis τ_1 and the old axis τ at point C is equal to the jet angle change from the point A to the point C. According to Fig. 2, the jet angle change from point A to point C may be defined as $\alpha(C) - \alpha(A) = \tan^{-1}(V_n/V_\tau)$,

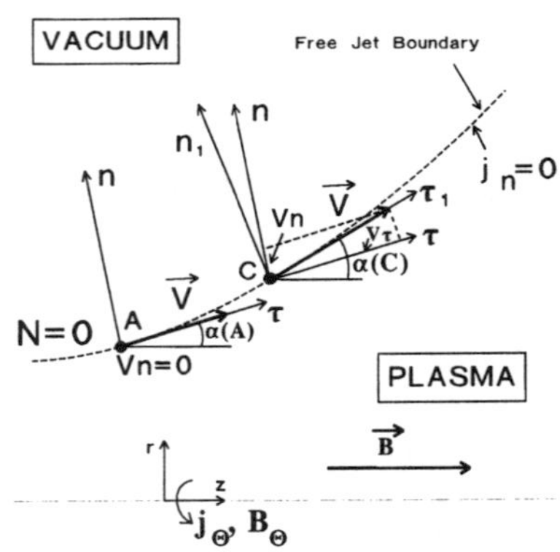

Fig. 2 Schematic presentation of the boundary conditions at the free plasma-jet boundary.

where $\alpha(A)$ is the jet angle at point A, $\alpha(C)$ is the jet angle at point C, and V_n and V_τ are the normal and tangential components of the velocity vector at point C in the coordinate system $\{n,\tau\}$. The change in the jet angle α can be obtained by differentiation of the preceding expression:

$$\frac{\partial\alpha}{\partial\tau} = \frac{1}{1+(V_n/V_\tau)^2} \cdot \frac{(\partial V_n/\partial\tau)V_\tau - (\partial V_\tau/\partial\tau)V_n}{V_\tau^2} \tag{12}$$

The velocity component V_n changes along axis τ from zero at point A to some nonzero value. Taking into account that $(\partial V_n/\partial\tau) \neq 0$ and $V_n = 0$ in point A, together with Eq. (12), yields the final expression for the jet angle changes at point A:

$$\frac{\partial\alpha}{\partial\tau} = \frac{\partial V_n}{\partial\tau}\frac{1}{V_\tau} \tag{13}$$

To solve Eq. (13), the expression for $\partial V_n/\partial\tau$, V_τ will be obtained from the n component of Eq. (1) (see Fig. 2).

The main theoretical problem in the plasma expansion is defining the boundary conditions for the density and electric potential at the plasma-vacuum interface. The condition for the density on the external surface of the plasma jet is $N = 0$ in the case of plasma expansion into a vacuum, as illustrated in Fig. 2.

To close this problem, the boundary conditions for the potential will be formulated. The present model is based on the premise that the current-carrying plasma flows from the cathode to the anode, which does not influence the flowfield. At the anode plane we set the anode potential. This procedure models the physical conditions presented by a grid anode and to some extent those presented by a large diameter thin ring anode. At the free plasma boundary the condition $J_n = 0$ is used for the electrical current.

The set of equations (5–13) with boundary conditions for the density, potential, and current fully determines the free boundary jet expansion problem.

IV. Results

A. Analytical Estimation

In this section a simple analytical model will be developed to obtain the plasma jet shape and to study the influence of current and boundary conditions at the starting plane on plasma expansion. For this purpose, a free plasma-jet expansion into the vacuum without an external magnetic field is studied. A plasma flow will be modeled assuming that the plasma parameters have no radial variation. This means that the kinetic force term in Eq. (1) may be neglected. Taking into account that in the vacuum-arc jet the axial velocity is larger than the radial velocity, we obtain the following simple equation from Eq. (5):

$$V_z \frac{dV_r}{dz} = -\frac{j_z B_\theta}{N m_i} \tag{14}$$

Assuming further that the axial velocity does not change substantially during expansion, we have from the mass conservation equation:

$$N(z) = N_0[R_0/R(z)]^2 \tag{15}$$

where N_0 and R_0 are the initial plasma density and jet radius and $R(z)$ is the local jet radius.

This equation should be supplemented by the equation for the jet shape. Equation (13) in the r-z plane is suited for analytical analyses. After the coordinate transformation ($n \rightarrow r, \tau \rightarrow z$), the equation may be written in the following form:

$$\frac{d^2R(z)}{dz^2} = \frac{1}{V_z}\frac{dV_r}{dz} \tag{16}$$

Equation (16) describes the plasma-jet shape. For this equation we use boundary conditions at the starting plane:

$$R(0) = R_0; \qquad dR/dz(0) = \gamma_0 = V_r/V_z(0)$$

Taking that the current density varies as $j_z(z) = [I/\pi R(z)^2]$ where I is the vacuum-arc current, the plasma jet was calculated for different γ_0 and I.

Results of this calculation are presented in Figs. 3 and 4 for arc current I = 200 A and 1000 A, respectively, where r_b is normalized by the R_0 local jet radius. In the 200-A case we find that the plasma jet does not expand, but rather contracts if the initial angle $\gamma_0 < 0.1$, whereas in the 1000-A case the jet contracts even at $\gamma_0 = 0.5$. These results were obtained assuming that there were no radial variations. The validity of these results must be established or rejected by comparison with a two-dimensional calculation, as presented in the next section.

B. Numerical Solution

Numerical Method

The equation system (5–11) was solved numerically in the following manner. An iterative scheme for obtaining the self-consistent steady-state plasma parameter distribution starts with an assumed initial potential distribution $\varphi^{(0)}(r,z)$. The ion velocity, current density, and density distribution are

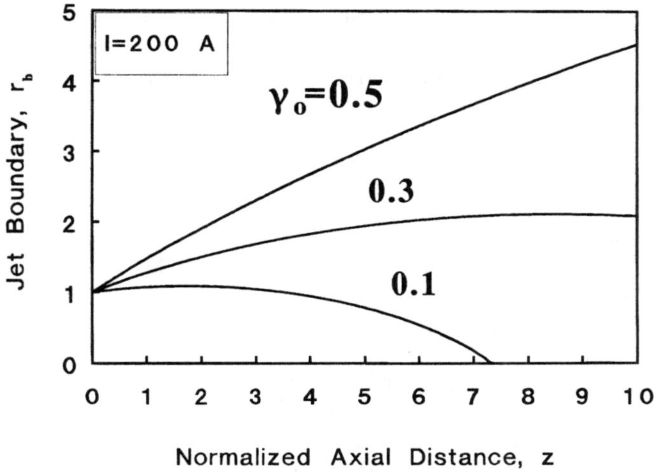

Fig. 3 Analytical one-dimensional calculation of the free plasma-jet boundary in the *r-z* plane with the radial to axial velocity ratio at the starting plane γ_0 as a parameter for I = 200 A.

then obtained by numerically calculating Eqs. (5–11) using the implicit two-layer method.[23]

The calculated velocity and density distribution are used for the second iteration from which a new potential distribution $\varphi^{(1)}(r,z)$ can be calculated by solving Eq. (4) for the potential using the relaxation method.[24,25] Gener-

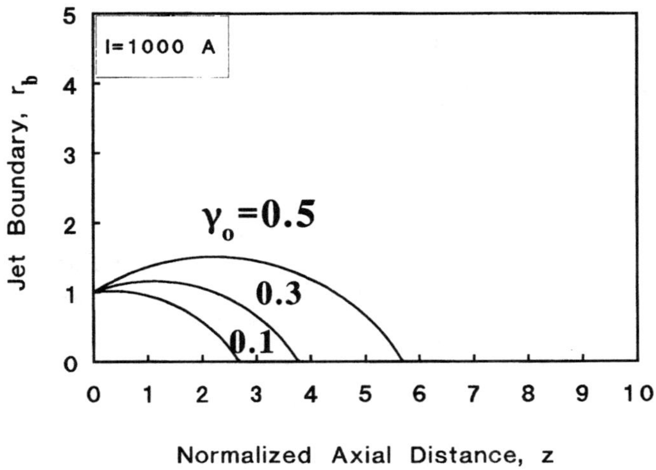

Fig. 4 Analytical one-dimensional calculation of the free plasma-jet boundary in the *r-z* plane with the radial to axial velocity ratio at the starting plane γ_0 as a parameter for I = 1000 A.

ally the whole procedure consists of finding $\varphi^{(n)}(r,z)$ from $\varphi^{(n-1)}(r,z)$. After a sufficient number of the iterations, we obtain a potential distribution coinciding with the self-consistent plasma parameter distribution with sufficient accuracy. In our case the solution converges within about 10 iterations.[27]

Results

The free plasma-jet expansion in the magnetic field was studied using Eqs. (5–13) with the preceding boundary conditions (see Section III). The model described here is related to the single- or multicathode spot vacuum-arc plasma-jet expansion. The problem was analyzed using dimensionless variables where all velocities are normalized by C_s, the density by its value at the starting boundary N_0, all coordinates by R_0, the electrical current density by eN_0C_s, and the electrical potential by the electron temperature (kT_e/e). The results presented in this work correspond to the *Ti* plasma jet.

The calculated plasma-jet boundary in the *r-z* plane is shown in Fig. 5 for different magnetic fields for the following conditions: initial Mach number $M = 5$, electron temperature $T_e = 2$ eV, $T_i/T_e = 0.5$, $Z_i = 1$, ion current fraction $f_i = 0.1$, and plasma density at the starting plane $N_0 = 10^{20}$ m^{-3}, which corresponds to the conditions in the external part of the cathode jet.[9] The radial to the axial velocity ratio at the starting plane will be introduced as parameter γ. We can see in Fig. 5 that the jet has an approximately conical shape with an angle that changes in the axial direction and depends on the magnetic field. The main jet shape change occurs over axial distances of $z \sim (3–4)R_0$. The influence of the self-magnetic field may be seen in Fig. 5 for the

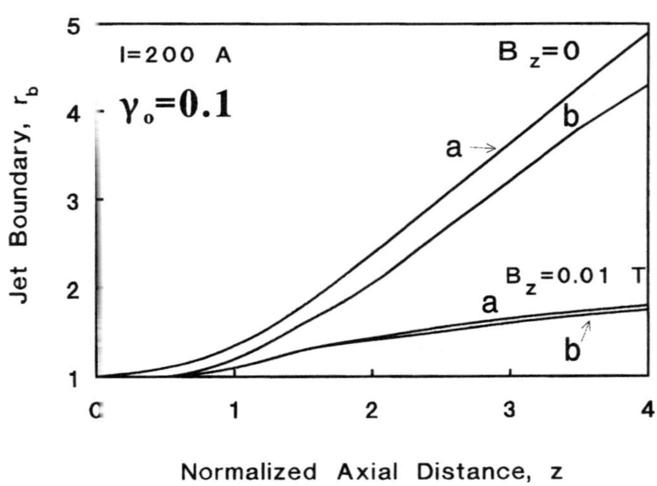

Fig. 5 The free plasma-jet boundary in the *r-z* plane with an external magnetic field as a parameter $\gamma_0 = 0.1$, $I = 200$, a) without considering a self-magnetic field and b) with a self-magnetic field.

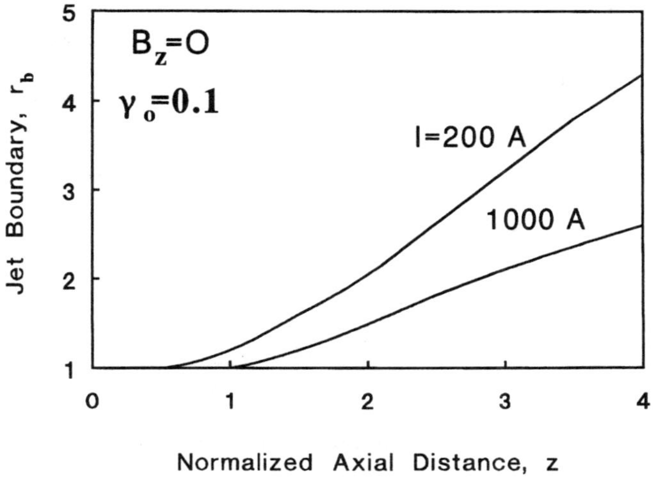

Fig. 6 The free plasma-jet boundary in the *r*-*z* plane with the current as a parameter $\gamma_0 = 0.1, B_z = 0$.

$\gamma_0 = 0.1, I = 200$ A case. The self-magnetic field decreases the jet radius by about 12% when no external field is applied and 6% when a 10 mT axial magnetic field is applied. The 200-A and 1000-A jet expansion are compared in Fig. 6, and we may see that the radial expansion of the 1-kA jet is half that of the 200-A jet. The difference is attributed to the increased importance of the self-magnetic field at higher arc currents. This calculation also shows the large role of the radial plasma density gradient on the jet expansion; in that expansion is predicted, in comparison with the results presented in Fig. 4, where the neglect of the radial gradients yielded the jet contraction. The radial plasma density distribution is plotted in Fig. 7 for 0.01 T and zero magnetic fields and shows that with a 0.01 T field the peak density is twice as high while the width is halved in comparison with the $B = 0$ case. The plasma density distribution is narrower when the self-magnetic field (curve b in Fig. 7) is taken into account in comparison to calculations in which the self-magnetic field was neglected (curve a in Fig. 7).

The axial velocity during expansion slightly increases because of the axial pressure gradient. The radial velocity increases approximately linearly with the radius starting from zero on the axis and reaching significant values at the plasma boundary, as a result of the expansion into the vacuum. The presence of a magnetic field, however, restricts the axial growth of the radial velocity, as can be seen in Fig. 8 where the radial to the axial velocity ratio at the plasma boundary is shown.

V. Discussion and Conclusions

In the present work an expansion of the plasma jet into a vacuum was analyzed in the low-density part using a MHD approach. The conditions at

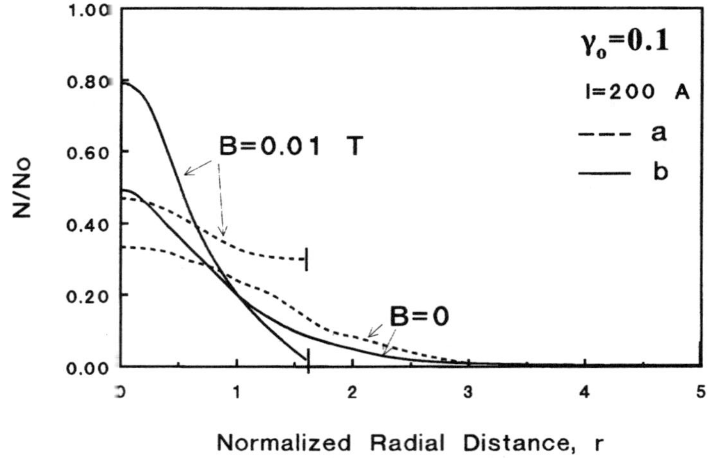

Fig. 7 Normalized radial density distribution for a different magnetic field at z = 4 with the magnetic field as a parameter I = 200; a), γ_0 = 0.1; a) without considering a self-magnetic field, and b) with a self-magnetic field. I is the plasma boundary position.

the free-jet boundary were formulated. The plasma density, velocity distribution, and jet shape were obtained by solving a two-dimensional expansion problem.

The expansion of the low-density part of a vacuum-arc plasma jet strongly depends on the boundary conditions in the plasma generation and flow

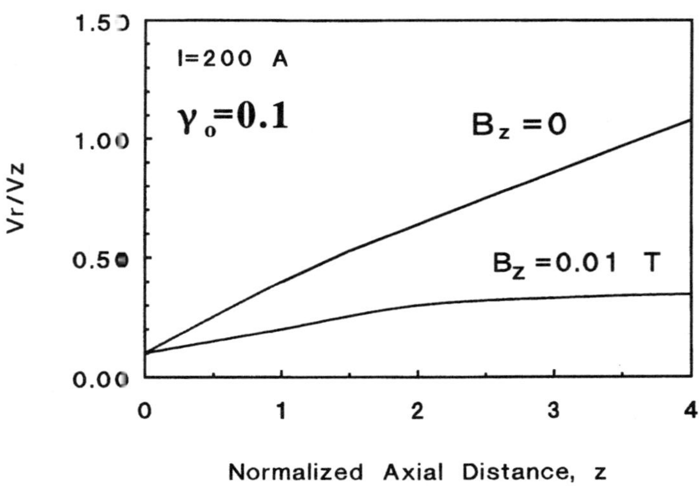

Fig. 8 A plot of the radial to the axial velocity ratio with the magnetic field as a parameter γ_0 = 0.1, I = 200 A.

formation region near the cathode, where the plasma is so dense that the self-magnetic field may be neglected and the plasma expands nearly with a conical shape. The low-density flow must be analyzed using a two-dimensional analysis that takes into account radial gradiants, in as much as one-dimensional analytic solutions erroneously yield plasma contraction, rather than expansion.

In the low-density portion of the jet, the discovery was made that the self-magnetic field generated by the current flow within the plasma restricts the radial expansion in the flow, especially for large currents (e.g., ~ kA). This influence is mainly near the boundary of the jet and is most significant when no external magnetic field is imposed.

An external magnetic field may be used to control the plasma-jet expansion. In the case of an axial magnetic field larger than 0.01 T, the influence of the self-magnetic field on the expansion is small.

Acknowledgments

The authors gratefully acknowledge the financial support from Friends of Tel Aviv University in France and the grant from the United States–Israel Binational Science Foundation (BSF), Jerusalem, Israel, No. 94-00290.

References

[1]Boxman, R. L., and Goldsmith, S., "Principle and Applications of Vacuum Arc Coating," *IEEE Transactions on Plasma Science*, Vol. 17, 1989, pp. 705–712.

[2]Dorodnov, A. M., "Technical Applications of Plasma Accelerators," *Soviet Physics—Technical Physics*, Vol. 23, No. 9, 1978, pp. 1058–1964.

[3]Plyutto, A. A., Ryzhkov, V. M., and Karpin, A. T., "High Speed Plasma Streams in Vacuum Arc," *Soviet Physics—JETR*, Vol. 20, 1965, pp. 328–337.

[4]Khoroshikh, V. M., Aksenov, I. I., and Konovalov, I. I., "Structure of Plasma Jets Generated by the Cathode Spot of a Vacuum Arc," *Soviet Physics—Technical Physics*, Vol. 33, 1988, pp. 723–724.

[5]Kimblin, C. V., "Vacuum Arc Ion Currents and Electrode Phenomena," *Proceedings of the IEEE*, Vol. 59, 1971, pp. 546–555.

[6]Kutzner, J., and Miller, H. C., "Integrated Ion Flux Emitted from the Cathode Spot Region of a Diffuse Vacuum Arc," *Journal of Physics D: Applied Physics*, Vol. 25, 1992, pp. 686–693.

[7]Cohen, Y. A., Boxman, R. L., and Goldsmith, S., "Angular Distribution of Ion Current Emerging from an Aperture Anode in a Vacuum Arc," *IEEE Transactions on Plasma Sciences*, Vol. 17, 1989, pp. 713–716.

[8]Beilis, I. I., Zektser, M. P., Lyubimov, G. A., "Analysis of the Formulation and Solution of the Problem of the Cathode Jet of a Vacuum Arc," *Soviet Physics—Technical Physics*, Vol. 33, 1988, pp. 1132–1137.

[9]Beilis, I. I., and Zektser, M. P., "Calculation of the Cathode Jet Parameters of the Arc Discharge," *High Temperature*, Vol. 29, 1991, pp. 501–504.

[10]Hantzsche, E., "Theory of the Expanding Plasma of Vacuum Arcs," *Journal of Physics D: Applied Physics*, Vol. 24, 1991, pp. 1339–1353.

[11]Morozov, A. I., "Focusing of Cold Quasineutral Beams in Electromagnetic Fields," *Soviet Physics—Doklady*, Vol. 10, 1966, pp. 775–777.

[12]Boxman, R. L., "Magnetic Constriction Effects in High-Current Vacuum Arcs Prior to the Release of Anode Vapor," *Journal of Applied Physics*, Vol. 48, No. 6, 1977, pp. 2338–2345.

[13]Izraeli, I., Boxman, R. L., and Goldsmith S., "The Current Distribution and the Magnetic Pressure Profile in a Vacuum Arc Subject to an Axial Magnetic Field," PS-15, 1987, pp. 502–505.

[14]Shohet, J. L., Pedrow, P. D., and Burrage, L. M., "Numerical Simulation of a Planar Vacuum Arc," *IEEE Transactions on Plasma Sciences*, PS-9, 1981, pp. 254–258.

[15]Gonzales, J. J. Gleizes, A., Proulx, P., and Boulos, M., "Mathematical Modeling of a Free-Burning Arc in the Presence of Metal Vapor," *Journal of Applied Physics*, Vol. 74, No. 5, 1993, pp. 3065–3069.

[16]Chen, D. M., and Pfender, E., "Modeling of the Anode Contraction Region of High Intensity Arcs," *IEEE Transactions on Plasma Sciences*, PS-8 1980, pp. 252–259.

[17]Huysmans, G. T. A., Goedbloed, J. P., and Kerner, W., "Free Boundary Resistive Modes," *Physics of Fluids B,* Vol. 5, No. 5, 1993, pp. 1545–1558.

[18]Kress, M. E., and Riedel, K. S., "Semi-Implicit Reduced Magnetohydrodynamics," *Journal of Computational Physics,* Vol. 83, 1989, pp. 237–239.

[19]Braginsky, S. I., *Reviews of the Plasma Physics*, edited by M. A. Leontovich, Vol. 1, Consultants Bureau, New York, 1965.

[20]Crank, J., *Free and Moving Boundary Problem*, Clarendon Press, Oxford, England, UK, 1984.

[21]Mitchell, A. R., and Griffiths, D. F., *The Finite Difference Method in Partial Differential Equations*, Wiley, New York, 1980.

[22]Kiedar, M., Beilis, I., Boxman, R. L., and Goldsmith, S., "Two-dimensional Expansion of the Low-Density Interelectrode Vacuum Arc Plasma Jet in an Axial Magnetic Field," *Journal of Physics D: Applied Physics* (to be published).

[23]Paskonov, V. M., "A Standard Program for the Solution of Boundary-Layer Problems," *Numerical Methods in Gas Dynamics,* edited by G. S. Roslyakov and L. A. Chudov, Israel Program for Scientific Translation, Jerusalem, Israel, 1966.

[24]Hockney, R. W., "The Potential Calculation and Some Application," *Methods in Computational Physics*, Academic, New York, 1970.

[25]Pedit, H., and Kuhn, S., "Kinetic Modeling of a Two-Dimensional, Collisionless Bounded Plasma," *Physics of Plasma*, Vol. 1, No. 1, 1994, pp. 13–31.

PROGRESS IN ASTRONAUTICS AND AERONAUTICS SERIES VOLUMES

***1. Solid Propellant Rocket Research (1960)**
Martin Summerfield
Princeton University

***2. Liquid Rockets and Propellants (1960)**
Loren E. Bollinger
Ohio State University
Martin Goldsmith
The Rand Corp.
Alexis W. Lemmon Jr.
Battelle Memorial Institute

***3. Energy Conversion for Space Power (1961)**
Nathan W. Snyder
Institute for Defense Analyses

***4. Space Power Systems (1961)**
Nathan W. Snyder
Institute for Defense Analyses

***5. Electrostatic Propulsion (1961)**
David B. Langmuir
Space Technology Laboratories, Inc.
Ernst Stuhlinger
NASA George C. Marshall Space Flight Center
J. M. Sellen Jr.
Space Technology Laboratories, Inc.

***6. Detonation and Two-Phase Flow (1962)**
S. S. Penner
California Institute of Technology
F. A. Williams
Harvard University

***7. Hypersonic Flow Research (1962)**
Frederick R. Riddell
AVCO Corp.

***8. Guidance and Control (1962)**
Robert E. Roberson
Consultant
James S. Farrior
Lockheed Missiles and Space Co.

***9. Electric Propulsion Development (1963)**
Ernst Stuhlinger
NASA George C. Marshall Space Flight Center

***10. Technology of Lunar Exploration (1963)**
Clifford I. Cumming
Harold R. Lawrence
Jet Propulsion Laboratory

***11. Power Systems for Space Flight (1963)**
Morris A. Zipkin
Russell N. Edwards
General Electric Co.

***12. Ionization in High-Temperature Gases (1963)**
Kurt E. Shuler, Editor
National Bureau of Standards
John B. Fenn,
Associate Editor
Princeton University

***13. Guidance and Control–II (1964)**
Robert C. Langford
General Precision Inc.
Charles J. Mundo
Institute of Naval Studies

***14. Celestial Mechanics and Astrodynamics (1964)**
Victor G. Szebehely
Yale University Observatory

***15. Heterogeneous Combustion (1964)**
Hans G. Wolfhard
Institute for Defense Analyses
Irvin Glassman
Princeton University
Leon Green Jr.
Air Force Systems Command

***16. Space Power Systems Engineering (1966)**
George C. Szego
Institute for Defense Analyses
J. Edward Taylor
TRW Inc.

***17. Methods in Astrodynamics and Celestial Mechanics (1966)**
Raynor L. Duncombe
U.S. Naval Observatory
Victor G. Szebehely
Yale University Observatory

***18. Thermophysics and Temperature Control of Spacecraft and Entry Vehicles (1966)**
Gerhard B. Heller
NASA George C. Marshall Space Flight Center

***19. Communication Satellite Systems Technology (1966)**
Richard B. Marsten
Radio Corporation of America

*Out of print.

***20. Thermophysics of Spacecraft and Planetary Bodies: Radiation Properties of Solids and the Electromagnetic Radiation Environment in Space (1967)**
Gerhard B. Heller
NASA George C. Marshall Space Flight Center

***21. Thermal Design Principles of Spacecraft and Entry Bodies (1969)**
Jerry T. Bevans
TRW Systems

***22. Stratospheric Circulation (1969)**
Willis L. Webb
Atmospheric Sciences Laboratory, White Sands, and University of Texas at El Paso

***23. Thermophysics: Applications to Thermal Design of Spacecraft (1970)**
Jerry T. Bevans
TRW Systems

***24. Heat Transfer and Spacecraft Thermal Control (1971)**
John W. Lucas
Jet Propulsion Laboratory

25. Communication Satellites for the 70's: Technology (1971)
Nathaniel E. Feldman
The Rand Corp.
Charles M. Kelly
The Aerospace Corp.

26. Communication Satellites for the 70's: Systems (1971)
Nathaniel E. Feldman
The Rand Corp.
Charles M. Kelly
The Aerospace Corp.

27. Thermospheric Circulation (1972)
Willis L. Webb
Atmospheric Sciences Laboratory, White Sands, and University of Texas at El Paso

28. Thermal Characteristics of the Moon (1972)
John W. Lucas
Jet Propulsion Laboratory

***29. Fundamentals of Spacecraft Thermal Design (1972)**
John W. Lucas
Jet Propulsion Laboratory

***30. Solar Activity Observations and Predictions (1972)**
Patrick S. McIntosh
Murray Dryer
Environmental Research Laboratories, National Oceanic and Atmospheric Administration

31. Thermal Control and Radiation (1973)
Chang-Lin Tien
University of California at Berkeley

***32. Communications Satellite Systems (1974)**
P. L. Bargellini
COMSAT Laboratories

33. Communications Satellite Technology (1974)
P. L. Bargellini
COMSAT Laboratories

***34. Instrumentation for Airbreathing Propulsion (1974)**
Allen E. Fuhs
Naval Postgraduate School
Marshall Kingery
Arnold Engineering Development Center

35. Thermophysics and Spacecraft Thermal Control (1974)
Robert G. Hering
University of Iowa

36. Thermal Pollution Analysis (1975)
Joseph A. Schetz
Virginia Polytechnic Institute
ISBN 0-915928-00-0

***37. Aeroacoustics: Jet and Combustion Noise; Duct Acoustics (1975)**
Henry T. Nagamatsu, Editor
General Electric Research and Development Center
Jack V. O'Keefe, Associate Editor
The Boeing Co.
Ira R. Schwartz, Associate Editor
NASA Ames Research Center
ISBN 0-915928-01-9

***38. Aeroacoustics: Fan, STOL, and Boundary Layer Noise; Sonic Boom; Aeroacoustics Instrumentation (1975)**
Henry T. Nagamatsu, Editor
General Electric Research and Development Center
Jack V. O'Keefe, Associate Editor
The Boeing Co.
Ira R. Schwartz, Associate Editor
NASA Ames Research Center
ISBN 0-915928-02-7

39. Heat Transfer with Thermal Control Applications (1975)
M. Michael Yovanovich
University of Waterloo
ISBN 0-915928-03-5

---
*Out of print.

***40. Aerodynamics of Base Combustion (1976)**
S. N. B. Murthy, Editor
J. R. Osborn, Associate Editor
Purdue University
A. W. Barrows
J. R. Ward, Associate Editors
Ballistics Research Laboratories
ISBN 0-915928-04-3

41. Communications Satellite Developments: Systems (1976)
Gilbert E. LaVean
Defense Communications Agency
William G. Schmidt
CML Satellite Corp.
ISBN 0-915928-05-1

***42. Communications Satellite Developments: Technology (1976)**
William G. Schmidt
CML Satellite Corp.
Gilbert E. LaVean
Defense Communications Agency
ISBN 0-915928-06-X

***43. Aeroacoustics: Jet Noise, Combustion and Core Engine Noise (1976)**
Ira R. Schwartz, Editor
NASA Ames Research Center
Henry T. Nagamatsu, Associate Editor
General Electric Research and Development Center
Warren C. Strahle, Associate Editor
Georgia Institute of Technology
ISBN 0-915928-07-8

***44. Aeroacoustics: Fan Noise and Control; Duct Acoustics; Rotor Noise (1976)**
Ira R. Schwartz, Editor
NASA Ames Research Center
Henry T. Nagamatsu, Associate Editor
General Electric Research and Development Center
Warren C. Strahle, Associate Editor
Georgia Institute of Technology
ISBN 0-915928-08-6

***45. Aeroacoustics: STOL Noise; Airframe and Airfoil Noise (1976)**
Ira R. Schwartz, Editor
NASA Ames Research Center
Henry T. Nagamatsu, Associate Editor
General Electric Research and Development Center
Warren C. Strahle, Associate Editor
Georgia Institute of Technology
ISBN 0-915928-09-4

***46. Aeroacoustics: Acoustic Wave Propagation; Aircraft Noise Prediction; Aeroacoustic Instrumentation (1976)**
Ira R. Schwartz, Editor
NASA Ames Research Center
Henry T. Nagamatsu, Associate Editor
General Electric Research and Development Center
Warren C. Strahle, Associate Editor
Georgia Institute of Technology
ISBN 0-915928-10-8

***47. Spacecraft Charging by Magnetospheric Plasmas (1976)**
Alan Rosen
TRW Inc.
ISBN 0-915928-11-6

48. Scientific Investigations on the Skylab Satellite (1976)
Marion I. Kent
Ernst Stuhlinger
NASA George C. Marshall Space Flight Center
Shi-Tsan Wu
University of Alabama
ISBN 0-915928-12-4

49. Radiative Transfer and Thermal Control (1976)
Allie M. Smith
ARO Inc.
ISBN 0-915928-13-2

***50. Exploration of the Outer Solar System (1976)**
Eugene W. Greenstadt
TRW Inc.
Murray Dryer
National Oceanic and Atmospheric Administration
Devrie S. Intriligator
University of Southern California
ISBN 0-915928-14-0

***51. Rarefied Gas Dynamics, Parts I and II (two volumes) (1977)**
J. Leith Potter
ARO Inc.
ISBN 0-915928-15-9

***52. Materials Sciences in Space with Application to Space Processing (1977)**
Leo Steg
General Electric Co.
ISBN 0-915928-16-7

*Out of print.

***53. Experimental Diagnostics in Gas Phase Combustion Systems (1977)**
Ben T. Zinn, Editor
Georgia Institute of Technology
Craig T. Bowman, Associate Editor
Stanford University
Daniel L. Hartley, Associate Editor
Sandia Laboratories
Edward W. Price, Associate Editor
Georgia Institute of Technology
James G. Skifstad, Associate Editor
Purdue University
ISBN 0-915928-18-3

54. Satellite Communication: Future Systems (1977)
David Jarett
TRW Inc.
ISBN 0-915928-18-3

***55. Satellite Communications: Advanced Technologies (1977)**
David Jarett
TRW Inc.
ISBN 0-915928-19-1

***56. Thermophysics of Spacecraft and Outer Planet Entry Probes (1977)**
Allie M. Smith
ARO Inc.
ISBN 0-915928-20-5

***57. Space-Based Manufacturing from Nonterrestrial Materials (1977)**
Gerald K. O'Neill, Editor
Brian O'Leary, Assistant Editor
Princeton University
ISBN 0-915928-21-3

***58. Turbulent Combustion (1978)**
Lawrence A. Kennedy
State University of New York at Buffalo
ISBN 0-915928-22-1

***59. Aerodynamic Heating and Thermal Protection Systems (1978)**
Leroy S. Fletcher
University of Virginia
ISBN 0-915928-23-X

***60. Heat Transfer and Thermal Control Systems (1978)**
Leroy S. Fletcher
University of Virginia
ISBN 0-915928-24-8

***61. Radiation Energy Conversion in Space (1978)**
Kenneth W. Billman
NASA Ames Research Center
ISBN 0-915928-26-4

***62. Alternative Hydrocarbon Fuels: Combustion and Chemical Kinetics (1978)**
Craig T. Bowman
Stanford University
Jorgen Birkeland
Department of Energy
ISBN 0-915928-25-6

***63. Experimental Diagnostics in Combustion of Solids (1978)**
Thomas L. Boggs
Naval Weapons Center
Ben T. Zinn
Georgia Institute of Technology
ISBN 0-915928-28-0

***64. Outer Planet Entry Heating and Thermal Protection (1979)**
Raymond Viskanta
Purdue University
ISBN 0-915928-29-9

***65. Thermophysics and Thermal Control (1979)**
Raymond Viskanta
Purdue University
ISBN 0-915928-30-2

***66. Interior Ballistics of Guns (1979)**
Herman Krier
University of Illinois at Urbana–Champaign
Martin Summerfield
New York University
ISBN 0-915928-32-9

***67. Remote Sensing of Earth from Space: Role of "Smart Sensors" (1979)**
Roger A. Breckenridge
NASA Langley Research Center
ISBN 0-915928-33-7

68. Injection and Mixing in Turbulent Flow (1980)
Joseph A. Schetz
Virginia Polytechnic Institute and State University
ISBN 0-915928-35-3

---

*Out of print.

***69. Entry Heating and Thermal Protection (1980)**
Walter B. Olstad
NASA Headquarters
ISBN 0-915928-38-8

***70. Heat Transfer, Thermal Control, and Heat Pipes (1980)**
Walter B. Olstad
NASA Headquarters
ISBN 0-915928-39-6

***71. Space Systems and Their Interactions with Earth's Space Environment (1980)**
Henry B. Garrett
Charles P. Pike
Hanscom Air Force Base
ISBN 0-915928-41-8

***72. Viscous Flow Drag Reduction (1980)**
Gary R. Hough
Vought Advanced Technology Center
ISBN 0-915928-44-2

***73. Combustion Experiments in a Zero-Gravity Laboratory (1981)**
Thomas H. Cochran
NASA Lewis Research Center
ISBN 0-915928-48-5

***74. Rarefied Gas Dynamics, Parts I and II (two volumes) (1981)**
Sam S. Fisher
University of Virginia
ISBN 0-915928-51-5

***75. Gasdynamics of Detonations and Explosions (1981)**
J. R. Bowen
University of Wisconsin at Madison
N. Manson
Universite de Poitiers
A. K. Oppenheim
University of California at Berkeley
R. I. Soloukhin
Institute of Heat and Mass Transfer, BSSR Academy of Sciences
ISBN 0-915928-46-9

76. Combustion in Reactive Systems (1981)
J. R. Bowen
University of Wisconsin at Madison
N. Manson
Universite de Poitiers
A. K. Oppenheim
University of California at Berkeley
R. I. Soloukhin
Institute of Heat and Mass Transfer, BSSR Academy of Sciences
ISBN 0-915928-47-7

***77. Aerothermodynamics and Planetary Entry (1981)**
A. L. Crosbie
University of Missouri-Rolla
ISBN 0-915928-52-3

***78. Heat Transfer and Thermal Control (1981)**
A. L. Crosbie
University of Missouri-Rolla
ISBN 0-915928-53-1

***79. Electric Propulsion and Its Applications to Space Missions (1981)**
Robert C. Finke
NASA Lewis Research Center
ISBN 0-915928-55-8

***80. Aero-Optical Phenomena (1982)**
Keith G. Gilbert
Leonard J. Otten
Air Force Weapons Laboratory
ISBN 0-915928-60-4

***81. Transonic Aerodynamics (1982)**
David Nixon
Nielsen Engineering & Research, Inc.
ISBN 0-915928-65-5

***82. Thermophysics of Atmospheric Entry (1982)**
T. E. Horton
University of Mississippi
ISBN 0-915928-66-3

***83. Spacecraft Radiative Transfer and Temperature Control (1982)**
T. E. Horton
University of Mississippi
ISBN 0-915928-67-1

84. Liquid-Metal Flows and Magneto-hydrodynamics (1983)
H. Branover
Ben-Gurion University of the Negev
P. S. Lykoudis
Purdue University
A. Yakhot
Ben-Gurion University of the Negev
ISBN 0-915928-70-1

*Out of print.

***85. Entry Vehicle Heating and Thermal Protection Systems: Space Shuttle, Solar Starprobe, Jupiter Galileo Probe (1983)**
Paul E. Bauer
McDonnell Douglas Astronautics Co.
Howard E. Collicott
The Boeing Co.
ISBN 0-915928-74-4

***86. Spacecraft Thermal Control, Design, and Operation (1983)**
Howard E. Collicott
The Boeing Co.
Paul E. Bauer
McDonnell Douglas Astronautics Co.
ISBN 0-915928-75-2

***87. Shock Waves, Explosions, and Detonations (1983)**
J. R. Bowen
University of Washington
N. Manson
Universite de Poitiers
A. K. Oppenheim
University of California at Berkeley
R. I. Soloukhin
Institute of Heat and Mass Transfer, BSSR Academy of Sciences
ISBN 0-915928-76-0

88. Flames, Lasers, and Reactive Systems (1983)
J. R. Bowen
University of Washington
N. Manson
Universite de Poitiers
A. K. Oppenheim
University of California at Berkeley
R. I. Soloukhin
Institute of Heat and Mass Transfer, BSSR Academy of Sciences
ISBN 0-915928-77-9

***89. Orbit-Raising and Maneuvering Propulsion: Research Status and Needs (1984)**
Leonard H. Caveny
Air Force Office of Scientific Research
ISBN 0-915928-82-5

90. Fundamentals of Solid-Propellant Combustion (1984)
Kenneth K. Kuo
Pennsylvania State University
Martin Summerfield
Princeton Combustion Research Laboratories, Inc.
ISBN 0-915928-84-1

91. Spacecraft Contamination: Sources and Prevention (1984)
J. A. Roux
University of Mississippi
T. D. McCay
NASA Marshall Space Flight Center
ISBN 0-915928-85-X

92. Combustion Diagnostics by Nonintrusive Methods (1984)
T. D. McCay
NASA Marshall Space Flight Center
J. A. Roux
University of Mississippi
ISBN 0-915928-86-8

93. The INTELSAT Global Satellite System (1984)
Joel Alper
COMSAT Corp.
Joseph Pelton
INTELSAT
ISBN 0-915928-90-6

94. Dynamics of Shock Waves, Explosions, and Detonations (1984)
J. R. Bowen
University of Washington
N. Manson
Universite de Poitiers
A. K. Oppenheim
University of California at Berkeley
R. I. Soloukhin
Institute of Heat and Mass Transfer, BSSR Academy of Sciences
ISBN 0-915928-91-4

95. Dynamics of Flames and Reactive Systems (1984)
J. R. Bowen
University of Washington
N. Manson
Universite de Poitiers
A. K. Oppenheim
University of California at Berkeley
R. I. Soloukhin
Institute of Heat and Mass Transfer, BSSR Academy of Sciences
ISBN 0-915928-92-2

96. Thermal Design of Aeroassisted Orbital Transfer Vehicles (1985)
H. F. Nelson
University of Missouri-Rolla
ISBN 0-915928-94-9

97. Monitoring Earth's Ocean, Land, and Atmosphere from Space—Sensors, Systems, and Applications (1985)
Abraham Schnapf
Aerospace Systems Engineering
ISBN 0-915928-98-1

*Out of print.

98. Thrust and Drag: Its Prediction and Verification (1985)
Eugene E. Covert
Massachusetts Institute of Technology
C. R. James
Vought Corp.
William F. Kimzey
Sverdrup Technology AEDC Group
George K. Richey
U.S. Air Force
Eugene C. Rooney
U.S. Navy Department of Defense
ISBN 0-930403-00-2

99. Space Stations and Space Platforms—Concepts, Design, Infrastructure, and Uses (1985)
Ivan Bekey
Daniel Herman
NASA Headquarters
ISBN 0-930403-01-0

100. Single- and Multi-Phase Flows in an Electromagnetic Field: Energy, Metallurgical, and Solar Applications (1985)
Herman Branover
Ben-Gurion University of the Negev
Paul S. Lykoudis
Purdue University
Michael Mond
Ben-Gurion University of the Negev
ISBN 0-930403-04-5

101. MHD Energy Conversion: Physiotechnical Problems (1986)
V. A. Kirillin
A. E. Sheyndlin
Soviet Academy of Sciences
ISBN 0-930403-05-3

102. Numerical Methods for Engine-Airframe Integration (1986)
S. N. B. Murthy
Purdue University
Gerald C. Paynter
Boeing Airplane Co.
ISBN 0-930403-09-6

103. Thermophysical Aspects of Re-Entry Flows (1986)
James N. Moss
NASA Langley Research Center
Carl D. Scott
NASA Johnson Space Center
ISBN 0-930430-10-X

***104. Tactical Missile Aerodynamics (1986)**
M. J. Hemsch
PRC Kentron, Inc.
J. N. Nielson
NASA Ames Research Center
ISBN 0-930403-13-4

105. Dynamics of Reactive Systems Part I: Flames and Configurations; Part II: Modeling and Heterogeneous Combustion (1986)
J. R. Bowen
University of Washington
J.-C. Leyer
Universite de Poitiers
R. I. Soloukhin
Institute of Heat and Mass Transfer, BSSR Academy of Sciences
ISBN 0-930403-14-2

106. Dynamics of Explosions (1986)
J. R. Bowen
University of Washington
J.-C. Leyer
Universite de Poitiers
R. I. Soloukhin
Institute of Heat and Mass Transfer, BSSR Academy of Sciences
ISBN 0-930403-15-0

***107. Spacecraft Dielectric Material Properties and Spacecraft Charging (1986)**
A. R. Frederickson
U.S. Air Force Rome Air Development Center
D. B. Cotts
SRI International
J. A. Wall
U.S. Air Force Rome Air Development Center
F. L. Bouquet
Jet Propulsion Laboratory, California Institute of Technology
ISBN 0-930403-17-7

***108. Opportunities for Academic Research in a Low-Gravity Environment (1986)**
George A. Hazelrigg
National Science Foundation
Joseph M. Reynolds
Louisiana State University
ISBN 0-930403-18-5

109. Gun Propulsion Technology (1988)
Ludwig Stiefel
U.S. Army Armament Research, Development and Engineering Center
ISBN 0-930403-20-7

*Out of print.

110. Commercial Opportunities in Space (1988)
F. Shahrokhi
K. E. Harwell
University of Tennessee Space Institute
C. C. Chao
National Cheng Kung University
ISBN 0-930403-39-8

111. Liquid-Metal Flows: Magnetohydrodynamics and Application (1988)
Herman Branover
Michael Mond
Yeshajahu Unger
Ben-Gurion University of the Negev
ISBN 0-930403-43-6

112. Current Trends in Turbulence Research (1988)
Herman Branover
Micheal Mond
Yeshajahu Unger
Ben-Gurion University of the Negev
ISBN 0-930403-44-4

113. Dynamics of Reactive Systems Part I: Flames; Part II: Heterogeneous Combustion and Applications (1988)
A. L. Kuhl
R&D Associates
J. R. Bowen
University of Washington
J.-C. Leyer
Universite de Poitiers
A. Borisov
USSR Academy of Sciences
ISBN 0-930403-46-0

114. Dynamics of Explosions (1988)
A. L. Kuhl
R & D Associates
J. R. Bowen
University of Washington
J.-C. Leyer
Universite de Poitiers
A. Borisov
USSR Academy of Sciences
ISBN 0-930403-47-9

115. Machine Intelligence and Autonomy for Aerospace (1988)
E. Heer
Heer Associates, Inc.
H. Lum
NASA Ames Research Center
ISBN 0-930403-48-7

116. Rarefied Gas Dynamics: Space Related Studies (1989)
E. P. Muntz
University of Southern California
D. P. Weaver
U.S. Air Force Astronautics Laboratory (AFSC)
D. H. Campbell
University of Dayton Research Institute
ISBN 0-930403-53-3

117. Rarefied Gas Dynamics: Physical Phenomena (1989)
E. P. Muntz
University of Southern California
D. P. Weaver
U.S. Air Force Astronautics Laboratory (AFSC)
D. H. Campbell
University of Dayton Research Institute
ISBN 0-930403-54-1

118. Rarefied Gas Dynamics: Theoretical and Computational Techniques (1989)
E. P. Muntz
University of Southern California
D. P. Weaver
U.S. Air Force Astronautics Laboratory (AFSC)
D. H. Campbell
University of Dayton Research Institute
ISBN 0-930403-55-X

119. Test and Evaluation of the Tactical Missile (1989)
Emil J. Eichblatt Jr.
Pacific Missile Test Center
ISBN 0-930403-56-8

120. Unsteady Transonic Aerodynamics (1989)
David Nixon
Nielsen Engineering & Research, Inc.
ISBN 0-930403-52-5

121. Orbital Debris from Upper-Stage Breakup (1989)
Joseph P. Loftus Jr.
NASA Johnson Space Center
ISBN 0-930403-58-4

122. Thermal-Hydraulics for Space Power, Propulsion and Thermal Management System Design (1990)
William J. Krotiuk
General Electric Co.
ISBN 0-930403-64-9

*Out of print.

123. Viscous Drag Reduction in Boundary Layers (1990)
Dennis M. Bushnell
Jerry N. Hefner
NASA Langley Research Center
ISBN 0-930403-66-5

***124. Tactical and Strategic Missile Guidance (1990)**
Paul Zarchan
Charles Stark Draper Laboratory, Inc.
ISBN 0-930403-68-1

125. Applied Computational Aerodynamics (1990)
P. A. Henne
Douglas Aircraft Company
ISBN 0-930403-69-X

126. Space Commercialization: Launch Vehicles and Programs (1990)
F. Shahrokhi
University of Tennessee Space Institute
J. S. Greenberg
Princeton Synergetics Inc.
T. Al-Saud
Ministry of Defense and Aviation Kingdom of Saudi Arabia
ISBN 0-930403-75-4

127. Space Commercialization: Platforms and Processing (1990)
F. Shahrokhi
University of Tennessee Space Institute
G. Hazelrigg
National Science Foundation
R. Bayuzick
Vanderbilt University
ISBN 0-930403-76-2

128. Space Commercialization: Satellite Technology (1990)
F. Shahrokhi
University of Tennessee Space Institute
N. Jasentuliyana
United Nations
N. Tarabzouni
King Abulaziz City for Science and Technology
ISBN 0-930403-77-0

129. Mechanics and Control of Large Flexible Structures (1990)
John L. Junkins
Texas A&M University
ISBN 0-930403-73-8

130. Low-Gravity Fluid Dynamics and Transport Phenomena (1990)
Jean N. Koster
Robert L. Sani
University of Colorado at Boulder
ISBN 0-930403-74-6

131. Dynamics of Deflagrations and Reactive Systems: Flames (1991)
A. L. Kuhl
Lawrence Livermore National Laboratory
J.-C. Leyer
Universite de Poitiers
A. A. Borisov
USSR Academy of Sciences
W. A. Sirignano
University of California
ISBN 0-930403-95-9

132. Dynamics of Deflagrations and Reactive Systems: Heterogeneous Combustion (1991)
A. L. Kuhl
Lawrence Livermore National Laboratory
J.-C. Leyer
Universite de Poitiers
A. A. Borisov
USSR Academy of Sciences
W. A. Sirignano
University of California
ISBN 0-930403-96-7

133. Dynamics of Detonations and Explosions: Detonations (1991)
A. L. Kuhl
Lawrence Livermore National Laboratory
J.-C. Leyer
Universite de Poitiers
A. A. Borisov
USSR Academy of Sciences
W. A. Sirignano
University of California
ISBN 0-930403-97-5

134. Dynamics of Detonations and Explosions: Explosion Phenomena (1991)
A. L. Kuhl
Lawrence Livermore National Laboratory
J.-C. Leyer
Universite de Poitiers
A. A. Borisov
USSR Academy of Sciences
W. A. Sirignano
University of California
ISBN 0-930403-98-3

---
*Out of print.

135. Numerical Approaches to Combustion Modeling (1991)
Elaine S. Oran
Jay P. Boris
Naval Research Laboratory
ISBN 1-56347-004-7

136. Aerospace Software Engineering (1991)
Christine Anderson
U.S. Air Force Wright Laboratory
Merlin Dorfman
Lockheed Missiles & Space Company, Inc.
ISBN 1-56347-005-0

137. High-Speed Flight Propulsion Systems (1991)
S. N. B. Murthy
Purdue University
E. T. Curran
Wright Laboratory
ISBN 1-56347-011-X

138. Propagation of Intensive Laser Radiation in Clouds (1992)
O. A. Volkovitsky
Yu. S. Sedenov
L. P. Semenov
Institute of Experimental Meteorology
ISBN 1-56347-020-9

139. Gun Muzzle Blast and Flash (1992)
Günter Klingenberg
Fraunhofer-Institut für Kurzzeitdynamik, Ernst-Mach-Institut
Joseph M. Heimerl
U.S. Army Ballistic Research Laboratory
ISBN 1-56347-012-8

140. Thermal Structures and Materials for High-Speed Flight (1992)
Earl. A. Thornton
University of Virginia
ISBN 1-56347-017-9

141. Tactical Missile Aerodynamics: General Topics (1992)
Michael J. Hemsch
Lockheed Engineering & Sciences Company
ISBN 1-56347-015-2

142. Tactical Missile Aerodynamics: Prediction Methodology (1992)
Michael R. Mendenhall
Nielsen Engineering & Research, Inc.
ISBN 1-56347-016-0

143. Nonsteady Burning and Combustion Stability of Solid Propellants (1992)
Luigi De Luca
Politecnico di Milano
Edward W. Price
Georgia Institute of Technology
Martin Summerfield
Princeton Combustion Research Laboratories, Inc.
ISBN 1-56347-014-4

144. Space Economics (1992)
Joel S. Greenberg
Princeton Synergetics, Inc.
Henry R. Hertzfeld
HRH Associates
ISBN 1-56347-042-X

145. Mars: Past, Present, and Future (1992)
E. Brian Pritchard
NASA Langley Research Center
ISBN 1-56347-043-8

146. Computational Nonlinear Mechanics in Aerospace Engineering (1992)
Satya N. Atluri
Georgia Institute of Technology
ISBN 1-56347-044-6

147. Modern Engineering for Design of Liquid-Propellant Rocket Engines (1992)
Dieter K. Huzel
David H. Huang
Rocketdyne Division of Rockwell International
ISBN 1-56347-013-6

148. Metallurgical Technologies, Energy Conversion, and Magnetohydrodynamic Flows (1993)
Herman Branover
Yeshajahu Unger
Ben-Gurion University of the Negev
ISBN 1-56347-019-5

149. Advances in Turbulence Studies (1993)
Herman Branover
Yeshajahu Unger
Ben-Gurion University of the Negev
ISBN 1-56347-018-7

150. Structural Optimization: Status and Promise (1993)
Manohar P. Kamat
Georgia Institute of Technology
ISBN 1-56347-056-X

*Out of print.

151. Dynamics of Gaseous Combustion (1993)
A. L. Kuhl
Lawrence Livermore National Laboratory
J.-C. Leyer
Universite de Poitiers
A. A. Borisov
USSR Academy of Sciences
W. A. Sirignano
University of California
ISBN 1-56347-060-8

152. Dynamics of Heterogeneous Gaseous Combustion and Reacting Systems (1993)
A. L. Kuhl
Lawrence Livermore National Laboratory
J.-C. Leyer
Universite de Poitiers
A. A. Borisov
USSR Academy of Sciences
W. A. Sirignano
University of California
ISBN 1-56347-058-6

153. Dynamic Aspects of Detonations (1993)
A. L. Kuhl
Lawrence Livermore National Laboratory
J.-C. Leyer
Universite de Poitiers
A. A. Borisov
USSR Academy of Sciences
W. A. Sirignano
University of California
ISBN 1-56347-057-8

154. Dynamic Aspects of Explosion Phenomena (1993)
A. L. Kuhl
Lawrence Livermore National Laboratory
J.-C. Leyer
Universite de Poitiers
A. A. Borisov
USSR Academy of Sciences
W. A. Sirignano
University of California
ISBN 1-56347-059-4

155. Tactical Missile Warheads (1993)
Joseph Carleone
Aerojet General Corporation
ISBN 1-56347-067-5

156. Toward a Science of Command, Control, and Communications (1993)
Carl R. Jones
Naval Postgraduate School
ISBN 1-56347-068-3

157. Tactical and Strategic Missile Guidance Second Edition (1994)
Paul Zarchan
Charles Stark Draper Laboratory, Inc.
ISBN 1-56347-077-2

158. Rarefied Gas Dynamics: Experimental Techniques and Physical Systems (1994)
Bernie D. Shizgal
University of British Columbia
David P. Weaver
Phillips Laboratory
ISBN 1-56347-079-9

159. Rarefied Gas Dynamics: Theory and Simulations (1994)
Bernie D. Shizgal
University of British Columbia
David P. Weaver
Phillips Laboratory
ISBN 1-56347-080-2

160. Rarefied Gas Dynamics: Space Sciences and Engineering (1994)
Bernie D. Shizgal
University of British Columbia
David P. Weaver
Phillips Laboratory
ISBN 1-56347-081-0

161. Teleoperation and Robotics in Space (1994)
Steven B. Skaar
University of Notre Dame
Carl F. Ruoff
Jet Propulsion Laboratory, California Institute of Technology
ISBN 1-56347-095-0

162. Progress in Turbulence Research (1994)
Herman Branover
Yeshajahu Unger
Ben-Gurion University of the Negev
ISBN 1-56347-099-3

163. Global Positioning System: Theory and Applications, Volume I (1996)
Bradford W. Parkinson
Stanford University
James J. Spilker Jr.
Stanford Telecom
Penina Axelrad,
Associate Editor
University of Colorado
Per Enge,
Associate Editor
Stanford University
ISBN 1-56347-107-8

164. Global Positioning System: Theory and Applications, Volume II (1996)
Bradford W. Parkinson
Stanford University
James J. Spilker Jr.
Stanford Telecom
Penina Axelrad,
Associate Editor
University of Colorado
Per Enge,
Associate Editor
Stanford University
ISBN 1-56347-106-X

*Out of print.

165. Developments in High-Speed Vehicle Propulsion Systems (1996)
S. N. B. Murthy
Purdue University
E. T. Curran
Wright Laboratory
ISBN 1-56347-176-0

166. Recent Advances in Spray Combustion: Spray Atomization and Drop Burning Phenomena, Volume I (1996)
Kenneth K. Kuo
Pennsylvania State University
ISBN 1-56347-175-2

167. Fusion Energy in Space Propulsion (1995)
Terry Kammash
University of Michigan
ISBN 1-56347-184-1

168. Aerospace Thermal Structures and Materials for a New Era (1995)
Earl A. Thornton
University of Virginia
ISBN 1-56347-182-5

169. Liquid Rocket Engine Combustion Instability (1995)
Vigor Yang
William E. Anderson
Pennsylvania State University
ISBN 1-56347-183-3

170. Tactical Missile Propulsion (1996)
G. E. Jensen
United Technologies Corporation
David W. Netzer
Naval Postgraduate School
ISBN 1-56347-118-3

171. Recent Advances in Spray Combustion: Spray Combustion Measurements and Model Simulation, Volume II (1996)
Kenneth K. Kuo
Pennsylvania State University
ISBN 1-56347-181-7

172. Future Aeronautical and Space Systems (1997)
Ahmed K. Noor
NASA Langley Research Center
Samuel L. Venneri
NASA Headquarters
ISBN 1-56347-188-4

173. Advances in Combustion Science: In Honor of Ya. B. Zel'dovich (1997)
William A. Sirignano
University of California
Alexander G. Merzhanov
Russian Academy of Sciences
Luigi De Luca
Politecnico di Milano
ISBN 1-56347-178-7

174. Fundamentals of High Accuracy Inertial Navigation (1997)
Averil B. Chatfield
ISBN 1-56347-243-0

175. Liquid Propellant Gun Technology (1997)
Günter Klingenberg
Fraunhofer-Institut für Kurzzeitdynamik, Ernst-Mach-Institut
John D. Knapton
Walter F. Morrison
Gloria P. Wren
U.S. Army Research Laboratory
ISBN 1-56347-196-5

176. Tactical and Strategic Missile Guidance Third Edition (1998)
Paul Zarchan
Charles Stark Draper Laboratory, Inc.
ISBN 1-56347-279-1

177. Orbital and Celestial Mechanics (1998)
John P. Vinti
Gim J. Der, Editor
TRW
Nino L. Bonavito, Editor
NASA Goddard Space Flight Center
ISBN 1-56347-256-2

178. Some Engineering Applications in Random Vibrations and Random Structures (1998)
Giora Maymon
RAFAEL
ISBN 1-56347-258-9

179. Conventional Warhead Systems Physics and Engineering Design (1998)
Richard M. Lloyd
Raytheon Systems Company
ISBN 1-56347-255-4

180. Advances in Missile Guidance Theory (1998)
Joseph Z. Ben-Asher
Isaac Yaesh
Israel Military Industries—Advanced Systems Division
ISBN 1-56347-275-9

181. Satellite Thermal Control for Systems Engineers (1998)
Robert D. Karam
ISBN 1-56347-276-7

182. Progress in Fluid Flow Research: Turbulence and Applied MHD
Herman Branover
Yeshajahu Unger
Ben Gurion University of the Negev
ISBN 1-56347-284-8

---
*Out of print.

Progress in Fluid Flow Research: Turbulence and Applied MHD